BASIC AND APPLIED THERMODYNAMICS

Second Edition

About the Author

P K Nag has been with the Indian Institute of Technology, Kharagpur, for about four decades, almost since his graduation. After retirement from IIT, he was an Emeritus Fellow of AICTE, New Delhi, stationed at Jadavpur University, Kolkata, till June 2005. He was a Visiting Professor in the Technical University of Nova Scotia (now Dalhousie University), Halifax, Canada, for two years during 1985–86 and 1993–94.

Dr Nag is the recipient of the President of India medal (1995) from the Institution of Engineers (India). He is a Fellow of the National Academy of Engineers (FNAE) and a Fellow of the Institution of Engineers (India). In his early days, he was also an NCC Officer at IIT, Kharagpur. Besides these, Dr Nag is also a Life Member of the Indian Society for Technical Education, Indian Society for Heat and Mass Transfer, and the Combustion Institute, USA (Indian section). He was formerly a member of the New York Academy of Sciences, USA.

His areas of interest include circulating fluidized bed boilers, combined cycle power generation, second law analysis of thermal systems, and waste heat recovery. He has authored four books, including this one, and more than 150 research papers in several national and international journals and proceedings. He has also attended international conferences in the US, Japan, Canada and Bangladesh where he has presented several papers.

Other books from the author

- *Engineering Thermodynamics*, 4/e
- *Power Plant Engineering*, 3/e
- *Heat Transfer*, 2/e

BASIC AND APPLIED THERMODYNAMICS

Second Edition

P K Nag

Former Professor
Department of Mechanical Engineering
Indian Institute of Technology
Kharagpur

McGraw Hill Education (India) Private Limited
CHENNAI

McGraw Hill Education Offices
Chennai New York St Louis San Francisco Auckland Bogotá Caracas
Kuala Lumpur Lisbon London Madrid Mexico City Milan Montreal
San Juan Santiago Singapore Sydney Tokyo Toronto

McGraw Hill Education (India) Private Limited

Published by McGraw Hill Education (India) Private Limited,
444/1, Sri Ekambara Naicker Industrial Estate, Alapakkam, Porur, Chennai 600116, Tamil Nadu, India.

28th reprint 2018
RAQZCRLEDDARB

ISBN (13 digits): 978-0-07-015131-4
ISBN (10 digits): 0-07-015131-8

29 30 31 32 322 SAI 22 21 20 19 18

Head—Higher Education (Publishing and Marketing): *Vibha Mahajan*
Senior Publishing Manager—SEM & Tech Ed.: *Shalini Jha*
Senior Production Manager: *P L Pandita*
General Manager—Production: *Rajender P Ghansela*
Manager—Production: *Reji Kumar*

Typeset at Mukesh Technologies Pvt. Ltd. #10, 100 Feet Road, Ellapillaichavadi,
Puducherry 605005, and
Printed and Bound in India at Sai Printo Pack Pvt. Ltd., A-102/4, Okhla Industrial Area, New Delhi 110020.

Visit us at: www.mheducation.co.in

To My Grandchildren

Madhurima Basu Thakur
and
Abhiroop Guha

with love and blessings

Contents

Preface *xv*
Content Highlights *xvii*

1. Introduction **1**
1.1 Macroscopic vs Microscopic Viewpoint *1*
1.2 Thermodynamic System and Control Volume *1*
1.3 Thermodynamic Properties, Processes and Cycles *2*
1.4 Homogeneous and Heterogeneous Systems *3*
1.5 Thermodynamic Equilibrium *3*
1.6 Quasi-Static Process *4*
1.7 Pure Substance *5*
1.8 Concept of Continuum *5*
1.9 Thermostatics *6*
1.10 Units and Dimensions *6*
1.11 History of Thermodynamics *10*
Solved Examples *15*
Review Questions *15*
Problems *16*

2. Temperature **17**
2.1 Zeroth Law of Thermodynamics *17*
2.2 Measurement of Temperature—The Reference Points *17*
2.3 Comparison of Thermometers *19*
2.4 Ideal Gas *19*
2.5 Gas Thermometers *20*
2.6 Ideal Gas Temperature *21*
2.7 Celsius Temperature Scale *22*
2.8 Electrical Resistance Thermometer *22*
2.9 Thermocouple *22*
2.10 International Practical Temperature Scale *23*
Solved Examples *24*
Review Questions *25*
Problems *26*

3. Work and Heat Transfer **27**
3.1 Work Transfer *27*
3.2 pdV–Work or Displacement Work *28*
3.3 Indicator Diagram *30*
3.4 Other Types of Work Transfer *32*

3.5 Free Expansion with Zero Work Transfer *35*
3.6 Net Work Done by a System *35*
3.7 Heat Transfer *36*
3.8 Heat Transfer—A Path Function *36*
3.9 Specific Heat and Latent Heat *37*
3.10 Points to Remember Regarding Heat Transfer and Work Transfer *38*
Solved Examples 38
Review Questions 42
Problems 42

4. First Law of Thermodynamics **45**
4.1 First Law for a Closed System Undergoing a Cycle *45*
4.2 First Law for a Closed System Undergoing a Change of State *46*
4.3 Energy—A Property of the System *47*
4.4 Different Forms of Stored Energy *47*
4.5 Specific Heat at Constant Volume *49*
4.6 Enthalpy *50*
4.7 Specific Heat at Constant Pressure *50*
4.8 Energy of an Isolated System *51*
4.9 Perpetual Motion Machine of the First Kind–PMM1 *51*
4.10 Limitations of the First Law *51*
Solved Examples 51
Review Questions 55
Problems 55

5. First Law Applied to Flow Processes **58**
5.1 Control Volume *58*
5.2 Steady Flow Process *59*
5.3 Mass Balance and Energy Balance in a Simple Steady Flow Process *59*
5.4 Some Examples of Steady Flow Processes *61*
5.5 Comparison of S.F.E.E. with Euler and Bernoulli Equations *64*
5.6 Variable Flow Processes *65*
5.7 Example of a Variable Flow Problem *66*
5.8 Discharging and Charging a Tank *68*
Solved Examples 69
Review Questions 77
Problems 77

6. Second Law of Thermodynamics **81**
6.1 Qualitative Difference between Heat and Work *81*
6.2 Cyclic Heat Engine *82*
6.3 Energy Reservoirs *83*
6.4 Kelvin–Planck Statement of Second Law *83*
6.5 Clausius' Statement of the Second Law *84*
6.6 Refrigerator and Heat Pump *84*
6.7 Equivalence of Kelvin–Planck and Clausius Statements *86*
6.8 Reversibility and Irreversibility *87*
6.9 Causes of Irreversibility *87*
6.10 Conditions for Reversibility *91*
6.11 Carnot Cycle *91*

6.12 Reversed Heat Engine *92*
6.13 Carnot's Theorem *93*
6.14 Corollary of Carnot's Theorem *93*
6.15 Absolute Thermodynamic Temperature Scale *94*
6.16 Efficiency of the Reversible Heat Engine *96*
6.17 Equality of Ideal Gas Temperature and Kelvin Temperature *97*
6.18 Types of Irreversibility *98*
Solved Examples 99
Review Questions 104
Problems 105

7. Entropy 109

7.1 Introduction *109*
7.2 Two Reversible Adiabatic Paths Cannot Intersect Each Other *109*
7.3 Clausius' Theorem *109*
7.4 The Property of Entropy *111*
7.5 The Inequality of Clausius *114*
7.6 Entropy Change in an Irreversible Process *115*
7.7 Entropy Principle *116*
7.8 Applications of Entropy Principle *117*
7.9 Entropy Transfer Mechanisms *122*
7.10 Entropy Generation in a Closed System *124*
7.11 Entropy Generation in an Open System *126*
7.12 First and Second Laws Combined *128*
7.13 Reversible Adiabatic Work in a Steady Flow System *129*
7.14 Entropy and Direction: The Second Law—A Directional Law of Nature *131*
7.15 Entropy and Disorder *131*
7.16 Absolute Entropy *132*
Solved Examples 132
Review Questions 143
Problems 144

8. Available Energy, Exergy and Irreversibility 150

8.1 Available Energy *150*
8.2 Available Energy Referred to a Cycle *150*
8.3 Quality of Energy *153*
8.4 Maximum Work in a Reversible Process *155*
8.5 Reversible Work by an Open System Exchanging Heat only with the Surroundings *156*
8.6 Useful Work *159*
8.7 Dead State *160*
8.8 Availability *161*
8.9 Availability in Chemical Reactions *162*
8.10 Irreversibility and Gouy–Stodola Theorem *163*
8.11 Availability or Exergy Balance *166*
8.12 Second Law Efficiency *168*
8.13 Comments on Exergy *173*
Solved Examples 175
Review Questions 193
Problems 194

9. Properties of Pure Substances **198**

9.1 p–v Diagram for a Pure Substance *198*
9.2 p–T Diagram for a Pure Substance *201*
9.3 p–v–T Surface *202*
9.4 T–s Diagram for a Pure Substance *203*
9.5 h–s Diagram or Mollier Diagram for a Pure Substance *204*
9.6 Quality or Dryness Fraction *206*
9.7 Steam Tables *208*
9.8 Charts of Thermodynamic Properties *209*
9.9 Measurement of Steam Quality *211*
Solved Examples 214
Review Questions 232
Problems 232

10. Properties of Gases and Gas Mixtures **236**

10.1 Avogadro's Law *236*
10.2 Equation of State of a Gas *236*
10.3 Ideal Gas *237*
10.4 Equations of State *248*
10.5 Virial Expansions *249*
10.6 Law of Corresponding States *250*
10.7 Other Equations of State *255*
10.8 Properties of Mixtures of Gases—Dalton's Law of Partial Pressures *257*
10.9 Internal Energy, Enthalpy and Specific Heats of Gas Mixtures *259*
10.10 Entropy of Gas Mixtures *260*
10.11 Gibbs Function of a Mixture of Inert Ideal Gases *261*
Solved Examples 262
Review Questions 274
Problems 275

11. Thermodynamic Relations, Equilibrium and Third Law **282**

11.1 Some Mathematical Theorems *282*
11.2 Maxwell's Equations *283*
11.3 *TdS* Equations *284*
11.4 Difference in Heat Capacities *285*
11.5 Ratio of Heat Capacities *286*
11.6 Energy Equation *287*
11.7 Joule-Kelvin Effect *290*
11.8 Clausius-Clapeyron Equation *292*
11.9 Evaluation of Thermodynamic Properties from an Equation of State *295*
11.10 General Thermodynamic Considerations on an Equation of State *297*
11.11 Mixtures of Variable Composition *298*
11.12 Conditions of Equilibrium of a Heterogeneous System *301*
11.13 Gibbs Phase Rule *302*
11.14 Types of Equilibrium *303*
11.15 Local Equilibrium Conditions *305*
11.16 Conditions of Stability *306*
Solved Examples 308
Review Questions 322
Problems 323

12. Vapour Power Cycles **328**

12.1 Simple Steam Power Cycle *328*
12.2 Rankine Cycle *329*
12.3 Actual Vapour Cycle Processes *332*
12.4 Comparison of Rankine and Carnot Cycles *333*
12.5 Mean Temperature of Heat Addition *334*
12.6 Reheat Cycle *335*
12.7 Ideal Regenerative Cycle *337*
12.8 Regenerative Cycle *338*
12.9 Reheat-Regenerative Cycle *341*
12.10 Feedwater Heaters *343*
12.11 Exergy Analysis of Vapour Power Cycles *345*
12.12 Characteristics of an Ideal Working Fluid in Vapour Power Cycles *346*
12.13 Binary Vapour Cycles *347*
12.14 Thermodynamics of Coupled Cycles *349*
12.15 Process Heat and By-Product Power *350*
12.16 Efficiencies in Steam Power Plant *352*
Solved Examples 354
Review Questions 372
Problems 372

13. Gas Power Cycles **378**

13.1 Carnot Cycle (1824) *378*
13.2 Stirling Cycle (1827) *379*
13.3 Ericsson Cycle (1850) *379*
13.4 Air Standard Cycles *380*
13.5 Otto Cycle (1876) *380*
13.6 Diesel Cycle (1892) *382*
13.7 Limited Pressure Cycle, Mixed Cycle or Dual Cycle *384*
13.8 Comparison of Otto, Diesel, and Dual Cycles *385*
13.9 Brayton Cycle *387*
13.10 Aircraft Propulsion *396*
13.11 Brayton-Rankine Combined Cycle *398*
Solved Examples 400
Review Questions 413
Problems 414

14. Refrigeration Cycles **419**

14.1 Refrigeration by Non-Cyclic Processes *419*
14.2 Reversed Heat Engine Cycle *419*
14.3 Vapour Compression Refrigeration Cycle *420*
14.4 Selection of a Refrigerant *434*
14.5 Absorption Refrigeration Cycle *436*
14.6 Electrolux Refrigerator *439*
14.7 Steam Jet Refrigeration System *440*
14.8 Heat Pump Systems *441*
14.9 Gas Cycle Refrigeration *442*
14.10 Application to Aircraft Refrigeration *444*
14.11 Liquefaction of Gases *446*
14.12 Production of Solid Ice *448*

Solved Examples 450
Review Questions 462
Problems 463

15. Psychrometrics and Air-conditioning Systems **467**

15.1 Properties of Atmospheric Air 467
15.2 Psychrometric Properties of Air 468
15.3 Psychrometric Chart 474
15.4 Psychrometric Processes 476
15.5 Simple Air-conditioning System 486
15.6 Load Calculations 489
15.7 Air-conditioning Systems 490
Solved Examples 491
Review Questions 504
Problems 505

16. Reactive Systems **507**

16.1 Degree of Reaction 507
16.2 Reaction Equilibrium 509
16.3 Law of Mass Action 510
16.4 Heat of Reaction 511
16.5 Temperature Dependence of the Heat of Reaction 512
16.6 Temperature Dependence of the Equilibrium Constant 512
16.7 Thermal Ionization of a Monatomic Gas 513
16.8 Gibbs Function Change 515
16.9 Fugacity and Activity 517
16.10 Displacement of Equilibrium Due to a Change in Temperature or Pressure 518
16.11 Heat Capacity of Reacting Gases in Equilibrium 519
16.12 Combustion 520
16.13 Enthalpy of Formation 521
16.14 First Law for Reactive Systems 522
16.15 Adiabatic Flame Temperature 523
16.16 Enthalpy and Internal Energy of Combustion: Heating Value 523
16.17 Absolute Entropy and the Third Law of Thermodynamics 525
16.18 Second Law Analysis of Reactive Systems 526
16.19 Chemical Exergy 526
16.20 Second Law Efficiency of a Reactive System 529
16.21 Fuel Cells 529
Solved Examples 531
Review Questions 544
Problems 544

17. Compressible Fluid Flow **548**

17.1 Velocity of Pressure Pulse in a Fluid 548
17.2 Stagnation Properties 550
17.3 One-Dimensional Steady Isentropic Flow 550
17.4 Critical Properties–Choking in Isentropic Flow 552
17.5 Normal Shocks 556
17.6 Adiabatic Flow with Friction and Diabatic Flow without Friction 560

Solved Examples *561*
Review Questions *566*
Problems *566*

18. Elements of Heat Transfer **568**

18.1 Basic Concepts *568*
18.2 Conduction Heat Transfer *568*
18.3 Convective Heat Transfer *574*
18.4 Heat Exchangers *579*
18.5 Radiation Heat Transfer *584*
Solved Examples *591*
Review Questions *599*
Problems *600*

19. Gas Compressors **603**

19.1 Compression Processes *603*
19.2 Work of Compression *603*
19.3 Single-Stage Reciprocating Air Compressor *605*
19.4 Volumetric Efficiency *607*
19.5 Multi-Stage Compression *609*
19.6 Air Motors *613*
19.7 Rotary Compressor *613*
19.8 Blowers and Fans *624*
Solved Examples *624*
Review Questions *646*
Problems *647*

20. Internal Combustion Engines **650**

20.1 Internal and External Combustion Engines *650*
20.2 Classification of IC Engines *650*
20.3 An Overview of Reciprocating Engines *652*
20.4 Description of an SI Engine *653*
20.5 Operating Principles *653*
20.6 The Two-Stroke Engine *658*
20.7 Engine Performance *659*
20.8 Air Standard Cycles *662*
20.9 Fuels for IC Engines *666*
20.10 Combustion in SI Engines *669*
20.11 Simple Float Carburetor *673*
20.12 Fuel Injection in CI Engines *676*
20.13 Cooling System of IC Engines *680*
20.14 Lubricating System *681*
20.15 Starting of Engine *685*
20.16 Performance Characteristics *685*
20.17 Supercharging *688*
Solved Examples *691*
Review Questions *705*
Problems *706*

21. Gas Tubines and Propulsion Systems **708**

21.1 Closed Cycle and Open Cycle Plants *708*
21.2 Advantages of a GT Plant *708*
21.3 Disadvantages of a GT Plant *710*
21.4 Analysis of GT Plant *711*
21.5 Closed-Cycle Gas Turbine *715*
21.6 Semi-Closed Cycle GT Plant *715*
21.7 Performance of Gas Turbine Power Plants *715*
21.8 Components of a Gas Turbine Plant *718*
21.9 Gas Turbine with Water Injection *722*
21.10 Gas Turbine Fuels *723*
21.11 Gas Turbine Materials *723*
21.12 Jet Propulsion System *724*
21.13 Propulsive Devices for Aircrafts and Missiles *726*
21.14 Turbojet Engine with Afterburner *732*
21.15 Turbo-Prop Engine *733*
21.16 Bypass Turbo-JET Engine *735*
21.17 Athodyds-Ramjets and Pulsejets *736*
21.18 Rocket Propulsion *739*
21.19 Nuclear Rockets *744*
Solved Examples 744
Review Questions 759
Problems 760

22. Transport Processes in Gases **762**

22.1 Mean Free Path and Collision Cross-Section *762*
22.2 Distribution of Free Paths *763*
22.3 Transport Properties *764*
Solved Examples 772
Review Questions 778
Problems 779

Appendix A **780**

A.1 Steam Tables *780*
A.2 Thermodynamic Properties of Refrigerant-12 (Dichlorodifluoromethane) *798*
A.3 Thermodynamic Properties of Refrigerant-22 (Monochlorodifluoromethane) *803*
A.4 Thermodynamic Properties of Refrigerant-134A (1,1,1,2–Tetrafluoroethane) *810*
A.5 Thermodynamic Properties of Ammonia *817*
A.6 Variation of $\bar{c}_p$ with Temperature for Various Ideal Gases *822*
A.7 Ideal Gas Specific Heats of Some Common Gases (kJ/kg.K) *822*
A.8 Logarithm to the Base 10 of the Equilibrium Constant K *824*
A.9 Enthalpy of Formation at 25°C, Ideal Gas Enthalpy, and Absolute Entropy at 0.1 MPa (1 bar) Pressure *826*
A.10 Gas Tables *829*
A.11 Critical Point Data *831*

Appendix B – Charts (*Fig. B.1 – Fig. B.6*)

Multiple-Choice Questions **833**

Bibliography **840**

Index **842**

Preface

The author feels greatly encouraged at the good response of the book *Engineering Thermodynamics*. The present book is a further updated and reorganized version, retaining the framework and style of presentation of the earlier book. The last four chapters of the book, viz., Statistical Thermodynamics, Irreversible Thermodynamics, Kinetic Theory of Gases and Transport Properties of Gases are of little relevance in undergraduate courses on Thermodynamics. Hence, the last four chapters of the book and also the topics not important for UG courses have been omitted in this edition and three new chapters on **Gas Compressors, Internal Combustion Engines** and **Gas Turbines and Propulsion Systems** have been included instead in this edition in conformity with the UG curricula of many colleges in Applied Thermodynamics and also the recommendation of the reviewers. In addition to contributing as a basic course, the book can be used for a further advanced course at a senior undergraduate or graduate level with topics selectively chosen.

The **first eight chapters** of the book are devoted to a thorough treatment of the basic principles and concepts of classical thermodynamics. The second law and entropy have been introduced using the concept of heat engine. **Chapters 9** and **10** present the properties of substances. **Chapter 11** gives the general thermodynamic relationships among properties. A detailed analysis of power and refrigeration cycles is given in **chapters 12** to **14**. **Chapter 15** deals with psychrometrics and air-conditioning systems, while reactive systems are analyzed in **Chapter 16**. **Chapters 17** and **18** deal with compressible fluid flow and heat transfer respectively. Gas compressors are discussed in **Chapter 19**. Internal combustion engines and their operating principles are explained in **Chapter 20**. **Chapter 21** is devoted to gas turbines and propulsion systems. Finally, **Chapter 22** discusses the transport processes in gases.

The style of presentation and inclusion of short questions, solved and exercise problems have been retained. Many illustrative examples are solved and many problems are provided in each chapter to aid comprehension and to stimulate the interest of the students. **Multiple Choice Questions** on the subject have been added at the end. Throughout the text, SI units have been used. Tables and charts given in the Appendices are also in SI units.

The book also has an accompanying website which can be accessed at http://www.mhhe.com/nag/bat2e.

The author hopes that the book will be useful to the students and teachers alike. The study on Steam like Steam Boilers, Steam Turbines and Condensers will be considered for inclusion in future depending on the demand after omitting and editing some of the existing material.

I am thankful to Ms Surabhi Shukla and Ms Devshree Lohchab for their arduous task of syllabi research and competitive analysis which helped me a great deal in preparing this revision. I would also like to thank Ms Sohini Mukherjee for her deft editorial services. Mr P L Pandita too deserves a special mention for his efficient handling of the production process. I am also indebted to the following reviewers for taking out time to review the book.

M K Das	*MNNIT, Allahabad, Uttar Pradesh*
Ajay Trehan	*Dr B R Ambedkar National Institute of Technology Jalandhar, Punjab*
K P Tyagi	*Krishna Institute of Engineering and Technology UPTU, Ghaziabad, Uttar Pradesh*
T K Jana	*Haldia Institute of Technology WBUT, East Midnapur, West Bengal*
P K Bardhan	*JIS College of Engineering WBUT, Nadia, West Bengal*
P S Utgikar	*JSPM's Rajarshi Shahu College of Engineering Pune, Maharashtra*
D A Warke	*J T Mahajan College of Engineering Jalgaon, Maharashtra*
R G Kapadia	*Shri Sad Vidya Mandal Institute of Technology Bharuch, Gujarat*
P Chitambarnathan	*Dr Sivanthi Adithanar College of Engineering Thiruchendur, Tamil Nadu*
A V S Gupta	*JNTU College of Engineering Hyderabad, Andhra Pradesh*
S Ramanathan	*DRK Institute of Science and Technology JNTU, Hyderabad, Andhra Pradesh*
Y S Varadarajan	*The National Institute of Engineering Mysore, Karnataka*
K Chandrashekara	*Sri Jayachamarajendra College of Engineering Mysore, Karnataka*
S Makam	*R V College of Engineering Bangalore, Karnataka*
M S Bhagya Shekar	*Don Bosco Institute of Technology Bangalore, Karnataka*
Sharanabasava C Pilli	*K L E Society's College of Engineering and Technology Belgaum, Karnataka*
V V Bongale	*Malnad College of Engineering Hassan, Karnataka*

Any constructive criticism as well as suggestions for improvement of the book will be welcome and gratefully acknowledged.

The present edition of the book is being dedicated to my grandchildren, Kumari Madhurima Basu Thakur and Sri Abhiroop Guha, with love and blessings.

P K Nag

CONTENT HIGHLIGHTS

New Chapters

New chapters on *Gas Compressors*, *Internal Combustion Engines* and *Gas Turbines and Propulsion Systems* included in the second edition would help engineers comprehend the role of compression processes, IC engines and propulsion systems in thermodynamics.

C H A P T E R 19

Gas Compressors

A gas compressor is a device in which work is done on the gas to raise its pressure, with an appreciable increase in its density. The compression of gases is an important process in many power plants, refrigeration plants, and industrial plants. Industrial uses occur in connection with compressed air motors for tools,

C H A P T E R 20

Internal Combustion Engines

20.1 INTERNAL AND EXTERNAL COMBUSTION ENGINES

C H A P T E R 21

Gas Turbines and Propulsion Systems

The economics of power generation by gas turbines is now quite attractive due to low capital cost and its high reliability and flexibility in operation. Another outstanding feature is its capability of quick starting and using a wide variety of fuels from natural gas to residual oil or powdered coal. Its consumption of lubricating oil is quite low because of the absence of rubbing and reciprocating parts, and the balancing does not pose a major problem. Due to better materials being made available and with the use of adequate blade cooling, the inlet gas temperature to the gas turbine (GT) blades can now exceed 1200°C as a result of which the overall efficiency of a GT plant can be about 35%, almost the same as that of a conventional steam power plant.

Because of its low weight per unit power, gas turbine is exclusively used to drive aviation system of all kinds of aircraft. It is also being increasingly used in land vehicles like buses and trucks and also to drive locomotives and marine ships. In oil and gas industries, the gas turbine in widely employed to drive auxiliaries like compressors blowers and pumps.

21.1 CLOSED CYCLE AND OPEN CYCLE PLANTS

The essential components of a gas turbine (GT) power plant are the compressor, combustion chamber and the turbine. The air standard cycle of a GT plant is the Brayton cycle.

A GT plant can either be open or closed. Figure 21.1 shows the arrangement of an open-cycle plant which is more common. The compressor takes in ambient air and raises its pressure. The temperature of air is increased when it flows through a combustion chamber where a fossil fuel is burned or a heat exchanger. (nuclear fuel being used as a source of energy) is present. The high-pressure, high-temperature working fluid, mostly a gas, enters a turbine where it expands to a low-pressure (equal to or a little above the atmospheric pressure) fluid. In an open unit, the gas is released from the turbine to the surroundings and in a closed unit, the working fluid is cooled in a cooler after the exhaust from the turbine and is returned to the compressor

New Topics

Coverage on **Chapter 14** is enhanced by the insertion of the topics of *Expansion Device, Multi-Evaporator Systems* and *Electrolux Refrigerator*. Two essential topics on *Air-Conditioning Systems* and *Air Washer* are included in **Chapter 15.**

14.3.5 Components in a Vapour Compression Plant

(a) Condenser It is in the condenser that heat is rejected in a vapour compression plant. Desuperheating of the vapour takes place in the discharge line and in the first few coils of the condenser. It is followed by the condensation of the vapour at the saturation temperature. Then, subcooling may take place near the bottom. However, the sensible heat of the desuperheating and subcooling processes is quite small compared to the latent heat of condensation.

The type of condenser is generally characterized by the cooling medium used. There are three types of condensers: (i) air-cooled condensers, (ii) water-cooled condensers, and (iii) evaporative condensers. An air-cooled condenser is used in small self-contained units below 5 TR because of high power consumption and large fan noise.

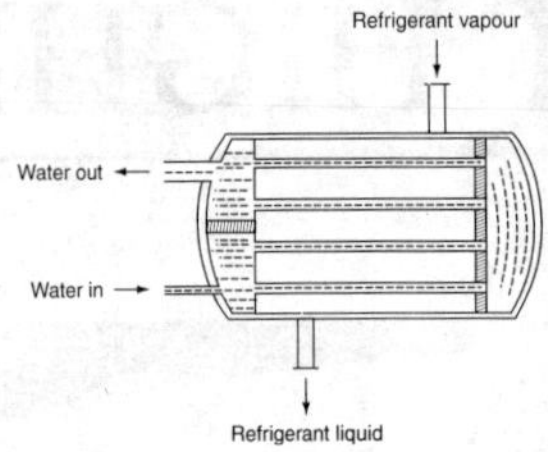

Fig. 14.17 *Schematic representation of a two-pass water-cooled shell and tube condenser*

Water-cooled condensers can be of three types, viz., shell and tube, shell and coil and double tube. The shell-and-tube type with water flowing through tubes and the refrigerant condensing in the shell is the most commonly used condenser (Fig. 14.17).

(b) Expansion Device An expansion device in a refrigeration system normally serves two functions, viz., (i) it reduces the pressure from the condenser to the evaporator, and (ii) it regulates the flow of the refrigerant to the evaporator depending on the load. It is essentially a restriction offering resistance to flow so that the pressure drops, ...

14.3.7 Multi-Evaporator Systems

A situation often arises when varied types of cooling loads are connected to the same refrigeration system. Each load may need an evaporator working at a different refrigeration temperature. A two-evaporator single-compressor system with individual expansion valves for each evaporator and one compressor is shown in Fig. 14.22(a) and the corresponding *p-h* diagram is shown is Fig. 14.22(b) The mass flow rates through evaporators 1 and 2 are

$$m_1 = \frac{Q_{e1}}{h_7 - h_5} \quad \text{and} \quad m_2 = \frac{Q_{e2}}{h_6 - h_4}$$

The enthalpy of liquid–vapour mixture entering the compressor is

$$h_1 = \frac{m_1 h_7 + m_2 h_6}{m_1 + m_2}$$

and the work input is

$$W = (m_1 + m_2)(h_2 - h_1)$$

14.3.8 Cascad...

Two important ...
of the two med...
To have reason...

14.6 ELECTROLUX REFRIGERATOR

An electrolux refrigerator works on three-fluid system. There is no circulation pump. The total pressure is the same throughout the system. The third fluid remains mainly in the evaporator, which reduces the partial pressure of refrigerant to enable it to evaporate at low pressure and hence low temperature.

The schematic diagram of an electrolux refrigerator working on $NH_3 - H_2O$ system with H_2 as the third fluid is shown in Fig. 14.30. Liquid NH_3 evaporates in the evaporator in the presence of H_2. Hydrogen is chosen because it is non-corrosive and insoluble in water. Thus a complete leakproof silent system is achieved.

A thermosiphon bubble pump is used to lift the weak aqua solution from the generator to the separator. The discharge tube from the generator is extended down below the liquid level in the generator. The bubbles rise and carry slugs of weak $NH_3 - H_2O$ solution into the separator. Two U-bend constrictions are provided as *vapour-locks* or *liquid-seal* to prevent H_2 from getting into the high side of the condenser.

The partial pressure of H_2 provides the pressure difference of NH_3 between the condenser and evaporator.

Accordingly, we have the following:

In the *condenser*, there is the pure NH_3 vapour pressure which is the total pressure.
In the *evaporator*, NH_3 vapour pressure = total pressure minus the partial pressure of H_2.

For example, let us consider the condenser temperatur... corresponding vapour pressure of NH_3 are

Condenser, $p_c = 20.33$ bar
Evaporator exit, $p_e = 2.36$ bar

The approximate pressures in various parts of the ...
The vapours leaving the generator top are assume... is 15.5 ... of NI...

Table 15.4
Heat of Respiration of Products in J/kg per 24 hours

Product	Storage temperature	
	0°C	15.6°C
Apples	312–1500	2300–8215
Cabbage	1248	4265
Carrots	2183	8420
Cauliflower	4680	10,500
Grapes	416	2290
Lemon	520	2390
Orange	728	2495
Potato	1350	1560
Tomato	1040	5820

15.7 AIR-CONDITIONING SYSTEMS

Among many applications of air conditioning, the important ones are

(a) Cold storage
(b) Industrial air conditioning
(c) Comfort air conditioning

15.4.13 Air Washer

An *air washer* is shown schematically in Fig. 15.22. It involves the flow of air through a spray of water. During the course of flow, the air may be cooled or heated, humidified or dehumidified, or simply adiabatically saturated, depending on the mean surface temperature of water spray. The water is, accordingly, cooled or heated externally, or simply recirculated by a pump. Make-up water is added for any loss due to humidification. Eliminator plates are provided to minimise the loss of water droplets.

The thermodynamic changes of state of air along paths 1–2 in an air washer are shown in Fig. 15.23. The following processes are possible:

1. *Process 1–2 A* Heating and humidification ($t_s > t_1$).
2. *Process 1–2B* Humidification ($t_s = t_1$).
3. *Process 1–2C* Cooling and humidification $\left(t_1' < t_s < t_1\right)$ where t' is the WBT and t_1 the DBT.
4. *Process 1–2D* Adiabatic saturation $\left(t_1' = t_s\right)$.
5. *Process 1–2E* Cooling and humidification $\left(t_d < t_s < t_1'\right)$ where t_d is the DPT.

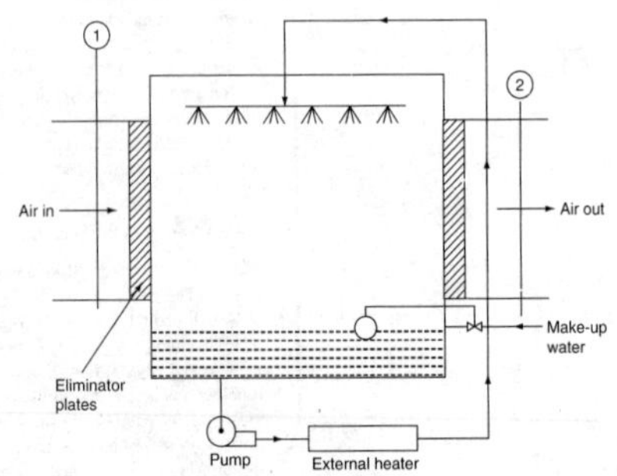

Fig. 15.22 *Air washer*

Solved Examples

Solved examples in various chapters at the end of the theory portion help the students understand the applicability of engineering calculations for thermodynamic applications.

24 *Basic and Applied Thermodynamics*

Solved Examples

Example 2.1

Two mercury-in-glass thermometers are made of identical materials and are accurately calibrated at 0°C and 100°C. One has a tube of constant diameter, while the other has a tube of conical bore, ten per cent greater in diameter at 100°C than at 0°C. Both thermometers have the length between 0 and 100 subdivided uniformly. What will be the straight bore thermometer read in a place where the conical bore thermometer reads 50°C?

Solution The volume of mercury in the tube at t°C, V_t, is given by

$$V_t = V_0 [1 + \beta (t - t_0)]$$

where V_0 is the volume of mercury at 0°C, β is the coefficient of volume expansion of mercury, and t_0 is the ice point temperature which is 0°C. The volume change of glass is neglected.

Therefore $\quad V_t - V_0 = \beta V_0 t$

The temperature t is thus a linear function of volume change of mercury $(V_t - V_0)$.

Therefore $\quad \Delta V_{0-100} = \beta V_0 \cdot 100$

$$\Delta V_{0-50} = \beta V_0 \cdot 50$$

$$\frac{\Delta V_{0-50}}{\Delta V_{0-100}} = \frac{1}{2}$$

100°C

50°C

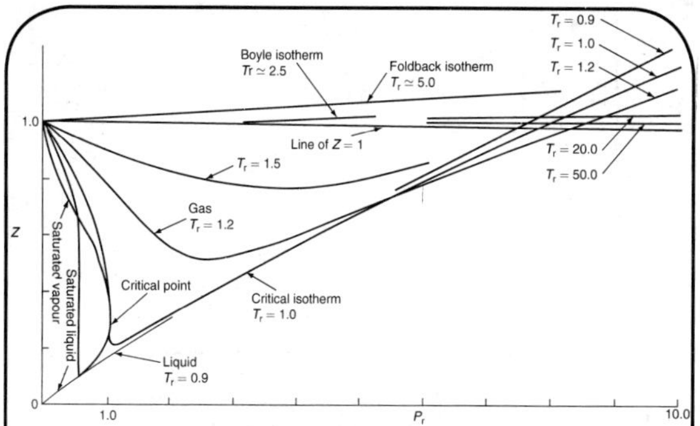

Fig. 10.5 *(a) Generalized compressibility chart*

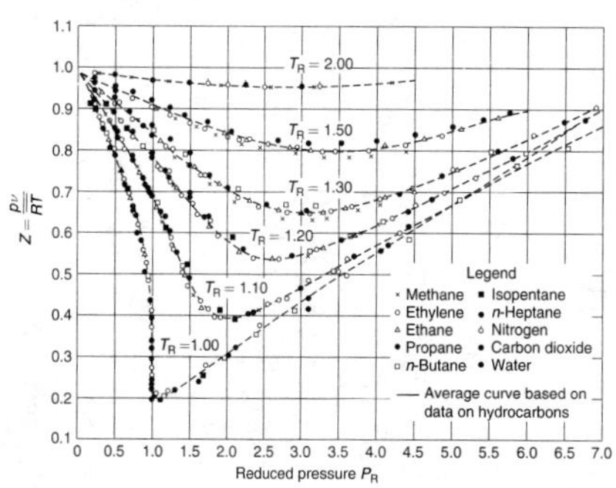

Fig. 10.5 *(b) Generalized compressibility chart*

Illustrations

Detailed illustrations are necessary for any book on engineering. Besides the existing ones, 133 new figures have been included in this edition to let students have a clear grasp of the contents.

Review Questions

Review questions are provided at the end of each chapter. They cover the contents of the entire chapter. These review questions are very helpful to teachers in setting up examination papers and class work assignments.

Review Questions

8.1 What do you understand by high grade energy and low grade energy?

8.2 What is available energy and unavailable energy?

8.3 Who propounded the concept of availability?

8.4 What is the available energy referred to a cycle?

8.5 Show that there is a decrease in available energy when heat is transferred through a finite temperature difference.

8.6 Deduce the expression for available energy from a finite energy source at temperature T when the environmental temperature is T_0.

8.7 What do you understand by exergy and energy?

8.8 What is meant by quality of energy?

8.9 Why is exergy of a fluid at a higher temperature more than that at a lower temperature?

8.10 How does the exergy value provide a useful measure of the quality of energy?

8.11 Why is the second law called the law of degradation of energy?

8.12 Energy is always conserved, but its quality is always degraded. Explain.

8.21 What do you understand by the dead state?

8.22 What is meant by availability?

8.23 Give expressions for availabilities of a closed system and a steady flow open system.

8.24 What are Helmholtz function and Gibbs function?

8.25 What is the availability in a chemical reaction if the temperature before and after the reaction is the same and equal to the temperature of the surroundings?

8.26 When is the availability of a chemical reaction equal to the decrease in the Gibbs function?

8.27 Derive the expression for irreversibility or exergy loss in a process executed by: (a) a closed system, (b) a steady flow system, in a given environment.

8.28 State and explain the Gouy-Stodola theorem.

8.29 How is heat transfer through a finite temperature difference equivalent to the destruction of its availability?

8.30 Considering the steady and adiabatic flow of an ideal gas through a pipe, show that the rate of decrease in availability or lost work is proportional to the pressure drop and the mass fl-

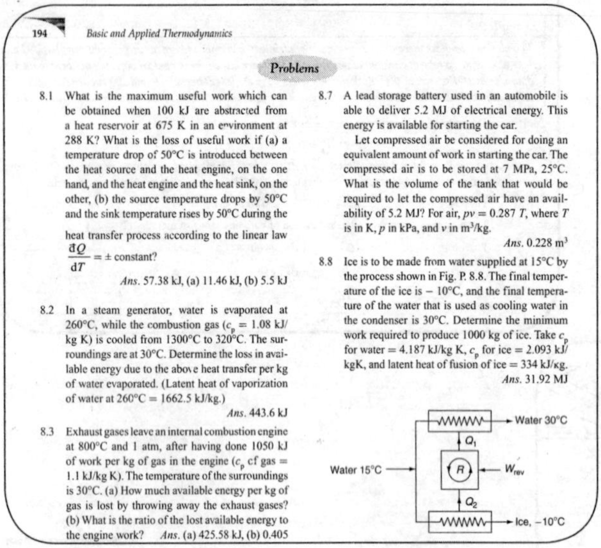

194 *Basic and Applied Thermodynamics*

Problems

8.1 What is the maximum useful work which can be obtained when 100 kJ are abstracted from a heat reservoir at 675 K in an environment at 288 K? What is the loss of useful work if (a) a temperature drop of 50°C is introduced between the heat source and the heat engine, on the one hand, and the heat engine and the heat sink, on the other, (b) the source temperature drops by 50°C and the sink temperature rises by 50°C during the heat transfer process according to the linear law $\frac{đQ}{dT} = \pm$ constant?

Ans. 57.38 kJ, (a) 11.46 kJ, (b) 5.5 kJ

8.2 In a steam generator, water is evaporated at 260°C, while the combustion gas (c_p = 1.08 kJ/kg K) is cooled from 1300°C to 320°C. The surroundings are at 30°C. Determine the loss in available energy due to the above heat transfer per kg of water evaporated. (Latent heat of vaporization of water at 260°C = 1662.5 kJ/kg.)

Ans. 443.6 kJ

8.3 Exhaust gases leave an internal combustion engine at 800°C and 1 atm, after having done 1050 kJ of work per kg of gas in the engine (c_p of gas = 1.1 kJ/kg K). The temperature of the surroundings is 30°C. (a) How much available energy per kg of gas is lost by throwing away the exhaust gases? (b) What is the ratio of the lost available energy to the engine work? *Ans.* (a) 425.58 kJ, (b) 0.405

8.7 A lead storage battery used in an automobile is able to deliver 5.2 MJ of electrical energy. This energy is available for starting the car.

Let compressed air be considered for doing an equivalent amount of work in starting the car. The compressed air is to be stored at 7 MPa, 25°C. What is the volume of the tank that would be required to let the compressed air have an availability of 5.2 MJ? For air, $pv = 0.287\ T$, where T is in K, p in kPa, and v in m³/kg.

Ans. 0.228 m³

8.8 Ice is to be made from water supplied at 15°C by the process shown in Fig. P. 8.8. The final temperature of the ice is – 10°C, and the final temperature of the water that is used as cooling water in the condenser is 30°C. Determine the minimum work required to produce 1000 kg of ice. Take c_p for water = 4.187 kJ/kg K, c_p for ice = 2.093 kJ/kgK, and latent heat of fusion of ice = 334 kJ/kg.

Ans. 31.92 MJ

Problems with Answers

A set of problems with answers is given in all the chapters. These help the students test their comprehension of the concepts.

Appendices

A total of 11 Appendices are given at the end of the book which contain relevant data, charts and tables related to thermodynamics.

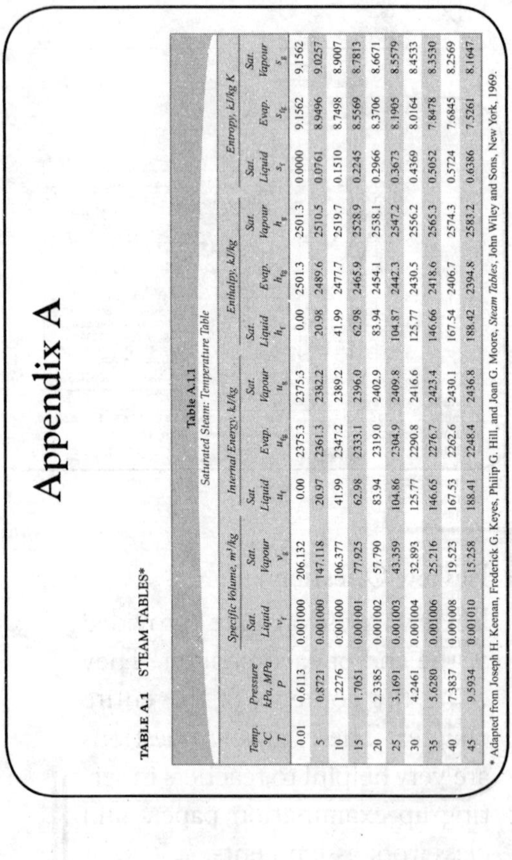

Appendix A

TABLE A.1 STEAM TABLES*

Table A.1.1
Saturated Steam: Temperature Table

Temp. °C T	Pressure kPa, MPa P	Specific Volume, m³/kg: Sat. Liquid v_f	Sat. Vapour v_g	Internal Energy, kJ/kg: Sat. Liquid u_f	Evap. u_{fg}	Sat. Vapour u_g	Enthalpy, kJ/kg: Sat. Liquid h_f	Evap. h_{fg}	Sat. Vapour h_g	Entropy, kJ/kg K: Sat. Liquid s_f	Evap. s_{fg}	Sat. Vapour s_g
0.01	0.6113	0.001000	206.132	0.00	2375.3	2375.3	0.00	2501.3	2501.3	0.0000	9.1562	9.1562
5	0.8721	0.001000	147.118	20.97	2361.3	2382.2	20.98	2489.6	2510.5	0.0761	8.9496	9.0257
10	1.2276	0.001000	106.377	41.99	2347.2	2389.2	41.99	2477.7	2519.7	0.1510	8.7498	8.9007
15	1.7051	0.001001	77.925	62.98	2333.1	2396.0	62.98	2465.9	2528.9	0.2245	8.5569	8.7813
20	2.3385	0.001002	57.790	83.94	2319.0	2402.9	83.94	2454.1	2538.1	0.2966	8.3706	8.6671
25	3.1691	0.001003	43.359	104.86	2304.9	2409.8	104.87	2442.3	2547.2	0.3673	8.1905	8.5579
30	4.2461	0.001004	32.893	125.77	2290.8	2416.6	125.77	2430.5	2556.2	0.4369	8.0164	8.4533
35	5.6280	0.001006	25.216	146.65	2276.7	2423.4	146.66	2418.6	2565.3	0.5052	7.8478	8.3530
40	7.3837	0.001008	19.523	167.53	2262.6	2430.1	167.54	2406.7	2574.3	0.5724	7.6845	8.2569
45	9.5934	0.001010	15.258	188.41	2248.4	2436.8	188.42	2394.8	2583.2	0.6386	7.5261	8.1647

* Adapted from Joseph H. Keenan, Frederick G. Keyes, Philip G. Hill, and Joan G. Moore, *Steam Tables*, John Wiley and Sons, New York, 1969.

Multiple-Choice Questions

Over 60 MCQs have been added at the end of the book to let students have a quick revision of the fundamental concepts.

Multiple-Choice Questions

1. Lead compounds were earlier used to be added to gasoline in S.I. engines so as to
   (a) reduce hydrocarbon emissions (b) reduce knocking
   (c) reduce exhaust gas temperature (d) increase power output
2. Decrease of air–fuel ratio in SI engines results in
   (a) an increase NO_x (b) a decrease of CO and unburnt hydrocarbon
   (c) an increase of CO and unburnt hydrocarbon (d) none of the above
3. If N is the rpm, the number of power strokes per minute in a four-stroke engine is
   (a) $2N$ (b) $N/2$ (c) N (d) $4N$
4. Volumetric efficiency is a measure of
   (a) speed of the engine (b) power of the engine
   (c) breathing capacity of the engine (d) pressure rise in the cylinder
5. If N is the rpm, the number of power strokes per minute in a two-stroke engine is
   (a) N (b) $N/2$ (c) $2N$ (d) $4N$
6. The volumetric efficiency of a well-designed engine is in the range of
   (a) 30 to 40% (b) 40 to 60% (c) 60 to 70% (d) 75 to 90%
7. If L is the stroke and N is the rpm, the mean piston speed of a two-stroke engine is
   (a) LN (b) $LN/2$ (c) $2LN$ (d) none of the abo

Bibliography

1. Adrian Bejan, *Advanced Engineering Thermodynamics*, John Wiley, New York, 1988.
2. Adrian Bejan, *Entropy Generation through Heat and Fluid Flow*, Wiley Int. Science, New York, 1982.
3. R.W. Benson, *Advanced Engineering Thermodynamics*, Pergamon Press, London, 1977.
4. H.B. Callen, *Thermodynamics*, John Wiley, New York, 1960.
5. Y.A. Cengel and M.A. Boles, *Thermodynamics: An Engineering Approach*, McGraw-Hill, New York, 1988.
6. G.N. Hatsopoulos and J.H. Keenan, *Principles of General Thermodynamics*, John Wiley, New York, 1965.
7. Jack P. Holman, *Heat Transfer*, McGraw-Hill, New York, 1978.
8. J.R. Howell and R.O. Buckius, *Fundamentals of Engineering Thermodynamics*, McGraw-Hill, New York, 1992.
9. J.S. Hsieh, *Principles of Thermodynamics*, McGraw-Hill, 1975.
10. J.B. Jones and R.E. Dugan, *Engineering Thermodynamics*, Prentice-Hall, India, 1998.
11. J.H. Keenan, *Thermodynamics*, John Wiley, New York, 1941.
12. J.F. Lay, *Thermodynamics*, Merril Books, 1963.
13. J.F. Lee and F.W. Sears, *Thermodynamics*, Addison-Wesely, Cambridge, Mass, 1962.
14. David A. Mooney, *Mechanical Engineering Thermodynamics*, Prentice-Hall, New Jersey, 1953.
15. M.J. Moran and H.N. Shapiro, *Fundamentals of Engineering Thermodynamics*, John Wiley, New York, 1988.
16. P.K. Nag, *Heat and Mass Transfer*, 2e, Tata McGraw-Hill, New Delhi, 2007.
17. E.F. Obert, *Concepts of Thermodynamics*, McGraw-Hill, New York, 1960.
18. W.C. Reynolds and H.C. Perkins, *Engineering Thermodynamics*, McGraw-Hill, New York, 1977.
19. Michel A. Saad, *Thermodynamics for Engineers*, Prentice-Hall, New Delhi, 1969.
20. A.H. Shapiro, *The Dynamics and Thermodynamics of Compressible Fluid Flow*, The Ronald Press, New York, 1953.

Bibliography

A comprehensive list of references in the form of textbooks and journals with the title of the article is given at the end of the book.

CHAPTER 1

Introduction

Thermodynamics is the science of energy transfer and its effect on the physical properties of substances. It is based upon observations of common experience which have been formulated into thermodynamic laws. These laws govern the principles of energy conversion. The applications of the thermodynamic laws and principles are found in all fields of energy technology, notably in steam and nuclear power plants, internal combustion engines, gas turbines, air conditioning, refrigeration, gas dynamics, jet propulsion, compressors, chemical process plants, and direct energy conversion devices.

1.1 MACROSCOPIC VS MICROSCOPIC VIEWPOINT

There are two points of view from which the behaviour of matter can be studied: the macroscopic and the microscopic. In the macroscopic approach, a certain quantity of matter is considered, without the events occurring at the molecular level being taken into account. From the microscopic point of view, matter is composed of myriads of molecules. If it is a gas, each molecule at a given instant has a certain position, velocity, and energy, and for each molecule these change very frequently as a result of collisions. The behaviour of the gas is described by summing up the behaviour of each molecule. Such a study is made in *microscopic* or *statistical thermodynamics*. *Macroscopic thermodynamics* is only concerned with the effects of the action of many molecules, and these effects can be perceived by human senses. For example, the macroscopic quantity, pressure, is the average rate of change of momentum due to all the molecular collisions made on a unit area. The effects of pressure can be felt. The macroscopic point of view is not concerned with the action of individual molecules, and the force on a given unit area can be measured by using, e.g. a pressure gauge. These macroscopic observations are completely independent of the assumptions regarding the nature of matter. All the results of classical or macroscopic thermodynamics can, however, be derived from the microscopic and statistical study of matter.

1.2 THERMODYNAMIC SYSTEM AND CONTROL VOLUME

A thermodynamic *system* is defined as a quantity of matter or a region in space upon which attention is concentrated in the analysis of a problem. Everything external to the system is called the *surroundings* or the *environment*. The system is separated from the surroundings by the system boundary (Fig. 1.1). The boundary may be either *fixed* or *moving*. A system and its surroundings together comprise a *universe*.

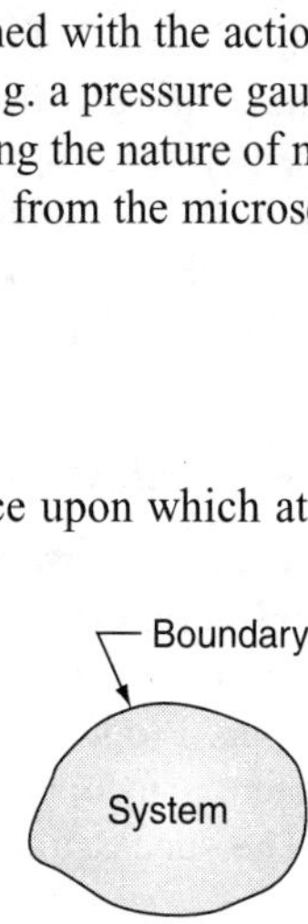

Fig. 1.1 *A thermodynamic system*

There are three classes of systems: (a) closed system, (b) open system and (c) isolated system. The *closed system* (Fig. 1.2) is a system of fixed mass. There is no mass transfer across the system boundary. There may be energy transfer into or out of the system.

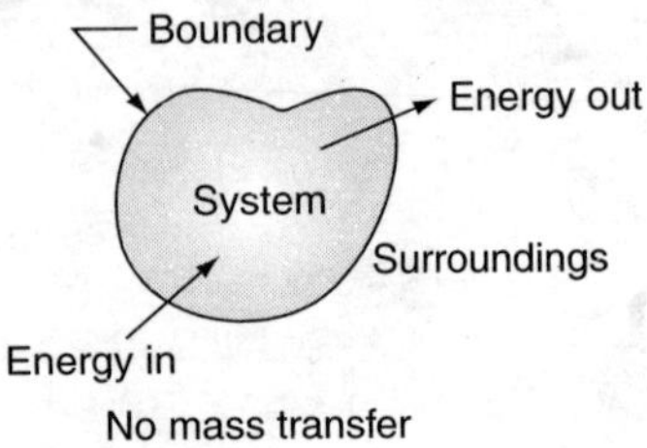

Fig. 1.2 *A closed system*

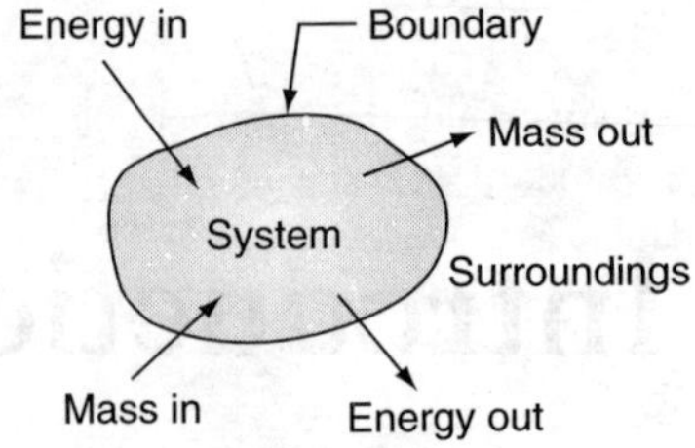

Fig. 1.3 *An open system*

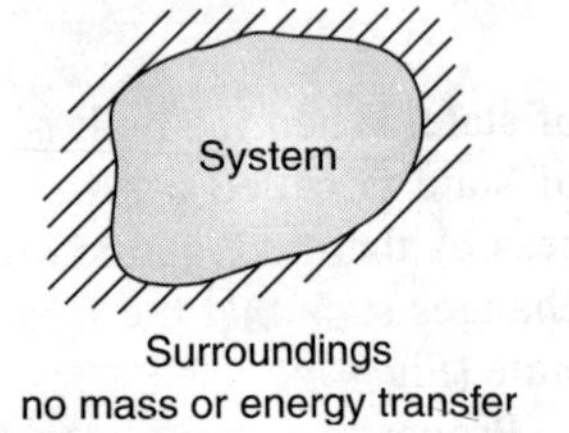

Fig. 1.4 *An isolated system*

A certain quantity of fluid in a cylinder bounded by a piston constitutes a closed system. The *open system* (Fig. 1.3) is one in which matter crosses the boundary of the system. There may be energy transfer also. Most of the engineering devices are generally open systems, e.g., an air compressor in which air enters at low pressure and leaves at high pressure and there are energy transfers across the system boundary. The *isolated system* (Fig. 1.4) is one in which there is no interaction between the system and the surroundings. It is of fixed mass and energy, and there is no mass or energy transfer across the system boundary.

If a system is defined as a certain quantity of matter, then the system contains the same matter and there can be no transfer of mass across its boundary. However, if a system is defined as a region of space within a prescribed boundary, then matter can cross the system boundary. While the former is called a closed system, the latter is an open system.

For thermodynamic analysis of an open system, such as an air compressor (Fig. 1.5), attention is focussed on a certain volume in space surrounding the compressor, known as the *control volume*, bounded by a surface called the *control surface*. Matter as well as energy crosses the control surface.

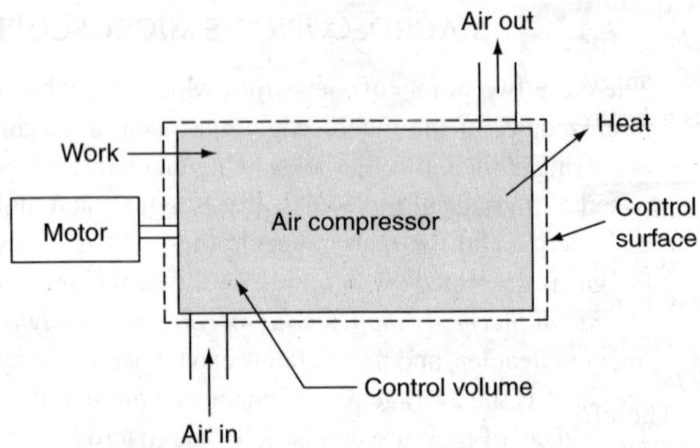

Fig. 1.5 *Control volume and control surface*

A closed system is a system closed to matter flow, though its volume can change against a flexible boundary. When there is matter flow, then the system is considered to be a volume of fixed identity, the control volume. There is thus no difference between an open system and a control volume.

1.3 THERMODYNAMIC PROPERTIES, PROCESSES AND CYCLES

Every system has certain characteristics by which its physical condition may be described, e.g., volume, temperature, pressure, etc. Such characteristics are called *properties* of the system. These are all *macroscopic* in nature. When all the properties of a system have definite values, the system is said to exist at a definite *state*. Properties are the coordinates to describe the state of a system. They are the state variables of the system. Any operation in which one or more of the properties of a system changes is called a *change of state*. The succession of states passed through during a change of state is called the *path* of the change of state. When the path is completely specified, the change of state is called a *process*, e.g., a constant pressure process. A thermodynamic *cycle* is defined as a series of state changes such that the final state is identical with the initial state (Fig. 1.6).

Properties may be of two types. *Intensive properties* are independent of the mass in the system, e.g. pressure, temperature, etc. *Extensive properties* are related to mass, e.g., volume, energy, etc. If mass is increased, the values of the extensive properties also increase. Specific extensive properties, i.e., extensive properties per unit mass, are intensive properties, e.g., specific volume, specific energy, density, etc.

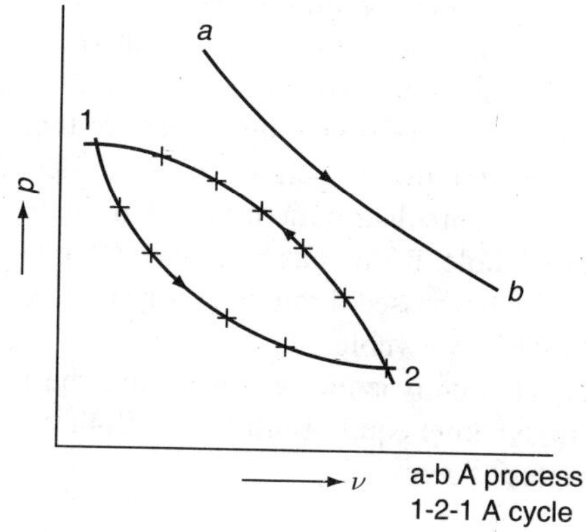

Fig. 1.6 *A process and a cycle*

1.4 HOMOGENEOUS AND HETEROGENEOUS SYSTEMS

A quantity of matter homogeneous throughout in chemical composition and physical structure is called a *phase*. Every substance can exist in any one of the three phases, viz., solid, liquid and gas. A system consisting of a single phase is called a *homogeneous system*, while a system consisting of more than one phase is known as a *heterogeneous system*.

1.5 THERMODYNAMIC EQUILIBRIUM

A system is said to exist in a state of *thermodynamic equilibrium* when no change in any macroscopic property is registered, if the system is isolated from its surroundings.

An isolated system always reaches in course of time a state of thermodynamic equilibrium and *can never depart from it spontaneously*.

Therefore, there can be no *spontaneous change in any macroscopic property* if the system exists in an equilibrium state. Thermodynamics studies mainly the properties of physical systems that are found in equilibrium states.

A system will be in a state of thermodynamic equilibrium, if the conditions for the following three types of equilibrium are satisfied:

(a) Mechanical equilibrium (b) Chemical equilibrium (c) Thermal equilibrium

In the absence of any unbalanced force within the system itself and also between the system and the surroundings, the system is said to be in a state of *mechanical equilibrium*. If an unbalanced force exists, either the system alone or both the system and the surroundings will undergo a change of state till mechanical equilibrium is attained.

If there is no chemical reaction or transfer of matter from one part of the system to another, such as diffusion or solution, the system is said to exist in a state of *chemical equilibrium*.

When a system existing in mechanical and chemical equilibrium is separated from its surroundings by a diathermic wall (diathermic means 'which allows heat to flow') and if there is no spontaneous change in any property of the system, the system is said to exist in a state of *thermal equilibrium*. When this is not satisfied, the system will undergo a change of state till thermal equilibrium is restored.

When the conditions for any one of the three types of equilibrium are not satisfied, a system is said to be in a *nonequilibrium state*. If the nonequilibrium of the state is due to an unbalanced force in the interior of a system or between the system and the surroundings, the pressure varies from one part of the system to another. There is no single pressure that refers to the system as a whole. Similarly, if the nonequilibrium is because of the temperature of the system being different from that of its surroundings, there is a nonuniform temperature distribution set up within the system and there is no single temperature that stands for the system as a whole. It can thus be inferred that when the conditions for thermodynamic equilibrium are not satisfied, the states passed through by a system cannot be described by thermodynamic properties which represent the system as a whole.

Thermodynamic properties are the macroscopic coordinates defined for, and significant to, only thermodynamic equilibrium states. Both classical and statistical thermodynamics study mainly the equilibrium states of a system.

1.6 QUASI-STATIC PROCESS

Let us consider a system of gas contained in a cylinder (Fig. 1.7). The system initially is in an equilibrium state, represented by the properties p_1, v_1, t_1. The weight on the piston just balances the upward force exerted by the gas. If the weight is removed, there will be an unbalanced force between the system and the surroundings, and under gas pressure, the piston will move up till it hits the stops. The system again comes to an equilibrium state, being described by the properties p_2, v_2, t_2. But the intermediate states passed through by the system are nonequilibrium states which cannot be described by thermodynamic coordinates. Figure 1.8 shows points 1 and 2 as the initial and final equilibrium states joined by a dotted line, which has got no meaning otherwise. Now if the single weight on the piston is made up of many very small pieces of weights (Fig. 1.9), and these weights are removed one by one very slowly from the top of the piston, at any instant of the upward travel of the piston, if the gas system is isolated, *the departure of the state of the system from the thermodynamic equilibrium state will be infinitesimally small*. So every state passed through by the system will be an equilibrium state. Such a process, which is but a locus of all the equilibrium points passed through by the system, is known as a *quasi-static process* (Fig. 1.10), 'quasi' meaning 'almost'. *Infinite slowness is the characteristic feature of a quasi-static process*. A quasi-static process is thus a succession of equilibrium states. A quasi-static process is also called a *reversible process*.

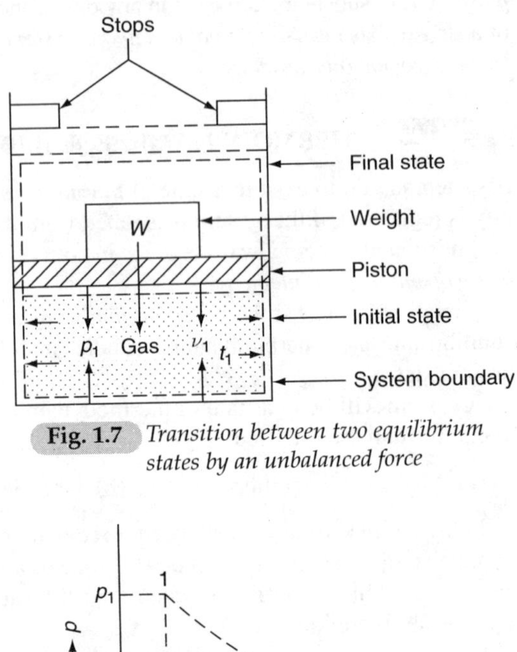

Fig. 1.7 *Transition between two equilibrium states by an unbalanced force*

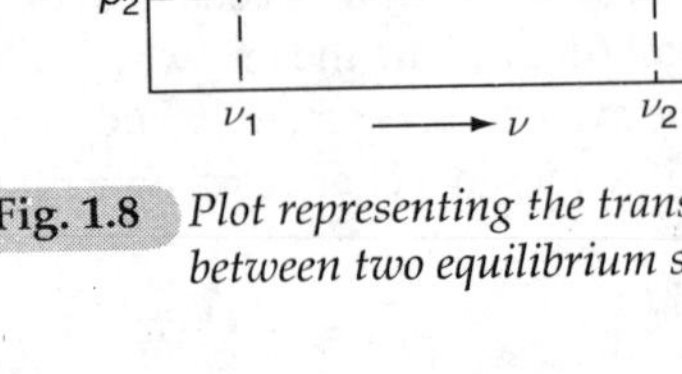

Fig. 1.8 *Plot representing the transition between two equilibrium states*

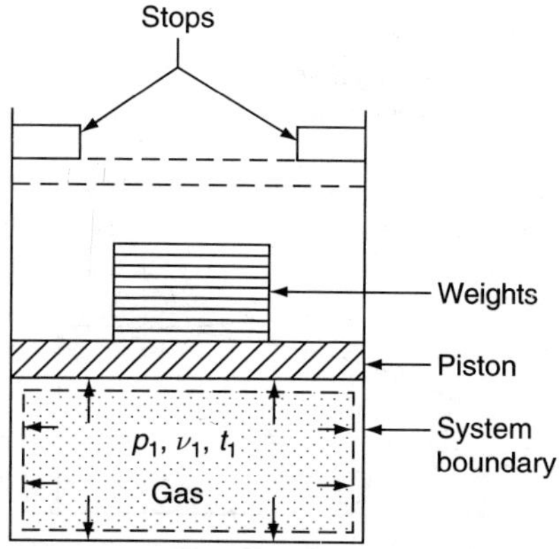

Fig. 1.9 *Infinitely slow transition of a system by infinitesimal force*

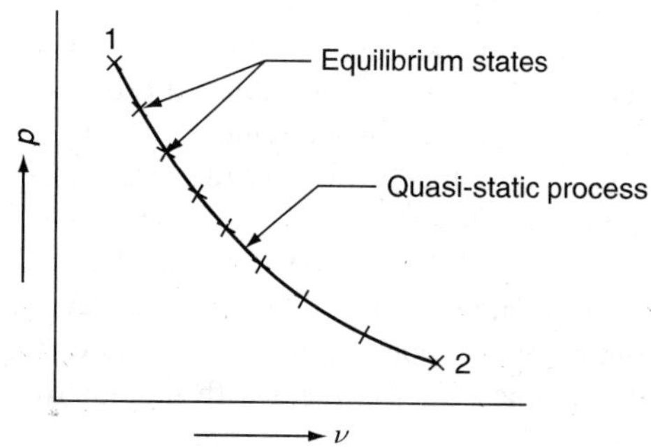

Fig. 1.10 *A quasi-static process*

1.7 PURE SUBSTANCE

A pure substance is defined as one that is homogeneous and invariable in chemical composition throughout its mass. The relative proportions of the chemical elements constituting the substance are also constant. Atmospheric air, steam-water mixture and combustion products of a fuel are regarded as pure substances. But the mixture of air and liquid air is not a pure substance, since the relative proportions of oxygen and nitrogen differ in the gas and liquid phases in equilibrium.

The state of a pure substance of given mass can be fixed by specifying two properties, provided the system is in equilibrium. This is known as the 'two-property rule'. The state can thus be represented as a point on thermodynamic property diagrams. Once any two properties of a pure substance are known, other properties can be determined from the available thermodynamic relations.

1.8 CONCEPT OF CONTINUUM

From the macroscopic viewpoint, we are always concerned with volumes which are very large compared to molecular dimensions. Even a very small volume of a system is assumed to contain a large number of molecules so that statistical averaging is meaningful and a property value can be assigned to it. Disregarding the behaviour of individual molecules, matter is here treated as continuous. Let us consider the mass δm in a volume δV surrounding the point P (Fig. 1.11). The ratio $\delta m/\delta V$ is the average mass density of the system within the volume δV. We suppose that at first δV is rather large, and is subsequently shrunk about the point P. If we plot $\delta m/\delta V$ against δV, the average density tends to approach an asymptote as δV increases (Fig. 1.12). However, when δV becomes so small as to contain relatively few molecules, the average density fluctuates substantially with time as molecules pass into and out of the volume in random motion, and so it is impossible to speak of a definite value of $\delta m/\delta V$. The smallest volume which may be regarded as continuous is $\delta V'$. The density ρ of the system at a point is thus defined as

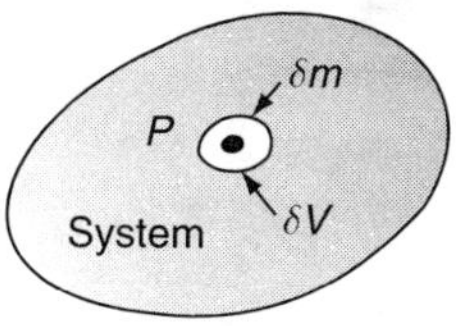

Fig. 1.11

$$\rho = \lim_{\delta V \to \delta V'} \frac{\delta m}{\delta V} \tag{1.1}$$

Similarly, the fluid velocity at a point P is defined as the instantaneous velocity of the centre of gravity of the smallest continuous volume $\delta V'$.

The concept of continuum loses validity when the mean free path of the molecules approaches the order of magnitude of the dimensions of the vessel, as, for instance, in highly rarefied gases encountered in high vacuum technology, in rocket flights at high altitudes and in electron tubes. In most engineering applications, however, the assumption of a continuum is valid and convenient, and goes hand in hand with the macroscopic point of view.

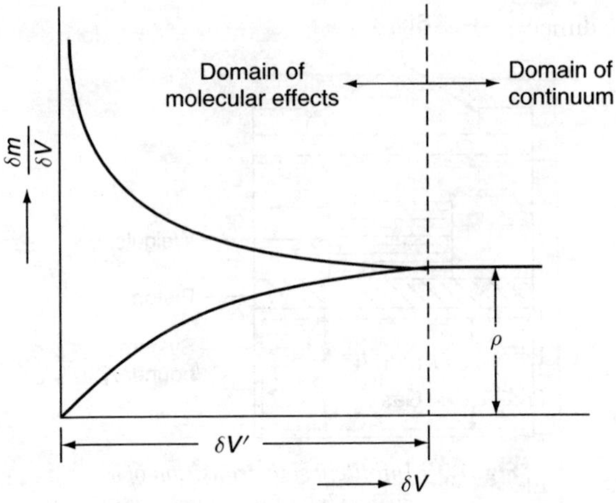

Fig.1.12 *Definition of the macroscopic property, density*

1.9 THERMOSTATICS

The science of thermodynamics deals with systems existing in thermodynamic equilibrium states which are specified by properties. Infinitely slow quasi-static processes executed by systems are only meaningful in thermodynamic plots. The name 'thermodynamics' is thus said to be a misnomer, since it does not deal with the dynamics of heat, which is nonquasi-static. The name 'thermostatics' then seems to be more appropriate. However, most of the real processes are dynamic and nonquasi-static, although the initial and final states of the system might be in equilibrium. Such processes can be successfully dealt with by the subject. Hence, the term 'thermodynamics' is not inappropriate.

1.10 UNITS AND DIMENSIONS

In the present text, the SI (System International) system of units has been used. The basic units in this system are given in Table 1.1.

Table 1.1
System: Basic Units

Quantity	*Unit*	*Symbol*
Length (L)	Metre	m
Mass (M)	Kilogramme	kg
Time (t)	Second	s
Amount of substance	Mole	mol
Temperature (T)	Kelvin	K
Electric current	Ampere	A
Luminous intensity	Candela	cd
Plane angle	Radian	rad
Solid angle	Steradian	sr

The dimensions of all other quantities are derived from these basic units which are given in Table 1.2.

Table 1.2
SI System: Derived Units

Quantity	*Unit*	*Symbol*	*Alternative unit*	*In basic units*
Force (F)	Newton	N		kg m/s^2
Energy (E)	Joule	J	Nm	kg m^2/s^2
Power	Watt	W	J/s	kg m^2/s^3
Pressure	Pascal	Pa	N/m^2	$kg/(ms^2)$
Frequency	Hertz	Hz		s^{-1}
Electric charge	Coulomb	C		As
Electric potential	Volt	V	W/A = J/C	kg $m^2/(s^3$ A)
Capacitance	Farad	F	C/V	s^4 A^2/(kg m^2)
Electrical resistance	Ohm	Ω	V/A	kg $m^2/(s^3$ $A^2)$
Magnetic flux	Weber	Wb	Vs	kg $m^2/(s^2$ A)
Magnetic flux density	Tesla	T	Wb/m^2	$kg/(s^2$ A)
Inductance	Henry	H	Wb/A	kg $m^2/(s^2$ $A^2)$

It is often convenient and desirable to use multiples of various units, the standard list of which is given in Table 1.3.

Table 1.3
SI System: Standard Multipliers

Factor	*Prefix*	*Factor*	*Prefix*
10^{12}	tera, T	10^{-3}	milli, m
10^{9}	giga, G	10^{-6}	micro, μ
10^{6}	mega, M	10^{-9}	nano, n
10^{3}	kilo, k	10^{-12}	pico, p

1.10.1 Force

The force acting on a body is defined by Newton's second law of motion. The unit of force is the newton (N). A force of one newton produces an acceleration of 1 ms^{-2} when applied to a mass of 1 kg.

$$1 \text{ N} = 1 \text{ kg m/s}^2$$

The weight of a body (W) is the force with which the body is attracted to the centre of the earth. It is the product of its mass (m) and the local gravitational acceleration (g), i.e. $W = mg$

The value of g at sea level is 9.80665 m/s^2. The mass of a substance remains constant with elevation, but its weight varies with elevation.

1.10.2 Pressure

Pressure is the normal force exerted by a system against unit area of the bounding surface. If δA is a small area and $\delta A'$ is the smallest area from continuum consideration, and δF_n is the component of force normal to δA (Fig. 1.13), the pressure p at a point on the wall is defined as

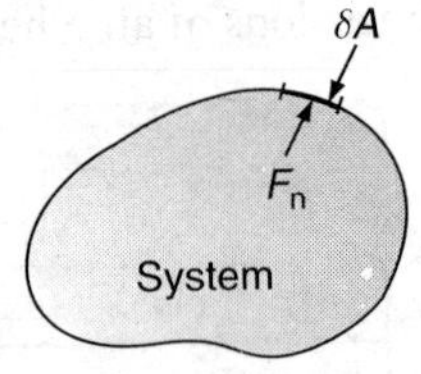

Fig. 1.13 *Definition of pressure*

$$p = \lim_{\delta A \to \delta A'} \frac{\delta F_n}{\delta A}$$

The pressure p at a point in a fluid in equilibrium is the same in all directions.

The unit for pressure in the SI system is the *pascal* (Pa), which is the force of one newton acting on an area of 1 m^2.

$$1 \text{ Pa} = 1 \text{ N/m}^2$$

The unit of pascal is very small. Very often kilo-pascal (kPa) or mega-pascal (MPa) is used.

Two other units, not within the SI system of units, continue to be widely used. These are the *bar*, where

$$1 \text{ bar} = 10^5 \text{ Pa} = 100 \text{ kPa} = 0.1 \text{ MPa}$$

and the standard atmosphere, where 1 atm = 101.325 kPa = 1.01325 bar

Most instruments indicate pressure relative to the atmospheric pressure, whereas the pressure of a system is its pressure above zero, or relative to a perfect vacuum. The pressure relative to the atmosphere is called *gauge pressure*. The pressure relative to a perfect vacuum is called *absolute pressure*.

$$\text{Absolute pressure} = \text{Gauge pressure} + \text{Atmospheric pressure}$$

When the pressure in a system is less than atmospheric pressure, the gauge pressure becomes negative, but is frequently designated by a positive number and called *vacuum*. For example, 16 cm vacuum will be

$$\frac{76 - 16}{76} \times 1.013 = 0.80 \text{ bar}$$

Figure 1.14 shows a few pressure measuring devices. Figure (a) shows the Bourdon gauge which measures the difference between the system pressure inside the tube and atmospheric pressure. It relies on the deformation of a bent hollow tube of suitable material which, when subjected to the pressure to be measured on the inside (and atmospheric pressure on the outside), tends to unbend. This moves a pointer through a suitable gear-and-lever mechanism against a calibrated scale. Figure (b) shows an open U-tube indicating gauge pressure, and Fig. (c) shows an open U-tube indicating vacuum. Figure (d) shows a closed U-tube indicating absolute pressure. If p is atmospheric pressure, this is a *barometer*. These are called U-tube manometers.

If Z is the difference in the heights of the fluid columns in the two limbs of the U-tube [Fig. (b) and Fig. (c)], ρ the density of the fluid and g the acceleration due to gravity, then from the elementary principle of hydrostatics, the gauge pressure p_g is given by

$$p_g = Z\rho g \left[m \cdot \frac{\text{kg}}{\text{m}^3} \frac{\text{m}}{\text{s}^2} \right] = Z\rho g \text{ N/m}^2$$

If the fluid is mercury having ρ = 13,616 kg/m^3, one metre head of mercury column is equivalent to a pressure of 1.3366 bar, as shown below

$$1 \text{ m Hg} = Z\rho g = 1 \times 13616 \times 9.81 = 1.3366 \times 10^5 \text{ N/m}^2 = 1.3366 \text{ bar}$$

The manometer is a sensitive, accurate and simple device, but it is limited to fairly small pressure differentials and, because of the inertia and friction of the liquid, is not suitable for fluctuating pressures, unless the rate of pressure change is small. A diaphragm-type pressure transducer along with a cathode ray oscilloscope can be used to measure rapidly fluctuating pressures.

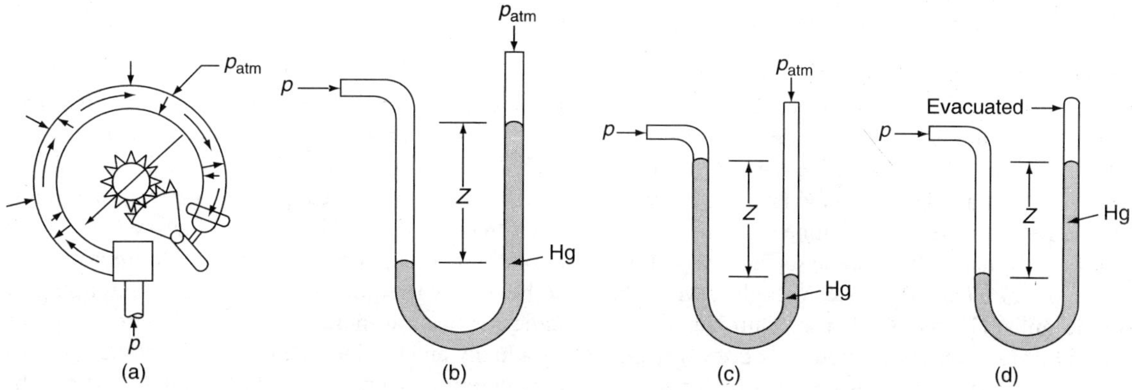

Fig. 1.14 *Pressure gauges (a) bourdon gauge (b) open U-tube indicating gauge pressure (c) open U-tube indicating vacuum (d) closed U-tube indicating absolute pressure*

1.10.3 Specific Volume and Density

Volume (V) is the space occupied by a substance and is measured in m^3. The specific volume (v) of a substance is defined as the volume per unit mass and is measured in m^3/kg. From continuum consideration the specific volume at a point is defined as $v = \lim_{\delta V \to \delta V'} \frac{\delta V}{\delta m}$

where $\delta V'$ is the smallest volume for which the system can be considered a continuum.

Density (ρ) is the mass per unit volume of a substance, which has been discussed earlier, and is given in kg/m^3.

$$\rho = \frac{m}{v}$$

In addition to m^3, another commonly used unit of volume is the litre (l).

$$1\,\text{l} = 10^{-3}\ \text{m}^3$$

The specific volume or density may be given either on the basis of mass or in respect of mole. A *mole* of a substance has a mass numerically equally to the molecular weight of the substance. One g mol of oxygen has a mass of 32 g and 1 kg mol (or kmol) of nitrogen has a mass of 28 kg. The symbol $\bar{v}$ is used for molar specific volume (m^3/kmol).

1.10.4 Energy

Energy is the capacity to exert a force through a distance, and manifests itself in various forms. Engineering processes involve the conversion of energy from one form to another, the transfer of energy from place to place, and the storage of energy in various forms, utilizing a working substance.

The unit of energy in the SI system is Nm or J (joule). The energy per unit mass is the specific energy, the unit of which is J/kg.

1.10.5 Power

The rate of energy transfer or storage is called power. The unit of power is watt (W), kilowatt (kW) or megawatt (MW).

$$1\ \text{W} = 1\ \text{J/s} = 1\ \text{Nm/s}$$

$$1\ \text{kW} = 1000\ \text{W}$$

1.11 HISTORY OF THERMODYNAMICS

The latter half of the eighteenth century ushered man into the modern world of machinery and manufacture, and brought about cataclysmic changes in the social, economic and political life of the people. The historians have called it the *Industrial Revolution.* It began in England due to a fortuitous combination of many factors. There was bustling creative activity in science and technology during this period in England, with the appearance of a galaxy of some brilliant individuals. The invention of the steam engine gave an impetus to this activity, and for the first time made man free from the forces of nature. The name of Savery, Newcomen and notably James Watt are associated with this invention. Watt brought about considerable improvement in the performance of the steam engine, which began to be widely used in coal mines, iron metallurgy and textile mills. George Stephenson introduced steam engine for rail transport, and Robert Fulton used it in steam boats. A variety of industries grew up, and man gradually entered into the modern machine age. The advent of steam engine also gave stimulus to the birth of thermodynamics. Thermodynamics is said to be the "daughter" of the steam engine. [D. Bradley, "Thermodynamics – the Daughter of Steam", Engineering Heritage, Vol. 2, I.Mech.E., London, 1966.]

There was once a young inventor who thought that he could produce energy out of nothing. "It is well known", said he, "that an electrical motor converts electrical energy into mechanical energy and that an electric generator converts mechanical energy into electrical energy. Why not then, use the motor to run the generator and the generator to run the motor, and create thereby an endless supply of energy"? But this is never to happen. A hypothetical device which creates energy from nothing is called a perpetual motion machine of the first kind, a PPMI. Like the proverbial touchstone which changes all metals into gold, man attempted to find such a PMMI for long long years, but it turned out to be a wild goose chase. In fact, the development of the principle of conservation of energy has been one of the most significant achievements in the evolution of physical science. The first recognition of this principle was made by Leibnitz in 1693, when he referred to the sum of kinetic energy and potential energy in a gravitational force field. Energy is neither created nor destroyed. Energy manifests in various forms and gets transformed from one form to another. Through gentle metabolic processes, a day labourer gradually transforms the chemical energy of the food he eats and the oxygen he breathes into heat, sound and useful work. Work was always considered a form of energy. The concept of heat was, however a very actively debated scientific topic. Until the middle of the nineteenth century, heat was regarded as an invisible colourless, weightless, odourless fluid that flowed from a body of higher calorie to a body of lower calorie. This was known as the caloric theory of heat, first proposed in 1789 by Antoine Lavoisier (1743–1794), the father of modern chemistry. When an object became full of caloric, it was then said to be saturated with it. This was the origin of the terms "saturated liquid", "saturated vapour" etc, that we use in thermodynamics today. The caloric was said to be conserved and it was indestructible. The caloric theory was, however, refuted and heat was confirmed as a form of energy in the middle of the nineteenth century leading to the formulation of the first law of thermodynamies. The names which stand out in the establishment of the first law were Benjamin Thompson (1753–1814), James Prescott Joule (1818–1889) and Julius Robert Mayer (1814–1878).

Benjamin Thompson, and American born in Massachusetts, did not support the revolt against the British during the American war of independence, and in 1775 he left for England where he took up government service. On a trip to Germany, he met the prince of Bavaria who offered him a job. He introduced many reforms in the government for which the title of Count von Rumford was conferred on him. While boring brass cannon hole, Count Rumford noticed that there was a continuous heat realease. How could the caloric fluid be conserved, when it was being produced continuously by mechanical friction? From the established principle of conservation of mass, a true fluid can be neither created nor destroyed, so heat could not be a fluid if it could be continuously created in an object by mechanical friction. Rumford conceived that heat was "a kind of motion" and the hotness of an object was due to the vibrating motion of the particles in the objects. On his return to England, he became a member of the Royal Society, and later founded the Royal Institution for the Advancement of Science. Rumford married the widow of Lavoisier and lived in Paris for the rest of his eventful life.

In the early forties of the nineteenth century, James P. Joule and Julius R. Mayer almost simultaneously set forth the idea that heat transfer and mechanical work were simply different forms of the same quantity, which we now recognize as energy in transit. In some modern treatments of engineering thermodynamies, Joule's name alone is attached to the establishment of the equivalence of "heat" and "work". The published record, however, shows that the idea of convertibility of heat into work was published independently by Mayer in May, 1842 and Joule in August, 1843. For an important aspect in the history of the first law, is the fact that both Mayer and Joule had difficulty in getting their papers published and in being taken seriously by their established contemporaries.

Robert Mayer was a doctor in a ship in the East Indies and from physiological observations, he believed in a principle of conservation of energy. He derived theoretically, the mechanical heat equivalent based on the calorimetric data of Joseph Black of Glasgow University. Mayer tried to publish his paper but remained unsuccessful for a long time. His despair was so great that he attempted suicide by jumping from a window, but he only broke his two legs. He was placed in an asylum for some time. In later years, however, he was given some measure or recognition and honoured equally with Joule in establishing the mechanical theory of heat.

Mayer argued that an amount of gas needs to be heated more at constant pressure than at constant volume, because at constant pressure it is free to dilate and do work against the atmosphere, which in today's notations becomes

$$mc_p\Delta T - mc_v\Delta T = P_{atm}\Delta V \tag{1.2}$$

Using the c_p and c_p/c_v constants that were known in his time, he estimated the left-hand side of the equation in calories, while the right-hand side was known in mechanical units. He thus established numerically the equivalence between these units. If the relation

$$Pv = RT \tag{1.3}$$

is used in Eq. (1.2), Mayer's argument reduces to

$$c_p - c_v = R \tag{1.4}$$

This classic relationship between the specific heats of an ideal gas is called Mayer's equation: while the ideal gas equation of state, Eq. (1.3), was first derived by Clapeyron [Bejan, 1988]*.

Joule was the ultimate experimentalist. His experiments seem to be the direct continuation of those of Rumford and the gap of some forty years between the two investigations appeared puzzling to some authors. Joule's first discovery from his measurements was that the flow of current in a resistance, is accompanied by the development of heat proportional to the resistance. He concluded that caloric was indeed created by the flow of current. He was firmly convinced that there existed some conservation law of a general nature and hence set out to investigate whether the conversion of the various forms of energy is governed by definite conversion factors. He considered the conversion of chemical, electric, caloric, and mechanical energy forms in all combinations. The determination of the mechanical equivalent of heat forms the central part of his experiments, the results of which can be summed up in the general relation:

$$W = JQ \tag{1.5}$$

where J is the mechanical equivalent of heat. Joule's experiments suggested that this relation may have universal validity with the same numerical value of J under all conditions.

Joule communicated the results of his experiments to the British Association for the Advancement of Science in 1843. It was received with entire incredulity and general silence. In 1844 a paper by Joule on the same subject was rejected by the Royal Society. To convince the skeptics, he produced a series of nakedly simple experiments whose message proved impossible to refute. From the point of view of mechanical engineers, the most memorable among these experiments was the heating of a pool of water by an array of paddle wheels driven by falling weights. He discussed in 1847, before the British Association at Oxford, his

---

* Adrian Bejan, "Research into the Origins of Engineering Thermodynamics", *Int. Comm. Heat Mass Transfer*, Vol. 15, No. 5, 1988, pp 571–580.

experimental results in which he suggested that the water at the bottom of the Niagara waterfall (160 feet high) should be warmer than at the top (by 0.2°F). From the thermal expansion of gases Joule deduced that there should be a "zero of temperature", 480°F below the freezing point of ice. This was the first suggestion of absolute zero. Although these results failed to provoke further discussion, it created interest in a young man who only two year ago had passed from the University of Cambridge with the highest honour. The young man was William Thomson, who later became Lord Kelvin. He somewhere stated that it was one of the most valuable recollections of his life. Michael Faraday was also present in the 1844 Oxford meeting, and he communicated Joule's paper "On the Mechanical Equivalent of Heat" to the Royal Society in 1849. The paper ultimately appeared in its Philosphical Transactions in 1850.

Even while Joule was perfecting the experimental basis of the energy law now called the Mayer-Joule principle, Herman Ludwig von Helmohltz (1821–1894) published in 1847, his famous essay on the conservation of force. In this work, he advanced the conservation of energy as a unifying principle extending over all branches of physics. Helmholtz, like Mayer, was a physician by profession and self-taught in Physics and Mathematics. He also faced great difficulties in getting his paper published in professional journals.

In the history of classical thermodynamics, one thinks of only the closed system formulations of the first law which were deliberated by the pioneers as stated above. In engineering thermodynamics, however, open system formulations are of prime interest. The first law for open systems was first stated by Gustave Zeuner, as part of the analysis of flow systems that operate in the steady state. Zeuner's formula for the heat transfer rate to a stream m in steady flow and without shaft work in present notations is given to be:

$$d\dot{Q} / \dot{m} = d(u + Pv + V^2 / 2 + gz) \tag{1.6}$$

The reference of this formula is found in Stodola's classic treatise on steam turbines, first published in the German language in 1903.

The first person to invent a theory simultaneously involving the ideas of conservation and conversion of energy was the young French military engineer Nicolas Leonard Sadi Carnot (1796–1832). The strikingly original ideas of Carnot's work make it among the most brilliant new departures in theoretical physics. Sadi Carnot was the son of Napoleon's general, Lazare Carnot, and was educated at the famous Ecole Polytechnique in Paris. Between 1794 and 1830, Ecole Polytechnique had such famous teachers as Lagrange, Fourier, Laplace, Ampere, Cauchy, Coriolis, Poisson. Gay-Lussac, and Poiseuille. After his formal education Carnot chose a career as an army officer. Britain was then a powerful military force, primarily as a result of the industrial revolution brought about by the steam engine. French technology was not developing as fast as Britain's. Carnot was convinced that France's inadequate utilization of steam power had made it militarily inferior. He began to study the fundamentals of steam engine technology, and in 1824 he published the results of his study in the form of a brochure "Reflection on the Motive Power of Heat and on Machines Fitted to Develop that Power". Carnot was trained in the basic principles of hydraulics, pumps and water wheels at the Ecole Polytechnique. During Carnot's time, caloric theory of heat was still persisting, and the water wheel as the major source of mechanical power was gradually getting replaced by the steam engine. Carnot conceived that the power of a steam engine was released, as the heat fluid or caloric fell from the high temperature of the boiler to the lower temperature of the condenser, in much the same way that water falls through a water wheel to produce mechanical shaft work output. Carnot stated, "The motive power of a water wheel depends on its height and the quantity of liquid. The motive power of heat also depends on the quantity of caloric used and on the height of its fall, i.e., the difference of temperatures of the bodies between which the exchange of caloric is made".

Till Carnot's time thermodynamics was developed primarily on an empricial basis provided by chemistry. Carnot approached an engineering problem, the efficiency of heat engines, in terms of entirely new concepts with the steam engine serving as the stimulus. Carnot observed that the existence of temperature differences is a necessary condition for producing mechanical work by means of a heat engine. He simplified the problem to its bare essentials and stipulates, that this system, consisting essentially of a working substance, should exchange heat with its surroundings only at two fixed temperatures. In order to conceptualize such a situation, he introduces the idea of heat reservoirs. Two important conclusions emerged from Carnot's work.

1. No one could build a water wheel that would produce a continuous work output unless water actually entered and exited the wheel. If water with a certain kinetic and potential energy entered the wheel, then the same amount of water with a lower energy must also exit the wheel. It is thus impossible to make a water wheel that converts all the energy of the inlet water into shaft work output. There must be an ourflow of water form the wheel.

If this idea is extended to a steam engine by replacing the water by heat fluid caloric, it can be concluded that when caloric at a certain energy level (temperature) enters a work producing heat engine, it must also exit the heat engine at a low energy level (temperature). Thus a continuously operating heat engine that converts all of its caloric (heat) input directly into work output is not possible. This is very close to the Kelvin-Planck statement of second law as it is known today.

2. The maximum efficiency of a water wheel was independent of the type of the liquid and depended only on the inlet and outlet flow energies. The maximum efficiency of the steam engine (or any heat engine) depends only on the temperatures of the high and low temperature thermal reservoirs of the engine and is independent of the working fluid. To achieve the maximum efficiency there must not be any mechanical friction or other losses of any kind.

Only at the age of 36, Sadi Carnot died of cholera following an attack of scarlet fever. The significance of Carnot's work was not recognized until 1850, when Rudolf Clausius (1822–1888) and William Thomson (1824–1907) worked out a clear formulation of the conservation of energy principle. Carnot's first conclusion was then called the second law of thermodynamics by Clausius, and Thomson used Carnot's second conclusion to develop the concept of absolute temperature scale. Thermodynamics is thus said to have originated from the "clumsy puffing of the early steam engines" and is often called "the daughter of steam engine".

Carnot's ideas were so revolutionary that they were largely ignored. Soon after Carnot's death, Emile Clapeyron (1799–1864), a French mining engineer, strengthened Carnot's ideas by using more precise mathematical derivation. Clapeyron constructed its thermodynamic cycle by deducing that it must be composed of two reversible isothermal processes and two reversible adiabatic processes. It is now known as Carnot's cycle. It was the first heat engine cycle to be conceptualized. No other heat engine can equal its efficiency.

Clapeyron was later able to derive a relation for the enthalpy change of the liquid to vapour phase (h_{fg}) in terms of pressure, temperature and specific volume. This provided the first equation, now called the Clausius-Clapeyron equation, representing the first order phase transition, which could be used to estimate a property that is nor directly measurable in terms of properties that are directly measurable. Clapeyron's equation is now most easily derived from one of Maxwell's equations.

William Thomson (1824–1907), who became a professor of natural philosophy at the University of Glasgow in 1848 at the age of 24 only, rejected the caloric theory of heat and for the first time used the terms "thermodynamics" and "mechanical energy". Apart from the deduction of the absolute temperature scale, Thomson worked with Joule from 1852 to 1862 in a series of experiments to measure the temperature of gas in a controlled expansion and propounded the Joule-Thomson effect for real gases.

Rudolf Julius Emanuel Clausius (1822–1888) realized that there were two distinct laws at work, the first law due to Joule and Mayer and the second law as expounded by Carnot. He defined the internal energy *U*. Although both Kelvin and Clausius used the function Q_{rev}/T for some years, Clausius recognized the value of this function and to describe it he coined the word "entropy" from the Greek work "tropee" meaning "transformation" and assigned it the symbol S. Clausius in 1865, summarized the first and second laws of thermodynamics in the following words:

"Die Energie der Welt ist konstant.
Die Entropie der Welt strebt einem Maximum zu"
which is translated as
"The energy of the world is constant.

The entropy of the world tends toward a maximum"

The world here means the universe, the system and the surroundings together.

These statements made a strong impression upon a young student, Max Karl Ernst Ludwig Planck (1858–1947). He was educated at the universities of Munich and Berlin. In his autobiography he stated, "One day I happened to come across the treatises of Rudolf Clausius, whose lucid style and enlightening clarity of reasoning made an enormous impression on me, and I became deeply absorbed in his articles, with an ever increasing enthusiasm. I appreciated especially his exact formulation of the two laws of thermodynamics, and the sharp distinction, which he was the first to establish between them". In 1897, Planck[1] demonstrated the close connection between the second law and the concept of reversibility. He stated the second law as the impossibility of a cyclic device which produces positive work and exchanges heat with a single reservoir. Similar statement was also made by Kelvin, and is now recognized as Kelvin-Planck statement of second law. Poincare[2] in 1908, extended the work of Planck and prescribed a complete structure of classical thermodynamics.

The property, entropy, plays a stellar role in thermodynamics. It was introduced via the concept of heat engines. In 1909, the Greek mathematician Caratheodory proved the existence of entropy function mathematically without the aid of Carnot engines and refrigerators. Caratheodory's statement of second law may be stated as: "In the neighbourhood of any arbitrary state P_0 of a physical system, there exist neighbouring states which are not accessible from P_0 along quasi-statie adiabatic paths". From the standpoint of the engineer and physicist it is entirely mathematical in form and devoid of physical insight.

William John Macquorn Rankine (1820–1872) defined the thermodynamic efficiency of a heat engine and showed the usefulness of p–v diagrams as related to work. He wrote the first text book on thermodynamics[3], and was the first to work out the thermodynamic cycle for the adiabatic cylinder steam engine, now known as Rankine cycle for a vapour power cycle.

In 1862, the cycle used in modern gasoline-powered I.C engines was proposed in a patent issued to Alphonse Beau de Rochas (1815–1893). The first practical engine was, however, built by Nikolous August Otto (1832–1891) which was demonstrated at the Paris Exposition in 1878. Otto fought many legal battles with Beau de Rochas for production of these engines, but finally lost to him.

Captain John Ericsson (1803–1889) was a Swedish engineer who marketed small solar-powered and coal-fired hot air engines. Rev. Robert Stirling (1790–1879), an English parish minister, patented a practical heat engine in 1816 that used air as the working fluid. In theory, the cycle used the Stirling engine approaches the ideal cycle later proposed by Carnot (1824).

George Bailey Brayton (1830–1892), an American engineer, marketed an I.C. engine with a separate combustion chamber, where combustion of fuel occurred at about constant pressure. This cycle later formed the basis for modern gas turbine plants.

Gottlieb Daimler (1834–1900) obtained a patent in 1879 for a multicylinder automotive engine, which was commercially successful. Dr. Rudolf Christian Karl Diesel (1858–1913) studied at Technische Hochschule in Munich. He designed large steam engines and boilers. He later developed in 1897 an I.C. engine with fuel injection which resembled the modern diesel engine. Failing health, continuing criticism and serious financial setbacks beset Diesel who in 1913 disappeared from a boat crossing the English channel in a moonlit night. Josiah Willard Gibbs (1839–1903) is often regarded as the most brilliant thermodynamicist produced in the USA. He received the first doctorate degree in engineering in the USA (Yale University). He contributed significantly in many areas of thermodynamics like heterogeneous systems, phase rule, physical chemistry and statistical thermodynamics. Some of his very important papers were published in obseure journals like Connecticut Academy of Sciences, and remained unknown to most scientists. Only after his death, these were discovered.

[1] M. Planck, *Treatise on Thermodynamics* (1897), translated by A. Ogg, Longman and Green; London, 1927.

[2] H. Poincare, *Thermodynamique,* Gauthier-Villars, Paris, 1908.

[3] W.J.M. Rankine, "*Manual of the Steam Engine and other Prime Movers*", 1859 going through 17 editions, as mentioned by Robert Balmer in "*Thermodynamics*", West Publishing Co., 1990, page 399.

Solved Examples

Example 1.1

The pressure of gas in a pipe line is measured with a mercury manometer having one limb open to the atmosphere (Fig. Ex. 1.1). If the difference in the height of mercury in the two limbs is 562 mm, calculate the gas pressure. The barometer reads 761 mm Hg, the acceleration due to gravity is 9.79 m/s², and the density of mercury is 13,640 kg/m³.

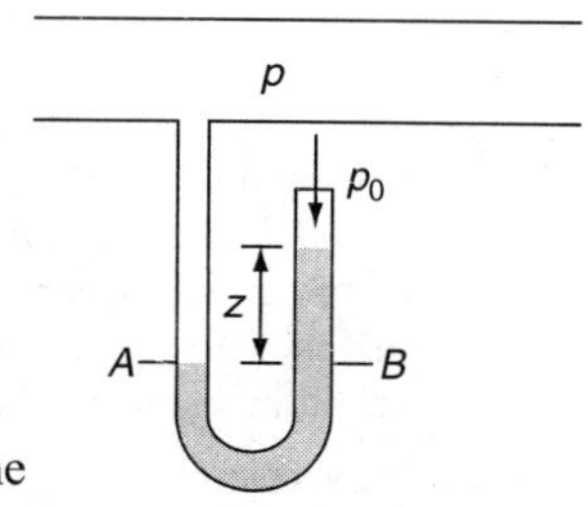

Fig. Ex. 1.1

Solution At the plane *AB*, we have $p = p_0 + \rho g z$

Now $p_0 = \rho g z_0$

where z_0 is the barometric height, ρ the density of mercury and p_0 the atmospheric pressure.

Therefore $p = \rho g\,(z + z_0)$

$= 13{,}640 \text{ kg/m}^3 \times 9.79 \text{ m/s}^2\,(0.562 + 0.761) \text{ m}$

$= 177 \times 10^3 \text{ N/m}^2 = 177 \text{ kPa}$

$= 1.77 \text{ bar} = 1.746 \text{ atm}$ *Ans.*

Example 1.2

A turbine is supplied with steam at a gauge pressure of 1.4 MPa. After expansion in the turbine the steam flows into a condenser which is maintained at a vacuum of 710 mm Hg. The barometric pressure is 772 mm Hg. Express the inlet and exhaust steam pressure in pascals (absolute). Take the density of mercury as 13.6 × 10³ kg/m³.

Solution *The atmospheric pressure* $p_0 = \rho g z_0 = 13.6 \times 10^3 \text{ kg/m}^3 \times 9.81 \text{ m/s}^2 \times 0.772 \text{ m}$

$= 1.03 \times 10^5 \text{ Pa}$

Inlet steam pressure $= [(1.4 \times 10^6) + (1.03 \times 10^5)] \text{ Pa}$

$= 15.03 \times 10^5 \text{ Pa} = 1.503 \text{ MPa}$ *Ans.*

Condenser pressure $= (0.772 - 0.710) \text{ m} \times 9.81 \text{ m/s}^2 \times 13.6 \times 10^3 \text{ kg/m}^3$

$= 0.827 \times 10^4 \text{ Pa} = 8.27 \text{ kPa}$ *Ans.*

Review Questions

1.1 What do you understand by macroscopic and microscopic viewpoints?

1.2 Is thermodynamics a misnomer for the subject?

1.3 How does the subject of thermodynamics differ from the concept of heat transfer?

1.4 What is the scope of classical thermodynamics?

1.5 What is a thermodynamic system?

1.6 What is the difference between a closed system and an open system?

1.7 An open system defined for a fixed region and a control volume are synonymous. Explain.

1.8 Define an isolated sysetm.

1.9 Distinguish between the terms 'change of state', 'path', and 'process'.

1.10 What is a thermodynamic cycle?

1.11 What are intensive and extensive properties?

1.12 What do you mean by homogeneous and heterogeneous systems?

1.13 Explain what you understand by thermodynamic equilibrium.

1.14 Explain mechanical, chemical and thermal equilibrium.

1.15 What is a quasi-static process/What is its characteristic feature?

1.16 What is the concept of continuum? How will you define density and pressure using this concept?

1.17 What is vacuum? How can it be measured?

1.18 What is a pressure transducer?

Problems

1.1 A pump discharges liquid into a drum at the rate of 0.0032 m^3/s. The drum, 1.50 m in diameter and 4.20 m in length, can hold 3000 kg of the liquid. Find the density of the liquid and the mass flow rate of the liquid handled by the pump.

1.2 The acceleration of gravity is given as a function of elevation above sea level by

$$g = 980.6 - 3.086 \times 10^{-6} H$$

where g is in cm/s^2, and H is in cm. If an aeroplane weighs 90,000 N at sea level, what is the gravity force upon it at 10,000 m elevation? What is the percentage difference from the sea-level weight?

1.3 Prove that the weight of a body at an elevation H above sea-level is given by

$$W = \frac{mg}{g_0}\left(\frac{d}{d+2H}\right)^2$$

where d is the diameter of the earth.

1.4 The first artificial earth satellite is reported to have encircled the earth at a speed of 28,840 km/h and its maximum height above the earth's surface was stated to be 916 km. Taking the mean diameter of the earth to be 12,680 km, and assuming the orbit to be circular, evaluate the value of the gravitational acceleration at this height.

The mass of the satellite is reported to have been 86 kg at sea-level. Estimate the gravitational force acting on the satellite at the operational altitude.

Ans. 8.9 m/s^2; 765 N

1.5 Convert the following readings of pressure to kPa, assuming that the barometer reads 760 mm Hg:

(a) 90 cm Hg gauge, (b) 40 cm Hg vacuum, (c) 1.2 m H_2O gauge, (d) 3.1 bar.

1.6 A 30 m high vertical column of a fluid of density 1878 kg/m^3 exists in a place where $g = 9.65$ m/s^2. What is the pressure at the base of the column.

Ans. 544 kPa

1.7 Assume that the pressure p and the specific volume v of the atmosphere are related according to the equation $pv^{1.4} = 2.3 \times 10^3$, where p is in N/m^2 abs and v is in m^3/kg. The acceleration due to gravity is constant at 9.81 m/s^2. What is the depth of atmosphere necessary to produce a pressure of 1.0132 bar at the earth's surface? Consider the atmosphere as a fluid column. *Ans.* 64.8 km

1.8 The pressure of steam flowing in a pipe line is measured with a mercury manometer, shown in Fig. P. 1.8. Some steam condenses into water. Estimate the steam pressure in kPa. Take the density of mercury as 13.6×10^3 kg/m^3, density of water as 10^3 kg/m^3, the barometer reading as 76.1 cm Hg, and g as 9.806 m/s^2.

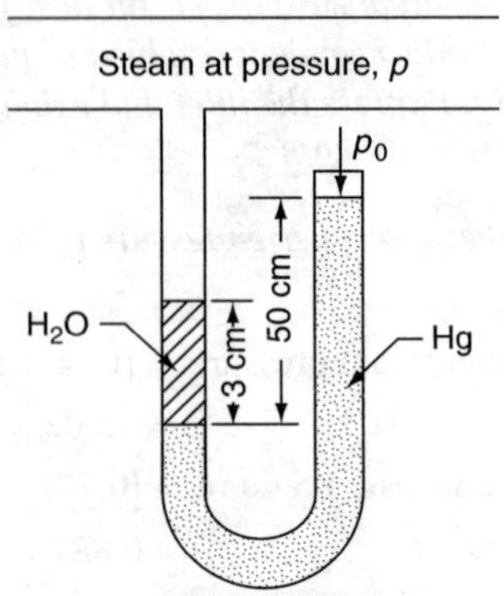

Fig. P.1.8

1.9 A vacuum gauge mounted on a condenser reads 0.66 m Hg. What is the absolute pressure in the condenser in kPa when the atmospheric pressure is 101.3 kPa? *Ans.* 8.8 kPa

1.10 The basis barometer can be used to measure the height of a building. If the barometric readings at the top and at the bottom of a building are 730 and 760 mm Hg, respectively, determine the height of the building. Assume an average air density of 1.18 kg/m^3.

C H A P T E R

2 Temperature

2.1 ZEROTH LAW OF THERMODYNAMICS

The property which distinguishes thermodynamics from other sciences is temperature. One might say that temperature bears as important a relation to thermodynamics as force does to statics or velocity does to dynamics. Temperature is associated with the ability to distinguish hot from cold. When two bodies at different temperatures are brought into contact, after some time they attain a common temperature and are then said to exist in thermal equilibrium.

When a body A is in thermal equilibrium with a body B, and also separately with a body C, then B and C will be in thermal equilibrium with each other.

This is known as the *zeroth law of thermodynamics*. It is the basis of temperature measurement.

In order to obtain a quantitative measure of temperature, a reference body is used, and a certain physical characteristic of this body which changes with temperature is selected. The changes in the selected characteristic may be taken as an indication of change in temperature. The selected characteristic is called the *thermometric property*, and the reference body which is used in the determination of temperature is called the *thermometer*. A very common thermometer consists of a small amount of mercury in an evacuated capillary tube. In this case the extension of the mercury in the tube is used as the thermometric property.

There are five different kinds of thermometer, each with its own thermometric property, as shown in Table 2.1.

Table 2.1
Thermometers and Thermometric Properties

	Thermometer	*Thermometric property*	*Symbol*
1.	Constant volume gas thermometer	Pressure	p
2.	Constant pressure gas thermometer	Volume	V
3.	Electrical resistance thermometer	Resistance	R
4.	Thermocouple	Thermal e.m.f.	ε
5.	Mercury-in-glass thermometer	Length	L

2.2 MEASUREMENT OF TEMPERATURE—THE REFERENCE POINTS

The temperature of a system is a property that determines whether or not a system is in thermal equilibrium with other systems. If a body is at, say, 70°C, it will be 70°C, whether measured by a mercury-in-glass thermometer, resistance thermometer or constant volume gas thermometer. If X is the thermometric property, let us arbitrarily choose for the temperature common to the thermometer and *to all systems in thermal equilibrium with it* the following *linear* function of X:

$$\theta(X) = aX, \text{ where } a \text{ is an arbitrary constant.}$$

If X_1 corresponds to $\theta(X_1)$, then X_2 will correspond to $\frac{\theta(X_1)}{X_1} \cdot X_2$

that is

$$\theta(X_2) = \frac{\theta(X_1)}{X_1} \cdot X_2 \tag{2.1}$$

Two temperatures on the linear X scale are to each other as the ratio of the corresponding X's.

2.2.1 Method Used Before 1954

The thermometer is first placed in contact with the system whose temperature $\theta(X)$ is to be measured, and then in contact with an arbitrarily chosen standard system in an easily reproducible state where the temperature is $\theta(X_1)$. Thus

$$\frac{\theta(X_1)}{\theta(X)} = \frac{X_1}{X} \tag{2.2}$$

Then the thermometer at the temperature $\theta(X)$ is placed in contact with another arbitrarily chosen standard system in another easily reproducible state where the temperature is $\theta(X_2)$. It gives

$$\frac{\theta(X_2)}{\theta(X)} = \frac{X_2}{X} \tag{2.3}$$

From Eqs (2.2) and (2.3)

$$\frac{\theta(X_1) - \theta(X_2)}{\theta(X)} = \frac{X_1 - X_2}{X}$$

or

$$\theta(X) = \frac{\theta(X_1) - \theta(X_2)}{X_1 - X_2} \cdot X \tag{2.4}$$

If we assign an arbitrary number of degrees to the temperature interval $\theta(X_1) - \theta(X_2)$, then $\theta(X)$ can be calculated from the measurements of X, X_1 and X_2.

An easily reproducible state of an arbitrarily chosen standard system is called a fixed point. Before 1954, there were two fixed points: (a) *the ice point*, the temperature at which pure ice coexisted in equilibrium with air-saturated water at one atmosphere pressure, and (b) *the steam point*, the temperature of equilibrium between pure water and pure steam at one atmosphere pressure. The temperature interval, $\theta(X_1) - \theta(X_2)$, between these two fixed points was chosen to be 100 degrees.

The use of two fixed points was found unsatisfactory and later abandoned, because of (a) the difficulty of achieving equilibrium between pure ice and air-saturated water (since when ice melts, it surrounds itself only with pure water and prevents intimate contact with air-saturated water), and (b) extreme sensitiveness of the steam point to the change in pressure.

2.2.2 Method in Use After 1954

Since 1954 only one fixed point has been in use, viz., *the triple point of water*, the state at which ice, liquid watei and water vapour coexist in equilibrium. The temperature at which this state exists is arbitrarily assigned the value of 273.16 degrees Kelvin, or 273.16 K (the reason for using Kelvin's name will be explained later). Designating the triple point of water by θ_t, and with X_t being the value of the thermometric property when the body, whose temperature θ is to be measured, is placed in contact with water at its triple point, it follows that

$$\theta_t = aX_t$$

$\therefore$

$$a = \frac{\theta_t}{X_t} = \frac{273.16}{X_t}$$

Therefore

$$\theta = aX$$

$$= \frac{273.16}{X_t} \cdot X$$

or

$$\theta = 273.16 \frac{X}{X_t} \tag{2.5}$$

The temperature of the triple point of water, which is an easily reproducible state, is now the *standard fixed point of thermometry*.

2.3 COMPARISON OF THERMOMETERS

Applying the above principle to the five thermometers listed in Table 2.1, the temperatures are given as:

(a) Constant volume gas thermometer $\theta(P) = 273.16 \dfrac{p}{p_t}$

(b) Constant pressure gas thermometer $\theta(V) = 273.16 \dfrac{V}{V_t}$

(c) Electric resistance thermometer $\theta(R) = 273.16 \dfrac{R}{R_t}$

(d) Thermocouple $\theta(\varepsilon) = 273.16 \dfrac{\varepsilon}{\varepsilon_t}$

(e) Liquid-in-glass thermometer $\theta(L) = 273.16 \dfrac{L}{L_t}$

If the temperature of a given system is measured simultaneously with each of the five thermometers, it is found that there is considerable difference among the readings. The smallest variation is, however, observed among different gas thermometers. That is why a gas is chosen as the standard thermometric substance.

2.4 IDEAL GAS

It has been established from experimental observations that the $p - v - T$ behaviour of gases at a low pressure is closely given by the following relation

$$p\bar{v} = \bar{R}T \tag{2.6}$$

where $\bar{R}$ is the universal gas constant, 8.3143 J/mol K and $\bar{v}$ is the molar specific volume, m^3/gmol. (see Sec. 10.3.). Dividing Eq. (2.6) by the molecular weight μ,

$$pv = RT \tag{2.7}$$

where v is specific volume, in m^3/kg, and R is the characteristic gas constant. Substituting $R = \bar{R}/\mu$ J/kg K, we get in terms of the total volume V of gas,

$$PV = n\bar{R}T$$

$$PV = mRT \tag{2.8}$$

where n is the number of moles and m is the mass of the gas. Equation (2.8) can be written for two states of the gas,

$$\frac{p_1V_1}{T_1} = \frac{p_2V_2}{T_2} \tag{2.9}$$

Equation (2.6), (2.7) or (2.8) is called the *ideal gas equation of state*. At very low pressure or density, all gases and vapour approach ideal gas behaviour.

2.5 GAS THERMOMETERS

A schematic diagram of a constant volume gas thermometer is given in Fig. 2.1. A small amount of gas is enclosed in bulb B which is in communication via the capillary tube C with one limb of the mercury manometer M. The other limb of the mercury manometer is open to the atmosphere and can be moved vertically to adjust the mercury levels so that the mercury just touches lip L of the capillary. The pressure in the bulb is used as a thermometric property and is given by

$$p = p_0 + \rho_M Zg$$

where p_0 is the atmospheric pressure ρ_M is the density of mercury.

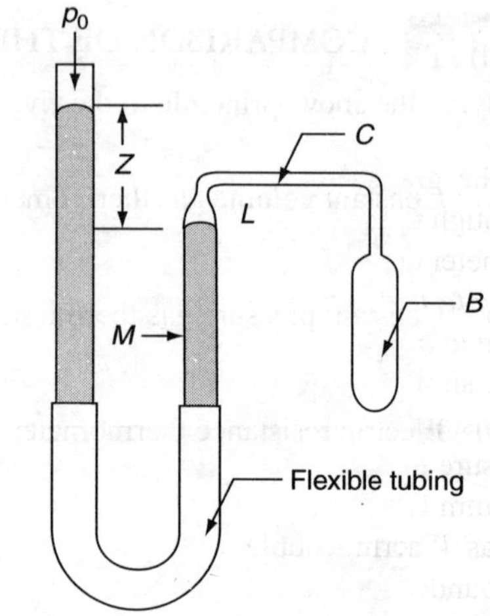

Fig. 2.1 *Constant volume gas thermometer*

When the bulb is brought in contact with the system whose temperature is to be measured, the bulb, in course of time, comes in thermal equilibrium with the system. The gas in the bulb expands, on being heated, pushing the mercury downward. The flexible limb of the manometer is then adjusted so that the mercury again touches the lip L. The difference in mercury level Z is recorded and the pressure p of the gas in the bulb is estimated. Since the volume of the trapped gas is constant, from the ideal gas equation,

$$\Delta T = \frac{V}{R}\Delta p \tag{2.10}$$

i.e. the temperature increase is proportional to the pressure increase.

In a constant pressure gas thermometer, the mercury levels have to be adjusted to keep Z constant, and the volume of gas V, which would vary with the temperature of the system, becomes the thermometric property.

$$\Delta T = \frac{p}{R}\Delta V \tag{2.11}$$

i.e. the temperature increase is proportional to the observed volume increase. The constant volume gas thermometer is, however, mostly in use, since it is simpler in construction and easier to operate.

2.6 IDEAL GAS TEMPERATURE

Let us suppose that the bulb of a constant volume gas thermometer contains an amount of gas such that when the bulb is surrounded by water at its triple point, the pressure p_t is 1000 mm Hg. Keeping the volume V constant, let the following procedure be conducted:

(a) Surround the bulb with steam condensing at 1 atm, determine the gas pressure p and calculate

$$\theta = 273.16 \frac{p}{1000}$$

(b) Remove some gas from the bulb so that when it is surrounded by water at its triple point, the pressure p_t is 500 mm Hg. Determine the new values of p and then θ for steam condensing at 1 atm.

$$\theta = 273.16 \frac{p}{500}$$

(c) Continue reducing the amount of gas in the bulb so that p_t and p have smaller and smaller values, e.g. p_t having, say, 250 mm Hg, 100 mm Hg, and so on. At each value of p_t calculate the corresponding θ.

(d) Plot θ vs. p_t and extrapolate the curve to the axis where $p_t = 0$. Read from the graph

$$\lim_{p_t \to 0} \theta$$

The graph, as shown in Fig. 2.2, indicates that although the readings of a constant volume gas thermometer depend upon the nature of the gas, *all gases indicate the same temperature as p_t is lowered and made to approach zero.*

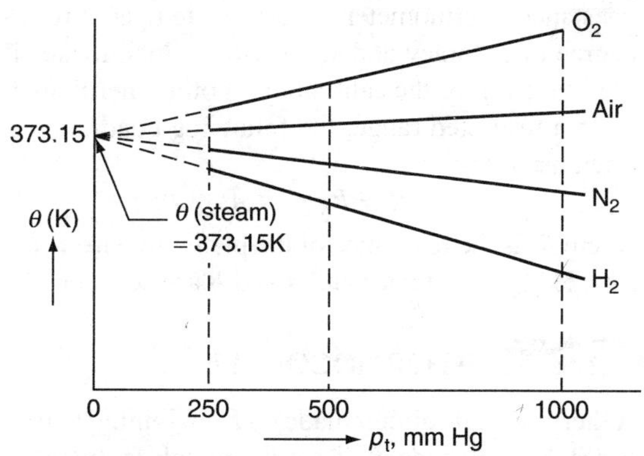

Fig. 2.2 *Ideal gas temperature for steam point*

A similar series of tests may be conducted with a constant pressure gas thermometer. The constant pressure may first be taken to be 1000 mm Hg, then 500 mm Hg, etc. and at each value of p, the volumes of gas V and V_t may be recorded when the bulb is surrounded by steam condensing at 1 atm and the triple point of water, respectively. The corresponding value of θ may be calculated from $\theta = 273.16 \dfrac{V}{V_t}$ and θ vs. p may be plotted, similar to Fig. 2.2. It is found from the experiments that all gases indicate the same value of θ as p approaches zero.

Since a real gas, as used in the bulb, behaves as an ideal gas as pressure approaches zero (which would be explained later in Chapter 10), the *ideal gas temperature T* is defined by either of the two equations

$$T = 273.16 \lim_{p_t \to 0} \frac{p}{p_t}$$

$$= 273.16 \lim_{p \to 0} \frac{V}{V_t}$$

where θ has been replaced by T to denote this particular temperature scale, the *ideal gas temperature scale.*

2.7 CELSIUS TEMPERATURE SCALE

The Celsius temperature scale employs a degree of the same magnitude as that of the ideal gas scale, but its zero point is shifted, so that the Celsius temperature of the triple point of water is 0.01 degree Celsius or 0.01°C. If t denotes the Celsius temperature, then $t = T - 273.15$

Thus the Celsius temperature t_s at which steam condenses at 1 atm. pressure

$$t_s = T_s - 273.15$$
$$= 373.15 - 273.15 = 100.00°C$$

Similar measurements for ice points show this temperature on the Celsius scale to be 0.00°C. The only Celsius temperature which is fixed by definition is that of the triple point.

2.8 ELECTRICAL RESISTANCE THERMOMETER

In the resistance thermometer (Fig. 2.3) the change in resistance of a metal wire due to its change in temperature is the thermometric property. The wire, frequently platinum, may be incorporated in a Wheatstone bridge circuit. The platinum resistance thermometer measures temperature to a high degree of accuracy and sensitivity, which makes it suitable as a standard for the calibration of other thermometers.

In a restricted range, the following quadratic equation is often used

$$R = R_0 (1 + At + Bt^2)$$

where R_0 is the resistance of the platinum wire when it is surrounded by melting ice and A and B are constants.

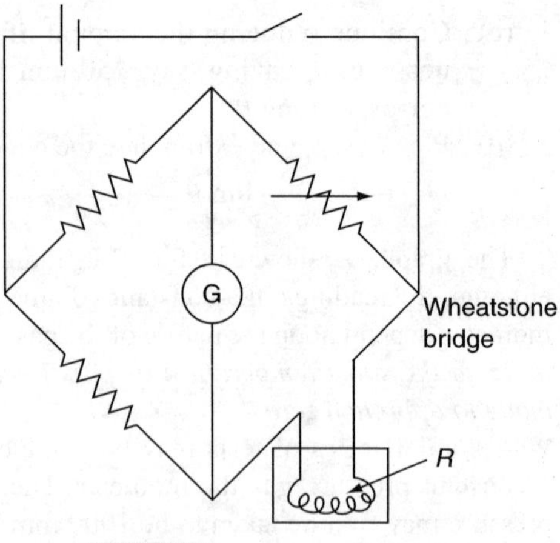

Fig. 2.3 *Resistance thermometer*

2.9 THERMOCOUPLE

A thermocouple circuit made up from joining two wires A and B made of dissimilar metals is shown in Fig. 2.4. Due to the Seebeck effect, a net e.m.f. is generated in the circuit which depends on the difference in temperature between the hot and cold junctions and is, therefore, a thermometric property of the circuit. This e.m.f. can be measured by a microvoltmeter to a high degree of accuracy. The choice of metals depends largely on the temperature range to be investigated, and copper-constantan, chromel-alumel and platinum-platinum-rhodium are typical combinations in use.

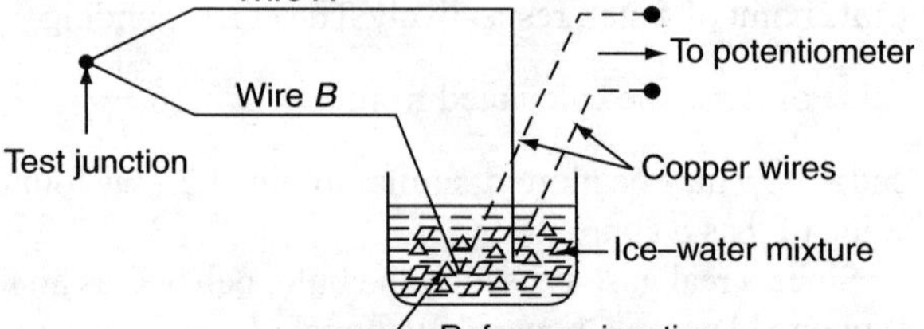

Fig. 2.4 *Thermocouple*

A thermocouple is calibrated by measuring the thermal e.m.f. at various known temperatures, the reference junction being kept at 0°C. The results of such measurements on most thermocouples can usually be represented by a cubic equation of the form $\varepsilon = a + bt + ct^2 + dt^3$

where ε is the thermal e.m.f. and the constants a, b, c and d are different for each thermocouple.

The advantage of a thermocouple is that it comes to thermal equilibrium with the system, whose temperature is to be measured, quite rapidly, because its mass is small.

2.10 INTERNATIONAL PRACTICAL TEMPERATURE SCALE

An international temperature scale was adopted at the Seventh General Conference on Weights and Measures held in 1927. It was not to replace the Celsius or ideal gas scales, but to provide a scale that could be easily and rapidly used to calibrate scientific and industrial instruments. Slight refinements were incorporated into the scale in revisions adopted in 1948, 1954, 1960 and 1968. The international practical scale agrees with the Celsius scale at the defining fixed points listed in Table 2.2. The temperature interval from the oxygen point to the gold point is divided into three main parts, as given below.

(a) From 0 to 660°C A platinum resistance thermometer with a platinum wire whose diameter must lie between 0.05 and 0.20 mm is used, and the temperature is given by the equation

$$R = R_0 (1 + At + Bt^2)$$

where the constants R_0, A, and B are computed by measurements at the ice point, steam point, and sulphur point.

(b) From - 190 to 0°C The same platinum resistance thermometer is used, and the temperature is given by

$$R = R_0 [1 + At + Bt^2 + C(t - 100) t^3]$$

where R_0, A and B are the same as before, and C is determined from a measurement at the oxygen point.

(c) From 660 to 1063°C A thermocouple, one wire of which is made of platinum and the other of an alloy of 90% platinum and 10% rhodium, is used with one junction at 0°C. The temperature is given by the formula

$$\varepsilon = a + bt + ct^2$$

where a, b, and c are computed from measurements at the antimony point, silver point, and gold point. The diameter of each wire of the thermocouple must lie between 0.35 and 0.65 mm.

An optical method is adopted for measuring temperatures higher than the gold point. The intensity of radiation of any convenient wavelength is compared with the intensity of radiation of the same wavelength emitted by a black body at the gold point. The temperature is then determined with the help of Planck's law of thermal radiation.

Table 2.2

Temperatures of Fixed Points

	Temperature °C
Normal boiling point of oxygen	−182.97
Triple point of water (standard)	+0.01
Normal boiling point of water	100.00
Normal boiling point of sulphur	444.60
(Normal melting point of zinc-suggested as an alternative to the sulphur point)	419.50
Normal melting point of antimony	630.50
Normal melting point of silver	960.80
Normal melting point of gold	1063.00

Solved Examples

Example 2.1

Two mercury-in-glass thermometers are made of identical materials and are accurately calibrated at 0°C and 100°C. One has a tube of constant diameter, while the other has a tube of conical bore, ten per cent greater in diameter at 100°C than at 0°C. Both thermometers have the length between 0 and 100 subdivided uniformly. What will be the straight bore thermometer read in a place where the conical bore thermometer reads 50°C?

Solution The volume of mercury in the tube at t°C, V_t, is given by

$$V_t = V_0 [1 + \beta (t - t_0)]$$

where V_0 is the volume of mercury at 0°C, β is the coefficient of volume expansion of mercury, and t_0 is the ice point temperature which is 0°C. The volume change of glass is neglected.

Therefore $$V_t - V_0 = \beta V_0 \, t$$

The temperature t is thus a linear function of volume change of mercury $(V_t - V_0)$.

Therefore $$\Delta V_{0-100} = \beta V_0 \cdot 100$$

$$\Delta V_{0-50} = \beta V_0 \cdot 50$$

$$\therefore \quad \frac{\Delta V_{0-50}}{\Delta V_{0-100}} = \frac{1}{2}$$

i.e., at 50°C, the volume of mercury will be half of that at 100°C, for the straight bore thermometer [Fig. Ex.2.1a]. But if the bore is conical [Fig. Ex. 2.1b], mercury will fill up the volume $ACDB$, which is less than half of the mercury volume at 100°C, i.e. volume $AEFB$. Let t be the true temperature when mercury rises half the length of the conical tube (the apparent temperature being 50°C). Let EA and FB be extended to meet at G. Let l represent the length of the thermometers and l' the vertical height of the cone ABG, as shown in the figure. Now,

$$\frac{l'}{l + l'} = \frac{d}{1.1d} = \frac{1}{1.1}$$

$$l' = 10$$

and $$\frac{l'}{l' + l/2} = \frac{d}{CD}$$

$$\therefore \quad CD = \frac{10.5}{10} d = 1.05d$$

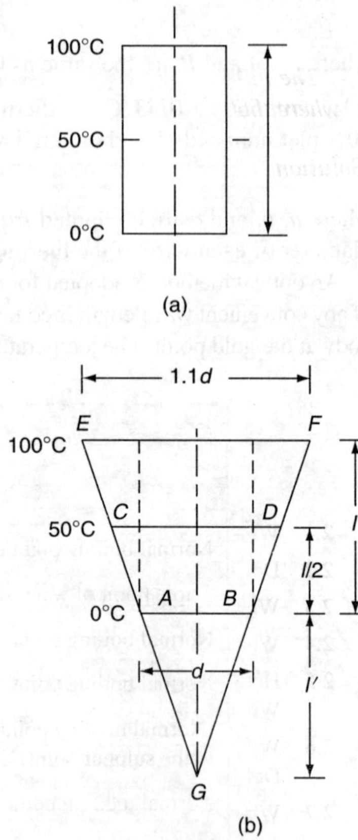

Fig. Ex. 2.1

Again $$\Delta V_{0-100} = V_0 \cdot \beta \cdot 100$$

$$\Delta V_{0-t} = V_0 \, \beta \, t$$

$$\frac{\Delta V_{0-t}}{\Delta V_{0-100}} = \frac{t}{100}$$

$$\text{or} \qquad \frac{\text{Volume } ACDB}{\text{Volume } AEFB} = \frac{t}{100}$$

$$\text{or} \qquad \frac{\frac{1}{3}\frac{\pi}{4}(1.05d)^2 \times 10.5l - \frac{1}{3}\frac{\pi}{4}d^2 \cdot 10l}{\frac{1}{3}\frac{\pi}{4}(1.1d)^2 \times 11l - \frac{1}{3}\frac{\pi}{4}d^2 \cdot 10l} = \frac{t}{100}$$

$$\text{or} \qquad \frac{1.05 \times 1.05 \times 10.5 - 10}{1.1 \times 1.1 \times 11 - 10} = \frac{t}{100}$$

$$\therefore \qquad t = \frac{1.58}{3.31} \times 100 = 47.7°\text{C} \qquad \textit{Ans.}$$

Example 2.2

The e.m.f. in a thermocouple with the test junction at t°C on gas thermometer scale and reference junction at ice point is given by

$$e = 0.20\,t - 5 \times 10^{-4}\,t^2 \text{ mV}$$

The millivoltmeter is calibrated at ice and steam points. What will this thermometer read in a place where the gas thermometer reads 50°C?

Solution At ice point, when $t = 0°C$, $\varepsilon = 0$ mV

At steam point, when $t = 100°C$, $\varepsilon = 0.20 \times 100 - 5 - 10^{-4} \times (100)^2 = 15$ mV

At $t = 50°C$, $\varepsilon = 0.20 \times 50 - 5 \times 10^{-4} (50)^2 = 8.75$ mV

When the gas thermometer reads 50°C, the thermocouple will read

$$\frac{100}{15} \times 8.75, \text{ or } 58.33°\text{C} \qquad \textit{Ans.}$$

Review Questions

2.1 What is the zeroth law of thermodynamics?

2.2 Define thermometric property.

2.3 What is a thermometer?

2.4 What is a fixed point?

2.5 How many fixed points were used prior to 1954? What are these?

2.6 What is the standard fixed point in thermometry? Define it.

2.7 Why is a gas chosen as the standard thermometric substance?

2.8 What is an ideal gas?

2.9 What is the difference between the universal gas constant and a characteristic gas constant?

2.10 What is a constant volume gas thermometer? Why is it preferred to a constant pressure gas thermometer?

2.11 What do you understand by the ideal gas temperature scale?

2.12 How can the ideal gas temperature for the steam point be measured?

2.13 What is the Celsius temperature scale?

2.14 What is the advantage of a thermocouple in temperature measurement?

2.15 How does the resistance thermometer measure temperature?

2.16 What is the need of the international practical temperature scale?

Problems

2.1 The limiting value of the ratio of the pressure of gas at the steam point and at the triple point of water when the gas is kept at constant volume is found to be 1.36605. What is the ideal gas temperature of the steam point?

2.2 In a constant volume gas thermometer the following pairs of pressures readings were taken at the boiling point of water and the boiling point of sulphur, respectively:

Water b.p.	50.0	100	200	300
Sulphur b.p.	96.4	193	387	582

The numbers are the gas pressures, mm Hg, each pair being taken with the same amount of gas in the thermometer, but the successive pairs being taken with different amounts of gas in the thermometer. Plot the ratio of $S_{b.p.} : H_2O_{b.p.}$ against the reading at the water boiling point, and extrapolate the plot to zero pressure at the water boiling point. This gives the ratio of $S_{b.p.} : H_2O_{b.p.}$ on a gas thermometer operating at zero gas pressure, i.e., an ideal gas thermometer. What is the boiling point of sulphur on the gas scale, from your plot?

Ans. 445°C

2.3 The resistance of a platinum wire is found to be 11.000 ohms at the ice point, 15.247 ohms at the steam point, and 28.887 ohms at the sulphur point. Find the constants A and B in the equation

$$R = R_0 (1 + At + Bt^2)$$

and plot R against t in the range 0 to 660°C.

2.4 When the reference junction of a thermocouple is kept at the ice point and the test junction is at the Celsius temperature t, and e.m.f. ε of the thermocouple is given by the equation

$$\varepsilon = at + bt^2$$

where $a = 0.20$ mV/deg, and $b = -5.0 \times 10^{-4}$ mV/deg^2

(a) Compute the e.m.f. when $t = -100$°C, 200°C, 400°C, and 500°C, and draw graph of ε against t in this range.

(b) Suppose the e.m.f. ε is taken as a thermometric property and that a temperature scale t^* is defined by the linear equation.

$$t^* = a'\varepsilon + b'$$

and that $t^* = 0$ at the ice point and $t^* = 100$ at the steam point. Find the numerical values of a' and b' and draw a graph of ε against t^*.

(c) Find the values of t^* when $t = -100$°C, 200°C, 400°C, and 500°C, and draw a graph of t^* against t.

(d) Compare the Celsius scale with the t^* scale.

2.5 The temperature t on a thermometric scale is defined in terms of a property K by the relation

$$t = a \ln K + b$$

where a and b are constants.

The values of K are found to be 1.83 and 6.78 at the ice point and the steam point, the temperatures of which are assigned the numbers 0 and 100 respectively. Determine the temperature corresponding to a reading of K equal to 2.42 on the thermometer.

Ans. 21.346°C

2.6 The resistance of the windings in a certain motor is found to be 80 ohms at room temperature (25°C). When operating at full load under steady state conditions, the motor is switched off and the resistance of the windings, immediately measured again, is found to be 93 ohms. The windings are made of copper whose resistance at temperature t°C is given by

$$R_t = R_0 [1 + 0.00393\, t]$$

where R_0 is the resistance at 0°C. Find the temperature attained by the coil during full load.

Ans. 70.41°C

2.7 A new scale N of temperature is divided in such a way that the freezing point of ice is 100°N and the boiling point is 400°N. What is the temperature reading on this new scale when the temperature is 150°C? At what temperature both the Celsius and the new temperature scale reading would be the same?

Ans. 550°N, – 50°C.

2.8 A platinum wire is used as a resistance thermometer. The wire resistance was found to be 10 ohm and 16 ohm at ice point and steam point respectively, and 30 ohm at sulphur boiling point of 444.6°C. Find the resistance of the wire at 500°C, if the resistance varies with temperature by the relation.

$$R = R_0 (1 + \alpha t + \beta t^2)$$

Ans. 31.3 ohm

C H A P T E R

3 Work and Heat Transfer

A closed system and its surroundings can interact in two ways: (a) by work transfer, and (b) by heat transfer. These may be called *energy interactions* and these bring about changes in the properties of the system. Thermodynamics mainly studies these energy interactions and the associated property changes of the system.

3.1 WORK TRANSFER

Work is one of the basic modes of energy transfer. In mechanics, the action of a force on a moving body is identified as work. A force is a means of transmitting an effect from one body to another. But a force itself never produces a physical effect except when coupled with motion and hence it is not a form of energy. An effect such as the raising of a weight through a certain distance can be performed by using a small force through a large distance or a large force through a small distance. The product of force and distance is the same to accomplish the same effect. In mechanics work is defined as:

The work is done by a force as it acts upon a body moving in the direction of the force.

The action of a force through a distance (or of a torque through an angle) is called *mechanical work* since other forms of work can be identified, as discussed later. The product of the force and the distance moved parallel to the force is the magnitude of mechanical work.

In thermodynamics, work transfer is considered as occurring between the system and the surroundings. *Work is said to be done by a system if the sole effect on things external to the system can be reduced to the raising of a weight.* The weight may not actually be raised, but the net effect external to the system would be the raising of a weight. Let us consider the battery and the motor in Fig. 3.1 as a system. The motor is driving a fan. The system is doing work upon the surroundings. When the fan is replaced by a pulley and a weight, as shown in Fig. 3.2, the weight may be raised with the pulley driven by the motor. The sole effect on things external to the system is then the raising of a weight.

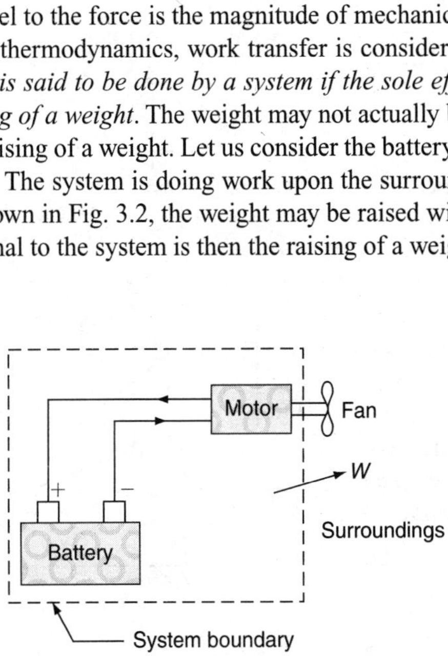

Fig. 3.1 *Battery-motor system driving a fan*

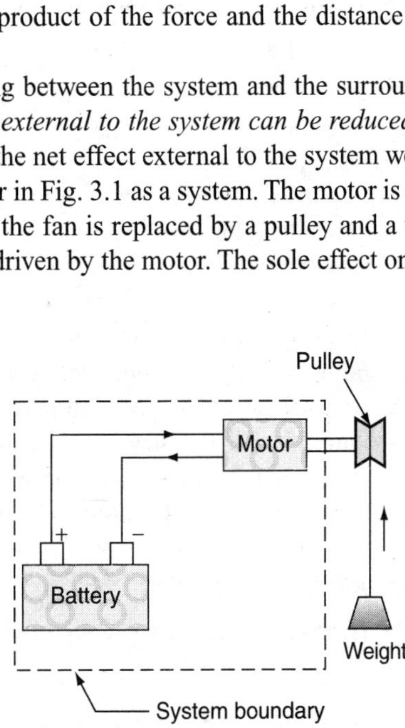

Fig. 3.2 *Work transfer from a system*

When work is done by a system, it is arbitrarily taken to be positive, and when work is done on a system, it is taken to be negative (Fig. 3.3). The symbol W is used for work transfer.

The unit of work is N.m or Joule [1 Nm = 1 Joule]. The rate at which work is done by, or upon, the system is known as *power*. The unit of power is J/s or watt.

Work is one of the forms in which a system and its surroundings can interact with each other. There are various types of work transfer which can get involved between them.

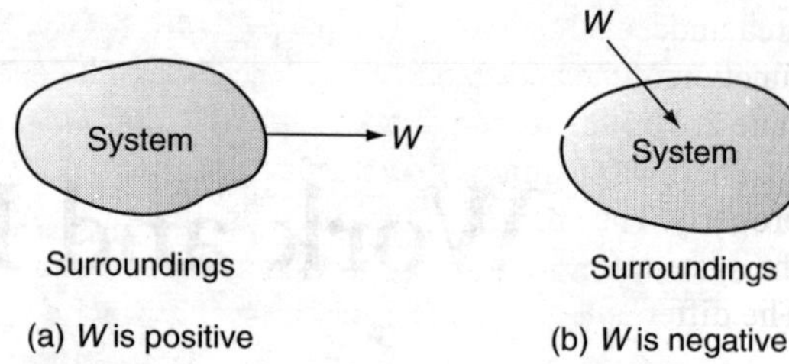

Fig. 3.3 *Work interaction between a system and the surroundings*

3.2 pdV–WORK OR DISPLACEMENT WORK

Let the gas in the cylinder (Fig. 3.4) be a system having initially the pressure p_1 and volume V_1. The system is in thermodynamic equilibrium, the state of which is described by the coordinates p_1, V_1. The piston is the only boundary which moves due to gas pressure. Let the piston move out to a new final position 2, which is also a thermodynamic equilibrium state specified by pressure p_2 and volume V_2. At any intermediate point in the travel of the piston, let the pressure be p and the volume V. This must also be an equilibrium state, since macroscopic properties p and V are significant only for equilibrium states. When the piston moves an infinitesimal distance dl, and if 'a' be the area of the piston, the force F acting on the piston $F = p.a.$ and the infinitesimal amount of work done by the gas on the piston

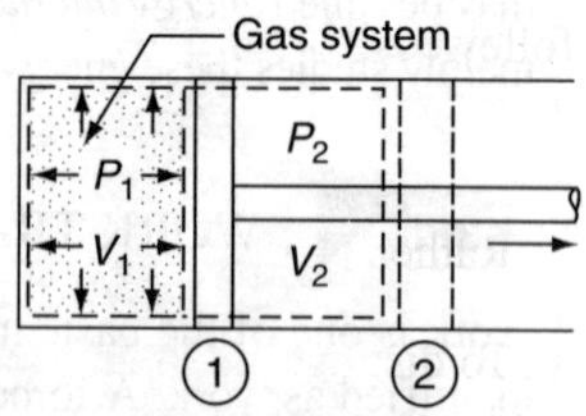

Fig. 3.4 *pdV work*

$$đ W = F \cdot dl = padl = pdV \tag{3.1}$$

where $dV = adl$ = infinitesimal displacement volume. The differential sign in dW with the line drawn at the top of it will be explained later.

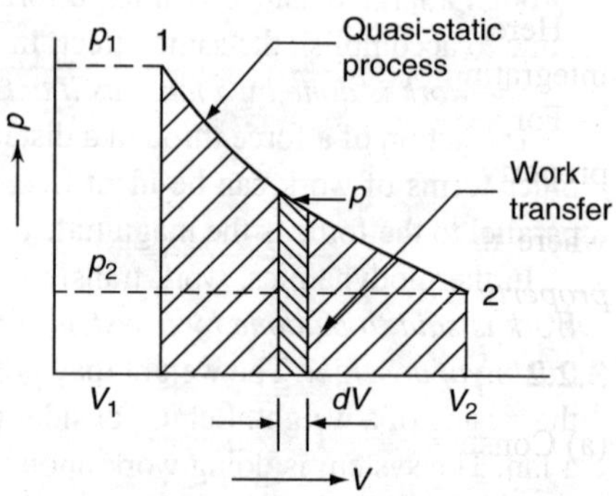

Fig. 3.5 *Quasi-static pdV work*

When the piston moves out from position 1 to position 2 with the volume changing from V_1 to V_2, the amount of work W done by the system will be

$$W_{1-2} = \int_{V_1}^{V_2} pdV$$

The magnitude of the work done is given by the area under the path 1–2, as shown in Fig. 3.5. Since p is at all times a thermodynamic coordinate, all the states passed through by the system as the volume changes from V_1 to V_2 must be equilibrium states, and the path 1–2 must be *quasi-static*. The piston moves infinitely slowly so that every state passed through is an equilibrium state. *The integration* $\int pdV$ *can be performed only on a quasi-static path.*

3.2.1 Path Function and Point Function

With reference to Fig. 3.6, it is possible to take a system from state 1 to state 2 along many quasi-static paths, such as A, B or C. Since the

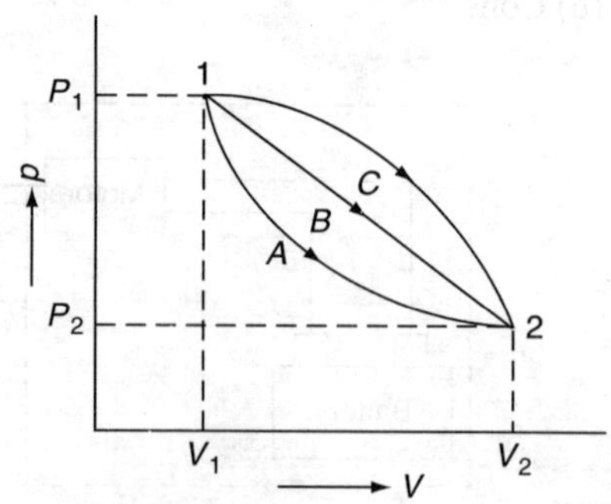

Fig. 3.6 *Work–a path function*

area under each curve represents the work for each process, the amount of work involved in each case is not a function of the end states of the process, and it depends on the path the system follows in going from state 1 to state 2. For this reason, work is called a *path function*, and đ *W is an inexact or imperfect differential.*

Thermodynamic properties are *point functions*, since for a given state, there is a definite value for each property. The change in a thermodynamic property of a system in a change of state is independent of the path the system follows during the change of state, and depends only on the initial and final states of the system. The differentials of point functions are *exact or perfect differentials*, and the integration is simply

$$\int_{V_1}^{V_2} dV = V_2 - V_1$$

The change in volume thus depends only on the end states of the system irrespective of the path the system follows.

On the other hand, work done in a quasi-static process between two given states depends on the path followed.

$$\int_1^2 đW \neq W_2 - W_1$$

Rather,
$$\int_1^2 đW = W_{1-2} \text{ or } {}_1W_2$$

To distinguish an inexact differential đW from an exact differential dV or dp the differential sign is being cut by a line at its top.

From Eq. (3.1),
$$dV = \frac{1}{p} đW \tag{3.2}$$

Here, $1/p$ is called the *integrating factor*. Therefore, an inexact differential đW when multiplied by an integrating factor $1/p$ becomes an exact differential dV.

For a cyclic process, the initial and final states of the system are the same, and hence, the change in any property is zero, i.e.
$$\oint dV = 0, \quad \oint dp = 0, \quad \oint dT = 0 \tag{3.3}$$

where the symbol $\oint$ denotes the cyclic integral for the closed path. Therefore, *the cyclic integral of a property is always zero.*

3.2.2 *p*d*V*-Work in Various Quasi-Static Processes

(a) Constant pressure process (Fig. 3.7) (isobaric or isopiestic process)

$$W_{1-2} = \int_{V_1}^{V_2} p dV = p(V_2 - V_1) \tag{3.4}$$

(b) Constant volume process (Fig. 3.8) (isochoric process)

$$W_{1-2} = \int p dV = 0 \tag{3.5}$$

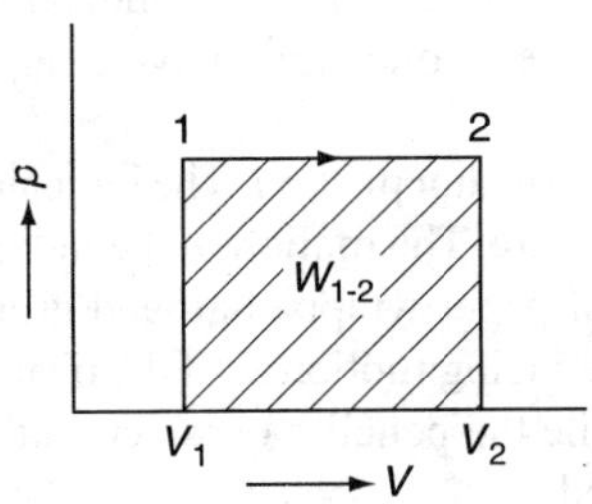

Fig. 3.7 *Constant pressure process*

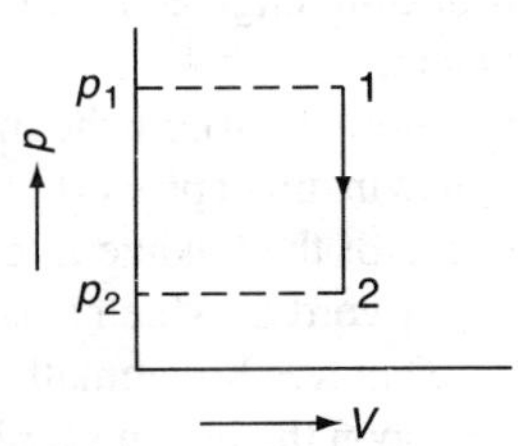

Fig. 3.8 *Constant volume process*

(c) Process in which $pV = C$ (Fig. 3.9)

$$\therefore \quad W_{1-2} = \int_{V_1}^{V_2} p\mathrm{d}V, \quad pV = p_1 V_1 = C$$

$$p = \frac{(p_1 V_1)}{V}$$

$$W_{1-2} = p_1 V_1 \int_{V_1}^{V_2} \frac{\mathrm{d}V}{V} = p_1 V_1 \ln \frac{V_2}{V_1}$$

$$= p_1 V_1 \ln \frac{p_1}{p_2} \tag{3.6}$$

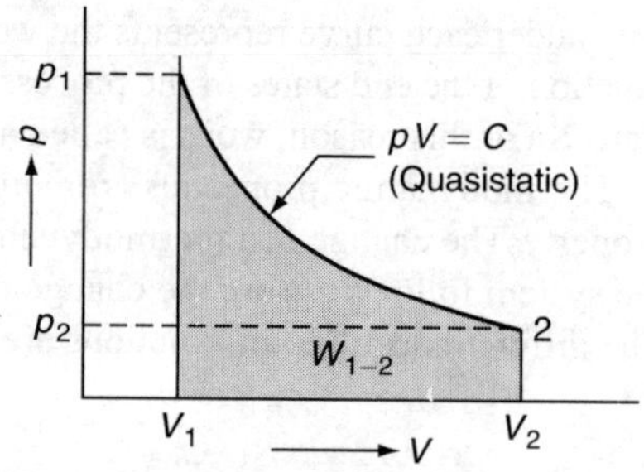

Fig. 3.9 *Process in which pV = constant*

(d) Process in which $pV^n = C$, where n is a constant (Fig. 3.10).

$$pV^n = p_1 V_1^n = p_2 V_2^n = C$$

$$\therefore \quad p = \frac{(p_1 V_1^n)}{V^n}$$

$$\therefore \quad W_{1-2} = \int_{V_1}^{V_2} p\mathrm{d}V$$

$$= \int_{V_1}^{V_2} \frac{p_1 V_1^n}{V^n} . \mathrm{d}V$$

$$= (p_1 V_1^n) \left[\frac{V^{-n+1}}{-n+1} \right]_{V_1}^{V_2}$$

$$= \frac{p_1 V_1^n}{1-n} (V_2^{1-n} - V_1^{1-n})$$

$$= \frac{p_2 V_2^n \times V_2^{1-n} - p_1 V_1^n \times V_1^{1-n}}{1-n}$$

$$= \frac{p_1 V_1 - p_2 V_2}{n-1} = \frac{p_1 V_1}{n-1} \left[1 - \left(\frac{p_2}{p_1} \right)^{n-1/n} \right] \tag{3.7}$$

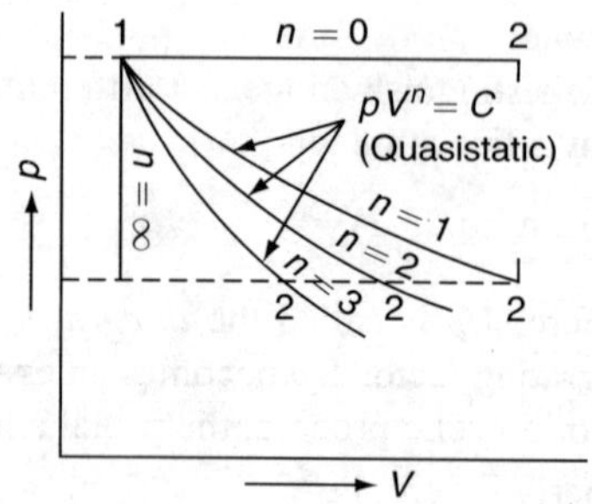

Fig. 3.10 *Process in which pV^n = constant*

3.3 INDICATOR DIAGRAM

An indicator diagram is a trace made by a recording pressure gauge, called the indicator, attached to the cylinder of a reciprocating engine. This represents the work done in one engine cycle. Figure 3.11 shows a typical engine indicator.

The same gas pressure acts on both the engine piston *P* and the indicator piston *I*. The indicator piston is loaded by a spring and it moves in direct proportion to the change in pressure. The motion of the indicator piston causes a pencil held at the end of the linkage *L* to move upon a strip of paper wrapped around drum *D*. The drum is rotated about its axis by cord *C*, which is connected through a reducing motion *R* to the piston *P* of the engine. The surface of drum *D* moves horizontally under the pencil while the pencil moves vertically over the surface and a plot of pressure upon the piston vs. piston travel is obtained.

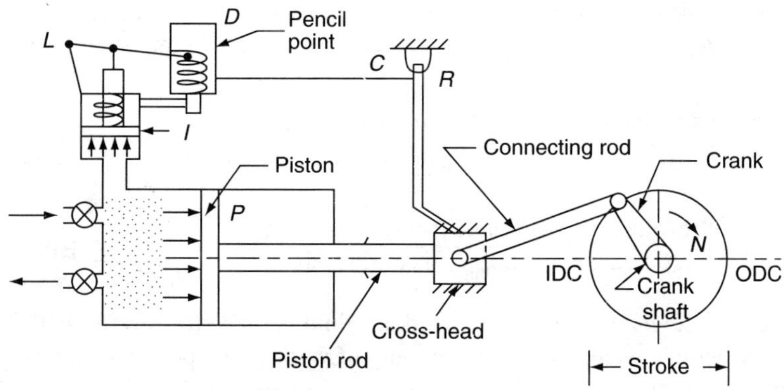

Fig. 3.11 *Engine indicator*

Before tracing the final indicator diagram, a pressure reference line is recorded by subjecting the indicator to the atmosphere and tracing a line at a constant pressure of one atmosphere.

The area of the indicator diagram represents the magnitude of the net work done by the system in one engine cycle. The area under the path 1–2 represents work done by the system and the area under the path 2–1 represents work done upon the system (Fig. 3.12). The area of the diagram, a_d, is measured by means of a planimeter, and the length of the diagram, l_d, is also measured. The *mean effective pressure* (m.e.p.) p_m is defined in the following way

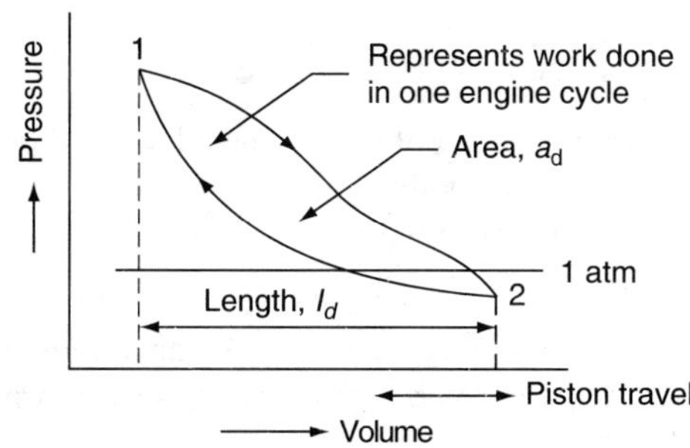

Fig. 3.12 *Indicator diagram*

$$p_m = \frac{a_d}{l_d} \times K$$

where K is the indicator spring constant (N/cm^2 × cm travel). Work done in one engine cycle

$$= (p_m \cdot A)\, L$$

where A = cross-sectional area of the cylinder = $\frac{\pi}{4} D^2$, where D is the cylinder diameter

and L = stroke of piston, or length of cylinder.

Let N be the revolutions per minute (r.p.m.) of the crankshaft. In a two stroke cycle, the engine cycle is completed in two strokes of the piston or in one revolution of the crankshaft. In a four-stroke cycle, the engine cycle is completed in four strokes of the piston or two revolutions of the crankshaft.

For a two-stroke engine, work done in one minute = $p_m ALN$, and for a four-stroke engine, work done in one minute = $p_m ALN/2$.

The power developed inside the cylinder of the engine is called *indicated power* (IP),

$$\therefore \qquad \text{IP} = \frac{p_m AL\left(N \text{ or } \frac{N}{2}\right) n}{60} \text{ kW} \qquad (3.8)$$

where p_m is in kPa and n is the number of cylinders in the engine.

The power available at the crankshaft is always less than this value (IP) due to friction, etc. and is called the *brake power* (BP) or *shaft power* (SP). If ω is the angular velocity of the crankshaft in radian/sec, then

$$\text{BP} = T\omega \tag{3.9}$$

where T is the torque transmitted to the crankshaft in mN.

$$\therefore \quad \text{BP} = \frac{2\pi TN}{60} \tag{3.10}$$

where N is the number of revolutions per minute (rpm).

The mechanical efficiency of the engine, η_{mech}, is defined as $$\eta_{\text{mech}} = \frac{\text{BP}}{\text{IP}} \tag{3.11}$$

An engine is said to be *double-acting*, if the working fluid is made to work on both sides of the piston. Such an engine theoretically develops twice the amount of work developed in a single-acting engine. Most reciprocating steam engines are double-acting, and so are many marine diesel engines. Internal combustion engines for road transport are always single-acting.

3.4 OTHER TYPES OF WORK TRANSFER

There are forms of work other than $p\mathrm{d}V$ or displacement work. The following are the additional types of work transfer which may get involved in system-surroundings interactions.

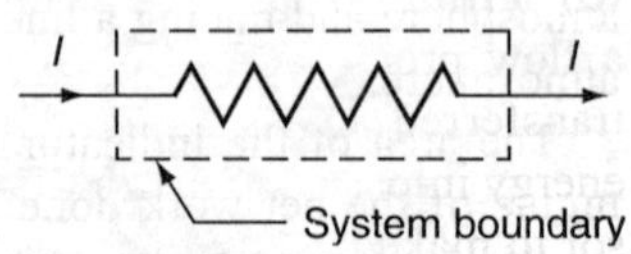

Fig. 3.13 *Electrical work*

(a) *Electrical Work* When a current flows through a resistor (Fig. 3.13), taken as a system, there is work transfer into the system. This is because the current can drive a motor, the motor can drive a pulley and the pulley can raise a weight.

The current flow, I, in amperes, is given by $I = \dfrac{\mathrm{d}C}{\mathrm{d}\tau}$

where C is the charge in coulombs and τ is time in seconds. Thus dC is the charge crossing a boundary during time dτ. If E is the voltage potential, the work is $đW = E \cdot \mathrm{d}C = EI\,\mathrm{d}\tau$

$$\therefore \quad W = \int_1^2 EI\,\mathrm{d}\tau \tag{3.12}$$

The electrical power will be $$\dot{W} = \lim_{\mathrm{d}\tau \to 0} \frac{\mathrm{d}W}{\mathrm{d}\tau} = EI \tag{3.13}$$

This is the rate at which work is transferred.

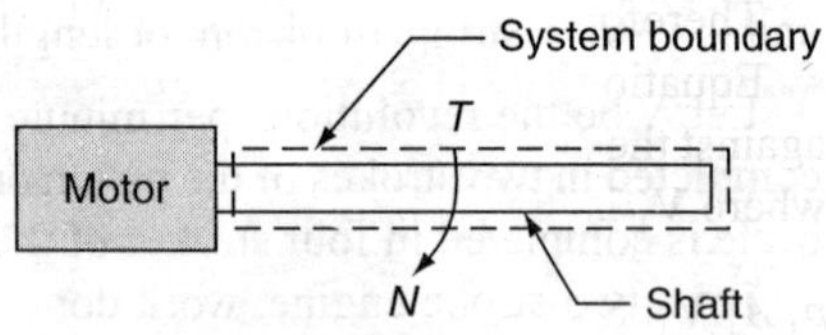

Fig. 3.14 *Shaft work*

(b) *Shaft Work* When a shaft, taken as the system (Fig. 3.14), is rotated by a motor, there is work transfer into the system. This is because the shaft can rotate a pulley which can raise a weight. If T is the torque applied to the shaft and dθ is the angular displacement of the shaft, the shaft work is $$W = \int_1^2 T\,\mathrm{d}\theta \tag{3.14}$$

and the shaft power is

$$\dot{W} = \int_1^2 T\frac{\mathrm{d}\theta}{\mathrm{d}\tau} = T\omega \tag{3.15}$$

where ω is the angular velocity and T is considered a constant in this case.

(c) *Paddle-Wheel Work or Stirring Work* As the weight is lowered, and the paddle wheel turns (Fig. 3.15), there is work transfer into the fluid system which gets stirred. Since the volume of the system remains constant, $\int p \mathrm{d}V = 0$. If m is the mass of the weight lowered through a distance dz and T is the torque transmitted by the shaft in rotating through an angle dθ, the differential work transfer to the fluid is given by

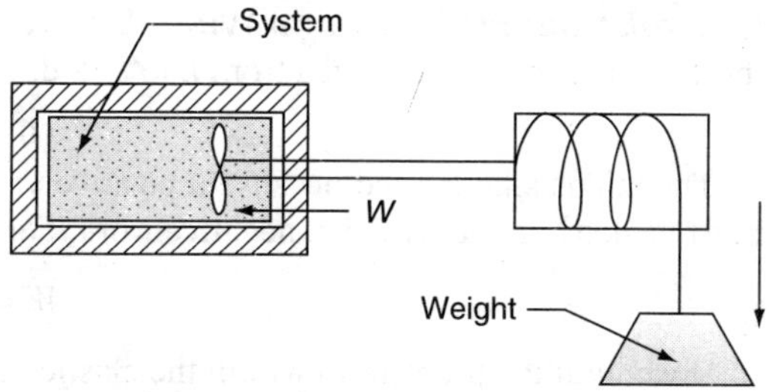

Fig. 3.15 *Paddle-wheel work*

$$\mathrm{d}W = mg\mathrm{d}z = T\mathrm{d}\theta$$

and the total work transfer is

$$W = \int_1^2 mg\mathrm{d}z = \int_1^2 W'\,\mathrm{d}z = \int_1^2 T\mathrm{d}\theta \tag{3.16}$$

where W' is the weight lowered.

(d) *Flow Work* The flow work, significant only in a flow process or an open system, represents the energy transferred across the system boundary as a result of the energy imparted to the fluid by a pump, blower or compressor to make the fluid flow across the control volume. Flow work is analogous to displacement work. Let p be the fluid pressure in the plane of the imaginary piston, which acts in a direction normal to it (Fig. 3.16). The work done on this imaginary piston by the external pressure as the piston moves forward is given by

$$\text{đ}\, W_{\text{flow}} = p\mathrm{d}V, \tag{3.17}$$

where dV is the volume of fluid element about to enter the system.

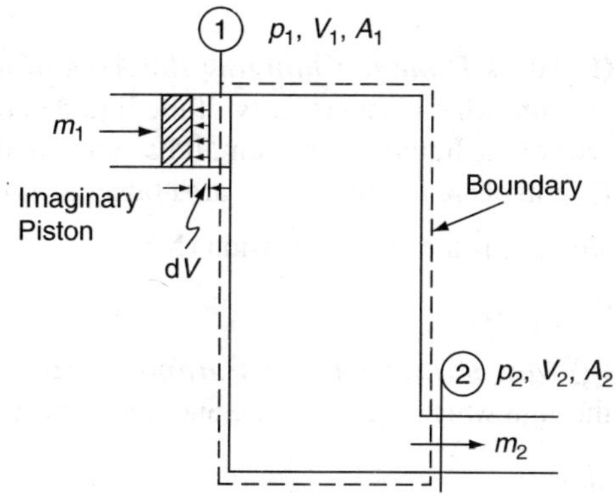

Fig. 3.16 *Flow work*

$$\therefore \qquad \text{đ}\, W_{\text{flow}} = pv\,\mathrm{d}m \tag{3.18}$$

where dV = v dm

Therefore, flow work at inlet (Fig. 3.16), $(\mathrm{d}W_{\text{flow}})_{\text{in}} = p_1 v_1\,\mathrm{d}m_1$ (3.19)

Equation (3.17) can also be derived in a slightly different manner. If the normal pressure p_1 is exerted against the area A_1, giving a total force $(p_1 A_1)$ against the piston, in time $d\tau$, this force moves a distance $\mathbf{V}_1 d\tau$, where $\mathbf{V}_1$ is the velocity of flow (piston). The work in the time $d\tau$ is $p_1 A_1 \mathbf{V}_1\, d\tau$, or the work per unit time is $p_1 A_1 \mathbf{V}_1$. Since the flow rate

$$w_1 = \frac{A_1 \mathbf{V}_1}{v_1} = \frac{dm_1}{d\tau}$$

the work done in time dτ becomes $(\text{đ}\, W_{\text{flow}})_{\text{in}} = p_1 v_1\,\mathrm{d}m_1$

Similarly, flow work of the fluid element leaving the system is

$$(\text{đ}\, W_{\text{flow}})_{\text{out}} = p_2 v_2\,\mathrm{d}m_2 \tag{3.20}$$

The flow work per unit mass is thus $W_{\text{flow}} = pv$ (3.21)

It is the displacement work done at the moving system boundary.

(e) Work Done in Stretching a Wire Let us consider a wire as the system. If the length of the wire in which there is a tension $\mathcal{F}$ is changed from L to $L + dL$, the infinitestimal amount of work that is done is equal to

$$đW = -\mathcal{F}\,dL$$

The minus sign is used because a positive value of dL means an expansion of the wire, for which work must be done on the wire, i.e. negative work. For a finite change of length,

$$W = -\int_1^2 \mathcal{F}\,dL \tag{3.22}$$

If we limit the problem to within the elastic limit, where E is the modulus of elasticity, s is the stress, ε is the strain, and A is the cross-sectional area, then

$$\mathcal{F} = sA = E\varepsilon A, \quad \text{since } \frac{s}{\varepsilon} = E$$

$$d\varepsilon = \frac{dL}{L}$$

$$đW = -\mathcal{F}\,dL = -E\varepsilon\, AL\, d\varepsilon$$

$$\therefore \quad W = -A\varepsilon L \int_1^2 \varepsilon\, d\varepsilon = -\frac{AEL}{2}(\varepsilon_2^2 - \varepsilon_1^2) \tag{3.23}$$

(f) Work Done in Changing the Area of a Surface Film A film on the surface of a liquid has a surface tension, which is a property of the liquid and the surroundings. The surface tension acts to make the surface area of the liquid a minimum. It has the unit of force per unit length. The work done on a homogeneous liquid film in changing its surface area by an infinitesimal amount dA is $đW = -\sigma dA$

where σ is the surface tension (N/m).

$$\therefore \quad W = -\int_1^2 \sigma dA \tag{3.24}$$

(g) Magnetization of a Paramagnetic Solid The work done per unit volume on a magnetic material through which the magnetic and magnetization fields are uniform is $đW = -HdI$

and

$$W_{1-2} = -\int_{I_1}^{I_2} HdI \tag{3.25}$$

where H is the field strength, and I is the component of the magnetization field in the direction of the field. The minus sign provides that an increase in magnetization (positive dI) involves negative work.

The following equations summarize the different forms of work transfer:

Displacement work

(compressible fluid) $\quad W = \int_1^2 pdV$

Electrical work $\quad W = \int_1^2 E\,dC = \int_1^2 EI\,d\tau$

Shaft work $\quad W = \int_1^2 T\,d\theta$

Surface film $\quad W = -\int_1^2 \sigma\,dA \quad$ (3.26)

Stretched wire $\quad W = -\int_1^2 \mathcal{F}\,dL$

Magnetised solid $\quad W = -\int_1^2 H\,dI$

It may be noted in the above expressions that the work is equal to the integral of the product of an intensive property and the change in its related extensive property. These expressions are valid only for infinitesimally slow quasi-static processes.

There are some other forms of work which can be identified in processes that are not quasi-static, for example, the work done by shearing forces in a process involving friction in a viscous fluid.

3.5 FREE EXPANSION WITH ZERO WORK TRANSFER

Work transfer is identified only at the boundaries of a system. It is a boundary phenomenon, and a form of energy in transit crossing the boundary. Let us consider a gas separated from the vacuum by a partition (Fig. 3.17). Let the partition be removed. The gas rushes to fill the entire volume. The expansion of a gas against vacuum is called *free expansion*. If we neglect the work associated with the removal of partition, and consider the gas and vacuum together as our system (Fig. 3.17a), there is no work transfer involved here, since no work crosses the system boundary, and hence

$$\int_1^2 đW = 0, \quad \text{although} \int_1^2 p\mathrm{d}V \neq 0$$

If only the gas is taken as the system (Fig. 3.17b), when the partition is removed there is a change in the volume of the gas, and one is tempted to calculate the work from the expression $\int_1^2 p\mathrm{d}V$. However, this is not a quasistatic process, although the initial and final end states are in equilibrium. Therefore, the work cannot be calculated from this relation. The two end states can be located on the p–V diagram and these are joined by a dotted line (Fig. 3.17c) to indicate that the process had occurred. However, if the vacuum space is divided into a large number of small volumes by partitions and the partitions are removed one by one slowly (Fig. 3.17d), then every state passed through by the system is an equilibrium state and the work done can then be estimated from the relation $\int_1^2 p\mathrm{d}V$ (Fig. 3.17e), Yet, in free expansion of a gas, thee is no resistance to the fluid at the system boundary as the volume of the gas increases to fill up the vacuum space. Work is done by a system to overcome some resistance. Since vacuum does not offer any resistance, there is no work transfer involved in free expansion.

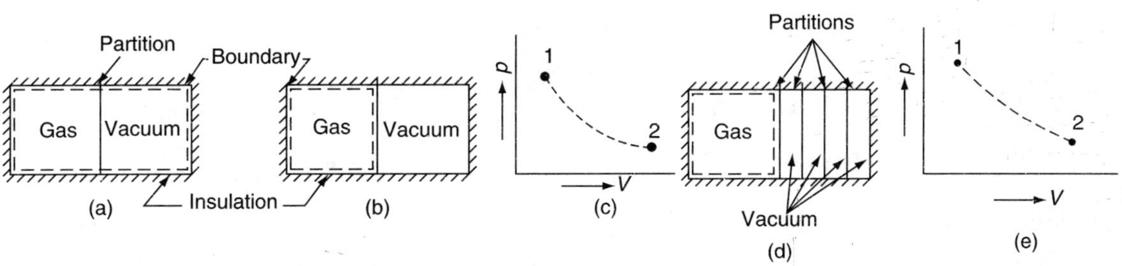

Fig. 3.17 *Free expansion*

3.6 NET WORK DONE BY A SYSTEM

Often different forms of a work transfer occur simultaneously during a process executed by a system. When all these work interactions have been evaluated, the total or net work done by the system would be equal to the algebraic sum of these as given below

$$W_{\text{total}} = W_{\text{displacement}} + W_{\text{shear}} + W_{\text{electrical}} + W_{\text{stirring}} + \cdots$$

3.7 HEAT TRANSFER

Heat is defined as the form of energy that is transferred across a boundary *by virtue of a temperature difference*. The temperature difference is the 'potential' or 'force' and heat transfer is the 'flux'.

The transfer of heat between two bodies in direct contact is called *conduction*. Heat may be transferred between two bodies separated by empty space or gases by the mechanism of *radiation* through electromagnetic waves. A third method of heat transfer is *convection* which refers to the transfer of heat between a wall and a fluid system in motion.

The direction of heat transfer is taken from the high temperature system to the low temperature system. *Heat flow into a system is taken to be positive, and heat flow out of a system is taken as negative* (Fig. 3.18). The symbol Q is used for heat transfer, i.e. the quantity of heat transferred within a certain time.

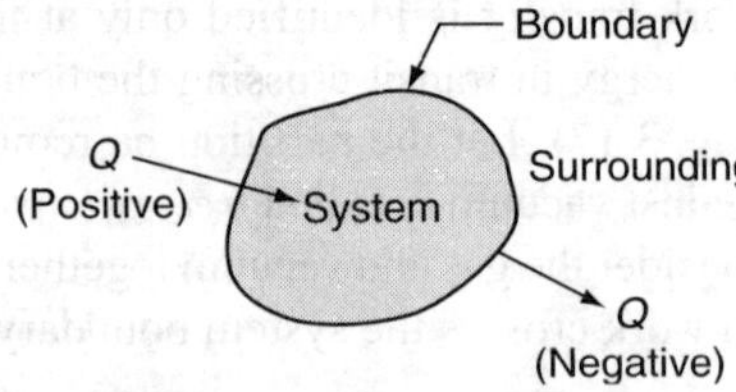

Fig. 3.18 *Direction of heat transfer*

Heat is a form of energy in transit (like work transfer). It is a boundary phenomenon, since it occurs only at the boundary of a system. Energy transfer by virtue of temperature difference only is called heat transfer. All other energy interactions may be termed as work transfer.

Heat is not that which inevitably causes a temperature rise. When heat is transferred to an ice-and-water mixture, the temperature does not rise until all the ice has melted. When a temperature rise in a system occurs, it may not be due to heat transfer, since a temperature rise may be caused by work transfer also. Heat, like work, is not a conserved quantity, and is not a property of a system.

A process in which no heat crosses the boundary of the system is called an *adiabatic process*. Thus, an adiabatic process is one in which there is only work interaction between the system and its surroundings.

A wall which is impermeable to the flow of heat is an *adiabatic wall*, whereas a wall which permits the flow of heat is a *diathermic wall*.

The unit of heat is Joule in S.I. units.

The rate of heat transfer or work transfer is given in kW or W.

3.8 HEAT TRANSFER—A PATH FUNCTION

Heat transfer is a *path function*, that is, the amount of heat transferred when a system changes from state 1 to state 2 depends on the intermediate states through which the system passes, i.e. its path. Therefore dQ is an inexəct differential, and we write $\int_1^2 đQ = Q_{1-2}$ or ${}_1Q_2$

The displacement work is given by $W_{1-2} = \int_1^2 đW = \int_1^2 p\mathrm{d}V$

It is valid for a quasi-static process, and the work transfer involved is represented by the area under the path on *p-v* diagram (Fig. 3.19a). Whenever there is a difference in pressure, there will be displacement work. The pressure difference is the cause and work transfer is the effect. The work transfer is equal to the integral of the product of the intensive property, p and the differential change in the extensive property, dV.

Likewise, whenever there is a difference in temperature, there will be heat flow. The temperature difference is the cause and heat transfer is the effect. Just like displacement work, the heat transfer can also be written as the integral of the product of the intensive property T and the differential change of an extensive property, say X (Fig. 3.19b).

$$Q_{1-2} = \int_1^2 đQ = \int_1^2 T\mathrm{d}X \tag{3.27}$$

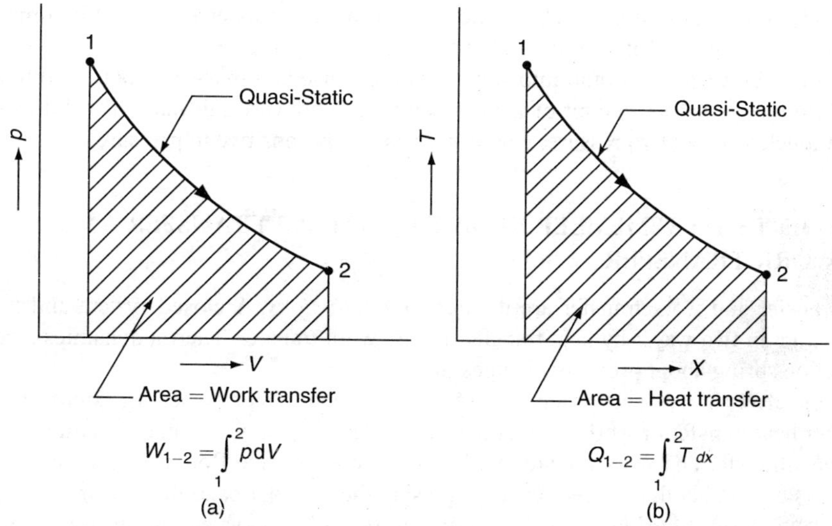

Fig. 3.19 *Representation of work transfer and heat transfer in quasi-static processes on p-v and T-x coordinates*

It must also be valid for a quasi-static process only, and the heat transfer involved is represented by the area under the path 1–2 in *T-X* plot (Fig. 3.19b). Heat transfer is, therefore, a path function, i.e. the amount of heat transferred when a system changes from a state 1 to a state 2 depends on the path the system follows (Fig. 3.19b). Therefore, $đQ$ is an inexact differential. Now, $đQ = T\,dX$ where X is an extensive property and dX is an exact differential.

$$\therefore \qquad dX = \frac{1}{T} đQ \tag{3.28}$$

To make $đQ$ integrable, i.e., an exact differential, it must be multiplied by an integrating factor which is, in this case, $1/T$. The extensive property X is yet to be defined. It has been introduced in Chapter 7 and it is called 'entropy'.

3.9 SPECIFIC HEAT AND LATENT HEAT

The *specific heat* of a substance is defined as the amount of heat required to raise a unit mass of the substance through a unit rise in temperature. The symbol c will be used for specific heat.

$$\therefore \qquad c = \frac{Q}{m \cdot \Delta t} \text{ J/kg K}$$

where Q is the amount of heat transfer (J), m, the mass of the substance (kg), and Δt, the rise in temperature (K).

Since heat is not a property, as explained later, so the specific heat is qualified with the process through which exchange of heat is made. For gases, if the process is at constant pressure, it is c_p, and if the process is at constant volume, it is c_v. For solids and liquids, however, the specific heat does not depend on the process. An elegant manner of defining specific heats, c_v and c_p, in terms of properties is given in Secs 4.5 and 4.6.

The product of mass and specific heat (mc) is called the *heat capacity* of the substance. The capital letter C, C_p or C_v, is used for heat capacity.

The latent heat is the amount of heat transfer required to cause a phase change in unit mass of a substance *at a constant pressure and temperature*. There are three phases in which matter can exist: solid, liquid, and vapour or gas. *The latent heat of fusion* (l_{fu}) is the amount of heat transferred to melt unit mass of solid into liquid, or to freeze unit mass of liquid to solid. The *latent heat of vaporization* (l_{vap}) is the quantity of heat required to vaporize unit mass of liquid into vapour, or condense unit mass of vapour into liquid. The *latent heat of sublimation* (l_{sub}) is the amount of heat transferred to convert unit mass of solid to vapour or vice versa. l_{fu} is not much affected by pressure, whereas l_{vap} is highly sensitive to pressure.

3.10 POINTS TO REMEMBER REGARDING HEAT TRANSFER AND WORK TRANSFER

(a) Heat transfer and work transfer are the *energy interactions*. A closed system and its surroundings can interact in two ways: by heat transfer and by work transfer. Thermodynamics studies how these interactions bring about property changes in a system.

(b) The same effect in a closed system can be brought about either by heat transfer or by work transfer. Whether heat transfer or work transfer has taken place depends on what constitutes the system.

(c) Both heat transfer and work transfer are boundary phenomena. Both are observed at the boundaries of the system, and both represent energy crossing the boundaries of the system.

(d) *It is wrong to say 'total heat' or 'heat content' of a closed system*, because heat or work is not a property of the system. Heat, like work, cannot be stored by the system. Both heat and work are the energy is transit.

(e) Heat transfer is the energy interaction due to temperature difference only. All other energy interactions may be termed as work transfer.

(f) Both heat and work are path functions and inexact differentials. The magnitude of heat transfer or work transfer depends upon the path the system follows during the change of state.

Solved Examples

Example 3.1

Gas from a bottle of compressed helium is used to inflate an inelastic flexible balloon, originally folded completely flat to a volume of 0.5 m^3. *If the barometer reads* 760 *mm Hg, what is the amount of work done upon the atmosphere by the balloon? Sketch the system before and after the process.*

Solution The firm line P_1 (Fig. Ex. 3.1) shows the boundary of the system before the process, and the dotted line P_2 shows the boundary after the process. The displacement work

$$W_d = \int_{\text{Balloon}} p\text{d}V + \int_{\text{Bottle}} p\text{d}V = p\Delta V + 0$$

$$= 101.325\ \frac{\text{kN}}{\text{m}^2} \times 0.5\ \text{m}^3 = 50.66\ \text{kJ}$$

This is positive because work is done by the system. Work done by the atmosphere is –50.66 kJ. Since the wall of the bottle is rigid, there is no pdV–work involved in it.

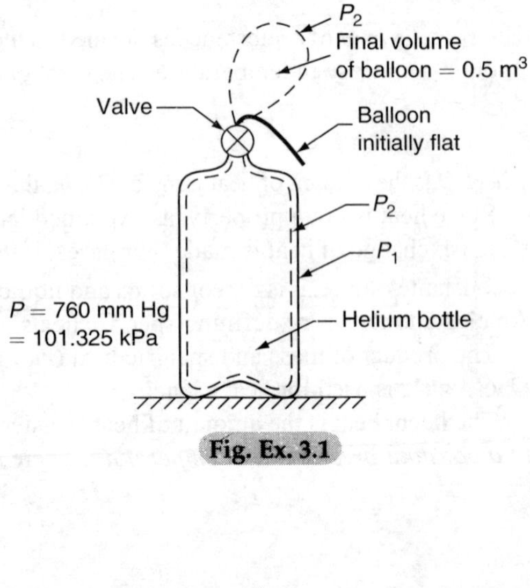

Fig. Ex. 3.1

It is assumed that the pressure in the balloon is atmospheric at all times, since the balloon fabric is light, inelastic and unstressed. If the balloon were elastic and stressed during the filling process, the work done by the gas would be greater than 50.66 kJ by an amount equal to the work done in stretching the balloon, although the displacement work done by the atmosphere is still –50.66 kJ. However, if the system includes both the gas and the balloon, the displacement work would be 50.66 kJ, as estimated above.

Example 3.2

When the valve of the evacuated bottle (Fig. *Ex.* 3.2) *is opened, atmospheric air rushes into it. If the atmospheric pressure is* 101.325 *kPa, and* 0.6 *m*^{3} *of air* (*measured at atmospheric conditions*) *enters into the bottle, calculate the work done by air.*

Solution The displacement work done by air

$$W_d = \int_{\text{Bottle}} p\,dV + \int_{\substack{\text{Free-air}\\\text{boundary}}} p\,dV$$

$$= 0 + p\Delta V$$

$$= 101.325 \text{ kN/m}^2 \times 0.6 \text{ m}^3 = 60.8 \text{ kJ}$$

Since the free-air boundary is contracting, the work done by the system is negative (ΔV being negative), and the surroundings do positive work upon the system.

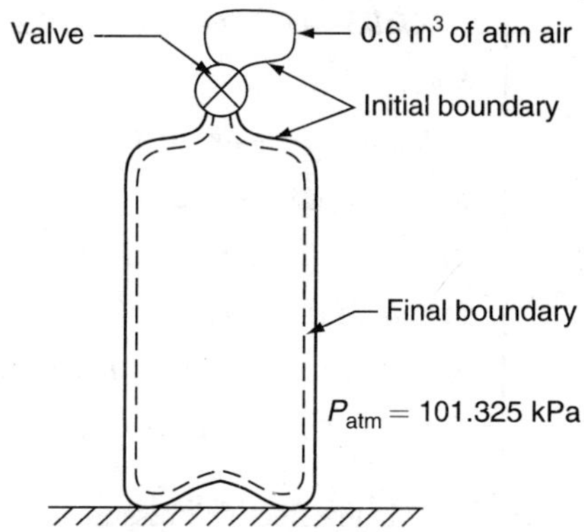

Fig. Ex. 3.2

Example 3.3

A piston and cylinder machine containing a fluid system has a stirring device in the cylinder (*Fig. Ex.* 3.3). *The piston is frictionless, and it is held down against the fluid due to the atmospheric pressure of* 101.325 *kPa. The stirring device is turned* 10,000 *revolutions with an average torque against the fluid of* 1.275 *mN. Meanwhile the piston of* 0.6 *m diameter moves out* 0.8 *m. Find the net work transfer for the system.*

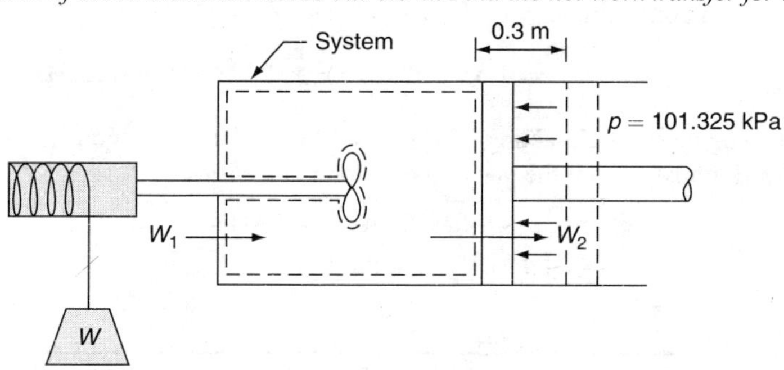

Fig. Ex. 3.3

Solution Work done by the stirring device upon the system (Fig. Ex. 3.3).

$$W_1 = 2\pi TN$$

$$= 2\pi \times 1.275 \times 10{,}000 \text{ Nm} = 80 \text{ kJ}$$

This is negative work for the system.

Work done by the system upon the surroundings

$$W_2 = (pA) \cdot L$$

$$= 101.325 \frac{\text{kN}}{\text{m}^2} \times \frac{\pi}{4} (0.6)^2 \text{ m}^2 \times 0.80 \text{ m} = 22.9 \text{ kJ}$$

This is positive work for the system. Hence, the net work transfer for the system

$$W = W_1 + W_2 = -80 + 22.9 = -57.1 \text{ kJ}$$

Example 3.4

The following data refer to a 12-cylinder, single-acting, two-stroke marine diesel engine:

Speed—150 rpm

Cylinder diameter—0.8 m

Stroke of piston—1.2 m

Area of indicator diagram—5.5 × 10^{-4} m²

Length of diagram—0.06 m

Spring value—147 MPa per m

Find the net rate of work transfer from the gas to the pistons in kW.

Solution Mean effective pressure, p_m, is given by

$$p_m = \frac{a_d}{l_d} \times \text{spring constant}$$

$$= \frac{5.5 \times 10^{-4} \text{ m}^2}{0.06} \times 147 \frac{\text{MPa}}{\text{m}} = 1.35 \text{ MPa}$$

One engine cycle is completed in two strokes of the piston or one revolution of the crank-shaft.

∴ Work done in one minute $= p_m LAN$

$$= 1.35 \times \frac{\pi}{4} (0.8)^2 \times 1.2 \times 150 = 122 \text{ MJ}$$

Since the engine is single-acting, and it has 12 cylinders, each contributing an equal power, the rate of work transfer from the gas to the piston is given by

$$\mathbf{W} = 122 \times 12 \text{ MJ/min} = 24.4 \text{ MJ/s}$$

$$= 24.4 \text{ MW} = 24{,}400 \text{ kW}$$ *Ans.*

Example 3.5

It is required to melt 5 tonnes/h of iron from a charge at 15°C to molten metal at 1650°C. The melting point is 1535°C, and the latent heat is 270 kJ/kg. The specific heat in solid state is 0.502 and in liquid state (29.93/atomic weight) kJ/kg K. If an electric furnace has 70% efficiency, find the kW rating needed. If the density in molten state is 6900 kg/m³ and the bath volume is three times the hourly melting rate, find the dimensions of the cylindrical furnace if the length to diameter ratio is 2. The atomic weight of iron is 56.

Solution Heat required to melt 1 kg of iron at 15°C to molten metal at 1650°C

= Heat required to raise the temperature from 15°C to 1535°C + Latent heat + Heat required to raise the temperature from 1535°C to 1650°C

= 0.502 (1535 – 15) + 270 + 29.93 (1650 – 1535)/56

= 763 + 270 + 61.5 = 1094.5 kJ/kg

Melting rate = 5×10^3 kg/h

So, the rate of heat supply required =($5 \times 10^3 \times 1094.5$) kJ/h

Since the furnace has 70% efficiency, the rating of the furnace would be

$$= \frac{\text{Rate of heat supply per second}}{\text{Furnace efficiency}}$$

$$= \frac{5 \times 10^3 \times 1094.5}{0.7 \times 3600} = 217 \times 10^3 \text{ kW}$$ *Ans.*

$$\text{Volume needed} = \frac{3 \times 5 \times 10^3}{6900} \text{ m}^3 = 2.18 \text{ m}^3$$

If d is the diameter and l the length of the furnace

$$\frac{\pi}{4} d^2 l = 2.18 \text{ m}^3$$

or $$\frac{\pi}{4} d^2 \times 2d = 2.18 \text{ m}^3$$

∴ $$d = 1.15 \text{ m}$$

and $$l = 2.30 \text{ m}$$ *Ans.*

Example 3.6

If it is desired to melt aluminium with solid state specific heat 0.9 kJ/kg K*, latent heat* 390 kJ/kg*, atomic weight* 27*, density in molten state* 2400 kg/m³ *and final temperature* 700°C*, find out how much metal can be melted per hour with the above kW rating. Other data are as in the above example. Also, find the mass of aluminium that the above furnace will hold. The melting point of aluminium is* 660°C.

Solution Heat required per kg of aluminium

$$= 0.9 (660 - 15) + 390 + \frac{29.93}{27} (700 - 660)$$

$$= 580.5 + 390 + 44.3 = 1014.8 \text{ kJ}$$

$$\text{Heat to be supplied} = \frac{1014.8}{0.7} = 1449.7 \text{ kJ/kg}$$

With the given power, the rate at which aluminium can be melted

$$= \frac{2.17 \times 10^3 \times 3600}{1449.7} \text{ kg/h} = 5.39 \text{ tonnes/h}$$ *Ans.*

Mass of aluminium that can be held in the above furnace

$$= 2.18 \times 2400 \text{ kg} = 5.23 \text{ tonnes}$$ *Ans.*

Review Questions

3.1 How can a closed system and its surroundings interact? What is the effect of such interactions on the system?

3.2 When is work said to be done by a system?

3.3 What are positive and negative work interactions?

3.4 What is displacement work?

3.5 Underwhatconditionsistheworkdoneequalto $\int_1^2 p\mathrm{d}V$?

3.6 What do you undersand by path function and point function? What are exact and inexact differentials?

3.7 Show that work is a path function, and not a property.

3.8 What is an indicator diagram?

3.9 What is mean effective pressure? How is it measured?

3.10 What are the indicated power and the brake power of an engine?

3.11 How does the current flowing through a resistor represent work transfer?

3.12 What do you understand by flow work? Is it different from displacement work?

3.13 Why does free expansion have zero work transfer?

3.14 What is heat transfer? What are its positive and negative directions?

3.15 What are adiabatic and diathermic walls?

3.16 What is an integrating factor?

3.17 Show that heat is a path function and not a property.

3.18 What is the difference between work transfer and heat transfer?

3.19 Does heat transfer inevitably cause a temperature rise?

Problems

3.1 (a) A pump forces 1 $\mathrm{m^3/min}$ of water horizontally from an open well to a closed tank where the pressure is 0.9 MPa. Compute the work the pump must do upon the water in an hour just to force the water into the tank against the pressure. Sketch the system upon which the work is done before and after the process.

Ans. 13.31 kJ

(b) If the work done as above upon the water had been used solely to raise the same amount of water vertically against gravity without change of pressure, how many metres would the water have been elevated?

(c) If the work done in (a) upon the water had been used solely to accelerate the water from zero velocity without change of pressure or elevation, what velocity would the water have reached? If the work had been used to accelerate the water from an initial velocity of 10 m/s, what would the final velocity have been?

3.2 The piston of an oil engine, of area 0.0045 $\mathrm{m^2}$, moves downwards 75 mm, drawing in 0.00028 $\mathrm{m^3}$ of fresh air from the atmosphere. The pressure in the cylinder is uniform during the process at 80 kPa, while the atmospheric pressure is 101.325 kPa, the difference being due to the flow resistance in the induction pipe and the inlet valve. Estimate the displacement work done by the air finally in the cylinder. *Ans.* 27 J

3.3 An engine cylinder has a piston of area 0.12 $\mathrm{m^2}$ and contains gas at a pressure of 1.5 MPa. The gas expands according to a process which is represented by a straight line on a pressure-volume diagram. The final pressure is 0.15 MPa. Calculate the work done by the gas on the piston if the stroke is 0.30 m. *Ans.* 29.7 kJ.

3.4 A mass of 1.5 kg of air is compressed in a quasi-static process from 0.1 MPa to 0.7 MPa for which pv = constant. The initial density of air is 1.16 $\mathrm{kg/m^3}$. Find the work done by the piston to compress the air. *Ans.* 251.62 kJ

3.5 A mass of gas is compressed in a quasi-static process from 80 kPa, 0.1 $\mathrm{m^3}$ to 0.4 MPa, 0.03 $\mathrm{m^3}$. Assuming that the pressure and volume are related by pv^n = constant, find the work done by the gas system. *Ans.* –11.83 kJ

3.6 A single-cylinder, double-acting, reciprocating water pump has an indicator diagram which is a rectangle 0.075 m long and 0.05 m high. The indicator spring constant is 147 MPa per m. The pump runs at 50 rpm. The pump cylinder diameter is 0.15 m and the piston stroke is 0.20 m. Find the rate in kW at which the piston does work on the water. *Ans*. 43.3 kW

3.7 A single-cylinder, single-acting, 4 stroke engine of 0.15 m bore develops an indicated power of 4 kW when running at 216 rpm. Calculate the area of the indicator diagram that would be obtained with an indicator having a spring constant of 25×10^6 N/m^3. The length of the indicator diagram is 0.1 times the length of the stroke of the engine.

Ans. 505 mm^2

3.8 A six-cylinder, 4-stroke gasoline engine is run at a speed of 2520 RPM. The area of the indicator card of one cylinder is 2.45×10^3 mm^2 and its length is 58.5 mm. The spring constant is 20×10^6 N/m^3. The bore of the cylinders is 140 mm and the piston stroke is 150 mm. Determine the indicated power, assuming that each cylinder contributes an equal power. *Ans*. 243.57 kW

3.9 A closed cylinder of 0.25 m diameter is fitted with a light frictionless piston. The piston is retained in position by a catch in the cylinder wall and the volume on one side of the piston contains air at a pressure of 750 kN/m^2. The volume on the other side of the piston is evacuated. A helical spring is mounted coaxially with the cylinder in this evacuated space to give a force of 120 N on the piston in this position. The catch is released and the piston travels along the cylinder until it comes to rest after a stroke of 1.2 m. The piston is then held in its position of maximum travel by a ratchet mechanism. The spring force increases linearly with the piston displacement to a final value of 5 kN. Calculate the work done by the compressed air on the piston. *Ans*. 3.07 kJ

3.10 A steam turbine drives a ship's propeller through an 8 : 1 reduction gear. The average resisting torque imposed by the water on the propeller is 750×10^3 N and the shaft power delivered by the turbine to the reduction gear is 15 MW. The turbine speed is 1450 rpm. Determine (a) the torque developed by the turbine, (b) the power delivered to the propeller shaft, and (c) the net rate of working of the reduction gear.

Ans. (a) $T = 98.84$ km N, (b) 14.235 MW, (c) 0.765 MW

3.11 A fluid, contained in a horizontal cylinder fitted with a frictionless leakproof piston, is continuously agitated by means of a stirrer passing through the cylinder cover. The cylinder diameter is 0.40 m. During the stirring process lasting 10 minutes, the piston slowly moves out a distance of 0.485 m against the atmosphere. The net work done by the fluid during the process is 2 kJ. The speed of the electric motor driving the stirrer is 840 rpm. Determine the torque in the shaft and the power output of the motor.

Ans. 0.08 mN, 6.92 W

3.12 At the beginning of the compression stroke of a two-cylinder internal combustion engine the air is at a pressure of 101.325 kPa. Compression reduces the volume to 1/5 of its original volume, and the law of compression is given by $pv^{1.2}$ = constant. If the bore and stroke of each cylinder is 0.15 m and 0.25 m, respectively, determine the power absorbed in kW by compression strokes when the engine speed is such that each cylinder undergoes 500 compression strokes per minute.

Ans. 17.95 kW

3.13 Determine the total work done by a gas system following an expansion process as shown in Fig. P. 3.13. *Ans*. 2952 MJ

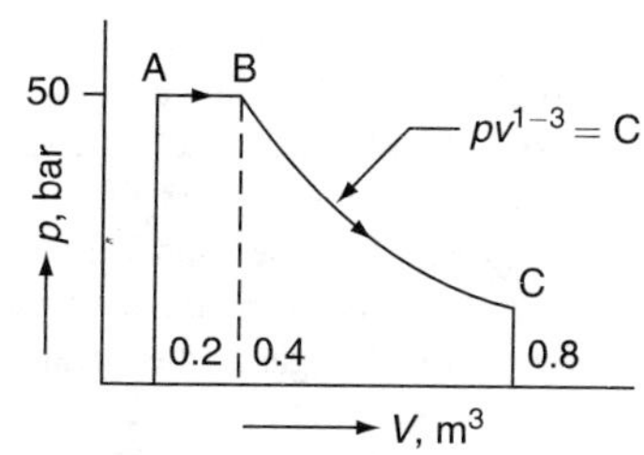

Fig. P. 3.13

3.14 A system of volume V contains a mass m of gas at pressure p and temperature T. The macroscopic properties of the system obey the following relationship:

$$\left(p + \frac{a}{V^2}\right)(V - b) = mRT$$

where a, b and R constants.

Obtain an expression for the displacement work done by the system during a constant-temperature expansion from volume V_1 to volume V_2. Calculate the work done by a system which contains 10 kg of this gas expanding from 1 m^3 to 10 m^3 at a temperature of 293 K. Use the values $a = 15.7 \times 10^4$ Nm^4, $b = 1.07 \times 10^{-2}$ m^3, and $R = 0.278$ kJ/kg–K. *Ans.* 1742.14 kJ

3.15 If a gas of volume 6000 cm^3 and at pressure of 100 kPa is compressed quasistatically according to pV^2 = constant until the volume becomes 2000 cm^3, determine the final pressure and the work transfer. *Ans.* 900 kPa, 1.2 kJ

3.16 The flow energy of 0.124 m^3/min of a fluid crossing a boundary to a system is 18 kW. Find the pressure at this point. *Ans.* 764 kPa

3.17 A milk chilling unit can remove heat from the milk at the rate of 41.87 MJ/h. Heat leaks into the milk from the surroundings at an average rate of 4.187 MJ/h. Find the time required for cooling a batch of 500 kg of milk from 45°C to 5°C. Take the c_p of milk to be 4.187 kJ/kgK. *Ans.* 2h 13 min

3.18 680 kg of fish at 5°C are to be frozen and stored at – 12°C. The specific heat of fish above freezing point is 3.182, and below freezing point is 1.717 kJ/kgK. The freezing point is – 2°C, and the latent heat of fusion is 234.5 kJ/kg. How much heat must be removed to cool the fish, and what per cent of this is latent heat? *Ans.* 186.28 MJ, 85.6%

3.19 A horizontal cylinder fitted with a sliding piston contains 0.1 m^3 of a gas at a pressure of 1 atm. The piston is restrained by a linear spring. In the initial state, the gas pressure inside the cylinder just balances the atmospheric pressure of 1 atm of the outside of the piston and the spring exerts no force on the piston. The gas is then heated reversibly untill its volume and presure become 0.16 m^2 and 2 atm, respectively. (a) Write the equation for the relation between the pressure and volume of the gas. (b) Calculate the work done by the gas. (c) Of the total work done by the gas, how much is done against the atmosphere? How much is done against the spring?

Ans. (a) p(N/m^2) = 2.026 × 10^{6}V–1.013 × 10^5

(b) 7,598 J, (c) 5,065 J, 2,533 J

3.20 An elastic sphere initially has a diameter of 1 m and contains a gas at a pressure of 1 atm. Due to heat transfer the diameter of the sphere increases to 1.1 m. During the heating process the gas pressure inside the sphere is proportional to the sphere diameter. Calculate the work done by the gas. *Ans.* 18.4 kJ

3.21 A piston-cylinder device contains 0.05 m^3 of a gas initially at 200 kPa. At this state, a linear spring having a spring constant of 150 kN/m is touching the piston but exerting no force on it. Now heat is transferred to the gas, causing the piston to rise and to compress the spring until the volume inside the cylinder doubles. If the cross-sectional area of the piston is 0.25 m^2, determine (a) the final pressure inside the cylinder, (b) the total work done by the gas, and (c) the fraction of this work done against the spring to compress it. *Ans.* (a) 320 kPa, (b) 13 kJ, (c) 3 kJ

3.22 A piston-cylinder device, whose piston is resting on a set of stops, initially contains 3 kg of air at 200 kPa and 27°C. The mass of the piston is such that a pressure of 400 kPa is required to move it. Heat is now transferred to the air until its volume doubles. Determine the work done by the air and the total heat transferred to the air. *Ans.* 516 kJ, 2674 kJ

C H A P T E R 4

First Law of Thermodynamics

Energy can be in two forms: (a) energy in *transit*, like heat and work transfer observed at the boundaries of a system, and (b) energy in *storage*, where energy is stored either *macroscopically* by virtue of motion, position or configuration of the system, or *microscopically* in the molecules or atoms constituting the system.

4.1 FIRST LAW FOR A CLOSED SYSTEM UNDERGOING A CYCLE

The transfer of heat and the performance of work may both cause the same effect in a system. Heat and work are different forms of the same entity, called energy, which is conserved. Energy which enters a system as heat may leave the system as work, or energy which enters the system as work may leave as heat.

Let us consider a closed system which consists of a known mass of water contained in an adiabatic vessel having a thermometer and a paddle wheel, as shown in Fig. 4.1. Let a certain amount of work W_{1-2} be done upon the system by the paddle wheel. The quantity of work can be measured by the fall of weight which drives the paddle wheel through a pulley. The system was initially at temperature t_1, the same as that of atmosphere, and after work transfer let the temperature rise to t_2. The pressure is always 1 atm. The process 1–2 undergone by the system is shown in Fig. 4.2 in generalized thermodynamic coordinates X, Y. Let the insulation now be removed. The system and the surroundings interact by heat transfer till the system returns to the original temperature t_1, attaining the condition of thermal equilibrium with the atmosphere. The amount of heat transfer Q_{2-1} from the system during this process, 2–1, shown in Fig. 4.2, can be estimated. The system thus executes a cycle, which consists of a definite amount of work input W_{1-2} to the system followed by the transfer of an amount of heat Q_{2-1} from the system. It has been found that this W_{1-2} is always proportional to the heat Q_{2-1}, and the constant of proportionality is called the Joule's equivalent or the *mechanical equivalent of heat*. In the simple example given here, there are only two energy transfer quantities as the system performs a thermodynamic cycle. If the cycle involves many more heat and work quantities, the same result will be found. Expressed algebraically.

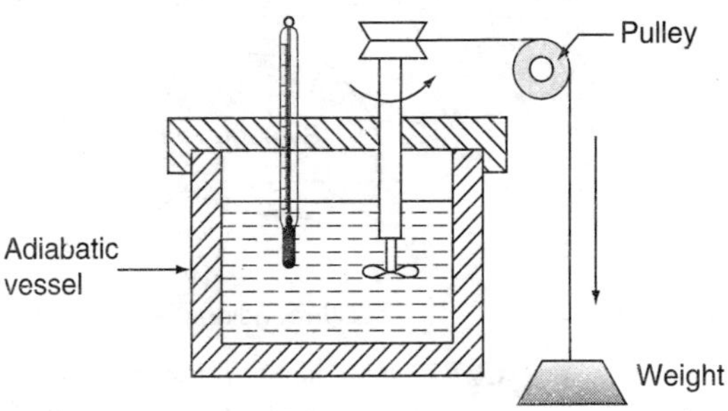

Fig. 4.1 *Adiabatic work*

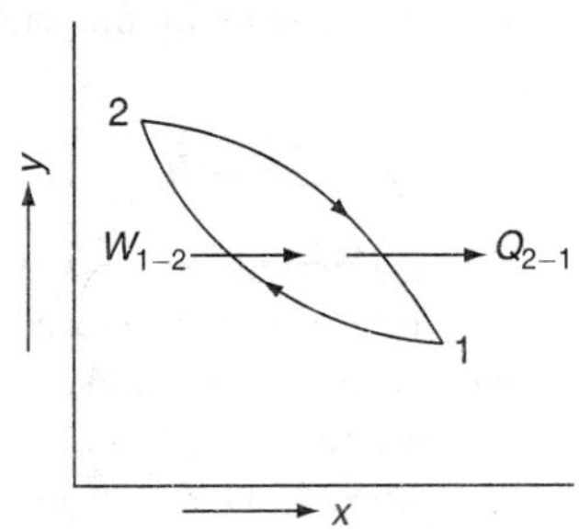

Fig. 4.2 *Cycle completed by a system with two energy interactions: adiabatic work transfer W_{1-2} followed by heat transfer Q_{2-1}*

$$(\Sigma W)_{\text{cycle}} = J(\Sigma Q)_{\text{cycle}} \tag{4.1}$$

where J is the Joule's equivalent. This is also expressed in the form $\oint dW = J \oint dQ$

where the symbol $\oint$ denotes the cyclic integral for the closed path. This is the *first law for a closed system undergoing a cycle*. It is accepted as a *general law of nature*, since no violation of it has ever been demonstrated.

In the S.I. system of units, both heat and work are measured in the derived unit of energy, the Joule. The constant of proportionality, J, is therefore unity ($J = 1$ Nm/J).

The first law of thermodynamics owes much to J.P. Joule who, during the period 1840–1849, carried out a series of experiments to investigate the equivalence of work and heat. In one of these experiments, Joule used an apparatus similar to the one shown in Fig. 4.1. Work was transferred to the measured mass of water by means of paddle wheel driven by the falling weight. The rise in the temperature of water was recorded. Joule also used mercury as the fluid system, and later a solid system of metal blocks which absorbed work by friction when rubbed against each other. Other experiments involved the supplying of work in an electric current. In every case, he found the same ratio (J) between the amount of work and the quantity of heat that would produce identical effects in the system.

Prior to Joule, heat was considered to be an invisible fluid flowing from a body of higher calorie to a body of lower calorie, and this was known as the *caloric theory of heat*. It was Joule who first established that heat is a form of energy, and thus laid the foundation of the first law of thermodynamics.

4.2 FIRST LAW FOR A CLOSED SYSTEM UNDERGOING A CHANGE OF STATE

The expression $(\Sigma W)_{\text{cycle}} = (\Sigma Q)_{\text{cycle}}$ applies only to systems undergoing cycles, and the algebraic summation of all energy transfer across system boundaries is zero. But if a system undergoes a change of state during which both heat transfer and work transfer are involved, the *net* energy transfer will be stored or accumulated within the system. If Q is the amount of heat transferred to the system and W is the amount of work transferred from the system during the process (Fig. 4.3), the net energy transfer $(Q - W)$ will be stored in the system. Energy in storage is neither heat nor work, and is given the name *internal energy* or simply, the *energy* of the system.

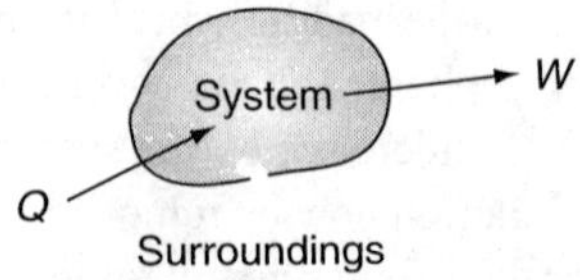

Fig. 4.3 *Heat and work interactions of a system with its surroundings in a process*

Therefore $Q - W = \Delta E$

where ΔE is the increase in the energy of the system

or
$$Q = \Delta E + W \tag{4.2}$$

Here Q, W, and ΔE are all expressed in the same units (in joules). Energy may be stored by a system in different modes, as explained in Article 4.4.

If there are more energy transfer quantities involved in the process, as shown in Fig. 4.4, the first law gives

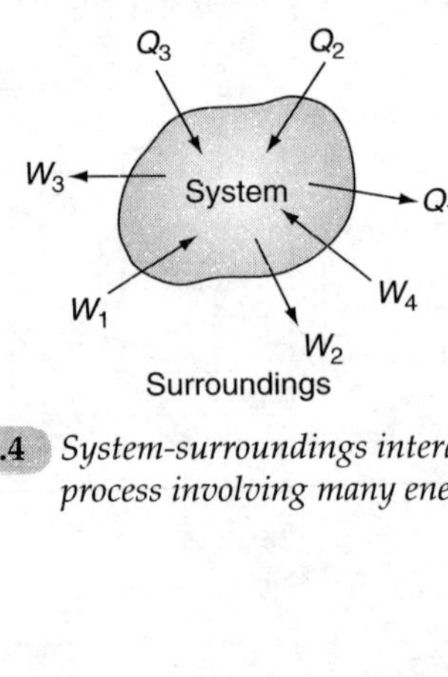

Fig. 4.4 *System-surroundings interaction in a process involving many energy fluxes*

$$(Q_2 + Q_3 - Q_1) = \Delta E + (W_2 + W_3 - W_1 - W_4)$$

Energy is thus conserved in the operation. The first law is a particular formulation of the principle of the conservation of energy. Equation (4.2) may also be considered as the definition of energy. This definition does not give an absolute value of energy E, but only the change of energy ΔE for the process. It can, however, be shown that the energy has a definite value at every state of a system and is, therefore, a property of the system.

4.3 ENERGY—A PROPERTY OF THE SYSTEM

Consider a system which changes its state from state 1 to state 2 by following the path A, and returns from state 2 to state 1 by following the path B (Fig. 4.5). So the system undergoes a cycle.

Writing the first law for path A $\quad Q_A = \Delta E_A + W_A \quad$ (4.3)

and for path B $\quad Q_B = \Delta E_B + W_B \quad$ (4.4)

The processes A and B together constitute a cycle, for which

$$(\Sigma W)_{cycle} = (\Sigma Q)_{cycle}$$

or

$$W_A + W_B = Q_A + Q_B$$

or

$$Q_A - W_A = W_B - Q_B \quad (4.5)$$

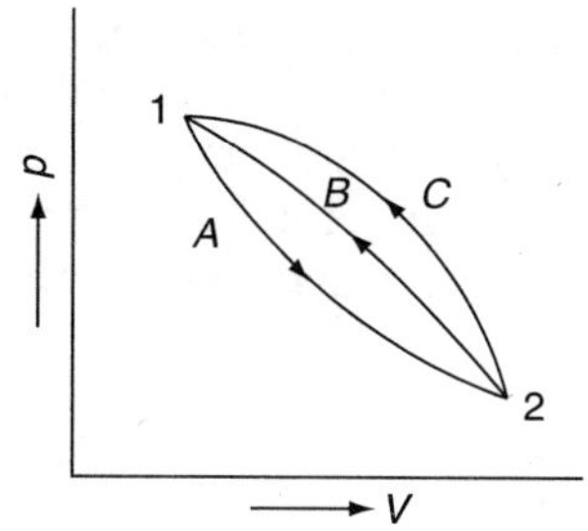

Fig. 4.5 *Energy–a property of a system*

From Eqs (4.3), (4.4), and (4.5), it yields $\quad \Delta E_A = -\Delta E_B \quad$ (4.6)

Similarly, had the system returned from state 2 to state 1 by following the path C instead of path B

$$\Delta E_A = -\Delta E_C \quad (4.7)$$

From Eqs (4.6) and (4.7) $\quad \Delta E_B = \Delta E_C \quad$ (4.8)

Therefore, it is seen that the change in energy between two states of a system is the same, whatever path the system may follow in undergoing that change of state. If some arbitrary value of energy is assigned to state 2, the value of energy at state 1 is fixed independent of the path the system follows. Therefore, energy has a definite value for every state of the system. Hence, it is a *point function and a property of the system.*

The energy E is an extensive property. The *specific energy*, $e = E/m$ (J/kg), is an intensive property.

The cyclic integral of any property is zero, because the final state is identical with the initial state. $\oint dE = 0$, $\oint dV = 0$, etc. So for a cycle, the Eq. (4.2) reduces to Eq. (4.1).

4.4 DIFFERENT FORMS OF STORED ENERGY

The symbol E refers to the total energy stored in a system. Basically there are two modes in which energy may be stored in a system:

(a) Macroscopic energy mode
(b) Microscopic energy mode

The macroscopic energy mode includes the macroscopic kinetic energy and potential energy of a system. Let us consider a fluid element of mass m having the centre of mass velocity $\bar{V}$ (Fig. 4.6). The macroscopic kinetic energy E_K of the fluid element by virtue of its motion is given by $E_K = \frac{m\bar{V}^2}{2}$

If the elevation of the fluid element from an arbitrary datum is z, then the macroscopic potential energy E_p by virtue of its position is given by $E_p = mgz$

The microscopic energy mode refers to the energy stored in the molecular and atomic structure of the system, which is called the *molecular internal energy* or *simply internal energy*, customarily denoted by the symbol U. Matter is composed of molecules. Molecules are in random thermal motion (for a gas) with an average velocity $\overline{v}$, constantly colliding with one another and with the walls (Fig. 4.6). Due to a collision, the molecules may be subjected to rotation as well as vibration. They can have translational kinetic energy, rotational kinetic energy, vibrational energy, electronic energy, chemical energy and nuclear energy (Fig. 4.7). If ε represents the energy of one molecule, then

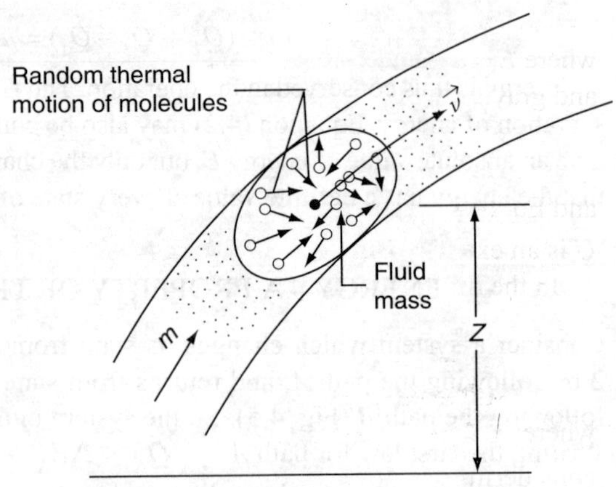

Fig. 4.6 *Macroscopic and microscopic energy*

$$\varepsilon = \varepsilon_{trans} + \varepsilon_{rot} + \varepsilon_{vib} + \varepsilon_{chem} + \varepsilon_{electronic} + \varepsilon_{nuclear} \tag{4.9}$$

If N is the total number of molecules in the system, then the total internal energy

$$U = N\varepsilon \tag{4.10}$$

In an ideal gas there are no intermolecular forces of attraction and repulsion, and the internal energy depends only on temperature. Thus

$$U = f(T) \text{ only} \tag{4.11}$$

for an ideal gas.

Other forms of energy which can also be possessed by a system are magnetic energy, electrical energy and surface (tension) energy. In the absence of these forms, the total energy E of a system is given by

$$E = \underbrace{E_K + E_P}_{\text{macro}} + \underbrace{U}_{\text{micro}} \tag{4.12}$$

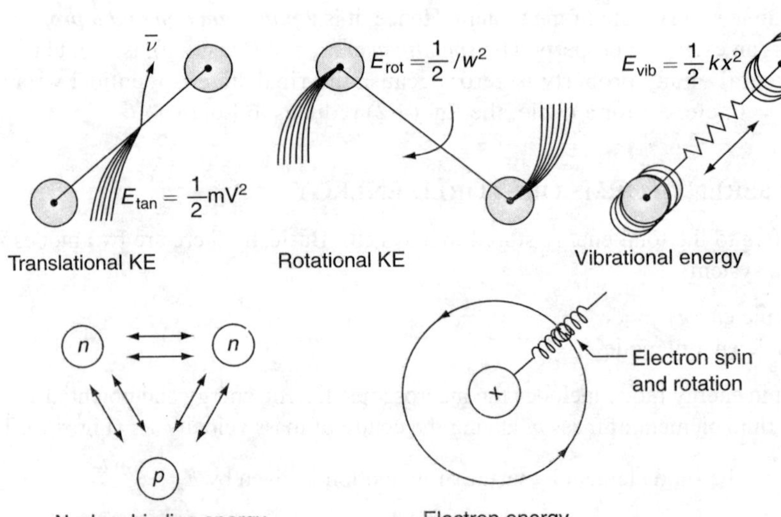

Fig. 4.7 *Various components of internal energy stored in a molecule*

where E_K, E_P, and U refer to the kinetic, potential and internal energy, respectively. In the absence of motion and gravity.

$$E_K = 0, \qquad E_P = 0$$
$$E = U$$

and Eq. (4.2) becomes

$$Q = \Delta U + W \tag{4.13}$$

U is an extensive property of the system. The specific internal energy u is equal to U/m and its unit is J/kg.

In the differential forms, Eqs (4.2) and (4.13) become

$$đQ = dE + đW \tag{4.14}$$

$$đQ = đU + đW \tag{4.15}$$

where

$$đW = đW_{pdV} + đW_{shaft} + đW_{electrical} + \ldots,$$

considering the different forms of work transfer which may be present. When only pdV work is present, the equations become

$$đQ = dE + pdV \tag{4.16}$$

$$đQ = dU + pdV \tag{4.17}$$

or, in the integral form

$$Q = \Delta E + \int pdV \tag{4.18}$$

$$Q = \Delta U + \int pdV \tag{4.19}$$

4.5 SPECIFIC HEAT AT CONSTANT VOLUME

The specific heat of a substance at constant volume c_V is defined as the rate of change of specific internal energy with respect to temperature when the volume is held constant, i.e.

$$c_v = \left(\frac{\partial u}{\partial T}\right)_v \tag{4.20}$$

For a constant-volume process

$$(\Delta u)_v = \int_{T_1}^{T_2} c_v \cdot dT \tag{4.21}$$

The first law may be written for a closed stationary system composed of a unit mass of a pure substance

$$Q = \Delta u + W$$

or

$$đQ = du + đW$$

For a process in the absence of work other than pdV work

$$đW = p\,đV$$

$\therefore$

$$đQ = du + pdv \tag{4.22}$$

When the volume is held constant

$$(Q)_v = (\Delta u)_v$$

$\therefore$

$$(Q)_v = \int_{T_1}^{T_2} c_v \, dT \tag{4.23}$$

Heat transferred at constant volume increases the internal energy of the system.

If the specific heat of a substance is defined in terms of heat transfer, then

$$c_v = \left(\frac{\partial Q}{\partial T}\right)_v$$

Since Q is not a property, this definition does not imply that c_v is a property of a substance. Therefore, this is not the appropriate method of defining the specific heat, although $(dQ)_v = du$.

Since u, T, and v are properties, c_v is a property of the system. The product $mc_v = C_v$ is called the *heat capacity at constant volume* (J/K).

4.6 ENTHALPY

The enthalpy of a substance, h, is defined as $$h = u + pv \tag{4.24}$$
It is an intensive property of a system (kJ/kg).

Internal energy change is equal to the heat transferred in a constant volume process involving no work other than pdV work. From Eq. (4.22), it is possible to drive an expression for the heat transfer in a constant pressure process involving no work other than pdV work. In such a process in a closed stationary system of unit mass of a pure substance

$$đQ = du + pdv$$

At constant pressure
$$pdv = d(pv)$$

∴
$$(đQ)_p = du + d(pv)$$

or
$$(đQ)_p = d(u + pv)$$

or
$$(đQ)_p = dh \tag{4.25}$$

where $h = u + pv$ is the *specific enthalpy*, a property of the system.

Heat transferred at constant pressure increases the enthalpy of a system.

For an ideal gas, the enthalpy becomes $$h = u + RT \tag{4.26}$$

Since the internal energy of an ideal gas depends only on the temperature (Eq. 4.11), the enthalpy of an ideal gas also depends on the temperature only, i.e.

$$h = f(T) \text{ only} \tag{4.27}$$

Total enthalpy $H = mh$

Also
$$H = U + pV$$

and
$$h = H/m \text{ (J/kg)}$$

4.7 SPECIFIC HEAT AT CONSTANT PRESSURE

The specific heat at constant pressure c_p is defined as the rate of change of enthalpy with respect to temperature when the pressure is held constant
$$c_p = \left(\frac{\partial h}{\partial T}\right)_p \tag{4.28}$$

Since h, T and p are properties, so c_p is a property of the system. Like c_v, c_p should not be defined in terms of heat transfer at constant pressure, although $(dQ)_p = dh$.

For a constant pressure process
$$(\Delta h)_p = \int_{T_1}^{T_2} c_p \cdot dT \tag{4.29}$$

The first law for a closed stationary system of unit mass

$$đQ = du + pdv$$

Again
$$h = u + pv$$

∴
$$dh = du + pdv + vdp$$
$$= dQ + vdp$$

∴
$$đQ = dh - vdp \tag{4.30}$$

$$\therefore \qquad (đQ)_p = dh \quad \text{or} \quad (Q)_p = (\Delta h)_p$$

$\therefore$ From Eqs (4.19) and (4.20)

$$(Q)_p = \int_{T_1}^{T_2} c_p \, dT$$

c_p is a property of the system, just like c_v. The *heat capacity at constant pressure* C_p is equal to mc_p (J/K).

4.8 ENERGY OF AN ISOLATED SYSTEM

An isolated system is one in which there is no interaction of the system with the surroundings. For an isolated system $đQ = 0$, $đW = 0$.

The first law gives $\quad dE = 0 \quad$ or $\quad E = \text{constant}$

The energy of an isolated system is always constant.

4.9 PERPETUAL MOTION MACHINE OF THE FIRST KIND–PMM1

The first law states the general principle of the conservation of energy. *Energy is neither created nor destroyed, but only gets transformed from one form to another.* There can be no machine which would continuously supply mechanical work without some other form of energy disappearing simultaneously (Fig. 4.8). Such a *fictitious machine* is called a *perpetual motion machine of the first kind*, or in brief, PMM1. *A PMM1 is thus impossible.*

The converse of the above statement is also true, i.e., there can be no machine which would continuously consume work without some other form of energy appearing simultaneously (Fig. 4.9).

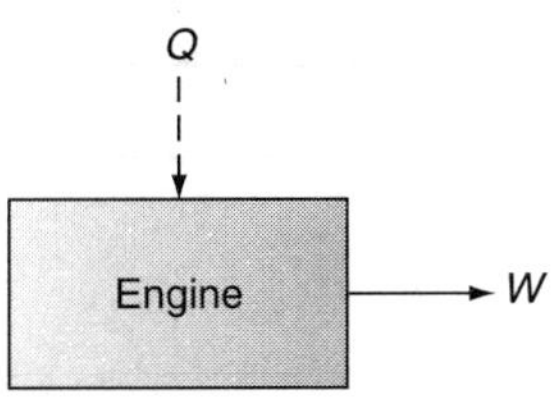

Fig. 4.8 *A PMMI*

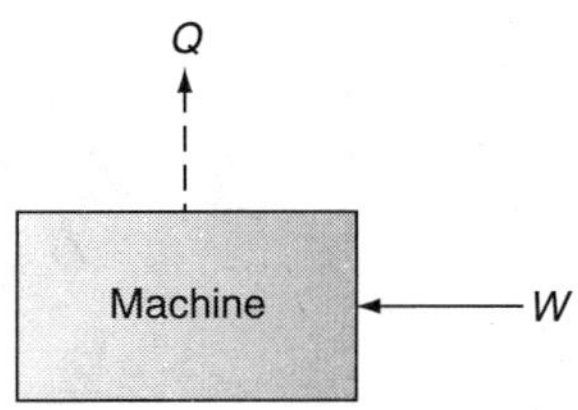

Fig. 4.9 *The converse of PMMI*

4.10 LIMITATIONS OF THE FIRST LAW

The first law deals with the amount of energy of various forms transferred between the system and its surroundings and with changes in the energy stroed in the system. It treats work and heat interactions as equivalent forms of energy in transit and does not indicate the possibility of a spontaneous process proceeding in a certain direction. It is the second law which assigns a quality to different forms of energy, and also indicates the direction of any spontaneous process.

Solved Examples

Example 4.1

A stationary mass of gas is compressed without friction from an initial state of 0.3 m^3 *and* 0.105 *MPa to a final state of* 0.15 m^3 *and* 0.105 *MPa, the pressure remaining constant during the process. There is a transfer of* 37.6 *kJ of heat from the gas during the process. How much does the internal energy of the gas change?*

Solution First law for a stationary system in a process gives

$$Q = \Delta U + W$$

or
$$Q_{1-2} = U_2 - U_1 + W_{1-2} \quad (1)$$

Here
$$W_{1-2} \int_{V_1}^{V_2} p\mathrm{d}V = p(V_2 - V_1) = 0.105\,(0.15 - 0.30)\text{ MJ} = -15.75\text{ kJ}$$

$$Q_{1-2} = -37.6\text{ kJ}$$

∴ Substituting in Eq. (1)
$$-37.6\text{ kJ} = U_2 - U_1 - 15.75\text{ kJ}$$

∴
$$U_2 - U_1 = -21.85\text{ kJ} \qquad \textit{Ans.}$$

The internal energy of the gas decreases by 21.85 kJ in the process.

Example 4.2

When a system is taken from state a tc state b, in Fig. Ex. 4.2, along path acb, 84 kJ of heat flow into the system, and the system does 32 kJ of work. (a) How much will the heat that flows into the system along path adb be, if the work done is 10.5 kJ? (b) When the system is returned from b to a along the curved path, the work done on the system is 21 kJ. Does the system absorb or liberate heat, and how much of the heat is absorbed or liberated? (c) If $U_a = 0$ and $U_d = 42$ kJ, find the heat absorbed in the processes ad and db.

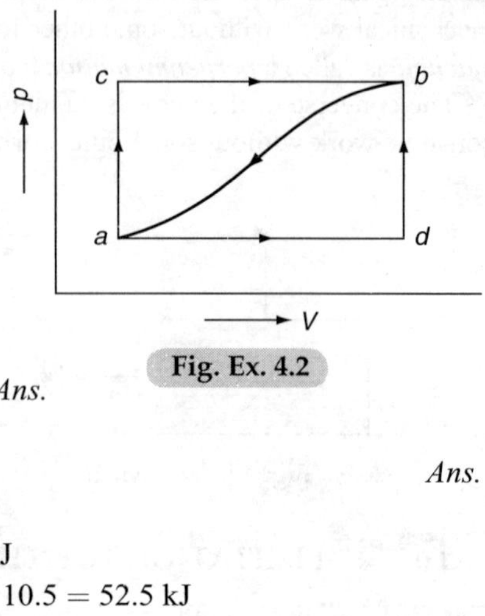

Fig. Ex. 4.2

Solution

$$Q_{\text{acb}} = 84\text{ kJ}$$
$$W_{\text{acb}} = 32\text{ kJ}$$

We have

$$Q_{\text{acb}} = U_{\text{b}} - U_{\text{a}} + W_{\text{acb}}$$

∴
$$U_{\text{b}} - U_{\text{a}} = 84 - 32 = 52\text{ kJ} \qquad \textit{Ans.}$$

(a)
$$Q_{\text{adb}} = U_{\text{b}} - U_{\text{a}} + W_{\text{adb}}$$
$$= 52 + 10.5 = 62.5\text{ kJ} \qquad \textit{Ans.}$$

(b)
$$Q_{\text{b-a}} = U_{\text{a}} - U_{\text{b}} + W_{\text{b-a}}$$
$$= -52 - 21 = -73\text{ kJ} \qquad \textit{Ans.}$$

The system liberates 73 kJ of heat.

(c)
$$W_{\text{adb}} = W_{\text{ad}} + W_{\text{db}} = W_{\text{ad}} = 10.5\text{ kJ}$$

∴
$$Q_{\text{ad}} = U_{\text{d}} - U_{\text{a}} + W_{\text{ad}} = 42 - 0 + 10.5 = 52.5\text{ kJ}$$

Now
$$Q_{\text{adb}} = 62.5\text{ kJ} = Q_{\text{ad}} + Q_{\text{db}}$$

∴
$$Q_{\text{db}} = 62.5 - 52.5 = 10\text{ kJ} \qquad \textit{Ans.}$$

Example 4.3

A piston and cylinder machine contains a fluid system which passes through a complete cycle of four processes. During a cycle, the sum of all heat transfers is – 170 kJ. The system completes 100 cycles per min. Complete the following table showing the method for each item, and compute the net rate of work output in kW.

Process	*Q (kJ/min)*	*W (kJ/min)*	*ΔE (kJ/min)*
a–b	0	2,170	—
b–c	21,000	0	—
c–d	–2,100	—	–36,600
d–a	—	—	—

Solution Process *a–b*:

$$Q = \Delta E + W$$
$$0 = \Delta E + 2170$$
$$\therefore \quad \Delta E = -2170 \text{ kJ/min}$$

Process *b–c*:

$$Q = \Delta E + W$$
$$21{,}000 = \Delta E + 0$$
$$\therefore \quad \Delta E = 21{,}000 \text{ kJ/min}$$

Process *c–d*:

$$Q = \Delta E + W$$
$$-2100 = -36{,}600 + W$$
$$\therefore \quad W = 34{,}500 \text{ kJ/min}$$

Process *d–a*:

$$\sum_{\text{cycle}} Q = -170 \text{ kJ}$$

The system completes 100 cycles/min.

$$\because \quad Q_{ab} + Q_{bc} + Q_{cd} + Q_{da} = -17{,}000 \text{ kJ/min}$$
$$0 + 21{,}000 - 2{,}100 + Q_{da} = -17{,}000$$
$$\therefore \quad Q_{da} = -35{,}900 \text{ kJ/min}$$

Now $\oint dE = 0$, since cyclic integral of any property is zero.

$$\therefore \quad \Delta E_{a-b} + \Delta E_{b-c} + \Delta E_{c-d} + \Delta E_{d-a} = 0$$
$$-2{,}170 + 21{,}000 - 36{,}600 + \Delta E_{d-a} = 0$$
$$\therefore \quad \Delta E_{d-a} = 17{,}770 \text{ kJ/min}$$
$$\therefore \quad W_{d-a} = Q_{d-a} - \Delta E_{d-a}$$
$$= -35{,}900 - 17{,}770 = -53{,}670 \text{ kJ/min}$$

The table becomes

Process	Q (kJ/min)	W (kJ/min)	ΔE (kJ/min)
a–b	0	2,170	– 2,170
b–c	21,000	0	21,000
c–d	–2,100	34,500	– 36,600
d–a	–35,900	– 53,670	17,770

Since

Rate of work output $\sum_{\text{cycle}} Q = \sum_{\text{cycle}} W$

$$= -17{,}000 \text{ kJ/min} = -283.3 \text{ kW}$$ *Ans.*

Example 4.4

The internal energy of a certain substance is given by the following equation

$$u = 3.56\, pv + 84$$

where u is given in kJ/kg, p is in kPa, and v is in m^3/kg.

A system composed of 3 *kg of this substance expands from an initial pressure of* 500 *kPa and a volume of* 0.22 m^3 *to a final pressure* 100 *kPa in a process in which pressure and volume are related by* $pv^{1.2}$ = *constant.*

(a) *If the expansion is quasi-static, find* Q, ΔU, *and* W *for the process.*

(b) *In another process the same system expands according to the same pressure-volume relationship as in part (a), and from the same initial state to the same final state as in part (a), but the heat transfer in this case is* 30 *kJ. Find the work transfer for this process.*

(c) *Explain the difference in work transfer in parts (a) and (b).*

Solution

(a)
$$u = 3.56\,pv + 84$$
$$\Delta u = u_2 - u_1 = 3.56\,(p_2v_2 - p_1v_1)$$
$\therefore$
$$\Delta U = 3.56\,(p_2V_2 - p_1V_1)$$
Now
$$p_1V_1^{\,1.2} = p_2V_2^{\,1.2}$$
$\therefore$
$$V_2 = V_1\left(\frac{p_1}{p_2}\right)^{1/1.2} = 0.2\left(\frac{5}{1}\right)^{1/1.2} = 0.22 \times 3.83 = 0.845\ \text{m}^3$$
$\therefore$
$$\Delta U = 356\,(1 \times 0.845 - 5 \times 0.22)\ \text{kJ}$$
$$= -\,356 \times 0.255 = -\,91\ \text{kJ}$$
Ans. (a)

For a quasi-static process
$$W = \int p\text{d}V = \frac{p_2V_2 - p_1V_1}{1-n}$$
$$= \frac{(1 \times 0.845 - 5 \times 0.22)\,100}{1-1.2} = 127.5\ \text{kJ}$$
$\therefore$
$$Q = \Delta U + W$$
$$= -\,91 + 127.5 = 36.5\ \text{kJ}$$
Ans. (a)

(b) Here $Q = 30$ kJ

Since the end states are the same, ΔU would remain the same as in (a).

$\therefore$
$$W = Q - \Delta U$$
$$= 30 - (-\,91) = 121\ \text{kJ}$$
Ans. (b)

(c) The work in (b) is not equal to $\int p\text{d}V$ since the process is not quasi-static.

Example 4.5

A fluid is confined in a cylinder by a spring-loaded, frictionless piston so that the pressure in the fluid is a linear function of the volume ($p = a + bV$). The internal energy of the fluid is given by the following equation

$$U = 34 + 3.15\,pV$$

where U is in kJ, p in kPa, and V in cubic metre. If the fluid changes from an initial state of 170 *kPa,* 0.03 *m*^{3} *to a final state of* 400 *kPa,* 0.06 *m*^{3}*, with no work other than that done on the piston, find the direction and magnitude of the work and heat transfer.*

Solution The change in the internal energy of the fluid during the process.
$$U_2 - U_1 = 3.15\,(p_2V_2 - p_1V_1)$$
$$= 315\,(4 \times 0.06 - 1.7 \times 0.03)$$
$$= 315 \times 0.189 = 59.5\ \text{kJ}$$

Now
$$p = a + bV$$
$$170 = a + b \times 0.03$$
$$400 = a + b \times 0.06$$

From these two equations
$$a = -\,60\ \text{kN/m}^2$$
$$b = 7667\ \text{kN/m}^5$$

Work transfer involved during the process

$$W_{1-2} = \int_{V_1}^{V_2} p\mathrm{d}V = \int_{V_1}^{V_2} (a + bV)\,\mathrm{d}V$$

$$= a(V_2 - V_1) + b\frac{V_2^2 - V_1^2}{2}$$

$$= (V_2 - V_1)\left[a + \frac{b}{2}(V_1 + V_2)\right]$$

$$= 0.03\ \mathrm{m}^3 \left[-60\ \mathrm{kN/m^2} + \frac{7667}{2}\frac{\mathrm{kN}}{\mathrm{m}^5} \times 0.09\,\mathrm{m}^3\right] = 8.55\ \mathrm{kJ}$$

Work is done by the system, the magnitude being 8.55 kJ.

∴ Heat transfer involved is given by

$$Q_{1-2} = U_2 - U_1 + W_{1-2} = 59.5 + 8.55 = 68.05\ \mathrm{kJ}$$

68.05 kJ of heat flow into the system during the process.

Review Questions

4.1 State the first law for a closed system undergoing a cycle.

4.2 What was the contribution of J.P. Joule in establishing the first law?

4.3 What is the caloric theory of heat? Why was it rejected?

4.4 Which is the property introduced by the first law?

4.5 State the first law for a closed system undergoing a change of state.

4.6 Show that energy is a property of a system.

4.7 What are the modes in which energy is stored in a system?

4.8 Define internal energy. How is energy stored in molecules and atoms?

4.9 What is the difference between the standard symbols of E and U?

4.10 What is the difference between heat and internal energy?

4.11 Define enthalpy. Why does the enthalpy of an ideal gas depend only on temperature?

4.12 Define the specific heats at constant volume and at constant pressure.

4.13 Why should specific heat not be defined in terms of heat transfer?

4.14 Which property of a system increases when heat is transferred: (a) at constant volume, (b) at constant pressure?

4.15 What is a PMM1? Why is it impossible?

Problems

4.1 An engine is tested by means of a water brake at 1000 rpm. The measured torque of the engine is 10000 mN and the water consumption of the brake is 0.5 $\mathrm{m^3/s}$, its inlet temperature being 20°C. Calculate the water temperature at exit, assuming that the whole of the engine power is ultimately transformed into heat which is absorbed by the cooling water. *Ans.* 20.5°C

4.2 In a cyclic process, heat transfers are + 14.7 kJ, – 25.2 kJ, – 3.56 kJ and + 31.5 kJ. What is the net work for this cycle process? *Ans.* 17.34 kJ

4.3 A slow chemical reaction takes place in a fluid at the constant pressure of 0.1 MPa. The fluid is surrounded by a perfect heat insulator during the reaction which begins at state 1 and ends at state 2. The insulation is then removed and 105 kJ of heat flow to the surroundings as the fluid goes to state 3. The following data are observed for the fluid at states 1, 2 and 3.

State	$V(m^3)$	t(°C)
1	0.003	20
2	0.3	370
3	0.06	20

For the fluid system, calculate E_2 and E_3, if $E_1 = 0$
Ans. $E_2 = -29.7$ kJ, $E_3 = -110.7$ kJ

4.4 During one cycle the working fluid in an engine engages in two work interactions: 15 kJ to the fluid and 44 kJ from the fluid, and three heat interactions, two of which are known: 75 kJ to the fluid and 40 kJ from the fluid. Evaluate the magnitude and direction of the third heat transfer. *Ans.* – 6 kJ

4.5 A domestic refrigerator is loaded with food and the door closed. During a certain period the machine consumes 1 kW h of energy and the internal energy of the system drops by 5000 kJ. Find the net heat transfer for the system. *Ans.* – 8.6 MJ

4.6 1.5 kg of liquid having a constant specific heat of 2.5 kJ/kg K is stirred in a well-insulated chamber causing the temperature to rise by 15°C. Find ΔE and W for the process.
Ans. $\Delta E = 56.25$ kJ, $W = -56.25$ kJ

4.7 The same liquid as in Problem 4.6 is stirred in a conducting chamber. During the process 1.7 kJ of heat are transferred from the liquid to the surroundings, while the temperature of the liquid is rising to 15°C. Find ΔE and W for the process.
Ans. $\Delta E = 56.25$ kJ, $W = 57.95$ kJ

4.8 The properties of a certain fluid are related as follows

$$u = 196 + 0.718\, t$$
$$pv = 0.287\,(t + 273)$$

where u is the specific internal energy (kJ/kg), t is in °C, p is pressure (kN/m^2), and v is specific volume (m^3/kg). For this fluid, find c_v and c_p *Ans.* 0.718, 1.005 kJ/kg K

4.9 A system composed of 2 kg of the above fluid expands in a frictionless piston and cylinder machine from an initial state of 1 MPa, 100°C to a final temperature of 30°C. If there is no heat transfer, find the net work for the process. *Ans.* 100.52 kJ

4.10 If all the work in the expansion of Problem 4.9 is done on the moving piston, show that the equation representing the path of the expansion in the pv-plane is given by $pv^{1.4}$ = constant.

4.11 A stationary system consisting of 2 kg of the fluid of Problem 4.8 expands in an adiabatic process according to $pv^{1.2}$ = constant. The initial conditions are 1 MPa and 200°C, and the final pressure is 0.1 MPa. Find W and ΔU for the process. Why is the work transfer not equal to $\int p dV$?
Ans. $W = 216.83$, $\Delta U = -216.83$ kJ, $\int p dV = 434.4$ kJ

4.12 A mixture of gases expands at constant pressure from 1 MPa, 0.03 m^3 to 0.06 m^3 with 84 kJ positive heat transfer. There is no work other than that done on a piston. Find ΔE for the gaseous mixture. *Ans.* 54 kJ
The same mixture expands through the same state path while a stirring device does 21 kJ of work on the system. Find ΔE, W, and Q for the process. *Ans.* 54 kJ, – 21 kJ, 33 kJ

4.13 A mass of 8 kg gas expands within a flexible container so that the p–v relationship is of the form $pv^{1.2}$ = const. The initial pressure is 1000 kPa and the initial volume is 1 m^3. The final pressure is 5 kPa. If specific internal energy of the gas decreases by 40 kJ/kg, find the heat transfer in magnitude and direction. *Ans.* + 2615 kJ

4.14 A gas of mass 1.5 kg undergoes a quasi-static expansion which follows a relationship $p = a + bV$, where a and b are constants. The initial and final pressures are 1000 kPa and 200 kPa respectively and the corresponding volumes are 0.20 m^3 and 1.20 m^3. The specific internal energy of the gas is given by the relation

$$u = 1.5\, pv - 85 \text{ kJ/kg}$$

where p is the kPa and v is in m^3/kg. Calculate the net heat transfer and the maximum internal energy of the gas attained during expansion.
Ans. 660 kJ, 503.3 kJ

4.15 The heat capacity at constant pressure of a certain system is a function of temperature only and may be expressed as

$$C_p = 2.093 \frac{41.87}{t + 100} \text{ J/K}$$

where t is the temperature of the system in °C. The system is heated while it is maintained at a pressure of 1 atmosphere until its volume increases from 2000 cm^3 to 2400 cm^3 and its temperature increases from 0°C to 100°C. (a) Find the magnitude of the heat interaction. (b) How much does the internal energy of the system increase?

Ans. (a) 238.32 J (b) 197.79 J

4.16 An imaginary engine receives heat and does work on a slowly moving piston at such rates that the cycle of operation of 1 kg of working fluid can be represented as a circle 10 cm in diameter on a p–v diagram on which 1 cm = 300 kPa and 1 cm = 0.1 m^3/kg. (a) How much work is done by each kg of working fluid for each cycle of operation? (b) The thermal efficiency of an engine is defined as the ratio of work done and heat input in a cycle. If the heat rejected by the engine in a cycle is 1000 kJ per kg of working fluid, what would be its thermal efficiency?

Ans. (a) 2356.19 kJ/kg, (b) 0.702

4.17 A gas undergoes a thermodynamic cycle consisting of three processes beginning at an initial state where p_1 = 1 bar, V_1 = 1.5 m^3 and U_1 = 512 kJ. The processes are as follows:

(i) Process 1–2: Compression with pV = constant to p_2 = 2 bar, U_2 = 690 kJ

(ii) Process 2–3: $W_{23} = 0$, $Q_{23} = -150$ kJ, and

(iii) Process 3–1: $W_{31} = +50$ kJ. Neglecting KE and PE changes, determine the heat interactions Q_{12} and Q_{31}. *Ans.* 74 kJ, 22 kJ

4.18 A gas undergoes a thermodynamic cycle consisting of the following processes: (i) Process 1–2: Constant pressure p = 1.4 bar, V_1 = 0.028 m^3, W_{12} = 10.5 kJ, (ii) Process 2–3: Compression with pV = constant, $U_3 = U_2$, (iii) Process 3–1: Constant volume, $U_1 - U_3 = -26.4$ kJ. There are no significant changes in KE and PE. (a) Sketch the cycle on a p-V diagram. (b) Calculate the net work for the cycle in kJ. (c) Calculate the heat transfer for process 1–2 (d) Show that $\sum_{\text{cycle}} Q = \sum_{\text{cycle}} W$.

Ans. (b) – 8.28 kJ, (c) 36.9 kJ

4.19 A certain gas of mass 4 kg is contained within a piston cylinder assembly. The gas undergoes a process for which $pV^{1.5}$ = constant. The initial state is given by 3 bar, 0.1 m^3. The change in internal energy of the gas in the process is $u_2 - u_1 = -4.6$ kJ/kg. Find the net heat transfer for the process when the final volume is 0.2 m^3. Neglect the changes in KE and PE. *Ans.* – 0.8 kJ

4.20 An electric generator coupled to a windmill produces an average electrical power output of 5 kW. The power is used to charge a storage battery. Heat transfer from the battery to the surroundings occurs at a constant rate of 0.6 kW. Determine the total amount of energy stored in the battery in 8*h* of operation. *Ans.* 1.27×10^5kJ

4.21 A gas in a piston-cylinder assembly undergoes two processes in series. From state 1 to state 2 there is energy transfer by heat to the gas of 500 kJ, and the gas does work on the piston amounting 800 kJ. The second process, from state 2 to state 3, is a constant pressure compression at 400 kPa, during which there is a heat transfer from the gas amounting 450 kJ. The following data are also known: U_1 = 2000 kJ and U_3 = 3500 kJ. Neglecting changes in KE and PE, calculate the change in volume of the gas during process 2–3.

Ans. –5.625 m^3

4.22 Air is contained in a rigid well-insulated tank with a volume of 0.2 m^3. The tank is fitted with a paddle wheel which transfers energy to the air at a constant rate of 4 W for 20 min. The initial density of the air is 1.2 kg/m^3. If no changes in KE or PE occur, determine (a) the specific volume at the final state, (b) the change in specific internal energy of the air.

Ans. (a) 0.833 m^3/kg, (b) 20 kJ/kg

C H A P T E R 5

First Law Applied to Flow Processes

5.1 CONTROL VOLUME

For any system and in any process, the first law can be written as

$$Q = \Delta E + W$$

where E represents all forms of energy stored in the system.

For a pure substance

$$E = E_K + E_P + U$$

where E_K is the K.E., E_P the P.E., and U the residual energy stored in the molecular structure of the substance.

$$Q = \Delta E_K + \Delta E_P + \Delta U + W \qquad (5.1)$$

When there is mass transfer across the system boundary, the system is called an open system. Most of the engineering devices are open systems involving the flow of fluids through them.

Equation (5.1) refers to a system having a particular mass of substance, and is free to move from place to place.

Consider a steam turbine (Fig. 5.1) in which steam enters at a high pressure, does work upon the turbine rotor, and then leaves the turbine at low pressure through the exhaust pipe.

If a certain mass of steam is considered as the thermodynamic system, then the energy equation becomes

$$Q = \Delta E_K + \Delta E_P + \Delta U + W$$

and in order to analyze the expansion process in turbine the moving system is to be followed as it travels through the turbine, taking into account the work and heat interactions all the way through. This method of analysis is similar to that of Langrange in fluid mechanics.

Although the system approach is quite valid, there is another approach which is found to be highly convenient. Instead of concentrating attention upon a certain quantity of fluid, which constitutes a moving system in flow process, attention is focussed upon a certain fixed region in space called a *control volume* through which the moving substance flows. This is similar to the analysis of Euler in fluid mechanics.

To distinguish the two concepts, it may be noted that while the system (closed) boundary usually changes shape, position and orientation relative to the observer, the control volume boundary remains fixed and unaltered. Again, while matter usually crosses the control volume boundary, no such flow occurs across the system boundary.

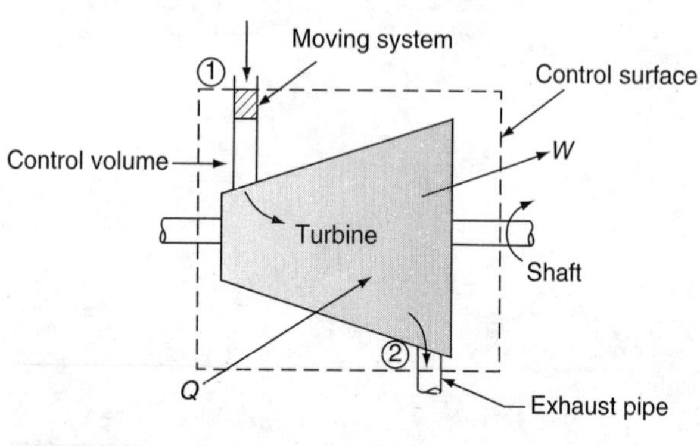

Fig. 5.1 *Flow process involving work and heat interactions*

The broken line in Fig. 5.1 represents the surface of the control volume which is known as the *control surface*. This is the same as the system boundary of the open system. The method of analysis is to inspect the control surface and account for all energy quantities transferred through this surface. Since there is mass transfer across the control surface, a mass balance also has to be made. Sections 1 and 2 allow mass transfer to take place, and Q and W are the heat and work interactions respectively.

5.2 STEADY FLOW PROCESS

As a fluid flows through a certain control volume, its thermodynamic properties may vary along the space coordinates as well as with time. If the rates of flow of mass and energy through the control surface change with time, the mass and energy within the control volume also would change with time.

'Steady flow' means that the rates of flow of mass and energy across the control surface are constant.

In most engineering devices, there is a constant rate of flow of mass and energy through the control surface, and the control volume in course of time attains a steady state. *At the steady state of a system, any thermodynamic property will have a fixed value at a particular location, and will not alter with time.* Thermodynamic properties may vary along space coordinates, but do not vary with time. 'Steady state' means that the state is steady or invariant with time.

5.3 MASS BALANCE AND ENERGY BALANCE IN A SIMPLE STEADY FLOW PROCESS

In Fig. 5.2, a steady flow system has been shown in which one stream of fluid enters and one another stream leaves the control volume. There is no accumulation of mass or energy within the control volume, and the properties at any location within the control volume are steady with time. Sections 1-1 and 2-2 indicate, respectively, the entrance and exit of the fluid across the control surface. The following quantities are defined with reference to Fig. 5.2.

A_1, A_2—cross-section of stream, m^2

w_1, w_2—mass flow rate, kg/s

p_1, p_2—pressure, absolute, N/m^2

v_1, v_2—specific volume, m^3/kg

u_1, u_2—specific internal energy, J/kg

$\mathbf{V}_1, \mathbf{V}_2$—velocity, m/s

Z_1, Z_2—elevation above an arbitrary datum, m

$\frac{đQ}{d\tau}$—net rate of heat transfer through the control surface, J/s

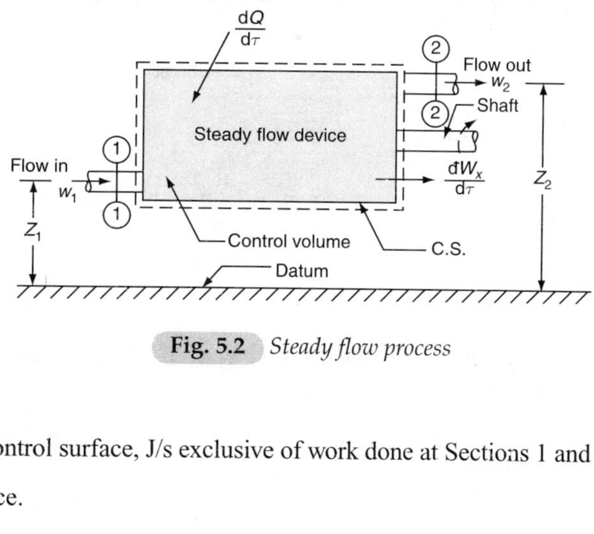

Fig. 5.2 *Steady flow process*

$\frac{đW_x}{d\tau}$—net rate of work transfer through the control surface, J/s exclusive of work done at Sections 1 and 2 in transferring the fluid through the control surface.

τ—time, s.

Subscripts 1 and 2 refer to the inlet and exit sections.

5.3.1 Mass Balance

By the conservation of mass, if there is no accumulation of mass within the control volume, the mass flow rate entering must equal the mass flow rate leaving, or

$$w_1 = w_2$$

or
$$\frac{A_1\mathbf{V}_1}{v_1} = \frac{A_2\mathbf{V}_2}{v_2} \tag{5.2}$$

This equation is known as the *equation of continuity.*

5.3.2 Energy Balance

In a flow process, the work transfer may be of two types: the *external work* and the *flow work*. The external work refers to all the work transfer across the control surface other than that due to normal fluid forces. In engineering thermodynamics the only kinds of external work of importance are *shear* (shaft or stirring) *work* and *electrical work*. In Fig. 5.2 the only external work occurs in the form of shaft work, W_x. The flow work, as discussed in Sec. 3.4, is the displacement work done by the fluid of mass dm_1 at the inlet Section 1 and that of mass dm_2 at the exit Section 2, which are $(-p_1v_1\, dm_1)$ and $(+p_2v_2\, dm_2)$ respectively. Therefore, the total work transfer is given by

$$W = W_x - p_1v_1dm_1 + p_2v_2dm_2 \tag{5.3}$$

In the rate form,

$$\frac{đW}{d\tau} = \frac{đW_x}{d\tau} - p_1v_1\frac{dm_1}{d\tau} + p_2v_2\frac{dm_2}{d\tau}$$

or
$$\frac{đW}{d\tau} = \frac{đW_x}{d\tau} - w_1p_1v_1 + w_2p_2v_2 \tag{5.4}$$

Since there is no accumulation of energy, by the conservation of energy, the total rate of flow of all energy streams entering the control volume must equal the total rate of flow of all energy streams leaving the control volume. This may be expressed in the following equation.

$$w_1e_1 + \frac{đQ}{d\tau} = w_2e_2 + \frac{đW}{d\tau}$$

Substituting for $\frac{đW}{d\tau}$ from Eq. (5.4)

$$w_1e_1 + \frac{đQ}{d\tau} = w_2e_2 + \frac{đW_x}{d\tau} - w_1p_1v_1 + w_2p_2v_2$$

$$w_1e_1 + w_1p_1v_1 + \frac{đQ}{d\tau} = w_2e_2 + w_2p_2v_2 + \frac{đW_x}{d\tau} \tag{5.5}$$

where e_1 and e_2 refer to the energy carried into or out of the control volume with unit mass of fluid.

The specific energy e is given by

$$e = e_k + e_p + u = \frac{\mathbf{V}^2}{2} + Zg + u \tag{5.6}$$

Substituting the expression for e in Eq. (5.5)

$$w_1\left(\frac{\mathbf{V}_1^2}{2} + Z_1g + u_1\right) + w_1p_1v_1 + \frac{đQ}{d\tau} = w_2\left(\frac{\mathbf{V}_2^2}{2} + Z_2g + u_2\right) + w_2p_2v_2 + \frac{đW_x}{d\tau}$$

or
$$w_1\left(h_1 + \frac{\mathbf{V}_1^2}{2} + Z_1g\right) + \frac{đQ}{d\tau} = w_2\left(h_2 + \frac{\mathbf{V}_2^2}{2} + Z_2g\right) + \frac{đW_x}{d\tau} \tag{5.7}$$

where $h = u + pv.$

And, since $w_1 = w_2$, let $w = w_1 = w_2 = \dfrac{dm}{d\tau}$

Dividing Eq. (5.7) by $\dfrac{dm}{d\tau}$

$$h_1 + \frac{V_1^2}{2} + Z_1 g + \frac{đQ}{dm} = h_2 + \frac{V_2^2}{2} + Z_2 g + \frac{đW_x}{dm} \tag{5.8}$$

Equations (5.7) and (5.8) are known as *steady flow energy equations* (S.F.E.E.), for a single stream of fluid entering and a single stream of fluid leaving the control volume. All the terms in Eq. (5.8) represent energy flow per unit mass of fluid (J/kg), whereas the terms in Eq. (5.7) represent energy flow per unit time (J/kg). The basis of energy flow per unit mass is usually more convenient when only a single stream of fluid enters and leaves a control volume. When more than one fluid stream is involved the basis of energy flow per unit time is more convenient.

Equation (5.8) can be written in the following form,

$$Q - W_x = (h_2 - h_1) + \frac{\bar{\mathbf{V}}_2^2 - \bar{\mathbf{V}}_1^2}{2} + g(Z_2 - Z_1) \tag{5.9}$$

where Q and W_x refer to energy transfer per unit mass. In the differential form, the SFEE becomes

$$đQ - đW_x = dh + \bar{\mathbf{V}} d\bar{\mathbf{V}} + gdZ \tag{5.10}$$

When more than one stream of fluid enters or leaves the control volume (Fig. 5.3), the mass balance and energy balance for steady flow are given below.

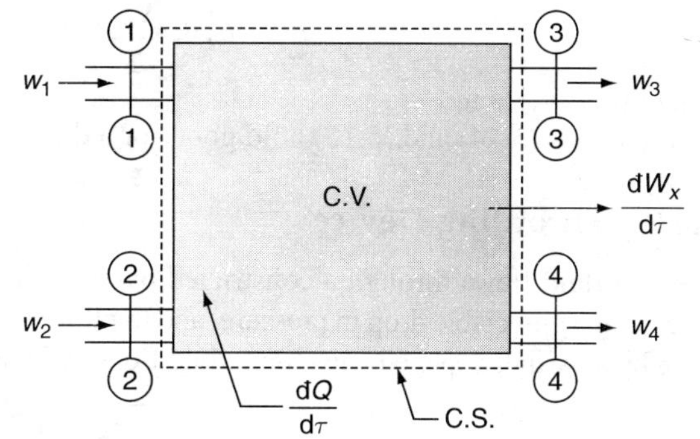

Fig. 5.3 *Steady flow process involving two fluid streams at the inlet and exit of the control volume*

Mass balance

$$w_1 + w_2 = w_3 + w_4 \tag{5.11}$$

$$\frac{A_1\mathbf{V}_1}{v_1} + \frac{A_2\mathbf{V}_2}{v_2} = \frac{A_3\mathbf{V}_3}{v_3} + \frac{A_4\mathbf{V}_4}{v_4} \tag{5.12}$$

Energy balance

$$w_1\left(h_1 + \frac{\mathbf{V}_1^2}{2} + Z_1 g\right) + w_2\left(h_2 + \frac{\mathbf{V}_2^2}{2} + Z_2 g\right) + \frac{đQ}{d\tau}$$

$$= w_3\left(h_3 + \frac{\mathbf{V}_3^2}{2} + Z_3 g\right) + w_4\left(h_4 + \frac{\mathbf{V}_4^2}{2} + Z_4 g\right) + \frac{đW_x}{d\tau} \tag{5.13}$$

The steady flow energy equation applies to a wide variety of processes like pipe line flows, heat transfer processes, mechanical power generation in engines and turbines, combustion processes, and flows through nozzles and diffusors. In certain problems, some of the terms in steady flow energy equation may be negligible or zero. But it is best to write the full equation first, and then eliminate the terms which are unnecessary.

5.4 SOME EXAMPLES OF STEADY FLOW PROCESSES

The following examples illustrate the applications of the steady flow energy equation in some of the engineering systems.

5.4.1 Nozzle and Diffusor

A nozzle is a device which increases the velocity or K.E. of a fluid at the expense of its pressure drop, whereas a diffusor increases the pressure of a fluid at the expense of its K.E. Figure 5.4 shows a nozzle which is insulated. The steady flow energy equation of the control surface gives

$$h_1 + \frac{\mathbf{V}_1^2}{2} + Z_1 g + \frac{đQ}{dm} = h_2 + \frac{\mathbf{V}_2^2}{2} + Z_2 g + \frac{đW_x}{dm}$$

Here $\frac{đQ}{dm} = 0$, $\frac{đW_x}{dm} = 0$, and the change in potential energy is zero. The equation reduces to

$$h_1 + \frac{\mathbf{V}_1^2}{2} = h_2 + \frac{\mathbf{V}_2^2}{2} \tag{5.14}$$

The continuity equation gives

$$w = \frac{A_1 \mathbf{V}_1}{v_1} = \frac{A_2 \mathbf{V}_2}{v_2} \tag{5.15}$$

When the inlet velocity or the 'velocity of approach' $\mathbf{V}_1$ is small compared to the exit velocity $\mathbf{V}_2$, Eq. (5.14) becomes

$$h_1 = h_2 + \frac{\mathbf{V}_2^2}{2} \quad \text{or} \quad \mathbf{V}_2 = \sqrt{2(h_1 - h_2)} \text{ m/s}$$

where $(h_1 - h_2)$ is in J/kg.

Equations (5.14) and (5.15) hold good for a diffusor as well.

5.4.2 Throttling Device

When a fluid flows through a constricted passage, like a partially opened valve, an orifice, or a porous plug, there is an appreciable drop in pressure, and the flow is said to be throttled. Figure 5.5 shows the process of throttling by a partially opened valve on a fluid flowing in an insulated pipe. In the steady-flow energy Eq. (5.8),

$$\frac{đQ}{dm} = 0, \quad \frac{đW_x}{dm} = 0$$

and the changes in P.E. are very small and ignored. Thus, the S.F.E.E. reduces to

$$h_1 + \frac{\mathbf{V}_1^2}{2} = h_2 + \frac{\mathbf{V}_2^2}{2}$$

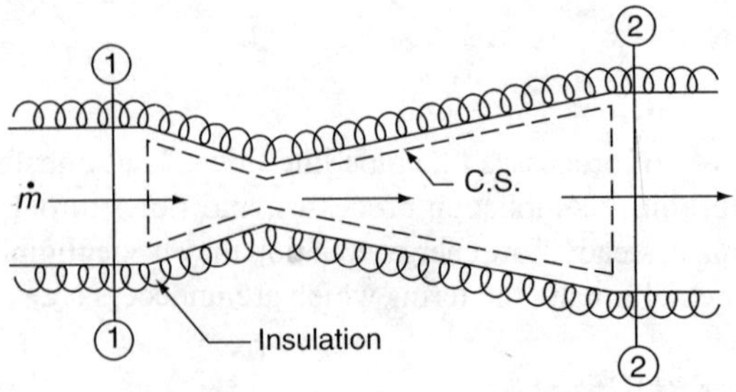

Fig. 5.4 *Steady flow process involving one fluid stream at the inlet and at exit of the control volume*

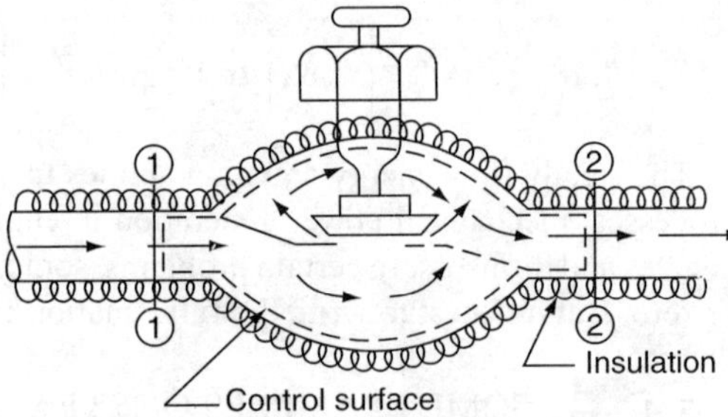

Fig. 5.5 *Flow through a valve*

Often the pipe velocities in throttling are so low that the K.E. terms are also negligible. So

$$h_1 = h_2 \tag{5.16}$$

or the enthalpy of the fluid before throttling is equal to the enthalpy of the fluid after throttling.

5.4.3 Turbine and Compressor

Turbines and engines give positive power output, whereas compressors and pumps require power input.

For a turbine (Fig. 5.6) which is well insulated, the flow velocities are often small, and the K.E. terms can be neglected. The S.F.E.E. then becomes

$$h_1 = h_2 + \frac{đW_x}{dm} \quad \text{or} \quad \frac{W_x}{m} = (h_1 - h_2)$$

It is seen that work is done by the fluid at the expense of its enthalpy.

Similarly, for an adiabatic pump or compressor, work is done upon the fluid and W is negative. So the S.F.E.E. becomes

$$h_1 = h_2 - \frac{W_x}{m} \quad \text{or} \quad \frac{W_x}{m} = h_2 - h_1$$

The enthalpy of the fluid increases by the amount of work input.

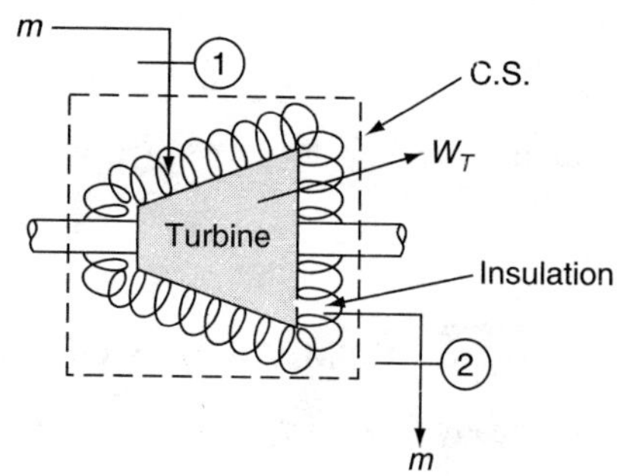

Fig. 5.6 *Flow through a turbine*

5.4.4 Heat Exchanger

A heat exchanger is a device in which heat is transferred from one fluid to another. Figure 5.7 shows a steam condenser, where steam condensers outside the tubes and cooling water flows through the tubes. The S.F.E.E. for the C.S. gives

$$w_c h_1 + w_s h_2 = w_c h_3 + w_s h_4 \quad \text{or} \quad w_s (h_2 - h_4) = w_c(h_3 - h_1)$$

Here the K.E. and P.E. terms are considered small, there is no external work done, and energy exchange in the form of heat is confined only between the two fluids, i.e., there is no external heat interaction or heat loss.

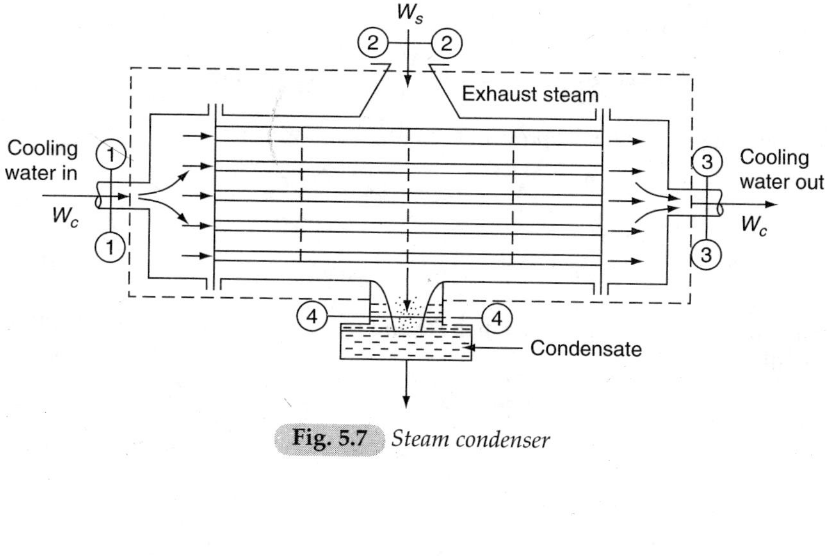

Fig. 5.7 *Steam condenser*

Figure 5.8 shows a steam desuperheater where the temperature of the superheated steam is reduced by spraying water. If w_1, w_2, and w_3 are the mass flow rates of the injected water, of the steam entering, and of the steam leaving, respectively, and h_1, h_2, and h_3 are the corresponding enthalpies, and if K.E. and P.E. terms are neglected as before, the S.F.E.E. becomes

$$w_1 h_1 + w_2 h_2 = w_3 h_3$$

and the mass balance gives

$$w_1 + w_2 = w_3$$

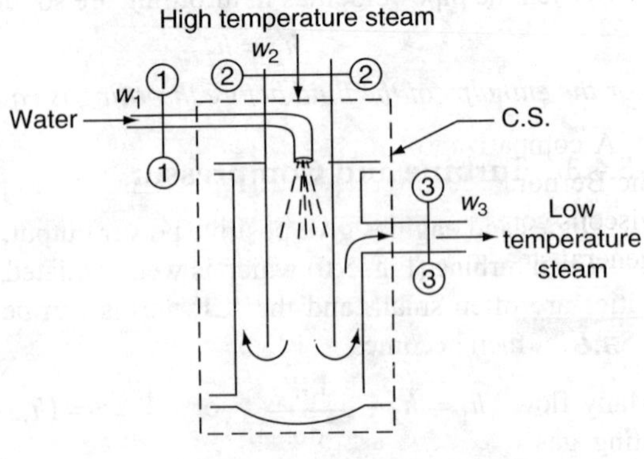

Fig. 5.8 *Steam desuperheater*

5.5 COMPARISON OF S.F.E.E. WITH EULER AND BERNOULLI EQUATIONS

The steady flow energy Eq. (5.8) can be written as

$$\frac{đQ}{dm} = (h_2 - h_1) + \frac{\mathbf{V}_2^2 - \mathbf{V}_1^2}{2} + (Z_2 - Z_1)\, g + \frac{dW_x}{dm}$$

In the differential form the S.F.E.E. becomes

$$đQ = dh + \bar{\mathbf{V}} d\bar{\mathbf{V}} + gdZ + dW_x \tag{5.17}$$

where $đQ$ and $đW_x$ refer to unit mass of the substance. Since $h = u + pv$ and $đQ = đu + pdv$ (for a quasi-static path involving only pdv-work), Eq. (5.17) can be written as

$$du + pd\mathbf{V} = du + pd\mathbf{V} + \mathbf{V}dp + \mathbf{V}d\mathbf{V} + gdZ + đW_x$$

For an inviscid frictionless fluid flowing through a pipe

$$\mathbf{V}dp + \mathbf{V}d\mathbf{V} + gdZ = 0 \tag{5.18}$$

This is the *Euler equation*. If we integrate between two sections 1 and 2 of the pipe

$$\int_1^2 \mathbf{v}dp + \int_1^2 \mathbf{V}d\mathbf{V} + \int_1^2 gdZ = 0$$

For an incompressible fluid, v = constant

$$\therefore \quad \mathbf{v}(p_2 - p_1) + \frac{\mathbf{V}_2^2}{2} - \frac{\mathbf{V}_1^2}{1} + g\,(Z_2 - Z_1) = 0 \tag{5.19}$$

Since the specific volume v is the reciprocal of the density ρ, we have

$$\frac{p_1}{\rho} + \frac{\mathbf{V}_1^2}{2} + Z_1 g = \frac{p_2}{\rho} + \frac{\mathbf{V}_2^2}{2} + Z_2 g \tag{5.20}$$

or

$$\frac{p}{\rho} + \frac{\mathbf{V}^2}{2} + Zg = \text{constant} \tag{5.21}$$

This is known as the *Bernoulli equation*, which is valid for an inviscid incompressible fluid. It can also be expressed in the following form

$$\Delta\left(p\mathbf{v} + \frac{\mathbf{V}^2}{2} + gZ\right) = 0 \tag{5.22}$$

where v is constant and Δ (...) means 'increase in ...

The S.F.E.E. as given by Eq. (5.18) or Eq. (5.17) can be written with $(u + pv)$ substituted for h, as follows:

$$Q - W_x = \Delta\left(u + p\mathbf{v} + \frac{\mathbf{V}^2}{2} + gZ\right) \tag{5.23}$$

A comparison of Eqs (5.22) and (5.23) shows that they have several terms in common. However, while the Bernoulli equation is restricted to frictionless incompressible fluids, the S.F.E.E. is not, and is valid for viscous compressible fluids as well. The Bernoulli equation is, therefore, a special limiting case of the more general steady flow energy equation.

5.6 VARIABLE FLOW PROCESSES

Many flow processes, such as filling up and evacuating gas cylinders, are not steady. Such processes can be analyzed by the control volume technique. Consider a device through which a fluid is flowing under non-steady state conditions (Fig. 5.9). The rate at which the mass of fluid within the control volume is accumulated is equal to the net rate of mass flow across the control surface, as given below

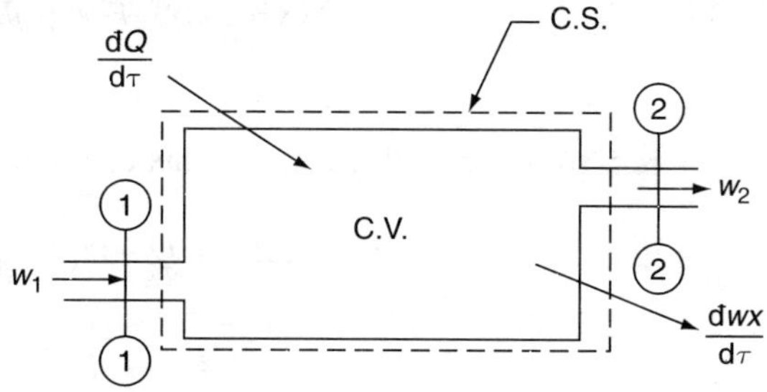

Fig. 5.9 *Variable flow process*

$$\frac{dm_V}{d\tau} = w_1 - w_2 = \frac{dm_1}{d\tau} - \frac{dm_2}{d\tau} \tag{5.24}$$

where m_V is the mass of fluid within the control volume at any instant.

Over any finite period of time

$$\Delta m_V = \Delta m_1 - \Delta m_2 \tag{5.25}$$

The rate of accumulation of energy within the control volume is equal to the net rate of energy flow across the control surface. If E_v is the energy of fluid within the control volume at any instant.

Rate of energy increase = Rate of energy inflow – Rate of energy outflow

$$\frac{dE_V}{d\tau} = w_1\left(h_1 + \frac{\mathbf{V}_1^2}{2} + Z_1 g\right) + \frac{đQ}{d\tau} - w_2\left(h_2 + \frac{\mathbf{V}_2^2}{2} + Z_2 g\right) - \frac{đW_x}{d\tau} \tag{5.26}$$

Now

$$E_V = \left(U + \frac{mV^2}{2} + mgZ\right)_V$$

where m is the mass of fluid in the control volume at any instant.

$$\therefore \quad \frac{dE_V}{d\tau} = \frac{d}{d\tau}\left(U + \frac{mV^2}{2} + mgZ\right)_V = \left(h_1 + \frac{\mathbf{V}_1^2}{2} + Z_1 g\right)\frac{dm_1}{d\tau} + \frac{đQ}{d\tau} - \left(h_2 + \frac{\mathbf{V}_2^2}{2} + Z_2 g\right)\frac{dm_2}{d\tau} - \frac{đW_x}{d\tau} \tag{5.27}$$

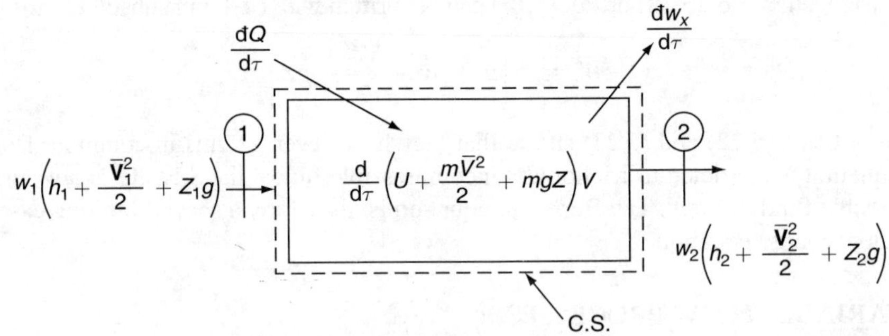

Fig. 5.10 *Energy fluxes in an unsteady system*

Figure 5.10 shows all these energy flux quantities. For any finite time interval, Eq. (5.27) becomes

$$\Delta E_V = Q - W_x + \int \left(h_1 + \frac{V_1^2}{2} + Z_1 g \right) dm_1 - \int \left(h_2 + \frac{V_2^2}{2} + Z_2 g \right) dm_2 \tag{5.28}$$

Equation (5.26) is the *general energy equation*. For steady flow,

$$\frac{dE_V}{d\tau} = 0,$$

and the equation reduces to Eq. (5.7). For a closed system $w_1 = 0$, $w_2 = 0$, then from Eq. (5.26),

$$\frac{dE_V}{d\tau} = \frac{đQ}{d\tau} - \frac{đW_x}{d\tau} \quad \text{or} \quad dE_V = đQ - đW_x \quad \text{or,} \quad đQ = dE + đW_x$$

as obtained earlier.

5.7 EXAMPLE OF A VARIABLE FLOW PROBLEM

Variable flow processes may be analyzed either by the system technique or the control volume technique, as illustrated below.

Consider a process in which a gas bottle is filled from a pipeline (Fig. 5.11). In the beginning the bottle contains gas of mass m_1 at state p_1, t_1, v_1, h_1 and u_1. The valve is opened and gas flows into the bottle till the mass of gas in the bottle is m_2 at state p_2, t_2, v_2, h_2 and u_2. The supply to the pipeline is very large so that the state of gas in the pipeline is constant at p_p, t_p, v_p, h_p, u_p, and V_p.

System Technique Assume an envelope (which is extensible) of gas in the pipeline and the tube which would eventually enter the bottle, as shown in Fig. 5.11.

Energy of the gas before filling

$$E_1 = m_1 u_1 + (m_2 - m_1) \left(\frac{V_p^2}{2} + u_p \right)$$

where $(m_2 - m_1)$ is the mass of gas in the pipeline and tube which would enter the bottle.

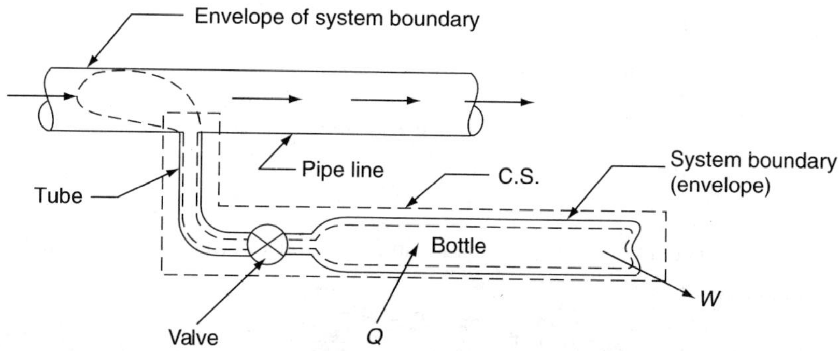

Fig. 5.11 *Bottle-filling process*

Energy of the gas after filling

$$E_2 = m_2u_2$$

$$\Delta E = E_2 - E_1 = m_2u_2 - \left[m_1u_1 + (m_2 - m_1)\left(\frac{\mathbf{V}_p^2}{2} + u_p\right)\right] \quad (5.29)$$

The P.E. terms are neglected: The gas in the bottle is not in motion, and so the K.E. terms have been omitted.

Now, there is a change in the volume of gas because of the collapse of the envelope to zero volume. Then the work done

$$W = p_p(V_2 - V_1) = p_p[0 - (m_2 - m_1)\nu_p] = -(m_2 - m_1)p_p\nu_p$$

∴ Using the first law for the process

$$Q = \Delta E + W$$

$$= m_2u_2 - m_1u_1 - (m_2 - m_1)\left(\frac{\mathbf{V}_p^2}{2} + u_p\right) - (m_2 - m_1)p_p\nu_p$$

$$= m_2u_2 - m_1u_1 - (m_2 - m_1)\left(\frac{\mathbf{V}_p^2}{2} + h_p\right) \quad (5.30)$$

which gives the energy balance for the process.

Control Volume Technique Assume a control volume bounded by a control surface, as shown in Fig. 5.11. Applying the energy Eq. (5.27) to this case, the following energy balance may be written on a time rate basis

$$\frac{dE_V}{d\tau} = \frac{đQ}{d\tau} + \left(h_p + \frac{\mathbf{V}_p^2}{2}\right)\frac{dm}{d\tau}$$

Since h_p and V_p are constant, the equation is integrated to give for the total process

$$\Delta E_V = Q + \left(h_p + \frac{\mathbf{V}_p^2}{2}\right)(m_2 - m_1)$$

Now $$\Delta E_V = U_2 - U_1 = m_2u_2 - m_1u_1$$

∴ $$Q = m_2u_2 - m_1u_1 - \left(h_p + \frac{\mathbf{V}_p^2}{2}\right)(m_2 - m_1)$$

This equation is the same as Eq. (5.30).

If $m_1 = 0$, i.e., the bottle is initially evacuated,

$$Q = m_2 u_2 - m_2 \left(h_p + \frac{V_p^2}{2} \right)$$

Again, if $Q = 0$ and $h_p \gg \frac{V_p^2}{2}$,

$$0 = m_2 u_2 - m_2 h_p$$

or $\quad u_2 = h_p = u_p + p_p v_p$

Thus, flow work ($p_p v_p$) is converted to an increase in molecular internal energy ($u_2 – u_p$).

If the gas is assumed ideal,

$$c_v T_2 = c_p T_p \quad \text{or} \quad T_2 = \gamma \, T_p$$

If $T_p = 27 + 273 = 300$ K, then for air

$$T_2 = 1.4 \times 300 = 420 \text{ K} \quad \text{or} \quad t_2 = 147°\text{C}$$

Therefore, in adiabatically filling a bottle with air at 27°C, the gas temperature rises to 147°C due to the flow work being converted to internal energy increase.

5.8 DISCHARGING AND CHARGING A TANK

Let us consider a tank discharging a fluid into a supply line (Fig. 5.12). Since $đW_x = 0$ and $dm_{in} = 0$, applying first law to the control volume,

$$dU_V = đQ + \left(h + \frac{V^2}{2} + gz \right)_{out} dm_{out} \tag{5.31}$$

Assuming K.E. and P.E. of the fluid to be small and $đQ = 0$

$$d(mu) = h dm$$

$$m du + u dm = u dm + pv\, dm$$

$$\frac{dm}{m} = \frac{du}{pv} \tag{5.32}$$

Again $\quad V = vm = \text{const.}$

$$v dm + m dv = 0$$

or

$$\frac{dm}{m} = -\frac{dv}{v} \tag{5.33}$$

From Eqs (5.32) and 5.33),

$$\frac{du}{pv} = -\frac{dv}{v}$$

$$d(u + pv) = 0 \quad \text{or} \quad đQ = 0$$

which shows that the process is adiabatic and quasi-static.

For charging the tank

$$\int (h dm)_{in} = \Delta U_V = m_2 u_2 - m_1 u_1 \tag{5.34}$$

$$m_p h_p = m_2 u_2 - m_1 u_1$$

where the subscript p refers to the constant state of the fluid in the pipeline. If the tank is initially empty, $m_1 = 0$.

$$m_p h_p = m_2 u_2$$

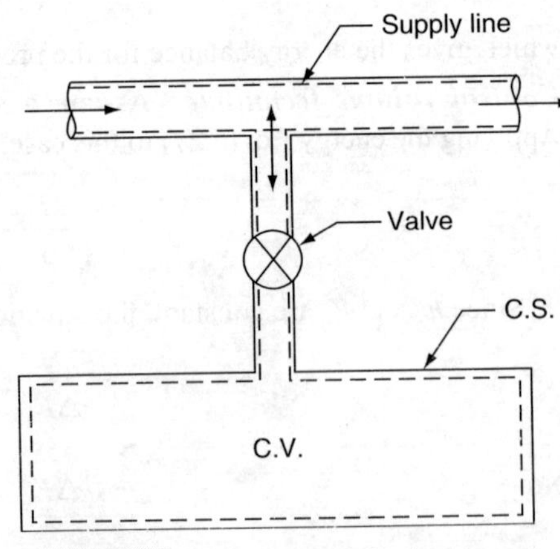

Fig. 5.12 *Charging and discharging a tank*

Since

$$m_p = m_2$$
$$h_p = u_2$$

Enthalpy is converted to internal energy.

If the fluid is an ideal gas, the temperature of the gas in the tank after it is charged is given by

$$c_p T_p = c_v T_2$$

or
$$T_2 = \gamma T_p \tag{5.35}$$

Solved Examples

Example 5.1

Air flows steadily at the rate of 0.5 kg/s through an air compressor, entering at 7 m/s velocity, 100 kPa pressure, and 0.95 m³/kg volume, and leaving at 5 m/s, 700 kPa, and 0.19 m³/kg. The internal energy of the air leaving is 90 kJ/kg greater than that of the air entering. Cooling water in the compressor jackets absorbs heat from the air at the rate of 58 kW. (a) Compute the rate of shaft work input to the air in kW. (b) Find the ratio of the inlet pipe diameter to outlet pipe diameter.

Solution Figure Ex. 5.1 shows the details of the problem.

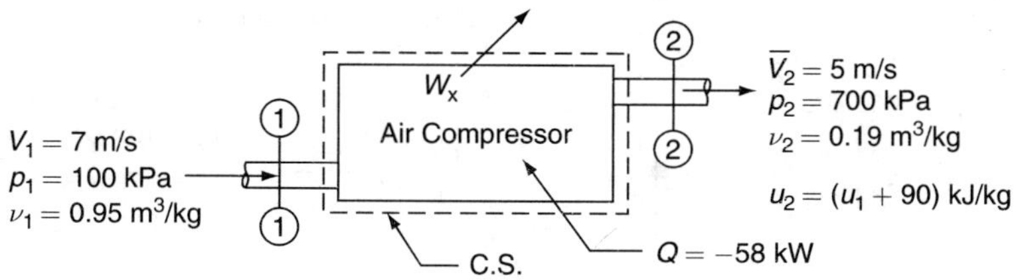

Fig. Ex. 5.1

(a) Writing the steady flow energy equation, we have

$$w\left(u_1 + p_1v_1 + \frac{\mathbf{V}_1^2}{2} + Z_1 g\right) + \frac{đQ}{d\tau} = w\left(u_2 + p_2v_2 + \frac{\mathbf{V}_2^2}{2} + Z_2 g\right) + \frac{đW_x}{d\tau}$$

$$\therefore \quad \frac{đW_x}{d\tau} = -w\left[(u_2 - u_1) + (p_2v_2 - p_1v_1) + \frac{\mathbf{V}_2^2 - \mathbf{V}_1^2}{2} + (Z_2 - Z_1)g\right] + \frac{đQ}{d\tau}$$

$$\therefore \quad \frac{đW_x}{d\tau} = -0.5\frac{\text{kg}}{\text{s}}\left[90\frac{\text{kJ}}{\text{kg}} + (7\times 0.19 - 1\times 0.95)\,100\frac{\text{kJ}}{\text{kg}} + \frac{(5^2 - 7^2)\times 10^{-3}}{2}\frac{\text{kJ}}{\text{kg}} + 0\right] - 58 \text{ kW}$$

$$= -0.5\,[90 + 38 - 0.012]\text{ kJ/s} - 58 \text{ kW}$$

$$= -122 \text{ kW}$$ *Ans.* (a)

Rate of work input is 122 kW.

(b) From mass balance, we have

$$w = \frac{A_1\mathbf{V}_1}{v_1} = \frac{A_2\mathbf{V}_2}{v_2}$$

$$\therefore \quad \frac{A_1}{A_2} = \frac{v_1}{v_2}\cdot\frac{\mathbf{V}_2}{\mathbf{V}_1} = \frac{0.95}{0.19}\times\frac{5}{7} = 3.57$$

$$\therefore \quad \frac{d_1}{d_2} = \sqrt{3.57} = 1.89$$

Ans. (b)

Example 5.2

In a steady flow apparatus, 135 kJ of work is done by each kg of fluid. The specific volume of the fluid, pressure, and velocity at the inlet are 0.37 m³/kg, 600 kPa, and 16 m/s. The inlet is 32 m above the floor, and the discharge pipe is at floor level. The discharge conditions are 0.62 m³/kg, 100 kPa, and 270 m/s. The total heat loss between the inlet and discharge is 9 kJ/kg of fluid. In flowing through this apparatus, does the specific internal energy increase or decrease, and by how much?

Solution Writing the steady flow energy equation for the control volume, as shown in Fig. Ex. 5.2.

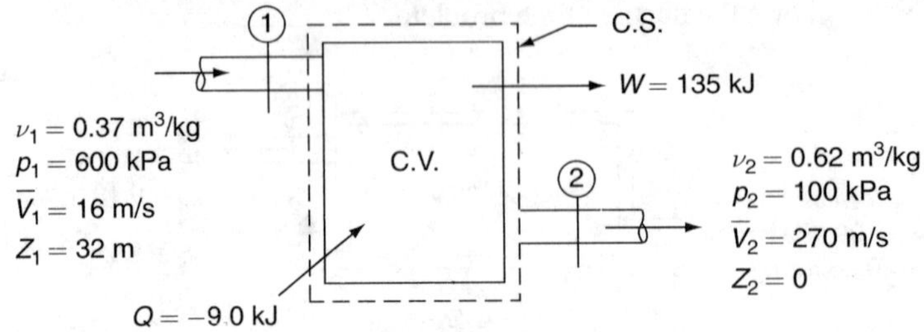

Fig. Ex. 5.2

$$u_1 + p_1v_1 + \frac{\mathbf{V}_1^2}{2} + Z_1g + \frac{đQ}{dm} = u_2 + p_2v_2 + \frac{\mathbf{V}_2^2}{2} + Z_2g + \frac{đW_x}{dm}$$

$$\therefore \quad u_1 - u_2 = (p_2v_2 - p_1v_1) + \frac{\mathbf{V}_2^2 - \mathbf{V}_1^2}{2} + (Z_2 - Z_1)g + \frac{đW_x}{dm} - \frac{đQ}{dm}$$

$$= (1\times 0.62 - 6\times 0.37)\times 10^2 + \frac{(270^2 - 16^2)\times 10^{-3}}{2}$$

$$+ (-32\times 9.81\times 10^{-3}) + 135 - (-9.0)$$

$$= -160 + 36.45 - 0.314 + 135 + 9 = 20.136 \text{ kJ/kg}$$

Specific internal energy decreases by 20.136 kJ.

Example 5.3

In a steam power station, steam flows steadily through a 0.2 m diameter pipeline from the boiler to the turbine. At the boiler end, the steam conditions are found to be: p = 4 MPa, t = 400°C, h = 3213.6 kJ/kg, and v = 0.073 m³/kg. At the turbine end, the conditions are found to be: p = 3.5 MPa, t = 392°C, h = 3202.6 kJ/kg, and v = 0.084 m³/kg. There is a heat loss of 8.5 kJ/kg from the pipeline. Calculate the steam flow rate.

Solution Writing the steady flow energy equation for the control volume as shown in Fig. Ex. 5.3

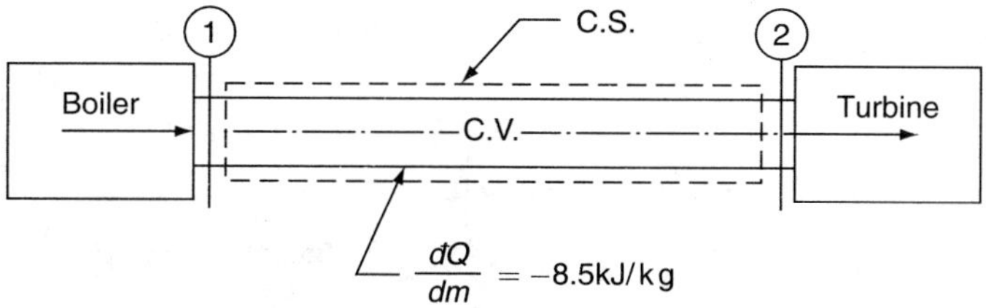

Fig. Ex. 5.3

$$h_1 + \frac{\mathbf{V}_1^2}{2} + Z_1 g + \frac{đQ}{dm} = h_2 + \frac{\mathbf{V}_2^2}{2} + Z_2 g + \frac{đW_x}{dm}$$

Here, there is no change in datum, so change in potential energy will be zero.

Now $$\frac{A_1\mathbf{V}_1}{v_1} = \frac{A_2\mathbf{V}_2}{v_2}$$

$\therefore$ $$\mathbf{V}_2 = \frac{A_1\mathbf{V}_1}{v_1} \cdot \frac{v_2}{A_2} = \frac{v_2}{v_1} \cdot \mathbf{V}_1 = \frac{0.084}{0.073}\mathbf{V}_1 = 1.15\,\mathbf{V}_1$$

and $$\frac{đW_x}{dm} = 0$$

$$h_1 + \frac{\mathbf{V}_1^2}{2} + \frac{đQ}{dm} = h_2 + \frac{\mathbf{V}_2^2}{2}$$

$$\therefore \frac{\left(\mathbf{V}_2^2 - \mathbf{V}_1^2\right) \times 10^{-3}}{2} = h_1 - h_2 + \frac{đQ}{dm} = 3213.6 - 3202.6 + (-8.5) = 2.5 \text{ kJ/kg}$$

$$\mathbf{V}_1^2 (1.15^2 - 1^2) = 5 \times 10^3$$

$$\mathbf{V}_1^2 = 15{,}650 \text{ m}^2/\text{s}^2$$

$\therefore$ $$\mathbf{V}_1 = 125.1 \text{ m/s}$$

$\therefore$ Mass flow rate $$w = \frac{A_1\mathbf{V}_1}{v_1} = \frac{\frac{\pi}{4} \times (0.2)^2 \text{ m}^2 \times 125.1 \text{ m/s}}{0.073 \text{ m}^3/\text{kg}}$$

$$= 53.8 \text{ kg/s}$$ *Ans.*

Example 5.4

A certain water heater operates under steady flow conditions receiving 4.2 *kg/s of water at* 75°C *temperature, enthalpy* 313.93 *kJ/kg. The water is heated by mixing with steam which is supplied to the heater at temperature* 100.2°C *and enthalpy* 2676 *kJ/kg. The mixture leaves the heater as liquid water at temperature* 100°C *and enthalpy* 419 *kJ/kg. How much steam must be supplied to the heater per hour?*

Solution By mass balance across the control surface (Fig. Ex. 5.4)

$$w_1 + w_2 = w_3$$

By energy balance

$$w_1\left(h_1+\frac{\mathbf{V}_1^2}{2}+Z_1 g\right)+\frac{đQ}{d\tau}+w_2\left(h_2+\frac{\mathbf{V}_2^2}{2}+Z_2 g\right)=w_3\left(h_3+\frac{\mathbf{V}_3^2}{2}+Z_3 g\right)+\frac{đW_x}{d\tau}$$

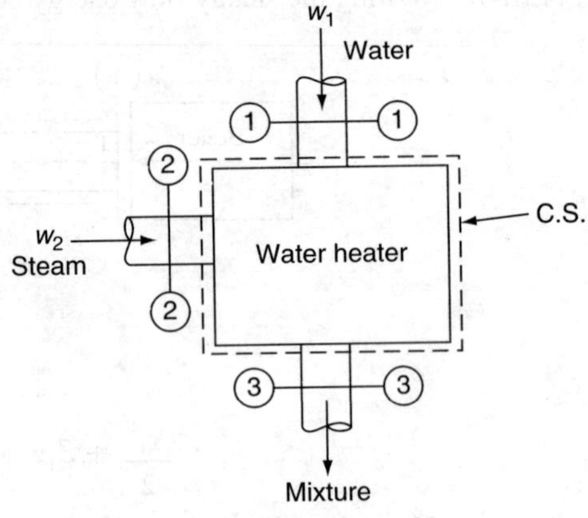

Fig. Ex. 5.4

By the nature of the process, there is no shaft work. Potential and kinetic energy terms are assumed to balance zero. The heater is assumed to be insulated. So the steady flow energy equation reduces to

$$w_1h_1 + w_2h_2 = w_3h_3$$

$$4.2 \times 313.93 + w_2 \times 2676 = (4.2 + w_2)\ 419$$

$$\therefore \quad w_2 = 0.196 \text{ kg/s}$$

$$= 705 \text{ kg/h}$$

Ans.

Example 5.5

Air at a temperature of 15°C passes through a heat exchanger at a velocity of 30 m/s where its temperature is raised to 800°C. It then enters a turbine with the same velocity of 30 m/s and expands until the temperature falls to 650°C. On leaving the turbine, the air is taken at a velocity of 60 m/s to a nozzle where it expands until the temperature has fallen to 500°C. If the air flow rate is 2 kg/s, calculate (a) the rate of heat transfer to the air in the heat exchanger, (b) the power output from the turbine assuming no heat loss, and (c) the velocity at exit from the nozzle, assuming no heat loss. Take the enthalpy of air as $h = c_p t$, where c_p is the specific heat equal to 1.005 kJ/kg K and t is the temperature.

Solution As shown in Fig. Ex. 5.5, writing the S.F.E.E. for the heat exchanger and eliminating the terms not relevant,

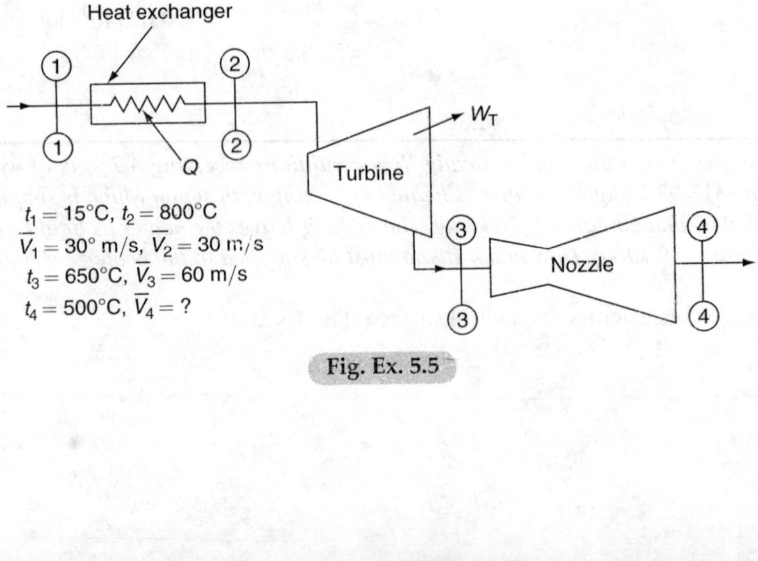

Fig. Ex. 5.5

$$w\left(h_1 + \frac{\mathbf{V}_1^2}{2} + Z_1 g\right) + Q_{1-2} = w\left(h_2 + \frac{\mathbf{V}_2^2}{2} + Z_2 g\right) + W_{1-2}$$

$$\therefore \quad wh_1 + Q_{1-2} = wh_2$$

$$\therefore \quad Q_{1-2} = w(h_2 - h_1) = wc_p(t_2 - t_1) = 2 \times 1.005\,(800 - 15) = 2.01 \times 785$$

$$= 1580 \text{ kJ/s}$$

Ans. (a)

Energy equation for the turbine gives

$$w\left(\frac{\mathbf{V}_2^2}{2} + h_2\right) = wh_3 + w\,\frac{\mathbf{V}_3^2}{2} + W_\text{T}$$

$$\frac{\mathbf{V}_2^2 - \mathbf{V}_3^2}{2} + (h_2 - h_3) = W_\text{T}/w$$

$$\frac{(30^2 - 60^2) \times 10^{-3}}{2} + 1.005\,(800 - 650) = W_\text{T}/w$$

$$\therefore \quad \frac{W_\text{T}}{w} = 1.35 + 150.75 = 149.4 \text{ kJ/kg}$$

$$\therefore \quad W_\text{T} = 149.4 \times 2 \text{ kJ/s}$$

$$= 298.8 \text{ kW}$$

Ans. (b)

Writing the energy equation for the nozzle

$$\frac{\mathbf{V}_3^3}{2} + h_3 = \frac{\mathbf{V}_4^2}{2} + h_4$$

$$\frac{\mathbf{V}_4^2 - \mathbf{V}_3^2}{2} = c_p\,(t_3 - t_4)$$

$$\mathbf{V}_4^2 - \mathbf{V}_3^2 = 1.005\,(650 - 500) \times 2 \times 10^3 = 301.50 \times 10^3 \text{ m}^2/\text{s}^2$$

$$\mathbf{V}_4^2 = 30.15 \times 10^4 + 0.36 \times 10^4 = 30.51 \times 10^4 \text{ m}^2/\text{s}^2$$

∴ Velocity at exit from the nozzle

$$\text{V}_4 = 554 \text{ m/s}$$

Ans. (c)

Example 5.6

The air speed of a turbojet engine in flight is 270 m/s. Ambient air temperature is – 15°C. Gas temperature at outlet of nozzle is 600°C. Corresponding enthalpy values for air and gas are respectively 260 and 912 kJ/kg. Fuel-air ratio is 0.0190. Chemical energy of the fuel is 44.5 MJ/kg. Owing to incomplete combustion, 5% of the chemical energy is not released in the reaction. Heat loss from the engine is 21 kJ/kg of air. Calculate the velocity of the exhaust jet.

Solution Energy equation for the turbojet engine (Fig. Ex. 5.6.) gives

$$w_\text{a}\left(h_\text{a} + \frac{\mathbf{V}_\text{a}^2}{2}\right) + w_\text{f} E_\text{f} + Q = w_\text{g}\left(h_\text{g} + \frac{\mathbf{V}_\text{g}^2}{2} + E_\text{g}\right)$$

$$\left(260 + \frac{270^2 \times 10^{-3}}{2}\right) + 0.0190 \times 44500 - 21$$

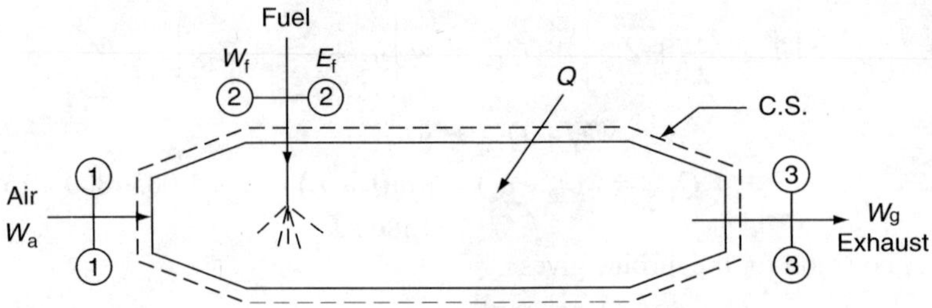

Fig. Ex. 5.6

$$= 1.0190\left(912 + \frac{\mathbf{V}_g^2 \times 10^{-3}}{2} + 0.05\frac{0.019}{1.019} \times 44500\right)$$

$$260 + 36.5 + 845 - 21 = 1.019\left(912 + \frac{\mathbf{V}_g^2 \times 10^{-3}}{2} + 42\right)$$

$$\therefore \qquad \frac{\mathbf{V}_g^2}{2} = 156 \times 10^3 \text{ m}^2/\text{s}^2$$

$$\mathbf{V}_g = \sqrt{3.12} \times 100 \text{ m/s}$$

Velocity of exhaust gas, $\mathbf{V}_g = 560$ m/s *Ans.*

Example 5.7

In a reciprocating engine, the mass of gas occupying the clearance volume is m_c kg at state p_1, u_1, v_1 and h_1. By opening the inlet valve, m_f kg of gas is taken into the cylinder, and at the conclusion of the intake process the state of the gas is given by p_2, u_2, v_2, h_2. The state of the gas in the supply pipe is constant and is given by p_p, u_p, v_p, h_p, $\mathbf{V}_p$. How much heat is transferred between the gas and the cylinder walls during the intake process?

Solution Let us consider the control volume as shown in Fig. Ex. 5.7. Writing the energy balance on a time rate basis

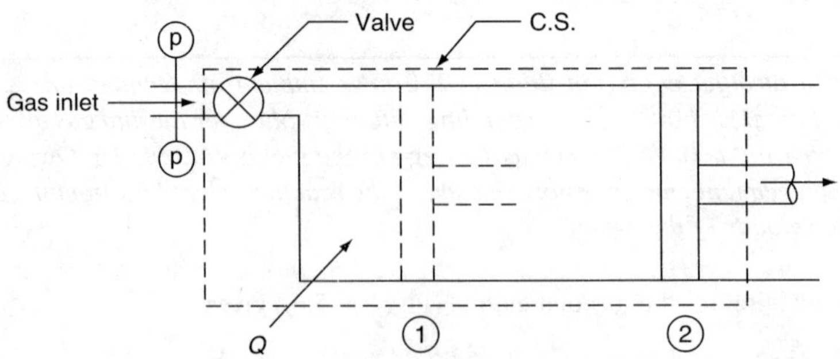

Fig. Ex. 5.7

$$\frac{dE_V}{d\tau} = \frac{đQ}{d\tau} - \frac{đW}{d\tau} + \left(h_p + \frac{\mathbf{V}_p^2}{2}\right)\frac{dm}{d\tau}$$

With h_p and $\mathbf{V}_p$ being constant, the above equation can be integrated to give for the total process

$$\Delta E_v = Q - W + \left(h_p + \frac{\mathbf{V}_p^2}{2}\right) m_f$$

Now $$\Delta E_v = U_2 - U_1 = (m_e + m_f)\, u_2 - m_e u_1$$

$\therefore$ $$Q = (m_c + m_f)u_2 - m_c u_1 - m_f \left(h_p + \frac{\mathbf{V}_p^2}{2}\right) + W \qquad \textit{Ans.}$$

Example 5.8

The internal energy of air is given by

$$u = u_0 + 0.718\, t$$

where u is in kJ/kg, u_0 is any arbitrary value of u at 0°C, kJ/kg, and t is the temperature in °C. Also for air, $p_v = 0.287\,(t + 273)$, where p is in kPa and v is in m³/kg.

A mass of air is stirred by a paddle wheel in an insulated constant volume tank. The velocities due to stirring make a negligible contribution to the internal energy of the air. Air flows out through a small valve in the tank at a rate controlled to keep the temperature in the tank constant. At a certain instant the conditions are as follows: tank volume 0.12 m³, pressure 1 MPa, temperature 150°C, and power to paddle wheel 0.1 kW. Find the rate of flow of air out of the tank at this instant.

Solution Writing the energy balance for the control volume as shown in Fig. Ex, 5.8

$$\frac{dE_V}{d\tau} = \frac{đW}{d\tau} - (h_p)\frac{dm}{d\tau}$$

Since there is no change in internal energy of air in the tank,

$$h_p \cdot \frac{dm}{d\tau} = \frac{đW}{d\tau}$$

where $$h_p = u + pv.$$

Let $u = 0$ at $t = 0\ K = -273$°C

$$u = u_0 + 0.718\, t$$
$$0 = u_0 + 0.718\,(-273)$$
$$u_0 = 0.718 \times 273 \text{ kJ/kg}$$

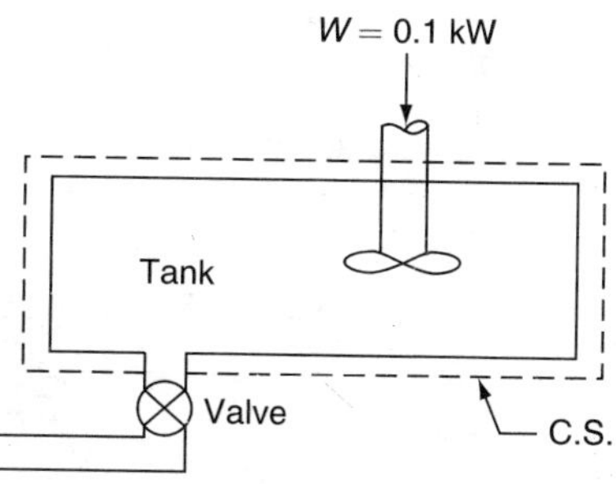

Fig. Ex. 5.8

At t°C

$$u = 0.718 \times 273 + 0.718\, t$$
$$= 0.718\,(t + 273) \text{ kJ/kg}$$
$$h_p = 0.718\,(t + 273) + 0.287\,(t + 273)$$

or $$h_p = 1.005\,(t + 273)$$

At 150°C

$$h_p = 1.005 \times 423 = 425 \text{ kJ/kg}$$

$\therefore$ $$\frac{dm}{d\tau} = \frac{1}{h_p}\frac{đW}{d\tau}$$

$$= \frac{0.1 \text{ kJ/s}}{425 \text{ kJ/kg}} = 0.236 \times 10^{-3} \text{ kg/s} = 0.845 \text{ kg/h}$$

This is the rate at which air flows out of the tank.

Example 5.9

A well-insulated vessel of volume V contains a gas at pressure p_0 and temperature t_0. The gas from a main at a uniform temperature t_1 is pumped into the vessel and the inflow rate decreases exponentially with time according to $\dot{m} = \dot{m}_0 e^{-a\tau}$, where a is a constant. Determine the pressure and temperature of the gas in the vessel as a function of time. Neglect the K.E. of the gas entering the vessel and assume that the gas follows the relation

$$pv = RT, \quad \text{where } T = t + 273$$

and its specific heats are constant.

(i) If the vessel was initially evacuated, show that the temperature inside the vessel is independent of time.

(ii) Determine the charging time when the pressure inside the vessel reaches that of the main.

Solution Since the vessel is well-insulated, $\dot{Q} = 0$ and there is no external work transfer, $W = C$. Therefore,

$$\frac{dE_V}{d\tau} = h_1 \frac{dm}{d\tau} = h_1 \dot{m}_o e^{-a\tau}$$

where h_1 is the enthalpy of the gas in the main.

On integration,

$$E = E_0 + \frac{h_1 \dot{m}_o}{a}(1 - e^{-a\tau})$$

where E_0 is the initial energy of the vessel at the beginning of the charging process, i.e. $E = E_0$ at $\tau = 0$. Neglecting K.E. and P.E. changes, by energy balance

$$Mu = M_0 u_0 + \frac{\dot{m}_o}{a}(1 - e^{-a\tau})(u_1 + p_1 v_1) \tag{1}$$

Again,

$$\frac{dm}{d\tau} = \dot{m}_o e^{-a\tau}$$

On integration,

$$M = M_0 + \frac{\dot{m}_o\left(1 - e^{-a\tau}\right)}{a} \tag{2}$$

where M_0 is the initial mass of the gas. Eliminating M from Eqs (1) and (2),

$$\left\{M_0 + \frac{\dot{m}_o}{a}\left(1 - e^{-a\tau}\right)\right\} u - M_0 u_0$$

$$= \frac{\dot{m}_o}{a}(1 - e^{-a\tau})(u_1 + RT_1)$$

$$M_0 c_v (T - T_0) = \frac{\dot{m}_o}{a}(1 - a^{-a\tau})\{c_v(T_1 - T) + RT_1\}$$

$$\therefore \quad T = \frac{M_0 c_v T_0 + \frac{\dot{m}_o}{a}\left(1 - e^{-a\tau}\right) c_p T_1}{\left\{M_0 + \frac{\dot{m}_o}{a}\left(1 - e^{-a\tau}\right)\right\} c_v}$$

$$p = \frac{MRT}{V} = \frac{R}{Vc_v}\left\{M_0 c_v T_0 + \frac{\dot{m}_0}{a}\left(1 - e^{-a\tau}\right)c_p T_1\right\}$$

$$= p_0 + \frac{\dot{m}_0 R}{aV}\ (1 - e^{-a\tau})\ \gamma T_1$$

The above two equations show the temperature and pressure of the gas in the vessel as functions of time.

(i) If $M_0 = 0$, $T = \gamma T_1$, i.e. the temperature inside the vessel becomes independent of time and is equal to γT_1 throughout the charging process.

(ii) The charging process will stop when pressure inside the vessel reaches that of the main. The charging time can be found by setting $p = p_1$ in the pressure relation

$$p_1 - p_0 = \frac{\dot{m}_0 R\gamma T_1}{aV} - \frac{\dot{m}_0 R}{aV} e^{-a\tau} g\, T_1$$

By rearrangement,

$$e^{a\tau} = \frac{\dot{m}_0 R\gamma T_1/(aV)}{\left(\frac{m_0 R\gamma T_1}{aV}\right) - (p_1 - p_0)}$$

$$\therefore \qquad \tau = -\frac{1}{a}\ln\left[1 - (p_1 - p_0)\frac{aV}{m_0 R\gamma T_1}\right]$$

Review Questions

5.1 Explain the system approach and the control volume approach in the analysis of a flow process.

5.2 What is a steady flow process? What is steady state?

5.3 Write the steady flow energy equation for a single stream entering and a single stream leaving a control volume and explain the various terms in it.

5.4 Give the differential form of the S.F.E.E.

5.5 Under what conditions does the S.F.E.E. reduce to Euler's equation?

5.6 How does Bernoulli's equation compare with S.F.E.E.?

5.7 What will be the velocity of a fluid leaving a nozzle, if the velocity of approach is very small?

5.8 Show that the enthalpy of a fluid before throttling is equal to that after throttling.

5.9 Write the general energy equation for a variable flow process.

5.10 What is the system technique in a bottle-filling process?

5.11 Explain the control volume technique in a variable flow process.

Problems

5.1 A blower handles 1 kg/s of air at 20°C and consumes a power of 15 kW. The inlet and outlet velocities of air are 100 m/s and 150 m/s respectively. Find the exit air temperature, assuming adiabatic conditions. Take c_p of air as 1.005 kJ/kg-K. *Ans.* 28.38°C

5.2 A turbine operates under steady flow conditions, receiving steam at the following state: pressure 1.2 MPa, temperature 188°C, enthalpy 2785 kJ/kg, velocity 33.3 m/s and elevation 3 m. The steam leaves the turbine at the following state: pressure 20 kPa, enthalpy 2512 kJ/kg, velocity 100 m/s, and elevation 0 m. Heat is lost to the surroundings at the rate of 0.29 kJ/s. If the rate of steam flow through the turbine is 0.42 kg/s, what is the power output of the turbine in kW? *Ans.* 112.51 kW

5.3 A nozzle is a device for increasing the velocity of a steadily flowing stream. At the inlet to a certain nozzle, the enthalpy of the fluid passing is 3000 kJ/kg and the velocity is 60 m/s. At the discharge end, the enthalpy is 2762 kJ/kg. The nozzle is horizontal and there is negligible heat loss from it. (a) Find the velocity at exist from the nozzle. (b) If the inlet area is 0.1 m^2 and the specific volume at inlet is 0.187 m^3/kg, find the mass flow rate. (c) If the specific volume at the nozzle exit is 0.498 m^3/kg, find the exit area of the nozzle.

Ans. (a) 692.5 m/s, (b) 32.08 kg/s (c) 0.023 m^2

5.4 In an oil cooler, oil flows steadily through a bundle of metal tubes submerged in a steady stream of cooling water. Under steady flow conditions, the oil enters at 90°C and leaves at 30°C, while the water enters at 25°C and leaves at 70°C. The enthalpy of oil at t°C is given by

$$h = 1.68\, t + 10.5 \times 10^{-4}\, t^2 \text{ kJ/kg}$$

What is the cooling water flow required for cooling 2.78 kg/s of oil? *Ans.* 1.473 kg/s

5.5 A thermoelectric generator consists of a series of semiconductor elements (Fig. P. 5.5), heated on one side and cooled on the other. Electric current flow is produced as a result of energy transfer as heat. In a particular experiment the current was measured to be 0.5 amp and the electrostatic potential at (1) was 0.8 volt above that at (2). Energy transfer as heat to the hot side of the generator was taking place at a rate of 5.5 watts. Determine the rate of energy transfer as heat from the cold side and the energy conversion efficiency. *Ans.* $Q_2 = 5.1$ watts, $\eta = 0.073$

5.6 A turbocompressor delivers 2.33 m^3/s of air at 0.276 MPa, 43°C which is heated at this pressure to 430°C and finally expanded in a turbine which delivers 1860 kW. During the expansion, there is a heat transfer of 0.09 MJ/s to the surroundings. Calculate the turbine exhaust temperature if changes in kinetic and potential energy are negligible. Take $c_p = 1.005$ kJ/kgK *Ans.* 157°C

5.7 A reciprocating air compressor takes in 2 m^3/min at 0.11 MPa, 20°C which it delivers at 1.5 MPa, 111°C to an aftercooler where the air is cooled at constant pressure to 25°C. The power absorbed by the compressor is 4.15 kW. Determine the heat transfer in (a) the compressor, and (b) the cooler. State your assumptions.

Ans. – 0.17 kJ/s, – 3.76 kJ/s.

5.8 In a water cooling tower air enters at a height of 1 m above the ground level and leaves at a height of 7 m. The inlet and outlet velocities are 20 m/s and 30 m/s respectively. Water enters at a height of 8 m and leaves at a height of 0.8 m. The velocity of water at entry and exit are 3 m/s and 1 m/s respectively. Water temperatures are 80°C and 50°C at the entry and exit respectively. Air

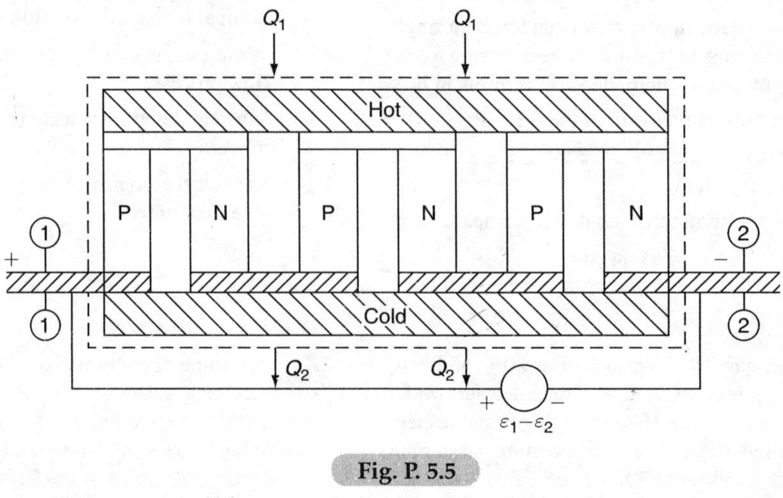

Fig. P. 5.5

temperatures are 30°C and 70°C at the entry and exit respectively. The cooling tower is well insulated and a fan of 2.25 kW drives the air through the cooler. Find the amount of air per second required for 1 kg/s of water flow. The values of c_p of air and water are 1.005 and 4.187 kJ/kg K respectively.

Ans. 3.16 kg

5.9 Air at 101.325 kPa, 20°C is taken into a gas turbine power plant at a velocity of 140 m/s through an opening of 0.15 m^2 cross-sectional area. The air is compressed heated, expanded through a turbine, and exhausted at 0.18 MPa, 150°C through an opening of 0.10 m^2 cross-sectional area. The power output is 375 kW. Calculate the net amount of heat added to the air in kJ/kg. Assume that air obeys the law $pv = 0.287\ (t + 273)$, where p is the pressure in kPa, v is the specific volume in m^3/kg, and t is the temperature in °C. Take $c_p = 1.005$ kJ/kg K.

Ans. 150.23 kJ/kg

5.10 A gas flows steadily through a rotary compressor. The gas enters the compressor at a temperature of 16°C, a pressure of 100 kPa, and an enthalpy of 391.2 kJ/kg. The gas leaves the compressor at a temperature of 245°C, a pressure of 0.6 MPa, and an enthalpy of 534.5 kJ/kg. There is no heat transfer to or from the gas as it flows through the compressor. (a) Evaluate the external work done per unit mass of gas assuming the gas velocities at entry and exit to be negligible. (b) Evaluate the external work done per unit mass of gas when the gas velocity at entry is 80 m/s and that at exit is 160 m/s. *Ans.* 143.3 kJ/kg, 152.9 kJ/kg

5.11 The steam supply to an engine comprises two streams which mix before entering the engine. One stream is supplied at the rate of 0.01 kg/s with an enthalpy of 2952 kJ/kg and a velocity of 20 m/s. The other stream is supplied at the rate of 0.1 kg/s with an enthalpy of 2569 kJ/kg and a velocity of 120 m/s. At the exit from the engine the fluid leaves as two streams, one of water at the rate of 0.001 kg/s with an enthalpy of 420 kJ/kg and the other of steam; the fluid velocities at the exit are negligible. The engine develops a shaft power of 25 kW. The heat transfer is negligible. Evaluate the enthalpy of the second exit stream.

Ans. 2402 kJ/kg

5.12 The stream of air and gasoline vapour, in the ratio of 14:1 by mass, enters a gasoline engine at a temperature of 30°C and leaves as combustion products at a temperature of 790°C. The engine has a specific fuel consumption of 0.3 kg/kWh. The net heat transfer rate from the fuel-air stream to the jacket cooling water and to the surroundings is 35 kW. The shaft power delivered by the engine is 26 kW. Compute the increase in the specific enthalpy of the fuel-air stream, assuming the changes in kinetic energy and in elevation to be negligible. *Ans.* – 1877 kJ/kg mixture

5.13 An air turbine forms part of an aircraft refrigerating plant. Air at a pressure of 295 kPa and a temperature of 58°C flows steadily into the turbine with a velocity of 45 m/s. The air leaves the turbine at a pressure of 115 kPa, a temperature of 2°C, and a velocity of 150 m/s. The shaft work delivered by the turbine is 54 kJ/kg of air. Neglecting changes in elevation, determine the magnitude and sign of the heat transfer per unit mass of air flowing. For air, take $c_p = 1.005$ kJ/kg K and the enthalpy $h = c_p\, t$.

Ans. + 7.96 kJ/kg

5.14 In a turbomachine handling an incompressible fluid with a density of 1000 kg/m^3 the conditions of the fluid at the rotor entry and exit are as given below

	Inlet	Exit
Pressure	1.15 MPa	0.05 MPa
Velocity	30 m/s	15.5 m/s
Height above datum	10 m	2 m

If the volume flow rate of the fluid is 40 m^3/s, estimate the net energy transfer from the fluid as work. *Ans.* 60.3 MW

5.15 A room for four persons has two fans, each consuming 0.18 kW power, and three 100 W lamps. Ventilation air at the rate of 80 kg/h enters with an enthalpy of 84 kJ/kg and leaves with an enthalpy of 59 kJ/kg. If each person puts out heat at the rate of 630 kJ/h determine the rate at which heat is to be removed by a room cooler, so that a steady state is maintained in the room

Ans. 1.92 kW

5.16 Air flows steadily at the rate of 0.4 kg/s through an air compressor, entering at 6 m/s with a pressure of 1 bar and a specific volume of 0.85 m^3/kg, and leaving at 4.5 m/s with a pressure of 6.9 bar and a specific volume of 0.16 m^3/kg. The internal energy of the air leaving is 88 kJ/kg greater than that of the air entering. Cooling water in a jacket, surrounding the cylinder absorbs heat from the air at the rate of 59 W. Calculate the power required to drive the compressor and the inlet and outlet cross-sectional areas.

Ans. 45.4 kW, 0.057 m^2, 0.0142 m^2

5.17 Steam flowing in a pipeline is at a steady state represented by p_p, t_p, u_p, v_p, h_p and $\mathbf{V}_p$. A small amount of the total flow is led through a small tube to an evacuated chamber which is allowed to fill slowly until the pressure is equal to the pipeline pressure. If there is no heat transfer, derive an expression for the final specific internal energy in the chamber, in terms of the properties in the pipeline.

5.18 The internal energy of air is given, at ordinary temperatures, by

$u = u_0 + 0.718\,t$

where u is in kJ/kg, u_0 is any arbitrary value of u at 0°C, kJ/kg, and t is temperature in °C.

Also for air,

$pv = 0.287\,(t + 273)$

where p is in kPa and v is in m^3/kg.

(a) An evacuated bottle is fitted with a valve through which air from the atmosphere, at 760 mm Hg and 25°C, is allowed to flow slowly to fill the bottle. If no heat is transferred to or from the air in the bottle, what will its temperature be when the pressure in the bottle reaches 760 mm Hg?

Ans. 144.2°C

(b) If the bottle initially contains 0.03 m^3 of air at 400 mm Hg and 25°C, what will the temperature be when the pressure in the bottle reaches 760 mm Hg?

Ans. 71.6°C

5.19 A pressure cylinder of volume V contains air at pressure p_0 and temperature T_0. It is to be filled from a compressed air line maintained at constant pressure p_1 and temperature T_1. Show that the temperature of the air in the cylinder after it has been charged to the pressure of the line is given by

$$T = \frac{\gamma T_1}{1 + \dfrac{p_0}{p_1}\left(\gamma \dfrac{T_1}{T_0} - 1\right)}$$

5.20 A small reciprocating vacuum pump having the rate of volume displacement V_d is used to evacuate a large vessel of volume V. The air in the vessel is maintained at a constant temperature T by energy transfer as heat. If the initial and final pressures are p_1 and p_2 respectively, find the time taken for the pressure drop and the necessary energy transfer as heat during evacuation. Assume that for air, $pV = mRT$, where m is the mass and R is a constant, and u is a function of T only.

$$\left[\textit{Ans. } t = \frac{V}{V_d}\ln\frac{p_1}{p_2};\quad Q = (p_1 - p_2)V\right]$$

[Hint: $dm = -p(V_d \cdot dt)/(RT) = V\,dp/(RT)$].

5.21 A tank containing 45 kg of water initially at 45°C has one inlet and one exit with equal mass flow rates. Liquid water enters at 45°C and a mass flow rate of 270 kg/h. A cooling coil immersed in the water removes energy at a rate of 7.6 kW. The water is well mixed by a paddle wheel with a power input of 0.6 kW. The pressures at inlet and exit are equal. Ignoring changes in KE and PE, find the variation of water temperature with time.

Ans. $T = 318 - 22\,[1 - \exp(-6t)]$

5.22 A rigid tank of volume 0.5 m^3 is initially evacuated. A tiny hole develops in the wall, and air from the surroundings at 1 bar, 21°C leaks in. Eventually, the pressure in the tank reaches 1 bar. The process occurs slowly enough that heat transfer between the tank and the surroundings keeps the temperature of the air inside the tank constant at 21°C. Determine the amount of heat transfer.

Ans. – 50 kJ

C H A P T E R 6

Second Law of Thermodynamics

6.1 QUALITATIVE DIFFERENCE BETWEEN HEAT AND WORK

The first law of thermodynamics states that a certain energy balance will hold when a system undergoes a change of state or a thermodynamic process. But it does not give any information on whether that change of state or the process is at all feasible or not. The first law cannot indicate whether a metallic bar of uniform temperature can spontaneously become warmer at one end and cooler at the other. All that the law can state is that if this process did occur, the energy gained by one end would be exactly equal to that lost by the other. *It is the second law of thermodynamics which provides the criterion as to the probability of various processes.*

Spontaneous processes in nature occur only in one direction. Heat always flows from a body at a higher temperature to a body at a lower temperature, water always flows downward, time always flows in the forward direction. The reverse of these never happens spontaneously. The spontaneity of the process is due to a finite driving potential, sometimes called the 'force' or the 'cause', and what happens is called the 'flux', the 'current' or the 'effect'. The typical forces like temperature gradient, concentration gradient, and electric potential gradient, have their respective conjugate fluxes of heat transfer, mass transfer, and flow of electric current. These transfer processes can never spontaneously occur from a lower to a higher potential. This directional law puts a limitation on energy transformation other than that imposed by the first law.

Joule's experiments (Article 4.1) amply demonstrate that energy, when supplied to a system in the form of work, can be completely converted into heat (work transfer → internal energy increase → heat transfer). But the complete conversion of heat into work in a cycle is not possible. *So heat and work are not completely interchangeable forms of energy.*

When work is converted into heat, we always have

$$W \underset{\rightarrow}{\equiv} Q$$

but when heat is converted into work in a complete closed cycle process

$$Q \underset{\rightarrow}{\geq} W.$$

The arrow indicates the direction of energy transformation. This is illustrated in Fig. 6.1. As shown in Fig. 6.1(a), a system is taken from state 1 to state 2 by work transfer W_{1-2}, and then by heat transfer Q_{2-1} the system is brought back from state 2 to state 1 to complete a cycle. It is always found that $W_{1-2} = Q_{2-1}$. But if the system is taken from state 1 to state 2 by heat transfer Q_{1-2}, as shown in Fig. 6.1(b), then the system cannot be brought back from state 2 to state 1 by work transfer W_{2-1}. Hence, heat cannot be converted completely and continuously into work in a cycle. Some heat has to be rejected. In Fig. 6.1(c), W_{2-3} is the work done and Q_{3-1} is the heat rejected to complete the cycle. This underlies the work of Sadi Carnot, a French military engineer, who

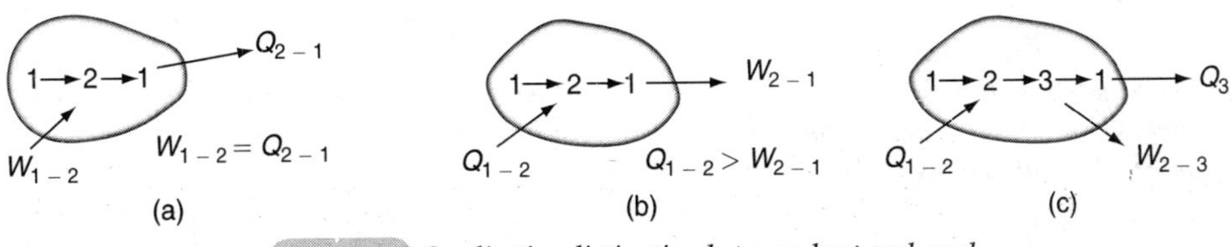

Fig. 6.1 *Qualitative distinction between heat and work*

first studied this aspect of energy transformation (1824). Work is said to be a *high grade energy* and heat a *low grade energy. The complete conversion of low grade energy into high grade energy in a cycle is impossible.*

6.2 CYCLIC HEAT ENGINE

For engineering purposes, the second law is best expressed in terms of the conditions which govern the production of work by a thermodynamic system operating in a cycle.

A heat engine cycle is a thermodynamic cycle in which there is a net heat transfer *to* the system and a net work transfer *from* the system. The system which executes a heat engine cycle is called a *heat engine*.

A heat engine may be in the form of mass of gas confined in a cylinder and piston machine (Fig. 6.2a) or a mass of water moving in a steady flow through a steam power plant (Fig. 6.2b).

In the cyclic heat engine, as represented in Fig. 6.2(a), heat Q_1 is transferred to the system, work W_E is done by the system, work W_c is done upon the system, and then heat Q_2 is rejected from the system. The system is brought back to the initial state through all these four successive processes which constitute a heat engine cycle. In Fig. 6.2(b) heat Q_1 is transferred from the furnace to the water in the boiler to form steam which then works on the turbine rotor to produce work W_T, then the steam is condensed to water in the condenser in which an amount Q_2 is rejected from the system, and finally work W_p is done on the system (water) to pump it to the boiler. The system repeats the cycle.

The net heat transfer in a cycle to either of the heat engines $\quad Q_{net} = Q_1 - Q_2 \quad$ (6.1)

and the net work transfer in a cycle $\quad W_{net} = W_T - W_p \quad$ (6.2)

$$(\text{or} \quad W_{net} = W_E - W_C)$$

By the first law of thermodynamics, we have

$$\sum_{cycle} Q = \sum_{cycle} W$$

$$\therefore \quad Q_{net} = W_{net}$$

or

$$Q_1 - Q_2 = W_T - W_P \tag{6.3}$$

Figure 6.3 represents a cyclic heat engine in the form of a block diagram indicating the various energy interactions during a cycle. Boiler (*B*), turbine (*T*), condenser (*C*), and pump (*P*), all four together constitute a heat engine. A heat engine is here a certain quantity of water undergoing the energy interactions, as shown, in cyclic operations to produce net work from a certain heat input.

The function of a heat engine cycle is to produce work continuously at the expense of heat input to the system. So the net work W_{net} and heat input Q_1 referred to the cycle are of primary interest. The efficiency of a heat engine or a heat engine cycle is defined as

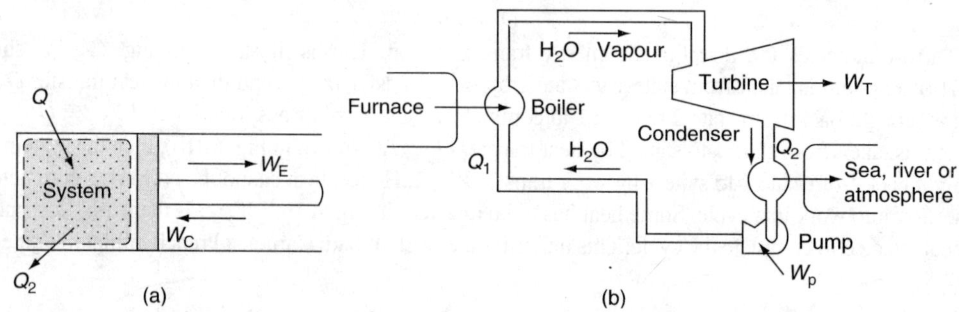

Fig. 6.2 *Cyclic heat engine: (a) Heat engine cycle performed by a closed system undergoing four successive energy interactions with the surroundings, (b) Heat engine cycle performed by a steady flow system interacting with the surroundings as shown*

$$\eta = \frac{\text{Net work output of the cycle}}{\text{Total heat input to the cycle}} = \frac{W_{net}}{Q_1} \quad (6.4)$$

From Eqs (6.1), (6.2), (6.3) and (6.4),

$$\eta = \frac{W_{net}}{Q_1} = \frac{W_T - W_P}{Q_1} = \frac{Q_1 - Q_2}{Q_1}$$

$$\eta = 1 - \frac{Q_2}{Q_1} \quad (6.5)$$

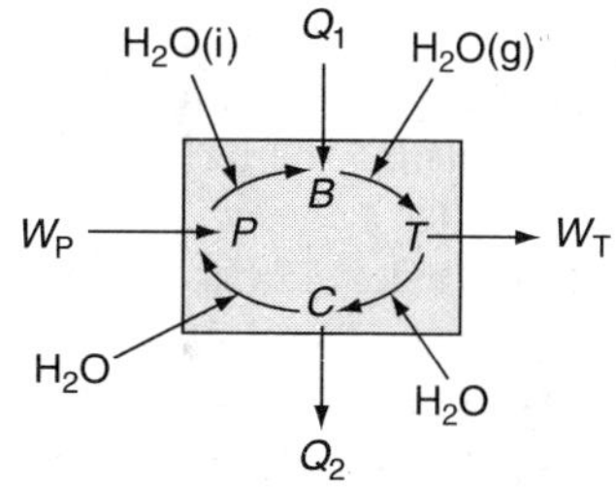

Fig. 6.3 *Cyclic heat engine with energy interactions represented in a block diagram*

This is also known as the *thermal efficiency* of a heat engine cycle. A heat engine is very often called upon to extract as much work (net) as possible from a certain heat input, i.e., to maximize the cycle efficiency.

6.3 ENERGY RESERVOIRS

A thermal energy reservoir (TER) is defined as a large body of infinite heat capacity, which is capable of absorbing or rejecting an unlimited quantity of heat without suffering appreciable changes in its thermodynamic coordinates. The changes that do take place in the large body as heat enters or leaves are so very slow and so very minute that all processes within it are quasi-static.

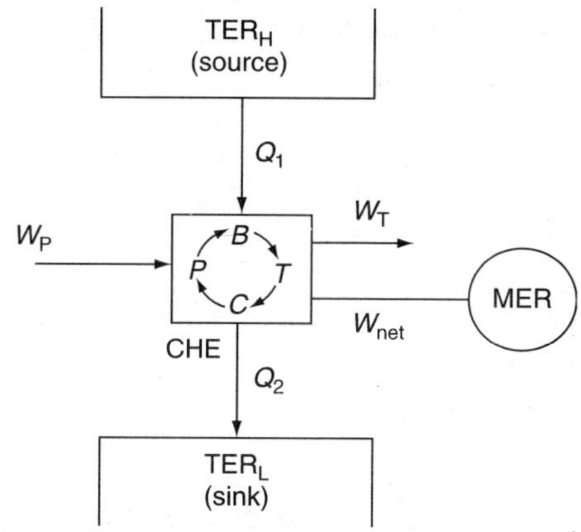

Fig. 6.4 *Cyclic heat engine (CHE) with source and sink*

The thermal energy reservoir TER_H from which heat Q_1 is transferred to the system operating in a heat engine cycle is called the *source*. The thermal energy reservoir TER_L to which heat Q_2 is rejected from the system during a cycle is the *sink*. A typical source is a constant temperature furnace where fuel is continuously burnt, and a typical sink is a river or sea or the atmosphere itself.

A mechanical energy reservoir (MER) is a large body enclosed by an adiabatic impermeable wall capable of storing work as potential energy (such as a raised weight or wound spring) or kinetic energy (such as a rotating flywheel). All processes of interest within an MER are essentially quasi-static. An MER receives and delivers mechanical energy quasi-statically.

Figure 6.4 shows a cyclic heat engine exchanging heat with a source and a sink and delivering W_{net} in a cycle to an MER.

6.4 KELVIN–PLANCK STATEMENT OF SECOND LAW

The efficiency of a heat engine is given by

$$\eta = \frac{W_{net}}{Q_1} = 1 - \frac{Q_2}{Q_1}$$

Experience shows that $W_{net} < Q_1$, since heat Q_1 transferred to a system cannot be completely converted to work in a cycle (Article 6.1). Therefore, η is less than unity. A heat engine can never be 100% efficient. Therefore, $Q_2 > 0$, i.e., there has always to be a heat rejection. To produce net work in a thermodynamic cycle, a heat engine has thus to exchange heat with two reservoirs, the source and the sink.

The *Kelvin-Planck statement* of the second law states: *It is impossible for a heat engine to produce net work in a complete cycle if it exchanges heat only with bodies at a single fixed temperature.*

If $Q_2 = 0$ (i.e., $W_{net} = Q_1$, or $\eta = 1.00$), the heat engine will produce net work in a complete cycle by exchanging heat with only one reservoir, thus violating the Kelvin-Planck statement (Fig. 6.5). Such a heat engine is called a *perpetual motion machine of the second kind*, abbreviated to PMM2. A PMM2 is impossible.

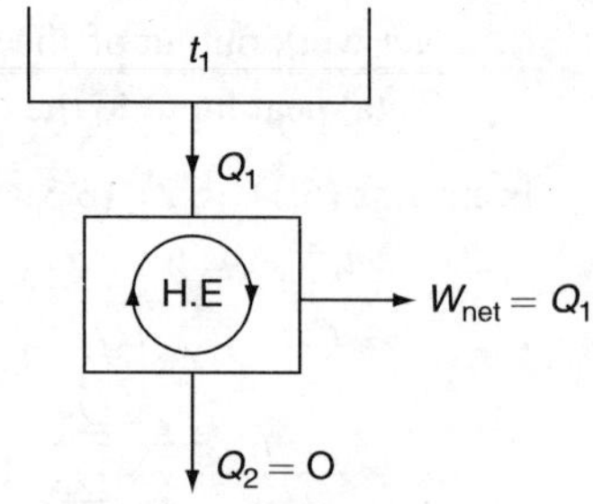

Fig. 6.5 *A PMM2*

A heat engine has, therefore, to exchange heat with two thermal energy reservoirs at two different temperatures to produce net work in a complete cycle (Fig. 6.6). So long as there is a difference in temperature, motive power (i.e. work) can be produced. If the bodies with which the heat engine exchanges heat are of finite heat capacities, work will be produced by the heat engine till the temperatures of the two bodies are equalized.

If the second law were not true, it would be possible to drive a ship across the ocean by extracting heat from the ocean or to run a power plant by extracting heat from the surrounding air. Neither of these impollibilities violates the first law of thermodynamics. Both the ocean and the surrounding air contain an enormous store of internal energy, which, in principle, may be extracted in the from of a flow of heat. There is nothing in the first law to preclude the possibility of converting this heat completely into work. The second law is, therefore, a separate law of nature, and not a deduction of the first law. The first law denies the possibility of creating or destroying energy; the second denies the possibility of utilizing energy in a particular way. The continual operation of a machine that creates its own energy and thus violates the first law is called the PMM1. The operation of a machine that utilizes the internal energy of only one TER, thus violating the second law, is called PMM2.

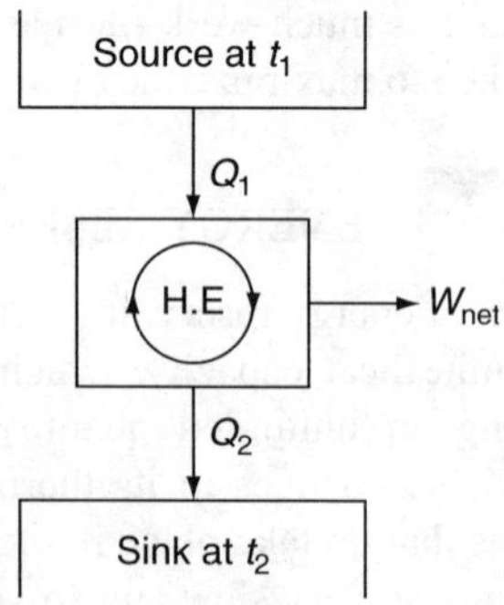

Fig. 6.6 *Heat engine producing net work in a cycle by exchanging heat at two different temperatures*

6.5 CLAUSIUS' STATEMENT OF THE SECOND LAW

Heat always flows from a body at a higher temperature to a body at a lower temperature. The reverse process never occurs spontaneously.

Clausius' statement of the second law gives: *It is impossible to construct a device which, operating in a cycle, will produce no effect other than the transfer of heat from a cooler to a hotter body.*

Heat cannot flow of itself from a body at a lower temperature to a body at a higher temperature. Some work must be expended to achieve this.

6.6 REFRIGERATOR AND HEAT PUMP

A *refrigerator* is a device which, operating in a cycle, maintains a body at a temperature lower than the temperature of the surroundings. Let the body *A* (Fig. 6.7) be maintained at t_2, which is lower than the ambient temperature t_1. Even though *A* is insulated, there will always be heat leakage Q_2 into the body from the surroundings by virtue of the temperature difference. In order to maintain, body *A* at the constant temperature t_2, heat has to be removed from the body at the same rate at which heat is leaking into the body. This heat (Q_2) is absorbed by a working fluid, called the refrigerant, which evaporates in the evaporator E_1 at a temperature lower than t_2 absorbing the latent heat of vaporization from the body *A* which is cooled or refrigerated (Process 4–1). The vapour is first compressed

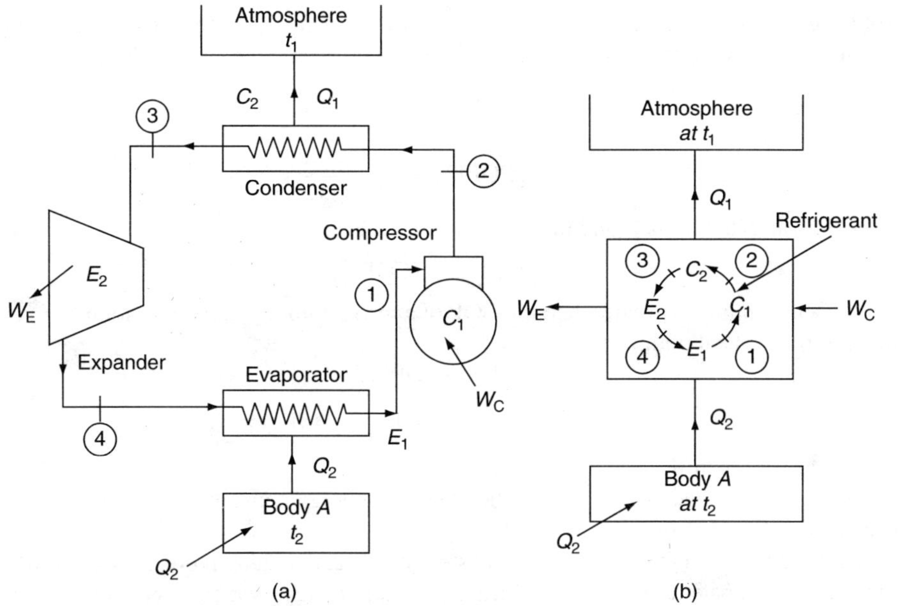

Fig. 6.7 *A cyclic refrigeration plant*

in the compressor C_1 driven by a motor which absorbs work W_C (Process 1–2), and is then condensed in the condenser C_2 rejecting the latent heat of condensation Q_1 at a temperature higher than that of the atmosphere (at t_1) for heat transfer to take place (Process 2–3). The condensate then expands adiabatically through an expander (an engine or turbine) producing work W_E, when the temperature drops to a value lower than t_2 such that heat Q_2 flows from the body A to make the refrigerant evaporate (Process 3–4). Such a cyclic device of flow through E_1–C_1–C_2–E_2 is called a *refrigerator*. In a refrigerator cycle, attention is concentrated on the body A. Q_2 and W are of primary interest. Just like efficiency in a heat engine cycle, there is a performance parameter in a refrigerator cycle, called the *coefficient of performance*, abbreviated to COP, which is defined as

$$\text{COP} = \frac{\text{Desired effect}}{\text{Work input}} = \frac{Q_2}{W}$$

$$\therefore \qquad [\text{COP}]_{\text{ref}} = \frac{Q_2}{Q_1 - Q_2} \tag{6.6}$$

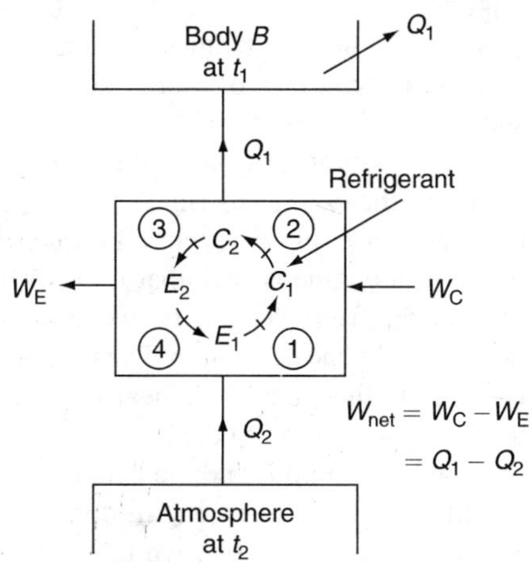

Fig. 6.8 *A cycle heat pump*

A *heat pump* is a device which, operating in a cycle, maintains a body, say B (Fig. 6.8), at a temperature higher than the temperature of the surroundings. By virtue of the temperature difference, there will be heat leakage Q_1 from the body to the surroundings. The body will be maintained at the constant temperature t_1, if heat is discharged into the body at the same rate at which heat leaks out of the body. The heat is extracted from the low temperature reservoir, which is nothing but the atmosphere, and discharged into the high temperature body B, with the expenditure of work W in a cyclic device called a heat pump. The working fluid operates in a cycle flowing through the evaporator E_1, compressor C_1, condenser C_2 and expander E_2, similar to a refrigerator, but the attention is here focussed on the high temperature body B. Here Q_1 and W are of primary interest, and the COP is defined as

$$\text{COP} = \frac{Q_1}{W}$$

$$\therefore \quad [\text{COP}]_{\text{H.P.}} = \frac{Q_1}{Q_1 - Q_2} \tag{6.7}$$

From Eqs (6.6) and (6.7), it is found that

$$[\text{COP}]_{\text{H.P.}} = [\text{COP}]_{\text{ref}} + 1 \tag{6.8}$$

The COP of a heat pump is greater than the COP of a refrigerator by unity. Equation (6.8) expresses a very interesting feature of a heat pump.

Since

$$Q_1 = [\text{COP}]_{\text{H.P.}}\, W$$

$$= [\text{COP}_{\text{ref}} + 1]\, W \tag{6.9}$$

Q_1 is always greater than W.

For an electrical resistance heater, if W is the electrical energy consumption, then the heat transferred to the space at steady state is W only, i.e., $Q_1 = W$.

A 1 kW electric heater can give 1 kW of heat at steady state and nothing more. In other words, 1 kW of work (high grade energy) dissipates to give 1 kW of heat (low grade energy), which is thermodynamically inefficient.

However, if this electrical energy W is used to drive the compressor of a heat pump, the heat supplied Q_1 will always be more than W, or $Q_1 > W$. Thus, a heat pump provides a thermodynamic advantage over direct heating.

For heat to flow from a cooler to a hotter body, W cannot be zero, and hence, the COP (both for refrigerator and heat pump) cannot be infinity. Therefore, $W > 0$, and COP $< \infty$.

6.7 EQUIVALENCE OF KELVIN-PLANCK AND CLAUSIUS STATEMENTS

At first sight, Kelvin-Planck's and Clausius' statements may appear to be unconnected, but it can easily be shown that they are virtually two parallel statements of the second law and are equivalent in all respects.

The equivalence of the two statements will be proved if it can be shown that the violation of one statement implies the violation of the second, and vice versa. (a) Let us first consider a cyclic heat pump P which transfers heat from a low temperature reservoir (t_2) to a high temperature reservoir (t_1) with no other effect, i.e., with no expenditure of work, violating Clausius statement (Fig. 6.9).

Let us assume a cyclic heat engine E operating between the same thermal energy reservoirs, producing W_{net} in one cycle. The rate of working of the heat engine is such that it draws an amount of heat Q_1 from the hot reservoir equal to that discharged by the heat pump. Then the hot reservoir may be eliminated and the heat Q_1 discharged by the heat pump is fed to the heat engine. So we see that the heat pump P and the heat engine E acting together constitute a heat engine operating in cycles and producing net work while exchanging heat only with one body at a single fixed temperature. This violates the Kelvin-Planck statement.

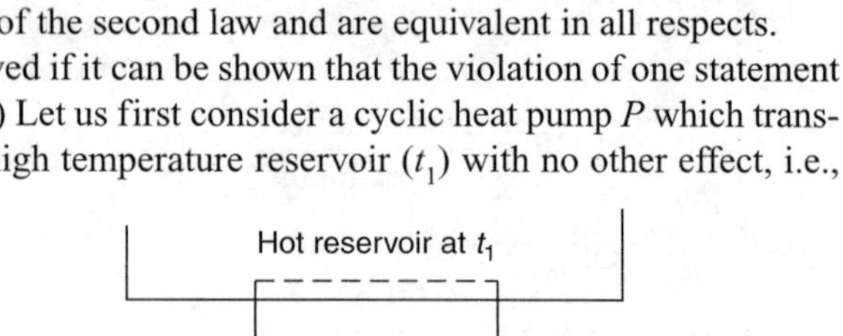

Fig. 6.9 *Violation of the Clausius statement*

(b) Let us now consider a perpetual motion machine of the second kind (E) which produces net work in a cycle by exchanging heat with only one thermal energy reservoir (at t_1) and thus violates the Kelvin-Planck statement (Fig. 6.10).

Let us assume a cyclic heat pump (P) extracting heat Q_2 from a low temperature reservoir at t_2 and discharging heat to the high temperature reservoir at t_1 with the expenditure of work W equal to what the PMM2 delivers in a complete cycle. So E and P together constitute a heat pump working in cycles and producing the sole effect of transferring heat from a lower to a higher temperature body, thus violating the Clausius statement.

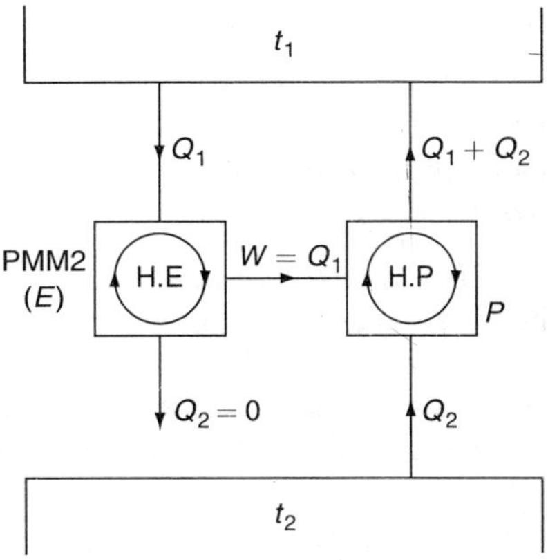

Fig. 6.10 *Violation of the Kelvin-Planck statement*

6.8 REVERSIBILITY AND IRREVERSIBILITY

The second law of thermodynamics enables us to divide all processes into two classes:

(a) Reversible or ideal process.
(b) Irreversible or natural process.

A *reversible* process is one which is performed in such a way that at the conclusion of the process, both the system and the surroundings may be restored to their initial states, without producing any changes in the rest of the universe. Let the state of a system be represented by A (Fig. 6.11), and let the system be taken to state B by following the path A–B. If the system and also the surroundings are restored to their initial states and no change in the universe is produced, then the process A–B will be a reversible process. In the reverse process, the system has to be taken from state B to A by following the same path B–A. A reversible process should not leave any trace or relic to show that the process had ever occurred.

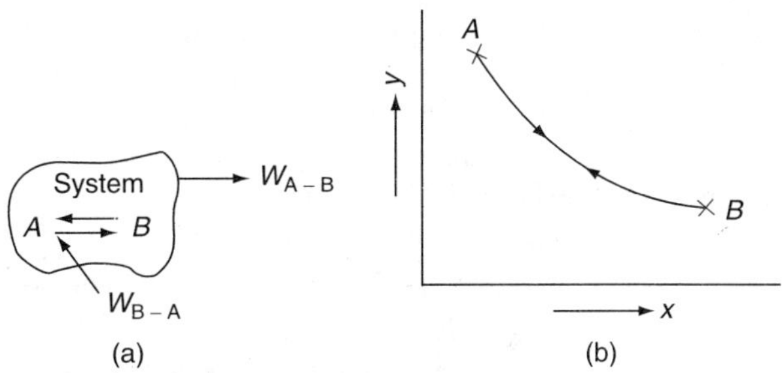

Fig. 6.11 *Reversible process*

A reversible process is carried out infinitely slowly with an infinitesimal gradient, so that every state passed through by the system is an equilibrium state. So a reversible process coincides with a quasi-static process.

Any natural process carried out with a finite gradient is an irreversible process. A reversible process, which consists of a *succession of equilibrium states*, is an idealized hypothetical process, approached only as a limit. It is said to be an *asymptote to reality*. All spontaneous processes are irreversible.

6.9 CAUSES OF IRREVERSIBILITY

Broken eggs, split milk, burnt boats, the wasted years of indolence that the locusts have eaten are merly proverbial metaphors for irreversibility.

The irreversibility of a process may be due to either one or both of the following:

(a) Lack of equilibrium during the process.
(b) Involvement of dissipative effects.

6.9.1 Irreversibility due to Lack of Equilibrium

The lack of equilibrium (mechanical, thermal or chemical) between the system and its surroundings, or between two systems, or two parts of the same system, causes a spontaneous change which is irreversible. The following are specific examples in this regard:

(a) Heat Transfer through a Finite Temperature Difference A heat transfer process approaches reversibility as the temperature difference between two bodies approaches zero. We define a reversible heat transfer process as one in which heat is transferred through an infinitesimal temperature difference. So to transfer a finite amount of heat through an infinitesimal temperature difference would require an infinite amount of time, or infinite area. All actual heat transfer processes are through a finite temperature difference and are, therefore, irreversible, and greater the temperature difference, the greater is the irreversibility.

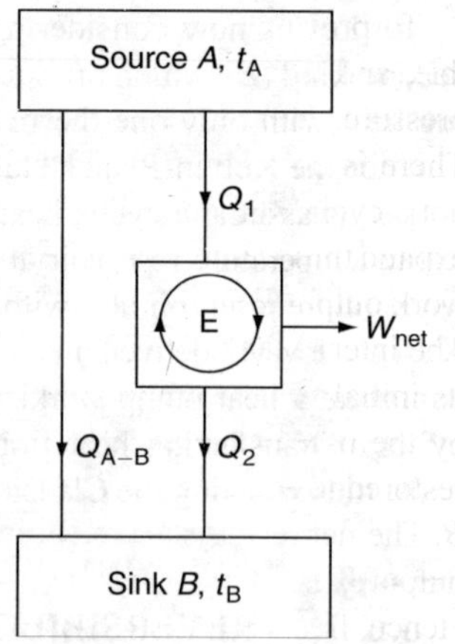

Fig. 6.12 *Heat transfer a finite temperature difference*

We can demonstrate by the second law that heat transfer through a finite temperature difference is irreversible. Let us assume that a source at t_A and a sink at t_B ($t_A > t_B$) are available, and let Q_{A-B} be the amount of heat flowing from A to B (Fig. 6.12). Let us assume an engine operating between A and B, taking heat Q_1 from A and discharging heat Q_2 to B. Let the heat transfer process be reversed, and Q_{B-A} be the heat flowing from B to A, and let the rate of working of the engine be such that $Q_2 = Q_{B-A}$.

(Fig. 6.13). Then the sink B may be eliminated. The net result is that E produces net work W in a cycle by exchanging heat only with A, thus violating the Kelvin-Planck statement. So the heat transfer process Q_{A-B} is irreversible, and Q_{B-A} is not possible.

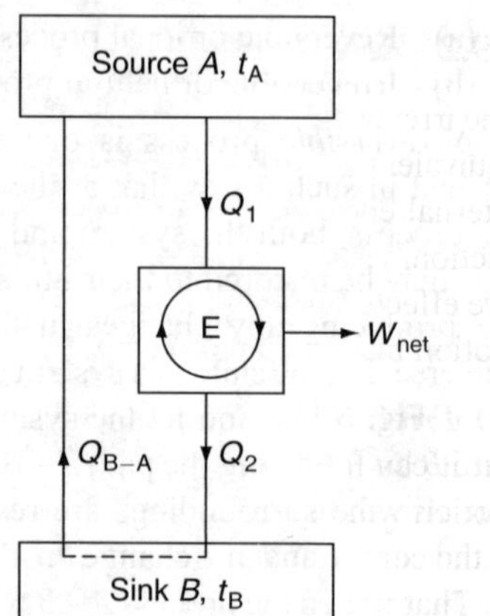

Fig. 6.13 *Heat transfer through a finite temperature difference is irreversible*

(b) Lack of Pressure Equilibrium within the Interior of the System or between the System and the Surroundings When there exists a difference in pressure between the system and the surroundings, or within the system itself, then both the system and its surroundings or the system alone, will undergo a change of state which will cease only when mechanical equilibrium is established. The reverse of this process is not possible spontaneously without producing any other effect. That the reverse process will violate the second law becomes obvious from the following illu^tration.

(c) Free Expansion Let us consider an insulated container (Fig. 6.14) which is divided into two compartments A and B by a thin diaphragm. Compartment A contains a mass of gas, while compartment B is completely evacuated. If the diaphragm is punctured, the gas in A will expand into B until the pressures in A and B become equal. This is known as free or unrestrained expansion. We can demonstrate by the second law, that the process of free expansion is irreversible.

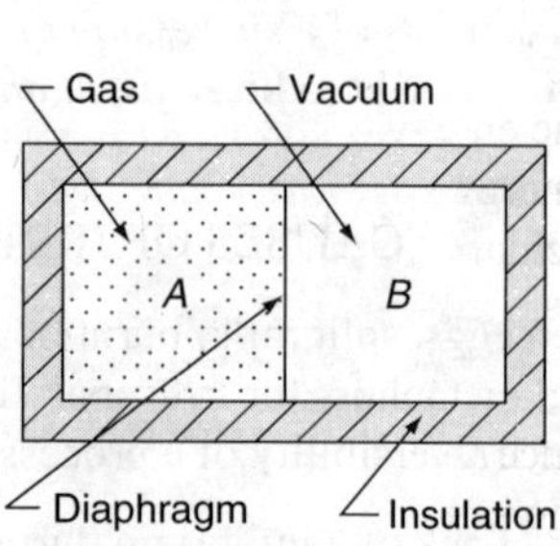

Fig. 6.14 *Free expansion*

To prove this, let us assume that free expansion is reversible, and that the gas in B returns into A with an increase in pressure, and B becomes evacuated as before (Fig. 6.15). There is no other effect. Let us install an engine (a machine, not a cyclic heat engine) between A and B, and permit the gas to expand through the engine from A to B. The engine develops a work output W at the expense of the internal energy of the gas. The internal energy of the gas (system) in B can be restored to its initial value by heat transfer Q $(= W)$ from a source. Now, by the use of the reversed free expansion, the system can be restored to the initial state of high pressure in A and vacuum in B. The net result is a cycle, in which we observe that net work output W is accomplished by exchanging heat with a single reservoir. This violates the Kelvin-Planck statement. Hence, free expansion is irreversible.

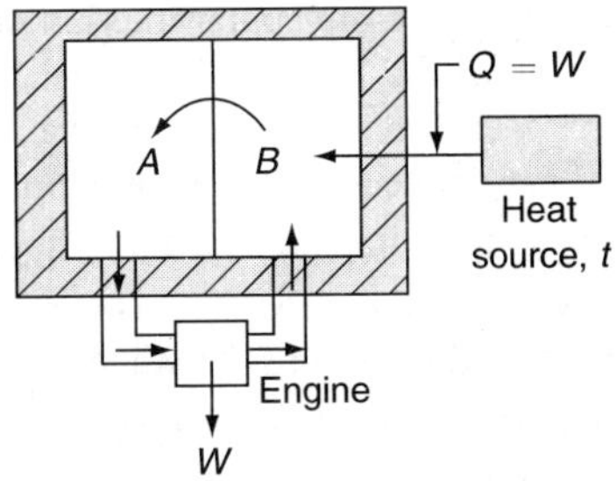

Fig. 6.15 *Second law demonstrates that free expansion is irreversible*

The same argument will hold if the compartment B is not in vacuum but at a pressure lower than that in compartment A (case b).

6.9.2 Irreversibility due to Dissipative Effects

The irreversibility of a process may be due to the *dissipative effects* in which work is done without producing an equivalent increase in the kinetic or potential energy of any system. The transformation of work into molecular internal energy either of the system or of the reservoir takes place through the agency of such phenomena as friction, viscosity, inelasticity, electrical resistance, and magnetic hysteresis. These effects are known as dissipative effects, and work is said to be dissipated. Dissipation of energy means the transition of ordered macroscopic motion into chaotic molecular motion, the reverse of which is not possible without violating second law.

(a) Friction Friction is always present in moving devices. Friction may be reduced by suitable lubrication, but it can never be completely eliminated. If this were possible, a movable device could be kept in continual motion without violating either of the two laws of thermodynamics. The continual motion of a movable device in the complete absence of friction is known as *perpetual motion of the third kind.*

That friction makes a process irreversible can be demonstrated by the second law. Let us consider a system consisting of a flywheel and a brake block (Fig. 6.16). The flywheel was rotating with a certain rpm, and it was brought to rest by applying the friction brake. The distance moved by the brake block is very small, so work transfer is very nearly equal to zero. If the braking process occurs very rapidly, there is little heat transfer. Using suffix 2 after braking and suffix 1 before braking, and applying the first law, we have

$$Q_{1-2} = E_2 - E_1 + W_{1-2} \qquad 0 = E_2 - E_1 + 0$$

$$\therefore \qquad E_2 = E_1 \tag{6.10}$$

The energy of the system (isolated) remains constant. Since the energy may exist in the forms of kinetic, potential, and molecular internal energy, we have

$$U_2 + \frac{mV_2^2}{2} + mZ_2g = U_1 + \frac{mV_1^2}{2} + mZ_1g$$

Since the wheel is brought to rest, $V_2 = 0$, and there is no change in P.E.

$$U_2 = U_1 + \frac{mV_1^2}{2} \tag{6.11}$$

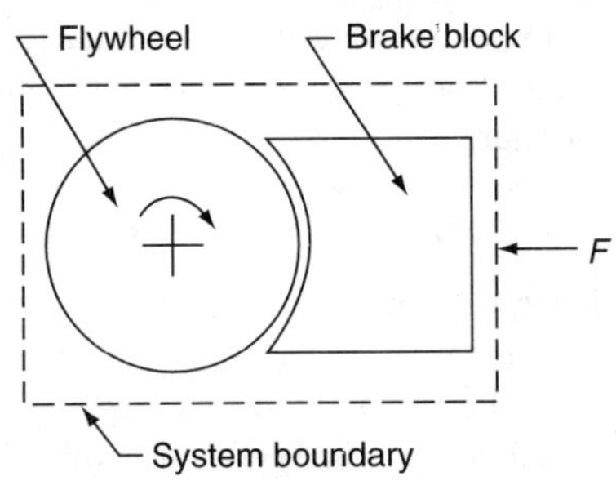

Fig. 6.16 *Irreversibility due to dissipative effect like friction*

Therefore, the molecular internal energy of the system (i.e., of the brake and the wheel) increases by the absorption of the K.E. of the wheel. The reverse process, i.e., the conversion of this increase in molecular internal energy into K.E. within the system to cause the wheel to rotate is not possible. To prove it by the second law, let us assume that it is possible, and imagine the following cycle with three processes:

Process A: Initially, the wheel and the brake are at high temperature as a result of the absorption of the K.E. of the wheel, and the flywheel is at rest. Let the flywheel now start rotating at a particular rpm at the expense of the internal energy of the wheel and brake, the temperature of which will then decrease.

Process B: Let the flywheel be brought to rest by using its K.E. in raising weights, with no change in temperature.

Process C: Now let heat be supplied from a source to the flywheel and the brake, to restore the system to its initial state.

Therefore, the processes *A*, *B*, and *C* together constitute a cycle producing work by exchanging heat with a single reservoir. This violates the Kelvin-Planck statement, and it will become a PMM2. So the braking process, i.e., the transformation of K.E. into molecular internal energy, is irreversible.

(b) Paddle-Wheel Work Transfer Work may be transferred into a system in an insulated container by means of a paddle wheel (Fig. 6.17) which is also known as stirring work. Here work transferred is dissipated adiabatically into an increase in the molecular internal energy of the system. To prove the irreversibility of the process, let us assume that the same amount of work is delivered by the system at the expense of its molecular internal energy, and the temperature of the system goes down (Fig. 6.18). The system is brought back to its initial state by heat transfer from a source. These two processes together constitute a cycle in which there is work output and the system exchanges heat with a single reservoir. It becomes a PMM2, and hence the dissipation of stirring work to internal energy is irreversible.

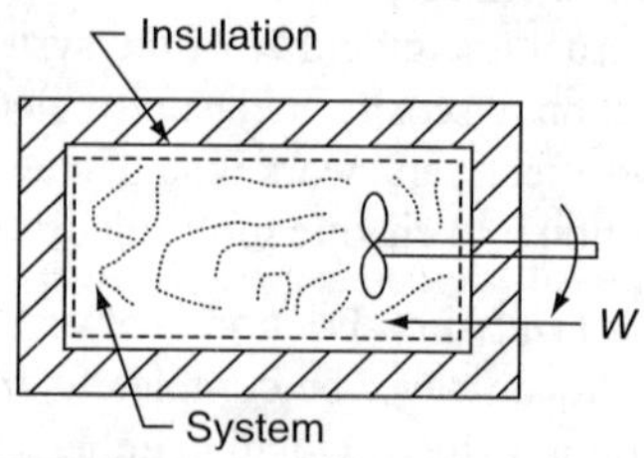

Fig. 6.17 *Adiabatic work transfer*

(c) Transfer of Electricity through a Resistor The flow of electric current through a wire represents work transfer, because the current can drive a motor which can raise a weight. Taking the wire or the resistor as the system (Fig. 6.19) and writing the first law

$$Q_{1-2} = U_2 - U_1 + W_{1-2}$$

Here both W_{1-2} and Q_{1-2} are negative.

$$\therefore \quad W_{1-2} = U_2 - U_1 + Q_{1-2} \quad (6.12)$$

A part of the work transfer is stored as an increase in the internal energy of the wire (to give an increase in its temperature), and the remainder leaves the system as heat. At steady state, the internal energy and hence the temperature of the resistor become constant with respect to time and

$$W_{1-2} = Q_{1-2} \quad (6.13)$$

The reverse process, i.e., the conversion of heat Q_{1-2} into electrical work W_{1-2} of the same magnitude is not possible. Let us assume that this is possible. Then heat Q_{1-2} will be absorbed and equal work W_{1-2} will be delivered. But this will become a PMM2. So the dissipation of electrical work into internal energy or heat is irreversible.

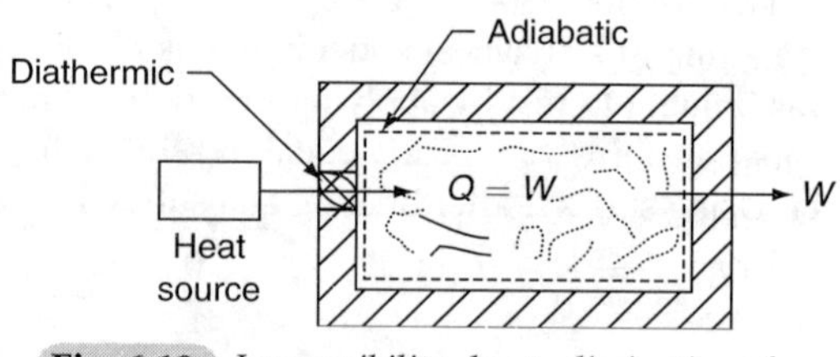

Fig. 6.18 *Irreversibility due to dissipation of stirring work into internal energy*

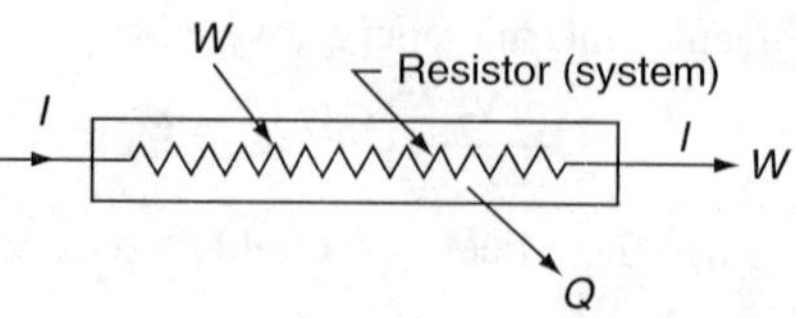

Fig. 6.19 *Irreversibility due to dissipation of electrical work into internal energy*

6.10 CONDITIONS FOR REVERSIBILITY

A natural process is irreversible because the conditions for mechanical, thermal and chemical equilibrium are not satisfied, and the dissipative effects, in which work is transformed into an increase in internal energy, are present. For a process to be reversible, it must not possess these features. If a process is performed quasi-statically, the system passes through states of thermodynamic equilibrium, which may be traversed as well in one direction as in the opposite direction. *If there are no dissipative effects, all the work done by the system during the performance of a process in one direction can be returned to the system during the reverse process.*

A process will be reversible when it is performed in such a way *that the system is at all times infinitesimally near a state of thermodynamic equilibrium and in the absence of dissipative effect of any form.* Reversible processes are, therefore, purely ideal, limiting cases of actual processes.

6.11 CARNOT CYCLE

A reversible cycle is an ideal hypothetical cycle in which all the processes constituting the cycle are reversible. Carnot cycle is a reversible cycle. For a stationary system, as in a piston and cylinder machine, the cycle consists of the following four successive processes (Fig. 6.20):

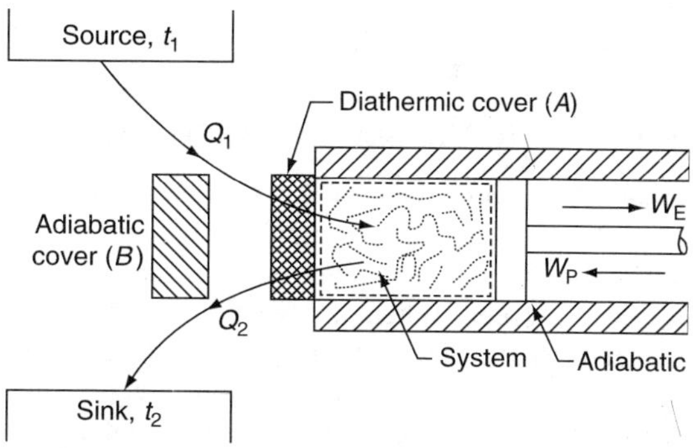

Fig. 6.20 *Carnot heat engine-stationary system*

(a) *A reversible isothermal process* in which heat Q_1 enters the system at t_1 reversibly from a constant temperature source at t_1 when the cylinder cover is in contact with the diathermic cover A. The internal energy of the system increases.

From First law, $Q_1 = U_2 - U_1 + W_{1-2}$ (6.14)

(for an ideal gas only, $U_1 = U_2$)

(b) *A reversible adiabatic process* in which the diathermic cover A is replaced by the adiabatic cover B, and work W_E is done by the system adiabatically and reversibly at the expense of its internal energy, and the temperature of the system decreases from t_1 to t_2.

Using the first law, $0 = U_3 - U_2 + W_{2-3}$ (6.15)

(c) *A reversible isothermal process* in which B is replaced by A and heat Q_2 leaves the system at t_2 to a constant temperature sink at t_2 reversibly, and the internal energy of the system further decreases.

From the first law, $-Q_2 = U_4 - U_3 - W_{3-4}$ (6.16)

only for an ideal gas, $U_3 = U_4$

(d) *A reversible adiabatic process* in which B again replaces A, and work W_p is done upon the system reversibly and adiabatically, and the internal energy of the system increases and the temperature rises from t_2 to t_1.

Applying the first law,

$$0 = U_1 - U_4 - W_{4-1} \quad (6.17)$$

Two reversible isotherms and two reversible adiabatics constitute a Carnot cycle, which is represented in p–v coordinates in Fig. 6.21.

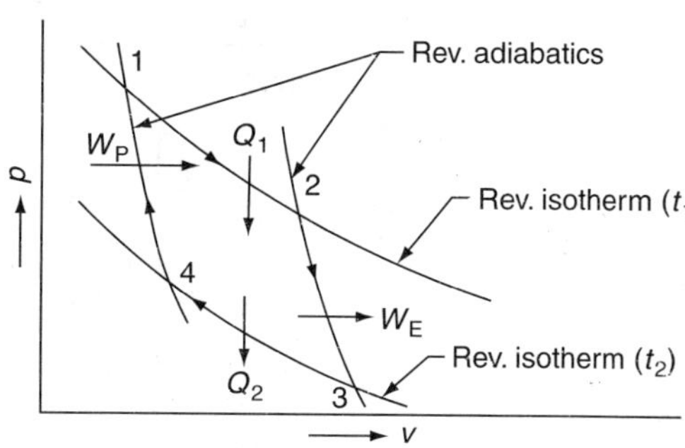

Fig. 6.21 *Carnot cycle*

Summing up Eqs (6.14) to (6.17),

$$Q_1 - Q_2 = (W_{1-2} + W_{2-3}) - (W_{3-4} + W_{4-1})$$

or

$$\sum_{\text{cycle}} Q_{\text{net}} = \sum_{\text{cycle}} W_{\text{net}}$$

A cyclic heat engine operating on the Carnot cycle is called a Carnot heat engine.

For a steady flow system, the Carnot cycle is represented as shown in Fig. 6.22. Here heat Q_1 is transferred to the system reversibly and isothermally at t_1 in the heat exchanger *A*, work W_T is done by the system reversibly and adiabatically in the turbine (*B*), then heat Q_2 is transferred from the system reversibly and isothermally at t_2 in the heat exchanger (*C*), and then work W_P is done upon the system reversibly and adiabatically by the pump (*D*). To satisfy the conditions for the Carnot cycle, there must not be any friction or heat transfer in the pipelines through which the working fluid flows.

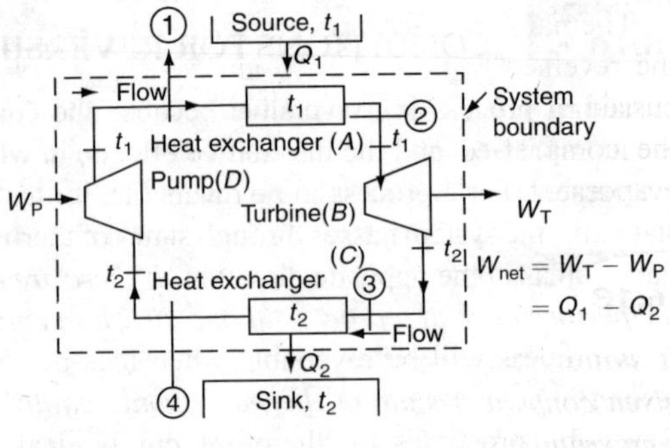

Fig. 6.22 *Carnot heat engine–steady flow system*

6.12 REVERSED HEAT ENGINE

Since all the processes of the Carnot cycle are reversible, it is possible to imagine that the processes are individually reversed and carried out in reverse order. When a reversible process is reversed, all the energy transfers associated with the process are reversed in direction, but remain the same in magnitude. The reversed Carnot cycle for a steady flow system is shown in Fig. 6.23. The reversible heat engine and the reversed Carnot heat engine are represented in block diagrams in Fig. 6.24. If *E* is a reversible heat engine (Fig. 6.24a), and if it is reversed (Fig. 6.24b), the quantities Q_1, Q_2 and *W* remain the same in magnitude, and only their directions are reversed. The reversed heat engine ∃ takes heat from a low temperature body, discharges heat to a high temperature body, and receives an inward flow of network.

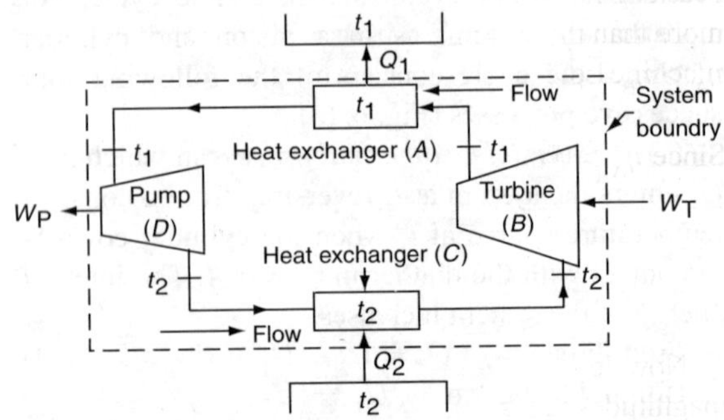

Fig. 6.23 *Reversed carnot heat engine-steady flow process*

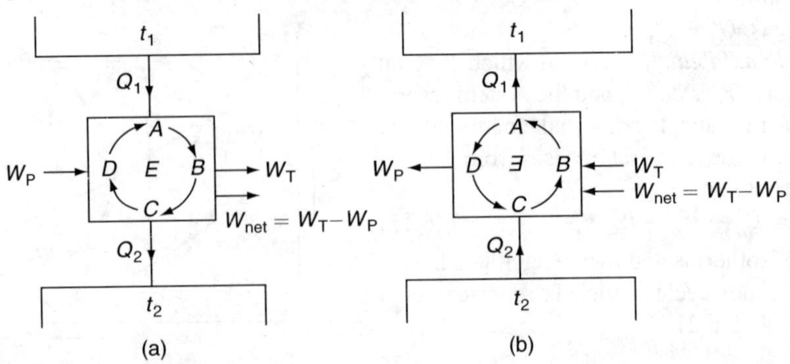

Fig. 6.24 *Carnot heat engine and reversed Carnot heat engine shown in block diagrams*

The names *heat pump* and *refrigerator* are applied to the reversed heat engine, which have already been discussed in Sec. 6.6, where the working fluid flows through the compressor (*B*), condenser (*A*), expander (*D*), and evaporator (*C*) to complete the cycle.

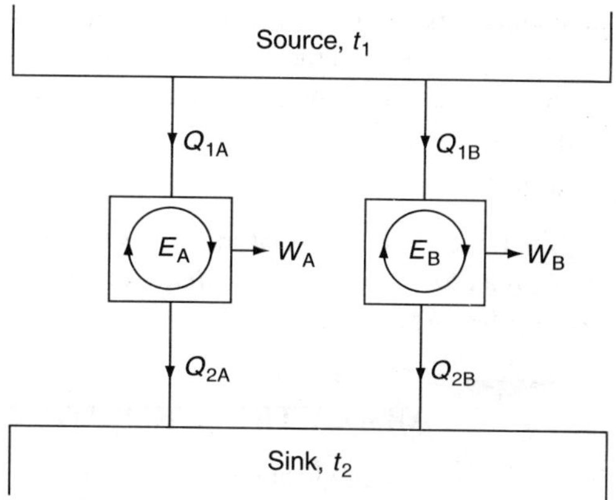

Fig. 6.25 *Two cyclic heat engines E_A and E_B operating between the same source and sink, of which E_B is reversible*

6.13 CARNOT'S THEOREM

It states that of all heat engines operating between a given constant temperature source and a given constant temperature sink, none has a higher efficiency than a reversible engine.

Let two heat engines E_A and E_B operate between the given source at temperature t_1 and the given sink at temperature t_2 as shown in Fig. 6.25.

Let E_A be any heat engine and E_B be *any reversible* heat engine. We have to prove that the efficiency of E_B is more than that of E_A. Let us assume that this is not true and $\eta_A > \eta_B$. Let the rates of working of the engines be such that

$$Q_{1A} = Q_{1B} = Q_1$$

Since $\eta_A > \eta_B$

$$\frac{W_A}{Q_{1A}} > \frac{W_B}{Q_{1B}}$$

$$\therefore \qquad W_A > W_B$$

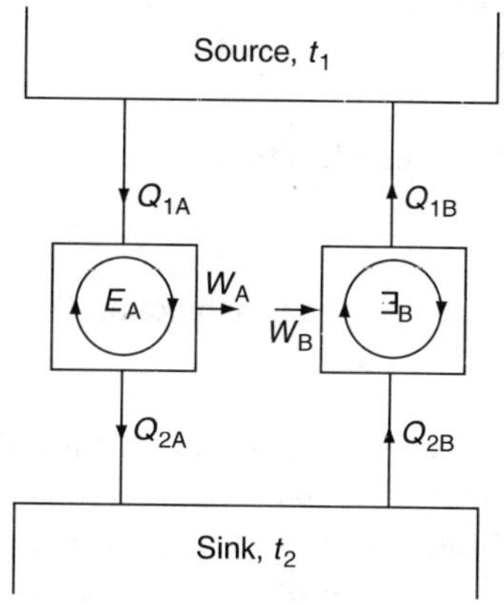

Fig. 6.26 E_B *is reversed*

Now, let E_B be reversed. Since E_B is a reversible heat engine, the magnitudes of heat and work transfer quantities will remain the same, but their directions will be reversed, as shown in Fig. 6.26. Since $W_A > W_B$, some part of W_A (equal to W_B) may be fed to drive the reversed heat engine $\exists_B$.

Since $Q_{1A} = Q_{1B} = Q_1$, the heat discharged by $\exists_B$ may be supplied to E_A. The source may, therefore, be eliminated (Fig. 6.27). The net result is that E_A and $\exists_B$ together constitute a heat engine which, operating in a cycle, produces net work $W_A - W_B$, while exchanging heat with a single reservoir at t_2. This violates the Kelvin-Planck statement of the second law. Hence the assumption that $\eta_A \not> \eta_B$ is wrong.

Therefore $$\eta_B \geq \eta_A$$

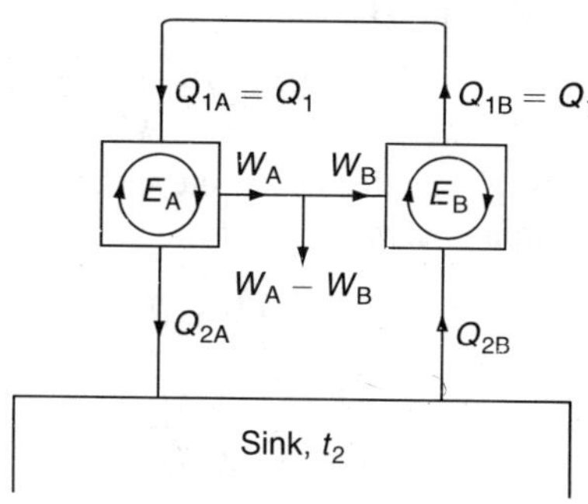

Fig. 6.27 *E_A and $\exists_B$ together violate the K-P statement*

6.14 COROLLARY OF CARNOT'S THEOREM

The efficiency of all reversible heat engines operating between the same temperature levels is the same.

Let both the heat engines E_A and E_B (Fig. 6.25) be reversible. Let us assume $\eta_A > \eta_B$. Similar to the procedure outlined in the preceding article, if E_B is reversed to run, say, as a heat pump using some part of the work output (W_A) of engine E_A, we see that the combined system of heat pump E_B and engine E_A, becomes a PMM2. So η_A cannot be greater than η_B. Similarly, if we assume $\eta_B > \eta_A$ and reverse the engine E_A, we observe that η_B cannot be greater than η_A.

Therefore $$\eta_A = \eta_B$$

Since the efficiencies of all reversible heat engines operating between the same heat reservoirs are the same, *the efficiency of a reversible engine is independent of the nature or amount of the working substance undergoing the cycle.*

6.15 ABSOLUTE THERMODYNAMIC TEMPERATURE SCALE

The efficiency of any heat engine cycle receiving heat Q_1 and rejecting heat Q_2 is given by

$$\eta = \frac{W_{net}}{Q_1} = \frac{Q_1 - Q_2}{Q_1} = 1 - \frac{Q_2}{Q_1} \tag{6.18}$$

By the second law, it is necessary to have a temperature difference ($t_1 - t_2$) to obtain work of any cycle. We know that the efficiency of all heat engines operating between the same temperature levels is the same, and it is independent of the working substance. Therefore, for a reversible cycle (Carnot cycle), the efficiency will depend solely upon the temperatures t_1 and t_2, at which heat is transferred, or

$$\eta_{rev} = f(t_1, t_2) \tag{6.19}$$

where f signifies some function of the temperatures. From Eqs (6.18) and (6.19)

$$1 - \frac{Q_2}{Q_1} = f(t_1, t_2)$$

In terms of a new function F

$$\frac{Q_1}{Q_2} = F(t_1, t_2) \tag{6.20}$$

If some functional relationship is assigned between t_1, t_2 and Q_1/Q_2, the equation becomes the definition of a temperature scale.

Let us consider two reversible heat engines, E_1 receiving heat from the source at t_1, and rejecting heat at t_2 to E_2 which, in turn, rejects heat to the sink at t_3 (Fig. 6.28).

Now $$\frac{Q_1}{Q_2} = F(t_1, t_2); \quad \frac{Q_2}{Q_3} = F(t_2, t_3)$$

E_1 and E_2 together constitute another heat engine E_3 operating between t_1 and t_3.

$$\therefore \quad \frac{Q_1}{Q_3} = F(t_1, t_3)$$

Now $$\frac{Q_1}{Q_2} = \frac{Q_1/Q_3}{Q_2/Q_3}$$

or $$\frac{Q_1}{Q_2} = F(t_1, t_2) = \frac{F(t_1, t_3)}{F(t_2, t_3)} \tag{6.21}$$

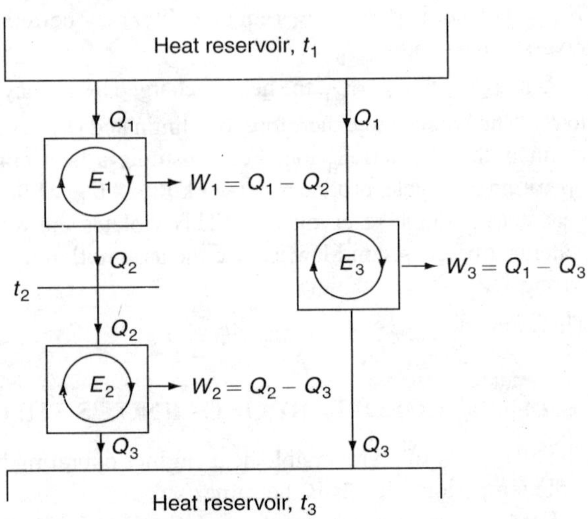

Fig. 6.28 *Three Carnot engines*

The temperatures t_1, t_2 and t_3 are arbitrarily chosen. The ratio Q_1/Q_2 depends only on t_1 and t_2, and is independent of t_3. So t_3 will drop out from the ratio on the right in Eq. (6.21). After it has been cancelled, the numerator can be written as $\phi(t_1)$, and the denominator as $\phi(t_2)$, where ϕ is another unknown function. Thus

$$\frac{Q_1}{Q_2} = F(t_1, t_2) = \frac{\phi(t_1)}{\phi(t_2)}$$

Since $\phi(t)$ is an arbitrary function, the simplest possible way to define the *absolute thermodynamic temperature* T is to let $\phi(t) = T$, as proposed by Kelvin. Then, by definition

$$\frac{Q_1}{Q_2} = \frac{T_1}{T_2} \tag{6.22}$$

The absolute thermodynamic temperature scale is also known as the *Kelvin scale*. Two temperatures on the Kelvin scale bear the same relationship to each other as do the heats absorbed and rejected respectively by a Carnot engine operating between two reservoirs at these temperatures. The Kelvin temperature scale is, therefore, independent of the peculiar characteristics of any particular substance.

The heat absorbed Q_1 and the heat rejected Q_2 during the two reversible isothermal processes bounded by two reversible adiabatics in a Carnot engine can be measured. In defining the Kelvin temperature scale also, the triple point of water is taken as the standard reference point. For a Carnot engine operating between reservoirs at temperatures T and T_t, T_t being the triple point of water (Fig. 6.29), arbitrarily assigned the value 273.16 K,

$$\frac{Q}{Q_t} = \frac{T}{T_t}$$

$$\therefore \quad T = 273.16 \frac{Q}{Q_t} \tag{6.23}$$

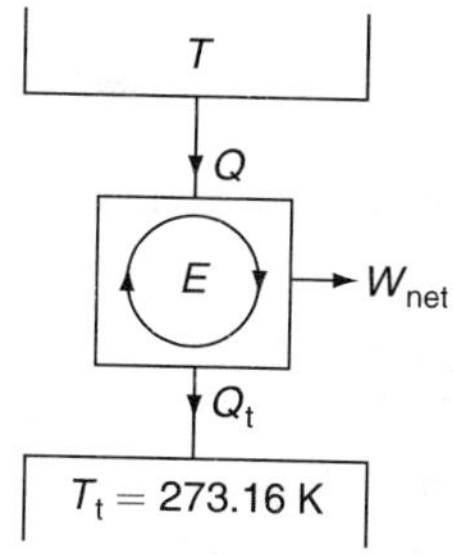

Fig. 6.29 *Carnot heat engine with sink at triple point of water*

If this equation is compared with the equations given in Article 2.3, it is seen that *in the Kelvin scale, Q plays the role of thermometric property*. The amount of heat supply Q changes with change in temperature, just like the thermal emf in a thermocouple.

It follows from the Eq. (6.23), $T = 273.16 \dfrac{Q}{Q_t}$

that the heat transferred isothermally between the given adiabatics decreases as the temperature decreases. Conversely, the smaller the value of Q, the lower the corresponding T. The smallest possible value of Q is zero, and the corresponding T is absolute zero. Thus, if a system undergoes a reversible isothermal process without transfer of heat, the temperature at which this process takes place is called the *absolute zero*. Thus, *at absolute zero, an isotherm and an adiabatic are identical.*

That the absolute thermodynamic temperature scale has a definite zero point can be shown by imagining a series of reversible engines, extending from a source at T_1 to lower temperatures (Fig. 6.30).

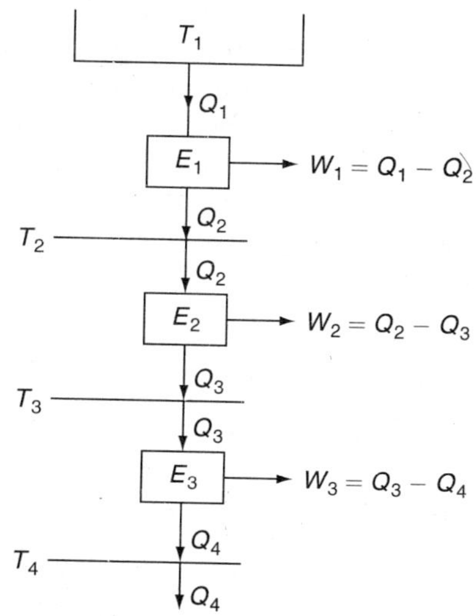

Fig. 6.30 *Heat engines operating in series*

Since $\qquad \dfrac{T_1}{T_2} = \dfrac{Q_1}{Q_2}$

$$\therefore \quad \frac{T_1 - T_2}{T_2} = \frac{Q_1 - Q_2}{Q_2}$$

or
$$T_1 - T_2 = (Q_1 - Q_2)\frac{T_2}{Q_2}$$

Similarly

$$T_2 - T_3 = (Q_2 - Q_3)\frac{T_3}{Q_3}$$

$$= (Q_2 - Q_3)\frac{T_2}{Q_2}$$

$$T_3 - T_4 = (Q_3 - Q_4)\frac{T_2}{Q_2} \quad \text{and so on.}$$

If $T_1 - T_2 = T_2 - T_3 = T_3 - T_4 = ...$, assuming equal temperature intervals

$$Q_1 - Q_2 = Q_2 - Q_3 = Q_3 - Q_4 = ...$$

or
$$W_1 = W_2 = W_3 = ...$$

Conversely, by making the work quantities performed by the engines in series equal ($W_1 = W_2 = W_3 = ...$), we will get
$$T_1 - T_2 = T_2 - T_3 = T_3 - T_4 = ...$$
at equal temperature intervals. A scale having one hundred equal intervals between the steam point and the ice point could be realized by a series of one hundred Carnot engines operating as in Fig. 6.30. Such a scale would be independent of the working substance.

If enough engines are placed in series to make the total work output equal to Q_1, then by the first law the heat rejected from the last engine will be zero. By the second law, however, the operation of a cyclic heat engine with zero heat rejection cannot be achieved, although it may be approached as a limit. When the heat rejected approaches zero, the temperature of heat rejection also approaches zero as a limit. *Thus it appears that a definite zero point exists on the absolute temperature scale but this point cannot be reached without a violation of the second law.*

Thus any attainable value of absolute temperature is always greater than zero. This is also known as the *Third Law of Thermodynamics* which may be stated as follows: *It is impossible by any procedure, no matter how idealized, to reduce any system to the absolute zero of temperature in a finite number of operations.*

This is what is called the Fowler-Guggenheim statement of the third law. The third law itself is an independent law of nature, and not an extension of the second law. The concept of heat engine is not necessary to prove the non-attainability of absolute zero of temperature by any system in a finite number of operations.

6.16 EFFICIENCY OF THE REVERSIBLE HEAT ENGINE

The efficiency of a reversible heat engine in which heat is received solely at T_1 is found to be

$$\eta_{rev} = \eta_{max} = 1 - \left(\frac{Q_2}{Q_1}\right)_{rev} = 1 - \frac{T_2}{T_1}$$

or
$$\eta_{rev} = \frac{T_1 - T_2}{T_1}$$

It is observed here that as T_2 decreases, and T_1 increases, the efficiency of the reversible cycle increases.

Since η is always less than unity, T_2 is always greater than zero and positive.

The COP of a refrigerator is given by

$$(\text{COP})_{\text{refr}} = \frac{Q_2}{Q_1 - Q_2} = \frac{1}{\dfrac{Q_1}{Q_2} - 1}$$

For a reversible refrigerator, using

$$\frac{Q_1}{Q_2} = \frac{T_1}{T_2}$$

$$[\text{COP}_{\text{refr}}]_{\text{rev}} = \frac{T_2}{T_1 - T_2} \tag{6.24}$$

Similarly, for a reversible heat pump

$$[\text{COP}_{\text{H.P}}]_{\text{rev}} = \frac{T_1}{T_1 - T_2} \tag{6.25}$$

6.17 EQUALITY OF IDEAL GAS TEMPERATURE AND KELVIN TEMPERATURE

Let us consider a Carnot cycle executed by an ideal gas, as shown in Fig. 6.31.

The two isothermal processes $a-b$ and $c-d$ are represented by equilateral hyperbolas whose equations are respectively

$$pV = nR\,\theta_1$$

and

$$pV = nR\,\theta_2$$

For any infinitesimal reversible process of an ideal gas, the first law may be written as

$$đQ = C_v\, d\theta + pdV$$

Applying this equation to the isothermal process $a-b$, the heat absorbed is found to be

$$Q_1 = \int_{V_a}^{V_b} pdV = \int_{V_a}^{V_b} \frac{nR\theta_1}{V} dV = nR\theta_1 \ln \frac{V_b}{V_a}$$

Similarly, for the isothermal process $c-d$, the heat rejected is

$$Q_2 = nR\theta_2 \ln \frac{V_c}{V_d}$$

$$\therefore \quad \frac{Q_1}{Q_2} = \frac{\theta_1 \ln \dfrac{V_b}{V_a}}{\theta_2 \ln \dfrac{V_c}{V_d}} \tag{6.26}$$

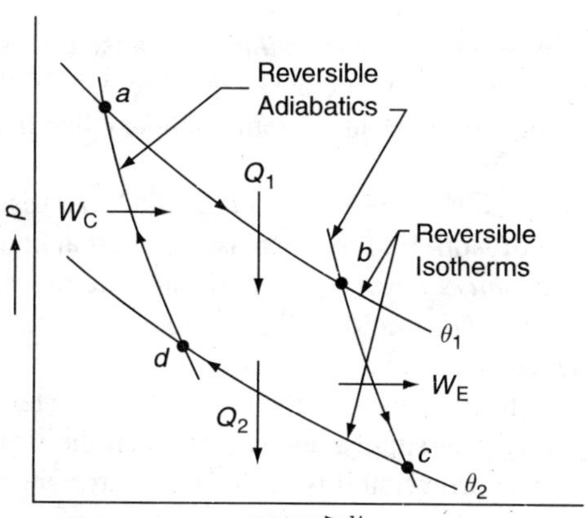

Fig. 6.31 Carnot cycle of an ideal gas

Since the process $b-c$ is adiabatic, the first law gives

$$-C_v\, d\theta = pdV = \frac{nR\theta}{V} dV \qquad \frac{1}{nR}\int_{\theta_2}^{\theta_1} C_v \frac{d\theta}{\theta} = \ln \frac{V_c}{V_b}$$

Similarly, for the adiabatic process $d-a$

$$\frac{1}{nR}\int_{\theta_2}^{\theta_1} C_v \frac{d\theta}{\theta} = \ln \frac{V_d}{V_a}$$

$$\therefore \qquad \ln \frac{V_c}{V_b} = \ln \frac{V_d}{V_a}$$

or
$$\frac{V_c}{V_b} = \frac{V_d}{V_a}$$

or
$$\frac{V_b}{V_a} = \frac{V_c}{V_d} \tag{6.27}$$

Equation (6.26) thus reduces to

$$\frac{Q_1}{Q_2} = \frac{\theta_1}{\theta_2} \tag{6.28}$$

Kelvin temperature was defined by Eq. (6.22)

$$\frac{Q_1}{Q_2} = \frac{T_1}{T_2}$$

If θ and T refer to any temperature, and θ_t and T_t refer to the triple point of water,

$$\frac{\theta}{\theta_t} = \frac{T}{T_t}$$

Since $\theta_t = T_t = 273.16$ K, it follows that

$$\theta = T \tag{6.29}$$

The Kelvin temperature is, therefore, numerically equal to the ideal gas temperature and may be measured by means of a gas thermometer.

6.18 TYPES OF IRREVERSIBILITY

It has been discussed in Sec. 6.9 that a process becomes irreversible if it occurs due to a finite potential gradient like the gradient in temperature or pressure, or if there is dissipative effect like friction, in which work is transformed into internal energy increase of the system. Two types of irreversibility can be distinguished:

(a) Internal irreversibility
(b) External irreversibility

The *internal irreversibility* is caused by the internal dissipative effects like friction, turbulence, electrical resistance, magnetic hysteresis, etc. within the system. The *external irreversibility* refers to the irreversibility occurring at the system boundary like heat interaction with the surroundings due to a finite temperature gradient.

Sometimes, it is useful to make other distinctions. If the irreversibility of a process is due to the dissipation of work into the increase in internal energy of a system, or due to a finite pressure gradient, it is called *mechanical irreversibility*. If the process occurs on account of a finite temperature gradient, it is *thermal irreversibility*, and if it is due to a finite concentration gradient or a chemical reaction, it is called *chemical irreversibility*.

A heat engine cycle in which there is a temperature difference (i) between the source and the working fluid during heat supply, and (ii) between the working fluid and the sink during heat rejection, exhibits external thermal irreversibility. If the real source and sink are not considered and hypothetical reversible processes for heat supply and heat rejection are assumed, the cycle can be reversible. With the inclusion of the actual source and sink, however, the cycle becomes externally irreversible.

Solved Examples

Example 6.1

A cyclic heat engine operates between a source temperature of 800°C and a sink temperature of 30°C. What is the least rate of heat rejection per kW net output of the engine?

Solution For a reversible engine, the rate of heat rejection will be minimum (Fig. Ex. 6.1).

$$\eta_{max} = \eta_{rev} = 1 - \frac{T_2}{T_1}$$

$$= 1 - \frac{30 + 273}{800 + 273}$$

$$= 1 - 0.282 = 0.718$$

Now $$\frac{W_{net}}{Q_1} = \eta_{max} = 0.718$$

∴ $$Q_1 = \frac{1}{0.718} = 1.392 \text{ kW}$$

Now $$Q_2 = Q_1 - W_{net} = 1.392 - 1$$

$$= 0.392 \text{ kW}$$

This is the least rate of heat rejection.

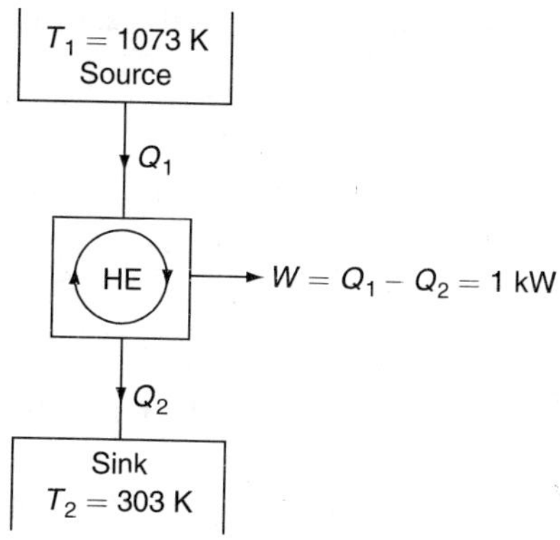

Fig. Ex. 6.1

Example 6.2

A domestic food freezer maintains a temperature of −15°C. The ambient air temperature is 30°C. If heat leaks into the freezer at the continuous rate of 1.75 kJ/s what is the least power necessary to pump this heat out continuously?

Solution Freezer temperature,

$$T_2 = -15 + 273 = 258 \text{ K}$$

Ambient air temperature,

$$T_1 = 30 + 273 = 303 \text{ K}$$

The refrigerator cycle removes heat from the freezer at the same rate at which heat leaks into it (Fig. Ex. 6.2).

For minimum power requirement

$$\frac{Q_2}{T_2} = \frac{Q_1}{T_1}$$

∴ $$Q_1 = \frac{1.75}{2.8} \times 303 = 2.06 \text{ kJ/s}$$

∴ $$W = Q_1 - Q_2$$

$$= 2.06 - 1.75 = 0.31 \text{ kJ/s} = 0.31 \text{ kW}$$

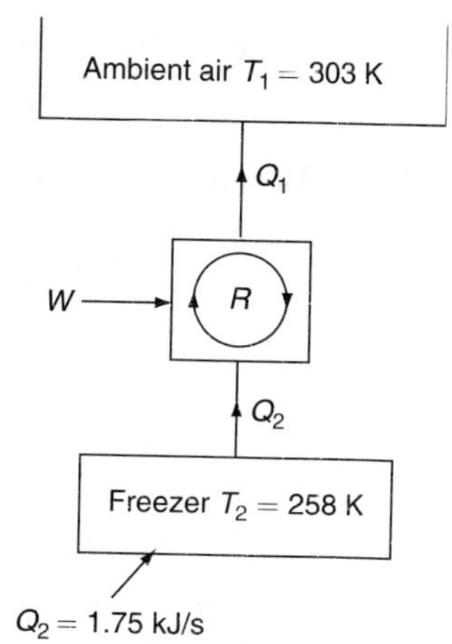

Fig. Ex. 6.2

Example 6.3

A reversible heat engine operates between two reservoirs at temperatures of 600°C and 40°C. The engine drives a reversible refrigerator which operates between reservoirs at temperatures of 40°C and –20°C. The heat transfer to the heat engine is 2000 kJ and the net work output of the combined engine refrigerator plant is 360 kJ.

(a) Evaluate the heat transfer to the refrigerant and the net heat transfer to the reservoir at 40°C.
(b) Reconsider (a) given that the efficiency of the heat engine and the COP of the refrigerator are each 40% of their maximum possible values.

Solution (a) Maximum efficiency of the heat engine cycle (Fig. Ex. 6.3) is given by

$$\eta_{max} = 1 - \frac{T_2}{T_1} = 1 - \frac{313}{873}$$

$$= 1 - 0.358 = 0.642$$

Again $\frac{W_1}{Q_1} = 0.642$

$\therefore$ $W_1 = 0.642 \times 2000 = 1284$ kJ

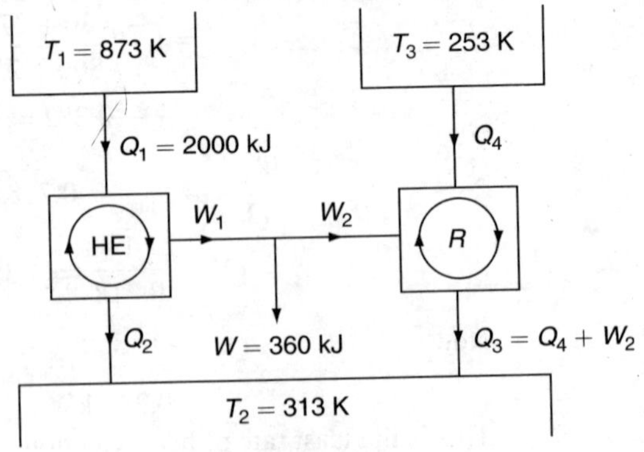

Fig. Ex. 6.3

Maximum COP of the refrigerator cycle

$$(\text{COP})_{max} = \frac{T_3}{T_2 - T_3} = \frac{253}{313 - 253} = 4.22$$

Also $\text{COP} = \frac{Q_4}{W_2} = 4.22$

Since $W_1 - W_2 = W = 360$ kJ

$\therefore$ $W_2 = W_1 - W = 1284 - 360 = 924$ kJ

$\therefore$ $Q_4 = 4.22 \times 924 = 3899$ kJ

$\therefore$ $Q_3 = Q_4 + W_2 = 924 + 3899 = 4823$ kJ

$Q_2 = Q_1 - W_1 = 2000 - 1284 = 716$ kJ

Heat rejection to the 40°C reservoir

$$= Q_2 + Q_3 = 716 + 4823 = 5539 \text{ kJ}$$ *Ans.* (a)

(b) Efficiency of the actual heat engine cycle

$$\eta = 0.4\,\eta_{max} = 0.4 \times 0.642$$

$\therefore$ $W_1 = 0.4 \times 0.642 \times 2000 = 513.6$ kJ

$\therefore$ $W_2 = 513.6 - 360 = 153.6$ kJ

COP of the actual refrigerator cycle

$$\text{COP} = \frac{Q_4}{W_2} = 0.4 \times 4.22 = 1.69$$

Therefore

$$Q_4 = 153.6 \times 1.69 = 259.6 \text{ kJ}$$

$$Q_3 = 259.6 + 153.6 = 413.2 \text{ kJ}$$ *Ans.* (b)

$$Q_2 = Q_1 - W_1 = 2000 - 513.6 = 1486.4 \text{ kJ}$$

Heat rejected to the 40°C reservoir

$$= Q_2 + Q_3 = 413.2 + 1486.4 = 1899.6 \text{ kJ}$$ *Ans.* (b)

Example 6.4

Which is the more effective way to increase the efficiency of a Carnot engine: to increase T_1, keeping T_2 constant; or to decrease T_2, keeping T_1 constant?

Solution The efficiency of a Carnot engine is given by

$$\eta = 1 - \frac{T_2}{T_1}$$

If T_2 is constant

$$\left(\frac{\partial \eta}{\partial T_1}\right)_{T_2} = \frac{T_2}{T_1^2}$$

As T_1 increases, η increases, and the slope $\left(\frac{\partial \eta}{\partial T_1}\right)_{T_2}$ decreases (Fig. Ex. 6.4.1).

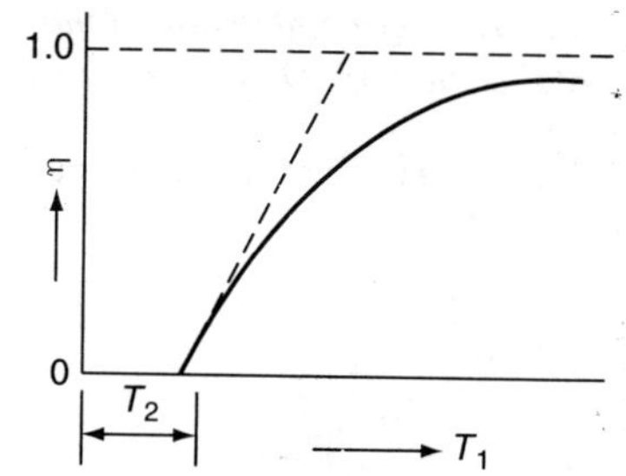

Fig. Ex. 6.4.1

If T_1 is constant,

$$\left(\frac{\partial \eta}{\partial T_2}\right)_{T_1} = -\frac{1}{T_1}$$

As T_2 decreases, η increases, but the slope $\left(\frac{\partial \eta}{\partial T_2}\right)_{T_1}$ remains constant (Fig. Ex. 6.4.2).

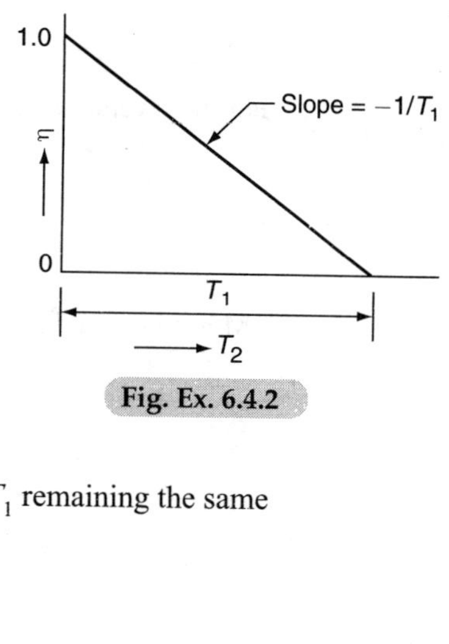

Fig. Ex. 6.4.2

Also $$\left(\frac{\partial \eta}{\partial T_1}\right)_{T_1} = \frac{T_2}{T_1^2} \text{ and } \left(\frac{\partial \eta}{\partial T_2}\right)_{T_1} = -\frac{T_1}{T_1^2}$$

Since $$T_1 > T_2, \left(\frac{\partial \eta}{\partial T_2}\right)_{T_1} > \left(\frac{\partial \eta}{\partial T_1}\right)_{T_2}$$

So, the more effective way to increase the efficiency is to decrease T_2. Alternatively, let T_2 be decreased by ΔT with T_1 remaining the same

$$\eta_1 = 1 - \frac{T_2 - \Delta T}{T_1}$$

If T_1 is increased by the same ΔT, T_2 remaining the same

$$\eta_2 = 1 - \frac{T_2}{T_1 + \Delta T}$$

Then

$$\eta_1 - \eta_2 = \frac{T_2}{T_1 + \Delta T} - \frac{T_2 - \Delta T}{T_1} = \frac{(T_1 - T_2)\Delta T + (\Delta T)^2}{T_1(T_1 + \Delta T)}$$

Since $$T_1 > T_2, (\eta_1 - \eta_2) > 0$$

The more effective way to increase the cycle efficiency is to decrease T_2.

Example 6.5

Kelvin was the first to point out the thermodynamic wastefulness of burning fuel for the direct heating of a house. It is much more economical to use the high temperature heat produced by combustion in a heat engine and then to use the work so developed to pump heat from outdoors up to the temperature desired in the house. In Fig. Ex. 6.5 a boiler furnishes heat Q_1 at the high temperature T_1. This heat is absorbed by a heat engine, which extracts work W and rejects the waste heat Q_2 into the house at T_2. Work W is in turn used to operate a mechanical refrigerator or heat pump, which extracts Q_3 from outdoors at temperature T_3 and reject Q'_2 (where $Q'_2 = Q_3 + W$) into the house. As a result of this cycle of operations, a total quantity of heat equal to $Q_2 + Q'_2$ is liberated in the house, against Q_1 which would be provided directly by the ordinary combustion of the fuel. Thus the ratio $(Q_2 + Q'_2)/Q_1$ represents the heat multiplication factor of this method. Determine this multiplication factor if $T_1 = 473$ K, $T_2 = 293$ K, and $T_3 = 273$ K.

Solution For the reversible heat engine (Fig. Ex. 6.5)

$$\frac{Q_2}{Q_1} = \frac{T_2}{T_1}$$

$$\therefore \quad Q_2 = Q_1\left(\frac{T_2}{T_1}\right)$$

Also

$$\eta = \frac{W}{Q_1} = \frac{T_1 - T_2}{T_1}$$

or

$$W = \frac{T_1 - T_2}{T_1} \cdot Q_1$$

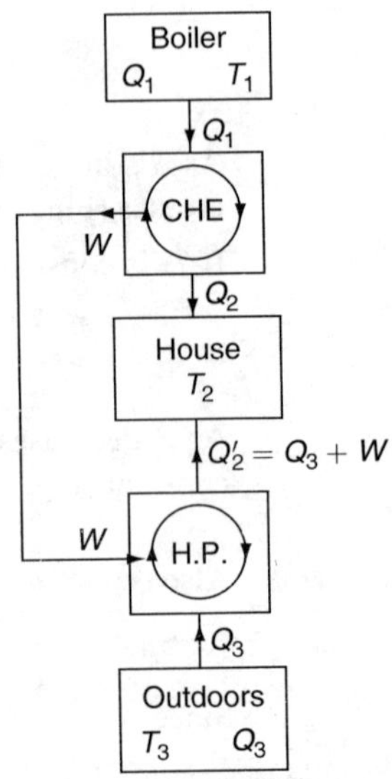

Fig. Ex. 6.5

For the reversible heat pump

$$\text{COP} = \frac{Q'_2}{W} = \frac{T_2}{T_2 - T_3}$$

$$\therefore \quad Q'_2 = \frac{T_2}{T_2 - T_3} \cdot \frac{T_1 - T_2}{T_1} \cdot Q_1$$

$\therefore$ Multiplication factor (M.F.)

$$= \frac{Q_2 + Q'_2}{Q_1} = \frac{Q_1 \dfrac{T_2}{T_1} + Q_1 \cdot \dfrac{T_2}{T_2 - T_3} \cdot \dfrac{T_1 - T_2}{T_1}}{Q_1}$$

or

$$\text{M.F.} = \frac{T_2^2 - T_2T_3 + T_2T_1 - T_2^2}{T_1(T_2 - T_3)}$$

or

$$\text{M.F.} = \frac{T_2(T_1 - T_3)}{T_1(T_2 - T_3)}$$

Here $T_1 = 473$ K, $T_2 = 293$ K and $T_3 = 273$ K

$$\therefore \quad \text{M.F.} = \frac{293(473 - 273)}{473(293 - 273)} = \frac{2930}{473} = 6.3 \qquad \textit{Ans.}$$

which means that every kg of coal burned would deliver the heat equivalent to over 6 kg. Of course, in an actual case, the efficiencies would be less than Carnot efficiencies, but even with a reduction of 50%, the possible savings would be quite significant.

Example 6.6

It is proposed that solar energy be used to warm a large collector plate. This energy would, in turn, be transferred as heat to a fluid within a heat engine, and the engine would reject energy as heat to the atmosphere. Experiments indicate that about 1880 kJ/m² h of energy can be collected when the plate is operating at 90°C. Estimate the minimum collector area that would be required for a plant producing 1 kW of useful shaft power. The atmospheric temperature may be assumed to be 20°C.

Solution The maximum efficiency for the heat engine operating between the collector plate temperature and the atmospheric temperature is

$$\eta_{max} = 1 - \frac{T_2}{T_1} = 1 - \frac{293}{363} = 0.192$$

The efficiency of any actual heat engine operating between these temperatures would be less than this efficiency.

$$\therefore \quad Q_{min} = \frac{W}{\eta_{max}} = \frac{1 \text{ kJ/s}}{0.192} = 5.21 \text{ kJ/s}$$

$$= 18{,}800 \text{ kJ/h}$$

∴ Minimum area required for the collector plate

$$= \frac{18{,}800}{1880} = 10 \text{ m}^2$$

Ans.

Example 6.7

A reversible heat engine in a satellite operates between a hot reservoir at T_1 and a radiating panel at T_2. Radiation from the panel is proportional to its area and to T_2^4. For a given work output and value of T_1 show that the area of the panel will be minimum when $\frac{T_2}{T_1} = 0.75$.

Determine the minimum area of the panel for an output of 1 kW if the constant of proportionality is 5.67×10^{-8} W/m² K⁴ and T_1 is 1000 K.

Solution For the heat engine (Fig. Ex. 6.7), the heat rejected Q_2 to the panel (at T_2) is equal to the energy emitted from the panel to the surroundings by radiation. If A is the area of the panel, $Q_2 \propto AT_2^4$, or $Q_2 = KAT_2^4$, where K is a constant.

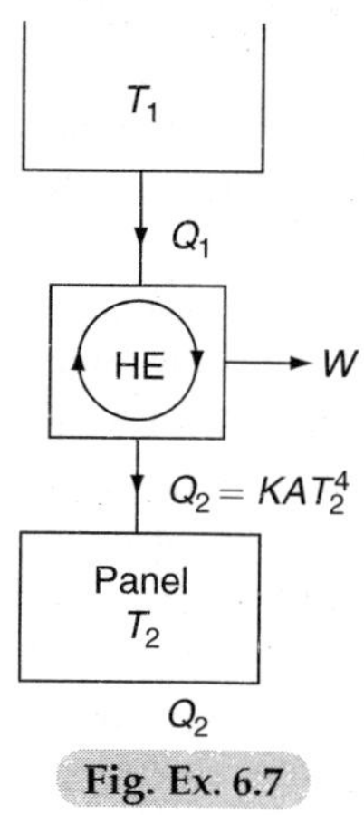

Fig. Ex. 6.7

Now

$$\eta = \frac{W}{Q_1} = \frac{T_1 - T_2}{T_1}$$

or

$$\frac{W}{T_1 - T_2} = \frac{Q_1}{T_1} = \frac{Q_2}{T_2} = \frac{KAT_2^4}{T_2}$$

$$= KAT_2^3$$

$$\therefore \quad A = \frac{W}{KT_2^3(T_1 - T_2)} = \frac{W}{K(T_1 T_2^3 - T_2^4)}$$

For a given W and T_1, A will be minimum when

$$\frac{dA}{dT_2} = -\frac{W}{K}(3T_1T_2^2 - 4T_2^3)\,.\,(T_1T_2^3 - T_2^4)^{-2} = 0$$

Since $\qquad (T_1T_2^3 - T_2^4)^{-2} \neq 0,\ 3T_1T_2^2 = 4T_2^3$

$$\therefore \qquad \frac{T_2}{T_1} = 0.75 \text{ Proved.}$$

$$\because \qquad A_{min} = \frac{W}{K(0.75)^3 T_1^3 (T_1 - 0.75T_1)}$$

$$= \frac{W}{K\dfrac{27}{256}T_1^4} = \frac{256W}{27KT_1^4}$$

Here $\qquad W = 1$ kW, $K = 5.67 \times 10^{-8}$ W/m² K⁴, and $T_1 = 1000$ K

$$\therefore \qquad A_{min} = \frac{256 \times 1\,\text{kW} \times \text{m}^2\text{K}^4}{27 \times 5.67 \times 10^{-8}\,\text{W} \times (1000)^4\,\text{K}^4}$$

$$= \frac{256 \times 10^3}{27 \times 5.67 \times 10^{-8} \times 10^{12}}\ \text{m}^2 = 0.1672\ \text{m}^2 \qquad \textit{Ans.}$$

Review Questions

6.1 What is the qualitative difference between heat and work? Why are heat and work not completely interchangeable forms of energy?

6.2 What is a cyclic heat engine?

6.3 Explain a heat engine cycle performed by a closed system.

6.4 Explain a heat engine cycle performed by a steady flow system.

6 5 Define the thermal efficiency of a heat engine cycle. Can this be 100%?

6.6 Draw a block diagram showing the four energy interactions of a cyclic heat engine.

6.7 What is a thermal energy reservoir? Explain the terms 'source' and 'sink'

6.8 What is a mechanical energy reservoir?

6.9 Why can all processes in a TER or an MER be assumed to be quasi-static?

6.10 Give the Kelvin-Planck statement of the second law.

6.11 To produce net work in a thermodynamic cycle, a heat engine has to exchange heat with two thermal reservoirs. Explain.

6.12 What is a PMM2? Why is it impossible?

6.13 Give the Clausius' statement of the second law.

6.14 Explain the operation of a cyclic refrigerator plant with a block diagram.

6.15 Define the COP of a refrigerator.

6.16 What is a heat pump? How does it differ from a refrigerator?

6.17 Can you use the same plant as a heat pump in winter and as a refrigerator in summer? Explain.

6.18 Show that the COP of a heat pump is greater than the COP of a refrigerator by unity.

6.19 Why is direct heating thermodynamically wasteful?

6.20 How can a heat pump upgrade low grade waste heat?

6.21 Establish the equivalence of Kelvin-Planck and Clausius statements.

6.22 What is a reversible process? A reversible process should not leave any evidence to show that the process had ever occurred. Explain.

6.23 How is a reversible process only a limiting process, never to be attained in practice?

6.24 All spontaneous processes are irreversible. Explain.

6.25 What are the causes of irreversibility of a process?

6.26 Show that heat transfer through a finite temperature difference is irreversible.

6.27 Demonstrate, using the second law, that free expansion is irreversible.

6.28 What do you understand by dissipative effects? When is work said to be dissipated?

6.29 Explain perpetual motion of the third kind.

6.30 Demonstrate using the second law how friction makes a process irreversible.

6.31 When a rotating wheel is brought to rest by applying a brake, show that the molecular internal energy of the system (of the brake and the wheel) increases.

6.32 Show that the dissipation of stirring work to internal energy is irreversible.

6.33 Show by second law that the dissipation of electrical work into internal energy or heat is irreversible.

6.34 What is a Carnot cycle? What are the four processes which constitute the cycle?

6.35 Explain the Carnot heat engine cycle executed by: (a) a stationary system, and (b) a steady flow system.

6.36 What is a reversed heat engine?

6.37 Show that the efficiency of a reversible engine operating between two given constant temperatures is the maximum.

6.38 Show that the efficiency of all reversible heat engines operating between the same temperature levels is the same.

6.39 Show that the efficiency of a reversible engine is independent of the nature or amount of the working substance going through the cycle.

6.40 How does the efficiency of a reversible cycle depend only on the two temperatures at which heat is transferred?

6.41 What is the absolute thermodynamic temperature scale? Why is it called absolute?

6.42 How is the absolute scale independent of the working substance?

6.43 How does Q play the role of thermometric property in the Kelvin Scale?

6.44 Show that a definite zero point exists on the absolute temperature scale but that this point cannot be reached without a violation of the second law.

6.45 Give the Fowler-Guggenheim statement of the third law.

6.46 Is the third law an extension of the second law? Is it an independent law of nature? Explain.

6.47 How does the efficiency of a reversible engine vary as the source and sink temperatures are varied? When does the efficiency become 100%?

6.48 For a given T_2, show that the COP of a refrigerator increases as T_1 decreases.

6.49 Explain how the Kelvin temperature can be measured with a gas thermometer.

6.50 Establish the equality of ideal gas temperature and Kelvin temperature.

6.51 What do you understand by internal irreversibility and external irreversibility?

6.52 Explain mechanical, thermal and chemical irreversibilities.

6.53 A Carnot engine with a fuel burning device as source and a heat sink cannot be treated as a reversible plant. Explain.

Problems

6.1 An inventor claims to have developed an engine that takes in 105 MKJ at a temperature of 400 K, rejects 42 MJ at a temperature of 200 K, and delivers 15 kWh of mechanical work. Would you advise investing money to put this engine in the market?

6.2 If a refrigerator is used for heating purposes in winter so that the atmosphere becomes the cold body and the room to be heated becomes the hot body, how much heat would be available for heating for each kW input to the driving motor? The COP of the refrigerator is 5, and the electromechanical efficiency of the motor is 90%. How does this compare with resistance heating?

Ans. 5.4 kW, 1 kW

6.3 Using an engine of 30% thermal efficiency to drive a refrigerator having a COP of 5, what is the heat input into the engine for each MJ removed from the cold body by the refrigerator?

Ans. 666.67 kJ

If this system is used as a heat pump, how many MJ of heat would be available for heating for each MJ of heat input to the engine? *Ans.* 1.8 MJ

6.4 An electric storage battery which can exchange heat only with a constant temperature atmosphere goes through a complete cycle of two processes. In process 1–2, 2.8 kWh of electrical work flow into the battery while 732 kJ of heat flow out to the atmosphere. During process 2–1, 2.4 kWh of work flow out of the battery. (a) Find the heat transfer in process 2–1. (b) If the process 1–2 has occurred as above, does the first law or the second law limit the maximum possible work of process 2–1? What is the maximum possible work? (c) If the maximum possible work were obtained in process 2–1, what will be the heat transfer in the process?

Ans. (a) –708 kJ (b) Second law, $W_{2-1} = 9348$ kJ (c) $Q_{2-1} = 0$

6.5 A household refrigerator is maintained at a temperature of 2°C. Every time the door is opened, warm material is placed inside, introducing an average of 420 kJ, but making only a small change in the temperature of the refrigerator. The door is opened 20 times a day, and the refrigerator operates at 15% of the ideal COP. The cost of work is 32 paise per kWh. What is the monthly bill for this refrigerator? The atmosphere is at 30°C. *Ans.* Rs. 15.20

6.6 A heat pump working on the Carnot cycle takes in heat from a reservoir at 5°C and delivers heat to a reservoir at 60°C. The heat pump is driven by a reversible heat engine which takes in heat from a reservoir at 840°C and rejects heat to a reservoir at 60°C. The reversible heat engine also drives a machine that absorbs 30 kW. If the heat pump extracts 17 kJ/s from the 5°C reservoir, determine (a) the rate of heat supply from the 840°C source, and (b) the rate of heat rejection to the 60°C sink.

Ans. (a) 47.61 kW; (b) 34.61 kW

6.7 A refrigeration plant for a food store operates with a COP which is 40% of the ideal COP of a Carnot refrigarator. The store is to be maintained at a temperature of – 5°C and the heat transfer from the store to the cycle is at the rate of 5 kW. If heat is transferred from the cycle to the atmosphere at a temperature of 25°C, calculate the power required to drive the plant and the heat discharged to the atmosphere. *Ans.* 4.4 kW, 6.4 kW

6.8 A heat engine is used to drive a heat pump. The heat transfers from the heat engine and from the heat pump are used to heat the water circulating through the radiators of a building. The efficiency of the heat engine is 27% and the COP of the heat pump is 4. Evaluate the ratio of the heat transfer to the circulating water to the heat transfer to the heat engine. *Ans.* 1.81

6.9 If 20 kJ are added to a Carnot cycle at a temperature of 100°C and 14.6 kJ are rejected at 0°C, determine the location of absolute zero on the Celsius scale. *Ans.* –270.37°C

6.10 Two reversible heat engines *A* and *B* are arranged in series, A rejecting heat directly to *B*. Engine *A* receives 200 kJ at a temperature of 421°C from a hot source, while engine *B* is in communication with a cold sink at a temperature of 4.4°C. If the work output of *A* is twice that of *B*, find (a) the intermediate temperature between *A* and *B*, (b) the efficiency of each engine, and (c) the heat rejected to the cold sink.

Ans. 143.4°C, 40% & 33.5%, 80 kJ

6.11 A heat engine operates between the maximum and minimum temperatures of 671°C and 60°C respectively, with an efficiency of 50% of the appropriate Carnot efficiency. It drives a heat pump which uses river water at 4.4°C to heat a block of flats in which the temperature is to be maintained at 21.1°C. Assuming that a temperature difference of 11.1°C exists between the working fluid and the river water, on the one hand, and the required room temperature on the other, and assuming the heat pump to operate on the reversed Carnot cycle, but with a COP of 50% of the ideal COP, find the heat input to the engine per unit heat output from the heat pump. Why is direct heating thermodynamically more wasteful?

Ans. 0.79 kJ/kJ heat input

6.12 An ice-making plant produces ice at atmospheric pressure and at 0°C from water at 0°C. The mean temperature of the cooling water circulating through the condenser of the refrigerating machine is 18°C. Evaluate the minimum electrical work in kWh required to produce 1 tonne of ice. (The enthalpy of fusion of ice at atmospheric pressure is 333.5 kJ/kg). *Ans.* 6.11 kWh

6.13 A reversible engine works between three thermal reservoirs, *A*, *B* and *C*. The engine absorbs an equal amount of heat from the thermal reservoirs *A* and *B* kept at temperatures T_A and T_B respectively, and rejects heat to the thermal reservoir *C* kept at temperature T_C. The efficiency of the engine is α times the efficiency of the reversible engine, which works between the two reservoirs *A* and *C*. Prove that

$$\frac{T_A}{T_B} = (2\alpha - 1) + 2(1 - \alpha)\frac{T_A}{T_C}$$

6.14 A reversible engine operates between temperatures T_1 and T ($T_1 > T$). The energy rejected from this engine is received by a second reversible engine at the same temperature *T*. The second engine rejects energy at temperature T_2 ($T_2 < T$). Show that (a) temperature *T* is the arithmetic mean of temperatures T_1 and T_2 if the engines produce the same amount of work output, and (b) temperature *T* is the geometric mean of temperatures T_1 and T_2 if the engines have the same cycle efficiencies.

6.15 Two Carnot engines *A* and *B* are connected in series between two thermal reservoirs maintained at 1000 K and 100 K respectively. Engine *A* receives 1680 kJ of heat from the high-temperature reservoir and rejects heat to the Carnot engine *B*. Engine *B* takes in heat rejected by engine *A* and rejects heat to the low-temperature reservoir. If engines *A* and *B* have equal thermal efficiencies, determine (a) the heat rejected by engine *B*, (b) the temperature at which heat is rejected by engine *A*, and (c) the work done during the process by engines *A* and *B* respectively. If engines *A* and *B* deliver equal work, determine (d) the amount of heat taken in by engine *B*, and (e) the efficiencies of engines *A* and *B*. *Ans.* (a) 168 kJ, b) 316.2 K, (c) 1148.7, 363.3 kJ, (d) 924 kJ, (e) 45%, 81.8%.

6.16 A heat pump is to be used to heat a house in winter and then reversed to cool the house in summer. The interior temperature is to be maintained at 20°C. Heat transfer through the walls and roof is estimated to be 0.525 kJ/s per degree temperature difference between the inside and outside. (a) If the outside temperature in winter is 5°C, what is the minimum power required to drive the heat pump? (b) If the power output is the same as in part (a), what is the maximum outer temperature for which the inside can be maintained at 20°C? *Ans.* (a) 403 W, (b) 35°C.

6.17 Consider an engine in outer space which operates on the Carnot cycle. The only way in which heat can be transferred from the engine is by radiation. The rate at which heat is radiated is proportional to the fourth power of the absolute temperature T_2 and to the area of the radiating surface. Show that for a given power output and a given T_1, the area of the radiator will be a minimum when

$$\frac{T_2}{T_1} = \frac{3}{4}$$

6.18 It takes 10 kW to keep the interior of a certain house at 20°C when the outside temperature is 0°C. This heat flow is usually obtained directly by burning gas or oil. Calculate the power required if the 10 kW heat flow were supplied by operating a reversible heat pump with the house as the upper reservoir and the outside surroundings as the lower reservoir. *Ans.* 0.6826 kW

6.19 Prove that the COP of a reversible refrigerator operating between two given temperatures is the maximum.

6.20 A house is to be maintained at a temperature of 20°C by means of a heat pump pumping heat from the atmosphere. Heat losses through the walls of the house are estimated at 0.65 kW per unit of temperature difference between the inside of the house and the atmosphere. (a) If the atmospheric temperature is – 10°C, what is the minimum power required to drive the pump? (b) It is proposed to use the same heat pump to cool the house in summer. For the same room temperature, the same heat loss rate, and the same power input to the pump, what is the maximum permissible atmospheric temperature? *Ans.* 2 kW, 50°C.

6.21 A solar-powered heat pump receives heat from a solar collector at T_h, rejects heat to the atmosphere at T_a, and pumps heat from a cold space at T_c. The three heat transfer rates are Q_h, Q_a, and Q_c respectively. Derive an expression for the minimum ratio Q_h/Q_c, in terms of the three temperatures.

If $T_h = 400$ K, $T_a = 300$ K, $T_c = 200$ K, $Q_c = 12$ kW, what is the minimum Q_h? If the collector captures 0.2 kW/m², what is the minimum collector area required? *Ans.* 24 kW, 120 m²

6.22 A heat engine operating between two reservoirs at 1000 K and 300 K is used to drive a heat pump which extracts heat from the reservoir at 300 K at a rate twice that at which the engine rejects heat to it. If the efficiency of the engine is 40% of the maximum possible and the COP of the heat pump is 50% of the maximum possible, what is the temperature of the reservoir to which the heat pump rejects heat? What is the rate of heat rejection from the heat pump if the rate of heat supply to the engine is 50 kW? *Ans.* 326.5 K, 86 kW

6.23 A reversible power cycle is used to drive a reversible heat pump cycle. The power cycle takes in Q_1 heat units at T_1 and rejects Q_2 at T_2. The heat pump abstracts Q_4 from the sink at T_4 and discharges Q_3 at T_3. Develop an expression for the ratio Q_4/Q_1 in terms of the four temperatures.

$$Ans.\ \frac{Q_4}{Q_1} = \frac{T_4(T_1 - T_2)}{T_1(T_3 - T_4)}$$

6.24 Prove that the following propositions are logically equivalent: (a) A PMM2 is impossible, (b) A weight sliding at constant velocity down a frictional inclined plane executes an irreversible process.

6.25 A heat engine receives half of its heat supply at 1000 K and half at 500 K while rejecting heat to a sink at 300 K. What is the maximum possible thermal efficiency of this heat engine? *Ans.* 0.55%

6.26 A heat pump provides 3 × 10⁴ kJ/h to maintain a dwelling at 23°C on a day when the outside temperature is 0°C. The power input to the heat pump is 4 kW. Determine the COP of the heat pump and compare it with the COP of a reversible heat pump operating between the reservoirs at the same two temperatures. *Ans.* 2.08, 12.87

6.27 A reversible power cycle receives energy Q_1 from a reservoir at temperature T_1 and rejects Q_2 to a reservoir at temperature T_2. The work developed by the power cycle is used to drive a reversible heat pump that removes energy Q'_2 from a reservoir at temperature T'_2 and rejects energy Q'_1 to a reservoir at temperature T'_1. (a) Determine an expression for the ratio Q'_1/Q_1 in terms of the four temperatures. (b) What must be the relationship of the temperatures T_1, T_2, T'_1 and T'_2, for Q'_1/Q_1 to exceed a value of unity?

$$Ans.\ (a)\ \frac{Q_1'}{Q_1} = \frac{T_1'(T_1 - T_2)}{T_1(T_1' - T_2')},\quad (b)\ \frac{T_2}{T_2'} < \frac{T_1}{T_1'}$$

6.28 When the outside temperature is – 10°C, a residential heat pump must provide 3.5 × 10⁶ kJ per day to a dwelling to maintain its temperature at 20°C. If electricity costs Rs. 2.10 per kWh, find the minimum theoretical operating cost for each day of operation. *Ans.* Rs. 208.83

CHAPTER 7

Entropy

7.1 INTRODUCTION

The first law of thermodynamics was stated in terms of cycles first and it was shown that the cyclic integral of heat is equal to the cyclic integral of work. When the first law was applied for thermodynamic processes, the existence of a property, the internal energy, was found. Similarly, the second law was also first stated in terms of cycles executed by systems. When applied to processes, the second law also leads to the definition of a new property, known as entropy. If the first law is said to be the law of internal energy, then second law may be stated to be the law of entropy. In fact, *thermodynamics is the study of three E's, namely, energy, equilibrium and entropy.*

7.2 TWO REVERSIBLE ADIABATIC PATHS CANNOT INTERSECT EACH OTHER

Let it be assumed that two reversible adiabatics *AC* and *BC* intersect each other at point *C* (Fig. 7.1). Let a reversible isotherm *AB* be drawn in such a way that it intersects the reversible adiabatics at *A* and *B*. The three reversible processes *AB*, *BC*, and *CA* together constitute a reversible cycle, and the area included represents the net work output in a cycle. But such a cycle is impossible, since net work is being produced in a cycle by a heat engine by exchanging heat with a single reservoir in the process *AB*, which violates the Kelvin-Planck statement of the second law. Therefore, the assumption of the intersection of the reversible adiabatics is wrong. *Through one point, there can pass only one reversible adiabatic.*

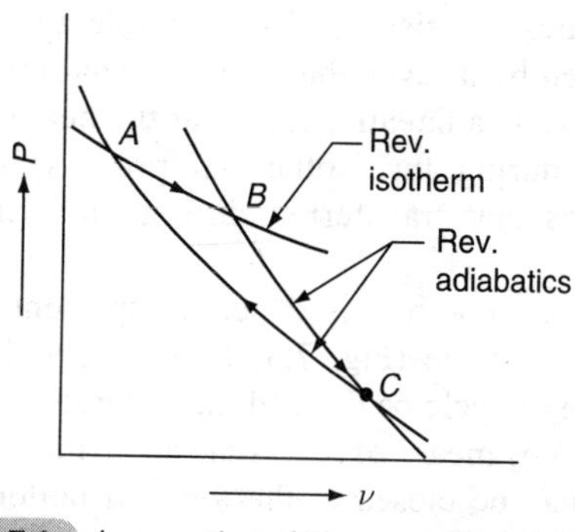

Fig. 7.1 *Assumption of two reversible adiabatics intersecting each other*

Since two constant property lines can never intersect each other, it is inferred that a reversible adiabatic path must represent some property, which is yet to be identified.

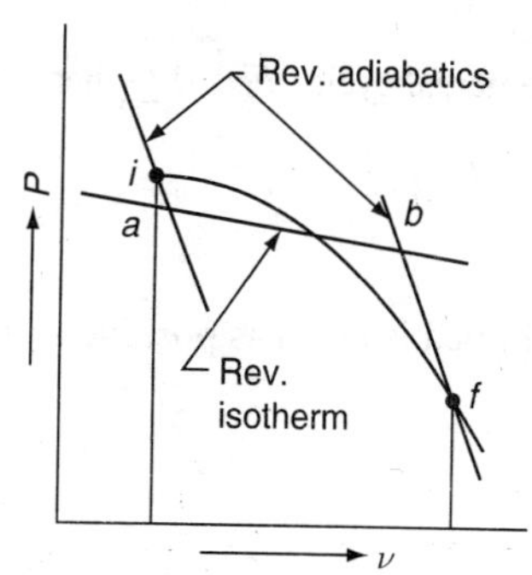

Fig. 7.2 *Reversible path substituted by two reversible adiabatics and a reversible isotherm*

7.3 CLAUSIUS' THEOREM

Let a system be taken from an equilibrium state *i* to another equilibrium state *f* by following the reversible path i–f (Fig. 7.2). Let a reversible adiabatic i–a be drawn through i and another reversible adiabatic b–f be drawn through f. Then a reversible isotherm a–b is drawn in such a way that the area under i–a–b–f is equal to the area under i–f. Applying the first law for

Process i–f

$$Q_{i-f} = U_f - U_i + W_{if} \tag{7.1}$$

Process i–a–b–f

$$Q_{iabf} = U_f - U_i + W_{iabf} \tag{7.2}$$

Since $\quad W_{if} = W_{iabf}$

∴ From Eqs (7.1) and (7.2)

$$Q_{if} = Q_{iabf} = Q_{ia} + Q_{ab} + Q_{bf}$$

Since $\quad Q_{ia} = 0$ and $Q_{bf} = 0$

$$Q_{if} = Q_{ab}$$

Heat transferred in the process i–f is equal to the heat transferred in the isothermal process a–b.

Thus any reversible path may be substituted by a reversible zigzag path, between the same end states, consisting of a reversible adiabatic followed by a reversible isotherm and then by a reversible adiabatic, such that the heat transferred during the isothermal process is the same as that transferred during the original process.

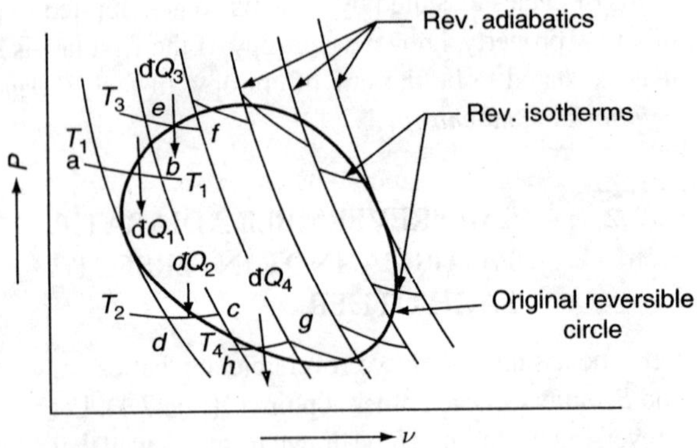

Fig. 7.3 *A reversible cycle split into a large number of carnot cycles*

Let a smooth closed curve representing a reversible cycle (Fig. 7.3) be considered. Let the closed cycle be divided into a large number of strips by means of reversible adiabatics. Each strip may be closed at the top and bottom by reversible isotherms. The original closed cycle is thus replaced by a zigzag closed path consisting of alternate adiabatic and isothermal processes, such that the heat transferred during all the isothermal processes is equal to the heat transferred in the original cycle. Thus the original cycle is replaced by a large number of Carnot cycles. If the adiabatics are close to one another and the number of Carnot cycles is large, the saw-toothed zigzag line will coincide with the original cycle.

For the elemental cycle *abcd* $đQ_1$ heat is absorbed reversibly at T_1, and $đQ_2$ heat is rejected reversibly at T_2

$$\frac{đQ_1}{T_1} = \frac{đQ_2}{T_2}$$

If heat supplied is taken as positive and heat rejected as negative

$$\frac{đQ_1}{T_1} + \frac{đQ_2}{T_2} = 0$$

Similarly, for the elemental cycle *efgh*

$$\frac{đQ_3}{T_3} + \frac{đQ_4}{T_4} = 0$$

If similar equations are written for all the elemental Carnot cycles, then for the whole original cycle

$$\frac{đQ_1}{T_1} + \frac{đQ_2}{T_2} + \frac{đQ_3}{T_3} + \frac{đQ_4}{T_4} + \ldots = 0$$

or

$$\oint_R \frac{đQ}{T} = 0 \tag{7.3}$$

The cyclic integral of $đQ/T$ for a reversible cycle is equal to zero. This is known as *Clausius' theorem*. The letter R emphasizes the fact that the equation is valid only for a reversible cycle.

7.4 THE PROPERTY OF ENTROPY

Let a system be taken from an initial equilibrium state i to a final equilibrium state f by following the reversible path R_1 (Fig. 7.4). The system is brought back from f to i by following another reversible path R_2. Then the two paths R_1 and R_2 together constitute a reversible cycle. From Clausius' theorem

$$\oint_{R_1R_2} \frac{đQ}{T} = 0$$

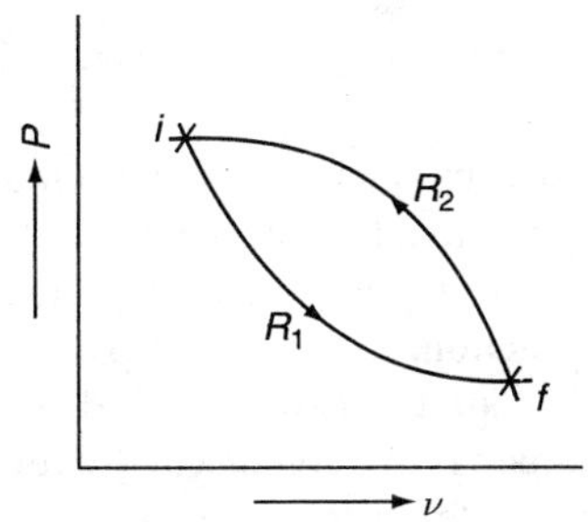

Fig. 7.4 *Two reversible paths R_1 and R_2 between two equilibrium states i and f*

The above integral may be replaced as the sum of two integrals, one for path R_1 and the other for path R_2

$$\int_{i\,R_1}^{f} \frac{đQ}{T} + \int_{f\,R_2}^{i} \frac{đQ}{T} = 0$$

or

$$\int_{i\,R_1}^{f} \frac{đQ}{T} = -\int_{f\,R_2}^{i} \frac{đQ}{T}$$

Since R_2 is a reversible path

$$\int_{i\,R_1}^{f} \frac{đQ}{T} = \int_{i\,R_2}^{f} \frac{đQ}{T}$$

Since R_1 and R_2 represent any two reversible paths, $\int_{i\,R}^{f} \frac{đQ}{T}$ is independent of the reversible path connecting i and f. Therefore, there exists a property of a system whose value at the final state f minus its value at the initial state *i* is equal to $\int_{i\,R}^{f} \frac{đQ}{T}$. This property is called *entropy*, and is denoted by *S*. If S_i is the entropy at the initial state *i*, and S_f is the entropy at the final state *f*, then

$$\int_{i\,R}^{f} \frac{đQ}{T} = S_f - S_i \tag{7.4}$$

When the two equilibrium states are infinitesimally near

$$\frac{đQ_R}{T} = dS \tag{7.5}$$

where d*S* is an *exact differential* because *S* is a point function and a property. The subscript *R* in $đQ$ indicates that heat $đQ$ is transferred *reversibly*.

The word 'entropy' was first used by Clausius, taken from the Greek word 'tropee' meaning 'transformation'. It is an extensive property, and has the unit J/K. The specific entropy

$$s = \frac{S}{m} \text{ J/kg K}$$

If the system is taken from an initial equilibrium state *i* to a final equilibrium state *f* by an *irreversible path*, since entropy is a point or state function, and the entropy change is independent of the path followed, the non-reversible path is to be replaced by a reversible path to integrate for the evaluation of entropy change in the irreversible process (Fig. 7.5).

$$S_f - S_i = \int_i^f \frac{đQ_{rev}}{T} = (\Delta S)_{\text{irrev path}} \tag{7.6}$$

Integration can be performed only on a reversible path.

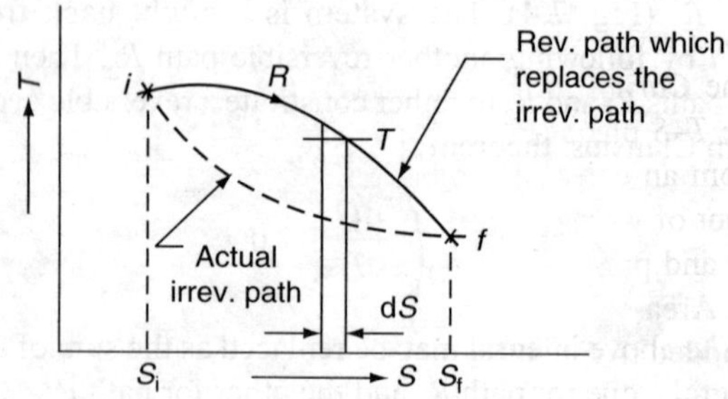

Fig. 7.5 *Integration can be done only on a reversible path*

7.4.1 Temperature-Entropy Plot

The infinitesimal change in entropy d*S* due to reversible heat transfer $đQ$ at temperature *T* is

$$dS = \frac{đQ_{rev}}{T}$$

If $đQ_{rev} = 0$, i.e., the process is reversible and adiabatic

$$dS = 0$$

and

$$S = \text{constant}$$

A reversible adiabatic process is, therefore, an isentropic process.

Now

$$đQ_{rev} = T dS$$

or

$$Q_{rev} = \int_i^f T dS$$

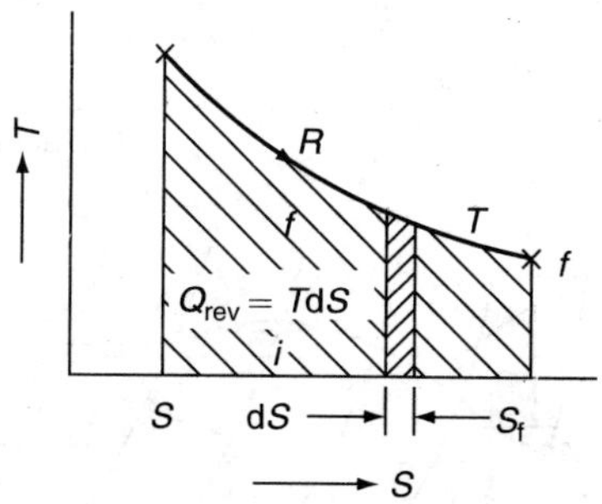

Fig. 7.6 *Area under a reversible path on the T-s plot represents heat transfer*

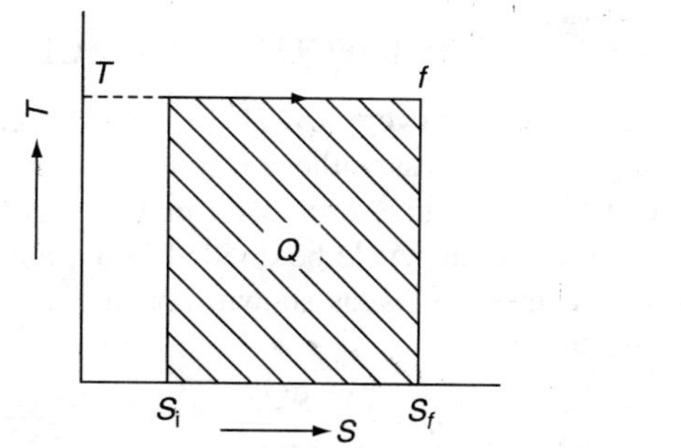

Fig. 7.7 *Reversible isothermal heat transfer*

The system is taken from *i* to *f* reversibly (Fig. 7.6). The area under the curve $\int_i^f TdS$ is equal to the heat transferred in the process.

For reversible isothermal heat transfer (Fig. 7.7), T = constant.

$$\therefore \qquad Q_{rev} = T\int_i^f dS = T(S_f - S_i)$$

For a reversible adiabatic process, $dS = 0$, $S = C$ (Fig. 7.8).

The *Carnot cycle* comprising two reversible isotherms and two reversible adiabatics forms a rectangle in the T–S plane (Fig. 7.9). Process 4–1 represents reversible isothermal heat addition Q_1 to the system at T_1 from an external source, process 1–2 is the reversible adiabatic expansion of the system producing W_E amount of work, process 2–3 is the reversible isothermal heat rejection from the system to an external sink at T_2, and process 3–4 represents reversible adiabatic compression of the system consuming W_c amount of work. Area 1 2 3 4 represents the net work output per cycle and the area under 4–1 indicates the quantity of heat added to the system Q_1.

$$\therefore \qquad \eta_{Carnot} = \frac{Q_1 - Q_2}{Q_1} = \frac{T_1(S_1 - S_4) - T_2(S_2 - S_3)}{T_1(S_1 - S_4)}$$

$$= \frac{T_1 - T_2}{T_1} = 1 - \frac{T_2}{T_1}$$

and

$$W_{net} = Q_1 - Q_2 = (T_1 - T_2)(S_1 - S_4)$$

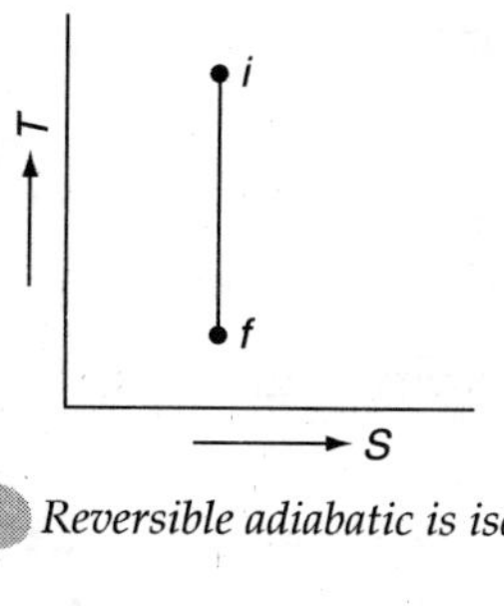

Fig. 7.8 *Reversible adiabatic is isentropic*

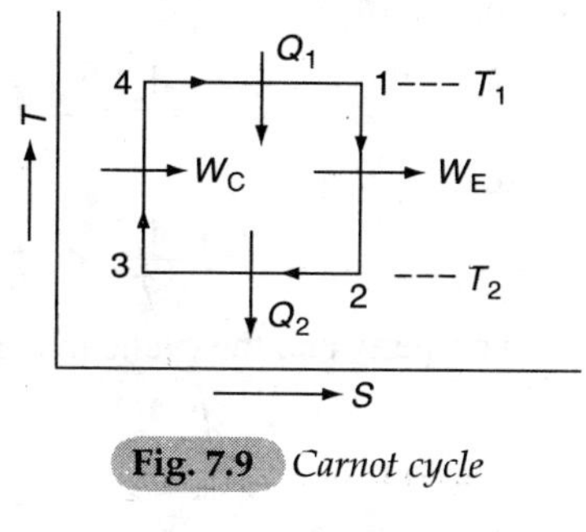

Fig. 7.9 *Carnot cycle*

7.5 THE INEQUALITY OF CLAUSIUS

Let us consider a cycle *ABCD* (Fig. 7.10). Let *AB* be a general process, either reversible or irreversible, while the other processes in the cycle are reversible. Let the cycle be divided into a number of elementary cycles, as shown. For one of these elementary cycles

$$\eta = 1 - \frac{\text{đ}Q_2}{\text{đ}Q}$$

where $\text{đ}Q$ is the heat supplied at T, and $\text{đ}Q_2$ the heat rejected at T_2.

Now, the efficiency of a general cycle will be equal to or less than the efficiency of a reversible cycle.

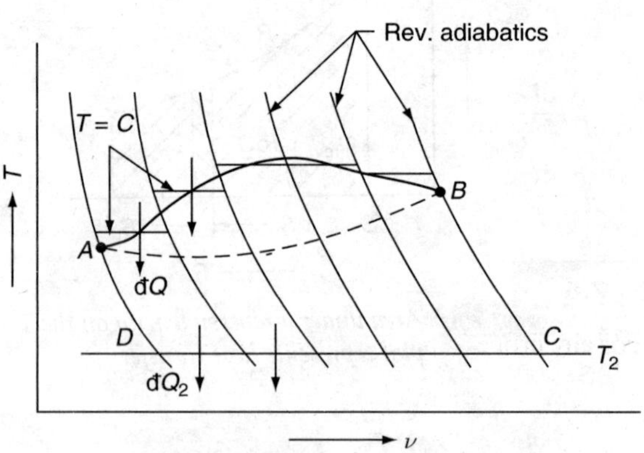

Fig. 7.10 *Inequality of clausius*

$$\therefore \qquad 1 - \frac{\text{đ}Q_2}{\text{đ}Q} \le \left(1 - \frac{\text{đ}Q_2}{\text{đ}Q}\right)_{\text{rev}}$$

or

$$\frac{\text{đ}Q_2}{\text{đ}Q} \ge \left(\frac{\text{đ}Q_2}{\text{đ}Q}\right)_{\text{rev}}$$

or

$$\frac{\text{đ}Q}{\text{đ}Q_2} \le \left(\frac{\text{đ}Q}{\text{đ}Q_2}\right)_{\text{rev}}$$

Since

$$\left(\frac{\text{đ}Q}{\text{đ}Q_2}\right)_{\text{rev}} = \frac{T}{T_2}$$

$$\therefore \qquad \frac{\text{đ}Q}{\text{đ}Q_2} \le \frac{T}{T_2}$$

or $\dfrac{\text{đ}Q}{T} \le \dfrac{\text{đ}Q_2}{T_2}$, for any process *AB*, reversible or irreversible.

For a reversible process

$$\text{d}s = \frac{\text{đ}Q_{\text{rev}}}{T} = \frac{\text{đ}Q_2}{T_2} \tag{7.7}$$

Hence, for any process *AB*

$$\frac{\text{đ}Q}{T} \le \text{d}s \tag{7.8}$$

Then for any cycle

$$\oint \frac{\text{đ}Q}{T} \le \oint \text{d}s$$

Since entropy is a property and the cyclic integral of any property is zero

$$\oint \frac{\text{đ}Q}{T} \le 0 \tag{7.9}$$

This equation is known as the *inequality of Clausius*. It *provides the criterion of the reversibility of a cycle.*

If $\oint \frac{đQ}{T} = 0$, the cycle is reversible,

$\oint \frac{đQ}{T} < 0$, the cycle is irreversible and possible

$\oint \frac{đQ}{T} > 0$, the cycle is impossible, since it violates the second law.

7.6 ENTROPY CHANGE IN AN IRREVERSIBLE PROCESS

For any process undergone by a system, we have from Eq. (7.8)

$$\frac{đQ}{T} \leq ds$$

or

$$ds \geq \frac{đQ}{T} \tag{7.10}$$

This is further clarified if we consider the cycles as shown in Fig. 7.11, where A and B are reversible processes and C is an irreversible process. For the reversible cycle consisting of A and B

$$\int_R \frac{đQ}{T} = \int_{\substack{1\\A}}^{2} \frac{đQ}{T} + \int_{\substack{2\\B}}^{1} \frac{đQ}{T} = 0$$

or

$$\int_{\substack{1\\A}}^{2} \frac{đQ}{T} = -\int_{\substack{2\\B}}^{1} \frac{đQ}{T} \tag{7.11}$$

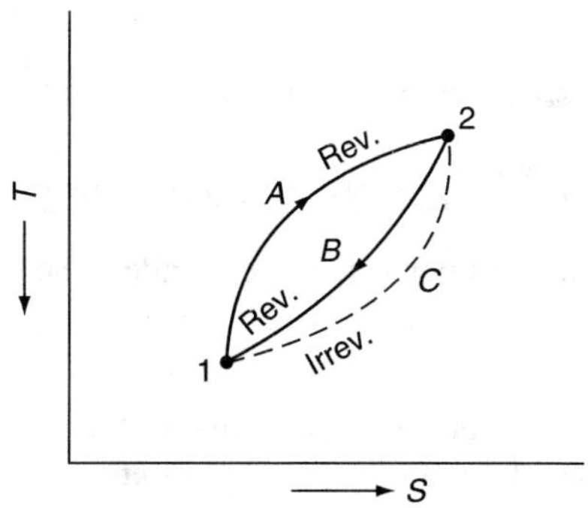

Fig. 7.11 *Entropy change in an irreversible process*

For the irreversible cycle consisting of A and C, by the inequality of Clausius,

$$\oint \frac{đQ}{T} = \int_{\substack{1\\A}}^{2} \frac{đQ}{T} + \int_{\substack{2\\C}}^{1} \frac{đQ}{T} < 0 \tag{7.12}$$

From Eqs (7.11) and (7.12),

$$-\int_{\substack{2\\B}}^{1} \frac{đQ}{T} + \int_{\substack{2\\C}}^{1} \frac{đQ}{T} < 0$$

$$\therefore \quad \int_{\substack{2\\B}}^{1} \frac{đQ}{T} > \int_{\substack{2\\C}}^{1} \frac{đQ}{T} \tag{7.13}$$

Since the path B is reversible,

$$\int_{\substack{2\\B}}^{1} \frac{đQ}{T} = \int_{\substack{2\\B}}^{1} dS \tag{7.14}$$

Since entropy is a property, entropy changes for the paths B and C would be the same. Therefore,

$$\underset{B}{\int_2^1} dS = \underset{C}{\int_2^1} dS \tag{7.15}$$

From Eqs (7.13) to (7.15),

$$\underset{C}{\int_2^1} dS > \underset{C}{\int_2^1} \frac{đQ}{T}$$

Thus, for any irreversible process,

$$dS > \frac{đQ}{T}$$

whereas for a reversible process

$$dS = \frac{đQ_{rev}}{T}$$

Therefore, for the general case, we can write

$$dS \geq \frac{đQ}{T}$$

or

$$S_2 - S_1 \geq \int_1^2 \frac{đQ}{T} \tag{7.16}$$

The equality sign holds good for a reversible process and the inequality sign for an irreversible process.

7.7 ENTROPY PRINCIPLE

For any infinitesimal process undergone by a system, we have from Eq. (7.10) for the total mass

$$dS \geq \frac{đQ}{T}$$

For an isolated system which does not undergo any energy interaction with the surroundings, $đQ = 0$. Therefore, for an isolated system

$$dS_{iso} \geq 0 \tag{7.17}$$

For a reversible process,

$$dS_{iso} = 0$$

or

$$S = \text{constant}$$

For an irreversible process

$$dS_{iso} > 0$$

It is thus proved that *the entropy of an isolated system can never decrease*. It always increases and remains constant only when the process is reversible. This is known as the *principle of increase of entropy*, or simply the *entropy principle*. It is the quantitative general statement of second law from the macroscopic viewpoint.

An isolated system can always be formed by including any system and its surroundings within a single boundary (Fig. 7.12). Sometimes the original system which is then only a part of the isolated system is called a 'subsystem'.

The system and the surroundings together (the universe or the isolated system) include everything which is affected by the process. For all possible processes that a system in the given surroundings can undergo

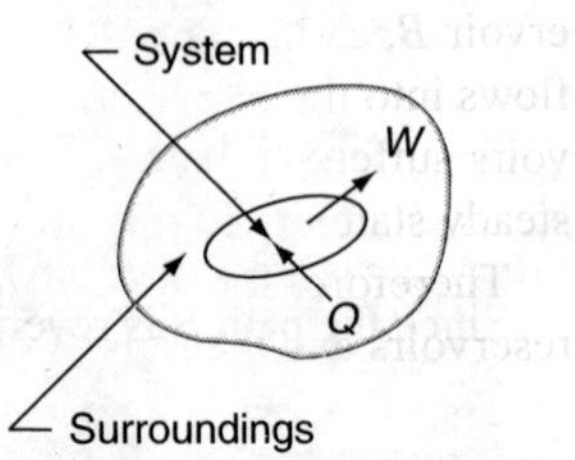

Fig. 7.12 *Isolated system*

$$dS_{univ} \geq 0$$

or

$$dS_{sys} + dS_{surr} \geq 0 \tag{7.18}$$

Entropy may decrease locally at some region within the isolated system, but it must be compensated by a greater increase of entropy somewhere within the system so that the net effect of an irreversible process is an entropy increase of the whole system. The entropy increase of an isolated system is a measure of the extent of irreversibility of the process undergone by the system.

Rudolf Clausius summarized the first and second laws of thermodynamics in the following words:

(a) Die Energie der Welt ist Costant.
(b) Die Entropie der Welt strebt einem Maximum zu.
[(a) The energy of the world (universe) is constant.
(b) The entropy of the world tends towards a maximum.]

The entropy of an isolated system always increases and becomes a maximum at the state of equilibrium. If the entropy of an isolated system varies with some parameter x, then there is a certain value of x_e which maximizes the entropy $\left(\text{when } \frac{dS}{dx} = 0\right)$ and represents the equilibrium state (Fig. 7.13). The system is then said to exist at the peak of the entropy hill, and $dS = 0$. *When the system is at equilibrium, any conceivable change in entropy would be zero.*

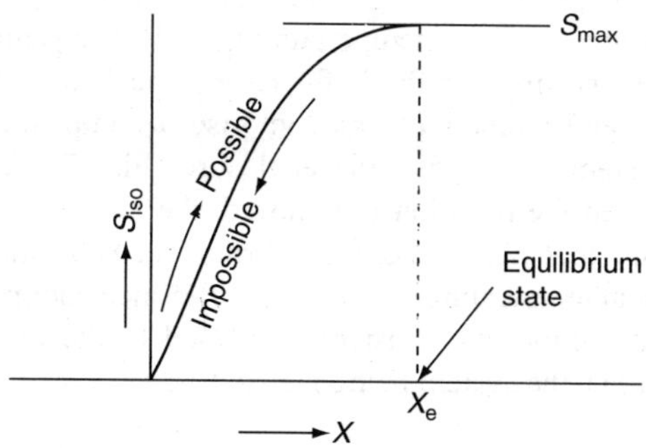

Fig. 7.13 *Equilibrium state of an isolated system*

7.8 APPLICATIONS OF ENTROPY PRINCIPLE

The principle of increase of entropy is one of the most important laws of physical science. It is the quantitative statement of the second law of thermodynamics. Every irreversible process is accompanied by entropy increase of the universe, and this entropy increase quantifies the extent of irreversibility of the process. The higher the entropy increase of the universe, the higher will be the irreversibility of the process. A few applications of the entropy principle are illustrated in the following.

7.8.1 Transfer of Heat through a Finite Temperature Difference

Let Q be the rate of heat transfer from reservoir A at T_1 to reservoir B at T_2, $T_1 > T_2$ (Fig. 7.14).

For reservoir A, $\Delta S_A = -Q/T_1$. It is negative because heat Q flows out of the reservoir. For reservoir B, $\Delta S_B = +Q/T_2$. It is positive because heat flows into the reservoir. The rod connecting the reservoirs suffers no entropy change because, once in the steady state, its coordinates do not change.

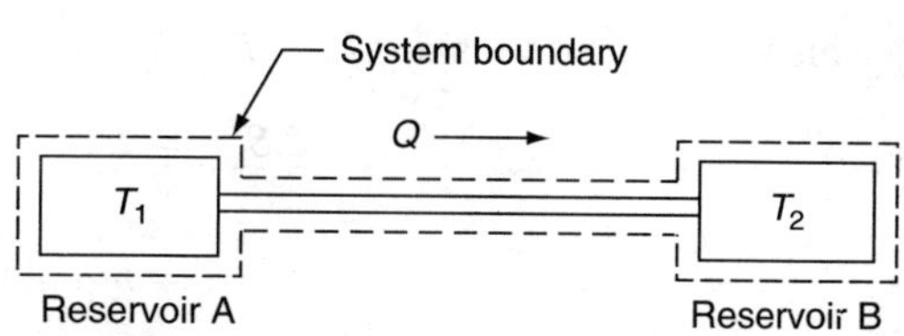

Fig. 7.14 *Heat transfer through a finite temperature difference*

Therefore, for the isolated system comprising the reservoirs and the rod, and since entropy is an additive property

$$S = S_A + S_B$$

$$\Delta S_{univ} = \Delta S_A + \Delta S_B$$

or
$$\Delta S_{\text{univ}} = -\frac{Q}{T_1} + \frac{Q}{T_2} = Q \cdot \frac{T_1 - T_2}{T_1 T_2}$$

Since $T_1 > T_2$, ΔS_{univ} is positive, and the process is irreversible and possible. If $T_1 = T_2$, ΔS_{univ} is zero, and the process is reversible. If $T_1 < T_2$, ΔS_{univ} is negative and the process is impossible.

7.8.2 Mixing of Two Fluids

Subsystem 1 having a fluid of mass m_1, specific heat c_1, and temperature t_1, and subsystem 2 consisting of a fluid of mass m_2, specific heat c_2, and temperature t_2, comprise a composite system in an adiabatic enclosure (Fig. 7.15). When the partition is removed, the two fluids mix together, and at equilibrium let t_f be the final temperature, and $t_2 < t_f < t_1$. Since energy interaction is exclusively confined to the two fluids, the system being isolated

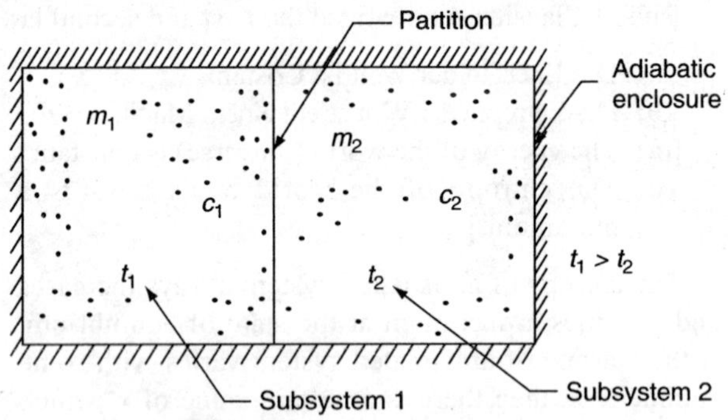

Fig. 7.15 *Mixing of two fluids*

$$m_1 c_1 (t_1 - t_f) = m_2 c_2 (t_f - t_2)$$

$$\therefore \qquad t_f = \frac{m_1 c_1 t_1 + m_2 c_2 t_2}{m_1 c_1 + m_2 c_2}$$

Entropy change for the fluid in subsystem 1

$$\Delta S_1 = \int_{T_1}^{T_f} \frac{đQ_{\text{rev}}}{T} = \int_{T_1}^{T_f} \frac{m_1 c_1 dT}{T} = m_1 c_1 \ln \frac{T_f}{T_1}$$

$$= m_1 c_1 \ln \frac{t_f + 273}{t_1 + 273}$$

This will be negative, since $T_1 > T_f$.

Entropy change for the fluid in subsystem 2

$$\Delta S_2 = \int_{T_1}^{T_f} \frac{m_2 c_2 dT}{T} = m_2 c_2 \ln \frac{T_f}{T_2} = m_2 c_2 \ln \frac{t_f + 273}{t_2 + 273}$$

This will be positive, since $T_2 < T_f$

$$\therefore \qquad \Delta S_{\text{univ}} = \Delta S_1 + \Delta S_2$$

$$= m_1 c_1 \ln \frac{T_f}{T_1} + m_2 c_2 \ln \frac{T_f}{T_2}$$

ΔS_{univ} will be positive definite, and the mixing process is irreversible.

Although the mixing process is irreversible, to evaluate the entropy change for the subsystems, the irreversible path was replaced by a reversible path on which the integration was performed.

If $m_1 = m_2 = m$ and $c_1 = c_2 = c$.

$$\Delta S_{\text{univ}} = mc \ln \frac{T_f^2}{T_1 \cdot T_2}$$

and
$$T_f = \frac{m_1c_1T_1 + m_2c_2T_2}{m_1c_1 + m_2c_2} = \frac{T_1 + T_2}{2}$$

$\therefore$
$$\Delta S_{univ} = 2\, mc \ln \frac{(T_1 + T_2)/2}{\sqrt{T_1 \cdot T_2}}$$

This is always positive, since the arithmetic mean of any two numbers is always greater than their geometric mean. This can also be proved geometrically. Let a semi-circle be drawn with $(T_1 + T_2)$ as diameter (Fig. 7.16).

Here, $AB = T_1$, $BC = T_2$ and $OE = (T_1 + T_2)/2$. It is known that $(DB)^2 = AB \cdot BC = T_1T_2$.

$\therefore$
$$DB = \sqrt{T_1T_2}$$

Now,
$$OE > DB$$

$$\frac{T_1 + T_2}{2} = \sqrt{T_1T_2}$$

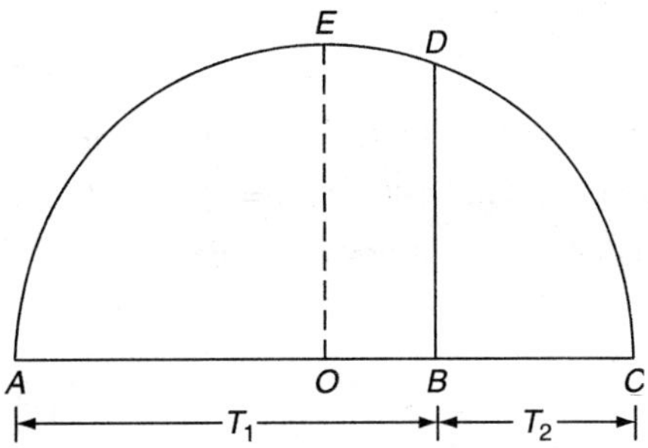

Fig. 7.16 *Geometrical proof to show that g.m. < a.m.*

7.8.3 Maximum Work Obtainable from Two Finite Bodies at Temperatures T_1 and T_2

Let us consider two identical finite bodies of constant heat capacity at temperatures T_1 and T_2 respectively, T_1 being higher than T_2. If the two bodies are merely brought together into thermal contact, delivering no work, the final temperature T_f reached would be the maximum

$$T_f = \frac{T_1 + T_2}{2}$$

If a heat engine is operated between the two bodies acting as thermal energy reservoirs (Fig. 7.17), part of the heat withdrawn from body 1 is converted to work W by the heat engine, and the remainder is rejected to body 2. The lowest attainable final temperature T_f corresponds to the delivery of the largest possible amount of work, and is associated with a reversible process.

As work is delivered by the heat engine, the temperature of body 1 will be decreasing and that of body 2 will be increasing. When both the bodies attain the final temperature T_f, the heat engine will stop operating. Let the bodies remain at constant pressure and undergo no change of phase.

Total heat withdrawn from body 1

$$Q_1 = C_p\,(T_1 - T_f)$$

where C_p is the heat capacity of the two bodies at constant pressure.

Total heat rejected to body 2

$$Q_2 = C_p\,(T_f - T_2)$$

$\therefore$ Amount of total work delivered by the heat engine

$$W = Q_1 - Q_2$$
$$= C_p\,(T_1 + T_2 - 2T_f) \qquad (7.19)$$

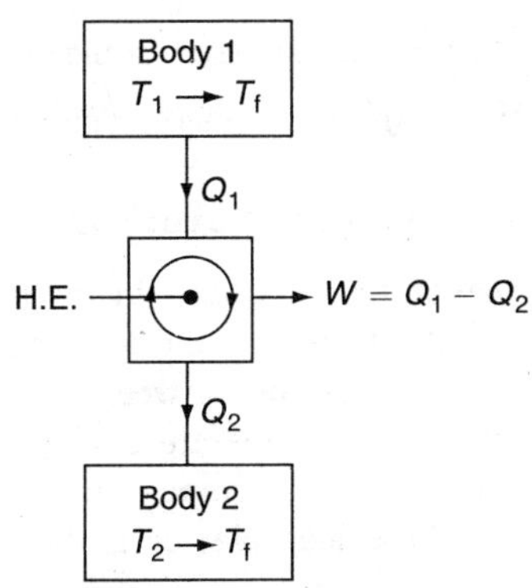

Fig. 7.17 *Maximum work obtainable from two finite bodies*

For given values of C_p, T_1 and T_2, the magnitude of work W depends on T_f. Work obtainable will be maximum when T_f is minimum.

Now, for body 1, entropy change ΔS_1 is given by

$$\Delta S_1 = \int_{T_1}^{T_f} C_p \frac{dT}{T} = C_p \ln \frac{T_f}{T_1}$$

For body 2, entropy change ΔS_2 would be

$$\Delta S_2 = \int_{T_2}^{T_f} C_p \frac{dT}{T} = C_p \ln \frac{T_f}{T_2}$$

Since the working fluid operating in the heat engine cycle does not undergo any entropy change, ΔS of the working fluid in heat engine $= \oint dS = 0$.

Applying the entropy principle

$$\Delta S_{univ} \geq 0$$

$$\therefore \quad C_p \ln \frac{T_f}{T_1} + C_p \ln \frac{T_f}{T_2} \geq 0$$

$$C_p \ln \frac{T_f^2}{T_1 T_2} \geq 0 \tag{7.20}$$

From Eq. (7.20), for T_f to be a minimum

$$C_p \ln \frac{T_f^2}{T_1 \cdot T_2} = 0$$

Since $C_p \neq 0$,

$$\ln \frac{T_f^2}{T_1 T_2} = 0 = \ln 1$$

$$\therefore \quad T_f = \sqrt{T_1 \cdot T_2} \tag{7.21}$$

For W to be a maximum, T_f will be $\sqrt{T_1 T_2}$. From Eq. (7.19)

$$W_{max} = C_p (T_1 + T_2 - 2\sqrt{T_1 T_2}) = C_p \left(\sqrt{T_1} - \sqrt{T_2}\right)^2$$

The final temperatures of the two bodies, initially at T_1 and T_2, can range from $(T_1 + T_2)/2$ with no delivery of work to $\sqrt{T_1 T_2}$ with maximum delivery of work.

7.8.4 Maximum Work Obtainable from a Finite Body and a TER

Let one of the bodies considered in the previous section be a thermal energy reservoir. The finite body has a thermal capacity C_p and is at temperature T and the TER is at temperature T_0, such that $T > T_0$. Let a heat engine operate between the two (Fig. 7.18). As heat is withdrawn from the body, its temperature decreases. The temperature of the TER would, however, remain unchanged at T_0. The engine would stop working, when the temperature of the body reaches T_0. During that period, the amount of work delivered is W, and the heat rejected to the TER is $(Q - W)$. Then

$$\Delta S_{Body} = \int_{T}^{T_0} C_p \frac{dT}{T} = C_p \ln \frac{T_0}{T}$$

$$\Delta S_{HE} = \oint dS = 0$$

$$\Delta S_{TER} = \frac{Q - W}{T_0}$$

$$\therefore \quad \Delta S_{univ} = C_p \ln \frac{T_0}{T} + \frac{Q - W}{T_0}$$

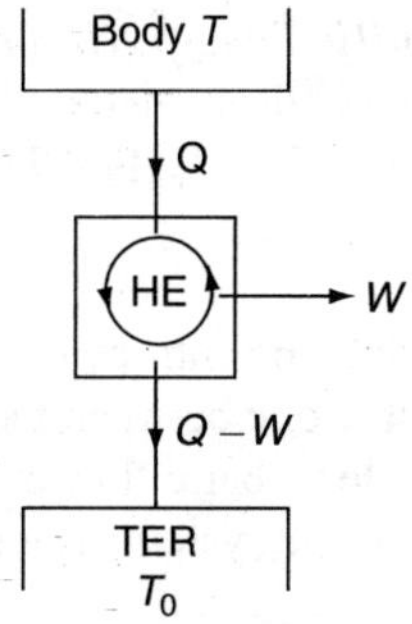

Fig. 7.18 *Maximum work obtainable when one of the bodies is a TER*

By the entropy principle,

$$\Delta S_{univ} \geq 0$$

$$C_p \ln \frac{T_0}{T} + \frac{Q - W}{T_0} \geq 0$$

or $$C_p \ln \frac{T_0}{T} \geq \frac{W - Q}{T_0}$$

or $$\frac{W - Q}{T_0} \leq C_p \ln \frac{T_0}{T}$$

or $$W \leq Q + T_0 C_p \ln \frac{T_0}{T}$$

$$\therefore \quad W_{max} = Q + T_0 C_p \ln \frac{T_0}{T}$$

or, $$W_{max} = C_p \left[(T - T_0) - T_0 \ln \frac{T}{T_0} \right] \tag{7.22}$$

7.8.5 Processes Exhibiting External Mechanical Irreversibility

(i) *Isothermal Dissipation of Work* Let us consider the isothermal dissipation of work through a system into the internal energy of a reservoir, as in the flow of an electric current *I* through a resistor in contact with a reservoir (Fig. 7.19). At steady state, the internal energy of the resistor and hence its temperature is constant. So, by first law

$$W = Q$$

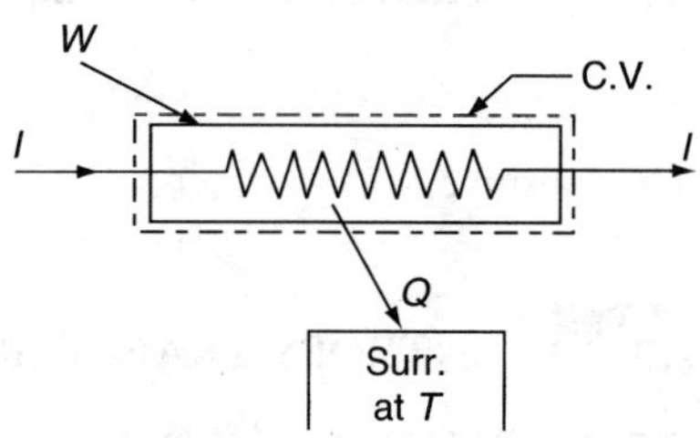

Fig. 7.19 *External mechanical irreversibility*

The flow of current represents work transfer. At steady state the work is dissipated isothermally into heat transfer to the surroundings. Since the surroundings absorb Q units of heat at temperature T,

$$\Delta S_{surr} = \frac{Q}{T} = \frac{W}{T}$$

At steady state, $$\Delta S_{sys} = 0$$

$$\therefore \quad \Delta S_{univ} = \Delta S_{sys} + \Delta S_{surr} = \frac{W}{T} \tag{7.23}$$

The irreversible process is thus accompanied by an entropy increase of the universe.

(ii) *Adiabatic Dissipation of Work* Let W be the stirring work supplied to a viscous thermally insulated liquid, which is dissipated adibatically into internal energy increase of the liquid, the temperature of which increases from T_i to T_f (Fig. 7.20). Since there is no flow of heat to or from the surroundings,

$$\Delta S_{surr} = 0$$

To calculate the entropy change of the system, the original irreversible path (dotted line) must be replaced by a reversible one between the same end states, *i* and *f*. Let us replace the irreversible performance of work by a reversible isobaric flow of heat from a series of reservoirs ranging from T_i to T_f to cause the same change in the state of the system. The entropy change of the system will be

$$\Delta S_{sys} = \int_{i \atop R}^{f} \frac{đQ}{T} = \int_{i \atop R}^{f} \frac{C_p dT}{T} = C_p \ln \frac{T_f}{T_i}$$

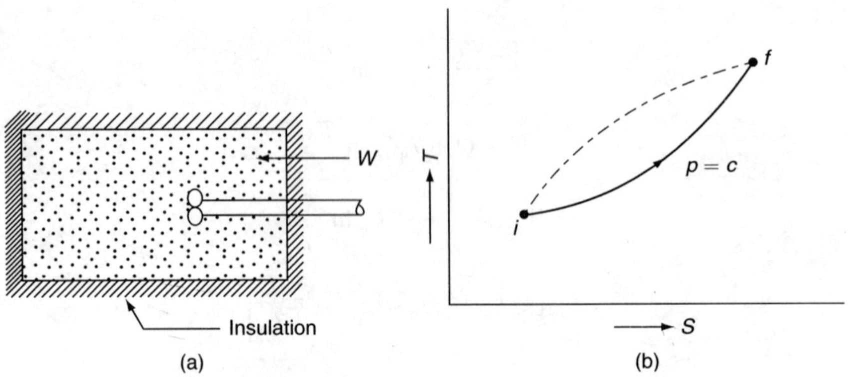

Fig. 7.20 *Adiabatic dissipation of work*

where C_p is the heat capacity of the liquid.

$$\Delta S_{univ} = \Delta S_{sys} + \Delta S_{surr} = C_p \ln \frac{T_f}{T_i} \tag{7.24}$$

which is positive.

7.9 ENTROPY TRANSFER MECHANISMS

Entropy can be *transferred* to or from a system in two forms: *heat transfer* and *mass flow*. In contrast, energy is transferred by work also. Entropy transfer is recognized at the system boundary as entropy crosses the boundary, and it represents the entropy gained or lost by a system during a process. The only form of entropy interaction associated with a fixed mass or closed system is heat transfer, and thus the entropy transfer for an adiabatic closed system is zero. It is being explained below in more detail:

(a) Heat Transfer Since $dS = \frac{đQ_{rev}}{T}$, when heat is added to a system, $đQ$ is positive, and the entropy of the system increases. When heat is removed from the system, $đQ$ is negative, and the entropy of the system decreases.

Heat transferred to the system of fixed mass increases the internal energy of the system, as a result of which the molecules (of a gas) move with higher kinetic energy and collide more frequently, and so the disorder in the system increases. Heat is thus regarded as disorganized or disordered energy transfer which increases molecular chaos (see Sec. 7.16). If heat Q flows reversibly from the system to the surroundings at T_0 (Fig. 7.21), the entropy increase of the surroundings is

$$\Delta S_{surr} = \frac{Q}{T_0}$$

The entropy of the system is reduced by

$$\Delta S_{sys} = -\frac{Q}{T_0}$$

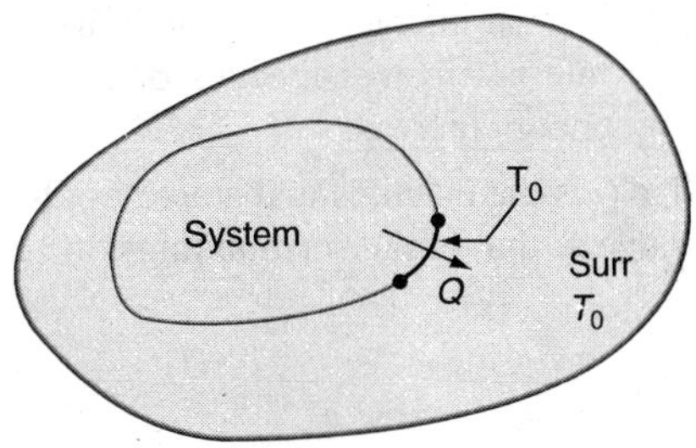

Fig. 7.21 *Entropy transfer along with heat flow*

The temperature of the boundary where heat transfer occurs is the constant temperature T_0. It may be said that the system has lost entropy to the surroundings. Alternatively, one may state that the surroundings have gained entropy from the system. Therefore, there is *entropy transfer* from the system to the surroundings along with heat flow. In other words, since the heat inflow increases the molecular disorder, there is flow or disorder along with heat. The sign of entropy transfer is the same as the sign of heat transfer: *positive*, if into the system, and *negative*, if out of the system.

On the other hand, *there is no entropy transfer associated with work*. In Fig. 7.22, the system delivers work to a flywheel, where energy is stored in a fully recoverable form. The flywheel molecules are simply put into rotation around the axis in a perfectly organized manner, and there is no dissipation and hence no entropy increase of the flywheel. The same can be said about work transfer in the compression of a spring or in the raising of a weight by a certain height. There is thus no entropy transfer along with work. If work is dissipated adiabatically into internal energy increase of the system (Subsection 7.9.5), there is an entropy increase in the system, but there is as such no entropy transfer to it.

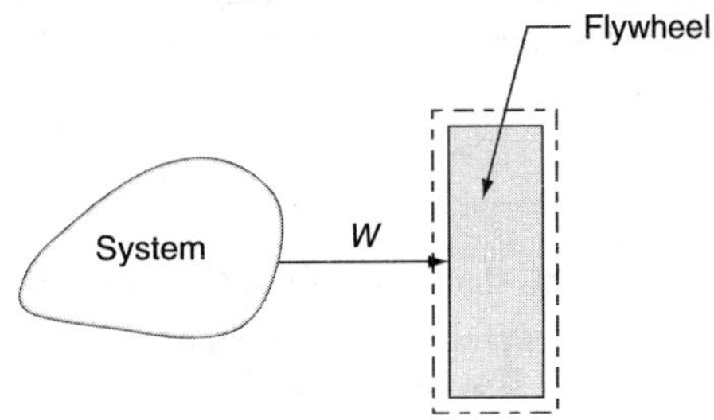

Fig. 7.22 *No entropy transfer along with work transfer*

Work is thus *entropy-free*, and no entropy is transferred with work. Energy is transferred with both heat and work, whereas entropy is transferred only with heat. The first law of thermodynamics makes no distinction between heat transfer and work. It considers them as *equals*. The distinction between heat transfer and work is brought about by the second law: *an energy interaction which is accompanied by entropy transfer is heat transfer, and an energy interaction which is not accompanied by entropy transfer is work. Thus, only energy is exchanged during work interaction, whereas both energy and entropy are exchanged during heat transfer.*

(b) Mass Flow Mass contains entropy as well as energy, and the entropy and energy of a system are proportional to the mass. When the mass of a system is doubled, so are the entropy and energy of the system. Both entropy and energy are carried into or out of a system by streams of matter, and the rates of entropy and energy transport into or out of a system are proportional to the mass flow rate. Closed systems do not involve any mass flow and thus any entropy transport. When an amount of mass m enters or leaves a system, an entropy of amount ms, s being the specific entropy, accompanies it. Therefore, the entropy of a system increases by ms when the mass of amount m enters it, and decreases by the same amount when it leaves it at the same state.

7.10 ENTROPY GENERATION IN A CLOSED SYSTEM

The entropy of any closed system can increase in two ways:

(a) by heat interaction in which there is entropy transfer
(b) internal irreversibilities or dissipative effects in which work (or K.E.) is dissipated into internal energy increase.

If $đQ$ is the infinitesimal amount of heat transferred to the system through its boundary at temperature T, the same as that of the surroundings, the entropy increase dS of the system can be expressed as

$$dS = d_eS + d_iS$$

$$= \frac{đQ}{T} + d_iS \tag{7.25}$$

where d_eS is the entropy increase due to external heat interaction and d_iS is the entropy increase due to internal irreversibility. From Eq. (7.25),

$$dS \geq \frac{đQ}{T}$$

$$\therefore \quad d_iS \geq 0 \tag{7.26}$$

The entropy increase due to internal irreversibility is also called entropy production or entropy generation, S_{gen}.

In other words, the entropy change of a system during a process is greater than the entropy transfer ($đQ/T$) by an amount equal to the entropy generated during the process within the system (d_iS), so that the *entropy balance* gives:

Entropy change = Entropy transfer + Entropy generation

$$\Delta S_{sytem} = \Delta S_{transfer} + \Delta S_{gen}$$

which is a verbal statement of Eq. (7.25) and illustrated in Fig. 7.23.

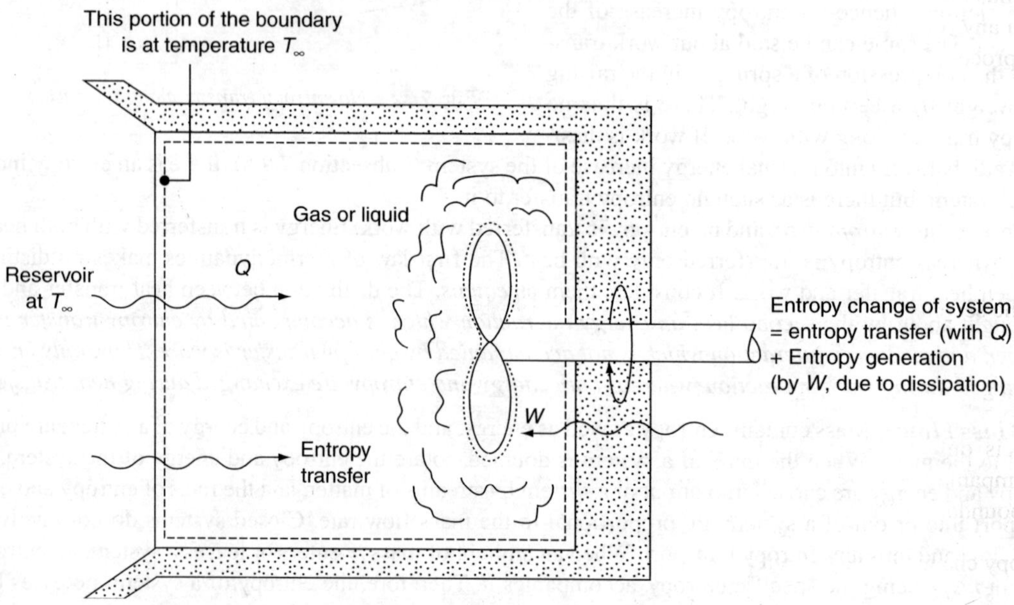

Fig. 7.23 *Illustration of the entropy transfer and entropy production concepts*

It may so happen that in a process (e.g. the expansion of a hot fluid in a turbine) the entropy decrease of the system due to heat loss to the surroundings $\left(-\int \frac{đQ}{T}\right)$ is equal to the entropy increase of the system due to internal irreversibilities such as friction, etc. $\left(\int \mathrm{d_i}S\right)$, in which case the entropy of the system before and after the process will remain the same $\left(\int \mathrm{d}S = 0\right)$. *Therefore, an isentropic process need not be adiabatic or reversible.*

But if the isentropic process is reversible, it must be adiabatic. Also, if the isentropic process is adiabatic, it cannot but be reversible. An adiabatic process need not be isentropic, since entropy can also increase due to friction etc. But if the process is adiabatic and reversible, it must be isentropic.

For an infinitesimal reversible process by a closed system,

$$đQ_R = \mathrm{d}U_R + p\mathrm{d}V$$

If the process is irreversible,

$$đQ_I = đU_I + đW$$

Since U is a property,

$$\mathrm{d}U_R = \mathrm{d}U_I$$

$\therefore$

$$đQ_R - p\mathrm{d}V = đQ_I - đW$$

or

$$\left(\frac{đQ}{T}\right)_R = \left(\frac{\mathrm{d}Q}{T}\right)_I + \frac{p\mathrm{d}V - đW}{T} \tag{7.27}$$

The difference $(p\mathrm{d}V - \mathrm{d}W)$ indicates the work that is lost due to irreversibility, and is called the *lost work* $đ(LW)$, which approaches zero as the process approaches reversibility as a limit. Equation (7.27) can be expressed in the form

$$\mathrm{d}S = \mathrm{d_e}S + \mathrm{d_i}S$$

Thus the entropy of a closed system increases due to heat addition $(\mathrm{d_e}S)$ and internal dissipation $(\mathrm{d_i}S)$.

In any process executed by a system, energy is always conserved, but entropy is produced internally. For any process between equilibrium states 1 and 2 (Fig. 7.24), the first law can be written as

$$\underbrace{\int_1^2 đQ - \int_1^2 đW}_{\text{Energy transfer}} = \underbrace{E_2 - E_1}_{\text{Energy change}}$$

or

$$Q_{1-2} = E_2 - E_1 + W_{1-2}$$

By the second law,

$$S_2 - S_1 \geq \int_1^2 \frac{đQ}{T}$$

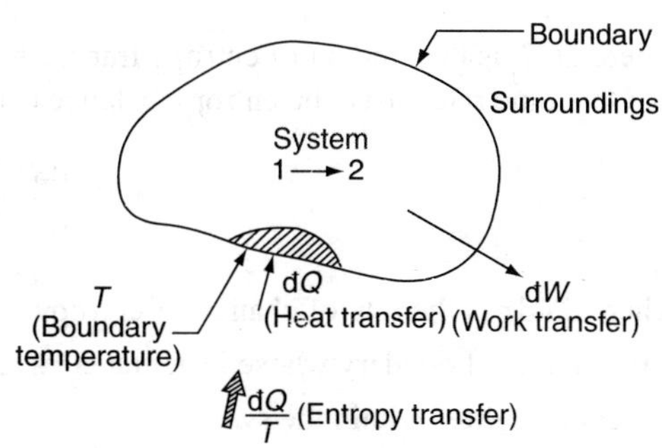

Fig. 7.24 *Schematic of a closed system interacting with its surroundings*

It is only the transfer of energy as heat which is accompanied by entropy transfer, both of which occur at the boundary where the temperature is T. Work interaction is not accompanied by any entropy transfer. The entropy change of the system $(S_2 - S_1)$ exceeds the entropy transfer $\int_1^2 \frac{đQ}{T}$. The difference is produced internally due to irreversibility. The amount of entropy generation S_{gen} is given by

$$S_2 - S_1 - \int_1^2 \frac{đQ}{T} = S_{gen} \qquad (7.28)$$

Entropy change — Entropy transfer — Entropy production

$$\therefore \qquad S_{gen} \geq 0$$

The second law states that, in general, *any thermodynamic process is accompanied by entropy generation.*

Process 1–2, which does not generate any entropy ($S_{gen} = 0$), is a reversible process (Fig. 7.25). Paths for which $S_{gen} > 0$ are considered irreversible. Like heat transfer and work transfer during the process 1–2, *the entropy generation also depends on the path the system follows.* S_{gen} is, therefore, not a thermodynamic property and $đS_{gen}$ is an inexact differential, although $(S_2 - S_1)$ depends only on the end states. In the differential form, Eq. (7.28) can be written as

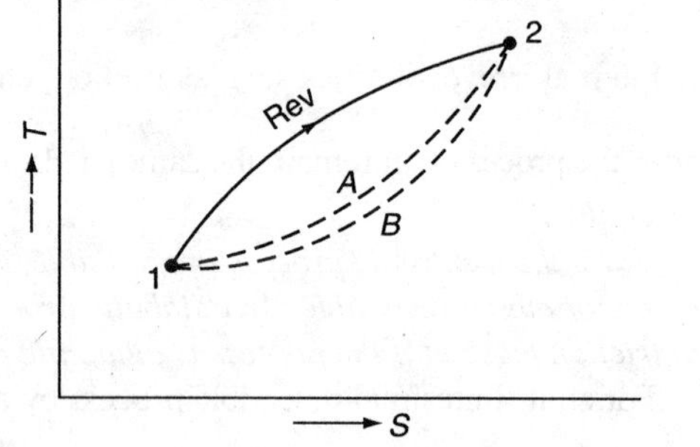

Fig. 7.25 *Entropy generation depends on the path*

$$dS_{gen} = dS - \frac{đQ}{T} \qquad (7.29)$$

The amount of entropy generation quantities the intrinsic irreversibility of the process. If the path *A* causes more entropy generation than path *B* (Fig. 7.25), i.e.

$$(S_{gen})_A > (S_{gen})_B$$

the path *A* is more irreversible than path *B* and involves more 'lost work'.

If heat transfer occurs at several locations on the boundary of a system, the entropy transfer term can be expressed as a sum, so Eq. (7.28) takes the form

$$S_2 - S_1 = \sum_j \frac{Q_j}{T_j} + S_{gen} \qquad (7.30)$$

where Q_j/T_j is the amount of entropy transferred through the portion of the boundary at temperature T_j.

On a time rate basis, the entropy balance can be written as

$$\frac{dS}{d\tau} = \sum_j \frac{\dot{Q}_j}{T_j} + \dot{S}_{gen} \qquad (7.31)$$

where $dS/d\tau$ is the rate of change of entropy of the system, $\dot{Q}_j/T_j$ is the rate of entropy transfer through the portion of the boundary whose instantaneous temperature is T_j, and $\dot{S}_{gen}$ is the rate of entropy generation due to irreversibilites within the system.

7.11 ENTROPY GENERATION IN AN OPEN SYSTEM

In an open system, there is transfer of three quantities: mass, energy and entropy. The control surface can have one or more openings for mass transfer (Fig. 7.26). It is rigid, and there is shaft work transfer across it.

The *continuity equation* gives

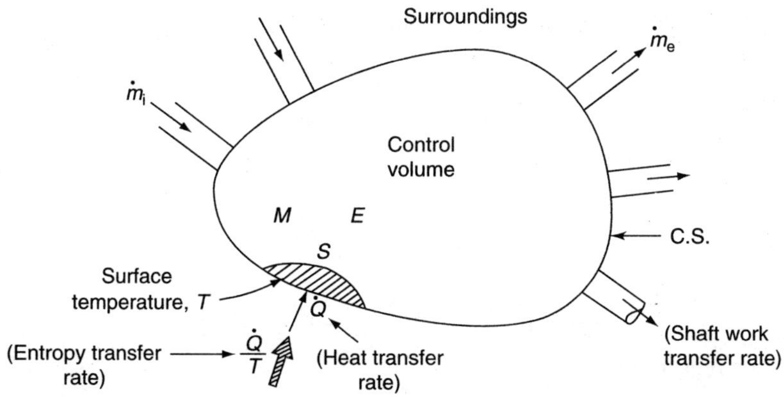

Fig. 7.26 *Schematic of an open system and its interaction with surroundings*

$$\underbrace{\sum_{i} \dot{m}_1 - \sum_{e} \dot{m}_e}_{\text{net mass transfer rate}} = \underbrace{\frac{\partial M}{\partial \tau}}_{\text{rate of mass accumulation in the CV}} \tag{7.32}$$

The *energy equation* gives

$$\underbrace{\sum_{i} \dot{m}_i \left(h + \frac{\mathbf{V}^2}{2} + gZ\right)_i - \sum_{e} \dot{m}_e \left(h + \frac{\mathbf{V}^2}{2} + gZ\right)_e}_{\text{net rate of energy transfer}} + \dot{Q} - \dot{W}_{sh} = \underbrace{\frac{\partial E}{\partial \tau}}_{\text{rate of energy accumulation in the CV}} \tag{7.33}$$

The *second law inequality* or the entropy principle gives

$$\underbrace{\sum_{i} \dot{m}_i s_i - \sum_{e} \dot{m}_e s_e + \frac{\dot{Q}}{T}}_{\text{net rate of entropy transfer}} \leq \underbrace{\frac{\partial S}{\partial \tau}}_{\text{rate of increase of entropy of the CV}} \tag{7.34}$$

Here $\dot{Q}$ represents the rate of heat transfer at the location of the boundary where the instantaneous temperature is T. The ratio $\dot{Q}/T$ accounts for the entropy transfer along with heat. The terms $\dot{m}_i s_i$ and $\dot{m}_e s_e$ account, respectively, for rates of entropy transfer into and out of the CV accompanying mass flow. The rate of entropy increase of the control volume exceeds, or is equal to, the net rate of entropy transfer into it. The difference is the entropy generated within the control volume due to irreversibility. Hence, the rate of entropy generation is given by

$$\dot{S}_{gen} = \frac{\partial S}{\partial \tau} - \sum_{i} \dot{m}_i s_i + \sum_{e} \dot{m}_e s_e - \frac{\dot{Q}}{T} \tag{7.35}$$

By the second law,

$$\dot{S}_{gen} \geq 0$$

If the process is reversible, $\dot{S}_{gen} = 0$. For an irreversible process, $\dot{S}_{gen} > 0$.

The magnitude of $\dot{S}_{gen}$ quantifies the irreversibility of the process. If systems *A* and *B* operate so that $(\dot{S}_{gen})_A > (\dot{S}_{gen})_B$ it can be said that the system *A* operates more irreversibly than system *B*. The unit of $\dot{S}_{gen}$ is W/K.

At *steady state*, the continuity equation gives

$$\sum_i \dot{m}_i = \sum_e \dot{m}_e \tag{7.36}$$

the energy equation becomes

$$0 = \dot{Q} - \dot{W}_{sh} + \sum_i \dot{m}_i \left(h + \frac{\mathbf{V}^2}{2} + gZ\right)_i - \sum_e \dot{m}_e \left(h + \frac{\mathbf{V}^2}{2} + gZ\right)_e \tag{7.37}$$

and the entropy equation reduces to

$$0 = \frac{\dot{Q}}{T} + \sum_i \dot{m}_i s_i - \sum_e \dot{m}_e s_e + \dot{S}_{gen} \tag{7.38}$$

These equations often must be solved simultaneously, together with appropriate property relations.

Mass and energy are conserved quantities, but entropy is not generally conserved. The rate at which entropy is transferred out must exceed the rate at which entropy enters the CV, the difference being the rate of entropy generated within the CV owing to irreversibilities.

For one-inlet and one-exit control volumes, the entropy equation becomes

$$0 = \frac{\dot{Q}}{T} + \dot{m}(s_1 - s_2) + \dot{S}_{gen}$$

$$\therefore \quad s_2 - s_1 = \frac{1}{\dot{m}}\left(\frac{\dot{Q}}{T}\right) + \frac{\dot{S}_{gen}}{\dot{m}} \tag{7.39}$$

7.12 FIRST AND SECOND LAWS COMBINED

By the second law

$$đQ_{rev} = TdS$$

and by the first law, for a closed non-flow system,

$$đQ = dU + pdV$$

$$TdS = dU + pdV \tag{7.40}$$

Again, the enthalpy

$$H = U + pV$$

$$dH = dU + pdV + Vdp$$

$$= TdS + Vdp$$

$$TdS = dH - Vdp \tag{7.41}$$

Equations (7.40) and (7.41) are the thermodynamic equations relating the properties of the system.

Let us now examine the following equations as obtained from the first and second laws:

(a) $đQ = dE + đW$—This equation holds good for any process, reversible or irreversible, and for any system.

(b) $đQ = dU + đW$—This equation holds good for any process undergone by a closed stationary system.

(c) $đQ = dU + pdV$—This equation holds good for a closed system when only pdV-work is present. This is true only for a reversible (quasi-static) process.
(d) $đQ = TdS$—This equation is true only for a reversible process.
(e) $TdS = dU + pdV$—This equation holds good for any process reversible or irreversible, undergone by a closed system, since it is a relation among properties which are independent of the path.
(f) $TdS = dH - Vdp$—This equation also relates only the properties of system. There is no path function term in the equation. Hence the equation holds good for any process.

The use of the term 'irreversible process' is doubtful, since no irreversible path or process can be plotted on thermodynamic coordinates. It is more logical to state that 'the change of state is irreversible' rather than say 'it is an irreversible process'. A natural process which is inherently irreversible is indicated by a dotted line connecting the initial and final states, both of which are in equilibrium. The dotted line has no other meaning, since it can be drawn in any way. To determine the entropy change for a real process, a known reversible path is made to connect the two end states, and integration is performed on this path using either equation (e) or equation (f), as given above. Therefore, the entropy change of a system between two identifiable equilibrium states is the same whether the intervening process is reversible or the change of state is irreversible.

7.13 REVERSIBLE ADIABATIC WORK IN A STEADY FLOW SYSTEM

In the differential form, the steady flow energy equation per unit mass is given by Eq. (5.11),

$$đQ = dh + \mathbf{V}d\mathbf{V} + gdZ + đW_x$$

For a reversible process, $đQ = Tds$

$$\therefore \qquad Tds = dh + \mathbf{V}d\mathbf{V} + gdZ + đW_x \tag{7.42}$$

Using the property relation, Eq. (7.41), per unit mass,

$$Tds = dh - vdp$$

in Eq. (7.42), we have

$$-vdp = \mathbf{V}d\mathbf{V} + gdZ + đW_x \tag{7.43}$$

On integration

$$-\int_1^2 vdp = \Delta\frac{\mathbf{V}^2}{2} + W_x \tag{7.44}$$

If the changes in K.E. and P.E. are neglected, Eq. (7.44) reduces to

$$W_x = -\int_1^2 vdp \tag{7.45}$$

If $đQ = 0$, implying $ds = 0$, the property relation gives

$$dh = vdp$$

or

$$h_2 - h_1 = \int_1^2 vdp \tag{7.46}$$

From Eqs (7.45) and (7.46),

$$W_x = h_1 - h_2 = -\int_1^2 vdp \tag{7.47}$$

The integral $-\int_1^2 vdp$ represents an area on the p–v plane (Fig. 7.27). To make the integration, one must have a relation between p and v such as pv^n = constant.

$$\therefore \qquad W_{1-2} = h_1 - h_2 = -\int_1^2 vdp$$

$$= \text{area } 12\, ab\, 1$$

Equation (7.47) holds good for a steady flow work-producing machine like an engine or turbine as well as for a work-absorbing machine like a pump or a compressor, when the fluid undergoes reversible adiabatic expansion or compression.

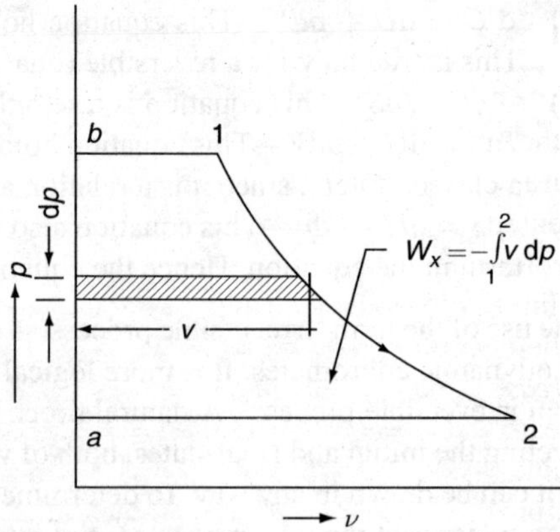

Fig. 7.27 *Reversible steady flow work interaction*

It may be noted that for a closed stationary system like a gas confined in a piston-cylinder machine (Fig. 7.28a), the reversible work done would be

$$W_{1-2} = \int_1^2 pdV = \text{Area } 12\, cd\, 1$$

The reversible work done by a steady flow system (Fig. 7.28b) would be

$$W_{1-2} = -\int_1^2 vdp = \text{Area } 12\, ab\, 1$$

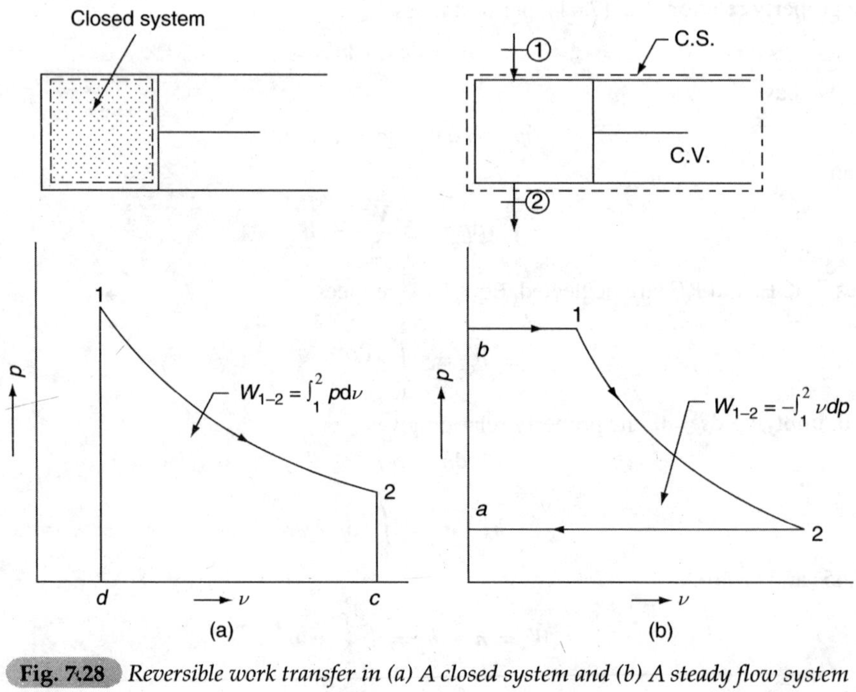

Fig. 7.28 *Reversible work transfer in (a) A closed system and (b) A steady flow system*

7.14 ENTROPY AND DIRECTION: THE SECOND LAW—A DIRECTIONAL LAW OF NATURE

Since the entropy of an isolated system can never decrease, it follows that only those processes are possible in nature which would give an entropy increase for the system and the surroundings together (the universe). All spontaneous processes in nature occur only in one direction from a higher to a lower potential, and these are accompanied by an entropy increase of the universe. When the potential gradient is infinitesimal (or zero in the limit), the entropy change of the universe is zero, and the process is reversible. The second law indicates the direction in which a process takes place. *A process always occurs in such a direction as to cause an increase in the entropy of the universe.* The macroscopic change ceases only when the potential gradient disappears and the equilibrium is reached when the entropy of the universe assumes a maximum value. To determine the equilibrium state of an isolated system it is necessary to express the entropy as a function of certain properties of the system and then render the function a maximum. At equilibrium, the system (isolated) exists at the peak of the entropy-hill, and $dS = 0$ (Fig. 7.13).

The natural direction of events in which entropy increases indicates the '*arrow of time*' which results from the universe not being in thermodynamic equilibrium. It undergoes a natural evolution, and inexorably approaches the state of quilibrium.

7.15 ENTROPY AND DISORDER

Work is a macroscopic concept. Work involves order or the orderly motion of molecules, as in the expansion or compression of a gas. The kinetic energy and potential energy of a system represent orderly forms of energy. The kinetic energy of a gas is due to the coordinated motion of all the molecules with the same average velocity in the same direction. The potential energy is due to the vantage position taken by the molecules or displacements of molecules from their normal positions. Heat or thermal energy is due to the random thermal motion of molecules in a completely disorderly fashion and the average velocity is zero. Orderly energy can be readily converted into disorderly energy, e.g. mechanical and electrical energies into internal energy (and then heat) by friction and Joule effect. Orderly energy can also be converted into one another. *But there are natural limitations on the conversion of disorderly energy into orderly energy, as delineated by the second law.* When work is dissipated into internal energy, the disorderly motion of molecules is increased. Two gases, when mixed, represent a higher degree of disorder than when they are separated. An irreversible process always tends to take the system (isolated) to a state of greater disorder. It is a tendency on the part of nature to proceed to a state of greater disorder. An isolated system always tends to a state of greater entropy. So there is a close link between entropy and disorder. It may be stated roughly that *the entropy of a system is a measure of the degree of molecular disorder existing in the system.* When heat is imparted to a system, the disorderly motion of molecules increases, and so the entropy of the system increases. The reverse occurs when heat is removed from the system.

Ludwig Boltzmann (1877) introduced statistical concepts to define disorder by attaching to each state a *thermodynamic probability*, expressed by the quantity W, which is greater the more disordered the state is. The increase of entropy implies that the system proceeds by itself from one state to another with a higher thermodynamic probability (or disorder number). An irreversible process goes on until the *most probable state* (equilibrium state when W is maximum) corresponding to the maximum value of entropy is reached. Boltzmann assumed a functional relation between S and W. While entropy is additive, probability is multiplicative. If the two parts A and B of a system in equilibrium are considered, the entropy is the sum

$$S = S_A + S_B$$

and the thermodynamic probability is the product

$$W = W_A \cdot W_B$$

Again, $$S = S(W),\ S_A = S(W_A), \text{ and } S_B = S(W_B)$$

$\therefore$ $$S(W) = S(W_A W_B) = S(W_A) + S(W_B)$$

which is a well-known functional equation for the logarithm. Thus the famous relation is reached

$$S = K \ln W \tag{7.48}$$

where K is a constant, known as Boltzmann constant. This is engraved upon Boltzmann's tombstone in Vienna.

When $W = 1$, which represents the greatest order, $S = 0$. This occurs only at $T = 0$ K. This state cannot be reached in a finite number of operations. This is the Nernst-Simon statement of third law of thermodynamics. In the case of a gas, W increases due to an increase in volume V or temperature T. In the reversible adiabatic expansion of a gas the increase in disorder due to an increase in volume is just compensated by the decrease in disorder due to a decrease in temperature, so that the disorder number or entropy remains constant.

7.16 ABSOLUTE ENTROPY

It is important to note that one is interested only in the amount by which the entropy of the system changes in going from an initial to final state, and not in the value of absolute entropy. In cases where it is necessary, a zero value of entropy of the system at an arbitrarily chosen standard state is assigned, and the entropy changes are calculated with reference to this standard state.

Solved Examples

Example 7.1

Water flows through a turbine in which friction causes the water temperature to rise from 35°C to 37°C. If there is no heat transfer, how much does the entropy of the water change in passing through the turbine? (Water is incompressible and the process can be taken to be at constant volume.)

Solution The presence of friction makes the process irreversible and causes an entropy increase for the system. The flow process is indicated by the dotted line joining the equilibrium states 1 and 2 (Fig. Ex. 7.1). Since entropy is a state property and the entropy change depends only on the two end states and is independent of the path the system follows, to find the entropy change, the irreversible path has to be replaced by a reversible path, as shown in the figure, because no integration can be made on a path other than a reversible path.

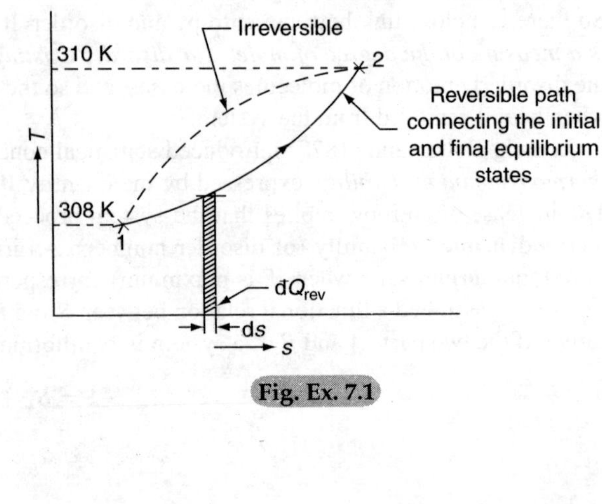

Fig. Ex. 7.1

$$T_2 = 37 + 273 = 310 \text{ K}$$

$$T_1 = 35 + 273 = 308 \text{ K}$$

We have $$đQ_{rev} = TdS$$

$$dS = \frac{mc_v dT}{T}$$

$$S_2 - S_1 = mc_v \ln \frac{T_2}{T_1} = 1 \times 4.187 \ln \frac{310}{308} = 0.0243 \text{ kJ/K}$$ *Ans.*

Example 7.2

(*a*) *One kg of water at 273 K is brought into contact with a heat reservoir at 373 K. When the water has reached 373 K, find the entropy change of the water, of the heat reservoir, and of the universe.*

(*b*) *If water is heated from 273 K to 373 K by first bringing it in contact with a reservoir at 323 K and then with a reservoir at 373 K, what will the entropy change of the universe be?*

(*c*) *Explain how water might be heated from 273 to 373 K with almost no change in the entropy of the universe.*

Solution (a) Water is being heated through a finite temperature difference (Fig. Ex. 7.2). The entropy of water would increase and that of the reservoir would decrease so that the net entropy change of the water (system) and the reservoir together would be positive definite. Water is being heated irreversibly, and to find the entropy change of water, we have to assume a reversible path between the end states which are at equilibrium.

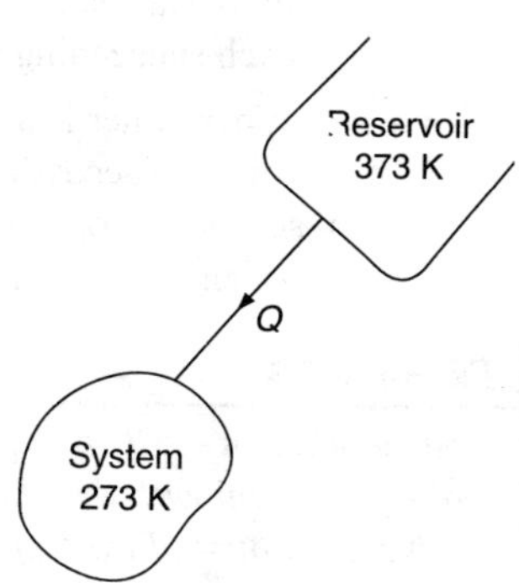

Fig. Ex. 7.2

$$(\Delta S)_{\text{water}} = \int \frac{đQ}{T} = \int_{T_1}^{T_2} \frac{mc dT}{T} = mc \ln \frac{T_2}{T_1}$$

$$= 1 \times 4.187 \ln \frac{373}{273} = 1.305 \text{ kJ/K}$$

The reservoir temperature remains constant irrespective of the amount of heat withdrawn from it.

Amount of heat absorbed by the system from the reservoir,

$$Q = 1 \times 4.187 \times (373 - 273) = 418.7 \text{ kJ}$$

∴ Entropy change of the reservoir

$$(\Delta S)_{\text{res}} = -\frac{Q}{T} = -\frac{418.7}{373} = -1.122 \text{ kJ/K}$$

∴ Entropy change of the universe

$$(\Delta S)_{\text{univ}} = (dS)_{\text{system}} + (\Delta S)_{\text{res}}$$

$$= 1.305 - 1.122 = 0.183 \text{ kJ/K}$$ *Ans.* (a)

(b) Water is being heated in two stages, first from 273 K to 323 K by bringing it in contact with a reservoir at 323 K, and then from 323 K to 373 K by bringing it in contact of a second reservoir at 373 K.

$$(\Delta S)_{\text{water}} = \int_{273\text{K}}^{323\text{K}} mc \frac{dT}{T} + \int_{323\text{K}}^{373\text{K}} mc \frac{dT}{T}$$

$$= 4.187 \left(\ln \frac{323}{273} + \ln \frac{373}{323} \right) = (0.1673 + 0.1441)\, 4.187$$

$$= 1.305 \text{ kJ/K}$$

$$(\Delta S)_{\text{res I}} = -\frac{1 \times 4.187 \times (323 - 273)}{323} = -0.647 \text{ kJ/K}$$

$$(\Delta S)_{\text{res II}} = \frac{1 \times 4.187 \times (373 - 323)}{373} = -0.56 \text{ kJ/K}$$

$$\therefore \quad (\Delta S)_{\text{uni V}} = (\Delta S)_{\text{water}} + (\Delta S)_{\text{res I}} + (\Delta S)_{\text{res II}}$$

$$= 1.305 - 0.647 - 0.56$$

$$= 0.098 \text{ kJ/K}$$ *Ans.*

(c) The entropy change of the universe would be less and less if the water is heated in more and more stages, by bringing the water in contact successively with more and more heat reservoirs, each succeeding reservoir being at a higher temperature than the preceding one.

When water is heated in infinite steps, by bringing it in contact with an infinite number of reservoirs in succession, so that at any instant the temperature difference between the water and the reservoir in contact is infinitesimally small, then the net entropy change of the universe would be zero, and the water would be *reversibly* heated.

Example 7.3

One kg of ice at – 5°C is exposed to the atmosphere which is at 20°C. The ice melts and comes into thermal equilibrium with the atmosphere. (a) *Determine the entropy increase of the universe.* (b) *What is the minimum amount of work necessary to convert the water back into ice at – 5°C? c_p of ice is 2.093 kJ/kg K and the latent heat of fusion of ice is 333.3 kJ/kg.*

Solution Heat absorbed by ice Q from the atmosphere (Fig. Ex. 7.3.1)

= Heat absorbed in solid phase + Latent heat + Heat absorbed in liquid phase

$$= 1 \times 2.093 \times [0 - (-5)] + 1 \times 333.3 + 1 \times 4.187 \times (20 - 0)$$

$$= 427.5 \text{ kJ}$$

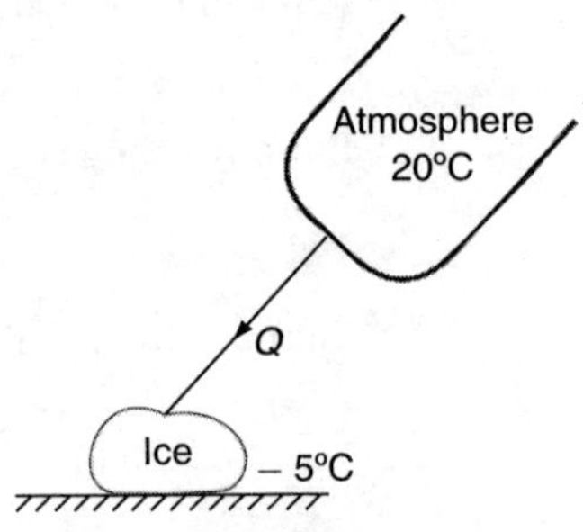

Fig. Ex. 7.3.1

Entropy change of the atmosphere.

$$(\Delta S)_{\text{atm}} = -\frac{Q}{T} = -\frac{427.5}{293} = -1.46 \text{ kJ/K}$$

Entropy change of the system (ice) as it gets heated from –5°C to 0°C

$$(\Delta S_{\text{I}})_{\text{system}} = \int_{268}^{293} mc_p \frac{dT}{T} = 1 \times 2.093 \ln \frac{273}{268} = 2.093 \times 0.0186$$

$$= 0.0389 \text{ kJ/K}$$

Entropy change of the system as ice melts at 0°C to become water at 0°C

$$(\Delta S_{II})_{system} = \frac{333.3}{273} = 1.22 \text{ kJ/K}$$

Entropy change of water as it gets heated from 0°C to 20°C

$$(\Delta S_{III})_{system} = \int_{273}^{293} mc_p \frac{dT}{T} = 1 \times 4.187 \ln \frac{293}{273} = 0.296 \text{ kJ/K}$$

Total entropy change of ice as it melts into water

$$(\Delta S)_{total} = \Delta S_I + \Delta S_{II} + \Delta S_{III}$$

$$= 0.0389 + 1.22 + 0.296 = 1.5549 \text{ kJ/K}$$

The entropy-temperature diagram for the system as ice at – 5°C converts to water at 20°C is shown in Fig. Ex. 7.3.2.

∴ Entropy increase of the universe

$$(\Delta S)_{univ} = (\Delta S)_{system} + (\Delta S)_{atm}$$

$$= 1.5549 - 1.46 = 0.0949 \text{ kJ/K}$$ *Ans.* (a)

(b) To convert 1 kg of water at 20°C to ice at – 5°C, 427.5 kJ of heat have to be removed from it, and the system has to be brought from state 4 to state 1 (Fig. 7.3.2). A refrigerator cycle, as shown in Fig. 7.3.3, is assumed to accomplish this.

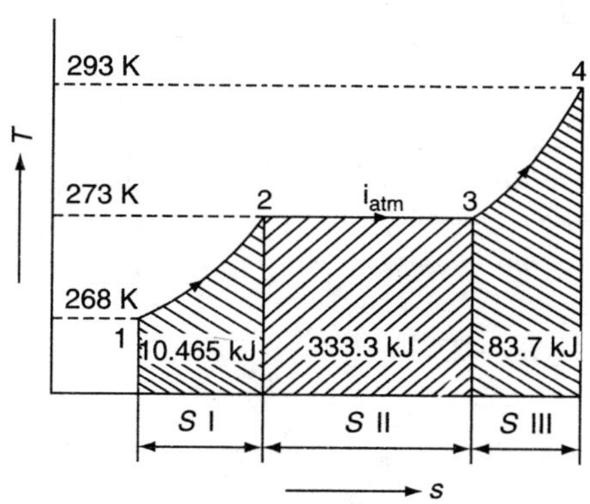

Fig. Ex. 7.3.2

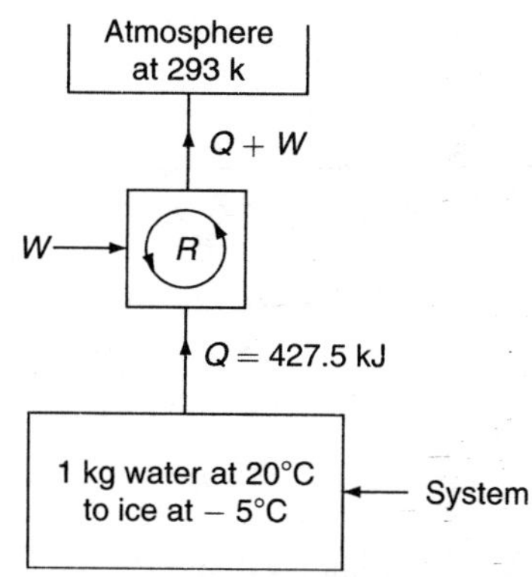

Fig. Ex. 7.3.3

The entropy change of the system would be the same, i.e. $S_4 - S_1$, with the only difference that its sign will be negative, because heat is removed from the system (Fig. Ex. 7.3.2).

$$(\Delta S)_{system} = S_1 - S_4$$

(negative)

The entropy change of the working fluid in the refrigerator would be zero, since it is operating in a cycle, i.e.,

$$(\Delta S)_{ref} = 0$$

The entropy change of the atmosphere (positive)

$$(\Delta S)_{\text{atm}} = \frac{Q+W}{T}$$

∴ Entropy change of the universe

$$(\Delta S)_{\text{univ}} = (\Delta S)_{\text{system}} + (\Delta S)_{\text{ref}} + (\Delta S)_{\text{atm}}$$

$$= (S_1 - S_4) + \frac{Q+W}{T}$$

By the principle of increase of entropy

$$(\Delta S)_{\text{univ or isolated system}} \geq 0$$

$$\therefore \quad \left[(S_1 - S_4) + \frac{Q+W}{T}\right] \geq 0$$

$$\therefore \quad \frac{Q+W}{T} \geq (S_4 - S_1)$$

$$W \geq T(S_4 - S_1) - Q$$

$$\therefore \quad W_{(\text{min})} = T(S_4 - S_1) - Q$$

Here $Q = 427.5$ kJ

$$T = 293 \text{ K}$$

$$S_4 - S_1 = 1.5549 \text{ kJ/K}$$

$$\therefore \quad W_{(\text{min})} = 293 \times 1.5549 - 427.5$$

$$= 28.5 \text{ kJ}$$

Ans. (b)

Example 7.4

Two identical bodies of constant heat capacity are at the same initial temperature T_i. A refrigerator operates between these two bodies until one body is cooled to temperature T_2. If the bodies remain at constant pressure and undergo no change of phase, show that the minimum amount of work needed to do this is

$$W_{(\text{min})} = C_p \left(\frac{T_i^2}{T_2} + T_2 - 2T_i\right)$$

Solution Both the finite bodies *A* and *B* are initially at the same temperature T_i. Body *A* is to be cooled to temperature T_2 by operating the refrigerator cycle, as shown in Fig. Ex. 7.4. Let T_2' be the final temperature of body *B*.

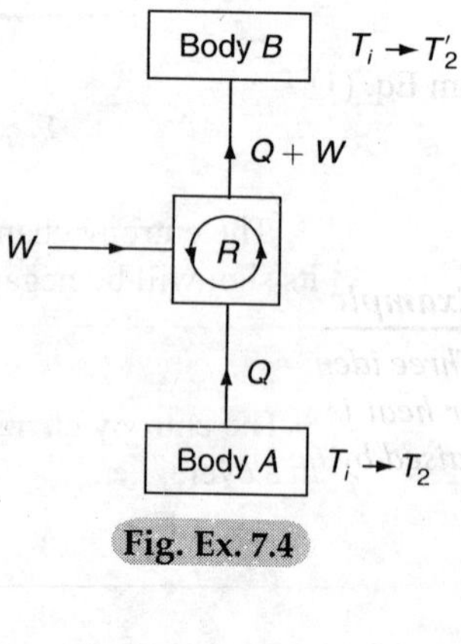

Fig. Ex. 7.4

Heat removed from body *A* to cool it from T_i to T_2

$$Q = C_p(T_i - T_2)$$

where C_p is the constant pressure heat capacity of the identical bodies *A* and *B*.

Heat discharged to body *B*

$$= Q + W = C_p(T_2' - T_i)$$

Work input, W

$$= C_p(T_2' - T_i) - C_p(T_i - T_2)$$
$$= C_p(T_2' + T_2 - 2T_i) \quad (1)$$

Now, the entropy change of body A

$$\Delta S_A = \int_{T_i}^{T_2} C_p \frac{dT}{T} = C_p \ln \frac{T_2}{T_i} \text{ (negative)}$$

The entropy change of body B

$$\Delta S_B = \int_{T_i}^{T_2'} C_p \frac{dT}{T} = C_p \ln \frac{T_2'}{T_i} \text{ (positive)}$$

Entropy change (cycle) of refrigerant = 0

∴ Entropy change of the universe

$$(\Delta S)_{univ} = \Delta S_A + \Delta S_B$$
$$= C_p \ln \frac{T_2}{T_i} + C_p \ln \frac{T_2'}{T_i}$$

By the entropy principle

$$(\Delta S)_{univ} \geq 0$$
$$\left(C_p \ln \frac{T_2}{T_i} + C_p \ln \frac{T_2'}{T_i}\right) \geq 0$$
$$C_p \ln \frac{T_2 T_2'}{T_i^2} \geq 0 \quad (2)$$

In Eq. (7.4.1) with C_p, T_2, and T_i being given, W will be a minimum when T_2' is a minimum. From Eq. (2), the minimum value of T_2' would correspond to

$$C_p \ln \frac{T_2 T_2'}{T_i^2} = 0 = \ln 1$$

$$\therefore \quad T_2' = \frac{T_i^2}{T_2}$$

From Eq. (1)

$$W_{(min)} = C_p \left(\frac{T_i^2}{T_2} + T_2 - 2T_i\right) \quad \text{Proved.}$$

***Example* 7.5**

Three identical finite bodies of constant heat capacity are at temperatures 300, 300, and 100 K. If no work or heat is supplied from outside, what is the highest temperature to which any one of the bodies can be raised by the operation of heat engines or refrigerators?

Solution Let the three identical bodies A, B, and C having the same heat capacity C be respectively at 300 K, 100 K and 300 K initially, and let us operate a heat engine and a refrigerator, as shown in Fig. Ex. 7.5. Let T_f be the final temperature of bodies A and B, and T_f' be the final temperature of body C. Now

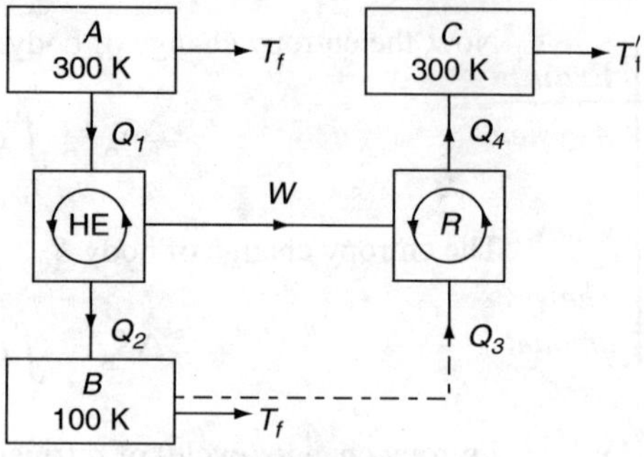

Fig. Ex. 7.5

$$(\Delta S)_A = C \ln \frac{T_f}{300}$$

$$(\Delta S)_B = C \ln \frac{T_f}{300}$$

$$(\Delta S)_C = C \ln \frac{T_f'}{300}$$

$$(\Delta S)_{H.E.} = 0$$

$$(\Delta S)_{ref} = 0$$

where C is the heat capacity of each of the three bodies.

Since
$$(\Delta S)_{univ} \geq 0$$

$$\left(C \ln \frac{T_f}{300} + C \ln \frac{T_f}{100} + C \ln \frac{T_f'}{300} \right) \geq 0$$

$$C \ln \frac{T_f^2 T_f'}{9,000,000} \geq 0$$

For minimum value of T_f, $C \ln \dfrac{T_f^2 T_f'}{9 \times 10^6} = 0 = \ln 1$

$\therefore$
$$T_f^2 T_f' = 9,000,000 \quad (1)$$

Now

$$Q_1 = C(300 - T_f)$$
$$Q_2 = C(T_f - 100)$$
$$Q_4 = C(T_f' - 300)$$

Again

$$Q_1 = \text{Heat removed from body } A$$
$$= \text{Heat discharged to bodies } B \text{ and } C$$
$$= Q_2 + Q_4$$

$\therefore$
$$C(300 - T_f) = C(T_f - 100) + C(T_f' - 300)$$

$\therefore$
$$T_f' = 700 - 2T_f \quad (2)$$

T_f' will be the highest value when T_f is the minimum.

From Eqs (1) and (2)

$$T_f^2 (700 - 2T_f) = 9,000,000$$

$\therefore$
$$2T_f^3 - 700\, T_f^2 + 9,000,000 = 0$$

or
$$T_f = 150 \text{ K}$$

From Eq. (2)

$$T'_f = (700 - 2 \times 150) \text{ K}$$
$$= 400 \text{ K} \qquad \textit{Ans.}$$

Example 7.6

A system has a heat capacity at constant volume

$$C_V = AT^2$$

where A = 0.042 J/K³.

The system is originally at 200 K, and a thermal reservoir at 100 K is available. What is the maximum amount of work that can be recovered as the system is cooled down to the temperature of the reservoir?

Solution Heat removed from the system (Fig. Ex.7.6)

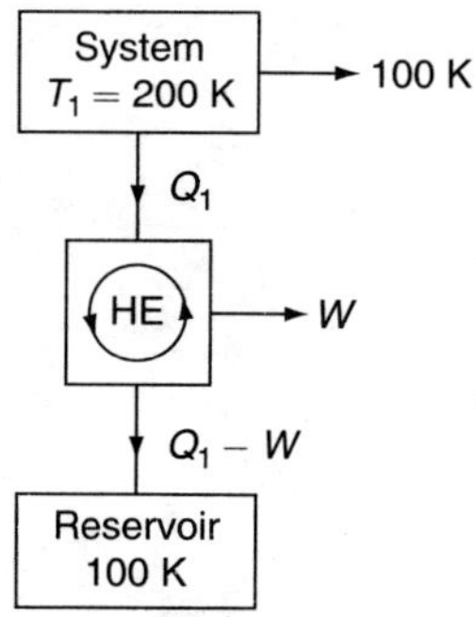

Fig. Ex. 7.6

$$Q_1 = \int_{T_1}^{T_2} C_V \, dT = \int_{T_1 = 200 \text{ K}}^{T_2 = 100 \text{ K}} 0.042 \, T^2 \, dT$$

$$= 0.042 \left[\frac{T^3}{3}\right]_{200 \text{ K}}^{100 \text{ K}}$$

$$= \frac{0.042}{3} \text{ J/K}^3 (100^3 - 200^3)\text{K}^3 = -98 \times 10^3 \text{ J}$$

$$(\Delta S)_{\text{system}} = \int_{200\text{K}}^{100\text{K}} C_V \frac{dT}{T} = \int_{200\text{K}}^{100\text{K}} 0.042 \, T^2 \frac{dT}{T}$$

$$= \frac{0.042}{2} \text{ J/K}^3 [100^2 - 200^2]\text{K}^2 = -630 \text{ J/K}$$

$$(\Delta S)_{\text{res}} = \frac{Q_1 - W}{T_{\text{res}}} = \frac{98 \times 10^3 - W}{100} \text{ J/K}$$

$$(\Delta S)_{\text{working fluid in H.E.}} = 0$$

$$\therefore \quad (\Delta S)_{\text{univ}} = (\Delta S)_{\text{system}} + (\Delta S)_{\text{res}} = -630 + \frac{98 \times 10^3 - W}{100}$$

Since

$$(\Delta S)_{\text{univ}} \geq 0$$

$$\therefore \quad -630 + \frac{98 \times 10^3 - W}{100} \geq 0$$

$$980 - \frac{W}{100} - 630 \geq 0$$

$$\frac{W}{100} \leq 350$$

$$W_{(\max)} = 35{,}000 \text{ J} = 35 \text{ kJ} \qquad \textit{Ans.}$$

Example 7.7

A fluid undergoes a reversible adiabatic compression from 0.5 MPa, 0.2 m³ to 0.05 m³ according to the law, $pv^{1.3}$ = constant. Determine the change in enthalpy, internal energy and entropy, and the heat transfer and work transfer during the process.

Solution

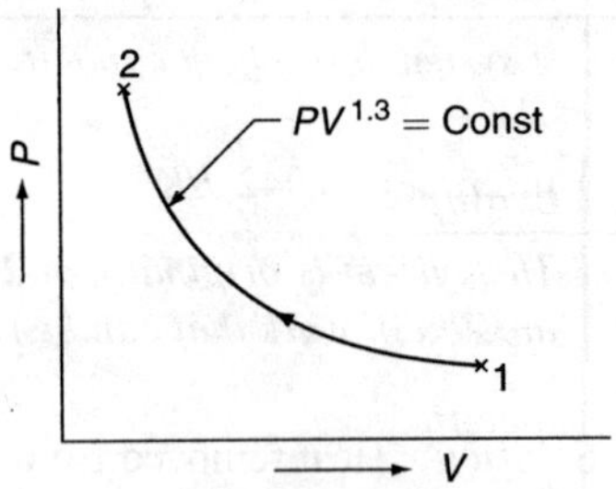

Fig. Ex. 7.7

$$TdS = dH - Vdp$$

For the reversible adiabatic process (Fig. Ex. 7.7)

$$dH = Vdp$$

$$p_1 = 0.5 \text{ MPa}, \quad V_1 = 0.2 \text{ m}^3$$

$$V_2 = 0.05 \text{ m}^3, \quad p_1 V_1^n = p_2 V_2^n$$

$$\therefore \quad p_2 = p_1 \left(\frac{V_1}{V_2}\right)^n$$

$$= 0.5 \times \left(\frac{0.20}{0.05}\right)^{1.3} \text{ MPa}$$

$$= 0.5 \times \left(\frac{0.20}{0.05}\right)^{1.3} \text{ MPa} = 0.5 \times 6.061 \text{ MPa} = 3.0305 \text{ MPa}$$

$$p_1 V_1^n = pV^n$$

$$\therefore \quad V = \left(\frac{p_1 V_1^n}{p}\right)^{1/n}$$

$$\int_{H_1}^{H_2} dH = \int_{p_1}^{p_2} Vdp$$

$$H_2 - H_1 = \int_{p_1}^{p_2} \left[\left(\frac{p_1 V_1^n}{p}\right)\right]^{1/n} dp$$

$$= (p_1 V_1^n)^{1/n} \left(\frac{p_1^{1-1/n} - p_1^{1-1/n}}{1 - 1/n}\right)$$

$$= \frac{n(p_2 V_2 - p_1 V_1)}{n - 1}$$

$$= \frac{1.3(3030.5 \times 0.05 - 500 \times 0.2)}{1.3 - 1} = 223.3 \text{ kJ}$$

$$H_2 - H_1 = (U_2 + p_2 V_2) - (U_1 + p_1 V_1)$$

$$= (U_2 - U_1) + (p_2 V_2 - p_1 V_1)$$

$$\therefore \quad U_2 - U_1 = (H_2 - H_1) - (p_2 V_2 - p_1 V_1)$$

$$= 223.3 - 51.53 = 171.77 \text{ kJ}$$ *Ans.*

$$S_2 - S_1 = 0$$ *Ans.*

$$Q_{1-2} = 0$$ *Ans.*

$$Q_{1-2} = U_2 - U_1 + W_{1-2}$$

$$\therefore \quad W_{1-2} = U_1 - U_2 = -171.77 \text{ kJ}$$ *Ans.*

Example 7.8

Air is flowing steadily in an insulated duct. The pressure and temperature measurements of the air at two stations A and B are given below. Establish the direction of the flow of air in the duct. Assume that for air, specific heat c_p is constant at 1.005 kJ/kg K, $h = c_p T$, and $\frac{v}{T} = \frac{0.287}{p}$, where p, v, and T are pressure (in kPa), volume (in m^3/kg) and temperature (in K) respectively.

	Station A	*Station B*
Pressure	130 kPa	100 kPa
Temperature	50°C	13°C

Solution From property relation

$$Tds = dh - vdp$$

$$ds = \frac{dh}{T} - v\frac{dp}{T}$$

For two states at A and B the entropy change of the system

$$\int_{s_A}^{s_B} ds = \int_{T_A}^{T_B} \frac{c_p dT}{T} - \int_{p_A}^{p_B} 0.287 \frac{dp}{p}$$

$$\therefore \quad s_B - s_A = 1.005 \ln \frac{T_B}{T_A} - 0.287 \ln \frac{p_B}{p_A}$$

$$= 1.005 \ln \frac{273 + 13}{273 + 50} - 0.287 \ln \frac{100}{130}$$

$$= -0.1223 + 0.0753 = -0.047 \text{ kJ/kg K}$$

$$(\Delta s)_{\text{system}} = -0.047 \text{ kJ/kg K}$$

Since the duct is insulated $(\Delta s)_{\text{surr}} = 0$

$$(\Delta s)_{\text{univ}} = -0.047 \text{ kJ/kg K}$$

This is impossible. So the flow must be from B to A.

Example 7.9

A hypothetical device is supplied with 2 kg/s of air at 4 bar, 300 K. Two separate streams of air leave the device, as shown in figure below. Each stream is at an ambient pressure of 1 bar, and the mass flow rate is the same for both streams. One of the exit streams is said to be at 330 K while the other is at 270 K. The ambient temperature is at 300 K. Determine whether such a device is possible.

Solution The entropy generation rate for the control volume (Fig. Ex. 7.9) is

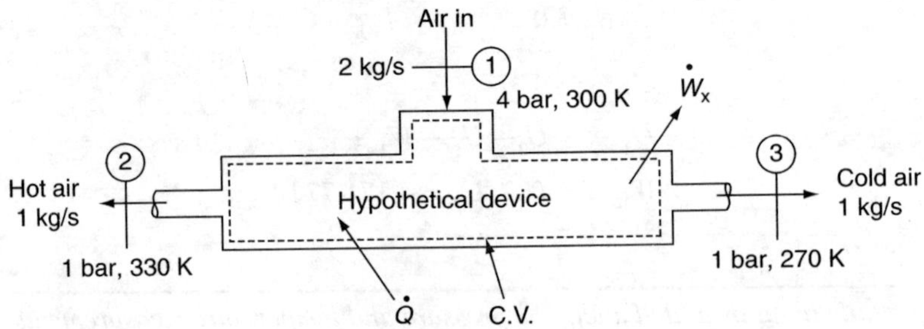

Fig. Ex. 7.9

$$\dot{S}_{gen} = \sum \dot{m}_e s_e - \sum \dot{m}_i s_i$$

$$= \dot{m}_2 s_2 + \dot{m}_3 s_3 - \dot{m}_1 s_1$$

$$= \dot{m}_2 s_2 + \dot{m}_3 s_3 - (\dot{m}_2 + \dot{m}_3) s_1$$

$$= \dot{m}_2 (s_2 - s_1) + \dot{m}_3 (s_3 - s_1)$$

Now,

$$s_2 - s_1 = c_p \ln \frac{T_2}{T_1} - R \ln \frac{p_2}{p_1}$$

$$= 1.005 \ln \frac{330}{300} - 0.287 \ln \frac{1}{4}$$

$$= 0.494 \text{ kJ/kg K}$$

$$s_3 - s_1 = c_p \ln \frac{T_3}{T_1} - R \ln \frac{p_3}{p_1}$$

$$= 1.005 \ln \frac{270}{300} - 0.287 \ln \frac{1}{4} = 0.292 \text{ kJ/kgK}$$

$$\dot{S}_{gen} = 1 \times 0.494 + 1 \times 0.292 = 0.786 \text{ kW/K}$$

Since $\dot{S}_{gen} > 0$, the device is possible. Such devices actually exist and are called *vortex tubes*. Although they have low efficiencies, they are suitable for certain applications like rapid cooling of soldered parts, electronic component cooling, cooling of machining operations and so on. The vortex tube is essentially a passive device with no moving parts. It is relatively maintenance free and durable.

Example 7.10

A room is maintained at 27°C while the surroundings are at 2°C. The temperatures of the inner and outer surfaces of the wall (k = 0.71 W/mK) are measured to be 21°C and 6°C, respectively. Heat flows steadily through the wall 5m × 7 m in cross-section and 0.32 m in thickness. Determine (a) the rate of heat transfer through the wall, (b) the rate of entropy generation in the wall, and (c) the rate of total entropy generation with this heat transfer process.

Solution

$$\dot{Q} = kA\frac{\Delta T}{L} = 0.71\,\frac{\text{W}}{\text{mK}} \times (5 \times 7)\,\text{m}^2 \times \frac{(21-6)\text{K}}{0.32\text{m}}$$

$$= 1164.84 \text{ W} \qquad \textit{Ans.} \text{ (a)}$$

Taking the wall as the system, the entropy balance in rate form gives:

$$\frac{dS_{\text{wall}}}{dt} = \dot{S}_{\text{transfer}} + \dot{S}_{\text{gen.wall}}$$

$$0 = \sum \frac{\dot{Q}}{T} + \dot{S}_{\text{gen.wall}}$$

$$0 = \frac{1164.84}{294} - \frac{1164.84}{279} + \dot{S}_{\text{gen.wall}}$$

Rate of entropy generation in the wall

$$\dot{S}_{\text{gen.wall}} = 4.175 - 3.962 = 0.213 \text{ W/K} \qquad \textit{Ans.} \text{ (b)}$$

The entropy change of the wall is zero during this process, since the state and hence the entropy of the wall does not change anywhere in the wall.

To determine the rate of total entropy generation during this heat transfer process, we extend the system to include the regions on both sides of the wall,

$$\frac{dS_{\text{wall}}}{dt} = \dot{S}_{\text{transfer}} + \dot{S}_{\text{gen.total}}$$

$$0 = \sum \frac{\dot{Q}}{T} + \dot{S}_{\text{gen.total}}$$

$$0 = \frac{1164.84}{300} - \frac{1164.84}{275} + \dot{S}_{\text{gen.total}}$$

$$\dot{S}_{\text{gen.total}} = 4.236 - 3.883 = 0.353 \text{ W/K} \qquad \textit{Ans.} \text{ (c)}$$

Review Questions

7.1 Show that through one point there can pass only one reversible adiabatic.

7.2 State and prove Clausius' theorem.

7.3 Show that entropy is a property of a system.

7.4 How is the entropy change of a reversible process estimated? Will it be different for an irreversible process between the same end states?

7.5 Why is the Carnot cycle on *T–s* plot a rectangle?

7.6 Establish the inequality of Clausius.

7.7 Give the criteria of reversibility, irreversibility and impossibility of a thermodynamic cycle.

7.8 What do you understand by the entropy principle?

7.9 When the system is at equilibrium, why would any conceivable change in entropy be zero?

7.10 Why is the entropy increase of an isolated system a measure of the extent of irreversibility of the process undergone by the system?

7.11 How did Rudolf Clausius summarize the first and second laws of thermodynamics?

7.12 Show that the transfer of heat through a finite temperature difference is irreversible.

7.13 Show that the adiabatic mixing of two fluids is irreversible.

7.14 What is the maximum work obtainable from two finite bodies at temperatures T_1 and T_2?

7.15 Determine the maximum work obtainable by using one finite body at temperature T and a thermal energy reservoir at temperature T_0, $T > T_0$.

7.16 What are the causes of entropy increase?

7.17 Why is an isentropic process not necessarily an adiabatic process?

7.18 What is the reversible adiabatic work for a steady flow system when K.E. and P.E. changes ae negligibly small? How is it different from that for a closed stationary system?

7.19 Under what conditions is the work done equal to (a) $\int pdv$, (b) $-\int vdp$?

7.20 Why are the equations

$$TdS = dU + pdV$$

$$TdS = dH - Vdp$$

valid for any process between two equilibrium end states?

7.21 Why is the second law called a directional law of nature?

7.22 How is entropy related to molecular disorder in a system?

7.23 Show that entropy varies logarithmically with the disorder number.

7.24 What do you understand by perfect order?

7.25 Give the Nernst-Simon statement of the third law of thermodynamics.

7.26 Why does entropy remain constant in a reversible adiabatic process?

7.27 What do you understand by 'lost work'?

7.28 The amount of entropy generation quantifies the intrinsic irreversibility of a process. Explain.

7.29 Show that S_{gen} is not a thermodynamic property.

7.30 Give the expression for the entropy generation rate for a control volume of a steady flow system.

7.31 What is the entropy generation in the isothermal dissipation of work?

7.32 What is the entropy generation in the adiabatic dissipation of work?

7.33 What do you understand by entropy transfer? Why is entropy transfer associated with heat transfer and not with work transfer?

Problems

7.1 On the basis of the first law fill in the blank spaces in the following table of imaginary heat engine cycles. On the basis of the second law classify each cycle as reversible, irreversible, or impossible.

Cycle Temperature		*Rate of Heat Flow*		*Rate of work*	*Efficiency*
Source	*Sink*	*Supply*	*Rejection*	*Output*	
(a) 327°C	27°C	420 kJ/s	230 kJ/s	...kW	
(b) 1000°C	100°C	...kJ/min	4.2 MJ/min	...kW	65%
(c) 750 K	300 K	... kJ/s	... kJ/s	26 kW	60%
(d) 700 K	300 K	3 kW	... kW	2 kW	—

Ans. (a) Irreversible, (b) Irreversible, (c) Reversible, (d) Impossible

7.2 The latent heat of fusion of water at 0°C is 335 kJ/kg. How much does the entropy of 1 kg of ice change as it melts into water in each of the following ways: (a) Heat is supplied reversibly to a mixture of ice and water at 0°C. (b) A mixture of ice and water at 0°C is stirred by a paddle wheel.
Ans. 1.2271 kJ/K

7.3 Two kg of water at 80°C are mixed adiabatically with 3 kg of water at 30°C in a constant pressure process of 1 atmosphere. Find the increase in the entropy of the total mass of water due to the mixing process (c_p of water = 4.187 kJ/kg K). *Ans.* 0.0592 kJ/K

7.4 In a Carnot cycle, heat is supplied at 350°C and rejected at 27°C. The working fluid is water which, while receiving heat, evaporates from liquid at 350°C to steam at 350°C. The associated entropy change is 1.44 kJ/kg K. (a) If the cycle operates on a stationary mass of 1 kg of water, how much is the work done per cycle, and how much is the heat supplied? (b) If the cycle operates in steady flow with a power output of 20 kW, what is the steam flow rate?
Ans. (a) 465.12, 897.12 kJ/kg, (b) 0.043 kg/s

7.5 A heat engine receives reversibly 420 kJ/cycle of heat from a source at 327°C, and rejects heat reversibly to sink at 27°C. There are no other heat transfers. For each of the three hypothetical amounts of heat rejected, in (a), (b), and (c) below, compute the cyclic integral of $đQ/T$. From these results show which case is irreversible, which reversible, and which impossible: (a) 210 kJ/cycle rejected, (b) 105 kJ/cycle rejected, (c) 315 kJ/cycle rejected.

Ans. (a) Reversible, (b) Impossible, (c) Irreversible

7.6 In Fig. P. 7.6, *abcd* represents a Carnot cycle bounded by two reversible adiabatics and two reversible isotherms at temperatures T_1 and T_2 ($T_1 > T_2$). The oval figure is a reversible cycle, where heat is absorbed at temperature less than, or equal to, T_1, and rejected at temperatures greater than, or equal to, T_2. Prove that the efficiency of the oval cycle is less than that of the Carnot cycle.

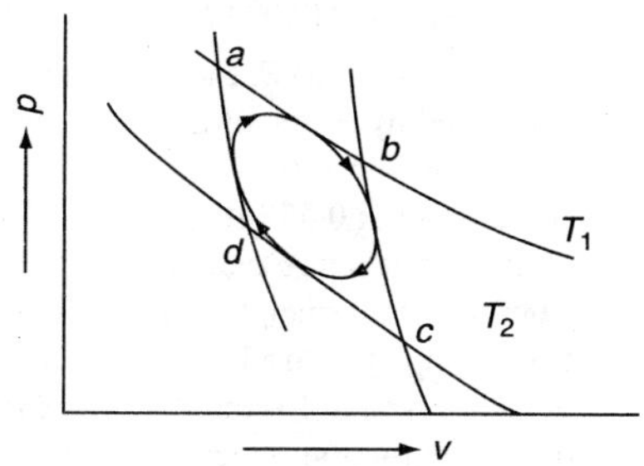

Fig. P. 7.6

7.7 Water is heated at a constant pressure of 0.7 MPa. The boiling point is 164.97°C. The initial temperature of water is 0°C. The latent heat of evaporation is 2066.3 kJ/kg. Find the increase of entropy of water, if the final state is steam.

Ans. 6.6967 kJ/kg K

7.8 One kg of air initially at 0.7 MPa, 20°C changes to 0.35 MPa, 60°C by the three reversible non-flow processes, as shown in Fig. P. 7.8. Process 1-*a*-2 consists of a constant pressure expansion followed by a constant volume cooling, process 1-*b*-2 an isothermal expansion followed by a constant pressure expansion, and process 1-*c*-2 an adiabatic expansion followed by a constant volume heating. Determine the changes of internal energy, enthalpy, and entropy for each process, and find the work transfer and heat transfer for each process. Take $c_p = 1.005$ and $c_v = 0.718$ kJ/kg, K and assume the specific heats to be constant. Also assume for air $pv = 0.287$ T, where p is the pressure in kPa, v the specific volume in m^3/kg, and T the temperature in K.

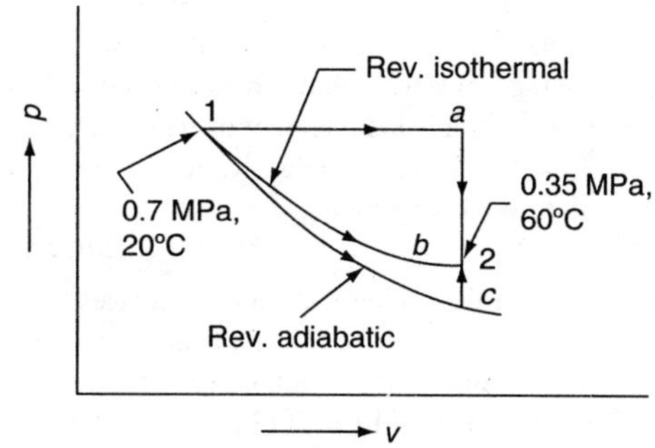

Fig. P. 7.8

7.9 Ten grammes of water at 20°C is converted into ice at –10°C at constant atmospheric pressure. Assuming the specific heat of liquid water to remain constant at 4.2 J/gK and that of ice to be half of this value, and taking the latent heat of fusion of ice at 0°C to be 335 J/g, calculate the total entropy change of the system.

Ans. 16.02 J/K

7.10 Calculate the entropy change of the universe as a result of the following processes:

(a) A copper block of 600 g mass and with C_p of 150 J/K at 100°C is placed in a lake at 8°C.

(b) The same block, at 8°C, is dropped from a height of 100 m into the lake.

(c) Two such blocks, at 100 and 0°C, are joined together.

Ans. (a) 6.63 J/K, (b) 2.095 J/K, (c) 3.64 J/K

7.11 A system maintained at constant volume is initially at temperature T_1, and a heat reservoir at the lower temperature T_0 is available. Show that the maximum work recoverable as the system is cooled to T_0 is

$$W = C_v \left[(T_1 - T_0) - T_0 \ln \frac{T_1}{T_0} \right]$$

7.12 A body of finite mass is originally at temperature T_1 which is higher than that of a reservoir at temperature T_2. Suppose an engine operates in a cycle between the body and the reservoir until it lowers the temperature of the body from T_1 to T_2, thus extracting heat Q from the body. If the engine does work W, then it will reject heat $Q–W$ to the reservoir at T_2. Applying the entropy principle, prove that the maximum work obtainable from the engine is

$$W_{(max)} = Q - T_2 (S_1 - S_2)$$

where $S_1 - S_2$ is the entropy decrease of the body. If the body is maintained at constant volume having constant volume heat capacity $C_v = 8.4$ kJ/K which is independent of temperature, and if $T_1 = 373$ K and $T_2 = 303$ K, determine the maximum work obtainable.

Ans. 58.96 kJ

7.13 Each of three identical bodies satisfies the equation $U = CT$, where C is the heat capacity of each of the bodies. Their initial temperatures are 200 K, 250 K, and 540 K. If $C = 8.4$ kJ/K, what is the maximum amount of work that can be extracted in a process in which these bodies are brought to a final common temperature?

Ans. 756 kJ

7.14 In the temperature range between 0°C and 100°C a particular system maintained at constant volume has a heat capacity.

$$C_V = A + 2BT$$

with $A = 0.014$ J/K and $B = 4.2 \times 10^{-4}$ J/K^2.

A heat reservoir at 0°C and a reversible work source are available. What is the maximum amount of work that can be transferred to the reversible work source as the system is cooled from 100°C to the temperature of the reservoir?

Ans. 4.508 J

7.15 A reversible engine, as shown in Fig. P. 7.15 during a cycle of operation draws 5 MJ from the 400 K reservoir and does 840 kJ of work. Find the amount and direction of heat interaction with other reservoirs.

Ans. $Q_2 = + 4.98$ MJ, $Q_3 = - 0.82$ MJ

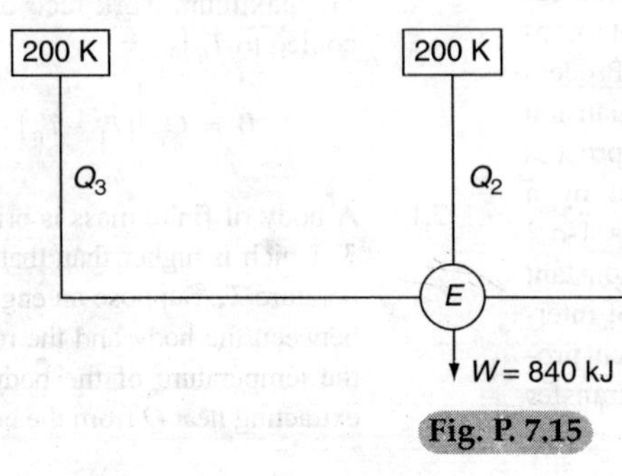

Fig. P. 7.15

7.16 For a fluid for which pv/T is a constant quantity equal to R, show that the change in specific entropy between two states A and B is given by

$$s_B - s_A = \int_{T_A}^{T_B} \frac{C_p}{T} dT - R \ln \frac{p_B}{p_A}$$

A fluid for which R is a constant and equal to 0.287 kJ/kg K, flows steadily through an adiabatic machine, entering and leaving through two adiabatic pipes. In one of these pipes the pressure and temperature are 5 bar and 450 K and in the other pipe the pressure and temperature are 1 bar and 300 K respectively. Determine which pressure and temperature refer to the inlet pipe. For the given temperature range, c_p is given by

$$c_p = a \ln T + b$$

where T is the numerical value of the absolute temperature and $a = 0.026$ kJ/kg K, $b = 0.86$ kJ/kg K.

Ans. $s_B - s_A = 0.0509$ kJ/kg K, A is the inlet pipe.

7.17 Two vessels, A and B, each of volume 3 m^3 may be connected by a tube of negligible volume. Vessel A contains air at 0.7 MPa, 95°C, while vessel B contains air at 0.35 MPa, 205°C. Find the change of entropy when A is connected to B by working from the first principles and assuming the mixing to be complete and adiabatic. For air take the relations as given in Example 7.8. *Ans.* 0.947 kJ/K

7.18 (a) An aluminium block ($c_p = 400$ J/kg K) with a mass of 5 kg is initially at 40°C in room air at 20°C. It is cooled reversibly by transferring heat to a completely reversible cyclic heat engine until the block reaches 20°C. The 20°C room air serves as a constant temperature sink for the engine. Compute (i) the change in entropy for the block, (ii) the change in entropy for the room air, (iii) the work done by the engine.

(b) If the aluminium block is allowed to cool by natural convection to room air, compute (i) the change in entropy for the block, (ii) the change in entropy for the room air (iii) the net change in entropy for the universe. *Ans.* (a) – 134.2 J/K, + 132 J/K, 1306 J, (b) – 132 J/K, + 136.5 J/K, 42.5 J/K

7.19 Two bodies of equal heat capacities C and temperatures T_1 and T_2 form an adiabatically closed system. What will the final temperature be if one lets this system come to equilibrium (a) freely? (b) reversibly? (c) What is the maximum work which can be obtained from this system?

7.20 A resistor of 30 ohms is maintained at a constant temperature of 27°C while a current of 10 amperes is allowed to flow for 1 sec. Determine the entropy change of the resistor and the universe.

Ans. $(\Delta S)_{\text{resistor}} = 0$, $(\Delta S)_{\text{univ}} = 10$ J/K

If the resistor initially at 27°C is now insulated and the same current is passed for the same time, determine the entropy change of the resistor and the universe. The specific heat of the resistor is 0.9 kJ/kg K and the mass of the resistor is 10 g.

Ans. $(\Delta S)_{\text{univ}} = 6.72$ J/K

7.21 An adiabatic vessel contains 2 kg of water at 25°C. By paddle-wheel work transfer, the temperature of water is increased to 30°C. If the specific heat of water is assumed constant at 4.187 kJ/kg K, find the entropy change of the universe. *Ans.* 0.139 kJ/K

7.22 A copper rod is of length 1 m and diameter 0.01 m. One end of the rod is at 100°C, and the other at 0°C. The rod is perfectly insulated along its length and the thermal conductivity of copper is 380 W/mK. Calculate the rate of heat transfer along the rod and the rate of entropy production due to irreversibility of this heat transfer. *Ans.* 2.985 W, 0.00293 W/K

7.23 A body of constant heat capacity C_p and at a temperature T_i is put in contact with a reservoir at a higher temperature T_f. The pressure remains constant while the body comes to equilibrium with the reservoir. Show that the entropy change of the universe is equal to

$$C_p \left[\frac{T_i - T_f}{T_f} - \ln\left(1 + \frac{T_i - T_f}{T_f}\right) \right]$$

Prove that this entropy change is positive.

Given:

$$\ln(1 + x) = x - \frac{x^2}{2} + \frac{x^3}{3} - \frac{x^4}{4} + \cdots -$$

where $x < 1$.

7.24 An insulated 0.75 kg copper calorimeter can containing 0.2 kg water is in equilibrium at a temperature of 20°C. An experimenter now places 0.05 kg of ice at 0°C in the calorimeter and encloses the latter with a heat insulating shield. (a) When all the ice has melted and equilibrium has been reached, what will be the temperature of water and the can? The specific heat of copper is 0.418 kJ/kg K and the latent heat of fusion of ice is 333 kJ/kg. (b) Compute the entropy increase of the universe resulting from the process. (c) What will be the minimum work needed by a stirrer to bring back the temperature of water to 20°C?

Ans. (a) 4.68°C, (b) 0.00276 kJ/K, (c) 20.84 kJ

7.25 Show that if two bodies of thermal capacities C_1 and C_2 at temperatures T_1 and T_2 are brought to the same temperature T by means of a reversible heat engine, then

$$\ln T = \frac{C_1 \ln T_1 + C_2 \ln T_2}{C_1 + C_2}$$

7.26 Two blocks of metal, each having a mass of 10 kg and a specific heat of 0.4 kJ/kg K, are at a temperature of 40°C. A reversible refrigerator receives heat from one block and rejects heat to the other. Calculate the work required to cause a temperature difference of 100°C between the two blocks. *Ans.* 32 kJ

7.27 A block of iron weighing 100 kg and having a temperature of 100°C is immersed in 50 kg of water at a temperature of 20°C. What will be the change of entropy of the combined system of iron and water? Specific heats of iron and water are 0.45 and 4.18 kJ/kg K respectively. *Ans.* 1.24 kJ/K

7.28 36 g of water at 30°C are converted into steam at 250°C at constant atmospheric pressure. The specific heat of water is assumed constant at 4.2 J/gK and the latent heat of vaporization at 100°C is 2260 J/g. For water vapour, assume $pV = mRT$ where $R = 0.4619$ kJ/kg K, and

$$\frac{c_p}{R} = a + bT + cT^2, \text{ where } a = 3.634,$$

$b = 1.195 \times 10^{-3}$ K^{-1} and $c = 0.135 \times 10^{-6}$ K^{-2}

Calculate the entropy change of the system.

Ans. 273.1 J/K

7.29 A 50 ohm resistor carrying a constant current of 1 A is kept at a constant temperature of 27°C by a stream of cooling water. In a time interval of 1 s, (a) what is the change in entropy of the resistor? (b) What is the change in entropy of the universe?

Ans. (a) 0, (b) 0.167 J/K

7.30 A lump of ice with a mass of 1.5 kg at an initial temperature of 260 K melts at the pressure of 1 bar as a result of heat transfer from the environment. After some time has elapsed the resulting water attains the temperature of the environment, 293 K. Calculate the entropy production associated with this process. The latent heat of fusion of ice is 333.4 kJ/kg, the specific heats of ice and water are 2.07 and 4.2 kJ/kg K respectively, and ice melts at 273.15 K. *Ans.* 0.1514 kJ/K

7.31 An ideal gas is compressed reversibly and adiabatically from state *a* to state *b*. It is then heated reversibly at constant volume to state *c*. After expanding reversibly and adiabatically to state *d* such that $T_b = T_d$, the gas is again reversibly heated at constant pressure to state *e* such that $T_e = T_c$. Heat is then rejected reversibly from the gas at constant volume till it returns to state *a*. Express T_a in terms of T_b and T_c. If $T_b = 555$ K and $T_c = 835$ K, estimate T_a. Take $\gamma = 1.4$.

Ans. $T_a = \dfrac{T_b^{\gamma+1}}{T_c^{\gamma}}$, 313.29 K

7.32 Liquid water of mass 10 kg and temperature 20°C is mixed with 2 kg of ice at – 5°C till equilibrium is reached at 1 atm pressure. Find the entropy change of the system. Given: c_p of water = 4.18 kJ/kg K, c_p of ice = 2.09 kJ/kg K and latent heat of fusion of ice = 334 kJ/kg. *Ans.* 104.9 J/K

7.33 A thermally insulated 50-ohm resistor carries a current of 1 A for 1 s. The initial temperature of the resistor is 10°C. Its mass is 5 g and its specific heat is 0.85 J/g K. (a) What is the change in entropy of the resistor? (b) What is the change in entropy of the universe?

Ans. (a) 0.173 J/K (b) 0.173 J/K

7.34 The value of c_p for a certain substance can be represented by $c_p = a + bT$. (a) Determine the heat absorbed and the increase in entropy of a mass *m* of the substance when its temperature is increased at constant pressure from T_1 to T_2. (b) Find the increase in the molal specific entropy of copper, when the temperature is increased at constant pressure from 500 to 1200 K. Given for copper: when $T = 500$ K, $c_p = 25.2 \times 10^3$ and when $T = 1200$ K, $c_p = 30.1 \times 10^3$ J/k mol K.

Ans. (a) $m \left[a(T_2 - T_1) + \frac{b}{2}(T_2^2 - T_1^2)\right] m \left[a \ln \frac{T_2}{T_1} + b(T_2 - T_1)\right]$

(b) 23.9 kJ/k mol K

7.35 An iron block of unknown mass at 85°C is dropped into an insulated tank that contains 0.1m³ of water at 20°C. At the same time a paddle-wheel driven by a 200 W motor is activated to stir the water. Thermal equilibrium is established after 20 min when the final temperature is 24°C. Determine the mass of the iron block and the entropy generated during the process. *Ans.* 52.2 kg, 1.285 kJ/K

7.36 A piston-cylinder device contains 1.2 kg of nitrogen gas at 120 kPa and 27°C. The gas is now compressed slowly in a polytropic process during which $pV^{1.3}$ = constant. The process ends when the volume is reduced by one-half. Determine the entropy change of nitrogen during this process. *Ans.* – 0.0615 kJ/K.

7.37 Air enters a compressor at ambient conditions of 96 kPa and 17°C with a low velocity and exits at 1 MPa, 327°C, and 120 m/s. The compressor is cooled by the ambient air at 17°C at a rate of 1500 kJ/min. The power input to the compressor is 300 kW. Determine (a) the mass flow rate of air and (b) the rate of entropy generation.

Ans. (a) 0.851 kg/s, (b) 0.144 kW/K

7.38 A gearbox operating at steady state receives 0.1 kW along the input shaft and delivers 0.095 kW along the output shaft. The outer surface of the gearbox is at 50°C. For the gearbox, determine (a) the rate of heat transfer, (b) the rate at which entropy is produced. *Ans.* (a) – 0.005 kW, (b) 1.54×10^{-5} kW/K

7.39 At steady state, an electric motor develops power along its output shaft at the rate of 2 kW while drawing 20 amperes at 120 volts. The outer surface of the motor is at 50°C. For the motor, determine the rate of heat transfer and the rate of entropy generation. *Ans.* – 0.4 kW, 1.24×10^{-3} kW/K

7.40 Show that the minimum theoretical work input required by a refrigeration cycle to bring two finite bodies from the same initial temperature to the final temperatures of T_1 and T_2 ($T_2 < T_1$) is given by

$$W_{min} = mc\,[2(T_1T_2)^{1/2} - T_1 - T_2]$$

7.41 A rigid tank contains an ideal gas at 40°C that is being stirred by a paddle wheel. The paddle wheel does 200 kJ of work on the ideal gas. It is observed that the temperature of the ideal gas remains constant during this process as a result of heat transfer between the system and the surroundings at 25°C. Determine (a) the entropy change of the ideal gas and (b) the total entropy generation.

Ans. (a) 0, (b) 0.671 kJ/K

7.42 A cylindrical rod of length L insulated on its lateral surface is initially in contact at one end with a wall at temperature T_1 and at the other end with a wall at a lower temperature T_2. The temperature within the rod initially varies linearly with position x according to:

$$T(x) = T_1 - \frac{T_1 - T_2}{L}x$$

The rod is insulated on its ends and eventually comes to a final equilibrium state where the temperature is T_f. Evaluate T_f and in terms of T_1 and T_2, and show that the amount of entropy generated is:

$$S_{gen} = mc\left[1 + \ln T_f + \frac{T_2}{T_1 - T_2}\ln T_2 - \frac{T_1}{T_1 - T_2}\ln T_1\right]$$

where c is the specific heat of the rod.

Ans. $T_f = [T_1 + T_2]/2$

7.43 Air flowing through a horizontal, insulated duct was studied by students in a laboratory. One student group measured the pressure, temperature, and velocity at a location in the duct as 0.95 bar, 67°C, 75 m/s. At another location the respective values were found to be 0.8 bar, 22°C, 310 m/s. The group neglected to note the direction of flow, however. Using the known data, determine the direction. *Ans*. Flow is from right to left

7.44 Nitrogen gas at 6 bar, 21°C enters an insulated control volume operating at steady state for which $W_{C.V.} = 0$. Half of the nitrogen exits the device at 1 bar, 82°C and the other half exits at 1 bar, – 40°C. The effects of KE and PE are negligible. Employing the ideal gas model, decide whether the device can operate as described.

Ans. Yes, the device can operate as described

CHAPTER 8

Available Energy, Exergy and Irreversibility

8.1 AVAILABLE ENERGY

The sources of energy can be divided into two groups, viz. high grade energy and low grade energy. The conversion of high grade energy to shaft work is exempted from the limitations of the second law, while conversion of low grade energy is subject to them.

The examples of two kinds of energy are:

High grade energy	*Low grade energy*
(a) Mechanical work	(a) Heat or thermal energy
(b) Electrical energy	(b) Heat derived from nuclear fission or fusion
(c) Water power	(c) Heat derived from combustion of fossil fuels
(d) Wind power	
(e) Kinetic energy of a jet	
(f) Tidal power	

The bulk of the high grade energy in the form of mechanical work or electrical energy is obtained from sources of low grade energy, such as fuels, through the medium of the cyclic heat engine. The complete conversion of low grade energy, heat, into high grade energy, shaft-work, is impossible by virtue of the second law of thermodynamics. That part of the low grade energy which is available for conversion is referred to as available energy, while the part which, according to the second law, must be rejected, is known as *unavailable energy*.

Josiah Willard Gibbs is accredited with being the originator of the availability concept. He indicated that environment plays an important part in evaluating the available energy.

8.2 AVAILABLE ENERGY REFERRED TO A CYCLE

The maximum work output obtainable from a certain heat input in a cyclic heat engine (Fig. 8.1) is called the *available energy* (A.E.), or the available part of the energy supplied. The minimum energy that has to be rejected to the sink by the second law is called the *unavailable energy* (U.E.), or the unavailable part of the energy supplied.

Therefore, $$Q_1 = \text{A.E.} + \text{U.E.} \tag{8.1}$$

or $$W_{max} = \text{A.E.} = Q_1 - \text{U.E.}$$

For the given T_1 and T_2,

$$\eta_{rev} = 1 - \frac{T_2}{T_1}$$

For a given T_1, η_{rev} will increase with the decrease of T_2. The lowest practicable temperature of heat rejection is the *temperature of the surroundings*, T_0

$$\therefore \quad \eta_{max} = 1 - \frac{T_0}{T_1}$$

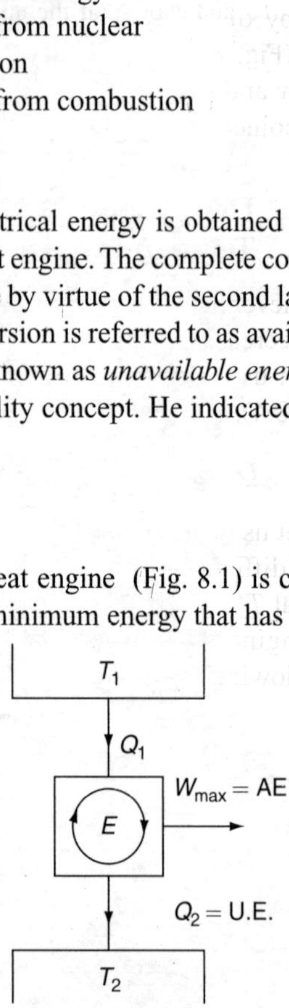

Fig. 8.1 *Available and unavailable energy in a cycle*

and $$W_{max} = \left(1 - \frac{T_0}{T_1}\right) Q_1$$

Let us consider a finite process x–y, in which heat is supplied reversibly to a heat engine (Fig. 8.2). Taking an elementary cycle, if $đQ_1$ is the heat received by the engine reversibly at T_1, then

$$đW_{max} = \frac{T_1 - T_0}{T_1} đQ_1 = đQ_1 - \frac{T_0}{T_1} đQ_1 = \text{A.E.}$$

For the heat engine receiving heat for the whole process x–y, and rejecting heat at T_0

$$\int_x^y đW_{max} = \int_x^y đQ_1 - \int_x^y \frac{T_0}{T_1} đQ_1$$

$\therefore$ $$W_{max} = \text{A.E.} = Q_{xy} - T_0(s_y - s_x) \quad (8.2)$$

or $$\text{U.E.} = Q_{xy} - W_{max}$$

or $$\text{U.E.} = T_0(s_y - s_x)$$

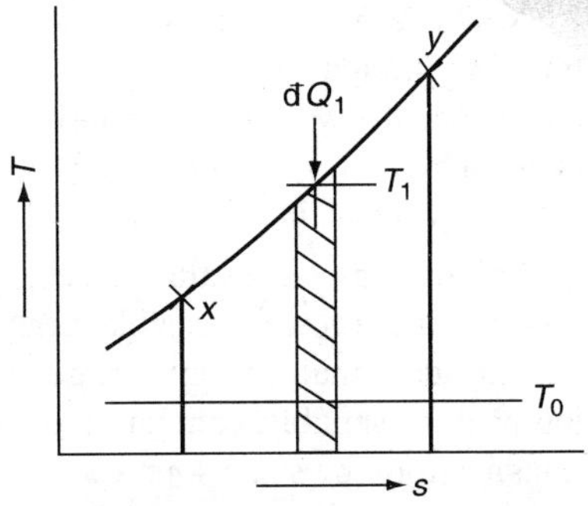

Fig. 8.2 Availability of energy

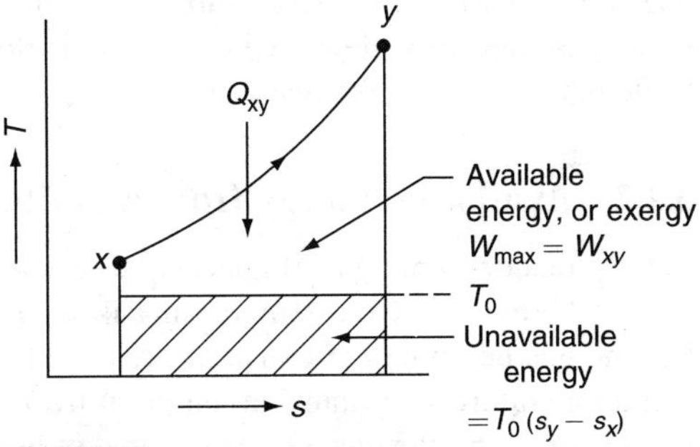

Fig. 8.3 Unavailable energy by the second law

The unavailable energy is thus the product of the lowest temperature of heat rejection, and the change of entropy of the system during the process of supplying heat (Fig. 8.3). The available energy is also known as *exergy* and the unavailable energy as *aenergy*, the words first coined by Rant (1956).

8.2.1 Decrease in Available Energy when Heat is Transferred through a Finite Temperature Difference

Whenever heat is transferred through a finite temperature difference, there is a decrease in the availability of energy so transferred.

Let us consider a reversible heat engine operating between T_1 and T_0 (Fig. 8.4). Then

$$Q_1 = T_1 \Delta s,\ Q_2 = T_0 \Delta s, \text{ and } W = \text{A.E.} = (T_1 - T_0)\Delta s$$

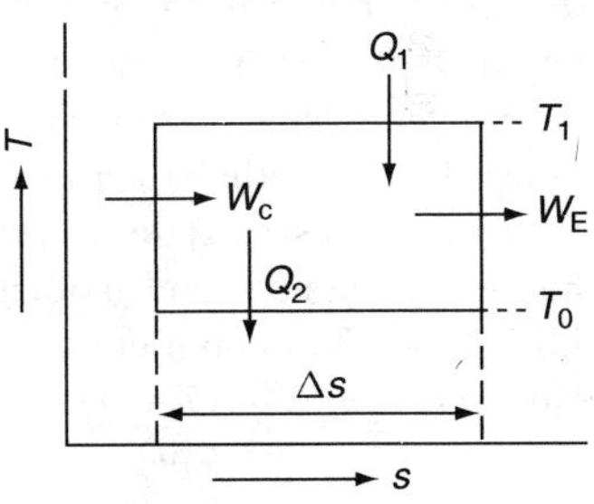

Fig. 8.4 Carnot cycle

Let us now assume that heat Q_1 is transferred through a finite temperature difference from the reservoir or source at T_1 to the engine absorbing heat at T'_1, lower than T_1 (Fig. 8.5). The availability of Q_1 as received by the engine at T'_1 lower than T_1 (Fig. 8.5). The availability of Q_1 as received by the engine at T'_1 can be found by allowing the engine to operate reversibly in a cycle between T'_1 and T_0, receiving Q_1 and rejecting Q_2.

Now $$Q_1 = T_1 \Delta s = T'_1 \Delta s'$$

Since $$T_1 > T'_1, \therefore \Delta s' > \Delta s \quad Q_2 = T_0 \Delta s \quad Q'_2 = T_0 \Delta s'$$

Since $$\Delta s' > \Delta s \quad \therefore Q'_2 > Q_2$$

$\therefore$ $$W' = Q_1 - Q'_2 = T'_1 \Delta s' - T_0 \Delta s'$$

and $$W = Q_1 - Q_2 = T'_1 \Delta s - T_0 \Delta s$$

$\therefore$ $$W' < W, \text{ because } Q'_2 > Q_2$$

Available energy or exergy lost due to irreversible heat transfer through finite temperature difference between the source and the working fluid during the heat addition process is given by $W - W' = Q'_2 - Q_2 = T_0\,(\Delta s' - \Delta s)$

or, decrease in A.E. $= T_0\,(\Delta s' - \Delta s)$

The decrease in available energy or exergy is thus the product of the lowest feasible temperature of heat rejection and the additional entropy change in the system while receiving heat irreversibly, compared to the case of reversible heat transfer from the same source.

The greater is the temperature difference $(T_1 - T'_1)$, the greater is the heat rejection Q'_2 and the greater will be the unavailable part of the energy supplied or anergy (Fig. 8.5). Energy is said to be degraded each time it flows through a finite temperature difference.

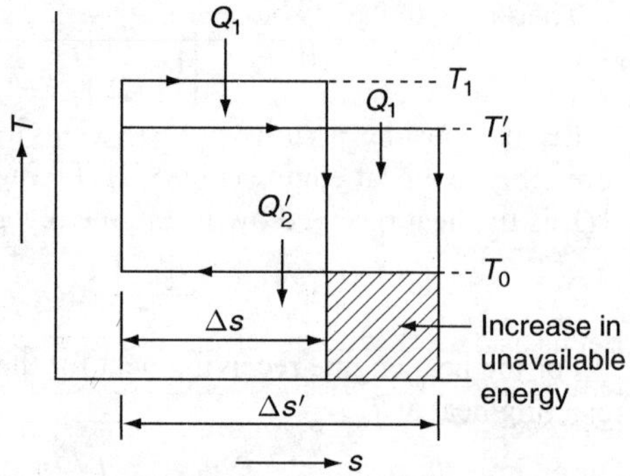

Fig. 8.5 *Increase in unavailable energy due to heat transfer through a finite temperature difference*

8.2.2 Available Energy from a Finite Energy Source

Let us consider a hot gas of mass m_g at temperature T when the environmental temperature is T_0 (Fig. 8.6). Let the gas be cooled at constant pressure from state 1 at temperature T to state 3 at temperature T_0 and the heat given up by the gas, Q_1, be utilized in heating up reversibly a working fluid of mass m_{wf} from state 3 to state 1 along the same path so that the temperature difference between the gas and the working fluid at any instant is zero and hence, the entropy increase of the universe is also zero. The working fluid expands reversibly and adiabatically in an engine or turbine from sate 1 to state 2 doing work W_E, and then rejects heat Q_2 reversibly and isothermally to return to the initial state 3 to complete a heat engine cycle.

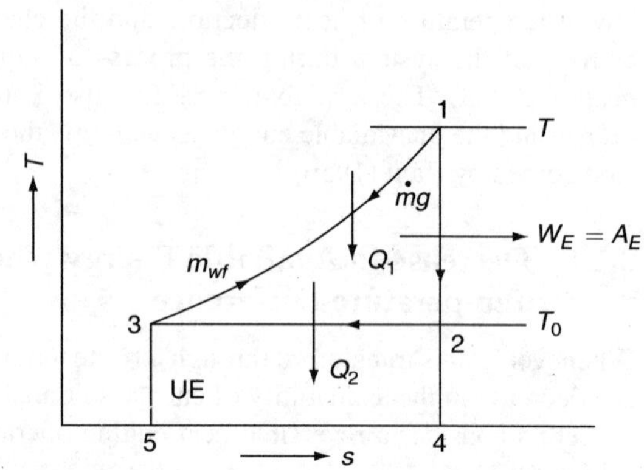

Fig. 8.6 *Available energy of a finite energy source*

Here, $Q_1 = m_g\, c_{p_g}\,(T - T_0) = m_{wf}\, c_{pwf}\,(T - T_0) =$ Area 1–4–5–3–1

$$\therefore \quad m_g\, c_{p_g} = m_{wf}\, c_{p_{wf}}$$

$$\Delta S_{gas} = \int_T^{T_0} m_g\, c_{p_g} \frac{dT}{T} = m_g\, c_{p_g} \ln \frac{T_0}{T} \text{ (negative)}$$

$$\Delta S_{wf} = \int_{T_0}^{T} m_{wf}\, c_{p_{wf}} \frac{dT}{T} = m_{wf}\, c_{p_{wf}} \ln \frac{T}{T_0} \text{ (positive)}$$

$$\therefore \quad \Delta S_{univ} = \Delta S_{gas} + \Delta S_{wf} = 0$$

$$Q_2 = T_0\, \Delta S_{wf} = T_0\, m_{wf}\, c_{p_{wf}} \ln \frac{T}{T_0} = \text{Area 2–4–5–3}$$

$$\therefore \quad \text{Available energy} = W_{max}$$

$$= Q_1 - Q_2 = m_g\, c_{p_g}\,(T - T_0) - T_0\, m_g\, c_{p_g} \ln \frac{T}{T_0}$$

$$= \text{Area 1–2–3–1}$$

Therefore, the available energy or exergy of a gas of mass m_g at temperature T is given by

$$AE = m_g\, c_{p_g} \left[(T - T_0) - T_0 \ln \frac{T}{T_0} \right] \tag{8.3}$$

This is similar to Eq. (7.22) derived from the entropy principle.

8.3 QUALITY OF ENERGY

Let us assume that a hot gas is flowing through a pipeline (Fig. 8.7). Due to heat loss to the surroundings, the temperature of the gas decreases continuously from inlet at state *a* to the exit at state *b*. Although the process is irreversible, let us assume a reversible isobaric path between the inlet and exit states of the gas (Fig. 8.8). For an infinitesimal reversible process at constant pressure,

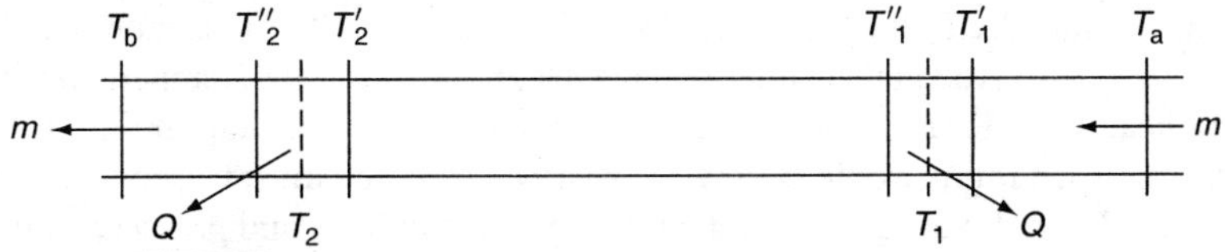

Fig. 8.7 *Heat loss from a hot gas flowing through a pipeline*

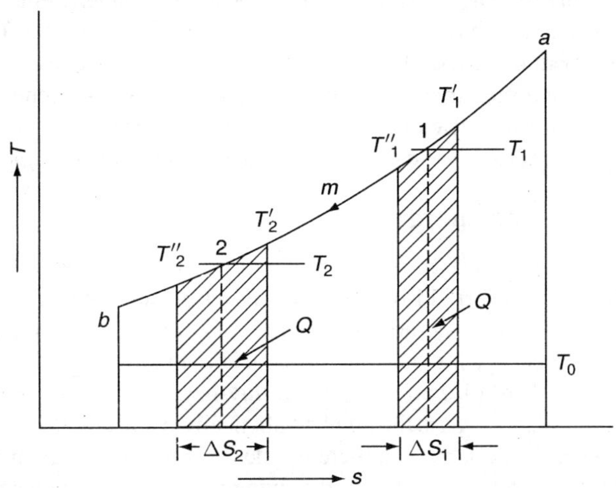

Fig. 8.8 *Energy quality at state 1 is superior to that at state 2*

$$dS = \frac{mc_p dT}{T}$$

or

$$\frac{dT}{dS} = \frac{T}{mc_p} \tag{8.4}$$

where m is the mass of gas flowing and c_p is its specific heat. The slope dT/dS depends on the gas temperature T. As T increases, the slope increases, and if T decreases the slope decreases.

Let us assume that Q units of heat are lost to the surroundings as the temperature of the gas decreases from T'_1 to T''_1, T_1 being the average of the two. Then,

Heat loss $$Q = mc_p (T'_1 - T''_1) = T_1\, \Delta S_1 \tag{8.5}$$

Exergy lost with this heat loss at temperature T_1 is

$$W_1 = Q - T_0\,\Delta S_1 \tag{8.6}$$

When the gas temperature has reached T_2 ($T_2 < T_1$), let us assume that the same heat loss Q occurs as the gas temperature decreases from T'_2 to T''_2, T_2 being the average temperature of the gas. Then

Heat loss
$$Q = mc_p\,(T'_2 - T''_2) = T_2\,\Delta S_2 \tag{8.7}$$

Exergy lost with this heat loss at temperature T_2 is

$$W_2 = Q - T_0\,\Delta S_2 \tag{8.8}$$

From Eqs (8.5) and (8.7), since $T_1 > T_2$

$$\Delta S_1 < \Delta S_2$$

Therefore, from Eqs (8.6) and (8.8),

$$W_1 > W_2 \tag{8.9}$$

The loss of exergy is more, when heat loss occurs at a higher temperature T_1 than when the same heat loss occurs at a lower temperature T_2. Therefore, a heat loss of 1 kJ at, say, 1000°C is more harmful than the same heat loss of 1 kJ at, say, 100°C. Adequate insulation must be provided for high temperature fluids ($T >> T_0$) to prevent the precious heat loss. This many not be so important for low temperature fluids ($T \sim T_0$), since the loss of available energy from such fluids would be low. (Similarly, insulation must be provided adequately for very low temperature fluids ($T << T_0$) to prevent heat gain from surroundings and preserve available energy.)

The available energy or exergy of a fluid at a higher temperature T_1 is more than that at a lower temperature T_2, and decreases as the temperature decreases. When the fluid reaches the ambient temperature, its exergy is zero.

The second law, therefore, affixes a quality to energy of a system at any state. The quality of energy of a gas at, say, 1000°C is superior to that at, say, 100°C, since the gas at 1000°C has the capacity of doing more work than the gas at 100°C, under the same environmental conditions. An awareness of this energy quality as of energy quantity is essential for the efficient use of our energy resources and for energy conservation. The concept of *available energy or exergy provides a useful measure of this energy quality.*

8.3.1 Law of Degradation of Energy

The available energy of a system decreases as its temperature or pressure decreases and approaches that of the surroundings. When heat is transferred from a system, its temperature decreases and hence the quality of its energy deteriorates. The degradation is more for energy loss at a higher temperature than that at a lower temperature. Quantity-wise the energy loss may be the same, but quality-wise the losses are different. *While the first law states that energy is always conserved quantity-wise, the second law emphasizes that energy always degrades quality-wise.* When a gas is throttled adiabatically from a high to a low pressure, the enthalpy (or energy per unit mass) remains the same, but there is a degradation of energy or available work. The same holds good for pressure drop due to friction of a fluid flowing through an insulated pipe. If the first law is the law of conservation of energy, the second law is called the *law of degradation of energy*. Energy is always conserved, but its quality is always degraded.

Article 8.2.1 which shows how energy gets degraded by thermal irreversibility and produces less useful work can be explained in little different way. Let two bodies 1 and 2 of constant heat capacities C_1 and C_2 be at temperatures T_1 and T_2 ($T_1 > T_2$). These are connected by a rod and a small quantity of head Q flows from 1 to 2. The total change of entropy is:

$$\Delta S = \Delta S_1 + \Delta S_2 = Q\left[\frac{1}{T_2} - \frac{1}{T_1}\right] > 0 \quad (\text{since } T_1 > T_2)$$

The entropy will continue to increase till thermal equilibrium is reached.

Let us now suppose that instead of allowing heat Q to flow from 1 to 2, we used it to operate a Carnot engine and obtain mechanical work, with T_0 as the sink temperature. The maximum work obtainable is:

$$W_1 = Q\left[1 - \frac{T_0}{T_1}\right]$$

If, however, we first allow Q to flow from 1 to 2 and then use it to operate the Carnot engine, we obtain:

$$W_2 = Q\left[1 - \frac{T_0}{T_2}\right] < W_1$$

Thus, in the course of the irreversible heat conduction the energy has become degraded to the extent that the useful work has been decreased by

$$\Delta W = W_1 - W_2 = T_0 \Delta S$$

The increase in entropy in an irreversible change is thus a measure of the extent to which energy becomes degraded in that change. Conversely, in order to extract the maximum work from a system, changes must be performed in a reversible manner so that total entropy ($\Delta S_{sys} + \Delta S_{sur}$) is conserved.

It is worth pointing that if the two bodies were allowed to reach thermal equilibrium (a) by heat conduction and (b) by operating a Carnot engine between them and extracting work, the final equilibrium temperatures would be different in the two cases. In the first, $U_1 + U_2$ is conserved and the final temperature is:

$$T_f^{(U)} = \frac{C_1 T_1 + C_2 T_2}{C_1 + C_2}$$

In the second case, $S_1 + S_2$ is conserved and W= $-\ \Delta U\,(-\Delta U_1 + U_2)$ ($\because$ d$W = Q -$ dU, $< T$ds $-$ dU so that d$W_{max} = -$dU). In the isentropic process, the final temperature is given by:

$$T_f^{(s)} = T_1^{C_1/(C_1+C_2)}\ T_2^{C_2/(C_1+C_2)} < T_f^{(U)}$$

If $\qquad C_1 = C_2 = C, T_f^{(U)} = \dfrac{T_1 + T_2}{2}$ and $T_f^{(s)} = (T_1 T_2)^{1/2}$

The difference in final temperature is due to the lower value of the total internal energy which results from work having been done at the expense of internal energy (see Art 7.9.3).

Similarly, it can be shown that due to mechanical irreversibility also, energy gets degraded so that the degradation of energy quality is a universal principle.

8.4 MAXIMUM WORK IN A REVERSIBLE PROCESS

Let us consider a closed stationary system undergoing a reversible process R from state 1 to state 2 interacting with the surroundings at p_0, T_0 (Fig. 8.9). Then by the first law,

$$Q_R = U_2 - U_1 + W_R \tag{8.10}$$

If the process were irreversible, as represented by the dotted line I, connecting the same equilibrium end states,

$$Q_I = U_2 - U_1 + W_I \tag{8.11}$$

Therefore, from Eqs (8.10) and (8.11),

$$Q_R - Q_I = W_R - W_I \tag{8.12}$$

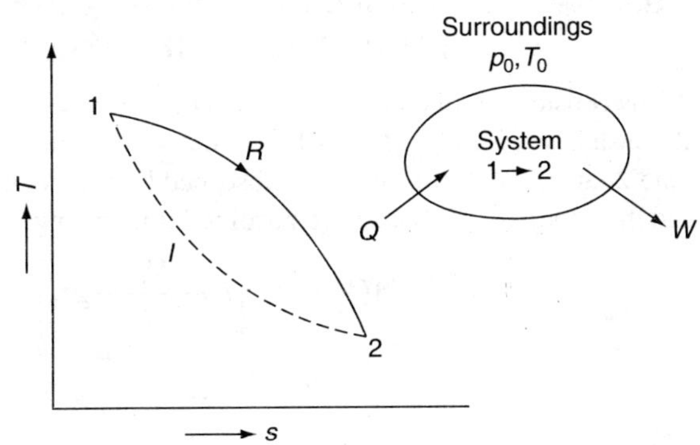

Fig. 8.9 Maximum work done by a closed system

Now,

$$\Delta S_{sys} = S_2 - S_1$$

and

$$\Delta S_{surr} = -\frac{Q}{T_0}$$

By the second law, $\qquad \Delta S_{univ} \geq 0$

For a reversible process, $\Delta S_{\text{univ}} = S_2 - S_1 - \frac{Q_R}{T_0} = 0$

$$\therefore \quad Q_R = T_0 (S_2 - S_1) \tag{8.13}$$

For an irreversible process,

$$\Delta S_{\text{univ}} > 0$$

$$\therefore \quad S_2 - S_1 - \frac{Q_I}{T_0} > 0$$

$$\therefore \quad Q_I < T_0 (S_2 - S_1) \tag{8.14}$$

From Eqs (8.13) and (8.14), $$Q_R > Q_I \tag{8.15}$$

Therefore, from Eqs (8.12) and (8.15), $$W_R > W_I \tag{8.16}$$

Therefore, the work done by a closed system by interacting only with the surroundings at p_0, T_0 in a reversible process is always more than that done by it in an irreversible process between the same end states.

8.4.1 Work Done in all Reversible Processes is the Same

Let us assume two reversible processes R_1 and R_2 between the same end states 1 and 2 undergone by a closed system by exchanging energy only with the surroundings (Fig. 8.10). Let one of the processes be reversed.

Then the system would execute a cycle 1–2–1 and produce network represented by the area enclosed by exchanging energy with only one reservoir, i.e. the surroundings. This violates the Kelvin-Planck statement. Therefore, *the two reversible processes must coincide and produce equal amounts of work.*

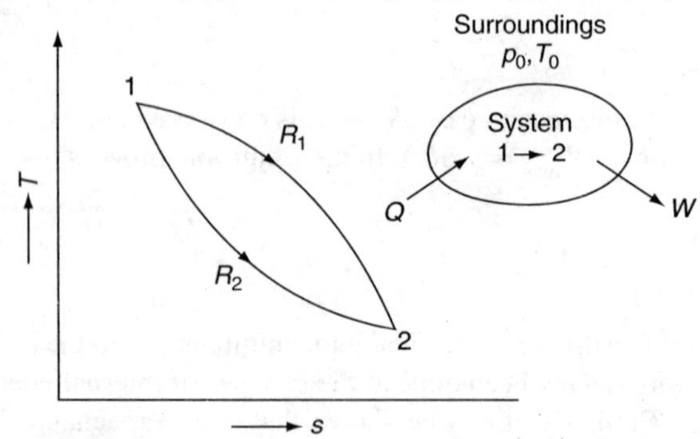

Fig. 8.10 *Equal work done in all reversible processes between the same end states*

8.5 REVERSIBLE WORK BY AN OPEN SYSTEM EXCHANGING HEAT ONLY WITH THE SURROUNDINGS

Let us consider an open system exchanging energy only with the surroundings at constant temperature T_0 and at constant pressure p_0 (Fig. 8.11). A mass dm_1 enters the system at state 1, a mass dm_2 leaves the system at state 2, an amount of heat $đQ$ is absorbed by the system, an amount of work $đW$ is delivered by the system, and the energy of the system (control volume) changes by an amount $d(E)_\sigma$. Applying the first law, we have

$$đQ + dm_1\left(h_1 + \frac{V_1^2}{2} + gz_1\right) - dm_2\left(h_2 + \frac{V_2^2}{2} + gz_2\right) - đW$$

$$= dE_\sigma = d\left[U + \frac{mV^2}{2} + mgz\right]_\sigma \tag{8.17}$$

For maximum work, the process must be entirely reversible. There is a temperature difference between the control volume and the surroundings. To make the heat transfer process reversible, let us assume a reversible heat engine E operating between the two. Again, the temperature of the fluid in the control volume may be different at different points. It is assumed that heat transfer occurs at points of the control surface σ where the temperature is T. Thus in an infinitesimal reversible process an amount of heat $đQ_0$ is absorbed by the engine E

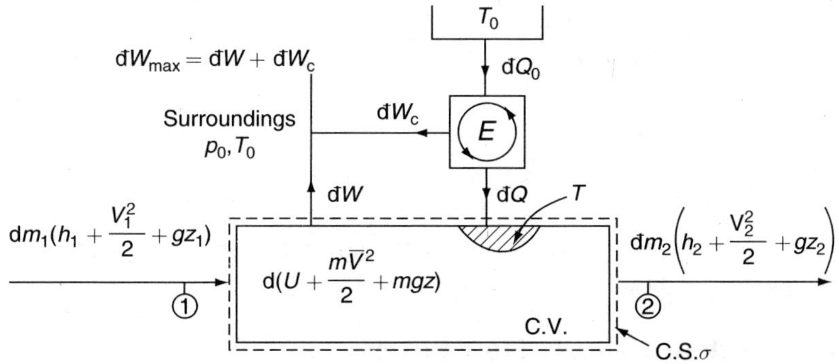

Fig. 8.11 *Reversible work done by an open system while exchanging heat only with the surroundings*

from the surroundings at temperature T_0, an amount of heat $đQ$ is rejected by the engine reversibly to the system where the temperature is T, and an amount of work $đW_c$ is done by the engine. For a reversible engine,

$$\frac{đQ_0}{T_0} = \frac{đQ}{T}$$

$$\therefore \quad đW_c = đQ_0 - đQ = đQ \cdot \frac{T_0}{T} - đQ$$

or

$$đW_c = đQ\left(\frac{T_0}{T} - 1\right) \tag{8.18}$$

The work $đW_c$ is always positive and is independent of the direction of heat flow. When $T_0 > T$, heat will flow from the surroundings to the system, $đQ$ is positive and hence $đW_c$ in Eq. (8.18) would be positive. Again, when $T_0 < T$, heat will flow from the system to the surroundings, $đQ$ is negative, and hence $đW_c$ would be positive.

Now, since the process is reversible, the entropy change of the system will be equal to the net entropy transfer, and $S_{gen} = 0$. Therefore,

$$\underset{\text{Entropy change}}{dS} = \underset{\text{Entropy transfer with heat}}{\frac{đQ}{T}} + \underset{\text{Entropy transfer with mass}}{dm_1 s_1 - dm_2 s_2}$$

$$\therefore \quad \frac{đQ}{T} = dS - dm_1 s_1 + dm_1 s_2 \tag{8.19}$$

Now, the maximum work is equal to the sum of the system work dW and the work dW_c of the reversible engine E, $đW_{max} = đW_{rev} = đW + đW_c$ (8.20)

From Eq. (8.18), $đW_{max} = dW + dQ\left(\frac{T_0}{T} - 1\right)$ (8.21)

Substituting Eq. (8.17) for $đW$ in Eq. (8.21),

$$đW_{max} = đQ + dm_1\left(h_1 + \frac{V_1^2}{2} + gz_1\right) - dm_2\left(h_2 + \frac{V_2^2}{2} + gz_2\right) - d\left[U + \frac{mV^2}{2} + mgz\right]_\sigma + đQ\left(\frac{T_0}{T} - 1\right)$$

$$= dm_1\left(h_1 + \frac{V_1^2}{2} + gz_1\right) - dm_2\left(h_2 + \frac{V_2^2}{2} + gz_2\right) - d\left[U + \frac{mV^2}{2} + mgz\right]_\sigma + \frac{đQ}{T}T_0 \tag{8.22}$$

On substituting the value of $đQ/T$ from Eq. (8.19),

$$đW_{max} = dm_1\left(h_1 + \frac{V_1^2}{2} + gz_1\right) - dm_2\left(h_2 + \frac{V_2^2}{2} + gz_2\right) - d\left[U + \frac{mV^2}{2} + mgz\right]_\sigma + T_0\,(dS - dm_1 s_1 + dm_2 s_2)$$

$$\therefore\ đW_{max} = dm_1\left(h_1 - T_0 s_1 + \frac{V_1^2}{2} + gz_1\right) - dm_2\left(h_2 - T_0 s_2 + \frac{V_2^2}{2} + gz_2\right) - d\left[U - T_0 S + \frac{mV^2}{2} + mgz\right]_\sigma \quad (8.23)$$

Equation (8.23) is the general expression for the maximum work of an open system which exchanges heat only with the surroundings at T_0, p_0.

8.5.1 Reversible Work in a Steady Flow Process

For a steady flow process $dm_1 = dm_2 = dm$

and
$$d\left[U - T_0 S + \frac{mV^2}{2} + mgz\right]_\sigma = 0$$

Equation (8.24) reduces to

$$đW_{max} = dm\left[\left(h_1 - T_0 s_1 + \frac{V_1^2}{2} + gz_1\right) - \left(h_2 - T_0 s_2 + \frac{V_2^2}{2} + gz_2\right)\right] \quad (8.24)$$

For total mass flow, the integral form of Eq. (8.24) becomes

$$W_{max} = \left(H_1 - T_0 S_1 + \frac{mV_1^2}{2} + mgz_1\right) - \left(H_2 - T_0 S_2 + \frac{mV_2^2}{2} + mgz_2\right) \quad (8.25)$$

The expression $(H - T_0 S)$ is called the *Keenan function*, *B*.

$$\therefore\quad W_{max} = \left(B_1 + \frac{mV_1^2}{2} + mgz_1\right) - \left(B_2 + \frac{mV_2^2}{2} + mgz_2\right)$$

$$= \Psi_1 - \Psi_2 \quad (8.26)$$

where Ψ is called the availability function of a steady flow process given by $\Psi = B + \dfrac{mV^2}{2} + mgz$

On a unit mass basis,

$$W_{max} = \left(h_1 - T_0 s_1 + \frac{V_1^2}{2} + gz_1\right) - \left(h_1 - T_0 s_2 + \frac{V_2^2}{2} + gz_2\right) = \left(b_1 + \frac{V_1^2}{2} + gz_1\right) - \left(b_2 + \frac{V_2^2}{2} + gz_2\right) \quad (8.27)$$

If K.E. and P.E. changes are neglected, Eqs (8.26) and (8.27) reduce to

$$W_{max} = B_1 - B_2 = (H_1 - T_0 S_1) - (H_2 - T_0 S_2) = (H_1 - H_2) - T_0(S_1 - S_2) \quad (8.28)$$

and per unit mass

$$W_{max} = b_1 - b_2 = (h_1 - h_2) - T_0\,(s_1 - s_2) \quad (8.29)$$

8.5.2 Reversible Work in a Closed System

For a *closed system*, $dm_1 = dm_2 = 0$

Equation (8.23) then becomes $dW_{max} = -\,d\left[U - T_0 S + \dfrac{mV^2}{2} + mgz\right]_\sigma = -\,d(E - T_0 S)_\sigma$

where $$E = U + \frac{mV^2}{2} + mgz$$

For a change of state of the system from the initial state 1 to the final state 2,

$$W_{max} = E_1 - E_2 - T_0 (S_1 - S_2)$$
$$= (E_1 - T_0S_1) - (E_2 - T_0S_2) \tag{8.30}$$

If the K.E. and P.E. changes are neglected, Eq. (8.30) reduces to

$$W_{max} = (U_1 - T_0S_1) - (U_2 - T_0S_2) \tag{8.31}$$

For unit mass of fluid,

$$W_{max} = (u_1 - u_2) - T_0 (s_1 - s_2)$$
$$= (u_1 - T_0s_1) - (u_2 - T_0s_2) \tag{8.32}$$

8.6 USEFUL WORK

All of the work W of the system with a flexible boundary would not be available for delivery, since a certain portion of it would be spent in pushing out the atmosphere (Fig. 8.12). The useful work is defined as the actual work delivered by a system less the work performed on the atmosphere. If V_1 and V_2 are the initial and final volume of the system and p_0 is the atmospheric pressure, then the work done on the atmosphere is $p_0 (V_2 - V_1)$. Therefore, the useful work W_u becomes

$$W_u = W_{act} - p_0 (V_2 - V_1) \tag{8.33}$$

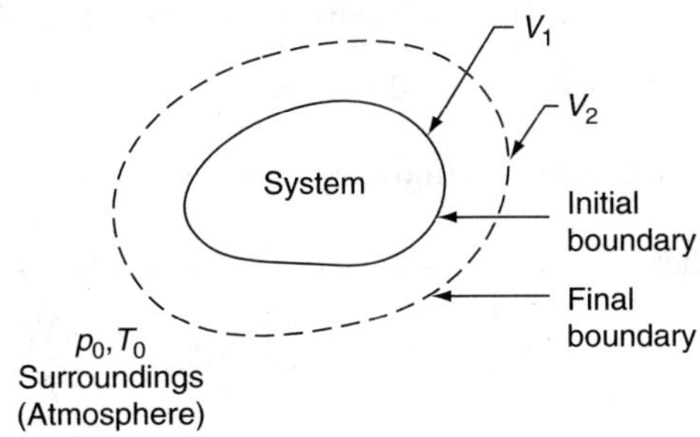

Fig. 8.12 *Work done by a closed system in pushing out the atmosphere*

Similarly, the *maximum useful work* will be

$$(W_u)_{max} = W_{max} - p_0 (V_2 - V_1) \tag{8.34}$$

In differential form

$$(dW_u)_{max} = dW_{max} - p_0 \, dV \tag{8.35}$$

In a steady flow system, the volume of the system does not change. Hence, the maximum useful work would remain the same, i.e., *no work is done on the atmosphere,* or

$$(đW_u)_{max} = dW_{max} \tag{8.36}$$

But in the case of an unsteady-flow open system or a closed system, the volume of the system changes. Hence, when a system exchanges heat only with the atmosphere, the maximum useful work becomes

$$(đW_u)_{max} = đW_{max} - p_0 \, dV$$

Substituting dW_{max} from Eq. (8.23),

$$(đW_u)_{max} = dm_1 \left(h_1 - T_0s_1 + \frac{V_1^2}{2} + gz_1\right) - dm_2 \left(h_2 - T_0s_2 + \frac{V_2^2}{2} + gz_2\right)$$
$$- d\left[U + p_0V - T_0S + \frac{mV^2}{2} + mgz\right]_\sigma \tag{8.37}$$

This is the maximum useful work for an *unsteady open system*. For the closed system, Eq. (8.37) reduces to

$$(dW_u)_{max} = -d\left[U + p_0V - T_0S + \frac{mV^2}{2} + mgz\right]_\sigma = -d[E + p_0V - T_0S]_\sigma \tag{8.38}$$

$$\therefore \quad (W_u)_{max} = E_1 - E_2 + p_0 (V_1 - V_2) - T_0(S_1 - S_2) \tag{8.39}$$

If K.E. and P.E. changes are neglected, Eq. (8.39) becomes

$$(W_u)_{max} = U_1 - U_2 + p_0 (V_1 - V_2) - T_0 (S_1 - S_2) \tag{8.40}$$

This can also be written in the following form

$$(W_u)_{max} = (U_1 + p_0V_1 - T_0S_1) - (U_2 + p_0V_2 - T_0S_2) = \phi_1 - \phi_2 \tag{8.41}$$

where ϕ is called the *availability function for a closed system* given by

$$\phi = U + p_0V - T_0S$$

The useful work per unit mass becomes

$$(W_u)_{max} = (u_1 + p_0v_1 - T_0s_1) - (u_2 + p_0v_2 - T_0s_2) \tag{8.42}$$

8.6.1 Maximum Useful Work Obtainable when the System Exchanges Heat with a Thermal Reservoir in Addition to the Atmosphere

If the open system discussed in Sec. 8.5 exchanges heat with a thermal energy reservoir at temperature T_R in addition to the atmosphere, the *maximum useful work* will be increased by $đQ_R\left(1-\frac{T_0}{T_R}\right)$, where $đQ_R$ is the heat received by the system. For a steady flow process,

$$(W_u)_{max} = W_{max} = \left(H_1 - T_0S_1 + \frac{mV_1^2}{2} + mgz_1\right) - \left(H_2 - T_0S_2 + \frac{mV_2^2}{2} + mgz_2\right) + Q_R\left(1-\frac{T_0}{T_R}\right)$$

$$= \psi_1 - \psi_2 + Q_R\left(1-\frac{T_0}{T_R}\right) \tag{8.43}$$

For a closed system

$$(W_u)_{max} = W_{max} - p_0(V_2 - V_1) + Q_R\left(1-\frac{T_0}{T_R}\right)$$

or

$$(W_u)_{max} = E_1 - E_2 + p_0(V_1 - V_2) - T_0(S_1 - S_2) + Q_R\left(1-\frac{T_0}{T_R}\right) \tag{8.44}$$

If K.E. and P.E. changes are neglected, then for a *steady flow process*:

$$(W_u)_{max} = (H_1 - H_2) - T_0 (S_1 - S_2) + Q_R\left(1-\frac{T_0}{T_R}\right) \tag{8.45}$$

and for a *closed system:*

$$(W_u)_{max} = U_1 + U_2 - p_0(V_1 - V_2) - T_0(S_1 - S_2) + Q_R\left(1-\frac{T_0}{T_R}\right) \tag{8.46}$$

8.7 DEAD STATE

If the state of a system departs from that of the surroundings, an opportunity exists for producing work (Fig. 8.13). However, as the system changes its state towards that of the surroundings, this opportunity diminishes, and it ceases to exist when the two are in equilibrium with each other. When the system is in equilibrium with the surroundings, it must be in pressure and temperature equilibrium with the surroundings, i.e., at p_0 and T_0. It must also be in chemical equilibrium with the surroundings, i.e., there should not be any chemical reaction or mass transfer. The system must have zero velocity and minimum potential energy. This state of the system is known as the *dead state*, which is designated by affixing subscript '0' to the properties. Any change in the state

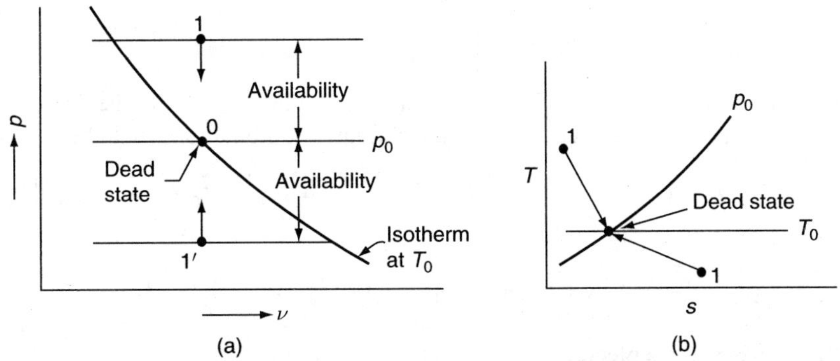

Fig. 8.13 *Available work of a system decreases as its state approaches P_0, T_0*

of the system from the dead state is a measure of the available work that can be extracted from it. Farther the initial point of the system from the dead state in terms of p, t either above or below it, higher will be the available energy or exergy of the system (Fig. 8.13). All spontaneous processes terminate at the dead state.

8.8 AVAILABILITY

Whenever useful work is obtained during a process in which a finite system undergoes a change of state, the process must terminate when the pressure and temperature of the system have become equal to the pressure and temperature of the surroundings, p_0 and T_0, i.e., when the system has reached the dead state. An air engine operating with compressed air taken from a cylinder will continue to deliver work till the pressure of air in the cylinder becomes equal to that of the surroundings, p_0. A certain quantity of exhaust gases from an internal combustion engine used as the high temperature source of a heat engine will deliver work until the temperature of the gas becomes equal to that of the surroundings, T_0.

The availability (A) of a given system is defined as *the maximum useful work* (*total work minus pdV work*) *that is obtainable in a process in which the system comes to equilibrium with its surroundings*. Availability is thus a composite property depending on the state of both the system and surroundings.

8.8.1 Availability in a Steady Flow Process

The reversible (maximum) work associated with a steady flow process for a single flow is given by Eq. (8.25)

$$W_{rev} = \left(H_1 - T_0 S_1 + \frac{m\mathrm{V}_1^2}{2} + mgz_1\right) - \left(H_2 - T_0 S_2 + \frac{m\mathrm{V}_2^2}{2} + mgz_2\right)$$

With a given state for the mass entering the control volume, the maximum useful work obtainable (i.e., the availability) would be when this mass leaves the control volume in equilibrium with the surroundings (i.e., at the dead state). Since there is no change in volume, no work will be done on the atmosphere. Let us designate the initial state of the mass entering the C.V. with parameters having *no subscript* and the final dead state of the mass leaving the C.V. with parameters having subscript 0. The maximum work or availability, A, would be

$$A = \left(H - T_0 S + \frac{m\mathrm{V}^2}{2} + mgz\right) - (H_0 - T_0 S_0 + mgz_0) = \psi - \psi_0 \quad (8.47)$$

where ψ is called the *availability function for a steady flow system* and $\mathrm{V}_0 = 0$. This is the availability of a system at any state as it enters a C.V. in a steady flow process. The availability per unit mass would be

$$a = \left(h - T_0 s + \frac{V^2}{2} + gz\right) - (h_0 - T_0 s_0 + gz) = \psi - \psi_0 \quad (8.48)$$

If subscripts 1 and 2 denote the states of a system entering and leaving a C.V., the decrease in availability or maximum work obtainable for the given system-surroundings combination would be

$$W_{max} = a_1 - a_2 = \psi_1 - \psi_2 = \left[\left(h_1 - T_0 s_1 + \frac{V_1^2}{2} + gz_1\right) - (h_0 - T_0 s_0 + gz_0)\right]$$

$$- \left[\left(h_2 - T_0 s_0 + \frac{V_2^2}{2} + gz_2\right) - (h_0 - T_0 s_0 + gz_0)\right] = (h_1 - h_2) - T_0(s_1 - s_2) + \frac{V_1^2 - V_2^2}{2} + g(z_1 - z_2) \quad (8.49)$$

If K.E. and P.E. changes are neglected,

$$W_{max} = (h_1 - T_0 s_1) - (h_2 - T_0 s_2) = b_1 - b_2$$

where b is the specific Keenan function.

If more than one flow into and out of the C.V. is involved.

$$W_{max} = \sum_i m_i \psi_i - \sum_i m_e \psi_e$$

8.8.2 Availability in a Nonflow Process

Let us consider a closed system and denote its initial state by parameters without any subscript and the final dead state with subscript '0'. The availability of the system A, i.e., the maximum useful work obtainable as the system reaches the dead state, is given by Eq. (8.39).

$$A = (W_u)_{max} = E - E_0 + p_0(V - V_0) - T_0(S - S_0)$$

$$= \left(U + \frac{mV^2}{2} + mgz\right) - (U_0 + mgz_0) + p_0(V - V_0) - T_0(S - S_0) \quad (8.50)$$

If K.E. and P.E. changes are neglected and for unit mass, the availability becomes

$$a = u - u_0 + p_0(v - v_0) - T_0(s - s_0) = (u + p_0 v - T_0 s) - (u_0 + p_0 v_0 - T_0 s_0) = \phi - \phi_0 \quad (8.51)$$

where ϕ is the availability function of the closed system.

If the system undergoes a change of state from 1 to 2, the decrease in availability will be

$$a = (\phi_1 - \phi_0) - (\phi_2 - \phi_0) = \phi_1 - \phi_2 = (u_1 - u_2) + p_0(v_1 - v_2) - T_0(s_1 - s_2) \quad (8.52)$$

This is the maximum useful work obtainable under the given surroundings.

8.9 AVAILABILITY IN CHEMICAL REACTIONS

In many chemical reactions the reactants are often in pressure and temperature equilibrium with the surroundings (before the reaction takes place) and so are the products after the reaction. An internal combustion engine can be cited as an example of such a process if we visualize the products being cooled to atmospheric temperature T_0 before being discharged from the engine.

(a) Let us first consider a system which is in temperature equilibrium with the surroundings before and after the process. The maximum work obtainable during a change of state is given by Eq. (8.30),

$$W_{max} = E_1 - E_2 - T_0(S_1 - S_2) = \left(U_1 + \frac{mV_1^2}{2} + mgz_1\right) - \left(U_2 + \frac{mV_2^2}{2} + mgz_2\right) - T_0(S_1 - S_2)$$

If K.E. and P.E. changes are neglected,

$$W_{max} = U_1 - U_2 - T_0(S_1 - S_2)$$

Since the initial and final temperatures of the system are the same as that of the surroundings, $T_1 = T_2 = T_0 = T$, say, then

$$(W_T)_{max} = (U_1 - U_2)_T - T(S_1 - S_2)_T \tag{8.53}$$

Let a property called *Helmholtz function F* be defined by the relation

$$F = U - TS \tag{8.54}$$

Then for two equilibrium states 1 and 2 at the same temperature T,

$$(F_1 - F_2)_T = (U_1 - U_2)_T - T(S_1 - S_2)_T \tag{8.55}$$

From Eqs (8.53) and (8.55),

$$(W_T)_{max} = (F_1 - F_2)_T \tag{8.56}$$

or

$$W_T \le (F_1 - F_2)_T \tag{8.57}$$

The work done by a system in any process between two equilibrium states at the same temperature during which the system exchanges heat only with the environment is equal to or less than the decrease in the Helmholtz function of the system during the process. The maximum work is done when the process is reversible and the equality sign holds. If the process is irreversible, the work is less than the maximum.

(b) Let us now consider a system which is in both pressure and temperature equilibrium with the surroundings before and after the process. When the volume of the system increases some work is done by the system against the surroundings ($p\mathrm{d}V$ work), and this is not available for doing useful work. The availability of the system, as defined by Eq. (8.50), neglecting the K.E. and P.E. changes, can be expressed in the form

$$A = (W_u)_{max} = (U + p_0 V - T_0 S) - (U_0 + p_0 V_0 - T_0 S_0) = \phi - \phi_0$$

The maximum work obtainable during a change of state is the decrease in availability of the system, as given by Eq. (8.52) for unit mass.

$$\therefore \quad (W_u)_{max} = A_1 - A_2 = \phi_1 - \phi_2 = (U_1 - U_2) + p_0(V_1 - V_2) - T_0(S_1 - S_2)$$

If the initial and final equilibrium states of the system are at the same pressure and temperature of the surroundings, say $p_1 = p_2 = p_0 = p$, and $T_1 = T_2 = T_0 = T$. Then,

$$(W_u)_{max} = (U_1 - U_2)_{p,T} + p(V_1 - V_2)_{p,T} - T(S_1 - S_2)_{p,T} \tag{8.58}$$

The *Gibbs function G* is defined as

$$G = H - TS = U + pV - TS \tag{8.59}$$

Then for two equilibrium states at the same pressure p and temperature T

$$(G_1 - G_2)_{p,T} = (U_1 - U_2)_{p,T} + p(V_1 - V_2)_{p,T} - T(S_1 - S_2)_{p,T} \tag{8.60}$$

From Eqs (8.58) and (8.60),

$$(W_u)_{\substack{max \\ p,T}} = (G_1 - G_2)_{p,T} \tag{8.61}$$

$$\therefore \quad (W_u)_{p,T} \le (G_1 - G_2)_{p,T} \tag{8.62}$$

The decrease in the Gibbs function of a system sets an upper limit to the work that can be performed, exclusive of $p\mathrm{d}V$ work, in any process between two equilibrium states at the same temperature and pressure, provided the system exchanges heat only with the environment which is at the same temperature and pressure as the end states of the system. If the process is irreversible, the useful work is less than the maximum.

8.10 IRREVERSIBILITY AND GOUY-STODOLA THEOREM

The actual work done by a system is always less than the idealized reversible work, and the difference between the two is called the irreversibility of the process.

$$I = W_{max} - W \tag{8.63}$$

This is also sometimes referred to as '*degradation*' or '*dissipation*'.

For a non-flow process between the equilibrium states, when the system exchanges heat only with the environment

$$I = [(U_1 - U_2) - T_0(S_1 - S_2)] - [(U_1 - U_2) + Q] = T_0(S_2 - S_1) - Q = T_0(\Delta S)_{\text{system}} + T_0(\Delta S)_{\text{surr}}$$

$$= T_0[(\Delta S)_{\text{system}} + (\Delta S)_{\text{surr}}] \tag{8.64}$$

$$\therefore \quad I \geq 0$$

Similarly, for the steady flow process

$$I = W_{\max} - W = \left[\left(B_1 + \frac{mV_1^2}{2} + mgZ_1\right) - \left(B_2 + \frac{mV^2}{2} + mgZ_2\right)\right]$$

$$- \left[\left(H_1 + \frac{mV_1^2}{2} + mgZ_1\right) - \left(H_2 + \frac{mV_2^2}{2} + mgZ_2\right) + Q\right] = T_0(S_2 - S_1) - Q = T_0(\Delta S)_{\text{system}} + T_0(\Delta S)_{\text{surr}}$$

$$= T_0(\Delta S_{\text{system}} + S_{\text{surr}}) = T_0 \Delta S_{\text{univ}} \tag{8.65}$$

The same expression for irreversibility applies to both flow and non-flow processes. The quantity $T_0(\Delta S_{\text{system}} + \Delta S_{\text{surr}})$ represents an increase in unavailable energy (or anergy).

The *Gouy-Stodola theorem* states that the rate of loss of available energy or exergy in a process is proportional to the rate of entropy generation, S_{gen}. If Eqs (8.64) and (8.65) are written in the rate form,

$$\dot{I} = \dot{W}_{\text{lost}} = T_0 \Delta \dot{S}_{\text{univ}} = T_0 \dot{S}_{\text{gen}} \tag{8.66}$$

This is known as the Gouy-Stodola equation. A thermodynamically efficient process would involve minimum exergy loss with minimum rate of entropy generation.

8.10.1 Applications of Gouy-Stodola Equation

(a) Heat Transfer through a Finite Temperature Difference If heat transfer $\dot{Q}$ occurs from the hot reservoir at temperature T_1 to the cold reservoir at temperature T_2 (Fig. 8.14a)

$$S_{\text{gen}} = \frac{\dot{Q}}{T_2} - \frac{\dot{Q}}{T_1} = \dot{Q}\,\frac{T_1 - T_2}{T_1 T_2} \quad \text{and} \quad \dot{W}_{\text{lost}} = \dot{Q}\left(1 - \frac{T_2}{T_1}\right) = \dot{Q}\,\frac{T_1 - T_2}{T_1}$$

$$\therefore \quad W_{\text{lost}} = T_2\, S_{\text{gen}}$$

If the heat transfer $\dot{Q}$ from T_1 to T_2 takes place through a reversible engine E, the entire work output is $\dot{W}$ dissipated in the brake, from which an equal amount of heat is rejected to the reservoir at T_2 (Fig. 8.14b). Heat transfer through a finite temperature difference is equivalent to the destruction of its availability.

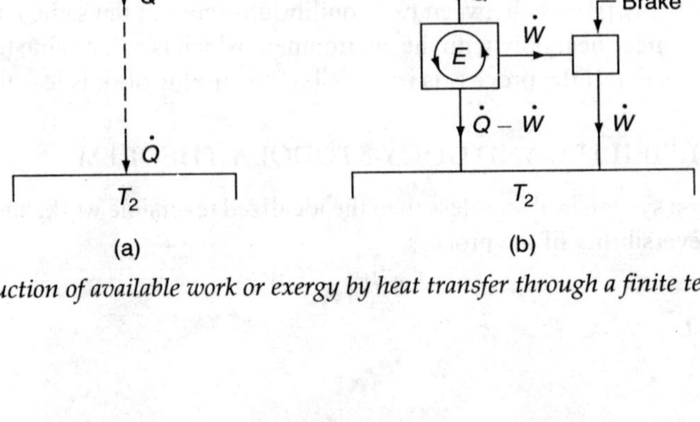

Fig. 8.14 *Destruction of available work or exergy by heat transfer through a finite temperature difference*

(b) *Flow with Friction* Let us consider the steady and adiabatic flow of an *ideal gas* through the segment of a pipe (Fig. 8.15a).

By the first law,

$$h_1 = h_2$$

and by the second law,

$$T\mathrm{d}s = \mathrm{d}h - v\mathrm{d}p$$

$$\int_1^2 \mathrm{d}s = \int_1^2 \frac{\mathrm{d}h}{T} - \int_1^2 \frac{v}{T}\,\mathrm{d}p = -\int_1^2 \frac{v}{T}\,\mathrm{d}p$$

$$\therefore \quad \dot{S}_{gen} = \int_1^2 \dot{m}\mathrm{d}s = -\int_{p1}^{p2} \dot{m}R\frac{\mathrm{d}p}{p} = -\dot{m}R\ln\frac{p_2}{p_1} = -\dot{m}R\ln\left(1-\frac{\Delta p}{p_1}\right) = -\dot{m}R\left(-\frac{\Delta p}{p_1}\right)$$

$$= \dot{m}R\frac{\Delta p}{p_1} \quad (8.67)$$

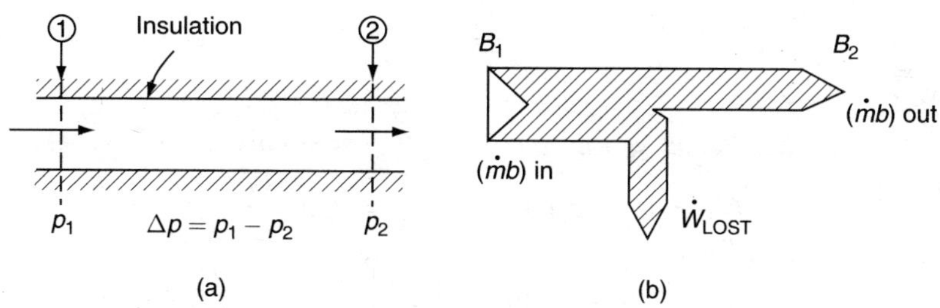

Fig. 8.15 *Irreversibility in a duct due to fluid friction*

where $\ln\left(1-\dfrac{\Delta p}{p_1}\right) \cong -\dfrac{\Delta p}{p_1}$, since $\dfrac{\Delta p}{p_1} < 1$

and higher terms are neglected.

$$\dot{W}_{lost} = \dot{B}_1 - \dot{B}_2 = \dot{m}[(h_1 - T_0 s_1) - (h_2 - T_0 s_2)] = \dot{m}T_0(s_2 - s_1) = T_0\,\dot{S}_{gen} = \dot{m}RT_0\frac{\Delta p}{p_1} \quad (8.68)$$

The decrease in availability or lost work is proportional to the pressure drop (Δp) and the mass flow rate ($\dot{m}$). It is shown on the right (Fig. 8.15b) by the *Grassmann diagram*, the width being proportional to the availability (or exergy) of the stream. It is an adaptation of the *Sankey diagram* used for energy transfer in a plant.

(c) *Mixing of Two Fluids* Two streams 1 and 2 of an incompressible fluid or an ideal gas mix adiabatically at constant pressure (Fig. 8.16).

Here, $\dot{m}_1 + \dot{m}_2 = \dot{m}_3 = \dot{m}$ (say)

Let $x = \dfrac{\dot{m}_1}{\dot{m}_1 + \dot{m}_2}$

By the first law,

$$\dot{m}_1 h_1 + \dot{m}_2 h_2 = (\dot{m}_1 + \dot{m}_2)h_3 \quad \text{or} \quad xh_1 + (1-x)h_2 = h_3$$

The preceding equation may be written in the following form, since enthalpy is a function of temperature.

$$xT_1 + (1-x)T_2 = T_3$$

$$\therefore \quad \frac{T_3}{T_1} = x + (1-x)\tau \quad (8.69)$$

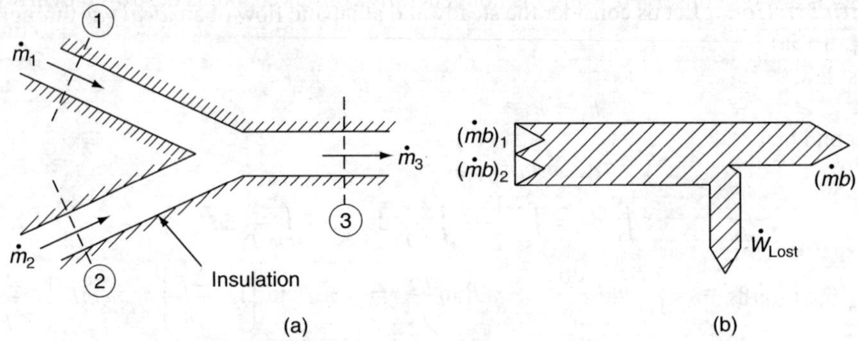

Fig. 8.16 *Irreversibility due to mixing*

where
$$\tau = \frac{T_2}{T_1}$$

By the second law,

$$\dot{S}_{gen} = \dot{m}_3 s_3 - \dot{m}_1 s_1 - \dot{m}_2 s_2 = \dot{m} s_3 - x \dot{m} s_1 - (1-x)\dot{m} s_2$$

$$\therefore \quad \frac{\dot{S}_{gen}}{\dot{m}} = (s_3 - s_2) + x(s_2 - s_1) = c_p \ln \frac{T_3}{T_2} + x c_p \ln \frac{T_2}{T_1}$$

or
$$\frac{\dot{S}_{gen}}{\dot{m} c_p} = \ln \left(\frac{T_3}{T_2}\right)\left(\frac{T_2}{T_1}\right)^{x}$$

or
$$N_s = \ln \left(\frac{T_3}{T_1}\right) \frac{T_1^{1-x}}{T_2^{1-x}}$$

or
$$N_s = \ln \frac{T_3/T_1}{(T_2/T_1)^{1-x}} \quad (8.70)$$

where N_s is a dimensionless quantity, called the *entropy generation number*, given by $\dot{S}_{gen}/\dot{m}c_p$.

Substituting T_3/T_1 from Eq. (8.69) in Eq. (8.70),

$$N_s = \ln \frac{x + \tau(1-x)}{\tau^{1-x}} \quad (8.71)$$

If $x = 1$ or $\tau = 1$, N_s becomes zero in each case. The magnitude of N_s depends on x and τ. The rate of loss of exergy due to mixing would be

$$\dot{W}_{lost} = \dot{I} = T_0 \dot{m} c_p \ln \frac{x + \tau(1-x)}{\tau^{1-x}} \quad (8.72)$$

8.11 AVAILABILITY OR EXERGY BALANCE

The availability or exergy is the maximum useful work obtainable from a system as it reaches the dead state (p_0, t_0). Conversely, availability or exergy can be regarded as the minimum work required to bring the closed system from the dead state to the given state. The value of exergy cannot be negative. If a closed system were at any state other than the dead state, the system would be able to change its state spontaneously toward the dead state. This tendency would stop when the dead state is reached. No work is done to effect such a spontaneous change. Since any change in state of the closed system to the dead state can be accomplished with zero work, the *maximum work* (*or exergy*) *cannot be negative.*

While energy is always conserved, exergy is not generally conserved, but is destroyed by irreversibilities. When the closed system is allowed to undergo a spontaneous change from the given state to the dead state, its exergy is completely destroyed without producing any useful work. The potential to develop work that exists originally at the given state is thus completely wasted in such a spontaneous process. Therefore, at steady state:

1. Energy in – Energy out = 0
2. Exergy in – Exergy out = Exergy destroyed

8.11.1 Exergy Balance for a Closed System

For a closed system, availability or exergy transfer occurs through heat and work interactions (Fig. 8.17).

1st law:
$$E_2 - E_1 = \int_1^2 đQ - W_{1-2} \tag{8.73}$$

2nd law:
$$S_2 - S_1 - \int_1^2 \left[\frac{đQ}{T}\right] = S_{gen}$$

or
$$T_0(S_2 - S_1) - T_0\int_1^2 \left[\frac{đQ}{T}\right]_\sigma = T_0 S_{gen} \tag{8.74}$$

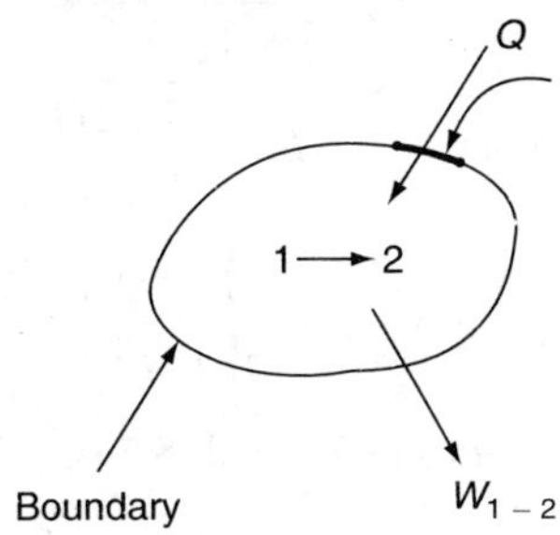

Fig. 8.17 *Exergy balance for a closed system*

Subtracting Eq. (8.74) from Eq. (8.73),

$$E_2 - E_1 - T_0(S_2 - S_1) = \int_1^2 đQ - W_{1-2} - T_0\int_1^2 \left[\frac{đQ}{T}\right] - T_0 S_{gen}$$

$$= \int_1^2 \left[1 - \frac{T_0}{T_\sigma}\right] đQ - W_{1-2} - T_0 S_{gen}$$

Since,

$$A_2 - A_1 = E_2 - E_1 + p_0(V_2 - V_1) - T_0(S_2 - S_1)$$

$$A_2 - A_1 = \int_1^2 \left[1 - \frac{T_0}{T_\sigma}\right] đQ - [W_{1-2} - p_0(V_2 - V_1)] - T_0 S_{gen} \tag{8.75}$$

Change in exergy; Exergy transfer with heat; Exergy transfer with work; Exergy destruction

In the form of the rate equation,

$$\frac{dA}{d\tau} = \sum_j \left[1 - \frac{T_0}{T_j}\right]\dot{Q}_j - \left[W - p_0\frac{dV}{d\tau}\right] - \dot{I} \tag{8.76}$$

Rate of change of exergy; Rate of exergy transfer with heat Q_j at the boundary where the instantaneous temperature is T_j; Rate of exergy transfer as work where dV/dt is the rate of change of system volume; Rate of exergy loss due to irreversibilities $(= T_0\dot{S}_{gen})$

For an isolated system, the exergy balance, Eq. (8.76), gives

$$\Delta A = -I \tag{8.77}$$

Since $I > 0$, the only processes allowed by the second law are those for which the exergy of the isolated system decreases. In other words,

The exergy of an isolated system can never increase.

It is the *counterpart of the entropy principle* which states that the entropy of an isolated system can never decrease.

The exergy balance of a system can be used to determine the locations, types and magnitudes of losses (waste) of the potential of energy resources (fuels) and find ways and means to reduce these losses for making the system more energy efficient and for more effective use of fuel.

8.11.2 Exergy Balance for a Steady Flow System

1st law: $$H_1 + \frac{mV_1^2}{2} + mgZ_1 + Q_{1-2}$$

$$= H_2 + \frac{mV_2^2}{2} + mgZ_2 + W_{1-2} \quad (8.78)$$

2nd law: $$S_1 + \int_1^2 \left[\frac{đQ}{T}\right] - S_2 = S_{gen}$$

or

$$T_0(S_1 - S_2) + T_0 \int_1^2 \left[\frac{đQ}{T}\right]_\sigma = T_0 S_{gen} = I \quad (8.79)$$

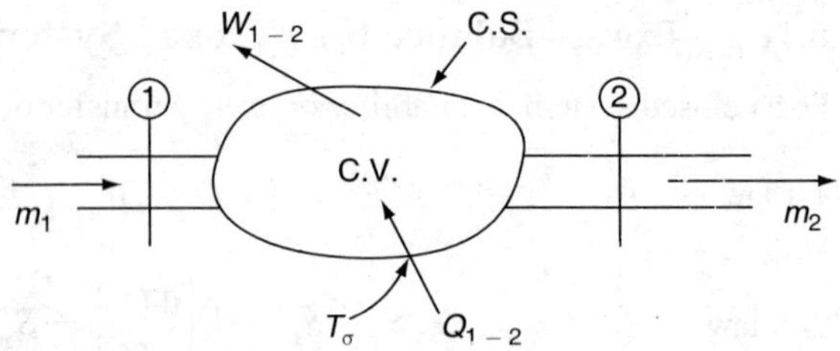

Fig. 8.18 *Exergy balance for a steady flow system*

From Eqs (8.78) and (8.79),

$$H_2 - H_1 - T_0(S_2 - S_1) + \frac{m(V_2^2 - V_1^2)}{2} + mg(Z_2 - Z_1) = \int_1^2 \left(1 - \frac{T_0}{T_\sigma}\right) đQ - W_{1-2} - I \quad (8.80)$$

or

$$A_2 - A_1 = \int_1^2 \left(1 - \frac{T_0}{T_\sigma}\right) đQ - W_{1-2} - I \quad (8.81)$$

In the form of rate equation at steady state:

$$\sum_j \left(1 - \frac{T_0}{T_j}\right) \dot{Q}_j - \dot{W}_{C.V.} + \dot{m}(a_{f_1} - a_{f_2}) - \dot{I}_{C.V.} = 0 \quad (8.82)$$

where $a_{f_1} - a_{f_2} = (h_1 - h_2) - T_0(s_1 - s_2) + \frac{V_1^2 - V_2^2}{2} + g(Z_1 - Z_2)$ and $[1 - T_0/T_j]\,\dot{Q}_j$ = time rate of exergy transfer along with heat $\dot{Q}_j$ occurring at the location on the boundary where the instantaneous temperature is T_j.

For a single stream entering and leaving, the exergy balance gives

$$\underbrace{\left[1 - \frac{T_0}{T_\sigma}\right]\frac{\dot{Q}}{\dot{m}} + a_{f_1}}_{\text{Exergy in}} = \underbrace{\frac{\dot{W}}{\dot{m}} - a_{f_2}}_{\text{Exergy out}} = \underbrace{\frac{\dot{I}}{\dot{m}}}_{\text{Exergy loss}} \quad (8.83)$$

8.12 SECOND LAW EFFICIENCY

A common measure on energy use efficiency is the first law efficiency, η_I. The first law efficiency is defined as the ratio of the output energy of a device to the input energy of the device. The first law is concerned only with the quantities of energy, and disregards the forms in which the energy exists. It does not also discriminate between the energies available at different temperatures. It is the second law of thermodynamics which provides a means of assigning a quality index to energy. The concept of available energy or exergy provides a useful measure of energy quality (Sec. 8.3).

With this concept it is possible to analyze means of minimizing the consumption of available energy to perform a given process, thereby ensuring the most efficient possible conversion of energy for the required task.

The second law efficiency, η_{II}, of a process is defined as the ratio of the minimum available energy (or exergy) which must be consumed to do a task divided by the actual amount of available energy (or exergy) consumed in performing the task.

$$\eta_{II} = \frac{\text{minimum exergy intake to perform the given task}}{\text{actual exergy intake to perform the same task}}$$

or

$$\eta_{II} = \frac{A_{min}}{A} \tag{8.84}$$

where A is the availability or exergy.

A power plant converts a fraction of available energy A or W_{max} to useful work W. For the desired output of W, $A_{min} = W$ and $A = W_{max}$. Here,

$$I = W_{max} - W \quad \text{and} \quad \eta_{II} = \frac{W}{W_{max}} \tag{8.85}$$

Now

$$\eta_I = \frac{W}{Q_1} = \frac{W}{W_{max}} \cdot \frac{W_{max}}{Q_1} = \eta_{II} \cdot \eta_{Carnot} \tag{8.86}$$

$\therefore$

$$\eta_{II} = \frac{\eta_I}{\eta_{Carnot}} \tag{8.87}$$

Since $W_{max} = Q_1\left(1 - \frac{T_0}{T}\right)$, Eq. (8.87) can also be obtained directly as follows

$$\eta_{II} = \frac{W}{Q_1\left(1 - \frac{T_0}{T}\right)} = \frac{\eta_I}{\eta_{Carnot}}$$

If work is involved, $A_{min} = W$ (desired) and if heat is involved, $A_{min} = Q\left(1 - \frac{T_0}{T}\right)$.

If solar energy Q_r is available at a reservoir storage temperature T_r and if quantity of heat Q_a is transferred by the solar collector at temperature T_a, then

$$\eta_I = \frac{Q_a}{Q_r}$$

and

$$\eta_{II} = \frac{\text{exergy output}}{\text{exergy input}} = \frac{Q_a\left(1 - \frac{T_0}{T_a}\right)}{Q_r\left(1 - \frac{T_0}{T_r}\right)} = \eta_I \frac{1 - \frac{T_0}{T_a}}{1 - \frac{T_0}{T_r}} \tag{8.88}$$

Table 8.1 shows availabilities, and both η_I and η_{II} expressions for several common thermal tasks.

Table 8.1

$T_p > T_a > T_0 > T_c$

Task	*Energy input*	
	Input shaft work, W_i	*Q from reservoir at T*
Produce work, W_0	$A = W_i$	$A = Q_r\left(1 - \frac{T_0}{T_r}\right)$
	$A_{min} = W_0$	$A_{min} = W_0$
	$\eta_I = \frac{W_0}{W_i}$	$\eta_I = \frac{W_0}{Q_r}$

Table 8.1 *(Continued)*

Task	*Energy input*	
	Input shaft work, W_i	*Q_r from reservoir at T_r*
	$\eta_{II} = \dfrac{A_{min}}{A}$	$\eta_{II} = \eta_I \cdot \dfrac{1}{1 - \dfrac{T_0}{T_r}}$
	$\eta_{II} = \eta_I$ (electric motor)	(heat engine)
Add heat Q_a to reservoir at T_a	$A = W_i$	$A = Q_r\left(1 - \dfrac{T_0}{T_r}\right)$
	$A_{min} = Q_a\left(1 - \dfrac{T_0}{T_a}\right)$	$A_{min} = Q_a\left(1 - \dfrac{T_0}{T_a}\right)$
	$^*\eta_I = \dfrac{Q_a}{W_i}$	$\eta_I = \dfrac{Q_a}{Q_r}$
	$^*\eta_{II} = \eta_I\left(1 - \dfrac{T_0}{T_a}\right)$	$\eta_{II} = \eta_I \dfrac{1 - \dfrac{T_0}{T_a}}{1 - \dfrac{T_0}{T_r}}$
	(heat pump)	(solar water heater)
Extract heat Q_c from cold reservoir at T_c (below ambient)	$A = W_i$	$A = Q_r\left(1 - \dfrac{T_0}{T_r}\right)$
	$A_{min} = Q_c\left(\dfrac{T_0}{T_c} - 1\right)$	$A_{min} = Q_c\left(\dfrac{T_0}{T_c} - 1\right)$
	$^*\eta_I = \dfrac{Q_c}{W_i}$	$^*\eta_I = \dfrac{Q_c}{Q_r}$
	$^*\eta_{II} = \eta_I\left(\dfrac{T_0}{T_c} - 1\right)$	$^*\eta_{II} = \eta_I\left(\dfrac{\dfrac{T_0}{T_c} - 1}{1 - \dfrac{T_0}{T_r}}\right)$
	(Refrigerator-electric motor driven)	(Refrigerator-heat operated)

* Strictly speaking, it is COP.

In the case of a heat pump, the task is to add heat Q_a to a reservoir to be maintained at temperature T_a and the input shaft work is W_i.

$$\text{COP} = \frac{Q_a}{W_i} = \eta_I, \text{ say } (\text{COP})_{max} = \frac{T_a}{T_a - T_0} = \frac{Q_a}{W_i} = \frac{Q_a}{A_{min}}$$

$$\therefore \quad A_{min} = Q_a\left(1 - \frac{T_0}{T_a}\right)$$

$$\therefore \quad \eta_{II} = \frac{A_{min}}{A} = \frac{Q_a\left(1 - \frac{T_0}{T_a}\right)}{W_i} \qquad \eta_{II} = \eta_I\left(1 - \frac{T_0}{T_a}\right) \tag{8.89}$$

Similarly, expressions of η_I and η_{II} can be obtained for other thermal tasks.

8.12.1 Matching End Use to Source

Combustion of a fuel releases the necessary energy for the tasks, such as space heating, process steam generation and heating in industrial furnaces. When the products of combustion are at a temperature much greater than that required by a given task, the end use is not well matched to the source and results in inefficient utilization of the fuel burned. To illustrate this, let us consider a closed system receiving a heat transfer $\dot{Q}_r$ at a source temperature T_r and delivering $\dot{Q}_a$ at a use temperature T_a (Fig. 8.19). Energy is lost to the surroundings by heat transfer at a rate $\dot{Q}_l$ across a portion of the surface at T_l. At a steady state the energy and availability rate balances become.

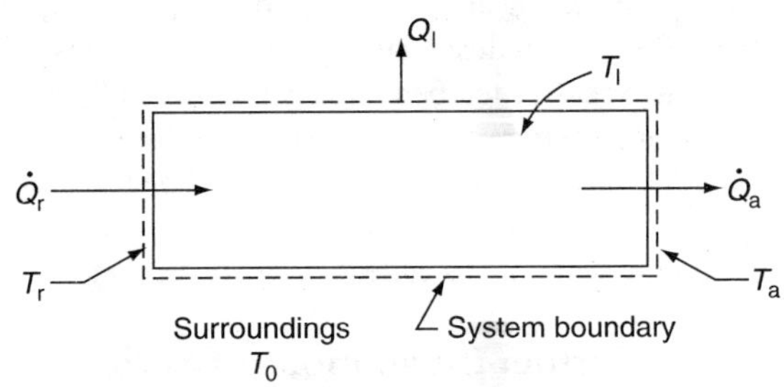

Fig. 8.19 *Efficient energy utilization from second law viewpoint*

$$\dot{Q}_r = \dot{Q}_a + \dot{Q}_l \tag{8.90}$$

$$\dot{Q}_r\left(1 - \frac{T_0}{T_r}\right) = \dot{Q}_a\left(1 - \frac{T_0}{T_a}\right) + \dot{Q}_l\left(1 - \frac{T_0}{T_l}\right) + \dot{I} \tag{8.91}$$

Equation (8.90) indicates that the energy carried in by heat transfer $\dot{Q}_r$ is either used, $\dot{Q}_a$, or lost to the surroundings, $\dot{Q}_l$. Then

$$\eta_I = \frac{\dot{Q}_a}{\dot{Q}_r} \tag{8.92}$$

The value of η_I can be increased by increasing insulation to reduce the loss. The limiting value, when $\dot{Q}_l = 0$, is $\eta_I = 1$ (100%).

Equation (8.91) shows that the availability carried into the system accompanying the heat transfer $\dot{Q}_r$ is either transferred from the system accompanying the heat transfer $\dot{Q}_a$ and $\dot{Q}_l$ or destroyed by irreversibilities within the system, $\dot{I}$. Therefore,

$$\eta_{II} = \frac{\dot{Q}_a\left(1 - \frac{T_0}{T_a}\right)}{\dot{Q}_r\left(1 - \frac{T_0}{T_r}\right)} = \eta_I \frac{1 - \frac{T_0}{T_a}}{1 - \frac{T_0}{T_r}} \tag{8.93}$$

Both η_I and η_{II} indicate how effectively the input is converted into the product. The parameter η_I does this on energy basis, whereas η_{II} does it on an availability or exergy basis.

For proper utilization of exergy, it is desirable to make η_I as close to unity as practical and also a good match between the source and use temperatures, T_r and T_a. Figure 8.20 demonstrates the second law efficiency against the use temperature T_a for an assumed source temperature $T_r = 2200$ K. It shows that η_{II} tends to unity (100%) as T_a approaches T_r. The lower the T_a, the lower becomes the value of η_{II}. Efficiencies for three applications, viz., space heating at $T_a = 320$ K, process steam generation at $T_a = 480$ K, and heating in industrial furnaces at $T_a = 700$ K, are indicated on the Figure. It suggests that fuel is used far more effectively in the high temperature use. An excessive temperatures gap between T_r and T_a causes a low η_{II} and an inefficient energy utilization. A fuel or any energy source is consumed efficiently when the first user temperature approaches the fuel temperature. This means that the fuel should first be used for high temperature applications. The heat rejected from these applications can then be cascaded to applications at lower temperatures, eventually to the task of, say, keeping a building warm. This is called *energy cascading* and ensures more efficient energy utilization.

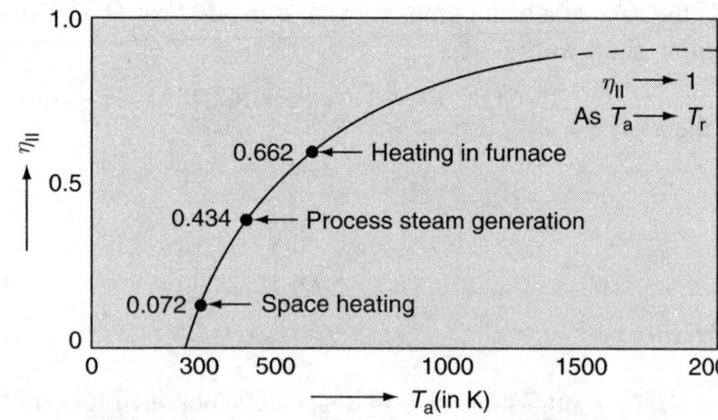

Fig. 8.20 *Effect of use temperature T_a on the second law efficiency ($T_r = 2200$ K, $T_0 = 300$ K, $n_l = 100\%$)*

8.12.2 Further Illustrations of Second Law Efficiencies

Second law efficiency of different components can be expressed in different forms. It is derived by using the exergy balance rate given below:

(a) Turbines The steady state exergy balance (Fig. 8.21) gives:

$$\frac{\dot{Q}}{\dot{m}}\left[1-\frac{T_0}{T_\sigma}\right] + a_{f_1} = \frac{\dot{W}}{\dot{m}} + a_{f_2} + \frac{\dot{I}}{\dot{m}}$$

If there is not heat loss,

$$a_{f_1} - a_{f_2} = \frac{\dot{W}}{\dot{m}} + \frac{\dot{I}}{\dot{m}} \tag{8.94}$$

The second law efficiency,
$$\eta_{II} = \frac{\dot{W}/\dot{m}}{a_{f_1} - a_{f_2}} \tag{8.95}$$

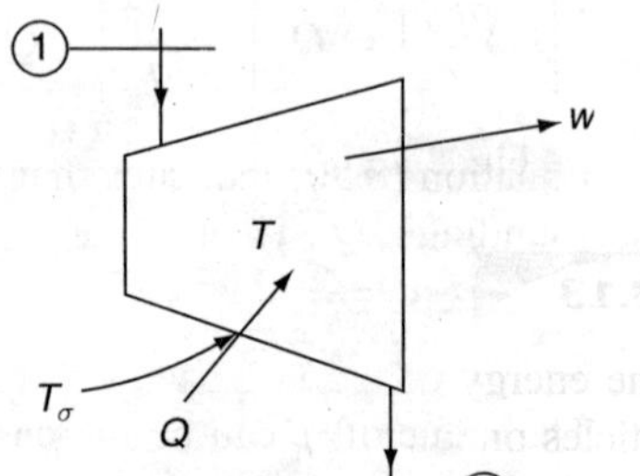

Fig. 8.21 *Exergy balance of a turbine*

(b) Compressor and Pump Similarly, for a compressor or a pump,

$$\frac{\dot{W}}{\dot{m}} = a_{f_2} - a_{f_1} + \frac{\dot{I}}{\dot{m}}$$

and
$$\eta_{II} = \frac{a_{f_2} - a_{f_2}}{-\dot{W}/\dot{m}} \tag{8.96}$$

(c) Heat Exchanger Writing the exergy balance for the heat exchanger, (Fig. 8.22)

$$\sum\left[1-\frac{T_0}{T_j}\right]\dot{Q}_j - \dot{W}_{C.V.} + [\dot{m}_h\, a_{f_1} + \dot{m}_c\, a_{f_3}] - [\dot{m}_h\, a_{f_2} + \dot{m}_c\, a_{f_4}] - \dot{I}_{C.V.} = 0$$

If there is no heat transfer and work transfer,

$$\dot{m}_h\,[a_{f_1} - a_{f_2}] = \dot{m}_c\,[a_{f_4} - a_{f_3}] + \dot{I} \tag{8.97}$$

$$\eta_{II} = \frac{\dot{m}_c\,[a_{f_4} - a_{f_3}]}{\dot{m}_h\,[a_{f_1} - a_{f_2}]} \tag{8.98}$$

(d) Mixing of Two Fluids Exergy balance for the mixer (Fig. 8.23) gives:

$$\left[1-\frac{T_0}{T_\sigma}\right]\dot{Q} + \dot{m}_1\, a_{f_1} + \dot{m}_2\, a_{f_2} = \dot{W}_{C.V.} + \dot{m}_3\, a_{f_3} + I_{C.V.}$$

If the mixing is adiabatic and since $\dot{W}_{C.V.} = 0$ and $\dot{m}_1 + \dot{m}_2 = \dot{m}_3$.

$$\dot{m}_1\,[a_{f_1} - a_{f_3}] = \dot{m}_2\,[a_{f_3} - a_{f_2}] + \dot{I} \tag{8.99}$$

and

$$\eta_{II} = \frac{\dot{m}_2\,[a_{f_3} - a_{f_2}]}{\dot{m}_1\,[a_{f_1} - a_{f_3}]} \tag{8.100}$$

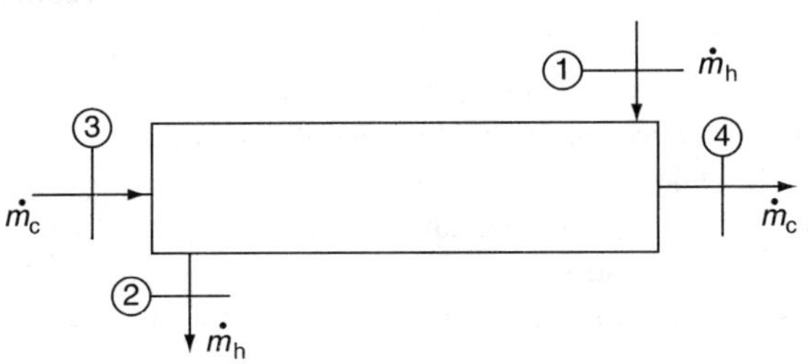

Fig. 8.22 *Exergy balance of a heat exchanger*

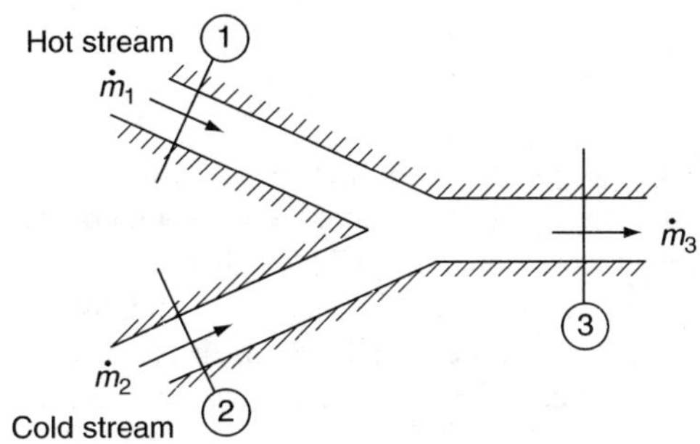

Fig. 8.23 *Exergy loss due to mixing*

8.13 COMMENTS ON EXERGY

The energy of the universe, like its mass, is constant. Yet at times, we are bombarded with speeches and articles on how to "conserve" energy. As engineers, we know that energy is always conserved. What is not conserved is the exergy, i.e., the useful work potential of the energy. Once the exergy is wasted, it can never be recovered. When we use energy (electricity) to heat our homes, we are not destroying any energy, we are merely converting it to a less useful form, a form of less exergy value.

The maximum useful work potential of a system at the specified state is called exergy which is a composite property depending on the state of the system and the surroundings. A system which is in equilibrium with its surroundings is said to be at the dead state having zero exergy.

The mechanical forms of energy such as KE and PE are entirely available energy or exergy. The exergy (W) of thermal energy (Q) of reservoirs (TER) is equivalent to the work output of a Carnot heat engine operating between the reservoir at temperature T and environment at T_0, i.e., $W = Q\left[1-\frac{T_0}{T}\right]$.

The actual work W during a process can be determined from the first law. If the volume of the system changes during a process, part of this work (W_{surr}) is used to push the surrounding medium at constant pressure p_0 and it cannot be used for any useful purpose. The difference between the actual work and the surrounding work is called *useful work*, W_u

$$W_u = W - W_{surr} = W - p_0(V_2 - V_1)$$

W_{surr} is zero for cyclic devices, for steady flow devices, and for system with fixed boundaries (rigid walls).

The maximum amount of useful work that can be obtained from a system as it undergoes a process between two specified states is called *reversible work*, W_{rev}. If the final state of the system is the dead state, the reversible work and the exergy become identical.

The difference between the reversible work and useful work for a process is called *irreversibility*.

$$I = W_{rev} - W_u = T_0 S_{gen} \qquad \dot{I} = T_0 \dot{S}_{gen}$$

For a total reversible process, $W_{rev} = W_u$ and $I = 0$.

The first law efficiency alone is not a realistic measure of performance for engineering devices. Consider two heat engines, having e.g., a thermal efficiency of, say, 30%. One of the engines (A) is supplied with heat Q from a source at 600 K and the other engine (B) is supplied with the same amount of heat Q from a source at 1000 K. Both the engines reject heat to the surroundings at 300 K.

$$(W_A)_{rev} = Q\left(1 - \frac{300}{600}\right) = 0.5Q, \text{ while } (W_A)_{act} = 0.3Q$$

Similarly,

$$(W_B)_{rev} = Q\left(1 - \frac{300}{1000}\right) = 0.7Q, \text{ and } (W_B)_{act} = 0.3Q$$

At first glance, both engines seem to convert the same fraction of heat, that they receive, to work, thus performing equally well from the viewpoint of the first law. However, in the light of second law, the engine B has a greater work potential (0.7 Q) available to it and thus should do a lot better than engine A. Therefore, it can be said that engine B is performing poorly relative to engine A, even though both have the same thermal efficiency.

To overcome the deficiency of the first law efficiency, a second law efficiency η_{II} can be defined as the ratio of actual thermal efficiency to the maximum possible thermal efficiency under the same conditions:

$$\eta_{II} = \frac{\eta_I}{\eta_{rev}}$$

So, for engine A, $\eta_{II} = 0.3/0.5 = 0.60$
and for engine B, $\eta_{II} = 0.3/0.7 = 0.43$

Therefore, the engine A is converting 60% of the available work potential (exergy) to useful work. This is only 43% for the engine B. Therefore,

$$\eta_{II} = \frac{\eta_{act}}{\eta_{rev}} = \frac{W_u}{W_{rev}} \text{ (for heat engines and other work producing devices)}$$

$$\eta_{II} = \frac{COP}{COP_{rev}} = \frac{W_{rev}}{W_u} \text{ (for refrigerators, heat pumps and other work absorbing devices)}$$

The exergies of a closed system (ϕ) and a flowing fluid system (ψ) are given on unit mass basis:

$$\phi = (u - u_0) - T_0(s - s_0) + p_0(v - v_0) \text{ kJ/kg}$$

$$\Psi = (h - h_0) - T_0(s - s_0) + \frac{V^2}{2} + gz \text{ kJ/kg}$$

Reversible Work Expressions

(a) Cyclic Devices $W_{rev} = \eta_{rev}\, Q_1$ (Heat engines)

$$-W_{rev} = \frac{Q_2}{(COP_{rev})_{Ref.}} \quad \text{(Refrigerators)}$$

$$-W_{rev} = \frac{Q_2}{(COP_{rev})_{HP}} \quad \text{(Heat pumps)}$$

(b) Closed System $W_{rev} = U_1 - U_2 - T_0(S_1 - S_2) + p_0(V_1 - V_2) = m(\phi_1 - \phi_2)$

(c) Steady Flow System (single stream)

$$W_{rev} = m\left[\left(h_1 + \frac{V_1^2}{2} + gz_1 - T_0 s_1\right) - \left(h_2 + \frac{V_2^2}{2} + gz_2 - T_0 s_2\right)\right] = m(\Psi_1 - \Psi_2)$$

When the system exchanges heat with another reservoir at temperature T_k other than the atmosphere,

$$\dot{W}_{rev} = \dot{m}(\Psi_1 - \Psi_2) + \dot{Q}_k\left(1 - \frac{T_0}{T_k}\right)$$

The first law efficiency is defined as the ratio of energy output and energy input, while their difference is the energy loss. Likewise, the second law efficiency is defined as the ratio of exergy output and exergy input and their difference is irreversibility. By reducing energy loss, first law efficiency can be improved. Similarly, by reducing irreversibilities, the second law efficiency can be enhanced.

Solved Examples

Example 8.1

In a certain process, a vapour, while condensing at 420°C, transfers heat to water evaporating at 250°C. The resulting steam is used in a power cycle which rejects heat at 35°C. What is the fraction of the available energy in the heat transferred from the process vapour at 420°C that is lost due to the irreversible heat transfer at 250°C?

Solution *ABCD* (Fig.Ex. 8.1) would have been the power cycle, if there was no temperature difference between the vapour condensing and the water evaporating and the area under *CD* would have been the unavailable energy. *EFGD* is the power cycle when the vapour condenses at 420°C and the water evaporates at 250°C. The unavailable energy becomes the area under *DG*. Therefore, the increase in unavailable energy due to irreversible heat transfer is represented by the area under *CG*.

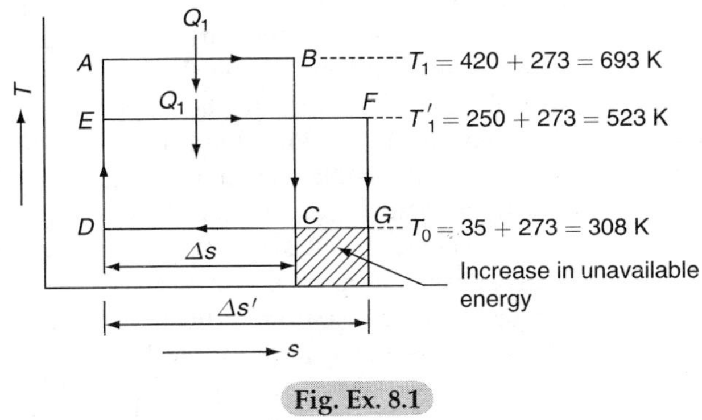

Fig. Ex. 8.1

Now $Q_1 = T_1\, \Delta S = T'_1\, \Delta S'$

$$\frac{\Delta S'}{\Delta S} = \frac{T_1}{T_1'}$$

$$W' = \text{work done in cycle } ABCD = (T_1 - T_0)\, \Delta S$$

$$W' = \text{Work done in cycle } EFGD = (T'_1 - T_0)\, \Delta S'$$

The fraction of energy that becomes unavailable due to irreversible heat transfer

$$= \frac{W - W'}{W} = \frac{T_0\,(\Delta S' - \Delta S)}{(T_1 - T_0)\,\Delta S} = \frac{T_0\left(\dfrac{\Delta S'}{\Delta S} - 1\right)}{(T_1 - T_0)}$$

$$= \frac{T_0\,(T_1 - T_1')}{T_1'\,(T_1 - T_0)} = \frac{308\,(693 - 523)}{523\,(693 - 308)}$$

$$= 0.26$$

Ans.

Example 8.2

In a steam boiler, hot gases from a fire transfer heat to water which vaporizes at constant temperature. In a certain case, the gases are cooled from 1100°C to 550°C while the water evaporates at 220°C. The specific heat of gases is 1.005 kJ/kgK, and the latent heat of water at 220°C, is 1858.5 kJ/kg. All the heat transferred from the gases goes to the water. How much does the total entropy of the combined system of gas and water increase as a result of the irreversible heat transfer? Obtain the result on the basis of 1 kg of water evaporated.

If the temperature of the surroundings is 30°C, find the increase in unavailable energy due to irreversible heat transfer.

Solution Gas ($\dot{m}_g$) is cooled from state 1 to state 2 (Fig. Ex. 8.2). For reversible heat transfer, the working fluid (w.f.) in the heat engine having the same c_p would have been heated along 2–1, so that at any instant, the temperature difference between gas and the working fluid is zero. Then 1–*b* would have been the expansion of the working fluid down to the lowest possible temperature T_0, and the amount of heat rejection would have been given by the area *abcd*.

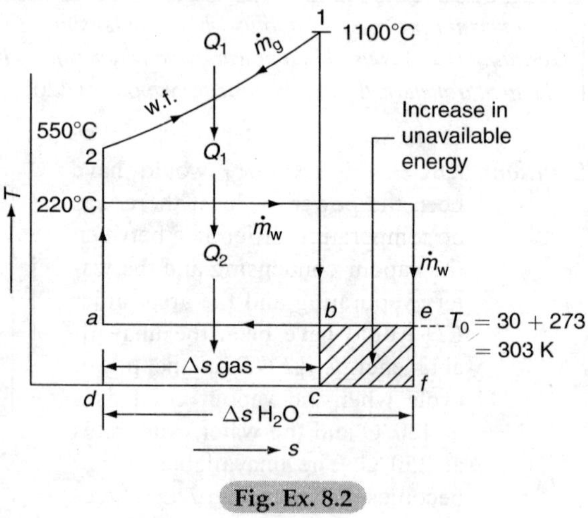

Fig. Ex. 8.2

When water evaporates at 220°C as the gas gets cooled from 1100°C to 550°C, the resulting power cycle has an unavailable energy represented by the area *aefd*. The increase in unavailable energy due to irreversible heat transfer is thus given by area *befc*.

Entropy increase of 1 kg water

$$(\Delta S)_{\text{water}} = \frac{\text{Latent heat absorbed}}{T} = \frac{1858.5}{(273 + 220)} = 3.77 \text{ kJ/kgK}$$

Q_1 = Heat transferred from the gas = Heat absorbed by water during evaporation

$= m_g c_{p_g} (1100 - 550) = 1 \times 1858.5$ kJ

$$\therefore \quad \dot{m}_g \, c_{p_g} = \frac{1858.5}{550} = 3.38 \text{ kJ/°C}$$

$$\Delta S_{\text{gas}} = \int_{T_{g1}}^{T_{g2}} \frac{đQ}{T} = \int_{T_{g1}}^{T_{g2}} \dot{m}_g \, c_{p_g} \frac{dT}{T}$$

$$= \dot{m}_g \, c_{p_g} \ln \frac{T_{g2}}{T_{g1}} = 3.38 \ln \frac{823}{1373}$$

$$= -3.38 \times 0.51 = -1.725 \text{ kJ/K}$$

$$\therefore \quad \Delta S_{\text{total}} = (\Delta S)_{\text{water}} + (\Delta S)_{\text{gas}} = 3.77 - 1.725 = 2.045 \text{ kJ/K}$$ *Ans.*

Increase in unavailable energy $= T_0 (\Delta S)_{\text{total}} = 303 \times 2.045 = 620$ kJ *Ans.*

Example 8.3

Calculate the available energy in 40 *kg of water at* 75°C *with respect to the surroundings at* 5°C, *the pressure of water being* 1 *atm.*

Solution If the water is cooled at a constant pressure of 1 atm from 75°C to 5°C (Fig. Ex. 8.3) the heat given up may be used as a source for a series of Carnot engines each using the surroundings as a sink. It is assumed that the amount of energy received by any engine is small relative to that in the source and the temperature of the source does not change while heat is being exchanged with the engine.

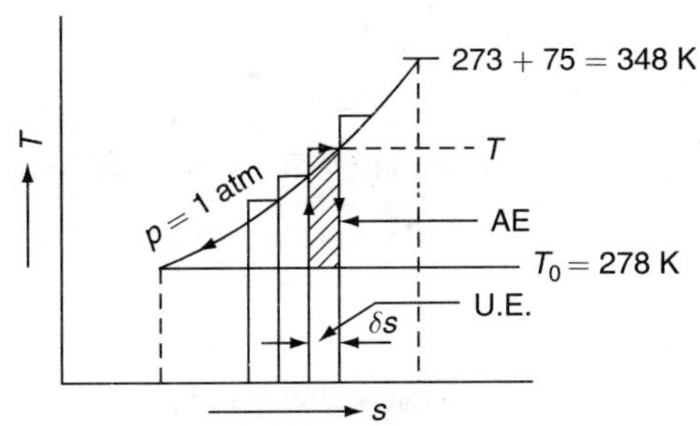

Fig. Ex. 8.3

Let us consider that the source has fallen to temperature T, at which level there operates a Carnot engine which takes in heat at this temperature and rejects heat at $T_0 = 278$ K. If δs is the entropy change of water, the work obtainable is

$$\delta W = -m (T - T_0) \delta s$$

where δs is negative.

$$\delta W = -40 (T - T_0) \frac{c_p \delta T}{T}$$

$$\therefore \quad = -40 \, c_p \left(1 - \frac{T_0}{T}\right) \delta T$$

With a very great number of engines in the series, the total work (maximum) obtainable when the water is cooled from 348 K to 278 K would be

$$W_{(\max)} = \text{A.E.} = -\lim \sum_{348}^{278} 40 c_p \left(1 - \frac{T_0}{T}\right) \delta T$$

$$= \int_{278}^{348} 40\, c_p \left(1 - \frac{T_0}{T}\right) dT = 40\, c_p \left[(348 - 278) - 278 \ln \frac{348}{278}\right] = 40 \times 4.2\,(70 - 62) = 1340 \text{ kJ} \quad \textit{Ans.}$$

$$Q_1 = 40 \times 4.2\,(348 - 278) = 11{,}760 \text{ kJ U.E.} = Q_1 - W_{(max)} = 11{,}760 - 1340 = 10{,}420 \text{ kJ}$$

Example 8.4

Calculate the decrease in available energy when 25 kg of water at 95°C mix with 35 kg of water at 35°C, the pressure being taken as constant and the temperature of the surroundings being 15°C (c_p of water = 4.2 kJ/kg K).

Solution The available energy of a system of mass m, specific heat c_p, and at temperature T, is given by

$$\text{A.E.} = mc_p \int_{T_0}^{T} \left(1 - \frac{T_0}{T}\right) dT$$

$$\therefore \quad (\text{A.E.})_{25} = \text{Available energy of 25 kg of water at 95°C} = 25 \times 4.2 \int_{273+15}^{273+95} \left(1 - \frac{288}{T}\right) dT$$

$$= 105 \left[(368 - 288) - 288 \ln \frac{368}{288}\right] = 987.49 \text{ kJ}$$

$$(\text{A.E.})_{35} = \text{Available energy of 35 kg of water at 35°C}$$

$$= 147 \left[(308 - 288) - 288 \ln \frac{308}{288}\right] = 97.59 \text{ kJ}$$

Total available energy

$$(\text{A.E.})_{\text{total}} = (\text{A.E.})_{25} + (\text{A.E.})_{35} = 987.49 + 97.59 = 1085.08 \text{ kJ}$$

After mixing, if t is the final temperature

$$25 \times 4.2\,(95 - t) = 35 \times 4.2\,(t - 35)$$

$$\therefore \quad t = \frac{25 \times 95 \times 35 \times 35}{25 + 35} = 60°\text{C}$$

Total mass after mixing = 25 + 35 = 60 kg

$$(\text{A.E.})_{60} = \text{Available energy of 60 kg of water at 60°C}$$

$$= 4.2 \times 60 \left[(333 - 288) - 288 \ln \frac{333}{288}\right] = 803.27 \text{ kJ}$$

∴ Decrease in available energy due to mixing

= Total available energy before mixing– Total available energy after mixing

= 1085.08 – 803.27 = 281.81 kJ *Ans.*

Example 8.5

The moment of inertia of a flywheel is 0.54 kg · m² and it rotates at a speed 3000 RPM in a large heat insulated system, the temperature of which is 15°C. If the kinetic energy of the flywheel is dissipated as frictional heat at the shaft bearings which have a water equivalent of 2 kg, find the rise in the temperature of the bearings when the flywheel has come to rest. Calculate the greatest possible amount of this heat which may be returned to the flywheel as high-grade energy, showing how much of the original kinetic energy is now unavailable. What would be the final RPM of the flywheel, if it is set in motion with this available energy?

Solution Initial angular velocity of the flywheel

$$\omega_1 = \frac{2\pi N_1}{60} = \frac{2\pi \times 3000}{60} = 314.2 \text{ rad/s}$$

Initial available energy of the flywheel $= (\text{K.E.})_{\text{initial}} = \frac{1}{2} I\omega_1^2 = 0.54 \text{ kg m}^2 \times (314.2)^2 \frac{\text{rad}^2}{\text{s}^2}$

$$= 2.66 \times 10^4 \text{ Nm} = 26.6 \text{ kJ}$$

When this K.E. is dissipated as frictional heat, if Δt is the temperature rise of the bearings, we have water equivalent of the bearings × rise in temperature = 26.6 kJ

$$\therefore \quad \Delta t = \frac{26.6}{2 \times 4.187} = 3.19°\text{C} \qquad \textit{Ans.}$$

∴ Final temperature of the bearings

$$t_f = 15 + 3.19 = 18.19°\text{C}$$

The maximum amount of energy which may be returned to the flywheel as high-grade energy is

$$\text{A.E.} = 2 \times 4.187 \int_{288}^{291.19} \left(1 - \frac{288}{T}\right) dT$$

$$= 2 \times 4.187 \left[(291.19 - 288) - 288 \ln \frac{291.19}{288}\right] = 0.1459 \text{ kJ}$$

The amount of energy rendered unavailable is

$$\text{U.E.} = (\text{A.E.})_{\text{initial}} - (\text{A.E.})_{\text{returnable as high energy}} = 26.6 - 0.1459 = 26.4541 \text{ kJ}$$

Since the amount of energy returnable to the flywheel is 0.146 kJ, if ω_2 is the final angular velocity, and the flywheel is set in motion with this energy

$$0.146 \times 10^3 = \frac{1}{2} \times 0.54\, \omega_2^2$$

$$\therefore \quad \omega_2^2 = \frac{146}{0.27} = 540.8$$

$$\therefore \quad \omega_2 = 23.246 \text{ rad/s}$$

If N_2 is the final RPM of the flywheel

$$\omega_2 = 23.246 = \frac{2\pi N_2}{60}$$

or

$$N_2 = \frac{23.246 \times 60}{2 \times \pi} = 222 \text{ RPM} \qquad \textit{Ans.}$$

Example 8.6

Two kg of air at 500 kPa, 80°C expands adiabatically in a closed system until its volume is doubled and its temperature becomes equal to that of the surroundings which is at 100 kPa, 5°C. For this process, determine (a) the maximum work, (b) the change in availability, and (c) the irreversibility. For air, take c_v = 0.718 kJ/kg K, $u = c_v T$ where c_v is contant, and $pV = mRT$ where p is pressure in kPa, V volume in m^3, m mass in kg, R a constant equal to 0.287 kJ/kg K, and T temperature in K.

Solution From the property relation

$$TdS = dU + pdV$$

the entropy change of air between the initial and final states is

$$\int_1^2 dS = \int_1^2 \frac{mc_v dT}{T} + \int_1^2 \frac{mR dV}{V}$$

or
$$S_2 - S_1 = mc_v \ln \frac{T_2}{T_1} + mR \ln \frac{V_2}{V_1}$$

From Eq. (8.32),
$$W_{max} = (U_1 - U_2) - T_0 (S_1 - S_2)$$

$$= m \left[c_v (T_1 - T_2 + T_0 \left(c_v \ln \frac{T_2}{T_1} + R \ln \frac{V_2}{V_1} \right) \right]$$

$$= 2 \left[0.718(80-5) + 278 \left(0.718 \ln \frac{278}{353} + 0.287 \ln \frac{2}{1} \right) \right]$$

$$= 2 [53.85 + 278 (-0.172 + 0.199)]$$

$$= 2 (53.85 + 7.51) = 122.72 \text{ kJ}$$ *Ans.* (a)

From Eq. (8.42), the change in availability $= \phi_1 - \phi_2$

$= (U_1 - U_2) - T_0(S_1 - S_2) + p_0(V_1 - V_2) = W_{max} + p_0(V_1 - V_2) = 122.72 + p_0(V_1 - 2V_1)$

$= 122.72 - 100 \times \dfrac{2 \times 0.287 \times 353}{500} = 82.2 \text{ kJ}$ *Ans.* (b)

The irreversibility
$$I = W_{max, useful} - W_{act}$$

From the first law,
$$W_{act} = Q - \Delta U = -\Delta U = U_1 - U_2$$

$$I = U_1 - U_2 - T_0(S_1 - S_2) - U_1 + U_2$$

$$= T_0(S_2 - S_1) = T_0(\Delta S)_{system}$$

For adiabatic process, $(\Delta S)_{surr} = 0$

$$I = T_0 \left[mc_v \ln \frac{T_2}{T_1} + mR \ln \frac{V_2}{V_1} \right]$$

$$= 278 \times 2 \left[0.718 \ln \frac{278}{353} + 0.287 \ln 2 \right]$$

$$= 278 \times 2 (-0.172 + 0.199) = 15.2 \text{ kJ}$$ *Ans.*(c)

Example 8.7

Air expands through a turbine from 500 kPa, 520°C to 100 kPa, 300°C. During expansion 10 kJ/kg of heat is lost to the surroundings which is at 98 kPa, 20°C. Neglecting the K.E. and P.E. changes, determine per kg of air (a) the decrease in availability, (b) the maximum work and (c) the irreversibility. For air, take $c_p = 1.005$ kJ/kg K, $h = c_p T$ where c_p is constant, and the p, V and T relation as in Example 8.6.

Solution From the property relation

$$TdS = dH - Vdp$$

the entropy change of air in the expansion process is

$$\int_1^2 dS = \int_1^2 \frac{mc_p dT}{T} - \int_1^2 \frac{mR dp}{p}$$

or $$S_2 - S_1 = mc_p \ln \frac{T_2}{T_1} - mR \ln \frac{p_2}{p_1}$$

For 1 kg of air, $$s_2 - s_1 = c_p \ln \frac{T_2}{T_1} - R \ln \frac{p_2}{p_1}$$

From Eq. (8.30), the change in availability

$$\Psi_1 - \Psi_2 = b_1 - b_2 = (h_1 - T_0 s_1) - (h_2 - T_0 s_2) = (h_1 - h_2) - T_0(s_1 - s_2)$$

$$= c_p (T_1 - T_2) - T_0 \left(R \ln \frac{p_2}{p_1} - c_p \ln \frac{T_2}{T_1} \right)$$

$$= 1.005\ (520 - 300) - 293 \left(0.287 \ln \frac{1}{5} - 1.005 \ln \frac{573}{793} \right)$$

$$= 1.005 \times 220 - 293\ (0.3267 - 0.4619) = 221.1 + 39.6$$

$$= 260.7 \text{ kJ/kg}$$ *Ans.* (a)

The maximum work is W_{max} = change in availability = $\Psi_1 - \Psi_2$

$$= 260.7 \text{ kJ/kg}$$ *Ans.* (b)

From S.F.E.E., $$Q + h_1 = W + h_2$$

$$W = (h_1 - h_2) + Q = c_p (T_1 - T_2) + Q$$

$$= 1.005\ (520 - 300) - 10 = 211.1 \text{ kJ/kg}$$

The irreversibility $$I = W_{max} - W = 260.7 - 211.1 = 49.6 \text{ kJ/kg}$$ *Ans.* (c)

Alternatively, $$I = T_0 (\Delta S_{system} + \Delta S_{surr})$$

$$= 293 \left(1.005 \ln \frac{573}{793} - 0.287 \ln \frac{1}{5} + \frac{10}{293} \right)$$

$$= 293 \times 0.1352 + 10 = 49.6 \text{ kJ/kg}$$ *Ans.* (d)

Example 8.8

An air preheater is used to cool the products of combustion from a furnace while heating the air to be used for combustion. The rate of flow of products is 12.5 kg/s and the products are cooled from 300 to 200°C, and for the products at this temperature c_p = 1.09 kJ/kg K. The rate of air flow is 11.5 kg/s, the initial air temperature is 40°C, and for the air c_p = 1.005 kJ/kg K. (a) Estimate the initial and final availability of the products. (b) What is the irreversibility for the process? (c) If the heat transfer from the products occurs reversibly through heat engines, what is the final temperature of the air? What is the power developed by the heat engine? Take T_0 = 300 K and neglect pressure drop for both the fluids and heat transfer to the surroundings.

Solution (a) Ψ_1 = initial availability of the products

$$= (h_1 - h_0) - T_0 (S_1 - S_0) = c_{p_g} (T_{g_1} - T_0) - T_0 c_{p_g} \ln \frac{T_{g_1}}{T_0}$$

$$= 1.09\ (573 - 300) - 300 \times 1.09 \ln \frac{573}{300} = 297.57 - 211.6 = 39.68 \text{ kJ/kg}$$

Ψ_2 = final availability of the products

$$= (h_2 - h_0) - T_0(s_2 - s_0) = 1.09\ (473 - 300) - 300 \times 1.09 \times \ln \frac{473}{300}$$

$$= 188.57 - 148.89 = 39.68 \text{ kJ/kg}$$

(b) Decrease in availability of the products $= \Psi_1 - \Psi_2 = (h_1 - h_2) - T_0(s_1 - s_2)$

$$= 1.09\ (573 - 473) - 300 \times 1.09 \ln \frac{573}{473}$$

$$= 109 - 62.72 = 46.28 \text{ kJ/kg}$$

By making an energy balance for the air preheater [Fig. Ex. 8.8(a)].

$$\dot{m}_g\, c_{p_g} (T_{g_1} - T_{g_2}) = \dot{m}_a\, c_{p_a} (T_{a_2} - T_{a_1})$$

$$12.5 \times 1.09\ (573 - 473) = 11.15 \times 1.005(T_{a_2} - 313)$$

$$\therefore \qquad T_{a_2} = \frac{12.5 \times 109}{11.5 \times 1.005} + 313 = 430.89 \text{ K}$$

Increase in availability for air $= \Psi_2 - \Psi_1 = (h_2 - h_1) - T_0(s_2 - s_1)$

$$= c_{p_a} (T_{a_2} - T_{a_1}) - T_0 c_{p_a} \ln \frac{T_{a_2}}{T_{a_1}}$$

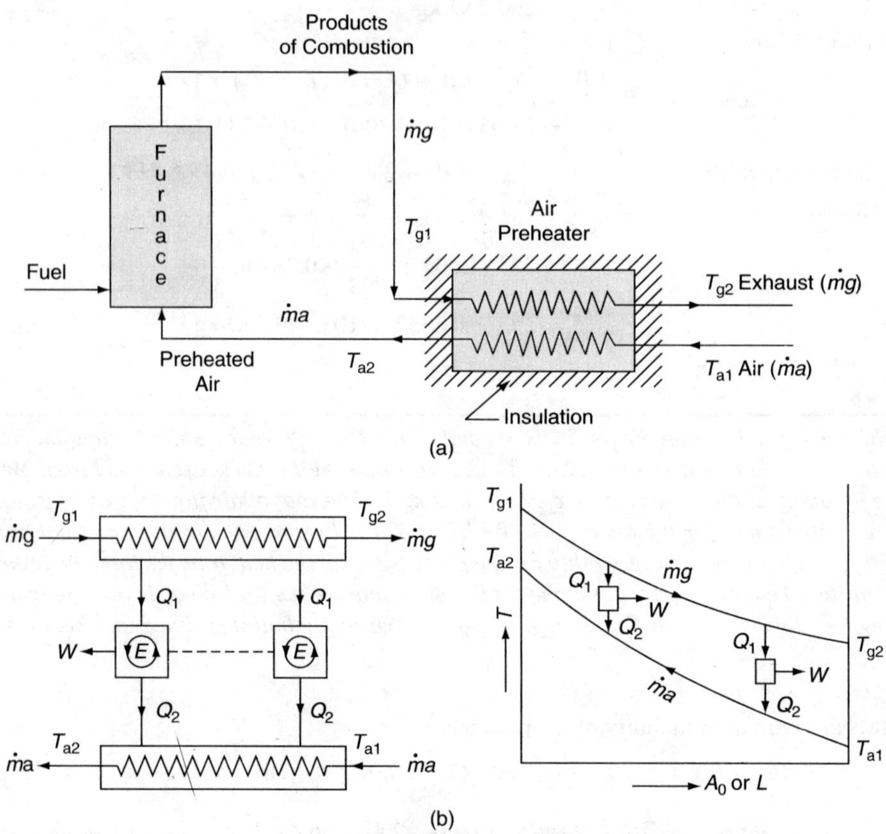

Fig. Ex. 8.8

$$= 1.005 \times (430.89 - 313) - 300 \times 1.005 \ln \frac{430.89}{313}$$

$$= 118.48 - 96.37 = 22.11 \text{ kJ/kg}$$

$\therefore$ Irreversibility of the process $= 12.5 \times 46.28 - 11.5 \times 22.11$

$$= 578.50 - 254.27 = 324.23 \text{ kW}$$

(c) Let us assume that heat transfer from the products to air occurred through heat engines reversibly as shown in Fig. Ex. 8.8(b).

For reversible heat transfer,

$$\Delta \dot{S}_{univ} = 0$$

$$\Delta \dot{S}_{sys} + \Delta \dot{S}_{surr} = 0$$

$$\Delta \dot{S}_{gas} + \Delta \dot{S}_{air} = 0$$

$$\Delta \dot{S}_{gas} = -\Delta \dot{S}_{air}$$

$$\dot{m}_g c_{p_g} \ln \frac{T_{g2}}{T_{g1}} = -\dot{m}_a c_{p_a} \ln \frac{T_{a2}}{T_{a1}}$$

$$12.5 \times 1.09 \ln \frac{473}{573} = -11.5 \times 1.005 \ln \frac{T_{a2}}{313}$$

$$\therefore \quad T_{a2} = 392.41 \text{ K}$$

Rate of heat supply from the gas to the working fluid in the heat engine,

$$\dot{Q}_1 = \dot{m}_g c_{pg} (T_{g1} - T_{g2}) = 12.5 \times 1.09 (573 - 473) = 1362.50 \text{ kW}$$

Rate of heat rejection from the working fluid in the heat engine to the air,

$$\dot{Q}_2 = \dot{m}_a c_{p_a} (T_{a2} - T_{a1}) = 11.5 \times 1.005 (392.41 - 313) = 917.78 \text{ kW}$$

Total power developed by the heat engine

$$\dot{W} = \dot{Q}_1 - \dot{Q}_2 = 1362.50 - 917.78 = 444.72 \text{ kW}$$

Example 8.9

A gas is flowing through a pipe at the rate of 2 *kg/s. Because of inadequate insulation the gas temperature decreases from* 800 *to* 790°C *between two sections in the pipe. Neglecting pressure losses, calculate the irreversibility rate (or rate of energy degradation) due to this heat loss. Take* $T_0 = 300$ *K and a constant* $c_p = 1.1$ *kJ/kg K.*

For the same temperature drop of 10°C *when the gas cools from* 80°C *to* 70°C *due to heat loss, what is the rate of energy degradation? Take the same values of* T_0 *and* c_p*. What is the inference you can draw from this example?*

Solution

$$\dot{S}_{gen} = \dot{S}_{sys} - \frac{\dot{Q}}{T_0}$$

$$= \dot{m}(s_2 - s_1) - \frac{\dot{m}c_p(T_2 - T_1)}{T_0}$$

Irreversibility rate = rate of energy degradation = rate of exergy loss

$$\dot{I} = T_0\dot{S}_{gen} = \dot{m}T_0(s_2 - s_1) - \dot{m}c_p(T_2 - T_1)$$

$$= \dot{m}c_p\left[(T_1 - T_2) - T_0 \ln\frac{T_1}{T_2}\right]$$

$$= 2 \times 1.1\left[(1073 - 1063) - 300 \ln\frac{1073}{1063}\right] = 15.818 \text{ kW}$$

When the same heat loss occurs at lower temperature

$$\dot{I} = 2 \times 1.1\left[(353 - 343) - 300 \ln\frac{353}{343}\right] = 3.036 \text{ kW}$$

It is thus seen that irreversibility rate or exergy destruction is more when the same heat loss occurs at higher temperature. Irreversibility rate decreases as the temperature of the gas decreases. Quantitatively, the heat loss may be the same, but qualitatively, it is different.

Example 8.10

An ideal gas is flowing through an insulated pipe at the rate of 3 kg/s. There is a 10% pressure drop from inlet to exit of the pipe. What is the rate of exergy loss because of the pressure drop due to friction? Take R = 0.287 kJ/kg K and T_0 = 300 K.

Solution Rate of entropy generation from Eq. (8.68),

$$\dot{S}_{gen} = \dot{m}R\frac{\Delta p}{p_1} = 3 \times 0.287 \frac{0.10\,p_1}{p_1} = 0.0861 \text{ kW/K}$$

Rate of exergy loss $\dot{I} = T_0\dot{S}_{gen} = 300 \times 0.0861 = 25.83$ kW

Example 8.11

Water at 90°C flowing at the rate of 2 kg/s mixes adiabatically with another stream of water at 30°C flowing at the rate of 1 kg/s. Estimate the entropy generation rate and the rate of exergy loss due to mixing. Take T_0 = 300 K.

Solution

$$\dot{m} = \dot{m}_1 + \dot{m}_2 = 2 + 1 = 3 \text{ kg/s}$$

Here

$$x = \frac{\dot{m}_1}{\dot{m}} = \frac{2}{3} = 0.67$$

$$\tau = \frac{T_2}{T_1} = \frac{303}{363} = 0.835$$

From Eq. (8.76),

$$\dot{S}_{gen} = \dot{m}c_p \ln\frac{x + \tau(1 - x)}{\tau^{1-x}}$$

$$= 3 \times 4.187 \ln \frac{0.67 \times 0.835 \times 0.33}{0.835^{0.33}} = 12.561 \ln \frac{0.94555}{0.94223} = 0.0442 \text{ kW/K}$$

Rate of exergy loss due to mixing

$$\dot{I} = T_0 \dot{S}_{gen} = 300 \times 0.0442 = 13.26 \text{ kW}$$

Alternatively,

Equilibrium temperature after mixing,

$$t = \frac{m_1 t_1 + m_2 t_2}{m_1 + m_2} = \frac{2 \times 90 + 1 \times 30}{2 + 1} = 70°\text{C}$$

$$\Delta \dot{S}_{univ} = \dot{S}_{gen} = 2 \times 4.187 \ln \frac{343}{363} + 1 \times 4.187 \ln \frac{343}{303}$$

$$= 0.0447 \text{ kW/K}$$

$$\therefore \quad \dot{I} = 300 \times 0.0447 = 13.41 \text{ kW}$$

Example 8.12

By burning a fuel the rate of heat release is 500 kW at 2000 K. What would be the first law and the second law efficiencies if (a) energy is absorbed in a metallurgical furnace at the rate of 480 kW at 1000 K, (b) energy is absorbed at the rate of 450 kW for generation of steam at 500 K, and (c) energy is absorbed in a chemical process at the rate of 300 kW at 320 K? Take $T_0 = 300$ K. (d) Had the energy absorption rate been equal to 450 kW in all these three cases, what would have been the second law efficiencies? What is the inference that you can draw from this example?

Solution If $\dot{Q}_r$ is the rate of heat release at temperature T_r and $\dot{Q}_a$ the rate of heat absorption at temperature T_a, then

$$\eta_I = \frac{\dot{Q}_a}{\dot{Q}_r} \quad \text{and} \quad \eta_{II} = \eta_I \frac{1 - \frac{T_0}{T_a}}{1 - \frac{T_0}{T_r}}$$

(a) Metallurgical furnace

$$\eta_I = \frac{480}{500} \times 100 = 96\%$$

$$\eta_{II} = 0.96 \frac{1 - \frac{300}{1000}}{1 - \frac{300}{2000}} \times 100 = 79\%$$

(b) Steam generation

$$\eta_I = \frac{450}{500} \times 100 = 90\%$$

$$\eta_{II} = 0.90 \frac{1 - \frac{300}{500}}{1 - \frac{300}{2000}} \times 100 = 42.3\%$$

(c) Chemical process

$$\eta_I = \frac{300}{500} \times 100 = 60\%$$

$$\eta_{II} = 0.60 \frac{1 - \frac{300}{320}}{1 - \frac{300}{2000}} \times 100 = 4.41\%$$

(d) In all the three cases, η_I would remain the same, where

$$\eta_I = \frac{450}{500} \times 100 = 0.90$$

$$\eta_{II\,(a)} = 0.90 \times \frac{1 - \frac{300}{1000}}{1 - \frac{300}{2000}} \times 100 = 74.11\%$$

$$\eta_{II\,(b)} = 0.90 \times \frac{1 - \frac{300}{500}}{1 - \frac{300}{2000}} \times 100 = 42.3\%$$

$$\eta_{II\,(c)} = 0.90 \times \frac{1 - \frac{300}{320}}{1 - \frac{300}{2000}} \times 100 = 6.61\%$$

It is seen that as the energy loss ($\dot{Q}_r - \dot{Q}_a$) increases, the first law efficiency decreases. For the same heat loss, however, as the temperature difference between the source and the use temperature increases, the second law efficiency decreases, or in other words, the rate of exergy loss increases.

Example 8.13

A system undergoes a power cycle while receiving energy Q_1 at temperature T_1 and discharging energy Q_2 at temperature T_2. There are no other heat transfers.

(a) Show that the thermal efficiency of the cycle can be expressed as:

$$\eta = 1 - \frac{T_2}{T_1} - \frac{T_2 I}{T_0 Q_1}$$

where T_0 is the ambient temperature and I is the irreversibility of the cycle.

(b) Obtain an expression for the maximum theoretical value for the thermal efficiency.

(c) Derive an expression for the irreversibility for which no network is developed by the cycle. What conclusion do you derive from it?

Solution An availability balance for the cycle gives (Fig. Ex. 8.13)

$$(\Delta A)_{\text{cycle}} = 0 = \left(1 - \frac{T_0}{T_1}\right) Q_1 - \left(1 - \frac{T_0}{T_2}\right) Q_2 - W - I$$

since each property is restored to its initial state.

Since $\quad Q_2 = Q_1 - W,$

$$0 = \left(1 - \frac{T_0}{T_1}\right) Q_1 - \left(1 - \frac{T_0}{T_2}\right) (Q_1 - W) - W - I$$

$$W = \frac{T_2}{T_0}\left[\left(1 - \frac{T_0}{T_1}\right) - \left(1 - \frac{T_0}{T_2}\right)\right] Q_1 - \frac{T_2 I}{T_0}$$

$$\eta = \frac{W}{Q_1} = 1 - \frac{T_2}{T_1} - \frac{T_2 I}{T_0 Q_1} \qquad \textit{Proved.}$$

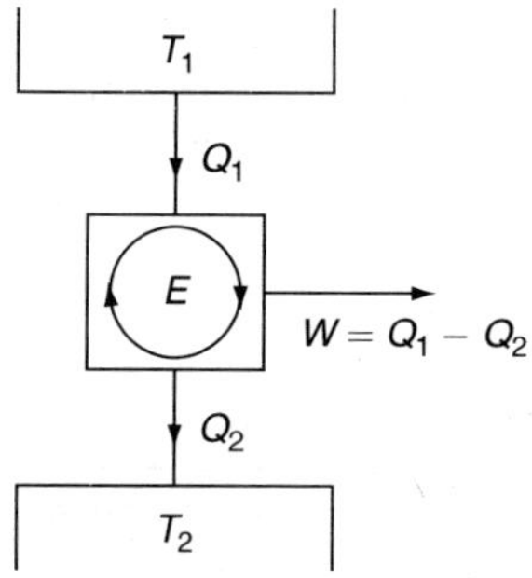

Fig. Ex. 8.13

(b) When $I = 0$, $\eta_{max} = 1 - \dfrac{T_2}{T_1}$

(c) When $W = 0$ $\quad \eta = 0 = 1 - \dfrac{T_2}{T_1} - \dfrac{T_2 I}{T_0 Q_1}$

$$I = T_0\left[\frac{1}{T_2} - \frac{1}{T_1}\right] Q_1 = T_0\left[\frac{Q_1}{T_2} - \frac{Q_1}{T_1}\right] = T_0 S_{gen}$$

The heat transfer Q_1 from T_1 to T_2 takes place through a reversible engine, and the entire work is dissipated in the brake, from which an equal amount of heat is rejected to the reservoir at T_2. Heat transfer through a finite temperature difference is thus equivalent to the destruction of its exergy. (See Art. 8.10.1(a)).

Example 8.14

A compressor operating at steady state takes in 1 kg/s of air at 1 bar and 25°C and compresses it to 8 bar and 160°C. Heat transfer from the compressor to its surroundings occurs at a rate of 100 kW. (a) Determine the power input in kW. (b) Evaluate the second law efficiency for the compressor. Neglect KE and PE changes. Take $T_0 = 25°C$ and $P_0 = 1$ bar.

Solution SFEE for the compressor (Fig. 8.14) gives:

$$\dot{W} = \dot{Q} + \dot{m}(h_1 - h_2) = -100 + 1 \times 1.005\,(25 - 160)$$

$$= -235.7 \text{ kW} \qquad \textit{Ans. } (a)$$

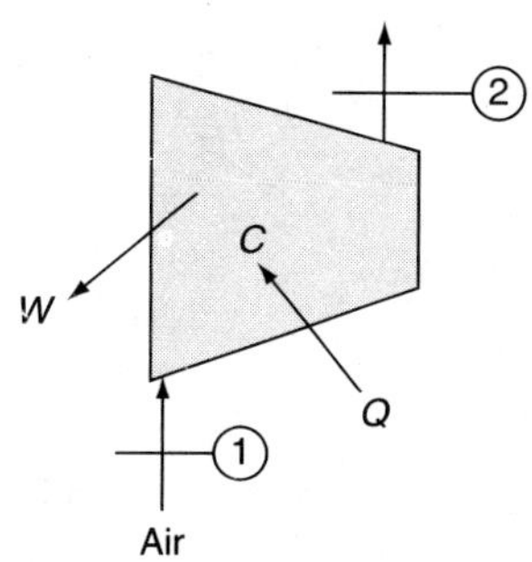

Fig. Ex. 8.14

Exergy balance for the compressor gives:

$$\dot{m} a_{f_1} + \dot{Q}\left(1 - \frac{T_0}{T}\right) - \dot{W} - \dot{m} a_{f_2} = \dot{I}$$

$$-\dot{W} = \dot{m}\,(a_{f_2} - a_{f_1}) - \dot{Q}\left(1 - \frac{T_0}{T}\right) + \dot{I}$$

$$\eta_{II} = \frac{\dot{m}\,(a_{f_2} - a_{f_1})}{\dot{W}}$$

$$a_{f_2} - a_{f_1} = h_2 - h_1 - T_0(s_2 - s_1)$$

$$= c_p(T_2 - T_1) - T_0\left(c_p \ln \frac{T_2}{T_1} - R \ln \frac{p_2}{p_1}\right) = 1.005\,(160 - 25) - 298\left(1.005 \ln \frac{433}{298} - 0.287 \ln 8\right)$$

$$= 200.95 \text{ kJ/kg}$$

$$\eta_{II} = \frac{200.95}{235.7} = 0.853 \text{ or, } 85.3\%$$ *Ans.* (b)

Example 8.15

Determine the exergy of 1 m³ *of complete vacuum.*

Solution $\phi = U - U_0 + p_0(V - V_0) - T_0(S - S_0) = H - H_0 - V(p - p_0) - T_0(S - S_0)$

Since a vacuum has zero mass,

$$U = 0,\ H_0 = 0, \text{ and } S = 0$$

If the vacuum was reduced to the dead state

$$U_0 = 0,\ H_0 = 0,\ S_0 = 0 \text{ and } V_0 = 0$$

The pressure p for the vacuum is zero.

But $p_0 = 1$ bar $= 100$ kPa and $V = 1$ m³ $\phi = p_0V = 100\ \frac{\text{kN}}{\text{m}^2} \times 1\ \text{m}^3 = 100$ kJ *Ans.*

If an air motor operates between the atmosphere and the vacuum, this is the maximum useful work obtainable. Therefore, the vacuum has an exergy or work potential.

Example 8.16

A mass of 1000 *kg of fish initially at* 1 *bar,* 300 *K is to be cooled to* – 20°C. *The freezing point of fish is* – 2.2°C, *and the specific heats of fish below and above the freezing point are* 1.7 *and* 3.2 *kJ/kg K respectively. The latent heat of fusion for the fish can be taken as* 235 *kJ/kg. Calculate the exergy produced in the chilling process. Take* $T_0 = 300$ *K and* $p_0 = 1$ *bar.*

Solution Exergy produced $= H_2 - H_1 - T_0(S_2 - S_1)$

With reference to Fig. Ex. 8.16,

$$H_1 - H_2 = 1000\,[1.7\,(270.8 - 253) + 235 + 3.2\,(300 - 270.8)]$$

$$= 1000\,[1.7 \times 17.8 + 235 + 3.2 \times 29.2]$$

$$= 1000\,[30.26 + 235 + 93.44] = 358.7 \text{ MJ}$$

$$H_2 - H_1 = -\,358.7 \text{ MJ}$$

$$S_1 - S_2 = 1000\left[1.7 \ln \frac{270.8}{253} + \frac{235}{270.8} + 3.2 \ln \frac{300}{270.8}\right]$$

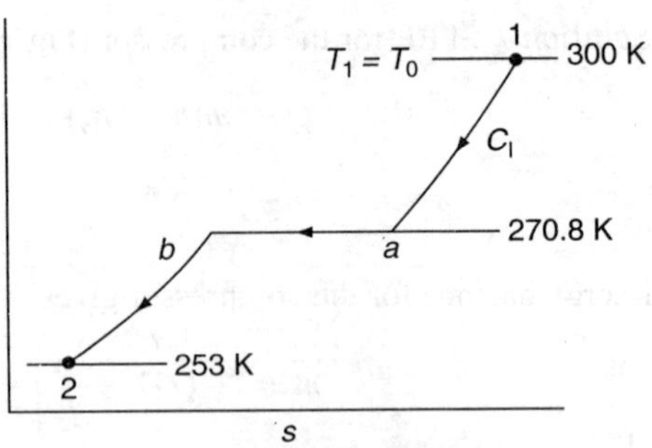

Fig. Ex. 8.16

$$= -\,1000\,[0.1156 + 0.8678 + 0.3277] = 1.311 \text{ MJ/K}$$

$$S_2 - S_1 = -\,1.311 \text{ MJ/K}$$

Exergy produced $= -\,358.7 + 300 \times 1.311 = -\,358.7 + 393.3 = 34.6$ MJ or 9.54 kWh *Ans.*

Example 8.17

A quantity of air initially at 1 bar, 300 K undergoes two types of interactions: (a) it is brought to a final temperature of 500 K adiabatically by paddle-wheel work transfer, (b) the same temperature rise is brought about by heat transfer from a thermal reservoir at 600 K. Take $T_0 = 300$ K, $p_0 = 1$ atm. Determine the irreversibility (in kJ/kg) in each case and comment on the results.

Solution *Case (a)*: As shown in the above Figures (Fig. Ex. 8.17),

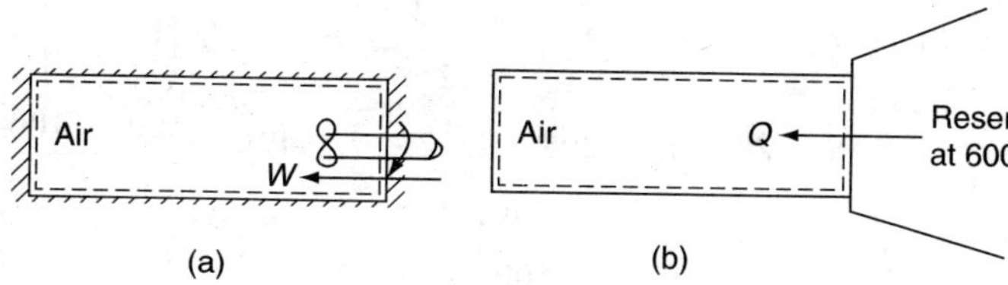

Fig. Ex. 8.17

$$\Delta s_{\text{univ}} = s_{\text{gen}} = c_v \ln \frac{T_2}{T_1} = 0.718 \ln \frac{500}{300} = 0.367 \text{ kJ/kg K}$$

$$I = 300 \times 0.367 = 110.1 \text{ kJ/kg}$$ *Ans.*

Case (b):
$$Q = m\, c_v (T_2 - T_1)$$

$$= 1 \times 0.718 (500 - 300) = 143.6 \text{ kJ/kg}$$

$$\Delta s_{\text{univ}} = s_2 - s_1 - \frac{Q_2}{T} = 0.367 - \frac{143.6}{600}$$

$$= 0.1277 \text{ kJ/kg K} I = 300 \times 0.1277 = 38.31 \text{ kJ/kg}$$

Comment:

The irreversibility in case (b) is less than in case (a).

$$I_a = T_0(s_2 - s_1), I_b = T_0(s_2 - s_1) - \frac{Q}{T}$$

$$I_a - I_b = \frac{Q}{T}$$

The irreversibility in case (b) is always less than in case (a) and the two values would approach each other only at high reservoir temperature, i.e.,

$$I_a \to I_b \text{ as } T \to \infty$$

Example 8.18

Steam enters a turbine at 30 bar, 400°C ($h = 3230$ kJ/kg, $s = 6.9212$ kJ/kg K) and with a velocity of 160 m/s. Steam leaves as saturated vapour at 100°C ($h = 2676.1$ kJ/kg, $s = 7.3549$ kJ/kg K) with a velocity of 100 m/s. At steady state the turbine develops work at a rate of 540 kJ/kg. Heat transfer between the turbine and its surroundings occurs at an average outer surface temperature of 500 K. Determine the irreversibility per unit mass. Give an exergy balance and estimate the second law efficiency of the turbine. Take $p_0 = 1$ atm, $T_0 = 298$ K and neglect PE effect.

Solution By exergy balance of the control volume (Fig. Ex. 8.18),

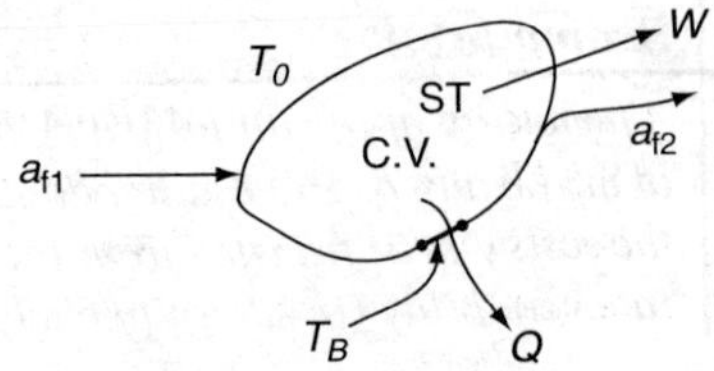

Fig. Ex. 8.18

$$a_{f_1} = W + Q\left(1 - \frac{T_0}{T_B}\right) + a_{f_2} + I$$

where a_f is the exergy transfer per unit mass.

$$I = a_{f_1} - a_{f_2} - W - Q\left(1 - \frac{T_0}{T_B}\right)$$

$$= (h_1 - h_2) - T_0(s_1 - s_2) + \frac{V_1^2 - V_2^2}{2} - W - Q\left(1 - \frac{T_0}{T_B}\right)$$

$$= (3230.9 - 2676.1) - 298(6.9212 - 7.3549) + \frac{160^2 - 100^2}{2}$$

$$\times 10^{-3} - 540 - Q\left(1 - \frac{298}{500}\right)$$

$$= 151.84 - Q(0.404) \quad (1)$$

By SFEE,

$$h_1 + \frac{V_1^2}{2} = W + h_2 + Q + \frac{V_2^2}{2}$$

$$Q = (h_1 - h_2) + \frac{V_1^2 - V_2^2}{2} - W$$

$$= (3230.9 - 2676.1) + \frac{160^2 - 100^2}{2} \times 10^{-3} - 540$$

$$= 22.6 \text{ kJ/kg.}$$

From Eq. (1),

$$I = 151.84 - 22.6 \times 0.404$$

$$= 142.71 \text{ kJ/kg} \qquad \textit{Ans.}$$

Net exergy transferred to turbine

$$a_{f_1} - a_{f_2} = 691.84 \text{ kJ/kg}$$

Work = 540 kJ/kg

Exergy destroyed = I = 142.71 kJ/kg

Exergy transferred out accompanying heat transfer

= 22.6 × 0.404 = 9.13 kJ/kg

Exergy Balance

Exergy transferred	*Exergy utilized*
691.84 kJ/kg	Work = 540 kJ/kg (78%)
	Destroyed = 142.71 kJ/kg (20.6%)
	Transferred with heat = 9.13 kJ/kg (1.3%)
	691.84 kJ/kg

$$\text{Second law efficiency, } \eta_{II} = \frac{540}{691.84} = 0.78 \text{ or } 78\% \qquad \textit{Ans.}$$

Example 8.19

A furnace is heated by an electrical resistor. At steady state, electrical power is supplied to the resistor at a rate of 8.5 kW per metre length to maintain it at 1500 K when the furnace walls are at 500 K. Let T_0 = 300 K (a) For the resistor as the system, determine the rate of availability transfer accompanying heat and the irreversibility rate, (b) For the space between the resistor and the walls as the system, evaluate the irreversibility rate.

Solution *Case (a):* At steady state for the resistor (Fig. Ex. 8.19).

$$\dot{Q} = \Delta\dot{U} + \dot{W} = \dot{W} = 8.5 \text{ kW}$$

Availability rate balance gives

$$\frac{dA}{dt} = \left(1 - \frac{T_0}{T}\right)\dot{Q} + \left(\dot{W} - p_0\frac{dV}{dt}\right) - \dot{I} = 0$$

$\dot{I}$ = Rate of irreversibility

$$= \left(1 - \frac{T_0}{T}\right)\dot{Q} + \dot{W} = \left(1 - \frac{300}{1500}\right)(-8.5) + 8.5$$

$$= 1.7 \text{ kW}$$ *Ans.* (a)

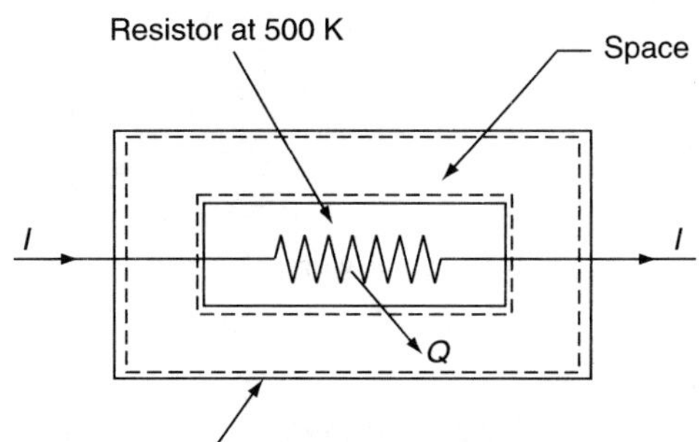

Fig. Ex. 8.19

Rate of availability transfer with heat

$$= \left(1 - \frac{T_0}{T}\right)\dot{Q} = \left(1 - \frac{300}{1500}\right)(-8.5) = -6.8 \text{ kW}$$ *Ans.* (a)

Case (b): Steady state,

$$\frac{dA}{dT} = \left(1 - \frac{T_0}{T}\right)\dot{Q} - \left(1 - \frac{T_0}{T_w}\right)\dot{Q} - \dot{W} - \dot{I} = 0$$

$$\dot{I} = \left(1 - \frac{300}{1500}\right)8.5 - \left(1 - \frac{300}{500}\right)8.5$$

$$= 6.8 - 3.4 = 3.4 \text{ kW}$$ *Ans.* (b)

Example 8.20

Air enters a compressor at 1 bar, 30°C, which is also the state of the environment. It leaves at 3.5 bar, 141°C and 90 m/s. Neglecting inlet velocity and P.E. effect, determine (a) whether the compression is adiabatic or polytropic, (b) if not adiabatic, the polytropic index, (c) the isothermal efficiency, (d) the minimum work input and irreversibility, and (d) the second law efficiency. Take c_p of air = 1.0035 kJ/kg K.

Solution (a) After isentropic compression

$$\frac{T_{2s}}{T_1} = \left[\frac{p_2}{p_1}\right]^{(\gamma-1)/\gamma}$$

$$T_{2s} = 303\,(3.5)^{0.286} = 433.6 \text{ K} = 160.6°\text{C}$$

Since this temperature is higher than the given temperature of 141°C, there is heat loss to the surroundings. The compression cannot be adiabatic. It must be polytropic.

(b)
$$\frac{T_2}{T_1} = \left[\frac{p_2}{p_1}\right]^{(n-1)/n}$$

$$\frac{141 + 273}{30 + 273} = 1.366 = \left[\frac{3.5}{1}\right]^{(n-1)/n}$$

$$\log 1.366 = \frac{n-1}{n} \log 3.5$$

$$1 - \frac{1}{n} = \frac{0.135}{0.544} = 0.248$$

$$n = 1.32978 = 1.33$$ *Ans.*

(c) Actual work of compression

$$W_a = h_1 - h_2 - \frac{V_2^2}{2} = 1.0035\,(30 - 141) - \frac{90^2}{2} \times 10^{-3} = -\,115.7 \text{ kJ/kg}$$

Isothermal work $W_T = \int_1^2 v\,dp - \frac{V_2^2}{2} = -\,RT_1 \ln \frac{p_2}{p_1} - \frac{V_2^2}{2}$

$$= -\,0.287 \times 303 \ln (3.5) - \frac{90^2}{2} \times 10^{-3} = -\,113 \text{ kJ/kg}$$

Isothermal efficiency:

$$\eta_T = \frac{W_T}{W_a} = \frac{113}{115.7} = 0.977 \text{ or } 97.7\%$$ *Ans.*

(d) Decrease in availability or exergy:

$$\Psi_1 - \Psi_2 = h_1 - h_2 - T_0(s_1 - s_2) + \frac{V_1^2 - V_2^2}{2}$$

$$= c_p(T_1 - T_2) - T_0\left[R \ln \frac{p_2}{p_1} - c_p \ln \frac{T_2}{T_1}\right] - \frac{V_2^2}{2}$$

$$= 1.0035\,(30 - 141)$$

$$-\,303\left[0.287 \ln 3.5 - 1.0035 \ln \frac{414}{303}\right] - \frac{90^2}{2000}$$

$$= -\,101.8 \text{ kJ/kg}$$

Minimum work input = – 101.8 kJ/kg *Ans.*

Irreversibility,
$$I = W_{rev} - W_a$$
$$= -\,101.8 - (-\,115.7)$$
$$= 13.9 \text{ kJ/kg}$$ *Ans.*

(e) Second law efficiency,

$$\eta_{II} = \frac{\text{Minimum work input}}{\text{Actual work input}} = \frac{101.8}{115.7}$$

$$= 0.88 \text{ or } 88\%$$ *Ans.*

Review Questions

8.1 What do you understand by high grade energy and low grade energy?

8.2 What is available energy and unavailable energy?

8.3 Who propounded the concept of availability?

8.4 What is the available energy referred to a cycle?

8.5 Show that there is a decrease in available energy when heat is transferred through a finite temperature difference.

8.6 Deduce the expression for available energy from a finite energy source at temperature T when the environmental temperature is T_0.

8.7 What do you understand by exergy and energy?

8.8 What is meant by quality of energy?

8.9 Why is exergy of a fluid at a higher temperature more than that at a lower temperature?

8.10 How does the exergy value provide a useful measure of the quality of energy?

8.11 Why is the second law called the law of degradation of energy?

8.12 Energy is always conserved, but its quality is always degraded. Explain.

8.13 Why is the work done by a closed system in a reversible process by interacting only with the surroundings the maximum?

8.14 Show that equal work is done in all reversible processes between the same end states of a system if it exchanges energy only with the surroundings.

8.15 Give the general expression for the maximum work of an open system which exchanges heat only with the surroundings.

8.16 What do you understand by Keenan function?

8.17 Give the expression for reversible work in a steady flow process under a given environment.

8.18 Give the expression for reversible work done by a closed system if it interacts only with the surroundings.

8.19 What do you understand by 'useful work'? Derive expressions for useful work for a closed system and a steady flow system which interact only with the surroundings.

8.20 What are the availability functions for a: (a) closed system, (b) steady flow system?

8.21 What do you understand by the dead state?

8.22 What is meant by availability?

8.23 Give expressions for availabilities of a closed system and a steady flow open system.

8.24 What are Helmholtz function and Gibbs function?

8.25 What is the availability in a chemical reaction if the temperature before and after the reaction is the same and equal to the temperature of the surroundings?

8.26 When is the availability of a chemical reaction equal to the decrease in the Gibbs function?

8.27 Derive the expression for irreversibility or exergy loss in a process executed by: (a) a closed system, (b) a steady flow system, in a given environment.

8.28 State and explain the Gouy-Stodola theorem.

8.29 How is heat transfer through a finite temperature difference equivalent to the destruction of its availability?

8.30 Considering the steady and adiabatic flow of an ideal gas through a pipe, show that the rate of decrease in availability or lost work is proportional to the pressure drop and the mass flow rate.

8.31 What do you understand by Grassman diagram?

8.32 What is entropy generation number?

8.33 Why is exergy always a positive value? Why cannot it be negative?

8.34 Why and when is exergy completely destroyed?

8.35 Give the exergy balance for a closed system.

8.36 Explain the statement: The exergy of an isolated system can never increase. How is it related to the principle of increase of entropy?

8.37 Give the exergy balance of a steady flow system.

8.38 Derive expressions for the irreversibility and second law efficiency of a:

(i) steam turbine (ii) compressor
(iii) heat exchanger, and (iv) mixer

8.39 What is the deficiency of the first law efficiency? How does the second law efficiency make up this deficiency?

8.40 How can you improve the first law efficiency and the second law efficiency?

Problems

8.1 What is the maximum useful work which can be obtained when 100 kJ are abstracted from a heat reservoir at 675 K in an environment at 288 K? What is the loss of useful work if (a) a temperature drop of 50°C is introduced between the heat source and the heat engine, on the one hand, and the heat engine and the heat sink, on the other, (b) the source temperature drops by 50°C and the sink temperature rises by 50°C during the heat transfer process according to the linear law $\frac{đQ}{dT} = \pm$ constant?

Ans. 57.38 kJ, (a) 11.46 kJ, (b) 5.5 kJ

8.2 In a steam generator, water is evaporated at 260°C, while the combustion gas (c_p = 1.08 kJ/kg K) is cooled from 1300°C to 320°C. The surroundings are at 30°C. Determine the loss in available energy due to the above heat transfer per kg of water evaporated. (Latent heat of vaporization of water at 260°C = 1662.5 kJ/kg.)

Ans. 443.6 kJ

8.3 Exhaust gases leave an internal combustion engine at 800°C and 1 atm, after having done 1050 kJ of work per kg of gas in the engine (c_p of gas = 1.1 kJ/kg K). The temperature of the surroundings is 30°C. (a) How much available energy per kg of gas is lost by throwing away the exhaust gases? (b) What is the ratio of the lost available energy to the engine work? *Ans.* (a) 425.58 kJ, (b) 0.405

8.4 A hot spring produces water at a temperature of 56°C. The water flows into a large lake, with a mean temperature of 14°C, at a rate of 0.1 m^3 of water per min. What is the rate of working of an ideal heat engine which uses all the available energy? *Ans.* 19.5 kW

8.5 0.2 kg of air at 300°C is heated reversibly at constant pressure to 2066 K. Find the available and unavailable energies of the heat added. Take T_0 = 30°C and c_p = 1.0047 kJ/kg K.

Ans. 211.9 and 78.1 kJ

8.6 Eighty kg of water at 100°C are mixed with 50 kg of water at 60°C, while the temperature of the surroundings is 15°C. Determine the decrease in available energy due to mixing. *Ans.* 240 kJ

8.7 A lead storage battery used in an automobile is able to deliver 5.2 MJ of electrical energy. This energy is available for starting the car.

Let compressed air be considered for doing an equivalent amount of work in starting the car. The compressed air is to be stored at 7 MPa, 25°C. What is the volume of the tank that would be required to let the compressed air have an availability of 5.2 MJ? For air, pv = 0.287 T, where T is in K, p in kPa, and v in m^3/kg.

Ans. 0.228 m^3

8.8 Ice is to be made from water supplied at 15°C by the process shown in Fig. P. 8.8. The final temperature of the ice is – 10°C, and the final temperature of the water that is used as cooling water in the condenser is 30°C. Determine the minimum work required to produce 1000 kg of ice. Take c_p for water = 4.187 kJ/kg K, c_p for ice = 2.093 kJ/kgK, and latent heat of fusion of ice = 334 kJ/kg.

Ans. 31.92 MJ

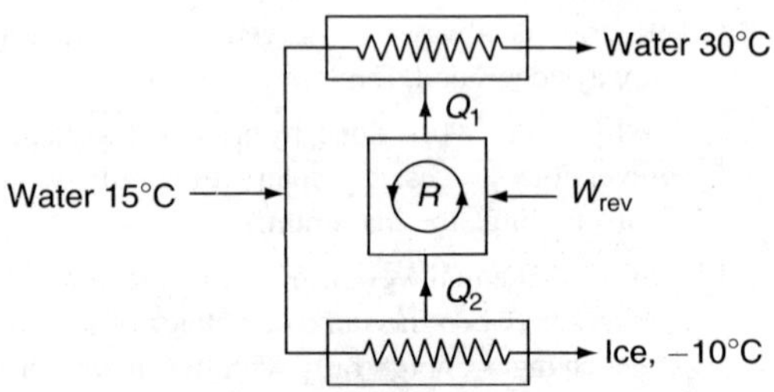

Fig. P. 8.8

8.9 A pressure vessel has a volume of 1 m^3 and contains air at 1.4 MPa, 175°C. The air is cooled to 25°C by heat transfer to the surroundings at 25°C. Calculate the availability in the initial and final states and the irreversibility of this process. Take p_0 = 100 kPa.

Ans. 135 kJ/kg, 114.6 kJ/kg, 222 kJ

8.10 Air flows through an adiabatic compressor at 2 kg/s. The inlet conditions are 1 bar and 310 K and the exit conditions are 7 bar and 560 K. Compute the net rate of availability transfer and the irreversibility. Take T_0 = 298 K.

Ans. 481.1 kW and 21.2 kW

8.11 An adiabatic turbine receives a gas (c_p = 1.09 and c_v = 0.838 kJ/kg K) at 7 bar and 1000°C and discharges at 1.5 bar and 665°C. Determine the second law and isentropic efficiencies of the turbine. Take T_0 = 298 K.

Ans. 0.956, 0.879

8.12 Air enters an adiabatic compressor at atmospheric conditions of 1 bar, 15°C and leaves at 5.5 bar. The mass flow rate is 0.01 kg/s and the efficiency of the compressor is 75%. After leaving the compressor, the air is cooled to 40°C in an aftercooler. Calculate (a) the power required to drive the compressor, and (b) the rate of irreversibility for the overall process (compressor and cooler).

Ans. (a) 2.42 kW, (b) 1 kW

8.13 In a rotary compressor, air enters at 1.1 bar, 21°C where it is compressed adiabatically to 6.6 bar, 250°C. Calculate the irreversibility and the entropy production for unit mass flow rate. The atmosphere is at 1.03 bar, 20°C. Neglect the K.E. changes. *Ans.* 19 kJ/kg, 0.064 kJ/kg K

8.14 In a steam boiler, the hot gases from a fire transfer heat to water which vaporizes at a constant temperature of 242.6°C (3.5 MPa). The gases are cooled from 1100 to 430°C and have an average specific heat, c_p = 1.046 kJ/kg K over this temperature range. The latent heat of vaporization of steam at 3.5 MPa is 1753.7 kJ/kg. If the steam generation rate is 12.6 kg/s and there is negligible heat loss from the boiler, calculate: (a) the rate of heat transfer, (b) the rate of loss of exergy of the gas, (c) the rate of gain of exergy of the steam, and (d) the rate of entropy generation. Take T_0 = 21°C.

Ans. (a) 22096 kW, (b) 15605.4 kW (c) 9501.0 kW, (d) 20.76 kW/K

8.15 An economizer, a gas-to-water finned tube heat exchanger, receives 67.5 kg/s of gas, c_p = 1.0046 kJ/kg K, and 51.1 kg/s of water, c_p = 4.186 kJ/kg. The water rises in temperature from 402 to 469 K, where the gas falls in temperature from 682 to 470 K. There are no changes of kinetic energy, and p_0 = 1.03 bar and T_0 = 289 K. Determine: (a) rate of change of availability of the water, (b) the rate of change of availability of the gas, and (c) the rate of entropy generation. *Ans.* (a) 4802.2 kW, (b) 7079.8 kW, (c) 7.73 kW/K

8.16 The exhaust gases from a gas turbine are used to heat water in an adiabatic counterflow heat exchanger. The gases are cooled from 260 to 120°C, while water enters at 65°C. The flow rates of the gas and water are 0.38 kg/s and 0.50 kg/s respectively. The constant pressure specific heats for the gas and water are 1.09 and 4.186 kJ/kg K respectively. Calculate the rate of exergy loss due to heat transfer. Take T_0 = 35°C.

Ans. 11.92 kW

8.17 The exhaust from a gas turbine at 1.12 bar, 800 K flows steadily into a heat exchanger which cools the gas to 700 K without significant pressure drop. The heat transfer from the gas heats an air flow at constant pressure, which enters the heat exchanger at 470 K. The mass flow rate of air is twice that of the gas and the surroundings are at 1.03 bar, 20°C. Determine: (a) the decrease in availability of the exhaust gases, and (b) the total entropy production per kg of gas. (c) What arrangement would be necessary to make the heat transfer reversible and how much would this increase the power output of the plant per kg of turbine gas? Take c_p for exhaust gas as 1.08 and for air as 1.05 kJ/kg K. Neglect heat transfer to the surroundings and the changes in kinetic and potential energy.

Ans. (a) 66 kJ/kg, (b) 0.0731 kJ/kg K, (c) 38.7 kJ/kg

8.18 An air preheater is used to heat up the air used for combustion by cooling the outgoing products of combustion from a furnace. The rate of flow of the products is 10 kg/s, and the products are cooled from 300°C to 200°C, and for the products at this temperature c_p = 1.09 kJ/kg K. The rate of air flow is 9 kg/s, the initial air temperature is 40°C, and for the air c_p = 1.005 kJ/kg K.

(a) What is the initial and final availability of the products?

(b) What is the irreversibility for this process?

(c) If the heat transfer from the products were to take place reversibly through heat engines, what would be the finally temperature of the air? What power would be developed by the heat engines? Take T_0 = 300 K.

Ans. (a) 85.97, 39.68 kJ/kg, (b) 256.5 kW, (c) 394.41 K, 353.65 kW

8.19 A mass of 2 kg of air in a vessel expands from 3 bar, 70°C to 1 bar, 40°C, while receiving 1.2 kJ of heat from a reservoir at 120°C. The environment is at 0.98 bar, 27°C. Calculate the maximum work and the work done on the atmosphere.

Ans. 177 kJ, 112.5 kJ

8.20 Air enters the compressor of a gas turbine at 1 bar, 30°C and leaves the compressor at 4 bar. The compressor has an efficiency of 82%. Calculate per kg of air (a) the work of compression, (b) the reversible work of compression, and (c) the irreversibility.

[For air, use $$\frac{T_{2s}}{T_1} = \left(\frac{p_2}{p_1}\right)^{\gamma - 1/\gamma}$$

where T_{2s} is the temperature of air after isentropic compression and $\gamma = 1.4$. The compressor efficiency is defined as $(T_{2s} - T_1)/(T_2 - T_1)$, where T_2 is the actual temperature of air after compression].

Ans. (a) 180.5 kJ/kg, (b) 159.5 kJ/kg (c) 21 kJ/kg

8.21 A mass of 6.98 kg of air is in a vessel at 200 kPa, 27°C. Heat is transferred to the air from a reservoir at 727°C. until the temperature of air rises to 327°C. The environment is at 100 kPa, 17°C. Determine (a) the initial and final availability of air, (b) the maximum useful work associated with the process.

Ans. (a) 103.5, 621.9 kJ (b) 582 kJ

8.22 Air enters a compressor in steady flow at 140 kPa, 17°C and 70 m/s and leaves it at 350 kPa, 127°C and 110 m/s. The environment is at 100 kPa, 7°C. Calculate per kg of air (a) the actual amount of work required, (b) the minimum work required, and (c) the irreversibility of the process.

Ans. (a) 114.4 kJ, (b) 97.3 kJ, (c) 17.1 kJ

8.23 Air expands in a turbine adiabatically from 500 kPa, 400 K and 150 m/s to 100 kPa, 300 K and 70 m/s. The environment is at 100 kPa, 17°C. Calculate per kg of air (a) the maximum work output, (b) the actual work output, and (c) the irreversibility.

Ans. (a) 159 kJ, (b) 109 kJ, (c) 50 kJ

8.24 Calculate the specific exergy of air for a state at 2 bar, 393.15 K when the surroundings are at 1 bar, 293.15 K. Take $c_p = 1$ and $R = 0.287$ kJ/kg K.

Ans. 72.31 kJ/kg

8.25 Calculate the specific exergy of CO_2 ($c_p = 0.8659$ and $R = 0.1889$ kJ/kg K) for a state at 0.7 bar, 268.15 K and for the environment at 1.0 bar and 293.15 K.

Ans. – 18.77 kJ/kg

8.26 A pipe carries a stream of brine with a mass flow rate of 5 kg/s. Because of poor thermal insulation the brine temperature increases from 250 K at the pipe inlet to 253 K at the exit. Neglecting pressure losses, calculate the irreversibility rate (or rate of energy degradation) associated with the heat leakage. Take $T_0 = 293$ K and $c_p = 2.85$ kJ/kg K.

Ans. 7.05 kW

8.27 In an adiabatic throttling process, energy per unit mass of enthalpy remains the same. However, there is a loss of exergy. An ideal gas flowing at the rate $\dot{m}$ is throttled from pressure p_1 to pressure p_2 when the environment is at temperature T_0. What is the rate of exergy loss due to throttling?

Ans. $\dot{I} = \dot{m} R T_0 \ln \frac{p_1}{p_2}$

8.28 Air at 5 bar and 20°C flows into an evacuated tank until the pressure in the tank is 5 bar. Assume that the process is adiabatic and the temperature of the surroundings is 20°C. (a) What is the final temperature of the air? (b) What is the reversible work produced between the initial and final states of the air? (c) What is the net entropy change of the air entering the tank? (d) Calculate the irreversibility of the process.

Ans. (a) 410.2 K, (b) 98.9 kJ/kg, (c) 0.3376 kJ/kg K, (d) 98.9 kJ/kg

8.29 A Carnot cycle engine receives and rejects heat with a 20°C temperature differential between itself and the thermal energy reservoirs. The expansion and compression processes have a pressure ratio of 50. For 1 kg of air as the working substance, cycle temperature limits of 1000 K and 300 K and $T_0 = 280$ K, determine the second law efficiency.

Ans. 0.965

8.30 Energy is received by a solar collector at the rate of 300 kW from a source temperature of 2400 K. If 60 kW of this energy is lost to the surroundings at steady state and if the user temperature remains constant at 600 K, what are the first law and the second law efficiencies? Take $T_0 = 300$ K.

Ans. 0.80, 0.457

8.31 For flow of an ideal gas through an insulated pipeline, the pressure drops from 100 bar to 95 bar. If the gas flows at the rate of 1.5 kg/s and has $c_p = 1.005$ and $c_v = 0.718$ kJ/kg K and if $T_0 = 300$ K, find the rate of entropy generation, and rate of loss of exergy.

Ans. 0.0215 kW/K, 6.46 kW

8.32 The cylinder of an internal combustion engine contains gases at 2500°C, 58 bar. Expansion takes place through a volume ratio of 9 according to $pv^{1.38}$ = const. The surroundings are at 20°C, 1.1 bar. Determine the loss of availability, the work transfer and the heat transfer per unit mass. Treat the gases as ideal having $R = 0.26$ kJ/kg K and $c_v = 0.82$ kJ/kg K.

Ans. 1144 kJ/kg, 1074 kJ/kg, – 213 kJ/kg

8.33 In a counterflow heat exchanger, oil ($c_p = 2.1$ kJ/kg K) is cooled from 440 to 320 K, while water ($c_p = 4.2$ kJ/kg K) is heated from 290 K to temperature T. The respective mass flow rates of oil and water are 800 and 3200 kg/h. Neglecting pressure drop, KE and PE effects and heat loss, determine (a) the temperature T, (b) the rate of exergy destruction, (c) the second law efficiency. Take $T_0 = 17$°C and $p_0 = 1$ atm.

Ans. (a) 305 K, (b) 41.4 MJ/h, (c) 10.9%

8.34 Oxygen enters a nozzle operating at steady state at 3.8 MPa, 387°C and 10 m/s. At the nozzle exit the conditions are 150 kPa, 37°C and 750 m/s. Determine (a) the heat transfer per kg and (b) the irreversibility. Assume oxygen as an ideal gas, and take $T_0 = 20$°C, $p_0 = 1$ atm.

Ans. (a) – 37.06 kJ/kg, (b) 81.72 kJ/kg

8.35 Argon gas expands adiabatically in a turbine from 2 MPa, 1000°C to 350 kPa. The mass flow rate is 0.5 kg/s and the turbine develops power at the rate of 120 kW. Determine (a) the temperature of argon at the turbine exit, (b) the irreversibility rate, and (c) the second law efficiency. Neglect KE and PE effects and take $T_0 = 20$°C, $p_0 = 1$ atm.

Ans. (a) 538.1°C (b) 18.78 kW, (c) 86.5%

8.36 In the boiler of a power plant are tubes through which water flows as it is brought from 0.8 MPa, 150°C (h = 632.6 kJ/kg, s = 1.8418 kJ/kg K) to 0.8 MPa, 250°C (h = 2950 kJ/kg, s = 7.0384 kJ/kg K). Combustion gases passing over the tubes cool from 1067°C to 547°C. These gases may be considered as air (ideal gas) having $c_p = 1.005$ kJ/kg K. Assuming steady state and neglecting any heat loss, and KE and PE effects, determine (a) the mass flow rate of combustion gases per kg of steam, (b) the loss of exergy per kg steam, and (c) the second law efficiency. Take $T_0 = 25$°C, $p_0 = 1$ atm.

Ans. (a) $m_g/m_w = 4.434$, (b) 802.29 kJ/kg steam, (c) 48.9%

8.37 Air enters a hair dryer at 22°C, 1 bar with a velocity of 3.7 m/s and exits at 83°C, 1 bar with a velocity of 9.1 m/s through an area of 18.7 cm². Neglecting any heat loss and PE effect and taking $T_0 = 22$°C, (a) evaluate the power required in kW, and (b) devise and evaluate a second law efficiency.

Ans. (a) – 1.02 kW, (b) 9%

8.38 An isolated system consists of two solid blocks. One block has a mass of 5 kg and is initially at 300°C. The other block has a mass of 10 kg and is initially at – 50°C. The blocks are allowed to come into thermal equilibrium. Assuming the blocks are incompressible with constant specific heats of 1 and 0.4 kJ/kg K, respectively, determine (a) the final temperature, (b) the irreversibility. Take $T_0 = 300$ K. *Ans.* (a) 417.4 K, (b) 277 kJ

8.39 Air flows into a heat engine at ambient conditions 100 kPa, 300 K. Energy is supplied as 1200 kJ per kg air from a 1500 K source and in some part of the process, a heat loss of 300 kJ/kg air happens at 750 K. The air leaves the engine at 100 kPa, 800 K. Find the first and the second law efficiencies. *Ans.* 0.315, 0.672

8.40 Consider two rigid containers each of volume 1 m³ containing air at 100 kPa, 400 K. An internally reversible Carnot heat pump is then thermally connected between them so that it heats one up and cools the other down. In order to transfer heat at a reasonable rate, the temperature difference between the working fluid inside the heat pump and the air in the containers is set to 20°C. The process stops when the air in the coldest tank reaches 300 K. Find the final temperature of the air that is heated up, the work input to the heat pump, and the overall second law efficiency.

Ans. 550 K, 31.2 kJ, 0.816

C H A P T E R 9

Properties of Pure Substances

A pure substance is a substance of constant chemical composition throughout its mass. It is a one-component system. It may exist in one or more phases.

9.1 p–v DIAGRAM FOR A PURE SUBSTANCE

Assume a unit mass of ice (solid water) at –10°C and 1 atm contained in a cylinder and piston machine (Fig. 9.1). Let the ice be heated slowly so that its temperature is always uniform. The changes which occur in the mass of water would be traced as the temperature is increased while the pressure is held constant. Let the state changes of water be plotted on p–v coordinates. The distinct regimes of heating, as shown in Fig. 9.2, are:

1–2 The temperature of ice increases from –10°C to 0°C. The volume of ice would increase, as would be the case for any solid upon heating. At state 2, i.e. 0°C, the ice would start melting.

2–3 Ice melts into water at a constant temperature of 0°C. At state 3, the melting process ends. *There is a decrease in volume, which is a peculiarity of water.*

3–4 The temperature of water increases, upon heating, from 0°C to 100°C. The volume of water increases because of thermal expansion.

4–5 The water starts boiling at state 4 and boiling ends at state 5. This phase change from liquid to vapour occurs at a constant temperature of 100°C (the pressure being constant at 1 atm). There is a large increase in volume.

5–6 The vapour is heated to, say, 250°C (state 6). The volume of vapour increases from v_5 to v_6.

Water existed in the solid phase between 1 and 2, in the liquid phase between 3 and 4, and in the gas phase beyond 5. Between 2 and 3, the solid changed into the liquid phase by absorbing the latent heat of fusion and between 4 and 5, the liquid changed into the vapour phase by absorbing the latent heat of vaporization, both at constant temperature and pressure.

The states 2, 3, 4 and 5 are known as *saturation states*. A saturation state is a state from which a change of phase may occur without a change of pressure or temperature. State 2 is a *saturated solid state* because a solid can change into liquid at constant pressure and temperature from state 2. States 3 and 4 are both saturated liquid states. In state 3, the liquid is saturated with respect to solidification, whereas in state 4, the liquid is saturated with respect to vaporization. State 5 is a *saturated vapour state*, because from state 5, the vapour can condense into liquid without a change of pressure or temperature.

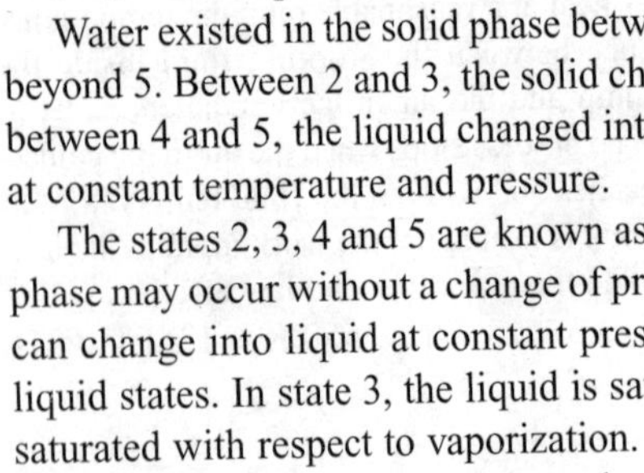

Fig. 9.1 *Heating of H_2O at a constant pressure of 1 atm*

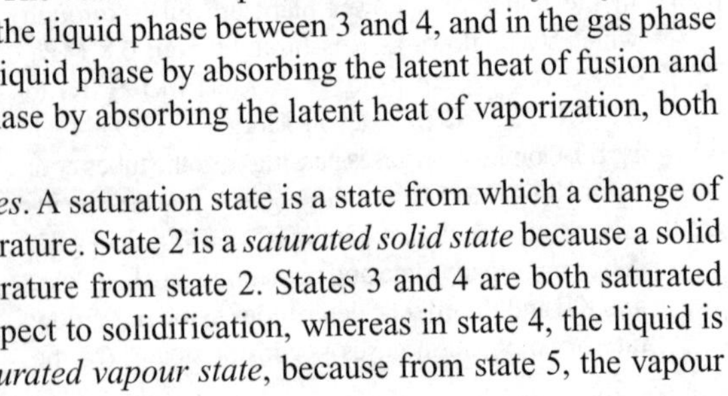

Fig. 9.2 *Changes in the volume of water during heating at constant pressure*

If the heating of ice at – 10°C to steam at 250°C were done at a constant pressure of 2 atm, similar regimes of heating would have been obtained with similar saturation states 2, 3, 4 and 5, as shown in Fig. 9.2. All the state changes of the system can similarly be plotted on the $p-v$ coordinates, when it is heated at different constant pressures. All the saturated solid states 2 at various pressures are joined by a line, as shown in Fig. 9.3.

Similarly, all the saturated liquid states 3 with respect to solidification, all the saturated liquid states 4 with respect to vaporization, and all the saturated vapour states 5, are joined together.

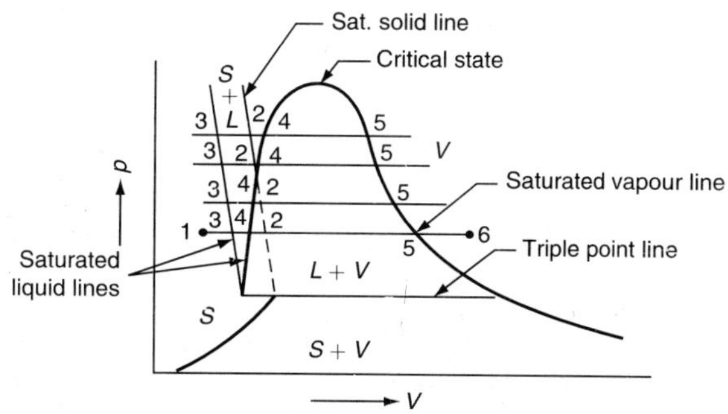

Fig. 9.3 *p–v diagram of water, whose volume decreases on melting*

Figure 9.4 shows state changes of a pure substance other than water whose volume increases on melting.

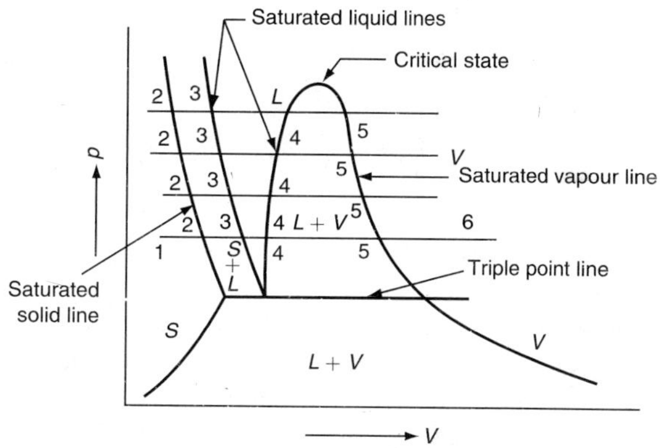

Fig. 9.4 *p–v diagram of a pure substance other than water, whose volume increases on melting*

The line passing through all the saturated solid states 2 (Figs. 9.3 and 9.4) is called the *saturated solid line*. The lines passing through all the saturated liquid states 3 and 4 with respect to solidification and vaporization respectively are known as the *saturated liquid lines*, and the line passing through all the saturated vapour states 5, is the *saturated vapour line*. The saturated liquid line with respect to vaporization and the saturated vapour line incline towards each other and form what is known as the *saturation* or *vapour dome*. The two lines meet at the *critical state*.

To the left of the saturated solid line is the *solid (S) region* (Fig. 9.4). Between the saturated solid line and saturated liquid line with respect to solidification there exists the *solid-liquid mixture* $(S + L)$ region. Between the two saturated liquid lines is the *compressed liquid region*. The *liquid-vapour mixture region* $(L + V)$ exists within the vapour dome between the saturated liquid and saturated vapour lines. to the right of the saturated vapour line is the *vapour region*. The *triple point* is a line on the $p-v$ diagram, where all the three phases, solid, liquid, and gas, exist in equilibrium. At a pressure below the triple point line, the substance cannot exist in the liquid phase, and the substance, when heated, transforms from solid to vapour (known as sublimation) by absorbing the latent heat of sublimation from the surroundings. The region below the triple point line is, therefore, the *solid-vapour* $(S + V)$ *mixture region*. Table 9.1 gives the triple point data for a number of substances.

Liquid is, most often, the working fluid in power cycles, etc. and interest is often confined to the liquid-vapour regions only. So to locate the state points, the solid regions from Figs 9.3 and 9.4 can be omitted. The $p-v$ diagram then becomes as shown in Fig. 9.5. If the vapour at state A is compressed slowly and isothermally, the pressure will rise until there is saturated vapour at point B. If the compression is continued, condensation takes place, the pressure remaining constant so long as the temperature remains constant. At any point between B and C, the liquid and vapour are in equilibrium. Since a very large increase in pressure is needed to compress the liquid, line CD is almost vertical. $ABCD$ is a typical *isotherm* of a pure substance

Table 9.1

Triple-point data

Substance	*Temperature, K*	*Pressure, mm Hg*
Acetylene, C_2H_2	192.4	962
Ammonia, NH_3	195.42	45.58
Argon, A	83.78	515.7
Carbon dioxide, CO_2	216.55	3885.1
Carbon monoxide, CO	68.14	115.14
Methane, C_2H_6	89.88	0.006
Ethylene, C_2H_4	104.00	0.9
Hydrogen, H_2	13.84	52.8
Methane, CH_4	90.67	87.7
Nitrogen, N_2	63.15	94.01
Oxygen, O_2	54.35	1.14
Water, H_2O	273.16	4.587

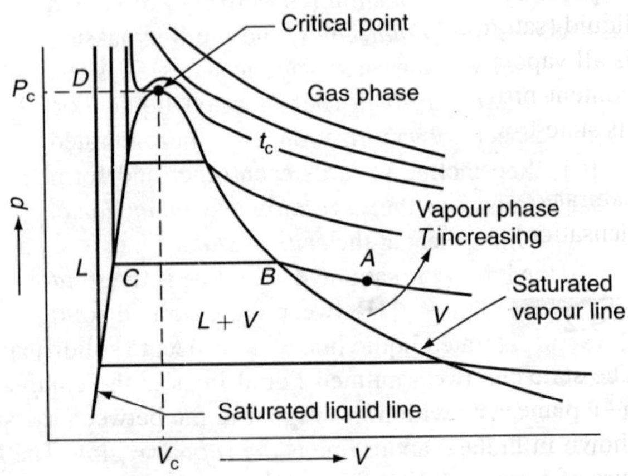

Fig 9.5 *Saturation curve on p–v diagram*

on a p–v diagram. Some isotherms are shown in Fig. 9.5. As the temperature increases, the liquid-vapour transition, as represented by *BC*, decreases, and becomes zero at the critical point. Below the critical point only, there is a liquid-vapour transition zone, where a saturated liquid, on heating, absorbs the latent heat of vaporization, and becomes saturated vapour at a constant pressure and temperature. Similarly, a saturated vapour, on cooling, releases the latent heat of condensation at constant pressure and temperature to become saturated liquid. Above the critical point, however, a liquid, upon heating, suddenly *flashes* into vapour, or a vapour, upon cooling, suddenly condenses into liquid. There is no distinct transition zone from liquid to vapour and vice versa. The isotherm passing through the critical point is called the *critical isotherm*, and the corresponding temperature is known as the *critical temperature* (t_c). The pressure and volume at the critical point are known as the *critical pressure* (p_c) and the *critical volume* (v_c) respectively. For water

$$p_c = 221.2 \text{ bar}$$
$$t_c = 374.15°\text{C}$$
$$v_c = 0.00317 \text{ m}^3/\text{kg}$$

The critical point data of certain substances are given in Appendix F. Above the critical point, the isotherms are continuous curves that at large volumes and low pressures approach equilateral hyperbolas.

When a liquid or solid is in equilibrium with its vapour at a given temperature, the vapour exerts a pressure that depends only on the temperature (Fig. 9.6). In general, the greater the temperature, the higher is the *vapour pressure*. The temperature at which the vapour pressure is equal to 760 mm Hg is called the *normal boiling point*.

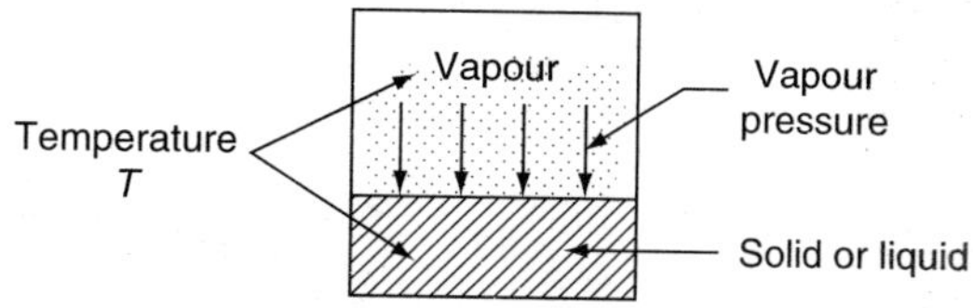

Fig. 9.6 *Vapour pressure*

Phase change occurs at constant pressure and temperature. A pure liquid at a given pressure will transform into vapour only at a particular temperature, known as *saturation temperature*, which is a function of pressure. Similarly, if the temperature is fixed, the liquid will boil (or condense) only at a particular pressure, called the *saturation pressure*, which is a function of temperature. In Fig. 9.7, if p_1 is the pressure, the corresponding saturation temperature is $(t_{sat})_1$, or if t_2 is the given temperature, the saturation pressure is $(p_{sat})_2$. As the pressure increases, the saturation temperature increases. Saturation states exist up to the critical point. At point *A*, the liquid starts boiling, and at point *B*, the boiling gets completed. At *A*, it is all liquid (saturated) and there is no vapour, while at *B*, it is all vapour (saturated) and there is no liquid. Vapour content progressively increases as the liquid changes its state from *A* towards *B*.

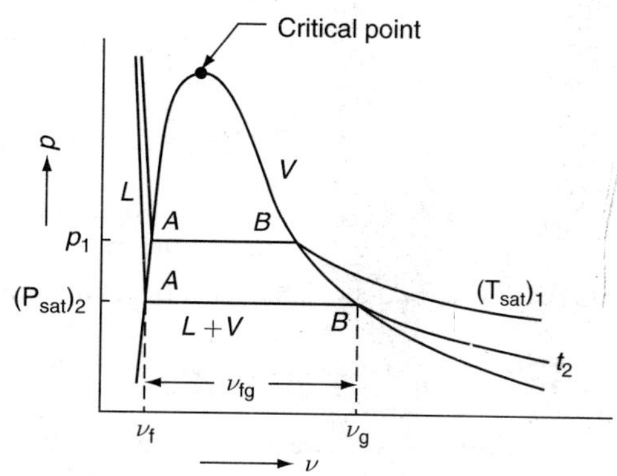

Fig. 9.7 *Saturation pressure and temperature*

If v_f is the specific volume of the saturated liquid at a given pressure, and v_g the specific volume of the saturated vapour, then $(v_g - v_f)$ or v_{fg} is the change in specific volume during phase transition (boiling or condensation) at the pressure. As pressure increases, v_{fg} decreases, and at the critical point v_{fg} becomes zero.

9.2 *p–T* DIAGRAM FOR A PURE SUBSTANCE

The state changes of a pure substance, upon slow heating at different constant pressures, are shown on the *p–v* plane, in Figs 9.2, 9.3, and 9.4. If these state changes are plotted on *p–T* coordinates, the diagram, as shown in Fig. 9.8, will be obtained. If the heating of ice at –10°C to steam at 250°C at the constant pressure of 1 atm is considered, 1–2 is the solid (ice) heating, 2–3 is the melting of ice at 0°C, 3–4 is the liquid heating, 4–5 is the vaporization of water at 100°C, and 5–6 is the heating in the vapour phase. The process will be reversed from state 6 to state 1 upon cooling. The curve passing through the 2, 3 points is called the *fusion curve*, and the curve passing through the 4, 5 points (which indicate the vaporization or condensation at different temperatures and pressures) is called the *vaporization curve*. If the vapour pressure of a solid is measured at different temperatures, and these are plotted, the *sublimation curve* will be obtained. The fusion curve, the vaporization curve, and the sublimation curve meet at the triple point.

The slopes of the sublimation and vaporization curves for all substances are positive. The slope of the fusion curve for most substances is positive, but for water, it is negative. The temperature at which a liquid boils is very sensitive to pressure, as indicated by the vaporization curve which gives the saturation temperatures at different pressures, but the temperature at which a solid melts is not such a strong function of pressure, as indicated by the small slope of the fusion curve.

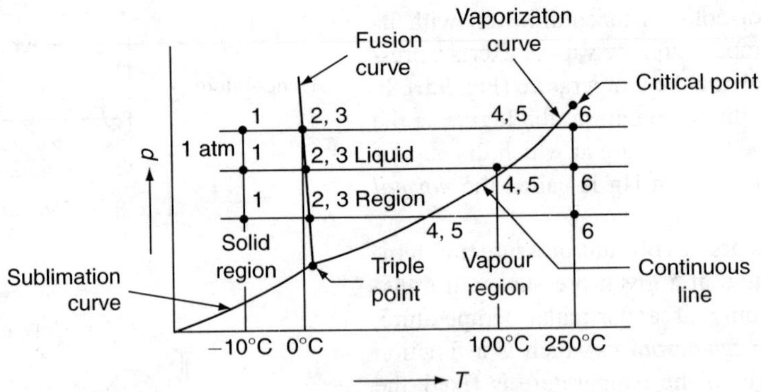

Fig. 9.8 *Phase equilibrium diagram on p–T coordinates*

The triple point of water is at 4.58 mm Hg and 273.16 K, whereas that of CO_2 is at 3885 mm Hg (about 5 atm) and 216.55 K. So when solid CO_2 ('dry ice') is exposed to 1 atm pressure, it gets transformed into vapour directly, absorbing the latent heat of sublimation from the surroundings, which gets cooled or 'refrigerated'.

9.3 *p–v–T* SURFACE

The relationships between pressure, specific volume, and temperature can be clearly understood with the aid of a three-dimensional *p–v–T* surface. Figure 9.9 illustrates a substance like water that expands upon freezing and Fig. 9.10 illustrates substances other than water which contract upon freezing. The projections on the *p–T* and *p–v* planes are also shown in these figures. Any point on the *p–v–T* surface represents an equilibrium state of the substance. The triple point line when projected to the *p–T* plane becomes a point. The critical isotherm has a point of inflection at the critical point.

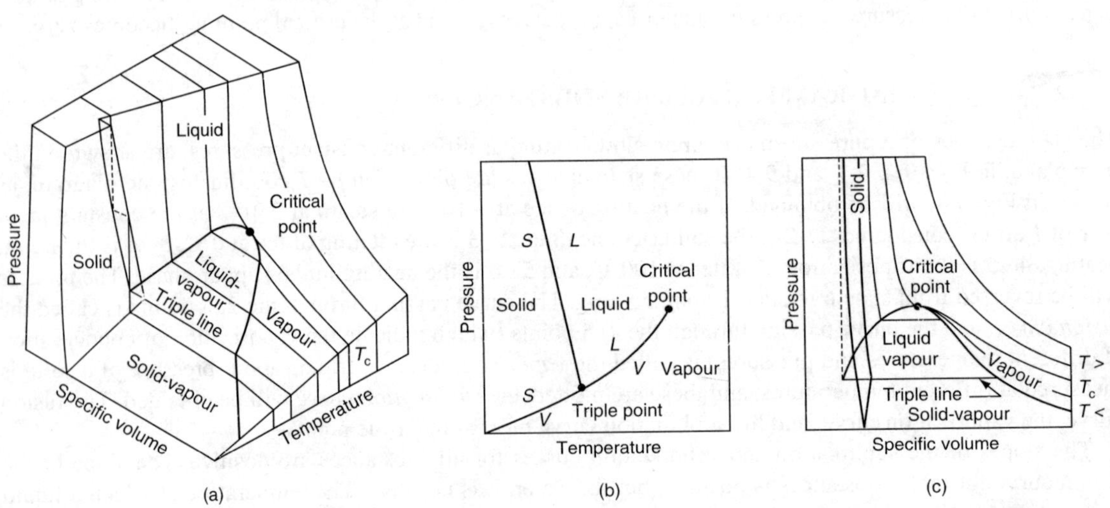

Fig. 9.9 *p–v–T surface and projections for a substance that expands on freezing (a) Three-dimensional view (b) Phase diagram (c) p–v diagram*

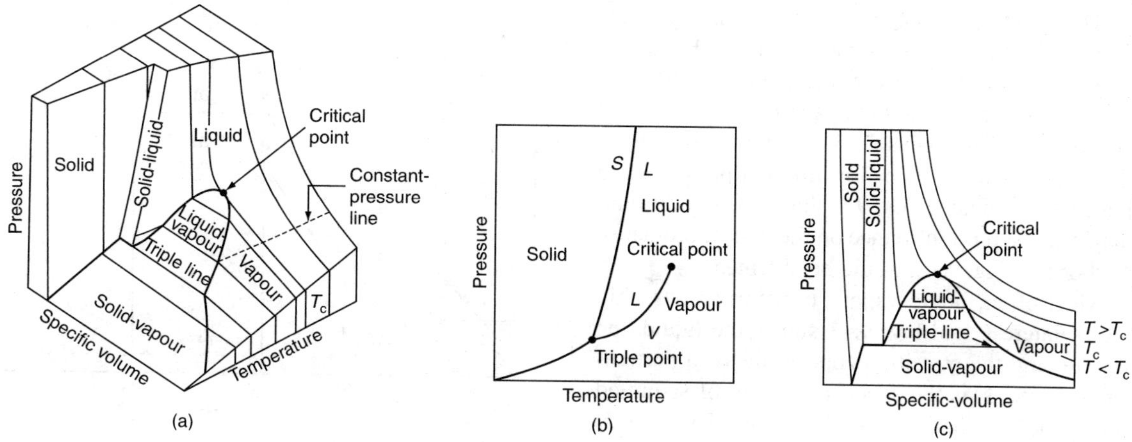

Fig. 9.10 *p–v–T surface and projections for a substance that contracts on freezing (a) Three-dimensional view (b) Phase diagram (c) p–v diagram*

9.4 T–s DIAGRAM FOR A PURE SUBSTANCE

The heating of the system of 1 kg of ice at –5°C to steam at 250°C is again considered, the pressure being maintained constant at 1 atm. The entropy increases of the system in different regimes of heating are given below.

1. The entropy increase of ice as it is heated from –5°C to 0°C at 1 atm. (c_{pice} = 2.093 kJ/kg K).

$$\Delta s_1 = s_2 - s_1 = \int \frac{đQ}{T} = \int_{T_1=268}^{T_2=273} \frac{mc_p dT}{T} = c_p \ln \frac{273}{268} = 2.093 \ln \frac{273}{268} = 0.0398 \text{ kJ/kg K.}$$

2. The entropy increase of ice as it melts into water at 0°C (latent heat of fusion of ice = 334.96 kJ/kg)

$$\Delta s_2 = s_3 - s_2 = \frac{334.96}{273} = 1.23 \text{ kJ/kg K}$$

3. The entropy increase of water as it is heated from 0°C to 100°C ($c_{\text{p water}}$ = 4.187 kJ/kg K)

$$\Delta s_3 = s_4 - s_3 = c_p \ln \frac{T_3}{T_2} = 4.187 \ln \frac{373}{273} = 1.305 \text{ kJ/kg K}$$

4. The entropy increase of water as it is vaporized at 100°C, absorbing the latent heat of vaporization (2257 kJ/kg)

$$\Delta s_4 = s_5 - s_4 = \frac{2257}{273} = 6.05 \text{ kJ/kg K}$$

5. The entropy increase of vapour as it is heated from 100°C to 250°C at 1 atm

$$\Delta s_5 = s_6 - s_5 = \int_{373}^{523} c_p \frac{dT}{T} = 2.093 \ln \frac{523}{373} = 0.706 \text{ kJ/kg K}$$

assuming the average specific heat of steam in the temperature range of 100°C to 250°C as 2.093 kJ/kg K.

These entropy changes are shown in Fig. 9.11. The curve 1-2-3-4-5-6 is the isobar of 1 atm. If, during the heating process, the pressure had been maintained constant at 2 atm, a similar curve would be obtained. The states 2, 3, 4, and 5 are saturation states. If these states for different pressures are joined, as in Figs 9.3 and 9.4, the phase equilibrium diagram of a pure substance on the T–s coordinates, as shown in Fig. 9.12, would be obtained.

Most often, liquid-vapour transformations only are of interest, and Fig. 9.13 shows the liquid, the vapour, and the transition zones only. At a particular pressure, s_f is the specific entropy of saturated water, and s_g is that of saturated vapour. The entropy change of the system during the phase change from liquid to vapour at that pressure is s_{fg} $(= s_g - s_f)$. The value of s_{fg} decreases as the pressure increases, and becomes zero at the critical point.

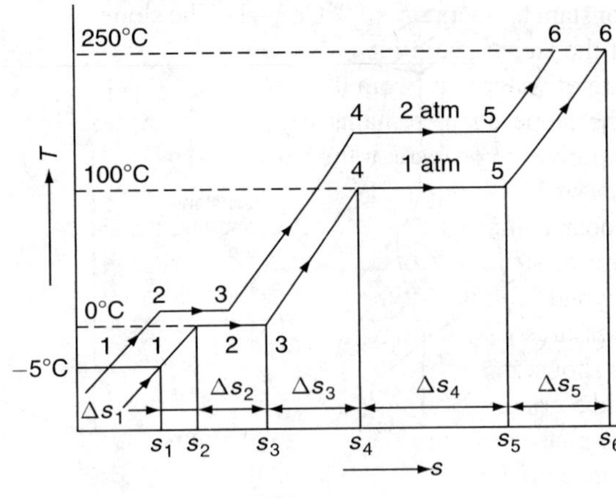

Fig. 9.11 *Isobars on T–s plot*

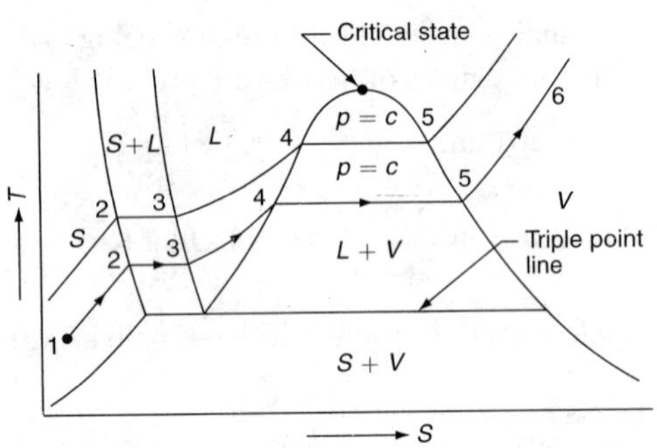

Fig. 9.12 *Phase equilibrium diagram on T–s coordinates*

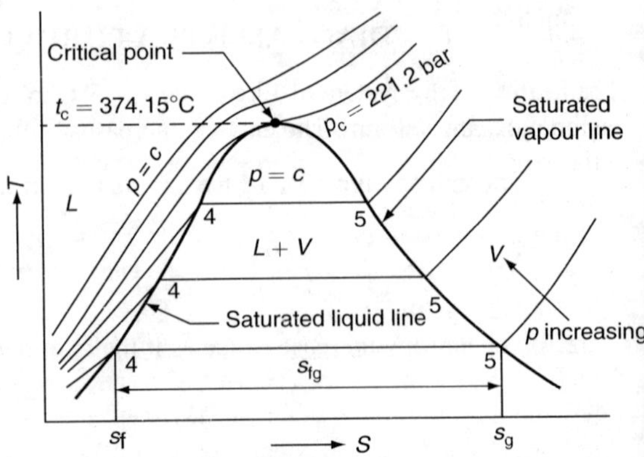

Fig. 9.13 *Saturation (or vapour) dome for water*

9.5 h–s DIAGRAM OR MOLLIER DIAGRAM FOR A PURE SUBSTANCE

From the first and second laws of thermodynamics, the following property relation was obtained.

$$Tds = dh - vdp$$

or

$$\left(\frac{\partial h}{\partial s}\right)_p = T \tag{9.1}$$

This equation forms the basis of the h–s diagram of a pure substance, also called the Mollier diagram. The slope of an isobar on the h–s coordinates is equal to the absolute saturation temperature $(t_{sat} + 273)$ at that pressure. If the temperature remains constant the slope will remain constant. If the temperature increases, the slope of the isobar will increase.

Consider the heating of a system of ice at −5°C to steam at 250°C, the pressure being maintained constant at 1 atm. The slope of the isobar of 1 atm on the h–s coordinates (Fig. 9.14) first increases as the temperature of the ice increases from −5°C to 0°C (1–2). Its slope then remains constant as ice melts into water at the constant temperature of 0°C (2–3). The slope of the isobar again increases as the temperature of water rises from 0°C to 100°C (3–4). The slope again remains constant as water vaporizes into steam at the constant temperature of 100°C (4–5). Finally, the slope of the isobar continues to increase as the temperature of steam increases to 250°C (5–6) and beyond. Similarly, the isobars of different pressures can be drawn on the $h-s$ diagram as shown in Figs 9.14 and 9.15. States 2, 3, 4 and 5 are saturation states. Figure 9.15 shows the phase equilibrium diagram of a pure substance on the $h-s$ coordinates, indicating the saturated solid line, saturated liquid lines and saturated vapour line, the various phases, and the transition (mixture) zones.

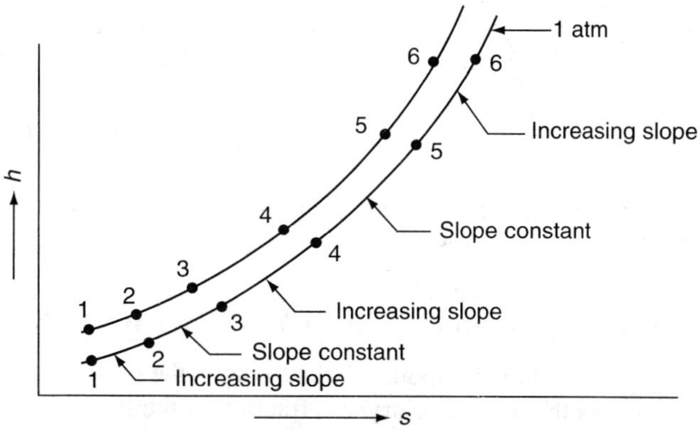

Fig. 9.14 *Isobars on h–s plot*

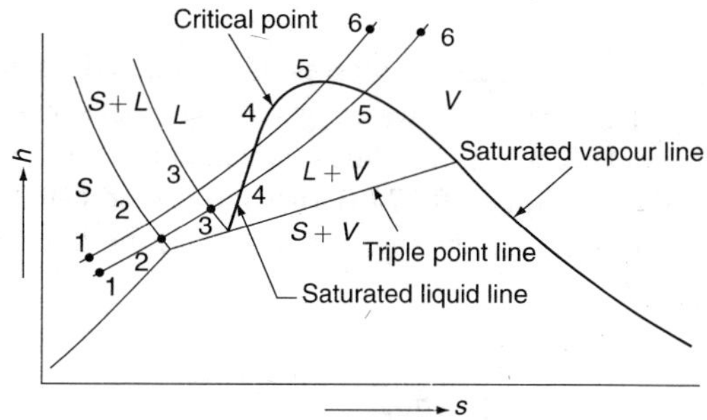

Fig. 9.15 *Phase equilibrium diagram on h–s coordinates (mollier diagram)*

Figure 9.16 is the $h-s$ or the Mollier diagram indicating only the liquid and vapour phases. As the pressure increases, the saturation temperature increases, and so the slope of the isobar also increases. Hence, the *constant pressure lines diverge from one another*, and the critical isobar is a tangent at the critical point, as shown. In the vapour region, the states of equal slopes at various pressures are joined by lines, as shown, which are the *constant temperature lines*. Although the slope of an isobar remains continuous beyond the saturated vapour line, the isotherm bends towards the right and its slope decreases asymptotically to zero, because in the ideal gas region it becomes horizontal and the constant enthalpy implies constant

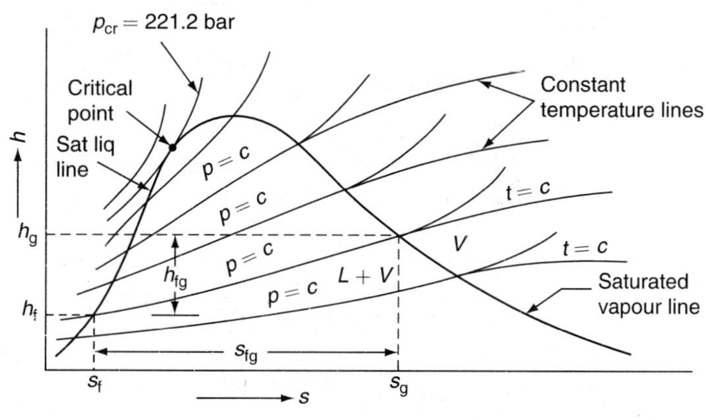

Fig. 9.16 *Enthalpy-entropy diagram of water*

temperature. $\left[\left(\frac{\partial h}{\partial s}\right)_T = T + v\left(\frac{\partial p}{\partial s}\right)_T = T - v\left(\frac{\partial T}{\partial v}\right)_p\right]$, by Maxwell's relation, Chapter 11. For an ideal gas, $\left(\frac{\partial h}{\partial s}\right)_T = T - v\frac{T}{v} = 0$. For $T = c$, $h = c$]. At a particular pressure, h_f is the specific enthalpy of saturated water, h_g is that of saturated vapour, and h_{fg} $(= h_g - h_f)$ is the latent heat of vaporization at that pressure. As the pressure increases, h_{fg} decreases, and at the critical pressure, h_{fg} becomes zero.

9.6 QUALITY OR DRYNESS FRACTION

If in 1 kg of liquid-vapour mixture, x kg is the mass of vapour and $(1 - x)$ kg is the mass of liquid, then x is known as the *quality* or dryness fraction of the liquid-vapour mixture. Therefore, quality indicates the mass fraction of vapour in a liquid vapour mixture, or

$$x = \frac{m_v}{m_v + m_l}$$

where m_v and m_l are the masses of vapour and liquid respectively in the mixture. The value of x varies between 0 and 1. For saturated water, when water just starts boiling, $x = 0$, and for saturated vapour, when vaporization is complete, $x = 1$, for which the vapour is said to be *dry saturated*.

Points m in Fig. 9.17 (a), (b), and (c) indicate the saturated liquid states with $x = 0$, and points n indicate the saturated vapour states with $x = 1$, the lines mn indicating the transition from liquid to vapour. Points

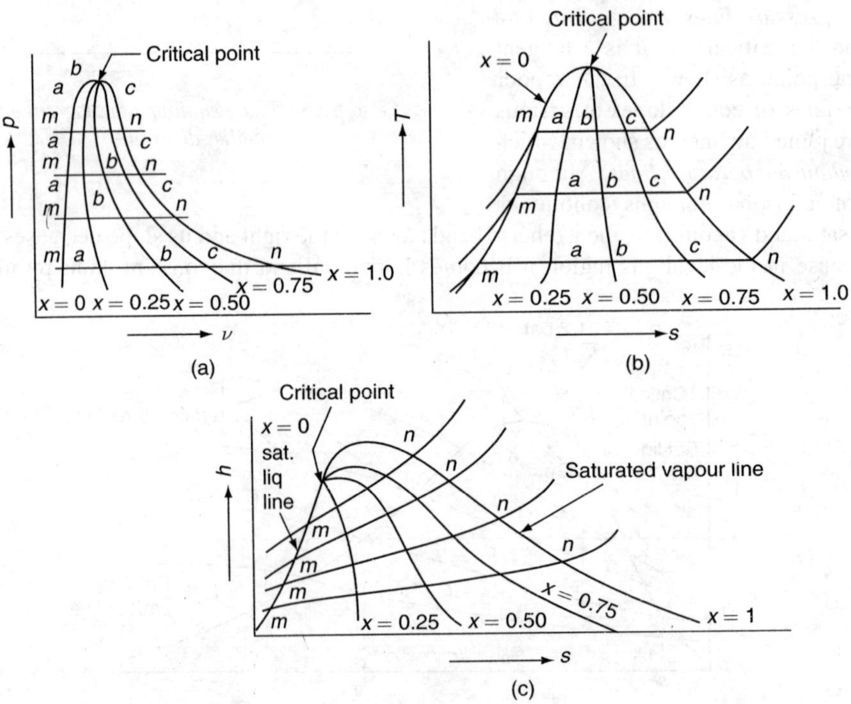

Fig. 9.17 *Constant quality lines on p–v, T–s and h–s diagrams*

a, b, and c at various pressures indicate the situations when the masses of vapour reached 25%, 50%, and 75% of the total mass, i.e. at points a, the mass of liquid is 75% and the mass of vapour is 25% of the total mass, at points b, the mixture consists of 50% liquid and 50% vapour by mass, and at points c, the mixture consists of 75% vapour and 25% liquid by mass. The lines passing through points a, b and c are the constant quality lines of 0.25, 0.50, and 0.75 respectively. Constant quality lines start from the critical point.

Let V be the total volume of a liquid vapour mixture of quality x, V_f the volume of the saturated liquid, and V_g the volume of the saturated vapour, the corresponding masses being m, m_f, and m_g respectively.

Now $$m = m_f + m_g$$

and $$V = V_f + V_g$$

$$mv = m_f v_f + m_g v_g = (m - m_g) v_f + m_g v_g$$

∴ $$v = \left(1 - \frac{m_g}{m}\right) v_f + \frac{m_g}{m} v_g$$

$$v = (1 - x) v_f + x v_g \tag{9.2}$$

where $x = \dfrac{m_g}{m}$, v_f = specific volume of saturated liquid, v_g = specific volume of saturated vapour, and v = specific volume of the mixture of quality x.

Similarly

$$s = (1 - x) s_f + x s_g \tag{9.3}$$

$$h = (1 - x) h_f + x h_g \tag{9.4}$$

$$u = (1 - x) u_f + x u_g \tag{9.5}$$

where s, h, and u refer to the mixture of quality x, the suffix f and suffix g indicate the conditions of saturated liquid and saturated vapour respectively.

From Eq. (9.2)

$$v = (1 - x) v_f + x v_g$$
$$= v_f + x(v_g - v_f)$$
$$= v_f + x\, v_{fg}$$

or $$v = v_f + x v_{fg} \tag{9.6}$$

Similarly

$$h = h_f + x h_{fg} \tag{9.7}$$

$$s = s_f + x s_{fg} \tag{9.8}$$

$$u = u_f + x u_{fg} \tag{9.9}$$

However, $$\rho \neq \rho_f + x \rho_{fg}$$

Volume fraction of vapour or voidage, $\alpha = v_g/v$

$$m = m_f + m_g$$

$$\rho v = \rho_f v_f + \rho_g v_g = \rho_f (v - v_g) + \rho_g v_g$$

$$\rho = (1 - \alpha) \rho_f + \alpha \rho_g \tag{9.9a}$$

9.7 STEAM TABLES

The properties of water are arranged in the steam tables as functions of pressure and temperature. Separate tables are provided to give the properties of water in the saturation states and in the liquid and vapour phases. The *internal energy of saturated water at the triple point* ($t = 0.01°C$) *is arbitrarily chosen to be zero*. Since $h = u + pv$, the enthalpy of saturated water at 0.01°C is slightly positive because of the small value of (pv) term. *The entropy of saturated water is also chosen to be zero at the triple point.*

9.7.1 Saturation States

When a liquid and its vapour are in equilibrium at a certain pressure and temperature, only the pressure or the temperature is sufficient to identify the saturation state. If the pressure is given, the temperature of the mixture gets fixed, which is known as the saturation temperature, or if the temperature is given, the saturation pressure gets fixed. Saturated liquid or the saturated vapour has only one independent variable, i.e. only one property is required to be known to fix up the state. Tables A.1 (a) and A.1 (b) in the appendix give the properties of saturated liquid and saturated vapour. In Table A.1 (a), the independent variable is temperature. At a particular temperature, the values of saturation pressure p, and v_f, v_g, h_f, h_{fg}, h_g, s_f and s_g are given, where v_f, h_f, and s_f refer to the saturated liquid states; v_g, h_g and s_g refer to the saturated vapour state; and v_{fg}, h_{fg}, and s_{fg} refer to the changes in the property values during evaporation (or condensation) at that temperature, where $v_{fg} = v_g - v_f$ and $s_{fg} = s_g - s_f$.

In Table A.1 (b), the independent variable is pressure. At a particular pressure, the values of saturation temperature t, and v_f, v_g, h_f, h_{fg}, h_g, s_f, and s_g are given. Depending upon whether the pressure or the temperature is given, either Table A.1 (a) or Table A.1 (b) can be conveniently used for computing the properties of saturation states.

If data are required for intermediate temperatures or pressures, *linear interpolation is normally accurate.* The reason for the two tables is to reduce the amount of interpolation required.

9.7.2 Liquid-vapour Mixtures

Let us consider a mixture of saturated liquid water and water vapour in equilibrium at pressure p and temperature t. The composition of the mixture by mass will be given by its quality x, and its state will be within the vapour dome (Fig. 9.18). The properties of the mixture are as given in Article 9.6, i.e.

$$v = v_f + xv_{fg}$$

$$u = u_f + xu_{fg}$$

$$h = h_f + xh_{fg}$$

$$s = s_f + xs_{fg}$$

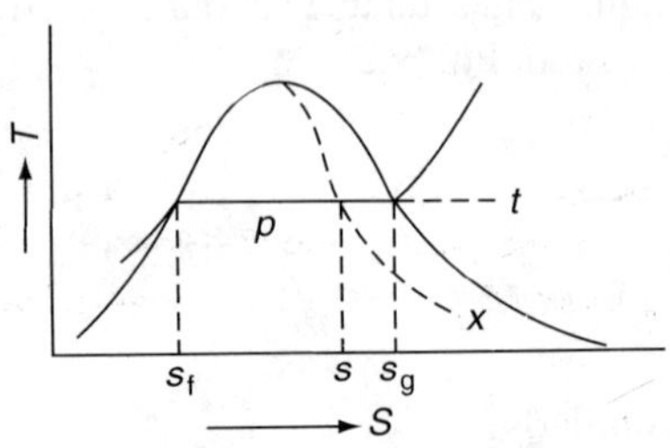

Fig. 9.18 *Property in two phase region*

where v_f, v_{fg}, u_f, u_{fg}, h_f, h_{fg}, s_f and s_{fg} are the saturation properties at the given pressure and temperature.

If p or t and the quality of the mixture are given, the properties of the mixture (v, u, h and s) can be evaluated from the above equations. Sometimes, instead of quality, one of the above properties, say, specific volume v, and pressure or temperature are given. In that case, the quality of the mixture x has to be calculated from the given v and p or t and then x being known, other properties are evaluated.

9.7.3 Superheated Vapour

When the temperature of the vapour is greater than the saturation temperature corresponding to the given pressure, the vapour is said to be *superheated* (state 1 in Fig. 9.19). The difference between the temperature of the superheated vapour and the saturation temperature at that pressure is called the *superheat* or the *degree of superheat*. As shown in Fig. 9.19, the difference ($t_1 - t_{sat}$) is the superheat.

In a superheated vapour at a given pressure, the temperature may have different values greater than the saturation temperature. Table A.2 in the appendix gives the values of the properties (volume, enthaly, and entropy) of superheated vapour for each tabulated pair of values of pressure and temperature, both of which are now independent. Interpolation or extrapolation is to be used for pairs of values of pressure and temperature not given.

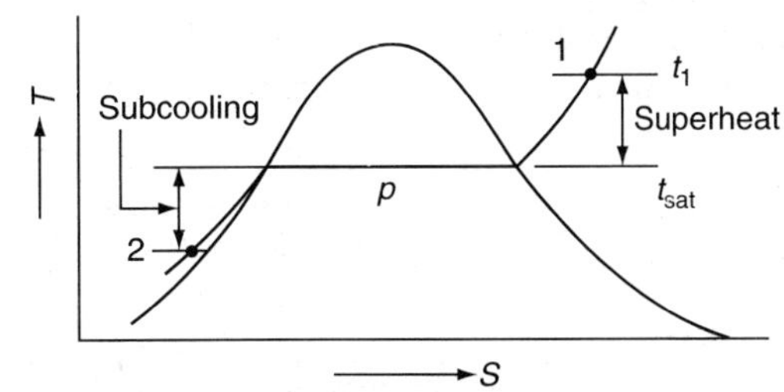

Fig. 9.19 *Superheat and subcooling*

9.7.4 Compressed Liquid

When the temperature of a liquid is less than the saturation temperature at the given pressure, the liquid is called *compressed liquid* (state 2 in Fig. 9.19). The pressure and temperature of compressed liquid may vary independently, and a table of properties like the superheated vapour table could be arranged to give the properties at any *p* and *t*. However, the *properties of liquids vary little with pressure.* Hence the properties are taken from the saturation tables at the temperature of the compressed liquid. When a liquid is cooled below its saturation temperature at a certain pressure it is said to be *subcooled*. The difference in saturation temperature and the actual liquid temperature is known as the degree of subcooling, or simply, subcooling (Fig. 9.17).

9.8 CHARTS OF THERMODYNAMIC PROPERTIES

The presentation of properties of substances in the form of a chart has certain obvious advantages. The manner of variation of properties is clearly demonstrated in the chart and there is no problem of interpolation. However, the precision is not as much as in steam tables.

The temperature-entropy plot and enthalpy-entropy plot (Fig. 9.20a) are commonly used. The temperature-entropy plot shows the vapour dome and the lines of constant pressure, constant volume, constant enthalpy, constant quality, and constant superheat. However, its scale is small and limited in use. The enthalpy-entropy plot or Mollier chart, has a larger scale to provide data suitable for many computations. It contains the same

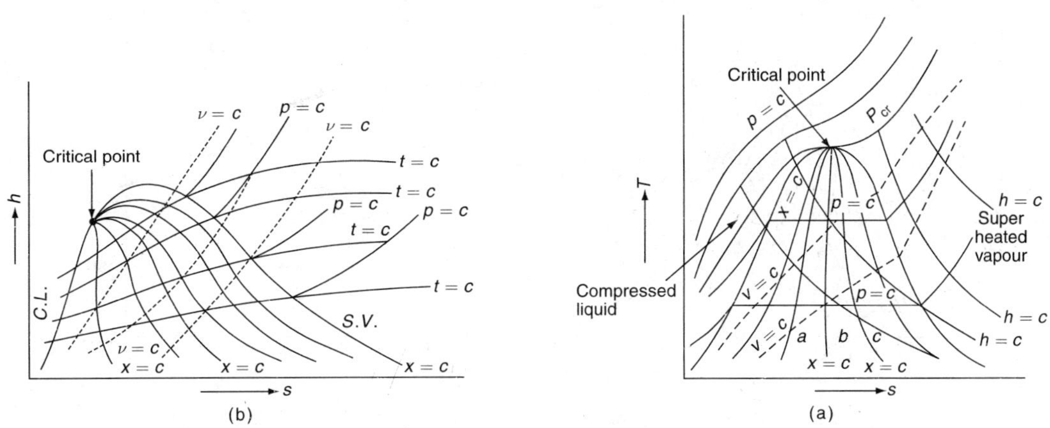

Fig. 9.20 *Constant property lines on (a) T–s plot and (b) Mollier diagram*

data as does the $T-s$ chart. The Mollier chart for water is given in Appendix F.I. The Mollier diagram for steam with data taken from Keenan et al. Steam Tables (John Willey, N.Y., 1969 is given in Fig. 9.21.)

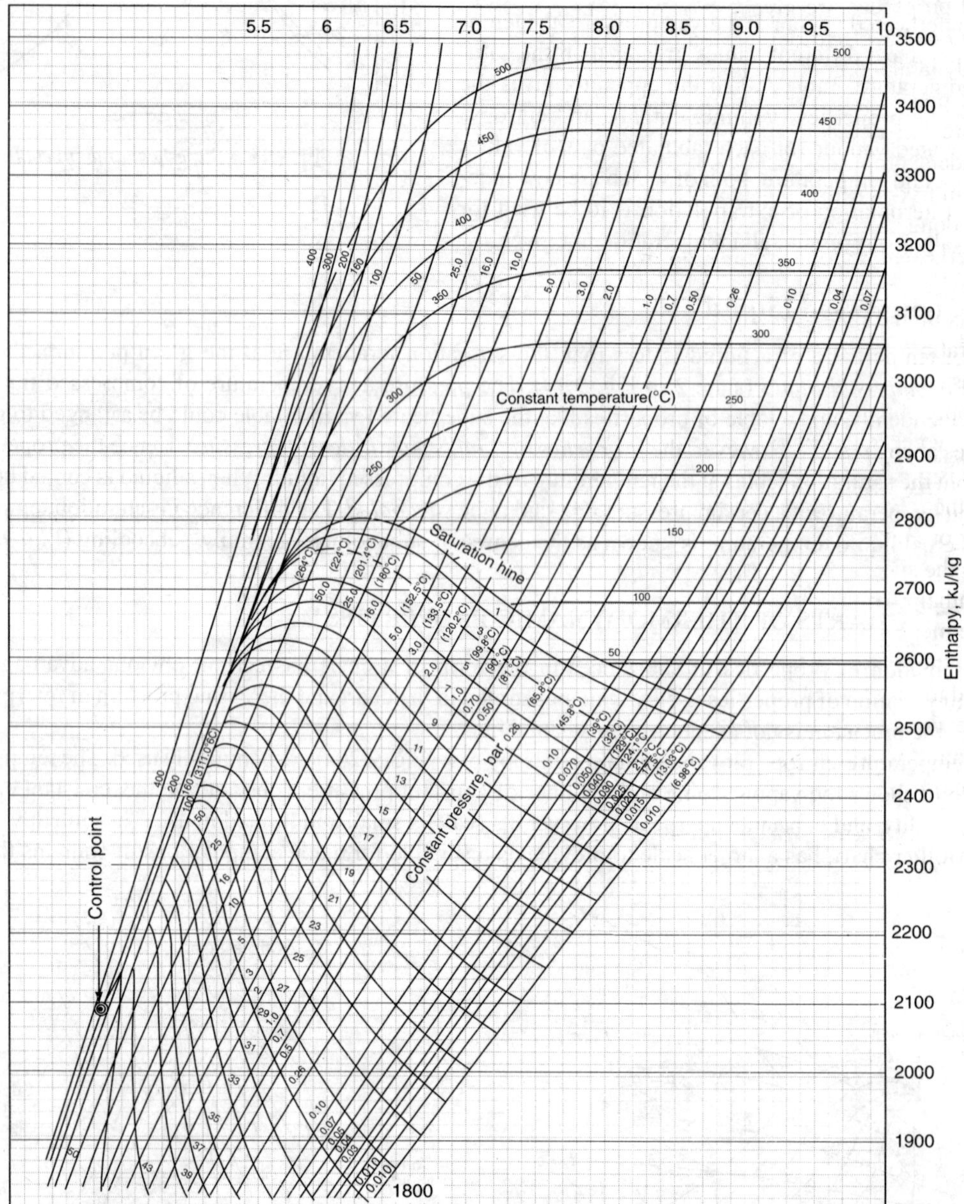

Fig. 9.21 *Mollier diagram for steam (Data taken from Keenan, J.H., F.G. keyes, P.C. Hill and J.g. Moore, Steam Tables, John Wiley, N.Y., 1969)*

9.9 MEASUREMENT OF STEAM QUALITY

The state of a pure substance gets fixed if two independent properties are given. A pure substance is thus said to have two degrees of freedom. Of all thermodynamic properties, it is easiest to measure the pressure and temperature of a substance. Therefore, whenever pressure and temperature are independent properties, it is the practice to measure them to determine that state of the substance. This is done in the compressed liquid region or the superheated vapour region (Fig. 9.22), where the measured values of pressure and temperature would fix up the state. But when the substance is in the saturation state or two-phase region (Fig. 9.22), the measured values of pressure and temperature could apply equally well to saturated liquid point f, saturated vapour point g, or to mixtures of any quality, points x_1, x_2 or x_3. Of the two properties, p and t, only one is independent; the other is a dependent property. If pressure is given, the saturation temperature gets automatically fixed for the substance. In order to fix up the state of the mixture, apart from either pressure or temperature, one more property, such as specific volume, enthalpy or composition of the mixture (quality) is required to be known. Since it is relatively difficult to measure the specific volume of a mixture, devices such as calorimeters are used for determining the quality or the enthalpy of the mixture.

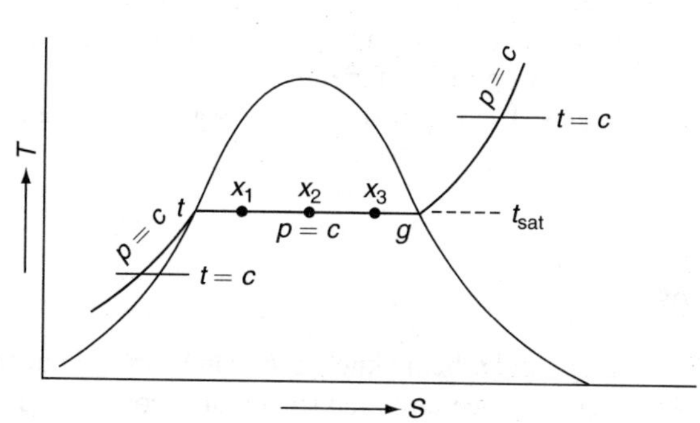

Fig. 9.22 *Quality of liquid-vapour mixture*

In the measurement of quality, the object is always to bring the state of the substance from the two-phase region to the single-phase or superheated region, where both pressure and temperature are independent, and measured to fix the state, either by adiabatic throttling or electric heating.

In the *throttling calorimeter*, a sample of wet steam of mass m and at pressure p_1 is taken from the steam main through a perforated sampling tube (Fig. 9.23). Then it is throttled by the partially-opened valve (or orifice) to a pressure p_2, measured by mercury manometer, and temperature t_2, so that after throttling

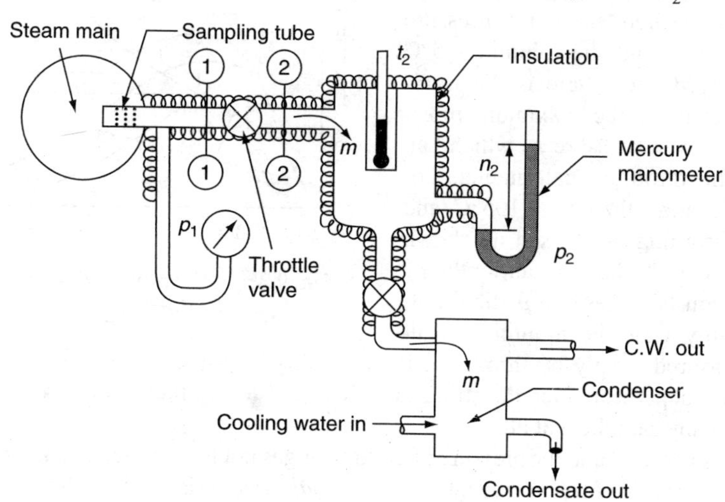

Fig. 9.23 *Throttling calorimeter*

the steam is in the superheated region. The process is shown on the T–s and h–s diagrams in Fig. 9.24. The steady flow energy equation gives the enthalpy after throttling as equal to enthalpy before throttling. The initial and final equilibrium states 1 and 2 are joined by a dotted line since throttling is irreversible (adiabatic but not isentropic) and the intermediate states are non-equilibrium states not describable by thermodynamic coordinates. The initial state (wet) is given by p_1 and x_1, and the final state by p_2 and t_2 (superheated). Now since

$$h_1 = h_2$$

$$h_{fp1} + x_1 h_{fgp1} = h_2$$

or

$$x_1 = \frac{h_2 - h_{fp1}}{h_{fgp1}}$$

With p_2 and t_2 being known, h_2 can be found out from the superheated steam table. The values of h_f and h_{fg} are taken from the saturated steam table corresponding to pressure p_1. Therefore, the quality of the wet steam x_1 can be calculated.

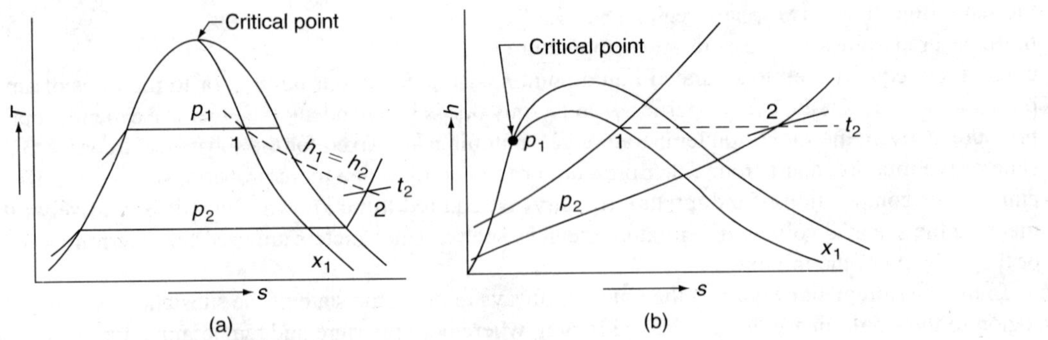

Fig. 9.24 *Throttling process on T–s and h–s plots*

To be sure that steam after throttling is in the single-phase or superheated region, a minimum of 5°C superheat is desired. So if the pressure after throttling is given and the minimum 5°C superheat is prescribed, then there is the minimum quality of steam (or the maximum moisture content) at the given pressure p_1 which can be measured by the throttling calorimeter. For example, if $p_2 = 1$ atm., then $t_2 = 105$°C and the state 2 after throttling gets fixed as shown in Fig. 9.25. From state 2, the constant enthalpy line intersects the constant pressure p_1 line at 1. Therefore, the quality x_1 is the minimum quality that can be measured simply by throttling. If the quality is, say, x'_1 less than x_1, then after throttling to $p_2 = 1$ atm., the superheat after throttling is less than 5°C. If the quality is x''_1, then throttling to 1 atm. does not give any superheat at all.

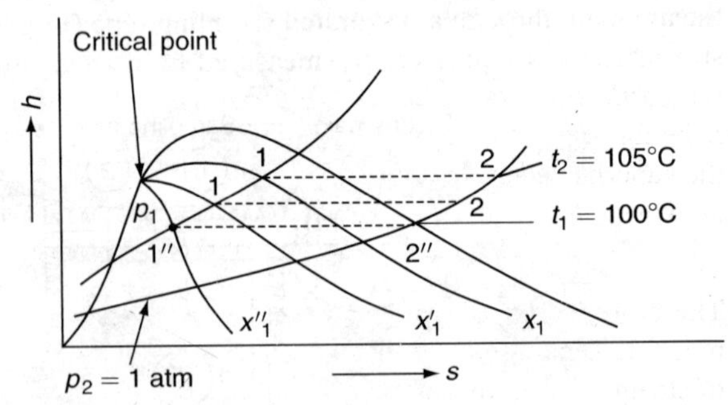

Fig. 9.25 *Minimum quality that can be measured only by throttling*

When the steam is very wet and the pressure after throttling is not low enough to take the steam to the superheated region, then a *combined separating* and *throttling calorimeter* is used for the measurement of quality. Steam from the main is first passed through a separator (Fig. 9.26), where some part of the moisture separates

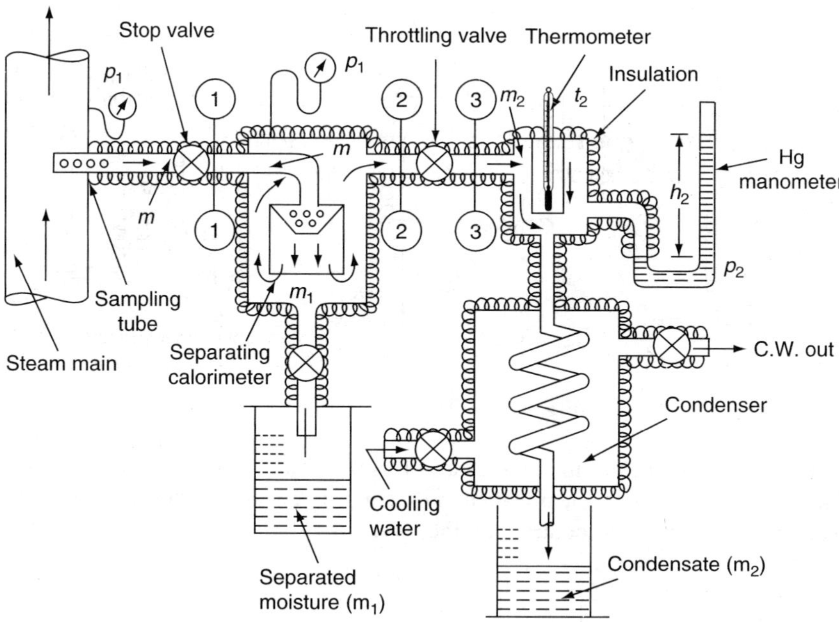

Fig. 9.26 *Separating and throttling calorimeter*

out due to the sudden change in direction and falls by gravity, and the partially dry vapour is then throttled and taken to the superheated region. In Fig. 9.27, process 1–2 represents the moisture separation from the wet sample of steam at constant pressure p_1 and process 2–3 represents throttling to pressure p_2. With p_2 and t_3 being measured, h_3 can be found out from the superheated steam table.

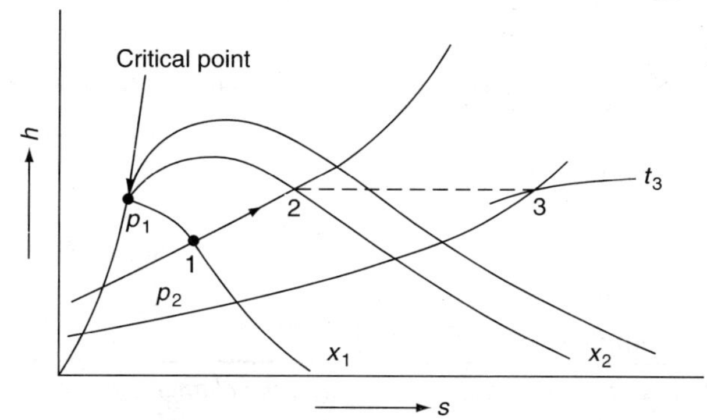

Fig. 9.27 *Separating and throttling processes on h-s plot*

Now,

$$h_3 = h_2 = h_{fp1} + x_2 h_{fgp1}$$

Therefore, x_2, the quality of steam after partial moisture separation, can be evaluated. If m kg of steam is taken through the sampling tube in t secs, m_1 kg of it is separated, and m_2 kg is throttled and then condensed to water and collected, then $m = m_1 + m_2$, and at state 2, the mass of dry vapour will be $x_2 m_2$. Therefore, the quality of the sample of steam at state 1, x_1 is given by

$$x_1 = \frac{\text{mass of dry vapour at state 1}}{\text{mass of liquid-vapour mixture at state 1}} = \frac{x_2 m_2}{m_1 + m_2}$$

The quality of wet steam can also be measured by an *electric calorimeter* (Fig. 9.28). The sample of steam is passed in steady flow through an electric heater, as shown. The electrical energy input Q should be sufficient to take the steam to the superheated region where pressure p_2 and temperature t_2 are measured. If I is

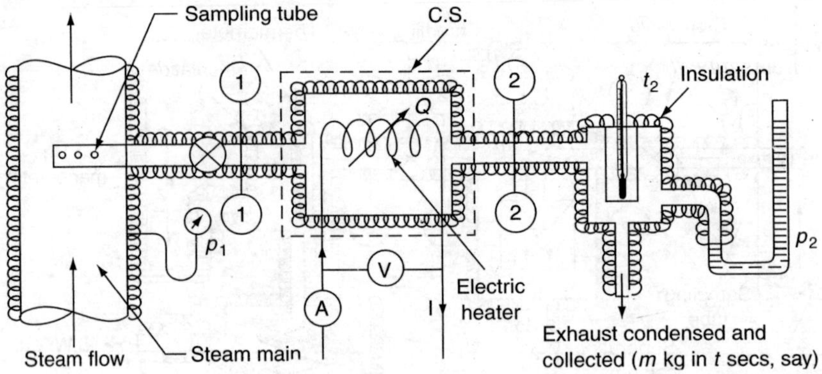

Fig. 9.28 *Electrical calorimeter*

the current flowing through the heater in amperes and V is the voltage across the coil, then at steady state $Q = VI \times 10^{-3}$ kW. If m is the mass of steam taken in t seconds under steady flow condition, then the steady flow energy equation for the heater (as control volume) gives

$$w_1 h_1 + Q = w_1 h_2$$

where w_1 is the steam flow rate in kg/s $\left(w_1 = \frac{m}{t} \text{ kg/s}\right)$

$$\therefore \quad h_1 + \frac{Q}{w_1} = h_2$$

With h_2, Q and w_1 being known, h_1 can be computed. Now

$$h_1 = h_{fp1} + x_1 h_{fgp1}$$

Hence x_1 can be evaluated.

Solved Examples

Example 9.1

Find the saturation temperature, the changes in specific volume and entropy during evaporation, and the latent heat of vaporization of steam at 1 MPa.

Solution At 1 MPa, from Table A.1(b) in the Appendix

$$t_{sat} = 179.91°\text{C}$$

$$v_f = 0.001127 \text{ m}^3/\text{kg} \qquad \textit{Ans.}$$

$$v_g = 0.19444 \text{ m}^3/\text{kg}$$

$$\therefore \quad v_{fg} = v_g - v_f = 0.1933 \text{ m}^3/\text{kg}$$

$$s_f = 2.1387 \text{ kJ/kg K}$$

$$s_g = 6.5865 \text{ kJ/kg K}$$

$$\therefore \quad s_{fg} = s_g - s_f = 4.4478 \text{ kJ/kg K} \qquad \textit{Ans.}$$

$$h_{fg} = h_g - h_f = 2015.3 \text{ kJ/kg} \qquad \textit{Ans.}$$

Example 9.2

Saturated steam has an entropy of 6.76 kJ/kg K. What are its pressure, temperature, specific volume, and enthalpy?

Solution In Table A.1(b), when $s_g = 6.76$ kJ/kg K

$$p = 0.6 \text{ MPa}, \qquad t = 158.85°\text{C}$$

$$v_g = 0.3156 \text{ m}^3/\text{kg}, \quad \text{and} \quad h_g = 2756.8 \text{ kJ/kg} \qquad \textit{Ans.}$$

Example 9.3

Find the enthalpy and entropy of steam when the pressure is 2 MPa and the specific volume is 0.09 m³/kg.

Solution In Table A.1(b), when $p = 2$ MPa, $v_f = 0.001177$ m³/kg and $v_g = 0.09963$ m³/kg. Since the given volume lies between v_f and v_g, the substance will be a mixture of liquid and vapour, and the state will be within the vapour dome. When in the two-phase region, the composition of the mixture or its quality has to be evaluated first. Now

$$v = v_f + x\, v_{fg}$$

$$0.09 = 0.001177 + x\,(0.09963 - 0.001177)$$

or

$$x = 0.904 \text{ or } 90.4\%$$

At

$$2 \text{ MPa}, h_f = 908.79 \quad \text{and} \quad h_{fg} = 1890.7 \text{ kJ/kg}$$

$$s_f = 2.4474 \quad \text{and} \quad s_{fg} = 3.8935 \text{ kJ/kg K}$$

$$h = h_f + x\, h_{fg} = 908.79 + 0.904 \times 1890.7 = 2618.79 \text{ kJ/kg} \qquad \textit{Ans.}$$

$$s = s_f + x\, s_{fg} = 2.4474 + 0.904 \times 3.8935 = 5.9534 \text{ kJ/kg K} \qquad \textit{Ans.}$$

Example 9.4

Find the enthalpy, entropy, and volume of steam at 1.4 MPa, 380°C.

Solution At $p = 1.4$ MPa, in Table A.1(b), $t_{sat} = 195.07°$C. Therefore, the state of steam must be in the superheated region. In Table A.2, for properties of superheated steam,

at 1.4 MPa, 350°C

$$v = 0.2003 \text{ m}^3/\text{kg}$$

$$h = 3149.5 \text{ kJ/kg}$$

$$s = 7.1360 \text{ kJ/kg K}$$

and at 1.4 MPa, 400°C

$$v = 0.2178 \text{ m}^3/\text{kg}$$

$$h = 3257.5 \text{ kJ/kg}$$

$$s = 7.3026 \text{ kJ/kg K}$$

∴ By interpolation

at 1.4 MPa, 380°C $v = 0.2108$ m³/kg

$h = 3214.3$ kJ/kg

$s = 7.2360$ kJ/kg K *Ans.*

Example 9.5

A vessel of volume 0.04 m³ contains a mixture of saturated water and saturated steam at a temperature of 250°C. The mass of the liquid present is 9 kg. Find the pressure, the mass, the specific volume, the enthalpy, the entropy, and the internal energy.

Solution From Table A.1(a), at 250°C $p_{sat} = 3.973$ MPa

$$v_f = 0.0012512 \text{ m}^3/\text{kg}, \quad v_g = 0.05013 \text{ m}^3/\text{kg}$$

$$h_f = 1085.36 \text{ kJ/kg}, \quad h_{fg} = 1716.2 \text{ kJ/kg}$$

$$s_f = 2.7927 \text{ kJ/kg K}, \quad s_{fg} = 3.2802 \text{ kJ/kg K}$$

Volume of liquid, $V_f = m_f v_f = 9 \times 0.0012512 = 0.01126$ m³

Volume of vapour, $V_g = 0.04 - 0.01126 = 0.02874$ m³

∴ Mass of vapour

$$m_g = \frac{V_g}{v_g} = \frac{0.02874}{0.05013} = 0.575 \text{ kg}$$

∴ Total mass of mixture,

$$m = m_f + m_g = 9 + 0.575 = 9.575 \text{ kg} \qquad \textit{Ans.}$$

Quality of mixture,

$$x = \frac{m_g}{m_f + m_g} = \frac{0.575}{9.575} = 0.06$$

∴ $v = v_f + xv_{fs} = 0.0012512 + 0.06\ (0.05013 - 0.0012512)$

$= 0.00418$ m³/kg *Ans.*

$h = h_f + xh_{fg} = 1085.36 + 0.06 \times 1716.2 = 1188.32$ kJ/kg *Ans.*

$s = s_f + xs_{fg} = 2.7927 + 0.06 \times 3.2802 = 2.9895$ kJ/kg K *Ans.*

$u = h - pv$

$= 1188.32 - 3.973 \times 10^3 \times 0.00418 = 1171.72$ kJ/kg *Ans.*

Also, at 250°C, $u_f = 1080.39$ and $u_{fg} = 1522.0$ kJ/kg

∴ $u = u_f + xu_{fg} = 1080.39 + 0.06 \times 1522 = 1071.71$ kJ/kg *Ans.*

Example 9.6

Steam initially at 0.3 MPa, 250°C is cooled at constant volume. (a) At what temperature will the steam become saturated vapour? (b) What is the quality at 80°C? What is the heat transferred per kg of steam in cooling from 250°C to 80°C?

Solution At 0.3 MPa, $t_{sat} = 133.55°C$

Since $t > t_{sat}$, the state would be in the superheated region (Fig. Ex. 9.6). From Table A.2, for properties of superheated steam, at 0.3 MPa, 250°C

$$v = 0.7964 \text{ m}^3/\text{kg}$$
$$h = 2967.6 \text{ kJ/kg}$$
$$\therefore \quad v_1 = v_3 = v_2 = 0.7964 \text{ m}^3/\text{kg}$$

In Table A.1

when $v_g = 0.8919, \quad t_{sat} = 120°C$

when $v_g = 0.7706, \quad t_{sat} = 125°C$

Therefore, when $v_g = 0.7964$, t_{sat}, by linear interpolation, would be 123.9°. Steam would become saturated vapour at $t = 123.9°C$ *Ans.* (a)

At 80°C, $v_f = 0.001029 \text{ m}^3/\text{kg}$, $v_g = 3.407 \text{ m}^3/\text{kg}$, $h_f = 334.91$ kJ/kg, $h_{fg} = 2308.8$ kJ/kg,

$$p_{sat} = 47.39 \text{ kPa } v_1 = v_2 = 0.7964 \text{ m}^3/\text{kg} = v_{f80°C} + x_2 v_{fg80°C}$$
$$= 0.001029 + x_2 (3.407 - 0.001029)$$
$$\therefore \quad x_2 = \frac{0.79539}{3.40597} = 0.234 \qquad \textit{Ans. (b)}$$
$$h_2 = 334.91 + 0.234 \times 2308.8 = 875.9 \text{ kJ/kg}$$
$$h_1 = 2967.6 \text{ kJ/kg}$$

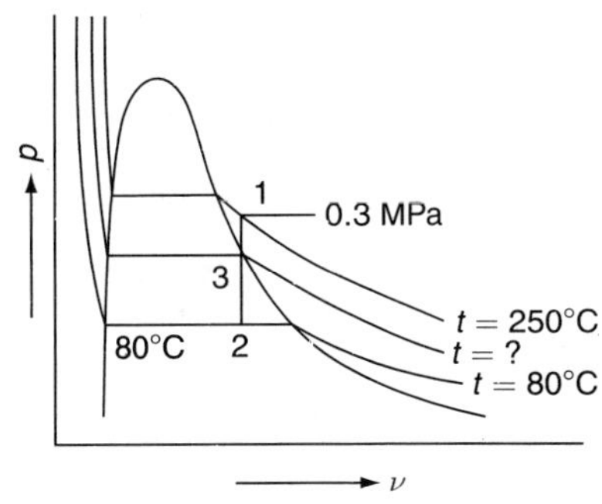

Fig. Ex. 9.6

From the first law of thermodynamics

$$đQ = du + pdv$$
$$\therefore \quad (đQ)_v = du$$
$$\text{or} \quad Q_{1-2} = u_2 - u_1 = (h_2 - p_2 v_2) - (h_1 - p_1 v_1)$$
$$= (h_2 - h_1) + v (p_1 - p_2)$$
$$= (875.9 - 2967.6) + 0.7964 (300 - 47.39)$$
$$= 2091.7 + 201.5$$
$$= -1890.2 \text{ kJ/kg}$$

Ans. (c)

Example 9.7

Steam initially at 1.5 MPa, 300°C expands reversibly and adiabatically in a steam turbine to 40°C. Determine the ideal work output of the turbine per kg of steam.

Solution The steady flow energy equation for the control volume, as shown in Fig. Ex. 9.7.1, gives (other energy terms being neglected)

$$h_1 = h_2 + W$$
$$\therefore \quad W = h_1 - h_2$$

Work is done by steam at the expense of a fall in its enthalpy value. The process is reversible and adiabatic, so it is isentropic. The process is shown on the T–s and h–s diagrams in Fig. Ex. 9.7.2.

From Table A.1(a), at 40°C

$$p_{sat} = 7.384 \text{ kPa}, \quad s_f = 0.5725, \text{ and } s_{fg} = 7.6845 \text{ kJ/kg K}$$

$h_f = 167.57$, and $h_{fg} = 2406.7$ kJ/kg

At p = 1.5 MPa, t = 300°C, from the tabulated properties of superheated steam (Table A.2)

$$s_1 = 6.9189 \text{ kJ/kg K}$$

$$h_1 = 3037.6 \text{ kJ/kg}$$

Since $s_1 = s_2$

$$6.9189 = s_f + x_2\, s_{fg\,40°C} = 0.5725 + x_2 \times 7.6845$$

$$x_2 = \frac{6.3464}{7.6845} = 0.826 \text{ or } 82.6\%$$

$$\therefore \quad h_2 = h_{f40°C} + x_2\, h_{fg\,40°C} = 167.57 + 0.826 \times 2406.7 = 2152.57 \text{ kJ/kg}$$

$$\therefore \quad W = h_1 - h_2 = 3037.6 - 2152.57 = 885.03 \text{ kJ/kg}$$ *Ans.*

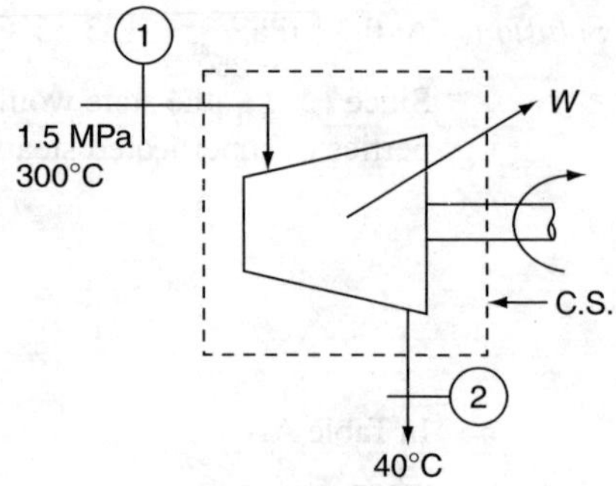

Fig. Ex. 9.7.1

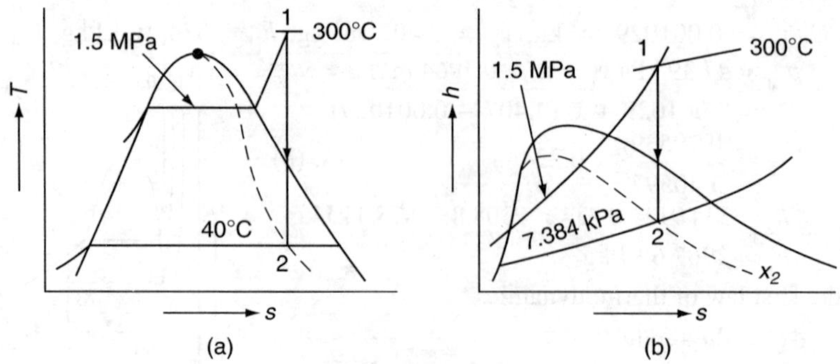

Fig. Ex. 9.7.2

Example 9.8

Steam at 0.8 MPa, 250°C and flowing at the rate of 1 kg/s passes into a pipe carrying wet steam at 0.8 MPa, 0.95 dry. After adiabatic mixing the flow rate is 2.3 kg/s. Determine the condition of steam after mixing.

The mixture is now expanded in a frictionless nozzle isentropically to a pressure of 0.4 MPa. Determine the velocity of the steam leaving the nozzle. Neglect the velocity of steam in the pipeline.

Solution Figure Ex. 1 gives the flow diagram.

$$w_2 = w_3 - w_1 = 2.3 - 1.0 = 1.3 \text{ kg/s}$$

The energy equation for the adiabatic mixing of the two streams gives

$$w_1 h_1 + w_2 h_2 = w_3 h_3 \tag{1}$$

At 0.8 MPa, 250°C, $h_1 = 2950.0$ kJ/kg

At 0.8 MPa, 0.95 dry

$$h_2 = h_f + 0.95\, h_{fg} = 721.11 + 0.95 \times 2048.0 = 2666.71 \text{ kJ/kg}$$

∴ From Eq. (1)

$$1 \times 2950 + 1.3 \times 2666.71 = 2.3 \times h_3$$

$$\therefore \quad h_3 = 2790 \text{ kJ/kg}$$

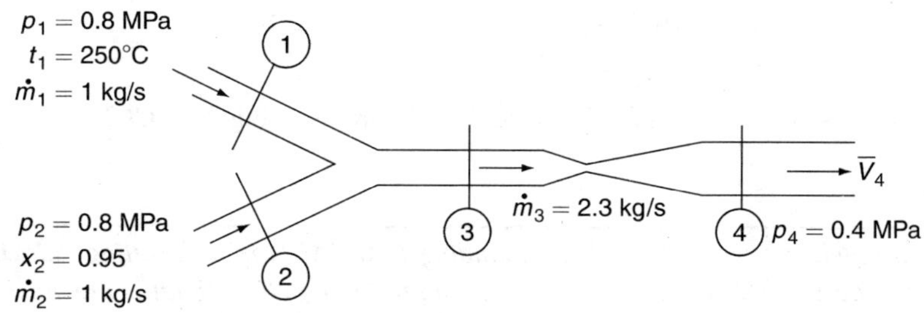

Fig. Ex. 9.8.1

Since $(h_g)_{0.8\text{ MPa}} = 2769.1$ kJ/kg

and $h_3 > h_g$, the state must be in the superheated region. From the steam tables, when $p = 0.8$ MPa, $t = 200$°C

$$h = 2839.3 \text{ kJ/kg}$$

When $$p = 0.8 \text{ MPa}, \quad t_{sat} = 170.43°\text{C}$$

$$h_g = 2769.1 \text{ kJ/kg}$$

By linear interpolation

$$t_3 = 179°\text{C}$$

$\therefore$ Degree of superheat = 179 − 170.33 = 8.57°C

$\therefore$ Condition of steam after mixing = 0.8 MPa, 179°C *Ans.*

The energy equation for the nozzle gives

$$h_3 = h_4 + \frac{V_4^2}{2}$$

since V_3 = – velocity of steam in the pipeline = 0

Steam expands isentropically in the nozzle to 0.4 MPa. By interpolation,

$$s_3 = 6.7087 \text{ kJ/kg K} = s_4$$

$\therefore$ $$6.7087 = 1.7766 + x_4 \times 5.1193$$

$$x_4 = 0.964$$

$\therefore$ $$h_4 = 604.74 + 0.964 \times 2133.8 = 2660 \text{ kJ/kg}$$

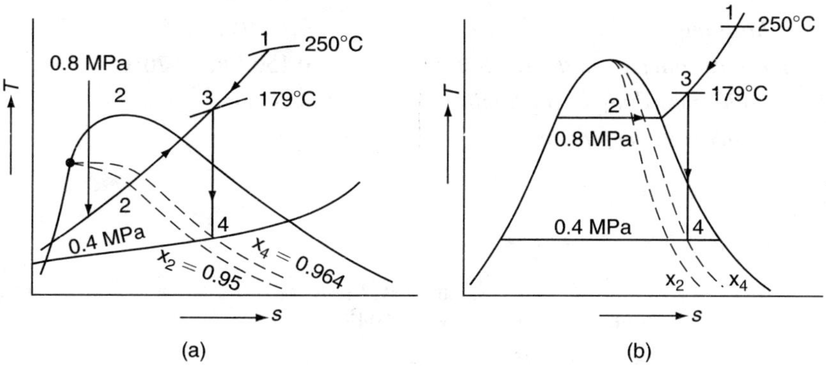

Fig. Ex. 9.8.2

$$V_4^2 \times 10^{-3} = 2(h_3 - h_4) = 2 \times 130 = 260$$

$$V_4 = \sqrt{26} \times 100 = 509.9 \text{ m/s}$$ *Ans.*

The processes are shown on the h–s and T–s diagrams in Fig. Ex. 9.8.2.

Example 9.9

Steam flows in a pipeline at 1.5 MPa. After expanding to 0.1 MPa in a throttling calorimeter, the temperature is found to be 120°C. Find the quality of steam in the pipeline. What is the maximum moisture at 1.5 MPa that can be determined with this set-up if at least 5°C of superheat is required after throttling for accurate readings?

Solution At state 2 (Fig. Ex. 9.9), when $p = 0.1$ MPa, $t = 120$°C by interpolation

$$h_2 = 2716.2 \text{ kJ/kg, and at } p = 1.5 \text{ MPa}$$

$$h_f = 844.89 \quad \text{and} \quad h_{fg} = 1947.3 \text{ kJ/kg}$$

Now, $$h_1 = h_2$$

or $$h_{f1.5MPa} + x_1 h_{fg1.5MPa} = h_2$$

$$844.89 + x_1 \times 1947.3 = 2716.2$$

$$x_1 = \frac{1871.3}{1947.3} = 0.963$$ *Ans.*

When $p = 0.1$ MPa and $t = 99.63 + 5 = 104.63$°C

$$h_3 = 2685.5 \text{ kJ/kg}$$

Since $$h_3 = h_4$$

$$2685.5 = 844.89 + x_4 \times 1947.3$$

$$\therefore \quad x_4 = \frac{1840.6}{1947.3} = 0.948$$

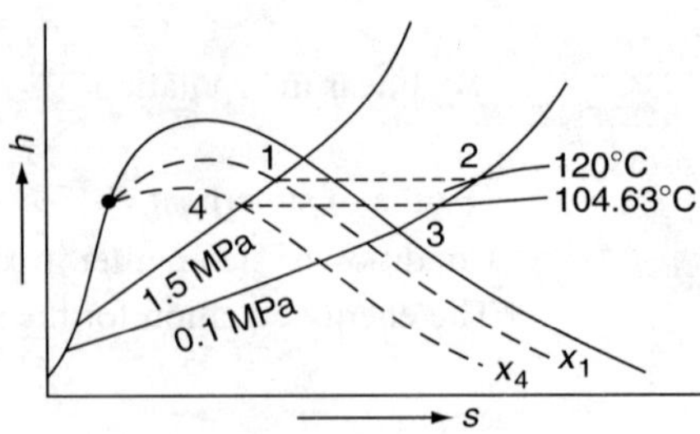

Fig. Ex. 9.9

The maximum moisture that can be determined with this set-up is only 5.2%. *Ans.*

Example 9.10

The following data were obtained with a separating and throttling calorimeter:

Pressure in pipeline	1.5 *MPa*
Condition after throttling	0.1 *MPa,* 110°*C*
During 5 min moisture collected in the separator	0.150 *litre at* 70°*C*
Steam condensed after throttling during 5 min	3.24 *kg*

Find the quality of steam in the pipeline

Solution As shown in Fig. Ex. 9.10

at 0.1 MPa, 110°C $$h_3 = 2696.2 \text{ kJ/kg}$$

Now $$h_3 = h_2 = h_{f1.5MPa} + x_2\, h_{fg1/5MPa}$$

or $$2696.2 = 844.89 + x_2 \times 1947.3$$

$$\therefore \quad x_2 = \frac{1851.31}{1947.3} = 0.955$$

If m_1 = mass of moisture collected in separator in 5 min and m_2 = mass of steam condensed after throttling in 5 min.

then $$x_1 = \frac{x_2 m_2}{m_1 + m_2}$$

At 70°C, $v_f = 0.001023 \text{ m}^3/\text{kg}$

$$m_1 = \frac{150 \times 10^{-6} \text{ m}^3}{1023 \times 10^{-6} \text{ m}^3/\text{kg}} = 0.1462 \text{ kg}$$

$$m_2 = 3.24 \text{ kg}$$

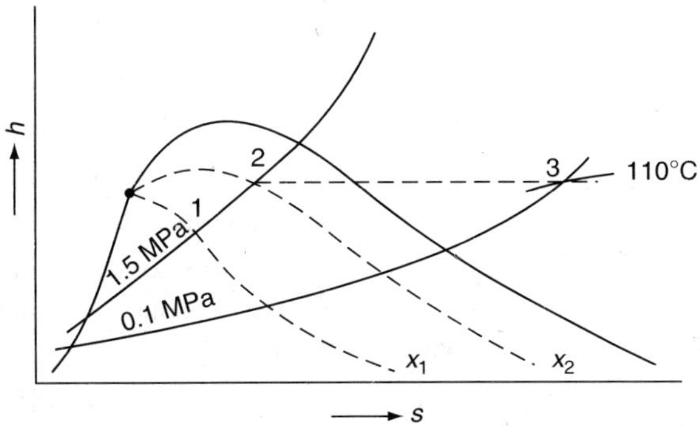

Fig. Ex. 9.10

$$\therefore \quad x_1 = \frac{0.955 \times 3.24}{0.1462 + 3.24} = \frac{3.1}{3.3862} = 0.915$$ *Ans.*

Example 9.11

A steam boiler initially contains 5 m³ of steam and 5 m³ of water at 1 MPa. Steam is taken out at constant pressure until 4 m³ of water is left. What is the heat transferred during the process?

Solution At 1 MPa,

$$v_f = 0.001127, \quad \text{and} \quad v_g = 0.1944 \text{ m}^3/\text{kg}$$
$$h_g = 2778.1 \text{ kJ/kg}$$
$$u_f = 761.68, \quad u_g = 2583.6 \text{ kJ/kg}$$
$$u_{fg} = 1822 \text{ kJ/kg}$$

The initial mass of saturated water and steam in the boiler (Fig. Ex. 9.11).

$$\frac{V_f}{v_f} + \frac{V_g}{v_g} = \frac{5}{0.001127} + \frac{5}{0.1944} = (4.45 \times 10^3 + 25.70) \text{ kg}$$

where suffix *f* refers to saturated water and suffix *g* refers to saturated vapour. Final mass of saturated water and steam

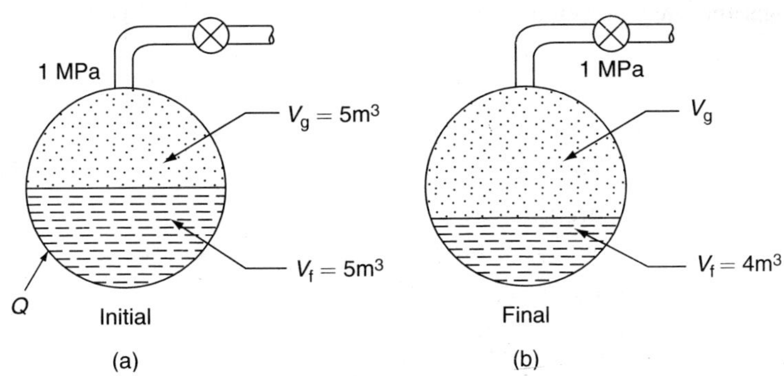

Fig. Ex. 9.11

$$= \frac{4}{0.001127} + \frac{6}{0.1944} = (3.55 \times 10^3 + 30.80) \text{ kg}$$

∴ Mass of steam taken out of the boiler (m_s)

$$= (4.45 \times 10^3 + 25.70) - (3.55 \times 10^3 + 30.80)$$
$$= 0.90 \times 10^3 - 5.1 = 894.9 \text{ kg}$$

Making an energy balance, we have: Initial energy stored in saturated water and steam + Heat transferred from the external source = Final energy stored in saturated water and steam + Energy leaving with the steam.

or $$U_1 + Q = U_f + m_s h_g$$

assuming that the steam taken out is dry ($x = 1$)

or $$4.45 \times 10^3 \times 761.68 + 25.70 \times 2583.6 + Q = 3.55 \times 10^3 \times 761.68 + 30.8 \times 2583.6 + 894.9 \times 2778.1$$

or $$Q = 894.9 \times 2778.1 - (0.90 \times 10^3) \times 761.68 + 5.1 \times 2583.6$$
$$= 2425000 - 685500 + 13176$$
$$= 1752{,}676 \text{ kJ} = 1752.676 \text{ MJ}$$ *Ans.*

Example 9.12

A 280 mm diameter cylinder fitted with a frictionless leakproof piston contains 0.02 kg of steam at a pressure of 0.6 MPa and a temperature of 200°C. As the piston moves slowly outwards through a distance of 305 mm, the steam undergoes a fully-resisted expansion during which the steam pressure p and the steam volume V are related by pV^n = constant, where n is a constant. The final pressure of the steam is 0.12 MPa. Determine (a) the value of n, (b) the work done by the steam, and (c) the magnitude and sign of heat transfer.

Solution Since the path of expansion (Fig. Ex. 9.12) follows the equation

$$pV^n = C$$
$$p_1 V_1^n = p_2 V_2^n$$

Taking logarithms and arranging the terms

$$n = \frac{\log \frac{p_1}{p_2}}{\log \frac{V_2}{V_1}}$$

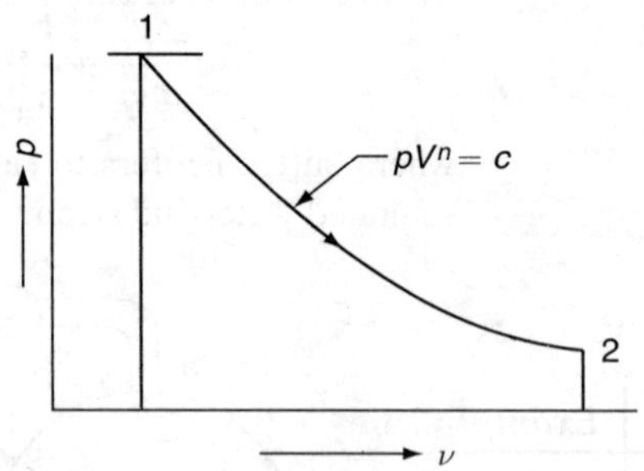

Fig. Ex. 9.12

Now, at 0.7 MPa, 200°C, from Tables A.2

$$v_1 = 0.352 \text{ m}^3/\text{kg}$$
$$h_1 = 2850.1 \text{ kJ/kg}$$

∴ Total volume, V_1, at state 1 = 0.352 × 0.02 = 0.00704 m^3

Displaced volume $= \frac{\pi}{4} d^2 \cdot l = \frac{\pi}{4} \times (0.28)^2 \times 0.305 = 0.0188 \text{ m}^3$

$\therefore$ Total volume V_2 after expansion $= 0.0188 + 0.00704 = 0.02584 \text{ m}^3$

$$n = \frac{\log \dfrac{0.6}{0.12}}{\log \dfrac{0.02584}{0.00704}} = \frac{\log 5}{\log 3.68} = 1.24 \qquad \textit{Ans. (a)}$$

Work done by steam in the expansion process

$$W_{1-2} = \int_{V_1}^{V_2} p\,dV = \frac{p_1V_1 - p_2V_2}{n-1}$$

$$= \frac{6 \times 10^5 \text{ N/m}^2 \times 0.00704 \text{ m}^2 - 1.2 \times 10^5 \text{ N/m}^2 \times 0.02584 \text{ m}^3}{1.24 - 1}$$

$$= \frac{4224 - 3100.8}{0.24} \text{ Nm} = 4680 \text{ Nm} = 4.68 \text{ kJ} \qquad \textit{Ans. (b)}$$

Now $\quad V_2 = 0.02584 \text{ m}^3$

$\therefore \quad v_2 = \dfrac{0.02584}{0.02} = 1.292 \text{ m}^3\text{/kg}$

Again $\quad v_2 = v_{f0.12\text{MPa}} + x_2 v_{fg0.12\text{MPa}}$

or $\quad 1.292 = 0.0010476 + x_2 \times 1.4271$

$\therefore \quad x_2 = \dfrac{1.291}{1.427} = 0.906$

At 0.12 MPa, $\quad u_f = 439.3 \text{ kJ/kg}, \quad u_g = 2512.0 \text{ kJ/kg}$

$\therefore \quad u_2 = 439.3 + 0.906\,(2512 - 439.3) = 2314.3 \text{ kJ/kg}$

Again $\quad h_1 = 2850.1 \text{ kJ/kg}$

$\therefore \quad u_1 = h_1 - p_1v_1 = 2850.1 - \dfrac{0.6 \times 10^6 \times 0.00704 \times 10^{-3}}{0.02}$

$\quad = 2850.1 - 211.2 = 2638.9 \text{ kJ/kg}$

By the first law

$$Q_{1-2} = U_2 - U_1 + W_{1-2} = m\,(u_2 - u_1) + W_{1-2}$$
$$= 0.02\,(2314.3 - 2638.5) + 4.68$$
$$= -\,6.484 + 4.68 = -\,1.804 \text{ kJ} \qquad \textit{Ans. (c)}$$

Example 9.13

A large insulated vessel is divided into two chambers, one containing 5 kg of dry saturated steam at 0.2 MPa and the other 10 kg of steam, 0.8 quality at 0.5 MPa. If the partition between the chambers is removed and the steam is mixed thoroughly and allowed to settle, find the final pressure, steam quality, and entropy change in the process.

Solution The vessel is divided into chambers, as shown in Fig. Ex. 9.13.1.

At 0.2 MPa, $\quad v_g = v_1 = 0.8857 \text{ m}^3\text{/kg}$

$\therefore \quad V_1 = m_1v_1 = 5 \times 0.8857 = 4.4285 \text{ m}^3$

At 0.5 MPa, $v_2 = v_f + x_2 v_{fg} = 0.001093 + 0.8 \times 0.3749 = 0.30101 \text{ m}^3/\text{kg}$

$\therefore$ $V_2 = m_2 v_2 = 10 \times 0.30101 = 3.0101 \text{ m}^3$

$\therefore$ Total volume, $V_m = V_1 + V_2 = 7.4386 \text{ m}^3$ (of mixture)

Total mass of mixture, $m_m = m_1 + m_2 = 5 + 10 = 15 \text{ kg}$

$\therefore$ Specific volume of mixture

$$v_m = \frac{V_m}{m_m} = \frac{7.4386}{15} = 0.496 \text{ m}^3/\text{kg}$$

By energy balance

$$m_1 u_1 + m_2 u_2 = m_3 u_3$$

At 0.2 MPa, $h_g = h_1 = 2706.7 \text{ kJ/kg}$

$u_1 = h_1 - p_1 v_1 \cong 2706.7 \text{ kJ/kg}$

At 0.5 MPa, $h_2 = h_f + x_2 h_{fg} = 640.23 + 0.8 \times 2108.5 = 2327.03 \text{ kJ/kg}$

$u_2 = h_2 - p_2 v_2 \cong h_2 = 2327.03$ kJ/kg

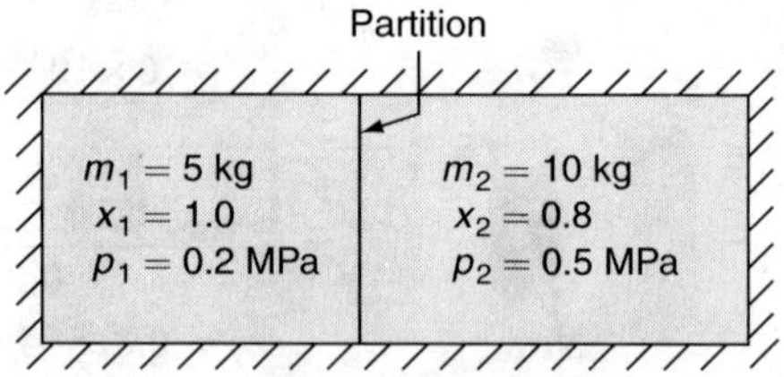

Fig. Ex. 9.13.1

$$h_3 = h_m = \frac{5 \times 2706.7 + 10 \times 2327.03}{15}$$

$\therefore$ $= 2453.6 \text{ kJ/kg} \cong u_3$

Now for the mixture

$$h_3 = 2453.6 \text{ kJ/kg} = u_3$$
$$v_3 = 0.496 \text{ m}^3/\text{kg}$$

From the Mollier diagram, with the given values of h and v, point 3 after mixing is fixed (Fig. Ex. 9.13.2).

$x_3 = 0.870$ *Ans.*

$s_3 = 6.29$ kJ/kg K

$p_3 = 3.5$ bar *Ans.*

$s_4 = s_{g\,0.2\text{MPa}} = 7.1271$ kJ/kg K

$s_2 = s_{f\,0.5\text{MPa}} + 0.8 s_{fg\,0.5\text{MPa}}$

$= 1.8607 + 0.8 \times 4.9606$

$- 5.8292$ kJ/kg K

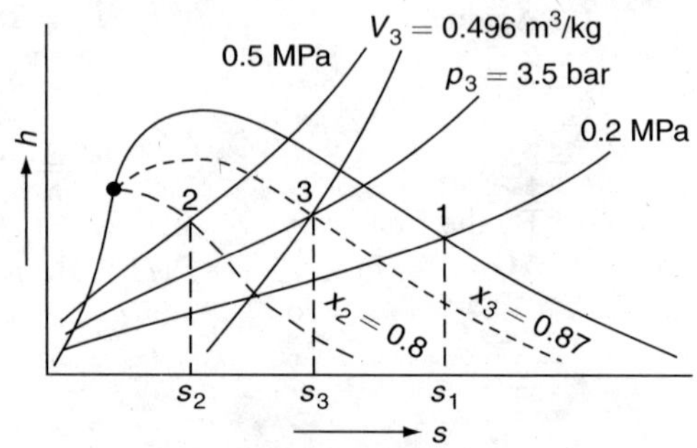

Fig. Ex. 9.13.2

Entropy change during the process

$= m_3 s_3 - (m_1 s_1 + m_2 s_2)$

$= 15 \times 6.298 - (5 \times 7.1271 + 10 \times 5.8292) = 0.43$ kJ/K *Ans.*

Example 9.14

Steam generated at a pressure of 6 MPa and a temperature of 400°C is supplied to a turbine via a throttle valve which reduces the pressure to 5 MPa. Expansion in the turbine is adiabatic to a pressure of 0.2 MPa, the isentropic efficiency (actual enthalpy drop/isentropic enthalpy drop) being 82%. The surroundings are at 0.1 MPa, 20°C. Determine the availability of steam before and after the throttle valve and at the turbine exhaust, and calculate the specific work output from the turbine. The K.E. and P.E. changes are negligible.

Solution Steady flow availability Ψ is given by

$$\Psi = (h - h_0) - T_0 (s - s_0) + \frac{1}{2} \mathbf{V}_1^2 + g(Z - Z_0)$$

where subscript 0 refers to the surroundings. Since the K.E. and P.E. changes are negligible

$$\Psi_1 = \text{Availability of steam before throttling}$$
$$= (h_1 - h_0) - T_0(s_1 - s_0)$$

At 6 MPa, 400°C (Fig. Ex. 9.14)

$$h_1 = 3177.2 \text{ kJ/kg}$$
$$s_1 = 6.5408 \text{ kJ/kg K}$$

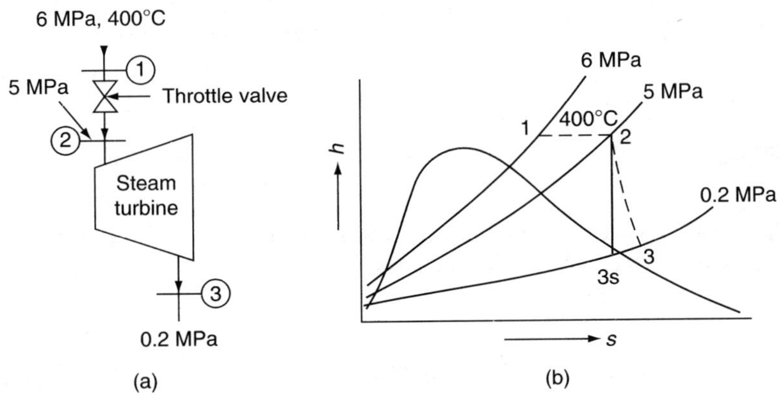

Fig. Ex. 9.14

At 20°C

$$h_0 = 83.96 \text{ kJ/kg}$$
$$s_0 = 0.2966 \text{ kJ/kg K}$$

$\therefore$ $$\Psi_1 = (3177.2 - 83.96) - 293\,(6.5408 - 0.2966)$$
$$= 3093.24 - 1829.54 = 1263.7 \text{ kJ/kg}$$ *Ans.*

Now $h_1 = h_2$, for throttling

At $h = 3177.2$ kJ/kg and $p = 5$ MPa, from the superheated steam table

$$\left.\begin{aligned} t_2 &= 390°\text{C} \\ s_2 &= 6.63 \text{ kJ/kg K} \end{aligned}\right\} \text{by linear interpolation}$$

$\therefore$ $$\Psi_2 = \text{Availability of steam after throttling}$$
$$= (h_2 - h_0) - T_0\,(s_2 - s_0) = (3177.2 - 83.96) - 293\,(6.63 - 0.2966)$$
$$= 3093.24 - 1855.69 = 1237.55 \text{ kJ/kg}$$

Decrease in availability due to throttling

$$= \Psi_1 - \Psi_2 = 1263.7 - 1237.55 = 26.15 \text{ kJ/kg}$$

Now

$$s_2 = s_{3s} = 6.63 = 1.5301 + x_{3s}\,(7.1271 - 1.5301)$$

$$\therefore \quad x_{3s} = \frac{5.10}{5.5970} = 0.9112$$

$$h_{3s} = 504.7 + 0.9112 \times 2201.9 = 2511.07 \text{ kJ/kg}$$

$$h_2 - h_{3s} = 3177.2 - 2511.07 = 666.13 \text{ kJ/kg}$$

$$\therefore \quad h_2 - h_3 = h_{is}(h_1 - h_{3s}) = 0.82 \times 666.13 = 546.2 \text{ kJ/kg}$$

$$\therefore \quad h_3 = 2631 \text{ kJ/kg} = 504.7 + x_2 \times 2201.7$$

$$\therefore \quad x_3 = \frac{2126.3}{2201.7} = 0.966$$

$$s_3 = 1.5301 + 0.966 \times 5.597 = 6.9368$$

$$\therefore \quad \Psi_3 = \text{Availability of steam at turbine exhaust}$$

$$= (h_3 - h_0) - T_0(s_3 - S_0) = (2631 - 83.96) - 293\,(6.9368 - 0.2966)$$

$$= 2547.04 - 1945.58 = 601.46 \text{ kJ/kg}$$

Specific work output from the turbine

$$= h_2 - h_3 = 3177.2 - 2631 = 546.2 \text{ kJ/kg}$$ *Ans.*

The work done is less than the loss of availability of steam between states 2 and 3, because of the irreversibility accounted for by the isentropic efficiency.

Example 9.15

A steam turbine receives 600 *kg/h of steam at* 25 *bar,* 350°C. *At a certain stage of the turbine, steam at the rate of* 150 *kg/h is extracted at* 3 *bar,* 200°C. *The remaining steam leaves the turbine at* 0.2 *bar,* 0.92 *dry. During the expansion process, there is heat transfer from the turbine to the surroundings at the rate of* 10 *kJ/s. Evaluate per kg of steam entering the turbine* (*a*) *the availability of steam entering and leaving the turbine,* (*b*) *the maximum work, and* (*c*) *the irreversibility. The atmosphere is at* 30°C.

Solution At 25 bar, 350°C

$$h_1 = 3125.87 \text{ kJ/kg}$$

$$s_1 = 6.8481 \text{ kJ/kg K}$$

At 30°C,

$$h_0 = 125.79 \text{ kJ/kg}$$

$$s_0 = s_{f\,30^\circ C} = 0.4369 \text{ kJ/kg K}$$

At 3 bar, 200°C

$$h_2 = 2865.5 \text{ kJ/kg}$$

$$s_2 = 7.3115 \text{ kJ/kg K}$$

At 0.2 bar (0.92 dry)

$$h_f = 251.4 \text{ kJ/kg}$$

$$h_{fg} = -2358.3 \text{ kJ/kg}$$

$$s_f = 0.8320 \text{ kJ/kg K}$$

$$s_g = 7.9085 \text{ kJ/kg K}$$

$$\therefore \quad h_3 = 251.4 + 0.92 \times 2358.3 = 2421.04 \text{ kJ/kg}$$

$$s_3 = 0.8320 + 0.92 \times 7.0765 = 7.3424 \text{ kJ/kg K}$$

The states of steam are shown in Fig. Ex. 9.15.

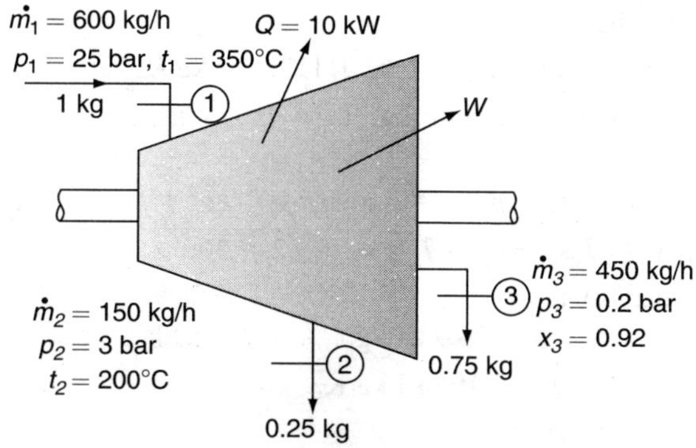

Fig. Ex. 9.15

(a) Availability of steam entering the turbine

$$\Psi_1 = (h_1 - h_0) - T_0 (s_1 - s_0)$$
$$= (3125.87 - 125.79) - 303 (6.8481 - 0.4369)$$
$$= 3000.08 - 1942.60 = 1057.48 \text{ kJ/kg}$$ *Ans.*

Availability of steam leaving the turbine at state 2,

$$\Psi_2 = (h_2 - h_0) - T_0(s_2 - s_0) = (2865.5 - 125.79) - 303 (7.3115 - 0.4369)$$
$$= 2739.71 - 2083.00 = 656.71 \text{ kJ/kg}$$ *Ans.*

Availability of steam leaving the turbine at state 3,

$$\Psi_3 = (h_3 - h_0) - T_0 (s_3 - s_0)$$
$$= (2421.04 - 125.79) - 303 (7.3524 - 0.4369)$$
$$= 199.85 \text{ kJ/kg}$$

(b) Maximum work per kg of steam entering the turbine

$$W_{\text{rev}} = \Psi_1 - \frac{m_2}{m_1} \Psi_2 - \frac{m_3}{m_1} \Psi_3 = 1057.48 - 0.25 \times 656.71 - 0.75 \times 199.85$$
$$= 743.41 \text{ kJ/kg}$$ *Ans.*

(c) Irreversibility

$$I = T_0(w_2 s_2 + w_3 s_3 - w_1 s_1) - Q$$
$$= 303 (150 \times 7.3115 + 450 \times 7.3424 - 600 \times 6.8481) - (-10 \times 3600)$$
$$= 303 (1096.73 + 3304.08 - 4108.86) + 36000 = 124{,}460.85 \text{ kJ/h}$$
$$= 124.461 \text{ MJ/h} = \frac{124.461 \times 103}{600} = 207.44 \text{ kJ/kg}$$ *Ans.*

***Example* 9.16**

Determine the exergy of (a) 3 kg of water at 1 bar and 90°C, (b) 0.2 kg of steam at 4 MPa, 500°C and (c) 0.4 kg of wet steam at 0.1 bar and 0.85 quality, (d) 3 kg of ice at 1 bar – 10°C. Assume a dead state of 1 bar and 300 K.

Solution At the dead state of 1 bar, 300 K,

$$u_0 = 113.1 \text{ kJ/kg}, \quad h_0 = 113.2 \text{ kJ/kg K.}$$
$$v_0 = 0.001005 \text{ m}^3\text{/kg}, \quad s_0 = 0.395 \text{ kJ/kg}$$

Exergy of the system:

$$\phi = m[(u + p_0v - T_0s) - (u_0 + p_0v_0 - T_0s_0)]$$

Now, $\quad u_0 + p_0v_0 - T_0s_0 = h_0 - T_0s_0 = 113.2 - 300 \times 0.395 = -5.3 \text{ kJ/kg}$

(a) For water at 1 bar, 90°C

$$u = 376.9 \text{ kJ/kg}, \quad h = 377 \text{ kJ/kg}, \quad v = 0.001035 \text{ m}^3\text{/kg}$$
$$s = 1.193 \text{ kJ/kg K.}$$

Since $\quad p = p_0,$

$$u + p_0v - T_0s = u + pv - T_0s = h - T_0s$$
$$= 377 - 300 \times 1.193 = 19.1 \text{ kJ/kg}$$

Hence, $\quad \phi = 3[19.1 - (-5.3)] = 3 \times 24.4 = 73.2 \text{ kJ}$ *Ans.*

(b) At $p = 4$ MPa, $t = 500$°C

$$u = 3099.8, \quad h = 3446.3 \text{ kJ/kg}, \quad v = 0.08637 \text{ m}^3\text{/kg}$$
$$s = 7.090 \text{ kJ/kg K}$$
$$u + p_0v - T_0s = 3099.8 + 100 \times 0.08637 - 300 \times 7.090 = 981.4 \text{ kJ/kg}$$
$$\phi = 0.2\,[981.4 - (-5.3)] = 197.34 \text{ kJ}$$ *Ans.*

(c) At 0.1 bar, 0.85 quality,

$$u = 192 + 0.85 \times 2245 = 2100.25 \text{ kJ/kg}$$
$$h = 192 + 0.85 \times 2392 = 2225.2 \text{ kJ/kg}$$
$$s = 0.649 + 0.85 \times 7.499 = 7.023 \text{ kJ/kg K}$$
$$v = 0.001010 + 0.85 \times 14.67 = 12.47 \text{ m}^3\text{/kg}$$
$$u + p_0v - T_0s = 2100.25 + 100 \times 12.47 - 300 \times 7.023 = 1240.4 \text{ kJ/kg}$$
$$\phi = 0.4\,[1240.4 - (-5.3)] = 498.3 \text{ kJ}$$ *Ans.*

(d) Since $p = p_0$,

$$\phi = U - U_0 + p_0(V - V_0) - T_0(S - S_0) = H - H_0 - V(p - p_0) - T_0(S - S_0)$$
$$= m[(h - h_0) - T_0(s - s_0)]$$

At 100 kPa, −10°C,

$$h = -354.1 \text{ kJ/kg and } s = -1.298 \text{ kJ/kg K}$$
$$\phi = 3\,[-354.1 - 113.2 - 300\,(-1.298 - 0.0395)] = 81.2 \text{ kJ.}$$ *Ans.*

Example 9.17

A flow of hot water at 90°C is used to heat relatively cold water at 25°C to a temperature of 50°C in a heat exchanger. The cold water flows at the rate of 1 kg/s. When the heat exchanger is operated in the parallel mode, the exit temperature of the hot water stream must not be less than 60°C. In the counterflow operation, the exit temperature of hot water can be as low as 35°C. Compare the second law efficiency and the rate of exergy destruction in the two modes of operation. Take $T_0 = 300$ K.

Solution Given: $t_{h_1} = 90°C$, $t_{c_1} = 25°C$, $t_{c_2} = 60°C$,

$$\dot{m}_c = 1 \text{ kg/s}, \quad T_0 = 300 \text{ K}.$$

The two modes of operation of (a) parallel flow and (b) counterflow are shown in Fig. Ex. 9.17.

In parallel flow mode (a), $t_{h_2} = 60°C$. Neglecting any heat loss,

$$\dot{m}_h c_h(t_{h_1} - t_{h_2}) = \dot{m}_c c_c(t_{c_2} - t_{c_1})$$

$$\dot{m}_h (90 - 60) = 1(50 - 25)$$

$$\dot{m}_h = 0.833 \text{ kg/s}$$

In counterflow mode, $t_{h_2} = 35°C$,

$$\dot{m}_h (90 - 35) = 1(50 - 25)$$

$$\dot{m}_h = \frac{25}{55} = 0.454 \text{ kg/s}$$

Thus, the counterflow arrangement uses significantly less hot water.

Assuming that the hot water stream on exit from the heat exchanger is simply dumped into the drain, the exergy flow rate of the hot water stream at entry is considered as the exergy input rate to the process.

$$a_{f1} = \dot{m}_h [(h_1 - h_0) - T_0(s_1 - s_0)]$$

At 300 K or 27°C, $h_0 = 113.2$ kJ/kg and $s_0 = 0.395$ kJ/kg K

At 90°C, $h_1 = 376.92$ kJ/kg, $s_1 = 1.1925$ kJ/kgK

$$a_{f_1} = 0.833 [(376.92 - 113.2) - 300 (1.1925 - 0.395)]$$
$$= 0.833 (263.72 - 239.25) = 20.38 \text{ kW}$$

Parallel flow:

At 60°C, $h_2 = 251.13$ kJ/kg, $s_2 = 0.8312$ kJ/kg K

At 25°C, $h_3 = 104.89$ kJ/kg, $s_3 = 0.3674$ kJ/kg K

At 50°C, $h_4 = 209.33$ kJ/kg, $s_4 = 0.7038$ kJ/kg K

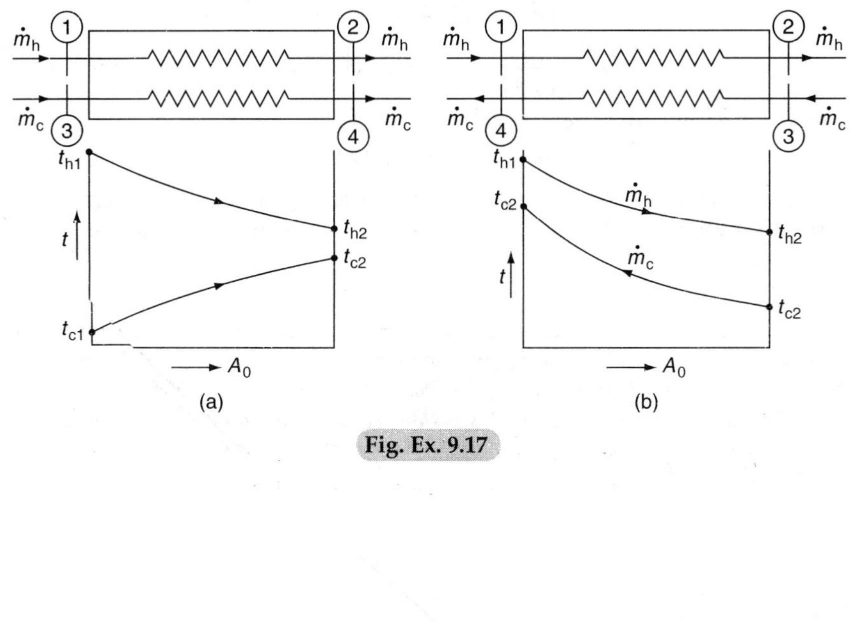

Fig. Ex. 9.17

Rate of exergy gain:

$$= \dot{m}_c [(h_4 - h_3) - T_0(s_4 - s_3)]$$
$$= 1 [(209.33 - 104.89) - 300(0.7038 - 0.3674)]$$
$$= 104.44 - 100.92 = 3.52 \text{ kW}$$
$$(\eta_{II})_P = \frac{3.52}{20.38} = 0.172 \text{ or } 17.2\%$$

Rate of exergy loss by hot water:

$$= \dot{m}_h [(h_1 - h_2) - T_0 (s_1 - s_2)]$$
$$= 0.833 [(376.92 - 251.13) - 300 (1.1925 - 0.8312)]$$
$$= 0.833 (125.79 - 108.39) = 14.494 \text{ kW}$$

Rate of irreversibility or exergy destruction:

$$= 14.494 - 3.52 = 10.974 \text{ kW}$$

If the hot water stream is not dumped to the drain,

$$\eta_{II, P} = \frac{3.52}{14.494} = 0.243 \quad \text{or} \quad 24.3\%$$ *Ans.*

Counterflow:

At 35°C, $h_2 = 146.68$ kJ/kg, $s_2 = 0.5053$ kJ/kg K

Rate of exergy gain of cold water $= \dot{m}_h [(h_4 - h_3) - T_0(s_4 - s_3)] = 3.52$ kW

(same as in parallel flow)

Rate of exergy input (if existing hot water is dumped to the surroundings)

$$= 0.454 (263.72 - 239.25) = 11.11 \text{ kW}$$
$$\eta_{II, C} = \frac{3.52}{11.11} = 0.3168 \text{ or } 31.68\%$$

Rate of exergy loss of hot water:

$$= \dot{m}_h [(h_1 - h_2) - T_0(s_1 - s_2)]$$
$$= 0.454 [(376.92 - 146.68) - 300 (1.1925 - 0.5053)]$$
$$= 0.454 (230.24 - 206.16) = 10.94 \text{ kW}$$
$$\eta_{II, C} = \frac{3.52}{10.94} = 0.3217 \quad \text{or} \quad 32.17\%$$ *Ans.*

Rate of irreversibility or exergy destruction:

$$= 10.94 - 3.52 = 7.42 \text{ kW}$$ *Ans.*

The second law efficiency for the counterflow arrangement is significantly higher and the rate of irreversibility is substantially lower compared to the parallel flow arrangement.

Example 9.18

A small geothermal well in a remote desert area produces 50 kg/h of saturated steam vapour at 150°C. The environment temperature is 45°C. This geothermal steam will be suitably used to produce cooling for homes at 23°C. The steam will emerge from this system as saturated liquid at 1 atm. Estimate the maximum cooling rate that could be provided by such a system.

Solution The energy balance of the control volume as shown in Fig. Ex. 9.18 gives:

$$\dot{Q} + wh_1 = \dot{Q}_0 + wh_2$$

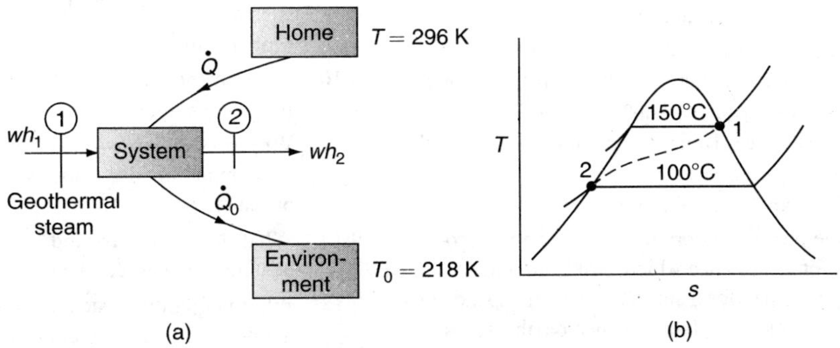

Fig. Ex. 9.18

The entropy balance is:

$$\dot{S}_{gen} = \left[\frac{\dot{Q}_0}{T_0} + ws_2\right] - \left[\frac{\dot{Q}}{T} + ws_1\right]$$

where T is the temperature maintained in the homes.

Solving for $\dot{Q}$,

$$\dot{Q} = \frac{w[(h_1 - T_0 s_1) - (h_2 - T_0 s_2)] - T_0 \dot{S}_{gen}}{(T_0/T) - 1}$$

By second law, $\dot{S}_{gen} > 0$.

Therefore, for a given discharge state 2, the maximum $\dot{Q}$ would be

$$\dot{Q}_{max} = \frac{w(b_1 - b_2)}{(T_0/T) - 1}$$

State-1: $T_1 = 150°C = 423$ K, saturated vapour

$h_1 = 2746.4$ kJ/kg

$s_1 = 6.8387$ kJ/kg K

State-2: $T_2 - 100°C = 373$ K, saturated liquid

$h_2 = 419.0$ kJ/kg

$s_2 = 1.3071$ kJ/kg K

So, since $T_0 = 318$ K,

$b_1 = h_1 - T_0 s_1 = 2746.4 - 318 \times 6.8387 = 571.7$ kJ/kg

$b_2 = h_2 - T_0 s_2 = 419.0 - 318 \times 1.3071 = 3.3$ kJ/kg

$$\dot{Q}_{max} = \frac{50 \times (571.7 - 3.3)}{(318/296) - 1} = 3.82 \times 10^5 \text{ kJ/h} = 106 \text{ kW}$$ *Ans.*

Review Questions

9.1 What is a pure substance?

9.2 What are saturation states?

9.3 What do you understand by triple point? Give the pressure and temperature of water at its triple point.

9.4 What is the critical state? Explain the terms critical pressure, critical temperature and critical volume of water?

9.5 What is normal boiling point.

9.6 Draw the phase equilibrium diagram on p–v coordinates for a substance which shrinks in volume on melting and then for a substance which expands in volume on melting. Indicate thereon the relevant constant property lines.

9.7 Draw the phase equilibrium diagram for a pure substance on p–T coordinates. Why does the fusion line for water have negative slope?

9.8 Draw the phase equilibrium diagram for a pure substance on T–s plot with relevant constant property lines.

9.9 Draw the phase equilibrium diagram for a pure substance on h–s plot with relevant constant property lines.

9.10 Why do the isobars on Mollier diagram diverge from one another?

9.11 Why do isotherms on Mollier diagram become horizontal in the superheated region at low pressures?

9.12 What do you understand by the degree of superheat and the degree of subcooling?

9.13 What is quality of steam? What are the different methods of measurement of quality?

9.14 Why cannot a throttling calorimeter measure the quality if the steam is very wet? How is the quality measured then?

9.15 What is the principle of operation of an electrical calorimeter?

Problems

9.1 Complete the following table of properties for 1 kg of water (liquid, vapour or mixture).

	p (bar)	*t* (°C)	*v* (m³/kg)	*x* (%)	*Super-heat* (°C)	*h* (kJ/kg)	*s* (kJ/kg K)
(a)	—	35	25.22	—	—	—	—
(b)	—	—	0.001044	—	—	419.04	—
(c)	—	212.42	—	90	—	—	—
(d)	1	—	—	—	—	—	6.104
(e)	10	320	—	—	—	—	—
(f)	5	—	0.4646	—	—	—	—
(g)	4	—	0.4400	—	—	—	—
(h)	—	500	—	—	—	3445.3	—
(i)	20	—	—	—	50	—	—
(j)	15	—	—	—	—	—	7.2690

9.2 (a) A rigid vessel of volume 0.86 m³ contains 1 kg of steam at a pressure of 2 bar. Evaluate the specific volume, temperature, dryness fraction, internal energy, enthalpy, and entropy of steam.

(b) The steam is heated to raise its temperature to 150°C. Show the process on a sketch of the p–v diagram, and evaluate the pressure, increase in enthalpy, increase in internal energy, increase in entropy of steam, and the heat transfer. Evaluate also the pressure at which the steam becomes dry saturated.

Ans. (a) 0.86 m³/kg, 120.23°C, 0.97, 2468.54 kJ/kg, 2640.54 kJ/kg, 6.9592 kJ/kg K (b) 2.3 bar, 126 kJ/kg, 106.6 kJ/kg, 0.2598 kJ/kg K, 106.6 kJ/K

9.3 Ten kg of water at 45°C is heated at a contant pressure of 10 bar until it becomes superheated vapour at 300°C. Find the changes in volume, enthalpy, internal energy and entropy.

Ans. 2.569 m^3, 28627.5 kJ, 26047.6 kJ, 64.842 kJ/K

9.4 Water at 40°C is continuously sprayed into a pipeline carrying 5 tonnes of steam at 5 bar, 300°C per hour. At a section downstream where the pressure is 3 bar, the quality is to be 95%. Find the rate of water spray in kg/h. *Ans.* 912.67 kg/h

9.5 A rigid vessel contains 1 kg of a mixture of saturated water and saturated steam at a pressure of 0.15 MPa. When the mixture is heated, the state passes through the critical point. Determine (a) the volume of vessel (b) the mass of liquid and of vapour in the vessel initially, (c) the temperature of the mixture when the pressure has risen to 3 MPa, and (d) the heat transfer required to produce the final state (c).

Ans. (a) 0.003155 m^3, (b) 0.9982 kg, 0.0018 kg, (c) 233.9°C, (d) 581.46 kJ/kg

9.6 A rigid closed tank of volume 3 m^3 contains 5 kg of wet steam at a pressure of 200 kPa. The tank is heated until the steam becomes dry saturated. Determine the final pressure and the heat transfer to the tank. *Ans.* 304 kPa, 3346 kJ

9.7 Steam flows through a small turbine at the rate of 5000 kg/h entering at 15 bar, 300°C and leaving at 0.1 bar with 4% moisture. The steam enters at 80 m/s at a point 3 m above the discharge and leaves at 40 m/s. Compute the shaft power assuming that the device is adiabatic but considering kinetic and potential energy changes. How much error would be made if these terms were neglected? Calculate the diameters of the inlet and discharge tubes.

Ans. 765.6 kW, 0.44%, 6.11 cm, 78.9 cm

9.8 A sample of steam from a boiler drum at 3 MPa is put through a throttling calorimeter in which the pressure and temperature are found to be 0.1 MPa, 120°C. Find the quality of the sample taken from the boiler. *Ans.* 0.951

9.9 It is desired to measure the quality of wet steam at 0.5 MPa. The quality of steam is expected to be not more than 0.9.

(a) Explain why a throttling calorimeter to atmospheric pressure will not serve the purpose.

(b) Will the use of a separating calorimeter, ahead of the throttling calorimeter, serve the purpose, if at best 5°C of superheat is desirable at the end of throttling? What is the minimum dryness fraction required at the exit of the separating calorimeter to satisfy this condition? *Ans.* 0.97

9.10 The following observations were recorded in an experiment with a combined separating and throttling calorimeter:

Pressure in the steam main—15 bar

Mass of water drained from the separator—0.55 kg

Mass of steam condensed after passing through the throttle valve —4.20 kg

Pressure and temperature after throttling—1 bar, 120°C

Evaluate the dryness fraction of the steam in the main, and state with reasons, whether the throttling calorimeter alone could have been used for this test. *Ans.* 0.85

9.11 Steam from an engine exhaust at 1.25 bar flows steadily through an electric calorimeter and comes out at 1 bar, 130°C. The calorimeter has two 1 kW heaters and the flow is measured to be 3.4 kg in 5 min. Find the quality in the engine exhaust. For the same mass flow and pressures, what is the maximum moisture that can be determined if the outlet temperature is at least 105°C? *Ans.* 0.944, 0.921

9.12 Steam expands isentropically in a nozzle from 1 MPa, 250°C to 10 kPa. The steam flow rate is 1 kg/s. Find the velocity of steam at the exit from the nozzle, and the exit area of the nozzle. Neglect the velocity of steam at the inlet to the nozzle.

The exhaust steam from the nozzle flows into a condenser and flows out as saturated water. The cooling water enters the condenser at 25°C and leaves at 35°C. Determine the mass flow rate of cooling water.

Ans. 1224 m/s, 0.0101 m^2. 47.81 kg/s

9.13 A reversible polytropic process, begins with steam at p_1 = 10 bar, t_1 = 200°C, and ends with p_2 = 1 bar. The exponent n has the value 1.15. Find the final specific volume, the final temperature, and the heat transferred per kg of fluid.

9.14 Two streams of steam, one at 2 MPa, 300°C and the other at 2 MPa, 400°C, mix in a steady flow adiabatic process. The rates of flow of the two streams are 3 kg/min and 2 kg/min respectively. Evaluate the final temperature of the emerging stream, if there is no pressure drop due to the mixing process. What would be the rate of increase in the entropy of the universe? This stream with a negligible velocity now expands adiabatically in a nozzle to a pressure of 1 kPa. Determine the exit velocity of the stream and the exit area of the nozzle.

Ans. 340°C, 0.042 kJ/K min, 1530 m/s, 53.77 cm^2

9.15 Boiler steam at 8 bar, 250°C, reaches the engine control valve through a pipeline at 7 bar, 200°C. It is throttled to 5 bar before expanding in the engine to 0.1 bar, 0.9 dry. Determine per kg of steam (a) the heat loss in the pipeline, (b) the temperature drop in passing through the throttle valve, (c) the work output of the engine, (d) the entropy change due to throttling and (e) the entropy change in passing through the engine.

Ans. (a) 105.3 kJ/kg, (b) 5°C, (c) 499.35 kJ/kg, (d) 0.1433 kJ/kg K, (e) 0.3657 kJ/kg K

9.16 Tank A (Fig. P 9.16) has a volume of 0.1 m^3 and contains steam at 200°C, 10% liquid and 90% vapour by volume, while tank *B* is evacuated. The valve is then opened, and the tanks eventually come to the same pressure, which is found to be 4 bar. During this process, heat is transferred such that the steam remains at 200°C. What is the volume of tank *B*?

Ans. 4.89 m^3

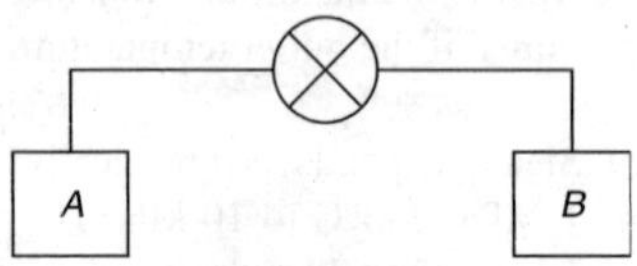

Fig. P 9.16

9.17 Calculate the amount of heat which enters or leaves 1 kg of steam initially at 0.5 MPa and 250°C, when it undergoes the following processes:

(a) It is confined by a piston in a cylinder and is compressed to 1 MPa and 300°C as the piston does 200 kJ of work on the steam.

(b) It passes in steady flow through a device and leaves at 1 MPa and 300°C while, per kg of steam flowing through it, a shaft puts in 200 kJ of work. Changes in K.E. and P.E. are negligible.

(c) It flows into an evacuated rigid container from a large source which is maintained at the initial condition of the steam. Then 200 kJ of shaft work is transferred to the steam, so that its final condition is 1 MPa and 300°C.

Ans. (a) – 130 kJ (b) – 109 kJ, and (c) – 367 kJ

9.18 A sample of wet steam from a steam main flows steadily through a partially open valve into a pipeline in which is fitted an electric coil. The valve and the pipeline are well insulated. The steam mass flow rates 0.008 kg/s while the coil takes 3.91 amperes at 230 volts. The main pressure is 4 bar, and the pressure and temperature of the steam downstream of the coil are 2 bar and 160°C respectively. Steam velocities may be assumed to be negligible.

(a) Evaluate the quality of steam in the main.

(b) State, with reasons, whether an insulated throttling calorimeter could be used for this test.

Ans. (a) 0.97, (b) not suitable

9.19 Two insulated tanks, *A* and *B*, are connected by a valve. Tank *A* has a volume of 0.70 m^3 and contains steam at 1.5 bar, 200°C. Tank *B* has a volume of 0.35 m^3 and contains steam at 6 bar with a quality of 90%. The valve is then opened, and the two tanks come to a uniform state. If there is no heat transfer during the process, what is the final pressure? Compute the entropy change of the universe.

Ans. 322.6 KPa, 0.1985 kJ/K

9.20 A spherical aluminium vessel has an inside diameter of 0.3 m and a 0.62 cm thick wall. The vessel contains water at 25°C with a quality of 1%. The vessel is then heated until the water inside is saturated vapour. Considering the vessel and water together as a system, calculate the heat transfer during this process. The density of aluminium is 2.7 g/cm^3 and its specific heat is 0.896 kJ/kg K.

Ans. 2682.82 kJ

9.21 Steam at 10 bar, 250°C flowing with negligible velocity at the rate of 3 kg/min mixes adiabatically with steam at 10 Bar, 0.75 quality, flowing also with negligible velocity at the rate of 5 kg/min. The combined stream of steam is throttled to 5 bar and then expanded isentropically in a nozzle to 2 bar. Determine (a) the state of steam after mixing, (b) the steam after throttling, (c) the increase in entropy due to throttling, (d) the velocity of steam at the exit from the nozzle, and (e) the exit area of the nozzle. Neglect the K.E. of steam at the inlet to the nozzle.

Ans. (a) 10 bar, 0.975 dry, (b) 5 bar, 0.894 dry, (c) 0.2669 kJ/kg K, (d) 540 m/s, (e) 1.864 cm^2

9.22 Steam of 65 bar, 400°C leaves the boiler to enter a steam turbine fitted with a throttle governor. At a reduced load, as the governor takes action, the pressure of steam is reduced to 59 bar by throttling before it is admitted to the turbine. Evaluate the availabilities of steam before and after the throttling process and the irreversibility due to it.

Ans. I = 21 kJ/kg

9.23 A mass of wet steam at temperature 165°C is expanded at constant quality 0.8 to pressure 3 bar. It is then heated at constant pressure to a degree of superheat of 66.5°C. Find the enthalpy and entropy changes during expansion and during heating. Draw the $T-s$ and $h-s$ diagrams.

Ans. – 59 kJ/kg, 0.163 kJ/kg K during expansion and 676 kJ/kg, 1.588 kJ/kg K during heating

9.24 Steam enters a turbine at a pressure of 100 bar and a temperature of 400°C. At the exit of the turbine the pressure is 1 bar and the entropy is 0.6 J/g K greater than that at inlet. The process is adiabatic and changes in KE and PE may be neglected. Find the work done by the steam in J/g. What is the mass flow rate of steam required to produce a power output of 1 kW? *Ans.* 625 J/g, 1.6 kg/s

9.25 One kg of steam in a closed system undergoes a thermodynamic cycle composed the following reversible processes: (1–2) The steam initially at 10 bar, 40% quality is heated at constant volume until the pressure rises to 35 bar; (2–3). It is then expanded isothermally to 10 bar; (3–1). It is finally cooled at constant pressure back to its initial state. Sketch the cycle on $T-s$ coordinates, and calculate the work done, the heat transferred, and the change of entropy for each of the three processes. What is the thermal efficiency of the cycle?

Ans. 0; 1364 kJ; 2.781 kJ/K, 367.5 kJ; 404.6 kJ; 0.639 kJ/K; – 209.1 kJ; – 1611 kJ; – 3.419 kJ/K 8.93%

9.26 Determine the exergy per unit mass for the steady flow of each of the following:

(a) steam at 1.5 MPa, 500°C
(b) air at 1.5 MPa, 500°C
(c) water at 4MPa, 300K
(d) air at 4 MPa, 300 K
(e) air at 1.5 MPa, 300 K

Ans. (a) 1220 kJ/kg, (b) 424 kJ/kg, (c) 3.85 kJ/kg; (d) 316 kJ/kg, (e) 232 kJ/kg

9.27 A liquid (c_p = 6 kJ/kg K) is heated at an approximately constant pressure from 298 K to 90°C by passing it through tubes immersed in a furnace. The mass flow rate is 0.2 kg/s. Determine (a) the heating load in kW. (b) the exergy production rate in kW corresponding to the temperature rise of the fluid.

Ans. (a) 78 kW, (b) 7.44 kW

9.28 A flow of hot water at 80°C is used to heat cold water from 20°C to 45°C in a heat exchanger. The cold water flows at the rate of 2 kg/s. When operated in parallel mode, the exit temperature of hot water stream cannot be less than 55°C, while in the counterflow mode, it can be as low as 30°C. Assuming the surroundings are at 300 K, compare the second law efficiencies for the two modes of operation.

9.29 Water at 90°C is flowing in a pipe. The pressure of the water is 3 bar, the mass flow rate is 10 kg/s, the velocity is 0.5 m/s and the elevation of the pipe is 200 m above the exit plane of the pipeline (ground level). Compute (a) the thermal exergy flux, (b) the pressure exergy flux, (c) the exergy flux from KE, (d) the exergy flux from PE, (e) total exergy flux of the stream.

Ans. (a) 260 kW, (b) 2.07 kW, (c) 1.25×10^{-3} kW, (d) 19.6 kW, (e) 282 kW

9.30 A cylinder fitted with a piston contains 2 kg steam at 500 kPa, 400°C. Find the entropy change and the work done when the steam expands to a final pressure of 200 kPa in each of the following ways: (a) adiabatically and reversibly, (b) adiabatically and irreversibly to an equilibrium temperature of 300°C.

Ans. (a) 0, 386.7 kJ, (b) 0.1976 kJ/K, 309.4 kJ

9.31 Steam expands isentropically in a nozzle from 1 MPa, 250°C to 10 kPa. The steam flow rate is 1 kg/s. Neglecting the KE of steam at inlet to the nozzle, find the velocity of steam at exit from the nozzle and the exit area of the nozzle.

Ans. 1223 m/s, 100 cm^2

9.32 Hot helium gas at 800°C is supplied to a steam generator and is cooled to 450°C while serving as a heat source for the generation of steam. Water enters the steam generator at 200 bar, 250°C and leaves as superheated steam at 200 bar, 500°C. The temperature of the surroundings is 27°C. For 1 kg helium, determine (a) the maximum work that could be produced by the heat removed from helium, (b) the mass of steam generated per kg of helium, (c) the actual work done in the steam cycle per kg of helium, (d) the net change for entropy of the universe, and (e) the irreversibility. Take the average c_p for helium as 5.1926 kJ/kg K and the properties of water at inlet to the steam generator as those of saturated water at 250°C.

Ans. (a) 1202.4 kJ/kg He, (b) 0.844 kg H_2O/kg He (c) 969.9 kJ/kg He, (d) 0.775 kJ/(kg He-K), (e) 232.5 kJ/kg He

CHAPTER 10

Properties of Gases and Gas Mixtures

10.1 AVOGADRO'S LAW

A mole of a substance has a mass numerically equal to the molecular weight of the substance.

One g mol of oxygen has a mass of 32 g, 1 kg mol of oxygen has a mass of 32 kg, 1 kg mol of nitrogen has a mass of 28 kg, and so on.

Avogadro's law states that the volume of a g mol of all gases at the pressure of 760 mm Hg and temperature of 0°C is the same, and is equal to 22.4 litres. Therefore, 1 g mole of a gas has a volume of 22.4×10^3 cm^3 and 1 kg mol of a gas has a volume of 22.4 m^3 at normal temperature and pressure (N.T.P.).

For a certain gas, if m is its mass in kg, and μ its molecular weight, then the number of kg moles of the gas, n, would be given by

$$n = \frac{m \text{ kg}}{\mu \dfrac{\text{kg}}{\text{kg mol}}} = \frac{m}{\mu} \text{ kg moles}$$

The molar volume, v, is given by

$$v = \frac{V}{n} \; m^3/\text{kg mol}$$

where V is the total volume of the gas in m^3.

10.2 EQUATION OF STATE OF A GAS

The functional relationship among the properties, pressure p, molar or specific volume v, and temperature T, is known as an *equation of state,* which may be expressed in the form,

$$f(p, v, T) = 0$$

If two of these properties of a gas are known, the third can be evaluated from the equation of state.

It was discussed in Chapter 2 that gas is the best-behaved thermometric substance because of the fact that the ratio of pressure p of a gas at any temperature to pressure p_t of the same gas at the triple point, as both p and p_t approach zero, approaches a value independent of the nature of the gas. The ideal gas temperature T of the system at whose temperature the gas exerts pressure p (Article 2.5) was defined as

$$T = 273.16 \lim_{p_t \to 0} \frac{p}{p_t} \qquad \text{(Const. } V\text{)}$$

$$T = 273.16 \lim_{p_p \to 0} \frac{V}{V_t} \qquad \text{(Const. } p\text{)}$$

The relation between pv and p of a gas may be expressed by means of a power series of the form

$$pv = A(1 + B'p + C'p^2 + ...) \qquad (10.1)$$

where A, B', C', etc., depend on the temperature and nature of the gas.

A fundamental property of gases is that $\lim_{p \to 0} (pv)$ is independent of the nature of the gas and depends only on T. This is shown in Fig. 10.1, where the product pv is plotted against p for four different gases in the bulb (nitrogen, air, hydrogen, and oxygen) at the boiling point of sulphur, at steam point and at the triple point of water. In each case, it is seen that as $p \to 0$, pv approaches the same value for all gases at the same temperature. From Eq. (10.1)

$$\lim_{p \to 0} pv = A$$

Therefore, the constant A is a function of temperature only and independent of the nature of the gas.

$$\lim \frac{p}{p_t} \text{ (Const. } V) = \lim \frac{pV}{p_t V} = \frac{\lim pv}{(pv)_t} = \frac{A}{A_t}$$

$$\lim \frac{V}{V_t} \text{ (Const. } p) = \lim \frac{pV}{pV_t} = \frac{\lim pv}{\lim (pv)_t} = \frac{A}{A_t}$$

The ideal gas temperature T, is thus

$$T = 273.16 \frac{\lim (pv)}{\lim (pv)_t}$$

$$\therefore \quad \lim (pv) = \left[\frac{\lim (pv)_t}{273.16}\right] T$$

The term within bracket is called the *universal gas constant* and is denoted by $\bar{R}$. Thus,

$$\bar{R} = \frac{\lim (pv)_t}{273.16} \tag{10.2}$$

The value obtained for $\lim_{p \to 0} (pv)_t$ is 22.4 $\frac{\text{litre-atm}}{\text{g mol}}$

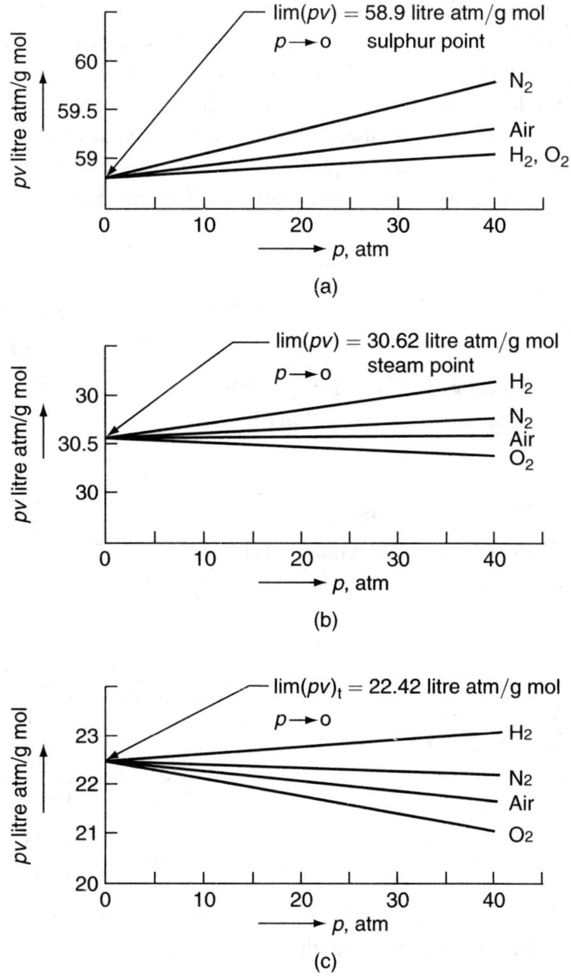

Fig. 10.1 *For any gas $\lim_{p \to 0} (pv)_T$ is independent of the nature of the gas and depends only on T*

$$\therefore \quad \bar{R} = \frac{22.4}{273.16} = 0.083 \frac{\text{litre-atm}}{\text{g mol K}}$$

The equation of state of a gas is thus

$$\lim_{p \to 0} p\bar{v} = \bar{R}T \tag{10.3}$$

where $\bar{v}$ is the molar volume.

10.3 IDEAL GAS

A hypothetical gas which obeys the law $p\bar{v} = \bar{R}T$ at all pressures and temperatures is called an *ideal gas.*

Real gases do not conform to this equation of state with complete accuracy. As $p \to 0$, or $T \to \infty$, the real gas approaches the ideal gas behaviour. In the equation $p\bar{v} = \bar{R}T$, as $T \to 0$, i.e. $t \to -273.15°C$, if $\bar{v}$ remains constant, $p \to 0$, or if p remains constant $\bar{v} \to 0$. Since negative volume or negative pressure is inconceivable, the lowest possible temperature is 0 K or – 273.15°C. T is, therefore, known as the absolute temperature.

There is no permanent or perfect gas. At atmospheric condition only, these gases exist in the gaseous state. They are subject to liquefication or solidification, as the temperatures and pressures are sufficiently lowered.

From Avogadro's law, when $p = 760$ mm Hg $= 1.013 \times 10^5$ N/m^2, $T = 273.15$ K, and $\bar{v} = 22.4$ m^3/kg mol

$$\bar{R} = \frac{1.013 \cdot 10^5 \cdot 22.4}{273.15}$$
$$= 8314.3 \text{ Nm/kg mol K}$$
$$= 8.3143 \text{ kJ/kg mol K}$$

Since $\bar{v} = V/n$, where V is the total volume and n the number of moles of the gas, the equation of state for an ideal gas may be written as

$$pV = n\bar{R}T \tag{10.4}$$

Also

$$n = \frac{m}{\mu}$$

where μ is the molecular weight

$\therefore$

$$pV = m \cdot \frac{\bar{R}}{\mu} \cdot T$$

or

$$pV = mRT \tag{10.5}$$

where R = characteristic gas constant $= \dfrac{\bar{R}}{\mu}$ (10.6)

For oxygen, e.g.,

$$R_{o2} = \frac{8.3143}{32} = 0.2598 \text{ kJ/kg K}$$

For air,

$$R_{air} = \frac{8.3143}{28.96} = 0.287 \text{ kJ/kg K}$$

There are 6.023×10^{23} molecules in a g mol of a substance.

This is known as Avogadro's number (A).

$\therefore$

$$A = 6.023 \times 10^{26} \text{ molecules/kg mol}$$

In n kg moles of gas, the total number of molecules, N, are

$$N = nA$$

or

$$n = N/A$$

$$pV = N\frac{\bar{R}}{A}T = NKT \tag{10.7}$$

where K = Boltzmann constant

$$= \frac{\bar{R}}{A} = \frac{8314.3}{6.023 \times 10^{26}} = 1.38 \times 10^{-23} \text{ J/molecule K}$$

Therefore, the equation of state of an ideal gas is given by

$$pV = mRT$$
$$= n\bar{R}T$$
$$= NKT$$

10.3.1 Specific Heats, Internal Energy, and Enthalpy of an Ideal Gas

An ideal gas not only satisfies the equation of state $pv = RT$, but its *specific heats are constant* also. For real gases, these vary appreciably with temperature, and little with pressure.

The properties of a substance are related by

$$Tds = du + pdv$$

or

$$ds = \frac{du}{T} + \frac{p}{T}\,dv \tag{10.8}$$

The internal energy u is assumed to be a function of T and v, i.e.

$$u = f(T, v)$$

or

$$du = \left(\frac{\partial u}{\partial T}\right)_v dT + \left(\frac{\partial u}{\partial v}\right)_T dv \tag{10.9}$$

From Eqs (10.8) and (10.9)

$$ds = \frac{1}{T}\left(\frac{\partial u}{\partial T}\right)_v dT + \frac{1}{T}\left[\left(\frac{\partial u}{\partial v}\right)_T + p\right] dv \tag{10.10}$$

Again, let

$$s = f(T, v)$$

$$ds = \left(\frac{\partial s}{\partial T}\right)_v dT + \left(\frac{\partial s}{\partial v}\right)_T dv \tag{10.11}$$

Comparing Eqs (10.10) and (10.11)

$$\left(\frac{\partial s}{\partial T}\right)_v = \frac{1}{T}\left(\frac{\partial u}{\partial T}\right)_v \tag{10.12}$$

$$\left(\frac{\partial s}{\partial v}\right)_T = \frac{1}{T}\left[\left(\frac{\partial u}{\partial v}\right)_T + p\right] \tag{10.13}$$

Differentiating Eq. (10.12) with respect to v when T is constant

$$\frac{\partial^2 s}{\partial T \partial v} = \frac{1}{T}\frac{\partial^2 u}{\partial T \partial v} \tag{10.14}$$

Differentiating Eq. (10.13) with respect to T when v is constant

$$\frac{\partial^2 s}{\partial v\, \partial T} = \frac{1}{T}\frac{\partial^2 u}{\partial v \partial T} + \frac{1}{T}\left(\frac{\partial p}{\partial T}\right)_v - \frac{1}{T^2}\left(\frac{\partial u}{\partial v}\right)_T - \frac{p}{T^2} \tag{10.15}$$

From Eqs (10.14) and (10.15)

$$\frac{1}{T}\frac{\partial^2 u}{\partial T \cdot \partial v} = \frac{1}{T}\frac{\partial^2 u}{\partial v \cdot \partial T} + \frac{1}{T}\left(\frac{\partial p}{\partial T}\right)_v - \frac{1}{T^2}\left(\frac{\partial u}{\partial v}\right)_T - \frac{p}{T^2}$$

or

$$\left(\frac{\partial u}{\partial v}\right)_T + p = T\left(\frac{\partial p}{\partial T}\right)_v \tag{10.16}$$

For an ideal gas

$$pv = RT$$

$$v\left(\frac{\partial p}{\partial T}\right)_v = R$$

$$\left(\frac{\partial p}{\partial T}\right)_{\text{v}} = \frac{R}{v} = \frac{p}{T} \tag{10.17}$$

From Eqs (10.16) and (10.17)

$$\left(\frac{\partial u}{\partial v}\right)_{\text{T}} = 0 \tag{10.18}$$

Therefore, u does not change when v changes at constant temperature.

Similarly, if $u = f(T, p)$, it can be shown that $\left(\frac{\partial u}{\partial p}\right)_{\text{T}} = 0$. Therefore, u does not change with p either, when T remains constant.

u does not change unless T changes.

Then

$$u = f(T) \tag{10.19}$$

only for an ideal gas. This is known as *Joule's law.*

If

$$u = f(T, v)$$

$$\text{d}u = \left(\frac{\partial u}{\partial T}\right)_{\text{v}} \text{d}T + \left(\frac{\partial u}{\partial v}\right)_{\text{T}} \text{d}v$$

Since the last term is zero by Eq. (10.18), and by definition

$$c_{\text{v}} = \left(\frac{\partial u}{\partial T}\right)_{\text{v}}$$

$$\text{d}u = c_{\text{v}}\, \text{d}T \tag{10.20}$$

The equation $\text{d}u = c_{\text{v}}\, \text{d}T$ holds good for an ideal gas for any process, whereas for any other substance it is true for a constant volume process only.

Since c_{v} is constant for an ideal gas,

$$\Delta u = c_{\text{v}}\, \Delta T$$

The enthalpy of any substance is given by

$$h = u + pv$$

For an ideal gas

$$h = u + RT$$

Therefore

$$h = f(T) \tag{10.21}$$

only for an ideal gas

Now

$$\text{d}h = \text{d}u + R\text{d}T$$

Since R is a constant

$$\begin{aligned} \Delta h &= \Delta u + R\Delta T \\ &= c_{\text{v}}\, \Delta T + R\Delta T \\ &= (c_{\text{v}} + R)\, \Delta T \end{aligned} \tag{10.22}$$

Since h is a function of T only, and by definition

$$c_{\text{p}} = \left(\frac{\partial h}{\partial T}\right)_{\text{p}}$$

$$\text{d}h = c_{\text{p}}\, \text{d}T \tag{10.23}$$

or

$$\Delta h = c_{\text{p}}\, \Delta T \tag{10.24}$$

From Eqs (10.22) and (10.23)

$$c_p = c_v + R$$

or
$$c_p - c_v = R \tag{10.25}$$

The Eq. $dh = c_p\, dT$ holds good for an ideal gas, even when pressure changes, but for any other substance, this is true only for a constant pressure change.

The ratio of c_p/c_v is of importance in ideal gas computations, and is designated by the symbol γ, i.e.

$$\frac{c_p}{c_v} = \gamma$$

or
$$c_p = \gamma\, c_v$$

From Eq. (10.25)

$$(\gamma - 1)\, c_v = R$$

and
$$\left.\begin{aligned} c_v &= \frac{R}{\gamma - 1} \\ c_p &= \frac{\gamma R}{\gamma - 1} \end{aligned}\right\} \text{kJ/kg K} \tag{10.26}$$

If $R = \dfrac{\bar{R}}{\mu}$ is substituted in Eq. (10.26)

and
$$\left.\begin{aligned} \bar{c}_v &= \mu c_v = (c_v)_{mol} = \frac{\bar{R}}{\gamma - 1} \\ \bar{c}_p &= \mu c_p = (c_p)_{mol} = \frac{\gamma \bar{R}}{\gamma - 1} \end{aligned}\right\} \text{kJ/(kg mol)(K)} \tag{10.27}$$

$\bar{c}_v$ and $\bar{c}_p$ are the *molar or molal specific heats* at constant volume and at constant pressure respectively.

It can be shown by the classical kinetic theory of gases that the values of γ are 5/3 for monatomic gases and 7/5 for diatomic gases. When the gas molecule contains more than two atoms (i.e. for polyatomic gases) the value of γ may be taken approximately as 4/3. The minimum value of γ is thus 1 and the maximum is 1.67.

The value of γ thus depends only on the molecular structure of the gas, i.e. whether the gas is monatomic, diatomic or polyatomic having one, two or more atoms in a molecule. It may be noted that c_p and c_v of an ideal gas depend only on γ and R, i.e. the number of atoms in a molecule and the molecular weight of the gas. They are independent of temperature or pressure of the gas.

10.3.2 Entropy Change of an Ideal Gas

From the general property relations

$$T ds = du + p dv$$

$$T ds = dh - v dp$$

and for an ideal gas, $du = c_v\, dT$, $dh = c_p\, dT$, and $pv = RT$, the entropy change between any two states 1 and 2 may be computed as given below

$$ds = \frac{du}{T} + \frac{p}{T}\, dv$$

$$= c_v \frac{dT}{T} + R \frac{dv}{v}$$

$\therefore$
$$s_2 - s_1 = c_v \ln \frac{T_2}{T_1} + R \ln \frac{v_2}{v_1} \tag{10.28}$$

Also
$$ds = \frac{dh}{T} - \frac{v}{T}\,dp = c_p \frac{dT}{T} - R\frac{dp}{p}$$

or
$$s_2 - s_1 = c_p \ln \frac{T_2}{T_1} - R \ln \frac{p_2}{p_1} \tag{10.29}$$

Since $R = c_p - c_v$, Eq. (10.29) may be written as

$$s_2 - s_1 = c_p \ln \frac{T_2}{T_1} - c_p \ln \frac{p_2}{p_1} + c_v \ln \frac{p_2}{p_1}$$

or
$$s_2 - s_1 = c_p \ln \frac{v_2}{v_1} + c_v \ln \frac{p_2}{p_1} \tag{10.30}$$

Any one of the three Eqs (10.28), (10.29), and (10.30), may be used for computing the entropy change between any two states of an ideal gas.

10.3.3 Reversible Adiabatic Process

The general property relations for an ideal gas may be written as

$$Tds = du + pdv = c_v\, dT + pdv$$

and
$$Tds = dh - vdp = c_p\, dT - vdp$$

For a reversible adiabatic change, $ds = 0$

$\therefore$
$$c_v\, dT = -\,pdv \tag{10.31}$$

and
$$c_p\, dT = vdp \tag{10.32}$$

By division

$$\frac{c_p}{c_v} = \gamma = -\frac{vdp}{pdv}$$

or
$$\frac{dp}{p} + \gamma \frac{dv}{v} = 0$$

or
$$d(\ln p) + \gamma d(\ln v) = d(\ln c)$$

where c is a constant.

$\therefore$
$$\ln p + \gamma \ln v = \ln c$$

$$pv^{\gamma} = c \tag{10.33}$$

Between any two states 1 and 2

$$p_1 v^{\gamma}{}_1 = p_2 v^{\gamma}{}_2$$

or
$$\frac{p_2}{p_1} = \left(\frac{v_1}{v_2}\right)^{\gamma}$$

For an ideal gas

$$pv = RT$$

From Eq. (10.33)

$$p = c \cdot v^{-\gamma}$$

$\therefore$
$$c \cdot v^{-\gamma} \cdot v = RT$$

$$c \cdot v^{1-\gamma} = RT$$

$$Tv^{\gamma-1} = \text{constant} \tag{10.34}$$

Between any two states 1 and 2, for a reversible adiabatic process in the case of an ideal gas

$$T_1 v_1^{\gamma-1} = T_2 v_2^{\gamma-1}$$

or

$$\frac{T_2}{T_1} = \left(\frac{v_1}{v_2}\right)^{\gamma-1} \tag{10.35}$$

Similarly, substituting from Eq. (10.33)

$$v = \left(\frac{c}{p}\right)^{1/\gamma} \text{ in the Eq. } pv = RT$$

$$p \cdot \frac{c'}{p^{1/\gamma}} = RT$$

$$p^{1-(1/\gamma)} \times c' = RT$$

$$\therefore \quad Tp^{(1-\gamma)/\gamma} = \text{constant} \tag{10.36}$$

Between any two states 1 and 2

$$T_1 p_1^{(1-\gamma)/\gamma} = T_2 p_2^{(1-\gamma)/\gamma}$$

$$\therefore \quad \frac{T_2}{T_1} = \left(\frac{p_2}{p_1}\right)^{\frac{\gamma-1}{\gamma}} \tag{10.37}$$

Equations (10.33), (10.34), and (10.36) give the relations among p, v, and T in a reversible adiabatic process for an ideal gas.

The *internal energy change* of an ideal gas for a reversible adiabatic process is given by

$$Tds = du + pdv = 0$$

or

$$\int_1^2 du = -\int_1^2 pdv = -\int_2^2 \frac{c}{v^\gamma}\, dv$$

where

$$pv^\gamma = p_1 v_1^\gamma = p_2 v_2^\gamma = c$$

$$\therefore \quad u_2 - u_1 = c\,\frac{v_2^{1-\gamma} - v_1^{1-\gamma}}{\gamma - 1} = \frac{p_2 v_2^\gamma \cdot v_2^{1-\gamma} - p_1 v_1^\gamma \cdot v_1^{1-\gamma}}{\gamma - 1}$$

$$= \frac{p_2 v_2 - p_1 v_1}{\gamma - 1}$$

$$= \frac{R(T_2 - T_1)}{\gamma - 1} = \frac{RT_1}{\gamma - 1}\left(\frac{T_2}{T_1} - 1\right)$$

$$= \frac{RT_1}{\gamma - 1}\left[\left(\frac{p_2}{p_1}\right)^{\gamma - 1/\gamma} - 1\right] \tag{10.38}$$

The enthalpy change of an ideal gas for a reversible adiabatic process may be similarly derived.

$$Tds = dh - vdp = 0$$

or

$$\int_1^2 dh = \int_1^2 vdp = \int_1^2 \frac{(c)^{1/\gamma}}{p^{1/\gamma}}$$

where $$p_1 v^{\gamma}_1 = p_2 v^{\gamma}_2 = c$$

$\therefore$
$$h_2 - h_1 = \frac{\gamma}{\gamma-1} c^{1/\gamma} [p_2^{(\gamma-1)/\gamma} - p_1^{(\gamma-1)/\gamma}]$$
$$= \frac{\gamma}{\gamma-1} (p_1 v_1^{\gamma})^{1/\gamma} \cdot (p_1)^{(\gamma-1)/\gamma} \left[\left(\frac{p_2}{p_1}\right)^{(\gamma-1)/\gamma} - 1\right]$$
$$= \frac{\gamma p_1 v_1}{\gamma-1} \left[\left(\frac{p_2}{p_1}\right)^{(\gamma-1)/\gamma} - 1\right]$$
$$= \frac{\gamma R T_1}{\gamma-1} \left[\left(\frac{p_2}{p_1}\right)^{(\gamma-1)/\gamma} - 1\right] \tag{10.39}$$

The work done by an ideal gas in a reversible adiabatic process is given by
$$đQ = dU + đW = 0$$
or
$$đW = -dU$$
i.e. work is done at the expense of the internal energy.

$\therefore$
$$W_{1-2} = U_1 - U_2 = m(u_1 - u_2)$$
$$= \frac{m(p_1 v_1 - p_2 v_2)}{\gamma-1} = \frac{mR(T_1 - T_2)}{\gamma-1} = \frac{mRT_1}{\gamma-1}\left[1 - \left(\frac{p_2}{p_1}\right)^{(\gamma-1)/\gamma}\right] \tag{10.40}$$

where m is the mass of gas.

In a steady flow process, where both flow work and external work are involved, we have from S.F.E.E.,
$$W_x + \Delta\frac{V^2}{2} + g\Delta z = h_1 - h_2 = c_p(T_1 - T_2) = \frac{\gamma R(T_1 - T_2)}{\gamma-1}$$
$$= \frac{\gamma(p_1 v_1 - p_2 v_2)}{\gamma-1} = \frac{\gamma}{\gamma-1} p_1 v_1 \left[1 - \left(\frac{p_2}{p_1}\right)^{\gamma-1/\gamma}\right] \tag{10.41}$$

If K.E. and P.E. changes are neglected,
$$W_x = \frac{\gamma}{\gamma-1} p_1 v_1 \left[1 - \left(\frac{p_2}{p_1}\right)^{\gamma-1/\gamma}\right] \tag{10.42}$$

10.3.4 Reversible Isothermal Process

When an ideal gas of mass m undergoes a reversible isothermal process from state 1 to state 2, the work done is given by
$$\int_1^2 đW = \int_{V_1}^{V_2} p\,dV$$
or
$$W_{1-2} = \int_{V_1}^{V_2} \frac{mRT}{V} dV = mRT \ln \frac{V_2}{V_1}$$
$$= mRT \ln \frac{p_1}{p_2} \tag{10.43}$$

The heat transfer involved in the process

$$Q_{1-2} = U_2 - U_1 + W_{1-2}$$
$$= W_{1-2} = mRT \ln V_2/V_1 = T(S_2 - S_1) \tag{10.44}$$

10.3.5 Polytropic Process

An equation of the form pv^n = constant, where n is a constant can be used approximately to describe many processes which occur in practice. Such a process is called a *polytropic process*. It is not adiabatic, but it can be reversible. It may be noted that γ is a property of the gas, whereas n is not. The value of n depends upon the process. It is possible to find the value of n which more or less fits the experimental results. For two statses on the process,

$$p_1 v^n{}_1 = p_2 v^n{}_2 \tag{10.45}$$

or

$$\left(\frac{v_2}{v_1}\right)^n = \frac{p_1}{p_2}$$

$$n = \frac{\log p_1 - \log p_2}{\log v_2 - \log v_1} \tag{10.46}$$

For known values of p_1, p_2, v_1 and v_2, n can be estimated from the above relation.

Two other relations of a polytropic process, corresponding to Eqs (10.35) and (10.37), are

$$\frac{T_2}{T_1} = \left(\frac{v_1}{v_2}\right)^{n-1} \tag{10.47}$$

$$\frac{T_2}{T_1} = \left(\frac{p_2}{p_1}\right)^{n-1/n} \tag{10.48}$$

(i) ***Entropy Change in a Polytropic Process*** In a reversible adiabatic process, the entropy remains constant. But in a reversible polytropic process, the entropy changes. Substituting Eqs (10.45), (10.47) and (10.48) in Eqs (10.28), (10.29) and (10.30), we have three expressions for entropy change as given below

$$s_2 - s_1 = c_v \ln \frac{T_2}{T_1} + R \ln \frac{v_2}{v_1}$$

$$= \frac{R}{\gamma - 1} \ln \frac{T_2}{T_1} + \frac{R}{n-1} \ln \frac{T_1}{T_2}$$

$$= \frac{n-\gamma}{(\gamma-1)(n-1)} R \ln \frac{T_2}{T_1} \tag{10.49}$$

Relations in terms of pressure and specific volume can be similarly derived. These are

$$s_2 - s_1 = \frac{n-\gamma}{n(\gamma-1)} R \ln \frac{p_2}{p_1} \tag{10.50}$$

and

$$s_2 - s_1 = -\frac{n-\gamma}{\gamma-1} R \ln \frac{v_2}{v_1} \tag{10.51}$$

It can be noted that when $n = \gamma$, the entropy change becomes zero. If $p_2 > p_1$, for $n \leq \gamma$, the entropy of the gas decreases, and for $n > \gamma$, the entropy of the gas increases. The increase of entropy may result from reversible heat transfer to the system from the surroundings. Entropy decrease is also possible if the gas is cooled.

(ii) *Heat and Work in a Polytropic Process* Using the first law to unit mass of an ideal gas,

$$Q - W = u_2 - u_1$$

$$= c_v (T_2 - T_1) = \frac{R(T_2 - T_1)}{\gamma - 1}$$

$$= \frac{p_2 v_2 - p_1 v_1}{\gamma - 1}$$

$$= \frac{p_1 v_1}{\gamma - 1}\left[\left(\frac{p_2}{p_1}\right)^{n-1/n} - 1\right] = \frac{p_1 v_1}{\gamma - 1}\left[\left(\frac{v_1}{v_2}\right)^{n-1} - 1\right] \tag{10.52}$$

For a steady flow system of unit mass of ideal gas, the S.F.E.E. Eq. (5.10), gives

$$Q = W_x - \Delta\left[\frac{\bar{V}^2}{2} + gz\right] = h_2 - h_1$$

$$= c_p (T_2 - T_1) = \frac{\gamma R\,(T_2 - T_1)}{\gamma - 1}$$

$$= \frac{\gamma}{\gamma - 1}(p_2 v_2 - p_1 v_1) \tag{10.53}$$

For a polytropic process,

$$Q - W_x = \Delta\left[\frac{\bar{V}^2}{2} + gz\right] = \frac{\gamma\, p_1 v_1}{\gamma - 1}\left[\left(\frac{p_2}{p_1}\right)^{n-1/n} - 1\right]$$

$$= \frac{\gamma\, p_1 v_1}{\gamma - 1}\left[\left(\frac{v_1}{v_2}\right)^{n-1} - 1\right] \tag{10.54}$$

Equations (10.52) and (10.54) can be used to determine heat and work quantities for a closed and a steady flow system respectively.

(iii) *Integral Property Relations in a Polytropic Process* In a pv^n = constant process,

$$\int_1^2 p\,dv = \int_1^2 \frac{p_1 v_1^n}{v^n}\,dv = \frac{p_1 v_1}{n - 1}\left[1 - \left(\frac{v_1}{v_2}\right)^{n-1}\right]$$

$$= \frac{p_1 v_1}{n - 1}\left[1 - \left(\frac{p_2}{p_1}\right)^{n-1/n}\right] \tag{10.55}$$

Similarly,

$$-\int_1^2 v\,dp = \frac{n p_1 v_1}{n - 1}\left[1 - \left(\frac{v_1}{v_2}\right)^{n-1}\right]$$

$$= \frac{n p_1 v_1}{n - 1}\left[1 - \left(\frac{p_2}{p_1}\right)^{n-1/n}\right] \tag{10.56}$$

The integral of Tds is obtained from the property relation

$$Tds = du + pdv$$

$$\therefore \qquad \int_1^2 Tds = \int_1^2 du + \int_1^2 pdv \quad = u_2 - u_1 + \int_1^2 pdv$$

Substituting from Eqs (10.50) and (10.53)

$$\int_1^2 Tds = \frac{\gamma - n}{(\gamma-1)(n-1)} p_1 v_1 \left[1 - \left(\frac{p_2}{p_1}\right)^{n-1/n}\right]$$

$$= \frac{\gamma - n}{(\gamma-1)(n-1)} p_1 v_1 \left[1 - \left(\frac{v_1}{v_2}\right)^{n-1}\right]$$

$$= \frac{\gamma - n}{(\gamma-1)(n-1)} R(T_1 - T_2) \tag{10.57}$$

Since $R/(\gamma - 1) = c_v$, and putting $\Delta T = T_2 - T_1$, the reversible heat transfer

$$Q_R = \int_1^2 Tds = c_v \frac{\gamma - n}{1 - n} \Delta T = c_n \Delta T \tag{10.58}$$

where $c_n = c_v (\gamma - n)/(1 - n)$ is called the polytropic specific heat. For $n > \gamma$ there will be positive heat transfer and gain in entropy. For $n < \gamma$, heat transfer will be negative and entropy of the gas would decrease.

Ordinarily both heat and work are involved in a polytropic process. To evaluate the heat transfer during such a process, it is necessary to first evaluate the work via either $\int pdv$ or $-\int vdp$, depending on whether it is a closed or an open steady flow system. The application of the first law will then yield the heat transfer.

The polytropic processes for various values of n are shown in Fig. 10.2 on the p–v and T–s diagrams.

$$pv^n = C$$

On differentiation,

$$v^n \, dp + pnv^{n-1} \, dv = 0$$

$$\frac{dp}{dv} = -n \frac{p}{v} \tag{10.59}$$

The slope of the curve increases in the negative direction with increase of n. The values of n for some familiar processes are given below

Isobaric process ($p = c$), $n = 0$

Isothermal process ($T = c$), $n = 1$

Isentropic process ($s = c$), $n = \gamma$

Isometric or isochoric process ($v = c$), $n = \infty$.

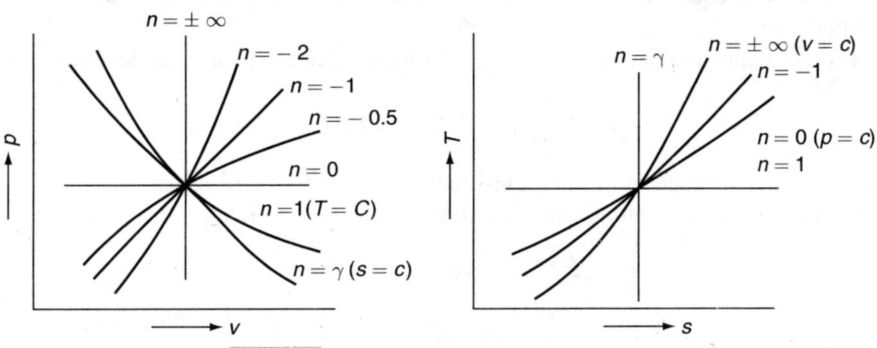

Fig. 10.2 *Process in which pv^n = constant*

10.4 EQUATIONS OF STATE

The ideal gas equation of state $p\bar{v} = \bar{R}T1$ can be established from the positulates of the kinetic theory of gases developed by Clerk Maxwell, with two important assumptions that there is little or no attraction between the molecules of the gas and that the volume occupied by the molecules themselves is negligibly small compared to the volume of the gas. When pressure is very small or temperature very large, the intermolecular attraction and the volume of the molecules compared to the total volume of the gas are nor of much significance, and the real gas obeys very closely the ideal gas equation. But as pressure increases, the intermolecular forces of attraction and repulsion increase, and also the volume of the molecules becomes appreciable compared to the gas volume. So then the real gases deviate considerably from the ideal gas equation. Van der Waals, by applying the laws of mechanics to individual molecules, introduced two correction terms in the equation of ideal gas, and his equation is given below.

$$\left(p + \frac{a}{v^2}\right)(v - b) = RT \tag{10.60}$$

The coefficient a was introduced to account for the existence of mutual attraction between the molecules. The term a/v^2 is called the *force of cohesion*. The coefficient b was introduced to account for the volumes of the molecules, and is known as *co-volume*.

Real gases conform more closely with the van der Waals equation of state than the ideal gas equation of state, particularly at higher pressures. But it is not obeyed by a real gas in all ranges of pressures and temperatures. Many more equations of state were later introduced, and notable among these are the equations developed by Berthelot, Dieterici, Beattie-Bridgeman, Kammerlingh Onnes, Hirshfelder-Bird-Spotz-McGee-Sutton, Wohl, Redlich-Kwong, and Martin-Hou.

Apart from the van der Waals equation, three two-constant equations of state are those of Berthelot, Dieterici, and Redlich-Kwong, as given below:

Berthelot:
$$p = \frac{RT}{v - b} - \frac{a}{Tv^2} \tag{10.61}$$

Dieterici:
$$p = \frac{RT}{v - b} . e^{-a/RT_v} \tag{10.62}$$

Redlich-Kwong:
$$p = \frac{RT}{v - b} - \frac{a}{T^{1/2} v(v + b)} \tag{10.63}$$

The constants, a and b are evaluated from the critical data, as shown for van der Waals equation in Article 10.6. The Berthelot and Dieterici equations of state, like the van der Waals equation, are of limited accuracy. But the Redlich-Kwong equation gives good results at high pressures and is fairly accurate for temperatures above the critical value.

Another two-constant equation which is again of limited accuracy is the Saha-Bose equation of state given as follows.

$$p = -\frac{RT}{2b} e^{-a/RT_v} \text{In}\left(\frac{v - 2b}{v}\right) \tag{10.64}$$

It is, however, quite accurate for densities less than about 0.8 times the critical density.

One more widely used equation of state with good accuracy is the Beattie-Bridgeman equation:

$$p = \frac{RT(1 - e)}{v^2}(v + B) - \frac{A}{v^2} \tag{10.65}$$

where

$$A = A_o\left(1 - \frac{a}{v}\right), B = B_o\left(1 - \frac{b}{v}\right), e = \frac{c}{vT^3}$$

There are five constants, A_0, B_0, a, b, and c, which have to be determined experimentally for each gas. The Beattie-Bridgeman equation does not give satisfactory results in the critical point region.

All these equations mentioned above reduce to the ideal gas equation for large volumes and temperatures and for small pressures.

10.5 VIRIAL EXPANSIONS

The relations between $p\bar{v}$ and p in the form of power series, as given in Eq. (10.1), may be expressed as

$$p\bar{v} = A(1 + B'p + C'p^2 + D'p^3 + ...)$$

For any gas, from Eq. (10.3)

$$\lim_{p \to 0} p\bar{v} = A = \bar{R}T$$

$$\therefore \quad \frac{p\bar{v}}{\bar{R}T} = 1 + B'p + C'p^2 + D'p^3 + ... \tag{10.66}$$

An alternative expression is

$$\frac{p\bar{v}}{\bar{R}T} = 1 + \frac{B}{\bar{v}} + \frac{C}{\bar{v}^2} + \frac{D}{\bar{v}^3} + \cdots \tag{10.67}$$

Both expressions in Eqs (10.66) and (10.67) are known as virial expansions or virial equations of state, first introduced by the Dutch physicist, Kammerlingh Onnes, B', C', B, C, etc. are called *virial coefficients*. B' and B are called second virial coefficients, C' and C are called third virial coefficients, and so on. For a given gas, these coefficients are functions of temperature only.

The ratio $p\bar{v}/\bar{R}T$ is called the compressibility factor, Z. For an ideal gas $Z = 1$. The magnitude of Z for a certain gas at a particular pressure and temperature gives an indication of the extent of deviation of the gas from the ideal gas behaviour. The virial expansions become

$$Z = 1 + B'p + C'p^2 + D'p^3 + ... \tag{10.68}$$

and

$$Z = 1 + \frac{B}{\bar{v}} + \frac{C}{\bar{v}^2} + \frac{D}{\bar{v}^3} + \cdots \tag{10.69}$$

The relations between B', C' and B, C, ... can be derived as given below

$$\begin{aligned}
\frac{pv}{\bar{R}T} &= 1 + B'p + C'p^2 + D'p^3 + ... \\
&= 1 + B'\left[\frac{\bar{R}T}{\bar{v}}\left(1 + \frac{B}{\bar{v}} + \frac{C}{\bar{v}^2} + ...\right)\right] \\
&\quad + C'\left[\left(\frac{\bar{R}T}{\bar{v}}\right)^2\left(1 + \frac{B}{\bar{v}} + \frac{C}{\bar{v}^2} + ... +\right)^2\right] + ... \\
&= 1 + \frac{B'\bar{R}T}{\bar{v}} + \frac{B'B\bar{R}T + C'(\bar{R}T)^2}{\bar{v}^2} \\
&\quad + \frac{B'\bar{R}TC + C'(\bar{R}T)^2 + D'(\bar{R}T)^3}{\bar{v}^3} + ...
\end{aligned} \tag{10.70}$$

Comparing this equation with Eq. (10.67) and rearranging

$$B' = \frac{B}{\bar{R}T}, \quad C' = \frac{C - B^2}{(\bar{R}T)^2},$$

$$D' = \frac{D - 3BC + 2B^3}{(\bar{R}T)^3}, \text{ and so on}$$

Therefore

$$Z = \frac{p\bar{v}}{\bar{R}T} = 1 + B'p + C'p^2 + ...$$

$$= 1 + \frac{B}{\bar{R}T}p + \frac{C - B^2}{(\bar{R}T)^2}p^2 + ... \quad (10.71)$$

The terms $B/\bar{v}$, $C/\bar{v}^2$ etc. of the virial expansion arise on account of molecular interactions. If no such interactions exist (at very low pressures) $B = 0$, $C = 0$, etc., $Z = 1$ and $p\bar{v} = \bar{R}T$.

10.6 LAW OF CORRESPONDING STATES

For a certain gas, the compressibility factor Z is a function of p and T [Eq. (10.71)], and so a plot can be made of lines of constant temperature on coordinates of p and Z (Fig. 10.3). From this plot Z can be obtained for any value of p and T, and the volume can then be obtained from the Equation $pv = ZRT$. The advantage of using Z instead of a direct plot of v is a smaller range of values in plotting.

For each substance, there is a compressibility factor chart. It would be very convenient if one chart could be used for all substances. The general shapes of the vapour dome and of the constant temperature lines on the $p-v$ plane are similar for all substances, although the scales may be different. This similarity can be exploited by using dimensionless properties called *reduced properties*. The reduced pressure is the ratio of the existing pressure to the critical pressure of the substance, and similarly for reduced temperature and reduced volume. Then

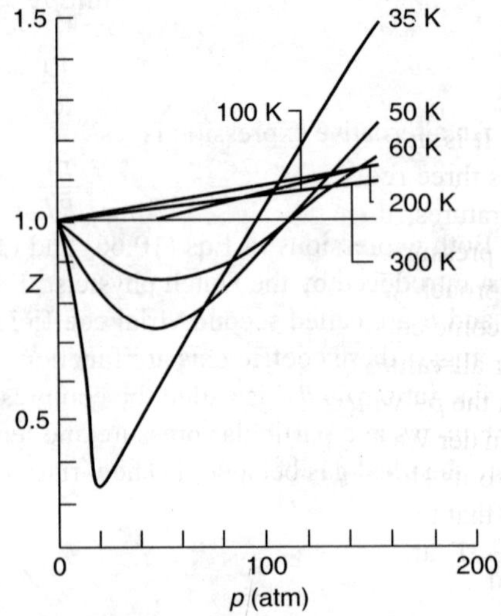

Fig. 10.3 *Variation of the compressibility factor of hydrogen with pressure at constant temperature*

$$p_r = \frac{p}{p_c},\ T_r = \frac{T}{T_c},\ v_r = \frac{v}{v_c}$$

where subscript r denotes the reduced property, and subscript c denotes the property at the critical state.

At the same pressure and temperature the specific or molal volumes of different gases are different. However, it is found from experimental data that at the same reduced pressure and reduced temperature, the reduced volumes of different gases are approximately the same. Therefore, for all substances

$$v_r = f(p_r, T_r) \quad (10.72)$$

Now,

$$v_r = \frac{v}{v_c} = \frac{ZRT\,p_c}{Z_c R T_c p} = \frac{Z}{Z_c} \cdot \frac{T_r}{p_r} \quad (10.73)$$

where $Z_c = \frac{p_c v_c}{RT_c}$. This is called the critical compressibility factor. Therefore from Eqs (10.72) and (10.73),

$$Z = f(p_r, T_r, Z_c) \quad (10.74)$$

Experimental values of Z_c for most substances fall within a narrow range 0.20–0.30. Therefore, Z_c may be taken to be a constant and Eq. (10.74) reduces

$$Z = f(p_r, T_r) \quad (10.75)$$

When T_r is plotted as a function of reduced pressure and Z, a single plot, known as *the generalized compressibility chart*, is found to be satisfactory for a great variety of substances. Although necessarily approximate, the plots are extremely useful in situations where detailed data on a particular gas are lacking but its critical properties are available.

The relation among the reduced properties, p_r, T_r, and v_r, is known as the law of *corresponding states*. It can be derived from the various equations of state, such as those of van der Waals, Berthelot, and Dieterici. For a van der Waals gas,

$$\left(p+\frac{a}{v^2}\right)(v-b)=RT$$

where a, b, and R are the characteristic constants of the particular gas.

$$\therefore \quad p=\frac{RT}{v-b}-\frac{a}{v^2}$$

or

$$pv^3-(pb+RT)v^2+av-ab=0$$

It is therefore a cubic in v and for given values of p and T has three roots of which only one need be real. For low temperatures, three positive real roots exist for a certain range of pressure. As the temperature increases the three real roots approach one another and at the critical temperature they become equal. Above this temperature only one real root exists for all values of p. The critical isotherm T_c at the critical state on the p–v plane (Fig. 10.4), where the three real roots of the van der Waals equation coincide, not only has a zero slope, but also its slope changes at the critical state (point of inflection), so that the first and second derivatives of p with respect to v at $T = T_c$ are each equal to zero. Therefore

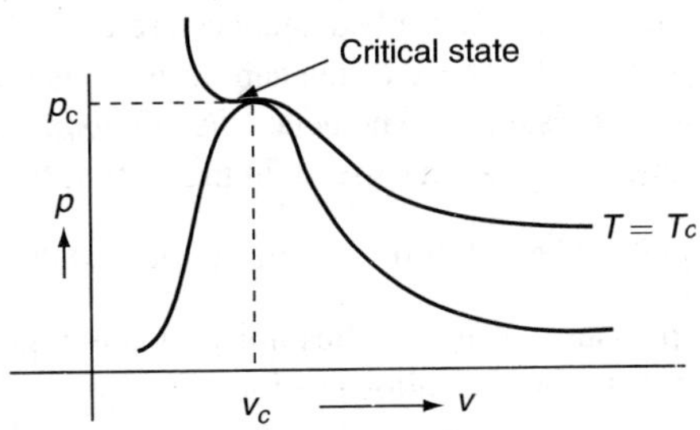

Fig. 10.4 *Critical properties on p–v diagram*

$$\left(\frac{\partial p}{\partial v}\right)_{T=T_c}=-\frac{RT_c}{(v_c-b)^2}+\frac{2a}{v_c^3}=0 \tag{10.76}$$

$$\left(\frac{\partial^2 p}{\partial v^2}\right)_{T=T_c}=\frac{2\cdot RT_c}{(v_c-b)^3}-\frac{6a}{v_c^4}=0 \tag{10.77}$$

From these two equations, by rearranging and dividing, $b=\frac{1}{3}v_c$.

Substituting the values of b in Eq. (10.76)

$$R=\frac{8a}{9T_c v_c}$$

Substituting the values of b and R in Eq. (10.60)

$$\left(p_c+\frac{a}{v_c^2}\right)\left(\frac{2}{3}v_c\right)=\frac{8a}{9T_c v_c}\cdot T_c$$

$$a=3p_c v_c^2$$

Therefore, the value of R becomes

$$R=\frac{8}{3}\frac{p_c v_c}{T_c}$$

The values of *a*, *b*, and *R* have thus been expressed in terms of critical properties. Substituting these in the van der Waals equation of state

$$\left(p+\frac{3p_c v_c^2}{v^2}\right)\left(v-\frac{1}{3}v_c\right)=\frac{8}{3}\frac{p_c v_c}{T_c}T$$

or,

$$\left(\frac{p}{p_c}+\frac{3v_c^2}{v^2}\right)\left(\frac{v}{v_c}-\frac{1}{3}\right)=\frac{8}{3}\frac{T}{T_c}$$

Using the reduced parameters,

$$\therefore \quad \left(p_r+\frac{3}{v_r^2}\right)(3v_r-1)=8T_r \tag{10.78}$$

In the reduced equation of state (10.78) the individual coefficients *a*, *b* and *R* for a particular gas have disappeared. So this equation is an expression of the *law of corresponding states* because it reduces the properties of all gases to one formula. It is a 'law' to the extent that real gases obey van der Waals equation. Two different substances are considered to be in 'corresponding states', if their pressures, volumes and temperatures are of the same fractions (or multiples) of the critical pressure, volume and temperature of the two substances. The *generalized compressibility* chart in terms of reduced properties is shown in Fig. 10.5(a) and (b). It is very useful in predicting the properties of substances for which more precise data are not available. The value of *Z* at the critical state of a van der Waals gas is 0.375 $\left(\text{since } R=\frac{8}{3}\frac{p_c v_c}{T_c}\right)$. At very low pressures *Z* approaches unity, as a real gas approaches the ideal gas behaviour. Equation (10.78) can also be written in the following form

$$\left(p_r v_r+\frac{3}{v_r}\right)(3v_r-1)=8T_r v_r$$

$$\therefore \quad p_r v_r=\frac{8T_r v_r}{3v_r-1}-\frac{3}{v_r} \tag{10.79}$$

Figure 10.6 shows the law of corresponding states in reduced coordinates, $(p_r v_r)$ vs. p_r. Differentiating Eq. (10.79) with respect to p_r and making it equal to zero, it is possible to determine the minima of the isotherms as shown below.

$$\frac{\partial}{\partial p_r}\left[\frac{8T_r v_r}{3v_r-1}-\frac{3}{v_r}\right]_{T_r}=0$$

or

$$\frac{\partial}{\partial v_r}\left[\frac{8T_r v_r}{3v_r-1}-\frac{3}{v_r}\right]_{T_r}\left[\frac{\partial v_r}{\partial p_r}\right]_{T_r}=0$$

Since

$$\left[\frac{\partial v_r}{\partial p_r}\right]_{T_r}\neq 0$$

$$\frac{\partial}{\partial v_r}\left[\frac{8T_r v_r}{3v_r-1}-\frac{3}{v_r}\right]_{T_r}=0$$

$$\therefore \quad \frac{8T_r}{(3v_r-1)^2}=\frac{3}{v_r^2}$$

or

$$\frac{3(3v_r-1)^2}{v_r^2}=8T_r=\left[p_r+\frac{3}{v_r^2}\right](3v_r-1)$$

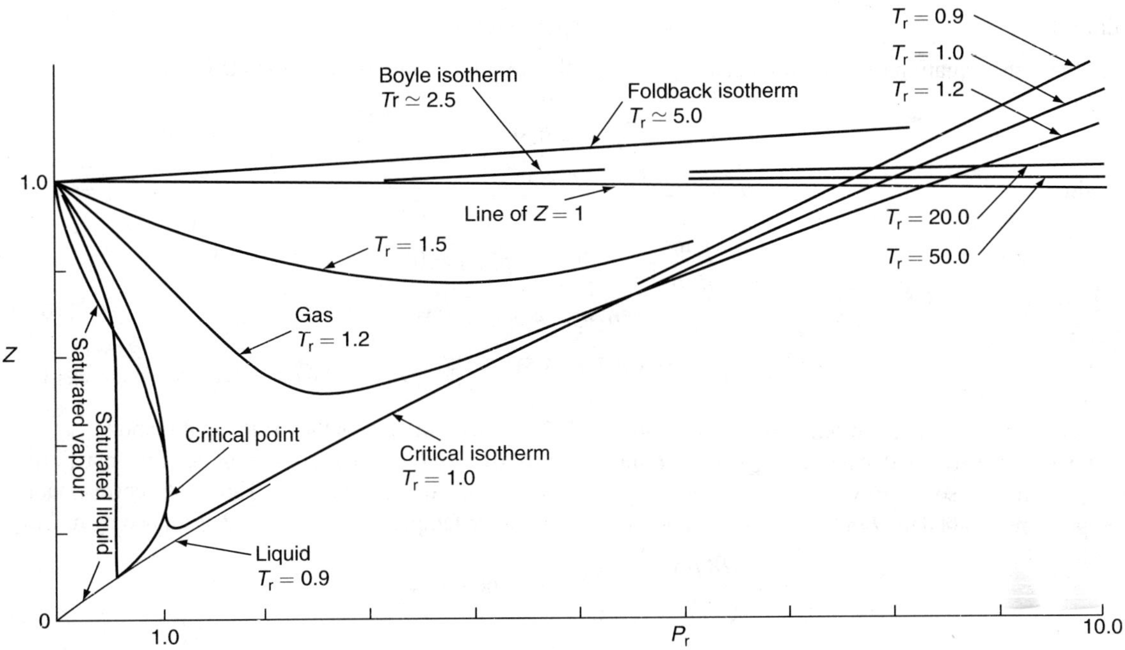

Fig. 10.5 *(a) Generalized compressibility chart*

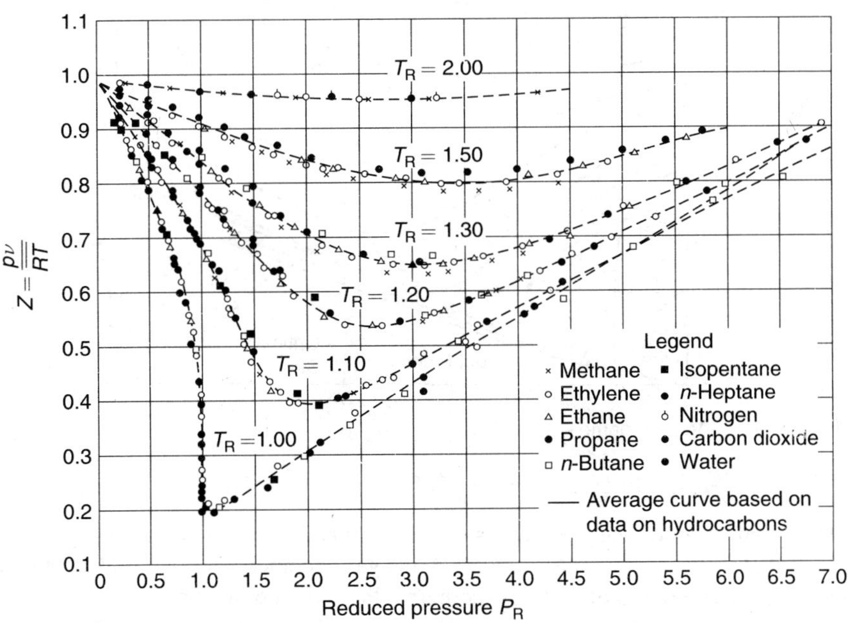

Fig. 10.5 *(b) Generalized compressibility chart*

Simplifying $$(p_r v_r)^2 - 9(p_r v_r) + 6p_r = 0$$

This is the equation of a parabola passing through the minima of the isotherms (Fig. 10.6).

When $$p_r = 0,$$
$$p_r v_r = 0, 9$$

Again $$p_r = \frac{9(p_r v_r) - (p_r v_r)^2}{6}$$

$$\frac{dp_r}{d(p_r v_r)} = 9 - 2\,(p_r\, v_r) = 0$$

$\therefore$ $$p_r\, v_r = 4.5$$

$$p_r = \frac{9 \times 4.5 \times (4.5)^2}{6} = 3.375$$

The parabola has the vertex at $p_r v_r = 4.5$ and $p_r = 3.375$, and it intersects the ordinate at 0 and 9.

Each isotherm up to that marked T_B has a minimum (Fig. 10.6). The T_B isotherm has an initial horizontal portion ($p_r v_r$ = constant) so that Boyle's law is obeyed fairly accurately up to moderate pressures. Hence, the corresponding temperature is called the *Boyle temperature* for that gas. The Boyle temperature T_B can be determined by making

$$\left[\frac{\partial(p_r v_r)}{\partial p_r}\right]_{T_r = T_B} = 0 \text{ when } p_r = 0$$

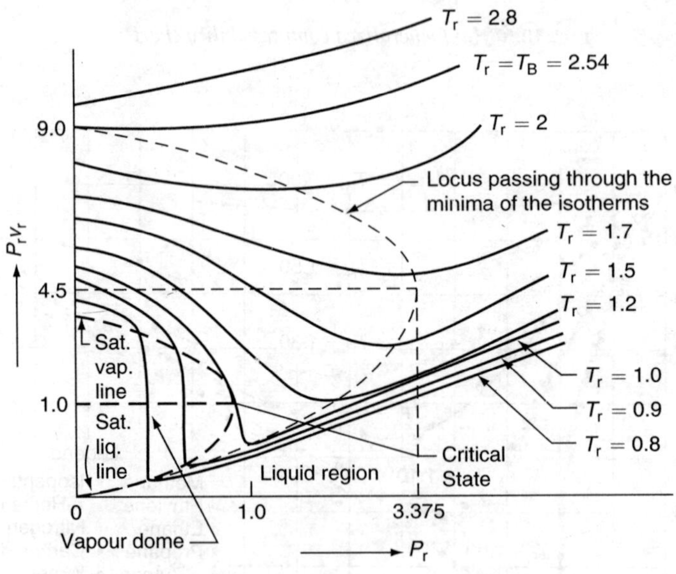

Fig. 10.6 *Law of corresponding states in reduced coordinates*

Above the Boyle temperature, the isotherms slope upward and show no minima.

As T_r is reduced below the critical (i.e. for $T_r < 1$), the gas becomes liquefied, and during phase transition isotherms are vertical. The minima of all these isotherms lie in the liquid zone.

Van der Waals equation of state can be expressed in the virial form as given below

$$\left(p + \frac{a}{v^2}\right)(v - b) = RT$$

$$\left(pv+\frac{a}{v}\right)\left(1-\frac{b}{v}\right)=RT$$

$$\therefore \qquad pv+\frac{a}{v}=RT\left(1-\frac{b}{v}\right)^{-1}$$

$$=RT\left[1+\frac{b}{v}+\frac{b^2}{v^2}+\frac{b^3}{v^3}+\cdots\right] \qquad \left(\text{where } \frac{b}{v}<1\right)$$

$$\therefore \qquad pv=RT\left[1+\left(b-\frac{a}{RT}\right)\frac{1}{v}+\frac{b^2}{v^2}+\frac{b^3}{v^3}+\cdots\right] \tag{10.80}$$

∴ The second virial coefficient $B = b - a/RT$, the third virial coefficient $C = b^2$, etc.

From Eq. (10.71), on mass basis

$$pv=RT\left(1+\frac{B}{\bar{R}T}p+\frac{C-B^2}{\bar{R}T^2}p^2+\ldots\right)$$

To determine Boyle temperature, T_B

$$\left[\frac{\partial(pv)}{\partial p}\right]_{\substack{T=C\\p=0}}=0=\frac{B}{\bar{R}T}$$

$$\therefore \qquad B=0$$

or $$T_B=\frac{a}{bR}, \text{ because } B=b-\frac{a}{RT}$$

The point at which B is equal to zero gives the Boyle temperature. The second virial coefficient is the most important. Since $\left[\frac{\partial(pv)}{\partial p}\right]_{p=0} = B$, when B is known, the behaviour of the gas at moderate pressures is completely determined. The terms which contain higher power (C/v^2, D/v^3, etc.) becomes significant only at very high pressures.

10.7 OTHER EQUATIONS OF STATE

van der Waals equation is one of the oldest equations of state introduced in 1899, where the constants a and b are related to the critical state properties as found earlier,

$$a=3p_c{v^2}_c=\frac{27}{64}\frac{R^2T_c^2}{p_c},\ b=\frac{1}{3}v_c=\frac{1}{8}\frac{RT_c}{p_c}$$

The Beattie-Bridgeman equation developed in 1928 is given in Eq. 10.79, which has five constants. It is essentially an empirical curve fit of data, and is reasonably accurate when values of specific vo!ume are greater than v_c.

Benedict, Webb, and Rubin extended the Beattie-Bridgeman equation of state to cover a broader range of states as given below:

$$p=\frac{\bar{R}T}{v}+\left(B\bar{R}T-A-\frac{c}{T^2}\right)\frac{1}{\bar{v}^2}+\frac{(b\bar{R}T-a)}{\bar{v}^3}+\frac{a}{\bar{v}^6}$$
$$+\frac{c}{\bar{v}^3T^2}\left(1+\frac{\gamma}{\bar{v}^2}\right)\exp\left(-\frac{\gamma}{\bar{v}^2}\right)$$

It has eight constants and is quite successful in predicting the $p-v-T$ behaviour of light hydrocarbons.

The Redlich-Kwong equation proposed in 1949 and given by Eq. 10.77 has the constants a and b in terms of critical properties as follows:

$$a = 0.4275 \frac{R^2 T_c^{2.5}}{p_c}, \quad b = 0.0867 \frac{R T_c}{p_c}$$

The values of the constants for the van der Waals, Redlich-Kwong and Benedict-Webb-Rubin equations of state are given in Table 10.1, while those for the Beattie-Bridgeman equation of state are given in Table 10.2. Apart from these, many other multi-constant equations of state have been proposed. With the advent of high speed computers, equations having 50 or more constants have been developed for representing the $p-v-T$ behaviour of different substances.

Table 10.1

Constants for the van der Waals, Redlich-Kwong, and Benedict-Webb-Rubin Equations of State

1. van der Waals and Redlich-Kwong: Constants for pressure in bars, specific volume in m³/k mol, and temperature in K

Substance	*van der Waals* a, $bar\left(\frac{m^3}{k\,mol}\right)^2$	*van der Waals* b, $\frac{m^3}{k\,mol}$	*Redlich-Kwong* a, $bar\left(\frac{m^3}{k\,mol}\right)^2 K^{1/2}$	*Redlich-Kwong* b, $\frac{m^3}{k\,mol}$
Air	1.368	0.0367	15.989	0.02541
Butane (C_4H_{10})	13.86	0.1162	289.55	0.08060
Carbon dioxide (CO_2)	3.647	0.0428	64.43	0.02963
Carbon monoxide (CO)	1.474	0.0395	17.22	0.02737
Methane (CH_4)	2.293	0.0428	32.11	0.02965
Nitrogen (N_2)	1.366	0.0386	15.53	0.02677
Oxygen (O_2)	1.369	0.0317	17.22	0.02197
Propane (C_3H_8)	9.349	0.0901	182.23	0.06242
Refrigerant 12	10.49	0.0971	208.59	0.06731
Sulfur dioxide (SO_2)	6.883	0.0569	144.80	0.03945
Water (H_2O)	5.531	0.0305	142.59	0.02111

Source: Calculated from critical data.

2. Benedict-Webb-Rubin: Constants for pressure in bars, specific volume in m³/k mol, and temperature in K

Substance	*a*	*A*	*b*	*B*	*c*	*C*	*α*	*γ*
C_4H_{10}	1.9073	10.218	0.039998	0.12436	3.206×10^5	1.006×10^6	1.101×10^{-3}	0.0340
CO_2	0.1386	2.7737	0.007210	0.04991	1.512×10^4	1.404×10^5	8.47×10^{-5}	0.00539
CO	0.0371	1.3590	0.002632	0.05454	1.054×10^3	8.676×10^3	1.350×10^{-4}	0.0060
CH_4	0.0501	1.8796	0.003380	0.04260	2.579×10^3	2.287×10^4	1.244×10^{-4}	0.0060
N_2	0.0254	1.0676	0.002328	0.04074	7.381×10^2	8.166×10^1	1.272×10^{-4}	0.0053

Source: H.W. Coopper, and J.C. Goldfrank, Hydrocarbon Processing, 45 (12); 141 (1967).

Table 10.2

(a) The Beattie-Bridgeman equation of state is

$$P = \frac{R_u T}{\bar{v}^2}\left(1 - \frac{c}{\bar{v}T^2}\right)\left(\bar{v} + B\right) - \frac{A}{\bar{v}}, \text{ where } A = A_0\left(1 - \frac{a}{\bar{v}}\right) \text{ and } B = B_0\left(1 - \frac{b}{\bar{v}}\right)$$

When P is in kPa, $\bar{v}$ is in m^3/k mol, T is in K, and $R_u = 8.314$ kPa · m^3/(k mol · K), the five constants in the Beattie-Bridgeman equation are as follows:

Gas	A_0	a	B_0	b	c
Air	131.8441	0.01931	0.04611	– 0.001101	4.34×10^4
Argon, Ar	130.7802	0.02328	0.03931	0.0	5.99×10^4
Carbon dioxide, CO_2	507.2836	0.07132	0.10476	0.07235	6.60×10^5
Helium, He	2.1886	0.05984	0.01400	0.0	40
Hydrogen, H_2	20.0117	– 0.00506	0.02096	– 0.04359	504
Nitrogen, N_2	136.2315	0.02617	0.05046	– 0.00691	4.20×10^4
Oxygen, O_2	151.0857	0.02562	0.04624	0.004208	4.80×10^4

Source: Gordon J. Van Wylen and Richard E. Sonntag, Fundamentals of Classical Thermodynamics, English/S.I. Version, 3d Ed., Wiley New York, 1986, p. 46. Table 3.3.

10.8 PROPERTIES OF MIXTURES OF GASES—DALTON'S LAW OF PARTIAL PRESSURES

Let us imagine a homogeneous mixture of inert ideal gases at a temperature T, a pressure p, and a volume V. Let us suppose there are n_1 moles of gas A_1, n_2 moles of gas A_2, ... and upto n_c moles of gas A_c (Fig. 10.7). Since there is no chemical reaction, the mixture is in a state of equilibrium with the equation of state

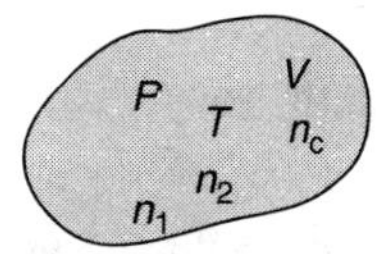

Fig. 10.7 *Mixture of gases*

$$pV = (n_1 + n_2 + \dots n_c)\ \bar{R}T$$

where $\bar{R} = 8.3143$ kJ/kg mol K

$$\therefore \quad p = \frac{n_1\bar{R}T}{V} + \frac{n_2\bar{R}T}{V} + \dots + \frac{n_c\bar{R}T}{V}$$

The expression $\frac{n_k\bar{R}T}{V}$ represents the pressure that the Kth gas would exert if it occupied the volume V alone at temperature T. This is called the *partial pressure* of the Kth gas and is denoted by p_K. Thus

$$p_1 = \frac{n_1\bar{R}T}{V},\ p_2 = \frac{n_2\bar{R}T}{V},\ \dots,\ p_c = \frac{n_c\bar{R}T}{V}$$

and

$$p = p_1 + p_2 + \dots + p_c \tag{10.81}$$

This is known as *Dalton's law of partial pressures* which states that the total pressure of a mixture of ideal gases is equal to the sum of the partial pressures.

Now

$$V = (n_1 + n_2 + \dots + n_c) \cdot \frac{\bar{R}T}{p}$$

$$= \Sigma n_K \cdot \frac{\bar{R}T}{p}$$

and the partial pressure of the Kth gas is

$$p_K = \frac{n_k \bar{R} T}{V}$$

Substituting the value of V

$$p_K = \frac{n_K \bar{R} T \cdot p}{\Sigma n_K \cdot \bar{R} T} = \frac{n_K \bar{R} T}{\Sigma n_K} \cdot p$$

Now

$$\Sigma n_K = n_1 + n_2 + \ldots + n_c$$

$$= \text{Total number of moles of gas}$$

The ratio $\frac{n_k}{\Sigma n_k}$ is called the *mole fraction of the Kth gas*, and is denoted by x_K.

Thus

$$x_1 = \frac{n_1}{\Sigma n_K},\ x_2 = \frac{n_2}{\Sigma n_K},\ \ldots,\ x_c = \frac{n_c}{\Sigma n_K}$$

and

$$p_1 = x_1 p,\ p_2 = x_2 p,\ \ldots,\ p_c = x_c p$$

or

$$p_K = x_K \times p \tag{10.82}$$

Also

$$x_1 + x_2 + \ldots + x_c = 1 \tag{10.83}$$

In a mixture of gases, if all but one mole fraction is determined, the last can be calculated from the above equation. Again, in terms of masses

$$p_1 V = m_1 R_1 T$$

$$p_2 V = m_2 R_2 T$$

$$\ldots\ldots\ldots\ldots\ldots\ldots$$

$$p_c V = m_c R_c T$$

Adding, and using Dalton's law

$$pV = (m_1 R_1 + m_2 R_2 + \ldots + m_c R_c) T \tag{10.84}$$

where

$$p = p_1 + p_2 + \ldots + p_c$$

For the gas mixture

$$pV = (m_1 + m_2 + \ldots + m_c) R_m T \tag{10.85}$$

where R_m is the gas constant for the mixture. From Eqs (10.84) and (10.85)

$$R_m = \frac{m_1 R_1 + m_2 R_2 + \ldots + m_c R_c}{m_1 + m_2 + \ldots + m_c} \tag{10.86}$$

The gas constant of the mixture is thus the weighted mean, on a mass basis, of the gas constants of the components.

The total mass of gas mixture m is

$$m = m_1 + \ldots + m_c$$

If μ denotes the *equivalent molecular* weight of the mixture having n total number of moles.

$$n\mu = n_1 \mu_1 + n_2 \mu_2 + \ldots + n_c \mu_c$$

$\therefore$

$$\mu = x_1 \mu_1 + x_2 \mu_2 + \ldots + x_c \mu_c$$

or

$$\mu = \Sigma x_K \mu_K \tag{10.87}$$

A quantity called the *partial volume* of a component of a mixture is the volume that the component alone would occupy at the pressure and temperature of the mixture. Designating the partial volumes by V_1, V_2, etc.

$$pV_1 = m_1R_1T, pV_2 = m_2R_2T, ..., pV_c = m_cR_cT$$

or
$$p(V_1 + V_2 + ... + V_c) = (m_1R_1 + m_2R_2 + ... + m_cR_c)T \tag{10.88}$$

From Eqs (10.84), (10.85), and (10.88)

$$V = V_1 + V_2 + ... + V_c \tag{10.89}$$

The total volume is thus equal to the sum of the partial volumes.

The *specific volume of the mixture*, v, is given by

$$v = \frac{V}{m} = \frac{V}{m_1 + m_2 + ... + m_c}$$

or
$$\frac{1}{v} = \frac{m_1 + m_2 + ... + m_c}{V}$$

$$= \frac{m_1}{V} + \frac{m_2}{V} + ... + \frac{m_c}{V}$$

or
$$\frac{1}{v} = \frac{1}{v_1} + \frac{1}{v_2} + ... + \frac{1}{v_c} \tag{10.90}$$

where v_1, v_2, ... denote specific volumes of the components, each component occupying the total volume.

Therefore, the density of the mixture $\rho = \rho_1 + \rho_2 + ... + \rho_c$ (10.91)

10.9 INTERNAL ENERGY, ENTHALPY AND SPECIFIC HEATS OF GAS MIXTURES

When gases at equal pressures and temperatures are mixed adiabatically without work, as by inter-diffusion in a constant volume container, the first law requires that the internal energy of the gaseous system remains constant, and experiments show that the temperature remains constant. Hence, the internal energy of a mixture of gases is equal to the sum of the internal energies of the individual components, each taken at the temperature and volume of the mixture (i.e. sum of the 'partial' internal energies). This is also true for any of the thermodynamic properties like H, C_v, C_p, S, F, and G, and is known as *Gibbs theorem*. Therefore, on a mass basis

$$mu_m = m_1u_1 + m_2u_2 + ... + m_cu_c$$

$$u_m = \frac{m_1u_1 + m_2u_2 + ... + m_cu_c}{m_1 + m_2 + ... + m_c} \tag{10.92}$$

which is the average specific internal energy of the mixture.

Similarly, the total enthalpy of a gas mixture is the sum of the 'partial' enthalpies

$$mh_m = m_1h_1 + m_2h_2 + ... + m_ch_c$$

and
$$h_m = \frac{m_1h_1 + m_2h_2 + ... + m_ch_c}{m_1 + m_2 + ... + m_c} \tag{10.93}$$

From the definitions of specific heats, it follows that

$$c_{vm} = \frac{m_1c_{v_1} + m_2c_{v_2} + ... + m_cc_{v_c}}{m_1 + m_2 + ... + m_c} \tag{10.94}$$

and
$$c_{pm} = \frac{m_1c_{p_1} + m_2c_{p_2} + ... + m_cc_{p_c}}{m_1 + m_2 + ... + m_c} \tag{10.95}$$

10.10 ENTROPY OF GAS MIXTURES

Gibbs theorem states that the total entropy of a mixture of gases is the sum of the partial entropies. The partial entropy of one of the gases of a mixture is the entropy that the gas would have if it occupied the whole volume alone at the same temperature. Let us imagine a number of inert ideal gases separated from one another by suitable partitions, all the gases being at the same temperature T and pressure p. The total entropy (initial)

$$S_i = n_1 s_1 + n_2 s_2 + \dots + n_c s_c \quad = \Sigma\, n_K s_K$$

From property relation

$$T ds = dh - v dp = c_p dT - v dp$$

$$\therefore \quad d\bar{s} = \bar{c}_p \frac{dT}{T} - \bar{R}\frac{dp}{p}$$

The entropy of 1 mole of the Kth gas at T and p

$$\bar{s}_K = \int c_{p_K} \frac{dT}{T} - \bar{R} \ln p + \bar{s}_{0K}$$

where s_{0K} is the constant of integration.

$$\therefore \quad S_i = \bar{R}\, \Sigma\, n_K \left(\frac{1}{\bar{R}} \int \bar{c}_{p_K} \frac{dT}{T} + \frac{\bar{s}_{0K}}{\bar{R}} - \ln p \right)$$

Let

$$\sigma_K = \frac{1}{\bar{R}} \int \bar{c}_{p_K} \frac{dT}{T} + \frac{\bar{s}_{0K}}{\bar{R}}$$

then

$$S_i = \bar{R}\, \Sigma\, n_k (\sigma_K - \ln p) \tag{10.96}$$

After the partitions are removed, the gases diffuse into one another at the same temperature and pressure, and by Gibbs theorem, the entropy of the mixture, S_f, is the sum of the partial entropies, with each gas exerting its respective partial pressure. Thus

$$S_f = \bar{R}\, \Sigma\, n_K (\sigma_K - \ln p_K)$$

Since

$$p_K = x_K\, p$$

$$S_f = \bar{R}\, \Sigma\, n_K (\sigma_K - \ln x_K - \ln p) \tag{10.97}$$

A change in entropy due to the diffusion of any number of inert ideal gases is

$$S_f - S_i = -\bar{R}\, \Sigma\, n_K \ln x_K \tag{10.98}$$

or

$$S_f - S_i = -\bar{R}\, (n_1 \ln x_1 + n_2 \ln x_2 + \dots + n_c \ln x_c)$$

Since the mole fractions are less than unity, $(S_f - S_i)$ is always positive, conforming to the Second Law.

Again

$$S_f - S_i = -\bar{R} \left(n_1 \ln \frac{p_1}{p} + n_2 \ln \frac{p_2}{p} + \dots + n_c \ln \frac{p_c}{p} \right) \tag{10.99}$$

which indicates that each gas undergoes in the diffusion process a *free expansion* from total pressure p to the respective partial pressure at constant temperature.

Similarly, on a mass basis, the entropy change due to diffusion

$$S_f - S_i = -\, \Sigma m_K R_K \ln \frac{p_K}{p}$$

$$= -\left(m_1 R_1 \ln \frac{p_1}{p} + m_2 R_2 \ln \frac{p_2}{p} + \dots + m_c R_c \ln \frac{p_1}{p} \right)$$

10.11 GIBBS FUNCTION OF A MIXTURE OF INERT IDEAL GASES

From the equations

$$d\bar{h} = \bar{c}_p \, dT$$

$$d\bar{s} = \bar{c}_p \frac{dT}{T} - \bar{R}\frac{dp}{p}$$

the enthalpy and entropy of 1 mole of an ideal gas at temperature T and pressure p are

$$\bar{h} = \bar{h}_0 + \int c_p \, dT$$

$$\bar{s} = \int \bar{c}_p \frac{dT}{T} + \bar{s}_0 - \bar{R} \ln p$$

Therefore, the molar Gibbs function

$$\bar{g} = \bar{h} - T\bar{s}$$

$$= \bar{h}_0 + \int \bar{c}_p dT - T\int \bar{c}_p \frac{dT}{T} - T\bar{s}_0 + \bar{R}T \ln p$$

Now $$\int d(uv) = \int u dv + \int v du = uv$$

Let $$u = \frac{1}{T}, \; v = \int c_p \, dT$$

Then $$\frac{1}{T} \int c_p \, dT = \int \frac{1}{T} \, c_p \, dT + \int c_p \, dT \left(-\frac{1}{T^2}\right) dT$$

$$= \int \frac{1}{T} \, c_p \, dT - \int \frac{\int c_p dT}{T^2} \, dT$$

$$\int c_p \, dT - T\int c_p \frac{dT}{T} = -T\int \frac{\int c_p dT}{T^2} \, dT$$

Therefore

$$\bar{g} = \bar{h}_0 - T\int \frac{\int \bar{c}_p dT}{T^2} \, dT - T\bar{s}_0 + \bar{R}T \ln p$$

$$= \bar{R}T \left(\frac{\bar{h}_0}{\bar{R}T} - \frac{1}{\bar{R}} \int \frac{\int \bar{c}_p dT}{T^2} dT - \frac{\bar{s}_0}{\bar{R}} + \ln p \right)$$

Let

$$\phi = \frac{\bar{h}_0}{\bar{R}T} - \frac{1}{\bar{R}} \int \frac{\int \bar{c}_p dT}{T^2} dT - \frac{\bar{s}_0}{\bar{R}} \tag{10.100}$$

Thus $$\bar{g} = \bar{R}T \, (\phi + \ln p) \tag{10.101}$$

where ϕ is a function of temperature only.

Let us consider a number of inert ideal gases separated from one another at the same T and p

$$G_i = \Sigma n_K g_K$$

$$= \bar{R}T \, \Sigma n_K(\phi_K + \ln p)$$

After the partitions are removed, the gases will diffuse, and the partial Gibbs function of a particular gas is the value of G, if that gas occupies the same volume at the same temperature exerting a partial pressure p_K. Thus

$$G_f = \bar{R}T \,\Sigma\, n_K(\phi_K + \ln p_K)$$
$$= \bar{R}T \,\Sigma\, n_K(\phi_K + \ln p + \ln x_K)$$

Therefore

$$G_f - G_i = \bar{R}T \,\Sigma\, n_K \ln x_K \quad (10.102)$$

Since $x_K < 1$, $(G_f - G_i)$ is negative because G decreases due to diffusion. Gibbs function of a mixture of ideal gases at T and p is thus

$$G = \bar{R}T\Sigma\, n_K(\phi_K + \ln p + \ln x_K) \quad (10.103)$$

Solved Examples

Example 10.1

Two vessels, A and B, both containing nitrogen, are connected by a valve which is opened to allow the contents to mix and achieve an equilibrium temperature of 27°C. Before mixing the following information is known about the gases in the two vessels.

Vessel A	*Vessel B*
$p = 1.5$ MPa	$p = 0.6$ MPa
$t = 50$°C	$t = 20$°C
Contents = 0.5 kg mol	*Contents* = 2.5 kg

Calculate the final equilibrium pressure, and the amount of heat transferred to the surroundings. If the vessel had been perfectly insulated, calculate the final temperature and pressure which would have been reached. Take $\gamma = 1.4$.

Solution For the gas in vessel A (Fig. Ex. 10.1)

$$P_A V_A = n_A \bar{R}\, T_A$$

where V_A is the volume of vessel A

$$1.5 \times 10^3 \times V_A = 0.5 \times 8.3143 \times 323$$
$$V_A = 0.895 \text{ m}^3$$

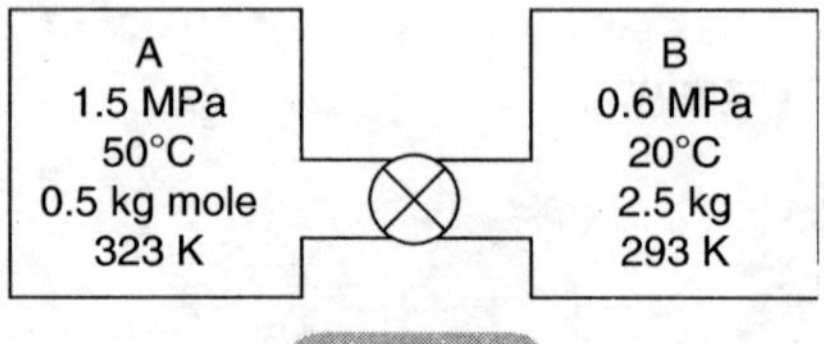

Fig. Ex. 10.1

The mass of gas in vessel A

$$m_A = n_A\, \mu_A$$
$$= 0.5 \text{ kg mol} \times 28 \text{ kg/kg mol} = 14 \text{ kg}$$

Characteristic gas constant R of nitrogen

$$R = \frac{8.3143}{28} = 0.297 \text{ kJ/kg K}$$

For the vessel B

$$p_B V_B = m_B\, RT_B$$
$$0.6 \times 10^3 \times V_B = 2.5 \times 0.297 \times 293$$
$$\therefore \quad V_B = 0.363 \text{ m}^3$$

Total volume of A and B

$$V = V_A + V_B = 0.895 + 0.363$$
$$= 1.258 \text{ m}^3$$

Total mass of gas

$$m = m_A + m_B = 14 + 2.5 = 16.5 \text{ kg}$$

Final temperature after mixing

$$T = 27 + 273 = 300 \text{ K}$$

For the final condition after mixing

$$pV = mRT$$

where p is the final equilibrium pressure

$$\therefore \quad p \times 1.258 = 16.5 \times 0.297 \times 300$$

$$\therefore \quad p = \frac{16.5 \times 0.297 \times 300}{1.258}$$
$$= 1168.6 \text{ kPa}$$
$$= 1.168 \text{ MPa}$$

$$c_v = \frac{R}{\gamma - 1} = \frac{0.297}{0.4} = 0.743 \text{ kJ/kg K}$$

Since there is no work transfer, the amount of heat transfer

$$Q = \text{change of internal energy}$$
$$= U_2 - U_1$$

Measuring the internal energy above the datum of absolute zero (at $T = 0$ K, $u = 0$ kJ/kg)

Initial internal energy U_1 (before mixing)

$$= m_A c_v T_A + m_B c_v T_B$$
$$= (14 \times 323 + 2.5 \times 293) \times 0.743$$
$$= 3904.1 \text{ kJ}$$

Final internal energy U_2 (after mixing)

$$= m c_v T$$
$$= 16.5 \times 0.743 \times 300$$
$$= 3677.9 \text{ kJ}$$

$$\therefore \quad Q = 3677.9 - 3904.1 = -226.2 \text{ kJ}$$ *Ans.*

If the vessels were insulated

$$U_1 = U_2$$
$$m_A c_v T_A + m_B c_v T_B = m c_v T$$

where T would have been the final temperature.

$$\therefore \quad T = \frac{m_A T_A + m_B T_B}{m}$$

$$= \frac{14 \times 323 + 2.5 \times 293}{16.5} = 318.5 \text{ K}$$

or $$t = 45.5°\text{C}$$ *Ans.*

The final pressure

$$p = \frac{mRT}{V} = \frac{16.5 \times 0.297 \times 318.5}{1.258}$$

$$= 1240.7 \text{ kPa}$$

$$= 1.24 \text{ MPa}$$

***Example* 10.2**

A certain gas has $c_p = 1.968$ and $c_v = 1.507$ kJ/kg K. Find its molecular weight and the gas constant.

A constant volume chamber of 0.3 m³ capacity contains 2 kg of this gas at 5°C. Heat is transferred to the gas until the temperature is 100°C. Find the work done, the heat transferred, and the changes in internal energy, enthalpy and entropy.

Solution Gas constant,

$$R = c_p - c_v = 1.968 - 1.507$$

$$= 0.461 \text{ kJ/kg K}$$ *Ans.*

Molecular weight,

$$\mu = \frac{\overline{R}}{R} = \frac{8.3143}{0.461} = 18.04 \text{ kg/kg mol}$$ *Ans.*

At constant volume

$$Q_{1-2} = mc_v (t_2 - t_1)$$

$$= 2 \times 1.507 (100 - 5)$$

$$= 286.33 \text{ kJ}$$ *Ans.*

Change in internal energy

$$W_{1-2} = \int_1^2 p\text{dv} = 0$$ *Ans.*

$$U_2 - U_1 = Q_{1-2} = 286.33 \text{ kJ}$$ *Ans.*

Change in enthalpy

$$H_2 - H_1 = mc_p (t_2 - t_1)$$

$$= 2 \times 1.968 (100 - 5) = 373.92 \text{ kJ}$$ *Ans.*

Change in entropy

$$S_2 - S_1 = mc_v \ln \frac{T_2}{T_1} = 2 \times 1.507 \ln \frac{373}{268}$$

$$= 0.886 \text{ kJ/K}$$ *Ans.*

***Example* 10.3**

(*a*) *The specific heats of a gas are given by $c_p = a + kT$ and $c_v = b + kT$, where a, b, and k are constants and T is in K. Show that for an isentropic expansion of this gas*

$$T^b v^{a-b} e^{kT} = \text{constant}$$

(*b*) *1.5 kg of this gas occupying a volume of 0.06 m³ at 5.6 MPa expands isentropically until the temperature is 240°C. If a = 0.946, b = 0.662, and k = 10^{-4}, calculate the work done in the expansion.*

Solution (a) $c_p - c_v = a + kT - b - kT$

$$= a - b = R$$

Now $$ds = c_v \frac{dT}{T} + R \frac{dv}{v}$$

$$= (b + kT) \frac{dT}{T} + (a - b) \frac{dv}{v} = b \frac{dT}{T} + k\,dT + (a - b) \frac{dv}{v}$$

For an isentropic process

$$b \ln T + kT + (a - b) \ln v = \text{constant}$$

$$\therefore \quad T^b \cdot v^{a-b} e^{kT} = \text{constant} \qquad \text{(Q.E.D.)}$$

(b)

$$R = a - b = 0.946 - 0.662 = 0.284 \text{ kJ/kg K}$$

$$T_2 = 240 + 273 = 513 \text{ K}$$

$$T_1 = \frac{p_1 V_1}{mR} = \frac{5.6\times10^3 \times 0.06}{1.5\times0.284} = 788.73 \text{ K} = 789 \text{ K}$$

$$TdS = dU + đW = 0$$

$$\therefore \quad W_{1-2} = -\int_{T_1}^{T_2} mc_v\, dT$$

$$= 1.5 \int_{513}^{789} (0.662 + 0.0001T)\, dT$$

$$= 1.5\,[0.662\,(789 - 513) + 10^{-4} \times 0.5\,\{(789)^2 - (513)^2\}]$$

$$= 1.5\,(182.71 + 19.97)$$

$$= 304 \text{ kJ}$$

Ans.

Example 10.4

Show that for an ideal gas, the slope of the constant volume line on the T–s diagram is more than that of the constant pressure line.

Solution We have, for 1 kg of ideal gas

$$Tds = du + pdv = c_v\, dT + pdv$$

$$\therefore \quad \left(\frac{\partial T}{\partial s}\right)_v = \frac{T}{c_v}$$

Also $$Tds = dh - vdp = c_p\, dT - vdp$$

$$\therefore \quad \left(\frac{\partial T}{\partial s}\right)_p = \frac{T}{c_p}$$

Since $$c_p > c_v, \quad \frac{T}{c_v} > \frac{T}{c_p}$$

$$\therefore \quad \left(\frac{\partial T}{\partial s}\right)_v > \left(\frac{\partial T}{\partial s}\right)_p$$

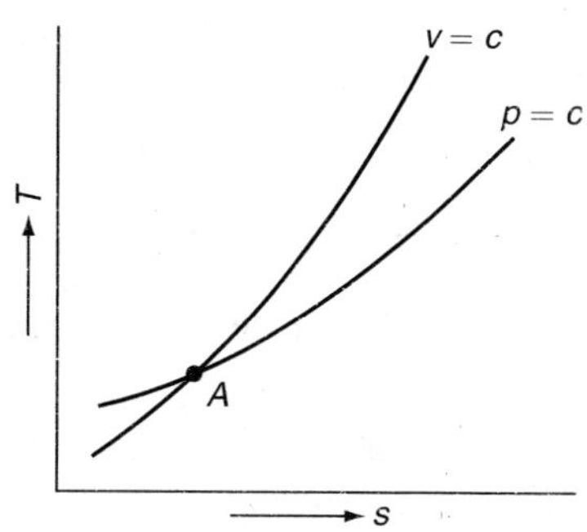

Fig. Ex. 10.4

This is shown in Fig. Ex. 10.4. The slope of the constant volume line passing through point A is steeper than that of the constant pressure line passing through the same point. (Q.E.D.)

Example 10.5

0.5 kg of air is compressed reversibly and adiabatically from 80 kPa, 60°C to 0.4 MPa, and is then expanded at constant pressure to the original volume. Sketch these processes on the p–v and T–s planes. Compute the heat transfer and work transfer for the whole path.

Solution The processes have been shown on the $p-v$ and $T-s$ planes in Fig. Ex. 10.5. At state 1

$$p_1V_1 = mRT_1$$

$\therefore$ V_1 = volume of air at state 1

$$= \frac{mRT_1}{p_1} = \frac{1\times0.287\times333}{2\times80} = 0.597 \text{ m}^3$$

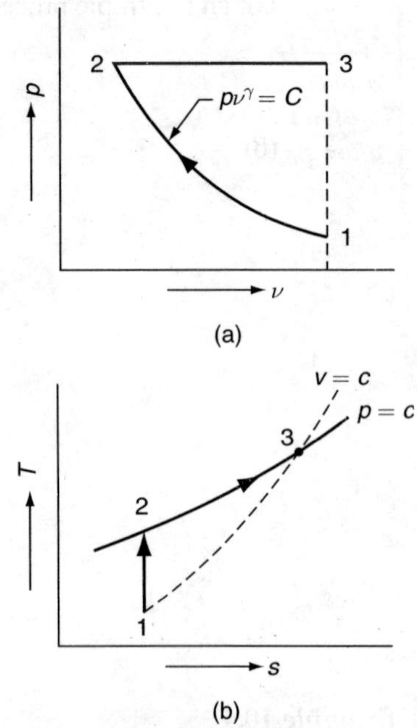

Fig. Ex. 10.5

Since the process 1–2 is reversible and adiabatic

$$\frac{T_2}{T_1} = \left(\frac{p_2}{p_1}\right)^{(\gamma-1)/\gamma}$$

$$\therefore \quad \frac{T_2}{T_1} = \left(\frac{400}{80}\right)^{(1.4-1)/1.4} = (5)^{2/7}$$

$$\therefore \quad T_2 = 333 \times (5)^{2/7} = 527 \text{ K}$$

For process 1–2, work done

$$W_{1-2} = \frac{p_1V_1 - p_2V_2}{\gamma-1} = \frac{mR(T_1-T_2)}{\gamma-1}$$

$$= \frac{1/2\times0.287\,(333-527)}{0.4}$$

$$= -\,69.6 \text{ kJ}$$

Again $\quad p_1 v_1^\gamma = p_2 v_2^\gamma$

$$\therefore \quad \left(\frac{v_2}{v_1}\right)^\gamma = \frac{p_1}{p_2} = \frac{80}{400} = \frac{1}{5}$$

$$\therefore \quad \frac{v_2}{v_1} = \left(\frac{1}{5}\right)^{1/1.4} = \frac{1}{3.162} = \frac{V_2}{V_1}$$

$$\therefore \quad V_2 = \frac{0.597}{3.162} = 0.189 \text{ m}^3$$

For process 2–3, work done

$$W_{2-3} = p_2(V_1 - V_2) = 400\,(0.597 - 0.189)$$

$$= 163.2 \text{ kJ}$$

$\therefore$ Total work transfer

$$W = W_{1-2} + W_{2-3}$$

$$= -\,69.6 + 163.2 = 93.6 \text{ kJ}$$ *Ans.*

For states 2 and 3

$$\frac{p_2V_2}{T_2} = \frac{p_3V_3}{T_3}$$

$$\therefore \qquad T_3 = T_2 \cdot \frac{V_3}{V_2} = 527 \times 3.162 = 1667 \text{ K}$$

Total heat transfer

$$Q = Q_{1-2} + Q_{2-3} = Q_{2-3} = mc_p (T_3 - T_2)$$
$$= 1/2 \times 1.005 (1667 - 527)$$
$$= 527.85 \text{ kJ}$$

Ans.

Example 10.6

A mass of air is initially at 260°C and 700 kPa, and occupies 0.028 m³. The air is expanded at constant pressure to 0.084 m³. A polytropic process with n = 1.50 is then carried out, followed by a constant temperature process which completes a cycle. All the processes are reversible. (a) Sketch the cycle in the p–v and T–s planes. (b) Find the heat received and heat rejected in the cycle. (c) Find the efficiency of the cycle.

Solution (a) The cycle is sketched on the $p-v$ and $T-s$ planes in Fig. Ex. 10.6.

Given $p_1 = 700$ kPa, $T_1 = 260 + 273 = 533 \text{ K} = T_3$

$V_1 = 0.028 \text{ m}^3$

$V_2 = 0.084 \text{ m}^3$

From the ideal gas equation of state

$$p_1V_1 = mRT_1$$

$$m = \frac{700 \times 0.028}{0.287 \times 533} = 0.128 \text{ kg}$$

Now $$\frac{T_2}{T_1} = \frac{p_2V_2}{p_1V_1} = \frac{0.0084}{0.028} = 3$$

$$\therefore \qquad T_2 = 3 \times 533 = 1599 \text{ K}$$

Again $$\frac{p_2}{p_2} = \left(\frac{T_2}{T_3}\right)^{n/(n-1)} = \left(\frac{1599}{533}\right)^{1.5/0.5} = (3)^3 = 27$$

Heat transfer in process 1–2

$$Q_{1-2} = mc_p (T_2 - T_1)$$
$$= 0.128 \times 1.005 (1599 - 533)$$
$$= 137.13 \text{ kJ}$$

Heat transfer in process 2–3

$$Q_{2-3} = \Delta U + \int p dv$$
$$= mc_v (T_3 - T_2) + \frac{mR (T_2 - T_3)}{n-1}$$
$$= mc_v \frac{n-\gamma}{n-1} (T_3 - T_2)$$
$$= 0.128 \times 0.718 \times \frac{1.5-1.4}{1.5-1} (533 - 1599)$$
$$= 0.128 \times 0.718 \times \frac{0.1}{0.5} (-1066) \qquad = -19.59 \text{ kJ}$$

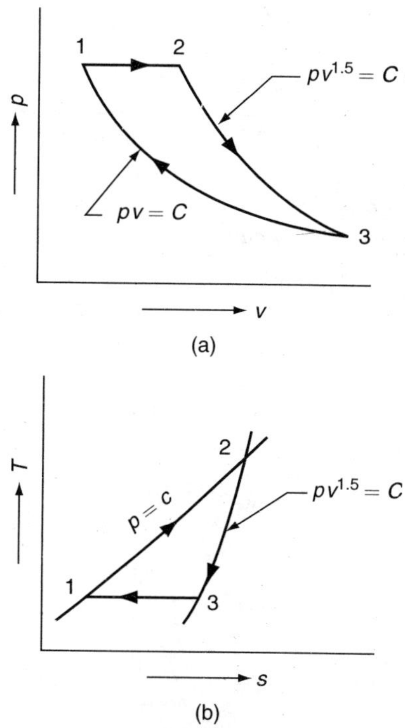

Fig. Ex. 10.6

For process 3–1

$$đQ = dU + đW = đW$$

$$\therefore \quad Q_{3-1} = W_{3-1} = \int_3^1 pdV = mRT_1 \ln \frac{V_1}{V_3}$$

$$= mRT_1 \ln \frac{p_3}{p_1} = 0.128 \times 0.287 \times 533 \ln \left(\frac{1}{27}\right)$$

$$= -0.128 \times 0.287 \times 533 \times 3.2959$$

$$= -64.53 \text{ kJ}$$

(b) Heat received in the cycle

$$Q_1 = 137.13 \text{ kJ}$$

Heat rejected in the cycle

$$Q_2 = 19.59 + 64.53 = 84.12 \text{ kJ}$$ *Ans.*

(c) The efficiency of the cycle

$$\eta_{cycle} = 1 - \frac{Q_2}{Q_1} = 1 - \frac{84.12}{137.13} = 1 - 0.61$$

$$= 0.39, \text{ or } 39\%$$ *Ans.*

Example 10.7

A mass of 0.25 kg of an ideal gas has a pressure of 300 kPa, a temperature of 80°C, and a volume of 0.07 m³. The gas undergoes an irreversible adiabatic process to a final pressure of 300 kPa and final volume of 0.10 m³, during which the work done on the gas is 25 kJ. Evaluate the c_p and c_v of the gas and the increase in entropy of the gas.

Solution From

$$p_1V_1 = mRT_1$$

$$R = \frac{300 \times 0.07}{0.25 \times (273 + 80)} = 0.238 \text{ kJ/kg K}$$

Final temperature

$$T_2 = \frac{p_2V_2}{mR} = \frac{300 \times 0.1}{0.25 \times 0.238} = 505 \text{ K}$$

Now

$$Q = (U_2 - U_1) + W = mc_v(T_2 - T_1) + W$$

$$0 = 0.25\, c_v\,(505 - 353) - 25$$

$$\therefore \quad c_v = \frac{25}{0.25 \times 152} = 0.658 \text{ kJ/kg K}$$

Now

$$c_p - c_v = R$$

$$c_p = 0.658 + 0.238 = 0.896 \text{ kJ/kg K}$$

Entropy change

$$S_2 - S_1 = mc_v \ln \frac{p_2}{p_1} + mc_p \ln \frac{v_2}{v_1}$$

$$= mc_p \ln \frac{V_2}{V_1} = 0.25 \times 0.896 \ln \frac{0.10}{0.07}$$

$$= 0.224 \times 0.3569 = 0.08 \text{ kJ/kg K}$$ *Ans.*

Example 10.8

A mixture of ideal gases consists of 3 kg of nitrogen and 5 kg of carbondioxide at a pressure of 300 kPa and a temperature of 20°C. Find (a) the mole fraction of each constituent, (b) the equivalent molecular weight of the mixture, (c) the equivalent gas constant of the mixture, (d) the partial pressures and the partial volumes, (e) the volume and density of the mixture, and (f) the c_p and c_v of the mixture.

If the mixture is heated at constant volume to 40°C, find the changes in internal energy, enthalpy and entropy of the mixture. Find the changes in internal energy, enthalpy and entropy of the mixture if the heating is done at constant pressure. Take γ for CO_2 and N_2 to be 1.286 and 1.4 respectively.

Solution (a) Since mole fraction $x_i = \frac{n_i}{\Sigma n_i}$

$$x_{N_2} = \frac{\frac{3}{28}}{\frac{3}{28}+\frac{5}{44}} = 0.485$$

$$x_{CO_2} = \frac{\frac{5}{44}}{\frac{3}{28}+\frac{5}{44}} = 0.515$$ *Ans.*

(b) Equivalent molecular weight of the mixture

$$M = x_1\,\mu_1 + x_2\,\mu_2$$
$$= 0.485 \times 28 + 0.515 \times 44$$
$$= 36.25 \text{ kg/kg mol}$$ *Ans.*

(c) Total mass,

$$m = m_{N_2} + m_{CO_2} = 3 + 5 = 8 \text{ kg}$$

Equivalent gas constant of the mixture

$$R = \frac{m_{N_2}R_{N_2} + m_{CO_2}R_{CO_2}}{m}$$

$$= \frac{3\times\frac{8.3143}{28}+5\times\frac{8.3143}{44}}{8} = \frac{0.89+0.94}{8}$$

$$= 0.229 \text{ kJ/kg K}$$ *Ans.*

(d)

$$p_{N_2} = x_{N_2}\cdot p = 0.485 \times 300 = 145.5 \text{ kPa}$$

$$p_{CO_2} = x_{CO_2}\cdot p = 0.515 \times 300 = 154.5 \text{ kPa}$$

$$V_{N_2} = \frac{m_{N_2}R_{N_2}T}{p} = \frac{3\times\frac{8.3143}{28}\times 293}{300} = 0.87 \text{ m}^3$$

$$V_{CO_2} = \frac{m_{CO_2}R_{CO_2}T}{p} = \frac{5\times\frac{8.3143}{44}\times 293}{300} = 0.923 \text{ m}^3$$ *Ans.*

(e) Total volume of the mixture

$$V = \frac{mRT}{p} = \frac{m_{N_2} R_{N_2} T}{p_{N_2}} = \frac{m_{CO_2} R_{CO_2} T}{p_{CO_2}}$$

$$\therefore \quad V = \frac{8 \times 0.229 \times 293}{300} = 1.79 \text{ m}^3$$

Density of the mixture

$$\rho = \rho_{N_2} + \rho_{CO_2} = \frac{m}{V} = \frac{8}{1.79}$$

$$= 4.46 \text{ kg/m}^3 \qquad \textit{Ans.}$$

(f)

$$c_{pN_2} - c_{vN_2} = R_{N_2}$$

$$\therefore \quad c_{vN_2} = \frac{R_{N_2}}{\gamma - 1} = \frac{8.3143}{28 \times (1.4 - 1)}$$

$$= 0.742 \text{ kJ/kg K}$$

$$\therefore \quad c_{pN_2} = 1.4 \times 0.742$$

$$= 1.039 \text{ kJ/kg K}$$

For CO_2,

$$\gamma = 1.286$$

$$\therefore \quad c_{vCO_2} = \frac{R_{CO_2}}{\gamma - 1} = \frac{8.3143}{44 \times 0.286} = 0.661 \text{ kJ/kg K}$$

$$c_{pCO_2} = 1.286 \times 0.661 = 0.85 \text{ kJ/kg K}$$

For the mixture

$$c_p = \frac{m_{N_2} c_{pN_2} + m_{CO_2} c_{pCO_2}}{m_{N_2} + m_{CO_2}}$$

$$= 3/8 \times 1.039 + 5/8 \times 0.85 = 0.92 \text{ kJ/kg K}$$

$$c_v = \frac{m_{N_2} c_{vN_2} + m_{CO_2} c_{vCO_2}}{m}$$

$$= 3/8 \times 0.742 + 5/8 \times 0.661 = 0.69 \text{ kJ/kg K} \qquad \textit{Ans.}$$

If the mixture is heated at constant volume

$$U_2 - U_1 = mc_v (T_2 - T_1)$$

$$= 8 \times 0.69 \times (40 - 20) = 110.4 \text{ kJ}$$

$$H_2 - H_1 = mc_p (T_2 - T_1)$$

$$= 8 \times 0.92 \times 20 = 147.2 \text{ kJ}$$

$$S_2 - S_1 = mc_v \ln \frac{T_2}{T_1} + mR \ln \frac{V_2}{V_1}$$

$$= mc_v \ln \frac{T_2}{T_1} = 8 \times 0.69 \times \ln \frac{313}{293}$$

$$= 0.368 \text{ kJ/kg K}$$

If the mixture is heated at constant pressure, ΔU and ΔH will remain the same. The change in entropy will be

$$S_2 - S_1 = mc_p \ln \frac{T_2}{T_1} - mR \ln \frac{p_2}{p_1}$$

$$= mc_p \ln \frac{T_2}{T_1} = 8 \times 0.92 \ln \frac{313}{293}$$

$$= 0.49 \text{ kJ/kg K}$$ *Ans.*

Example 10.9

Find the increase in entropy when 2 kg of oxygen at 60°C are mixed with 6 kg of nitrogen at the same temperature. The initial pressure of each constituent is 103 kPa and is the same as that of the mixture.

Solution

$$x_{O_2} = \frac{p_{O_2}}{p} = \frac{\frac{2}{32}}{\frac{2}{32}+\frac{6}{28}} = 0.225$$

$$x_{N_2} = \frac{p_{N_2}}{p} = 0.775$$

Entropy increase due to diffusion

$$\Delta S = -m_{O_2} R_{O_2} \ln \frac{p_{O_2}}{p} - m_{N_2} R_{N_2} \ln \frac{p_{N_2}}{p}$$

$$= -2\left(\frac{8.3143}{32}\right) \ln 0.225 - 6\left(\frac{8.3143}{28}\right) \ln 0.775$$

$$= 1.2314 \text{ kJ/kg K}$$ *Ans.*

Example 10.10

The gas neon has a molecular weight of 20.183 and its critical temperature, pressure and volume are 44.5 K, 2.73 MPa and 0.0416 m³/kg mol. Reading from a compressibility chart for a reduced pressure of 2 and a reduced temperature of 1.3, the compressibility factor Z is 0.7. What are the corresponding specific volume, pressure, temperature, and reduced volume?

Solution At $p_r = 2$ and $T_r = 1.3$ from chart (Fig. Ex. 10.13),

$$z = 0.7$$

$$p = 2 \times 2.73 = 5.46 \text{ MPa}$$ *Ans.*

$$\frac{T}{T_c} = 1.3$$

$$T = 1.3 \times 44.5 = 57.85 \text{ K}$$ *Ans.*

$$p_v = ZRT$$

$$\therefore \quad v = \frac{0.7 \times 8.3143 \times 57.85}{20.183 \times 5.46 \times 10^3}$$

$$= 3.05 \times 10^{-3} \text{ m}^3/\text{kg}$$

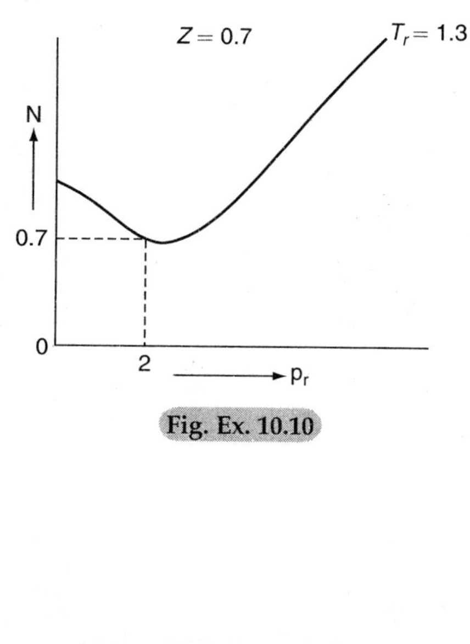

Fig. Ex. 10.10

$$\therefore \quad v_r = \frac{v}{v_c} = \frac{3.05\times10^{-3}\times20.183}{4.16\times10^{-2}}$$

$$= 1.48$$ *Ans.*

Example 10.11

For the Berthelot equation of state

$$p = \frac{RT}{v-b} - \frac{a}{Tv^2}$$

show that (a) $\lim_{\substack{p\to0\\T\to\infty}} (RT - pv) = 0$

(*b*) $\lim_{T\to\infty} \frac{v}{T} = \frac{R}{p}$

(*c*) *Boyle temperature,* $T_B = \sqrt{\frac{a}{bR}}$,

(*d*) *Critical properties* $p_c = \frac{1}{12b}\sqrt{\frac{2aR}{3b}}$, $v_c = 3b$, $T_c = \sqrt{\frac{8a}{27bR}}$,

(*e*) *Law of corresponding states*

$$\left(p_r + \frac{3}{T_r \cdot v_r^2}\right)(3v_r - 1) = 8T_r$$

Solution (a)

$$p = \frac{RT}{v-b} - \frac{a}{Tv^2}$$

$$\therefore \quad RT = \left(p + \frac{a}{Tv^2}\right)(v-b)$$

or

$$\frac{RT}{p} = v + \frac{a}{pvT} - b - \frac{ab}{pv^2T}$$

$$\therefore \quad RT - pv = \frac{a}{vT} - bp - \frac{ab}{v^2T}$$

$$\therefore \quad \lim_{\substack{p\to0\\T\to\infty}} (RT - pv) = 0$$ Proved (a)

(b) Now

$$v = \frac{RT}{p} - \frac{a}{pvT} + b + \frac{ab}{pv^2T}$$

$$\therefore \quad \frac{v}{T} = \frac{R}{p} - \frac{a}{pvT^2} + \frac{b}{T} + \frac{ab}{pv^2T^2}$$

$$\therefore \quad \lim_{T\to\infty} \frac{v}{T} = \frac{R}{p}$$ Proved (b)

(c)

$$pv = RT - \frac{a}{vT} + bp + \frac{ab}{v^2T}$$

The last three terms of the equation are very small, except at very high pressures and small volume. Hence substituting $v = RT/p$

$$pv = RT - \frac{ap}{RT^2} + bp + \frac{abp^2}{R^2T^3}$$

$$\therefore \quad \left[\frac{\partial (pv)}{\partial p}\right]_T = -\frac{a}{RT^2} + b + \frac{2abp}{R^2T^3} = 0$$

when $p = 0$, $T = T_B$, the Boyle temperature

$$\therefore \quad \frac{a}{RT_B^2} = b$$

or

$$T_B = \sqrt{\frac{a}{bR}}$$

Proved (c)

(d)

$$p = \frac{RT}{v-b} - \frac{a}{Tv^2}$$

$$\left(\frac{\partial p}{\partial v}\right)_{T=T_c} = -\frac{RT_c}{(v_c - b)^2} + \frac{2a}{T_c \cdot v_c^3} = 0$$

$$\left(\frac{\partial^2 p}{\partial v^2}\right)_{T=T_c} = \frac{2RT_c}{(v_c - b)^3} - \frac{6a}{T_c \cdot v_c^4} = 0$$

$$\left(p_c + \frac{a}{T_c v_c^2}\right)(v_c - b) = RT_c$$

By solving the three equations, as was done in the case of van der Waals equation of state in Article 10.7

$$p_c = \frac{1}{12b}\sqrt{\frac{2aR}{3b}},\ v_c = 3b, \text{ and } T_c = \sqrt{\frac{8a}{27bR}}$$

Proved (d)

(e) Solving the above three equations

$$a = \frac{8v_c^3 p_c^2}{R} = 3p_c \cdot v_c^2 \cdot T_c$$

$$b = \frac{v_c}{3},\ R = \frac{8}{3}\frac{p_v v_c}{T_c} \quad (\text{so that } Z_c = 3/8)$$

Substituting in the equation

$$\left(p + \frac{a}{Tv^2}\right)(v - b) = RT$$

$$\left(p + \frac{3p_c v_c^2 T_c}{Tv^2}\right)\left(v - \frac{v_c}{3}\right) = \frac{8p_c v_c}{3T_c} \cdot T$$

$$\left(p_r + \frac{3}{T_r v_r^2}\right)(3v_r - 1) = 8T_r$$

This is the law of corresponding states.

Proved (e)

Review Questions

10.1 What is a mole?

10.2 What is Avogadro's law?

10.3 What is an equation of state?

10.4 What is the fundamental property of gases with respect to the product pv?

10.5 What is universal gas constant?

10.6 Define an ideal gas.

10.7 What is the characteristic gas constant?

10.8 What is Boltzmann constant?

10.9 Why do the specific heats of an ideal gas depend only on the atomic structure of the gas?

10.10 Show that for an ideal gas the internal energy depends only on its temperature.

10.11 Show that the enthalpy of an ideal gas is a function of temperature only.

10.12 Why is there no temperature change when an ideal gas is throttled?

10.13 Show that for an ideal gas, $c_p - c_v = R$.

10.14 Derive the equations used for computing the entropy change of an ideal gas.

10.15 Show that for a reversible adiabatic process executed by an ideal gas, the following relations hold good: (i) pv^γ = constant, (ii) $Tv^{\gamma-1}$ = constant, and (iii) $Tp^{\frac{1-\gamma}{\gamma}}$ = constant.

10.16 Express the changes in internal energy and enthalpy of an ideal gas in a reversible adiabatic process in terms of the pressure ratio.

10.17 Derive the expression of work transfer for an ideal gas in a reversible isothermal process.

10.18 What is a polytropic process? What are the relations among p, v and T of an ideal gas in a polytropic process?

10.19 Show that the entropy change between states 1 and 2 in a polytropic process, pv^n = constant, is given by the following relations:

(i) $s_2 - s_1 = \dfrac{n-\gamma}{(\gamma-1)(n-1)} R \ln \dfrac{T_2}{T_1}$

(ii) $s_2 - s_1 = \dfrac{n-\gamma}{n(\gamma-1)} R \ln \dfrac{p_2}{p_1}$

(iii) $s_2 - s_1 = -\dfrac{n-\gamma}{\gamma-1} R \ln \dfrac{v_2}{v_1}$

10.20 What are the expressions of work transfer for an ideal gas in a polytropic process, if the gas is: (i) a closed system, and (ii) a steady flow system?

10.21 Derive the expression of heat transfer for an ideal gas in a polytropic process. What is the polytropic specific heat? What would be the direction of heat transfer if (a) $n > \gamma$, and (b) $n < \gamma$?

10.22 Why is the external work supplied to a compressor equal to $-\int_{p_1}^{p_2} v\, dp$?

10.23 Write down the van der Waals equation of state. How does it differ from the ideal gas equation of state. What is the force of cohesion? What is co-volume?

10.24 What are the two-constant equations of state?

10.25 Give the virial expansions for pv in terms of p and v.

10.26 What are virial coefficients? When do they become zero?

10.27 What is the compressibility factor?

10.28 What are reduced properties?

10.29 What is the generalized compressibility chart?

10.30 What is the law of corresponding states?

10.31 Express the van der Waals constants in terms of critical properties.

10.32 Draw the diagram representing the law of corresponding states in reduced coordinates indicating the isotherms and the liquid and vapour phases.

10.33 Define Boyle temperature? How is it computed?

10.34 State Dalton's law of partial pressures.

10.35 How is the partial pressure in a gas mixture related to the mole fraction?

10.36 How are the characteristic gas constant and the molecular weight of a gas mixture computed?

10.37 What is Gibb's theorem?

10.38 Show that in a diffusion process a gas undergoes a free expansion from the total pressure to the relevant partial pressure.

10.39 Show that in a diffusion process at constant temperature the entropy increases and the Gibbs function decreases.

Problems

10.1 What is the mass of air contained in a room 6 m × 9 m × 4 m if the pressure is 101.325 kPa and the temperature is 25°C? *Ans.* 256 kg

10.2 The usual cooking gas (mostly methane) cylinder is about 25 cm in diameter and 80 cm in height. It is charged to 12 MPa at room temperature (27°C). (a) Assuming the ideal gas law, find the mass of gas filled in the cylinder. (b) Explain how the actual cylinder contains nearly 15 kg of gas. (c) If the cylinder is to be protected against excessive pressure by means of a fusible plug, at what temperature should the plug melt to limit the maximum pressure to 15 MPa?

10.3 A certain gas has $c_p = 0.913$ and $c_v = 0.653$ kJ/kg K. Find the molecular weight and the gas constant R of the gas.

10.4 From an experimental determination the specific heat ratio for acetylene (C_2H_2) is found to 1.26. Find the two specific heats.

10.5 Find the molal specific heats of monatomic, diatomic, and polyatomic gases, if their specific heat ratios are respectively 5/3, 7/5 and 4/3.

10.6 A supply of natural gas is required on a site 800 m above storage level. The gas at – 150°C, 1.1 bar from storage is pumped steadily to a point on the site where its pressure is 1.2 bar, its temperature 15°C, and its flow rate 1000 m³/hr. If the work transfer to the gas at the pump is 15 kW, find the heat transfer to the gas between the two points. Neglect the change in K.E. and assume that the gas has the properties of methane (CH_4) which may be treated as an ideal gas having $\gamma = 1.33$ ($g = 9.75$ m/s²). *Ans.* 63.9 kW

10.7 A constant volume chamber of 0.3 m³ capacity contains 1 kg of air at 5°C. Heat is transferred to the air until the temperature is 100°C. Find the work done, the heat transferred, and the changes in internal energy, enthalpy and entropy.

10.8 One kg of air in a closed system, initially at 5°C and occupying 0.3 m³ volume, undergoes a constant pressure heating process to 100°C. There is no work other than pdv work. Find (a) the work done during the process, (b) the heat transferred, and (c) the entropy change of the gas

10.9 0.1 m³ of hydrogen initially at 1.2 MPa, 200°C undergoes a reversible isothermal expansion to 0.1 MPa. Find (a) the work done during the process, (b) the heat transferred, and (c) the entropy change of the gas.

10.10 Air in a closed stationary system expands in a reversible adiabatic process from 0.5 MPa, 15°C to 0.2 MPa. Find the final temperature, and per kg of air, the change in enthalpy, the heat transferred, and the work done.

10.11 If the above process occurs in an open steady flow system, find the final temperature, and per kg of air, the change in internal energy, the heat transferred, and the shaft work. Neglect velocity and elevation changes.

10.12 The indicator diagram for a certain water-cooled cylinder and piston air compressor shows that during compression $pv^{1.3}$ = constant. The compression starts at 100 kPa, 25°C and ends at 600 kPa. If the process is reversible, how much heat is transferred per kg of air?

10.13 An ideal gas of molecular weight 30 and $\gamma = 1.3$ occupies a volume of 1.5 m³ at 100 kPa and 77°C. The gas is compressed according to the law $pv^{1.25}$ = constant to a pressure of 3 MPa. Calculate the volume and temperature at the end of compression and heating, work done, heat transferred, and the total change of entropy.

10.14 Calculate the change of entropy when 1 kg of air changes from a temperature of 330 K and a volume of 0.15 m³ to a temperature of 550 K and a volume of 0.6 m³. If the air expands according to the law, pv^n = constant, between the same end states, calculate the heat given to, or extracted from, the air during the expansion, and show that it is approximately equal to the change of entropy multiplied by the mean absolute temperature.

10.15 0.5 kg of air, initially at 25°C, is heated reversibly at constant pressure until the volume is doubled, and is then heated reversibly at constant volume until the pressure is doubled. For the total path, find the work transfer, the heat transfer, and the change of entropy.

10.16 An ideal gas cycle of three processes uses Argon (Mol. wt. 40) as a working substance. Process 1–2 is a reversible adiabatic expansion from 0.014 m³, 700 kPa, 280°C to 0.056 m³. Process 2–3 is a reversible isothermal process. Process 3–1 is a constant pressure process in which heat transfer is zero. Sketch the cycle in the *p-v* and *T-s* planes, and find (a) the work transfer in process 1–2, (b) the work transfer in process 2–3, and (c) the net work of the cycle. Take $\gamma = 1.67$.

Ans. (a) 8.85kJ (b) 8.96kJ (c) 5.82kJ

10.17 A gas occupies 0.024m^3 at 700 kPa and 95°C. It is expanded in the non-flow process according to the law $pv^{1.2}$ = constant to a pressure of 70 kPa after which it is heated at constant pressure back to its original temperature. Sketch the process on the *p-v* and *T-s* diagrams, and calculate for the whole process the work done, the heat transferred, and the change of entropy. Take c_p = 1.047 and c_v = 0.775 kJ/kg K for the gas.

10.18 0.5 kg of air at 600 kPa receives an addition of heat at constant volume so that its temperature rises from 110°C to 650°C. It then expands in a cylinder polytropically to its original temperature and the index of expansion is 1.32. Finally, it is compressed isothermally to its original volume. Calculate (a) the change of entropy during each of the three stages, (b) the pressures at the end of constant volume heat addition and at the end of expansion. Sketch the processes on the *p-v* and *T-s* diagrams.

10.19 0.5 kg of helium and 0.5 kg of nitrogen are mixed at 20°C and at a total pressure of 100 kPa. Find (a) the volume of the mixture, (b) the partial volumes of the components, (c) the partial pressures of the components, (d) the mole fractions of the components, (e) the specific heats c_p and c_v of the mixture, and (f) the gas constant of the mixture.

Ans. (a) 3.481 m^3 (b) 3.045, 0.436 m^3 (c) 87.5, 12.5 kPa (d) 0.875, 0.125 (e) 3.11, 1.921 kJ/k (f) 1.189 kJ/kgK.

10.20 A gaseous mixture consists of 1 kg of oxygen and 2 kg of nitrogen at a pressure of 150 kPa and a temperature of 20°C. Determine the changes in internal energy, enthalpy and entropy of the mixture when the mixture is heated to a temperature of 100°C (a) at constant volume, and (b) at constant pressure.

10.21 A closed rigid cylinder is divided by a diaphragm into two equal compartments, each of volume 0.1 m^3. Each compartment contains air at a temperature of 20°C. The pressure in one compartment is 2.5 MPa and in the other compartment is 1 MPa. The diaphragm is ruptured so that the air in both the compartments mixes to bring the pressure to a uniform value throughout the cylinder which is insulated. Find the net change of entropy for the mixing process.

10.22 A vessel is divided into three compartments (a), (b), and (c) by two partitions. Part (a) contains oxygen and has a volume of 0.1 m^3, (b) has a volume of 0.2 m^3 and contains nitrogen, while (c) is 0.05 m^3 and holds CO_2. All three parts are at a pressure of 2 bar and a temperature of 13°C. When the partitions are removed and the gases mix, determine the change of entropy of each constituent, the final pressure in the vessel and the partial pressure of each gas. The vessel may be taken as being completely isolated from its surroundings.

Ans. 0.0875, 0.0783, 0.0680 kJ/K; 2 bar; 0.5714, 1.1329, 0.2857 bar.

10.23 A Carnot cycle uses 1 kg of air as the working fluid. The maximum and minimum temperatures of the cycle are 600 K and 300 K. The maximum pressure of the cycle is 1 MPa and the volume of the gas doubles during the isothermal heating process. Show by calculation of net work and heat supplied that the efficiency is the maximum possible for the given maximum and minimum temperatures.

10.24 An ideal gas cycle consists of three reversible processes in the following sequence: (a) constant volume pressure rise, (b) isentropic expansion to *r* times the initial volume, and (c) constant pressure decrease in volume. Sketch the cycle on the *p-v* and *T-s* diagrams. Show that the efficiency of the cycle is

$$\eta_{\text{cycle}} = \frac{r^\gamma - 1 - \gamma(r-1)}{r^\gamma - 1}$$

Evaluate the cycle efficiency when $\gamma = \frac{4}{3}$ and $r = 8$.

Ans. (η = 0.378)

10.25 Using the Dieterici equation of state

$$p = \frac{RT}{v-b} \cdot \exp\left(-\frac{a}{RTv}\right)$$

(a) Show that

$$p_c = \frac{a}{4e^2b^2},\ v_c = 2b,\ T_c = \frac{a}{4Rb}$$

(b) expand in the form

$$pv = RT\left(1 + \frac{B}{v} + \frac{C}{v^2} + \cdots\right)$$

(c) show that $T_B = \dfrac{a}{bR}$

10.26 The number of moles, the pressures, and the temperatures of gases *a*, *b*, and *c* are given below as follows

Gas	m (kg mol)	p (kPa)	t (°C)
N_2	1	350	100
CO	3	420	200
O_2	2	700	300

If the containers are connected, allowing the gases to mix freely, find (a) the pressure and temperature of the resulting mixture at equilibrium, and (b) the change of entropy of each constituent and that of the mixture.

10.27 Calculate the volume of 2.5 kg moles of steam at 236.4 atm. and 776.76 K with the help of compressibility factor versus reduced pressure graph. At this volume and the given pressure, what would the temperature be in K, if steam behaved like a van der Waals gas?

The critical pressure, volume, and temperature of steam are 218.2 atm, 57 cm³/g mole, and 647.3 K respectively.

10.28 Two vessels, *A* and *B*, each of volume 3 m³ may be connected together by a tube of negligible volume. Vessel *A* contains air at 7 bar, 95°C while *B* contains air at 3.5 bar, 205°C. Find the change of entropy when *A* is connected to *B*. Assume the mixing to be complete and adiabatic.

Ans. (0.975 kJ/kg K)

10.29 An ideal gas at temperature T_1 is heated at constant pressure to T_2 and then expanded reversibly, according to the law pv^n = constant, until the temperature is once again T_1. What is the required value of *n*, if the changes of entropy during the separate processes are equal?

Ans. $\left(n = \frac{2\gamma}{\gamma + 1}\right)$

10.30 A certain mass of sulphur dioxide (SO_2) is contained in a vessel of 0.142 m³ capacity, at a pressure and temperature of 23.1 bar and 18°C respectively. A valve is opened momentarily and the pressure falls immediately to 6.9 bar. Sometime later the temperature is again 18°C and the pressure is observed to be 9.1 bar. Estimate the value of specific heat ratio. *Ans.* 1.29

10.31 A gaseous mixture contains 21% by volume of nitrogen, 50% by volume of hydrogen, and 29% by volume of carbon-dioxide. Calculate the molecular weight of the mixture, the characteristic gas constant *R* for the mixture and the value of the reversible adiabatic index γ. (At 10°C, the c_p values of nitrogen, hydrogen, and carbon dioxide are 1.039, 14.235, and 0.828 kJ/kg K respectively.)

A cylinder contains 0.085 m³ of the mixture at 1 bar and 10°C. The gas undergoes a reversible non-flow process during which its volume is reduced to one-fifth of its original value. If the law of compression is $pv^{1.2}$ = constant, determine the work and heat transfers in magnitude and sense and the change in entropy.

Ans. 19.64 kg/kg mol, 0.423 kJ/kg K, 1.365, – 16 kJ, – 7.24 kJ, – 0.31 kJ/kg K

10.32 Two moles of an ideal gas at temperature *T* and pressure *p* are contained in a compartment. In an adjacent compartment is one mole of an ideal gas at temperature 2*T* and pressure *p*. The gases mix adiabatically but do not react chemically when a partition separating the compartments is withdrawn. Show that the entropy increase due to the mixing process is given by

$$R\left(\ln\frac{27}{4} + \frac{\gamma}{\gamma - 1}\ln\frac{32}{27}\right)$$

provided that the gases are different and that the ratio of specific heat γ is the same for both gases and remains constant.

What would the entropy change be, if the mixing gases were of the same species?

10.33 n_1 moles of an ideal gas at pressure p_1 and temperature *T* are in one compartment of an insulated container. In an adjoining compartment, separated by a partition, are n_2 moles of an ideal gas at pressure p_2 and temperature *T*. When the partition is removed, calculate (a) the final pressure of the mixture, (b) the entropy change when the gases are identical, and (c) the entropy change when the gases are different. Prove that the entropy change in (c) is the same as that produced by two independent free expansions.

10.34 Assume that 20 kg of steam are required at a pressure of 600 bar and a temperature of 750°C in order to conduct a particular experiment. A 140-litre heavy duty tank is available for storage.

Predict if this is an adequate storage capacity using:

(a) the ideal gas theory,
(b) the compressibility factor chart,

(c) the van der Waals equation with $a = 5.454$ (litre)^2 atm/(g mol)^2, $b = 0.03042$ litres/g mol for steam,
(d) the Mollier chart
(e) the steam tables.
Estimate the error in each.

Ans. (a) 157.75l, (b) 132.5l, (c) 124.94l, (d) 137.29l

10.35 Estimate the pressure of 5 kg of CO_2 gas which occupies a volume of 0.70 m^3 at 75°C, using the Beattie-Bridgeman equation of state.

Compare this result with the value obtained using the generalized compressibility chart. Which is more accurate and why?

For CO_2 with units of atm, litres/g mol and K, $A_0 = 5.0065$, $a = 0.07132$, $B_0 = 0.10476$, $b = 0.07235$, $C \times 10^{-4} = 66.0$.

10.36 Measurements of pressure and temperature at various stages in an adiabatic air turbine show that the states of air lie on the line $pv^{1.25}$ = constant. If kinetic and gravitational potential energy are neglected, prove that the shaft work per kg as a function of pressure is given by the following relation

$$W = 3.5\, p_1 v_1 \left[1 - \left(\frac{p_2}{p_1}\right)^{1/5}\right]$$

Take γ for air as 1.4.

10.37 Air flows steadily into a compressor at a temperature of 17°C and a pressure of 1.05 bar and leaves at a temperature of 247°C and a pressure of 6.3 bar. There is no heat transfer to or from the air as it flows through the compressor; changes in elevation and velocity are negligible. Evaluate the external work done per kg of air, assuming air as an ideal gas for which $R = 0.287$ kJ/kg K and $\gamma = 1.4$. Evaluate the minimum external work required to compress the air adiabatically from the same initial state to the same final pressure and the isentropic efficiency of the compressor.

Ans. – 225kJ/kg, – 190kJ/kg, 84.4%

10.38 A slow-speed reciprocating air compressor with a water jacket for cooling approximates a quasi-static compression process following a path $pv^{1.3}$ = const. If air enters at a temperature of 20°C and a pressure of 1 bar, and is compressed to 6 bar at a rate of 1000 kg/h, determine the discharge temperature of air, the power required and the heat transferred per kg. *Ans.* 443k, 51.82 kW, 36 kJ/kg

10.39 A single-acting two-stage reciprocating air compressor with complete intercooling delivers 6 kg/min at 15 bar pressure. Assume an intake condition of 1 bar and 15°C and that the compression and expansion processes are polytropic with $n = 1.3$. Calculate: (a) the power required, (b) the isothermal efficiency.

Ans. (a) 26.15 kW (b) 85.6%

10.40 A two-stage air compressor receives 0.238 m^3/s of air at 1 bar and 27°C and discharges it at 10 bar. The polytropic index of compression is 1.35. Determine (a) the minimum power necessary for compression, (b) the power needed for single-stage compression to the same pressure, (c) the maximum temperature for (a) and (b), and (d) the heat removed in the intercooler. *Ans.* (a) 63.8 kW, (b) 74.9 kW, (c) 404.2 k, 544.9 k, (d) 28.9 kW

10.41 A mass of an ideal gas exists initially at a pressure of 200 kPa, temperature 300 K, and specific volume 0.5 m^3/kg. The value of γ is 1.4. (a) Determine the specific heats of the gas. (b) What is the change in entropy when the gas is expanded to pressure 100 kPa according to the law $pv^{1.3}$ = const? (c) What will be the entropy change if the path is $pv^{1.5}$ = const. (by the application of a cooling jacket during the process)? (d) What is the inference you can draw from this example?

Ans. (a) 1.166, 0.833 kJ/kg K, (b) 0.044 kJ/kg K (c) – 0.039 kJ/kg K (d) Entropy increases when $n < \gamma$ and decreases when $n > \gamma$

10.42 (a) A closed system of 2 kg of air initially at pressure 5 atm and temperature 227°C, expands reversibly to pressure 2 atm following the law $pv^{1.25}$ = const. Assuming air as an ideal gas, determine the work done and the heat transferred. *Ans.* 193 kJ, 72 kJ

(b) If the system does the same expansion in a steady flow process,what is the work done by the system? *Ans.* 241 kJ

10.43 Air contained in a cylinder fitted with a piston is compressed reversibly according to the law $pv^{1.25}$ = const.Themassofairinthecylinderis0.1kg. The initial pressure is 100 kPa and the initial temperature 20°C. The final volume is 1/8 of the initial volume. Determine the work and the heat transfer. *Ans.* – 22.9 kJ, – 8.7 kJ

10.44 Air is contained in a cylinder fitted with a frictionless piston. Initially the cylinder contains 0.5 m^3 of air at 1.5 bar, 20°C. The air is then compressed reversibly according to the law pv^n = constant until the final pressure is 6 bar, at which point the temperature is 120°C. Determine: (a) the polytropic index n, (b) the final volume of air, (c) the work done on the air and the heat transfer, and (d) the net change in entropy.

Ans. (a) 1.2685, (b) 0.1676 m^3 (c) – 95.3 kJ, – 31.5 kJ, (d) 0.0153 kJ/K

10.45 The specific heat at constant pressure for air is given by

$$c_p = 0.9169 + 2.577 \times 10^{-4} T - 3.974 \times 10^{-8} T^2 \text{ kJ/kg K}$$

Determine the change in internal energy and that in entropy of air when it undergoes a change of state from 1 atm and 298 K to a temperature of 2000 K at the same pressure.

Ans. 1470.4 kJ/kg, 2.1065 kJ/kg K

10.46 A closed system allows nitrogen to expand reversibly from a volume of 0.25 m^3 to 0.75 m^3 along the path $pv^{1.32}$ = const. The original pressure of the gas is 250 kPa and its initial temperature is 100°C. (a) Draw the p-v and T-s diagrams. (b)What are the final temperature and the final pressure of the gas? (c) How much work is done and how much heat is transferred? (d) What is the entropy change of nitrogen?

Ans. (b) 262.44 K, 58.63 kPa, (c) 57.89 kJ, 11.4 kJ, (d) 0.0362 kJ/K

10.47 Methane has a specific heat at constant pressure given by c_p = 17.66 + 0.06188 T kJ/kg mol K when 1 kg of methane is heated at constant volume from 27 to 500°C. If the initial pressure of the gas is 1 atm, calculate the final pressure, the heat transfer, the work done and the change in entropy.

Ans. 2.577 atm, 1258.5 kJ/kg, 0, 2.3838 kJ/kg K

10.48 Air is compressed reversibly according to the law $pv^{1.25}$ = const. from an initial pressure of 1 bar and volume of 0.9 m^3 to a final volume of 0.6 m^3. Determine the final pressure and the change of entropy per kg of air.

Ans. 1.66 bar, – 0.0436 kJ/kg K

10.49 In a heat engine cycle, air is isothermally compressed. Heat is then added at constant pressure, after which the air expands isentropically to its original state. Draw the cycle on p-v and T-s coordinates. Show that the cycle efficiency can be expressed in the following form

$$\eta = 1 - \frac{(\gamma - 1)\ln r}{\gamma[r^{\gamma-1/\gamma} - 1]}$$

where r is the pressure ratio, p_2/p_1. Determine the pressure ratio and the cycle efficiency if the initial temperature is 27°C and the maximum temperature is 327°C. *Ans.* 13.4, 32.4%

10.50 What is the minimum amount of work required to separate 1 mole of air at 27°C and 1 atm pressure (assumed composed of 1/5 O_2 and 4/5 N_2) into oxygen and nitrogen each at 27°C and 1 atm pressure? *Ans.* 1250 J

10.51 A closed adiabatic cylinder of volume 1 m^3 is divided by a partition into two compartments 1 and 2. Compartment 1 has a volume of 0.6 m^3 and contains methane at 0.4 MPa, 40°C, while compartment 2 has a volume of 0.4 m^3 and contains propane at 0.4 MPa, 40°C. The partition is removed and the gases are allowed to mix. (a) When the equilibrium state is reached, find the entropy change of the universe. (b) What are the molecular weight and the specific heat ratio of the mixture?

The mixture is now compressed reversibly and adiabatically to 1.2 MPa. Compute (c) the final temperature of the mixture, (d) the work required per unit mass, and (e) the specific entropy change for each gas. Take c_p of methane and propane as 35.72 and 74.56 kJ/kg mol K respectively.

Ans. (a) 0.8609 kJ/K, (b) 27.2, 1.193 (c) 100.9°C, (d) 396 kJ, (e) 0.255 kJ/kg K

10.52 An ideal gas cycle consists of the following reversible processes: (i) isentropic compression, (ii) constant volume heat addition, (iii) isentropic expansion, and (iv) constant pressure heat rejection. Show that the efficiency of this cycle is given by

$$\eta = 1 - \frac{1}{r_k^{\gamma-1}}\left[\frac{\gamma(a^{1/\gamma} - 1)}{a - 1}\right]$$

where r_k is the compression ratio and a is the ratio of pressures after and before heat addition.

An engine operating on the above cycle with a compression ratio of 6 starts the compression with air at 1 bar, 300 K. If the ratio of pressures after and before heat addition is 2.5, calculate the efficiency and the m.e.p. of the cycle. Take $\gamma = 1.4$ and $c_v = 0.718$ kJ/kg K. *Ans.* 0.579, 2.5322 bar

10.53 The relation between *v, p* and *v* for many gases is of the form $u = a + bpv$ where *a* and *b* are constants. Show that for a reversible adiabatic process pv^{γ} = constant, where $\gamma = (b + 1)/b$.

10.54 (a) Show that the slope of a reversible adiabatic process on *p-v* coordinates is

$$\frac{dp}{dv} = -\frac{1}{kv}\frac{c_p}{c_v} \text{ where } k = -\frac{1}{v}\left(\frac{\partial v}{\partial p}\right)_T$$

(b) Hence, show that for an ideal gas, pv^{γ} = constant, for a reversible adiabatic process.

10.55 A certain gas obeys the Clausius equation of state $p(v - b) = RT$ and has its internal energy given by $u = c_v T$. Show that the equation for a reversible adiabatic process is $p(v - b)^{\gamma}$ = constant, where $\gamma = c_p/c_v$.

10.56 (a) Two curves, one representing a reversible adiabatic process undergone by an ideal gas and the other an isothermal process by the same gas, intersect at the same point on the *p-v* diagram. Show that the ratio of the slope of the adiabatic curve to the slope of the isothermal curve is equal to γ. (b) Determine the ratio of work done during a reversible adiabatic process to the work done during an isothermal process for a gas having $\gamma = 1.6$. Both processes have a pressure ratio of 6.

10.57 Two containers *p* and *q* with rigid walls contain two different monatomic gases with masses m_p and m_q, gas constants R_p and R_q, and initail temperatures T_p and T_q respectively, are brought in contact with each other and allowed to exchange energy until equilibrium is achieved. Determine: (a) the final temperature of the two gases and (b) the change of entropy due to this energy exchange.

10.58 The pressure of a certain gas (photon gas) is a function of temperature only and is related to the energy and volume by $p(T) = (1/3)\ (U/V)$. A system consisting of this gas confined by a cylinder and a piston undergoes a Carnot cycle between two pressures p_1 and p_2. (a) Find expressions for work and heat of reversible isothermal and adiabatic processes. (b) Plot the Carnot cycle on *p-v* and *T-s* diagrams. (c) Determine the efficiency of the cycle in terms of pressures. (d) What is the functional relation between pressure and temperature?

10.59 The gravimetric analysis of dry air is approximately: oxygen = 23%, nitrogen = 77%. Calculate: (a) the volumetric analysis, (b) the gas constant, (c) the molecular weight, (d) the respective partial pressures, (e) the specific volume at 1 atm, 15°C, and (f) How much oxygen must be added to 2.3 kg air to produce a mixture which is 50% oxygen by volume?

Ans. (a) 21% O_2, 79% N_2, (b) 0.288 kJ/kg K, (d) 21 kPa for O_2, (e) 0.84 m³/kg, (f) 1.47 kg

10.60 A vessel of volume 2*V* is divided into two equal compartments. These are filled with the same ideal gas, the temperature and pressure on one side of the partition being (p_1, T_1) and on the other (p_2, T_2). Show that if the gases on the two sides are allowed to mix slowly with no heat entering, the final pressure and temperature will be given by

$$p = \frac{p_1 + p_2}{2},\ T = \frac{T_1 T_2 (p_1 + p_2)}{p_1 T_2 + p_2 T_1}$$

Further, show that the entropy gain is

$$\Delta S = V\left[\left(\frac{c_p}{R}\right)\left\{\frac{p_1}{T_2}\ln\frac{T}{T_1} + \frac{p_2}{T_2}\ln\frac{T}{T_2}\right\} - \frac{p_1}{T_1}\ln\frac{p}{p_1} - \frac{p_2}{T_2}\ln\frac{p}{p_2}\right]$$

10.61 An ideal gas with a constant volume of c_p = 29.6 J/gmol-K is made to undergo a cycle consisting of the following reversible processes in a closed system:

Process 1–2: The gas expands adiabatically from 5 MPa, 550 K to 1 MPa;

Process 2–3: The gas is heated at constant volume until 550 K;

Process 3–1: The gas is compressed isothermally back to its initial condition.

Calculate the work, the heat and the change of entropy of the gas for each of the three processes. Draw the *p-v* and *T-s* diagrams.

Ans. W_{1-2} = 4260 J/gmol, $Q_{1-2} = 0$, $\Delta s_{1-2} = 0$, $W_{2-3} = 0$, Q_{2-3} = 4260 J/gmol, Δs_{2-3} = 9.62 J/gmol-K, W_{3-1} = – 5290 J/mol $= Q_{3-1}$, Δs_{3-1} = – 9.62 J/gmol-K, $W_{net} - Q_{net}$ = – 1030 Jg/gmol, $\oint dS = 0$.

10.62 Air in a closed system expands reversibly and adiabatically from 3 MPa. 200°C to two times its initial volume, and then cools at constant volume until the pressure drops to 0.8 MPa. Calculate the work done and heat transferred per kg of air. Use $c_p = 1.017$ and $c_v = 0.728$ kJ/kgK.

Ans. 82.7 kJ/kg, – 78.1 kJ/kg

10.63 A vessel is divided into three compartments (a), (b) and (c) by two partitions. Part (a) contains hydrogen and has a volume of 0.1 m^3, part (b) contains nitrogen and has a volume of 0.2 m^3 and part (c) contains carbon dioxide and has a volume of 0.05 m^3. All the three parts are at a pressure of 2 bar and a temperature of 13°C. The partitions are removed and the gases are allowed to mix. Determine (a) the molecular weight of the mixture, (b) the characteristics gas constant for the mixture, (c) the partial pressures of each gas, (d) the reversible adiabatic index γ, and (e) the entropy change due to diffusion. The specific heats of hydrogen, nitrogen and carbon dioxide are 14.235, 1.039 and 0.828 kJ/kg K respectively.
The above gas mixture is then reversibly compressed to a pressure of 6 bar according to the law $pv^{1.2}$ = constant, (f) Determine the work and heat interactions in magnitude and sense, and (g) the change in entropy.
Ans. (a) 22.8582 (b) 0.3637 kJ/kg K (c) $p_{H_2} = 0.5714$, $p_{N_2} = 1.1428$, $p_{CO_2} = 0.2858$ bar (d) 1.384 (e) 0.3476 kJ/kgK (f) – 70.455 kJ, – 33.772 kJ (g) – 0.1063 kJ/K.

10.64 A four cylinder single-stage air compressor has a bore of 200 mm and a stroke of 300 mm and runs at 400 rpm. At a working pressure of 721.3 kPa it delivers 3.1 m^3 of air per min at 270°C. Calculate (a) the mass flow rate, (b) the free air delivery (FAD) (c) effective swept volume, (d) volumetric efficiency. Take the inlet condition as that of the free air at 101.3 kPa, 21°C. *Ans.* (a) 0.239 kg/s (b) 0.199 m^3/s (c) 0.299 m^3, (d) 79.2%

10.65 Predict the pressure of nitrogen gas at $T = 175$ K and $v = 0.00375$ m^3/kg on the basis of (a) the ideal gas equation of state, (b) the van der Waals equation of state, (c) the Beattie-Bridgeman equation of state and (d) the Benedict-Webb-Rubin equation of state. Compare the values obtained with the experimentally determined value of 10,000 kPa. *Ans.* (a) 13,860 kPa (b) 9468 kPa (c) 10,110 kPa (d) 10,000 kPa

10.66 The pressure in an automobile tyre depends on the temperature of the air in the tyre. When the air temperature is 25°C, the pressure gauge reads 210 kPa. If the volume of the tyre is 0.025 m^3, determine the pressure rise in the tyre when the air temperature in the tyre rises to 50°C. Also find the amount of air that must be bled off to restore pressure to its original value at this temperature. Take atmospheric pressure as 100 kPa.

10.67 Two tanks are connected by a valve. One tank contains 2 kg of CO gas at 77°C and 0.7 bar. The other tank holds 8 kg of the same gas at 27°C and 1.2 bar. The valve is opened and the gases are allowed to mix while receiving energy by heat transfer from the surroundings. The final equilibrium temperature is 42°C. Using the ideal gas model, determine (a) the final equilibrium pressure, (b) the heat transfer for the process.
Ans. (a) 1.05 bar (b) 37.25 kJ

C H A P T E R 11

Thermodynamic Relations, Equilibrium and Third Law

11.1 SOME MATHEMATICAL THEOREMS

Theorem 1 If a relation exists among the variables *x*, *y*, and *z*, then *z* may be expressed as a function of *x* and *y*, or

$$dz = \left(\frac{\partial z}{\partial x}\right)_y dx + \left(\frac{\partial z}{\partial y}\right)_x dy$$

If $\left(\frac{\partial z}{\partial x}\right)_y = M$, and $\left(\frac{\partial z}{\partial y}\right)_x = N$

then $$dz = M\,dx + N\,dy,$$

where *z*, *M* and *N* are functions of *x* and *y*. Differentiating *M* partially with respect to *y*, and *N* with respect to *x*

$$\left(\frac{\partial M}{\partial y}\right)_x = \frac{\partial^2 z}{\partial x \cdot \partial y}$$

$$\left(\frac{\partial N}{\partial x}\right)_y = \frac{\partial^2 z}{\partial y \cdot \partial x}$$

$$\therefore \quad \left(\frac{\partial M}{\partial y}\right)_x = \left(\frac{\partial N}{\partial x}\right)_y \tag{11.1}$$

This is the *condition of exact* (*or perfect*) *differential.*

Theorem 2 If a quantity *f* is a function of *x*, *y*, and *z*, and a relation exists among *x*, *y* and *z*, then *f* is a function of any two of *x*, *y*, and *z*. Similarly, any one of *x*, *y*, and *z* may be regarded to be a function of *f* and any one of *x*, *y*, and *z*. Thus, if

$$x = x\,(f, y)$$

$$dx = \left(\frac{\partial x}{\partial f}\right)_y df + \left(\frac{\partial x}{\partial y}\right)_f dy$$

Similarly, if

$$y = y(f, z)$$

$$dy = \left(\frac{\partial y}{\partial f}\right)_z df + \left(\frac{\partial y}{\partial z}\right)_f dz$$

Substituting the expression of d*y* in the preceding equation

$$dx = \left(\frac{\partial x}{\partial f}\right)_y df + \left(\frac{\partial x}{\partial y}\right)_f \left[\left(\frac{\partial y}{\partial f}\right)_z df + \left(\frac{\partial y}{\partial z}\right)_f dz\right] = \left[\left(\frac{\partial x}{\partial f}\right)_y + \left(\frac{\partial x}{\partial y}\right)_f \left(\frac{\partial y}{\partial f}\right)_z\right] df + \left(\frac{\partial x}{\partial y}\right)_f \left(\frac{\partial y}{\partial z}\right)_f dz$$

Again $$dx = \left(\frac{\partial x}{\partial f}\right)_z df + \left(\frac{\partial x}{\partial z}\right)_f dz$$

$$\therefore \quad \left(\frac{\partial x}{\partial z}\right)_f = \left(\frac{\partial x}{\partial y}\right)_f \left(\frac{\partial y}{\partial z}\right)_f$$

$$\therefore \quad \left(\frac{\partial x}{\partial y}\right)_f \left(\frac{\partial y}{\partial z}\right)_f \left(\frac{\partial z}{\partial x}\right)_f = 1 \tag{11.2}$$

Theorem 3 Among the variables x, y, and z, any one variable may be considered as a function of the other two. Thus

$$x = x(y, z)$$

$$dx = \left(\frac{\partial x}{\partial y}\right)_z dy + \left(\frac{\partial x}{\partial z}\right)_y dz$$

Similarly,

$$dz = \left(\frac{\partial z}{\partial x}\right)_y dx + \left(\frac{\partial z}{\partial y}\right)_x dy$$

$$\therefore \quad dx = \left(\frac{\partial x}{\partial y}\right)_z dy + \left(\frac{\partial x}{\partial z}\right)_y \left[\left(\frac{\partial z}{\partial x}\right)_y dx + \left(\frac{\partial z}{\partial y}\right)_x dy\right]$$

$$= \left[\left(\frac{\partial x}{\partial y}\right)_z + \left(\frac{\partial x}{\partial z}\right)_y \left(\frac{\partial z}{\partial y}\right)_x\right] dy + \left(\frac{\partial x}{\partial z}\right)_y \left(\frac{\partial z}{\partial x}\right)_y dx$$

$$= \left[\left(\frac{\partial x}{\partial y}\right)_z + \left(\frac{\partial x}{\partial z}\right)_y \left(\frac{\partial z}{\partial y}\right)_x\right] dy + dx$$

$$\therefore \quad \left(\frac{\partial x}{\partial y}\right)_z + \left(\frac{\partial x}{\partial z}\right)_y \left(\frac{\partial z}{\partial y}\right)_x = 0$$

or

$$\left(\frac{\partial x}{\partial y}\right)_z \left(\frac{\partial z}{\partial x}\right)_y \left(\frac{\partial y}{\partial z}\right)_x = -1 \tag{11.3}$$

Among the thermodynamic variables p, V and T, the following relation holds good

$$\left(\frac{\partial p}{\partial V}\right)_T \left(\frac{\partial V}{\partial T}\right)_p \left(\frac{\partial T}{\partial p}\right)_V = -1$$

11.2 MAXWELL'S EQUATIONS

A pure substance existing in a single phase has only two independent variables. Of the eight quantities p, V, T, S, U, H, F (Helmholtz function), and G (Gibbs function) *any one may be expressed as a function of any two others.*

For a pure substance undergoing an infinitesimal reversible process

(a) $dU = TdS - pdV$
(b) $dH = dU + pdV + Vdp = TdS + Vdp$
(c) $dF = dU - TdS - SdT = -pdV - SdT$
(d) $dG = dH - TdS - SdT = Vdp - SdT$

Since U, H, F and G are thermodynamic properties and exact differentials of the type

$$dz = M\,dx + N\,dy, \text{ then}$$

$$\left(\frac{\partial M}{\partial y}\right)_x = \left(\frac{\partial N}{\partial x}\right)_y$$

Applying this to the four equations

$$\left(\frac{\partial T}{\partial V}\right)_S = -\left(\frac{\partial p}{\partial S}\right)_V \tag{11.4}$$

$$\left(\frac{\partial T}{\partial p}\right)_S = \left(\frac{\partial V}{\partial S}\right)_p \tag{11.5}$$

$$\left(\frac{\partial p}{\partial T}\right)_V = \left(\frac{\partial S}{\partial V}\right)_T \tag{11.6}$$

$$\left(\frac{\partial V}{\partial T}\right)_p = -\left(\frac{\partial S}{\partial p}\right)_T \tag{11.7}$$

These four equations are known as *Maxwell's equations.*

11.3 *TdS* EQUATIONS

Let entropy S be imagined as a function of T and V. Then

$$dS = \left(\frac{\partial S}{\partial T}\right)_V dT + \left(\frac{\partial S}{\partial V}\right)_T dV$$

$$\therefore \quad TdS = T\left(\frac{\partial S}{\partial T}\right)_V dT + T\left(\frac{\partial S}{\partial V}\right)_T dV$$

Since $T\left(\frac{\partial S}{\partial T}\right)_V = C_v$, heat capacity at constant volume, and

$\left(\frac{\partial S}{\partial T}\right)_T = \left(\frac{\partial p}{\partial T}\right)_V$, Maxwell's third equation,

$$TdS = C_v\,dT + T\left(\frac{\partial p}{\partial T}\right)_V dV \tag{11.8}$$

This is known as the *first TdS equation.*

If $S = S(T, p)$

$$dS = \left(\frac{\partial S}{\partial T}\right)_p dT + \left(\frac{\partial S}{\partial p}\right)_T dp$$

$$\therefore \quad TdS = T\left(\frac{\partial S}{\partial T}\right)_p dT + T\left(\frac{\partial S}{\partial p}\right)_T dp$$

Since $\quad T\left(\frac{\partial S}{\partial T}\right)_p = C_p$, and $\left(\frac{\partial S}{\partial p}\right)_T = -\left(\frac{\partial V}{\partial T}\right)_p$

then
$$TdS = C_p dT - T\left(\frac{\partial V}{\partial T}\right)_p dp \tag{11.9}$$

This is known as the *second TdS equation*.

11.4 DIFFERENCE IN HEAT CAPACITIES

Equating the first and second TdS equations

$$TdS = C_p dT - T\left(\frac{\partial V}{\partial T}\right)_p dp = C_v\, dT + T\left(\frac{\partial p}{\partial T}\right)_V dV$$

$$(C_p - C_v)\, dT = T\left(\frac{\partial p}{\partial T}\right)_V dV + T\left(\frac{\partial V}{\partial T}\right)_p dp$$

$\therefore$
$$dT = \frac{T\left(\frac{\partial p}{\partial T}\right)_V}{C_p - C_v} dV + \frac{T\left(\frac{\partial V}{\partial T}\right)_p}{C_p - C_v} dp$$

Again
$$dT = \left(\frac{\partial T}{\partial V}\right)_p dV + \left(\frac{\partial T}{\partial p}\right)_v dp$$

$\therefore$
$$\frac{T\left(\frac{\partial p}{\partial T}\right)_V}{C_p - C_v} = \left(\frac{\partial T}{\partial V}\right)_p \quad \text{and} \quad \frac{T\left(\frac{\partial V}{\partial T}\right)_p}{C_p - C_v} = \left(\frac{\partial T}{\partial p}\right)_V$$

Both these equations give

$$C_p - C_v = T\left(\frac{\partial p}{\partial T}\right)_v \left(\frac{\partial V}{\partial T}\right)_p$$

But
$$\left(\frac{\partial p}{\partial T}\right)_v \left(\frac{\partial T}{\partial V}\right)_p \left(\frac{\partial V}{\partial p}\right)_T = -1$$

$\therefore$
$$C_p - C_v = -T\left(\frac{\partial V}{\partial T}\right)_p^2 \left(\frac{\partial p}{\partial V}\right)_T \tag{11.10}$$

This is a very important equation in thermodynamics. It indicates the following important facts.

(a) Since $\left(\frac{\partial V}{\partial T}\right)_p^2$ is always positive, and $\left(\frac{\partial p}{\partial V}\right)_T$ for any substance is negative, $(C_p - C_v)$ is always positive. Therefore, C_p is always greater than C_v.

(b) As $T \to 0\,K$, $C_p \to C_v$ or at absolute zero, $C_p = C_v$.

(c) When $\left(\frac{\partial V}{\partial T}\right)_p = 0$ (e.g., for water at 4°C, when density is maximum, or specific volume minimum), $C_p = C_v$.

(d) For an ideal gas, $pV = mRT$

$$\left(\frac{\partial V}{\partial T}\right)_p = \frac{mR}{p} = \frac{V}{T}$$

and

$$\left(\frac{\partial p}{\partial V}\right)_T = -\frac{mRT}{V^2}$$

$\therefore$

$$C_p - C_v = mR$$

or

$$c_p - c_v = R$$

Equation (11.10) may also be expressed in terms of volume expansivity (β), defined as

$$\beta = \frac{1}{V}\left(\frac{\partial V}{\partial T}\right)_p$$

and isothermal compressibility (k_T), defined as

$$k_T = -\frac{1}{V}\left(\frac{\partial V}{\partial T}\right)_T$$

$$C_p - C_v = \frac{TV\left[\frac{1}{V}\left(\frac{\partial V}{\partial T}\right)_p\right]^2}{-\frac{1}{V}\left(\frac{\partial V}{\partial p}\right)_T}$$

$\therefore$

$$C_p - C_v = \frac{TV\beta^2}{k_T} \tag{11.11}$$

11.5 RATIO OF HEAT CAPACITIES

At constant S, the two TdS equations become

$$C_p dT_s = T\left(\frac{\partial V}{\partial T}\right)_p dp_s$$

$$C_v dT_s = -T\left(\frac{\partial p}{\partial T}\right)_V dV_s$$

$\therefore$

$$\frac{C_p}{C_v} = -\left(\frac{\partial V}{\partial T}\right)_p \left(\frac{\partial T}{\partial p}\right)_V \left(\frac{\partial p}{\partial V}\right)_S = \frac{\left(\frac{\partial p}{\partial V}\right)_S}{\left(\frac{\partial p}{\partial V}\right)_T} = \gamma$$

Since $\gamma > 1$,

$$\left(\frac{\partial p}{\partial V}\right)_S > \left(\frac{\partial p}{\partial V}\right)_T$$

Therefore, the slope of an isentrope is greater than that of an isotherm on $p - v$ diagram (Fig. 11.1). For reversible and adiabatic compression, the work done is

$$W_s = h_{2s} - h_1 = \int_1^{2S} v\,dp = \text{Area } 1\text{–}2S\text{–}3\text{–}4\text{–}1$$

For reversible and isothermal compression, the work done would be

$$W_T = h_{2T} - h_1 = \int_1^{2T} v\,dp = \text{Area } 1\text{–}2T - 3 - 4 - 1$$

$$\therefore \qquad W_T < W_S$$

For polytropic compression with $1 < n < \gamma$, the work done will be between these two values. So, isothermal compression requires minimum work. (See Sec. 10.4)

The adiabatic compressibility (k_s) is defined as

$$k_s = -\frac{1}{V}\left(\frac{\partial V}{\partial p}\right)_S$$

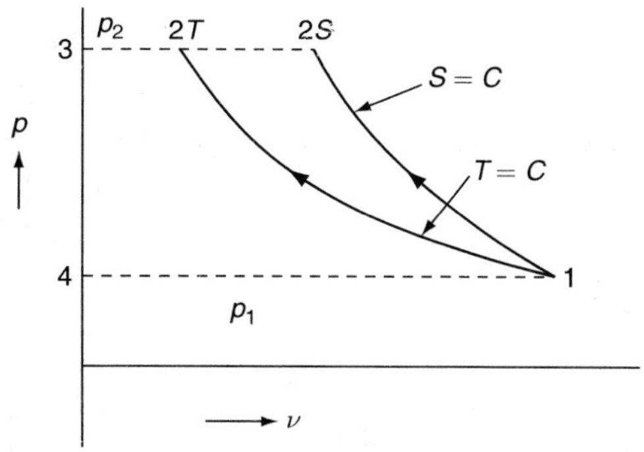

Fig. 11.1 *Compression work in different reversible processes*

$$\therefore \qquad \frac{C_p}{C_v} = \frac{-\dfrac{1}{V}\left(\dfrac{\partial V}{\partial p}\right)_T}{-\dfrac{1}{V}\left(\dfrac{\partial V}{\partial p}\right)_s} = \gamma$$

or

$$\gamma = \frac{k_T}{k_s} \tag{11.12}$$

11.6 ENERGY EQUATION

For a system undergoing an infinitesimal reversible process between two equilibrium states, the change of internal energy is

$$dU = TdS - pdV$$

Substituting the first TdS equation

$$dU = C_v dT + T\left(\frac{\partial p}{\partial T}\right)_V dV - pdV$$

$$= C_v dT + \left[T\left(\frac{\partial p}{\partial T}\right)_V - p\right] dV \tag{11.13}$$

If

$$U = U(T, V)$$

$$dU = \left(\frac{\partial U}{\partial T}\right)_V dT + \left(\frac{\partial U}{\partial V}\right)_T dV$$

$$\left(\frac{\partial U}{\partial V}\right)_T = T\left(\frac{\partial p}{\partial T}\right)_V - p \tag{11.14}$$

This is known as the *energy equation*. Two applications of the equation are given below:

(a) For an ideal gas,

$$p = \frac{n\bar{R}T}{V}$$

$$\therefore \qquad \left(\frac{\partial p}{\partial T}\right)_V = \frac{n\bar{R}}{V} = \frac{p}{T}$$

$$\therefore \qquad \left(\frac{\partial U}{\partial V}\right)_T = T \cdot \frac{p}{T} - p = 0$$

U does not change when V changes at $T = C$.

$$\left(\frac{\partial U}{\partial p}\right)_T \left(\frac{\partial p}{\partial V}\right)_T \left(\frac{\partial V}{\partial U}\right)_T = 1$$

$$\therefore \qquad \left(\frac{\partial U}{\partial p}\right)_T \left(\frac{\partial p}{\partial V}\right)_T = \left(\frac{\partial U}{\partial V}\right)_T = 0$$

Since
$$\left(\frac{\partial p}{\partial V}\right)_T \neq 0, \left(\frac{\partial U}{\partial p}\right)_T = 0$$

U does not change either when p changes at $T = C$. So the internal energy of an ideal gas is a function of temperature only, as shown earlier in Chapter 10.

Another important point to note is that in Eq. (11.13), for an ideal gas

$p V = n\bar{R}T$ and
$$T\left(\frac{\partial p}{\partial T}\right)_V - p = 0$$

Therefore
$$\mathrm{d}U = C_v \, \mathrm{d}T$$

holds good for an ideal gas in any process (even when the volume changes). But for any other substance

$$\mathrm{d}U = C_v \, \mathrm{d}T$$

is true only when the volume is constant and $\mathrm{d}V = 0$.

Similarly
$$\mathrm{d}H = T\mathrm{d}S + V\mathrm{d}p$$

and
$$T\mathrm{d}S = C_p \mathrm{d}T - T\left(\frac{\partial V}{\partial T}\right)_p \mathrm{d}p$$

$$\therefore \qquad \mathrm{d}H = C_p \, \mathrm{d}T + \left[V - T\left(\frac{\partial V}{\partial T}\right)_p\right]\mathrm{d}p \tag{11.15}$$

$$\therefore \qquad \left(\frac{\partial H}{\partial p}\right)_T = V - T\left(\frac{\partial V}{\partial T}\right)_p \tag{11.16}$$

As shown for internal energy, it can be similarly proved from Eq. (11.16) that the enthalpy of an ideal gas is not a function of either volume or pressure

$$\left[\text{i.e.} \left(\frac{\partial H}{\partial p}\right)_T = 0 \text{ and } \left(\frac{\partial H}{\partial V}\right)_T = 0\right]$$

but a function of temperature alone.

Since for an ideal gas, $p V = n\bar{R}T$

and
$$V - T\left(\frac{\partial V}{\partial T}\right)_p = 0$$

the relation $\mathrm{d}H = C_p \, \mathrm{d}T$ is true for any process (even when the pressure changes). However, for any other substance the relation $\mathrm{d}H = C_p \, \mathrm{d}T$ holds good only when the pressure remains constant or $\mathrm{d}p = 0$.

(b) Thermal radiation in equilibrium with the enclosing walls possesses an energy that depends only on the volume and temperature. The energy density (u), defined as the ratio of energy to volume, is a function of temperature only, or

$$u = \frac{U}{V} = f(T) \text{ only.}$$

The electromagnetic theory of radiation states that radiation is equivalent to a photon gas and it exerts a pressure, and that the pressure exerted by the black-body radiation in an enclosure is given by

$$p = \frac{u}{3}$$

Black-body radiation is thus specified by the pressure, volume, and temperature of the radiation.

Since $$U = uV \text{ and } p = \frac{u}{3}$$

$$\left(\frac{\partial U}{\partial V}\right)_T = u \text{ and } \left(\frac{\partial p}{\partial T}\right)_V = \frac{1}{3}\frac{du}{dT}$$

By substituting in the energy Eq. (11.13)

$$u = \frac{T}{3}\frac{du}{dT} - \frac{u}{3}$$

$\therefore$ $$\frac{du}{u} = 4\frac{dT}{T}$$

or $$\ln u = \ln T^4 + \ln b$$

or $$u = bT^4$$

where b is a constant. This is known as the *Stefan-Boltzmann Law*.

Since $$U = uV = VbT^4$$

$$\left(\frac{\partial U}{\partial T}\right)_V = C_v = 4VbT^3$$

and $$\left(\frac{\partial p}{\partial T}\right)_V = \frac{1}{3}\frac{du}{dT} = \frac{4}{3}bT^3$$

from the first TdS equation

$$TdS = C_v\,dT + T\left(\frac{\partial p}{\partial T}\right)_V dV = 4VbT^3\,dT + \frac{4}{3}bT^4 \cdot dV$$

For a reversible isothermal change of volume, the heat to be supplied reversibly to keep temperature constant

$$Q = \frac{4}{3}bT^4\,\Delta V$$

For a reversible adiabatic change of volume

$$\frac{4}{3}bT^4\,dV = -4VbT^3\,dT$$

$\therefore$ $$\frac{dV}{V} = -3\frac{dT}{T}$$

or $$VT^3 = \text{const.}$$

If the temperature is one-half the original temperature, the volume of black-body radiation is to be increased adiabatically eight times its original volume so that the radiation remains in equilibrium with matter at that temperature.

11.7 JOULE-KELVIN EFFECT

A gas is made to undergo continuous throttling process by a valve, as shown in Fig. 11.2. The pressures and temperatures of the gas in the insulated pipe upstream and downstream of the valve are measured with suitable manometers and thermometers.

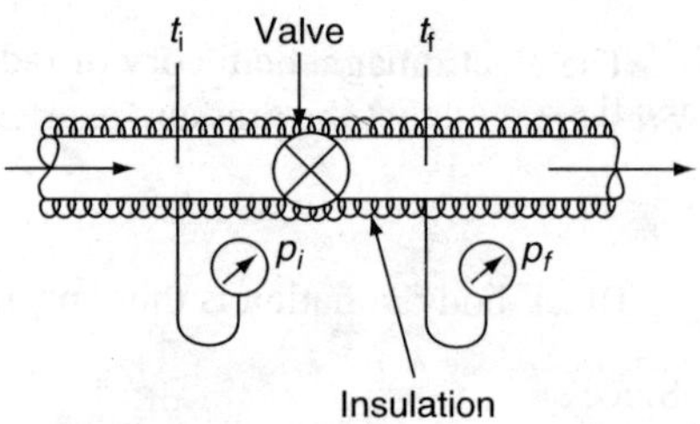

Fig. 11.2 *Joule-Thomson expansion*

Let p_i and T_i be the arbitrarily chosen pressure and temperature before throttling and let them be kept constant. By operating the valve manually, the gas is throttled successively to different pressures and temperatures p_{f1}, T_{f1}; p_{f2}, T_{f2}; p_{f3}, T_{f3} and so on. These are then plotted on the $T-p$ coordinates as shown in Fig. 11.3. All the points represent equilibrium states of some constant mass of gas, say, 1 kg, at which the gas has the *same enthalpy*.

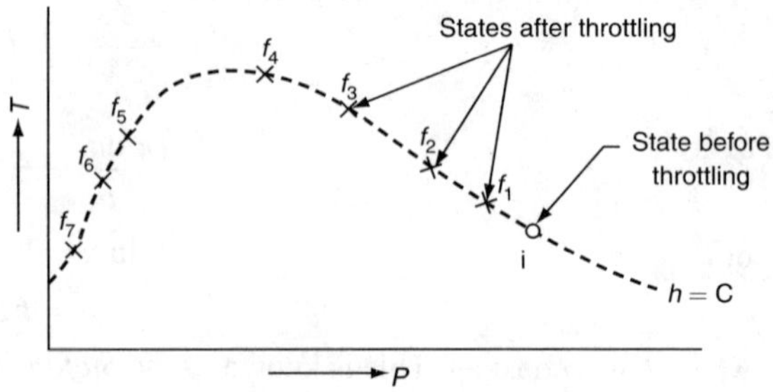

Fig. 11.3 *Isenthalpic states of a gas*

The curve passing through all these points is an isenthalpic curve or an *isenthalpe.* It is not the graph of a throttling process, but the graph through points of equal enthalpy.

The initial temperature and pressure of the gas (before throttling) are then set to new values, and by throttling to different states, a *family of isenthalpes* is obtained for the gas, as shown in Figs. 11.4 and 11.5. The curve passing through the maxima of these isenthalpes is called the *inversion curve.*

The numerical value of the slope of an isenthalpe on a $T-p$ diagram at any point is called the *Joule-Kelvin coefficient* and is denoted by μ_J. Thus the locus of all points at which μ_J is zero is the inversion curve. The region inside the inversion curve where μ_J is positive is called the *cooling region* and the region outside where μ_J is negative is called the *heating region*. So,

$$\mu_J = \left(\frac{\partial T}{\partial p}\right)_h$$

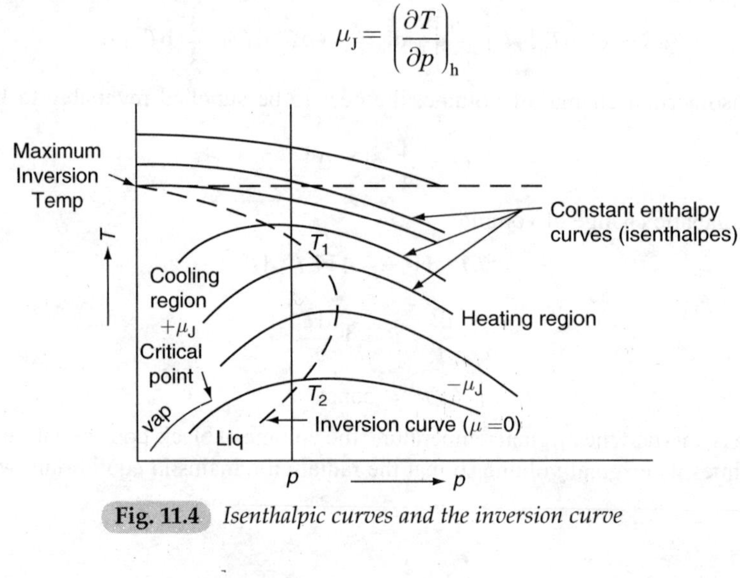

Fig. 11.4 *Isenthalpic curves and the inversion curve*

The difference in enthalpy between two neighbouring equilibrium states is

$$dh = Tds + vdp$$

and the second TdS equation (per unit mass)

$$Tds = c_p\, dT - T\left(\frac{\partial v}{\partial T}\right)_p dp$$

$$\therefore \quad dh = c_p\, dT - \left[T\left(\frac{\partial v}{\partial T}\right)_p - v\right] dp$$

The second term in the above equation stands only for a real gas, because for an ideal gas, $dh = c_p\, dT$.

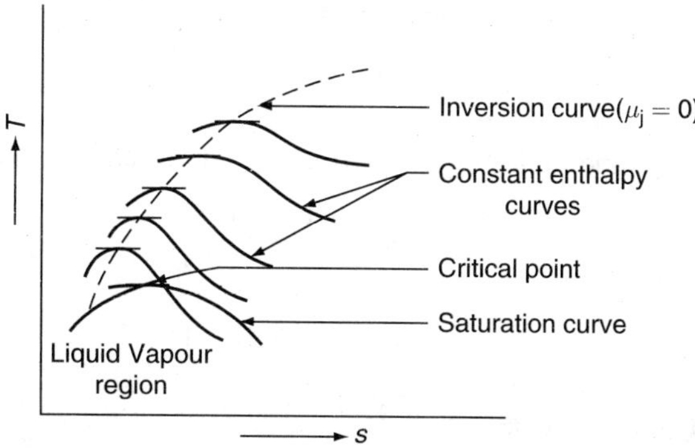

Fig. 11.5 *Inversion and saturation curves on T – s plot*

$$\mu_J = \left(\frac{\partial T}{\partial p}\right)_h = \frac{1}{c_p}\left[T\left(\frac{\partial v}{\partial T}\right)_p - v\right] \tag{11.17}$$

For an ideal gas

$$pv = RT$$

$$\therefore \quad \left(\frac{\partial v}{\partial T}\right)_p = \frac{R}{p} = \frac{v}{T}$$

$$\therefore \quad \mu_J = \frac{1}{c_p}\left(T \cdot \frac{v}{T} - v\right) = 0$$

There is no change in temperature when an ideal gas is made to undergo a Joule-Kelvin expansion (i.e. throttling).

For achieving the effect of cooling by Joule-Kelvin expansion, the initial temperature of the gas must be below the point where the inversion curve intersects the temperature axis, i.e. below the *maximum inversion temperature*. For nearly all substances, the maximum inversion temperature is above the normal ambient temperature, and hence cooling can be obtained by the Joule-Kelvin effect. In the case of hydrogen and helium, however, the gas is to be precooled in heat exchangers below the maximum inversion temperature before it is throttled. For liquefaction, the gas has to be cooled below the critical temperature.

Let the initial state of gas before throttling be at A (Fig. 11.6). The change in temperature may be positive, zero, or negative, depending upon the final pressure after throttling. If the final pressure lies between A and B, there will be a rise in temperature or heating effect. If it is at C, there will be no change in temperature. If the final pressure is below p_C, there will be a cooling effect, and if the final pressure is p_D, the temperature drop will be $(T_A - T_D)$.

Maximum temperature drop will occur if the initial state lies on the inversion curve. In Fig. 11.6, it is $(T_B - T_D)$.

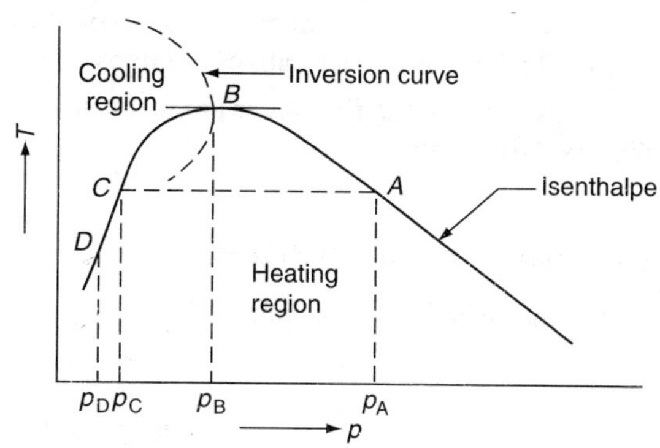

Fig. 11.6 *Maximum cooling by Joule-Kelvin expansion*

The volume expansivity is

$$\beta = \frac{1}{v}\left(\frac{\partial v}{\partial T}\right)_p$$

So the Joule-Kelvin coefficient μ_J is given by, from Eq. (11.17)

$$\mu_J = \frac{1}{c_p}\left[T\left(\frac{\partial v}{\partial T}\right)_p - v\right]$$

or

$$\mu_J = \frac{v}{c_p}\,[\beta T - 1]$$

For an ideal gas, $\beta = \frac{1}{T}$ and $\mu_J = 0$

There are two inversion temperatures for each pressure, e.g. T_1 and T_2 at pressure p (Fig. 11.4).

11.8 CLAUSIUS-CLAPEYRON EQUATION

During phase transitions like melting, vaporization and sublimation, the temperature and pressure remain constant, while the entropy and volume change. If x is the fraction of initial phase i which has been transformed into final phase f, then

$$s = (1 - x)\, s^{(i)} + x s^{(f)}$$

$$v = (1 - x)\, v^{(i)} + x v^{(f)}$$

where s and v are linear function of x.

For reversible phase transition, the heat transferred per mole (or per kg) is the latent heat, given by

$$l = T\{s^{(f)} - s^{(i)}\}$$

which indicates the change in entropy.

Now

$$dg = -s\,dT + v\,dp$$

or

$$s = -\left(\frac{\partial g}{\partial T}\right)_p$$

and

$$v = \left(\frac{\partial g}{\partial p}\right)_T$$

A *phase change of the first order* is known as any phase change that satisfies the following requirements:

(a) There are changes of entropy and volume.

(b) The first-order derivatives of Gibbs function change discontinuously.

Let us consider the first-order phase transition of one mole of a substance from phase i to phase f. Using the first TdS equation

$$TdS = c_v\, dT + T\left(\frac{\partial p}{\partial T}\right)_V dv$$

for the phase transition which is reversible, isothermal and isobaric, and integrating over the whole change of phase, and since $\left(\frac{\partial p}{\partial T}\right)_V$ is independent of v

$$T\{s^{(f)} - s^{(i)}\} = T\frac{dp}{dT}\cdot\{v^{(f)} - v^{(i)}\}$$

$$\therefore \quad \frac{dp}{dT} = \frac{s^{(f)} - s^{(i)}}{v^{(f)} - v^{(i)}} = \frac{l}{T\{v^{(f)} - v^{(i)}\}} \tag{11.18}$$

The above equation is known as the *Clausius-Clapeyron equation.*

The Clausius-Clapeyron equation can also be derived in another way.

For a reversible process at constant T and p, the Gibbs function remains constant. Therefore, for the first-order phase change at T and p

$$g^{(i)} = g^{(f)}$$

and for a phase change at $T + dT$ and $p + dp$ (Fig. 11.7)

$$g^{(i)} + dg^{(i)} = g^{(f)} + dg^{(f)}$$

Subtracting

$$dg^{(i)} = dg^{(f)}$$

or

$$-s^{(i)}\,dT + v^{(i)}\,dp = -s^{(f)}\,dT + v^{(f)}\,dp$$

$\therefore$

$$\frac{dp}{dT} = \frac{s^{(f)} - s^{(i)}}{v^{(f)} - v^{(i)}} = \frac{l}{T\left[v^{(f)} - v^{(i)}\right]}$$

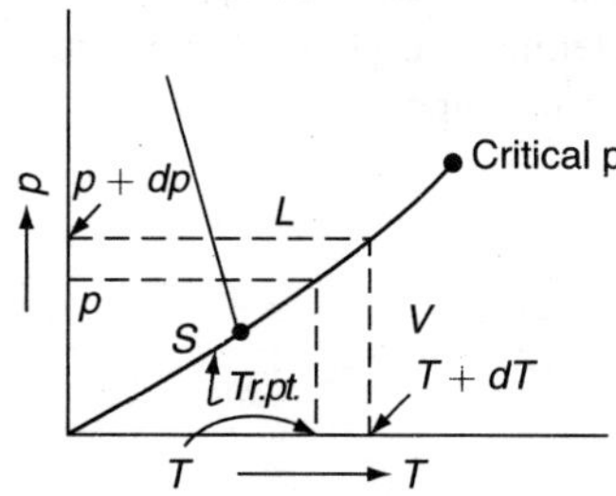

Fig. 11.7 *First order phase transition*

For *fusion*

$$\frac{dp}{dT} = \frac{l_{fu}}{T(v'' - v')}$$

where l_{fu} is the latent heat of fusion, the first prime indicates the saturated solid state, and the second prime the saturated liquid state. The slope of the fusion curve is determined by $(v'' - v')$, since l_{fu} and T are positive. If the substance expands on melting, $v'' > v'$, the slope is positive. This is the usual case. Water, however, contracts on melting and has the fusion curve with a negative slope (Fig. 11.8).

For *vaporization*

$$\frac{dp}{dT} = \frac{l_{vap}}{T(v''' - v'')}$$

where l_{vap} is the latent heat of vaporization, and the third prime indicates the saturated vapour state.

$\therefore$

$$l_{vap} = T\,\frac{dp}{dT}\,(v''' - v'')$$

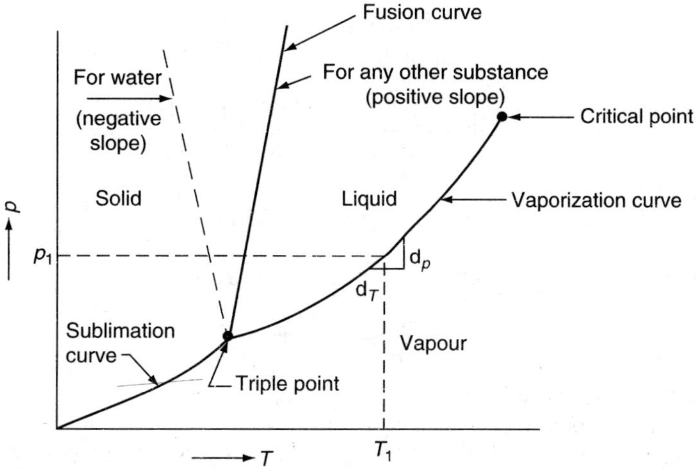

Fig. 11.8 *Phase diagram for water and any other substance on p-T coordinates*

At temperature considerably below the critical temperature, $v''' >> v''$ and using the ideal gas equation of state for vapour

$$v''' = \frac{RT}{p}$$

$$l_{vap} \cong T \cdot \frac{dp}{dT}\frac{RT}{p}$$

or

$$l_{vap} = \frac{RT^2}{p}\frac{dp}{dT} \tag{11.19}$$

If the slope dp/dT at any state (e.g. point p_1, T_1 in Fig. 11.8) is known, the latent heat of vaporization can be computed from the above equation.

The vapour pressure curve is of the form

$$\ln p = A + \frac{B}{T} + C \ln T + DT$$

where A, B, C and D are constants. By differentiating with respect to T

$$\frac{1}{p}\frac{dp}{dT} = -\frac{B}{T^2} + \frac{C}{T} + D \tag{11.20}$$

Equations (11.19) and (11.20) can be used to estimate the latent heat of vaporization.

Clapeyron's equation can also be used to estimate approximately the vapour pressure of a liquid at any arbitrary temperature in conjunction with a relation for the latent heat of a substance, known as *Trouton's rule*, which states that

$$\frac{\bar{h}_{fg}}{T_B} \cong 88 \text{ kJ/kgmol K}$$

where $\bar{h}_{fg}$ is the latent heat of vaporization in kJ/kgmol and ΔT_B is the boiling point at 1.013 bar. On substituting this into Eq. (11.19)

$$\frac{dp}{dT} = \frac{88T_B}{\bar{R}T^2}p$$

or

$$\int_{101.325}^{p} \frac{dp}{p} = \frac{88T_B}{\bar{R}}\int_{T_B}^{T}\frac{dT}{T^2}$$

$$\ln \frac{p}{101.325} = -\frac{88T_B}{\bar{R}}\left(\frac{1}{T} - \frac{1}{T_B}\right)$$

$$\therefore \qquad p = 101.325 \exp\left[\frac{88}{\bar{R}}\left(1 - \frac{T_B}{T}\right)\right] \tag{11.21}$$

This gives the vapour pressure p in kPa at any temperature T.

For *sublimation*

$$\frac{dp}{dT} = \frac{l_{sub}}{T(v''' - v')}$$

where l_{sub} is the latent heat of sublimation.

Since $v''' >> v'$, and vapour pressure is low, $v''' = \dfrac{RT}{p}$

$$\frac{dp}{dT} = \frac{l_{sub}}{T \cdot \dfrac{RT}{p}}$$

or

$$l_{sub} = -2.303\, R \frac{d(\log p)}{d(1/T)}$$

the slope of log p vs. $1/T$ curve is negative, and if it is known, l_{sub} can be estimated.

At the triple point (Fig. 9.12),

$$l_{sub} = l_{vap} + l_{fus} \tag{11.22}$$

$$\left(\frac{dp}{dT}\right)_{vap} = \frac{p_{tp} l_{vap}}{\bar{R} T_{tp}^2}$$

$$\left(\frac{dp}{dT}\right)_{sub} = \frac{p_{tp} l_{sub}}{\bar{R} T_{tp}^2}$$

Since $l_{sub} > l_{vap}$, at the triple point

$$\left(\frac{dp}{dT}\right)_{sub} > \left(\frac{dp}{dT}\right)_{vap}$$

Therefore, the slope of the sublimation curve at the triple point is greater than that of the vaporization curve (Fig. 11.8).

11.9 EVALUATION OF THERMODYNAMIC PROPERTIES FROM AN EQUATION OF STATE

Apart from calculating pressure, volume, or temperature, an equation of state can also be used to evaluate other thermodynamic properties such as internal energy, enthalpy and entropy. The property relations to be used are:

$$du = c_v\, dT + \left[T\left(\frac{\partial p}{\partial T}\right)_v - p\right] dv \tag{11.23}$$

$$dh = c_p\, dT + \left[v - T\left(\frac{\partial v}{\partial T}\right)_p\right] dp \tag{11.24}$$

$$ds = \frac{1}{T}\left[c_v\, dT + T\left(\frac{\partial p}{\partial T}\right)_v dv\right]$$

$$= \frac{1}{T}\left[c_p\, dT - T\left(\frac{\partial v}{\partial T}\right)_p dv\right] \tag{11.25}$$

Integrations of the differential relations of the properties p, v and T in the above equations are carried out with the help of an equation of state. The changes in properties are independent of the path and depend only on the end states. Let us consider that the change in enthalpy per unit mass of a gas from a reference state 0 at p_0, T_0 having enthalpy, h_0 to some other state B at p, T with enthalpy h is to be calculated (Fig. 11.9). The reversible path $0B$ may be replaced for convenience by either path 0–a–B or path 0–b–B, both also being reversible.

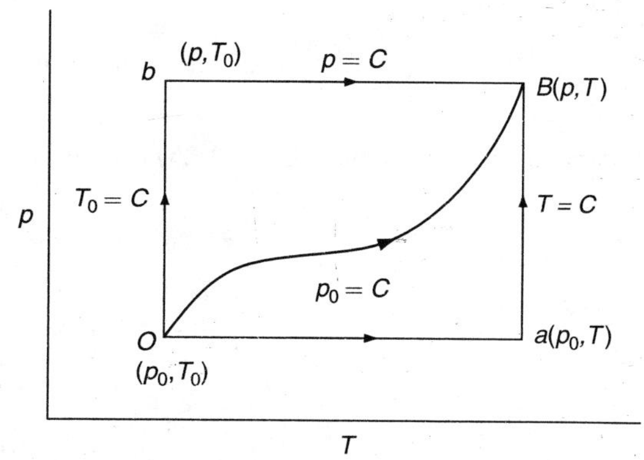

Fig. 11.9 *Processes connecting states (p_0, T_0) and (p, T)*

Path 0–*a*–*B*:
From Eq. 11.24,

$$h_a - h_0 = \int_{T_0}^{T} c_p \, dT$$

$$h - h_a = \int_{p_0}^{p} \left[v - T\left(\frac{\partial v}{\partial T}\right)_p \right]_T dp$$

On addition,

$$h - h_0 = \left[\int_{T_0}^{T} c_p dT\right]_{p_0} + \left\{\int_{p_0}^{p} \left[v - T\left(\frac{\partial v}{\partial T}\right)_p\right] dp\right\}_T \tag{11.26}$$

Similarly, for
Path 0-*b*-*B*:

$$h - h_0 = \left\{\int_{p_0}^{p} \left[v - T\left(\frac{\partial v}{\partial T}\right)_p\right] dp\right\}_{T_0} + \left[\int_{T_0}^{T} c_p \, dT\right]_p \tag{11.27}$$

Equation (11.26) is preferred to Eq. (11.27) since c_p at lower pressure can be conveniently measured.

Now,

$$\int_{p_0 v_a}^{pv} d\,(pv) = \int_{v_a}^{v} p \, dv + \int_{p_0}^{p} v \, dp$$

$$\int_{p_0}^{p} v \, dp = pv - p_0 v_a - \left[\int_{v_a}^{v} p dv\right]_T \tag{11.28}$$

Again,

$$\left[\frac{\partial v}{\partial T}\right]_p \left[\frac{\partial T}{\partial p}\right]_v \left[\frac{\partial p}{\partial v}\right]_T = -1$$

$$\therefore \quad \left[\frac{\partial v}{\partial T}\right]_p = -\left[\frac{\partial p}{\partial T}\right]_v \left[\frac{\partial v}{\partial p}\right]_T$$

Substituting in Eq. (11.26),

$$h - h_0 = \left[\int_{T_0}^{T} c_p dT\right]_{p_0} + pv - p_0 v_a - \left[\int_{v_a}^{v} p dv\right]_T + \left\{\int_{p_0}^{p} T\left(\frac{\partial p}{\partial T}\right)_p \left(\frac{\partial v}{\partial p}\right)_T dp\right\}_T$$

$$= \left[\int_{T_0}^{T} c_p dT\right]_{p_0} + pv - p_0 v_a - \left\{\int_{v_a}^{v} \left[p - T\left(\frac{\partial p}{\partial T}\right)_v\right] dv\right\}_T \tag{11.29}$$

To find the entropy change, Eq. (11.25) is integrated to yield:

$$s - s_0 = \left[\int_{T_0}^{T} c_p \frac{dT}{T}\right]_{p_0} - \left[\int_{p_0}^{p} \left(\frac{\partial v}{\partial T}\right)_p dp\right]_T$$

$$= \left[\int_{T_0}^{T} c_p \frac{dT}{T}\right]_{p_0} - \left[-\int_{p_0}^{p} \left(\frac{\partial p}{\partial T}\right)_u \left(\frac{\partial v}{\partial p}\right)_T dp\right]_T$$

$$= \left[\int_{T_0}^{T} c_p \frac{dT}{T}\right]_{p_0} + \left[\int_{v_a}^{v} \left(\frac{\partial p}{\partial T}\right)_v dv\right]_T \tag{11.30}$$

11.10 GENERAL THERMODYNAMIC CONSIDERATIONS ON AN EQUATION OF STATE

Certain general characteristics are common to all gases. These must be clearly observed in the developing and testing of an equation of state. It is edifying to discuss briefly some of the more important ones:

(i) Any equation of state must reduce to the ideal gas equation as pressure approaches zero at any temperature. This is clearly seen in a generalized compressibility factor chart in which all isotherms converge to the point $z = 1$ at zero pressure. Therefore,

$$\lim_{p\to 0}\left[\frac{pv}{RT}\right] = 1 \text{ at any temperature}$$

Also, as seen from Fig. 10.6, the reduced isotherms approach the line $z = 1$ as the temperature approaches infinity, or:

$$\lim_{T\to\infty}\left[\frac{pv}{RT}\right] = 1 \text{ at any pressure.}$$

(ii) The critical isotherm of an equation of state should have a point of inflection at the critical point on p–v coordinates, or

$$\left[\frac{\partial p}{\partial v}\right]_{T=T_c} = 0 \text{ and } \left[\frac{\partial^2 p}{\partial v^2}\right]_{T=T_c} = 0$$

(iii) The isochores of an equation of state on a p–T diagram should be essentially straight, or:

$$\left[\frac{\partial p}{\partial T}\right]_v = \text{constant}, \quad \left[\frac{\partial^2 p}{\partial T^2}\right]_v = 0 \text{ as } p \to 0, \text{ or as } T \to \infty.$$

An equation of state can predict the slope of the critical isochores of a fluid. This slope is identical with the slope of the vaporization curve at the critical point. From the Clapegron equation, $dp/dT = \Delta s/\Delta v$, the slope of the vaporization curve at the critical point becomes:

$$\frac{dp}{dT} = \left[\frac{\partial s}{\partial v}\right]_{T_c} = \left[\frac{\partial p}{\partial T}\right]_{v_c} \quad \text{(by Maxwell's equation)}$$

Therefore, the vapour-pressure slope at the critical point, dp/dT, is equal to the slope of the critical isochore $(\partial p/\partial T)_{vc}$ (Fig. 11.10).

(iv) The slopes of the isotherms of an equation of state on a Z–p compressibility factor chart as p approaches zero should be negative at lower temperatures and positive at higher temperatures. At the Boyle temperature, the slope is zero as p approaches zero, or

$$\lim_{p\to 0}\left[\frac{\partial z}{\partial p}\right]_T = 0 \text{ at } T = T_B$$

An equation of state should predict the Boyle temperature which is about 2.54 T_c for many gases.

An isotherm of maximum slope on the Z–p plot as p approaches zero, called the *foldback isotherm*, which is about 5T_c for many gases, should be predicted by an equation of state, for which:

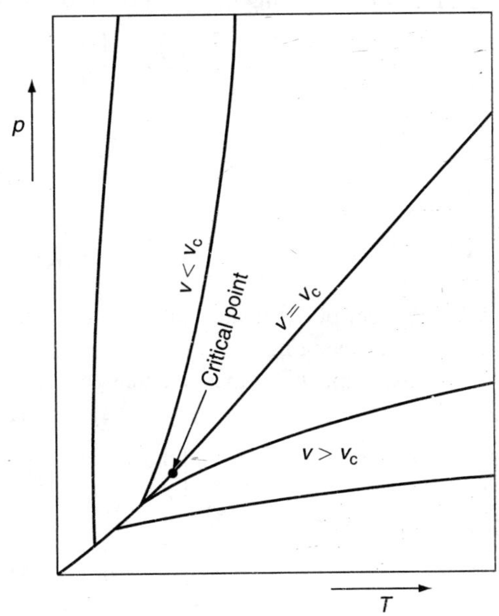

Fig. 11.10 *Pressure-temperature diagram with isochoric lines*

$$\lim_{p \to 0} \left[\frac{\partial^2 z}{\partial T \cdot \partial p} \right] = 0 \text{ at } T = T_F$$

where T_F is the foldback temperature (Fig. 10.10 a). As temperature increases beyond T_F the slope of the isotherm decreases, but always remains positive.

(v) An equation of state should predict the Joule-Thomson coefficient, which is

$$\mu_J = \frac{1}{c_p} \left[T \left(\frac{\partial v}{\partial T} \right)_p - v \right] = \frac{RT^2}{pc_p} \left(\frac{\partial z}{\partial T} \right)_p$$

For the inversion curve, $\mu_J = 0$,

or,
$$\left(\frac{\partial z}{\partial T} \right)_p = 0$$

11.11 MIXTURES OF VARIABLE COMPOSITION

Let us consider a system containing a mixture of substances 1, 2, 3 ... K. If some quantities of a substance are added to the system, the energy of the system will increase. Thus for a system of variable composition, the internal energy depends not only on S and V, but also on the number of moles (or mass) of various constituents of the system.

Thus
$$U = U(S, V, n_1, n_2, ..., n_K)$$

where $n_1, n_2, ..., n_K$ are the number of moles of substances 1, 2, ..., K. The composition may change not only due to addition or subtraction, but also due to chemical reaction and inter-phase mass transfer. For a small change in U, assuming the function to be continuous.

$$dU = \left(\frac{\partial U}{\partial S} \right)_{V, n_1, n_2, ..., n_K} dS + \left(\frac{\partial U}{\partial V} \right)_{S, n_1, n_2, ..., n_K} dV + \left(\frac{\partial U}{\partial n_1} \right)_{S, V, n_2, ..., n_K} dn_1 + \left(\frac{\partial U}{\partial n_2} \right)_{S, V, n_1, n_3, ..., n_K} dn_2$$

$$+ ... + \left(\frac{\partial U}{\partial n_K} \right)_{S, V, n_1, n_2, ..., n_{K-1}} dn_K$$

or
$$dU = \left(\frac{\partial U}{\partial S} \right)_{V, n_i} dS + \left(\frac{\partial U}{\partial V} \right)_{S, n_i} dV + \sum_{i=1}^{K} \left(\frac{\partial U}{\partial n_i} \right)_{S, V, n_j} dn_i$$

where subscript i indicates any substance and subscript j any other substance except the one whose number of moles is changing.

If the composition does not change

$$dU = TdS - pdV$$

$$\therefore \quad \left(\frac{\partial U}{\partial S} \right)_{V, n_i} = T, \text{ and } \left(\frac{\partial U}{\partial V} \right)_{S, n_i} = -p$$

$$\therefore \quad dU = TdS - pdV + \sum_{i=1}^{K} \left(\frac{\partial U}{\partial n_i} \right)_{S, V, n_j} dn_i \tag{11.31}$$

Molal *chemical potential*, μ, of component i is defined as

$$\mu_i = \left(\frac{\partial U}{\partial n_i} \right)_{S, V, n_j}$$

signifying the change in internal energy per unit mole of component *i* when *S*, *V*, and the number of moles of all other components are constant.

$$\therefore \qquad dU = TdS - pdV + \sum_{i=1}^{K} \mu_i \, dn_i$$

or
$$TdS = dU + pdV - \sum_{i=1}^{K} \mu_i \, dn_i \tag{11.32}$$

This is known as *Gibbs entropy equation*.

In a similar manner

$$G = G(p, T, n_1, n_2, ..., n_K)$$

or
$$dG = \left(\frac{\partial G}{\partial p}\right)_{T, n_i} dp + \left(\frac{\partial G}{\partial T}\right)_{p, n_i} dT + \sum_{i=1}^{K} \left(\frac{\partial G}{\partial n_i}\right)_{T, p, n_j} dn_i$$

$$= Vdp - SdT + \sum_{i=1}^{K} \left(\frac{\partial G}{\partial n_i}\right)_{T, p, n_j} dn_i \tag{11.33}$$

Since
$$G = U + pV - TS$$

$$d(U + pV - TS) = VdP - SdT + \sum_{i=1}^{K} \left(\frac{\partial G}{\partial n_i}\right)_{T, p, n_j} dn_i$$

or
$$dU + pdV + Vdp - TdS - SdT = Vdp - SdT + \sum_{i=1}^{K} \left(\frac{\partial G}{\partial n_i}\right)_{T, p, n_j} dn_i$$

or
$$dU = TdS - pdV + \sum_{i=1}^{K} \left(\frac{\partial G}{\partial n_i}\right)_{T, p, n_j} dn_i$$

Comparing this equation with Eq. (11.31)

$$\left(\frac{\partial U}{\partial n_i}\right)_{S, V, n_j} = \left(\frac{\partial G}{\partial n_i}\right)_{T, p, n_j} = \mu_i$$

$\therefore$ Equation (11.33) becomes

$$dG = Vdp - SdT + \sum_{i=1}^{K} \mu_i \, dn_i$$

Similar equations can be obtained for changes in *H* and *F*.

Thus

$$dU = TdS - pdV + \sum_{i=1}^{K} \mu_i \, dn_i$$

$$dG = Vdp - SdT + \sum_{i=1}^{K} \mu_i \, dn_i$$

$$dH = TdS + Vdp + \sum_{i=1}^{K} \mu_i \, dn_i \tag{11.34}$$

$$dF = -SdT - pdV + \sum_{i=1}^{K} \mu_i \, dn_i$$

where

$$\mu_i = \left(\frac{\partial U}{\partial n_i}\right)_{S,V,n_j} = \left(\frac{\partial G}{\partial n_i}\right)_{T,p,n_j} = \left(\frac{\partial H}{\partial n_i}\right)_{S,p,n_j} = \left(\frac{\partial F}{\partial n_i}\right)_{T,V,n_j} \tag{11.35}$$

Chemical potential is an intensive property.

Let us consider a homogeneous phase of a multi-component system, for which

$$dU = TdS - pdV + \sum_{i=1}^{K} \mu_i \, dn_i$$

If the phase is enlarged in size, U, S, and V will increase, whereas T, p and μ will remain the same. Thus

$$\Delta U = T\Delta S - p\Delta V + \Sigma \mu_i \, \Delta n_i$$

Let the system be enlarged to K-times the original size. Then

$$\Delta U = KU - U = (K-1)U$$

$$\Delta S = KS - S = (K-1)S$$

$$\Delta V = (K-1)V$$

$$\Delta n_i = (K-1)\, n_i$$

Substituting

$$(K-1)\, U = T(K-1)\, S - p\,(K-1)\, V + \Sigma \mu_i \,(K-1)\, n_i$$

$$\therefore \quad U = TS - pV + \Sigma\, \mu_i n_i$$

$$\therefore \quad G_{T,p} = \Sigma\, \mu_i n_i \tag{11.36}$$

Let us now find a relationship if there is a simultaneous change in intensive property. Differentiating Eq. (11.36)

$$dG = \Sigma\, n_i \, d\mu_i + \Sigma\, \mu_i dn_i \tag{11.37}$$

at constant T and p, with only μ changing.

When T and p change

$$dG = -SdT + Vdp + \Sigma \mu_i dn_i \tag{11.38}$$

Combining Eqs (11.37) and (11.38)

$$-SdT + Vdp - \Sigma\, n_i \, d\,\mu_i = 0 \tag{11.39}$$

This is known as *Gibbs-Duhem equation*, which shows the necessary relationship for simultaneous changes in T, p, and μ.

Now

$$G_{T\cdot p} = \Sigma\, \mu_i n_i = \mu_1 n_1 + \mu_2 n_2 \dots + \mu_K n_K$$

For a phase consisting of only one constituent

$$G = \mu n$$

or

$$\mu = \frac{G}{n} = g$$

i.e. the chemical potential is the molar Gibbs function and is a function of T and p only.

For a single phase, multi-component system, μ_i is a function of T, p, and the mole fraction x_i.

11.12 CONDITIONS OF EQUILIBRIUM OF A HETEROGENEOUS SYSTEM

Let us consider a heterogeneous system of volume V, in which several homogeneous phases ($\phi = a, b, ..., r$) exist in equilibrium. Let us suppose that each phase consists of i (= 1, 2, ..., C) constituents and that the number of constituents in any phase is different from the others.

Within each phase, a change in internal energy is accompanied by a change in entropy, volume and composition, according to

$$dU_\phi = T_\phi\, dS_\phi - p_\phi\, dV_\phi + \sum_{i=1}^{C} (\mu_i\, dn_i)_\phi$$

A change in the internal energy of the entire system can, therefore, be expressed as

$$\sum_{\phi=a}^{r} dU_\phi = \sum_{\phi=a}^{r} T_\phi\, dS_\phi - \sum_{\phi=a}^{r} p_\phi\, dV_\phi + \sum_{\phi=a}^{r} \sum_{i=1}^{c} (\mu_i\, dn_i)_\phi \tag{11.40}$$

Also, a change in the internal energy of the entire system involves changes in the internal energy of the constituent phases.

$$dU = dU_a + dU_b + ... + dU_r = \sum_{\phi=a}^{r} dU_\phi$$

Likewise, changes in the volume, entropy, or chemical composition of the entire system result from contributions from each of the phases

$$dV = dV_a + dV_b + ... + dV_r = \sum_{\phi=a}^{r} dV_\phi$$

$$dS = dS_a + dS_b + ... + dS_r = \sum_{\phi=a}^{r} dS_\phi$$

$$dn = dn_a + dn_b + ... + dn_r = \sum_{\phi=a}^{r} dn_\phi$$

In a closed system in equilibrium, the internal energy, volume, entropy, and mass are constant.

$$\therefore \quad dU = dV = dS = dn = 0$$

or

$$dU_a = -(dU_b + ... + dU_r) = -\sum_j dU_j$$

$$dV_a = -\sum_j dV_j$$

$$dS_a = -\sum_j dS_j \tag{11.41}$$

$$dn_a = -\sum_j dn_j$$

where subscript j includes all phases except phase a.

Equation (11.40) can be written in terms of j independent variables and the dependent variable a Eq. (11.41).

$$\left[T_a\, dS_a + \sum_j T_j\, dS_j\right] - \left[p_a\, dV_a + \sum_j p_j\, dV_j\right] + \left[\sum_i (\mu_i\, dn_i)_a + \sum_j \sum_j (\mu_i\, dn_i)_j\right] = 0$$

Substituting from Eq. (11.41)

$$\left[-T_a \sum_j dS_j + \sum_j T_j\, dS_j\right] - \left[-p_a \sum_j dV_j + \sum_j p_j \times dV_j\right] + \left[-\sum_j \sum_t \mu_{ia}\, dn_{ij} + \sum_j \sum_i (\mu_i\, dn_i)_j\right] = 0$$

where subscript $j\ a$ refers to component i of phase a.

Rearranging and combining the coefficients of the independent variables, dS_j, dV_j and dn_j, gives

$$\sum_j (T_j - T_a) dS_j - \sum_j (p_j - p_a) dV_j + \sum_j \sum_i (\mu_{ij} - \mu_{ia})\, dn_{ij} = 0$$

But since dS_j, dV_j, and dn_j are independent, their coefficients must each be equal to zero.

$$\therefore \quad T_j = T_a, \quad p_j = p_a, \quad \mu_{ij} = \mu_{ia} \tag{11.42}$$

These equations represent conditions that exist when the system is in thermal, mechanical, and chemical equilibrium. The temperature and pressure of phase a must be equal to those of all other phases, and the chemical potential of the ith component in phase a must be equal to the chemical potential of the same component in all other phases.

11.13 GIBBS PHASE RULE

Let us consider a heterogeneous system of C chemical constituents which do not combine chemically with one another. Let us suppose that there are ϕ phases, and every constituent is present in each phase. The constituents are denoted by subscripts and the phases by superscripts. The Gibbs function of the whole heterogeneous system is

$$G_{T,p} = \sum_{i=1}^{C} n_i^{(1)} \mu_i^{(1)} + \sum_{i=1}^{C} n_i^{(2)} \mu_i^{(2)} + \ldots + \sum_{i=1}^{C} n_i^{(\phi)} \mu_i^{(\phi)}$$

G is a function of T, p, and the n's of which there are $C\phi$ in number. Since there are no chemical reactions, the only way in which the n's may change is by the transport of the constituents from one phase to another. In this case the total number of moles of each constituent will remain constant.

$$n_1^{(1)} + n_1^{(2)} + \ldots + n_1^{(\phi)} = \text{constant}$$

$$n_2^{(1)} + n_2^{(2)} + \ldots + n_2^{(\phi)} = \text{constant}$$

$$\cdots\cdots\cdots\cdots\cdots\cdots$$

$$n_C^{(1)} + n_C^{(2)} + \ldots + n_C^{(\phi)} = \text{constant}$$

These are the *equations of constraint.*

At chemical equilibrium, G will be rendered a minimum at constant T and p, subject to these equations of constraint. At equilibrium, from Eq. (11.42).

$$\mu_{ij} = \mu_{ia}$$

$$\mu_1^{(1)} = \mu_1^{(2)} = \ldots = \mu_1^{(\phi)}$$

$$\mu_2^{(1)} = \mu_2^{(2)} = \ldots = \mu_2^{(\phi)} \tag{11.43}$$

$$\cdots\cdots\cdots\cdots\cdots\cdots$$

$$\mu_c^{(1)} = \mu_c^{(2)} = \ldots = \mu_c^{(\phi)}$$

These are known as the *equations of phase equilibrium*. The equations of the phase equilibrium of one constituent are $(\phi - 1)$ in number. Therefore, for C constituents, there are $C(\phi - 1)$ such equations.

When equilibrium has been reached, there is no transport of matter from one phase to another. Therefore, in each phase, $\Sigma x = 1$. For ϕ phases, there are ϕ such equations available.

The state of the system at equilibrium is determined by the temperature, pressure, and $C\phi$ mole fractions. Therefore

Total number of variables $= C\phi + 2$.

Among these variables, there are $C(\phi - 1)$ equations of phase equilibrium and ϕ equations of $\Sigma x = 1$ type. Therefore

Total number of equations $= C(\phi - 1) + \phi$

If the number of variables is equal to the number of equations, the system is *nonvariant*. If the number of variables exceeds the number of equations by one, then the system is called *monovariant* and is said to have a variance of 1.

The excess of variables over equations is called the *variance*, *f*. Thus

$$f = (C\phi + 2) - [C(\phi - 1) + \phi]$$

or

$$f = C - \phi + 2 \tag{11.44}$$

This is known as the *Gibbs Phase Rule* for a non-reactive system. The variance '*f*' is also known as the *degree of freedom*.

For a pure substance existing in a single phase, $C = 1$, $\phi = 1$, and therefore, the variance is 2. There are two properties required to be known to fix up the state of the system at equilibrium.

If $C = 1$, $\phi = 2$, then $f = 1$, i.e. only one property is required to fix up the state of a single-component two-phase system.

If $C = 1$, $\phi = 3$, then $f = 0$. The state is thus unique for a substance; and refers to the triple point where all the three phases exist in equilibrium.

11.14 TYPES OF EQUILIBRIUM

The thermodynamic potential which controls equilibrium in a system depends on the particular *constraints* imposed on the system. Let $\text{đ}Q$ be the amount of heat transfer involved between the system and the reservoir in an infinitesimal irreversible process (Fig. 11.11). Let dS denote the entropy change of the system and dS_0 the entropy change of the reservoir. Then, from the entropy principle

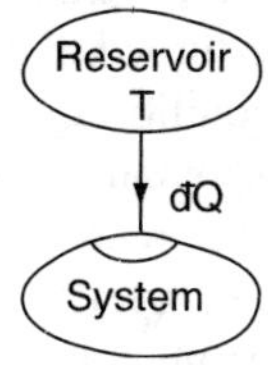

Fig. 11.11 *Heat interaction between a system and its surroundings*

$$dS_0 + dS > 0$$

Since

$$dS_0 = -\frac{\text{đ}Q}{T} \quad -\frac{\text{đ}Q}{T} + dS > 0$$

or

$$\text{đ}Q - TdS < 0$$

During the infinitesimal process, the internal energy of the system changes by an amount dU, and an amount of work pdV is performed. So, by the first law

$$\text{đ}Q = dU + pdV$$

Thus the inequality becomes

$$dU + pdV - TdS < 0 \tag{11.45}$$

If the constraints are constant U and V, then the Eq. (11.46) reduces to

$$dS > 0$$

The condition of constant U and V refers to an isolated system. Therefore, entropy is the critical parameter to determine the state of thermodynamic equilibrium of an isolated system. The entropy of an isolated system always increases and reaches a maximum value when equilibrium is reached.

If the constraints imposed on the system are constant T and V, the Eq. (11.45) reduces to

$$dU - d(TS) < 0$$

or

$$d(U - TS) < 0$$
$$dF < 0$$

which expresses that the Helmholtz function decreases, becoming a minimum at the final equilibrium state.

If the constraints are constant T and p, the Eq. (11.46) becomes

$$dU + d(pV) - d(TS) < 0$$
$$d(U + pV - TS) < 0$$
$$dG < 0$$

The Gibbs function of a system at constant T and p decreases during an irreversible process, becoming a minimum at the final equilibrium state. For a system constrained in a process to constant T and p, G is the critical parameter to determine the state of equilibrium.

The thermodynamic potential and the corresponding constrained variables are shown below.

S	U	V
H		F
P	G	T

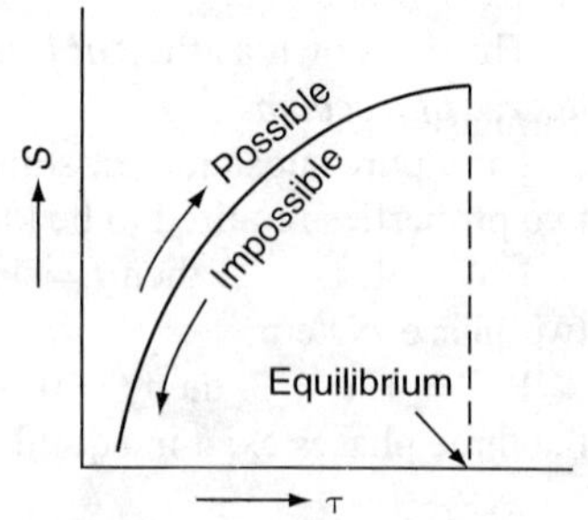

Fig. 11.12 Possible process for an isolated system

This trend of G, F, or S establishes four types of equilibrium, namely (a) stable, (b) neutral, (c) unstable, and (d) metastable.

A system is said to be in a state of *stable* equilibrium if, when the state is perturbed, the system returns to its original state. A system is not in equilibrium if there is a spontaneous change in the state. If there is a spontaneous change in the system, the entropy of the system increases and reaches a maximum when the equilibrium condition is reached (Fig. 11.12). Both A and B (Fig. 11.13) are assumed to be at the same temperature T. Let there be some spontaneous change; the temperature of A rises to $T + dT_1$, and that of B decreases to $T - dT_2$. For simplicity, let the heat capacities of the bodies be the same, so that $dT_1 = dT_2 = dT$. If $đQ$ is the heat interaction involved, then the entropy change

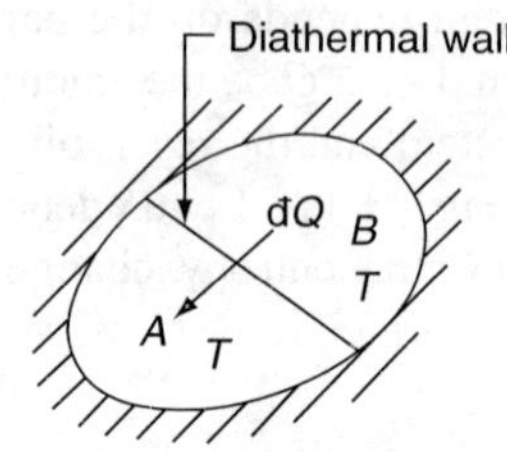

Fig. 11.13 Spontaneous changes in A and B due to heat interaction

$$dS_A = \frac{đQ}{T + dT}, \quad dS_B = -\frac{đQ}{T - dT}$$

$$\therefore \quad dS = dS_A + dS_B = đQ\left[\frac{1}{T + dT} - \frac{1}{T - dT}\right] = -\frac{2 \cdot dT}{T^2} \cdot đQ$$

So there is a decrease in entropy for the isolated system of A and B together. It is thus clear that the variation in temperature dT cannot take place. The system, therefore, exists in a stable equilibrium condition. Perturbation of the state produces an absurd situation and the system must revert to the original stable state. It may be observed:

If for all the possible variations in state of the isolated system, there is a negative change in entropy, then the system is in stable equilibrium.

$$\begin{aligned} &(dS)_{U,V} > 0 && \text{Spontaneous change} \\ &(dS)_{U,V} = 0 && \text{Equilibrium} \\ &(dS)_{U,V} < 0 && \text{Criterion of stability} \end{aligned} \qquad (11.46)$$

Similarly

$$(dG)_{p,T} < 0, (dF)_{T,V} < 0 \quad \text{Spontaneous change}$$
$$(dG)_{p,T} = 0, (dF)_{T,V} = 0 \quad \text{Equilibrium} \qquad (11.47)$$
$$(dG)_{p,T} > 0, (dF)_{T \times V} > 0 \quad \text{Criterion of stability}$$

A system is in a state of *stable equilibrium* if, for any finite variation of the system at constant T and p, G increases, i.e. the stable equilibrium state corresponds to the minimum value of G.

A system is said to be in a state of *neutral equilibrium* when the thermodynamic criterion of equilibrium (G, F, S, U, or H) remains at constant value for all possible variations of finite magnitude. If perturbed, the system does not revert to the original state.

For a system at constant T and p, the criterion of neutral equilibrium is

$$\delta\, G_{T,p} = 0$$

Similarly

$$\delta\, F_{T,V} = 0,\ \delta\, H_{S,p} = 0,\ \delta\, U_{S,V} = 0,\ \delta\, S_{U,V} = 0$$

A system is in a state of *unstable equilibrium* when the thermodynamic criterion is neither an extremum nor a constant value for all possible variations in the system. If the system is in unstable equilibrium, there will be a spontaneous change accompanied by

$$\delta G_{T,p} < 0,\ \delta F_{T,V} < 0,\ \delta U_{S,V} < 0,\ \delta H_{S,p} < 0,\ \delta S_{U,V} > 0$$

A system is in a state of *metastable equilibrium* if it is stable to small but not to large disturbances. A mixture of oxygen and hydrogen is in a metastable equilibrium. A little spark may start a chemical reaction. Such a mixture is not in its most stable state, even though in the absence of a spark it appears to be stable.

Figure 11.14 shows different types of equilibrium together with their mechanical analogies. S has been used as the criterion for equilibrium.

11.15 LOCAL EQUILIBRIUM CONDITIONS

Let an arbitrary division of an isolated system be considered, such that

$$S = S_1 + S_2,\ U = U_1 + U_2$$

Then for equilibrium, it must satisfy the condition

$$(\delta S)_{U,V} = 0$$

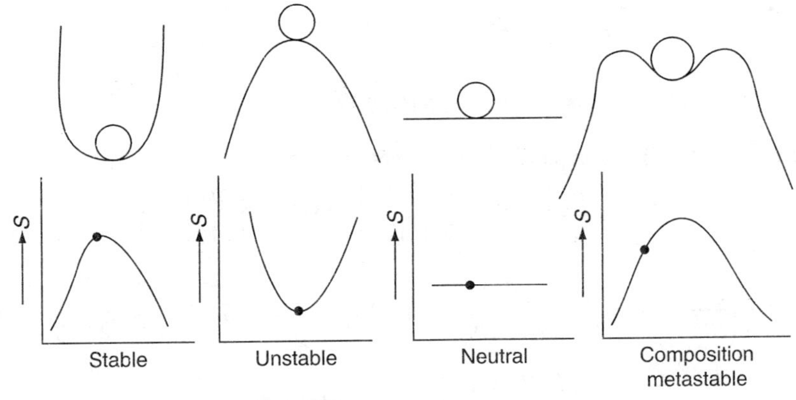

Fig. 11.14 *Types of equilibrium*

to first order in small displacements (otherwise δS could be made positive because of higher order terms). Now to the first order in a very small change

$$\delta_1 S = \left(\frac{\partial S}{\partial U_1}\right)_V \delta U_1 + \left(\frac{\partial S}{\partial U_2}\right)_V \delta U_2 + \left(\frac{\partial S}{\partial U_1}\right)_u \delta V_1 + \left(\frac{\partial S}{\partial V_2}\right)_u \delta V_2$$

Now

$$TdS = dU + pdV$$

$$\therefore \quad \left(\frac{\partial S}{\partial U}\right)_V = \frac{1}{T}, \left(\frac{\partial S}{\partial V}\right)_U = \frac{p}{T}$$

$$\therefore \quad \delta_1 S = \frac{1}{T_1}\,\delta U_1 + \frac{1}{T_2}\,\delta U_2 + \frac{p_1}{T_1}\,\delta V_1 + \frac{p_2}{T_2}\,\delta V_2$$

Again

$$\delta U_1 = -\,\delta U_2 \text{ and } \delta V_1 = -\,\delta V_2$$

$$\therefore \quad \delta_1 S = \left(\frac{1}{T_1} - \frac{1}{T_2}\right)\delta U_1 + \left(\frac{p_1}{T_1} - \frac{p_2}{T_2}\right)\delta V_1 + \text{Second order terms}$$

When

$$\delta_1 S = 0, \text{ at equilibrium}$$

$$T_1 = T_2, p_1 = p_2$$

11.16 CONDITIONS OF STABILITY

At equilibrium, $S = S_{max}$, $F = F_{min}$, $G = G_{min}$, and $SS = 0$, $\delta F = 0$; $\delta G = 0$; these are necessary but not sufficient conditions for equilibrium. To prove that S is a maximum, and G or F a minimum, it must satisfy

$$\delta^2 S < 0, \delta^2 F > 0, \delta^2 G > 0$$

If the system is perturbed, and for any infinitesimal change of the system

$$(\delta S)_{U,V} < 0, (\delta G)_{p,T} > 0, (\delta F)_{T,V} > 0$$

it represents the stability of the system. The system must revert to the original state.

For a spontaneous change, from Eq. (11.46)

$$\delta U + p\delta V - T\delta S < 0$$

For stability

$$\delta U + p\delta V - T\delta S > 0$$

Let us choose $U = U(S, V)$ and expand δU in powers of δV and δS.

$$\delta U = \left(\frac{\partial U}{\partial S}\right)_V \delta S + \frac{1}{2}\left(\frac{\partial^2 U}{\partial S^2}\right)_V (\delta S)^2 + \left(\frac{\partial U}{\partial V}\right)_S \delta V$$

$$+\frac{1}{2}\left(\frac{\partial^2 U}{\partial V^2}\right)_S (\delta V)^2 + \frac{\partial^2 U}{\partial V \cdot \partial S} \times \delta V \cdot \delta S + \ldots = T\delta S - p\delta V + \frac{1}{2}\left(\frac{\delta^2 U}{\delta S^2}\right)_V (\delta S)^2$$

$$+\frac{1}{2}\left(\frac{\partial^2 U}{\partial V^2}\right)_S (\delta V)^2 + \frac{\partial^2 U}{\partial V \cdot \partial S}\,\delta V \cdot \delta S + \ldots$$

The third order and higher terms are neglected.

Since $\delta U + p\delta v - T\delta S > 0$, it must satisfy the conditions given below

$$\left(\frac{\partial^2 U}{\partial S^2}\right)_V > 0, \left(\frac{\partial^2 U}{\partial V^2}\right)_S > 0, \frac{\partial^2 U}{\partial V \cdot \partial S} > 0$$

These inequalities indicate how the signs of some important physical quantities become restricted for a system to be stable.

Since
$$\left(\frac{\partial U}{\partial S}\right)_V = T$$

$$\left(\frac{\partial^2 U}{\partial S^2}\right)_V = \left(\frac{\partial T}{\partial S}\right)_V = \frac{T}{C_v}$$

$\therefore$
$$\frac{T}{C_v} > 0$$

Since $T > 0$ K

$\therefore$
$$C_v > 0 \tag{11.48}$$

which is the condition of *thermal stability*.

Also
$$\left(\frac{\partial U}{\partial V}\right)_S = -p$$

$$\left(\frac{\partial^2 U}{\partial V^2}\right)_S = -\left(\frac{\partial p}{\partial V}\right)_S$$

$\therefore$
$$\left(\frac{\partial p}{\partial V}\right)_S < 0 \tag{11.49}$$

i.e. the adiabatic bulk modulus must be negative.

Similarly, if $F = F(T \cdot V)$, then by Taylor's expansion, and using appropriate substitution

$$\delta F = -S\delta T - p\delta V + \frac{1}{2}\left(\frac{\partial^2 F}{\partial V^2}\right)_T (\delta V)^2 + \frac{1}{2}\left(\frac{\partial^2 F}{\partial T^2}\right)_S (\delta T)^2 + \frac{\partial^2 \cdot F}{\partial V \cdot \partial T} \cdot \delta V \cdot \delta T + \ldots$$

For stability

$$\delta F + S\delta T + p\delta V > 0$$

$$\left(\frac{\partial^2 F}{\partial V^2}\right)_T > 0$$

Again
$$\left(\frac{\partial F}{\partial V}\right)_T = -p$$

$\therefore$
$$\left(\frac{\partial^2 F}{\partial V^2}\right)_T = -\left(\frac{\partial p}{\partial V}\right)_T$$

$$\therefore \qquad \left(\frac{\partial p}{\partial V}\right)_T < 0 \qquad (11.50)$$

which is known as the condition of *mechanical stability*. The isothermal bulk modulus must also be negative.

Solved Examples

Example 11.1

(a) Derive the equation

$$\left(\frac{\partial C_p}{\partial p}\right)_T = -T\left(\frac{\partial^2 V}{\partial T^2}\right)_p$$

(b) Prove that C_p *of an ideal gas is a function of T only.*

(c) In the case of a gas obeying the equation of state

$$\frac{pv}{RT} = 1 + B'p$$

where B' is a function of T only, show that

$$C_p = -\bar{R}\,Tp\frac{d^2}{dT^2}(B'T) + (C_p)_0$$

where $(C_p)_0$ is the value at very low pressures.

Solution

(a)

$$C_p = T\left(\frac{\partial S}{\partial T}\right)_p$$

$$\left(\frac{\partial C_p}{\partial p}\right)_p = T\left(\frac{\partial^2 S}{\partial T \cdot \partial p}\right)$$

Now $\left(\frac{\partial S}{\partial p}\right)_T = -\left(\frac{\partial V}{\partial T}\right)_p$, by Maxwell's relation

$$\therefore \qquad \frac{\partial^2 S}{\partial p \cdot \partial T} = -\left(\frac{\partial^2 V}{\partial T^2}\right)_p$$

$$\therefore \qquad \left(\frac{\partial C_p}{\partial p}\right)_T = -T\left(\frac{\partial^2 V}{\partial T^2}\right)_p$$

Proved.

(b) For an ideal gas

$$V = \frac{n\bar{R}T}{p}$$

$$\left(\frac{\partial V}{\partial T}\right)_p = \frac{n\bar{R}}{p} \text{ and } \left(\frac{\partial^2 V}{\partial T^2}\right)_p = 0$$

$$\therefore \qquad \left(\frac{\partial \bar{C}_p}{\partial p}\right)_T = 0, \text{ i.e. } C_p \text{ is a function of } T \text{ alone.}$$

(c)
$$\frac{p\bar{v}}{\bar{\bar{R}}T} = 1 + B'p$$

$$\therefore \qquad B'p = \frac{p\bar{v}}{\bar{\bar{R}}T} - 1$$

$$B'T = \frac{T}{p}\left(\frac{p\bar{v}}{\bar{\bar{R}}T} - 1\right) = \left(\frac{\bar{v}}{\bar{\bar{R}}} - \frac{T}{p}\right)$$

$$\left[\frac{\partial}{\partial T}(B'T)\right]_p = +\frac{1}{\bar{\bar{R}}}\left(\frac{\partial \bar{v}}{\partial T}\right)_p - \frac{1}{p}$$

$$\left[\frac{\partial^2 (B'T)}{\partial T^2}\right]_p = \frac{1}{\bar{\bar{R}}}\left(\frac{\partial^2 \bar{v}}{\partial T^2}\right)_p = -\frac{1}{\bar{\bar{R}}T}\left(\frac{\partial \bar{C}_p}{\partial p}\right)_T$$

$\therefore$ On integration

$$\bar{C}_p = -\bar{R}Tp\frac{d^2}{dT^2}(B'T) + \bar{C}_{p0}$$

where $\bar{C}_{p0}$ (integration constant) is the value of $\bar{C}_p$ at very low values of pressure.

***Example* 11.2**

The Joule-Kelvin coefficient μ_J is a measure of the temperature change during a throttling process. A similar measure of the temperature change produced by an isentropic change of pressure is provided by the coefficient μ_s, where

$$\mu_s = \left(\frac{\partial T}{\partial p}\right)_S$$

Prove that

$$\mu_s - \mu_J = \frac{V}{C_p}$$

Solution *The Joule-Kelvin coefficient, μ_J, is given by*

$$\frac{T\left(\frac{\partial V}{\partial T}\right)_p - V}{C_p}$$

Since $C_p = T\left(\frac{\partial S}{\partial T}\right)_p$ and by Maxwell's relation

$$\left(\frac{\partial V}{\partial T}\right)_p = -\left(\frac{\partial S}{\partial p}\right)_T$$

$$\mu_j = \frac{-T\left(\frac{\partial S}{\partial p}\right)_T}{T\left(\frac{\partial S}{\partial T}\right)_p} - \frac{V}{C_p}$$

$$\therefore \quad \mu_j = -\left(\frac{\partial S}{\partial p}\right)_T \left(\frac{\partial T}{\partial S}\right)_p - \frac{V}{C_p}$$

Since
$$\left(\frac{\partial S}{\partial p}\right)_T \left(\frac{\partial T}{\partial S}\right)_p \left(\frac{\partial p}{\partial T}\right)_S = -1$$

$$\therefore \quad \mu_j = +\mu_s - \frac{V}{C_p}$$

$$\therefore \quad \mu_s - \mu_j = \frac{V}{C_p}$$ Proved.

Alternative method:

From the second *TdS* equation

$$TdS = C_p \, dT - T\left(\frac{\partial V}{\partial T}\right)_p dp$$

$$\left(\frac{\partial T}{\partial p}\right)_s = \mu_s = \frac{T}{C_p}\left(\frac{\partial V}{\partial T}\right)_p$$

Now
$$\mu_j = \frac{T}{C_p}\left[T\left(\frac{\partial V}{\partial T}\right)_p - V\right]$$

$$\therefore \quad \mu_s - \mu_j = \frac{V}{C_p}$$ Proved.

Example 11.3

If the boiling point of benzene at 1 atm pressure is 353 K, estimate the approximate value of the vapour pressure of benzene at 303 K.

Solution Using Clapeyron's equation and Trouton's rule, Eq. (11.21),

$$p = 101.325 \exp\left\{\frac{88}{\bar{R}}\left(1 - \frac{T_B}{T}\right)\right\}$$

$$= 101.325 \exp\left\{\frac{88}{8.3143}\left(1 - \frac{353}{303}\right)\right\}$$

$$= 17.7 \text{ kPa}$$ *Ans.*

Example 11.4

The vapour pressure, in mm of mercury, of solid ammonia is given by

$$\ln p = 23.03 - \frac{3754}{T}$$

and that of liquid ammonia by

$$\ln p = 19.49 - \frac{3063}{T}$$

(a) What is the temperature of the triple point? What is the pressure? (b) What are the latent heats of sublimation and vaporization? (c) What is the latent heat of fusion at the triple point?

Solution At the triple point, the saturated solid and saturated liquid lines meet.

$$23.03 - \frac{3754}{T} = 19.49 - \frac{3063}{T}$$

$$\therefore \quad T = 195.2 \text{ K} \qquad \textit{Ans. (a)}$$

$$\ln p = 23.03 - \frac{3754}{195.2}$$

$$\ln p = 3.80$$

$$p = 44.67 \text{ mm Hg} \qquad \textit{Ans.}$$

With the assumptions, $v''' >> v'$ and $v''' \sim \frac{\bar{R}T}{p}$

Clausius-Clapeyron equation reduces to

$$\frac{dp}{dT} = \frac{p}{\bar{R}T^2} \cdot l_{sub}$$

where l_{sub} is the latent heat of sublimation.

The vapour pressure of solid ammonia is given by

$$\ln p = 23.03 - \frac{3754}{T}$$

$$\therefore \quad \frac{1}{p} \cdot \frac{dp}{dT} = \frac{3754}{T^2}$$

$$\therefore \quad \frac{dp}{dT} = 3754 \frac{p}{T^2} = \frac{p}{\bar{R}T^2} \cdot l_{sub}$$

$$\therefore \quad l_{sub} = 3754 \times 8.3143 = 31{,}200 \text{ kJ/kg mol}$$

The vapour pressure of liquid ammonia is given by

$$\ln p = 19.49 - \frac{3063}{T}$$

$$\therefore \qquad \frac{dp}{dT} = 3063\,\frac{p}{T^2} = \frac{p}{\bar{R}T^2}\cdot l_{vap}$$

where l_{vap} is the latent heat of vaporization.

$$\therefore \qquad l_{vap} = 3063 \times 8.3143 = 25{,}500 \text{ kJ/kg mol} \qquad \textit{Ans. (b)}$$

At the triple point

$$l_{sub} = l_{vap} + l_{fu}$$

where l_{fu} is the latent heat of fusion.

$$\therefore \qquad l_{fu} = l_{sub} - l_{vap} = 31{,}200 - 25{,}500 = 5{,}700 \text{ kJ/kg mol}$$

Example 11.5

Explain why the specific heat of a saturated vapour may be negative.

Solution *As seen in Fig. Ex. 11.5, if heat is transferred along the saturation line, there is a decrease in temperature. The slope of the saturated vapour line is negative, i.e. when dS is positive, dT is negative. Therefore, the specific heat at constant saturation*

$$C'''_{sat} = T\left(\frac{dS'''}{dT}\right)$$

is negative. From the second TdS equation

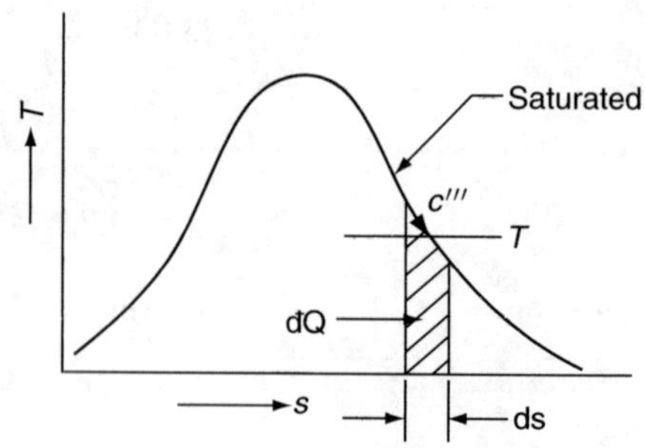

Fig. Ex.11.5

$$TdS = C_p dT - T\left(\frac{\partial V}{\partial T}\right)_p dp$$

$$T\,\frac{dS''}{dT} = C_p - T\left(\frac{\partial V'''}{\partial T}\right)_p\left(\frac{dp}{dT}\right)_{sat}$$

$$= C_p - T\cdot\frac{n\bar{R}}{p}\cdot\frac{l_{vap}}{T\,(V''' - V'')} \qquad \text{[using } pV''' = n\bar{R}\,T \text{ and Clapeyron's equation]}$$

$$C'''_{sat} = C_p\,\frac{V'''}{T}\cdot\frac{l_{vap}}{V'''} \qquad [\because V''' >> V'']$$

$$\therefore \qquad C'''_{sat} = C_p - \frac{l_{vap}}{T}$$

Now the value of l_{vap}/T for common substances is about 83.74 J/g mol K (*Trouton's rule*), where C_p is less than 41.87 J/g mol K. Therefore, C'''_{sat} can be negative. Proved.

Example 11.6

(*a*) *Establish the condition of equilibrium of a closed composite system consisting of two simple systems separated by a movable diathermal wall that is impervious to the flow of matter.*

(b) *If the wall were rigid and diathermal, permeable to one type of material, and impermeable to all others, state the condition of equilibrium of the composite system.*

(c) *Two particular systems have the following equations of state.*

$$\frac{1}{T_1} = \frac{3}{2}\bar{R}\frac{N_1}{U_1}, \frac{p_1}{T_1} = \frac{N_1}{V_1}\bar{R}$$

and

$$\frac{1}{T_2} = \frac{3}{2}\bar{R}\frac{N_2}{U_2}, \frac{p_2}{T_2} = \bar{R}\frac{N_2}{V_2}$$

where $\bar{R} = 8.3143$ *kJ/kg mol K, and the subscripts indicate systems* 1 *and* 2. *The mole number of the first system is* $N_1 = 0.5$, *and that of the second is* $N_2 = 0.75$. *The two systems are contained in a closed adiabatic cylinder, separated by a movable diathermal piston. The initial temperatures are* $T_1 = 200$ *K and* $T_2 = 300$ *K, and the total volume is* $0.02\ m^3$. *What is the energy and volume of each system in equilibrium? What is the pressure and temperature?*

Solution For the composite system, as shown in Fig. Ex. 11.6(a)

$$U_1 + U_2 = \text{constant}$$

$$V_1 = V_2 = \text{constant}$$

The values of U_1, U_2, V_1, and V_2 would change in such a way as to maximize the value of entropy. Therefore, when the equilibrium condition is achieved

$$dS = 0$$

for the whole system. Since

$$S = S_1 + S_2 = S_1(U_1, V_1, ..., N_{k1} ...) + S_2(U_2, V_2, ..., N_{k2} ...)$$

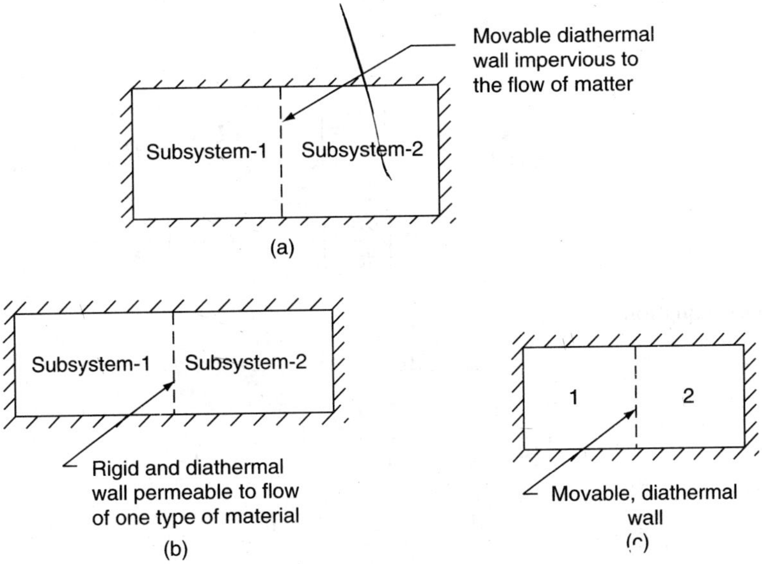

Fig. Ex.11.6

$$\therefore \quad dS = \left(\frac{\partial S_1}{\partial U_1}\right)_{V_1 \ldots, N_{k1} \ldots} dU_1 + \left(\frac{\partial S_1}{\partial V_1}\right)_{U_1 \ldots, N_{k1} \ldots,} dV_1$$

$$+ \left(\frac{\partial S_2}{\partial U_2}\right)_{V_2 \ldots, N_{k2} \ldots,} dU_2 + \left(\frac{\partial S_2}{\partial V_2}\right)_{V_2 \ldots, N_{k2} \ldots} dV_2$$

$$= \frac{1}{T_1} dU_1 + \frac{p_1}{T_1} dV_1 + \frac{1}{T_2} dU_2 + \frac{p_2}{T_2} dV_2$$

Since $dU_1 + dU_2 = 0$ and $dV_1 + dV_2 = 0$

$$\therefore \quad dS = \left(\frac{1}{T_1} - \frac{1}{T_2}\right) dU_1 + \left(\frac{p_1}{T_1} - \frac{p_2}{T_2}\right) dV_1 = 0$$

Since the expression must vanish for arbitrary and independent values of dU_1 and dV_1

$$\frac{1}{T_1} - \frac{1}{T_2} = 0 \text{ and } \frac{p_1}{T_1} - \frac{p_2}{T_2} = 0$$

or
$$p_1 = p_2 \text{ and } T_1 = T_2$$

$\therefore$ These are the conditions of mechanical and thermal equilibrium.

(b) We will consider the equilibrium state of two simple subsystems (Fig. Ex. 11.6(b)) connected by a rigid and diathermal wall, permeable to one type of material (N_1) and impermeable to all others ($N_2, N_3, \ldots N_r$). We thus seek the equilibrium values of U_1 and of U_2, and of N_{1-1} and N_{1-2} (i.e. material N_1 in subsystems 1 and 2 respectively.)

At equilibrium, an infinitesimal change in entropy is zero

$$dS = 0$$

Now
$$dS = dS_1 + dS_2$$

$$= \left(\frac{\partial S_1}{\partial U_1}\right)_{V_1, N_{1-1}, \ldots} dU_1 + \left(\frac{\partial S_1}{\partial N_{1-1}}\right)_{U_1, V_1, N_{1-2}, \ldots} dN_{1-1}$$

$$+ \left(\frac{\partial S_2}{\partial U_2}\right)_{V_2, N_{1-2}} dU_2 + \left(\frac{\partial S_2}{\partial N_{1-2}}\right)_{U_2, V_2, N_{2-2}} dN_{1-2}$$

From the equation

$$TdS = dU + p\,dV - \mu dN$$

$$\therefore \quad \left(\frac{\partial S}{\partial U}\right)_{V, N, \ldots} = \frac{1}{T}, \left(\frac{\partial S}{\partial N}\right)_{U, V} = -\frac{\mu}{T}$$

and
$$dN_{1-1} + dN_{1-2} = 0$$

$$dU_1 + dU_2 = 0$$

$$dS = \left(\frac{1}{T_1} - \frac{1}{T_2}\right) dU_1 - \left(\frac{\mu_{1-1}}{T_1} - \frac{\mu_{1-2}}{T_2}\right) dN_{1-1} = 0$$

As dS must vanish for arbitrary values of both dU_1 and dN_{1-1}

$$T_1 = T_2$$

$$\mu_{1-1} = \mu_{1-2}$$

which are the conditions of thermal and chemical equilibrium.

(c)

$$N_1 = 0.5 \text{ g mol}, N_2 = 0.75 \text{ g mol}$$

$$T_{1-1} = 200 \text{ K}, T_{1-2} = 300 \text{ K}$$

$$V = V_1 + V_2 = 0.02 \text{ m}^3$$

$$U_1 + U_2 = \text{constant}$$

$$\Delta U_1 + \Delta U_2 = 0$$

Let T_f be the final temperature (Fig. Ex. 11.6(c))

$$(U_{f-1} - U_{i-1}) = -(U_{f-2} - U_{i-2})$$

$$\frac{3}{2}\bar{R}N_1 (T_f - T_{i-1}) = -\frac{3}{2}\bar{R}N_2 (T_f - T_{i-2})$$

$$0.5 (T_f - 200) = -0.75 (T_f - 300)$$

$\therefore$

$$1.25\, T_f = 325$$

or

$$T_f = 260 \text{ K}$$ *Ans.*

$$U_{f-1} = \frac{3}{2}\bar{R}\, N_1 T_f = \frac{3}{2} \times 8.3143 \times 0.5 \times 10^{-3} \times 260 = 1.629 \text{ kJ}$$

$$U_{f-2} = \frac{3}{2} \times 8.3143 \times 0.75 \times 10^{-3} \times 260 = 2430 \text{ kJ}$$

$$V_{f-2} = \frac{\bar{R}N_1 T_{f-1}}{p_{f-1}}$$ At equilibrium $p_{f-1} = p_{f-2} = p_f$ *Ans.*

$$V_{f-2} = \frac{\bar{R}N_2 T_{f-2}}{p_{f-2}}$$ $T_{f-1} = T_{f-2} = T_f$

$$V_{f-1} + V_{f-2} = \frac{\bar{R}T_f}{p_f}(N_1 + N_2) = 0.02 \text{ m}^3$$

$$\frac{8.3143 \times 260}{p_f} \times 1.25 \times 10^{-3} = 0.02 \text{ m}^3$$

$\therefore$

$$p_f = \frac{8.3143 \times 260 \times 1.25 \times 10^{-3}}{0.02} \text{ kN/m}^2$$

$$= 135 \text{ kN/m}^2 = 1.35 \text{ bar}$$

$$\therefore \quad V_{f-1} = \frac{8.3143 \times 0.5 \times 10^{-3} \times 260}{135} = 0.008 \text{ m}^3$$

$$\therefore \quad V_{f-2} = 0.02 - 0.008 = 0.012 \text{ m}^3$$

Ans.

Example 11.7

Show that for a van der Waals' gas

(a) $\left(\frac{\partial c_v}{\partial v}\right)_T = 0$

(b) $(s_2 - s_1)_T = R \ln \frac{v_2 - b}{v_1 - b}$

(c) $T(v-b)^{R/c_v}$ = *constant, for an isentropic*

(d) $c_p - c_v = \frac{R}{1 - 2a(v-b)^2 / RTv^3}$

(e) $(h_2 - h_1)_T = (p_2 v_2 - p_1 v_1) + a\left(\frac{1}{v_1} - \frac{1}{v_2}\right)$

Solution (a) From the energy Eq. (11.13)

$$\left(\frac{\partial U}{\partial V}\right)_T = T\left(\frac{\partial p}{\partial T}\right)_V - p$$

$$\frac{\partial^2 U}{\partial V \cdot \partial T} = T\left(\frac{\partial^2 p}{\partial T^2}\right)_V + \left(\frac{\partial p}{\partial T}\right)_V - \left(\frac{\partial p}{\partial T}\right)_V$$

$$\frac{\partial^2 U}{\partial V \cdot \partial T} = T\left(\frac{\partial^2 p}{\partial T^2}\right)_V$$

$$C_v = \left(\frac{\partial U}{\partial T}\right)_V$$

$$\left(\frac{\partial C_v}{\partial V}\right)_T = \frac{\partial^2 U}{\partial T \cdot \partial V} = T\left(\frac{\partial^2 p}{\partial T^2}\right)_V = \left(\frac{\partial c_v}{\partial v}\right)_T$$

For a van der Waals' gas

$$\left(p + \frac{a}{v^2}\right)(v - b) = RT$$

$$p = \frac{RT}{v-b} - \frac{a}{v^2}$$

$$\therefore \quad \left(\frac{\partial^2 p}{\partial T^2}\right)_V = 0$$

$$\therefore \qquad \left(\frac{\partial c_v}{\partial v}\right)_T = 0 \qquad \text{Proved (a)}$$

$\therefore$ c_v is independent of volume.

(b) From the first *Tds* Eq. (11.8)

$$Tds = c_v\, dT + T\left(\frac{\partial p}{\partial T}\right)_V dv$$

and energy Eq. (11.13), $\left(\frac{\partial U}{\partial V}\right)_T = T\left(\frac{\partial p}{\partial T}\right)_V - p$

$$ds = c_v \frac{dT}{T} + \frac{1}{T}\left[p + \left(\frac{\partial U}{\partial V}\right)_T\right] dv$$

For van der Waals' gas

$$\left(\frac{\partial U}{\partial V}\right)_T = \frac{a}{v^2}$$

$$\therefore \qquad ds = c_v \frac{dT}{T} + \frac{1}{T}\left(p + \frac{a}{v^2}\right) dv = c_v \frac{dT}{T} + \frac{R}{v-b}\, dv$$

$$\therefore \qquad (s_2 - s_1)_T = R \ln \frac{v_2 - b}{v_1 - b} \qquad \text{Proved (b)}$$

(c) At constant entropy

$$c_v \frac{dT}{T} + \frac{R}{v-b}\, dv = 0$$

or

$$\frac{dT}{T} + \frac{R}{c_v} \frac{dv}{v-b} = 0$$

by integration, $T(v-b)^{R/C_v} = \text{constant}$ Proved (c)

(d)

$$c_p - c_v = T\left(\frac{\partial p}{\partial T}\right)_V \left(\frac{\partial v}{\partial T}\right)_p = \left[\left(\frac{\partial U}{\partial V}\right)_T + p\right]\left(\frac{\partial v}{\partial T}\right)_p$$

$$= \left(\frac{a}{v^2} + p\right)\left(\frac{\partial v}{\partial T}\right)_p = \left(\frac{RT}{v-b}\right)\left(\frac{\partial v}{\partial T}\right)_p$$

From the equation

$$\left(p + \frac{a}{v^2}\right)(v-b) = RT$$

$$\therefore \qquad (v-b)(-2av^{-3})\left(\frac{\partial v}{\partial T}\right)_p + \left(p + \frac{a}{v^2}\right)\left(\frac{\partial v}{\partial T}\right)_p = R$$

$$\left(\frac{\partial v}{\partial T}\right)_p = \frac{R/(v-b)}{\dfrac{RT}{(v-b)} - \dfrac{2a}{v^3}}$$

$$\therefore \qquad c_p - c_v = \frac{R}{1-2a(v-b)^2/RT\,v^3}$$ Proved (d)

(e)
$$\left(\frac{\partial U}{\partial V}\right)_T = T\left(\frac{\partial p}{\partial T}\right)_V - p = \frac{a}{v^2}$$

$$du_T = \frac{a}{v^2}\,dv_T$$

$$\therefore \qquad (u_2 - u_1)_T = a\left(\frac{1}{v_1} - \frac{1}{v_2}\right)$$

$$\therefore \qquad (h_2 - h_1)_T = (p_2v_2 - p_1v_1) + a\left(\frac{1}{v_1} - \frac{1}{v_2}\right)$$ Proved (e)

Example 11.8

The virial equation of state of a gas is given by

$$pv = RT(1 + B'p + C'p^2 + \ldots)$$

Show that

$$\lim_{p\to 0}\left[T\left(\frac{\partial v}{\partial T}\right)_p - v\right] = RT^2\frac{dB'}{dT}$$

Hence, prove that the inversion temperature of a van der Waals' gas is twice the Boyle temperature.

Solution

$$pv = RT\,(1 + B'p + C'p^2 + \ldots)$$

$$v = \frac{RT}{p} + RTB' + RTpC' + \ldots$$

$$\left(\frac{\partial v}{\partial T}\right)_p = \frac{R}{p} + RT\frac{dB'}{dT} + RB' + RTp\frac{dC'}{dT} + RpC' + \ldots$$

$$\therefore \qquad T\left(\frac{\partial v}{\partial T}\right)_p = \frac{RT}{p} + RT^2\frac{dB'}{dT} + RTB' + RT^2p\frac{dC'}{dT} + RTpC' + \ldots$$

$$\therefore \qquad T\left(\frac{\partial v}{\partial T}\right) - v = RT^2\frac{dB'}{dT} + RT^2p\frac{dC'}{dT} + \ldots$$

$$\mu_J = \frac{1}{c_p}\left[T\left(\frac{\partial v}{\partial T}\right)_p - v\right]$$

$$\therefore \qquad = \frac{RT^2}{c_p}\left[\frac{dB'}{dT} + p\frac{dC'}{dT} + \ldots\right]$$

$$\lim_{p \to 0} \mu_J = \frac{RT^2}{c_p} \frac{dB'}{dT}$$

or
$$\lim_{p \to 0} \left[T\left(\frac{dv}{dT}\right)_p - v \right] = RT^2 \frac{dB'}{dT}$$
Proved

For a van der Waals' gas, to find Boyle temperature T_B,

$$B = b - \frac{a}{RT} = 0$$

$$\therefore \quad T_B = \frac{a}{bR}$$

$$B' = \frac{B}{RT} = \frac{b}{RT} - \frac{a}{R^2T^2}$$

$$\frac{dB'}{dT} = -\frac{b}{RT^2} + \frac{2a}{R^2T^3}$$

$$\therefore \quad \lim_{p \to 0} \mu_J = \frac{RT^2}{c_p}\left(-\frac{b}{RT^2} + \frac{2a}{R^2T^3}\right) = 0$$

$$\therefore \quad \frac{b}{RT^2} = \frac{2a}{R^2T^3}$$

$$\therefore \quad T_i = \frac{2a}{bR}$$

$$\therefore \quad T_1 = 2T_B$$

or Inversion temperature = 2 × Boyle temperature Proved

Example 11.9

Over a certain range of pressures and temperatures, the equation of a certain substance is given by the relation

$$v = \frac{RT}{p} - \frac{C}{T^3}$$

where C is a constant. Derive an expression for: (a) the change of enthalpy and (b) the change of entropy, of this substance in an isothermal process.

Solution (a) From Eq. (11.15)

$$dh = c_p\, dT + \left[v - T\left(\frac{\partial v}{\partial T}\right)_p \right] dp$$

$$\therefore \quad (h_2 - h_1)_T = \int_1^2 \left[v - T\left(\frac{\partial v}{\partial T}\right)_p \right] dp$$

Now,
$$v = \frac{RT}{p} - \frac{C}{T^3}$$

$$\left(\frac{\partial v}{\partial T}\right)_p = \frac{R}{p} + \frac{3C}{T^4}$$

$$\therefore \quad T\left(\frac{\partial v}{\partial T}\right)_p = \frac{RT}{p} + \frac{3C}{T^3}$$

$$\therefore \quad v - T\left(\frac{\partial v}{\partial T}\right)_p = \frac{RT}{p} - \frac{C}{T^3} - \frac{RT}{p} - \frac{3C}{T^3} = \frac{4C}{T^3}$$

On substitution,
$$(h_2 - h_1)_T = \int_{p_1}^{p_2} -\frac{4C}{T^3}\, dp = \frac{4C}{T^3}\,(p_1 - p_2)_T$$

(b) Using second *Tds* equation
$$TdS = c_p\, dT - T\left(\frac{\partial v}{\partial T}\right)_p dp$$

$$\therefore \quad dS_T = -\left(\frac{\partial v}{\partial T}\right)_p dp_T = -\left(\frac{R}{p} + \frac{3C}{T^4}\right) dp_T$$

$$\therefore \quad (s_2 - s_1)_T = R \ln \frac{p_1}{p_2} + \frac{3C}{T^4}\,(p_1 - p_2)_T$$

Example 11.10

Agron gas is compressed reversibly and isothermally at the rate of 1.5 kg/s from 1 atm, 300 K to 400 atm. Calculate the power required to run the compressor and the rate at which heat must be removed from the compressor. The gas is assumed to obey the Redlich-Kwong equation of state, for which the constants are:
$$a = 0.42748\frac{\overline{R}^2 T_c^{2.5}}{p_c} \text{ and } b = 0.08664\frac{\overline{R}^2 T_c}{p_c}$$

For argon, $T_c = 151$ K and $p_c = 48$ atm. Take $\overline{R} = 0.082$ litre-atm/gmol-K.

Solution Substituting the values of p_c, T_c and R,
$$a = 0.42748\frac{(82)^2 (151)^{2.5}}{48} = 16.8\times10^6 \,\frac{\text{atm K}^{1/2}\text{cm}^6}{(\text{g mol})^2}$$

$$b = 0.08664\frac{(82)(151)}{48} = 22.4\frac{\text{cm}^3}{\text{g mol}}$$

Substituting the numerical values of p_2, T_2, a, b and R into the Redlich-Kwong equation

$$p = \frac{RT}{v-b} - \frac{a}{T^{1/2}v(v+b)}$$

$$v_2^3 - 49.24\, v_2^2 + 335.6\, v_2 - 43,440 = 0$$

from which we obtain

$$v_2 = 56.8 \text{ cm}^3/\text{g mol}$$

Since $p_1 = 1$ atm, the volume of the gas at the initial state can be obtained from the ideal gas equation:

$$v_1 = \frac{RT_1}{p_1} = \frac{82 \times 300}{1} = 24,600 \text{ cm}^3/\text{g mol}$$

For isothermal compression,

$$\Delta h_{12} = h_2 - h_1 = \left\{ \int_{p_1}^{p_2} \left[v - T\left(\frac{\partial v}{\partial T}\right)_p \right] dp \right\}_T$$

Now, $d(pv) = p\, dv + v\, dp$

$$\int_{p1}^{p_2} v\, dp = p_2 v_2 - p_1 v_1 - \left[\int_{v_1}^{v_2} p\, dv \right]_T$$

Since
$$\left(\frac{\partial v}{\partial T}\right)_p = -\left(\frac{\partial v}{\partial p}\right)_T \left(\frac{\partial p}{\partial T}\right)_v$$

we have
$$\left[\int_{p_1}^{p_2} \left(\frac{\partial v}{\partial T}\right)_p dp \right]_T = -\left[\int_{v_1}^{v_2} \left(\frac{\partial v}{\partial T}\right)_v dv \right]_T$$

Hence,
$$\Delta h_{12} = h_2 - h_1 = p_2 v_2 - p_1 v_1 - \left\{ \int_{v_1}^{v_2} \left[p - T\left(\frac{\partial p}{\partial T}\right)_v \right] dv \right\}_T$$

According to Redlich-Kwong equation, we have, $\left(\frac{\partial p}{\partial T}\right) = \frac{R}{v-b} + \frac{a}{2T^{3/2}v(v+b)}$

Thus, $\Delta h_{12} = h_2 - h_1 = (p_2 v_2 - p_1 v_1) - \left\{ \int_{v_1}^{v_2} \left[\frac{-3a}{2T^{1/2}v(v+b)} \right]_v dv \right\}_T$

$$= (p_2 v_2 - p_1 v_1) - \frac{1.5a}{T_1^{1/2}} \frac{1}{b} \ln \frac{(v_2 + b)/v_2}{(v_1 + b)/v_1}$$

Substituting the numerical values,

$$h_2 - h_1 = -1,790 \text{ J/g mol}$$

$$\Delta s_{12} = s_2 - s_1 = \left[\int_{p_1}^{p_2} \left(\frac{\partial v}{\partial T}\right)_p dp\right]_T = \left[\int_{v_1}^{v_2} \left(\frac{\partial p}{\partial T}\right)_v dv\right]_T$$

For the Redlich-Kwong equation this becomes,

$$s_2 - s_1 = \left\{\int_{v_1}^{v_2} \left[\frac{R}{v-b} + \frac{a}{2T^{3/2}v(v+b)}\right] dv\right\}_T$$

$$= R \ln \frac{v_2 - b}{v_1 - b} - \frac{a}{2bT_1^{3/2}} \ln \frac{(v_2+b)/v_2}{(v_1+b)/v_1}$$

Substituting the numerical values,

$$s_2 - s_1 = -57 \text{ J/gmol-K}$$

$$\dot{Q}_{12} = \dot{m}T_1(s_2 - s_1)$$

$$= \frac{10^5 \text{ g/h}}{39.8 \text{ g/gmol}} \times 300 \text{ K} \times (-57) \frac{\text{J}}{\text{g mol K}}$$

$$= -4.29 \times 10^7 \text{ J/h}$$

$$= -11.917 \text{ kW (heat removed from the gas)}$$ *Ans.*

$$\dot{W}_{12} = \dot{Q}_{12} + \dot{m}(h_1 - h_2)$$

$$= -4.29 \times 10^7 + \frac{10^5}{39.9} \times 1790$$

$$= -3.84 \times 10^7 \text{ J/h}$$

$$= -10.67 \text{ kW(Work is done on the gas)}$$ *Ans.*

Review Questions

11.1 What is the condition for exact differential?

11.2 Derive Maxwell's equations.

11.3 Write down the first and second *TdS* equations, and derive the expression for the difference in heat capacities, C_p and C_v. What does the expression signify?

11.4 Define volume expansivity and isothermal compressibility.

11.5 Show that the slope of an isentrope is greater than that of an isotherm on *p–v* plot. How is it meaningful for estimating the work of compression?

11.6 What is the energy equation? How does this equation lead to the derivation of the Stefan-Boltzman law of thermal radiation?

11.7 Show that the internal energy and enthalpy of an ideal gas are functions of temperature only.

11.8 Why are $dU = C_v\, dT$ and $dH = C_p\, dT$ true for an ideal gas in any process, whereas these are true for any other substance only at constant volume and at constant pressure respectively?

11.9 Explain Joule-Kelvin effect. What is inversion temperature?

11.10 What is Joule-Thomson coefficient? Why is it zero for an ideal gas?

11.11 Why does the hydrogen gas need to be precooled before being throttled to get the cooling effect?

11.12 Why does the maximum temperature drop occur if the state before throttling lies on the inversion curve?

11.13 Why does the Gibbs function remain constant during phase transition?

11.14 What are the characteristics of the first order phase transition?

11.15 Write down the representative equation for phase transition. Why does the fusion line for water have negative slope on the p–T diagram?

11.16 Why is the slope of the sublimation curve at the triple point on the p–T diagram greater than that of the vaporization curve at the same point?

11.17 Explain how thermodynamic properties are evaluated from an equation of state.

11.18 Illustrate how enthalpy change and entropy change of a gas can be estimated with the help of an equation of state.

11.19 State the important thermodynamic criteria which an equation of state should satisfy.

11.20 Explain how the Boyle temperature is yielded when:

$$\lim_{p\to 0}\left(\frac{\partial z}{\partial p}\right)_T = 0$$

11.21 What is foldback temperature?

11.22 Show that for an inversion curve $\left(\dfrac{\partial z}{\partial T}\right)_p = 0$.

11.23 Define chemical potential of a component in terms of U, H, and G.

11.24 What is the use of the Gibbs entropy equation?

11.25 Explain the significance of the Gibbs-Duhem equation.

11.26 State the conditions of equilibrium of a heterogeneous system.

11.27 What do you understand by phase equilibrium?

11.28 Give the Gibbs phase rule for a nonreactive system. Why is the triple point of a substance nonvariant?

11.29 What are the four types of equilibrium? What is stable equilibrium?

11.30 State the conditions of spontaneous change, equilibrium and criterion of stability for: (a) a system having constant U and V (i.e., isolated), and (b) a system having constant T and p.

11.31 What do you understand by neutral and unstable equilibrium?

11.32 What is metastable equilibrium?

11.33 Show that for a system to be stable, these conditions are satisfied

(a) $C_v > 0$ (thermal stability)

(b) $\left(\dfrac{\partial p}{\partial V}\right)_T < 0$ (mechanical stability)

Problems

11.1 Derive the following equations

(a) $U = F - T\left(\dfrac{\partial F}{\partial T}\right)_V = -T^2\left(\dfrac{\partial F/T}{\partial T}\right)_V$

(b) $C_v = -T\left(\dfrac{\partial^2 F}{\partial T^2}\right)_v$

(c) $H = G - T\left(\dfrac{\partial G}{\partial T}\right)_p = -T^2\left(\dfrac{\partial G/T}{\partial T}\right)_p$

(d) $C_p = -T\left(\dfrac{\partial^2 G}{\partial T^2}\right)_p$

11.2 (a) Derive the equation

$$\left(\frac{\partial c_v}{\partial v}\right)_T = T\left(\frac{\partial^2 p}{\partial T^2}\right)_V$$

(b) Prove that c_v of an ideal gas is a function of T only.

(c) In the case of a gas obeying the equation of state

$$\frac{pv}{RT} = 1 + \frac{B''}{v}$$

where B'' is a function of T only, show that

$$c_v = -\frac{RT}{v}\frac{d^2}{dT^2}(B''T) + (c_v)_0$$

where $(c_v)_0$ is the value at very large volumes.

11.3 Derive the third TdS equation

$$TdS = C_v\left(\frac{\partial T}{\partial p}\right)_v dp + C_p\left(\frac{\partial T}{\partial V}\right)_p dV$$

and show that the three TdS equations may be written as

(a) $TdS = C_v\,dT + \dfrac{\beta T}{k}\,dV$

(b) $TdS = C_p\,dT - V\beta\,Tdp$

(c) $TdS = \dfrac{C_v}{\beta}\,k\,dp + \dfrac{C_v}{\beta V}\,dV$

11.4 Derive the equations

(a) $C_p = T\left(\dfrac{\partial V}{\partial T}\right)_p\left(\dfrac{\partial p}{\partial T}\right)_s$

(b) $\left(\dfrac{\partial p}{\partial T}\right)_s = \dfrac{C_p}{V\beta T}$

(c) $\dfrac{(\partial p/\partial T)_s}{(\partial p/\partial T)_v} = \dfrac{\gamma}{\gamma-1}$

11.5 Derive the equations

(a) $C_v = -T\left(\dfrac{\partial p}{\partial T}\right)_v\left(\dfrac{\partial V}{\partial T}\right)_s$

(b) $\left(\dfrac{\partial V}{\partial T}\right)_s = -\dfrac{C_v k}{\beta T}$

(c) $\dfrac{(\partial V/\partial T)_s}{(\partial V/\partial T)_p} = \dfrac{1}{1-\gamma}$

11.6 (a) Prove that the slope of a curve on a Mollier diagram representing a reversible isothermal process is equal to

$$T - \frac{1}{\beta}$$

(b) Prove that the slope of a curve on a Mollier diagram representing a reversible isochoric process is equal to

$$T + \frac{\gamma - 1}{\beta}$$

11.7 (a) Show that

$$\mu_J\, c_p = T^2\left(\frac{\partial V/T}{\partial T}\right)_p$$

For 1 mole of a gas, in the region of moderate pressures, the equation of sate may be written as

$$\frac{p\bar{v}}{\bar{R}T} = 1 + B'p + C'p^2$$

where B' and C' are functions of temperature only.

(b) Show that as $p \to 0$

$$\mu_J\, c_p \to \bar{R}T^2\,\frac{dB'}{dT}$$

(c) Show that the equation of the inversion curve is

$$p = -\frac{dB'/dT}{dC'/dT}$$

11.8 Prove the following functional relationship of the reduced properties for the inversion curve of a van der Waals' gas

$$T_r = \frac{3(3v_r - 1)^2}{4v_r^2} \text{ and } p_r = \frac{9(2v_r - 1)}{v_r^2}$$

Hence, show that

$$\frac{\text{Maximum inversion temperature}}{\text{Critical temperature}} = 6.75$$

and $$\frac{\text{Minimum inversion temperature}}{\text{Critical temperature}} = 0.75$$

11.9 Estimate the maximum inversion temperature of hydrogen if it is assumed to obey the equation of state

$$pV = RT + B_1 p + B_2 p^2 + B_3 p^3 + \ldots$$

For hydrogen, $B_1 \times 10^5 = a + 10^{-2}\,bT + 10^2\,c/T$

where $a = 166$, $b = -7.66$, $c = -172.33$

11.10 The vapour pressure of mercury at 399 K and 401 K is found to be 0.988 mm and 1.084 mm of mercury respectively. Calculate the latent heat of vaporization of liquid mercury at 400 K.

Ans. 61,634.96 kJ/kg mol

11.11 In the vicinity of the triple point, the vapour pressure of liquid ammonia (in atmospheres) is represented by

$$\ln p = 15.16 - \frac{3063}{T}$$

This is the equation of the liquid-vapour boundary curve in a p–T diagram. Similarly, the vapour pressure of solid ammonia is

$$\ln p = 18.70 - \frac{3754}{T}$$

(a) What is the temperature and pressure at the triple point?

(b) What are the latent heats of sublimation and vaporization?

(c) What is the latent heat of fusion at the triple point? *Ans.* 195.2 K, 0.585 atm., 1498 kJ/kg, 1836 kJ/kg, 338 kJ/kg

11.12 It is found that a certain liquid boils at a temperature of 95°C at the top of a hill, whereas it boils at a temperature of 105°C at the bottom. The latent heat is 4.187 kJ/g mole. What is the approximate height of the hill? Assume $T_0 = 300$ K. *Ans.* 394 m

11.13 Show that for an ideal gas in a mixture of ideal gases

$$d\mu_k = \frac{\mu_k - h_k}{T}\, dT + v_k\, dp + RTd \ln x_k$$

11.14 Compute μ_j for a gas whose equation of state is

$$p(v - b) = RT \qquad \textit{Ans. } \mu_J = -\,b/c_p$$

11.15 Show that

(a) $$\left(\frac{\partial \beta}{\partial p}\right)_T = -\left(\frac{\partial k}{\partial T}\right)_p$$

(b) $$\left(\frac{\partial u}{\partial p}\right)_T = -T\left(\frac{\partial v}{\partial T}\right)_p - p\left(\frac{\partial v}{\partial p}\right)_T$$

11.16 Two particular systems have the following equations of state

$$\frac{1}{T^{(1)}} = \frac{3}{2}\bar{R}\frac{N^{(1)}}{U^{(1)}} \text{ and } \frac{1}{T^{(2)}} = \frac{5}{2}\bar{R}\frac{N^{(2)}}{U^{(2)}}$$

where $\bar{R} = 8.3143$ kJ/kg mol K. The mole number of the first system is $N^{(1)} = 2$, and that of the second is $N^{(2)} = 3$. The two systems are separated by a diathermal wall, and the total energy in the composite system is 25.120 kJ. What is the internal energy of each system in equilibrium? *Ans.* 7.2 kJ, 17.92 kJ

11.17 Two systems with the equations of state given in Problem 11.16 are separated by a diathermal wall. The respective mole numbers are $N^{(1)} = 2$ and $N^{(2)} = 3$. The initial temperatures are $T^{(1)} = 250$ K and $T^{(2)} = 350$ K. What are the values of $U^{(1)}$ and $U^{(2)}$ after equilibrium has been established? What is the equilibrium temperature? *Ans.* 8.02 kJ, 20.04 kJ, 321.4 K

11.18 Show that the change in latent heat L with temperature is given by the following relation

$$\left(\frac{dL}{dT}\right) = (C_p''' - C_p'') + \frac{L}{T} - \frac{v'''\beta''' - v''\beta''}{v''' - v''}\, L$$

11.19 Show that for a van der Waals' gas, the Joule-Thomson coefficient is given by

$$\mu_j = \frac{v}{c_p}\left[\frac{2a(v-b)^2 - RTbv^2}{RTv^3 - 2a(v-b)^2}\right]$$

11.20 At 273.15 K the specific volumes of water and ice are 0.001 and 0.001091 m³/kg and the latent heat of fusion of ice is 334 kJ/kg. Determine the melting point increase due to increase of pressure by 1 atm (101.325 kPa). *Ans.* –0.00753 K

11.21 Calculate the latent heat of vaporization of steam formed by boiling water under a pressure of 101.325 kPa. At a pressure near this, a rise of temperature of 1 K causes an increase of vapour pressure of 3.62 kPa. *Ans.* 2257 kJ/kg

11.22 It is known that radiation exerts a pressure $p = 1/3\, u$, where u is the energy per unit volume.

(a) Show that $du = TdS + \frac{1}{V}\left(Ts - \frac{4}{3}u\right) dV$

where s is the entropy per unit volume.

(b) Assuming u and s as functions of temperature only, show that

(i) $u = As^{4/3}$

(ii) $s = \frac{4}{3}\, aT^3$

(iii) $u = aT^4$

where A is the constant of integration and $a = 81/256\, A^3$.

(c) Show that the average time radiation remains in a spherical enclosure of radius r is given by

$$t = \frac{4r}{3c}$$

where c is the speed of radiation.

(d) If E_B is the energy emitted per unit area of spherical surface per unit time, show that

$$E_B = \sigma T^4$$

where $\sigma = ac/4$ and T is the temperatur of the surface.

11.23 Show that the inversion temperature of a van der Waals' gas is given by $T_i = 2a/bR$.

11.24 Show that:

(a) $$\left(\frac{\partial u}{\partial v}\right)_T = T^2\left[\frac{\partial(p/T)}{\partial T}\right]_v$$

(b) $\left(\frac{\partial h}{\partial p}\right)_T = -T^2\left[\frac{\partial(v/T)}{\partial T}\right]_p$

11.25 Show that for a van der Waals' gas at low pressures, a Joule-Thomson expansion from pressure p_1 to p_2 produces a temperature change which can be found from the solution of

$$p_1 - p_2 = \frac{c_p}{b}(T_1 - T_2) + T_i \ln \frac{T_2 - T_i}{T_1 - T_i}$$

where T_i is the inversion temperature.

11.26 Using the Redlick-Kwong equation of state, develop expressions for the changes in entropy and internal energy of a gas in an isothermal process.

Ans. $(s_2 - s_1)_T = R \ln \frac{v_2 - b}{v_1 - b} + \frac{a}{2bT^{3/2}} \ln \left[\frac{v_2(v_1 + b)}{v_1(v_2 + b)}\right]$

$$(u_2 - u_1)_T = \frac{a}{2bT^{1/2}} \ln \left[\frac{v_2(v_1 + b)}{v_1(v_2 + b)}\right]$$

11.27 Find the change of entropy of a gas following Clausius equation of state at constant temperature

$p(v - b) = RT$ *Ans.* $R \ln \frac{v_2 - b}{v_1 - b}$

11.28 (a) Show that for a van der Waals' gas

$$\beta = \frac{Rv^2(v-b)}{RTv^3 - 2a(v-b)^2}$$

$$k_T = \frac{v^2(v-b)^2}{RTv^3 - 2a(v-b)^2}$$

(b) What is the value of k_T/β expressed in its simplest form?

(c) What do the above relations become when $a = 0$, $b = 0$ (ideal gas)?

11.29 (a) Show that

(i) $\left(\frac{\partial u}{\partial p}\right)_v = \frac{k}{\beta} \cdot c_v$ (ii) $\left(\frac{\partial u}{\partial v}\right)_p = \frac{c_p}{v_B} - p$

(b) Hence show that the slope of a reversible adiabatic process on p–v coordinates is

$$\frac{dp}{dv} = -\frac{\gamma}{kv}$$

where k is the isothermal compressibility.

11.30 According to Berthelot, the temperature effect of the second virial coefficient is given by

$$B'(T) = \frac{b}{T} - \frac{a}{T^3}$$

where a and b are constants. Show that according to Berthelot,

$$T_{inv}/T_B = \sqrt{3}$$

11.31 The following expressions for the equation of state and the specific heat c_p are obeyed by a certain gas:

$$v = \frac{RT}{p} + \alpha T^2 \quad \text{and} \quad c_p = A + BT + Cp$$

where α, A, B, C are constants. Obtain an expression for (a) the Joule-Thomson coefficient, and (b) the specific heat c_v.

Ans. (a) $\mu_J = \frac{\alpha T^2}{A + BT + Cp}$

(b) $c_v = A + BT + \frac{CRT}{v - \alpha T^2} - R\left(\frac{v + \alpha T^2}{v - \alpha T^2}\right)$

11.32 Determine the maximum Joule-Thomson inversion temperature in terms of the critical temperature T_c predicted by the

(a) van der Waals equations
(b) Redlich–Kwong equation
(c) Dieterici equation

Ans. (a) 6.75 T_c, (b) 5.34 T_c (c) 8T_c

11.33 From the virial form of the equation of state of a gas

$$v = \frac{RT}{p} + RTB'(T) + RTC'(T)p + \ldots$$

show that the Joule-Thomson coefficient is

$$\mu_J = \frac{RT^2}{c_p}\left[\frac{dB'}{dT} + \frac{dC'}{dT}p + \ldots\right]$$

(b) For a van der Waals gas

$$B'(T) = \frac{bRT - a}{R^2T^2}$$

Show that the limiting value of μ_J at low pressure is

$$\mu_J = \frac{1}{c_p}\left(\frac{2a}{RT} - b\right)$$

11.34 Show that $k_T - k_s = \frac{Tv\beta^2}{c_p}$

11.35 For a simple compressible system, show that

(a) $\left[\frac{\partial u}{dv}\right]_T = T^2\left[\frac{\partial p/T}{\partial T}\right]_v$

(b) $\left[\frac{\partial h}{\partial p}\right]_T = -T^2\left[\frac{\partial v/T}{\partial T}\right]_p$

11.36 The liquid-vapour equilibrium curve for nitrogen over the range from the triple point to the normal boiling point may be expressed by the relation:

$$\log p = A - BT - \frac{C}{T}$$

where p is the vapour pressure in mm Hg, T is the temperature in K, and $A = 7.782$, $B = 0.006265$, and $C = 341.6$.

(a) Derive an expression for the enthalpy of vaporization h_{fg} in terms of A, B, C, T and v_{fg}.

(b) Calculate h_{fg} for nitrogen at 71.9 K with v_{fg} = 11,530 cm^3/gmol. *Ans.* 5,790 J/gmol

11.37 For a gas obeying the van der Waals equation of state, show that:

(a) $c_p - c_v = \dfrac{R}{1 - 2a\,(v-b)^2/RTv^3}$

(b) $\left[\frac{\partial c_v}{\partial v}\right]_T = T\left[\frac{\partial^2 p}{\partial T^2}\right]_v = 0$ to prove that c_v is a function of temperature only.

(c) $\left[\frac{\partial c_p}{\partial p}\right]_T = -T\left[\frac{\partial^2 v}{\partial T^2}\right]_p$

$$= R^2T\left[\frac{2av^3 - 6abv^{-4}}{\left(p - av^{-2} + 2abv^{-3}\right)^3}\right]$$

to prove that c_p for a van der Waals gas is not a function of temperature only.

(d) The relation between T and v is given by:

$$T\left(v-b\right)^{R/c_v} = \text{constant}$$

(e) The relation between p and v is given by:

$$\left[p + \frac{a}{v^2}\right]\left(v-b\right)^{1+R/c_v} = \text{constant.}$$

11.38 Nitrogen at a pressure of 250 atm and a temperature of 400 K expands reversibly and adiabatically in a turbine to an exhaust pressure of 5 atm. The flow rate is 1 kg/s. Calculate the power output if nitrogen obeys the Redlich–Kwong equation of state. For nitrogen at 1 atm take.

$$c_p = 6.903 - 0.3753 \times 10^{-3}T + 1.930 \times 10^{-6}T^2 - 6.861 \times 10^{-9}T^3$$

where c_p is in cal/gmol-K and T is in K.

For nitrogen, $T_c = 126.2$ K,

$p_c = 33.5$ atm. *Ans.* 272 kW

Hints: See Fig. P-11.34

$h_1 - h_2 = (h_1 - h_4) + (h_4 - h_3) + (h_3 - h_2)$ and

$s_1 - s_2 = 0 = (s_1 - s_4) + (s_4 - s_3) + (s_3 - s_2)$

$a = 15.4 \times 10^6$ atm/K$^{1/2}$ cm^6/(g mol)^2, b = 26.8 cm^3/gmol

By trial-and-error, v_1 = 143 cm^3/g mol, v_4 = 32,800 cm^3/g mol

$T_2 = 124$ K, $h_1 - h_2 = 7.61$ kJ/g mol.

CHAPTER 12

Vapour Power Cycles

12.1 SIMPLE STEAM POWER CYCLE

A power cycle continuously converts heat (energy released by the burning of fuel) into work (shaft work), in which a working fluid repeatedly performs a succession of processes. In the vapour power cycle the working fluid, which is water, undergoes a change of phase. Figure 12.1 gives the schematic of a simple steam power plant working on the vapour power cycle. Heat is transferred to water in the boiler from an external source (furnace, where fuel is continuously burnt) to raise steam, the high pressure, high temperature steam leaving the boiler expands in the turbine to produce shaft work, the steam leaving the turbine condenses into water in the condenser (where cooling water circulates), rejecting heat, and then the water is pumped back to the boiler. Figure 12.2 shows how a unit mass of the working fluid, sometimes in the

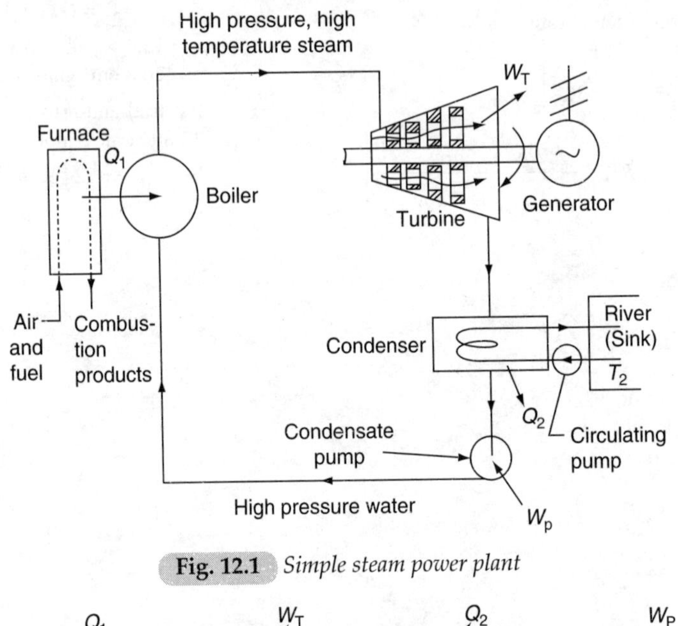

Fig. 12.1 *Simple steam power plant*

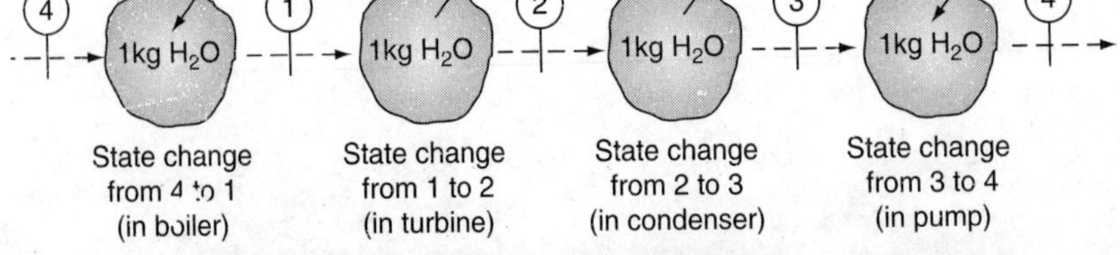

Fig. 12.2 *One kg H_2O executing a heat engine cycle*

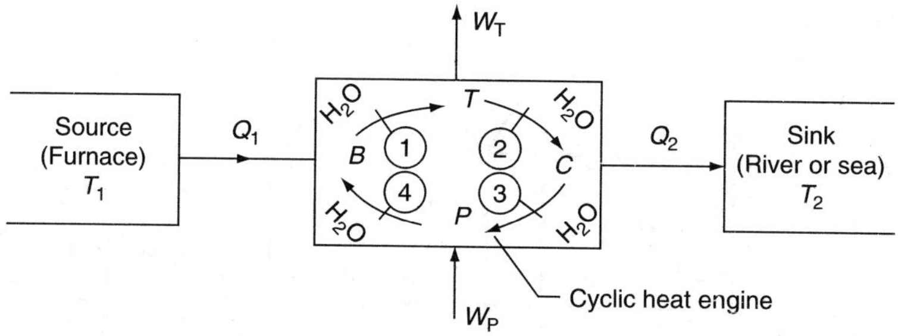

Fig. 12.3 *Cyclic heat engine with water as the working fluid*

liquid phase and sometimes in the vapour phase, undergoes various external heat and work interactions in executing a power cycle. Since the fluid is undergoing a cyclic process, there will be no net change in its internal energy over the cycle, and consequently the net energy transferred to the unit mass of the fluid as heat during the cycle must equal the net energy transfer as work from the fluid. Figure 12.3 shows the cyclic heat engine operating on the vapour power cycle, where the working substance, water, follows along the B-T-C-P (Boiler-Turbine-Condenser-Pump) path, interacting externally as shown, and converting net heat input to net work output continuously. By the first law

$$\sum_{\text{cycle}} Q_{\text{net}} = \sum_{\text{cycle}} W_{\text{net}}$$

or

$$Q_1 - Q_2 = W_T - W_P$$

where Q_1 = heat transferred to the working fluid (kJ/kg)
Q_2 = heat rejected from the working fluid (kJ/kg)
W_T = work transferred from the working fluid (kJ/kg)
W_P = work transferred into the working fluid (kJ/kg)

The efficiency of the vapour power cycle would be given by

$$\eta_{\text{cycle}} = \frac{W_{\text{net}}}{Q_1} = \frac{W_T - W_P}{Q_1} = \frac{Q_1 - Q_2}{Q_1} = 1 - \frac{Q_2}{Q_1} \qquad (12.1)$$

12.2 RANKINE CYCLE

For each process in the vapour power cycle, it is possible to assume a hypothetical or ideal process which represents the basic intended operation and involves no extraneous effects. For the steam boiler, this would be a reversible constant pressure heating process of water to form steam, for the turbine the ideal process would be a reversible adiabatic expansion of steam, for the condenser it would be a reversible constant pressure heat rejection as the steam condenses till it becomes saturated liquid, and for the pump, the ideal process would be the reversible adiabatic compression of this liquid ending at the initial pressure. When all these four processes are ideal, the cycle is an ideal cycle, called a *Rankine cycle*. This is a reversible cycle. Figure 12.4 shows the flow diagram of the Rankine cycle, and in Fig. 12.5, the cycle has been

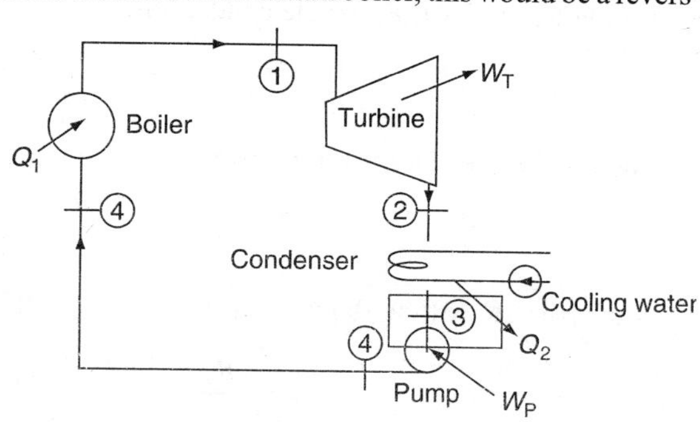

Fig. 12.4 *A simple steam plant*

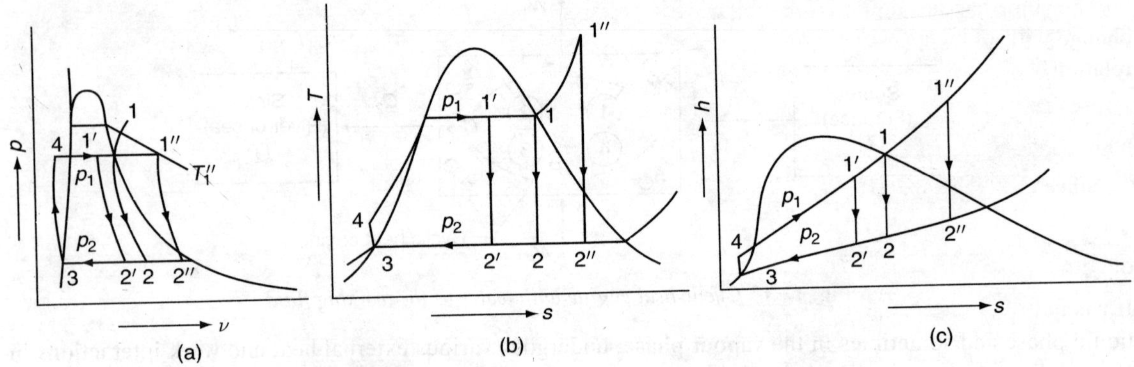

Fig. 12.5 *Rankine cycle on p–v, T–s and h–s diagrams*

plotted on the *p–v*, *T–s*, and *h–s* planes. The numbers on the plots correspond to the numbers on the flow diagram. For any given pressure, the steam approaching the turbine may be dry saturated (state 1) wet (state 1′), or superheated (state 1″), but the fluid approaching the pump is, in each case, saturated liquid (state 3). Steam expands reversibly and adiabatically in the turbine from state 1 to state 2 (or 1′ to 2′, or 1″ to 2″), the steam leaving the turbine condenses to water in the condenser reversibly at constant pressure from state 2 (or 2′, or 2″) to state 3, the water at state 3 is then pumped to the boiler at state 4 reversibly and adiabatically, and the water is heated in the boiler to form steam reversibly at constant pressure from state 4 to state 1 (or 1′ or 1″).

For purposes of analysis the Rankine cycle is assumed to be carried out in a steady flow operation. Applying the steady flow energy equation to each of the processes on the basis of unit mass of fluid, and neglecting changes in kinetic and potential energy, the work and heat quantities can be evaluated in terms of the properties of the fluid.

For 1 kg fluid

The S.F.E.E. for the boiler (control volume) gives

$$h_4 + Q_1 = h_1$$

$$\therefore \quad Q_1 = h_1 - h_4 \tag{12.2}$$

The S.F.E.E. for the turbine as the control volume gives

$$h_1 = W_T + h_2$$

$$\therefore \quad W_T = h_1 - h_2 \tag{12.3}$$

Similarly, the S.F.E.E. for the condenser is

$$h_2 = Q_2 + h_3$$

$$\therefore \quad Q_2 = h_2 - h_3 \tag{12.4}$$

and the S.F.E.E. for the pump gives

$$h_3 + W_p = h_4$$

$$\therefore \quad W_p = h_4 - h_3 \tag{12.5}$$

The efficiency of the Rankine cycle is then given by

$$\eta = \frac{W_{net}}{Q_1} = \frac{W_T - W_P}{Q_1} = \frac{(h_1 - h_2) - (h_4 - h_3)}{h_1 - h_4} \tag{12.6}$$

The pump handles liquid water which is incompressible, i.e. its density or specific volume undergoes little change with an increase in pressure. For reversible adiabatic compression, by the use of the general property relation

$$T\mathrm{d}s = \mathrm{d}h - v\mathrm{d}p;\ \mathrm{d}s = 0$$

and

$$\mathrm{d}h = v\mathrm{d}p$$

Since change in specific volume is negligible

$$\Delta h = v\Delta p$$

or

$$h_4 - h_3 = v_3(p_1 - p_2)$$

If v is in m^3/kg and p is in bar

$$h_4 - h_3 = v_3(p_1 - p_2) \times 10^5 \text{ J/kg} \quad (12.7)$$

The *work ratio is* defined as the ratio of net work output to positive work output.

$$\therefore \quad \text{work ratio} = \frac{W_{\text{net}}}{W_{\text{T}}} = \frac{W_{\text{T}} - W_{\text{P}}}{W_{\text{T}}}$$

Usually, the pump work is quite small compared to the turbine work and is sometimes neglected. Then $h_4 = h_3$, and the cycle efficiency approximately becomes

$$\eta \equiv \frac{h_1 - h_2}{h_1 - h_4}$$

The efficiency of the Rankine cycle is presented graphically in the T–s plot in Fig. 12.6. Thus Q_1 is proportional to area 1564, Q_2 is proportional to area 2563, and W_{net} $(= Q_1 - Q_2)$ is proportional to area 1 2 3 4 enclosed by the cycle.

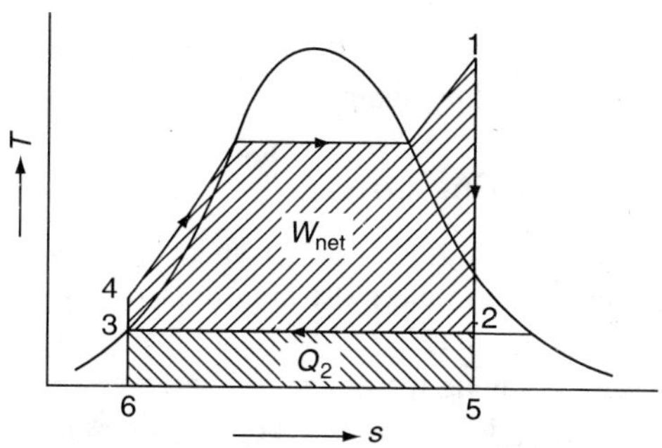

Fig. 12.6 Q_1, W_{net} *and* Q_2 *are proportional to areas*

The capacity of a steam plant is often expressed in terms of *steam rate*, which is defined as the rate of steam flow (kg/h) required to produce unit shaft output (1 kW). Therefore

$$\text{Steam rate} = \frac{1}{W_{\text{T}} - W_{\text{P}}}\frac{\text{kg}}{\text{kJ}} \cdot \frac{1\,\text{kJ/s}}{1\,\text{kW}}$$

$$= \frac{1}{W_{\text{T}} - W_{\text{P}}}\frac{\text{kg}}{\text{kWs}} = \frac{3600}{W_{\text{T}} - W_{\text{P}}}\frac{\text{kJ}}{\text{kWh}} \quad (12.8)$$

The cycle efficiency is sometimes expressed alternatively as heat rate which is the rate of heat input (Q_1) required to produce unit work output (1 kW)

$$\text{Heat rate} = \frac{3600\,Q_1}{W_{\text{T}} - W_{\text{P}}} = \frac{3600}{\eta_{\text{cycle}}}\frac{\text{kJ}}{\text{kWh}} \quad (12.9)$$

From the equation $W_{\text{rev}} = -\int_1^2 v\,\mathrm{d}p$, it is obvious that the reversible steady-flow work is closely associated with the specific volume of fluid flowing through the device. The larger the specific volume, the larger the reversible work produced or consumed by the steady-flow device. Therefore, every effort should be made to keep the specific volume of a fluid as small as possible during a compression process to minimize the work input and as large as possible, during an expansion process to maximize the work output.

In steam or gas power plants (Chapter 13), the pressure rise in the pump or compressor is equal to the pressure drop in the turbine if we neglect the pressure losses in various other components. In steam power plants, the pump handles liquid, which has a very small specific volume, and the turbine handles vapour, whose specific volume is many times larger. Therefore, the work output of the turbine is much larger than the work input to the pump. This is one of the reasons for the overwhelming popularity of steam power plants in electric power generation.

If we were to compress the steam exiting the turbine back to the turbine inlet pressure before cooling it first in the condenser in order to "save" the heat rejected, we would have to supply all the work produced by the turbine back to the compressor. In reality, the required work input would be still greater than the work output of the turbine because of the irreversibilities present in both processes (see Example 12.1).

12.3 ACTUAL VAPOUR CYCLE PROCESSES

The processes of an actual cycle differ from those of the ideal cycle. In the actual cycle conditions might be as indicated in Figs 12.7 and 12.8, showing the various losses. The thermal efficiency of the cycle is

$$\eta_{th} = \frac{W_{net}}{Q_1}$$

where the work and heat quantities are the measured values for the actual cycle, which are different from the corresponding quantities of the ideal cycle.

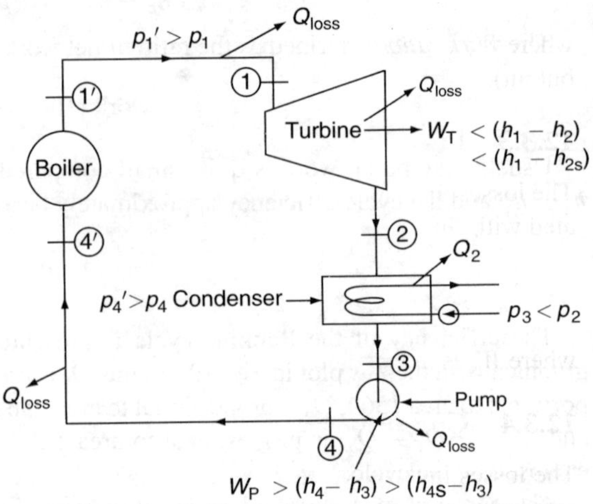

Fig. 12.7 *Various losses in a steam plant*

12.3.1 Piping Losses

Pressure drop due to friction and heat loss to the surroundings are the most important piping losses. States 1′ and 1 (Fig. 12.8) represent the states of the steam leaving the boiler and entering the turbine respectively, 1′ – 1″ represents the frictional losses, and 1″ – 1 shows the constant pressure heat loss to the surroundings. Both the pressure drop and heat transfer reduce the availability of steam entering the turbine.

A similar loss is the pressure drop in the boiler and also in the pipeline from the pump to the boiler. Due to this pressure drop, the water entering the boiler must be pumped to a much higher pressure than the desired steam pressure leaving the boiler, and this requires additional pump work.

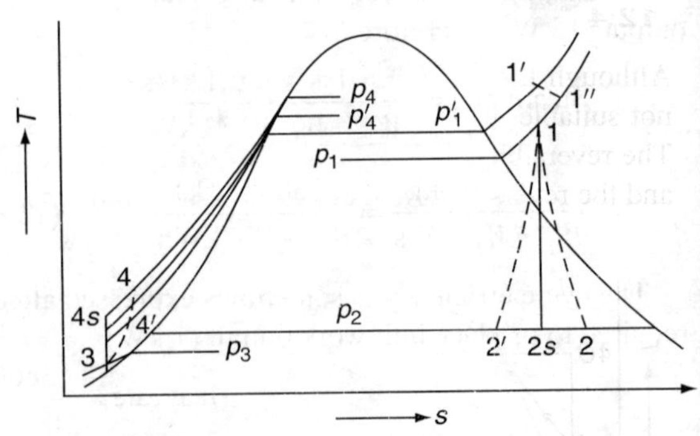

Fig. 12.8 *Various losses on T–s plot*

12.3.2 Turbine Losses

The losses in the turbine are those associated with frictional effects and heat loss to the surroundings. The steady flow energy equation for the turbine in Fig. 12.7 gives

$$h_1 = h_2 + W_T + Q_{loss}$$

$$\therefore \qquad W_T = h_1 - h_2 - Q_{loss} \tag{12.10}$$

For the reversible adiabatic expansion, the path will be 1 – 2*s*. For an ordinary real turbine the heat loss is small, and W_T is $h_1 - h_2$, with Q_2 equal to zero. Since actual turbine work is less than the reversible ideal work output, h_2 is greater than h_{2s}. However, if there is heat loss to the surroundings, h_2 will decrease, accompanied by a decrease in entropy. If the heat loss is large, the end state of steam from the turbine may be 2′. It may so happen that the entropy increase due to frictional effects just balances the entropy decrease due to heat loss, with the result that the initial and final entropies of steam in the expansion process are equal, *but the expansion is neither adiabatic nor reversible*. Except for very small turbines, heat loss from turbines is generally negligible. The isentropic efficiency of the turbine is defined as

$$\eta_T = \frac{W_T}{h_1 - h_{2s}} = \frac{h_1 - h_2}{h_1 - h_{2s}} \tag{12.11}$$

where W_T is the actual turbine work, and $(h_1 - h_{2s})$ is the isentropic enthalpy drop in the turbine (i.e. the ideal output).

12.3.3 Pump Losses

The losses in the pump are similar to those of the turbine, and are primarily due to the irreversibilities associated with fluid friction. Heat transfer is usually negligible. The pump efficiency is defined as

$$\eta_P = \frac{h_{4s} - h_3}{W_P} = \frac{h_{4s} - h_3}{h_4 - h_3} \tag{12.12}$$

where W_P is the actual pump work.

12.3.4 Condenser Losses

The losses in the condenser are usually small. These include the loss of pressure and the cooling of condensate below the saturation temperature.

12.4 COMPARISON OF RANKINE AND CARNOT CYCLES

Although the Carnot cycle has the maximum possible efficiency for the given limits of temperature, it is not suitable in steam power plants. Figure 12.9 shows the Rankine and Carnot cycles on the *T–s* diagram. The reversible adiabatic expansion in the turbine, the constant temperature heat rejection in the condenser, and the reversible adiabatic compression in the pump, are similar characteristic features of both the Rankine

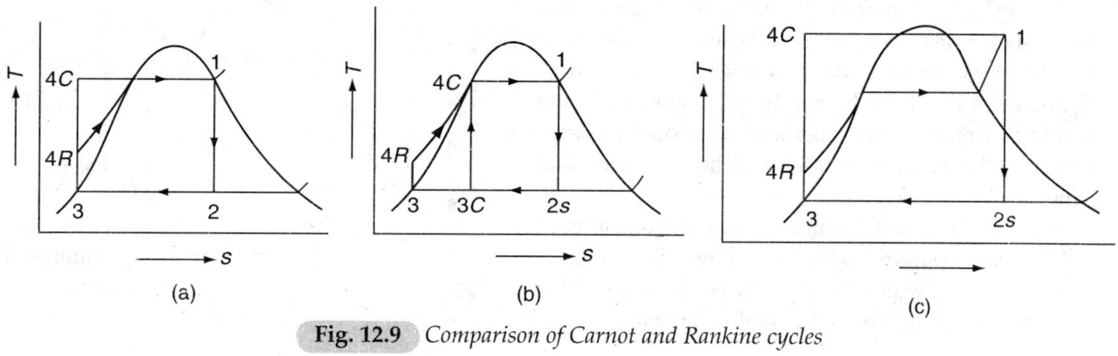

Fig. 12.9 *Comparison of Carnot and Rankine cycles*

and Carnot cycles. But whereas the heat addition process in the Rankine cycle is reversible and at constant pressure, in the Carnot cycle it is reversible and isothermal. In Figs 12.9(a) and 12.9(c), Q_2 is the same in both the cycles, but since Q_1 is more, η_{Carnot} is greater than η_{Rankine}. The two Carnot cycles in Fig. 12.9(a) and 12.9(b) have the same thermal efficiency. Therefore, in Fig. 12.9(b) also, $\eta_{\text{Carnot}} > \eta_{\text{Rankine}}$. But the Carnot cycle cannot be realized in practice because the pump work [in all the three cycles (a), (b), and (c)] is very large. Whereas in (a) and (c) it is impossible to add heat at infinite pressures and at constant temperature from state 4*c* to state 1, in (b), it is difficult to control the quality at 3*c*, so that insentropic compression leads to saturated liquid state.

12.5 MEAN TEMPERATURE OF HEAT ADDITION

In the Rankine cycle, heat is added reversibly at a constant pressure, but at infinite temperatures. If T_{m1} is the mean temperature of heat addition, as shown in Fig. 12.10, so that the area under 4*s* and 1 is equal to the area under 5 and 6, then heat added

$$Q_1 = h_1 - h_{4s} = T_{m1}(s_1 - s_{4s})$$

$$\therefore \quad T_{m1} = \text{Mean temperature of heat addition} = \frac{h_1 - h_{4s}}{s_1 - s_{4s}}$$

Since $\quad Q_2 = \text{Heat rejected} = h_{2s} - h_3$

$$= T_2(s_1 - s_{4s})$$

$$\eta_{\text{Rankine}} = 1 - \frac{Q_2}{Q_1} = 1 - \frac{T_2(s_1 - s_{4s})}{T_{m1}(s_1 - s_{4s})}$$

$$\therefore \quad \eta_{\text{Rankine}} = 1 - \frac{T_2}{T_{m1}} \tag{12.13}$$

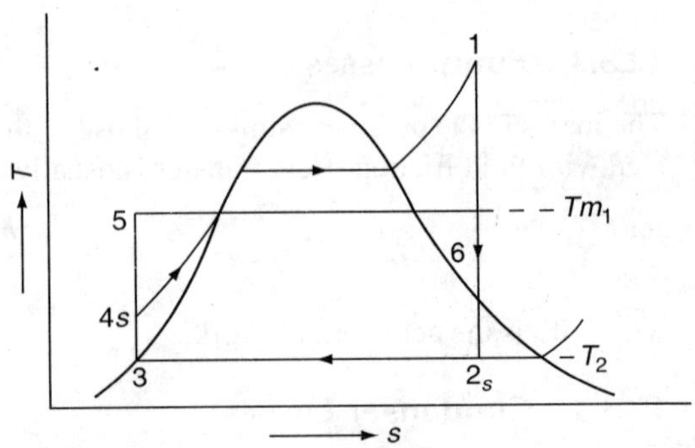

Fig. 12.10 *Mean temperature of heat addition*

where T_2 is the temperature of heat rejection. The lower is the T_2 for a given T_{m1}, the higher will be the efficiency of the Rankine cycle. But the lowest practicable temperature of heat rejection is the temperature of the surroundings (T_0). This being fixed,

$$\eta_{\text{Rankine}} = f(T_{m1}) \text{ only} \tag{12.14}$$

The higher the mean temperature of heat addition, the higher will be the cycle efficiency.

The effect of increasing the initial temperature at constant pressure on cycle efficiency is shown in Fig. 12.11. When the initial state changes from 1 to 1′, T_{m1} between 1 and 1′ is higher than T_{m1} between 4s and 1. So an increase in the superheat at constant pressure increases the mean temperature of heat addition and hence the cycle efficiency.

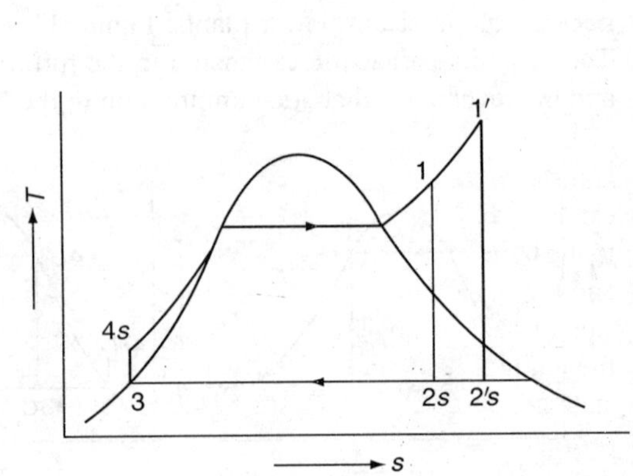

Fig. 12.11 *Effect of superheat on mean temperature of heat addition*

The maximum temperature of steam that can be used is *fixed from metallurgical considerations* (i.e. the materials used for the manufacture of the components which are subjected to high-pressure, high-temperature steam like the superheaters, valves, pipelines, inlet stages of turbine, etc.). When the maximum temperature is fixed, as the operating steam pressure at which heat is added in the boiler increases from p_1 to p_2 (Fig. 12.12), the mean temperature of heat addition increases, since T_{m1} between 7*s* and 5 is higher than that between 4s and 1. But when the turbine inlet pressure increases from p_1 to p_2, the ideal expansion line shifts to the left and the moisture content at the turbine exhaust increases (because $x_{6s} < x_{2s}$). If the moisture content of steam in the later stages of the turbine is higher, the entrained water particles along with the vapour coming out from the nozzles with high velocity strike the blades and erode their surfaces, as a result of which the longevity of the blades decreases. From a consideration of the erosion of blades in the later stages of the turbine, the maximum moisture content at the turbine exhaust is not allowed to exceed 15%, or the quality to fall below 85%. It is desirable that most of the turbine expansion should take place in the single phase or vapour region.

Therefore, with the maximum steam temperature at the turbine inlet, the minimum temperature of heat rejection, and the minimum quality of steam at the turbine exhaust being fixed, the maximum steam pressure at the turbine inlet also gets fixed (Fig. 12.13). The vertical line drawn from 2*s*, fixed by T_2 and x_{2s}, intersects the T_{max} line, fixed by material, at 1, which gives the maximum steam pressure at the turbine inlet. The irreversibility in the expansion process has, however, not been considered.

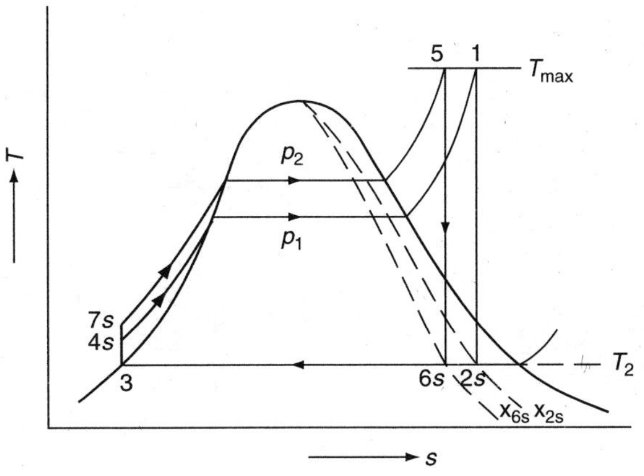

Fig. 12.12 *Effect of increase of pressure on Rankine cycle*

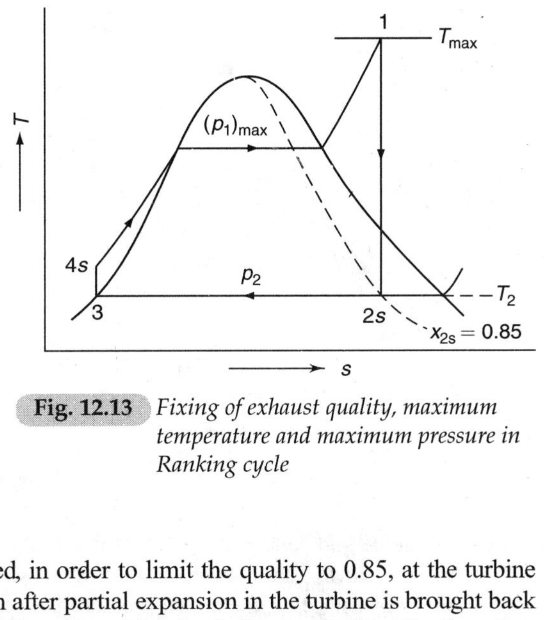

Fig. 12.13 *Fixing of exhaust quality, maximum temperature and maximum pressure in Ranking cycle*

12.6 REHEAT CYCLE

If a steam pressure higher than $(p_1)_{max}$ (Fig. 12.13) is used, in order to limit the quality to 0.85, at the turbine exhaust, reheat has to be adopted. In that case all the steam after partial expansion in the turbine is brought back to the boiler, reheated by combustion gases and then fed back to the turbine for further expansion. The flow, *T–s*, and *h–s* diagrams for the ideal Rankine cycle with reheat are shown in Fig. 12.14. In the reheat cycle the expansion of steam from the initial state 1 to the condenser pressure is carried out in two or more steps, depending upon the number of reheats used. In the first step, steam expands in the high pressure (H.P.) turbine from the initial state to approximately the saturated vapour line (process 1–2*s* in Fig. 12.14). The steam is then resuperheated (or reheated) at constant pressure in the boiler (process 2*s*–3) and the remaining expansion (process 3–4*s*) is carried out in the low pressure (L.P.) turbine. In the case of use of two reheats, steam is resuperheated twice at two different constant pressures. To protect the reheat tubes, steam is not allowed to expand deep into the two-phase region before it is taken for reheating, because in that case the moisture particles in steam while evaporating

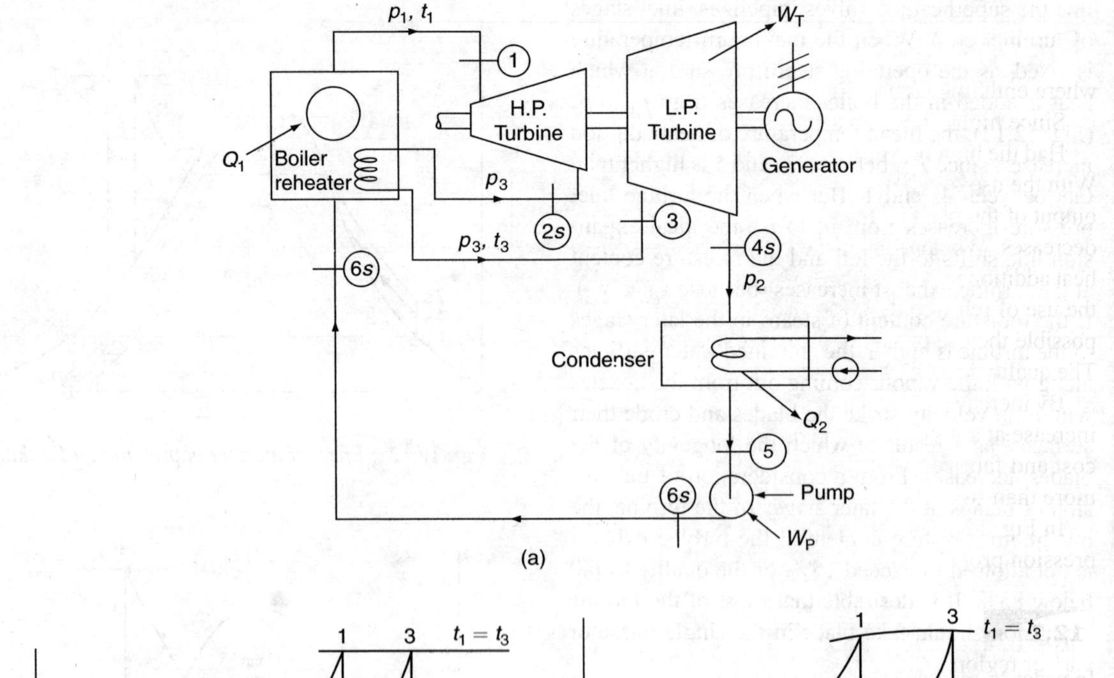

Fig. 12.14 *Reheat cycle*

would leave behind solid deposits in the form of scale which is difficult to remove. Also, a low reheat pressure may bring down T_{m1} and hence, cycle efficiency. Again, a high reheat pressure increases the moisture content at turbine exhaust. Thus, the reheat pressure is optimized. The optimum reheat pressure for most of the modern power plants is about 0.2 to 0.25 of the initial steam pressure. For the cycle in Fig. 12.14, for 1 kg of steam

$$Q_1 = h_1 - h_{6s} + h_3 - h_{2s}$$

$$Q_2 = h_{4s} - h_5$$

$$W_T = h_1 - h_{2s} + h_3 - h_{4s}$$

$$W_P = h_{6s} - h_5$$

$$\eta = \frac{W_T - W_P}{Q_1} = \frac{(h_1 - h_{2s} + h_3 - h_{4s}) - (h_{6s} - h_5)}{h_1 - h_{6s} + h_3 - h_{2s}} \tag{12.15}$$

$$\text{Steam rate} = \frac{3600}{(h_1 - h_{2s} + h_3 - h_{4s}) - (h_{6s} - h_5)} \text{ kg/kWh} \tag{12.16}$$

where enthalpy is in kJ/kg.

Since higher pressures are used in a reheat cycle, pump work may be appreciable.

Had the high pressure p_1 been used without reheat, the ideal Rankine cycle would have been $1 - 4's - 5 - 6s$. With the use of reheat, the area $2s - 3 - 4s - 4's$ has been added to the basic cycle. It is obvious that net work output of the plant increases with reheat, because $(h_3 - h_{4s})$ is greater than $(h_{2s} - h_{4's})$, and hence the steam rate decreases. Whether the cycle efficiency improves with reheat depends upon whether the mean temperature of heat addition in process $2s - 3$ is higher than the mean temperature of heat addition in process $6s - 1$. In practice, the use of reheat only gives a small increase in cycle efficiency, but it increases the net work output by making possible the use of higher pressures, keeping the quality of steam at turbine exhaust within a permissible limit. The quality improves from $x_{4's}$ to x_{4s} by the use of reheat.

By increasing the number of reheats, still higher steam pressures could be used, but the mechanical stresses increase at a higher proportion than the increase in pressure, because of the prevailing high temperature. The cost and fabrication difficulties will also increase. In that way, the maximum steam pressure gets fixed, and more than two reheats have not yet been used so far.

In Fig. 12.14, only ideal processes have been considered. The irreversibilities in the expansion and compression processes have been considered in the example given later.

12.7 IDEAL REGENERATIVE CYCLE

In order to increase the mean temperature of heat addition (T_{m1}), attention was so far confined to increasing the amount of heat supplied at high temperatures, such as increasing superheat, using higher pressure and temperature of steam, and using reheat. The mean temperature of heat addition can also be increased by decreasing the amount of heat added at low temperatures. In a saturated steam Rankine cycle (Fig. 12.15), a considerable part of the total heat supplied is in the liquid phase when heating up water from 4 to 4′, at a temperature lower than T_1, the maximum temperature of the cycle. For maximum efficiency, all heat should be supplied at T_1, and feedwater should enter the boiler at state 4′. This may be accomplished in what is known as an ideal regenerative cycle, the flow diagram of which is shown in Fig. 12.16 and the corresponding T–s diagram in Fig. 12.17.

The unique feature of the ideal regenerative cycle is that the condensate, after leaving the pump circulates around the turbine casing, counterflow to the direction of vapour flow in the turbine (Fig. 12.16).

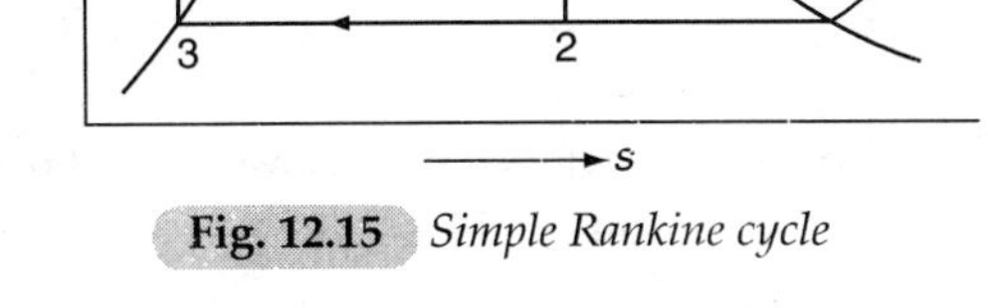

Fig. 12.15 *Simple Rankine cycle*

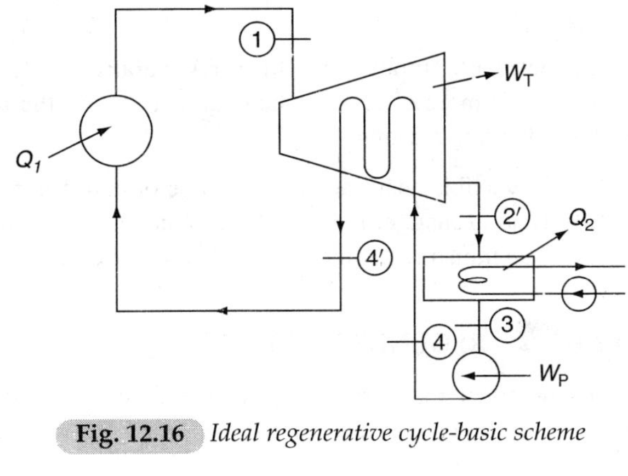

Fig. 12.16 *Ideal regenerative cycle-basic scheme*

Thus it is possible to transfer heat from the vapour as it flows through the turbine to the liquid flowing around the turbine. Let us assume that this is a reversible heat transfer, i.e. at each point the temperature of the vapour is only infinitesimally higher than the temperature of the liquid. The process 1–2′ (Fig. 12.17) thus represents reversible expansion of steam in the turbine with reversible heat rejection. For any small step in the process of heating the water,

$$\Delta T \text{ (water)} = -\Delta T \text{ (steam)}$$

and

$$\Delta s \text{ (water)} = -\Delta s \text{ (steam)}$$

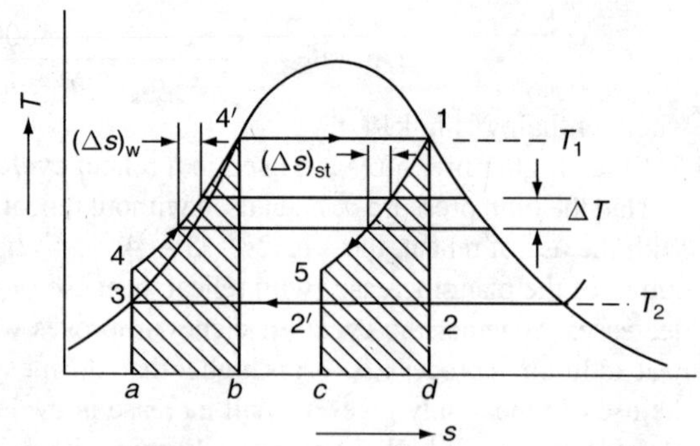

Fig. 12.17 *Ideal regenerative cycle on T–s plot*

Then the slopes of lines 1–2′ and 4′–3 (Fig. 12.17) will be identical at every temperature and the lines will be identical in contour. Areas 4–4′–*b*–*a*–4 and 2′–1–*d*–*c*–2′ are not only equal but congruous. Therefore, all the heat added from an external source (Q_1) is at the constant temperature T_1, and all the heat rejected (Q_2) is at the constant temperature T_2, both being reversible.

Then

$$Q_1 = h_1 - h_{4'} = T_1(s_1 - s_{4'})$$
$$Q_2 = h_{2'} - h_3 = T_2(s_{2'} - s_3)$$

Since

$$s_{4'} - s_3 = s_1 - s_{2'}$$

or

$$s_1 - s_{4'} = s_{2'} - s_3$$

$$\therefore \quad \eta = 1 - \frac{Q_2}{Q_1} = 1 - \frac{T_2}{T_1}$$

The efficiency of the ideal regenerative cycle is thus equal to the Carnot cycle efficiency.

Writing the steady flow energy equation for the turbine

$$h_1 - W_T - h_{2'} + h_4 - h_{4'} = 0$$

or

$$W_T = (h_1 - h_{2'}) - (h_{4'} - h_4) \tag{12.17}$$

The pump work remains the same as in the Rankine cycle, i.e.

$$W_P = h_4 - h_3$$

The net work output of the ideal regenerative cycle is thus less, and hence its steam rate will be more, although it is more efficient, when compared with the Rankine cycle. However, the cycle is not practicable for the following reasons:

(a) Reversible heat transfer cannot be obtained in finite time.
(b) Heat exchanger in the turbine is mechanically impracticable.
(c) The moisture content of the steam in the turbine will be high.

12.8 REGENERATIVE CYCLE

In the practical regenerative cycle, the feedwater enters the boiler at a temperature between 4 and 4′ (Fig. 12.17), and it is heated by steam extracted from intermediate stages of the turbine. The flow diagram of

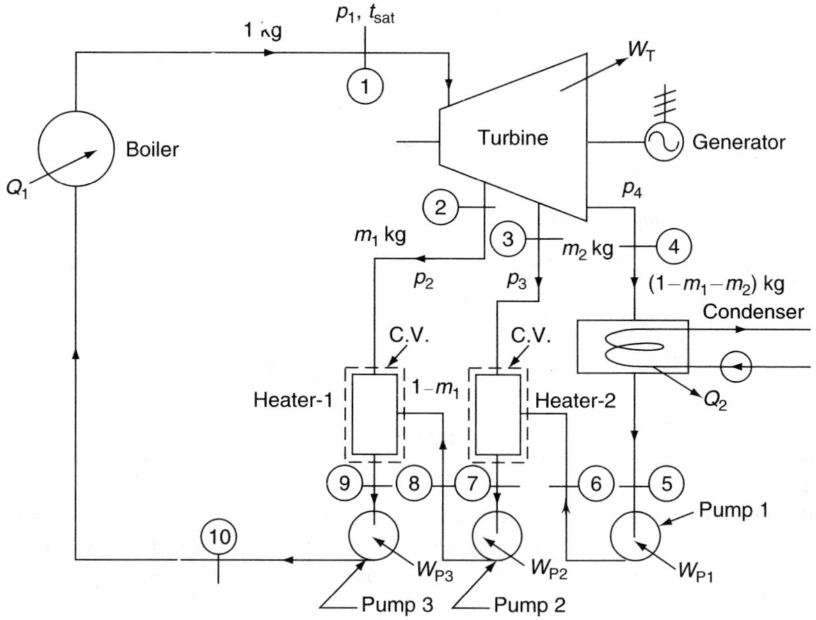

Fig. 12.18 *Regenerative cycle flow diagram with two feedwater heaters*

the regenerative cycle with saturated steam at the inlet to the turbine, and the corresponding T–s diagram are shown in Figs 12.18 and 12.19 respectively. For every kg of steam entering the turbine, let m_1 kg steam be extracted from an intermediate stage of the turbine where the pressure is p_2, and it is used to heat up feedwater [$(1 - m_1)$ kg at state 8] by mixing in heater 1. The remaining $(1 - m_1)$ kg of steam then expands in the turbine from pressure p_2 (state 2) to pressure p_3 (state 3) when m_2 kg of steam is extracted for heating feedwater in heater 2. So $(1 - m_1 - m_2)$ kg of steam then expands in the remaining stages of the turbine to pressure p_4, gets condensed into water in the condenser, and then pumped to heater 2, where it mixes with m_2 kg of steam extracted at pressure p_3. Then $(1 - m_1)$ kg of water is pumped to heater 1 where it mixes with m_1 kg

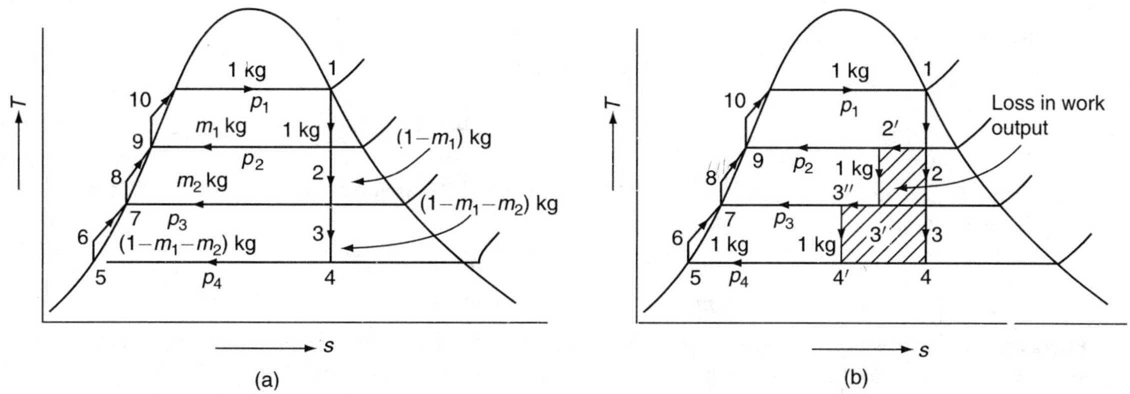

Fig. 12.19 *(a) Regenerative cycle on T–s plot with decreasing mass of fluid*
(b) Regenerative cycle on T–s plot for unit mass of fluid

of steam extracted at pressure p_2. The resulting 1 kg of steam is then pumped to the boiler where heat from an external source is supplied. Heaters 1 and 2 thus operate at pressures p_2 and p_3 respectively. The amounts of steam m_1 and m_2 extracted from the turbine are such that at the exit from each of the heaters, the state is saturated liquid at the respective pressures. The heat and work transfer quantities of the cycle are

$$W_T = 1\,(h_1 - h_2) + (1 - m_1)\,(h_2 - h_3) + (1 - m_1 - m_2)\,(h_3 - h_4) \text{ kJ/kg} \tag{12.18}$$

$$\begin{aligned} W_P &= W_{P1} + W_{P2} + W_{P3} \\ &= (1 - m_1 - m_2)\,(h_6 - h_5) + (1 - m_1)\,(h_8 - h_7) + 1\,(h_{10} - h_9) \text{ kJ/kg} \end{aligned} \tag{12.19}$$

$$Q_1 = 1\,(h_1 - h_{10}) \text{ kJ/kg} \tag{12.20}$$

$$Q_2 = (1 - m_1 - m_2)\,(h_4 - h_5) \text{ kJ/kg} \tag{12.21}$$

Cycle efficiency,

$$\eta = \frac{Q_1 - Q_2}{Q_1} = \frac{W_T - W_P}{Q_1}$$

$$\text{Steam rate} = \frac{3600}{W_T - W_P} \text{ kg/kW h}$$

In the Rankine cycle operating at the given pressures, p_1 and p_4, the heat addition would have been from state 6 to state 1. By using two stages of regenerative feedwater heating, feedwater enters the boiler at state 10, instead of state 6, and heat addition is, therefore, from state 10 to state 1. Therefore

$$(T_{m_1})_{\text{with regeneration}} = \frac{h_1 - h_{10}}{s_1 - s_{10}} \tag{12.22}$$

and

$$(T_{m_1})_{\text{without regeneration}} = \frac{h_1 - h_6}{s_1 - s_6} \tag{12.23}$$

Since

$$(T_{m_1})_{\text{with regeneration}} > (T_{m_1})_{\text{without regeneration}}$$

the efficiency of the regenerative cycle will be higher than that of the Rankine cycle.

The energy balance for heater 2 gives

$$m_1 h_2 + (1 - m_1)\,h_8 = 1 h_9$$

$$m_1 = \frac{h_9 - h_8}{h_2 - h_8} \tag{12.24}$$

The energy balance for heater 1 gives

$$m_2 h_3 + (1 - m_1 - m_2)\,h_6 = (1 - m_1)\,h_7$$

or

$$m_2 = (1 - m_1)\,\frac{h_7 - h_6}{h_3 - h_6} \tag{12.25}$$

From Eqs (12.24) and (12.25), m_1 and m_2 can be evaluated. Equations (12.24) and (12.25) can also be written alternatively as

$$(1 - m_1)\,(h_9 - h_8) = m_1\,(h_2 - h_9)$$

$$(1 - m_1 - m_2)\,(h_7 - h_6) = m_2\,(h_3 - h_7)$$

Energy gain of feedwater = Energy given off by vapour in condensation. Heaters have been assumed to be adequately insulated, and there is no heat gain from, or heat loss to, the surroundings.

Path 1–2–3–4 in Fig. 12.19 represents the states of a decreasing mass of fluid. For 1 kg of steam, the states would be represented by the path 1–2′–3′–4′. From Equation (12.18),

$$W_T = (h_1 - h_2) + (1 - m_1)(h_2 - h_3) + (1 - m_1 - m_2)(h_3 - h_4)$$
$$= (h_1 - h_2) + (h_{2'} - h_{3'}) + (h_{3'} - h_{4'}) \tag{12.26}$$

where
$$(1 - m_1)(h_2 - h_3) = 1(h_{2'} - h_{3'}) \tag{12.27}$$
$$(1 - m_1 - m_2)(h_3 - h_4) = 1(h_{3''} - h_{4'}) \tag{12.28}$$

The cycle 1–2–2′–3′–3″–4′–5–6–7–8–9–10–1 represents 1 kg of working fluid. The heat released by steam condensing from 2 to 2′ is utilized in heating up the water from 8 to 9.

∴
$$1(h_2 - h_{2'}) = 1(h_9 - h_8) \tag{12.29}$$

Similarly
$$1(h_{3'} - h_{3''}) = 1(h_7 - h_6) \tag{12.30}$$

From Eqs (12.26), (12.29) and (12.30)
$$W_T = (h_1 - h_{4'}) - (h_2 - h_{2'}) - (h_{3'} - h_{3''})$$
$$= (h_1 - h_{4'}) - (h_9 - h_8) - (h_7 - h_6) \tag{12.31}$$

The similarity of Eqs (12.17) and (12.31) can be noticed. It is seen that the stepped cycle 1–2′–3′–4′–5–6–7–8–9–10 approximates the ideal regenerative cycle in Fig. 12.17, and that a greater number of stages would give a closer approximation (Fig. 12.20). Thus the heating of feedwater by steam 'bled' from the turbine, known as regeneration, *carnotizes* the Rankine cycle.

The heat rejected Q_2 in the cycle decreases from $(h_4 - h_5)$ to $(h_{4'} - h_5)$. There is also loss in work output by the amount (Area under 2–2′ + Area under 3′–3″ – Area under 4–4′), as shown by the hatched area in Fig. 12.19(b). So the steam rate increases by regeneration, i.e. more steam has to circulate per hour to produce unit shaft output.

The enthalpy-entropy diagram of a regenerative cycle is shown in Fig. 12.21.

12.9 REHEAT-REGENERATIVE CYCLE

The reheating of steam is adopted when the vaporization pressure is high. The effect of reheat alone on the thermal efficiency of the cycle is very small. Regeneration or the heating up of feedwater by steam extracted from the turbine has a marked effect on cycle efficiency. A modern steam power plant is equipped with both. Figures 12.22 and 12.23 give the flow and *T–s* diagrams of a steam plant with reheat and three stages of feedwater heating. Here

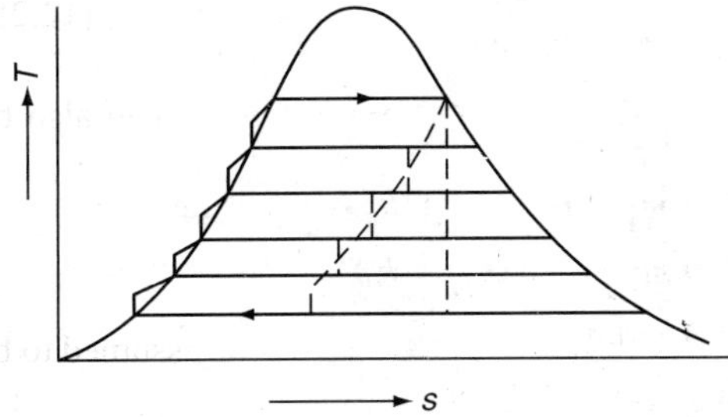

Fig. 12.20 *Regenerative cycle with many stages of feedwater heating*

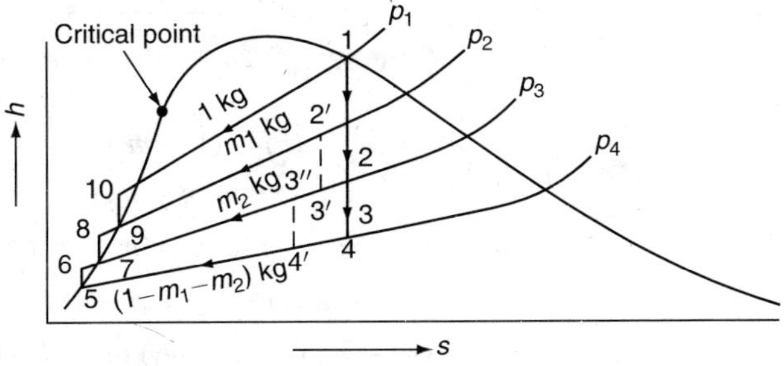

Fig. 12.21 *Regenerative cycle on h–s diagram*

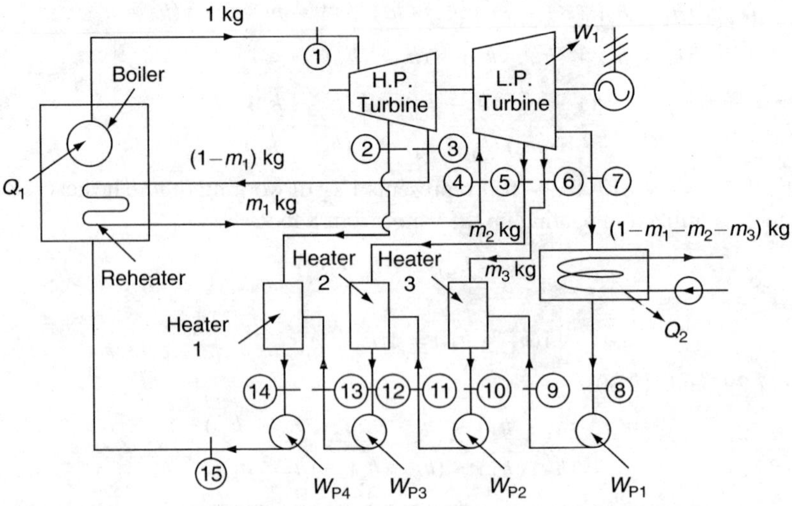

Fig. 12.22 *Reheat-regenerative cycle flow diagram*

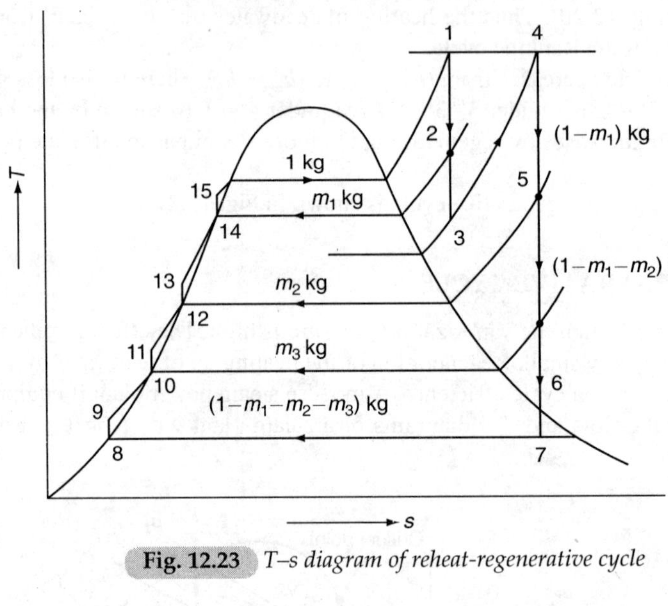

Fig. 12.23 *T–s diagram of reheat-regenerative cycle*

$$W_T = (h_1 - h_2) + (1 - m_1)(h_2 - h_3) + (1 - m_1)(h_4 - h_5) + (1 - m_1 - m_2)(h_5 - h_6) + (1 - m_1 - m_2 - m_3)(h_6 - h_7) \text{ kJ/kg}$$

$$W_P = (1 - m_1 - m_2 - m_3)(h_9 - h_8) + (1 - m_1 - m_2)(h_{11} - h_{10}) + (1 - m_1)(h_{13} - h_{12}) + 1(h_{15} - h_{14}) \text{ kJ/kg}$$

$$Q_1 = (h_1 - h_{15}) + (1 - m_1)(h_4 - h_3) \text{ kJ/kg}$$

and

$$Q_2 = (1 - m_1 - m_2 - m_3)(h_7 - h_8) \text{ kJ/kg}$$

The energy balance of heaters 1, 2, and 3 give

$$m_1 h_2 + (1 - m_1)h_{13} = 1 \times h_{14}$$

$$m_2 h_5 + (1 - m_1 - m_2)h_{11} = (1 - m_1)h_{12}$$

$$m_3 h_6 + (1 - m_1 - m_2 - m_3)h_9 = (1 - m_1 - m_2)h_{10}$$

from which m_1, m_2, and m_3 can be evaluated.

12.10 FEEDWATER HEATERS

Feedwater heaters are of two types, viz., open heaters and closed heaters. In an open or contact-type heater, the extracted or bled steam is allowed to mix with feedwater, and both leave the heater at a common temperature, as shown in Figs 12.18 and 12.22. In a closed heater, the fluids are kept separate, and not allowed to mix together (Fig. 12.24). The feedwater flows through the tubes in the heater and the extracted steam condenses on the outside of the tubes in the shell. The heat released by condensation is transferred to the feedwater through the walls of the tubes. The condensate (saturated water at the steam extraction pressure), sometimes called the heater-drip, then passes through a trap into the next lower pressure heater. This, to some extent, reduces the steam required by that heater. The trap passes only liquid and no vapour. The drip from the lowest pressure heater could similarly be trapped to the condenser, but this would be throwing away energy to the condenser cooling water. To avoid this waste, a drip pump feeds the drip directly into the feedwater stream.

Figure 12.25 shows the *T–s* plot corresponding to the flow diagram in Fig. 12.24. The temperature of the feedwater (at '*l*' or '*o*') leaving a particular heater is always less than the saturation temperature at the steam extraction pressure (at *e* or *g*), the difference being known as the *terminal temperature difference* of the heater.

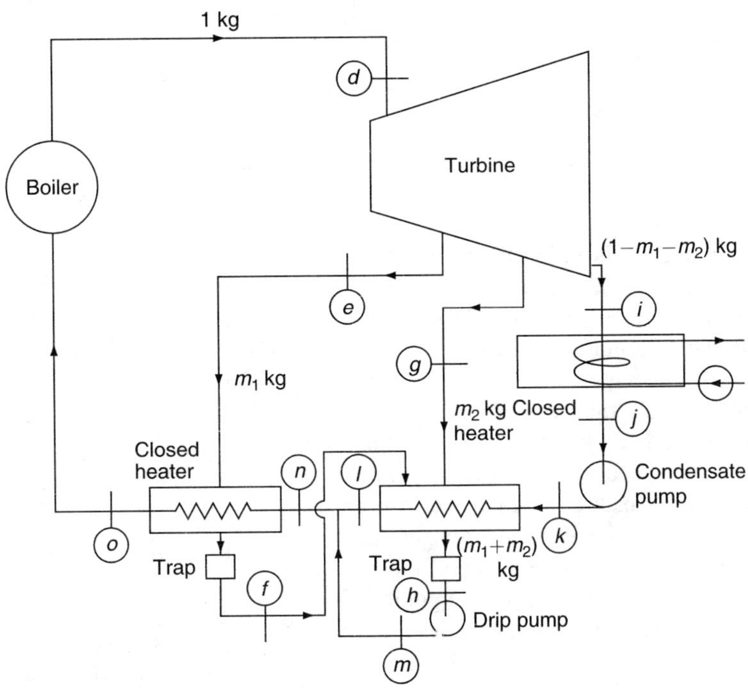

Fig. 12.24 *Regenerative cycle flow diagram with closed feedwater heaters*

The advantages of the open heater are simplicity, lower cost, and high heat transfer capacity. The disadvantage is the necessity of a pump at each heater to handle the large feedwater stream.

A closed heater system requires only a single pump for the main feedwater stream regardless of the number of heaters. The drip pump, if used, is relatively small. Closed heaters are costly and may not give as high a feedwater temperature as do open heaters. In most steam power plants, closed heaters are favoured, but at least one open heater is used, primarily for the purpose of feedwater deaeration. The open heater in such a system is called the *deaerator*.

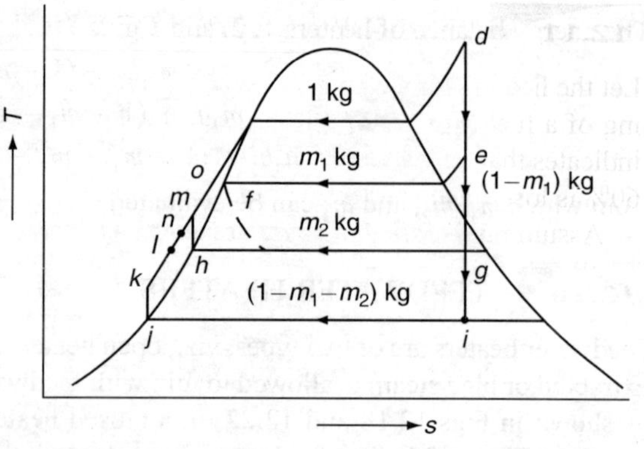

Fig. 12.25 *T–s diagram of regenerative cycle with closed feedwater heaters*

The higher the number of heaters used, the higher will be the cycle efficiency. If n heaters are used, the greatest gain in efficiency occurs when the overall temperature rise is about $n/(n + 1)$ times the difference between the condenser and boiler saturation temperatures. (See *Analysis of Engineering Cycles* by R.W. Haywood, Pergamon Press, 1973).

If $(\Delta f)_0 = t_{\text{boiler sat}} - t_{\text{cond}}$ and $(\Delta t)_{\text{fw}}$ = temperature rise of feedwaters, it is seen that.

$$n = 0 \quad (\Delta t)_{fw} = 0$$

$$n = 1, \quad (\Delta t)_{fw} = \frac{1}{2}(\Delta t)_0$$

$$\text{Gain} = \frac{1}{6}(\Delta t)_0$$

$$n = 2, \quad (\Delta t)_{fw} = \frac{2}{3}(\Delta t)_0$$

$$\text{Gain} = \frac{1}{12}(\Delta t)_0$$

$$n = 3, \quad (\Delta t)_{fw} = \frac{3}{4}(\Delta t)_0$$

$$\text{Gain} = \frac{1}{20}(\Delta t)_0$$

$$n = 4, \quad (\Delta t)_{fw} = \frac{4}{2}(\Delta t)_0$$

Since the cycle efficiency is proportional to $(\Delta t)_{\text{fw}}$ the efficiency gain follows *the law of diminishing return* with the increase in the number of heaters. The greatest increment in efficiency occurs by the use of the first heater (Fig. 12.26). The increments for each additional heater thereafter successively diminish. The number of heaters is fixed up by the energy balance of the whole plant when it is found that the cost of adding another does not justify the saving in Q_1 or the marginal increase in cycle efficiency. An increase in feedwater temperature may, in some cases, cause a reduction in boiler efficiency. So, the number of heaters gets optimized. Five points of extraction are often used in practice. Some cycles use as many as nine.

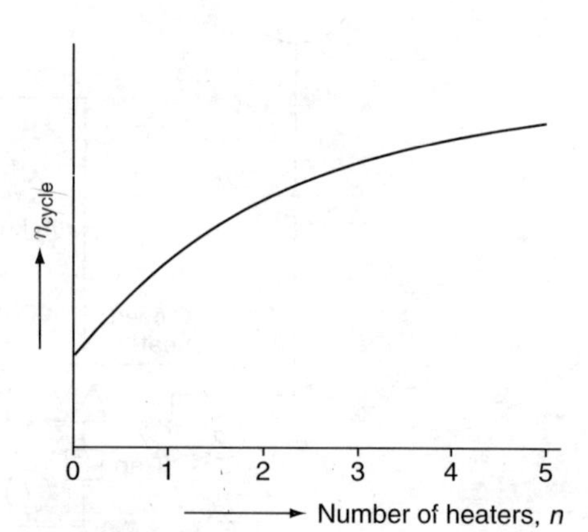

Fig. 12.26 *Effect of the use of number of heaters on cycle efficiency*

12.11 EXERGY ANALYSIS OF VAPOUR POWER CYCLES

Let the heating for steam generation in the boiler unit be provided by a stream of hot gases produced by burning of a fuel (Fig. 12.27). The distribution of input energy is shown in the Sankey diagram 12.27 (b) which indicates that only about 30% of the input energy to the simple ideal plant is converted to shaft work and about 60% is lost to the condenser. The exergy analysis, however, gives a different distribution as discussed below.

Assuming that the hot gases are at atmospheric pressure, the exergy input is

$$a_{f_1} = w_g c_{p_g}\left[T_i - T_0 - T_0 \ln \frac{T_i}{T_0}\right]$$

$$= w_g c_{p_g} T_0 \left[\frac{T_i}{T_0} - 1 - \ln \frac{T_i}{T_0}\right]$$

Similarly, the exergy loss rate with the exhaust stream is:

$$a_{f_2} = w_g c_{p_g} T_0 \left[\frac{T_e}{T_0} - 1 - \ln \frac{T_e}{T_0}\right]$$

Net exergy input rate in the steam generation process:

$$a_{i_1} = a_{f_1} - a_{f_2}$$

The exergy utilization rate in the steam generation is:

$$a_{f_u} = w_s[(h_1 - h_4) - T_0(s_1 - s_4)]$$

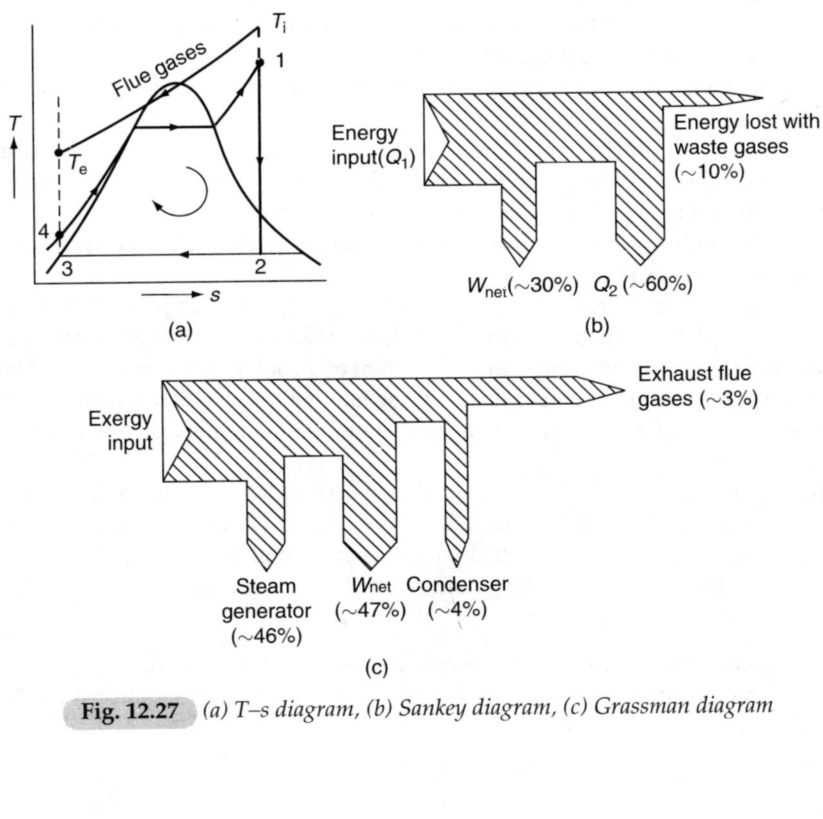

Fig. 12.27 *(a) T–s diagram, (b) Sankey diagram, (c) Grassman diagram*

Rate of exergy loss in the steam generator:

$$I = a_{i_1} - a_{f_u}$$

The useful mechanical power output:

$$= W_{net} = w_s[(h_1 - h_2) - (h_4 - h_3)]$$

Exergy flow rate of the wet steam to the condenser:

$$a_{f_c} = w_s[(h_2 - h_3) - T_0(s_2 - s_3)]$$

Second law efficiency,

$$\eta_{II} = \frac{W_{net}}{a_{f_1} - a_{f_2}}$$

Exergy flow or Grassman diagram is shown in Fig. 12.27 (c). The energy disposition diagram (b) shows that the major energy loss (~60%) takes place in the condenser. This energy rejection, however, occurs at a temperature close to the ambient temperature, and, therefore, corresponds to a very low exergy value (~4%). The major exergy destruction due to irreversibilities takes place in the steam generation. To improve the performance of the steam plant the finite source temperatures must be closer to the working fluid temperatures to reduce thermal irreversibility.

12.12 CHARACTERISTICS OF AN IDEAL WORKING FLUID IN VAPOUR POWER CYCLES

There are certain drawbacks with steam as the working substance in a power cycle. The maximum temperature that can be used in steam cycles consistent with the best available material is about 600°C, while the critical temperature of steam is 375°C, which necessitates large superheating and permits the addition of only an infinitesimal amount of heat at the highest temperature.

High moisture content is involved in going to higher steam pressures in order to obtain higher mean temperature of heat addition (T_{m1}). The use of reheat is thus necessitated. Since reheater tubes are costly, the use of more than two reheats is hardly recommended. Also, as pressure increases, the metal stresses increase, and the thicknesses of the walls of boiler drums, tubes, pipe lines, etc. increase not in proportion to pressure increase, but much faster, because of the prevalence of high temperature.

It may be noted that high T_{m1} is only desired for high cycle efficiency. High pressures are only forced by the characteristics (weak) of steam.

If the lower limit is now considered, it is seen that at the heat rejection temperature of 40°C, the saturation pressure of steam is 0.075 bar, which is considerably lower than atmospheric pressure. The temperature of heat rejection can be still lowered by using some refrigerant as a coolant in the condenser. The corresponding vacuum will be still higher, and to maintain such low vacuum in the condenser is a big problem.

It is the low temperature of heat rejection that is of real interest. The necessity of a vacuum is a disagreeable characteristic of steam.

The saturated vapour line in the T–s diagram of steam is sufficiently inclined, so that when steam is expanded to lower pressures (for higher turbine output as well as cycle efficiency), it involves more moisture content, which is not desired from the consideration of the erosion of turbine blades in later stages.

The desirable characteristics of the working fluid in a vapour power cycle to obtain best thermal efficiency are given below:

(a) The fluid should have a high critical temperature so that the saturation pressure at the maximum permissible temperature (metallurgical limit) is relatively low. It should have a large enthalpy of evaporation at that pressure.

(b) The saturation pressure at the temperature of heat rejection should be above atmospheric pressure so as to avoid the necessity of maintaining vacuum in the condenser.

(c) The specific heat of liquid should be small so that little heat transfer is required to raise the liquid to the boiling point.

(d) The saturated vapour line of the *T–s* diagram should be steep, very close to the turbine expansion process so that excessive moisture does not appear during expansion.

(e) The freezing point of the fluid should be below room temperature, so that it does not get solidified while flowing through the pipelines.

(f) The fluid should be chemically stable and should not contaminate the materials of construction at any temperature.

(g) The fluid should be nontoxic, noncorrosive, not excessively viscous, and low in cost.

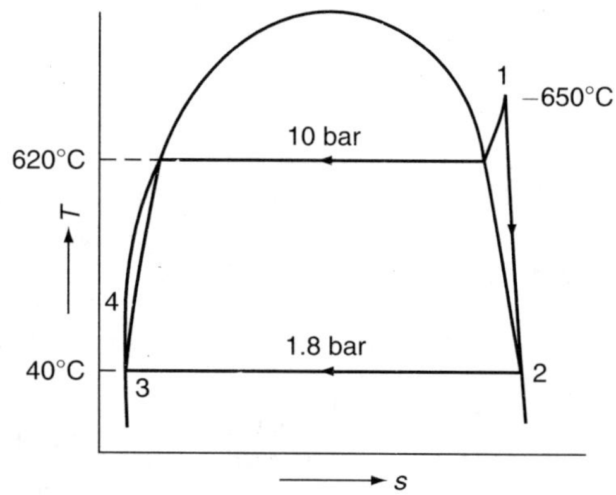

Fig. 12.28 *T–s diagram of an ideal working fluid for a vapour power cycle*

The characteristics of such an ideal fluid are approximated in the *T-s* diagram as shown in Fig. 12.28. Some superheat is desired to reduce piping losses and improve turbine efficiency. The thermal efficiency of the cycle is very close to the Carnot efficiency.

12.13 BINARY VAPOUR CYCLES

No single fluid can meet all the requirements as mentioned above. Although in the overall evaluation, water is better than any other working fluid, however, in the high temperature range, there are a few better fluids, and notable among them are (a) diphenyl ether, $(C_6H_5)_2O$, (b) aluminium bromide, Al_2Br_6, and (c) mercury and other liquid metals like sodium or potassium. From among these, only mercury has actually been used in practice. Diphenyl ether could be considered, but it has not yet been used because, like most organic substances, it decomposes gradually at high temperatures. Aluminium bromide is a possibility and yet to be considered.

When $p = 12$ bar, the saturation temperature for water, aluminium bromide, and mercury are 187°C, 482.5°C, and 560°C respectively. Mercury is thus a better fluid in the high temperature range, because at high temperature, its vaporization pressure is relatively low. Its critical pressure and temperature are 1080 bar and 1460°C respectively.

But in the low temperature range, mercury is unsuitable, because its saturation pressure becomes exceedingly low and it would be impractical to maintain such a high vacuum in the condenser. At 30°C, the saturation pressure of mercury is only 2.7×10^{-4} cm Hg. Its specific volume at such a low pressure is very large, and it would be difficult to accommodate such a large volume flow.

For this reason, mercury vapour leaving the mercury turbine is condensed at a higher temperature, and the heat released during the condensation of mercury is utilized in evaporating water to form steam to operate on a conventional turbine.

Thus in the binary (or two-fluid) cycle, two cycles with different working fluids are coupled in series, the heat rejected by one being utilized in the other.

The flow diagram of mercury-steam binary cycle and the corresponding *T–s* diagram are given in Figs 12.29 and 12.30 respectively. The mercury cycle, *a–b–c–d*, is a simple Rankine type of cycle using saturated vapour. Heat is supplied to the mercury in process *d–a*. The mercury expands in a turbine (process *a–b*) and is then condensed in process *b–c*. The feed pump process, *c–d*, completes the cycle.

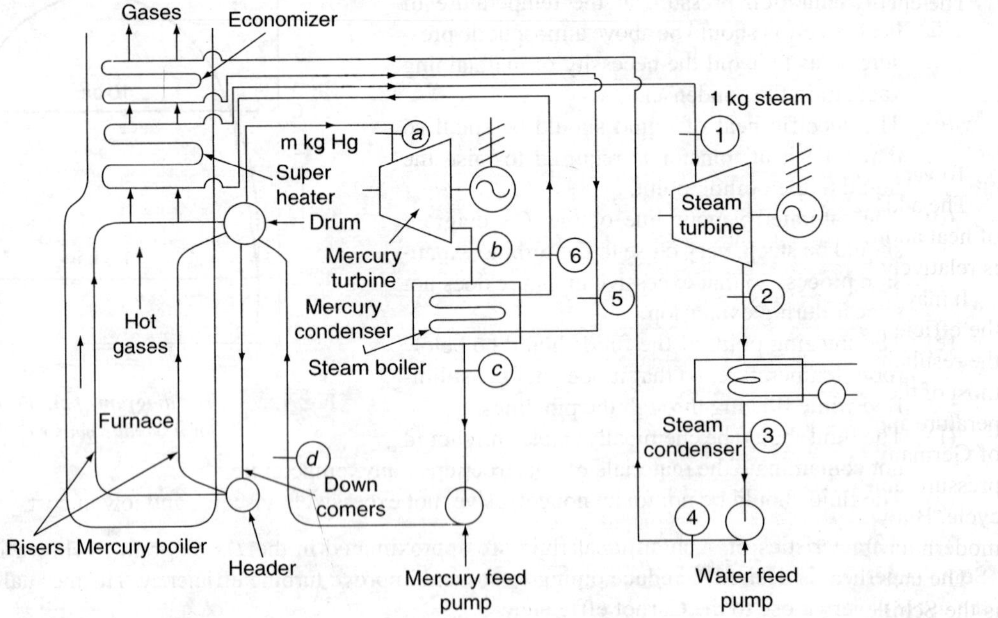

Fig. 12.29 *Mercury-steam plant flow diagram*

The heat rejected by mercury during condensation is transferred to boil water and form saturated vapour (process 5–6). The saturated vapour is heated from the external source (furnace) in the superheater (process 6–1). Superheated steam expands in the turbine (process 1–2) and is then condensed (process 2–3). The feedwater (condensate) is then pumped (process 3–4), heated till it is saturated liquid in the economizer (process 4–5) before going to the mercury condenser-steam boiler, where the latent heat is absorbed. In an actual plant the steam cycle is always a regenerative cycle, but for the sake of simplicity, this complication has been omitted.

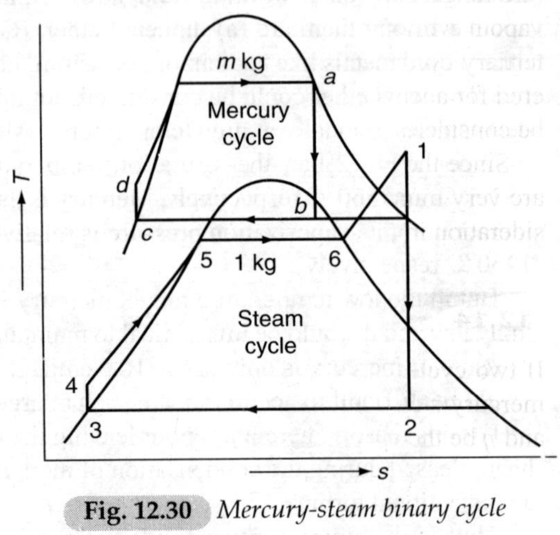

Fig. 12.30 *Mercury-steam binary cycle*

Let m represent the flow rate of mercury in the mercury cycle per kg of steam circulating in the steam cycle. Then for 1 kg of steam

$$Q_1 = m(h_a - h_d) + (h_1 - h_6) + (h_5 - h_4)$$

$$Q_2 = h_2 - h_3$$

$$W_T = m(h_a - h_b) + (h_1 - h_2)$$

$$W_P = m(h_d - h_c) + (h_4 - h_3)$$

$$\eta_{\text{cycle}} = \frac{Q_1 - Q_2}{Q_1} = \frac{W_T - W_P}{Q_1}$$

$$\text{and steam rate} = \frac{3600}{W_T - W_P} \text{ kg/kWh}$$

The energy balance of the mercury condenser-steam boiler gives

$$m(h_b - h_c) = h_6 - h_5$$

$$\therefore \quad m = \frac{h_6 - h_5}{h_b - h_c} \text{ kg Hg/kg } H_2O$$

To vaporize one kg of water, seven to eight kg of mercury must condense.

The addition of the mercury cycle to the steam cycle results in a marked increase in the mean temperature of heat addition to the plant as a whole and consequently the efficiency is increased. The maximum pressure is relatively low.

It may be interesting to note that the concept of the binary vapour cycle evolved from the need of improving the efficiency of the reciprocating steam engine. When steam expands up to, say, atmospheric temperature, the resultant volume flow rate of steam becomes too large for the steam engine cylinder to accommodate. So most of the early steam engines are found to be non-condensing. The binary cycle with steam in the high temperature and ammonia or sulphur dioxide in the low temperature range, was first suggested by Professor Josse of Germany in the middle of the nineteenth century. Steam exhausted from the engine at a relatively higher pressure and temperature was used to evaporate ammonia or sulphur dioxide which operated on another cycle. But with the progress in steam turbine design, such a cycle was found to be of not much utility, since modern turbines can cope efficiently with a large volume flow of steam.

The mercury-steam cycle has been in actual commercial use for more than three decades. One such plant is the Schiller Station in the USA. But it has never attained wide acceptance because there has always been the possibility of improving steam cycles by increasing pressure and temperature, and by using reheat and regeneration. Over and above, mercury is expensive, limited in supply, and highly toxic.

The mercury-steam cycle represents the two-fluid cycles. The mercury cycle is called the *topping cycle* and the steam cycle is called the *bottoming cycle*. If a sulphur dioxide cycle is added to it in the low temperature range, so that the heat released during the condensation of steam is utilized in forming sulphur dioxide vapour which expands in another turbine, then the mercury-steam-sulphur dioxide cycle is a three-fluid or tertiary cycle. Similarly, other liquid metals, apart from mercury, like sodium or potassium, may be considered for a working fluid in the topping cycle. Apart from SO_2 other refrigerants (ammonia, freons, etc.) may be considered as working fluids for the bottoming cycle.

Since the possibilities of improving steam cycles are diminishing, and the incentives to reduce fuel cost are very much increasing, coupled cycles, like the mercury-steam cycle, may receive more favourable consideration in the near future.

12.14 THERMODYNAMICS OF COUPLED CYCLES

If two cycles are coupled in series where heat lost by one is absorbed by the other (Fig. 12.31), as in the mercury-steam binary cycle, let η_1 and η_2 be the efficiencies of the topping and bottom cycles respectively, and η be the overall efficiency of the combined cycle.

$$\eta_1 = 1 - \frac{Q_2}{Q_1} \quad \text{and} \quad \eta_2 = 1 - \frac{Q_3}{Q_2}$$

or

$$Q_2 = Q_1(1 - \eta_1) \quad \text{and} \quad Q_3 = Q_2(1 - \eta_2)$$

Now

$$\eta = 1 - \frac{Q_3}{Q_1} = 1 - \frac{Q_2(1-\eta_2)}{Q_1}$$

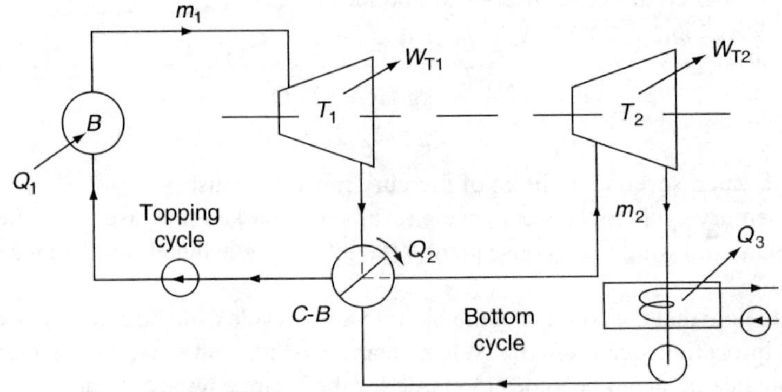

Fig. 12.31 *Two vapour cycles coupled in series*

$$= 1 - \frac{Q_1(1-\eta_1)(1-\eta_2)}{Q_1}$$

$$= 1 - (1 - \eta_1)(1 - \eta_2)$$

If there are n cycles coupled in series, the overall efficiency would be given by

$$\eta = 1 - \prod_{i=1}^{n}(1-\eta_i)$$

i.e.
$$\eta = 1 - (1 - \eta_1)(1 - \eta_2)(1 - \eta_3)\dots(1 - \eta_n)$$

or
$$1 - \eta = (1 - \eta_1)(1 - \eta_2)(1 - \eta_3)\dots(1 - \eta_n)$$

$\therefore$ Total loss = Product of losses in all the cycles.

For two cycles coupled in series

$$\eta = 1 - (1 - \eta_1)(1 - \eta_2)$$

$$= 1 - (1 - \eta_1 - \eta_2 + \eta_1\eta_2)$$

$$= \eta_1 + \eta_2 - \eta_1\eta_2$$

or
$$\eta = \eta_1 + \eta_2 - \eta_1\eta_2$$

This shows that the overall efficiency of two cycles coupled in series equals the sum of the individual efficiencies minus their product.

By combining two cycles in series, even if individual efficiencies are low, it is possible to have a fairly high combined efficiency, which cannot be attained by a single cycle.

For example, if $\eta_1 = 0.50$ and $\eta_2 = 0.40$

$$\eta = 0.5 + 0.4 - 0.5 \times 0.4 = 0.70$$

It is almost impossible to achieve such a high efficiency in a single cycle.

12.15 PROCESS HEAT AND BY-PRODUCT POWER

There are several industries, such as paper mills, textile mills, chemical factories, dying plants, rubber manufacturing plants, sugar factories, etc., where saturated steam at the desired temperature is required for heating, drying, etc. For constant temperature heating (or drying), steam is a very good medium, since isothermal condition can be maintained by allowing saturated steam to condense at that temperature and utilizing the latent heat released for heating purposes. Apart from the process heat, the factory also needs power to drive various machines, for lighting, and for other purposes.

Formerly it was the practice to generate steam for power purposes at a moderate pressure and to generate separately saturated steam for process work at a pressure which gave the desired heating temperature. Having two separate units for process heat and power is wasteful, for of the total heat supplied to the steam for power purposes, a greater part will normally be carried away by the cooling water in the condenser.

By modifying the initial steam pressure and exhaust pressure, it is possible to generate the required power and make available for process work the required quantity of exhaust steam at the desired temperature. In Fig. 12.32, the exhaust steam from the turbine is utilized for process heating, the process heater replacing the condenser of the ordinary Rankine cycle. The pressure at exhaust from the turbine is the saturation pressure corresponding to the temperature desired in the process heater. Such a turbine is called *a back pressure turbine*. A plant producing both power and process heat is sometimes known as a *cogeneration plant*. When the process steam is the basic need, and the power is produced incidentally as a by-product, the cycle is sometimes called a by-product power cycle. Figure 12.33 shows the *T–s* plot of such a cycle. If W_T is the turbine output in kW, Q_H the process heat required in kJ/h, and w is the steam flow rate in kg/h

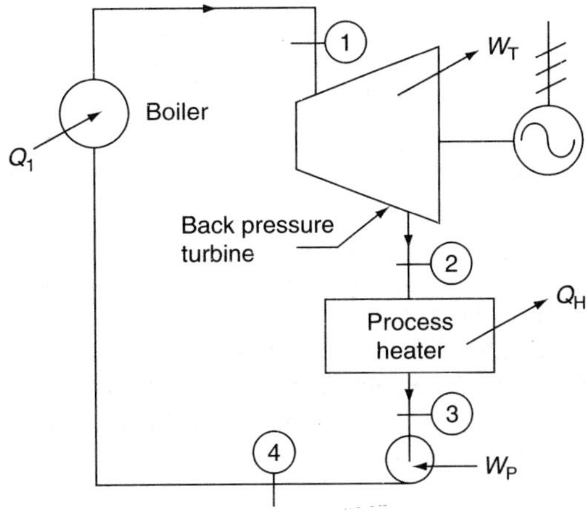

Fig. 12.32 *Back pressure turbine*

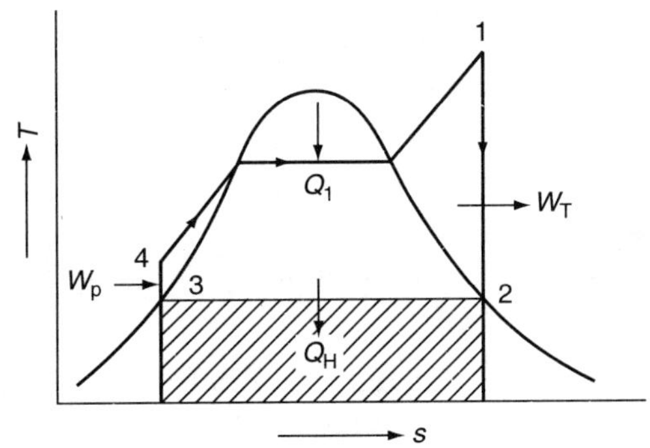

Fig. 12.33 *By-product power cycle*

$$W_T \times 3600 = w(h_1 - h_2)$$

and $$w(h_2 - h_3) = Q_H$$

$\therefore$ $$W_T \times 3600 = \frac{Q_H}{h_2 - h_3}(h_1 - h_2)$$

or $$Q_H = \frac{W_T \times 3600 \times (h_2 - h_3)}{h_1 - h_2} \text{ kJ/h}$$

Of the total energy input Q_1 (as heat) to the by-product cycle, W_T part of it only is converted into shaft work (or electricity). The remaining energy ($Q_1 - W_T$), which would otherwise have been a waste, as in the Rankine cycle (by the Second Law), is utilized as process heat.

Fraction of energy (Q_1) utilized in the form of work (W_T), and process heat (Q_H) in a by-product power cycle

$$= \frac{W_T + Q_H}{Q_1}$$

Condenser loss, which is the biggest loss in a steam plant, is here zero, and the fraction of energy utilized is very high.

In many cases the power available from the back pressure turbine through which the whole of the heating steam flows is appreciably less than that required in the factory. This may be due to relatively high back pressure, or small heating requirement, or both. *Pass-out turbines* are employed in these cases, where a certain quantity of steam is continuously extracted for heating purposes at the desired temperature and pressure. (Figs 12.34 and 12.35).

$$Q_1 = w(h_1 - h_8) \text{ kJ/h}$$
$$Q_2 = (w - w_1)(h_3 - h_4) \text{ kJ/h}$$
$$Q_H = w_1(h_2 - h_6) \text{ kJ/h}$$
$$W_T = w(h_1 - h_2) + (w - w_1)(h_2 - h_3) \text{ kJ/h}$$
$$W_P = (w - w_1)(h_5 - h_4) + w_1(h_7 - h_6) \text{ kJ/h}$$
$$w_1 h_7 + (w - w_1)h_5 = w \times h_8$$

where w is the boiler capacity (kg/h) and w_1 is the steam flow rate required (kg/h) at the desired temperature for process heating.

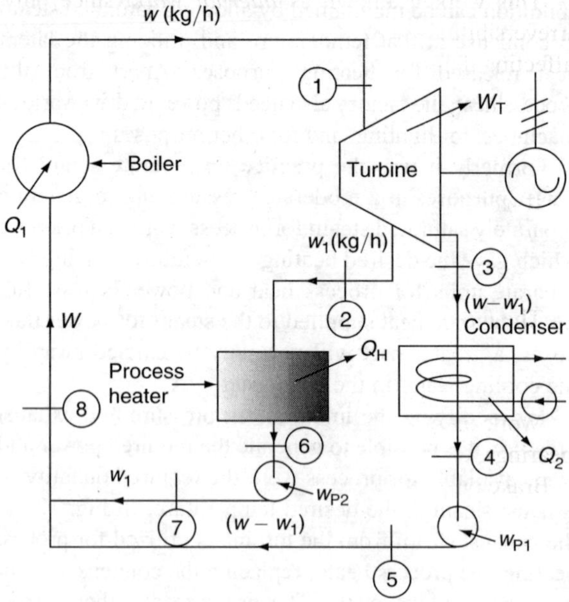

Fig. 12.34 *Pass-out turbine*

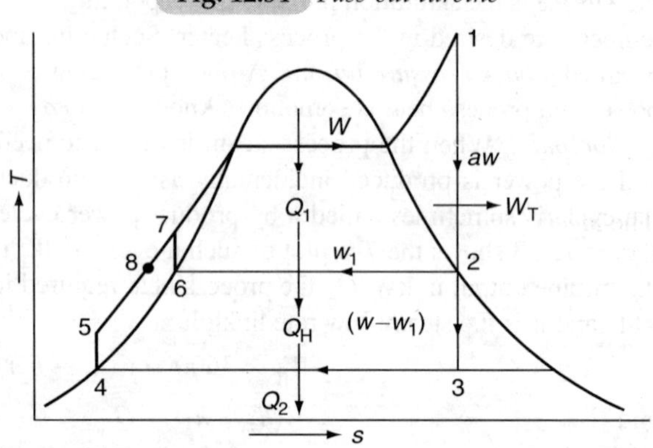

Fig. 12.35 *T–s diagram of power and process heat plant*

12.16 EFFICIENCIES IN STEAM POWER PLANT

For the steady flow operation of a turbine, neglecting changes in K.E. and P.E. (Figs 12.36 and 12.37).

Maximum or ideal work output per unit mass of steam

$$(W_T)_{max} = (W_T)_{ideal} = h_1 - h_{2s}$$
$$= \text{Reversible and adiabatic enthalpy drop in turbine}$$

This work is, however, not obtainable, since no real process is reversible. The expansion process is accompanied by irreversibilities. The actual final state 2 can be defined, since the temperature, pressure, and quality can be found by actual measurement. The actual path 1–2 is not known and its nature is immaterial, since the work output is here being expressed in terms of the change of a property, enthalpy. Accordingly, the work done by the turbine in irreversible adiabatic expansion from 1 to 2 is

$$(W_T)_{actual} = h_1 - h_2$$

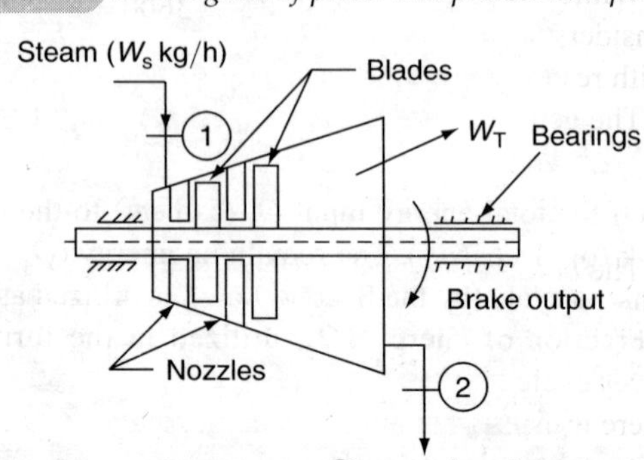

Fig. 12.36 *Efficiencies in a steam turbine*

This work is known as *internal work*, since only the irreversibilities within the flow passages of turbine are affecting the state of steam at the turbine exhaust.

$\therefore$ Internal output = Ideal output – Friction and other losses within the turbine casing

If w_s is the steam flow rate in kg/h

$$\text{Internal output} = w_s(h_1 - h_2)\ \text{kJ/h}$$

$$\text{Ideal output} = w_s(h_1 - h_{2s})\ \text{kJ/h}$$

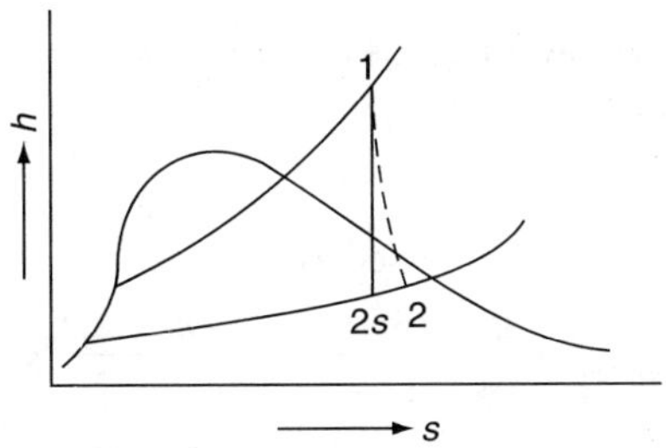

Fig. 12.37 *Internal efficiency of a steam turbine*

The *internal efficiency* of turbine is defined as

$$\eta_{\text{internal}} = \frac{\text{Internal output}}{\text{Ideal output}} = \frac{h_1 - h_2}{h_1 - h_{2s}}$$

Work output available at the shaft is less than the internal output because of the external losses in the bearings, etc.

$\therefore$ Brake output or shaft output

= Internal output External losses

= Ideal output – Internal and External losses

= (kW × 3600 kJ/h)

The *brake efficiency* of turbine is defined as

$$\eta_{\text{brake}} = \frac{\text{Brake output}}{\text{Ideal output}}$$

$$= \frac{\text{kW} \times 3600}{w_s(h_1 - h_{2s})}$$

The *mechanical efficiency* of turbine is defined as

$$\eta_{\text{mech}} = \frac{\text{Brake output}}{\text{Internal output}}$$

$$= \frac{\text{kW} \times 3600}{w_s(h_1 - h_2)}$$

$$\therefore \quad \eta_{\text{brakc}} = \eta_{\text{internal}} \times \eta_{\text{mech}}$$

While the internal efficiency takes into consideration the internal losses, and the mechanical efficiency considers only the external losses, the brake efficiency takes into account both the internal and external losses (with respect to turbine casing).

The generator (or *alternator*) efficiency is defined as

$$\eta_{\text{generator}} = \frac{\text{Output at generator terminals}}{\text{Brake output of turbine}}$$

The *boiler efficiency* is defined as

$$\eta_{\text{boiler}} = \frac{\text{Energy utilized}}{\text{Energy supplied}} = \frac{w_s(h_1 - h_4)}{w_f \times \text{C.V.}}$$

where w_f is the fuel burning rate in the boiler (kg/h) and C.V. is the calorific value of the fuel (kJ/kg), i.e. the heat energy released by the complete combustion of unit mass of fuel.

The power plant is an *energy converter* from fuel to electricity (Fig. 12.38), and the *overall efficiency* of the plant is defined as

$$\eta_{\text{overall}} = \eta_{\text{plant}} = \frac{\text{kW} \times 3600}{w_f \times \text{C.V.}}$$

This may be expressed as

$$\eta_{\text{overall}} = \frac{\text{kW} \times 3600}{w_f \times \text{C.V.}} = \frac{w(h_1 - h_4)}{w_f \times \text{C.V.}} \times \frac{w(h_1 - h_2)}{w_s(h_1 - h_4)} \times \frac{\text{Brake output}}{w_s(h_1 - h_2)} \times \frac{\text{kW} \times 3600}{\text{Brake output}}$$

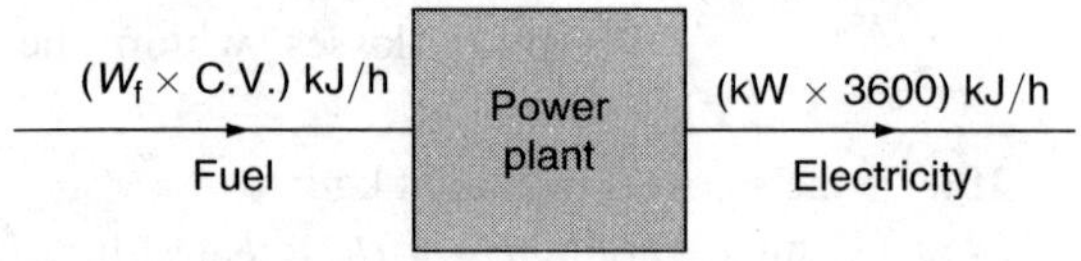

Fig. 12.38 *Power plant–an energy converter from fuel to electricity*

or $\eta_{\text{overall}} = \eta_{\text{boiler}} \times \eta_{\text{cycle}} \times \eta_{\text{turbine (mech)}} \times \eta_{\text{generator}}$

where pump work has been neglected in the expression for cycle efficiency.

Solved Examples

Example 12.1

Determine the work required to compress steam isentropically from 1 *bar to* 10 *bar, assuming that at the initial state the steam exists as* (*a*) *saturated liquid and* (*b*) *saturated vapour. Neglect changes in kinetic and potential energies. What conclusion do you derive from this example?*

Solution The compression processes are shown in Fig. Ex. 12.1

(a) Steam is a saturated liquid initially, and its specific volume is:

$$v_1 = (v_f)_{1\text{bar}} = 0.001043 \text{ m}^3/\text{kg}$$

Since liquid is incompressible, v_1 remains constant.

$$W_{\text{rev}} = -\int_1^2 v\text{d}p = v_1(p_1 - p_2)$$

$$= 0.001043\,(1 - 10) \times 10^2 = -\,0.9387 \text{ kJ/kg}.$$

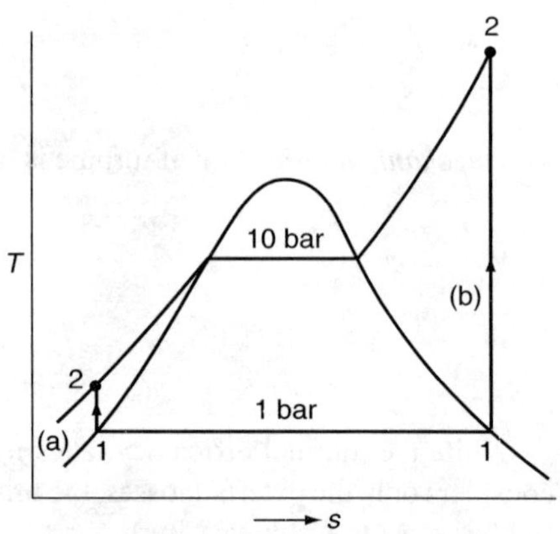

Fig. Ex. 12.1 *Compression of steam isentropically*

(b) Steam is a saturated vapour initially and remains a vapour during the entire compression process. Since the specific volume of a gas changes considerably during a compression process, we need to know how v varies with p to perform the integration $-\int v\,\text{d}p$. This relation is not readily available. But for an isentropic process, it is easily obtained from the property relation

$$T\,\text{d}s = \text{d}h - v\text{d}p = 0$$

or

$$v\,\text{d}p = \text{d}h$$

$$W_{\text{rev}} = -\int_1^2 v\,\text{d}p = -\int_1^2 \text{d}h = h_1 - h_2$$

From steam tables,

$$h_1 = (h_g)_{1\text{bar}} = 2675.5 \text{ kJ/kg}$$

$$s_1 = (s_g)_{1\text{bar}} = 7.3594 \text{ kJ/kg K} = s_2$$

For $p = 10$ bar $= 1$ MPa and $s = 7.3594$ kJ/kg K, by interpolation

$$h_2 = 3195.5 \text{ kJ/kg}$$

$$W_{rev} = 2675.5 - 3195.5 = -520 \text{ kJ/kg}$$

It is thus observed that compressing steam in vapour form would require over 500 times more work than compressing it in liquid form for the same pressure rise.

Example 12.2

Steam at 20 bar, 360°C is expanded in a steam turbine to 0.08 bar. It then enters a condenser, where it is condensed to saturated liquid water. The pump feeds back the water into the boiler. (a) Assuming ideal processes, find per kg of steam the net work and the cycle efficiency. (b) If the turbine and the pump have each 80% efficiency, find the percentage reduction in the net work and cycle efficiency.

Solution The property values at different state points (Fig. Ex. 12.2) found from the steam tables are given below.

$$h_1 = 3159.3 \text{ kJ/kg} \qquad s_1 = 6.9917 \text{ kJ/kg K}$$

$$h_3 = h_{fp2} = 173.88 \text{ kJ/kg} \qquad s_3 = s_{fp2} = 0.5926 \text{ kJ/kg K}$$

$$h_{fgp2} = 2403.1 \text{ kJ/kg} \qquad s_{gp2} = 8.2287 \text{ kJ/kg K}$$

$$v_{fp2} = 0.001008 \text{ m}^3\text{/kg} \qquad \therefore s_{fgp2} = 7.6361 \text{ kJ/kg K}$$

Now $\quad s_1 = s_{2s} = 6.9917 = s_{fp2} + x_{2s}\, s_{fgp2} = 0.5926 + x_2 \cdot 7.6361$

$$\therefore \quad x_{2s} = \frac{6.3991}{7.6361} = 0.838$$

$$\therefore \quad h_{2s} = h_{fp2} + x_{2s}\, h_{fgp2} = 173.88 + 0.838 \times 2403.1$$

$$= 2187.68 \text{ kJ/kg}$$

(a) $W_P = h_{4s} - h_3 = v_{fp2}(p_1 - p_2)$

$$= 0.001008 \frac{\text{m}^3}{\text{kg}} \times 19.92 \times 100 \frac{\text{kN}}{\text{m}^2}$$

$$= 2.008 \text{ kJ/kg}$$

$$h_{4s} = 175.89 \text{ kJ/kg}$$

$$W_T = h_1 - h_{2s}$$

$$= 3159.3 - 2187.68 = 971.62 \text{ kJ/kg}$$

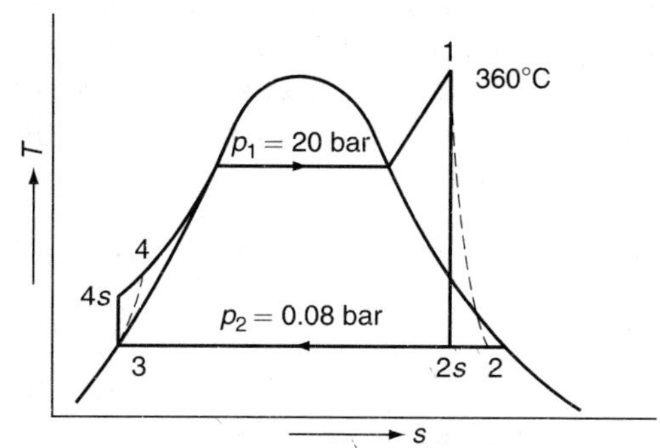

Fig. Ex. 12.2

$$\therefore W_{net} = W_T - W_P = 969.61 \text{ kJ/kg}$$

$$Q_1 = h_1 - h_{4s} = 3159.3 - 175.89$$

$$= 2983.41 \text{ kJ/kg}$$ *Ans.*

$$\therefore \eta_{cycle} = \frac{W_{net}}{Q_1} = \frac{969.61}{2983.41} = 0.325, \text{ or } 32.5\%$$ *Ans.*

(b) If $\eta_P = 80\%$, and $\eta_T = 80\%$

$$W_P = \frac{2.008}{0.8} = 2.51 \text{ kJ/kg}$$

$$W_T = 0.8 \times 971.62 = 777.3 \text{ kJ/kg}$$

$$\therefore W_{net} = W_T - W_P = 774.8 \text{ kJ/kg}$$

∴ % Reduction in work output

$$= \frac{969.61 - 774.8}{969.61} \times 100 = 20.1\%$$ *Ans.*

$$h_{4s} = 173.88 + 2.51 = 176.39 \text{ kJ/kg}$$

$$\therefore Q_1 = 3159.3 - 176.39 = 2982.91 \text{ kJ/kg}$$

$$\therefore \eta_{cycle} = \frac{774.8}{2982.91} = 0.2597, \text{ or } 25.97\%$$

∴ % Reduction in cycle efficiency

$$= \frac{0.325 - 0.2597}{0.325} \times 100 = 20.1\%$$ *Ans.*

Example 12.3

A cyclic steam power plant is to be designed for a steam temperature at turbine inlet of 360°C and an exhaust pressure of 0.08 bar. After isentropic expansion of steam in the turbine, the moisture content at the turbine exhaust is not to exceed 15%. Determine the greatest allowable steam pressure at the turbine inlet, and calculate the Rankine cycle efficiency for these steam conditions. Estimate also the mean temperature of heat addition.

Solution As state 2*s* (Fig. Ex. 12.3), the quality and pressure are known.

$$\therefore s_{2s} = s_f + x_{2s} s_{fg} = 0.5926 + 0.85\,(8.2287 - 0.5926)$$

$$= 7.0833 \text{ kJ/kg K}$$

Since $s_1 = s_{2s}$

$$\therefore s_1 = 7.0833 \text{ kJ/kg K}$$

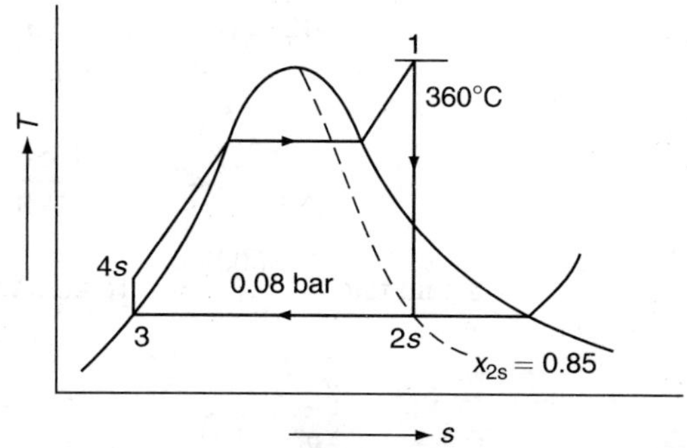

Fig. Ex. 12.3

At state 1, the temperature and entropy are thus known. At 360°C, s_g = 5.0526 kJ/kg K, which is less than s_1. So from the table of superheated steam, at t_1 = 360°C and s_1 = 7.0833 kJ/kg K, the pressure is found to be 16.832 bar (by interpolation).

∴ The greatest allowable steam pressure is

$$p_1 = 16.832 \text{ bar}$$
$$h_1 = 3165.54 \text{ kJ/kg}$$
$$h_{2s} = 173.88 + 0.85 \times 2403.1 = 2216.52 \text{ kJ/kg}$$
$$h_3 = 173.88 \text{ kJ/kg}$$
$$h_{4s} - h_3 = 0.001 \times (16.83 - 0.08) \times 100 = 1.675 \text{ kJ/kg}$$
$$\therefore \quad h_{4s} = 175.56 \text{ kJ/kg}$$
$$Q_1 = h_1 - h_{4s} = 3165.54 - 175.56 = 2990 \text{ kJ/kg}$$
$$W_T = h_1 - h_{2s} = 3165.54 - 2216.52 = 949 \text{ kJ/kg}$$
$$W_P = 1.675 \text{ kJ/kg}$$
$$\eta_{\text{cycle}} = \frac{W_{\text{net}}}{Q_1} = \frac{247.32}{2990} = 0.3168 \text{ or } 31.68\%$$ *Ans.*

Mean temperature of heat addition

$$T_{m1} = \frac{h_1 - h_{4s}}{s_1 - s_{4s}} = \frac{2990}{7.0833 - 0.5926} = 460.66 \text{ K} = 187.51°\text{C}.$$

Example 12.4

A steam power station uses the following cycle:

Steam at boiler outlet—150 bar, 550°C

Reheat at 40 bar to 550°C

Condenser at 0.1 bar.

Using the Mollier chart and assuming ideal processes, find the (a) quality at turbine exhaust, (b) cycle efficiency, and (c) steam rate.

Solution The property values at different states (Fig. Ex. 12.4) are read from the Mollier chart.

$$h_1 = 3465,\ h_{2s} = 3065,\ h_3 = 3565,$$
$$h_{4s} = 2300 \text{ kJ/kg } x_{4s} = 0.88,\ h_5(\text{steam table}) = 191.83 \text{ kJ/kg}$$

Quality at turbine exhaust = 0.88 *Ans.* (a)

$$W_P = v\,\Delta p = 10^{-3} \times 150 \times 10^2 = 15 \text{ kJ/kg}$$
$$\therefore \quad h_{6s} = 206.83 \text{ kJ/kg}$$
$$Q_1 = (h_1 - h_{6s}) + (h_3 - h_{2s}) = (3465 - 206.83) + (3565 - 3065) = 3758.17 \text{ kJ/kg}$$

$$W_T = (h_1 - h_{2s}) + (h_3 - h_{4s})$$
$$= (3465 - 3065) + (3565 - 2300) = 1665 \text{ kJ/kg}$$
$$\therefore \quad W_{net} = W_T - W_P = 1665 - 15 = 1650 \text{ kJ/kg}$$
$$\therefore \quad \eta_{cycle} = \frac{W_{net}}{Q_1} = \frac{1650}{3758.17} = 0.4390, \text{ or } 43.9\%$$ *Ans.* (b)

$$\text{Steam rate} = \frac{3600}{1650} = 2.18 \text{ kg/kW h}$$ *Ans.* (c)

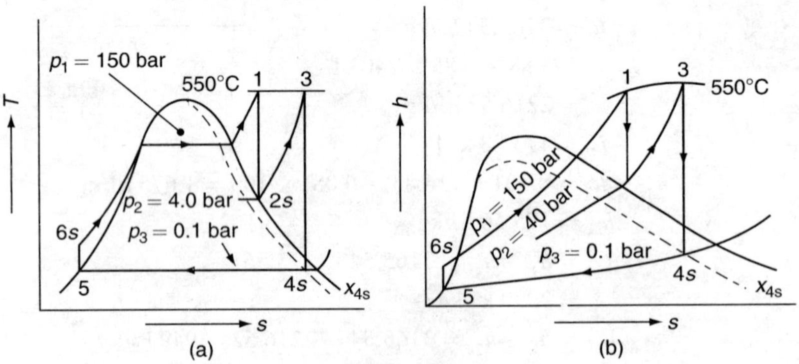

Fig. Ex. 12.4

Example 12.5

In a single-heater regenerative cycle the steam enters the turbine at 30 bar, 400°C and the exhaust pressure is 0.10 bar. The feed water heater is a direct-contact type which operates at 5 bar. Find (a) the efficiency and the steam rate of the cycle and (b) the increase in mean temperature of heat addition, efficiency and steam rate, as compared to the Rankine cycle (without regeneration). Neglect pump work.

Solution Figure Ex. 12.5 gives the flow, *T–s*, and *h–s* diagrams. From the steam tables, the property values at various states have been obtained.

$$h_1 = 3230.9 \text{ kJ/kg}$$
$$s_1 = 6.9212 \text{ kJ/kg K} = s_2 = s_3$$
$$s_g \text{ at 5 bar} = 6.8213 \text{ kJ/kg K}$$

Since $s_2 > s_g$, the state 2 must lie in the superheated region. From the table for superheated steam $t_2 = 172°\text{C}$, $h_2 = 2796$ kJ/kg

$$s_3 = 6.9212 = s_{f0.1\text{ bar}} + x_{3s}\, s_{fg0.1\text{ bar}}$$
$$= 0.6493 + x_3 7.5009$$
$$\therefore \quad x_3 = \frac{6.2719}{7.5009} = 0.836$$
$$\therefore \quad h_3 = 191.83 + 0.836 \times 2392.8 = 2192.2 \text{ kJ/kg}$$

Since pump work is neglected

$$h_4 = 191.83 \text{ kJ/kg} = h_5$$
$$h_6 = 640.23 \text{ kJ/kg} = h_7$$

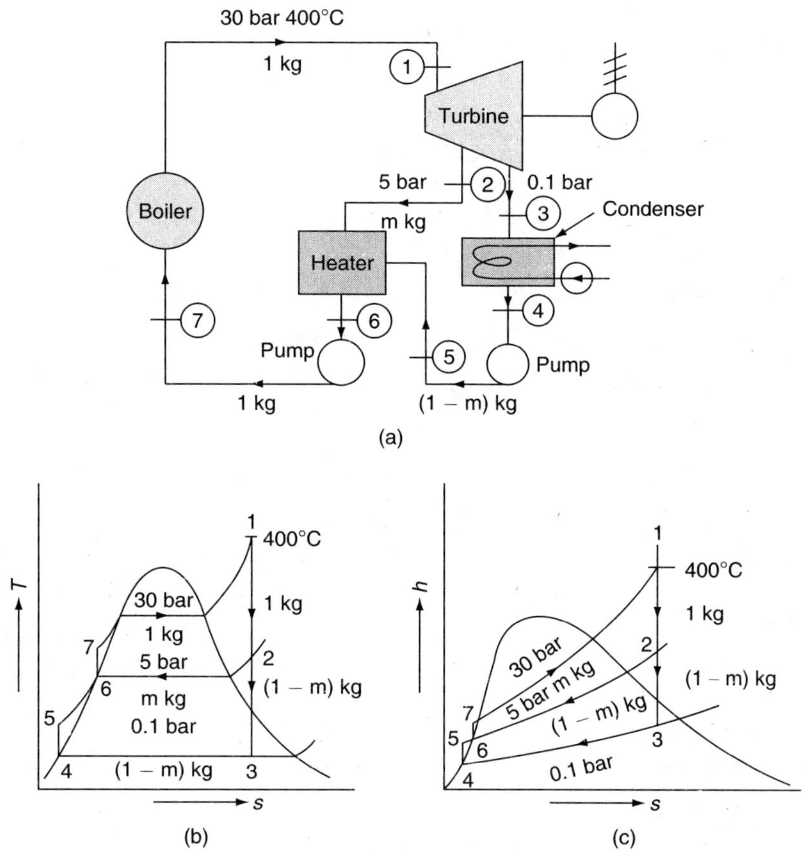

Fig. Ex. 12.5

Energy balance for the heater gives

$$m(h_2 - h_6) = (1 - m)(h_6 - h_5)$$

$$m(2796 - 640.23) = (1 - m)(640.23 - 191.83)$$

$$2155.77\, m = 548.4 - 548.4\, m$$

$$\therefore \quad m = \frac{548.4}{2704.17} = 0.203 \text{ kg}$$

$$\therefore \quad W_T = (h_1 - h_2) + (1 - m)(h_2 - h_3)$$

$$= (3230.9 - 2796) + 0.797\,(2796 - 2192.2) = 916.13 \text{ kJ/kg}$$

$$Q_1 = h_1 - h_6 = 3230.9 - 640.23 = 2590.67 \text{ kJ/kg}$$

$$\eta_{\text{cycle}} = \frac{916.13}{2590.67} = 0.3536, \text{ or } 35.36\%$$ *Ans.* (a)

$$\text{Steam rate} = \frac{3600}{916.13} = 3.93 \text{ kg/kW h}$$ *Ans.* (b)

$$T_{m1} = \frac{h_1 - h_7}{s_1 - s_7} = \frac{2590.67}{6.9212 - 1.8607} = 511.95 \text{ K} = 238.8°\text{C}$$

$$T_{m1} \text{ (without regeneration)} = \frac{h_1 - h_4}{s_1 - s_4}$$

$$= \frac{3039.07}{6.9212 - 0.6493}$$

$$= 484.55 \text{ K}$$

$$= 211.4\ °\text{C}$$

Increase in T_{m1} due to regeneration = 238.8 – 211.4 = 27.4°C *Ans.* (b)

W_T (without regeneration) = $h_1 - h_3$ = 3230.9 – 2192.2 = 1038.7 kJ/kg

$$\text{Steam rate (without regeneration)} = \frac{3600}{1038.7} = 3.46 \text{ kg/kW h}$$

∴ Increase in steam rate due to regeneration

= 3.93 – 3.46 = 0.47 kg/kW h *Ans.* (b)

$$\eta_{cycle} \text{ (without regeneration)} = \frac{h_1 - h_3}{h_1 - h_4} = \frac{1038.7}{3039.07} = 0.3418 \text{ or } 34.18\%$$

∴ Increase in cycle efficiency due to regeneration

= 35.36 – 34.18 = 1.18% *Ans.* (c)

Example 12.6

In a steam power plant the condition of steam at inlet to the steam turbine is 20 *bar and* 300°*C and the condenser pressure is* 0.1 *bar. Two feedwater heaters operate at optimum temperature. Determine: (a) the quality of steam at turbine exhaust, (b) net work per kg of steam, (c) cycle efficiency, and (d) the stream rate. Neglect pump work.*

Solution From Fig. 12.19 (a),

$$h_1 = 3023.5 \text{ kJ/kg}$$

$$s_1 = 6.7664 \text{ kJ/kg K} = s_2 = s_3 = s_4$$

$$t_{sat} \text{ at 20 bar} \cong 212°\text{C}$$

$$t_{sat} \text{ at 0.1 bar} \cong 46°\text{C}$$

$$\Delta t_{OA} = 212 - 46 = 166°\text{C}$$

$$\therefore \text{ Temperature rise per heater} = \frac{166}{3} = 55°\text{C}$$

∴ Temperature at which the first heater operates

= 212 – 55 = 157°C ≅ 150°C (assumed)

Temperature at which the second heater operates = 157 – 55 = 102°C ≅ 100°C (assumed)

At 0.1 bar,

$$h_f = 191.83,\ h_{fg} = 2392.8,\ s_f = 0.6493$$

$$s_g = 8.1502$$

At 100°C,

$$h_f = 419.04,\ h_{fg} = 2257.0,\ s_f = 1.3069,\ s_g = 7.3549$$

At 150°C,

$$h_f = 632.20,\ h_{fg} = 2114.3,\ s_f = 1.8418,\ s_g = 6.8379$$

$$6.7664 = 1.8418 + x_2 \times 4.9961$$

$$\therefore\ x_2 = 0.986$$

$$\therefore\ h_2 = 632.2 + 0.986 \times 2114.3 = 2716.9 \text{ kJ/kg}$$

$$6.7664 = 1.3069 + x_3 \times 6.0480$$
$$x_3 = 0.903$$
$$\therefore \quad h_3 = 419.04 + 0.903 \times 2257.0 \quad \text{or} \quad h_3 = 2457.1 \text{ kJ/kg}$$
$$6.7664 = 0.6493 + x_4 \times 7.5010$$
$$x_4 = 0.816$$
$$\therefore \quad h_4 = 191.83 + 0.816 \times 2392.8 = 2144.3 \text{ kJ/kg}$$

Since pump work is neglected, $h_{10} = h_9$, $h_8 = h_7$, $h_6 = h_5$. By making an energy balance for the *hp* heater

$$(1 - m_1)(h_9 - h_8) = m_1(h_2 - h_9)$$

Rearranging

$$m_1 = \frac{h_9 - h_7}{h_2 - h_7} = \frac{213.16}{2297.86} = 0.093 \text{ kg}$$

By making an energy balance for the *lp* heater,

$$(1 - m_1 - m_2)(h_7 - h_6) = m_2(h_3 - h_7)$$
$$(1 - 0.093 - m_2)(419.04 - 191.83) = m_2(2457.1 - 419.04)$$
$$\therefore \quad m_2 = 0.091 \text{ kg}$$
$$\therefore \quad W_T = 1(h_1 - h_2) + (1 - m_1)(h_2 - h_3) + (1 - m_1 - m_2)(h_3 - h_4)$$
$$= (3023.5 - 2716.9) + (1 - 0.093)(2716.9 - 2457.1)$$
$$+ (1 - 0.093 - 0.091)(2457.1 - 2144.3) = 797.48 \text{ kJ/kg}$$
$$Q_1 = h_1 - h_9 = 3023.5 - 632.2 = 2391.3 \text{ kJ/kg}$$
$$\therefore \quad \eta_{\text{cycle}} = \frac{W_T - W_P}{Q_1} = \frac{797.48}{2391.3} = 0.3334 \text{ or } 33.34\%$$

$$\text{Steam rate} = \frac{3600}{W_{\text{net}}} = \frac{3600}{797.48} = 4.51 \text{ kJ/kW h}$$

Example 12.7

Dry saturated steam at 40 *bar expands in a turbine isentropically to the condenser pressure of* 0.075 *bar. Hot gases available at* 2000 *K, and* 1 *atm pressure are used for steam generation and then exhausted at* 450 *K to the ambient atmosphere which is at* 300 *K and* 1 *atm. The heating rate provided by the gas stream is* 100 *MW. Assuming* c_p *of hot gases as* 1.1 *kJ/kg K, give an exergy balance of the plant and compare it with the energy balance, and find the second law efficiency.*

Solution

$$\dot{Q}_1 = w_g c_{p_g}(T_i - T_e) = 100 \text{ MW}$$
$$w_g = \text{mass flow rate of hot gas}$$
$$= \frac{100 \times 10^3}{1.1 \times (2000 - 450)} = 58.7 \text{ kg/s}$$

Exergy flow rate of inlet gas

$$a_{f_1} = w_g c_{p_g} T_0 \left[\frac{T_i}{T_0} - 1 - \ln\frac{T_i}{T_0}\right] = 58.7 \times 1.1 \times 300 \left[\frac{2000}{300} - 1 - \ln\frac{2000}{300}\right] = 73 \text{ MW}$$

Exergy flow rate of exhaust gas stream

$$a_{f_2} = 58.7 \times 1.1 \times 300 \left[\frac{450}{300} - 1 - \ln\frac{450}{300}\right] = 1.83 \text{ MW}$$

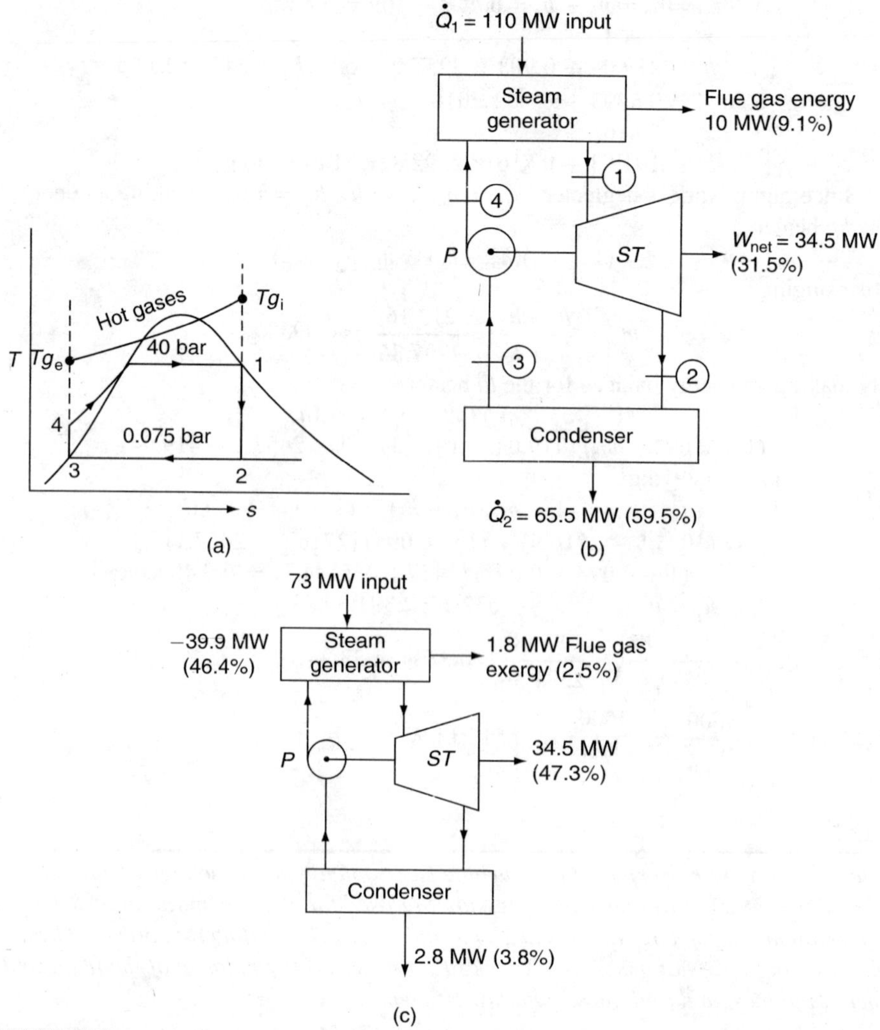

Fig. Ex. 12.7 *(a) T–s diagram, (b) Energy distribution diagram, (c) Exergy distribution diagram*

The exergy loss rate is only about $\left[\dfrac{1.83}{73}\times 100\right]$ or 2.5% of the initial exergy of the source gas.

The rate of exergy decrease of the gas stream,

$$a_{f_i} = \text{Exergy input rate} = 73 - 1.83 = 71.17 \cong 71.2 \text{ MW}$$

The rate of exergy increase of steam = Exergy utilization rate

$$a_{fu} = w_s\,[h_1 - h_4 - T_0(s_1 - s_4)]$$

Now,

$$h_1 = (h_g)_{40\text{ bar}} = 2801 \text{ kJ/kg}, \quad h_3 = 169 \text{ kJ/kg}$$

$$s_3 = s_4 = 0.576 \text{ kJ/kg K}, \quad h_4 = 172.8 \text{ kJ/kg}$$

$$s_1 = s_2 = 6.068 \text{ kJ/kg K}, \quad h_2 = 1890.2 \text{ kJ/kg}$$

$$W_T = h_1 - h_2 = 2801 - 1890.2 = 910.8 \text{ kJ/kg}$$

$$W_P = h_4 - h_3 = 172.8 - 169 = 3.8 \text{ kJ/kg}$$
$$Q_1 = h_1 - h_4 = 2801 - 172.8 = 2628 \text{ kJ/kg}$$
$$Q_2 = h_2 - h_3 = 1890.2 - 169 = 1721 \text{ kJ/kg}$$
$$W_{net} = W_T - W_P = Q_1 - Q_2 = 907 \text{ kJ/kg}$$
$$Q_1 = w_s \times 2628 = 100 \times 10^3 \text{ kW}$$
$$w_s = 38 \text{ kg/s}$$
$$a_{f_u} = 38\,[2801 - 172.8 - 300\,(6.068 - 0.576)] = 37.3 \text{ MW}$$

Rate of exergy destruction in the steam generator

= Rate of exergy decrease of gases – Rate of exergy increase of steam.

$$\dot{I} = a_{f_i} - a_{f_u} = 71.2 - 37.3 = 33.9 \text{ MW}$$

Rate of useful mechanical power output

$$\dot{W}_{net} = 38 \times 907 = 34.5 \text{ MW}$$

Exergy flow rate of wet steam to the condenser

$$a_{f_c} = w_s\,[h_2 - h_3 - T_0(s_2 - s_3)]$$
$$= 38\,[1890 - 169 - 300\,(6.068 - 0.576)] = 2.8 \text{ MW}$$

This is the exergy loss to the surroundings.

The energy and exergy balances are shown in Fig. Ex. 12.7 (b) and (c). The second law efficiency is given by

$$\eta_{II} = \frac{\text{Useful exergy output}}{\text{Exergy input}} = \frac{34.5}{73} = 0.473 \text{ or } 47.3\%$$ *Ans.*

Example 12.8

In a steam power plant, the condition of steam at turbine inlet is 80 *bar,* 500°*C and the condenser pressure is* 0.1 *bar. The heat source comprises a steam of exhaust gases from a gas turbine discharging at* 560°*C and* 1 *atm pressure. The minimum temperature allowed for the exhaust gas stream is* 450 *K. The mass flow rate of the hot gases is such that the heat input rate to the stream cycle is* 100 *MW. The ambient condition is given by* 300 *K and* 1 *atm. Determine* η_I*, work ratio and* η_{II} *of the following cycles: (a) basic Rankine cycle, without superheat, (b) Rankine cycle with superheat, (c) Rankine cycle with reheat such that steam expands in the h.p. turbine until it exits as dry saturated vapour, (d) ideal regenerative cycle, with the exit temperature of the exhaust gas steam taken as* 320°*C, because the saturation temperature of steam at* 80 *bar is close to* 300°*C.*

Solution For the first law analysis of each cycle, knowledge of the h values at each of the states indicated in Fig. Ex. 12.8 is required.

(a) *Basic Rankine cycle* (*Fig.* 12.8*a*):

By usual procedure with the help of steam tables,

$$h_1 = 2758,\ h_2 = 1817,\ h_3 = 192 \text{ and } h_4 = 200 \text{ kJ/kg}$$
$$W_T = h_1 - h_2 = 941 \text{ kJ/kg},\ W_p = h_4 - h_3 = 8 \text{ kJ/kg}$$
$$Q_1 = h_1 - h_4 = 2558 \text{ kJ/kg},\ W_{net} = 933 \text{ kJ/kg}$$
$$\eta_I = \frac{W_{net}}{Q_1} = \frac{933}{2558} = 0.365 \text{ or } 36.5\%$$ *Ans.*

$$\text{Work ratio} = \frac{W_T - W_P}{W_T} = \frac{933}{941} = 0.991$$ *Ans.*

$$\text{Power output} = \eta_I\, Q_1 = 0.365 \times 100 = 36.5 \text{ MW}$$

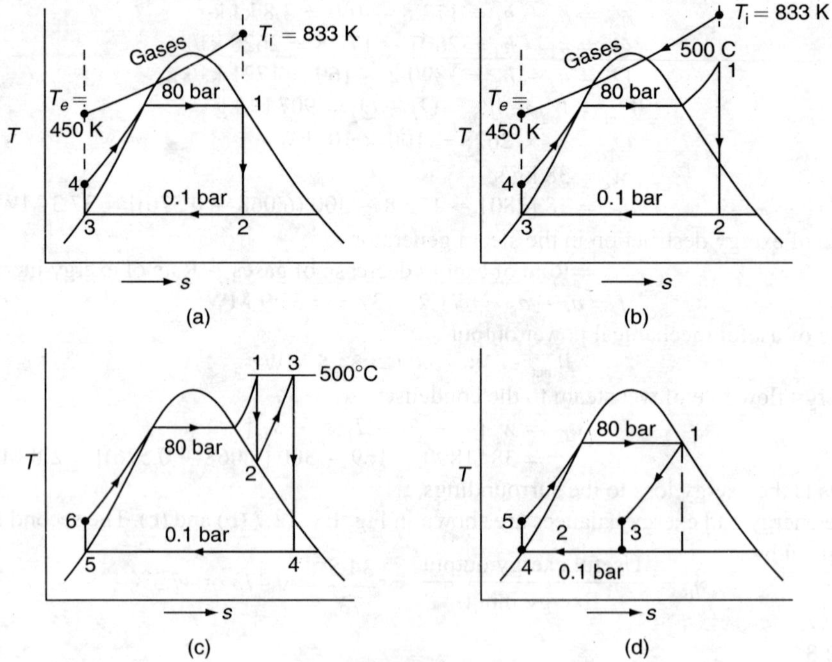

Fig. Ex. 12.8

$$\text{Exergy input rate} = w_g \, c_{p_g} \left[(T_i - T_0) - T_0 \ln \frac{T_i}{T_0} \right]$$

$$= \frac{100 \times 1000}{833 - 450} \left[(833 - 300) - 300 \ln \frac{833}{300} \right] = 59.3 \text{ MW}$$

$$\eta_{II} = \frac{36.5}{59.3} = 0.616 \text{ or } 61.6\% \qquad \textit{Ans.}$$

(b) *Rankine cycle with superheat (Fig. 12.8b):*

$$h_1 = 3398, h_2 = 2130, h_3 = 192 \text{ and } h_4 = 200 \text{ kJ/kg}$$

$$W_T = 1268 \text{ kJ/kg}, W_P = 8 \text{ kJ/kg}, Q_1 = 3198 \text{ kJ/kg}$$

$$\eta_I = \frac{1260}{3198} = 0.394 \text{ or } 39.4\% \qquad \textit{Ans.}$$

$$\text{Work ratio} = \frac{1260}{1268} = 0.994 \qquad \textit{Ans.}$$

$$\text{Exergy input rate} = 59.3 \text{ MW}, W_{net} = Q_1 \times \eta_I = 39.4 \text{ MW}$$

$$\eta_{II} = \frac{36.5}{59.3} = 0.664 \text{ or } 66.4\% \qquad \textit{Ans.}$$

Improvements in both first law and second law efficiencies are achieved with superheating. The specific work output is also increased. Therefore, conventional vapour power plants are almost always operated with some superheat.

(c) *Rankine cycle with reheat* (*Fig.* 12.8*c*):

$h_1 = 3398, h_2 = 2761, h_3 = 3482, h_4 = 2522, h_5 = 192$ and $h_6 = 200$ kJ/kg

$$W_{T1} = 637 \text{ kJ/kg}, W_{T2} = 960 \text{ kJ/kg}$$
$$W_T = 637 + 960 = 1597 \text{ kJ/kg}, W_P = 8 \text{ kJ/kg}$$
$$W_{net} = 1589 \text{ kJ/kg}, Q_1 = 3198 + 721 = 3919 \text{ kJ/kg}$$
$$\eta_I = \frac{1589}{3919} = 0.405 \text{ or } 40.5\% \quad \textit{Ans.}$$
$$\text{Work ratio} = \frac{W_{net}}{W_T} = \frac{1589}{1597} = 0.995 \quad \textit{Ans.}$$

Mechanical power output = 100 × 0.405 = 40.5 MW

Exergy input rate = 59.3 MW

$$\eta_{II} = \frac{40.5}{59.3} = 0.683 \text{ or } 68.3\% \quad \textit{Ans.}$$

Compared with basic Rankine cycle, the second law efficiency for the reheat cycle shows an increase of about 11% [(0.683 – 0.616)/0.616]. Therefore, most of the large conventional steam power plants in use today operate on the Rankine cycle with reheat.

(d) *Rankine cycle with complete regeneration* (*Fig.* 12.8*d*)

t_{sat} at 0.1 bar = 45.8°C = 318.8 K and

t_{sat} at 80 bar = 295°C = 568 K

$$\eta_I = \eta_{Carnot} = 1 - \frac{T_3}{T_1} = 1 - \frac{318.8}{568.0} = 0.439 \text{ or, } 43.9\% \quad \textit{Ans.}$$
$$Q_1 = h_1 - h_6 = 2758 - 1316 = 1442 \text{ kJ/kg}$$
$$W_{net} = Q_1 \times \eta_I = 1442 \times 0.439 = 633 \text{ kJ/kg}$$
$$W_P = 8 \text{ kJ/kg} \quad W_T = 641 \text{ kJ/kg}$$
$$\text{Work ratio} = \frac{633}{641} = 0.988 \quad \textit{Ans.}$$

Power output = 0.439 × 100 = 43.9 MW

$$\text{Exergy input rate} = \frac{100 \times 1000}{833 - 593}\left[(833 - 300) - 300 \ln \frac{833}{300}\right] = 94.583 \text{ MW} \cong 94.6 \text{ MW}$$
$$\eta_{II} = \frac{43.9}{94.6} = 0.464 \text{ or } 46.4\% \quad \textit{Ans.}$$

The second law efficiency is lower for regeneration because of the more substantial loss of exergy carried by the effluent gas stream at 593 K.

***Example* 12.9**

A certain chemical plant requires heat from process steam at 120°C *at the rate of* 5.83 *MJ/s and power at the rate of* 1000 *kW from the generator terminals. Both the heat and power requirements are met by a back pressure turbine of* 80% *brake and* 85% *internal efficiency, which exhausts steam at* 120°C *dry saturated. All the latent heat released during condensation is utilized in the process heater. Find the pressure and temperature of steam at the inlet to the turbine. Assume* 90% *efficiency for the generator.*

Solution At 120°C, $h_{fg} = 2202.6 \text{ kJ/kg} = h_2 - h_3$ (Fig. Ex. 12.9)

$$Q_H = w_s (h_2 - h_3) = 5.83 \text{ MJ/s}$$
$$\therefore \quad w_s = \frac{5830}{2202.6} = 2.647 \text{ kg/s}$$

$$W_{net} = \frac{1000}{0.9} \text{ kJ/s} = \text{Brake output}$$

Now
$$\eta_{brake} = \frac{\text{Brake output}}{\text{Ideal output}} = \frac{(1000)/0.9}{w_s (h_1 - h_{2s})} = 0.80$$

$\therefore$
$$h_1 - h_{2s} = \frac{1000}{0.9 \times 0.8 \times 2.647} = 524.7 \text{ kJ/kg}$$

Again
$$\eta_{internal} = \frac{h_1 - h_2}{h_1 - h_{2s}} = 0.85$$

$\therefore$
$$h_1 - h_2 = 0.85 \times 524.7 = 446 \text{ kJ/kg}$$

$$h_2 = h_g \text{ at } 120°\text{C} = 2706.3 \text{ kJ/kg}$$

$\therefore$
$$h_1 = 3152.3 \text{ kJ/kg}$$

$$h_{2s} = h_1 - 524.7 = 2627.6 \text{ kJ/kg} = h_f + x_{2s} h_{fg} = 503.71 + x_{2s} \times 2202.6$$

$\therefore$
$$x_{2s} = \frac{2123.89}{2202.6} = 0.964$$

$\therefore$
$$s_{2s} = s_f + x_{2s} s_{fg} = 1.5276 + 0.964 \times 5.6020 = 6.928 \text{ kJ/kg K}$$

At state 1,
$$h_1 = 3152.3 \text{ kJ/kg}$$
$$s_1 = 6.928 \text{ kJ/kg K}$$

From the Mollier chart
$$p_1 = 22.5 \text{ bar}$$
$$t_1 = 360°\text{C}$$
Ans.

Example 12.10

A certain factory has an average electrical load of 1500 kW and requires 3.5 MJ/s for heating purposes. It is proposed to install a single-extraction passout steam turbine to operate under the following conditions:

Initial pressure 15 bar.
Initial temperature 300°C.
Condenser pressure 0.1 bar.

Steam is extracted between the two turbine sections at 3 bar, 0.96 dry, and is isobarically cooled without subcooling in heaters to supply the heating load. The internal efficiency of the turbine (in the L.P. Section) is 0.80 and the efficiency of the boiler is 0.85 when using oil of calorific value 44 MJ/kg.

If 10% of boiler steam is used for auxiliaries calculate the oil consumption per day. Assume that the condensate from the heaters (at 3 bar) and that from the condenser (at 0.1 bar) mix freely in a separate vessel (hot well) before being pumped to the boiler. Neglect extraneous losses.

Solution Let w_s be the flow rate of steam (kg/h) entering the turbine, and w the amount of steam extracted per hour for process heat (Fig. Ex. 12.10).

$$h_1 = 3037.3 \text{ kJ/kg}$$
$$h_2 = 561.47 + 0.96 \times 2163.8 = 2638.7 \text{ kJ/kg}$$
$$s_2 = 1.6718 + 0.96 \times 5.3201 = 6.7791 \text{ kJ/kg K} = s_{3s}$$
$$s_{3s} = 6.7791 = 0.6493 + x_{3s} \times 7.5009$$

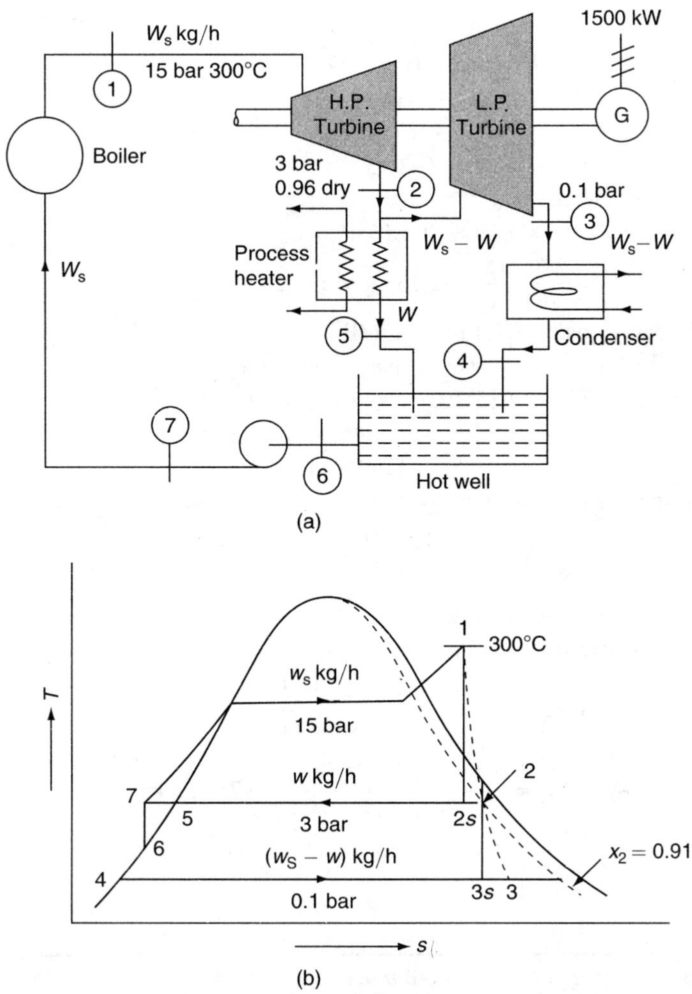

Fig. Ex. 12.10

$$x_{3s} = \frac{6.1298}{7.5009} = 0.817$$

$$h_{3s} = 191.83 + 0.817 \times 2392.8 = 2146.75 \text{ kJ/kg}$$

$$h_2 - h_{3s} = 2638.7 - 2146.75 = 491.95 \text{ kJ/kg}$$

$$h_2 - h_3 = 0.8 \times 491.95 = 393.56 \text{ kJ/kg}$$

$$\therefore \quad h_3 = 2638.7 - 393.56 = 2245.14 \text{ kJ/kg}$$

$$h_5 = 561.47 \text{ kJ/kg}, h_4 = 191.83 \text{ kJ/kg}$$

$$Q_H = w(h_2 - h_5) = w(2638.7 - 561.47) = 3.5 \text{ MJ/s}$$

$$\therefore \quad w = \frac{3500}{2077.23} = 1.685 \text{ kg/s}$$

Now

$$W_T = w_s(h_1 - h_2) + (w_s - w)(h_2 - h_3)$$
$$= w_s(3037.3 - 2638.7) + (w_s - 1.685) \times 393.56$$
$$= w_s \times 398.6 + w_s \times 393.56 - 663.15$$
$$= 792.16\, w_s - 663.15$$

Neglecting pump work

$$W_T = 1500 \text{ kJ/s} = 792.16\, w_s - 663.15$$

$$\therefore \quad w_s = \frac{2163.15}{792.16} = 2.73 \text{ kg/s} = 9828 \text{ kg/h}$$

By making energy balance for the hot well

$$(w_s - w)h_4 + wh_5 = w_s h_6$$
$$(2.73 - 1.685)191.83 + 1.685 \times 561.47 = 2.73 \times h_6$$
$$200.46 + 946.08 = 2.73\, h_6$$
$$\therefore \quad h_6 = 419.98 \text{ kJ/kg} \cong h_7$$

Steam raising capacity of the boiler = 1.1 w_s kg/h, since 10% of boiler steam is used for auxiliaries.

$$\therefore \quad \eta_{\text{boiler}} = \frac{1.1\, w_s\,(h_1 - h_7)}{w_f \times \text{C.V.}}$$

where w_f = fuel burning rate (kg/h)
and C.V. = calorific value of fuel = 44 MJ/kg

$$\therefore \quad 0.85 = \frac{1.1 \times 9828 \times (3037.3 - 419.98)}{w_f \times 44000}$$

or

$$w_f = \frac{1.1 \times 9828 \times 2617.32}{0.85 \times 44000} = 756.56 \text{ kg/h}$$

$$= \frac{756.56 \times 24}{1000} = 18.16 \text{ tonnes/day}$$ *Ans.*

Example 12.11

A steam turbine gets its supply of steam at 70 bar and 450°C. After expanding to 25 bar in high pressure stages, it is reheated to 420°C at the constant pressure. Next, it is expanded in intermediate pressure stages to an appropriate minimum pressure such that part of the steam bled at this pressure heats the feedwater to a temperature of 180°C. The remaining steam expands from this pressure to a condenser pressure of 0.07 bar in the low pressure stage. The isentropic efficiency of the h.p. stage is 78.5%, while that of the intermediate and l.p. stages is 83% each. From the above data (a) determine the minimum pressure at which bleeding is necessary, and sketch a line diagram of the arrangement of the plant, (b) sketch on the T–s diagram all the processes, (c) determine the quantity of steam bled per kg of flow at the turbine inlet, and (d) calculate the cycle efficiency. Neglect pump work.

Solution Figure Ex. 12.11 gives the flow and *T–s* diagrams of the plant. It would be assumed that the feedwater heater is an open heater. Feedwater is heated to 180°C. So p_{sat} at 180°C ≅ 10 bar is the pressure at which the heater operates.

Therefore, the pressure at which bleeding is necessary is 10 bar. *Ans.* (a).

From the Mollier chart

$$h_1 = 3285,\ h_{2s} = 3010,\ h_3 = 3280,\ h_{4s} = 3030 \text{ kJ/kg}$$

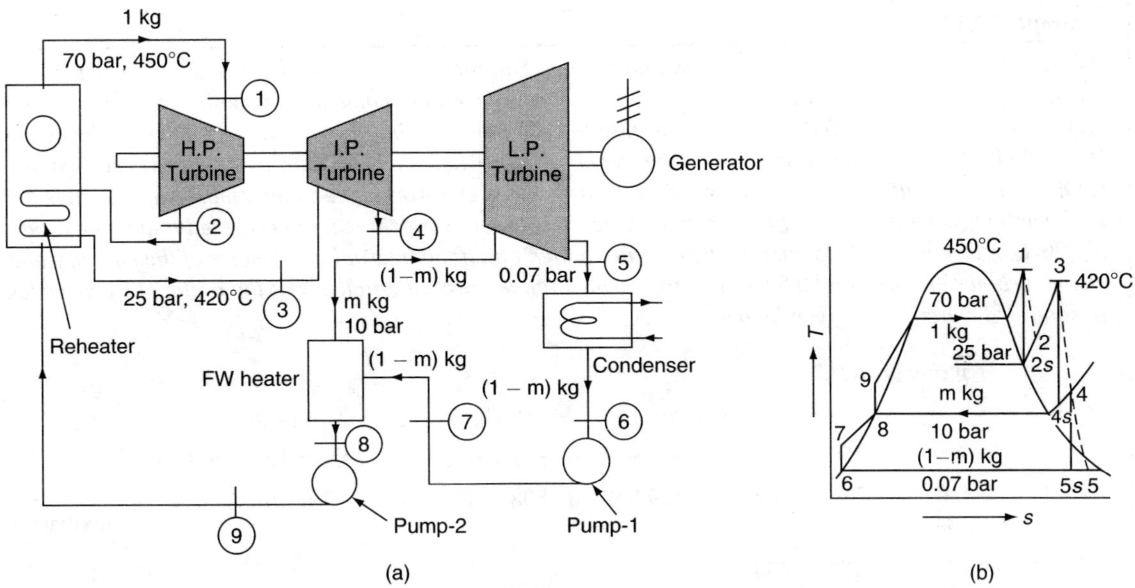

Fig. Ex. 12.11

$$h_3 - h_4 = 0.83\,(h_3 - h_{4s}) = 0.83 \times 250 = 207.5 \text{ kJ/kg}$$

$$h_4 = 3280 - 207.5 = 3072.5 \text{ kJ/kg}$$

$$h_{5s} = 2225 \text{ kJ/kg}$$

$$h_4 - h_5 = 0.83(h_4 - h_{5s}) = 0.83 \times 847.5 = 703.4 \text{ kJ/kg}$$

$$\therefore \quad h_5 = 3072.5 - 703.4 = 2369.1 \text{ kJ/kg}$$

$$h_6 = 162.7 \text{ kJ/kg}$$

$$h_8 = 762.81 \text{ kJ/kg}$$

$$h_1 - h_2 = 0.785\,(h_1 - h_{2s}) = 0.785 \times 275 = 215.9 \text{ kJ/kg}$$

$$h_2 = 3285 - 215.9 = 3069.1 \text{ kJ/kg}$$

Energy balance for the heater gives

$$m \times h_4 + (1 - m)h_7 = 1 \times h_8$$

$$m \times 3072.5 + (1 - m) \times 162.7 = 1 \times 762.81$$

$$m = \frac{600.11}{2909.8} = 0.206 \text{ kg/kg steam flow at turbine inlet.}$$ *Ans.* (c)

$$\therefore \quad \eta_{\text{cycle}} = \frac{(h_1 - h_2) + (h_3 - h_4) + (1 - m)(h_4 - h_5)}{(h_1 - h_8) + (h_3 - h_2)}$$

$$= \frac{215.9 + 207.5 + 0.794 \times 703.4}{2522.2 + 210.9}$$

$$= \frac{981.9}{2733.1}\ 0.3592 \text{ or } 35.92\%$$ *Ans.* (d)

Example 12.12

A binary-vapour cycle operates on mercury and steam. Saturated mercury vapour at 4.5 bar is supplied to the mercury turbine, from which it exhausts at 0.04 bar. The mercury condenser generates saturated steam at 15 bar which is expanded in a steam turbine to 0.04 bar. (a) Find the overall efficiency of the cycle. (b) If 50,000 kg/h of steam flows through the steam turbine, what is the flow through the mercury turbine? (c) Assuming that all processes are reversible, what is the useful work done in the binary vapour cycle for the specified steam flow? (d) If the steam leaving the mercury condenser is superheated to a temperature of 300°C in a superheater located in the mercury boiler, and if the internal efficiencies of the mercury and steam turbines are 0.85 and 0.87 respectively, calculate the overall efficiency of the cycle. The properties of saturated mercury are given below

p (bar)	*t*(° C)	h_f	h_g	s_f	s_g	v_f	v_g
		(kJ/kg)		(kJ/kg K)		(m³/kg)	
4.5	450	62.93	355.98	0.1352	0.5397	79.9×10^{-6}	0.068
0.04	216.9	29.98	329.85	0.0808	0.6925	76.5×10^{-6}	5.178

Solution The cycle is shown in Fig. Ex. 12.12.

For the mercury cycle, $h_a = 355.98$ kJ/kg

$$s_a = 0.5397 \text{ kJ/kg K} = s_b = s_f + x_b s_{fg}$$
$$= 0.0808 + x_b (0.6925 - 0.0808)$$

$$\therefore \quad x_b = \frac{0.4589}{0.6117} = 0.75$$

$$h_b = 29.98 + 0.75 \times 299.87$$
$$= 254.88 \text{ kJ/kg}$$

$$(W_T)_m = h_a - h_b$$
$$= 355.98 - 254.88 = 101.1 \text{ kJ/kg}$$

$$(W_P)_m = h_d - h_c$$
$$= 76.5 \times 10^{-6} \times 4.46 \times 100$$
$$= 3.41 \times 10^{-2} \text{ kJ/kg}$$

$$\therefore W_{net} = 101.1 \text{ kJ/kg}$$

$$Q_1 = h_a - h_d$$
$$= 355.98 - 29.98 = 326 \text{ kJ/kg}$$

$$\therefore \eta_m = \frac{W_{net}}{Q_1} = \frac{101.1}{326} = 0.31 \text{ or } 31\%$$

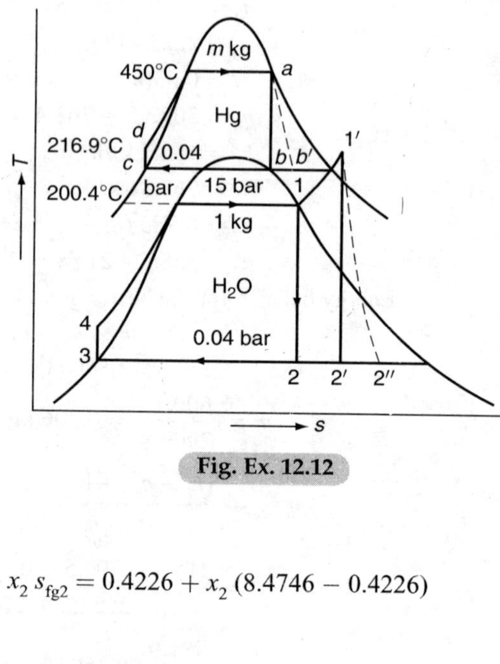

Fig. Ex. 12.12

For the steam cycle

$$h_1 = 2792.2 \text{ kJ/kg}$$

$$s_1 = 6.4448 \text{ kJ/kg K} = s_2 = s_f + x_2 s_{fg2} = 0.4226 + x_2 (8.4746 - 0.4226)$$

$$x_2 = \frac{6.0222}{8.0520} = 0.748$$

$$h_2 = 121.46 + 0.748 \times 2432.9 = 1941.27 \text{ kJ/kg}$$

$$(W_T)_{St} = h_1 - h_2 = 2792.2 - 1941.27 = 850.93 \text{ kJ/kg}$$
$$(W_P)_{St} = h_4 - h_3 = 0.001 \times 14.96 \times 100 = 1.496 \text{ kJ/kg} \cong 1.5 \text{ kJ/kg}$$
$$h_4 = 121.46 + 1.5 = 122.96 \text{ kJ/kg}$$
$$Q_1 = h_1 - h_4 = 2792.2 - 122.96 = 2669.24 \text{ kJ/kg}$$
$$(W_{net})_{St} = 850.93 - 1.5 = 849.43 \text{ kJ/kg}$$
$$\therefore \quad \eta_{st} = \frac{W_{net}}{Q_1} = \frac{849.43}{2669.24} = 0.318 \text{ or } 31.8\%$$

Overall efficiency of the binary cycle would be

$$\eta_{overall} = \eta_m + \eta_{St} - \eta_m \cdot \eta_{St} = 0.31 + 0.318 - 0.31 \times 0.318$$
$$= 0.5294 \text{ or } 52.94\%$$ *Ans.* (a)

$\eta_{overall}$ can also be determined in, the following way:
By writing the energy balance for a mercury condenser-steam boiler

$$m(h_b - h_c) = 1(h_1 - h_4)$$

where m is the amount of mercury circulating for 1 kg of steam in the bottom cycle.

$$\therefore \quad m = \frac{h_1 - h_4}{h_b - h_c} = \frac{2669.24}{254.88 - 29.88} = \frac{2669.24}{224.90} = 11.87 \text{ kg}$$
$$(Q_1)_{total} = m(h_a - h_d) = 11.87 \times 326 = 3869.6 \text{ kJ/kg}$$
$$(W_T)_{total} = m(h_a - h_b) + (h_1 - h_2) = 11.87 \times 101.1 - 850.93 = 2051 \text{ kJ/kg}$$

$(W_P)_{total}$ may be neglected

$$\therefore \quad \eta_{overall} = \frac{W_{net}}{Q} = \frac{2051}{3869.6} = 0.53 \text{ or } 53\%$$

If 50,000 kg/h of steam flows through the steam turbine, the flow rate of mercury w_m would be

$$w_m = 50{,}000 \times 11.87 = 59.35 \times 10^4 \text{ kg/h}$$ *Ans.* (b)

$$(W_T)_{total} = 2051 \times 50{,}000 = 10255 \times 10^4 \text{ kJ/h} = 0.2849 \times 10^5 \text{ kW} = 28.49 \text{ MW}$$ *Ans.* (c)

Considering the efficiencies of turbines

$$(W_T)_m = h_a - h'_b = 0.85 \times 101.1 = 84.95 \text{ kJ/kg}$$
$$\therefore \quad h'_b = 355.98 - 85.94 = 270.04 \text{ kJ/kg}$$
$$\therefore \quad m'(h'_b - h'_c) = (h_1 - h_4)$$
$$\therefore \quad m' = \frac{2669.24}{240.06} = 11.12 \text{ kg}$$
$$(Q_1)_{total} = m'(h_a - h_d) + 1(h'_1 - h_1)$$
$$= 11.12 \times 326 + (3037.3 - 2792.2) = 3870.22 \text{ kJ/kg}$$
$$s'_1 = 6.9160 = 0.4226 + x'_2 (8.4746 - 0.4226)$$
$$x'_2 = \frac{6.4934}{8.0520} = 0.806$$
$$h'_2 = 121.46 + 0.806 \times 2432.9 = 2082.38 \text{ kJ/kg}$$
$$(W_T)_{St} = h_i - h_2'' = 0.87(3037.3 - 2082.38) = 830.78 \text{ kJ/kg}$$
$$(W_T)_{total} = 11.12 \times 85.94 + 830.78 = 1786.43 \text{ kJ/kg}$$

Pump work is neglected.

$$\eta_{overall} = \frac{1786.43}{3870.22} = 0.462 \text{ or } 46.2\%$$ *Ans.* (d)

Review Questions

12.1 What are the four basic components of a steam power plant?

12.2 What is the reversible cycle that represents the simple steam power plant? Draw the flow, p–v, T–s and h–s diagrams of this cycle.

12.3 What do you understand by steam rate and heat rate? What are their units?

12.4 Why is Carnot cycle not practicable for a steam power plant?

12.5 What do you understand by the mean temperature of heat addition?

12.6 For a given T_2, show how the Rankine cycle efficiency depends on the mean temperature of heat addition.

12.7 What is metallurgical limit?

12.8 Explain how the quality at turbine exhaust gets restricted.

12.9 How are the maximum temperature and maximum pressure in the Rankine cycle fixed?

12.10 When is reheating of steam recommended in a steam power plant? How does the reheat pressure get optimized?

12.11 What is the effect of reheat on (a) the specific output, (b) the cycle efficiency, (c) steam rate, and (d) heat rate, of a steam power plant?

12.12 Give the flow and T–s diagrams of the ideal regenerative cycle. Why is the efficiency of this cycle equal to Carnot efficiency? Why is this cycle not practicable?

12.13 What is the effect of regeneration on the (a) specific output, (b) mean temperature of heat addition, (c) cycle efficiency, (d) steam rate and (e) heat rate of a steam power plant?

12.14 How does the regeneration of steam carnotize the Rankine cycle?

12.15 What are open and closed heaters? Mention their merits and demerits.

12.16 Why is one open feedwater heater used in a steam plant? What is it called?

12.17 How are the number of heaters and the degree of regeneration get optimized?

12.18 Draw the T–s diagram of an ideal working fluid in a vapour power cycle.

12.19 Discuss the desirable characteristics of a working fluid in a vapour power cycle.

12.20 Mention a few working fluids suitable in the high temperature range of a vapour power cycle.

12.21 What is a binary vapour cycle?

12.22 What are topping and bottoming cycles?

12.23 Show that the overall efficiency of two cycles coupled in series equals the sum of the individual efficiencies minus their product.

12.24 What is a cogeneration plant? What are the thermodynamic advantages of such a plant?

12.25 What is a back pressure turbine? What are its applications?

12.26 What is the biggest loss in a steam plant? How can this loss be reduced?

12.27 What is a pass-out turbine? When is it used?

12.28 Define the following: (a) internal work, (b) internal efficiency, (c) brake efficiency (d) mechanical efficiency, and (e) boiler efficiency.

12.29 Express the overall efficiency of a steam plant as the product of boiler, turbine, generator and cycle efficiencies.

Problems

12.1 For the following steam cycles, find (a) W_T in kJ/kg (b) W_P in kJ/kg, (c) Q_1 in kJ/kg, (d) cycle efficiency, (e) steam rate in kg/kW h, and (f) moisture at the end of the turbine process. Show the results in tabular form with your comments.

Boiler outlet	*Condenser pressure*	*Type of cycle*
10 bar, saturated	1 bar	Ideal Rankine cycle
-do-	-do-	Neglect W_P
-do-	-do-	Assume 75% pump and turbine efficiency
-do-	0.1 bar	Ideal Rankine cycle

(Continued)

10 bar, 300°C	-do-	-do-
150 bar, 600°C	-do-	-do-
-do-	-do-	Reheat to 600°C at maximum intermediate pressure to limit end moisture to 15%
-do-	-do-	-do- but with 85% turbine efficiency
10 bar, saturated	0.1 bar	Isentropic pump process ends on saturated liquid line
-do-	-do-	-do- but with 80% machine efficiencies
-do-	-do-	Ideal regenerative cycle
-do-	-do-	Single open heater at 110°C
-do-	-do-	Two open heaters at 90°C and 135°C
-do-	-do-	-do- but the heaters are closed heaters

12.2 A geothermal power plant utilizes steam produced by natural means underground. Steam wells are drilled to tap this steam supply which is available at 4.5 bar and 175°C. The steam leaves the turbine at 100 mm Hg absolute pressure. The turbine isentropic efficiency is 0.75. Calculate the efficiency of the plant. If the unit produces 12.5 MW, what is the steam flow rate?

12.3 A simple steam power cycle uses solar energy for the heat input. Water in the cycle enters the pump as a saturated liquid at 40°C, and is pumped to 2 bar. It then evaporates in the boiler at this pressure, and enters the turbine as saturated vapour. At the turbine exhaust the conditions are 40°C and 10% moisture. The flow rate is 150 kg/h. Determine (a) the turbine isentropic efficiency, (b) the net work output (c) the cycle efficiency, and (d) the area of solar collector needed if the collectors pick up 0.58 kW/m^2. *Ans.* (a) 0.767, (b) 15.51 kW, (c) 14.7%, (d) 182.4 m^2

12.4 In a reheat cycle, the initial steam pressure and the maximum temperature are 150 bar and 550°C respectively. If the condenser pressure is 0.1 bar and the moisture at the condenser inlet is 5%, and assuming ideal processes, determine (a) the reheat pressure, (b) the cycle efficiency, and (c) the steam rate. *Ans.* 13.5 bar, 43.6%, 2.05 kg/kW h

12.5 In a nuclear power-plant heat is transferred in the reactor to liquid sodium. The liquid sodium is then pumped to a heat exchanger where heat is transferred to steam. The steam leaves this heat exchanger as saturated vapour at 55 bar, and is then superheated in an external gas-fired superheater to 650°C. The steam then enters the turbine, which has one extraction point at 4 bar, where steam flows to an open feedwater heater. The turbine efficiency is 75% and the condenser temperature is 40°C. Determine the heat transfer in the reactor and in the superheater to produce a power output of 80 MW.

12.6 In a reheat cycle, steam at 500°C expands in a h.p. turbine till it is saturated vapour. It is reheated at constant pressure to 400°C and then expands in a l.p. turbine to 40°C. If the maximum moisture content at the turbine exhaust is limited to 15%, find (a) the reheat pressure, (b) the pressure of steam at the inlet to the h.p. turbine, (c) the net specific work output, (d) the cycle efficiency, and (e) the steam rate. Assume all ideal processes.

What would have been the quality, the work output, and the cycle efficiency without the reheating of steam? Assume that the other conditions remain the same.

12.7 A regenerative cycle operates with steam supplied at 30 bar and 300°C, and condenser pressure of 0.08 bar. The extraction points for two heaters (one closed and one open) are at 3.5 bar and 0.7 bar respectively. Calculate the thermal efficiency of the plant, neglecting pump work.

12.8 The net power output of the turbine in an ideal reheat-regenertive cycle is 100 MW. Steam enters the high-pressure (H.P.) turbine at 90 bar, 550°C. After expansion to 7 bar, some of the steam goes to an open heater and the balance is reheated to 400°C, after which it expands to 0.07 bar. (a) What is the steam flow rate to the H.P. turbine? (b) What is the total pump work? (c) Calculate the cycle efficiency. (d) If there is a 10°C rise in the temperature of the cooling water, what is the rate of flow of the cooling water in the condenser? (e) If the velocity of the steam flowing from the turbine to the condenser is limited to a maximum of 130 m/s, find the diameter of the connecting pipe.

12.9 A mercury cycle is superposed on the steam cycle operating between the boiler outlet condition of 40 bar, 400°C and the condenser temperature of 40°C. The heat released by mercury condensing at 0.2 bar is used to impart the latent heat of vaporization to the water in the steam cycle. Mercury enters the mercury turbine as saturated vapour at 10 bar. Compute (a) kg of mercury circulated per kg of water, and (b) the efficiency of the combined cycle.

The property values of saturated mercury are given below

p (*bar*)	*t* (°C)	h_f	h_g	s_f	s_g	v_f	v_g
		(kJ/kg)		(kJ/kg K)		(m³/kg)	
10	515.5	72.23	363.0	0.1478	0.5167	80.9×10^{-6}	0.0333
0.2	277.3	38.35	336.55	0.0967	0.6385	77.4×10^{-6}	1.163

12.10 In an electric generating station, using a binary vapour cycle with mercury in the upper cycle and steam in the lower, the ratio of mercury flow to steam flow is 10:1 on a mass basis. At an evaporation rate of 1,000,000 kg/h for the mercury, its specific enthalpy rises by 356 kJ/kg in passing through the boiler. Superheating the steam in the boiler furnace adds 586 kJ to the steam specific enthalpy. The mercury gives up 251.2 kJ/kg during condensation, and the steam gives up 2003 kJ/kg in its condenser. The overall boiler efficiency is 85%. The combined turbine mechanical and generator efficiencies are each 95% for the mercury and steam units. The steam auxiliaries require 5% of the energy generated by the units. Find the overall efficiency of the plant.

12.11 A sodium-mercury-steam cycle operates between 1000°C and 40°C. Sodium rejects heat at 670°C to mercury. Mercury boils at 24.6 bar and rejects heat at 0.141 bar. Both the sodium and mercury cycles are saturated. Steam is formed at 30 bar and is superheated in the sodium boiler to 350°C. It rejects heat at 0.08 bar. Assume isentropic expansions, no heat losses, and no regeneration and neglect pumping work. Find (a) the amounts of sodium and mercury used per kg of steam, (b) the heat added and rejected in the composite cycle per kg steam, (c) the total work done per kg steam. (d) the efficiency of the composite cycle, (e) the efficiency of the corresponding Carnot cycle, and (f) the work, heat added, and efficiency of a supercritical pressure steam (single fluid) cycle operating at 250 bar and between the same temperature limits.

For mercury, at 24.6 bar, $h_g = 366.78$ kJ/kg
$s_g = 0.48$ kJ/kg K and at 0.141 bar, $s_f = 0.09$
and $s_g = 0.64$ kJ/kg K, $h_f = 36.01$
and $h_g = 330.77$ kJ/kg
For sodium, at 1000°C, $h_g = 4982.53$ kJ/kg
At turbine exhaust, $h = 3914.85$ kJ/kg
At 670°C, $h_f = 745.29$ kJ/kg

For a supercritical steam cycle, the specific enthalpy and entropy at the turbine inlet may be computed by extrapolation from the steam tables.

12.12 A textile factory requires 10,000 kg/h of steam for process heating at 3 bar saturated and 1000 kW of power, for which a back pressure turbine of 70% internal efficiency is to be used. Find the steam condition required at the inlet to the turbine.

12.13 A 10,000 kW steam turbine operates with steam at the inlet at 40 bar, 400°C and exhausts at 0.1 bar. Ten thousand kg/h of steam at 3 bar are to be extracted for process work. The turbine has 75% isentropic efficiency throughout. Find the boiler capacity required.

12.14 A 50 MW steam plant built in 1935 operates with steam at the inlet at 60 bar, 450°C and exhausts at 0.1 bar, with 80% turbine efficiency. It is proposed to scrap the old boiler and put in a new boiler and a topping turbine of efficiency 85% operating with inlet steam at 180 bar, 500°C. The exhaust from the topping turbine at 60 bar is reheated to 450°C and admitted to the old turbine. The flow rate is just sufficient to produce the rated output from the old turbine. Find the improvement in efficiency with the new set up. What is the additional power developed?

12.15 A steam plant operates with an initial pressure at 20 bar and temperature 400°C, and exhausts to a heating system at 2 bar. The condensate from the heating system is returned to the boiler plant at 65°C, and the heating system utilizes for its intended purpose 90% of the energy transferred from the steam it receives. The turbine efficiency is 70%. (a) What fraction of the energy supplied to the steam plant serves a useful purpose? (b) If two separate steam plants had been set up to produce the same useful energy, one to generate heating steam at 2 bar, and the other to generate power through a cycle working between 20 bar, 400°C and 0.07 bar, what fraction of the energy supplied would have served a useful purpose? *Ans*. 91.2%, 64.5%

12.16 In a nuclear power plant saturated steam at 30 bar enters a h.p. turbine and expands isentropically to a pressure at which its quality is 0.841. At this pressure the steam is passed through a moisture separator which removes all the liquid. Saturated vapour leaves the separator and is expanded isentropically to 0.04 bar in l.p. turbine, while the saturated liquid leaving the separator is returned via a feed pump to the boiler. The condensate leaving the condenser at 0.04 bar is also returned to the boiler via a second feed pump. Calculate the cycle efficiency and turbine outlet quality taking into account the feed pump term. Recalculate the same quantities for a cycle with the same boiler and condenser pressures but without moisture separation. *Ans*. 35.5%, 0.824; 35%; 0.716

12.17 The net power output of an ideal regenerative-reheat steam cycle is 80 MW. Steam enters the h.p. turbine at 80 bar, 500°C and expands till it becomes saturated vapour. Some of the steam then goes to an open feedwater heater and the balance is reheated to 400°C, after which it expands in the l.p. turbine to 0.07 bar. Compute (a) the reheat pressure, (b) the steam flow rate to the h.p. turbine, and (c) the cycle efficiency. Neglect pump work. *Ans*. 6.5 bar, 58.4 kg/s, 43.7%

12.18 Figure P 12.18 shows the arrangement of a steam plant in which steam is also required for an industrial heating process. The steam leaves boiler *B* at 30 bar, 320°C and expands in the H.P. turbine to 2 bar, the efficiency of the H.P. turbine being 75%. At this point one half of the steam passes to the process heater *P* and the remainder enters separator *S* which removes all the moisture. The dry steam enters the L.P. turbine at 2 bar and expands to the condenser pressure 0.07 bar, the efficiency of the L.P. turbine being 70%. The drainage from the separator mixes with the condensate from the process heater and the combined flow enters the hotwell *H* at 80°C. Traps are provided at the exist from *P* and *S*. A pump extracts the condensate from condenser *C* and this enters

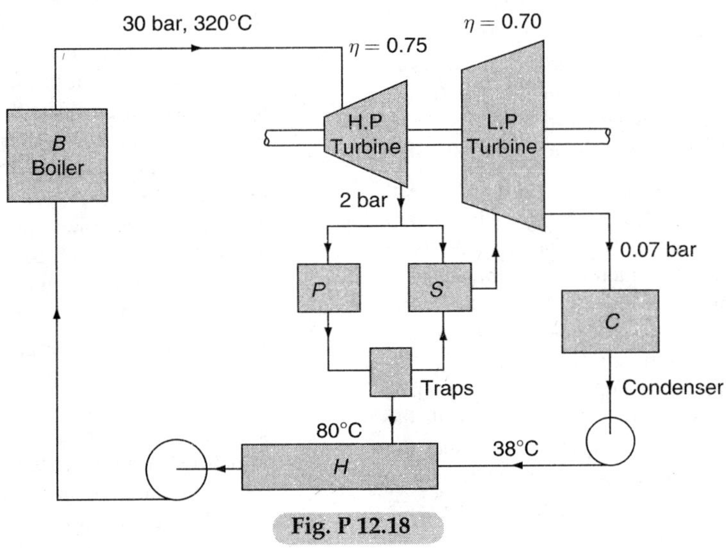

Fig. P 12.18

the hotwell at 38°C. Neglecting the feed pump work and radiation loss, estimate the temperature of water leaving the hotwell which is at atmospheric pressure. Also calculate, as percentage of heat transferred in the boiler, (a) the heat transferred in the process heater, and (b) the work done in the turbines.

12.19 In a combined power and process plant the boiler generates 21,000 kg/h of steam at a pressure of 17 bar, and temperature 230°C. A part of the steam goes to a process heater which consumes 132.56 kW, the steam leaving the process heater 0.957 dry at 17 bar being throttled to 3.5 bar. The remaining steam flows through a H.P. turbine which exhausts at a pressure of 3.5 bar. The exhaust steam mixes with the process steam before entering the L.P. turbine which develops 1337.5 kW. At the exhaust the pressure is 0.3 bar, and the steam is 0.912 dry. Draw a line diagram of the plant and determine (a) the steam quality at the exhaust from the H.P. turbine, (b) the power developed by the H.P. turbine, and (c) the isentropic efficiency of the H.P. turbine.

Ans. (a) 0.96, (b) 1125 kW, (c) 77%

12.20 In a cogeneration plant, the power load is 5.6 MW and the heating load is 1.163 MW. Steam is generated at 40 bar and 500°C and is expanded isentropically through a turbine to a condenser at 0.06 bar. The heating load is supplied by extracting steam from the turbine at 2 bar which condensed in the process heater to saturated liquid at 2 bar and then pumped back to the boiler. Compute (a) the steam generation capacity of the boiler in tonnes/h, (b) the heat input to the boiler in MW, and (c) the heat rejected to the condenser in MW.

Ans. (a) 19.07 t/h, (b) 71.57 MW, and (c) 9.607 MW

12.21 Steam is supplied to a pass-out turbine at 35 bar, 350°C and dry saturated process steam is required at 3.5 bar. The low pressure stage exhausts at 0.07 bar and the condition line may be assumed to be straight (the condition line is the locus passing through the states of steam leaving the various stages of the turbine). If the power required is 1 MW and the maximum process load is 1.4 kW, estimate the maximum steam flow through the high and low pressure stages. Assume that the steam just condenses in the process plant.

Ans. 1.543 and 1.182 kg/s

12.22 Geothermal energy from a natural geyser can be obtained as a continuous supply of steam 0.87 dry at 2 bar and at a flow rate of 2700 kg/h. This is utilized in a mixed-pressure cycle to augment the superheated exhaust from a high pressure turbine of 83% internal efficiency, which is supplied with 5500 kg/h of steam at 40 bar and 500°C. The mixing process is adiabatic and the mixture is expanded to a condenser pressure of 0.10 bar in a low pressure turbine of 78% internal efficiency. Determine the power output and the thermal efficiency of the plant. *Ans.* 1745 kW, 35%

12.23 In a study for a space project it is thought that the condensation of a working fluid might be possible at −40°C. A binary cycle is proposed, using Refrigerant-12 as the low temperature fluid, and water as the high temperature fluid. Steam is generated at 80 bar, 500°C and expands in a turbine of 81% isentropic efficiency to 0.06 bar, at which pressure it is condensed by the generation of dry saturated refrigerant vapour at 30°C from saturated liquid at −40°C. The isentropic efficiency of the R-12 turbine is 83%. Determine the mass ratio of R-12 to water and the efficiency of the cycle. Neglect all losses. *Ans.* 10.86; 44.4%

12.24 Steam is generated at 70 bar, 500°C and expands in a turbine to 30 bar with an isentropic efficiency of 77%. At this condition it is mixed with twice its mass of steam at 30 bar, 400°C. The mixture then expands with an isentropic efficiency of 80% to 0.06 bar. At a point in the expansion where the pressure is 5 bar, steam is bled for feedwater heating in a direct contact heater, which raises the feedwater to the saturation temperature of the bled steam. Calculate the mass of steam bled per kg of high pressure steam and the cycle efficiency. Assume that the L.P. expansion condition line is straight. *Ans.* 0.53 kg; 31.9%

12.25 An ideal steam power plant operates between 70 bar, 550°C and 0.075 bar. It has seven feedwater heaters. Find the optimum pressure and temperature at which each of the heaters operate.

12.26 In a reheat cycle steam at 550°C expands in an h.p. turbine till it is saturated vapour. It is reheated at constant pressure to 400°C and then expands in a l.p. turbine to 40°C. If the moisture content at turbine exhaust is given to be 14.67%, find (a) the reheat pressure, (b) the pressure of steam at inlet to the h.p. turbine, (c) the net work output per kg, and (d) the cycle efficiency. Assume all processes to be ideal.

Ans. (a) 20 bar, (b) 200 bar, (c) 1604 kJ/kg, (d) 43.8%

12.27 In a reheat steam cycle, the maximum steam temperature is limited to 500°C. The condenser pressure is 0.1 bar and the quality at turbine exhaust is 0.8778. Had there been no reheat, the exhaust quality would have been 0.7592. Assuming ideal processes, determine (a) reheat pressure, (b) the boiler pressure, (c) the cycle efficiency, and (d) the steam rate. *Ans.* (a) 30 bar, (b) 150 bar, (c) 50.51%, (d) 1.9412 kg/kWh

12.28 In a cogeneration plant, steam enters the h.p. stage of a two-stage turbine at 1 MPa, 200°C and leaves it at 0.3 MPa. At this point some of the steam is bled off and passed through a heat exchanger which it leaves as saturated liquid at 0.3 MPa. The remaining steam expands in the l.p. stage of the turbine to 40 kPa. The turbine is required to produce a total power of 1 MW and the heat exchanger to provide a heating rate of 500 KW. Calculate the required mass flow rate of steam into the h.p. stage of the turbine. Assume (a) steady condition throughout the plant, (b) velocity and gravity terms to be negligible, (c) both turbine stages are adiabatic with isentropic efficiencies of 0.80. *Ans.* 2.457 kg/s

12.29 A steam power plant is designed to operate on the basic Rankine cycle. The heat input to the boiler is at the rate of 50 MW. The H_2O exits the condenser as saturated liquid and exits the boiler as saturated vapour. The pressure of steam at boiler exit is 120 bar and the condenser pressure is 0.04 bar. The heat input to the boiler is provided by a steady stream of hot gases initially at 2200 K and 1 atm. The hot gases exhaust at 600 K and 1 atm to the surroundings which are at 600 K and 1 atm. Taking the c_p of hot gases as 1.1 kJ/kgK, determine (a) the cycle efficiency, (b) the work output, (c) the power output (in MW), (d) the required mass flow rate of steam (in kg/h), (e) the specific steam consumption (in kg/kWh), (f) the mass flow rate (in kg/h) of the stream of hot gases, (g) the exergy flux (in MW) of the inlet gases, (h) the exergy loss rate (in MW) with the exhaust gases, (i) the exergy consumption (in MW) in the steam generation process, (j) the exergy consumption (in MW) in the condensation process, (i) the second law efficiency.

Ans. (f) 1.20×10^5 kg/h, (g) 41.5 MW, (h) 3.14 MW, (i) 16.9 MW, (j) 1.40 MW, (k) 48.3%

12.30 In a cogeneration plant the steam generator provides 10^6 kg/h of steam at 80 bar, 480°C, of which 4×10^5 kg/h is extracted between the first and second turbine stages at 10 bar and diverted to a process heating load. Condensate returns from the process heating load at 9.5 bar, 120ºC and is mixed with liquid exiting the lower-pressure pump at 9.5 bar. The entire flow is then pumped to the steam generator pressure. Saturated liquid at 0.08 bar leaves the condenser. The turbine stages and the pumps operate with isentropic efficiencies of 86% and 80%, respectively. Determine (a) the heating load, in kJ/h, (b) the power developed by the turbine, in kW, (c) the rate of heat transfer in the steam generator, in kJ/h. *Ans.* (a) 9.529×10^8 kJ/h, (b) 236,500 kW, (c) 3.032×10^9 kJ/h

CHAPTER 13

Gas Power Cycles

Here gas is the working fluid. It does not undergo any phase change. Engines operating on gas cycles may be either cyclic or non-cyclic. Hot air engines using air as the working fluid operate on a closed cycle. Internal combustion engines where the combustion of fuel takes place inside the engine cylinder are non-cyclic heat engines.

13.1 CARNOT CYCLE (1824)

The Carnot cycle (Fig. 13.1) has been discussed in Chapters 6 and 7. It consists of:

Two reversible isotherms and two reversible adiabatics. If an ideal gas is assumed as the working fluid. Then for 1 kg of gas,

$$Q_{1-2} = RT_1 \ln \frac{v_2}{v_1}; \quad W_{1-2} = RT_1 \ln \frac{v_2}{v_1} \quad Q_{2-3} = 0; \quad W_{2-3} = -c_v (T_3 - T_2)$$

$$Q_{3-4} = RT_2 \ln \frac{v_4}{v_3}; \quad W_{3-4} = RT_2 \ln \frac{v_4}{v_3} \quad Q_{4-1} = 0; \quad W_{4-1} = -c_v (T_1 - T_4)$$

$$\therefore \quad \sum_{\text{cycle}} đQ = \sum_{\text{cycle}} đW$$

Now

$$\frac{v_2}{v_3} = \left(\frac{T_2}{T_1}\right)^{1/(\gamma-1)} \quad \text{and} \quad \frac{v_1}{v_4} = \left(\frac{T_2}{T_1}\right)^{1/(\gamma-1)}$$

$$\therefore \quad \frac{v_2}{v_3} = \frac{v_1}{v_4} \quad \text{or} \quad \frac{v_2}{v_1} = \frac{v_3}{v_4}$$

Therefore

$$Q_1 = \text{Heat added} = RT_1 \ln \frac{v_2}{v_1} \quad W_{\text{net}} = Q_1 - Q_2 = R \ln \frac{v_2}{v_1} \cdot (T_1 - T_2)$$

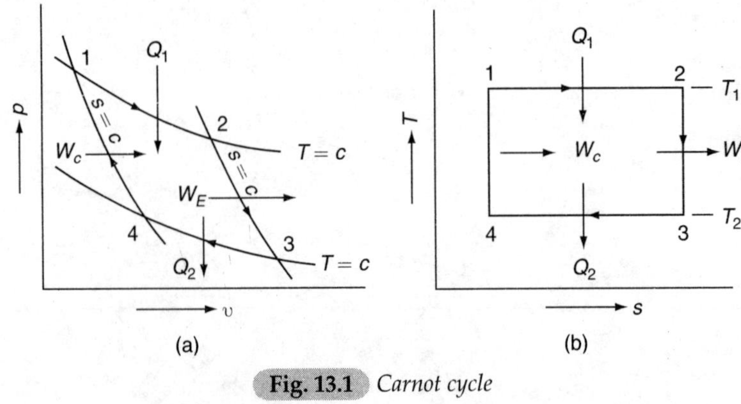

Fig. 13.1 *Carnot cycle*

$$\therefore \qquad \eta_{cycle} = \frac{W_{net}}{Q_1} = \frac{T_1 - T_2}{T_1} \tag{13.1}$$

The large back work ($W_c = W_{4-1}$) is a big drawback for the Carnot gas cycle, as in the case of the Carnot vapour cycle.

13.2 STIRLING CYCLE (1827)

The Stirling cycle (Fig. 13.2) consists of:
Two reversible isotherms and two reversible isochores. For 1 kg of ideal gas

$$Q_{1-2} = W_{1-2} = RT_1 \ln \frac{v_2}{v_1} \qquad Q_{2-3} = -c_v(T_2 - T_1);\ W_{2-3} = 0$$

$$Q_{3-4} = W_{3-4} = -RT_2 \ln \frac{v_3}{v_4} \qquad Q_{4-1} = c_v(T_1 - T_2);\ W_{4-1} = 0$$

Due to heat transfers at constant volume processes, the efficiency of the Stirling cycle is less than that of the Carnot cycle. However, if a regenerative arrangement is used such that $Q_{2-3} = Q_{4-1}$, i.e. the area under 2 – 3 is equal to the area under 4 – 1, then the cycle efficiency becomes

$$\eta = \frac{RT_1 \ln \frac{v_2}{v_1} - RT_2 \ln \frac{v_3}{v_4}}{RT_1 \ln \frac{v_2}{v_1}} = \frac{T_1 - T_2}{T_1} \tag{13.2}$$

So, the regenerative Stirling cycle has the same efficiency as the Carnot cycle.

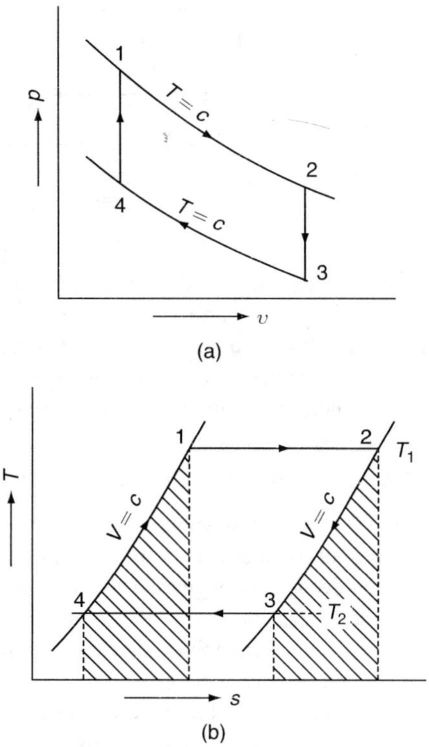

Fig. 13.2 *Stirling cycle*

13.3 ERICSSON CYCLE (1850)

The Ericsson cycle (Fig. 13.3) is made up of:

Two reversible isotherms and two reversible isobars.

For 1 kg of ideal gas

$$Q_{1-2} = W_{1-2} = RT_1 \ln \frac{p_1}{p_2} \qquad Q_{2-3} = c_p(T_2 - T_1);\ W_{2-3} = p_2(v_3 - v_2) = R(T_2 - T_1)$$

$$Q_{3-4} = W_{3-4} = -RT_2 \ln \frac{p_1}{p_2} \qquad Q_{4-1} = c_p(T_1 - T_4);\ W_{4-1} = p_1(v_1 - v_4) = R(T_1 - T_2)$$

Since part of the heat is transferred at constant pressure and part at constant temperature, the efficiency of the Ericsson cycle is less than that of the Carnot cycle. But with ideal regeneration, $Q_{2-3} = Q_{4-1}$ so that all the heat is added from the external source at T_1 and all the heat is rejected to an external sink at T_2, the efficiency of the cycle becomes equal to the Carnot cycle efficiency, since

$$\eta = 1 - \frac{Q_2}{Q_1} = 1 - \frac{RT_2 \ln \frac{p_1}{p_2}}{RT_1 \ln \frac{p_1}{p_2}} = 1 - \frac{T_2}{T_1} \tag{13.3}$$

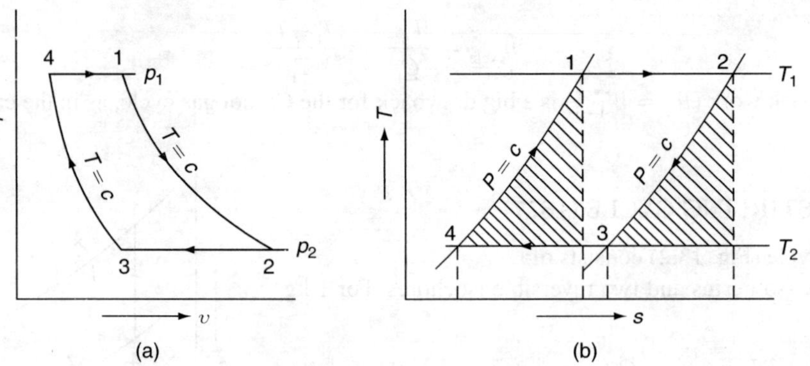

Fig. 13.3 *Ericsson cycle*

The regenerative, Stirling and Ericsson cycles have the same efficiency as the Carnot cycle, but much less back work. Hot air engines working on these cycles have been successfully operated. But it is difficult to transfer heat to a gas at high rates, since the gas film has a very low thermal conductivity. So there has not been much progress in the development of hot air engines. However, since the cost of internal combustion engine fuels is getting excessive, these may find a field of use in the near future.

13.4 AIR STANDARD CYCLES

Internal combustion engines (Fig. 13.4) in which the combustion of fuel occurs in the engine cylinder itself are non-cyclic heat engines. The temperature due to the evolution of heat because of the combustion of fuel inside the cylinder is so high that the cylinder is cooled by water circulation around it to avoid rapid deterioration. The working fluid, the fuel-air mixture, undergoes permanent chemical change due to combustion, and the products of combustion after doing work are thrown out of the engine, and a fresh charge is taken. So the working fluid does not undergo a complete thermodynamic cycle.

To simplify the analysis of I.C. engines, *air standard cycles* are conceived. In an air standard cycle, a certain mass of air operates in a complete thermodynamic cycle, where heat is added and rejected with external heat reservoirs, and all the processes in the cycle are reversible. Air is assumed to behave as an ideal gas, and its specific heats are assumed to be constant. These air standard cycles are so conceived that they correspond to the operations of internal combustion engines.

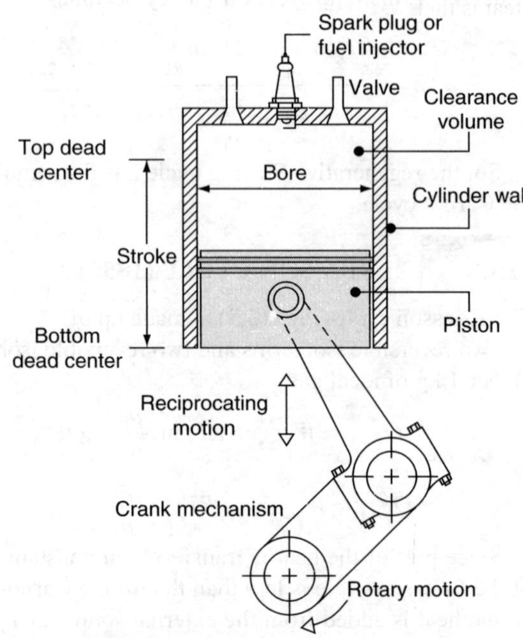

Fig. 13.4 *Nomenclature for reciprocating piston-cylinder engines*

13.5 OTTO CYCLE (1876)

One very common type of internal combustion engines is the *Spark Ignition (S.I.) engine* used inautomobiles. The Otto cycle is the air standard cycle of such an engine. The sequence of precesses in the elementary operation of the S.I. engine is given below, whith reference to Fig. 13.5(a,b) where the sketches of the engine and the indicatior diagram are given.

Process 1–2, *Intake*. The inlet valve is open, the piston moves to the right, admitting fuel-air mixture into the cylinder at constant pressure.

Process 2–3, *Compression*. Both the valves are closed, the piston compresses the combustible mixture to the minimum volume.

Process 3–4, *Combustion*. The mixture is then ignited by means of a spark, combustion takes place, and there is an increase in temperature and pressure.

Process 4–5, *Expansion*. The products of combustion do work on the piston which moves to the right, and the pressure and temperature of the gases decrease.

Process 5–6, *Blow-down*. The exhaust valve opens, and the pressure drops to the initial pressure.

Process 6–1, *Exhaust*. With the exhaust valve open, the piston moves inwards to expel the combustion products from the cylinder at constant pressure.

The series of processes as described above constitute a *mechanical cycle*, and not a thermodynamic cycle. The cycle is completed in four strokes of the piston.

Figure 13.5 (c) shows the air standard cycle (Otto cycle) corresponding to the above engine. It consists of: Two reversible adiabatics and two reversible isochores.

Air is compressed in process 1 – 2 reversibly and adiabatically. Heat is then added to air reversibly at constant volume in process 2 – 3. Work is done by air in expanding reversibly and adiabatically in process 3 – 4. Heat is then rejected by air reversibly at constant volume in process 4 – 1, and the system (air) comes back to

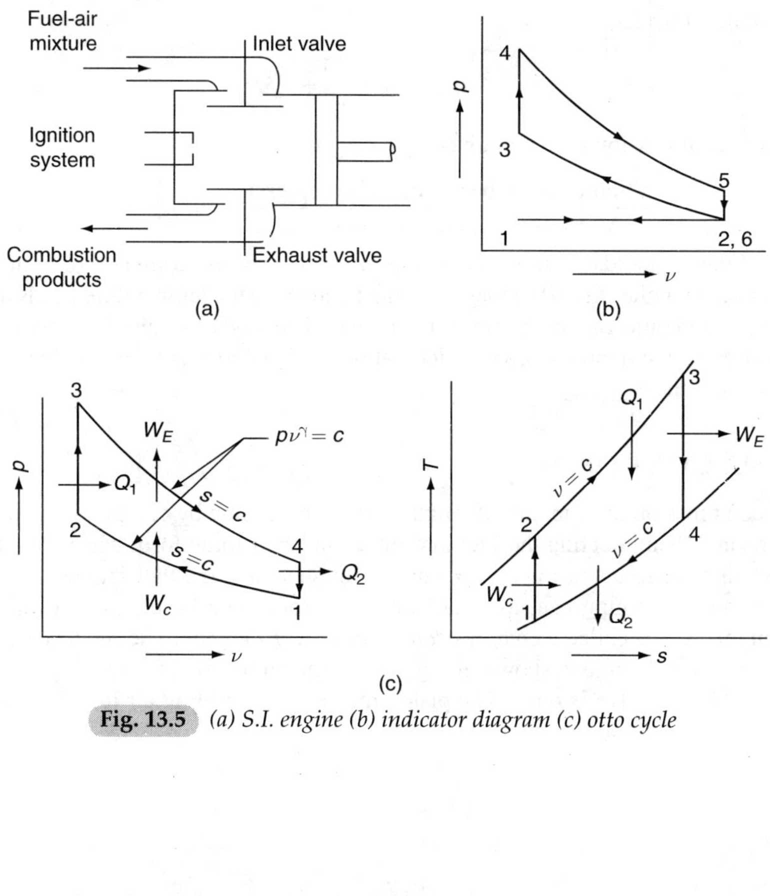

Fig. 13.5 *(a) S.I. engine (b) indicator diagram (c) otto cycle*

its initial state. Heat transfer processes have been substituted for the combustion and blow-down processes of the engine. The intake and exhaust processes of the engine cancel each other.

Let m be the fixed mass of air undergoing the cycle of operations as described above.

Heat supplied $Q_1 = Q_{2-3} = mc_v\,(T_3 - T_2)$

Heat rejected $Q_2 = Q_{4-1} = mc_v\,(T_4 - T_1)$

$$\text{Efficiency } \eta = 1 - \frac{Q_2}{Q_1} = 1 - \frac{mc_v(T_4 - T_1)}{mc_v(T_3 - T_2)} = 1 - \frac{T_4 - T_1}{T_3 - T_2} \tag{13.4}$$

Process 1–2,
$$\frac{T_2}{T_1} = \left(\frac{v_1}{v_2}\right)^{\gamma-1}$$

Process 3–4,
$$\frac{T_3}{T_4} = \left(\frac{v_4}{v_3}\right)^{\gamma-1} = \left(\frac{v_1}{v_2}\right)^{\gamma-1}$$

$$\therefore \quad \frac{T_2}{T_1} = \frac{T_3}{T_4} \quad \text{or} \quad \frac{T_3}{T_2} = \frac{T_4}{T_1}$$

$$\frac{T_3}{T_2} - 1 = \frac{T_4}{T_1} - 1$$

$$\therefore \quad \frac{T_4 - T_1}{T_3 - T_2} = \frac{T_1}{T_2} = \left(\frac{v_2}{v_1}\right)^{\gamma-1}$$

$\therefore$ From Eq. (13.4), $\eta = 1 - \left(\dfrac{v_2}{v_1}\right)^{\gamma-1}$

or
$$\eta_{\text{otto}} = 1 - \frac{1}{r_k^{\gamma-1}} \tag{13.5}$$

where r_k is called the compression ratio and given by

$$r_k = \frac{\text{Volume at the beginning of compression}}{\text{Volume at the end of compression}} = \frac{V_1}{V_2} = \frac{v_1}{v_2}$$

The efficiency of the air standard Otto cycle is thus a function of the compression ratio only. The higher the compression ratio, the higher the efficiency. It is independent of the temperature levels at which the cycle operates. The compression ratio cannot, however, be increased beyond a certain limit, because of a noisy and destructive combustion phenomenon, known as detonation. It also depends upon the fuel, the engine design, and the operating conditions.

13.6 DIESEL CYCLE (1892)

The limitation on compression ratio in the S.I. engine can be overcome by compressing air alone, instead of the fuel-air mixture, and then injecting the fuel into the cylinder in spray form when combustion is desired. The temperature of air after compression must be high enough so that the fuel sprayed into the hot air burns spontaneously. The rate of burning can, to some extent, be controlled by the rate of injection of fuel. An engine operating in this way is called a *compression ignition (C.I.) engine*. The sequence of processes in the elementary operation of a C.I. engine, shown in Fig. 13.6, is given below.

Process 1–2, *intake*. The air valve is open. The piston moves out admitting air into the cylinder at constant pressure.

Process 2–3, *Compression*. The air is then compressed by the piston to the minimum volume with all the valves closed.

Process 3–4, *Fuel injection and combustion*. The fuel valve is open, fuel is sprayed into the hot air, and combustion takes place at constant pressure.

Process 4–5, *Expansion*. The combustion products expand, doing work on the piston which moves out to the maximum volume.

Process 5–6, *Blow-down*. The exhaust valve opens, and the pressure drops to the initial pressure.

Pressure 6–1, *Exhaust*. With the exhaust valve open, the piston moves towards the cylinder cover driving away the combustion products from the cylinder at constant pressure.

The above processes constitute an engine cycle, which is completed in four strokes of the piston or two revolution of the crank shaft.

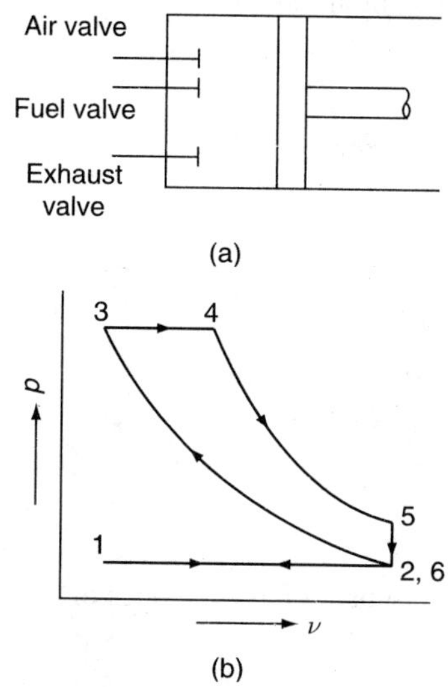

Fig. 13.6 *(a) C.I. engine, (b) indicator diagram*

Figure 13.7 shows the air standard cycle, called the *Diesel cycle*, corresponding to the C.I. engine, as described above. The cycle is composed of:

Two reversible adiabatics, one reversible isobar, and one reversible isochore.

Air is compressed reversibly and adiabatically in process 1 – 2. Heat is then added to it from an external source reversibly at constant pressure in process 2 – 3. Air then expands reversibly and adiabatically in process 3 – 4. Heat is rejected reversibly at constant volume in process 4 – 1, and the cycle repeats itself.

For m kg of air in the cylinder, the efficiency analysis of the cycle can be made as given below.

Heat supplied, $$Q_1 = Q_{2-3} = mc_p (T_3 - T_2)$$

Heat rejected, $$Q_2 = Q_{4-1} = mc_v (T_4 - T_1)$$

Efficiency $$\eta = 1 - \frac{Q_2}{Q_1} = 1 - \frac{mc_v (T_4 - T_1)}{mc_p (T_3 - T_2)}$$

$$\therefore \quad \eta = 1 - \frac{T_4 - T_1}{\gamma (T_3 - T_2)} \tag{13.6}$$

The efficiency may be expressed in terms of any two of the following three ratios

Compression ratio, $$r_k = \frac{V_1}{V_2} = \frac{v_1}{v_2}$$

Expansion ratio, $$r_e = \frac{V_4}{V_3} = \frac{v_4}{v_3}$$

Cut-off ratio, $$r_c = \frac{V_3}{V_2} = \frac{v_3}{v_2}$$

It is seen that $$r_k = r_e \cdot r_c$$

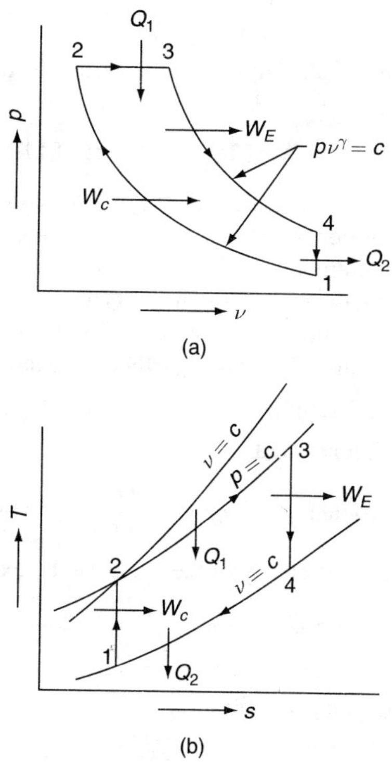

Fig. 13.7 *Diesel cycle*

Process 3–4

$$\frac{T_4}{T_3} = \left(\frac{v_3}{v_4}\right)^{\gamma-1} = \frac{1}{r_e^{\gamma-1}} \qquad T_4 = T_3 \frac{r_c^{\gamma-1}}{r_k^{\gamma-1}}$$

Process 2–3

$$\frac{T_2}{T_3} = \frac{p_2 v_2}{p_3 v_3} = \frac{v_2}{v_3} = \frac{1}{r_c}$$

$$\therefore \qquad T_2 = T_3 \cdot \frac{1}{r_c}$$

Process 1–2

$$\frac{T_1}{T_2} = \left(\frac{v_2}{v_1}\right)^{\gamma-1} = \frac{1}{r_k^{\gamma-1}}$$

$$\therefore \qquad T_1 = T_2 \cdot \frac{1}{r_k^{\gamma-1}} = \frac{T_3}{r_c} \cdot \frac{1}{r_k^{\gamma-1}}$$

Substituting the values of T_1, T_2 and T_4 in the expression of efficiency (Eq. 13.6)

$$\eta = 1 - \frac{T_3 \cdot \dfrac{r_c^{\gamma-1}}{r_k^{\gamma-1}} - \dfrac{T_3}{r_c} \dfrac{1}{r_k^{\gamma-1}}}{\gamma\left(T_3 - T_3 \cdot \dfrac{1}{r_c}\right)} \qquad \eta_{\text{Diesel}} = 1 - \frac{1}{\gamma} \cdot \frac{1}{r_k^{\gamma-1}} \cdot \frac{r_c^{\gamma} - 1}{r_c - 1} \tag{13.7}$$

As $r_c > 1$, $\dfrac{1}{\gamma}\left(\dfrac{r_c^{\gamma}-1}{r_c-1}\right)$ is also greater than unity. Therefore, the efficiency of the Diesel cycle is less than that of the Otto cycle for the same compression ratio.

13.7 LIMITED PRESSURE CYCLE, MIXED CYCLE OR DUAL CYCLE

The air standard Diesel cycle does not simulate exactly the pressure-volume variation in an actual compression ignition engine, where the fuel injection is started before the end of compression stroke. A closer approximation is the limited pressure cycle in which some part of heat is added to air at constant volume, and the remainder at constant pressure.

Figure 13.8 shows the $p - v$ and $T - s$ diagrams of the dual cycle. Heat is added reversibly, partly at constant volume (2 – 3) and partly at constant pressure (3 – 4).

Heat supplied $\qquad Q_1 = mc_v (T_3 - T_2) + mc_p (T_4 - T_3)$

Heat rejected $\qquad Q_2 = mc_v (T_5 - T_1)$

Efficiency
$$\eta = 1 - \frac{Q_2}{Q_1} = 1 - \frac{mc_v (T_5 - T_1)}{mc_v (T_3 - T_2) + mc_v (T_4 - T_3)} = 1 - \frac{T_5 - T_1}{(T_3 - T_2) + \gamma (T_4 - T_3)} \tag{13.8}$$

The efficiency of the cycle can be expressed in terms of the following ratios

Compression ratio, $\qquad r_k = \dfrac{V_1}{V_2}$

Expansion ratio, $\qquad r_e = \dfrac{V_5}{V_4}$

Cut-off ratio, $\qquad r_c = \dfrac{V_4}{V_3}$

Constant volume pressure ratio, $$r_p = \frac{p_3}{p_2}$$

It is seen, as before that

$$r_k = r_c \cdot r_e$$

or

$$r_e = \frac{r_k}{r_c}$$

Process 3 – 4

$$r_c = \frac{V_4}{V_3} = \frac{T_4 p_3}{p_4 T_3} = \frac{T_4}{T_3}$$

$$T_3 = \frac{T_4}{r_c}$$

Process 2 – 3

$$\frac{p_2 V_2}{T_2} = \frac{p_3 V_3}{T_3}$$

$$T_2 = T_3 \frac{p_2}{p_3} = \frac{T_4}{r_p \cdot r_c}$$

Process 1 – 2

$$\frac{T_1}{T_2} = \left(\frac{v_2}{v_1}\right)^{\gamma-1} = \frac{1}{r_k^{\gamma-1}}$$

$\therefore$

$$T_1 = \frac{T_4}{r_p \cdot r_c \cdot r_k^{\gamma-1}}$$

Process 4 – 5

$$\frac{T_5}{T_4} = \left(\frac{v_4}{v_5}\right)^{\gamma-1} = \frac{1}{r_e^{\gamma-1}}$$

$\therefore$

$$T_5 = T_4 \cdot \frac{r_c^{\gamma-1}}{r_k^{\gamma-1}}$$

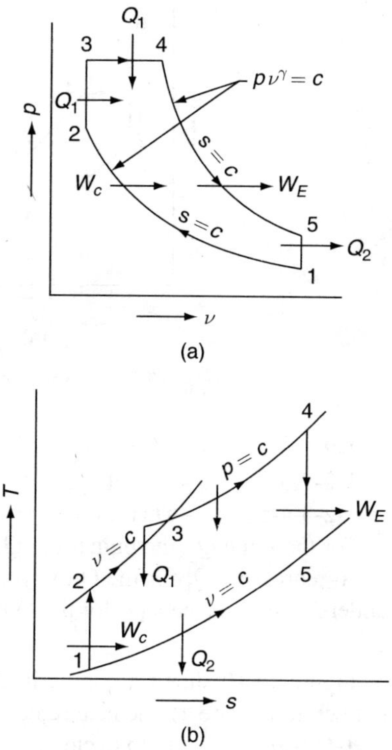

Fig. 13.8 *Limited pressure cycle, mixed cycle or dual cycle*

Substituting the values of T_1, T_2, T_3, and T_5 in the expression of efficiency (Eq. 13.8).

$$\eta = 1 - \frac{T_4 \cdot \dfrac{r_c^{\gamma-1}}{r_k^{\gamma-1}} - \dfrac{T_4}{r_p \cdot r_c \cdot r_k^{\gamma-1}}}{\left(\dfrac{T_4}{r_c} - \dfrac{T_4}{r_p \cdot r_c}\right) + \gamma\left(T_4 - \dfrac{T_4}{r_c}\right)}$$

$\therefore$

$$\eta_{\text{Dual}} = 1 - \frac{1}{r_k^{\gamma-1}} \frac{r_p \cdot r_c^{\gamma} - 1}{r_p - 1 + \gamma r_p (r_c - 1)} \quad (13.9)$$

13.8 COMPARISON OF OTTO, DIESEL, AND DUAL CYCLES

The three cycles can be compared on the basis of either the same compression ratio or the same maximum pressure and temperature.

Figure 13.9 shows the comparison of Otto, Diesel, and Dual cycles for the same compression ratio and heat rejection. Here

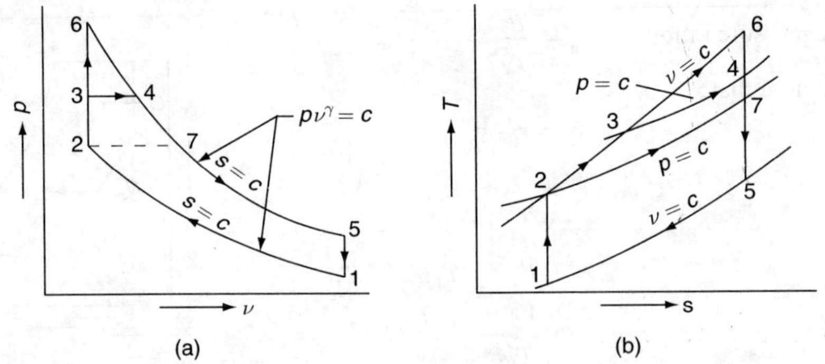

Fig. 13.9 *Comparison of otto, diesel and dual cycles for the same compression ratio*

1-2-6-5 —Otto cycle
1-2-7-5 —Diesel cycle
1-2-3-4-5 —Dual cycle

For the same Q_2, the higher the Q_1, the higher is the cycle efficiency. In the $T - s$ diagram, the area under $2 - 6$ represents Q_1 for the Otto cycle, the area under $2 - 7$ represents Q_1 for the Diesel cycle, and the area under 2-3-4 represents Q_1 for the Dual cycle. Therefore, for the same r_k and Q_2

$$\eta_{\text{Otto}} > \eta_{\text{Dual}} > \eta_{\text{Diesel}}$$

Figure 13.10 shows a comparison of the three air standard cycles for the same maximum pressure and temperature (state 4), the heat rejection being also the same. Here

1-6-4-5 —Otto cycle
1-7-4-5 —Diesel cycle
1-2-3-4-5 —Dual cycle

Q_1 is represented by the area under 6–4 for the Otto cycle, by the area under 7–4 for the Diesel cycle and by the area under 2-3-4 for the Dual cycle in the T–s plot, Q_2 being the same.

$$\therefore \qquad \eta_{\text{Diesel}} > \eta_{\text{Dual}} > \eta_{\text{Otto}}$$

This comparison is of greater significance, since the Diesel cycle would definitely have a higher compression ratio than the Otto cycle.

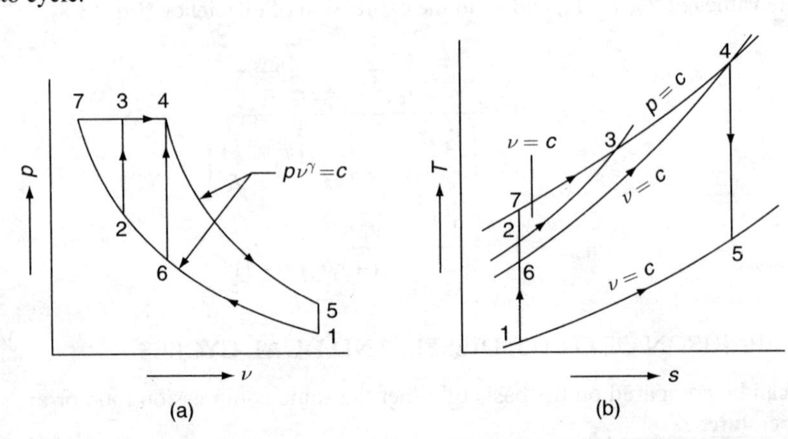

Fig. 13.10 *Comparison of otto, diesel and dual cycles for the same maximum pressure and temperature*

13.9 BRAYTON CYCLE

A simple gas turbine power plant is shown in Fig. 13.11. Air is first compressed adiabatically in process $a - b$, it then enters the combustion chamber where fuel is injected and burned essentially at constant pressure in process $b - c$, and then the products of combustion expand in the turbine to the ambient pressure in process $c - d$ and are thrown out to the surroundings. The cycle is open. The state diagram on the $p - v$ coordinates is shown in Fig. 13.12. Open cycles are used in aircraft, automotive (buses and trucks) and industrial gas turbine installations.

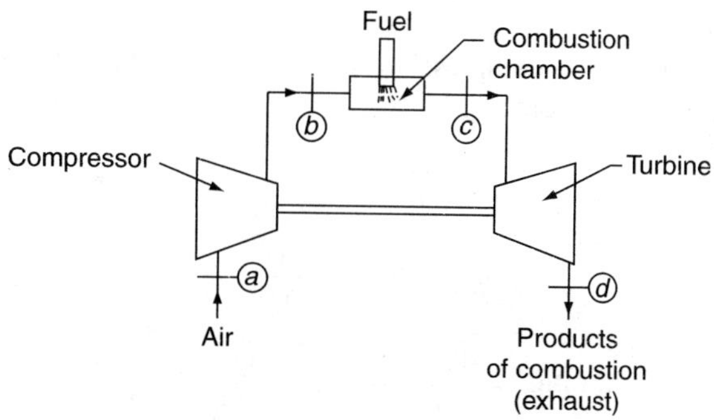

Fig. 13.11 *A simple gas turbine plant*

The Brayton cycle is the air standard cycle for the gas turbine power plant. Here air is first compressed reversibly and adiabatically, heat is added to it reversibly at constant pressure, air expands in the turbine reversibly and adiabatically, and heat is then rejected from the air reversibly at constant pressure to bring it to the initial state. The Brayton cycle, therefore, consists of:

Two reversible isobars and two reversible adiabatics.

The flow, $p - v$, and $T - s$ diagrams are shown in Fig. 13.13. For m kg of air

$$Q_1 = \text{heat supplied} = mc_p(T_3 - T_2)$$

$$Q_2 = \text{heat rejected} = mc_p(T_4 - T_1)$$

$\therefore$ Cycle efficiency,

$$\eta = 1 - \frac{Q_2}{Q_1}$$

$$= 1 - \frac{T_4 - T_1}{T_3 - T_2} \qquad (13.10)$$

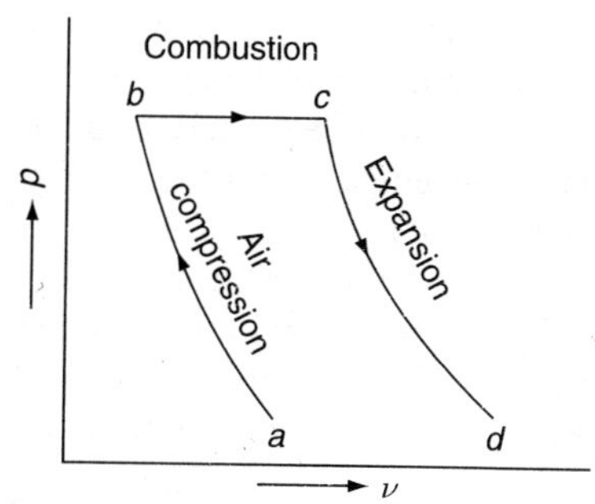

Fig. 13.12 *State diagram of a gas turbine plant on p–v plot*

Now

$$\frac{T_2}{T_1} = \left(\frac{p_2}{p_1}\right)^{(\gamma-1)/\gamma} = \frac{T_3}{T_4} \quad (\text{Since } p_2 = p_3, \text{ and } p_4 = p_1)$$

$\therefore$

$$\frac{T_4}{T_1} - 1 = \frac{T_3}{T_2} - 1$$

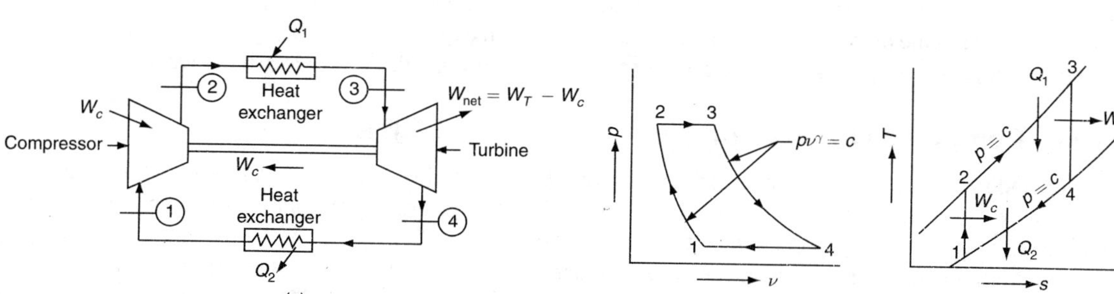

Fig. 13.13 *(a–c) Brayton cycle, (d) comparison of Rankine cycle and braytom cycle, both operating between the same pressures p_1 and p_2*

or
$$\frac{T_4 - T_1}{T_3 - T_2} = \frac{T_1}{T_2} = \left(\frac{p_1}{p_2}\right)^{(\gamma-1)/\gamma} = \left(\frac{v_2}{v_1}\right)^{\gamma-1}$$

If r_k = compression ratio = v_1/v_2 the efficiency becomes (from Eq. 13.10)

$$\eta = 1 - \left(\frac{v_2}{v_1}\right)^{\gamma-1}$$

or
$$\eta_{\text{Brayton}} = 1 - \frac{1}{r_k^{\gamma-1}} \tag{13.11}$$

$$\text{Work ratio} = \frac{W_T - W_C}{W_T} = \frac{Q_1 - Q_2}{W_T}$$

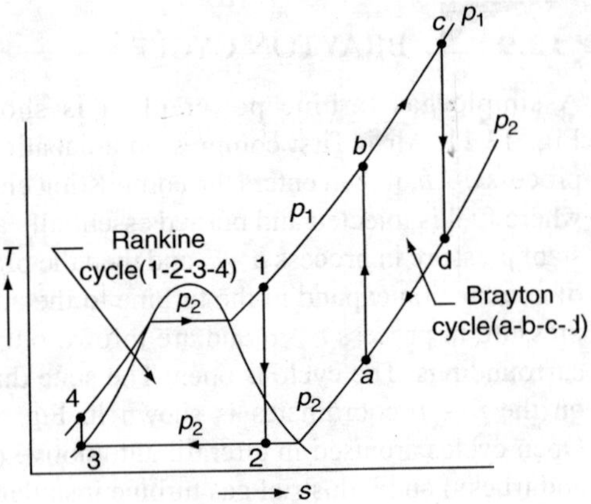

Fig. 13.13 *(Continued)*

If r_p = pressure ratio = p_2/p_1 the efficiency may be expressed in the following form also

$$\eta = 1 - \left(\frac{p_1}{p_2}\right)^{(\gamma-1)/\gamma}$$

or
$$\eta_{\text{Brayton}} = 1 - \frac{1}{(r_p)^{(\gamma-1)/\gamma}} \tag{13.12}$$

The efficiency of the Brayton cycle, therefore, depends upon either the compression ratio or the pressure ratio. For the same compression ratio, the Brayton cycle efficiency is equal to the Otto cycle efficiency.

A closed cycle gas turbine plant (Fig. 13.13) is used in a gas-cooled nuclear reactor plant, where the source is a high temperature gas-cooled reactor (HTGR) supplying heat from nuclear fission directly to the working fluid (a gas).

Both Rankine cycle and Brayton cycle consist of two reversible isobars and two reversible adiabatics (Fig. 13.13(d)). While in Rankine cycle, the working fluid undergoes phase change, in Brayton cycle the working fluid always remains in the gaseous phase. Both the pump and the steam turbine in the case of Rankine cycle, and the compressor and the gas turbine in the case of Brayton cycle operate through the same pressure difference of p_1 and p_2. All are steady-flow machines and the work transfer is given by $-\int_{p_1}^{p_2} v\, dp$. For Brayton cycle, the average specific volume of air handled by the compressor is less than the same of gas in the gas turbine (since the gas temperature is much higher), the work done by the gas turbine is more than the work input to the compressor, so that there is W_{net} available to deliver. In the case of Rankine cycle, the specific volume of water in the pump is much less than that of the steam expanding in the steam turbine, so $W_T >> W_P$. Therefore, steam power plants are more popular than the gas turbine plants for electricity generation.

13.9.1 Comparison between Brayton Cycle and Otto Cycle

Brayton and Otto cycles are shown superimposed on the $p - v$ and $T - s$ diagrams in Fig. 13.14. For the same r_k and work capacity, the Brayton cycle (1-2-5-6) handles a larger range of volume and a smaller range of pressure and temperature than does the Otto cycle (1-2-3-4).

In the reciprocating engine field, the Brayton cycle is not suitable. A reciprocating engine cannot efficiently handle a large volume flow of low pressure gas, for which the engine size ($\pi/4D^2 L$) becomes large, and the friction losses also become more. So the Otto cycle is more suitable in the reciprocating engine field.

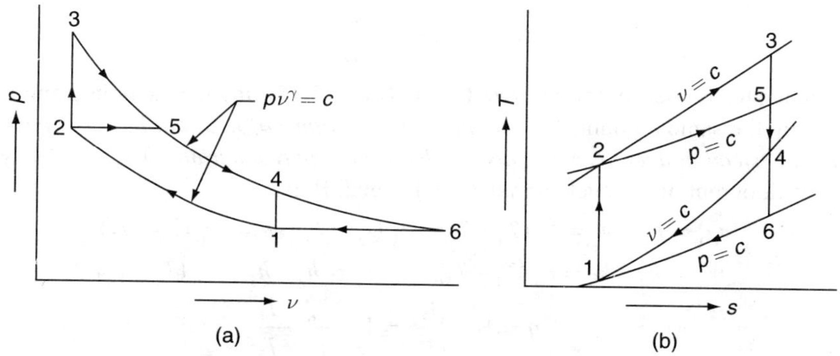

Fig. 13.14 *Comparison of otto and brayton cycles*

In turbine plants, however, the Brayton cycle is more suitable than the Otto cycle. An internal combustion engine is exposed to the highest temperature (after the combustion of fuel) only for a short while, and it gets time to become cool in the other processes of the cycle. On the other hand, a gas turbine plant, a steady flow device, is always exposed to the highest temperature used. So to protect material, the maximum temperature of gas that can be used in a gas turbine plant cannot be as high as in an internal combustion engine. Also, in the steady flow machinery, it is more difficult to carry out heat transfer at constant volume than at constant pressure. Moreover, a gas turbine can handle a large volume flow of gas quite efficiently. So we find that the Brayton cycle is the basic air standard cycle for all modern gas turbine plants.

13.9.2 Effect of Regeneration on Brayton Cycle Efficiency

The efficiency of the Brayton cycle can be increased by utilizing part of the energy of the exhaust gas from the turbine in heating up the air leaving the compressor in a heat exchanger called a *regenerator*, thereby reducing the amount of heat supplied from an external source and also the amount of heat rejected. Such a cycle is illustrated in Fig. 13.15. The temperature of air leaving the turbine at 5 is higher than that of air leaving the compressor at 2. In the regenerator, the temperature of air leaving the compressor is raised by heat transfer from the turbine exhaust. The maximum temperature to which the cold air at 2 could be heated is the temperature of the hot air leaving the turbine at 5. This is possible only in an infinite heat exchanger. In the real case, the temperature at 3 is less than that at 5. The ratio of the actual temperature rise of air to the maximum possible rise is called the *effectiveness* of the regenerator. For this case illustrated

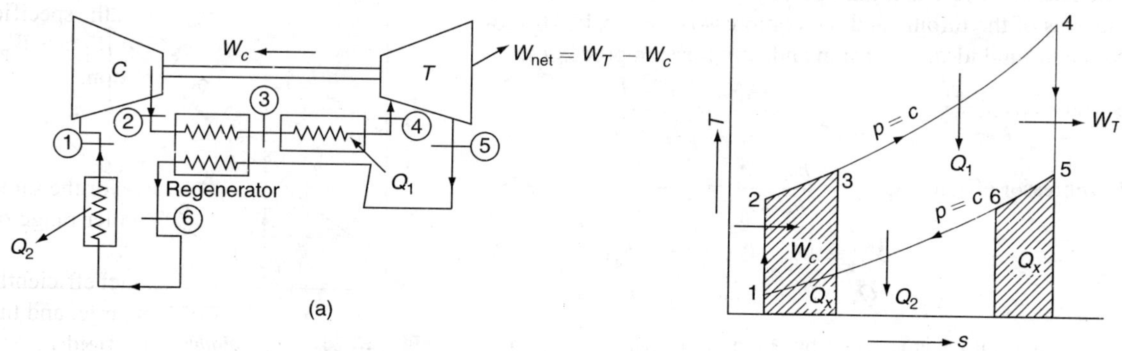

Fig. 13.15 *Effect of regeneration on Brayton cycle*

$$\text{Effectiveness} = \frac{t_3 - t_2}{t_5 - t_2}$$

When the regenerator is used in the idealized cycle (Fig. 13.15), the heat supplied and the heat rejected are each reduced by the same amount, Q_x. *The mean temperature of heat addition increases and the mean temperature of heat rejection decreases because of the use of the regenerator*. The efficiency is increased as a result, but the work output of the cycle remains unchanged. Here,

$$Q_1 = h_4 - h_3 = c_p (T_4 - T_3) \qquad Q_2 = h_6 - h_1 = c_p (T_6 - T_1)$$

$$W_T = h_4 - h_5 = c_p (T_4 - T_5) \qquad W_c = h_2 - h_1 = c_p (T_2 - T_1)$$

$$\eta = 1 - \frac{Q_2}{Q_1} = 1 - \frac{T_6 - T_1}{T_4 - T_3}$$

In practice the regenerator is costly, heavy and bulky, and causes pressure losses which bring about a decrease in cycle efficiency. These factors have to be balanced against the gain in efficiency to decide whether it is worthwhile to use the regenerator.

Above a certain pressure ratio (p_2/p_1) the addition of a regenerator causes a loss in cycle efficiency when compared to the original Brayton cycle. In this situation the compressor discharge temperature (T_2) is higher than the turbine exhaust gas temperature (T_5) (Fig. 13.18). The compressed air will thus be cooled in the regenerator and the exhaust gas will be heated. As a result both the heat supply and heat rejected are increased. However, the compressor and turbine works remain unchanged. So, the cycle efficiency (W_{net}/Q_1) decreases.

Let us now derive an expression for the ideal regenerative cycle when the compressed air is heated to the turbine exhaust temperature in the regenerator so that $T_3 = T_5$ and $T_2 = T_6$ (Fig. 13.15). Therefore,

$$\eta = 1 - \frac{Q_2}{Q_1} = 1 - \frac{T_6 - T_1}{T_4 - T_3} = 1 - \frac{T_1}{T_4}\left[\frac{(T_2/T_1) - 1}{1 - (T_5/T_4)}\right] = 1 - \frac{T_1}{T_4} \cdot \frac{T_2}{T_1}\left[\frac{1 - (T_1/T_2)}{1 - (T_5/T_4)}\right]$$

Since $$\frac{T_2}{T_1} = \left(\frac{p_2}{p_1}\right)^{\gamma - 1/\gamma} = \frac{T_4}{T_5} \qquad \eta = 1 - \frac{T_1}{T_4} r_p^{\gamma - 1/\gamma} \tag{13.17}$$

For a fixed ratio of (T_1/T_4), the cycle efficiency drops with increasing pressure ratio.

13.9.3 Effect of Irreversibilities in Turbine and Compressor

The Brayton cycle is highly sensitive to the real machine efficiencies of the turbine and the compressor. Figure 13.16 shows the actual and ideal expansion and compression processes.

$$\text{Turbine efficiency,} \quad \eta_T = \frac{h_3 - h_4}{h_3 - h_{4s}} = \frac{T_3 - T_4}{T_3 - T_{4s}}$$

$$\text{Compressor efficiency,} \quad \eta_C = \frac{h_{2s} - h_1}{h_2 - h_1} = \frac{T_{2s} - T_1}{T_2 - T_1}$$

$$W_{net} = W_T - W_C = (h_3 - h_4) - (h_2 - h_1)$$

$$Q_1 = h_3 - h_2 \text{ and } Q_2 = h_4 - h_1$$

The net output is reduced by the amount $[(h_4 - h_{4s}) + (h_2 - h_{2s})]$, and the heat supplied is reduced by the amount $(h_2 - h_{2s})$.

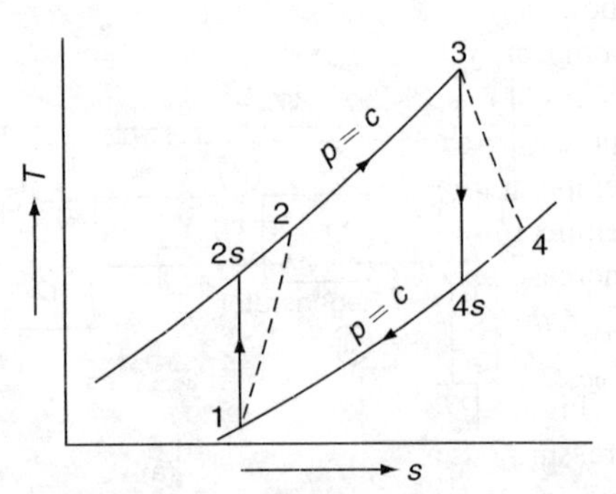

Fig. 13.16 *Effect of machine efficiencies on Brayton cycle*

The efficiency of the cycle will thus be less than that of the ideal cycle. As η_T and η_C decrease, η_{cycle} also decreases. The cycle efficiency may approach zero even when η_T and η_C are of the order of 60 to 70%. This is the main drawback of a gas turbine plant. The machines have to be highly efficient to obtain reasonable cycle efficiency.

13.9.4 Effect of Pressure Ratio on the Brayton Cycle

The efficiency of the Brayton cycle is a function of the pressure ratio as given by the equation

$$\eta = 1 - \frac{1}{(r_p)^{(\gamma-1)/\gamma}}$$

The more the pressure ratio, the more will be the efficiency.

Let T_1 = the lowest temperature of the cycle, which is the temperature of the surroundings (T_{min}), and T_3 = the maximum or the highest temperature of the cycle limited by the characteristics of the material available for burner and turbine construction (T_{max}).

Since the turbine, a steady-flow machine, is always exposed to the highest temperature gas, the maximum temperature of gas at the inlet to the turbine is limited to about 800°C by using a high air-fuel ratio. With turbine blade cooling, however, the maximum gas inlet temperature can be 1100°C or even higher.

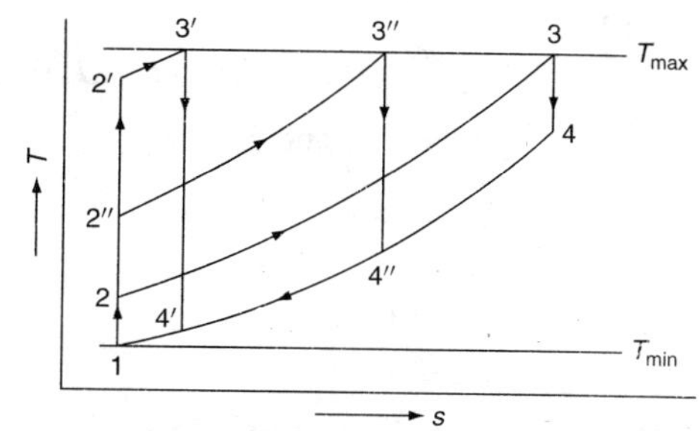

Fig. 13.17 *Effect of pressure ratio on brayton cycle*

Figure 13.17 shows the Brayton cycles operating between the same T_{max} and T_{min} at various pressure ratios. As the pressure ratio changes, the cycle shape also changes. For the cycle 1-2-3-4 of low pressure ratio r_p, since the average temperature of heat addition

$$T_{m1} = \frac{h_3 - h_2}{s_3 - s_2}$$

is only a little greater than the average temperature of heat rejection

$$T_{m2} = \frac{h_4 - h_1}{s_4 - s_1}$$

the efficiency will be low. *At the lower limit of unity pressure ratio, both work output and efficiency will be zero.*

As the pressure ratio is increased, the efficiency steadily increases, because T_{m1} increases and T_{m2} decreases. The mean temperature of heat addition T_{m1} approaches T_{max} and the mean temperature of heat rejection T_{m2} approaches T_{min}, with the increase in r_p. *In the limit when the compression process ends at T_{max}, the Carnot efficiency is reached, r_p has the maximum value (r_{pmax}), but the work capacity again becomes zero.*

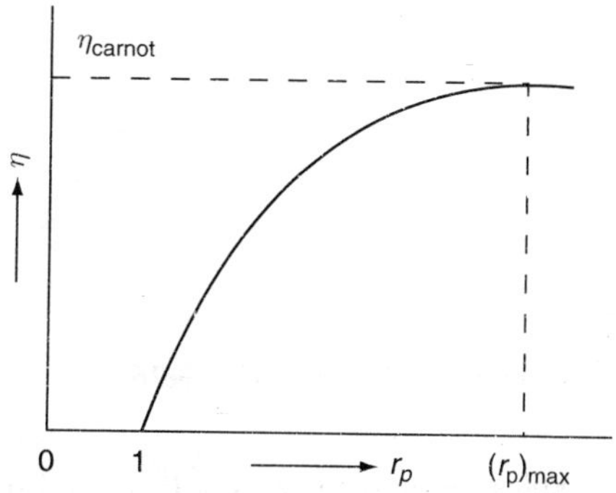

Fig. 13.18 *Effect of pressure ratio on brayton cycle efficiency*

Figure 13.18 shows how the cycle efficiency varies with the pressure ratio, with r_p varying between two limiting values of 1 and $(r_p)_{max}$ when the Carnot efficiency is reached. When $r_p = (r_p)_{max}$, the cycle efficiency is given by

$$\eta = 1 - \frac{1}{((r_p)_{max})^{\gamma-1/\gamma}} = \eta_{Carnot} = 1 - \frac{T_{min}}{T_{max}}$$

$$\therefore \quad ((r_p)_{max})^{(\gamma-1)/\gamma} = \frac{T_{max}}{T_{min}}$$

$$\therefore \quad (r_p)_{max} = \left(\frac{T_{max}}{T_{min}}\right)^{\gamma/(\gamma-1)} \tag{13.14}$$

From Fig. 13.17 it is seen that the work capacity of the cycle, operating between T_{max} and T_{min}, is zero when $r_p = 1$ passes through a maximum, and then again becomes zero when the Carnot efficiency is reached. There is an optimum value of pressure ratio $(r_p)_{opt}$ at which work capacity becomes a maximum, as shown in Fig. 13.19.

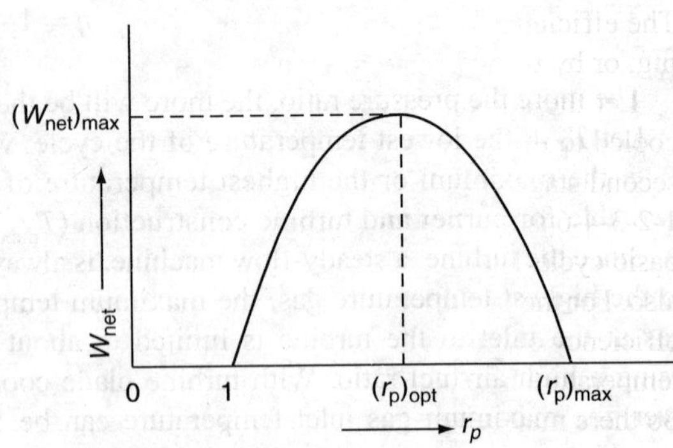

Fig. 13.19 *Effect of pressure ratio on net output*

For 1 kg,

$$W_{net} = c_p\,[(T_3 - T_4) - (T_2 - T_1)]$$

where $T_3 = T_{max}$ and $T_1 = T_{min}$

Now $\dfrac{T_3}{T_4} = (r_p)^{(\gamma-1)/\gamma}$

$$\therefore \quad T_4 = T_3 \cdot r_p^{-(\gamma-1)/\gamma}$$

Similarly $T_2 = T_1 \cdot r_p^{(\gamma-1)/\gamma}$

Substituting in the expression for W_{net}

$$W_{net} = c_p\,[T_3 - T_3 \cdot (r_p)^{-(\gamma-1)/\gamma} - T_1\, r_p^{(\gamma-1)/\gamma} + T_1] \tag{13.15}$$

To find $(r_p)_{opt}$

$$\frac{dW_{net}}{dr_p} = c_p\left[-T_3\left(-\frac{\gamma-1}{\gamma}\right) r_p^{(-1+(1/\gamma)-1)} - T_1\left(\frac{\gamma-1}{\gamma}\right) r_p^{(1-(1/\gamma)-1)}\right] = 0$$

$$\therefore \quad T_3\left(\frac{\gamma-1}{\gamma}\right) r_p^{(1/\gamma)-2} = T_1\left(\frac{\gamma-1}{\gamma}\right) \cdot r_p^{-1/\gamma}$$

$$\therefore \quad r_p^{-(1/\gamma)-(1/\gamma)+2} = \frac{T_3}{T_1}$$

or

$$(r_p)_{opt} = \left(\frac{T_3}{T_1}\right)^{\gamma/2(\gamma-1)}$$

$$\therefore \quad (r_p)_{opt} = \left(\frac{T_{max}}{T_{min}}\right)^{\gamma/2(\gamma-1)} \tag{13.16}$$

From Eqs. (13.14) and (13.26)

$$(r_p)_{opt} = \sqrt{(r_p)_{max}} \tag{13.17}$$

Substituting the values of $(r_p)_{opt}$ in Eq. (13.15)

$$W_{net} = (W_{net})_{max} = c_p\left[T_3 - T_3\left(\frac{T_1}{T_3}\right)^{\frac{\gamma}{2(\gamma-1)}\cdot\frac{\gamma-1}{\gamma}} - T_1\left(\frac{T_3}{T_1}\right)^{\frac{\gamma}{2(\gamma-1)}\cdot\frac{\gamma-1}{\gamma}} + T_1\right] = c_p[T_3 - 2\sqrt{T_1 T_3} + T_1]$$

or

$$(W_{net})_{max} = c_p(\sqrt{T_{max}} - \sqrt{T_{min}})^2 \tag{13.18}$$

$$\eta_{cycle} = 1 - \frac{1}{r_p^{\gamma-1/\gamma}} = 1 - \sqrt{\frac{T_{min}}{T_{max}}} \tag{13.19}$$

Considering the cycles 1-2′-3′-4′ and 1-2″-3″-4″ (Fig. 13.17), it is obvious that to obtain a reasonable work capacity, a certain reduction in efficiency must be accepted.

13.9.5 Effect of Intercooling and Reheating on Brayton Cycle

The efficiency of the Brayton cycle may often be increased by the use of staged compression with intercooling, or by using staged heat supply, called reheat.

Let the compression process be divided into two stages. Air, after being compressed in the first stage, is cooled to the initial temperature in a heat exchanger, called an intercooler, and then compressed further in the second stage (Fig. 13.20). 1-2′-5-6 is the ideal cycle without intercooling, having a single-stage compression, 1-2-3-4-6 is the cycle with intercooling, having a two-stage compression. The cycle 2-3-4-2′ is thus added to the basic cycle 1-2′-5-6. There is more work capacity, since the included area is more. There is more heat supply also. For the cycle 4-2′-2-3, T_{m1} is lower and T_{m2} higher (lower r_p) than those of the basic cycle 1-2′-5-6. So the efficiency of the cycle reduces by staging the compression and intercooling. But if a regenerator is used, the low temperature heat addition (4–2′) may be obtained by recovering energy from the exhaust gases from the turbine. So there may be a net gain in efficiency when intercooling is adopted in conjunction with a regenerator.

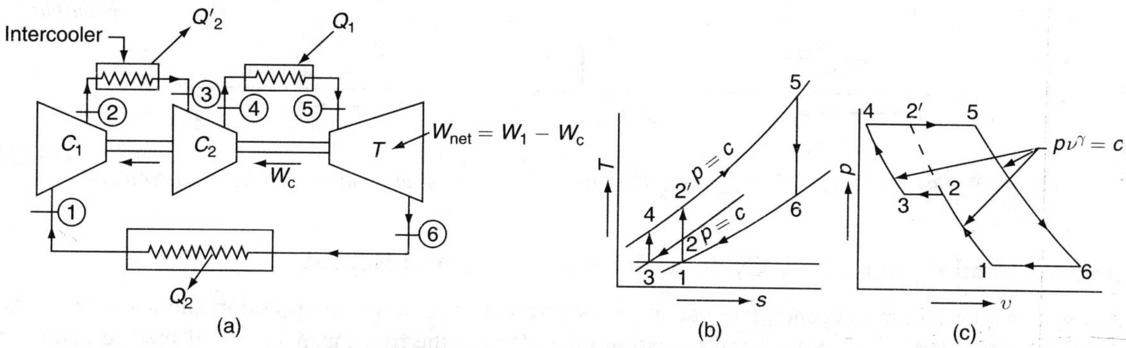

Fig. 13.20 *Effect of intercooling on Brayton cycle*

Similarly, let the total heat supply be given in two stages and the expansion process be divided in stages in two turbines (T_1 and T_2) with intermediate reheat, as shown in Fig. 13.21. 1-2-3-4′ is the cycle with a single-stage heat supply having no reheat, with total expansion in one turbine only. 1-2-3-4-5-6 is the cycle with a single-stage reheat, having the expansion divided into two stages. With the basic cycle, the cycle 4-5-6-4 is added because of reheat. The work capacity increases, but the heat supply also increases.

In the cycle 4-5-6-4′, r_p is lower than in the basic cycle 1-2-3-4′, so its efficiency is lower. Therefore, the efficiency of the cycle decreases with the use of reheat. But T_6 is greater than T'_4. Therefore, if regeneration is employed, there is more energy that can be recovered from the turbine exhaust gases. So when regeneration is employed in conjunction with reheat, there may be a net gain in cycle efficiency.

If in one cycle, several stages of intercooling and several stages of reheat are employed, a cycle as shown in Fig. 13.22 is obtained. When the number of such stages is large the cycle reduces to the Ericsson cycle with two reversible isobars and two reversible isotherms. With ideal regeneration the cycle efficiency becomes equal to the Carnot efficiency.

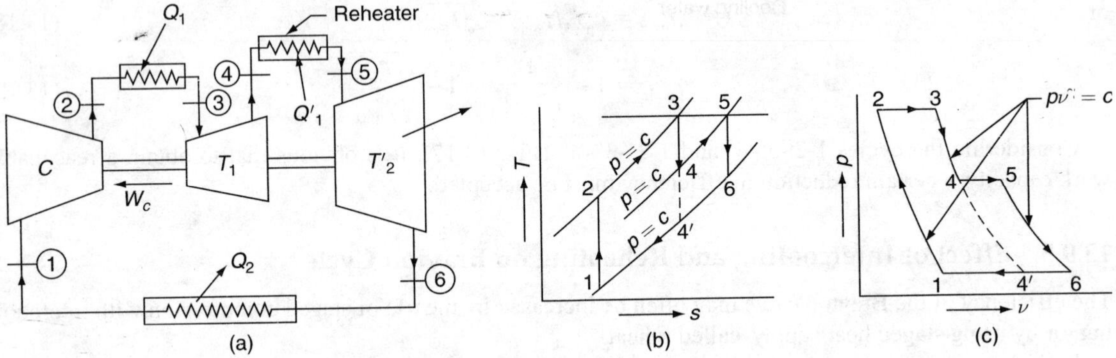

Fig. 13.21 *Effect of reheat on Brayton cycle*

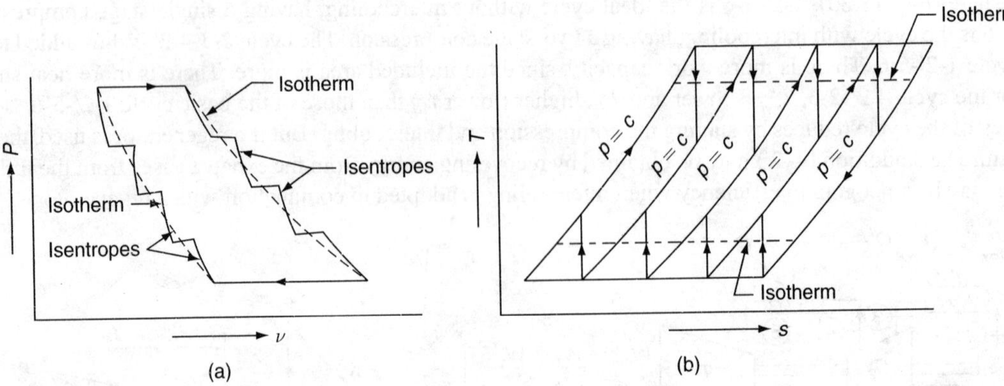

Fig. 13.22 *Brayton cycle with many stages of intercooling and reheating approximates to ericsson cycle*

13.9.6 Ideal Regenerative Cycle with Intercooling and Reheat

Let us consider an ideal regenerative gas turbine cycle with two-stage compression and a single reheat. It assumes that both intercooling and reheating take place at the root mean square of the high and low pressure in the cycle, so that $p_3 = p_2 = p_7 = p_8 = \sqrt{p_1 p_4} = \sqrt{p_6 p_9}$ (Fig. 13.23). Also, the temperature after intercooling is equal to the compressor inlet temperature ($T_1 = T_3$) and the temperature after reheat is equal to the temperature entering the turbine initially ($T_6 = T_8$).

Here, $$Q_1 = c_p(T_6 - T_5) + c_p(T_8 - T_7)$$

Since $\dfrac{p_6}{p_7} = \dfrac{p_8}{p_9}$ and $T_6 = T_8$, it follows that $T_5 = T_7 = T_9$.

$\therefore$ $$Q_1 = 2c_p(T_6 - T_7)$$

Again, $$Q_2 = c_p(T_{10} - T_1) + c_p(T_2 - T_3)$$

but $$\frac{p_2}{p_1} = \frac{p_4}{p_3} \quad \text{and} \quad T_3 = T_1$$

so that $$T_2 = T_4 = T_{10}$$

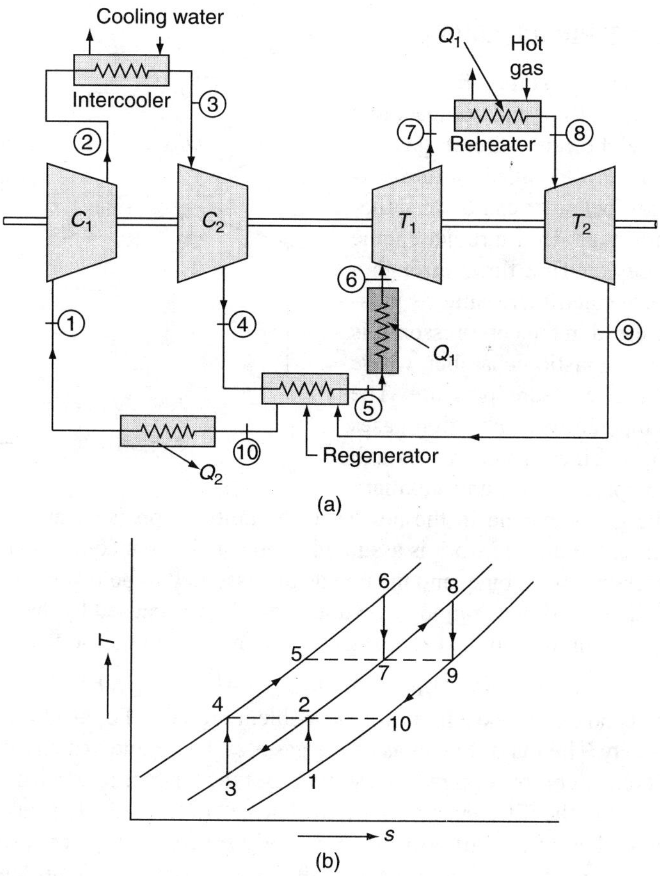

Fig. 13.23 *Ideal regenerative gas turbine cycle with two-stage compression and reheat*

$$\therefore \qquad Q_2 = 2c_p(T_2 - T_1)$$

$$W_{net} = Q_1 - Q_2 = 2c_p[(T_6 - T_7) - (T_2 - T_1)]$$

$$\therefore \eta_{cycle} = \frac{W_{net}}{Q_1} = 1 - \frac{T_2 - T_1}{T_6 - T_7} = 1 - \frac{T_1}{T_6}\left[\frac{(T_2/T_1) - 1}{1 - (T_7/T_6)}\right] = 1 - \frac{T_1}{T_6} \times \frac{T_2}{T_1}\left[\frac{1 - (T_1/T_2)}{1 - (T_7/T_6)}\right] = 1 - \frac{T_1}{T_6} \times \left(\frac{p_2}{p_1}\right)^{\gamma-1/\gamma}$$

But
$$p_2 = \sqrt{p_1 p_4},$$

$$\therefore \qquad \eta_{cycle} = 1 - \frac{T_1}{T_6}\left(\frac{p_4}{p_1}\right)^{(\gamma-1)/2\gamma} \tag{13.20}$$

13.9.7 Free-shaft Turbine

So far only a single shaft has been shown in flow diagrams, on which were mounted all the compressors and the turbines. Sometimes, for operating convenience and part-load efficiency, one turbine is used for driving the compressor only on one shaft, and a separate turbine is used on another shaft, known as free-shaft, for supplying the load, as shown in Fig. 13.24.

13.10 AIRCRAFT PROPULSION

Gas turbines are particularly suited for aircraft propulsion because they are light and compact and have a high power-to-weight ratio. Aircraft gas turbines operate on an open cycle called "*jet propulsion cycle*", as explained below. It can be of turbojet, turbofan or turborop type. In a turbojet engine (Fig. 13.25), high velocity air first flows through a diffuser where it is decelerated increasing its pressure. Air is then compressed in the compressor. It is mixed with fuel in the combustion chamber, where the mixture is burned at constant pressure. The high pressure, high temperature combustion gases partially expand in the turbine, producing enough power to drive the compressor and any auxiliary equipment. Finally, the gases expand in the nozzle to the ambient pressure and leave the aircraft at high velocity. In the ideal case, the turbine work is assumed to be equal to the compressor work. The processes in the diffuser, the compressor, the turbine, and the nozzle are assumed to be reversible and adiabatic.

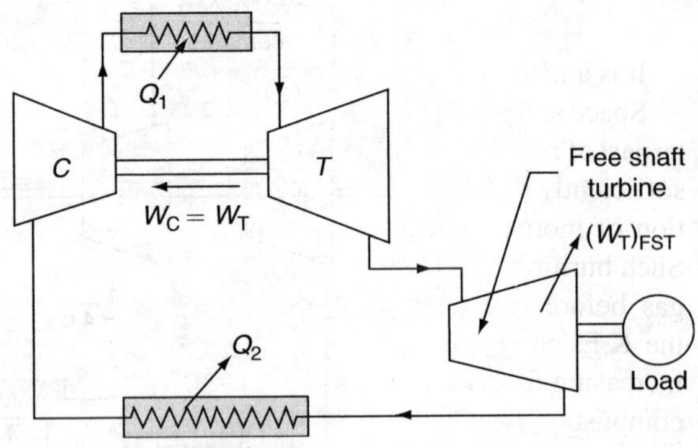

Fig. 13.24 *Free shaft turbine*

The thrust developed in a turbojet engine is the unbalanced force caused by the difference in the momentum of the air entering the engine and the exhaust gases leaving the engine, so that,

$$F = (\dot{m}\overline{V})_{\text{exit}} - (\dot{m}\overline{V})_{\text{inlet}} = \dot{m}(\overline{V}_{\text{exit}} - \overline{V}_{\text{inlet}}) \tag{13.21}$$

The pressures at inlet and exit of the engine are the ambient pressure. For an aircraft cruising in still air, $\overline{V}$ inlet is the aircraft velocity. The mass flow rates of the gases at the engine exit and the inlet are different, the difference being equal to the combustion rate of the fuel. But the air-fuel ratio used in jet propulsion engines is usually very high, making the difference very small. Thus $\dot{m}$ in Eq. (13.21) is taken as the mass flow rate of air through the engine. For an aircraft cruising at a steady speed, the thrust is used to overcome the fluid drag, and the net force acting on the body of the aircraft is zero. Commercial airplanes save fuel by flying at higher altitudes during long trips since the air at higher altitudes is of less density and exerts a smaller drag force on the aircraft.

The power developed from the thrust of the engine is called the propulsive power, W_p, given by:

$$\dot{W}_p = F\overline{V}_{\text{aircraft}} = \dot{m}(\overline{V}_{\text{exit}} - \overline{V}_{\text{inlet}})\overline{V}_{\text{aircraft}} \tag{13.22}$$

The propulsive efficiency, η_p, is defined by:

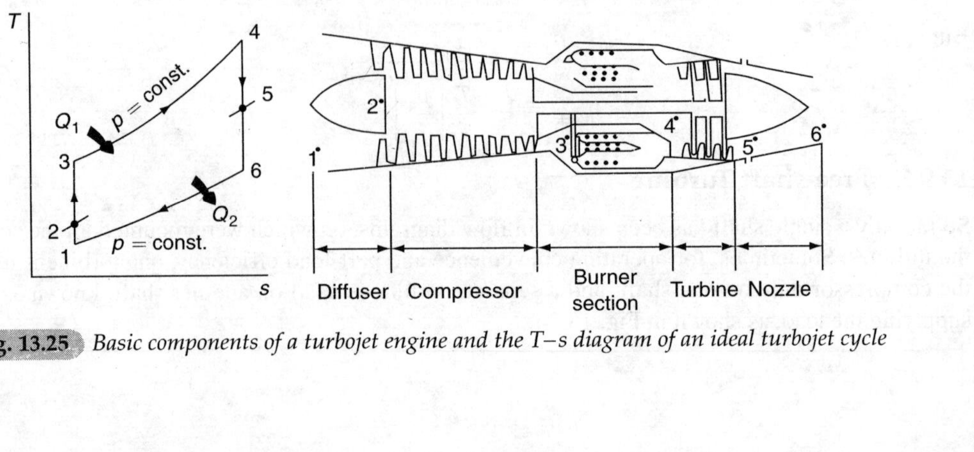

Fig. 13.25 *Basic components of a turbojet engine and the T–s diagram of an ideal turbojet cycle*

$$\eta_p = \frac{\text{Propulsive power}}{\text{Energy input rate}} = \frac{\dot{W}_p}{\dot{Q}_{in}} \tag{13.23}$$

It is a measure of how efficiently the energy released during combustion is converted to propulsive power.

Space and weight limitations prohibit the use of regenerators and intercoolers on aircraft engines. The counterpart of reheating is afterburning. The air-fuel ratio in a jet engine is so high that the turbine exhaust gases are sufficiently rich in oxygen to support the combustion of more fuel in an afterburner (Fig. 13.26). Such burning of fuel raises the temperature of the gas before it expands in the nozzle, increasing the K.E. change in the nozzle and consequently increasing the thrust. In the air-standard case, the combustion is replaced by constant pressure heat addition.

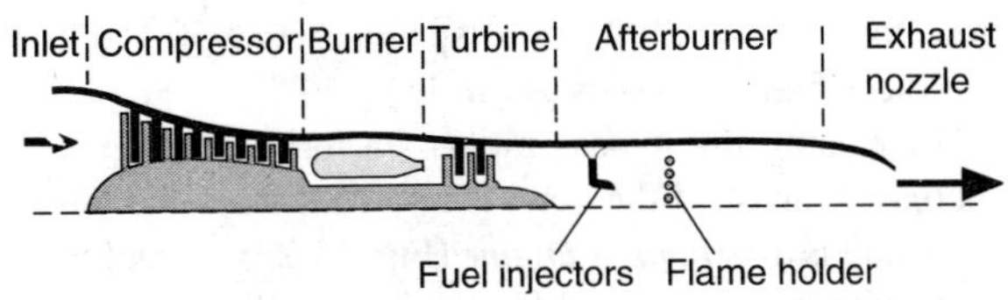

Fig. 13.26 *Turbojet engine with afterburner*

The most widely used engine in aircraft propulsion is the *turbofan engine* wherein a large fan driven by the turbine forces a considerable amount of air through a duct (cowl) surrounding the engine (Figs. 13.27 and 13.28). The fan exhaust leaves the duct at a higher velocity, enhancing the total thrust of the engine significantly. Some of the air entering the engine flows through the compressor, combustion chamber and turbine, and the rest passes through the fan into a duct and is either mixed with the exhaust gases or is discharged separately. It improves the engine performance over a broad operating range. The ratio of the mass flow rates of the two streams is called the *bypass ratio.*

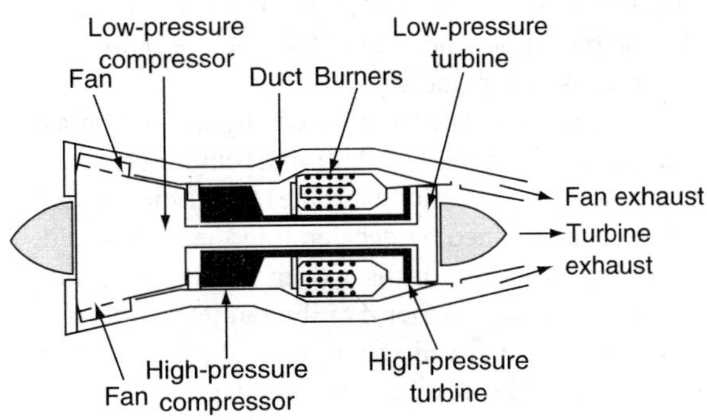

Fig. 13.27 *Turbofan engine*

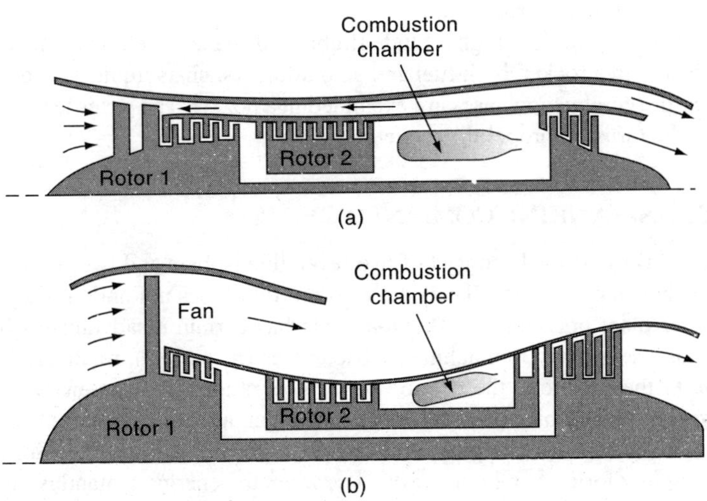

Fig. 13.28 *Turbofan or bypass jet engines*

$$\text{Bypass ratio} = \frac{\dot{m}_{\text{total}} - \dot{m}_{\text{turbine}}}{\dot{m}_{\text{turbine}}}$$

The bypass ratio can be varied in flight by various means. Turbofan engines deserve most of the credit for the success of jumbo jets, which weigh almost 400,000 kg and are capable of carrying 400 passengers for up to 10,000 km at speeds over 950 km/h with less fuel per passenger mile.

Increasing the bypass ratio of a turbofan engine increases thrust. If the cowl is removed from the fan the result is a *turboprop engine* (Fig. 13.29). Turbofan and turboprop engines differ mainly in their bypass ratio 5 or 6 for turbofans and as high as 100 for turboprop. In general, propellers are more efficient than jet engines, but they are limited to low-speed and low-altitude operation since their efficiency decreases at high speeds and altitudes.

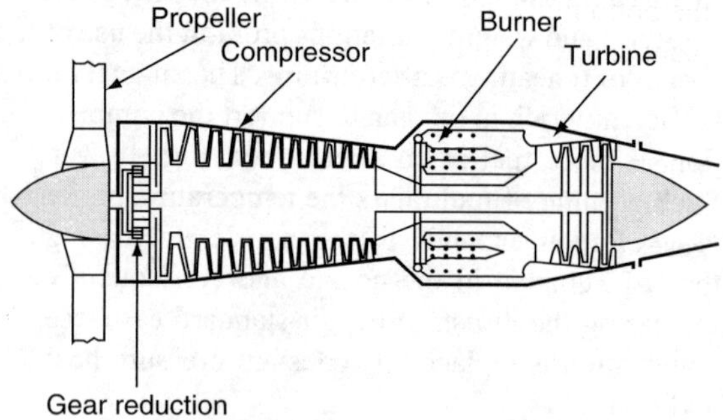

Fig. 13.29 *Turboprop engine*

A particularly simple type of engine known as a *ramjet* is shown in Fig. 13.30. This engine requires neither a compressor nor a turbine. A sufficient pressure rise is obtained by decelerating the high speed incoming air in the diffuser (ram effect) on being rammed against a barrier. For the ramjet to operate, the aircraft must already be in flight at a sufficiently great speed. The combustion products exiting the combustor are expanded through a nozzle to produce the thrust.

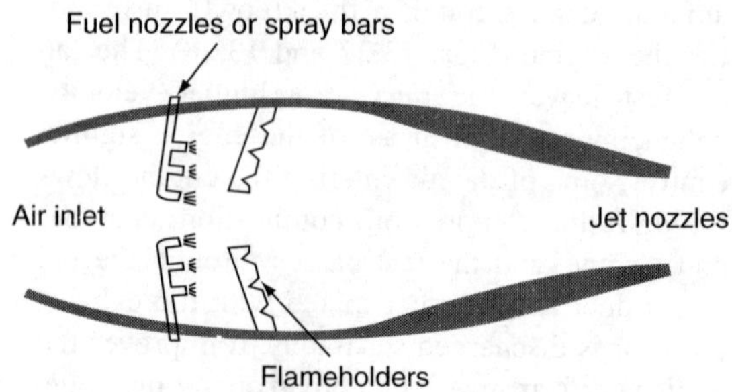

Fig. 13.30 *Ramjet engine*

In each of the engines mentioned so far, combustion of the fuel is supported by air brought into the engines from the atmosphere. For very high altitude flights and space travel, where this is no longer possible, rockets may be employed. In a rocket, both fuel and an oxidizer (such as liquid oxygen) are carried on board of the craft. High pressure combustion gases are expanded in a nozzle. The gases leave the rocket at very high velocities, producing the thrust to propel the rocket.

13.11 BRAYTON-RANKINE COMBINED CYCLE

Both Rankine cycle and Brayton cycle consist of two reversible isobars and two reversible adiabatics. While the former is a phase change cycle, in the latter the working fluid does not undergo any phase change.

A gas turbine power plant operating on Brayton cycle has certain disadvantages like large compressor work, large exhaust loss, sensitivity to machine inefficiencies (η_T and η_C), relatively lower cycle efficiency and costly fuel. Due to these factors, the cost of power generation by a stationary gas turbine in a utility system is high. However, a gas turbine plant offers certain advantages also, such as less installation cost, less installation time, quick starting and stopping, and fast response to load changes. So, a gas turbine plant is often used as a *peaking unit* for certain hours of the day, when the energy demand is high. To utilize the high temperature exhaust and to raise its plant efficiency a gas turbine may be used in conjunction with a steam turbine plant to offer the gas turbine advantages of quick starting and stopping and permit flexible operation of the combined plant over a wide range of loads.

Let us consider two cyclic power plants coupled in series, the topping plant operating on Brayton cycle and the bottoming one operating on Rankine cycle (Fig. 13.31). Helium may be the working fluid in the topping plant and water in the bottoming plant. The overall efficiency of the combined plant is:

$$\eta = \eta_1 + \eta_2 - \eta_1 \eta_2$$

where η_1 and η_2 are the efficiencies of the Brayton cycle and Rankine cycle respectively.

For capacity augmentation often supplementary firing is used (Fig. 13.32). For expansion of combustion gases in the gas turbine

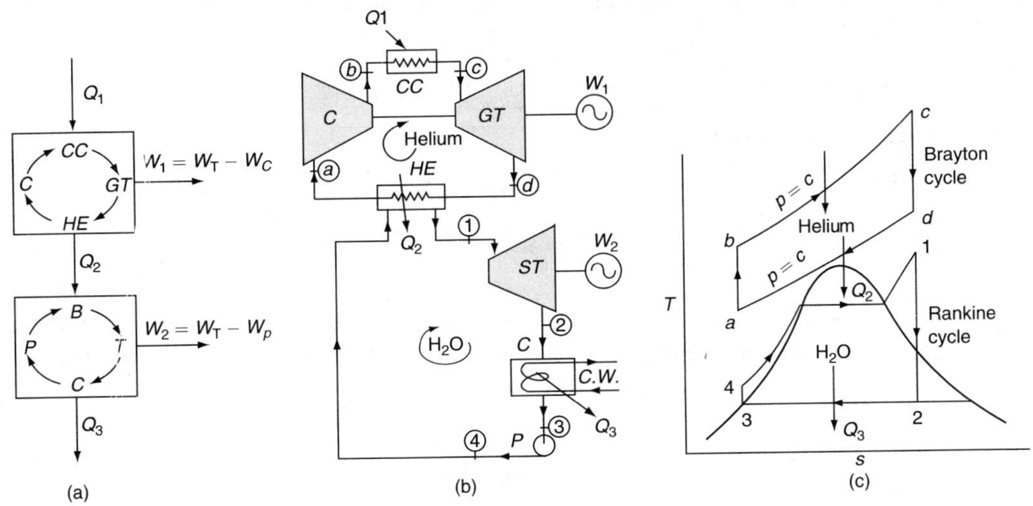

Fig. 13.31 *Brayton-Rankine combined cycle plant*

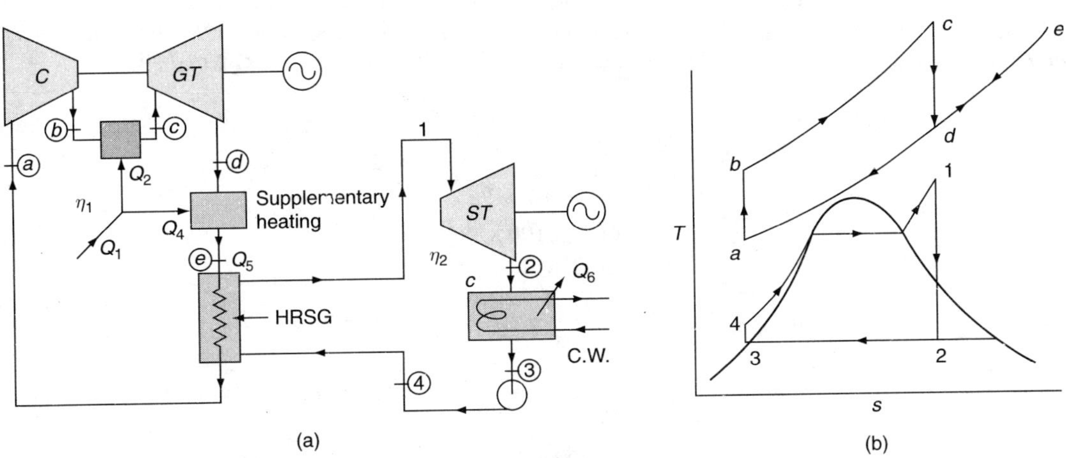

Fig. 13.32 *Brayton/Rankine cyclic plants with supplementary heating*

$$T_c/T_d = (p_2/p_1)^{(\gamma - 1)/\gamma}$$

where $\gamma = 1.3$ (assumed).

For the compressor, $$T_b/T_a = (p_2/p_1)^{(\gamma' - 1)/\gamma'}$$

where $$\gamma' = 1.4.$$

$$W_{GT} = w_a\,[c_{pg}\,(T_c - T_d) - c_{pa}\,(T_b - T_a)]$$

neglecting the mass of fuel (for a high air-fuel ratio), and w_a being the mass flow of air.

$$W_{ST} = w_s\,(h_1 - h_2)$$

where w_s is the steam flow rate. The pump work is neglected. By energy balance,

$$w_a\,c_{pg}\,(T_c - T_f) = w_s\,(h_1 - h_4)$$

Now, $$Q_1 = w_a\,c_{pg}\,[(T_c - T_b) + (T_e - T_d)]$$

The overall efficiency of the plant is:

$$\eta = \frac{W_{GT} + W_{ST}}{Q_1}$$

Again, $$Q_1 = w_f \times \text{C.V.}$$

where w_f is the fuel burning rate.

High overall efficiency, low investment cost, less water requirement, large operating flexibility, phased installation, and low environmental impact are some of the advantages of combined gas-steam cycles.

Solved Examples

Example 13.1

An engine working on the Otto cycle is supplied with air at 0.1 MPa, 35°C. The compression ratio is 8. Heat supplied is 2100 kJ/kg. Calculate the maximum pressure and temperature of the cycle, the cycle efficiency, and the mean effective pressure. (For air, $c_p = 1.005$, $c_v = 0.718$, and $R = 0.287$ kJ/kg K).

Solution From Fig. Ex. 13.1

$$T_1 = 273 + 35 = 308 \text{ K}$$
$$p_1 = 0.1 \text{ MPa} = 100 \text{ kN/m}^2$$
$$Q_1 = 2100 \text{ kJ/kg}$$
$$r_k = 8,\ \gamma = 1.4$$

$$\therefore \quad \eta_{cycle} = 1 - \frac{1}{r_k^{\gamma-1}} = 1 - \frac{1}{8^{0.4}} = 1 - \frac{1}{2.3} = 0.565 \text{ or } 56.5\% \qquad \textit{Ans.}$$

$$\frac{v_1}{v_2} = 8,\ v_1 = \frac{RT_1}{p_1} = \frac{0.287 \times 308}{100} = 0.884 \text{ m}^3\text{/kg}$$

$$\therefore \quad v_2 = \frac{0.884}{8} = 0.11 \text{ m}^3\text{/kg}$$

$$\frac{T_2}{T_1} = \left(\frac{v_1}{v_2}\right)^{\gamma-1} = (8)^{0.4} = 2.3$$

$\therefore \quad T_2 = 2.3 \times 308 = 708.4 \text{ K}$

$Q_1 = c_v (T_3 - T_2) = 2100 \text{ kJ/kg}$

$$\therefore \quad T_3 - 708.4 = \frac{2100}{0.718} = 2925 \text{ K}$$

or $\quad T_3 = T_{max} = 3633 \text{ K}$ *Ans.*

$$\frac{p_2}{p_1} = \left(\frac{v_1}{v_2}\right)^{\gamma} = (8)^{1.4} = 18.37$$

$\therefore \quad p_2 = 1.873 \text{ MPa}$

Again

$$\frac{p_3 v_3}{T_3} = \frac{p_2 v_2}{T_2}$$

$$\therefore \quad p_3 = p_{max} = 1.837 \times \frac{3633}{708} = 9.426 \text{ MPa} \quad \textit{Ans.}$$

$W_{net} = Q_1 \times \eta_{cycle} = 2100 \times 0.565$

$= 1186.5 \text{ kJ/kg}$

$= p_m(v_1 - v_2) = p_m(0.884 - 0.11)$

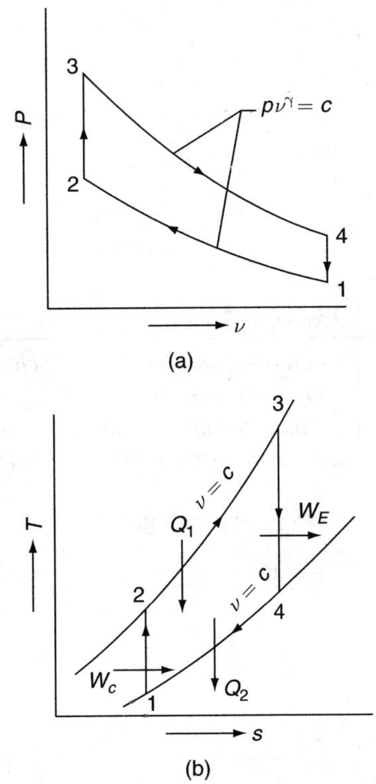

Fig. Ex. 13.1

$$\therefore \quad p_m = \text{m.e.p.} = \frac{1186.5}{0.774} = 1533 \text{ kPa} = 1.533 \text{ MPa} \quad \textit{Ans.}$$

Example 13.2

A Diesel engine has a compression ratio of 14 *and cut-off takes place at* 6% *of the stroke. Find the air standard efficiency.*

Solution From Fig. Ex. 13.2

$$r_k = \frac{v_1}{v_2} = 14$$

$v_3 - v_2 = 0.06\,(1\, v_1 - v_2)$

$= 0.06\,(14\, v_2 - v_2)$

$= 0.78\, v_2$

$v_3 = 1.78\, v_2$

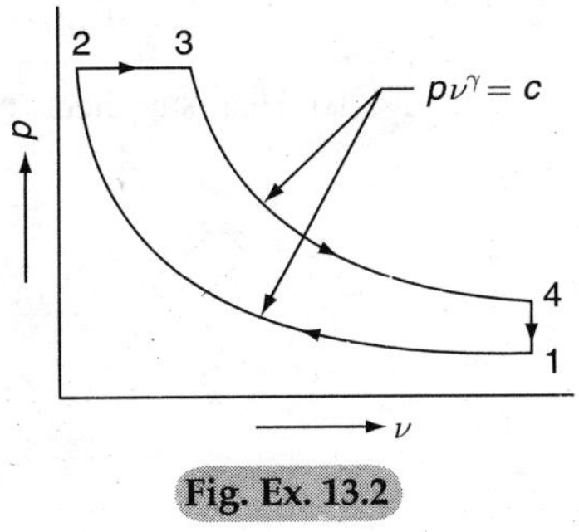

Fig. Ex. 13.2

$\therefore$ Cut-off ratio, $r_c = \dfrac{v_1}{v_2} = 1.78$

$$\eta_{\text{Diesel}} = 1 - \frac{1}{\gamma} \cdot \frac{1}{r_k^{\gamma-1}} \cdot \frac{r_c^\gamma - 1}{r_c - 1} = 1 - \frac{1}{1.4} \cdot \frac{1}{(14)^{0.4}} \cdot \frac{(1.78)^{1.4} - 1}{1.78 - 1}$$

$$= 1 - 0.248 \cdot \frac{1.24}{0.78} = 0.605, \text{ i.e., } 60.5\%$$ *Ans.*

Example 13.3

In an air standard Diesel cycle, the compression ratio is 16, and at the beginning of isentropic compression, the temperature is 15°C and the pressure is 0.1 MPa. Heat is added until the temperature at the end of the constant pressure process is 1480°C. Calculate (a) the cut-off ratio, (b) the heat supplied per kg of air, (c) the cycle efficiency, and (d) the m.e.p.

Solution From Fig. Ex. 13.3

$$r_k = \frac{v_1}{v_2} = 16$$

$$T_1 = 273 + 15 = 288 \text{ K}$$

$$p_1 = 0.1 \text{ MPa} = 100 \text{ kN/m}^2$$

$$T_3 = 1480 + 273 = 1753 \text{ K}$$

$$\frac{T_2}{T_1} = \left(\frac{v_1}{v_2}\right)^{\gamma-1} = (16)^{0.4} = 3.03$$

$\therefore$ $$T_2 = 288 \times 3.03 = 873 \text{ K}$$

$$\frac{p_2 v_2}{T_2} = \frac{p_3 v_3}{T_3}$$

(a) Cut-off ratio,

$$r_c = \frac{v_3}{v_2} = \frac{T_3}{T_2} = \frac{1753}{873} = 2.01$$ *Ans.*

(b) Heat supplied, $Q_1 = c_p (T_3 - T_2)$

$$= 1.005 (1753 - 873)$$

$$= 884.4 \text{ kJ/kg}$$

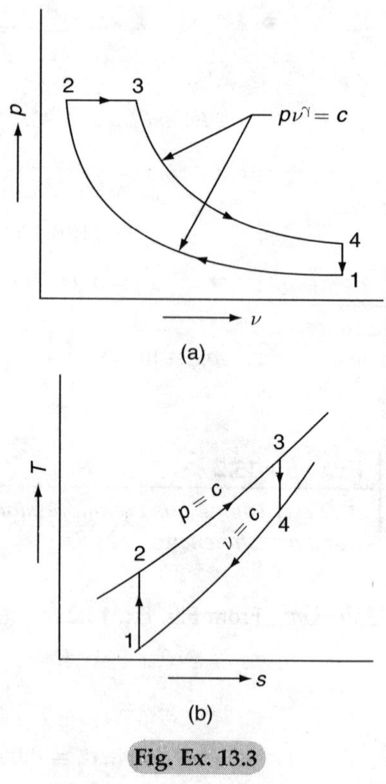

Fig. Ex. 13.3

$$\frac{T_3}{T_4} = \left(\frac{v_4}{v_3}\right)^{\gamma-1} = \left(\frac{v_1}{v_2} \times \frac{v_2}{v_3}\right)^{\gamma-1} = \left(\frac{16}{2.01}\right)^{0.4} = 2.29$$

$$\therefore \quad T_4 = \frac{1753}{2.29} = 766 \text{ K}$$

Heat rejected, $Q_2 = c_v (T_4 - T_1) = 0.718 (766 - 288) = 343.2$ kJ/kg

(c) Cycle efficiency $= 1 - \frac{Q_2}{Q_1} = 1 - \frac{343.2}{884.4} = 0.612$ or 61.2% *Ans.*

It may also be estimated from the equation

$$\eta_{cycle} = 1 - \frac{1}{\gamma} \cdot \frac{1}{r_k^{\gamma-1}} \cdot \frac{r_c^{\gamma} - 1}{r_c - 1} = 1 - \frac{1}{1.4} \frac{1}{(16)^{0.4}} \cdot \frac{(2.01)^{1.4} - 1}{2.01 - 1}$$

$$= 1 - \frac{1}{1.4} \cdot \frac{1}{3.03} \cdot 1.64 = 0.612 \text{ or } 61.2\%$$ *Ans.*

$$W_{net} = Q_1 \times \eta_{cycle} = 884.4 \times 0.612 = 541.3 \text{ kJ/kg}$$

$$v_1 = \frac{RT_1}{p_1} = \frac{0.287 \times 288}{100} = 0.827 \text{ m}^3\text{/kg}$$

$$v_2 = \frac{0.827}{16} = 0.052 \text{ m}^3\text{/kg}$$

$$\therefore \quad v_1 - v_2 = 0.827 - 0.052 = 0.775 \text{ m}^3\text{/kg}$$

(d) m.e.p. $= \frac{W_{net}}{v_1 - v_2} = \frac{541.3}{0.775} = 698.45$ kPa *Ans.*

Example 13.4

An air standard dual cycle has a compression ratio of 16, and compression begins at 1 bar, 50°C. The maximum pressure is 70 bar. The heat transferred to air at constant pressure is equal to that at constant volume. Estimate (a) the pressures and temperatures at the cardinal points of the cycle, (b) the cycle efficiency, and (c) the m.e.p. of the cycle, $c_v = 0.718$ kJ/kg K, $c_p = 1.005$ kJ/kg K.

Solution Given: (Fig. Ex. 13.4)

$$T_1 = 273 + 50 = 323 \text{ K}$$

$$\frac{T_2}{T_1} = \left(\frac{v_1}{v_2}\right)^{\gamma-1} = (16)^{0.4}$$

$$\therefore \quad T_2 = 979 \text{ K}$$

$$p_2 = p_1 \left(\frac{v_1}{v_2}\right)^{\gamma} = 1.0 \times (16)^{1.4} = 48.5 \text{ bar}$$

$$T_3 = T_2 \cdot \frac{p_3}{p_2} = 979 \times \frac{70}{48.5} = 1413 \text{ K}$$

$$Q_{2-3} = c_v (T_3 - T_2) = 0.718 (1413 - 979) = 312 \text{ kJ/kg}$$

Now

$$Q_{2-3} = Q_{3-4} = c_p (T_4 - T_3)$$

$$\therefore \quad T_4 = \frac{312}{1.005} + 1413 = 1723 \text{ K}$$

$$\frac{v_4}{v_3} = \frac{T_4}{T_3} = \frac{1723}{1413} = 1.22$$

$$\therefore \quad \frac{v_5}{v_4} = \frac{v_1}{v_2} \times \frac{v_3}{v_4} = \frac{16}{1.22} = 13.1$$

$$\therefore \quad T_5 = T_4 \left(\frac{v_4}{v_5}\right)^{\gamma-1} = 1723 \times \frac{1}{(13.1)^{0.4}} = 615 \text{ K}$$

$$p_5 = p_1 \left(\frac{T_5}{T_1}\right) = 1.0 \times \frac{615}{323} = 1.9 \text{ bar}$$

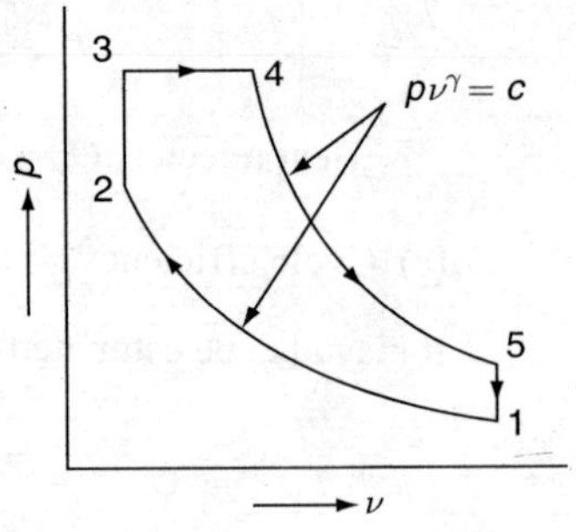

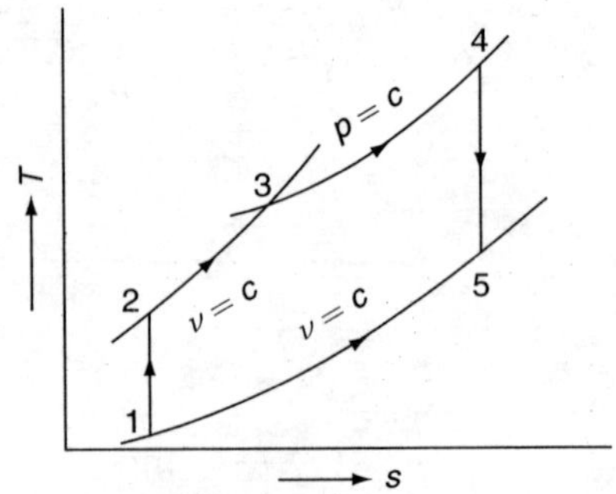

Fig. Ex. 13.4

$$\eta_{\text{cycle}} = 1 - \frac{Q_2}{Q_1} = 1 - \frac{c_v(T_5 - T_1)}{c_v(T_3 - T_2) + c_v(T_4 - T_3)}$$

$$= 1 - \frac{0.718(615 - 323)}{312 - 312}$$

$$= 1 - \frac{0.718 \times 292}{624} = 0.665 \text{ or } 66.5\% \qquad \textit{Ans. (b)}$$

$$v_1 = \frac{RT_1}{p_1} = \frac{0.287 \text{ kJ/kg K} \times 323 \text{ K}}{10^2 \text{ kN/m}^2} = 0.927 \text{ m}^3\text{/kg}$$

$$v_1 - v_2 = v_1 - \frac{v_1}{16} = \frac{15}{16} v_1$$

$$W_{\text{net}} = Q_1 \times \eta_{\text{cycle}} = 0.665 \times 624 \text{ kJ/kg}$$

$$\therefore \quad \text{m.e.p.} = \frac{W_{\text{net}}}{v_1 - v_2} = \frac{0.665 \times \text{kJ/kg}}{\frac{15}{16} \times 0.927 \text{ m}^3\text{/kg}} = 476 \text{ kN/m}^2 = 4.76 \text{ bar} \qquad \textit{Ans. (c)}$$

Example 13.5

In a gas turbine plant, working on the Brayton cycle with a regenerator of 75% effectiveness, the air at the inlet to the compressor is at 0.1 MPa, 30°C, the pressure ratio is 6, and the maximum cycle temperature is 900°C. If the turbine and compressor have each an efficiency of 80%, find the percentage increase in the cycle efficiency due to regeneration.

Solution Given: (Fig. Ex. 13.5)

$$p_1 = 0.1 \text{ MPa}$$
$$T_1 = 303 \text{ K}$$
$$T_3 = 1173 \text{ K}$$
$$r_p = 6, \eta_T = \eta_C = 0.8$$

Without a regenerator

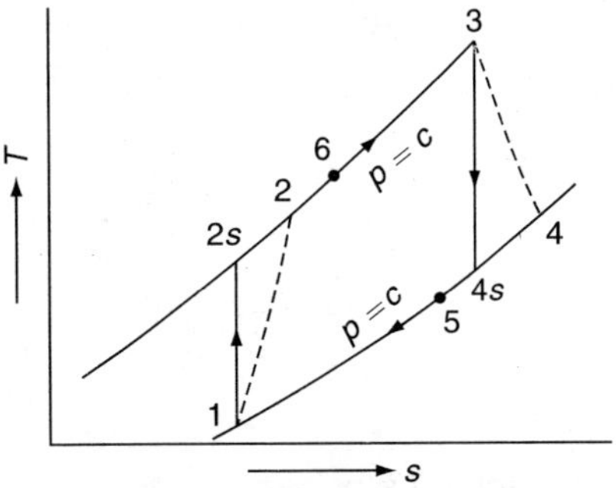

Fig. Ex. 13.5

$$\frac{T_{2s}}{T_1} = \left(\frac{p_2}{p_1}\right)^{(\gamma-1)/\gamma} = \frac{T_3}{T_{4s}} = (6)^{0.4/1.4} = 1.668$$

$$T_{2s} = 303 \times 1.668 = 505 \text{ K}$$

$$T_{4s} = \frac{1173}{1.668} = 705 \text{ K}$$

$$T_2 - T_1 = \frac{T_{2s} - T_1}{\eta_C} = \frac{505 - 303}{0.8} = 252 \text{ K}$$

$$T_3 - T_4 = \eta_T (T_3 - T_{4s}) = 0.8 (1173 - 705) = 375 \text{ K}$$

$$W_T = h_3 - h_4 = c_p (T_3 - T_4)$$

$$= 1.005 \times 375 = 376.88 \text{ kJ/kg}$$

$$W_C = h_2 - h_1 = c_p (T_2 - T_1) = 1.005 \times 252 = 253.26 \text{ kJ/kg}$$

$$T_2 = 252 + 303 = 555 \text{ K}$$

$$Q_1 = h_3 - h_2 = c_p (T_3 - T_2) = 1.005 (1173 - 555) = 621.09 \text{ kJ/kg}$$

$$\therefore \quad \eta = \frac{W_T - W_C}{Q_1} = \frac{376.88 - 253.26}{621.09} = 0.199 \text{ or } 19.9\%$$

With regenerator

$$T_4 = T_3 - 375 = 1173 - 375 = 798 \text{ K}$$

$$\text{Regenerator effectiveness} = \frac{T_6 - T_2}{T_4 - T_2} = 0.75$$

$$\therefore \quad T_6 - 555 = 0.75 (798 - 555)$$

or $$T_6 = 737.3 \text{ K}$$

$$\therefore \quad Q_1 = h_3 - h_6 = c_p (T_3 - T_6) = 1.005 (1173 - 737.3) = 437.88 \text{ kJ/kg}$$

W_{net} remains the same.

$$\therefore \quad \eta = \frac{W_{net}}{Q_1} = \frac{123.62}{437.9} = 0.2837 \text{ or } 28.37\%$$

$\therefore$ Percentage increase due to regeneration

$$= \frac{0.2837 - 0.199}{0.199} = 0.4256, \text{ or } 42.56\%$$

Example 13.6

A gas turbine plant operates on the Brayton cycle between T_{min} = 300 K and T_{max} = 1073 K. Find the maximum work done per kg of air, and the corresponding cycle efficiency. How does this efficiency compare with the Carnot cycle efficiency operating between the same two temperatures?

Solution

$$(W_{net})_{max} = c_p \left(\sqrt{T_{max}} - \sqrt{T_{min}}\right)^2 = 1.005 \left(\sqrt{1073} - \sqrt{300}\right)^2$$

$$= 1.005 (15.43)^2 = 239.28 \text{ kJ/kg}$$ *Ans.*

$$\eta_{cycle} = 1 - \frac{1}{(r_p)^{(\gamma-1)/\gamma}} = 1 - \sqrt{\frac{T_{min}}{T_{max}}} = 1 - \sqrt{\frac{300}{1073}} = 0.47 \text{ or } 47\% \qquad \textit{Ans.}$$

$$\eta_{Carnot} = 1 - \frac{T_{min}}{T_{max}} = 1 - \frac{300}{1073} = 0.721 \text{ or } 72.1\%$$

$$\therefore \quad \frac{\eta_{Brayton}}{\eta_{Carnot}} = \frac{0.47}{0.721} = 0.652 \qquad \textit{Ans.}$$

Example 13.7

In an ideal Brayton cycle, air from the atmosphere at 1 atm, 300 K is compressed to 6 atm and the maximum cycle temperature is limited to 1100 K by using a large air-fuel ratio. If the heat supply is 100 MW, find (a) the thermal efficiency of the cycle, (b) work ratio, (c) power output, (d) exergy flow rate of the exhaust gas leaving the turbine.

Solution The cycle efficiency,

$$\eta_{cycle} = 1 - \frac{1}{r_p^{(\gamma-1)/\gamma}} = 1 - \frac{1}{6^{0.4/1.4}} = 0.401 \text{ or } 40.1\% \qquad \textit{Ans. (a)}$$

$$T_2/T_1 = (r_p)^{(\gamma-1)/\gamma} = 1.67$$

$$T_2 = 501 \text{ K}$$

$$T_3/T_4 = 1.67, \ T_4 = 1100/1.67 = 658.7 \text{ K}$$

$$W_C = 1.005\,(501 - 300) = 202 \text{ kJ/kg}$$

$$W_T = 1.005\,(1100 - 658.7) = 443.5 \text{ kJ/kg}$$

$$\text{Work ratio} = \frac{W_T - W_C}{W_T} = \frac{241.5}{443.5} = 0.545 \qquad \textit{Ans. (b)}$$

$$\text{Power output} = 100 \times 0.401 = 40.1 \text{ MW} \qquad \textit{Ans. (c)}$$

$$Q_1 = \dot{m}\, c_p\, (T_3 - T_2) = 100{,}000 \text{ kW}$$

$$\dot{m} = 166.1 \text{ kg/s}$$

Exergy flow rate of the exhaust gas stream

$$= \dot{m}\, c_p\, T_0 \left(\frac{T_4}{T_0} - 1 - \ln\frac{T_4}{T_0}\right) = 166.1 \times 1.005 \times 300 \left(\frac{658}{300} - 1 - \ln\frac{658.7}{300}\right)$$

$$= 20.53 \text{ MW} \qquad \textit{Ans.}$$

Example 13.8

The following refer to a stationary gas turbine:

Compressor inlet temperature = 311 K

Compressor pressure ratio = 8

Combustion chamber pressure drop = 5% of inlet pressure

Turbine inlet temperature = 1367 *K*
Turbine exit and compressor inlet pressures are atmospheric.

There exists a facility to take air from the compressor exit for use in cooling the turbine. Find the percentage of air that may be taken from the compressor for this purpose so that the overall cycle efficiency drops by 5% from that of the case of no usage of compressed air for cooling of turbine. For simplicity, assume the following: (a) Take properties of gas through the turbine as those of air, (b) Addition of cooling air to the turbine and addition of fuel to the combustion chamber do not affect the turbine power, (c) Compressor and turbine efficiencies are 0.87 *and* 0.90 *respectively.*

Solution Given: $\eta_C = 0.87$, $\eta_T = 0.9$, $T_1 = 311$ K, $p_2/p_1 = 8$,

$p_3 = 0.95p_2$, $T_3 = 1367$ K, $p_4 = p_1 = 1$ atm, $\gamma = 1.4$

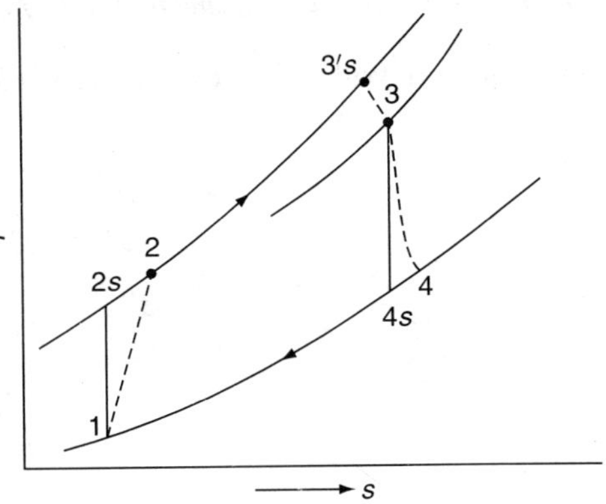

Fig. Ex. 13.8

Case-1: No cooling

$$W_C = \frac{\dot{m}_a c_{pa}(T_{2s} - T_1)}{\eta_c}$$

$$W_T = \dot{m}_g c_{pg}(T_3 - T_{4s})\,\eta_T$$

$$Q_1 = \dot{m}_c p_c (T_3 - T_2)$$

Here, $\dot{m}_a = \dot{m}_g = \dot{m}_c = \dot{m}$

$$c_{pa} = c_{pg} = c_{pc} = c_p$$

$$(T_{2s}/T_1 = p_{2s}/p_1)^{(\gamma-1)/\gamma} = 8^{0.4/1.4} = 1.181$$

$$T_{2s} = 563.3 \text{ K}$$

$$\frac{T_{2s} - T_1}{T_2 - T_1} = 0.87, \; T_2 = 601 \text{ K}$$

$$T_3/T_{4s} = (p_3/p_{4s})^{(\gamma-1)/\gamma} = \left(\frac{0.95p_2}{p_1}\right)^{0.4/1.4} = 1.785$$

$$T_{4s} = 765.83 \text{ K}$$

$$W_C = 290\ \dot{m}c_p,\; W_T = 541.06\ \dot{m}\,c_p \text{ and } Q_1 = 766\ \dot{m}\,c_p$$

$$\eta_{cycle} = \frac{541.06 - 290}{766} = 0.328$$

Case-2: With cooling

$$\eta_{cycle} = 0.328 - 0.05 = 0.278$$

Since the extraction of compressed air for turbine cooling does not contribute to turbine work or burner fuel flow, it can be treated as an increment x added to the compressor mass flow.

$$\frac{541.06 - 290(1+x)}{766} = 0.278$$

$\therefore$ $x = 0.13$

$$\% \text{ of compressor delivery air flow} = \frac{0.13}{1.13} \times 100 = 11.6\%$$ *Ans.*

Example 13.9

In a gas turbine plant the ratio of T_{max}/T_{min} is fixed. Two arrangements of components are to be investigated: (a) single-stage compression followed by expansion in two turbines of equal pressure ratios with reheat to the maximum cycle temperature, and (b) compression in two compressors of equal pressure ratios, with intercooling to the minimum cycle temperature, followed by single-stage expansion. If η_C and η_T are the compressor and turbine efficiencies, show that the optimum specific output is obtained at the same overall pressure ratio for each arrangement.

If η_C is 0.85 and η_T is 0.9, and T_{max}/T_{min} is 3.5, determine the above pressure ratio for optimum specific output and show that with arrangement (a) the optimum output exceeds that of arrangement (b) by about 11%.

Solution (a) With reference to Fig. Ex. 13.9(a)

$$T_1 = T_{min},\ T_3 = T_5 = T_{max},\ \frac{p_2}{p_4} = \frac{p_4}{p_1}$$

$$\therefore \quad p_4 = \sqrt{p_1 p_2}$$

$$\frac{p_{2s}}{p_1} = r, \text{ pressure ratio}$$

$$\therefore \quad p_{2s} = p_2 = r p_1$$

$$\therefore \quad p_4 = \sqrt{r} \cdot p_1$$

$$\frac{T_{2s}}{T_1} = \left(\frac{p_2}{p_1}\right)^{(\gamma-1)/\gamma} = r^x$$

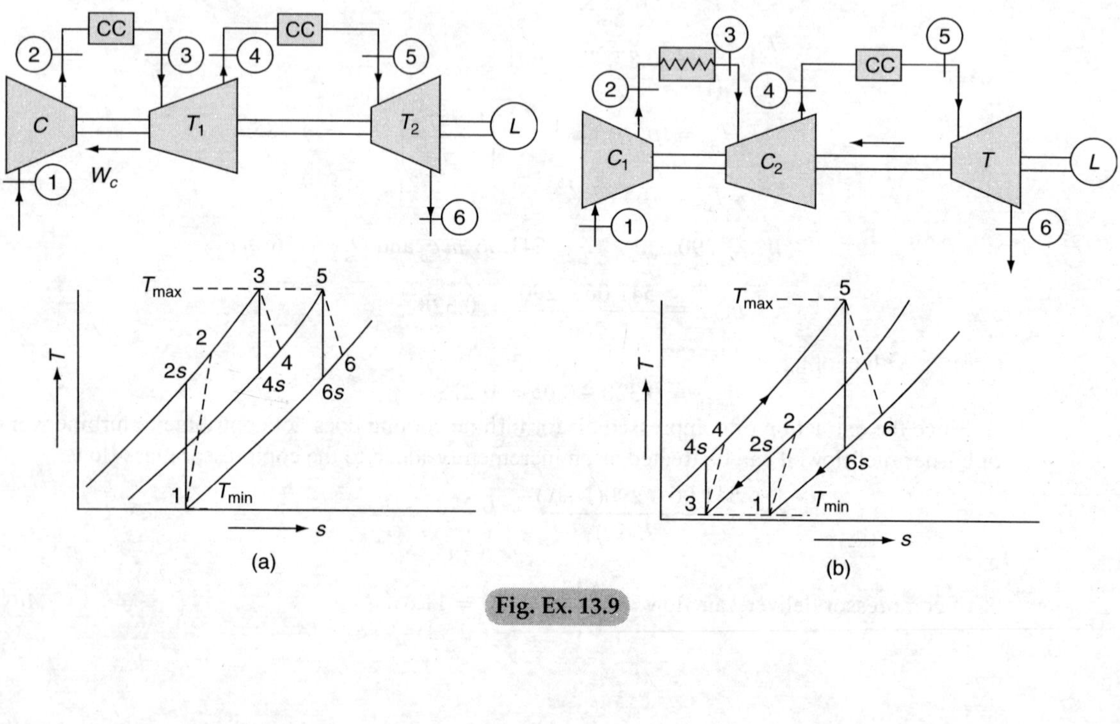

Fig. Ex. 13.9

where $$x = \frac{\gamma - 1}{\gamma}$$

$$\therefore \quad T_{2s} = T_{min}\, r^{x}$$

$$(\Delta T_s)_{comp} = T_{2s} - T_1 = T_{min}\, r^x - T_{min} = T_{min}\,(r^x - 1)$$

$$\therefore \quad (\Delta T)_{comp} = \frac{T_{min}\left(r^x - 1\right)}{\eta_C}$$

$$\frac{T_3}{T_{4s}} = \left(\frac{p_3}{p_4}\right)^{(\gamma-1)/\gamma} = \left(\frac{rp_1}{\sqrt{r}\cdot p_1}\right)^{x} r^{x/2}$$

$$\therefore \quad T_{4s} = T_3\, r^{-x/2} = T_{max}\, r^{-x/2}$$

$$(\Delta T_s)_{turb} = T_3 - T_{4s} = T_{max} - T_{max} \cdot r^{-x/2} = T_{max}\,(1 - r^{-x/2})$$

$$\therefore \quad (\Delta T)_{turb\ 1} = \eta_T\, T_{max}\,(1 - r^{-x/2}) = (\Delta T)_{turb\ 2}$$

$$\therefore \quad W_{net} = c_p\left[2\eta_T T_{max}\left(1 - r^{-x/2}\right) - \frac{T_{min}}{\eta_C}\left(r^x - 1\right)\right]$$

$$\frac{dW_{net}}{dr} = c_p\left[2\eta_T T_{max}\,\frac{x}{2}\cdot r^{-x/2-1} - \frac{T_{min}}{\eta_C}\,x\cdot r^{x-1}\right] = 0$$

On simplification

$$r^{3x/2} = \eta_T\,\eta_C\,\frac{T_{max}}{T_{min}}$$

$$\therefore \quad r_{opt} = \left(\eta_T \eta_C\,\frac{T_{max}}{T_{min}}\right)^{2\gamma/3(\gamma-1)}$$

(b) With reference to Fig. Ex. 13.9(b)

$$\frac{T_{2s}}{T_{min}} = \left(\frac{p_2}{p_1}\right)^{(\gamma-1)/\gamma} = \left(\sqrt{r}\right)^{x} = r^{x/2}$$

$$(\Delta T_s)_{comp\ 1} = T_{2s} - T_1 = T_{min}\,(r^{x/2} - 1)$$

$$(\Delta T)_{comp1} = \frac{T_{min}\left(r^{x/2} - 1\right)}{\eta_C} = (\Delta T)_{comp\ 2}$$

$$\frac{T_{max}}{T_{6s}} = \left(\frac{p_5}{p_6}\right)^{(\gamma-1)/\gamma} = r^{x}$$

$$\therefore \quad T_{6s} = T_{max} \cdot r^{-x}$$

$$(\Delta T_s)_{turb} = T_{max} - T_{6s} = T_{max}\,(1 - r^{-x})$$

$$(\Delta T)_{turb} = \eta_T\, T_{max}\,(1 - r^{-x})$$

$$W_{net} = c_p \left[\eta_T T_{max} \left(1 - r^{-x}\right) - \frac{2T_{min}\left(r^{x/2} - 1\right)}{\eta_C} \right]$$

$$\frac{dW_{net}}{dr} = c_p \left[\eta_T T_{max}\, x \cdot r^{-(x+1)} - \frac{2T_{min}}{\eta_C} \frac{x}{2} \cdot r^{(x/2)-1} \right] = 0$$

on simplification $$r_{opt} = \left(\eta_T \eta_C \frac{T_{max}}{T_{min}} \right)^{2\gamma/3(\gamma-1)}$$

This is the same as in (a).

If $$\eta_C = 0.85,\ \eta_T = 0.9$$

$$\frac{1}{x} = \frac{\gamma}{\gamma - 1} = \frac{1.4}{0.4} = \frac{7}{2}$$

$$\frac{T_{max}}{T_{min}} = 3.5$$

$\therefore$ $$r_{opt} = (0.85 \times 0.9 \times 3.5)^{2/3 \times 7/2} = 9.933$$ *Ans.*

$$W_{net}\,(a) = c_p \left[2\eta_T T_{max} \left(1 - r^{-x/2}\right) - \frac{T_{min}}{\eta_C}\left(r^{x} - 1\right) \right]$$

$$= c_p \left[2 \times 0.9 \times T_{max} \left(1 - 9.933^{-0.143}\right) - \frac{T_{min}}{0.85}\left(9.933^{0.286} - 1\right) \right]$$

$$= c_p \cdot T_{min} \left[2 \times 0.9 \times 3.5 \left(1 - \frac{1}{1.388}\right) - T\, 1.178\,(0.928) \right]$$

$$= 0.670\, c_p \cdot T_{min}$$

$$W_{net}\,(b) = c_p\, T_{min} \left[0.9 \times 3.5 \left(1 - 9.933^{-0.286}\right) - \frac{2}{0.85}\left(9.933^{0.143} - 1\right) \right]$$

$$= c_p\, T_{min}\,(1.518 - 0.914) = 0.604\, c_p\, T_{min}$$

$$\frac{W_{net}(a) - W_{net}(b)}{W_{net}(a)} \times 100$$

$$= \frac{0.670 - 0.604}{0.670} \times 100 = 10.9\%$$ Proved.

Example 13.10

A turbojet aircraft flies with a velocity of 300 m/s at an altitude where the air is at 0.35 bar and – 40°C. The compressor has a pressure ratio of 10, and the temperature of the gases at the turbine inlet is 1100°C. Air enters the compressor at a rate of 50 kg/s. Estimate (a) the temperature and pressure of the gases at the turbine exit, (b) the velocity of gases at the nozzle exit, and (c) the propulsive efficiency of the cycle.

Solution (a) For isentropic flow of air in the diffusor (Fig. Ex. 13.10)

$$\cancelto{0}{Q_{1-2}} - \cancelto{0}{W_{1-2}} = h_2 - h_1 + \frac{\cancelto{0}{V_2^2} - V_1^2}{2}$$

$$0 = c_p (T_2 - T_1) - \frac{V_1^2}{2}$$

$$T_2 = T_1 + \frac{V_1^2}{2c_p} = 233 + \frac{300^2}{2 \times 1.005} \times 10^{-3}$$

$$= 277.78 \text{ K}$$

$$p_2 = p_1 (T_2/T_1)^{\gamma/(\gamma - 1)}$$

$$= 35 \frac{\text{kN}}{\text{m}^2} \left(\frac{277.78}{233}\right)^{1.4/0.4} = 64.76 \text{ kPa}$$

$$p_3 = r_p p_2 = 10 \times 64.76 = 647.6 \text{ kPa}$$

$$T_3 = \left(\frac{p_3}{p_2}\right)^{(\gamma-1)/\gamma} T_2 = 277.78\,(10)^{0.4/1.4}$$

$$= 536.66 \text{ K}$$

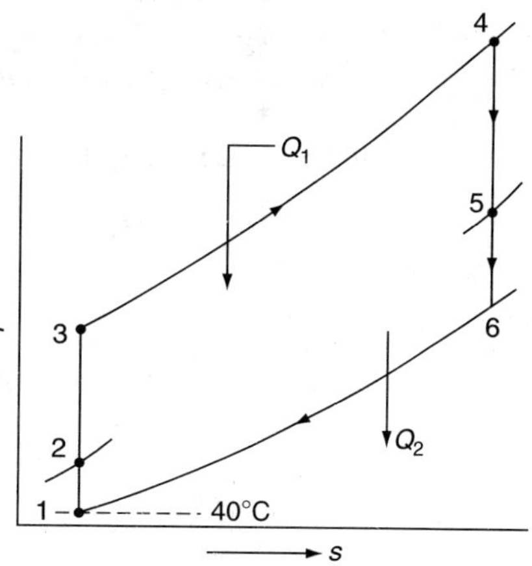

Fig. Ex. 13.10

$$W_C = W_T$$

$$h_3 - h_2 = h_4 - h_5$$

or, $$T_3 - T_2 = T_4 - T_5$$

$$T_5 = T_4 - T_3 + T_2 = 1373 - 536.66 + 277.78 = 1114.12 \text{ K}$$

$$p_5 = \left(\frac{T_5}{T_4}\right)^{\gamma/(\gamma-1)} p_4 = 647.6 \left(\frac{1114.12}{1373}\right)^{3.5} = 311.69 \text{ K}$$ *Ans.* (a)

(b) For isentropic expansion of gases in the nozzle,

$$T_6 = T_5 \left(\frac{p_6}{p_5}\right)^{(\gamma-1)/\gamma} = 1114.12 \left(\frac{35}{311.69}\right)^{0.286} = 596.12 \text{ K}$$

Neglecting the K.E. of gas at nozzle inlet,

$$V_6 = [2c_p (T_5 - T_6) \times 1000]^{1/2} = [2 \times 1.005\,(1114.12 - 596.12) \times 1000]^{1/2}$$

$$= 1020.4 \text{ m/s}$$ *Ans.* (b)

(c) The propulsive efficiency of a turbojet engine is the ratio of the propulsive power developed $\dot{W}_P$ to the total heat transfer to the fluid.

$$\dot{W}_P = w[V_{exit} - V_{inlet}]\, V_{aircraft} = 50\,[1020.4 - 300] \times 300 \frac{\text{kg}}{\text{s}} \times \frac{\text{m}^2}{\text{s}^2} = 10.806 \text{ MW}$$

$$Q_1 = w\,(h_4 - h_3) = 50 \times 1.005\,(1373 - 536.66) = 42.026 \text{ MW}$$

$$\eta_P = \frac{10.806}{42.026} = 0.257 \text{ or } 25.7\%$$ *Ans.*

Example 13.11

In a combined GT – ST plant, the exhaust gas from the GT is the supply gas to the steam generator at which a further supply of fuel is burned in the gas. The pressure ratio for the GT is 8, the inlet air temperature is 15°C and the maximum cycle temperature is 800°C.

Combustion in the steam generator raises the gas temperature to 800°C and the gas leaves the generator at 100°C. The steam condition at supply is at 60 bar, 600°C and the condenser pressure is 0.05 bar. Calculate the flow rates of air and steam required for a total power output of 190 MW and the overall efficiency of the combined plant. What would be the air-fuel ratio? Assume ideal processes. Take $c_{p_g} = 1.11$ *and* $c_{p_a} = 1.005$ *kJ/kg K,* $\gamma_g = 1.33$, $\gamma_a = 1.4$, *C.V. of fuel = 43.3 MJ/kg. Neglect the effect of fuel flow on the total mass flow of gas expanding in the gas turbine.*

Solution With reference to Fig. 13.32.

$$T_b = T_a (p_b/p_a)^{(\gamma-1)/\gamma} = 288 \times 8^{1.4/0.4} = 522 \text{ K}$$

$$T_d = \frac{T_C}{r_p^{(\gamma-1)/\gamma}} = \frac{1073}{8^{0.33/1.33}} = \frac{1073}{1.682} = 638 \text{ K}$$

$$W_{GT} = c_{p_g}(T_c - T_d) - c_{p_a}(T_b - T_a)$$

$$= 1.11 (1073 - 638) - 1.005 (522 - 288) = 249 \text{ kJ/kg}$$

$$Q_1 = c_{p_g}(T_c - T_b) = 1.11 (1073 - 522) = 612 \text{ kJ/kg}$$

$$Q'_1 = c_{p_g}(T_e - T_d) = 1.11(1073 - 638)$$

$$= 483 \text{ kJ/kg}$$

$$h_1 = 3775,\ h_2 = 2183,\ h_3 = 138 = h_4, \text{ all in kJ/kg}$$

$$(Q_1)_{St} = h_1 - h_3 = 3775 - 138 = 3637 \text{ kJ/kg}$$

$$Q''_{fe} = 1.11(800 - 100) = 777 \text{ kJ/kg}$$

By energy balance of the steam generator,

$$w_a \times 777 = w_s \times 3637$$

$$w_a/w_s = 4.68$$

$$W_{ST} = h_1 - h_2 = 3775 - 2183 = 1592 \text{ kJ/kg}$$

$$w_a \times 249 + w_s \times 1592 = 190 \times 10^3 \text{ kW}$$

$$w_s (4.68 \times 249 + 1592) = w_s \times 2.757 \times 10^3 = 190 \times 10^3$$

$$w_s = 68.9 \text{ kg/s and } w_a = 322.5 \text{ kg/s}$$

Now,

$$w_a(612 + 483) = w_f \times 43{,}300$$

$$w_a/w_f = \text{A/F ratio} = \frac{43{,}300}{1095} = 39.5 \qquad \textit{Ans.}$$

$$\text{Fuel energy input} = \frac{322.5}{39.5} \times 43{,}300 = 353525 \text{ kW} = 353.5 \text{ MW}$$

$$\eta_{0A} = \frac{190}{353.5} = 0.537 \text{ or } 53.7\% \qquad \textit{Ans.}$$

Review Questions

13.1 What are cyclic and non-cyclic heat engines? Give examples.

13.2 What are the four processes which constitute the Stirling cycle? Show that the regenerative Stirling cycle has the same efficiency as the Carnot cycle.

13.3 State the four processes that constitute the Ericsson cycle. Show that the regenerative Ericsson cycle has the same efficiency as the Carnot cycle.

13.4 Mention the merits and demerits of the Stirling and Ericsson cycles.

13.5 What is an air standard cycle? Why are such cycles conceived?

13.6 What is a spark ignition engine? What is the air standard cycle of such an engine? What are its four processes?

13.7 Show that the efficiency of the Otto cycle depends only on the compression ratio.

13.8 How is the compression ratio of an SI engine fixed?

13.9 What is a compression ignition engine? Why is the compression ratio of such an engine more than that of an SI engine?

13.10 State the four processes of the Diesel cycle.

13.11 Explain the mixed or dual cycle.

13.12 For the same compression ratio and heat rejection, which cycle is most efficient: Otto, Diesel or Dual? Explain with p–v and T–s diagrams.

13.13 With the help of p–v and T–s diagrams, show that for the same maximum pressure and temperature of the cycle and the same heat rejection,

$$\eta_{\text{Diesel}} > \eta_{\text{Dual}} > \eta_{\text{Otto}}$$

13.14 What are the three basic components of a gas turbine plant? What is the air standard cycle of such a plant? What are the processes it consists of?

13.15 Show that the efficiency of the Brayton cycle depends only on the pressure ratio.

13.16 What is the application of the closed cycle gas turbine plant?

13.17 Discuss the merits and demerits of Brayton and Otto cycles applied to reciprocating and rotating plants.

13.18 What is the effect of regeneration on Brayton cycle efficiency? Define the effectiveness of a regenerator.

13.19 What is the effect of irreversibilities in turbine and compressor on Brayton cycle efficiency?

13.20 Explain the effect of pressure ratio on the net output and efficiency of a Brayton cycle.

13.21 Drive the expression of optimum pressure ratio for maximum net work output in an ideal Brayton cycle. What is the corresponding cycle efficiency.

13.22 Explain the effects of: (a) intercooling, and (b) reheating, on Brayton cycle.

13.23 What is a free shaft turbine?

13.24 With the help of flow and T–s diagrams explain the air standard cycle for a jet propulsion plant.

13.25 With the help of a neat sketch explain the operation of a turbojet engine. How is the thrust developed in this engine? Why does a commercial airplane fly at high altitudes?

13.26 Define propulsive power and propulsive efficiency.

13.27 Why are regenerators and intercoolers not used in aircraft engines? What is after-burning? Why is it used?

13.28 Explain the working of a turbofan engine with the help of a neat sketch. Define "bypass ratio". How does it influence the engine thrust?

13.29 How does a turboprop engine differ from a turbofan engine?

13.30 What is a ramjet? How is the thrust produced here?

13.31 What is a rocket? How is it propelled?

13.32 Explain the advantages and disadvantages of a gas turbine plant for a utility system.

13.33 What are the advantages of a combined gas turbine-steam turbine power plant?

13.34 With the help of flow and T–s diagrams explain the operation of a combined GT–ST plant. Why is supplementary firing often used?

Problems

13.1 In a Stirling cycle the volume varies between 0.03 and 0.06 m^3, the maximum pressure is 0.2 MPa, and the temperature varies between 540°C and 270°C. The working fluid is air (an ideal gas). (a) Find the efficiency and the work done per cycle for the simple cycle. (b) Find the efficiency and the work done per cycle for the cycle with an ideal regenerator, and compare with the Carnot cycle having the same isothermal heat supply process and the same temperature range. *Ans.* (a) 27.7%, 53.7 kJ/kg, (b) 32.2%

13.2 An Ericsson cycle operating with an ideal regenerator works between 1100 K and 288 K. The pressure at the beginning of isothermal compression is 1.013 bar. Determine (a) the compressor and turbine work per kg of air, and (b) the cycle efficiency. *Ans.* (a) $W_T = 465$ kJ/kg, $W_C = 121.8$ kJ/kg (b) 0.738

13.3 Plot the efficiency of the air standard Otto cycle as a function of the compression ratio for compression ratios from 4 to 16.

13.4 Find the air standard efficiencies for Otto cycles with a compression ratio of 6 using ideal gases having specific heat ratios 1.3, 1.4 and 1.67. What are the advantages and disadvantages of using helium as the working fluid?

13.5 An engine equipped with a cylinder having a bore of 15 cm and a stroke of 45 cm operates on an Otto cycle. If the clearance volume is 2000 cm^3, compute the air standard efficiency. *Ans.* 47.4%

13.6 In an air standard Otto cycle the compression ratio is 7, and compression begins at 35°C, 0.1 MPa. The maximum temperature of the cycle is 1100°C. Find (a) the temperature and pressure at the cardinal points of the cycle, (b) the heat supplied per kg of air, (c) the work done per kg of air, (d) the cycle efficiency, and (e) the m.e.p. of the cycle.

13.7 An engine working on the Otto cycle has an air standard cycle efficiency of 56% and rejects 544 kJ/kg of air. The pressure and temperature of air at the beginning of compression are 0.1 MPa and 60°C respectively. Compute (a) the compression ratio of the engine, (b) the work done per kg of air, (c) the pressure and temperature at the end of compression, and (d) the maximum pressure in the cycle.

13.8 For an air standard Diesel cycle with a compression ratio of 15, plot the efficiency as a function of the cut-off ratio for cut-off ratios from 1 to 4. Compare with the results of Problem 13.3.

13.9 In an air standard Diesel cycle, the compression ratio is 15. Compression begins at 0.1 MPa, 40°C. The heat added is 1.675 MJ/kg. Find (a) the maximum temperature of the cycle, (b) the work done per kg of air, (c) the cycle efficiency, (d) the temperature at the end of the isentropic expansion, (e) the cut-off ratio, (f) the maximum pressure of the cycle, and (g) the m.e.p. of the cycle.

13.10 Two engines are to operate on Otto and Diesel cycles with the following data:
Maximum temperature 1400 K, exhaust temperature 700 K. State of air at the beginning of compression 0.1 MPa, 300 K.
Estimate the compression ratios, the maximum pressures, efficiencies, and rate of work outputs (for 1 kg/min of air) of the respective cycles.
Ans. Otto—$r_k = 5.656$, $p_{max} = 2.64$ MPa, $W =$ 2872 kJ/kg, $\eta = 50\%$
Diesel—$r_k = 7.456$, $p_{max} = 1.665$ MPa, $W = 446.45$ kJ/kg, $\eta = 60.8\%$

13.11 An air standard limited pressure cycle has a compression ratio of 15 and compression begins at 0.1 MPa, 40°C. The maximum pressure is limited to 6 MPa and the heat added is 1.675 MJ/kg. Compute (a) the heat supplied at constant volume per kg of air, (b) the heat supplied at constant pressure per kg of air, (c) the work done per kg of air, (d) the cycle efficiency, (e) the temperature at the end of the constant volume heating process, (f) the cut-off ratio, and (g) the m.e.p. of the cycle. *Ans.* (a) 235 kJ/kg, (b) 1440 kJ/kg, (c) 1014 kJ/kg, (d) 60.5%, (e) 1252 K, (f) 2.144, (g) 1.21 MPa

13.12 In an ideal cycle for an internal combustion engine the pressure and temperature at the beginning of adiabatic compression are respectively 0.11 MPa and 115°C, the compression ratio being 16. At the end of compression heat is added to the working fluid, first, at constant volume, and then at constant pressure reversibly. The working fluid is then expanded adiabatically and reversibly to the original volume.

If the working fluid is air and the maximum pressure and temperature are respectively 6 MPa and 2000°C, determine, per kg of air (a) the pressure, temperature, volume, and entropy of the air at the five cardinal points of the cycle (take s_1 as the entropy of air at the beginning of compression), and (b) the work output and efficiency of the cycle.

13.13 Show that the air standard efficiency for a cycle comprising two constant pressure processes and two isothermal processes (all reversible) is given by

$$\eta = \frac{(T_1 - T_2)\ \text{In}\left(r_p\right)^{(\gamma-1)\gamma}}{T_1\left[1 + \text{In}\left(r_p\right)^{(\gamma-1)/\gamma} - T_2\right]}$$

where T_1 and T_2 are the maximum and minimum temperatures of the cycle, and r_p is the pressure ratio.

13.14 Obtain an expression for the specific work done by an engine working on the Otto cycle in terms of the maximum and minimum temperatures of the cycle, the compression ratio r_k, and constants of the working fluid (assumed to be an ideal gas).

Hence show that the compression ratio for maximum specific work output is given by

$$r_k = \left(\frac{T_{min}}{T_{max}}\right)^{1/2(1-\gamma)}$$

13.15 A dual combustion cycle operates with a volumetric compression ratio $r_k = 12$, and with a cut-off ratio 1.615. The maximum pressure is given by $p_{max} = 54p_1$, where p_1 is the pressure before compression. Assuming indices of compression and expansion of 1.35, show that the m.e.p. of the cycle

$$p_m = 10\, p_1$$

Hence evaluate (a) temperatures at cardinal points with $T_1 = 335$ K, and (b) cycle efficiency.

Ans. (a) $T_2 = 805$ K, $p_2 = 29.2\, p_1$, $T_3 = 1490$ K, $T_4 = 2410$ K, $T_5 = 1200$ K, (b) $\eta = 0.67$

13.16 Recalculate (a) the temperatures at the cardinal points, (b) the m.e.p., and (c) the cycle efficiency when the cycle of Problem 13.15 is a Diesel cycle with the same compression ratio and a cut-off ratio such as to give an expansion curve coincident with the lower part of that of the dual cycle of Problem 13.15. *Ans.* (a) $T_2 = 805$ K, $T_3 = 1970$ K, $T_4 = 1142$ K (b) $6.82\, p_1$, (c) $\eta = 0.513$

13.17 In an air standard Brayton cycle the compression ratio is 7 and the maximum temperature of the cycle is 800°C. The compression begins at 0.1 MPa, 35°C. Compare the maximum specific volume and the maximum pressure with the Otto cycle of Problem 13.6. Find (a) the heat supplied per kg of air, (b) the net work done per kg of air, (c) the cycle efficiency, and (d) the temperature at the end of the expansion process.

13.18 A gas turbine plant operates on the Brayton cycle between the temperatures 27°C and 800°C. (a) Find the pressure ratio at which the cycle efficiency approaches the Carnot cycle efficiency, (b) find the pressure ratio at which the work done per kg of air is maximum, and (c) compare the efficiency at this pressure ratio with the Carnot efficiency for the given temperatures.

13.19 In a gas turbine plant working on the Brayton cycle the air at the inlet is at 27°C, 0.1 MPa. The pressure ratio is 6.25 and the maximum temperature is 800°C. The turbine and compressor efficiencies are each 80%. Find (a) the compressor work per kg of air, (b) the turbine work per kg of air, (c) the heat supplied per kg of air, (d) the cycle efficiency, and (e) the turbine exhaust temperature.

Ans. (a) 259 kJ/kg, (b) 351.68 kJ/kg, (c) 569.43 kJ/kg, (d) 16.2%, (e) 723 K

13.20 Solve Problem 13.19 if a regenerator of 75% effectiveness is added to the plant.

13.21 Solve Problem 13.19 if the compression is divided into two stages, each of pressure ratio 2.5 and efficiency 80%, with intercooling to 27°C.

13.22 Solve Problem 13.21 if a regenerator of 75% effectiveness is added to the plant.

13.23 Solve Problem 13.19 if a reheat cycle is used. The turbine expansion is divided into two stages, each of pressure ratio 2.5 and efficiency 80%, with reheat to 800°C.

13.24 Solve Problem 13.23 if a regenerator of 75% effectiveness is added to the plant.

13.25 Solve Problem 13.24 if the staged compression of Problem 13.21 is used in the plant.

13.26 Find the inlet condition for the free-shaft turbine if a two-shaft arrangement is used in Problem 13.19.

13.27 A simple gas turbine plant operating on the Brayton cycle has air inlet temperature 27°C, pressure ratio 9, and maximum cycle temperature 727°C.

What will be the improvement in cycle efficiency and output if the turbine process is divided into two stages each of pressure ratio 3, with intermediate reheating to 727°C? *Ans.* – 18.3%, 30.6%

13.28 Obtain an expression for the specific work output of a gas turbine unit in terms of pressure ratio, isentropic efficiencies of the compressor and turbine, and the maximum and minimum temperatures, T_3 and T_1. Hence show that the pressure ratio r_p for maximum power is given by

$$r_p = \left(\eta_T \eta_C \frac{T_3}{T_1}\right)^{\gamma/2(\gamma-1)}$$

If $T_3 = 1073$ K, $T_1 = 300$ K, $\eta_C = 0.8$, $\eta_T = 0.8$ and $\gamma = 1.4$ compute the optimum value of pressure ratio, the maximum net work output per kg of air, and corresponding cycle efficiency. *Ans.* 4.263, 100 kJ/kg, 17.2%

13.29 A gas turbine plant draws in air at 1.013 bar, 10°C and has a pressure ratio of 5.5. The maximum temperature in the cycle is limited to 750°C.

Compression is conducted in an uncooled rotary compressor having an isentropic efficiency of 82%, and expansion takes place in a turbine with an isentropic efficiency of 85%. A heat exchanger with an efficiency of 70% is fitted between the compressor outlet and combustion chamber. For an air flow of 40 kg/s, find (a) the overall cycle efficiency, (b) the turbine output, and (c) the air-fuel ratio if the calorific value of the fuel used in 45.22 MJ/kg. *Ans.* (a) 30.4%, (b) 4272 kW, (c) 115

13.30 A gas turbine for use as an automotive engine is shown in Fig. P 13.30. In the first turbine, the gas expands to just a low enough pressure p_5, for the turbine to drive the compressor. The gas is then expanded through a second turbine connected to the drive wheels. Consider air as the working fluid, and assume that all processes are ideal. Determine (a) pressure p_5, (b) the net work per kg and mass flow rate, (c) temperature T_3 and cycle thermal efficiency, and (d) the *T-s* diagram for the cycle.

13.31 Repeat Problem 13.30 assuming that the compressor has an efficiency of 80%, both the turbines have efficiencies of 85%, and the regenerator has an efficiency of 72%.

13.32 An ideal air cycle consists of isentropic compression, constant volume heat transfer, isothermal expansion to the original pressure, and constant pressure heat transfer to the original temperature. Deduce an expression for the cycle efficiency in terms of volumetric compression ratio r_k, and isothermal expansion ratio, r_e. In such a cycle, the pressure and temperature at the start of compression are 1 bar and 40°C, the compression ratio is 8, and the maximum pressure is 100 bar. Determine the cycle efficiency and the m.e.p.

Ans. 51.5%, 3.45 bar

13.33 For a gas turbine jet propulsion unit, shown in Fig. 13.24, the pressure and temperature entering the compressor are 1 atm and 15°C respectively. The pressure ratio across the compressor is 6 to 1 and the temperature at the turbine inlet is 1000°C. On leaving the turbine the air enters the nozzle and expands to 1 atm. Determine the pressure at the nozzle inlet and the velocity of the air leaving the nozzle.

13.34 Repeat Problem 13.33, assuming that the efficiency of the compressor and turbine are both 85%, and that the nozzle efficiency is 95%.

13.35 Develop expressions for work output per kg and the efficiency of an ideal Brayton cycle with regeneration, assuming maximum possible regeneration. For fixed maximum and minimum temperatures, how do the efficiency and work outputs vary with the pressure ratio? What is the optimum pressure ratio?

13.36 For an air standard Otto cycle with fixed intake and maximum temperatures, T_1 and T_3, find the compression ratio that renders the net work per cycle a maximum. Derive the expression for cycle efficiency at this compression ratio.

If the air intake temperature, T_1, is 300 K and the maximum cycle temperature, T_3, is 1200 K, compute the compression ratio for maximum net work, maximum work output per kg in a cycle, and the corresponding cycle efficiency.

Find the changes in work output and cycle efficiency when the compression ratio is increased from this optimum value to 8. Take $c_v = 0.718$ kJ/kg K.

Ans. 5.65, 215 kJ/kg, 50%, $\Delta W_{net} = -9$ kJ/kg, $\Delta\eta = +6.4\%$

13.37 Show that the mean effective pressure, p_m, for the Otto cycle is given by

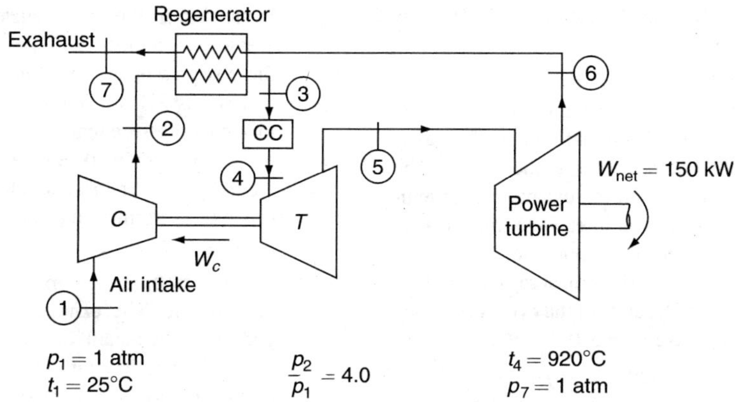

Fig. P. 13.30

$$p_m = \frac{\left(p_3 - p_1 r_k^{\gamma}\right)\left(1 - \frac{1}{r_k^{\gamma-1}}\right)}{(\gamma - 1)(r_k - 1)}$$

where $p_3 = p_{max}$, $p_1 = p_{min}$ and r_k is the compression ratio.

13.38 A gas turbine plant operates on the Brayton cycle using an optimum pressure ratio for maximum net work output and a regenerator of 100% effectiveness.

Derive expressions for net work output per kg of air and corresponding efficiency of the cycle in terms of the maximum and the minimum temperatures.

If the maximum and minimum temperatures are 800°C and 30°C respectively, compute the optimum value of pressure ratio, the maximum net work output per kg and the corresponding cycle efficiency.

$$Ans.\ (W_{net})_{max} = c_p \left(\sqrt{T_{max}} - \sqrt{T_{min}}\right)^2$$

$$(\eta_{cycle})_{max} = 1 - \sqrt{\frac{T_{min}}{T_{max}}},\ (r_p)_{opt} = 9.14,$$

$$(W_{net})_{max} = 236.79 \text{ kJ/kg};\ \eta_{cycle} = 0.469$$

13.39 A gas turbine unit is to provide peaking power for an electrical utility with a net power output of 10 MW. The pressure ratio across the compressor is 7, the efficiency of the compressor 80%, and the efficiency of the turbine is 92%. In order to conserve fuel, a regenerator with an effectiveness of 85% is used. The maximum temperature of the cycle is 1200 K. The air at compressor inlet is at 20°C, 1.1 bar. Assume the working fluid to be air which behaves as an ideal gas with $c_p = 1.005$ kJ/kg K and $\gamma = 1.4$. Neglect pressure drops in the combustion chamber and regenerator. Determine the required air flow and the fuel flow rates for a fuel heating value of 42 MJ/kg, and the power plant efficiency.

13.40 Show that for the Stirling cycle with all the processes occurring reversibly but where the heat rejected is not used for regenerative heating, the efficiency is given by

$$\eta = 1 - \frac{\left(\frac{T_1}{T_2} - 1\right) + (\gamma - 1)\ln r}{\left(\frac{T_1}{T_2} - 1\right) + (\gamma - 1)\frac{T_1}{T_2}\ln r}$$

where r is the compression ratio and T_1/T_2 the maximum to minimum temperature ratio.

Determine the efficiency of this cycle using hydrogen ($R = 4.307$ kJ/kg K, $c_p = 14.50$ kJ/kg K) with a pressure and temperature prior to isothermal compression of 1 bar and 300 K respectively, a maximum pressure of 2.55 MPa and heat supplied during the constant volume heating of 9300 kJ/kg. If the heat rejected during the constant volume cooling can be utilized to provide the constant volume heating, what will be the cycle efficiency? Without altering the temperature ratio, can the efficiency be further improved in the cycle?

13.41 Helium is used as the working fluid in an ideal Brayton cycle. Gas enters the compressor at 27ºC and 20 bar and is discharged at 60 bar. The gas is heated to 1000ºC before entering the turbine. The cooler returns the hot turbine exhaust to the temperature of the compressor inlet. Determine: (a) the temperatures at the end of compression and expansion, (b) the heat supplied, the heat rejected and the net work per kg of He, and (c) the cycle efficiency and the heat rate. Take $c_p = 5.1926$ kJ/kg K.

Ans. (a) 465.5, 820.2 K, (b) 4192.5, 2701.2, 1491.3 kJ/kg, (c) 0.3557, 10,121 kJ/kWh

13.42 In an air standard cycle for a gas turbine jet propulsion unit, the pressure and temperature entering the compressor are 100 kPa and 290 K, respectively. The pressure ratio across the compressor is 6 to 1 and the temperature at the turbine inlet is 1400 K. On leaving the turbine the air enters the nozzle and expands to 100 kPa. Assuming that the efficiency of the compressor and turbine are both 85% and that the nozzle efficiency is 95%, determine the pressure at the nozzle inlet and the velocity of the air leaving the nozzle. *Ans.* 285 kPa, 760 m/s

13.43 A stationary gas turbine power plant operates on the Brayton cycle and delivers 20 MW to an electric generator. The maximum temperature is 1200 K and the minimum temperatures is 290 K. The minimum pressure is 95 kPa and the maximum pressure is 380 kPa. If the isentropic efficiencies of the turbine and compressor are 0.85 and 0.80 respectively, find (a) the mass flow rate of air to the compressor, (b) the volume flow rate of air to the compressor, (c) the fraction of the turbine work output needed to drive the compressor, (d) the cycle efficiency.

If a regenerator of 75% effectiveness is added to the plant, what would be the changes in the cycle efficiency and the net work output? *Ans.* (a) 126.37 kg/s, (b) 110.71 m^3/s, (c) 0.528, (d) 0.2146, $\Delta\eta = 0.148$ $\Delta W_{net} = 0$

13.44 Air enters the compressor of a gas turbine operating on Brayton cycle at 1 bar, 27ºC. The pressure ratio in the cycle is 6. Calculate the maximum temperature in the cycle and the cycle efficiency. Assume $W_T = 2.5\ W_C$ and $\gamma = 1.4$. *Ans.* 1251.4 K, 40%

13.45 In an ideal jet propulsion cycle, air enters the compressor at 1 atm, 15°C. The pressure of air leaving the compressor is 5 atm and the maximum temperature is 870°C. The air expands in the turbine to such a pressure that the turbine work is equal to the compressor work. On leaving the turbine the air expands reversibly and adiabatically in a nozzle to 1 atm. Determine the velocity of air leaving the nozzle. *Ans.* 710.3 m/s

13.46 In a gas turbine the compressor is driven by the h.p. turbine. The exhaust from the h.p. turbine goes to a free-shaft l.p. turbine which runs the load. The air flow rate is 20 kg/s, and the minimum and maximum temperatures are respectively 300 K and 1000 K. The compressor pressure ratio is 4. Calculate the pressure ratio of the l.p. turbine and the temperature of the exhaust gases from the unit. The compressor and turbine are isentropic. Take c_p of air and exhaust gases as 1 kJ/kg K and $\gamma = 1.4$. *Ans.* 2.3, 673 K

13.47 A regenerative gas turbine with intercooling and reheat operates at steady state. Air enters the compressor at 100 kPa, 300 K with a mass flow rate of 5.807 kg/s. The pressure ratio across the two-stage compressor as well as the turbine is 10. The intercooler and reheater each operate at 300 kPa. At the inlets to the turbine stages, the temperature is 1400 K. The temperature at inlet to the second compressor stage is 300 K. The efficiency of each compressor and turbine stage is 80%. The regenerator effectiveness is 80%. Determine (a) the thermal efficiency, (b) the back work ratio, W_C/W_T, (c) the net power developed. *Ans.* (a) 0.443, (b) 0.454, (c) 2046 kW

13.48 In a regenerative gas turbine power plant air enters the compressor at 1 bar, 27ºC and is compressed to 4 bar. The isentropic efficiency of the compressor is 80% and the regenerator effectiveness is 90%. All of the power developed by the h.p. turbine is used to drive the compressor and the l.p. turbine provides the net power output of 97 kW. Each turbine has an isentropic efficiency of 87% and the temperature at inlet to the h.p. turbine is 1200 K. Determine (a) the mass flow rate of air into the compressor, (b) the thermal efficiency, (c) the temperature of the air at the exit of the regenerator. *Ans.* (a) 0.562 kg/s, (b) 0.432, (c) 523.2 K

CHAPTER 14

Refrigeration Cycles

14.1 REFRIGERATION BY NON-CYCLIC PROCESSES

Refrigeration is the cooling of a system below the temperature of its surroundings.

The melting of ice or snow was one of the earliest methods of refrigeration and is still employed. Ice melts at 0°C. So when ice is placed in a given space warmer than 0°C, heat flows into the ice and the space is cooled or refrigerated. The latent heat of fusion of ice is supplied from the surroundings, and the ice changes its state from solid to liquid.

Another medium of refrigeration is solid carbon dioxide or dry ice. At atmospheric pressure CO_2 cannot exist in a liquid state, and consequently, when solid CO_2 is exposed to atmosphere, it sublimates, i.e. it goes directly from solid to vapour, by absorbing the latent heat of sublimation (620 kJ/kg at 1 atm, −78.5°C from the surroundings (Fig. 14.1). Thus dry ice is suitable for low temperature refrigeration.

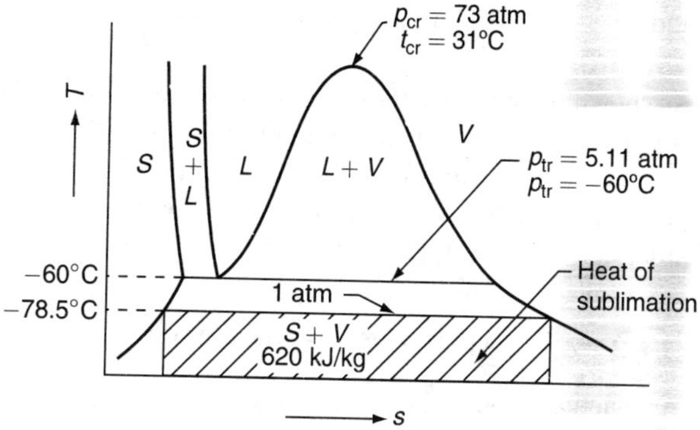

Fig. 14.1 *T-s diagram of CO_2*

In these two examples it is observed that the refrigeration effect has been accomplished by non-cyclic processes. Of greater importance, however, are the methods in which the cooling substance is not consumed and discarded, but used again and again in a thermodynamic cycle.

14.2 REVERSED HEAT ENGINE CYCLE

A reversed heat engine cycle, as explained in Sec. 6.12, is visualized as an engine operating in the reverse way, i.e. receiving heat from a low temperature region, discharging heat to a high temperature region, and receiving a net inflow of work (Fig. 14.2). Under such conditions the cycle is called a *heat pump cycle* or a *refrigeration cycle* (see Sec. 6.6). For a heat pump

$$(\text{COP})_{\text{H.P.}} = \frac{Q_1}{W} = \frac{Q_1}{Q_1 - Q_2}$$

and for a refrigerator

$$(\text{COP})_{\text{ref}} = \frac{Q_2}{W} = \frac{Q_2}{Q_1 - Q_2}$$

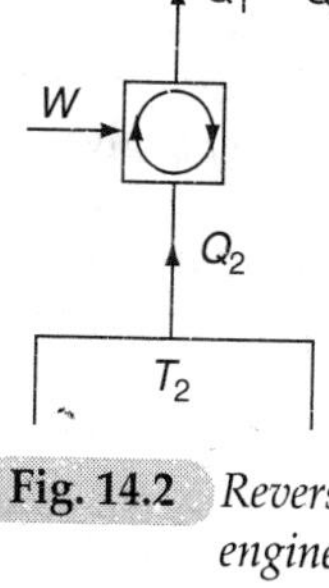

Fig. 14.2 *Reversed heat engine cycle*

The working fluid in a refrigeration cycle is called a *refrigerant.* In the reversed Carnot cycle (Fig. 14.3), the refrigerant is first compressed reversibly and adiabatically in process 1–2 where the work input per kg of refrigerant is W_c, then it is condensed reversibly in process 2–3 where the heat rejection is Q_1, the refrigerant then expands reversibly and adiabatically in process 3–4 where the work output is W_E, and finally it absorbs heat Q_2 reversibly by evaporation from the surroundings in process 4–1.

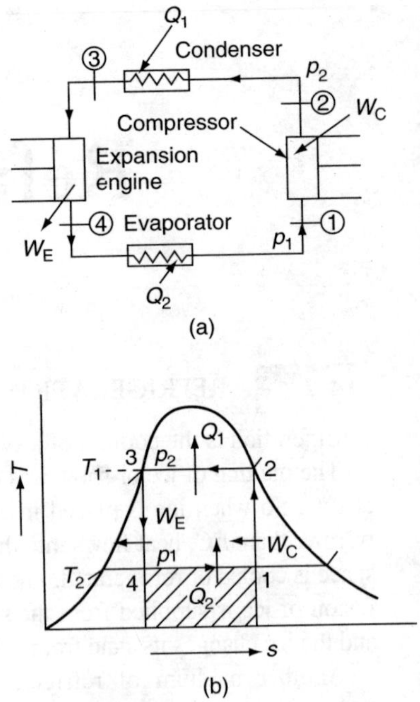

Fig. 14.3 *Reversed carnot cycle*

Here $$Q_1 = T_1(s_2 - s_3),\ Q_2 = T_2(s_1 - s_4)$$

and $$W_{net} = W_C - W_E = Q_1 - Q_2 = (T_1 - T_2)(s_1 - s_4)$$

where T_1 is the temperature of heat rejection and T_2 the temperature of heat absorption.

$$(\text{COP}_{\text{ref}})_{\text{rev}} = \frac{Q_2}{W_{net}} = \frac{T_2}{T_1 - T_2}$$

and $$(\text{COP}_{\text{H.P.}})_{\text{rev}} = \frac{Q_1}{W_{net}} = \frac{T_1}{T_1 - T_2} \tag{14.1}$$

As shown in Sec. 6.16, these are the maximum values for any refrigerator or heat pump operating between T_1 and T_2. It is important to note that for the same T_2 or T_1, the COP increases with the decrease in the temperature difference $(T_1 - T_2)$, i.e. the closer the temperatures T_1 and T_2, the higher the COP.

14.3 VAPOUR COMPRESSION REFRIGERATION CYCLE

In an actual vapour refrigeration cycle, an expansion engine, as shown in Fig. 14.3, is not used, since power recovery is small and does not justify the cost of the engine. A throttling valve or a capillary tube is used for expansion in reducing the pressure from p_2 to p_1. The basic operations involved in a vapour compression refrigeration plant are illustrated in the flow diagram, Fig. 14.4, and the property diagrams, Fig. 14.5.

The operations represented are as follows for an idealized plant:

1. *Compression* A reversible adiabatic process 1–2 or 1′–2′ either starting with saturated vapour (state 1), called *dry compression*, or starting with wet vapour (state 1′), called *wet compression.* Dry compression (1–2) is always preferred to wet compression (1′–2′), because with wet compression there is a danger of the liquid refrigerant being trapped in the head of the cylinder by the rising piston which may damage the valves or the cylinder head, and the droplets of liquid refrigerant may wash away the lubricating oil from the walls of the cylinder, thus accelerating wear.

It is, therefore, desirable to have compression with vapour initially dry saturated at 1, as shown in Fig. 14.5(b) or even slightly superheated. The state of the vapour at the end of dry compression will be at 2 at pressure p_2 which is the saturation pressure of the refrigerant corresponding to the condensing temperature t_1, instead of being at t_2'', for the Carnot cycle. The increased work of the cycle due to the dry compression instead of wet compression is given by the area 2–2′–2″, known as *superheat horn* (Fig. 14.5(b))

2. *Cooling and Condensing* A reversible constant pressure process, 2–3, first desuperheated and then condensed, ending with saturated liquid. Heat Q_1 is transferred out.

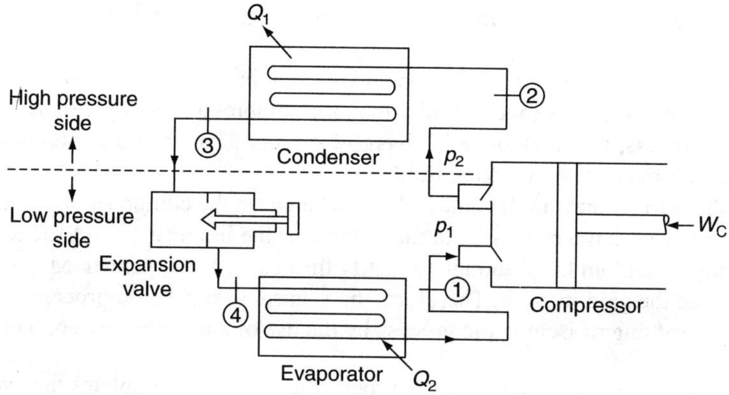

Fig. 14.4 *Vapour compression refrigeration plant-flow diagram*

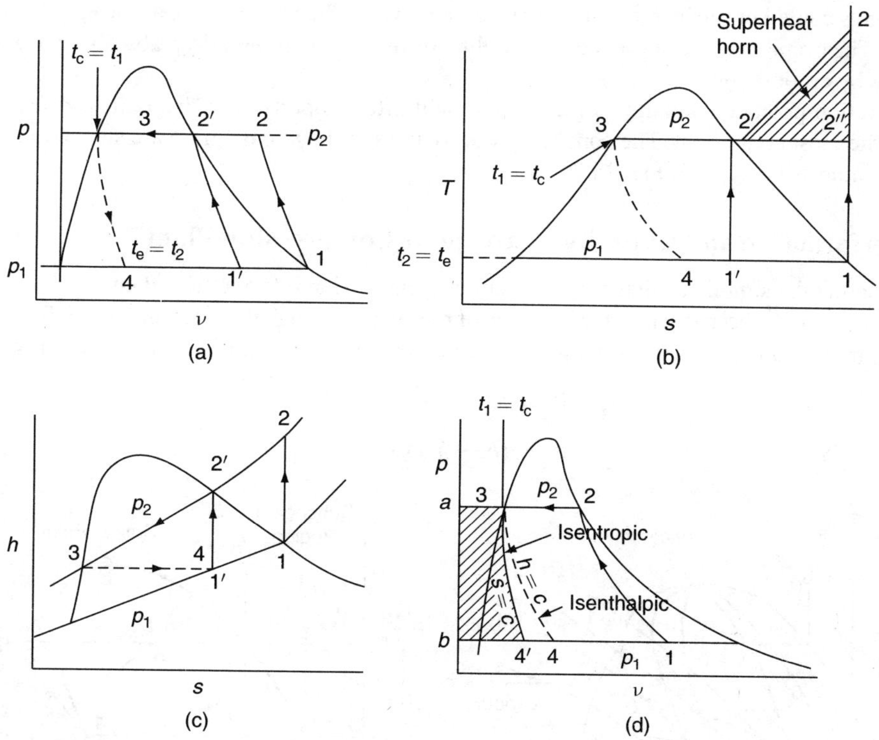

Fig. 14.5 *Vapour compression refrigeration cycle: Throttling vs Isentropic expansion*

3. *Expansion* An adiabatic throttling process 3–4, for which enthalpy remains unchanged. States 3 and 4 are equilibrium points. Process 3–4 is adiabatic (then only $h_3 = h_4$ by S.F.E.E.), but not isentropic.

$$T\mathrm{d}s = \mathrm{d}h - v\mathrm{d}p, \quad \text{or} \quad s_4 - s_3 = -\int_{p_1}^{p_2} \frac{v\mathrm{d}p}{T}$$

Hence it is irreversible and cannot be shown in property diagrams. States 3 and 4 have simply been joined by a dotted line.

The positive work recovered during the isentropic expansion in the expansion engine is given by the area $3 - a - b - 4'$ (Fig. 14.5(d)), which is much smaller than the compressor work given by the area $1 - 2 - a - b - 1$. In a steady flow process, the work done is given by $-\int v dp$. For the same pressure difference dp, the work depends on the volume v of the working fluid.

In the expander, the refrigerant is in the liquid phase, whereas in the compressor, it is in the gas phase. The volume of the vapour is very large compared to the volume of the liquid($v_g >> v_f$). Hence, the positive work of isentropic expansion is seldom large enough to justify the cost of the expander. Moreover, the friction and other losses may exceed the gain in work. Therefore, the isentropic expansion process of the Carnot cycle is replaced by a simple throttling or isenthalpic process by the use of a throttle valve or a capillary tube.

4. *Evaporation* A constant pressure reversible process, 4–1, which completes the cycle. The refrigerant is throttled by the expansion valve to a pressure, the saturation temperature at this pressure being below the temperature of the surroundings. Heat then flows, by virtue of temperature difference, from the surroundings, which gets cooled or refrigerated, to the refrigerant, which then evaporates, absorbing the latent heat of evaporation. The evaporator thus produces the cooling or the *refrigerating effect,* absorbing heat Q_2 from the surroundings by evaporation.

In refrigeration practice, enthalpy is the most sought-after property. The diagram in p–h coordinates is found to be the most convenient. The constant property lines in the p–h diagram are shown in Fig. 14.6, and the vapour compression cycle in Fig. 14.7.

14.3.1 Performance and Capacity of a Vapour Compression Plant

Figure 14.8 shows the simplified diagram of a vapour compression refrigeration plant.

When steady state has been reached, for 1 kg flow of refrigerant through the cycle, the steady flow energy equations (neglecting K.E. and P.E. changes) may be written for each of the components in the cycle as given below.

Compressor $$h_1 + W_c = h_2$$

$\therefore$ $$W_c = (h_2 - h_1) \text{ kJ/kg}$$

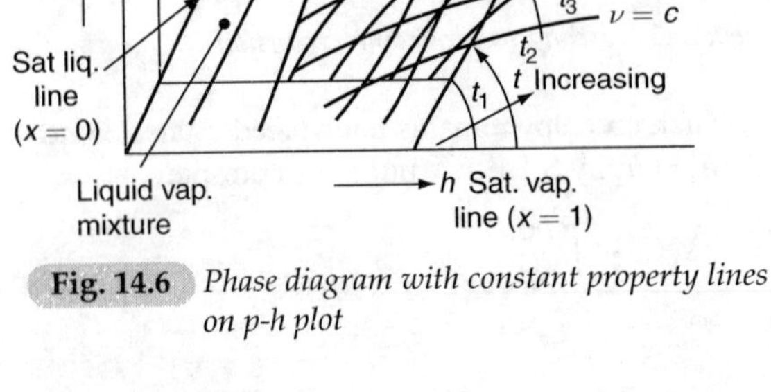

Fig. 14.6 *Phase diagram with constant property lines on p-h plot*

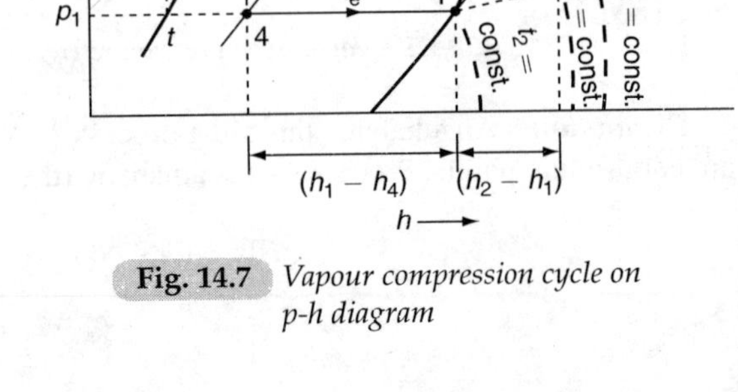

Fig. 14.7 *Vapour compression cycle on p-h diagram*

Condenser $h_2 = Q_1 + h_3$

$\therefore$ $Q_1 = (h_2 - h_3)$ kJ/kg

Expansion valve $h_3 = h_4$

or $(h_f)p_1 = (h_f)p_2 + x_4 (h_{fg})p_2$

$$\therefore \quad x_4 = \frac{(h_f)\,p_1 - (h_f)\,p_2}{(h_{fg})\,p_2}$$

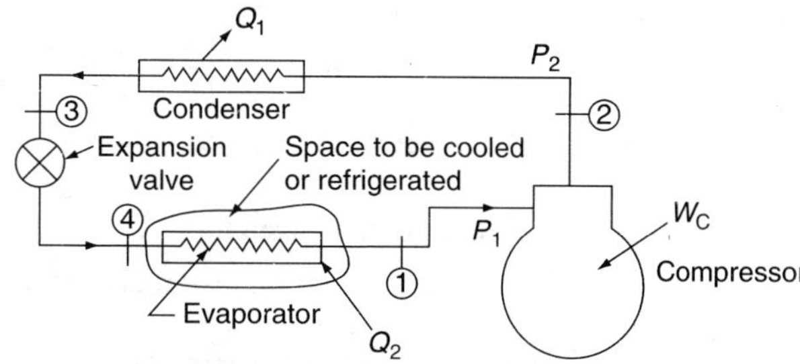

Fig. 14.8 *Vapour compression plant*

This is the quality of the refrigerant at the inlet to the evaporator (mass fraction of vapour in liquid-vapour mixture).

Evaporator $h_4 = Q_2 = h_1$

$\therefore$ $Q_2 = (h_1 - h_4)$ kJ/kg

This is known as the *refrigerating effect,* the amount of heat removed from the surroundings per unit mass flow of refrigerant.

If the *p-h* chart for a particular refrigerant is available with the given parameters, it is possible to obtain from the chart the values of enthalpy at all the cardinal points of the cycle. Then for the cycle

$$\text{COP} = \frac{Q_2}{W_c} = \frac{h_1 - h_4}{h_2 - h_1} \tag{14.2}$$

If w is the mass flow of refrigerant in kg/s, then the rate of heat removal from the surroundings

$$= w(h_1 - h_4) \text{ kJ/s} = w(h_1 - h_4) \times 3600 \text{ kJ/h}$$

One tonne of refrigeration is defined as the rate of heat removal from the surroundings equivalent to the heat required for melting 1 tonne of ice in one day. If the latent heat of fusion of ice is taken as 336 kJ/kg, then 1 tonne is equivalent to heat removal at the rate of $(1000 \times 336)/24$ or 14,000 kJ/h

$\therefore$ Capacity of the refrigerating plant

$$\frac{w(h_1 - h_4) \times 3600}{14{,}000} \text{ tonnes}$$

One ton or ton or tonne of refrigeration is SI units is often taken approximately equivalent to the heat removal rate of 3.5 kW or 210 kJ/min or 12,600 kJ/h.

The rate of heat removal in the condenser

$$Q_1 = w(h_2 - h_3) \text{ kJ/s}$$

If the condenser is water-cooled, $\dot{m}_c$ the flow-rate of cooling water in kg/s, and $(t_{c2} - t_{c1})$ the rise in temperature of water, then

$$Q_1 = w(h_2 - h_3) = \dot{m}_c c_c (t_{c2} - t_{c1}) \text{ kJ/s}$$

provided the heat transfer is confined only between the refrigerant and water, and there is no heat interaction with the surroundings.

The rate of work input to the compressor

$$W_c = w(h_2 - h_1) \text{ kJ/s}$$

Expressing the power consumption per ton of refrigeration as *unit power consumption*, we have mass flow rate and power consumption per ton refrigeration.

$$\omega = \frac{3.5}{Q_2} \text{ kg/(SATR)}$$

$$W_e = 3.5\left(\frac{h_2 - h_1}{h_1 - h_4}\right)\text{kW/TR} = \frac{3.5}{\text{COP}}\text{kW/TR}$$

Theoretical piston displacement of the compressor or suction volume flow rate

$$\dot{V} = \omega \upsilon_1 \text{ m}^3/\text{s}$$

where υ_1 is the specific volume at suction:

$$\text{Actual piston displacement} = \frac{\omega \upsilon_1}{\eta_{vol}} \text{ m}^3/\text{s}$$

where η_{vol} is the volumetric efficiency, defined later. Suction volume requirement per ton is

$$\dot{V} = \frac{210}{Q_2}\upsilon_1 \text{ m}^3/(\text{min} \times \text{TR})$$

The isentropic discharge temperature t_2 can be found by one of the following three methods:

(i) Graphically, from the p–h diagram of the refrigerant by drawing the isentropic line from point 1 (at pressure p_1) to pressure p_2 or by iteration, finding t_2 corresponding to $s_2 = s_1$.

(ii) Using saturation properties and the specific heat of vapour c_p,

$$s_1 = s_2 = s_2' + c_p \ln \frac{T_2}{T_2'}$$

(iii) Using superheat tables and interpolating for the degree of superheat $(T_2 - T_2')$ corresponding to the entropy difference $(s_2 - s_2')$ which is known.

14.3.2 Actual Vapour Compression Cycle

In order to ascertain that there is no droplet of liquid refrigerant being carried over into the compressor, some superheating of vapour is recommended after the evaporator.

A small degree of subcooling of the liquid refrigerant after the condenser is also used to reduce the mass of vapour formed during expansion, so that too many vapour bubbles do not impede the flow of liquid refrigerant through the expansion valve.

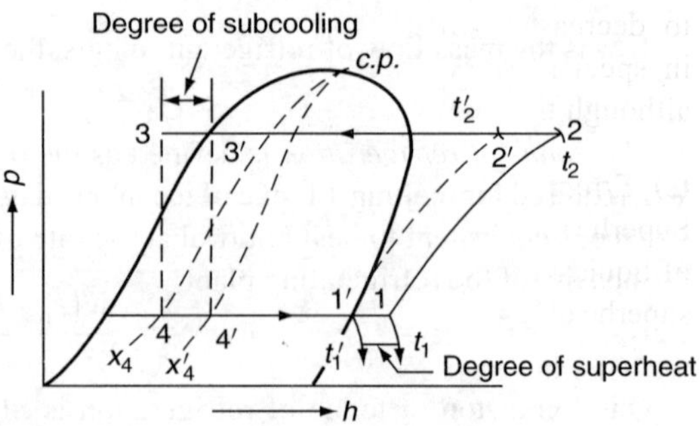

Fig. 14.9 *Superheat and subcooling in a vapour compression cycle*

Both the superheating of vapour at the evaporator outlet and the subcooling of liquid at the condenser outlet contribute to an increase in the refrigerating effect, as shown in Fig. 14.9. The compressor discharge temperature, however, increases, due to superheat, from t_2' to t_2, and the load on the condenser also increases.

Sometimes, a liquid-line heat exchanger is used in the plant, as shown in Fig. 14.10. The liquid is subcooled in the heat exchanger, reducing the load on the condenser and improving the COP. For 1 kg flow

$$Q_2 = h_6 - h_5,\ Q_1 = h_2 - h_3$$

$$W_c = h_2 - h_1$$

and

$$h_1 - h_6 = h_3 - h_4$$

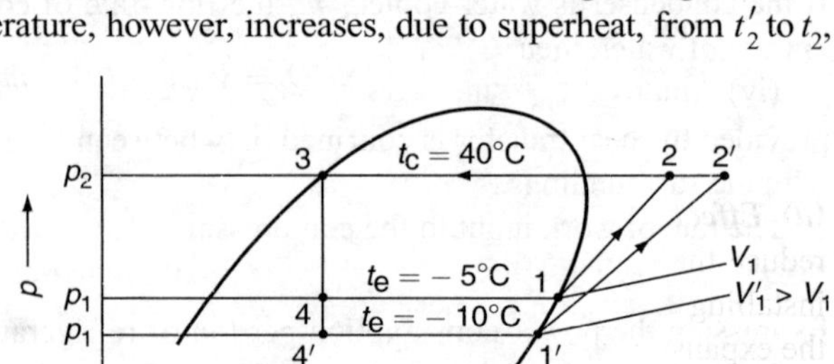

Fig. 14.10 *Effect of evaporator pressure*

14.3.3 Effect of Change in Operating Conditions on the Performance of Vapour Compression Cycle

(a) Effect of Evaporator Pressure For a single stage, freon compressor, let $t_{ev} = -5°C$, $t_{cond} = 40°C$. Let there be a decrease of evaporator temperature to –10°C. (Fig. 14.10). It is found that there is

(i) a decrease in refrigerating effect from $(h_1 - h_4)$ to $(h_1' - h_4')$

(ii) an increase in specific volume of suction vapour from v_1 to v_1' $(v_1' > v_1)$.

(iii) a decrease in volumetric efficiency, due to increase in the pressure ratio

(iv) an increase in compressor work from $(h_2 - h_1)$ to $(h_2' - h_1')$.

(b) Effect of Condenser Pressure An increase in condensing pressure results in a decrease of the refrigerating capacity and an increase in power consumption, as seen from the cycle 1–2–3–4 changed to cycle 1–2′–3′–4′ for $t_c' = 45°C$ from $t_c = 40°C$, as seen in Fig. 14.11. The decrease in refrigerating capacity is due to the decrease in refrigerating effect and volumetric efficiency. The increase in power consumption is due to the increased mass flow due to decreased refrigerating effect and an increase in specific work due to increased pressure ratio, although the isentropic line remains the same.

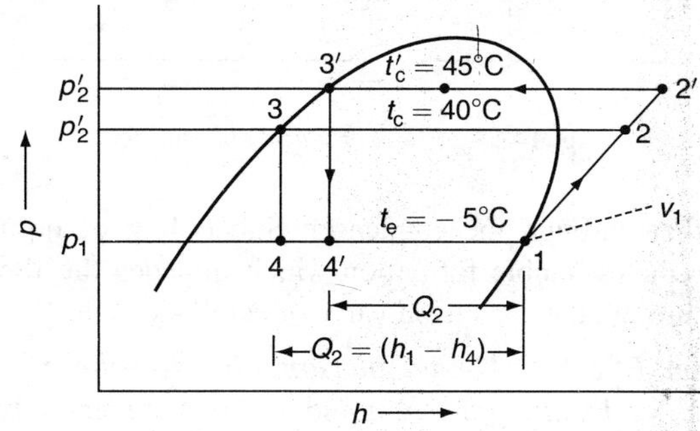

Fig. 14 1 *Effect of condenser pressure*

(c) Effect of Suction Vapour Superheat Superheating of suction vapour is useful, as mentioned earlier, because it ensures complete vaporization of liquid in the evaporator before it enters the compressor. It can be seen from Fig. 14.12, that the effect of superheating of the vapour from $t_1 = t_e$ to t_1' is as follows:

(i) increase in specific volume of suction vapour from v_1 to v_1'

(ii) increase in refrigerating effect from $(h_1 - h_4)$ to $(h_1' - h_4)$.

(iii) increase in specific work from $(h_2 - h_1)$ to $(h_2' - h_1')$.

(iv) increase in condenser load from $(h_2 - h_3)$ to $(h_2' - h_3)$.

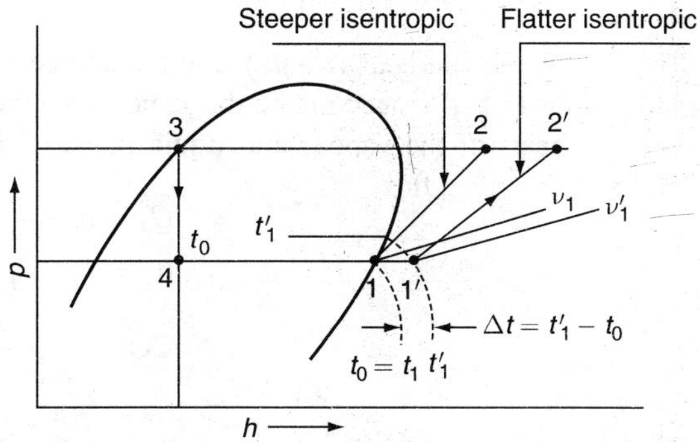

Fig. 14.12 *Effect of suction vapour superheat*

(d) Effect of Liquid Subcooling It is possible to reduce the temperature of the liquid refrigerant by installing a *subcooler* between the condenser and the expansion valve. The effect of subcooling of the liquid from $t_3 = t_4$ to t_3' (Fig. 14.13), it is seen that subcooling reduces flashing of the liquid during expansion $(x_4' < x_4)$ and increases the refrigerating effect. Cooling water usually first passes through the subcooler and

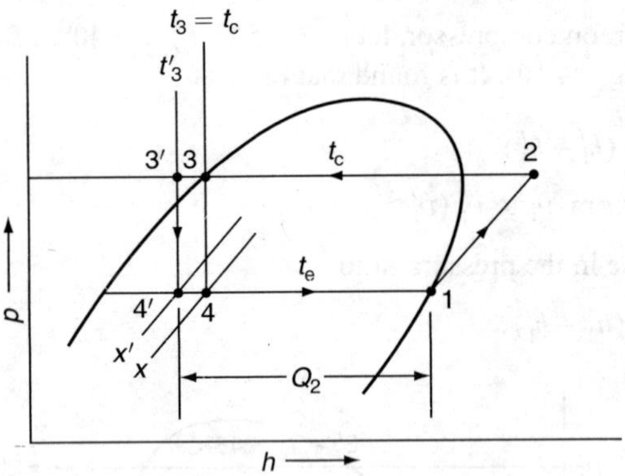

Fig. 14.13 *Effect of liquid subcooling*

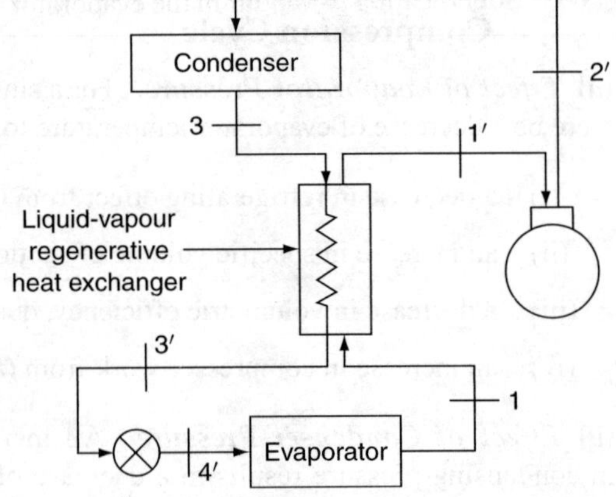

Fig. 14.14 *Effect of using suction line regenerative heat exchange*

then through the condenser. Subcooling of liquid reduces bubble formation which impedes the flow through the expansion valve or capillary tube.

(e) Effect of Using Suction Line Regenerative Heat Exchanger A liquid – vapour regenerative heat exchanger may be installed as shown in Fig. 14.14. In this, the refrigerant from the evaporator is superheated in the regenerator with consequent sub-cooling of the liquid from the condenser (Fig. 14.15). Since the mass flow rate of the liquid and vapour is the same, by energy balance we have

$$h'_1 - h_1 = h_3 - h'_3$$

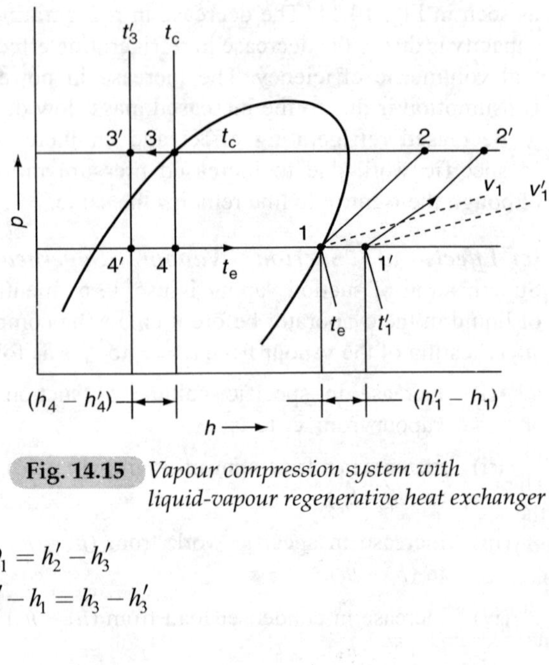

Fig. 14.15 *Vapour compression system with liquid-vapour regenerative heat exchanger*

The degree of superheat $(t'_1 - t_e)$ and the degree of subcooling $(t_c - t'_3)$ need not be the same, since the specific heats of the vapour and liquid phases are different. For 1-kg flow,

$$Q_2 = h_1 - h_4, \quad Q_1 = h'_2 - h'_3$$

$$W_c = h'_2 - h'_1, \quad h'_1 - h_1 = h_3 - h'_3$$

$$\text{COP} = \frac{h_1 - h'_4}{h'_2 - h'_1}.$$

14.3.4 Actual Vapour Compression Cycle

As the refrigerant flows through the condenser, evaporator and piping, there will be drops in pressure. There will also be heat losses or gains depending on the temperature difference between the refrigerant and the surroundings. Further, compression will be polytropic with friction and heat transfer instead of isentropic. The actual vapour compression cycle may have some or all of the items of departure from the simple saturation cycle as enlisted below and shown on the *p-h* diagram as shown in Fig. 14.16.

(i) Superheating of vapour in the evaporator, $1d - 1c$
(ii) Heat gain and superheating of the vapour in the suction line, $1c - 1b$
(iii) Pressure drop in the suction line, $1b - 1a$
(iv) Pressure drop due to throttling in the compressor–suction valve, $1a - 1$
(v) Polytropic compression, $1 - 2$
(vi) Pressure drop at the compressor–discharge valve, $2 - 2a$
(vii) Pressure drop in the delivery line, $2a - 2b$
(viii) Heat loss and desuperheating of vapour in the delivery line, $2b - 2c$
(ix) Pressure drop in the condenser, $2b - 3$
(x) Subcooling of the liquid in the condenser, $3 - 3a$
(xi) Heat gain in the liquid line, $3a - 3b$
(xii) Pressure drop in the evaporator, $4 - 1d$

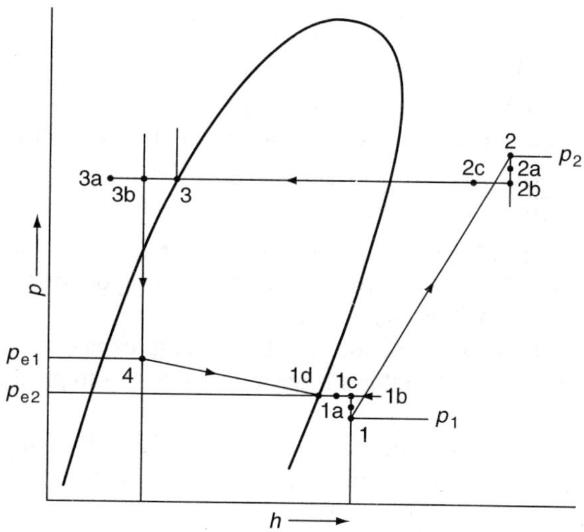

Fig. 14.16 *Actual vapour compression cycle on p-h diagram*

It may be observed that the pressure drop in the evaporator is large, while that in the condenser is not significant.

14.3.5 Components in a Vapour Compression Plant

(a) Condenser It is in the condenser that heat is rejected in a vapour compression plant. Desuperheating of the vapour takes place in the discharge line and in the first few coils of the condenser. It is followed by the condensation of the vapour at the saturation temperature. Then, subcooling may take place near the bottom. However, the sensible heat of the desuperheating and subcooling processes is quite small compared to the latent heat of condensation.

The type of condenser is generally characterized by the cooling medium used. There are three types of condensers: (i) air-cooled condensers, (ii) water-cooled condensers, and (iii) evaporative condensers. An air-cooled condenser is used in small self-contained units below 5 TR because of high power consumption and large fan noise.

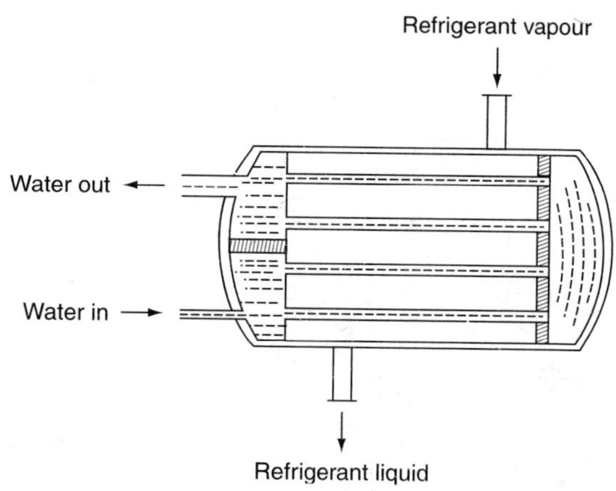

Fig. 14.17 *Schematic representation of a two-pass water-cooled shell and tube condenser*

Water-cooled condensers can be of three types, viz., shell and tube, shell and coil and double tube. The shell-and-tube type with water flowing through tubes and the refrigerant condensing in the shell is the most commonly used condenser (Fig. 14.17).

(b) Expansion Device An expansion device in a refrigeration system normally serves two functions, viz., (i) it reduces the pressure from the condenser to the evaporator, and (ii) it regulates the flow of the refrigerant to the evaporator depending on the load. It is essentially a restriction offering resistance to flow so that the pressure drops, resulting in a throttling process. There are two types of such devices, viz.,

(i) Variable restriction type
(ii) Constant restriction type

In the variable restriction type, the extent of opening or area of flow keeps on changing depending on the control required. There are two types of such control devices:

(i) Automatic expansion valve
(ii) Thermostatic expansion valve

The constant restriction type device is the *capillary tube* which is a long narrow tube of small diameter. Once the size and length are fixed, the evaporator pressure, etc., gets fixed. No modification in the operating conditions is possible.

The basic function of an automatic expansion valve (A.E.V.)* is to maintain constant pressure in the evaporator. A bellows is acted on by forces to push a needle to open or close the orifice with the variation of cooling load.

A thermostatic expansion valve, (T.E.V.) maintains a constant degree of superheat in the evaporator. A sensing bulb at the evaporator outlet maintains the constant temperature of the refrigerant entering the compressor (Fig. 14.18).

(c) Compressor Compressors may be of three types: (1) reciprocating, (b) rotary, and (c) centrifugal. When the volume flow rate of the refrigerant is large, centrifugal compressors are used. Rotary compressors are used for small units. Reciprocating compressors are used in plants up to 100 tonnes capacity. For plants of higher capacities, centrifugal compressors are employed.

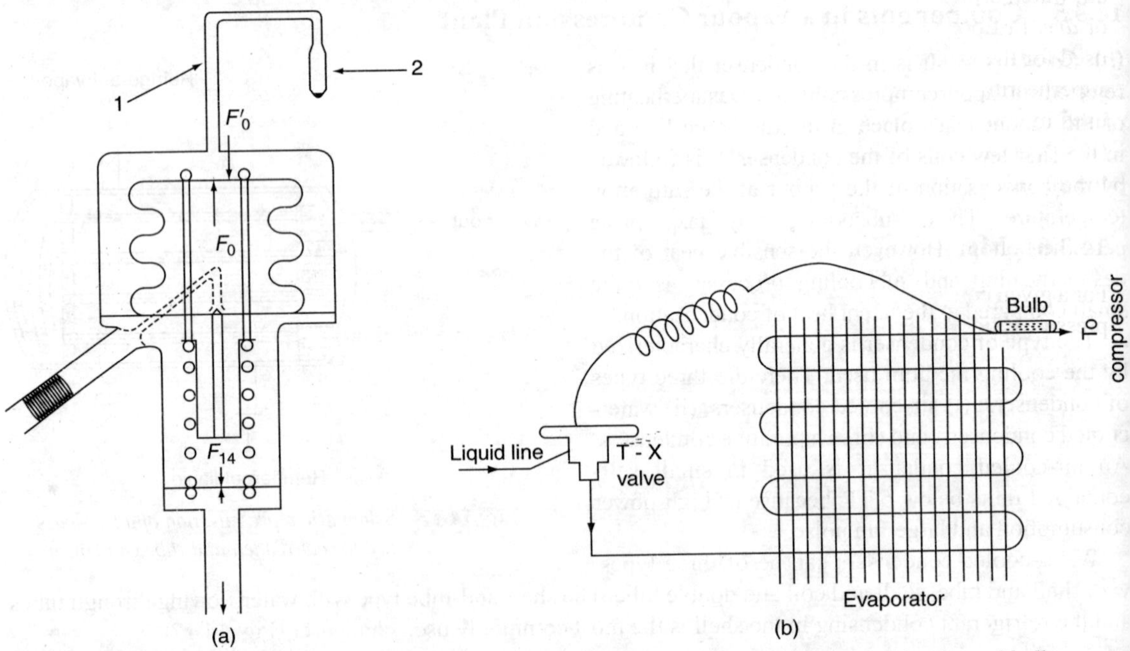

Fig. 14.18 *Arrangement showing installation of thermostatic expansion valve and its thermal bulb*

*To understand the mechanisms and functions of the A.E.V. and T.E.V. See *Refrigeration and Air Conditioning* by C.P. Arora. The T.E.V. meets the varying load requirements better than the A.E.V.

In reciprocating compressors, which may be single-cylinder or multi-cylinder ones, because of clearance, leakage past the piston and valves, and throttling effects at the suction and discharge valves, the actual volume of gas drawn into the cylinder is less than the volume displaced by the piston. This is accounted for in the term *volumetric efficiency,* which is defined as

$$\eta_{\text{vol}} = \frac{\text{Actual volume of gas drawn at evaporator pressure and temperature}}{\text{Piston displacement}}$$

∴ Volume of gas handled by the compressor

$$= w \cdot v_1 (\text{m}^3/\text{s}) = \left(\frac{\pi}{4} D^2 L \frac{N}{60} n\right) \times \eta_{\text{vol}}$$

where w is the refrigerant flow rate,

v_1 is the specific volume of the refrigerant at the compressor inlet,

D and L are the diameter and stroke of the compressor,

n is the number of cylinders in the compressor, and

N is the r.p.m.

The clearance volumetric efficiency is given by Eq. (18.13)

$$\eta_{\text{vol}} = 1 + C - C\left(\frac{p_2}{p_1}\right)^{1/n}$$

where C is the clearance. For details see the Chapter 19.

(d) Evaporator The evaporator is the component of a refrigeration system in which heat is removed from air, water, brine, etc, required to be cooled by the evaporating refrigerant. Evaporators are classified as *flooded* or *dry*. In flooded evaporators, the liquid refrigerant covers the entire heat transfer surface. A float valve is used for the expansion of the refrigerant. In dry evaporators, a part of the heat transfer surface is used for the superheating of vapour. A thermostatic expansion valve for large units or a capillary tube for small units is used in conjunction with a dry evaporator. The refrigerant flows inside the tubes in dry or direct expansion (D – X) evaporators, whereas it flows outside the tubes in flooded evaporators. One D – X coil with fins on the air side is shown in Fig. 14.19 with 4 rows of tubing and a T.E.V.

14.3.6 Multi-Stage Vapour Compression Systems

For a given condensation temperature, the lower the evaporator temperature, the higher becomes the compressor pressure ratio. For a reciprocating compressor, a high pressure ratio across a single stage means low volumetric

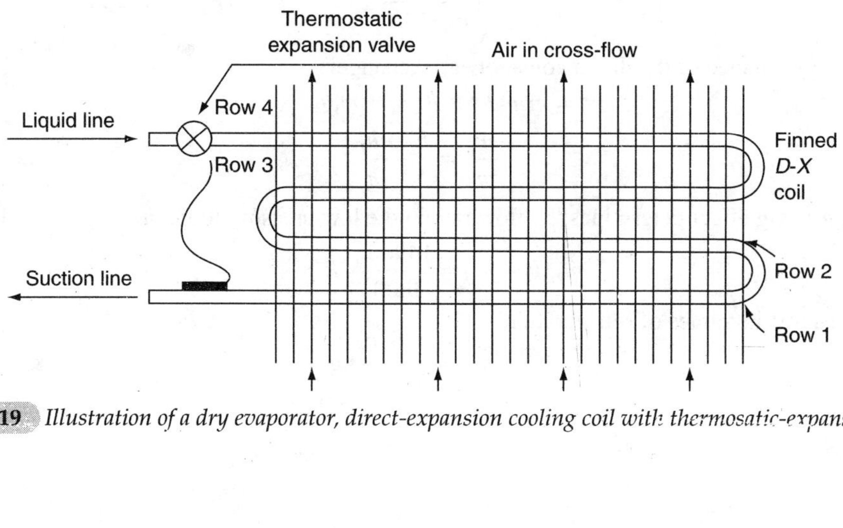

Fig. 14.19 *Illustration of a dry evaporator, direct-expansion cooling coil with thermosatic-expansion valve*

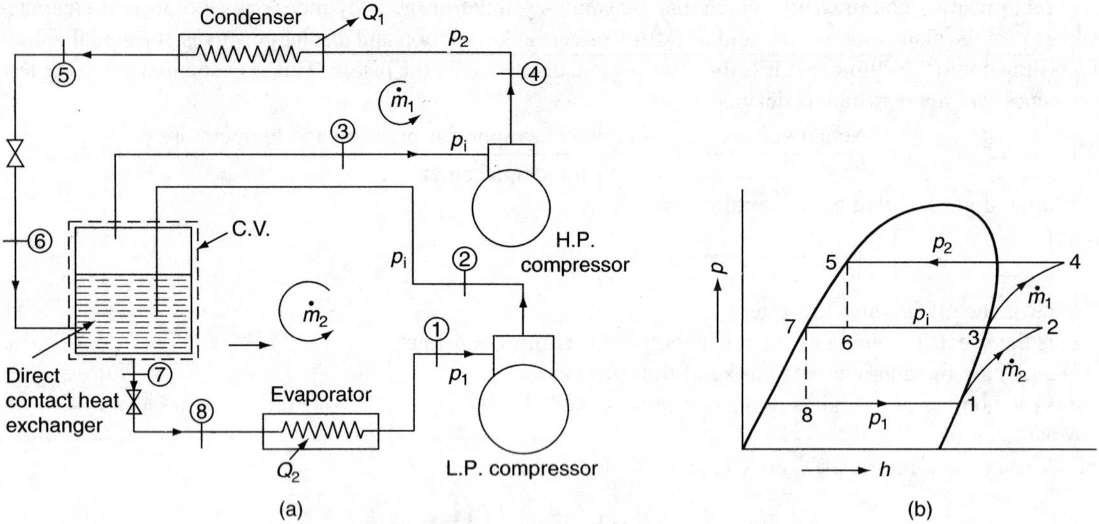

Fig. 14.20 *Two-stage vapour compression system*

efficiency. Also, with dry compression the high pressure ratio results in high compressor discharge temperature which may damage the refrigerant. To reduce the work of compression and improve the COP, multistage compression with intercooling may be adopted. Since the intercooler temperature may be below the temperature of available cooling water used for the condenser, the refrigerant itself may be used as the intercooling medium. Figure 14.20 shows a two-stage compression system with a direct contact heat exchanger.

As shown later in Sec. 14.5, for minimum work, the intercooler pressure p_i is the geometric mean of the evaporator and condenser pressures, p_1 and p_2, or

$$p_i = \sqrt{p_1 \cdot p_2}$$

However, this condition is true for complete intercooling to the initial temperature. In refrigeration systems, it is not possible. It is seen that the discharge temperature of the low-stage is much lower than that of the high stage. To reduce the discharge temperature of the higher stage, therefore the pressure ratio of the lower stage can be increased, with a corresponding reduction in the pressure ratio of the higher stage. The expression recommended for determining the intermediate pressure in refrigeration systems is

$$P_i = \sqrt{P_c P_e^{T_c/T_e}}.$$

By making an energy balance of the direct contact heat exchanger,

$$\dot{m}_2 h_2 + \dot{m}_1 h_6 = \dot{m}_2 h_7 + \dot{m}_1 h_3$$

$$\therefore \quad \frac{\dot{m}_1}{\dot{m}_2} = \frac{h_2 - h_7}{h_3 - h_6}$$

The desired refrigerating effect determines the flow rate in the low pressure loop, $\dot{m}_2$, as given below

$$\dot{m}_2 (h_1 - h_8) = \frac{14000}{3600} \times P$$

where P is the capacity, in tonnes of refrigeration.

$$\dot{m}_2 = \frac{3.89P}{h_1 - h_8} \text{ kg/s}$$

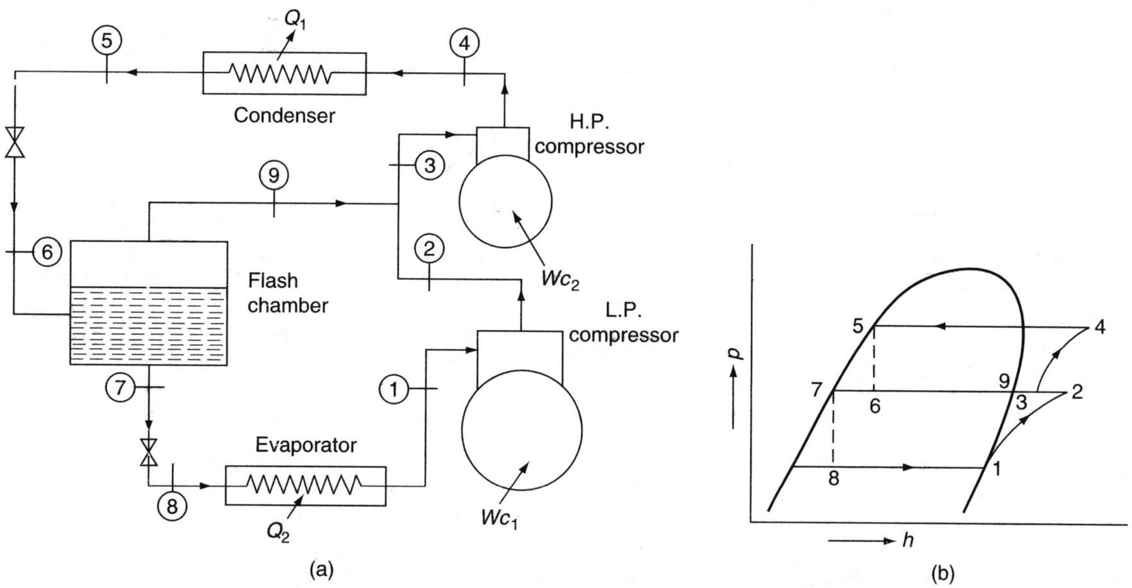

Fig. 14.21 *Two-stage vapour compression system with a flash intercooler*

Figure 14.21 shows a two-stage vapour compression system with a flash chamber intercooler, where the vapour from the flash chamber (state 9) mixes with the vapour from the LP compressor (state 2) to form vapour at state 3, which enters the HP compressor.

14.3.7 Multi-Evaporator Systems

A situation often arises when varied types of cooling loads are connected to the same refrigeration system. Each load may need an evaporator working at a different refrigeration temperature. A two-evaporator single-compressor system with individual expansion valves for each evaporator and one compressor is shown in Fig. 14.22(a) and the corresponding *p-h* diagram is shown is Fig. 14.22(b) The mass flow rates through evaporators 1 and 2 are

$$m_1 = \frac{Q_{e1}}{h_7 - h_5} \quad \text{and} \quad m_2 = \frac{Q_{e2}}{h_6 - h_4}$$

The enthalpy of liquid–vapour mixture entering the compressor is

$$h_1 = \frac{m_1 h_7 + m_2 h_6}{m_1 + m_2}$$

and the work input is

$$W = (m_1 + m_2)(h_2 - h_1)$$

14.3.8 Cascade Systems

Two important parameters that need to be considered in the selection of a refrigerant are the temperatures of the two media (the refrigerated space and the environment) with which the refrigerant exchanges heat. To have reasonable heat transfer rate, a temperature difference of 5°C to 10°C should be maintained between the refrigerant and the medium. If a space is to be maintained at –10°C e.g., the refrigerant should evaporate at –20°C (Fig. 14. 23), the saturation pressure at which must be above the atmospheric pres-

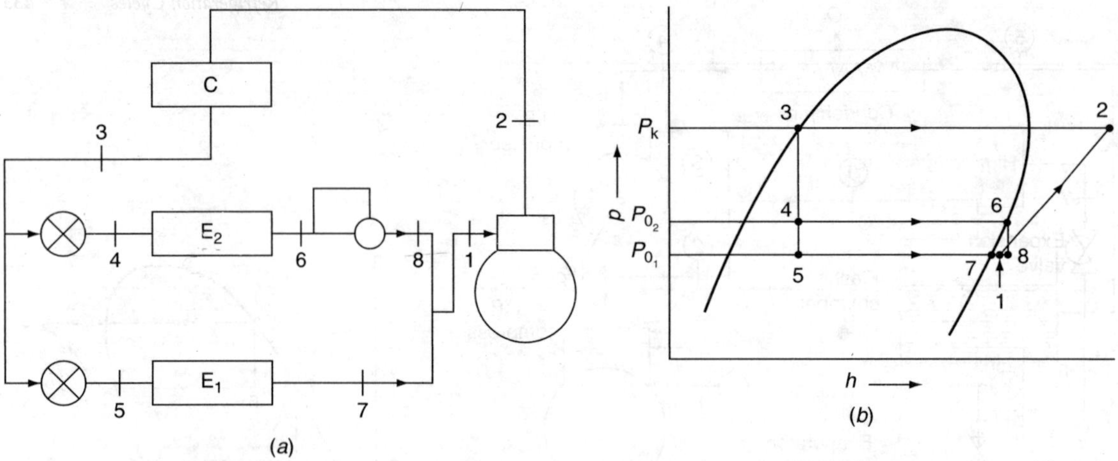

Fig. 14.22 *(a) System with two evaporators and single compressor, with individual expansion valves (b) Thermodynamic cycle for the system of Fig. 14.22(a)*

sure to prevent any air leakage into the system. Again, the temperature of the refrigerant in the condenser should be above the cooling medium by about 10°C, as shown in the figure, the saturation pressure at which must be below the critical pressure of the refrigerant. If a single refrigerant cannot meet the temperature requirements in the range from –20°C to 50°C, two cycles with two different refrigerants can be used in series, as shown in Fig. 14.24. Such a coupled cycle makes a *cascade refrigeration system*. Two independent vapour compression systems combine together in such a way that the evaporator of the high-temperature system becomes the condenser of the low-temperature system. But the working media of the two systems are separated from each other. The intermediate heat exchanger is also called a *cascade heat exchanger or condenser*. There are many industrial and medical applications where very low temperatures are required such as liquefaction of petroleum vapours, liquefaction of gases, dry ice manufacture etc. The blood storage needs as low as –80°C and precipitation hardening of special alloy steels needs around –90°C. At such low temperatures, the saturation pressures are very low prompting air leakage and the piston displacements become very large. For low-temperature side, R-13 can be used for which at atmospheric pressure the saturation temperature is below –80°C and for the high temperature side, R-12 or R-22 can be used.

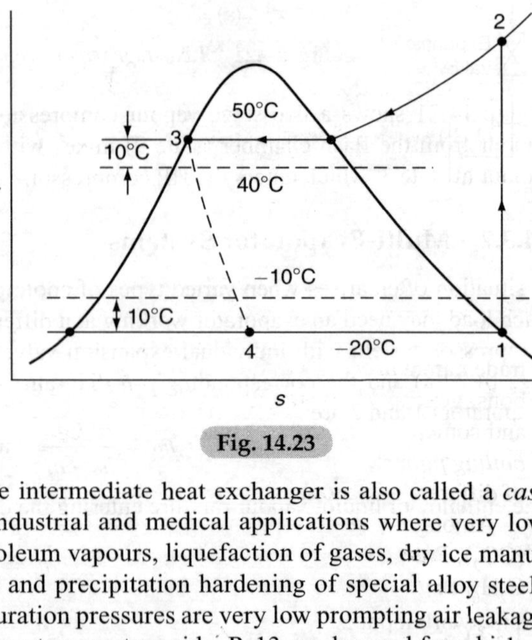

Fig. 14.23

14.3.9 Refrigerants

In the earlier days, ethyl chloride (C_2H_5Cl), sulphur dioxide (SO_2), methylchloride (CH_3Cl), and carbon dioxide (CO_2), etc., were used as refrigerants. A great breakthrough occurred with the development of freons (a

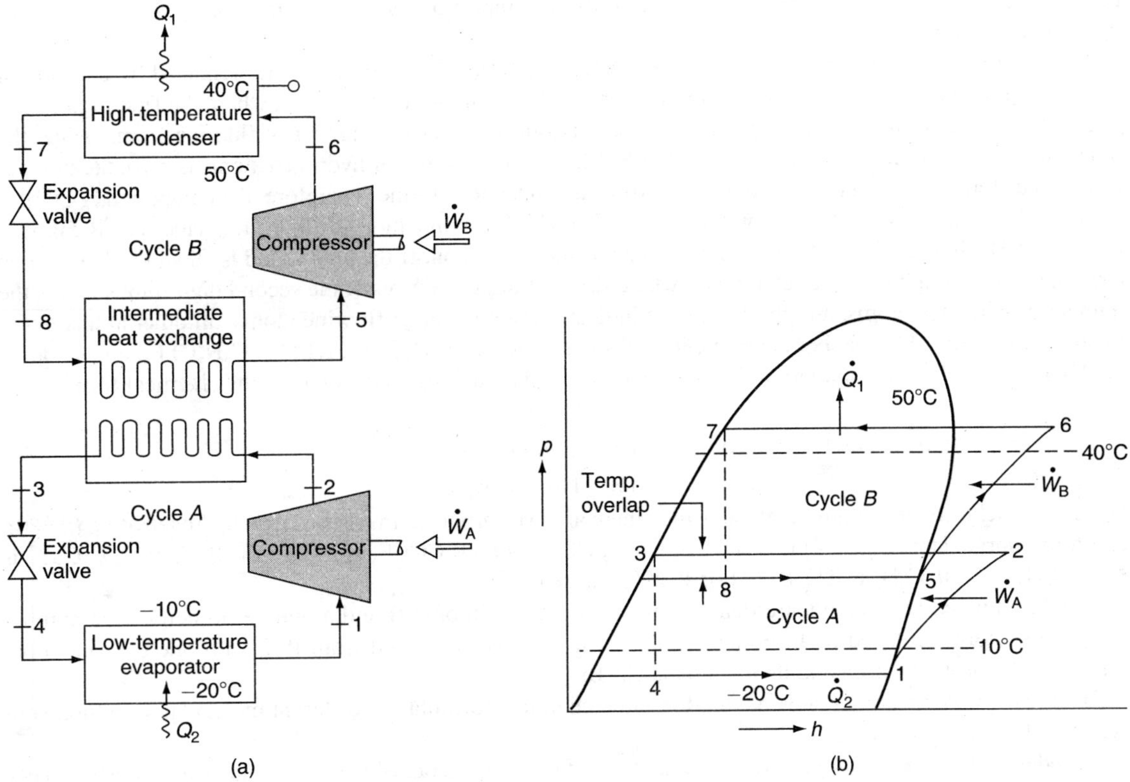

Fig. 14.24 *Cascade refrigeration cycle*

trade name) in 1930s by E.I. du Pont de Nemours and Co, USA. Freons are a series of fluorinated hydrocarbons, generally known as *fluorocarbons*, derived from methane and ethane as bases. With fluorine, chlorine and sometimes bromine in their molecule, these form a series of refrigerants with a wide range of *normal boiling points* (NBP) (saturation temperature at atmospheric pressure), which satisfy the varied requirements of different refrigerators.

Among the most common inorganic refrigerants are ammonia (NH_3), water (H_2O), and carbon dioxide (CO_2). Presently, the most commonly used organic refrigerants are the chloro-fluoro derivatives of CH_4 and C_2H_6, as given below:

1. *CFCs*: Fully halogenated ones with chlorine in their molecule are *chloro-fluoro carbons*, referred to as CFCs.

2. *HCFCs*: These contain H atoms in the molecule along with Cl and F, called *hydro-chloro-fluoro carbons*.

3. *HCs*: These have no chlorine atoms in the molecule, called *hydrocarbons*. Thus, we have HCs, HFCs, HCFCs and CFCs.

The *F-atom* in the molecule of these substances makes the substances physically more favourable. The *Cl-atom* in the molecule is considered responsible for the depletion of the ozone layer in the upper atmosphere (stratosphere), thus allowing harmful ultra-violet radiation to penetrate through the atmosphere and reach the earth's surface. The *H-atoms* in the molecule impart a degree of flammability to the substance depending upon the number of these atoms.

Freons are either methane-based or ethane-based, where the hydrogen atoms are replaced by chlorine or fluorine atoms. Methane-based compounds are denoted by a number of two digits, where the first digit minus one is the number of hydrogen atoms and the second digit indicates the number of fluorine atoms, while the other atoms are chlorine. For Refrigerant-12 or R-12, e.g., the number of hydrogen atoms is zero, the number of fluorine atoms is two, and hence the other two atoms must be chlorine. Therefore, the compound is CCl_2F_2, dichloro-difluoro methane. Similarly, for R-22, it is $CHClF_2$, monochloro-difluoro methane. For R-50, it is methane, CH_4. For R-10, it is CCl_4, carbon tetrachloride, and so on. If the compound is ethane-based, a three digit number is assigned to the refrigerant, where the first digit is always 1, the second digit minus one is the number of hydrogen atoms, and the third digit indicates the number of fluorine atoms, all other atoms in the hydrocarbon being chlorine. For example, R 110 is C_2Cl_6, R-113 is $C_2Cl_3F_3$, R142 is $C_2H_3Cl\,F_2$, and so on.

Alternatively, for a compound derived from a saturated hydrocarbon denoted by the chemical formula

$$C_m H_n F_p Cl_g$$

in which $(n + p + q) = 2m + 2$, the complete designation is

$$R - (m - 1)(n + 1)(p)$$

Hence, for dichloro-tetrafluoro-ethane in which there are two carbon atoms ($m = 2$), no hydrogen atom ($n = 0$), and four fluorine atoms ($p = 4$) is R-114. Similary, CCl_3F is R11, CCl_2F_2 is R-12, $CHClF_2$ is R22, $C_2Cl_3F_3$ is R113, CH_4 is R-50, C_2H_6 is R170, C_3H_8 is R-290, and so on.

The brominated refrigerants are denoted by putting an additional B and a number to denote the number of chlorine atoms replaced by bromine atoms. Thus, R13 B1 is derived from R13 with one chlorine atom replaced by a bromine atom and its formula is CF_3Br.

In case of *isomers*, compounds having the same chemical formula but different molecular structure, subscripts a, b, etc., are used.

In addition, there are *azeotropes*, which are mixtures of refrigerants, but which behave like pure substances. They are given arbitrary designations, e.g., R502 for a mixture of 48.8% R-22, an HCFC, and 51.2% R-115, a CFC.

In the case of common inorganic refrigerants, numerical designations are given according to their molecular weight added to 700. Thus ammonia is noted by 717, water by 718 and carbon dioxide by 744.

14.4 SELECTION OF A REFRIGERANT

Refrigerants have to be physiologically nontoxic and nonflammable. There is no ideal refrigerant as such. A refrigerant which is suitable in a certain application may not suit at all in other applications. The selection of a refrigerant depends on certain thermodynamic, chemical and physical properties:

14.4.1 Thermodynamic Properties

(i) Evaporation and Condensing Pressure The pressure in the evaporator should be positive and a little above the ambient pressure so that the atmospheric air and moisture do not leak into the system. The pressure in the condenser should not be high so as to necessitate heavy construction of equipment and high cost.

(ii) Critical Pressure and Temperature Fluids have a better heat-transfer rate at temperatures below critical temperature. So the critical temperature should be as high as possible. This would give a high COP. The critical pressure should be low so as to give lower condensing pressure.

(iii) Freezing Point The freezing-point temperature should be as low as possible so that the refrigerant can operate freely at higher temperatures in the cycle.

(iv) Latent Heat and Specific Heat A higher latent heat of vaporization at the working pressure would permit a greater amount of heat extraction in the evaporator. The specific heat of the liquid should be small so that less liquid flashes into vapour during expansion and there is higher refrigerating effect.

(v) Liquid and Vapour Density Refrigerants must not have a high specific volume at evaporator pressure so as to have smaller suction and discharge lines and the displacement volumes.

(vi) COP and kW per TR The refrigerant should have a high COP and low power input to compressor per ton of refrigeration.

14.4.2 Chemical Properties

(i) Inflammability The refrigerant should not be inflammable and explosive as far as possible.

(ii) Toxicity Toxic refrigerants should not be used in domestic refrigerating systems and comfort air conditioning.

(iii) Solubility in Water Refrigerants should have poor affinity for water, since the presence of moisture would lead to formation of ice and choking of the capillary tube.

(iv) Action on Material of Construction Refrigerants must not react chemically or carrode materials of construction. If a refrigerant, e.g., attacks copper and copper alloys, then pipes, fittings, valves, etc. are to be made with ferrous materials.

14.4.3 Physical Properties

(i) Thermal Conductivity For better heat-transfer rate, the refrigerant should have a high value of thermal conductivity.

(ii) Viscosity A low viscosity is desirable for low pumping power and high heat-transfer rate.

(iii) Leak Tendency The tendency for leakage of the refrigerant should be low and the detection of leak should be easy.

Besides these, the refrigerant should be cheap in cost and readily available. It should be chemically stable at all conditions of operating pressures and temperatures. It should not allow the lubricating oil to be carried over into the evaporator and condenser to casue fouling and reduce the rate of heat transfer.

14.4.4 Ozone Depletion Potential of CFC Refrigerants

The earth's *ozone layer* in the upper atmosphere or stratosphere absorbs harmful ultraviolet rays from the sun and protects us from skin cancer. Scientists were shocked to find in 1985 a gaping hole in the ozone layer above the Antartic through which ultraviolet rays directly reach the earth. CFCs are linked to the depletion of this ozone layer. They have varying degrees of *ozone depletion potential* (ODP). In addition, they also act as *greenhouse gases*, and hence they have *global warming potential* (GWP). According to an international agreement, called the Montreal Protocol, the use of fully halogenated CFCs (no hydrogen in the molecule), that are considered to have high ODP, viz., R-11, R-12, R-113 and R-114, have to be phased out by the year 2000. HCFCs like R-22 have ODP of 5% of that of R-12. However, because of their GWP, they will be phased out by 2030.

CFCs being heavier than N_2 and O_2 migrate to the upper atmosphere by molecular diffusion due to partial pressure difference. Chlorine atoms from the CFC molecule would be split off by the action of sunlight and the free chlorine atoms react with ozone in the stratosphere as follows:

$$CCl_2F_2 \xrightarrow{\text{sunlight}} CClF_2 + Cl$$

$$O_3 + Cl \xrightarrow{\text{sunlight}} ClO + O_2$$

Thus, O_3 will be depleted to O_2. Because of chain reaction, a single atom of Cl reacts taking out 100,000 O_3 molecules. Therefore, even a small concentration of CFC becomes very significant.

HCs and HFCs substitute for CFCs, since they do not contain any Cl atom and have zero ODP. Even HCFCs do contain Cl atoms, but in association with H-atoms have much reduced ODP. The two most common refrigerants having very high ODP are being replaced by R-134a (Tetra fluoro ethane), the most preferred substitute.

14.5 ABSORPTION REFRIGERATION CYCLE

The absorption refrigeration system is a *heat operated unit* which uses a refrigerant that is *alternately absorbed and liberated from the absorbent.* In the basic absorption system, the compressor in the vapour compression cycle is *replaced by an absorber-generator assembly* involving less mechanical work. Figure 14.25 gives the basic absorption refrigeration cycle, in which *ammonia is the refrigerant and water is the absorbent.* This is known as the *aqua-ammonia absorption* system.

Ammonia vapour is vigorously absorbed in water. So when low-pressure ammonia vapour from the evaporator comes in contact in the absorber with the weak solution (the concentration of ammonia in water is low) coming from the generator, it is readily absorbed, releasing the latent heat of condensation. The temperature of the solution tends to rise, while the absorber is cooled by the circulating water, absorbing the *heat of solution* (Q_A), and maintaining a constant temperature. Strong solution, rich in ammonia, is pumped to the generator where heat (Q_G) is supplied from an external source (steam, electricity, gas flame, etc.). Since the boiling point of ammonia is less than that of water, the ammonia vapour is given off from the aqua-ammonia solution at high pressure, and the weak solution returns to the absorber through a pressure reducing valve.

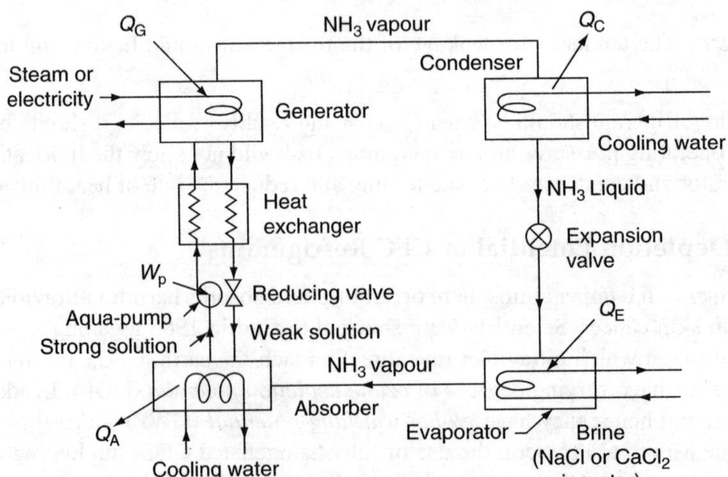

Fig. 14.25 *Vapour absorption refrigeration plant-flow diagram*

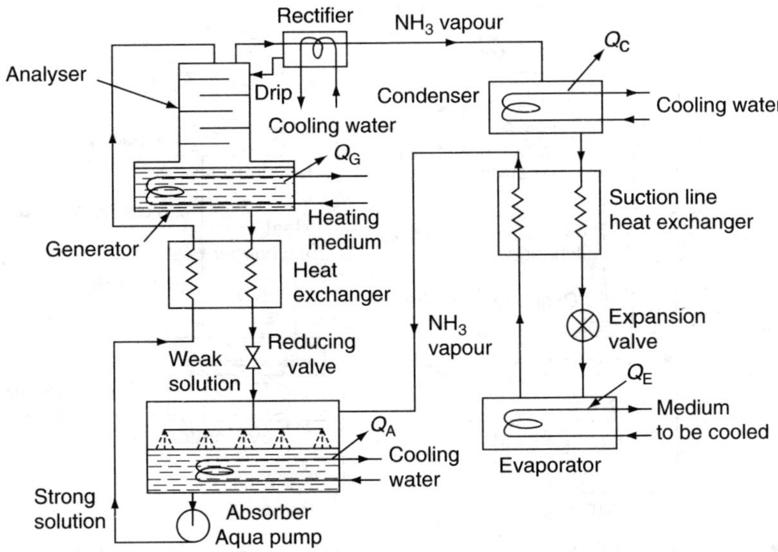

Fig. 14.26 *Actual vapour absorption refrigeration plant with analyzer and rectifier*

The heat exchanger preheats the strong solution and precools the weak solution, reducing both Q_G and Q_A, the heat to be supplied in the generator and the heat to be removed in the absorber respectively. The ammonia vapour then condenses in the condenser, is throttled by the expansion valve, and then evaporates, absorbing the heat of evaporation from the surroundings or the brine to be chilled.

In driving the ammonia vapour out of the solution in the generator, it is impossible to avoid evaporating some of the water. This water vapour going to the condenser along with the ammonia vapour, after condensation, may get frozen to ice and block the expansion valve. So an *analyzer-rectifier combination* (Fig. 14.26) is used to eliminate water vapour from the ammonia vapour going into the condenser.

The analyzer is a direct-contact heat exchanger consisting of a series of trays mounted above the generator. The strong solution from the absorber flows downward over the trays to cool the outgoing vapours. Since the saturation temperature of water is higher than that of ammonia at a given pressure, it is the water vapour which condenses first. As the vapour passes upward through the analyzer, it is cooled and enriched by ammonia, and the liquid is heated. Thus the vapour going to the condenser is lower in temperature and richer in ammonia, and the heat input to the generator is decreased.

The final reduction in the percentage of water vapour in the ammonia going to the condenser occurs in the rectifier which is a water-cooled heat exchanger which condenses water vapour and returns it to the generator through the drip line, as shown in Fig. 14.26. The use of a suction-line heat exchanger is to reduce Q_A and increase Q_E, thus achieving a double benefit. In the absorber the weak solution is sprayed to expose a larger surface area so as to accelerate the rate of absorption of ammonia vapour.

There is another absorption refrigeration system, namely, lithium bromide-water vapour absorption (Fig. 14.27). Here the refrigerant is water and the absorbent is the solution of lithium bromide salt in water. Since water cannot be cooled below 0°C, it can be used as a refrigerant in air conditioning units. Lithium bromide solution has a strong affinity for water vapour because of its very low vapour pressure. It absorbs water vapour as fast as it is released in the evaporator.

While the vapour compression refrigeration system requires the expenditure of 'high-grade' energy in the form of shaft work to drive the compressor with the concommitant disadvantage of vibration and noise, the

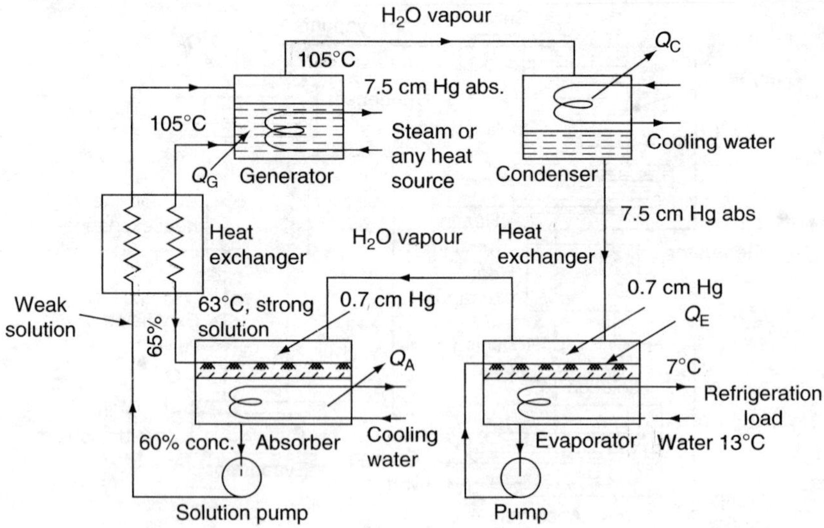

Fig. 14.27 *Lithium bromide-water absorption refrigeration plant*

absorption refrigeration system requires only 'low-grade' energy in the form of heat to drive it, and it is relatively silent in operation and subject to little wear. Although the COP $= Q_E/Q_G$ is low, the absorption units are usually built when waste heat is available, and they are built in relatively bigger sizes. One current application of the absorption system that may grow in importance is the *utilization of solar energy* for the generator heat source of a refrigerator for food preservation and perhaps for comfort cooling.

14.5.1 Theoretical COP of an Absorption System

Let us assume that an absorption refrigeration plant uses heat Q_G from a source at T_1, provides refrigeration Q_E for a region at T_R, and rejects heat (Q_A+Q_C) to a sink (atmosphere) at T_2, as shown in Fig 14.28.

By the first law

$$Q_E + Q_G = Q_C + Q_A \tag{14.3}$$

By the second law

$$(\Delta S)_{\text{source}} + (\Delta S)_{\text{sink}} + (\Delta S)_{\text{region}} \geq 0$$

$$\therefore \quad -\frac{Q_G}{T_1} + \frac{Q_E + Q_G}{T_2} - \frac{Q_E}{T_R} \geq 0$$

From Eq. (14.3)

$$-\frac{Q_G}{T_1} + \frac{Q_E + Q_G}{T_2} - \frac{Q_E}{T_R} \geq 0$$

$$\therefore \quad \frac{T_1 - T_2}{T_1 T_2} Q_G + \frac{T_R - T_2}{T_2 T_R} Q_E \geq 0$$

$$\therefore \quad \frac{T_2 - T_R}{T_2 T_R} Q_E \leq \frac{T_1 - T_2}{T_1 T_2} Q_G$$

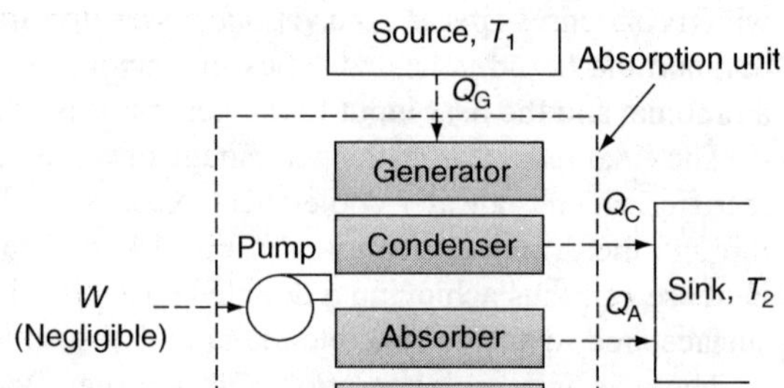

Fig. 14.28 *Energy fluxes in vapour absorption plant*

$$\therefore \quad \frac{Q_E}{Q_G} \le \frac{(T_1 - T_2)T_R}{(T_2 - T_R)T_1}$$

or

$$\text{COP} \le \frac{(T_1 - T_2)T_R}{(T_2 - T_R)T}$$

$$\therefore \quad (\text{COP})_{\max} = \frac{(T_1 - T_2)T_R}{(T_2 - T_R)T_1} \qquad (14.4)$$

or

$$(\text{COP})_{\max} = \frac{T_R}{T_2 - T_R} \times \frac{T_1 - T_2}{T_1}$$

Therefore, the maximum possible COP is the product of the ideal COP of a refrigerator working between T_R and T_2, and the ideal thermal efficiency of an engine working between T_1 and T_2. The cyclic heat engine and the refrigerator together, as shown in Fig. 14.29, is equivalent to the absorption cycle.

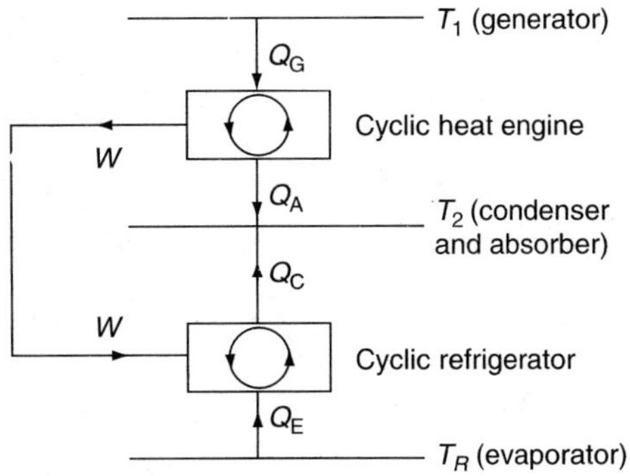

Fig. 14.29 *An absorption cycle as equivalent to a cyclic heat engine and a cyclic refrigerator*

14.6 ELECTROLUX REFRIGERATOR

An electrolux refrigerator works on three-fluid system. There is no circulation pump. The total pressure is the same throughout the system. The third fluid remains mainly in the evaporator, which reduces the partial pressure of refrigerant to enable it to evaporate at low pressure and hence low temperature.

The schematic diagram of an electrolux refrigerator working on $NH_3 - H_2O$ system with H_2 as the third fluid is shown in Fig. 14.30. Liquid NH_3 evaporates in the evaporator in the presence of H_2. Hydrogen is chosen because it is non-corrosive and insoluble in water. Thus a complete leakproof silent system is achieved.

A thermosiphon bubble pump is used to lift the weak aqua solution from the generator to the separator. The discharge tube from the generator is extended down below the liquid level in the generator. The bubbles rise and carry slugs of weak $NH_3 - H_2O$ solution into the separator. Two U-bend constrictions are provided as *vapour-locks* or *liquid-seal* to prevent H_2 from getting into the high side of the condenser.

The partial pressure of H_2 provides the pressure difference of NH_3 between the condenser and evaporator.

Accordingly, we have the following:

In the *condenser*, there is the pure NH_3 vapour pressure which is the total pressure.
In the *evaporator*, NH_3 vapour pressure = total pressure minus the partial pressure of H_2.

For example, let us consider the condenser temperature as 50°C, and evaporator temperature as –15°C. The corresponding vapour pressure of NH_3 are

Condenser, p_c = 20.33 bar
Evaporator exit, p_e = 2.36 bar

The approximate pressures in various parts of the system are given below:

The vapours leaving the generator top are assumed to be in equilibrium with the rich solution entering at 40°C, at which the saturation pressure of NH_3 is 15.54 bar. The temperature at the evaporator inlet is assumed to be at –25°C at which the saturation pressure of NH_3 is 1.516 bar.

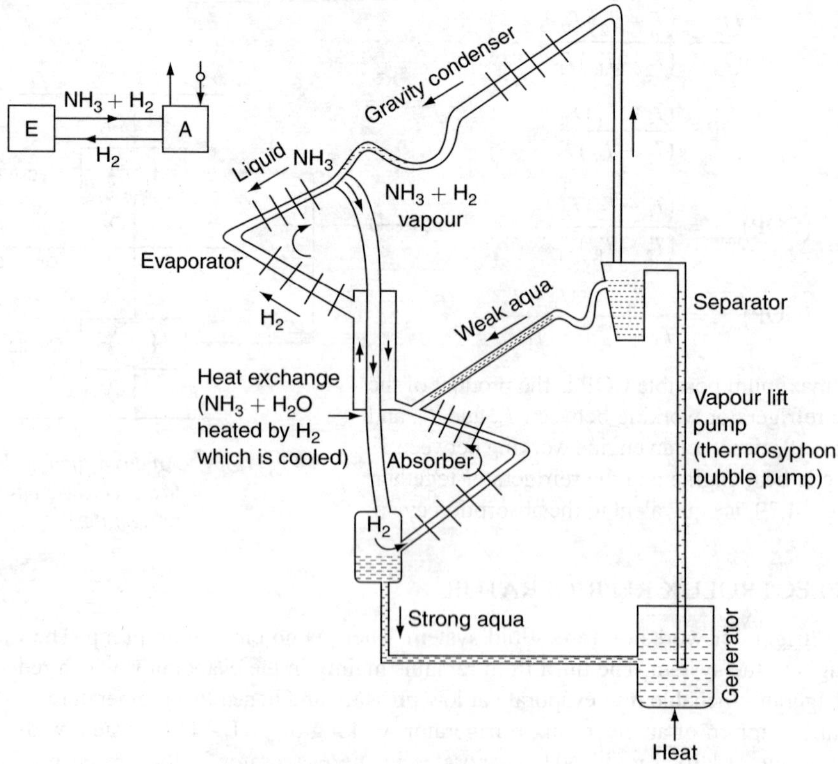

Fig. 14.30 *Electrolux refrigerator*

Table 14.1
Partial pressures in bar in the electrolux refrigerator

Component	NH_3	H_2O	H_2	*Total*
Condenser	20.33	0	0	20.33
Evaporator inlet	1.516	0	18.814	20.33
Evaporator exit	2.36	0	17.97	20.33
Generator top	15.54	4.79	0	20.33

14.7 STEAM JET REFRIGERATION SYSTEM

A simple steam-jet refrigeration system and its thermodynamic cycle is shown in Figs.14.31 and 14.32. Water at state 9 expands to 11 through the throttle valve into the flash chiller. Due to flashing (evaporation) of some liquid, the remaining water gets chilled to the desired temperature T_e. The pressure in the flash chamber is maintained at the corresponding saturation temperature T_e. The chilled water at 7 is recirculated after taking up the load in the cooling coil. The amount of water vaporized is compensated by make-up water at 6 entering through a throttle valve. Water vapour at 2 is compressed to 4 by the ejector driven by motive steam at 1. The compressed water vapour at 4 is then condensed to 5 and pumped back to the boiler.

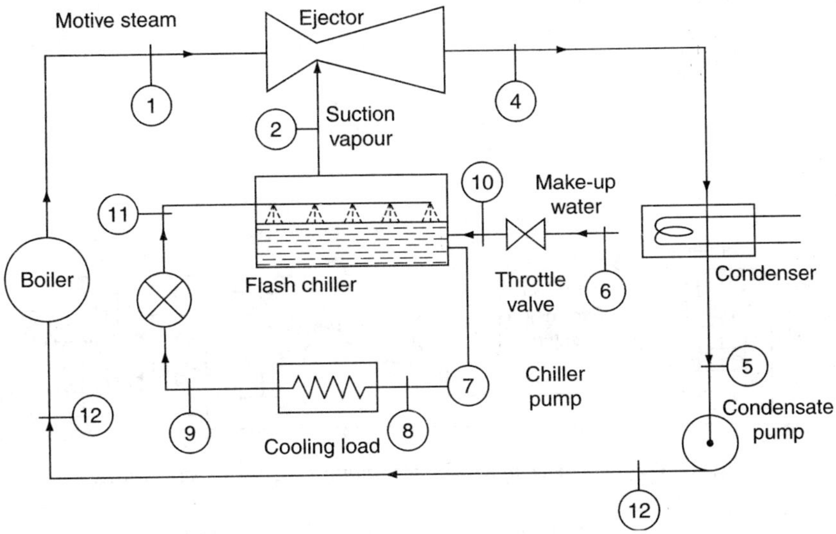

Fig. 14.31 *Water-vapour refrigeration system*

Though the overall COP of the system is low, the steam-jet refrigeration finds application in air conditioning for comfort where steam is available and safety is the prime concern. It is also suitably used for the concentration or drying of heat-sensitive foods and chemicals. It has certain limitations like

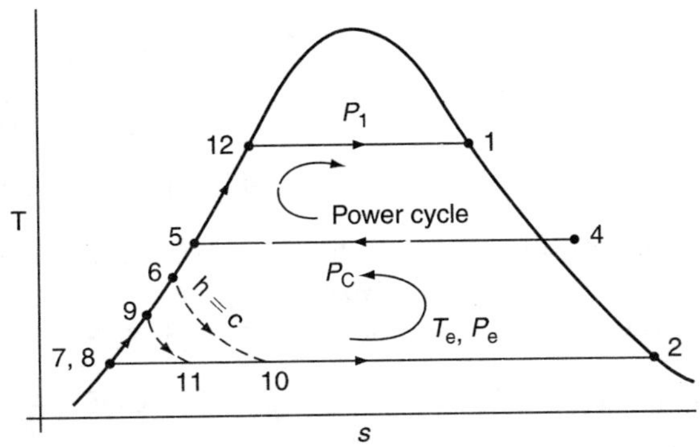

Fig. 14.32 *Thermodynamic cycle of water-vapour refrigeration system*

(i) Enormous volume to be handled, e.g., at 8°C, 665 m^3/(hr) × (TR) are required, whereas for R-12, it is only 7 m^3/h × TR. Centrifugal compressors can be used, but they are expensive. Ejector compression is the most economical to use with water as refrigerant.

(ii) High operating vacuum, e.g., at 8°C evaporator, the pressure is 0.01072 bar and the same at 35°C condenser, it is 0.0595 bar.

(iii) High freezing point, which can be used for refrigeration above 0°C only, such as is required for air conditioning

14.8 HEAT PUMP SYSTEMS

The heat pump is a cyclic device which is able to extract energy at a low temperature heat source and upgrade it to a high temperature heat source, enabling it to be used more effectively. Low grade reject heat available at a low temperature may be upgraded to a high temperature heat source by a heat pump. While a refrigerator is meant for the removal of heat and to achieve cooling, a heat pump is used to supply heat at a high temperature.

Heat pump systems have many features in common with the refrigeration systems and may be of the vapour-compression or the absorption type. A typical vapour-compression heat pump for space heating

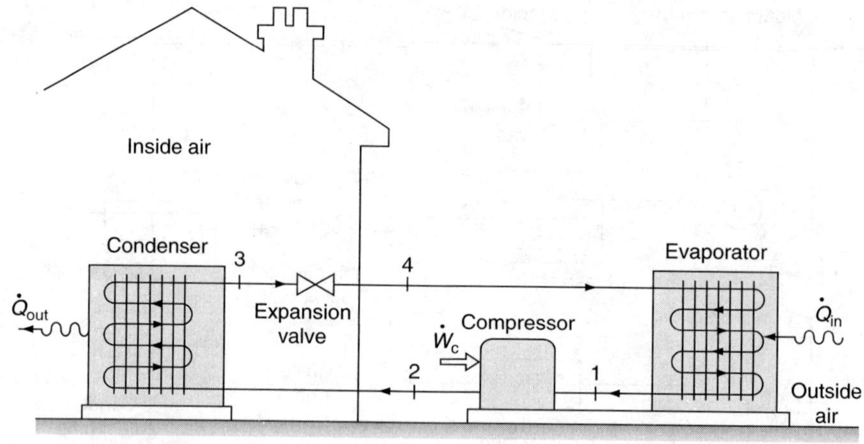

Fig. 14.33 *Vapour-compression heat pump system for space heating*

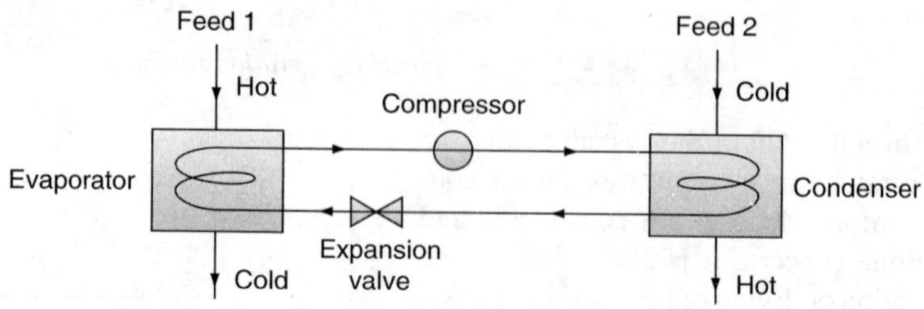

Fig. 14.34 *Heating and cooling of two process streams*

(Fig. 14.33) has the same basic components as the vapour-compression refrigeration system: compressor, condenser, expansion valve, and evaporator. The heat $\dot{Q}_{in}$ comes from the surroundings (cold air), and $\dot{Q}_{out}$ is directed to the dwelling as the desired effect, with the COP given by:

$$\text{COP} = \frac{\dot{Q}_{out}}{W_C} = \frac{h_2 - h_3}{h_2 - h_1}$$

Many possible sources are available for heat transfer $\left(\dot{Q}_{in}\right)$ to the refrigerant passing through the evaporator. These include the outside air, the ground, and water from lakes, rivers, or wells. Liquid circulated through a solar collector and stored in an insulated tank also can be used as a source for a heat pump. Industrial heat pumps employ waste heat or warm liquid or gas streams as the low temperature heat source and upgrade it.

An air-air heat pump can be used for year-round air conditioning, to achieve heating during winter and cooling during summer, by using a reversing valve with the evaporator and condenser executing opposite duties. A heat pump in industry will be very effective if both condenser and evaporator are utilized for heating and cooling respectively, e.g., for cooling a process stream and heating another (Fig. 14.34).

14.9 GAS CYCLE REFRIGERATION

Refrigeration can also be accomplished by means of a gas cycle. In the gas cycle, an expander replaces the throttle valve of a vapour compression system, because the drop in temperature by throttling a real gas is very

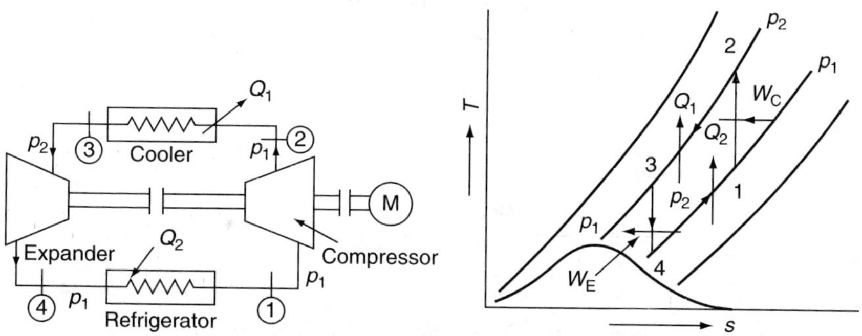

Fig. 14.35 *Gas refrigeration cycle*

small. For an ideal gas, enthalpy is a function of temperature only, and since in throttling enthalpy remains unchanged, there would not be any change in temperature also. Work output obtained from the expander is used as an aid in compression, thus decreasing the net work input. The ideal gas-refrigeration cycle is the same as the *reversed Brayton cycle*. The flow and *T-s* diagrams of the cycle are shown in Fig. 14.35. *Since there is no* phase change, the condenser and evaporator in a vapour compression system are here called the cooler and refrigerator respectively. At 1, the air is isentropically compressed from (1 to 2) from p_1 to p_2 after which it is cooled at constant pressure p_2 to p_3. The air, since it is presently used only in aircraft, is finally expanded isentropically to cool to 4. The COP of the refrigeration cycle, assuming the gas to be ideal, is given by

$$\text{COP} = \frac{Q_2}{W_{\text{net}}} = \frac{h_1 - h_4}{(h_2 - h_1) - (h_3 - h_4)}$$

$$= \frac{T_1 - T_4}{(T_2 - T_1) - (T_3 - T_4)} = \frac{T_1 - T_4}{T_1\left[\dfrac{T_2}{T_1} - 1\right] - T_4\left[\dfrac{T_3}{T_4} - 1\right]}$$

For isentropic compression and expansion

$$\frac{T_2}{T_1} = \left(\frac{p_2}{p_1}\right)^{\frac{\gamma-1}{\gamma}} = \frac{T_3}{T_4}$$

$$\therefore \qquad \text{COP} = \frac{T_1 - T_4}{(T_1 - T_4)\left[\dfrac{T_3}{T_4} - 1\right]} = \frac{T_4}{T_3 - T_4}$$

Also

$$\text{COP} = \frac{1}{\left[\dfrac{p_2}{p_1}\right]^{(\gamma-1)/\gamma} - 1} \qquad (14.5)$$

where p_2 is the pressure after compression and p_1 is the pressure before compression.

The COP of a gas-cycle refrigeration system is low. The power requirement per unit capacity is high. Its prominent application is in aircraft and missiles, where the vapour compression refrigeration system becomes heavy and bulky. Figure 14.36 shows the open cycle aircraft cabin cooling.

The compressed air is available and is a small percentage of the amount handled by the compressor of a turbojet or a supercharged aircraft engine. Large amounts of cool ambient air are available for cooling the compressed

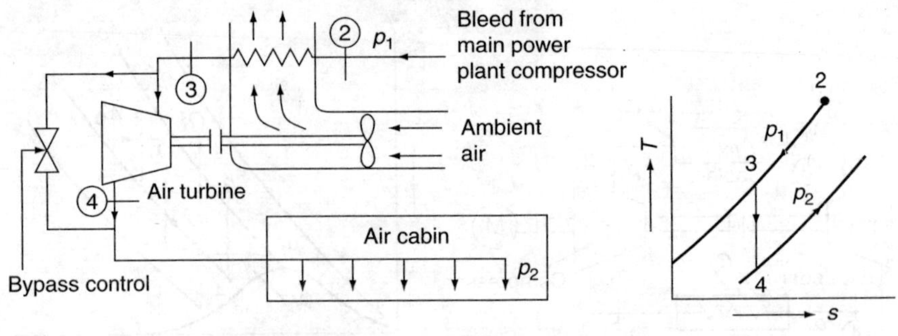

Fig. 14.36 *Open cycle aircraft cabin cooling*

air. In addition to cooling, the replacement of stale air in the cabin is possible. At high altitudes the pressurization of cabin air is also possible. Because of these considerations, air cycle refrigeration is favoured in aircrafts.

The flow and T-s diagrams of gas cycle refrigeration are shown in Fig. 14.37.

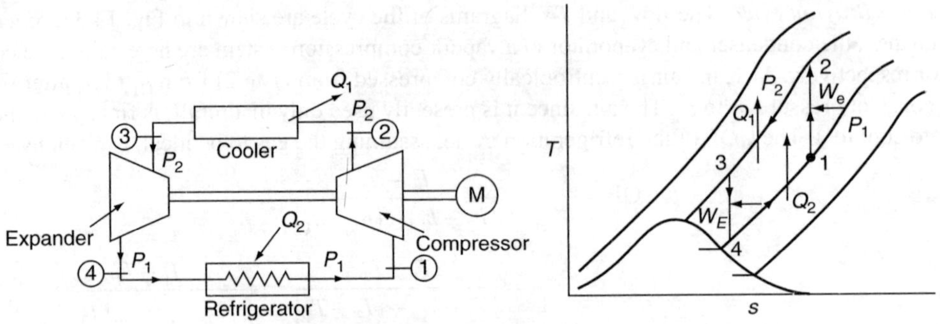

Fig. 14.37 *Gas refrigeration cycle*

14.9.1 Polytropic and Multistage Compression

Polytropic compression with cooling would reduce the net work input with the compression index being n instead of γ. The expression for compressor work becomes

$$W_c = \frac{n}{n-1}(p_2 v_2 - p_1 v_1) = \frac{n}{n-1}\frac{\gamma-1}{\gamma}c_p(T_2 - T_1)$$

Net work is

$$W_{net} = W_c - W_e = c_p\left[\frac{n}{n-1}\frac{\gamma-1}{\gamma}(T_2 - T_1) - (T_3 - T_4)\right]$$

$$\therefore \quad \text{COP} = \frac{T_1 - T_4}{\frac{n}{n-1}\frac{\gamma-1}{\gamma}(T_2 - T_1) - (T_3 - T_4)}$$

14.10 APPLICATION TO AIRCRAFT REFRIGERATION

The gas cycle is exclusively used in air conditioning systems of military and commercial aircrafts, though refrigerated cargo aircrafts generally use dry ice. It is appropriate to call it *air-cycle refrigeration*, since only air is the working substance in this cycle. The air cycle can work as an open cycle or a closed cycle. The main considerations for using air-cycle refrigeration in aircrafts are weight, space and operating power. Other advantages are the following:

(i) Small amounts of leakages are tolerable with air as the refrigerant.
(ii) In its simplest form as an open cycle it requires only one heat exchanger.
(iii) Availability of the refrigerant in mid air is of no concern.
(iv) Cabin pressurization and air conditioning can be combined into one operation.
(v) Initial compression of the air can be obtained by ram effect.

A simple aircraft refrigeration cycle is shown in Fig. 14.38(a) and 14.38(b). The ram effect is shown by the line 1–2. Point 2′ denotes the state after isentropic compression to pressure p_2' and temperature T_2'. The energy equation for the diffusion process:

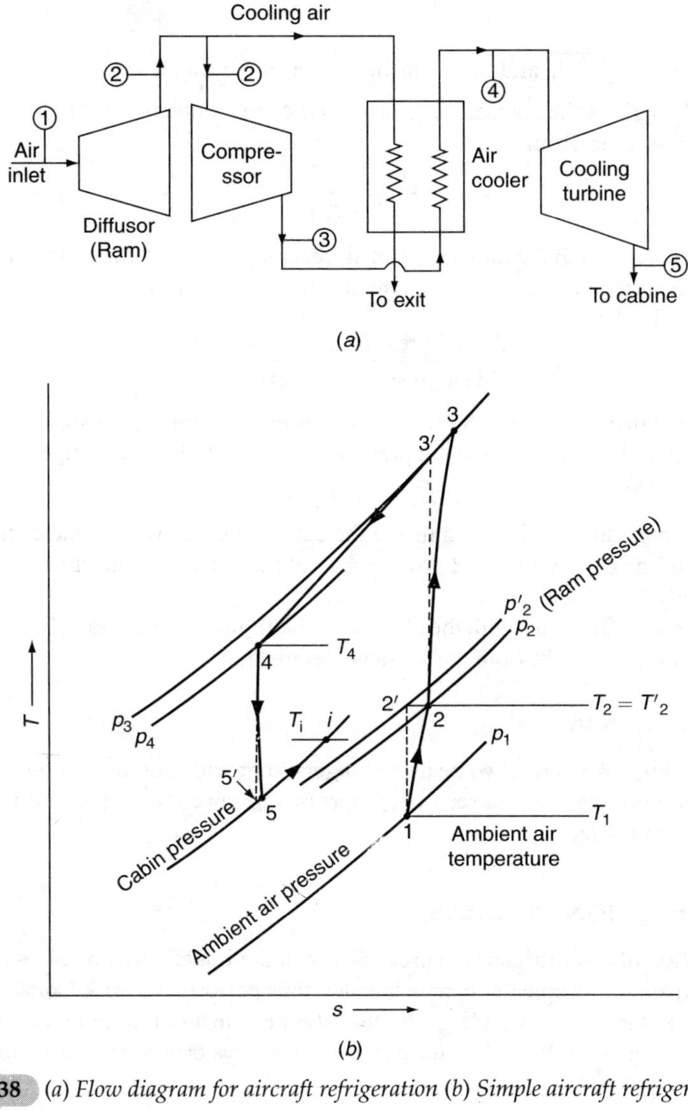

Fig. 14.38 *(a) Flow diagram for aircraft refrigeration (b) Simple aircraft refrigeration cycle*

$$h_2 = h_2' = h_1 + \frac{v^2}{2}$$

where v is the velocity of the aircraft.

$$c_p T_2 = c_p T_2' = c_p T_1 + \frac{v^2}{2}$$

$$T_2 = T_2' = T_1 + \frac{v^2}{2c_p}$$

$$\frac{T_2'}{T_1} = 1 + \frac{v^2}{\frac{2\gamma R}{\gamma - 1} T_1} = 1 + \frac{\gamma - 1}{2}\frac{v^2}{c^2} = 1 + \frac{\gamma - 1}{2} M^2$$

where c = sonic velocity = $\sqrt{\gamma R T_1}$ and M = Mach number of the flight.

The temperature $T_2 = T_2'$ is the stagnation temperature of the ambient air. The stagnation pressure after isentropic diffusion, p_2' is given from

$$\frac{p_2'}{p_1} = \left(\frac{T_2'}{T_1}\right)^{\frac{\gamma}{\gamma - 1}}$$

The irreversible compression in the ram results in air reaching the point 2 instead of the point 2′, i.e., at the same stagnation temperature but at a reduced stagnation pressure p_2, which is obtained from the knowledge of ram efficiency η_R defined by

$$\eta_R = \frac{\text{Actual pressure recovery}}{\text{Ideal pressure recovery}} = \frac{p_2 - p_1}{p_2' - p_1}$$

The rest of the cycle is shown in Fig. 14.38. The pressure after expansion p_5 is slightly above the cabin pressure. It is also higher than the ambient air static pressure at that altitude. The refrigerating effect produced is

$$Q_2 = c_p (T_i - T_5)$$

where T_i is the room temperature maintained inside the cabin. The net work of the cycle is the difference of work for process 2–3 of the compressor and process 4–5 of the expander plus the ram air work, $c_p(T_2 - T_1)$, which is derived directly from the engine.

The ambient air temperature varies with the altitude of the flight of the aircraft. Generally, the temperature drops by 0.64°C per 100 m of height from the sea-level temperature.

14.10.1 Bootstrap System

The bootstrap system (Fig. 14.39) has two heat exchangers to provide additional cooling capacity whenever required. The expansion turbine drives a second compressor to raise the pressure of air before it enters the cooling turbine via the air cooler.

14.11 LIQUEFACTION OF GASES

An important application of gas refrigeration processes is in the liquefaction of gases. A gas may be cooled either by making it expand isentropically in an expander, thus performing work (sometimes known as *external-work method*), or making the gas undergo Joule-Kelvin expansion through a throttle valve (sometimes called the *internal-work method*). While the former method always brings about a temperature decrease, the

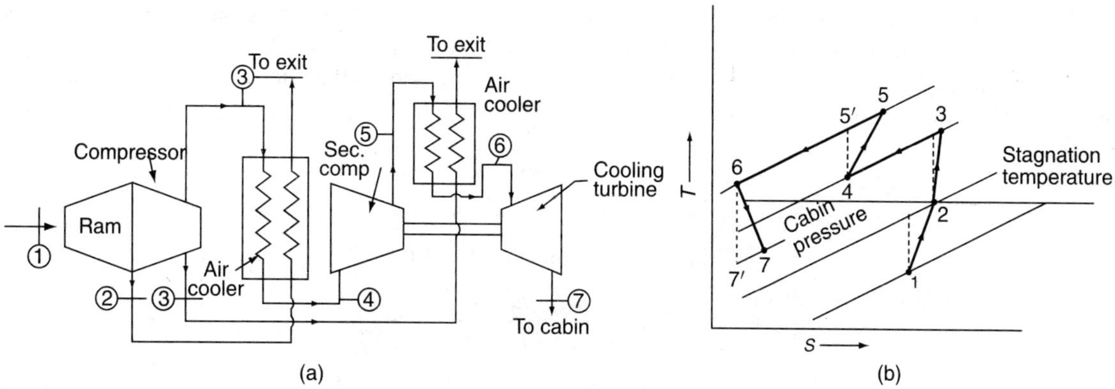

Fig. 14.39 *Thermodynamic cycle for bootstrap system*

expansion through the throttle valve may yield a temperature decrease only when the temperature before throttling is below the maximum inversion temperature.

14.11.1 Linde–Hampson System for Liquefaction of Air

In this system, Joule-Kelvin effect is utilized for cooling, and ultimately, liquefying the air. The schematic diagram and the *T-s* diagram are shown in Fig. 14.40. Ideally, the compression would be isothermal as shown on the *T-s* diagram. A two-stage compressor with intercooling and aftercooling is shown. The *yield, Y,* of the system is defined as the ratio of the mass of liquid produced to the mass of gas compressed. The energy required per unit mass of liquid produced is known as the *specific work* consumption, *W*.

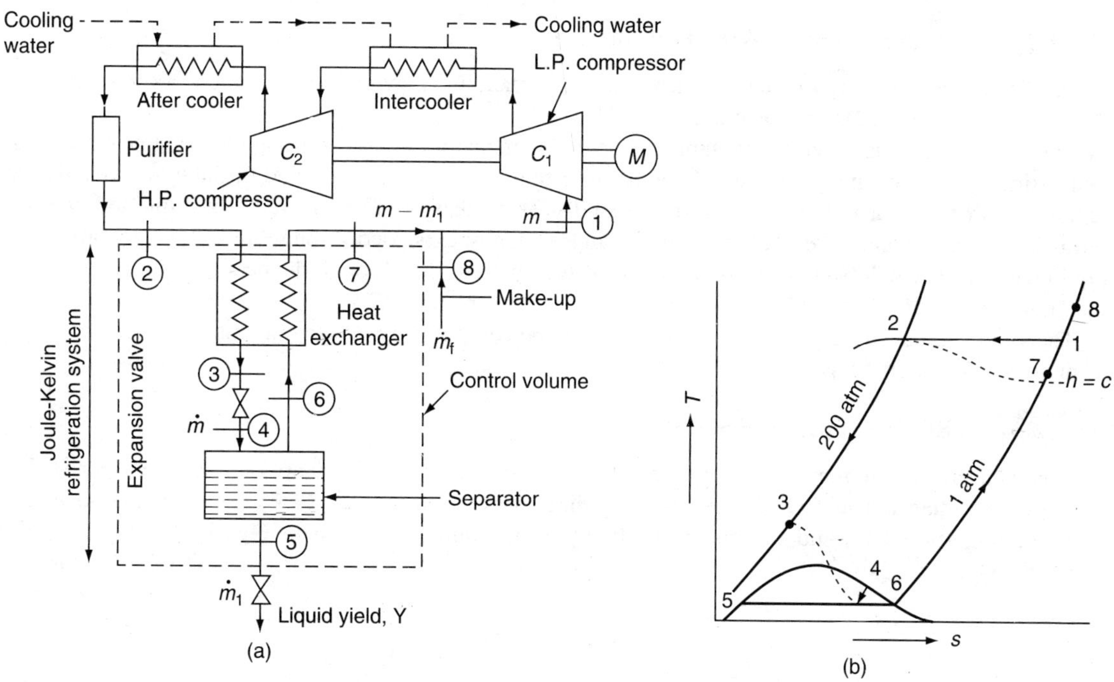

Fig. 14.40 *Linde-hampson cycle for air liquefaction*

The theoretical yield, assuming perfect insulation, can be determined by a mass and energy balance for the control volume (Joule-Kelvin refrigeration system) as shown in Fig. 14.40. Let $\dot{m}_f$ be the rate at which liquid air is produced (the same given to the system as make-up), and $\dot{m}$ the rate at which air is compressed and then expanded. Then the yield is

$$Y = \frac{\dot{m}_f}{\dot{m}}$$

and the energy balance gives

$$\dot{m}h_2 - \dot{m}_f h_5 - (\dot{m} - \dot{m}_f)h_7 = 0$$

$$\therefore \quad \dot{m}(h_2 - h_7) - \dot{m}_f(h_5 - h_7) = 0$$

$$Y = \frac{\dot{m}_f}{\dot{m}} = \frac{h_2 - h_7}{h_5 - h_7}$$

$$\therefore \quad Y = \frac{h_7 - h_2}{h_7 - h_5} \tag{14.6}$$

No yield is thus possible unless h_7 is greater than h_2. The energy balance for the compressor gives

$$\dot{m}h_1 + W_c = \dot{m}h_2 + Q_R$$

Where Q_R is the heat loss to the surroundings from the compressor

$$\therefore \quad \frac{W_c}{\dot{m}} = T_1(s_1 - s_2) - (h_1 - h_2)$$

This is the *minimum work requirement.*

Specific work consumption, W

$$= \frac{W_c}{\dot{m}} \times \frac{\dot{m}}{\dot{m}_f} = \frac{W_c}{\dot{m}}\frac{1}{Y} = \frac{h_7 - h_5}{h_7 - h_2}\left[T_1(s_1 - s_2) - (h_1 - h_2)\right]$$

14.11.2 Claude System of Air Liquefaction

In the Claude system, energy is removed from the gas stream by allowing it to do some work in an expander. The flow and *T-s* diagrams are given in Fig. 14.41.

The gas is first compressed to pressures of about 40 atm and then passed through the first heat exchanger. Approximatly 80% of the gas is then diverted from the main stream, expanded through an expander, and reunited with the return stream below the second heat exchanger. The stream to be liquefied continues through the second and third heat exchangers, and is finally expanded through an expansion valve to the liquid receiver. The cold vapour from the liquid receiver is returned through the heat exchangers to cool the incoming gas.

The yield and the specific work consumption may be computed by making the mass and energy balance as in the Linde-Hampson system.

14.12 PRODUCTION OF SOLID ICE

Dry ice is used for low temperature refrigeration, such as to preserve ice cream and other perishables. The property diagram of CO_2 on the *p-h* coordinates is given in Fig. 14.42. The schematic diagram of producing solid CO_2 and the corresponding *p-h* diagram are shown in Fig. 14.43 and Fig. 14.44 respectively.

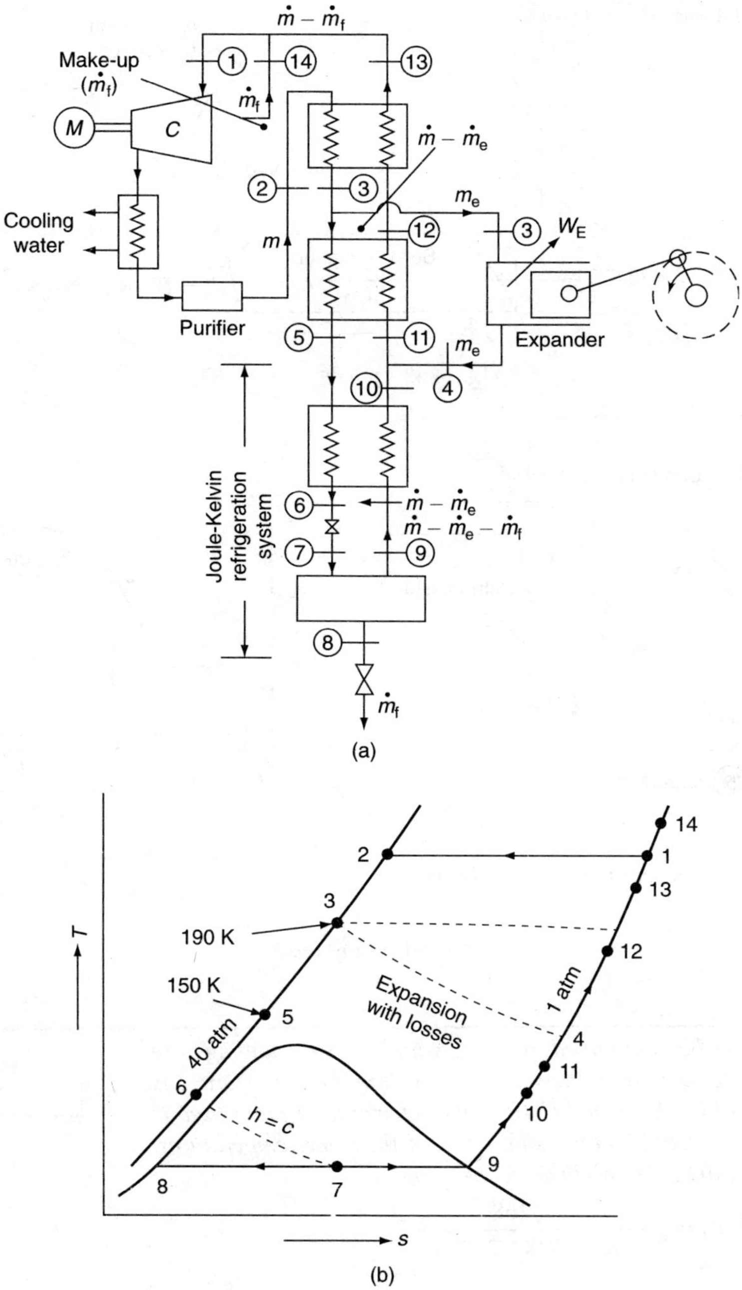

Fig. 14.41 *Claude cycle for air liquefaction*

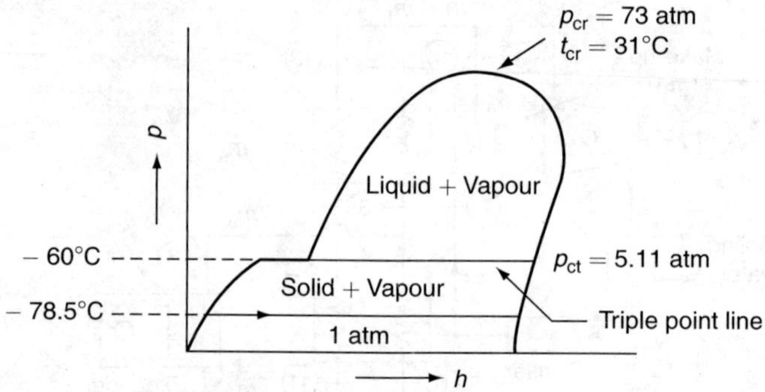

Fig. 14.42 *p-h diagram of CO_2*

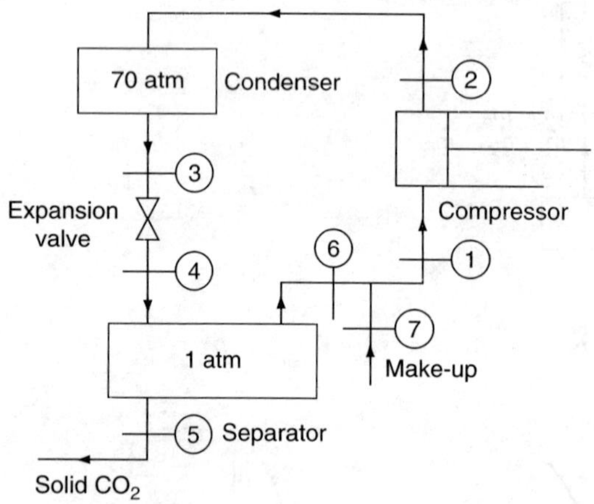

Fig. 14.43 *Production of dry ice-flow diagram*

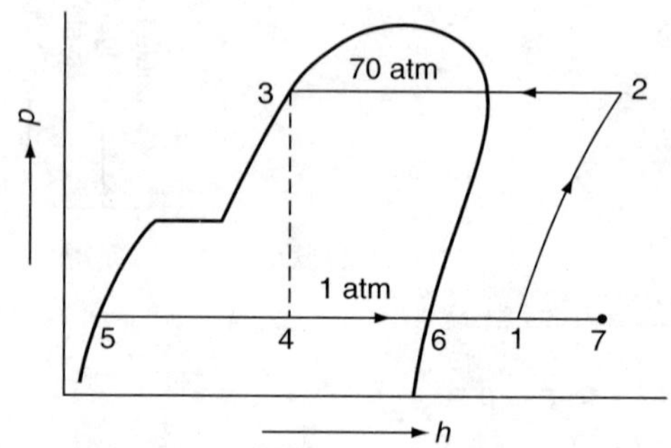

Fig. 14.44 *Refrigeration cycle of a dry ice plant on p-h plot*

Solved Examples

Example 14.1

A cold storage is to be maintained at – 5°C while the surroundings are at 35°C. The heat leakage from the surroundings into the cold storage is estimated to be 29 kW. The actual COP of the refrigeration plant used is one-third that of an ideal plant working between the same temperatures. Find the power required (in kW) to drive the plant.

Solution $\text{COP (Ideal)} = \dfrac{T_2}{T_1 - T_2} = \dfrac{268}{308 - 268} = 6.7$

$\therefore$ $\text{Actual COP} = 1/3 \times 6.7 = 2.23 = \dfrac{Q_2}{W}$

$\therefore$ Power required to drive the plant (Fig. Ex. 14.1)

$$W = \frac{Q_2}{2.23} = \frac{29}{2.23} = \mathbf{13\ kW}$$

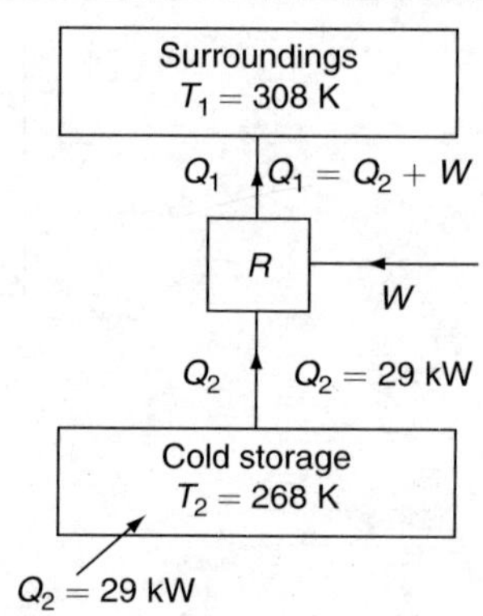

Fig. Ex. 14.1

Example 14.2

A refrigerator uses R-134a as the working fluid and operates on an ideal vapour compression cycle between 0.14 MPa and 0.8 MPa. If the mass flow rate of the refrigerant is 0.06 kg/s, determine (a) the rate of heat removal from the refrigerated space, (b) the power input to the compressor, (c) the heat rejection rate in the condenser, and (d) the COP.

Solution From the R-134*a* tables, the enthalpies at the four states (Fig. Ex. 14.2) are:

$$h_1 = 236.04 \text{ kJ/Kg}$$

$$s_1 = 0.9322 \text{ kJ/kg K} = s_2$$

For $$p_2 = 0.8 \text{ MPa}, s_2 = 0.9322 \text{ kJ/kgK},$$

$$h_2 = 272.05 \text{ kJ/kg}, h_3 = h_4 = 93.42 \text{ kJ/kg}$$

$$Q_2 = 0.06\,(236.04 - 93.42) = \mathbf{8.56\ kW}$$

$$W_c = 0.06\,(272.05 - 236.04) = \mathbf{2.16\ kW}$$

$$Q_1 = 0.06\,(272.05 - 93.42) = \mathbf{10.72\ kW}$$

$$\text{COP} = \frac{Q_2}{W_c} = \frac{8.56}{2.16} = \mathbf{3.963}$$

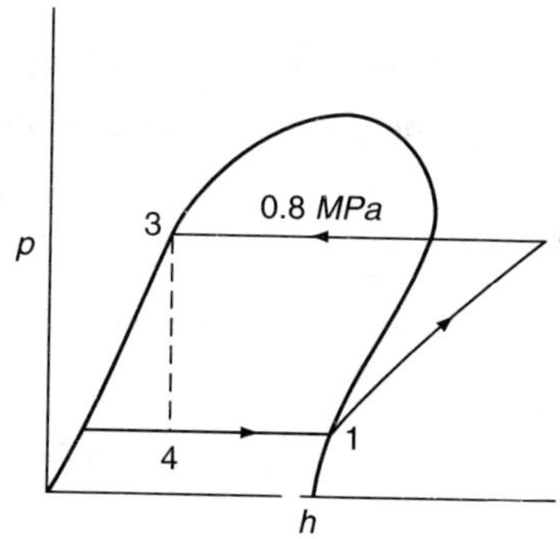

Fig. Ex. 14.2

Example 14.3

A simple R-12 plant is to develop 5 tonnes of refrigeration. The condenser and evaporator temperatures are to be 40°C and –10°C respectively. Determine (a) the refrigerant flow rate in kg/s, (b) the volume flow rate handled by the compressor in m³/s. (c) the compressor discharge temperature, (d) the pressure ratio, (e) the heat rejected to the condenser in kW, (f) the flash gas percentage after throttling, (g) the COP, and (h) the power required to drive the compressor.

How does this COP compare with that of a Carnot refrigerator operating between 40°C and –10°C?

Solution From the table of the thermodynamic properties of a saturated Refrigerant-12 (Fig. Ex. 14.3), $p_1 = (p_{sat})_{-10°C} = 2.1912$ bar, $h_1 = 183.19$ kJ/kg, $s_1 = 0.7019$ kJ/kg K, $v_1 = 0.077 \text{m}^3\text{/kg}$,

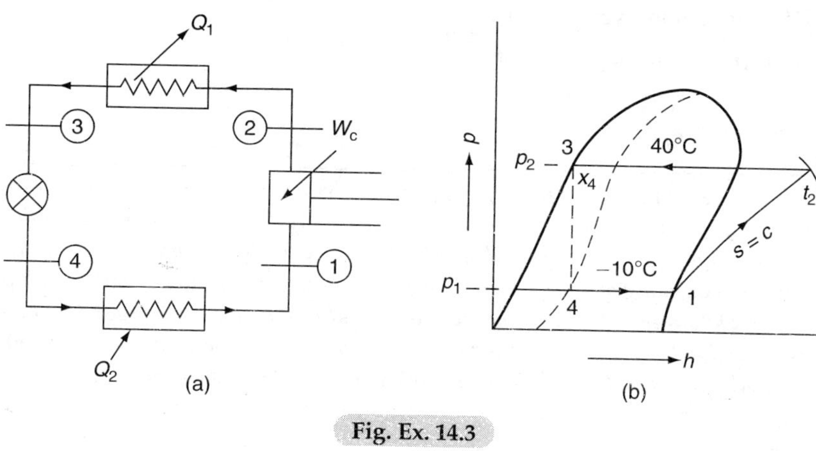

Fig. Ex. 14.3

$p_2 = (p_{sat})_{40°c} = 9.6066$ bar, and $h_3 = 74.59$ kJ/kg $= h_4$. From the superheated table of R-12, when $p_2 = 9.6066$ bar, and $s_2 = s_1 = 0.7019$ kJ/kg K, by interpolation, $t_2 = 48°C$, and $h_2 = 209.41$ kJ/kg. The capacity of the plant $= 5 \times 14{,}000 = 70{,}000$ kJ/h. If w is the refrigerant flow rate in kg/s

$$w(h_1 - h_4) = \frac{70.000}{3600} = 19.44 \text{ kW}$$

$$\therefore \quad w = \frac{19.44}{183.19 - 74.59} = \mathbf{0.18 \text{ kg/s}}$$

Volume flow rate $= w \cdot v_1 = 0.18 \times 0.077 = \mathbf{0.0139\ m^3/s}$

Compressor discharge temperature $=$ **48°C**

$$\text{Pressure ratio} = \frac{p_2}{p_1} = \frac{9.6066}{2.1912} = \mathbf{4.39}$$

$$\text{Heat rejected to the condenser} = w(h_2 - h_3)$$
$$= 0.18\,(209.41 - 74.59) = \mathbf{24.27\ kW}$$

$$h_4 = h_f + x_4 h_{fg} = 26.87 + x_4 \times 156.31 = 74.59$$

$$x_4 = \frac{47.72}{156.31} = 0.305$$

$$\therefore \quad \text{Flash gas percentage} = \mathbf{30.5\%}$$

$$\text{COP} = \frac{h_1 - h_4}{h_2 - h_1} = \frac{183.19 - 74.59}{209.41 - 183.19}$$

$$= \frac{108.60}{26.22} = \mathbf{4.14}$$

Power required to drive the compressor

$$= w(h_2 - h_1) = 0.18 \times 26.22 = \mathbf{4.72\ kW}$$

$$\text{COP (Reversible)} = \frac{T_2}{T_1 - T_2} = \frac{263}{50} = 5.26$$

$$\therefore \quad \frac{\text{COP(Vap. Comp. cycle)}}{\text{COP(Carnot cycle)}} = \frac{4.14}{5.26} = \mathbf{0.787}$$

Example 14.4

*A Refrigerant-*12 *vapour compression plant producing* 10 *tonnes of refrigeration operates with condensing and evaporating temperatures of* 35°C *and* –10°C *respectively. A suction line heat exchanger is used to subcool the saturated liquid leaving the condenser. Saturated vapour leaving the evaporator is superheated in the suction line heat exchanger to the extent that a discharge temperature of* 60°C *is obtained after isentropic compression. Determine (a) the subcooling achieved in the heat exchanger, (b) the refrigerant flow rate in kg/s, (c) the cylinder dimensions of the two-cylinder compressor, if the speed is* 900 *rpm, stroke-to-bore ratio is* 1.1*, and the volumetric efficiency is* 80% *(d) the COP of the plant, and (e) the power required to drive the compressor in kW.*

Solution From the p-h chart of R-12, the property values at the states, as shown in Fig. Ex. 14.4

$$h_3 = 882,\ h_2 = 1034$$
$$h_6 = 998,\ h_1 = 1008 \text{ kJ/kg}$$
$$v_1 = 0.084 \text{ m}^3\text{/kg}$$

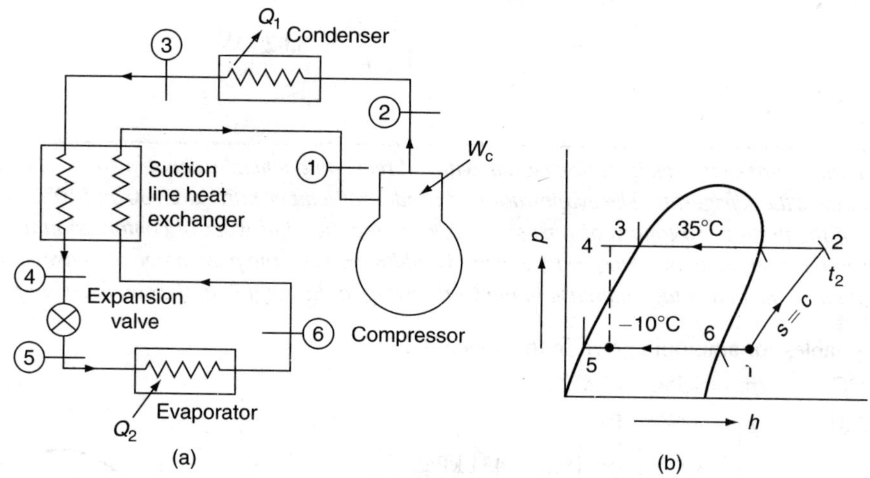

Fig. Ex. 14.4

$$h_3 - h_4 = h_1 - h_6$$
$$882 - h_4 = 1008 - 998 = 10$$
$$\therefore \quad h_4 = 872 \text{ kJ/kg}$$
$$\therefore \quad t_4 = 25°\text{C}$$

So 10°C subcooling is achieved in the heat exchanger.

Refrigeration effect $= h_6 - h_5 = 998 - 872 =$ **126 kJ/kg**

$$\therefore \quad \text{Refrigerant flow rate } \frac{10 \times 14000}{126} = \mathbf{1110 \text{ kg/h} = 0.31 \text{ kg/s}}$$

$$\text{Volume flow rate} = w \cdot v_1 = 1110 \times 0.084 = 93 \text{ m}^3\text{/h}$$

$$\text{Compressor displacement} = \frac{93}{0.8} = 116 \text{ m}^3\text{/h} = 1.94 \text{ m}^3\text{/min}$$

This is equal to $\frac{\pi}{4} D^2$ LNn

Where D = diameter

L = stroke

N = rpm

n = number of cylinders of the compressor.

$$\frac{\pi}{4} D^2 \times 1.1D \times 900 \times 2 = 1.94 \text{ m}^3\text{/min}$$

$$D^3 = 1250 \text{ cm}^3$$

or $$D = \mathbf{10.8 \text{ cm}} \quad \text{and} \quad L = \mathbf{11.88 \text{ cm}}$$

$$\text{COP} = \frac{h_6 - h_5}{h_2 - h_1} = \frac{126}{1034 - 1008} = \mathbf{4.85}$$

Power required to drive the compressor

$$= w(h_2 - h_1) = \frac{1110 \times 26}{3600} = \mathbf{8.02 \text{ kW}}$$

Example 14.5

A two-stage vapour compression refrigeration system with a direct contact heat exchanger (flash chamber) operates with ammonia as the refrigerant. The evaporator and condenser temperatures are –30 and 40°C respectively. If the capacity of the plant is 30 tonnes of refrigeration, estimate the total work of compression and the COP. Had the compression been done in a single stage, what would have been the percentage increase in the work of compression? What is the percentage increase in the COP owing to the staging of the compression process?

Solution Using tables for ammonia given in the appendix,

at 40°C, $p_2 = 1554.3$ kPA

at – 30°C, $p_1 = 119.5$ kPa

$\therefore \quad p_i = \sqrt{1554.3 \times 119.5} = 431$ kPa

$\therefore \quad h_1 = 1404.6 \; s_2 = s_1, h_2 = 1574.3$

$h_3 = 1443.5 \; s_4 = s_3, h_4 = 1628.1$

$h_5 = 371.7 = h_6, h_7 = 181.5$ hg

$$\therefore \quad \dot{m}_2 = \frac{3.89 \times 30}{1404.6 - 181.5} = \frac{116.7}{1223.1} = 0.0954 \text{ kg/s}$$

$$\dot{m}_1 = \dot{m}_2 \frac{h_2 - h_7}{h_3 - h_6} = 0.0954 \times \frac{1392.8}{1071.8} = 0.124 \text{ kg/s}$$

$$\dot{W}_c = \dot{m}_c (h_2 - h_1) + \dot{m}_1 (h_4 - h_3)$$

$$= 0.0954 \times 169.7 + 0.124 \times 184.6$$

$$= 16.19 + 22.89 = 39.08 \text{ kW}$$

$$\text{COP} = \frac{30 \times 3.89}{39.08} = 2.986$$

Single stage

$$h_1 = 1404.6 \quad h_2 = 1805.1$$

$$h_3 = 371.7 = h_4$$

$$\dot{m} = \frac{30 \times 3.89}{1404.6 - 371.7} = \frac{116.7}{1032.9} = 0.113 \text{ kg/s}$$

$$W_c = \dot{m}(h_2 - h_1) = 0.113 \times 400.5 = 45.26 \text{ kW}$$

$$\therefore \quad \text{COP} = \frac{116.7}{45.26} = 2.578$$

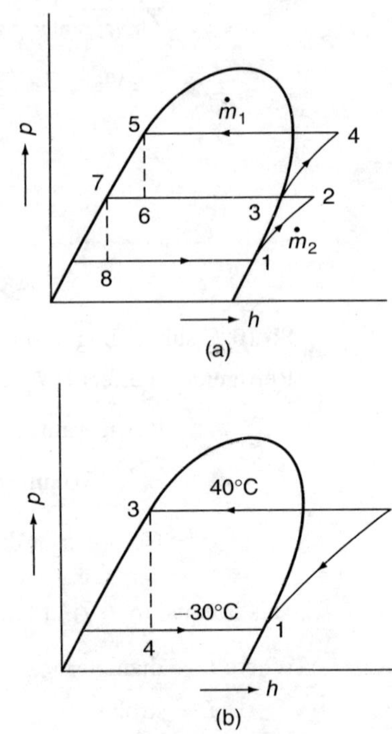

Fig. Ex. 14.5

Increase in work of compression (for single stage)

$$= \frac{45.26 - 39.08}{39.08} \times 100 = \mathbf{15.81\%}$$

Increase in COP for 2-stage compression

$$= \frac{2.986 - 2.578}{2.578} \times 100 = \mathbf{15.82\%}$$

***Example* 14.6**

A vapour compression plant uses R-134a as refrigerant. The evaporator temperature is –10°C. The condenser pressure is 7.675 bar and there is no subcooling of the condensate. The flash chamber pressure is 4.139 bar and the dry saturated vapour bled off from the flash chamber is mixed with refrigerant from the LP compressor before entering the HP compressor. The liquid from the flash chamber is throttled before entering the evaporator. Assuming isentropic compression, determine (a) the COP of the plant, (b) the mass flow rate of refrigerant in the evaporator when the power input to the compressor is 100 kW.

Solution From property tables of R-134a, the enthalpy of saturated liquid leaving the condenser at 30°C (p_{sat} = 7.675 bar) is

$$h_7 = h_8 = h_f + x_8 h_{fg}$$

$$\therefore \quad x_8 = \frac{140.96 - 113.29}{300.5 - 113.29} = 0.1478 \text{ kg/kg refrigerant}$$

Now, h_1 = 288.86 kJ/kg, s_1 = 1.7189 kJ/kg. K = s_2

By interpolation, h_2 = 303.468 kJ/kg

By energy balance,

$$h_5 = (1 - m_1)\, h_2 + m_1 h_g$$
$$= (1 - 0.1478) \times 303.468 + 0.1478 \times 300.5$$
$$= 303.0295 \text{ kJ/kg}$$

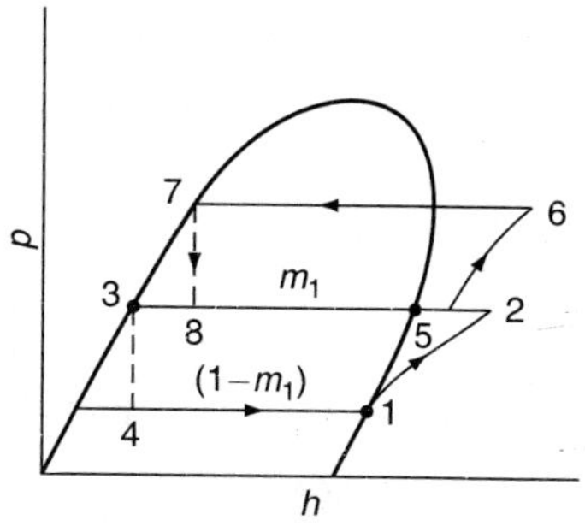

Fig. Ex. 14.6

By interpolation, $s_5 = s_6 = 1.7174$ kJ/kgK

$\therefore \quad h_6 = 315.79$ kJ/kg

Since work input = 100 kW,

$$\dot{m}(h_6 - h_5) + (1 - m_1)\dot{m}(h_2 - h_1) = 100$$

$$\therefore \quad \dot{m}(315.79 - 303.0295) + (1 - 0.1478)\dot{m}(303.468 - 288.86) = 100$$

$$\therefore \quad \dot{m} = 3.966 \text{ kg/s}$$

Mass flow rate of refrigerant through the evaporator

$$= (1 - m_1)\dot{m} = (1 - 0.1478) \times 3.966$$
$$= 3.38 \text{ kg/s.}$$

Ans. (*b*)

$$\text{COP} = \frac{3.38(288.96 - 113.29)}{100} = 5.93$$

Ans. (*a*)

Example 14.7

In an aqua-ammonia absorption refrigerator system, heat is supplied to the generator by condensing steam at 0.2 MPa, 90% quality. The temperature to be maintained in the refrigerator is –10°C, and the ambient temperature is 30°C. Estimate the maximum COP of the refrigerator.

If the actual COP is 40% of the maximum COP and the refrigeration load is 20 tonnes, what will the required steam flow rate be?

Solution At 0.2 MPa, from the steam table

$$t_{sat} = 120.2°C, h_{fg} = 2201.9 \text{ kJ/kg}$$

The maximum COP of the absorption refrigeration system is given by Eq. (14.4)

$$(\text{COP})_{max} = \frac{(T_1 - T_2)T_R}{(T_2 - T_R)T_1}$$

where T_1 = generator temperature

$= 120.2 + 273 = 393.2$ K

T_2 = condenser and absorber temperature

$= 30 + 273 = 303$ K

T_R = evaporator temperature

$= -10 + 273 = 263$ K

$$\therefore \quad (\text{COP})_{max} = \frac{(393.2 - 303) \times 263}{(303 - 263) \times 393.2} = \frac{90.2 \times 263}{40 \times 393.2} = 1.5$$

$\therefore$ Actual COP $= 1.5 \times 0.4 = 0.60$

Since $\text{COP} = \frac{Q_E}{Q_G}$

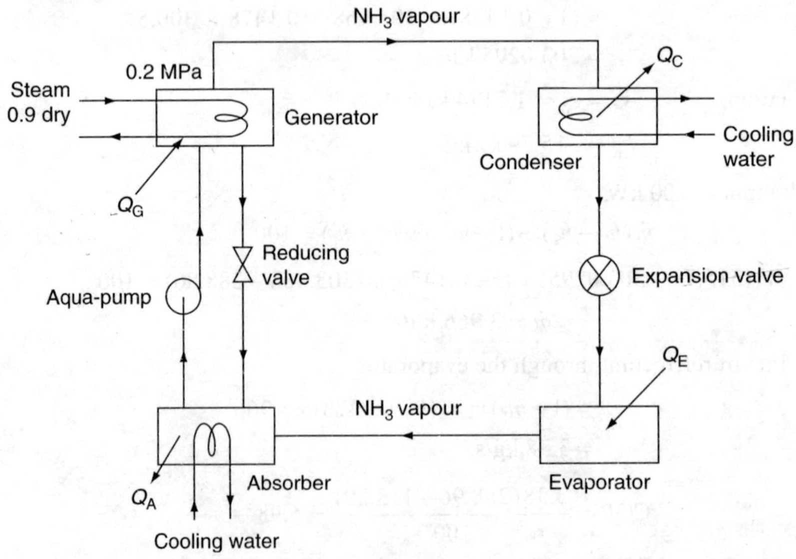

Fig. Ex. 14.7

$$Q_G = \frac{Q_E}{COP} = \frac{20 \times 14000}{0.60 \times 3600} = 129.6 \text{ kW}$$

Heat transferred by 1 kg of steam on condensation

$$= (h_f + xh_{fg}) - h_f = 0.9 \times 2201.9 = 1981.71 \text{ kJ/kg}$$

∴ Steam flow rate required

$$= \frac{129.6}{1981.71} = \mathbf{0.0654 \text{ kg / s}}$$

***Example* 14.8**

In a steam-jet refrigeration system, the flash chamber is maintained at 5°C. The quality of water vapour leaving the evaporator is 0.98 and the chilled water returns to the flash chamber at 14°C and the make-up water is available at 30°C. Estimate the COP of the system if the mass-flow rate of the chilled water is 12 kg/s, compressor efficiency is 0.80 and the condenser pressure is 0.10 bar.

Solution From steam tables, at 14°C, $h_f = 58.62$ kJ/kg

at 5°C, $h_f = 20.93$ kJ/kg

at 30°C, $h_f = 125.73$ kJ/kg

Enthalpy at water vapour at 5°C and 0.98 quality

$$h_v = 0.98 \times 2510.7 = 2460.49 \text{ kJ/kg}$$

$$\text{Refrigerating capacity} = \dot{m}_w (h_{w_i} - h_{w_o})$$

$$= 12(58.62 - 20.93)$$

$$= 452.28 \text{ kJ/s} / 3.5 = 129.22 \text{ TR}$$

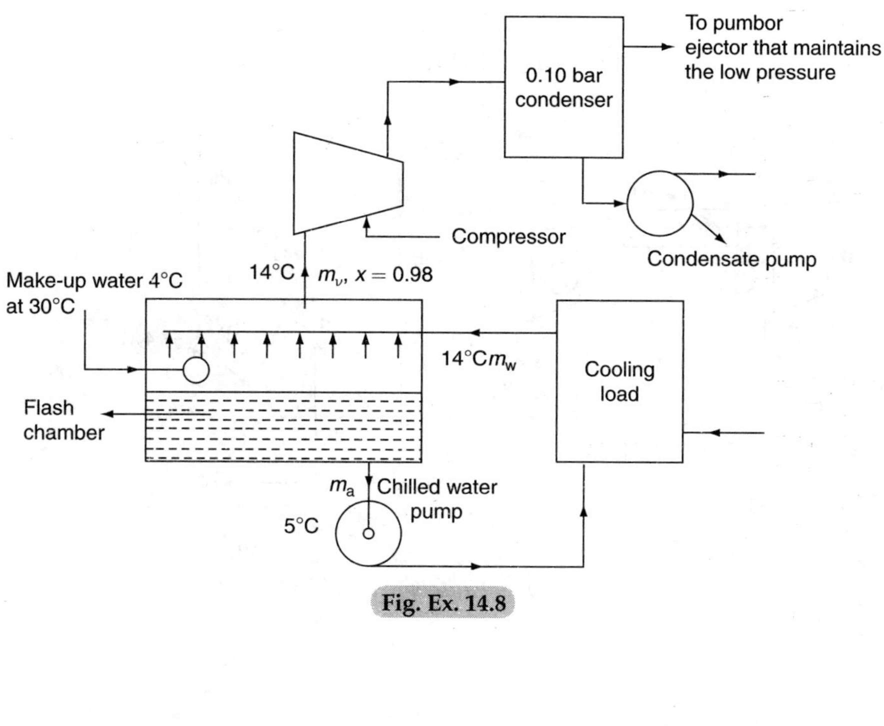

Fig. Ex. 14.8

By making an energy balance,

$$\dot{m}_w (h_{w_i} - h_{w_o}) = \dot{m}_\upsilon (h_{\upsilon_i} - h_{m,k})$$

where $h_{m,k}$ is the enthalpy of make-up water and $\dot{m}_\upsilon$ is the mass of vapour discharged from the evaporator/flash chamber.

$$\dot{m}_\upsilon = \frac{452.28}{2460.49 - 125.73} = 0.1937 \text{ kg/s}$$

For isentropic compression, the enthalpy of vapour leaving the compressor at 0.10 bar, $h_g = 2800$ kJ/kg

$$\text{Work input} = 0.1937\,(2800 - 2460.49)/0.8 = 82.204 \text{ kW}$$

$$\therefore \quad \text{COP} = \frac{452.28}{82.204} = 5.5 \qquad \textit{Ans.}$$

Example 14.9

In an aircraft cooling system, air enters the compressor at 0.1 MPa, 4°C, and is compressed to 0.3 MPa with an insentropic efficiency of 72%. After being cooled to 55°C at constant pressure in a heat exchanger the air then expands in a turbine to 0.1 MPa with an isentopic efficiency of 78%. The low temperature air absorbs a cooling load of 3 tonnes of refrigeration at constant pressure before re-entering the compressor which is driven by the turbine. Assuming air to be an ideal gas, determine the COP of the refrigerator, the driving power required, and the air mass flow rate.

Solution Given: (Fig. Ex. 14.9)

$$T_1 = 277 \text{ K}, T_3 = 273 + 55 = 328 \text{ K}$$

$$\frac{T_{2s}}{T_1} = \left(\frac{p_2}{p_1}\right)^{(\gamma-1)/\gamma}$$

$$\therefore \quad T_{2s} = 277(3)^{0.4/1.4} = 379 \text{ K}$$

$$T_{2s} - T_1 = 102 \text{ K}$$

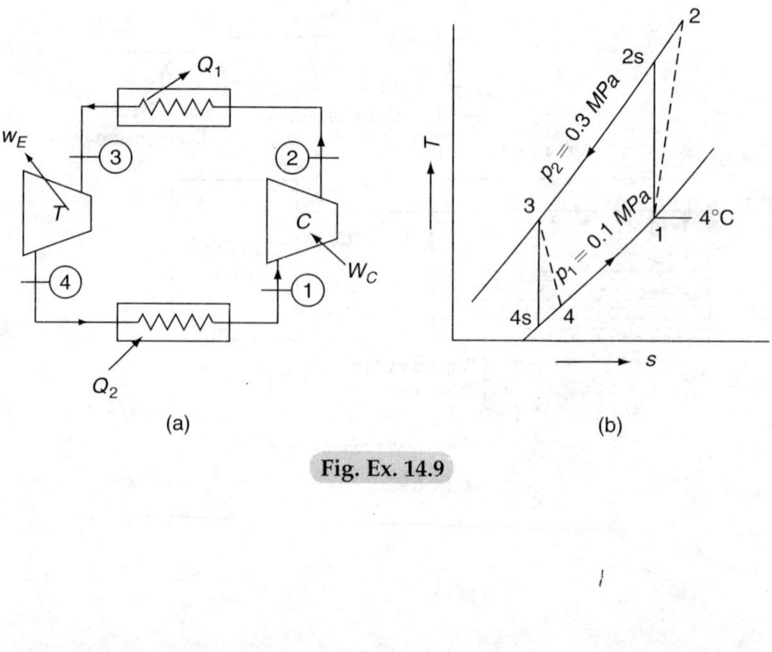

Fig. Ex. 14.9

$$\therefore \quad T_2 - T_1 = \frac{102}{0.72} = 141.8 \text{ K}$$

$$\frac{T_{4s}}{T_3} = \left(\frac{p_2}{p_1}\right)^{(\gamma-1)/\gamma}$$

$$\therefore \quad T_{4s} = 328(3)^{0.4/1.4} = \frac{328}{1.368} = 240 \text{ K}$$

$$T_3 - T_{4s} = 88 \text{ K}$$

$$\therefore \quad T_3 - T_4 = 0.78 \times 88 = 68.6 \text{ K}$$

$$\therefore \quad T_4 = 259.4 \text{ K}$$

Refrigerating effect $= c_p (T_1 - T_4) = 17.6\, c_p$ kJ/kg

Net work input $= c_p[(T_2 - T_1) - (T_3 - T_4)]$

$= c_p[141.8 - 68.6] = 73.2\, c_p$ kJ/kg

$$\text{COP} = \frac{17.6\, c_p}{73.2\, c_p} = \mathbf{0.24}$$

$$\text{Driving power required} = \frac{3 \times 14000}{0.24 \times 3600} = \mathbf{48.6\ kW}$$

$$\text{Mass flow rate of air} = \frac{3 \times 14000}{1.005 \times 17.6} = 2374.5 \text{ kg/h} = \mathbf{0.66\ kg/s}$$

Example 14.10

A vapour-compression heat pump system uses R-12 as the working fluid. The refrigerant enters the compressor at 2.4 bar. 0°C with a volumetric flow rate of 0.6 m³/min. Compression is adiabatic to 9 bar, 60°C and the saturated liquid exits the condenser at 9 bar. Determine (a) the power input to the compressor, (b) the heating capacity of the system, (c) the coefficient of performance, (d) the isentropic compressor efficiency.

Solution

At $p_1 = 2.4$ bar, $T_1 = 0$°C

$h_1 = 188.99$ kJ/kg

$s_1 = 0.7177$ kJ/kg K

$v_1 = 0.0703$ m³/kg

At $p_2 = 9$ bar, $T_2 = 60$°C

$h_2 = 219.37$ kJ/kg

$h_2s = 213.27$ kJ/kg

At $p_2 = 9$ bar, sat. liquid

$h_3 = 71.93$ kJ/kg $= h_4$

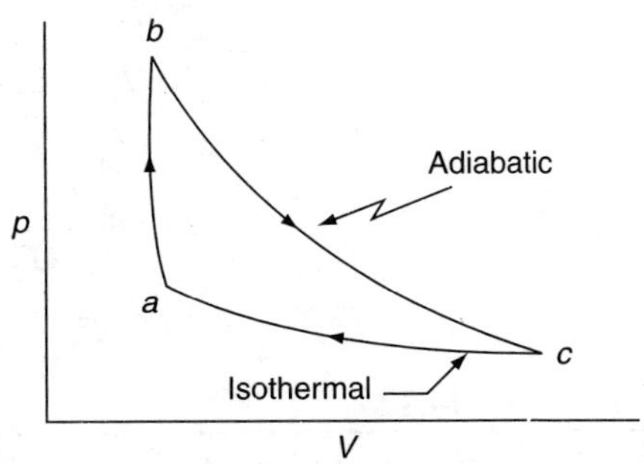

Fig. Ex. 14.10

$$\dot{m} = \frac{A_1 V_1}{v_1} = \frac{0.6}{60}\frac{m^3}{s} \times \frac{1}{0.0703} \text{kg m}^3 = 0.1422 \text{ kg/s}$$

(a) Power input $\dot{W}_c = \dot{m}(h_2 - h_1)$

$= 0.1422\ (219.37 - 188.99) = 4.32$ kW *Ans.*

(b) Heating capacity, $\dot{Q}_1 = \dot{m}(h_2 - h_3)$

$= 0.1422\ (219.37 - 71.93) = 20.97$ kW $= 5.963$ tonnes *Ans.*

(c) $\text{COP} = \frac{\dot{Q}_1}{\dot{W}_c} = \frac{20.97}{4.32} = 4.854$ *Ans.*

(d) $\eta_{is} = \frac{h_{2s} - h_1}{h_2 - h_1} = \frac{213.27 - 188.99}{219.37 - 188.99} = 0.799$ or 79.9% *Ans.*

Example 14.11

In air-refrigeration system working on reversed Brayton cycle, the temperature of air at entrance to compressor (pressure ratio = 4, efficiency = 0.8) is 275 K and the inlet pressure is 1 bar. The pressure loss in the cooler is 0.1bar and in the cold chamber it is 0.08 bar. The temperature of air at turbine (efficiency = 0.85) inlet is 310 K. Estimate the pressure ratio for the turbine and the COP of the cycle.

Solution $T_1 = 275$ K, $T_3 = 310$ K

$\frac{p_2}{p_1} = 4, p_1 = 1$ bar

$T_{2s} = T_1\,(p_2/p_1) = 275 \times 40.286 = 408.81$ K

$T_2 = T_1 + (T_{2s} - T_1)\,\eta_c = 442.26$ K

$p_3 = p_2 - pr$ loss in the cooler $= 4 - 0.1 = 3.9$ bar

$p_4 = p_1 + pr$ loss in the cold chamber $= 1 + 0.08$

$= 1.08$ bar

$\therefore$ Pressure ratio for the turbine $= \frac{3.9}{1.08} = 3.611$ *Ans.*

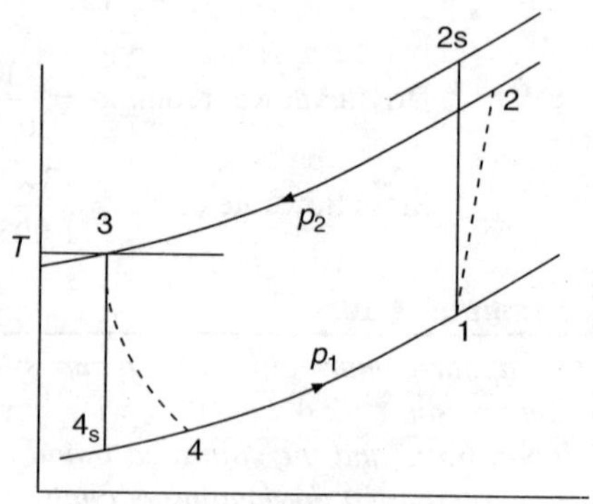

Fig. Ex. 14.11

$$\frac{T_3}{T_{4s}} = (p_3/p_{4s})^{0.286} = (3.611)^{0.286} = 1.4437$$

$$T_{4s} = 214.72 \text{ K}$$

$$T_4 = T_3 - (T_3 - T_{4s})\eta_t = 310 - (310 - 214.72) \times 0.85 = 229.015 \text{ K}$$

$$\text{COP} = \frac{T_1 - T_4}{(T_2 - T_3) - (T_1 - T_4)} = \frac{275 - 229.015}{(442.26 - 310) - (275 - 229.015)}$$

$= 0.533$ *Ans.*

Example 14.12

The cabin of a jet plane having a cooling load of 60 kW is to be maintained at 1 bar, 20°C. The aircraft is flying at a speed of 90 km/h in air where the pressure is 0.35 bar and temperature 255 K. The air flows through the diffuser of efficiency 85% and a compressor of pressure ratio 6 and efficiency 85% before entering the air cooler having an effectiveness of 0.9. The pressure drop in the air cooler is 0.08 bar. The air should leave the turbine having an efficiency of 88% at 1.08 bar. Determine the mass flow rate of air flowing through the cooling system and the COP.

Solution With reference to Fig. 14.37,

$$p_1 = 0.35 \text{ bar}, T_1 = 255 \text{ K}$$

Speed of aircraft, $$V = \frac{900{,}000}{36{,}00} = 250 \text{ m/s}$$

Stagnation temperature, $$T_2' = T_2 = T_1 + \frac{V^2}{2c_p}$$

$\therefore$ $$T_2' = T_2 = 255 + \frac{250^2}{2 \times 1.005} = 286.1 \text{ K}$$

Stagnation pressure, $$p_2' = p_1 \left(\frac{T_2'}{T_1}\right)^{\frac{\gamma}{\gamma-1}} = 0.35 \left(\frac{286.1}{2.55}\right)^{3.5} = 0.523 \text{ bar}$$

Actual pressure, $$p_2 = p_1 + 0.85\,(p_2' - p_1) = 0.497 \text{ bar}$$

$$p_3 = 6p_2 = 6 \times 0.497 = 2.982 \text{ bar}$$

$$T_3' = T_2 \left(\frac{p_3}{p_2}\right)^{\frac{\gamma-1}{\gamma}} = 286.1\,(6)^{0.286} = 477.6 \text{ K}$$

$$T_3 = T_2 + \frac{T_3' - T_2}{\eta_c} = 286.1 + \frac{477.6 - 286.1}{0.85} = 511.28 \text{ K}$$

Power required for compressor work per kg

$$= c_p(T_3 - T_2) = 1.005\,(511.28 – 286.1) = 226.3 \text{ kJ/kg}$$

Pressure at turbine inlet, $p_4 = p_3 -$ losses in cooler

$$= 2.982 - 0.08 = 2.902 \text{ bar}$$

Since the air is cooled by the air coming out of the diffuser, the effectiveness

$$\varepsilon = \frac{T_3 - T_4}{T_3 - T_2}$$

$\therefore$ $$T_4 = T_3 - \varepsilon\,(T_3 - T_2)$$

$$= 511.28 - 0.9\,(511.28 - 286.1) = 308.6 \text{ K}$$

$$T_3' = T_4 / (p_4 / p_5)^{0.286} = 308.6 / (2.902 / 1.08)^{0.286} = 232.6 \text{ K}$$

$$T_5 = T_4 - \frac{T_4 - T_5'}{\eta_t} = 308.6 - \frac{308.6 - 232.6}{0.88} = 241.75 \text{ K}$$

Refrigerating (cooling) effect $= c_p\,(T_{\text{cabin}} - T_5)$

$$= 1.005\,(293 - 241.75) = 51.5 \text{ kJ/kg}$$

Cooling load $= 60 \text{ kW} = \dot{m} \times 51.5$

where $\dot{m}$ = mass flow rate of air through the cooling system $= \dfrac{60}{51.5} = 1.165$ kg/s *Ans.*

Power required to drive the compressor

$$= \dot{m}\,(h_3 - h_2) = 1.165 \times 226.3 = 263.55 \text{ kW}$$

$$\therefore \quad \text{COP} = \frac{60}{263.55} = 0.23 \qquad \textit{Ans.}$$

Review Questions

14.1 What is refrigeration ? How is (a) ice and (b) dry ice used for the purpose of refrigeration?

14.2 Explain the vapour compression cycle with the help of flow, *T–s* and *p–h* diagrams. Can this cycle be reversible? If not, why?

14.3 What is superheat horn?

14.4 What do you understand by dry and wet compression? Which is preferred and why?

14.5 What is refrigerating effect?

14.6 Explain the effect of evaporator pressure on the performance of a vapour compression refrigeration plant.

14.7 Estimate the effect of suction vapour superheat on the vapour compression cycle.

14.8 What is the effect of liquid subcooling on the vapour compression cycle?

14.9 What is a tonne of refrigeration?

14.10 Explain the effect of superheat and subcooling on the vapour compression cycle.

14.11 How does the actual vapour compression cycle differ from the ideal one?

14.12 What is a suction line heat exchanger? When is it used?

14.13 What are the expansion devices used in vapour compression plant? When are they used?

14.14 What are the different types of compressors used in vapour compression plants and what are their applications?

14.15 What is a multistage vapour compression plant? When is it used?

14.16 What is a flash chamber? What is its advantage?

14.17 What are the most widely used refrigerants?

14.18 How are refrigerants numbered?

14.19 Why is the use of halogenated hydrocarbons as refrigerants now discouraged?

14.20 What are the effects of CFCs on the environment? How do they affect the ozone layer?

14.21 What is ODP? Why has R-22 less ODP than R-12?

14.22 What are the parameters to be considered in the selection of a refrigerant?

14.23 What do you understand by cascade refrigeration system? Explain it with the help of flow and *T-s* diagrams.

14.24 Evaluate ammonia as a refrigerant.

14.25 What is an absorption refrigeration cycle? How does it differ from a vapour compression cycle?

14.26 How is the refrigerant liberated from the absorbent in an aqua-ammonia system?

14.27 What are the functions of the analyzer and the rectifier?

14.28 What is the advantage of using a suction line heat exchanger in an absorption refrigeration system?

14.29 In a lithium bromide-water absorption system, which is the refrigerant? What is its limitation?

14.30 Derive the expression for the maximum COP of an absorption refrigeration system.

14.31 With a neat sketch explain the mechanism of operation of an electrolux refrigerator.

14.32 Evaluate water as a refrigerant. Explain the principle of steam-jet refrigeration with a neat sketch. What are its applications?

14.33 How does a heat pump upgrade low grade reject heat?

14.34 How can a heat pump be used for (a) space heating (b) year-round airconditioning?

14.35 How is a reversed Brayton cycle used for refrigeration?

14.36 Why is the COP of a gas cycle refrigeration system low?

14.37 Explain the flow and *T-s* diagrams of a simple aircraft refrigeration cycle.

14.38 What is a bootstrap system? How is it different from the simple aircraft refrigeration cycle?

14.39 Why is gas cycle refrigeration preferred in aircraft?

14.40 What is the principle of the Linde-Hampson system for liquefaction of air?

14.41 Derive the expressions of liquid yield and the minimum work requirement in a Linde-Hampson system.

14.42 How does Claude cycle differ from a Linde-Hampson cycle in the context of the liquefaction of air?

14.43 With the help of flow and *p–h* diagrams, explain how dry ice is produced.

Problems

14.1 A refrigerator using R-134*a* operates on an ideal vapour compression cycle between 0.12 and 0.7 MPa. The mass flow of refrigerant is 0.05 kg/s. Determine (a) the rate of heat removal from the refrigerated space, (b) the power input to the compressor, (c) the heat rejection to the environment, and (d) the COP.

Ans. (a) 7.35 kW, (b) 1.85 kW, (c) 9.20 kW, (d) 3.97

14.2 A Refrigerant-12 vapour compression cycle has a refrigeration load of 3 tonnes. The evaporator and condenser temperatures are –20°C and 40°C respectively. Find (a) the refrigerant flow rate in kg/s, (b) the volume flow rate handled by the compressor in m^3/s, (c) the work input to the compressor in kW. (d) the heat rejected in the condenser in kW, and (e) the isentropic discharge temperature.

If there is 5 C deg. of superheating of vapour before it enters the compressor, and 5 C deg. subcooling of liquid before it flows the through the expansion valve, determine the above quantities.

14.3 A 5 tonne R-12 plant maintains a cold store at –15°C. The refrigerant flow rate is 0.133kg/s. The vapour leaves the evaporator with 5 C deg. superheat. Cooling water is available in plenty at 25°C. A suction line heat exchanger subcools the refrigerant before throttling. Find (a) the compressor discharge temperature, (b) the COP, (c) the amount of subcooling in C deg., and (d) the cylinder dimensions of the compressor, if the speed is 900 rpm, stroke-to-bore ratio is 1.2, and volumetric efficiency is 95 %.

Allow approximately 5°C temperature difference in the evaporator and condenser.

Ans. (a) 66°C, (b) 4.1 (c) 125°C, (d) 104.5 mm, 125mm

14.4 A vapour compression refrigeration system uses R-12 and operates between pressure limits of 0.745 and 0.15 MPa. The vapour entering the compressor has a temperature of –10°C and the liquid leaving the condenser is at 28°C. A refrigerating load of 2 kW is required. Determine the COP and the swept volume of the compressor if it has a volumetric efficiency of 76% and runs at 600 rpm.

Ans. 4.15, 243 cm^3

14.5 A food-freezing system requires 20 tonnes of refrigeration at an evaporator temperature of –35°C and a condenser temperature of 25°C. The refrigerant, R-12, is subcooled 4°C before entering the expansion valve, and the vapour is superheated 5°C before leaving the evaporator. A six-cylinder single-acting compressor with stroke equal to bore is to be used, operating at 1500 rpm. Determine (a) the refrigerating effect, (b) the refrigerant flow rate, (c) the theoretical piston displacement per sec, (d) the theoretical power required in kW, (e) the COP, the (f) the heat removed in the condenser, and (g) the bore and stroke of the compressor.

14.6 A R-12 vapour compression refrigeration system is operation at a condenser pressure of 9.6 bar and an evaporator pressure of 2.19 bar. Its refrigeration capacity is 15 tonnes. The values of enthalpy at the inlet and outlet of the evaporator are 64.6 and 195.7 kJ/kg. The specific volume at inlet to the reciprocating compressor is 0.082 m^3/kg. The index of compression for the compressor is 1.13. Determine: (a) the power input in kW required for the compressor, and (b) the COP. Take 1 tonne of refrigeration as equivalent to heat removal at the rate of 3.517 kW.

Ans. (a) 11.57 kW, (b) 4.56

14.7 A refrigeration plant produces 0.139 kg/s of the ice at –5°C from water at 30°C. If the power required to drive the plant is 22 kW, determine the capacity of the ice plant in tones and the actual COP (c_p of ice = 2.1 kJ/kg K).

14.8 An air conditioning unit using R-12 as a refrigerant is mounted in the window of a room. During steady operation 1.5 kW of heat is transferred from the air in the room to evaporator coils of R-12. If this air is at 22°C and the temperature of R-12 in the evaporator is 15°C, determine (a) the refrigerant flow rate, and (b) the minimum power required to drive the compressor if the outside air is at 43°C and the temperature of the refrigerant during condensation is 50°C.

14.9 In a solar energy operated aqua-ammonia absorption refrigeration system, water is cooled at the rate of 10 kg/s from 38°C to 13°C. If the incident solar energy is 640 W/m^2 and the COP of the system is 0.32, estimate the area of the solar collector needed.

14.10 A gas refrigerating system using air as a refrigerant is to work between –12°C and 27°C using an ideal reversed Brayton cycle of pressure ratio 5 and minimum pressure 1 atm, and to maintain a load of 10 tonnes. Find (a) the COP, (b) the air flow rate in kg/s, (c) the volume flow rate entering the compressor in m^3/s, and (d) the maximum and minimum temperatures of the cycle.

14.11 An open cycle (Brayton) aircraft cabin cooler expands air at 27°C through a turbine which is 30% efficient from 2 to 1 atm. The cabin temperature is not to exceed 24°C. Estimate the mass flow rate of air required (kg/s) for each tonne of cooling.

14.12 Determine the ideal COP of an absorption refrigerating system in which the heating, cooling, and refrigeration take place at 197°C, 17°C, and –3°C respectively. Ans. 5.16

14.13 Heat pump is to use an R-12 cycle to operate between outdoor air at –1°C and air in a domestic heating system at 40°C. The temperature difference in the evaporator and the condenser is 8°C. The compressor efficiency is 80%, and the compression begins with saturated vapour. The expansion begins with saturated liquid. The combined efficiency of the motor and belt drive is 75%. If the required heat supply to the warm air is 43.6 kW, what will be the electrical load in kW?

14.14 An ideal (Carnot) refrigeration system operates between the temperature limits of –30°C and 25°C. Find the ideal COP and the power required from an external source to absorb 3.89 kW at high temperature.

14.15 A heat pump that operates on the ideal vapour compression cycle with R-134*a* is used to heat a house and maintain it at 20°C, using underground water at 10°C as the heat source. The house is losing heat at a rate of 75 MJ/h. The evaporator and condenser pressures are 320 and 800 kPa respectively. Determine the power input to the heat pump and the electric power saved by using a heat pump instead of a resistance heater.

Ans. 2.27 kW, 18.56 kW

14.16 An ammonia-absorption system has an evaporator temperature of –12°C and a condenser temperature of 50°C. The generator temperature is 150°C. In this cycle, 0.42 kJ is transferred to the ammonia in the evaporator for each kJ transferred to the ammonia solution in the generator from the high temperature source.

It is desired to compare the performance of this cycle with the performance of a similar vapour compression cycle. For this, it is assumed that a reservoir is available at 150°C, and that heat is transferred from this reservoir to a reversible engine which rejects heat to the surroundings at 25°C. This work is then used to drive an ideal vapour compression system with ammonia as the refrigerant. Compare the amount of refrigeration that can be achieved per kJ from the high temperature source in this case with the 0.42 kJ that can be achieved in the absorption system.

14.17 An R-12 plant is to cool milk from 30°C to 1°C involving a refrigeration capacity of 10 tonnes. Cooling water for the condenser is available at 25°C and 5 C deg. rise in its temperature is allowable. Determine the suitable condensing and evaporating temperatures, providing a minimum of 5 C deg. differential, and calculate the theoretical power required in kW and the cooling water requirement in kg/s. Also, find the percentage of flash gas at the end of the throttling. Assume a 2 C deg. subcooling in the liquid refrigerant leaving the condenser.

14.18 The following data pertain to an air cycle refrigeration system for an aircraft.

Capacity 5 tonnes

Cabin air inlet temperature 15°C and outlet temperature 25°C

Pressure ratio across the compressor 5

The aircraft is flying at 0.278 km/s where the ambient conditions are 0°C and 80 kPa. Find the COP and the cooling effectiveness of the exchanger. The cabin is at 0.1 MPa, and the cooling turbine powers the circulating fans.

14.19 A water cooler supplies chilled water at 7°C when water is supplied to it at 27°C at a rate of 0.7 litres/min., while the power consumed amounts to 200 watts. Compare the COP of this refrigeration plant with that of the ideal refrigeration cycle for a similar situation.

14.20 A refrigerating plant of 8 tonnes capacity has an evaporation temperature of – 8°C and condenser temperature of 30°C. The refrigerant, R-12, is subcooled 5°C before entering the expansion valve and the vapour is superheated 6°C before leaving the evaporator coil. The compression of the refrigerant is isentropic. If there is a suction pressure drop of 20 kPa through the valves, and discharge pressure drop of 10 kPa through the valves, determine the COP of the plant, theoretical piston displacement per sec. and the heat removal rate in the condenser.

14.21 An ultra-low-temperature freezer system employs a coupling of two vapour compression cycles of R-12 and R-13, as shown in Fig. P. 14.21. The states and properties of both cycles are shown on the *T–s* plot. Determine the ratio of the circulation rates of the two refrigerants, w_1/w_2 and the overall COP. How does this COP compare with the Carnot COP operating between 42°C and –70°C?

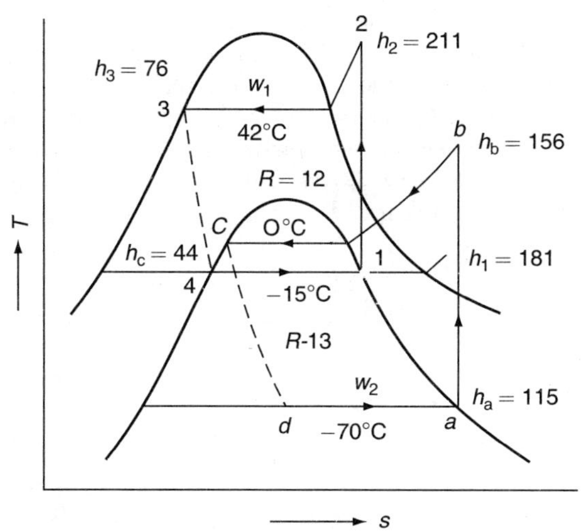

Fig. P. 14.21

14.22 Derive an expression for the COP of an ideal gas refrigeration cycle with a regenerative heat exchanger. Express the result in terms of the minimum gas temperature during heat rejection (T_h) maximum gas temperature during heat absorption (T_1), and pressure ratio for the cycle (p_2/p_1).

$$\text{Ans. COP} = \frac{T_1}{T_h r_p^{(\gamma-1)/\gamma} - T_1}$$

14.23 Large quantities of electrical power can be transmitted with relatively little loss when the transmission cable is cooled to a superconducting temperature. A regenerated gas refrigeration cycle operating with helium is used to maintain an electrical cable at 15 K. If the pressure ratio is 10 and heat is rejected directly to the atmosphere at 300 K, determine the COP and the performance ratio with respect to the Carnot cycle.

Ans. 0.02, 0.38

14.24 A 100 tonne refrigerating plant using R-12 has a condensing temp. of 35°C and an evaporating temp. of 5°C. Calculate the power requirement of the compressor in kW, the volume flow rate of compressor and the compressor displacement volume if the volumetric effy. is 75% and the mechanical efficiency is 80%.

If a liquid suction heat exchanger is installed in the above plant which subcools the condensed refrigerant to 30°C, what would be the refrigeration capacity of the plant and the power required by the compressor?

14.25 A heat pump installation is proposed for a home heating unit with an output rated at 30 kW. The evaporator temperature is 10°C and the condenser pressure is 0.5 bar. Using an ideal vapour cycle, estimate the power required to drive the compressor if steam/water mixture is used as the working fluid, the COP and the mass flow rate of the fluid. Assume saturated vapour at compressor inlet and saturated liquid at condenser outlet. Ans. 8.0 kW, 3.77, 0.001012 kg/s

14.26 A 100 tonne low temperature R-12 system is to operate on a 2-stage vapour compression refrigeration cycle with a flash chamber, with the refrigerant evaporating at –40°C, an intermediate pressure of 2.1912 bar, and condensation at 30°C. Saturated vapour enters both the compressors and saturated liquid enters each expansion valve. Consider bothe stages of compression to be isentropic. Determine:

(a) The flow rate of refrigerant handled by each compressor. (b) the total power required to drive the compressor, (c) the piston displacement of each compressor, if the clearance is 2.5% for each machine, and (d) the COP of the system, (e) What would have been the refrigerant flow rate, the total work of compression, the piston displacement in each compressor and the compressor and the COP, if the compression had occurred in a single stage?

Ans. (a) 2.464, 3.387 kg/s, (b) 123 kW, (c) 0.6274, 0.314 m^3/s, (d) 2.86, (e) 3.349 kg/s, 144.54 kW, 1.0236 m^3/s, 2.433

14.27 A vapour compression plant uses R-134a as refrigerant. The evaporator temperature is −10°C. The condenser pressure is 7.675 bar and there is no subcooling of the condensate. The flash chamber pressure is 4.139 bar and the dry saturated vapour bled off from the flash chamber is mixed with the refrigerant from the LP compressor before entering the HP compressor. The liquid from the flash chamber is throttled before entering the evaporator. Assuming isentropic compression, calculate (a) COP of the plant (b) mass flow of refrigerant in the evaporator when the power input to the plant is 100 kW. Ans. (a) 5.93, (b) 3.38 kg/s

14.28 The working fluid in a heat pump installation is ammonia. The ammonia after evaporation to a dry saturated state at 2°C is compressed to a pressure of 12.38 bar at which it is cooled and condensed to a saturated liquid state. It then passes through a throttle valve and returns to the evaporator. Calculate the COP assuming that the isentropic efficiency of the compressor is 0.85. Ans. 7.88

14.29 (a) A R-12 simple saturation cycle operates at temperatures of 35ºC and –15ºC for the condenser and evaporator respectively. Determine the COP and power input per TR.

(b) If a liquid-vapour heat exchanger is installed in the system, with the temperature of the vapour leaving the heat exchanger at 15ºC, what is the change in COP and power input per TR.

Ans. (a) 4.09, 0.865 kW/TR, (b) +2.56%, –2.5%

14.30 The cabin of a jet plane has an air-cooling system operating on a simple cycle. It is maintained at 25ºC. The ambient air condition at the altitude of flight is 0.35 bar, –15ºC. The pressure ratio of the compressor is 3. The plane is flying at a speed of 1000 km/h. The pressure drop in the air cooler is 0.1 bar. Air leaves the cooling turbine at 1.06 bar. The pressure in the cabin is 101.325 kpa. The cooling load of the cabin is 58 kW. Determine (a) the stagnation temperature and pressure of the air entering the compressor, (b) mass-flow rate of air flowing through the cooling system, (c) volume flow rate of air handled by the compressor and expander, (d) net power delivered by the engine to the cooling system, (e) the COP of the system.

Ans. (a) 296.5 K, 0.57 bar, (b) 5950 kg/h, (c) 9050 m^3/h, 430 m^3/h, (d) kW, (e) 0.3

14.31 A simple R-134a heat pump for space heating operates between temperature limits of 15 and 50ºC. The heat required to be pumped is 100 MJ/h. Determine (a) the quality of refrigerant entering the evaporator, (b) the discharge temperature from the compressor, assuming the specific heat of vapour as 0.996 kJ/kgK, (c) the theoretical piston displacement of the compressor, given the specific volume at suction as 0.04185 m^3/kg, (d) the theoretical power input, (e) the COP. Ans. (a) 0.2675, (b) 327.15 K, (c) 7.156×10^{-3} m^3/s, (d) 3.56 kW, (e) 8.1

CHAPTER 15

Psychrometrics and Air-Conditioning Systems

The singlemost important application of refrigeration is in air conditioning which means simultaneous control of the dry bulb temperature, humidity, cleanliness, ventilation and distribution of air and air motion in a place besides the control of contaminated gases.

15.1 PROPERTIES OF ATMOSPHERIC AIR

The working substance in air conditioning is the atmospheric or *moist air* which is a mixture of two gases, viz., *dry air* which is a mixture of a number of gases like N_2, O_2, A, CO_2 etc., as given in Table 15.1 and *water vapour* which may exist in a saturated or superheated state. A stated in the first chapter, a pure substance is homogeneous and invariable in chemical composition. Both dry air and water vapour are individually pure substances. But moist air is not a pure substance when condensation or evaporation of moisture can take place which varies the composition of the mixture. Thus, moist air consists of two parts: one comprising dry air, considered as the *fixed part*, and the other, solely of water vapour, considered as the *variable part*.

Completely dry air does not exist in nature. Water vapour in varying amounts is diffused through it. Both dry air and water vapour can be considered as ideal gases since both exist in the atmosphere at low pressures. Hence, ideal gas laws can be applied to them individually. If p_a and p_w are the partial pressures of dry air and water vapour respectively then by *Dalton's law of partial pressures*

$$p = p_a + p_w$$

where p is the atmospheric pressure.

Mole fraction of dry air, x_a

$$x_a = \frac{p_a}{p} = p_a, \text{ since } p = 1 \text{ atm.}$$

and mole fraction of water vapour, x_w

Table 15.1
Composition of Dry Air

Component	*Molecular mass*	*Part by volume*	*Part by mass*
N_2	28.02	0.7803	0.7547
O_2	32.0	0.2099	0.2319
Ar	39.91	0.0094	0.0129
CO_2	44.0	0.0003	0.0005
H_2	2.02	0.0001	0.0000

$$x_w = \frac{p_w}{p} = p_w$$

Let a dry air–water vapour mixture of mass m be contained in a chamber of volume V at the total pressure p and temperature T. If V_1 is the volume of dry air at p and T of the mixture, and V_2 is the volume of water vapour at the same p and T then by *Amagat's law of partial volumes*

$$V = V_1 + V_2$$

where $V_1 = \frac{n_1 \bar{R} T}{p}$ and $V_2 = \frac{n_2 \bar{R} T}{p}$ are the partial volumes of dry air and water vapour respectively, n_1 and n_2 being the respective mole numbers. Therefore, the mole fraction of dry air is

$$x_1 = p_1/p = V_1/V = n_1/n.$$

Now, if M is the molecular mass of the mixture

$$mM = M_1 m_1 + M_2 m_2$$

$$\therefore M = x_1 M_1 + x_2 M_2$$

The *Gibb's theorem* states that the internal energy of a mixture of gases is equal to the sum of the internal energies of individual gases, taken each at the temperature and volume of the mixture. Thus, we have

$$mu = m_1 u_1 + m_2 u_2,\ u = \frac{m_1 u_1 + m_2 u_2}{m_1 + m_2}$$

Similary,

$$mh = m_1 h_1 + m_2 h_2,\ h = \frac{m_1 h_1 + m_2 h_2}{m_1 + m_2}$$

$$mc = m_1 c_1 + m_2 c_2,\ c = \frac{m_1 c_1 + m_2 c_2}{m_1 + m_2}$$

for enthalpy and specific heat respectively.

The molecular mass of atmospheric air is

$$M = x_{N_2} M_{N_2} + x_{O_2} M_{O_2} + x_{Ar} M_{Ar} = 0.78 \times 28 + 0.21 \times 32 + 0.94 \times 40 = 28.96 \text{ kg/kg mol.}$$

15.2 PSYCHROMETRIC PROPERTIES OF AIR

Dry air and water vapour form a binary mixture. From Gibbs phase rule (Eq. 11.45), $f = C - \phi + 2$, where $C = 2$, the number of components, and ϕ = number of phases = 1; therefore, $f = 3$. Thus, three properties are required to completely define its thermodynamic state, unlike a pure substance which requires only two. One of the three properties can be the composition. The properties of moist air are called *psychrometric properties,* and the subject which deals with the behaviour of moist air is known as *psychrometry* or *psychrometrics.*

In air-conditioning practice, all calculations are based on the dry-air part (fixed), since the water-vapour part is continuously variable. For defining and calculating the relevant psychrometric properties, a certain volume V of moist air at pressure p and temperature T is considered (Fig. 15.1). It contains G kg of dry air and m_w kg of water vapour. The actual temperature t of moist air is called the *dry bulb temperature* (DBT). The total pressure p which is equal to the *barometric pressure* is constant.

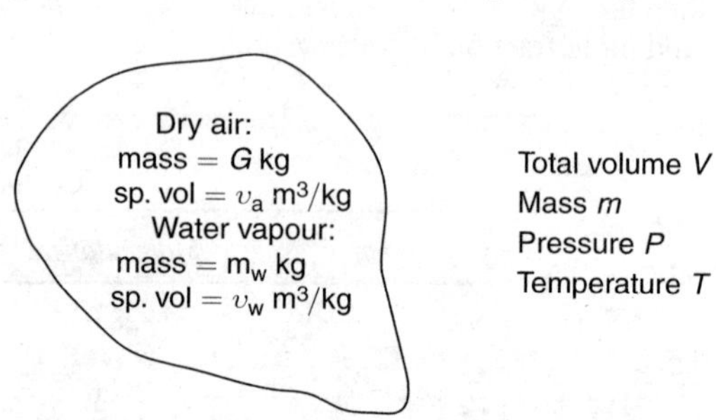

Fig. 15.1 *A mixture of dry air and water vapour (moist air)*

15.2.1 Specific Humidity or Humidity Ratio, W

The specific humidity of humidity ratio is defined as the mass of water vapour or moisture per unit mass of dry air in a mixture of air and water vapour. If G = mass of dry air and m = mass of water vapour, then

$$W = \frac{m}{G} = \frac{V/v_w}{V/v_a} = \frac{v_a}{v_w} \tag{15.1}$$

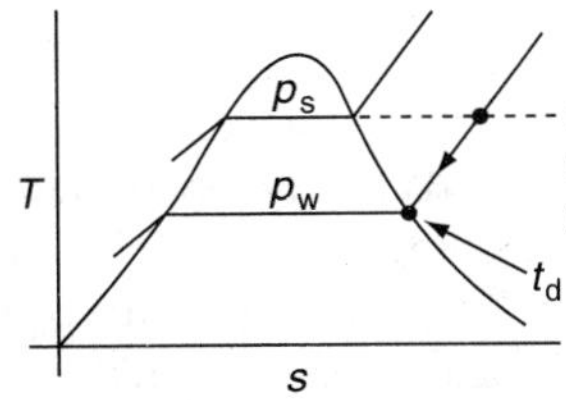

Fig. 15.2 *State of water vapour in moist air*

Since p_w is very small, the saturation temperature at p_w is less than atmospheric temperature t (Fig. 15.2). So the water vapour in air exists in *superheated state*, and air is said to be *unsaturated*.

Since both dry air and water vapour behave individually as ideal gases,

$$p_a v_a = \frac{\bar{R}}{\bar{M}_a} T, \quad p_a v = G \frac{\bar{R}}{\bar{M}_a} T$$

$$p_w v_w = \frac{\bar{R}}{M_w} T, \quad p_w v = m_w \frac{\bar{R}}{M_w} T$$

where subscripts 'a' and 'w' refer to dry air and water vapour respectively and M is the molecular weight.

Substituting in Eq. (15.1), we obtain

$$W = \frac{M_w p_w}{M_a p_a} = \frac{18.016}{28.96} \frac{p_w}{p_a} = 0.622 \frac{p_w}{p_a} \tag{15.2}$$

where W is in kg of water vapour per kg of dry air

Since

$$p = p_a + p_w,$$

$$\therefore \ W = 0.622 \frac{p_w}{p - p_w} \tag{15.3}$$

Since the total atmospheric pressure p remains constant, we see that only

$$W = f(p_w) \tag{15.4}$$

If we take 1 kg of dry air, Viz.,

$$G = 1 \text{ kg}$$

then the mass of water vapour associated with this dry air is

$$m_w = W \text{ kg}.$$

then the total mass of this volume of moist air is

$$m = (1 + W) \text{ kg} \tag{15.5}$$

The specific humidity thus is not a mass fraction of water vapour which would have been $W/(1+W)$, but a ratio of the mass of water vapour to that of dry air in a certain volume of the mixture.

15.2.2 Dew Point Temperature

If a sample of unsaturated moist air containing superheated water vapour is cooled at constant pressure p_w, the mixture will eventually reach the saturation temperature t_d of water vapour corresponding to its partial pressure p_w (Fig. 15.2) at which it starts condensing and dews start appearing. This temperature t_d is called the *dew point temperature* (DPT). It is therefore, the temperature to which moist air must be cooled at constant pressure before condensation of moisture takes place. Moisture can thus be removed from humid air by bringing the air in contact with a cold surface or cooling coil whose temperature is below its dew point temperature. During the process of cooling, the partial pressure p_w of water vapour and specific humidity W remain constant until the vapour starts condensing.

Now, unsaturated moist air with superheated water vapour in a control volume V (Fig. 15.1) at dry bulb temperature T and partial pressure p_w (Fig.15.3) is considered. If now more water vapour is added to this control volume V at this temperature T, the partial pressure p_w will go on increasing, until it reaches a value p_s corresponding to state 2, after which it cannot increase further, since p_s is the saturation pressure or the maximum possible pressure of vapour at temperature T. The water vapour at state 2 is now saturated and the air is said to be *saturated*. In this state, the air is holding the maximum amount of water vapour at temperature T and W is maximum.

$$\therefore W_{max} = W_s = 0.622 \frac{p_s}{p - p_s} \tag{15.6}$$

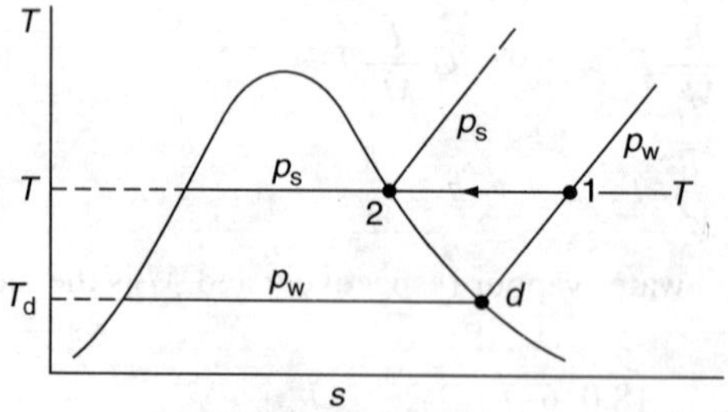

Fig. 15.3 *Isothermal saturation process of moist air*

The ratio of actual specific humidity W to the specific humidity W_s of saturated air at temperature T is called the *degree of saturation* denoted by μ. Thus, from Eqs (15.3) and (15.6),

$$\mu = \frac{W}{W_s} = \frac{p_w}{p_s}\left[\frac{1 - p_s/p}{1 - p_w/p}\right] \tag{15.7}$$

The degree of saturation μ is a measure of the capacity of air to absorb moisture. For dry air, $p_w = 0$ and $\mu = 0$ and for saturated air, $p_w = p_s$, $\mu = 1$, Thus μ varies between 0 and 1.

15.2.3 Relative Humidity (Denoted by ϕ or RH)

Relative humidity is defined as the ratio of the mass of water vapour m_w in a certain volume of moist air at a given temperature to the mass of water vapour $(m_w)_s$ in the same volume of saturated air at the same temperature. Thus if v_w and v_s are the specific volumes of water vapour in the moist air and saturated air respectively at temperature T and in volume V, viz., at points 1 and 2 respectively in Fig. 15.3, we have

$$\phi = \frac{m_w}{(m_w)_s} = \frac{p_w V/\bar{R}T}{p_s V/\bar{R}T} = \frac{p_w}{p_s} \tag{15.8}$$

Also,

$$\phi = \frac{V/v_w}{V/v_s} = \frac{v_s}{v_w} \tag{15.9}$$

Using ideal-gas relationship between states 1 and 2,

$$p_1 v_1 = p_2 v_2, \quad \text{or} \quad p_w v_w = p_s v_s,$$

We have

$$\phi = \frac{p_w}{p_s} = \frac{v_s}{v_w}$$

The relative humidity can thus be defined as the ratio of partial pressure of water vapour in a given unsaturated moist air at a certain temperature T to the saturation pressure of water vapour at the same temperature T. It is usually measured in percentage. When $p_w = p_s$, ϕ is equal to unity, and air is saturated with 100% RH.

From Eq. (15.3),

$$W = 0.622 \frac{p_w}{p - p_w} = 0.622 \frac{p_w}{p_a}$$

$$\therefore \phi = \frac{p_w}{p_s} = \frac{W}{0.622} \frac{p_a}{p_s} \tag{15.10}$$

Also, from Eqs (15.6) and (15.7),

$$\mu = \phi \left[\frac{1 - p_s/p}{1 - p_w/p} \right] \tag{15.11}$$

$$\text{or, } \phi = \frac{\mu}{1 - (1 - \mu)\, p_s/p} \tag{15.12}$$

15.2.4 Enthalpy of Moist Air

According to the Gibbs theorem, the enthalpy of a mixture of ideal gases is equal to the sum of the enthalpies of the constituent gases. Thus, the enthalpy of moist air is equal to the sum of the enthalpies of dry air and associated water vapour.

$$h = h_a + W h_w \tag{15.13}$$

per kg of dry air.

The enthalpy of the dry-air part above a datum of 0°C is expressed as

$$h_a = c_{p_a} t = 1.005t \text{ kJ/kg}, \tag{15.14}$$

t being the dry-bulb temperature in °C.

Taking the reference state enthalpy as zero for saturated water at 0°C (Fig. 15.4), the enthalpy of the water-vapour part at point p is expressed as

$$h_w = h_p = c_{p_w} t_d + (h_{fg})_{t_d} + c_{p_v} (t - t_d) \text{ kJ/kg} \tag{15.15}$$

$$= 4.187 t_d + (h_{fg})_{t_d} + 1.88(t - t_d) \tag{15.16}$$

where the specific heat of liquid water is taken as 4.187 kJ/kg K and that of superheated vapour as 1.88 kJ/kg K.

To simplify the above expression for h_w, since the pressure is low and the enthalpy is a function of temperature only,

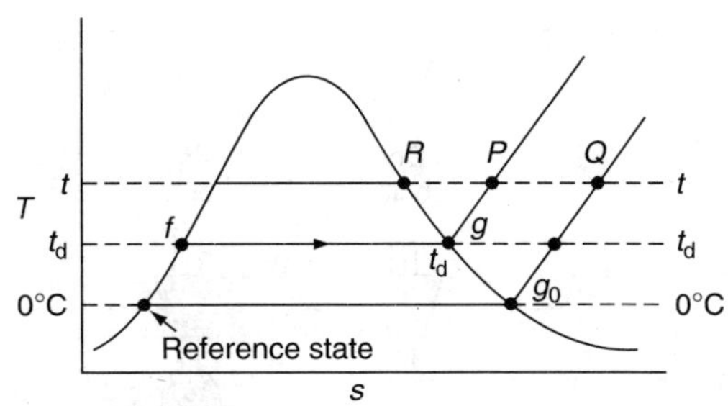

Fig. 15.4 *Enthalpy of water-vapour part*

$$h_P = h_Q = h_R = h_{g_0} + c_{p_v}(t - 0) = 2500 + 1.88t \tag{15.17}$$

where the enthalpy of saturated vapour at 0°C is taken as 2500 kJ/kg. Therefore, the enthalpy of moist air is

$$h = 1.005t + W(2500 + 1.88t) \text{ kJ/kg d.a.} \tag{15.18}$$

15.2.5 Wet-Bulb Temperature

The wet-bulb temperature (WBT) is the temperature recorded by a thermometer when the bulb is enveloped by a cotton wick saturated with water. As the air stream flows past it, some water evaporates, taking the latent heat from the water-soaked wick, thus decreasing its temperature. Energy is then transferred from the air to the wick. When the equilibrium condition is reached, there is a balance between energy removed from the water film by evaporation and energy supplied to the wick by heat transfer, and the minimum temperature recorded is the wet-bulb temperature.

A *Psychrometer* is an instrument which measures both the dry-bulb and wet-bulb temperatures of air. Figure 15.5(a) shows a continuous psychrometer with a fan for drawing air over the thermometer bulbs. A *sling psychrometer* (Fig.15.5(b)) has the two thermometers mounted on a frame with a handle. The handle is rotated fast (150–300 rpm) so that there is good air motion. The wet bulb temperature is the *lowest temperature* recorded by the moistened bulb.

The difference between the DBT and the WBT is called the *wet-bulb depression* (WBD).

Thus,

$$\text{WBD} = \text{DBT} - \text{WBT} = t - t'$$

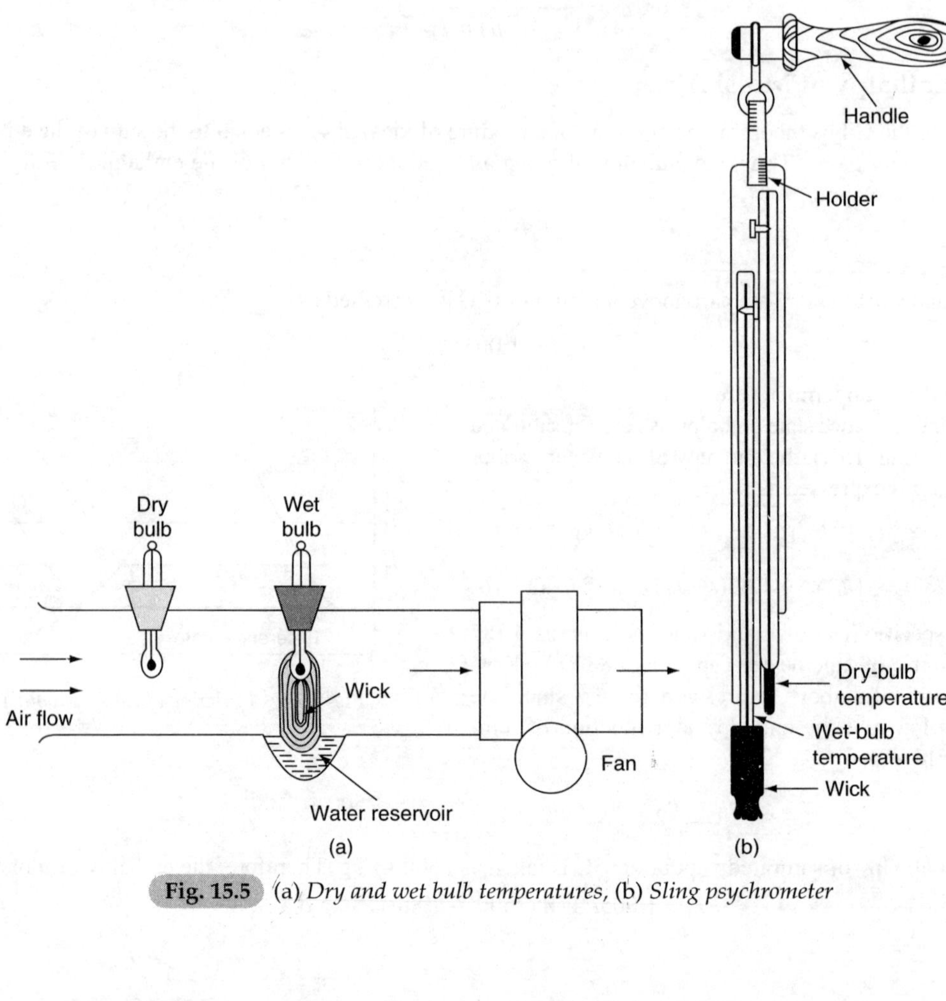

Fig. 15.5 (a) *Dry and wet bulb temperatures,* (b) *Sling psychrometer*

If the ambient air is saturated and RH is 100%, there will be no evaporation of water on the bulb, and hence WBT and DBT will be equal. The wet-bulb depression will be zero. The WBT is an indirect measure of the dryness of air.

The wet-bulb temperature is essentially not a thermodynamic property. It is the temperature of equilibrium reached by heat transfer from air to water in the wick due to temperature difference $(t - t')$ causing the evaporation of water and consequent diffusion of water vapour into air due to partial pressure difference $(p'_w - p_w)$, where p'_w is the saturation water vapour-pressure at temperature t' (Fig. 15.6). At any DBT, the greater the depression, the smaller is the amount of water vapour held in the mixture.

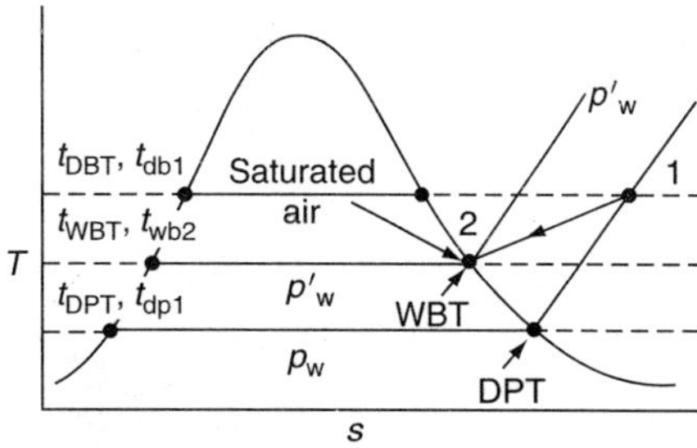

Fig. 15.6 *Wet bulb temperature and adiabatic saturation process*

15.2.6 Adiabatic Saturation Process and Thermodynamic Wet-Bulb Temperature

When unsaturated air flows over a long sheet of water (Fig. 15.7) in an insulated chamber, the water evaporates, and the specific humidity of the air increases. Both the air and water are cooled as evaporation takes place. The process continues until the energy transferred from the air to the water is equal to the energy required to vaporize the water. When this point is reached, thermal equilibrium exists with respect to the water, air and water vapour, and consequently the air is saturated. The equilibrium temperature is called the *adiabatic saturation temperature* or the *thermodynamic wet-bulb temperature*. The make-up water is introduced at this temperature to make the water level constant.

The 'adiabatic cooling' process is shown in Fig. 15.6 on *T-s* plot for the vapour in the air-water vapour mixture. Although the total pressure of the mixture is constant, the partial pressure of the vapour increases from p_w to p'_w, and in the saturation state corresponds to the adiabatic saturation temperature. The vapour is initially at t_{db1} and is adiabatically cooled to t_{wb2} which is equal to the adiabatic saturation temperature. The adiabatic saturation temperature and the wet-bulb temperature are taken to be equal for all practical purposes. The WBT lies between DBT and DPT.

Since the system is insulated and no work is done, the first law yields

$$Gh_{a1} + m_1 h_{w1} + (m_2 - m_1)h_{f2} = Gh_{a2} + m_2 h_{w2} \tag{15.19}$$

where $(m_2 - m_1)$ is the mass of water added, h_{f2} is the enthalpy of liquid water at $t_2 (= t_{wb2})$, h_a is the specific enthalpy of dry air, and h_w is the specific enthalpy of water vapour in air. Dividing by G, and since $h_{w2} = h_{g2}$

$$h_{a1} + W_1 h_{w1} + (W_2 - W_1)h_{f2} = h_{g2} + W_2 h_{g2} \tag{15.20}$$

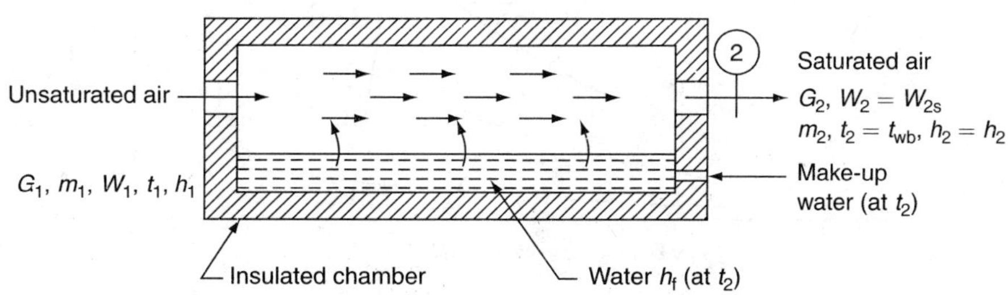

Fig. 15.7 *Adiabatic saturation process*

Solving for W_1,

$$W_1 = \frac{(h_{a2} - h_{a1}) + W_2(h_{g2} - h_{f2})}{h_{f2} - h_{w1}}$$

$$= \frac{c_{pa}(T_2 - T_1) + W_2 h_{fg2}}{h_{f2} - h_{w1}} \tag{15.21}$$

where $$W_2 = \frac{m_2}{G} = \frac{m_s}{G} = 0.622\frac{p_s}{p - p_s}.$$

The enthalpy of the air–water vapour mixture is given by

$$Gh = Gh_a + mh_w$$

where h is the enthalpy of the mixture per kg of dry air.

$$h = h_a + Wh_w$$

which was stated earlier in Eq. (15.13).

15.3 PSYCHROMETRIC CHART

The psychrometric chart is a graphical plot (Fig. 15.8) with specific humidity and partial pressure of water vapour as ordinate, and dry bulb temperature as abscissa. The volume of the mixture (m^3/kg dry air), wet-bulb temperature, relative humidity, and enthalpy of the mixture appear as parameters. The chart is plotted for one atmospheric pressure, usually 760 mm Hg.

The constant WBT line represents the adiabatic saturation process. It also coincides with the constant enthalpy line. To show this, let us consider the energy balance for the adiabatic saturation process (Fig. 15.7).

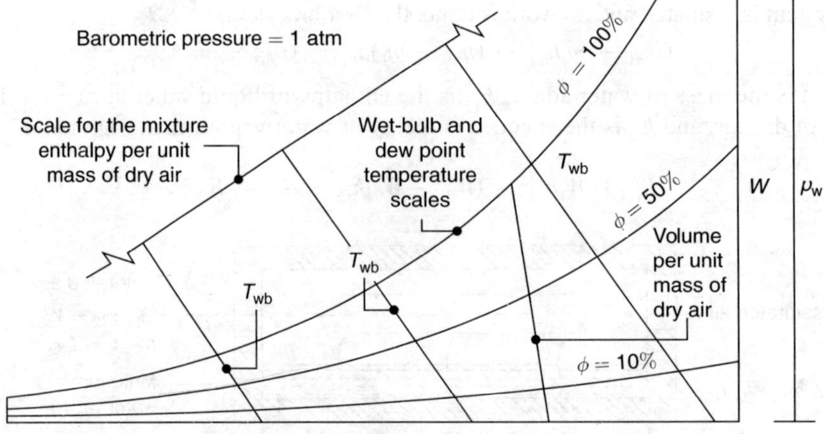

(a)

Fig. 15.8 *(Continued)*

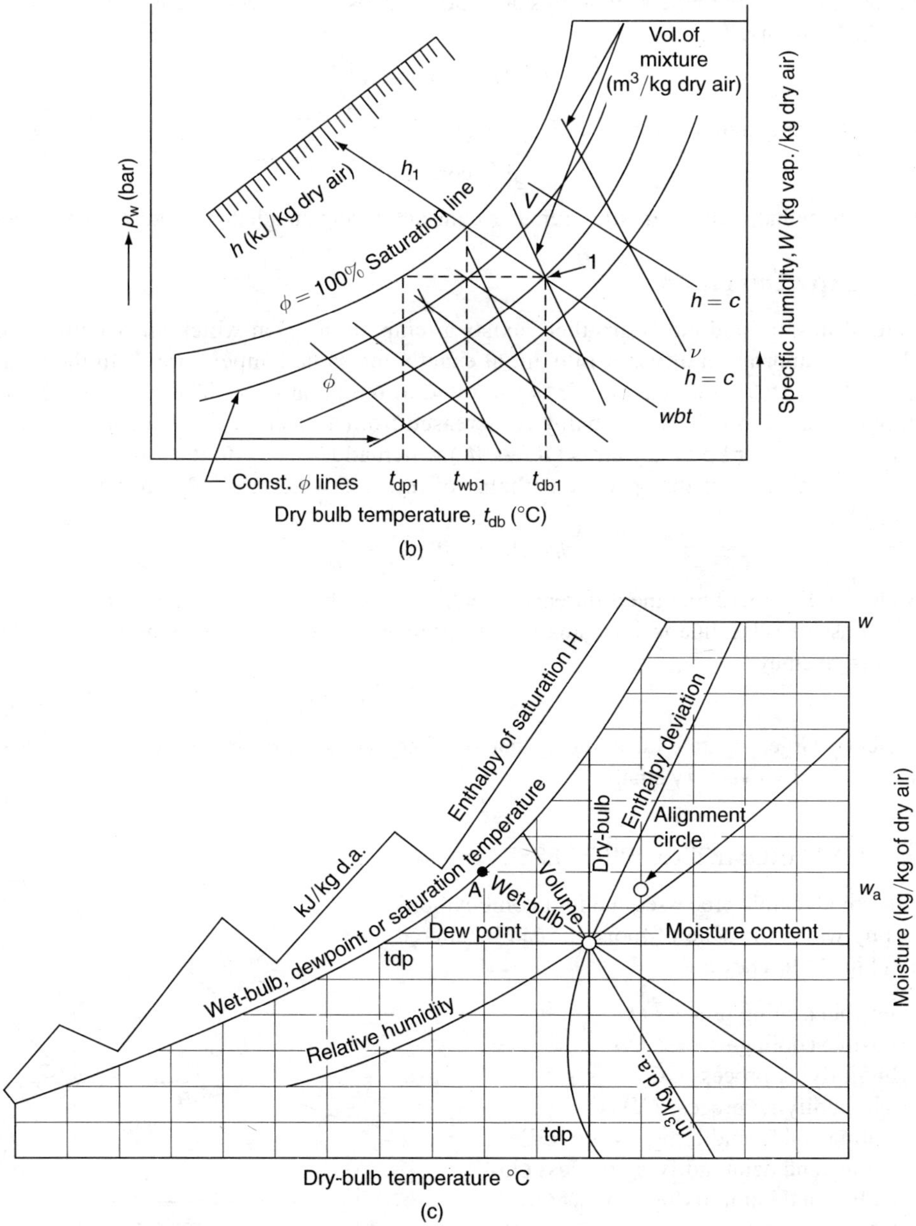

Fig. 15.8 *Psychrometric chart*

$$h_{a1} + W_1 h_{w1} + (W_2 - W_1) h_{f2} = h_{a2} + W_2 h_{w2}$$

Since $h = h_a + W h_w$ kJ/kg d.a.,

$$h_1 - W_1 h_{f2} = h_2 - W_2 h_{f2}$$

where the subscript 2 refers to the saturation state, and the subscript 1 denotes any state along the adiabatic saturation path. Therefore,

$$h - Wh_{f2} = \text{constant} \tag{15.22}$$

Since Wh_f is small compared to h (of the order of 1 or 2%)

$$h = \text{constant}$$

indicating that the enthalpy of the mixture remains constant during an adiabatic saturation process.

15.3.1 Enthalpy Deviation

For any state of unsaturated moist air, there exists a temperature t^* at which air becomes adiabatically saturated by the evaportation of water into air, at exactly the same temperature t^*. In the adiabatic saturation process (Fig. 15.7), the leaving air is saturated at temperature t^*. The specific humidity is correspondingly increased to W^*. The enthalpy is increased from a given initial value h to the value h^*. The weight of water added per kg of dry air is $(W^* - W)$ which adds energy to the moist air of amount equal to $(W^* - W)\ h_f^*$ where h_f^* is the specific enthalpy of injected water at t^*. By energy balance, since the process is adiabatic,

$$h + (W^* - W)\, h_f^* = h^*$$

It was stated, in Eq. 15.12 that the difference of enthalpies at the saturation point A (15.8c) and any other point on the constant WBT line is very small. The difference denoted by the symbol D is called *enthalpy deviation* and is given by

$$D = h - h^* = (W - W^*)h_f^* \tag{15.23}$$

If it is sufficiently large, separate lines of constnat WBT and constnat h can be shown on the psychrometric chart. However, usually D is very small.

15.4 PSYCHROMETRIC PROCESSES

Four basic thermodynamic processes and four combinations of processes by which the state of moist air can be altered are shown in Fig. 15.9. They are

(i) Sensible heating process OA
(ii) Sensible cooling process OB
(iii) Humidifying process OC
(iv) Dehumidifying process OD
(v) Heating and humidifying process OE
(vi) Cooling and dehumidifying process OF
(vii) Cooling and humidifying process OG
(viii) Heating and dehumidifying process OH

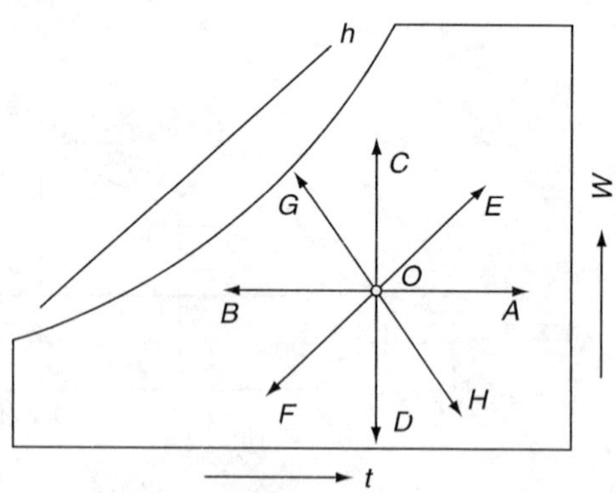

Fig. 15.9 *Basic psychrometric processes*

Sensible heating and cooling processes involve only a change in the dry bulb temperature, whereas the processes of humidifying and dehumidifying involve a change in the specific humidity. Thus, when the state of air moves from O to A or to B, there is no change in the moisture content of the air. If the state changes from O to C or to D, the DBT remains constant. The last four fundamental processes listed above involve both changes in temperature as well as in humidity.

15.4.1 Sensible Heating or Cooling (at W = constant)

When the state of moist air is altered along the W = constant line such as 1–2 in Fig. 15.10 with only DBT changing, we have

Balance of

Dry air $G_1 = G_2$

Moisture $m_1 = m_2$

Energy $G_1 h_1 + Q_{1-2} = G_2 h_2$

$$\therefore Q_{1-2} = G(h_2 - h_1) = G\,[h_{a2} + W_2 h_{w2} - (h_1 + W_1 h_w)]$$

$$= G[1.005(t_2 - t_1) + W_2\,[h_{g2} + 1.88(t_2 - t_{d2})] - W_1[h_{g1} + 1.88(t_1 - t_{d1})] = Q_s \qquad (15.24)$$

where the C_p of water vapour in the superheated state is assumed to be 1.88 kJ/kgK and t_d is the dew point temperature.

Since $$W_1 = W_2,\ h_{g2} = h_{g1},\ t_{d2} = t_{d1},\ Q_{1-2} = G(1.005 + 1.88\,W)(t_2 - t_1) = Q_s \qquad (15.25)$$

Similar equations will hold good for sensible cooling at constant W. Simple heating of moist air can be done to any desired temperature, simple cooling can be done only up to the dew-point temperature.

If a building to be air conditioned receives or loses heat due to transmission through the walls, it is supposed to have *sensible heat load* (Q_c). Heat gain in buildings will require the conditioning of air to lower temperatures, causing a *cooling load* on the air-conditioning equipment. However, heat loss in buildings will require the heating of air causing a *heating load* on the equipment.

15.4.2 Latent Heat Load: Humidification or Dehumidification

When the state of air is altered along the DBT= constant line, such as 3–4 in Fig. 15.11, moisture in the form of water vapour is transferred to change the specific humidity of the air. Thus the amount of moisture transfer is

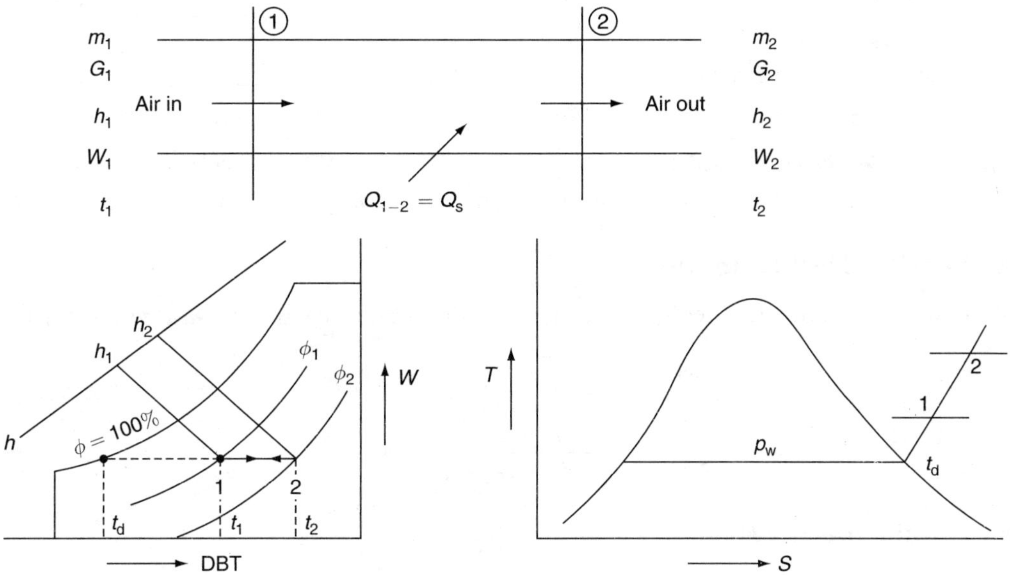

Fig. 15.10 *Sensible heating*

$$m = G(W_3 - W_4)$$

Because of this change in specific humidity, there is also a change in enthalpy of the air given by $(h_3 - h_4)$. In air-conditioning practice, this change of enthalpy due to the change in W is considered to cause *latent heat transfer* given by

$$Q_L = G(h_3 - h_4) = G[(c_p t_3 + h_{fgo} W_3) - (c_p t_4 + h_{fgo} W_4)]$$

$$= G h_{fgo} (W_3 - W_4) = m h_{fgo} = 2500 \text{ m kJ} \quad (15.26)$$

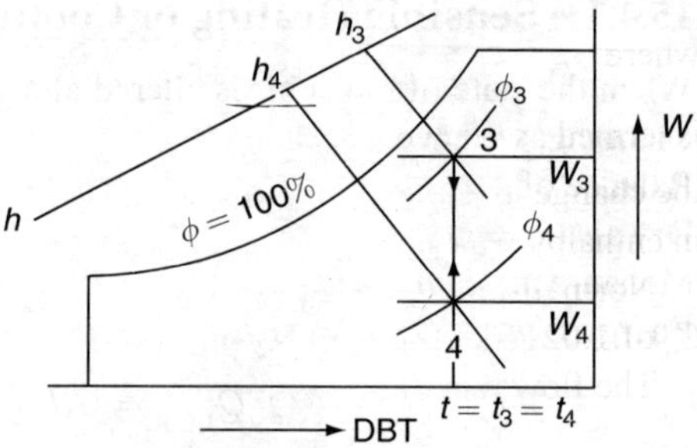

Fig. 15.11 *Latent heat process*

The latent heat load is proportional to the latent heat of vaporization h_{fgo}. Accordingly, if a building gains or loses moisture, it is supposed to have a *latent heat load.* A gain of moisture will require the condensation of moisture (water vapour) for the dehumidification of air in the conditioning apparatus, and hence a cooling load. On the other hand, a loss of moisture will require the evaporation of water for the humidification of air in the apparatus and hence a heating load.

15.4.3 Total Heat Load

Let us now consider a change in the state of air along the path 1–3 as shown in Fig. 15.12. This involves both a change in temperature as well as in the humidity ratio. The change in temperature causes a sensible heat load (SHL)given by

$$Q_S = Q_{1-2} = G(h_2 - h_1) = G c_p (t_2 - t_1)$$

The change in the humidity ratio causes a moisture transfer given by

$$m = G(W_3 - W_1)$$

and a latent heat load (LHL) given by

$$Q_L = Q_{2-3} = G(h_3 - h_2) = G h_{fgo} (W_3 - W_1)$$

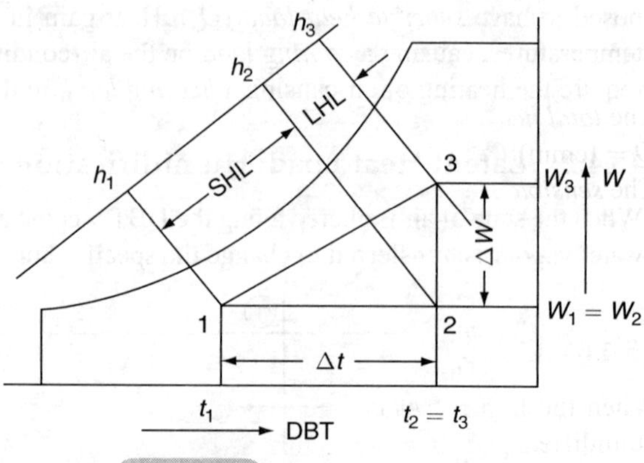

Fig. 15.12 *Total heat load process*

The *total heat load* is then expressed as

$$Q = Q_s + Q_L = G(h_3 - h_1) = G[c_p(t_3 - t_1) + h_{fgo}(W_3 - W_1)] \quad (15.27)$$

15.4.4 Sensible Heat Factor (SHF)

The ratio of the sensible heat transfer to the total heat transfer is called the *sensible heat factor.* Thus

$$\text{SHF} = \frac{Q_s}{Q_s + Q_L} = \frac{Q_s}{Q}$$

$$= \frac{h_2 - h_1}{(h_2 - h_1) + (h_3 - h_2)} = \frac{h_2 - h_1}{h_3 - h_1} \quad (15.28)$$

15.4.5 Humid Specific Heat

The enthalpy of moist air can be written in the form

$$h = (c_{pa} + W c_{pw}) t + W (h_{fg})_{0°C} = c_p t + W (h_{fg})_{0°C}$$

where $$c_p = c_{pa} + Wc_{pw} = (1.005 + 1.88c_{pw})\,\text{kJ/(kg d.a.)(K)}$$

is termed as the *humid specific heat*. It is the specific heat of moist air $(1+W)$ kg/kg d.a. The term $C_p t$ governs the change in enthalpy of moist air with temperature at constant W, and the term $W(h_{fg})_{0°C}$ governs the change in enthalpy with the change in W, i.e. due to addition or removal of moisture in air.

Normally, the second term 1.88 W is very small compared to the first term 1.005. An approximate value of C_p of 1.0216 kJ/(kg d.a.)(K) may be taken for all practical purposes in air-conditioning calculations.

The flow rate of air is measured in terms of m³/min or cmm. Then the mass flow rate of dry air is

$$G = \dot{Q}_v \rho / 60 \text{ kg d.a./s}$$

where Q_v is the volume flow rate of dry air (m³/min) and for standard air at 20°C and 50% RH, the density ρ is about 1.2 kg/m³ d.a.

Sensible heat load (SHL) is then

$$\dot{Q}_s = \frac{cmm \times 1.2 \times 1.0216}{60} \Delta t = 0.0204(cmm)\ \Delta t \text{ kW} \tag{15.29}$$

The *latent heat load* (LHL) is from Eq. (15.16)

$$\dot{Q}_L = 2500\ \dot{m} \text{ kW} = \frac{2500(cmm) \times 1.2 \times 2500}{60} \Delta W = 50(cmm)\ \Delta W \text{ kW} \tag{15.30}$$

The *total heat load* (THL)

$\dot{Q} = (\text{cmm})(0.0204)\Delta t + 50(\text{cmm})\ \Delta W$

The *sensible heat factor* (SHF) is from

$$\text{SHF} = \frac{\dot{Q}_s}{\dot{Q}} = \frac{0.0204(cmm)\Delta t}{(cmm)(0.0204)\Delta t + 50(cmm)\Delta W} = \frac{0.0204\Delta t}{0.02\Delta h} \tag{15.31}$$

15.4.6 Cooling and Dehumidification

When the humidity radio of air decreases, air is said to be dehumidified, and when it increases, air is humidified. Air may be cooled and dehumidified (a) by placing the evaporator coil across the air flow (Fig. 15.13(b)), (b) by circulating chilled water or brine (NaCl or $CaCl_2$) in a tube placed across the air flow (Fig.15.13(c)), or (c) by spraying chilled water to air in the form of mist (15.13(d)) to expose a large surface area. The temperature of the cooling surface or the spraywater must be below the dew point at state 1, t_d (Fig. 15.13(a)). If the cooling surface or the spray shower is of large magnitude, the air may come out at the saturation state 2s, known as the *apparatus dew point* (adp).

Balance gives $$G_1 = G_2 = G$$

If L is the amount of water vapour removed,

$$m_1 = m_2 + L, \quad L = m_1 - m_2 = G(W_1 - W_2).$$

The energy equation gives $$G_1 h_1 = G_2 h_2 + Q_{1-2} + L\,h_{f2}$$

where h_{f2} is the specific enthalpy of water at temperature t_2.

$$\therefore Q_{1-2} = G[(h_1 - h_2) - (W_1 - W_2)h_{f2}]$$

If $(W_1 - W_2)h_{f2}$ is small, the amount of heat removed is

$$Q_{1-2} = G(h_1 - h_2) = \text{Total heat load on the cooling coil (kJ/h)}.$$

Cooling and dehumidification of air is common in summer air conditioning.

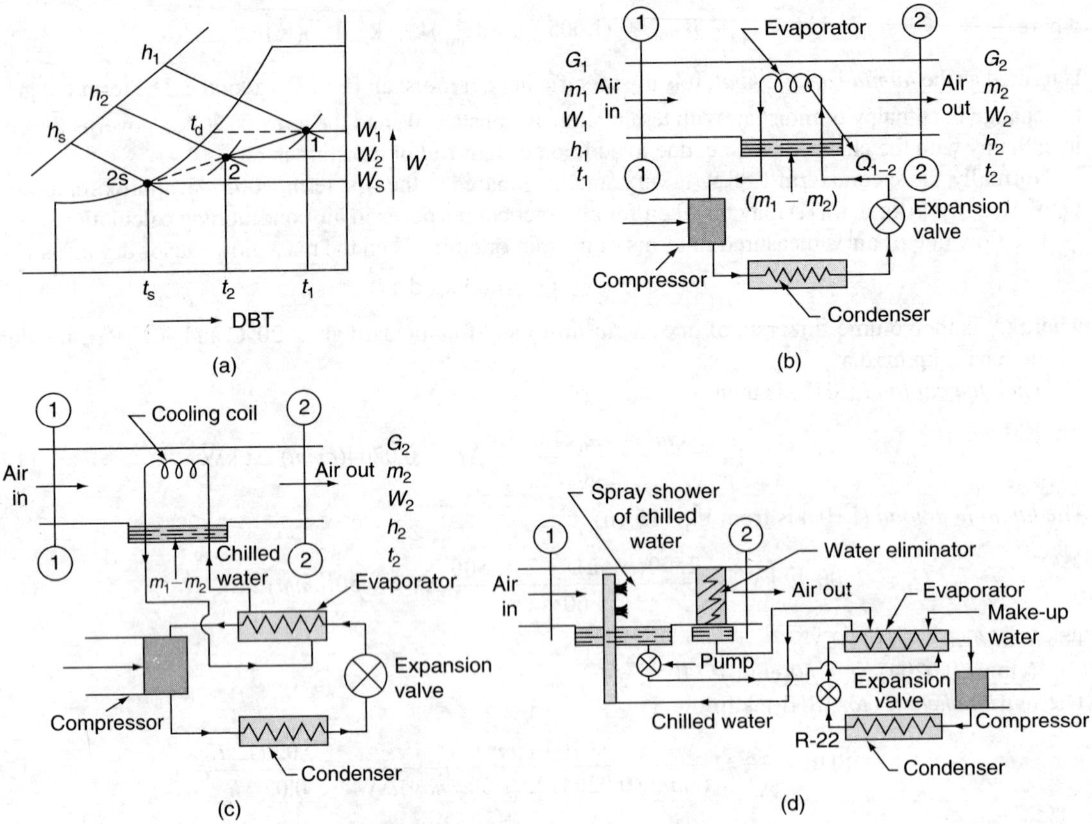

Fig. 15.13 *Cooling and dehumidification*

15.4.7 Heating and Humidification

The addition of heat and moisture to air is a problem for winter air conditioning (Fig. 15.14). The water added may be liquid or vapour. The following equations apply

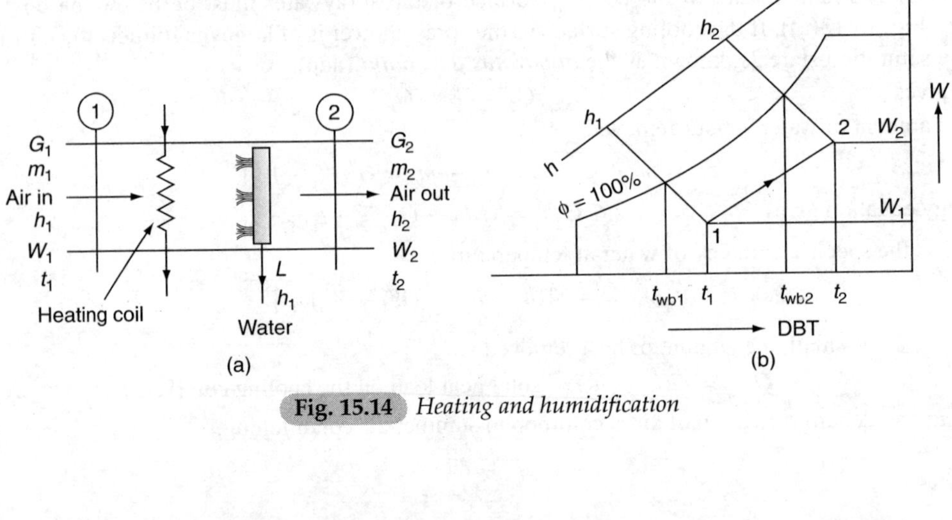

Fig. 15.14 *Heating and humidification*

$$G_1 = G_2 = G$$
$$m_1 + L = m_2$$
$$L = m_2 - m_1 = G(W_2 - W_1)$$
$$G_1 h_1 + L\,h_f + Q_{1-2} = G_2 h_2$$
$$Q_{1-2} = G(h_2 - h_1) - G(W_2 - W_1)h_f$$
$$\text{Or,} \quad Q_{1-2} = G[(h_2 - h_1) - (W_2 - W_1)h_f]$$

15.4.8 Dehumidification and Heating

In winter with cold and dry ambient air, it is often required to dehumidify and heat it, as shown in Fig. 15.15.

15.4.9 Adiabatic Mixing of Two Streams

This is a common problem in air conditioning, where ventilation air and some recirculated room air are mixed prior to processing it to the desired state, and supplying it to the conditioned space (Fig. 15.16).
The following equations hold good:

$$G_1 + G_2 = G_3$$
$$G_1\,W_1 + G_2\,W_2 = G_3\,W_3$$
$$G_1\,h_1 + G_2\,h_2 = G_3\,h_3$$

Combining these equations and rearranging,

$$\frac{G_1}{G_2} = \frac{h_3 - h_2}{h_1 - h_3} = \frac{W_3 - W_2}{W_1 - W_3}$$

The points 1, 2 and 3 fall in a straight line and the division of the line is inversely proportional to the ratio of the mass flow rates.

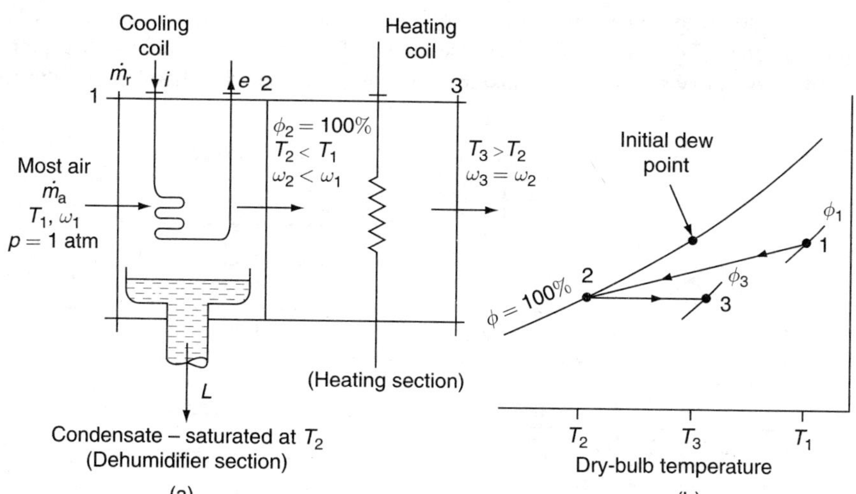

Fig. 15.15 *Dehumidification with heating* (a) *equapment schematic* (b) *psychrometric chart representation*

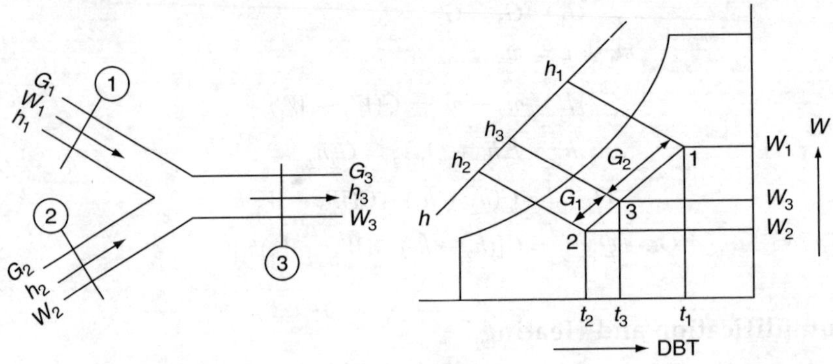

(a) (b)

Fig. 15.16 *Adiabatic mixing of two air streams*

15.4.10 Chemical Dehumidification

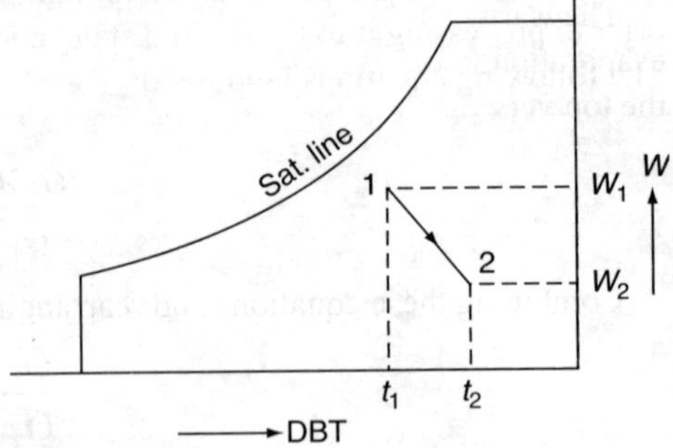

Fig. 15.17 *Chemical dehumidification*

Some substances like silica gel (product of fused sodium silicate and sulphuric acid) and activated alumina have great affinity with water vapour. They are called *adsorbents*. When moist air passes through a bed of silica gel, water vapour molecules get adsorbed on its surface. Latent heat of condensation is released. So the DBT of air increases. The process is shown in Fig. 15.17.

15.4.11 Adiabatic Evaporative Cooling

A large quantity of water is constantly circulated through a spray chamber. The air-vapour mixture is passed through the spray and, in doing so, evaporates some of the circulated water. The air may leave at a certain humidity ratio or in a saturation state (Fig. 15.18). The increase in specific humidity is equal to the quantity of water evaporated per unit mass of dry air. No heat transfer takes place between the chamber and the surroundings.

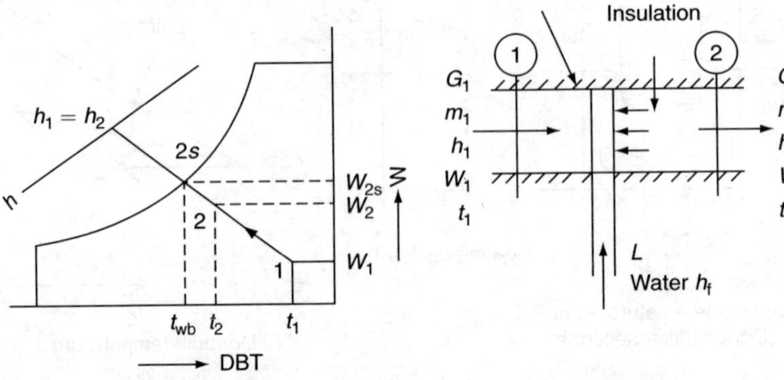

(a) (b)

Fig. 15.18 *Adiabatic evaporative cooling*

Therefore, the energy required for evaporation is supplied by the air, and so the DBT is lowered. After the process has been in operation for a sufficiently longtime, the circulating water approaches the WBT of the air. The following equations apply to the system.

$$G_1 = G_2 = G$$

$$G_1 W_1 + L = G_2 W_2$$

$$L = G (W_2 - W_1)$$

$$G_1 h_1 + L h_f = G_2 h_2$$

$$G(h_1 - h_2) + G (W_2 - W_1) h_f = 0 \quad \text{or,} \quad h_1 - W_1 h_f = h_2 - W_2 h_f$$

It is the same as Eq. 15.22 derived earlier. Since Wh_f is small, enthalpy of the mixture remains constant in adiabatic evaporative cooling.

The *cooling tower* utilizes the phenomenon of evaporative cooling to cool warm water below the dbt of the air. However, the water never reaches the minimum temperature, i.e., the wbt, since an excessively large cooling tower would then be required. Also, since warm water is continuously introduced to the tower (Fig. 15.19), the equilibrium conditions are not achieved, and the dbt of the air is increased. Hence, while the water is cooled, the air is heated and humidified.

The warm water is introduced at the top of the tower in the form of spray to expose a large surface area for evaporation to take place. The more the water evaporates, the more is the effect of cooling. Air leaves from the top very nearly saturated. The following equations apply

$$G_1 = G_2 = G$$

$$G_1 W_1 + m_{w3} = G_2 W_2 + m_{w4}$$

$$m_{w3} - m_{w4} = G (W_2 - W_1)$$

$$G_1 h_1 + m_{w3} h_{w3} = G_2 h_2 + m_{w4} h_{w4}$$

$$G (h_1 - h_2) + m_{w3} h_{w3} = m_{w4} h_{w4}$$

The difference in temperature of the cooled-water temperature and the wet bulb temperature of the entering air is known as the *approach*. The *range* is the temperature difference between the inlet and exit states of water. Cooling towers are rated in terms of approach and range.

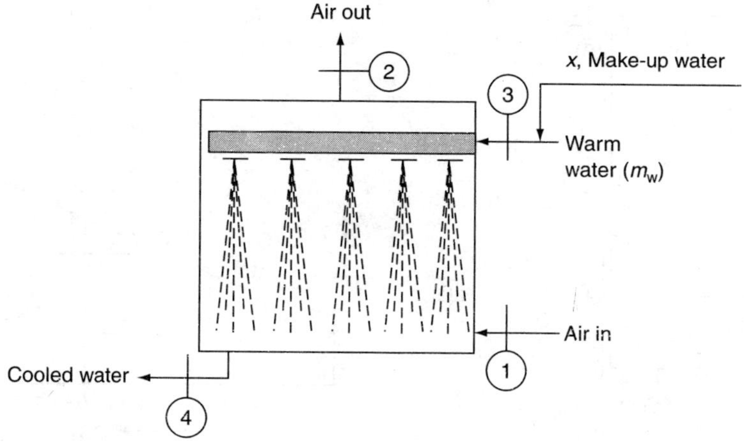

Fig. 15.19 *Cooling tower*

If x is the make-up water supplied (Fig. 15.19), then

$$x = G(W_2 - W_1) \text{ and } m_{w3} = m_{w4} = m_w$$

By energy balance,

$$G_1 h_1 + m_{w3} h_{w3} + x h_w = G_2 h_2 + m_{w3} h_{w4}$$

$$m_w (h_{w3} - h_{w4}) = G(h_2 - h_1) - G(W_2 - W_1) h_w$$

$$h_{w3} - h_{w4} = \frac{G}{m_w}\left[(h_2 - h_1) - (W_2 - W_1) h_w\right]$$

$$\therefore \text{Range} = t_{w3} - t_{w4} = \frac{G}{m_w c_{pw}}\left[(h_2 - h_1) - (W_2 - W_1) h_w\right]$$

where c_{pw} is the specific heat of water and h_w is the enthalpy of the make-up water.

$$\text{Approach} = t_{w3} - t_{wb1}$$

The active portion of the tower in which energy exchange occurs is filled with a packing which breaks up the flow and exposes a large surface area of water in contact with air.

The schematic layout of a cooling tower is shown in Fig. 15.20.

15.4.12 Bypass Factor

Figure 15.21 shows the process that the moist air undergoes while flowing over a surface. The air enters at 1 and leaves at 2 when the surface is maintained at s. In the transfer of heat and water of vapour in any conditioning

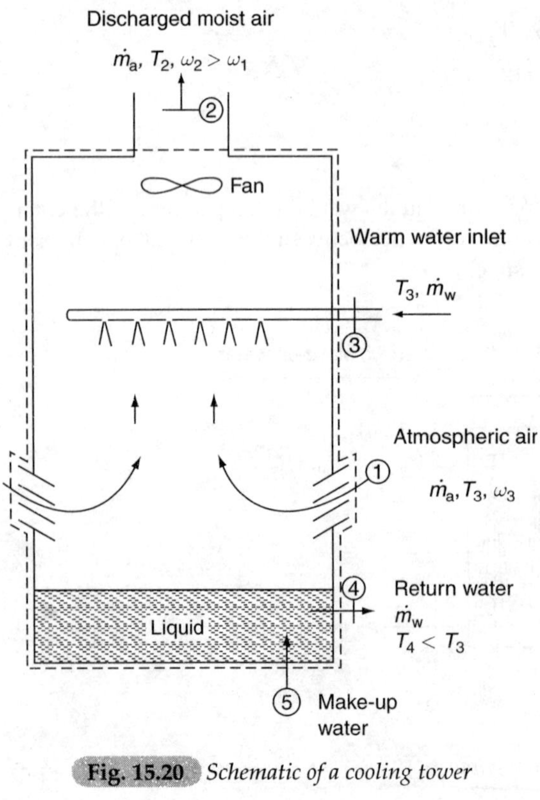

Fig. 15.20 *Schematic of a cooling tower*

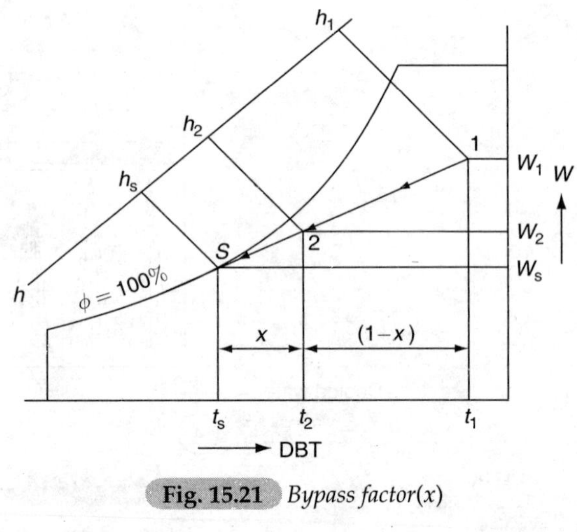

Fig. 15.21 *Bypass factor(x)*

process, the particles of air stream come in contact with the surface. The state of the *contacted air* is that of the saturated air at t_s. The *uncontacted air* remains in the entering state. The end state 2 of the air is produced by the mixing of contacted and uncontacted particles. The *bypass factor* (BPF) is defined as the fraction of the uncontacted air in terms of the states 1, 2 and *s*, as

$$\text{BPF}, X = \frac{t_2 - t_s}{t_1 - t_s} = \frac{W_2 - W_s}{W_1 - W_s} = \frac{h_2 - h_s}{h_1 - h_s}$$

The *contact factor* $(1 - x)$ represents the fraction of the contacted air.

15.4.13 Air Washer

An *air washer* is shown schematically in Fig. 15.22. It involves the flow of air through a spray of water. During the course of flow, the air may be cooled or heated, humidified or dehumidified, or simply adiabatically saturated, depending on the mean surface temperature of water spray. The water is, accordingly, cooled or heated externally, or simply recirculated by a pump. Make-up water is added for any loss due to humidification. Eliminator plates are provided to minimise the loss of water droplets.

The thermodynamic changes of state of air along paths 1–2 in an air washer are shown in Fig. 15.23. The following processes are possible:

1. *Process* 1–2*A* Heating and humidification ($t_s > t_1$).
2. *Process* 1–2*B* Humidification ($t_s = t_1$).
3. *Process* 1–2*C* Cooling and humidification $\left(t_1' < t_s < t_1\right)$ where t' is the WBT and t_1 the DBT.
4. *Process* 1–2*D* Adiabatic saturation $\left(t_1' = t_s\right)$.
5. *Process* 1–2*E* Cooling and humidification $\left(t_d < t_s < t_1'\right)$ where t_d is the DPT.

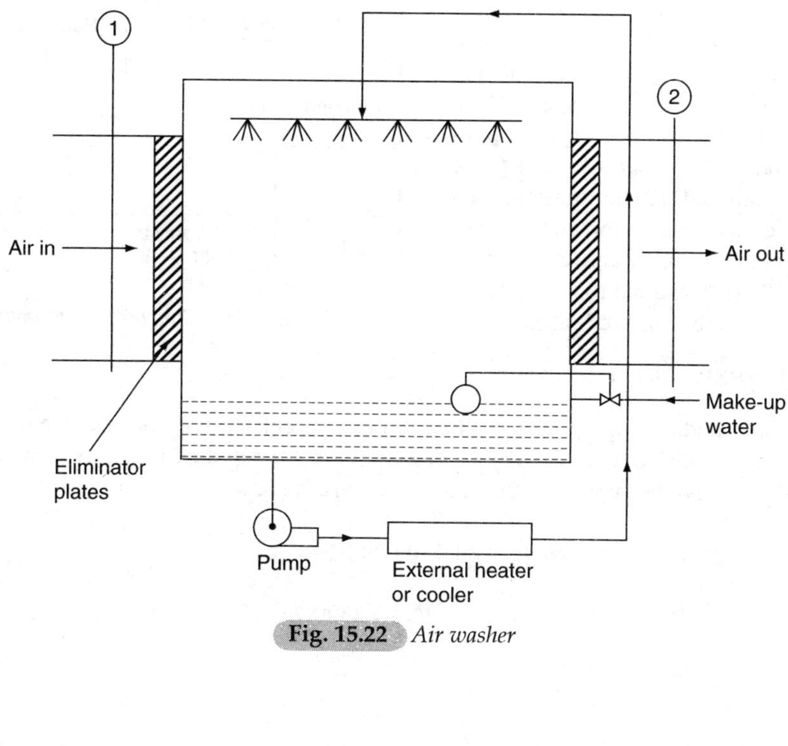

Fig. 15.22 *Air washer*

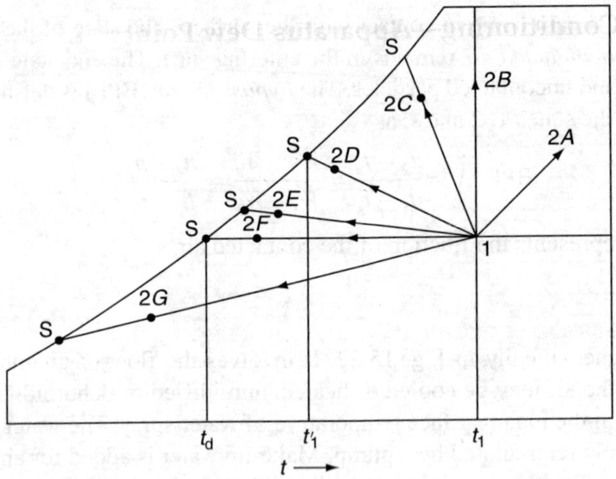

Fig. 15.23 *Range of psychrometric processes with an air washer*

6. *Process* 1–2*F* Cooling ($t_s = t_d$).
7. *Process* 1–2*G* Cooling and dehumidification ($t_s < t_d$).

It is thus seen that the air washer can be used for *year-round air conditioning.*

15.5 SIMPLE AIR-CONDITIONING SYSTEM

The problem of air conditioning a space is to pick up its sensible and latent heat loads and to calculate the state and mass rate of flow of air to be supplied to the space. For a simple air-conditioning system (Fig. 15.24), let us consider a space which is to be maintained at the room or inside conditions of DBT t_i and humidity ratio W_i. Let Q_s be the sensible heat gain and m, the moisture gain $G(W_2 - W_1)$ of the room from internal and external sources. The room sensible heat (RSH) and the room latent heat (RLH) make the room total heat (RTH) which requires to be removed. Thus

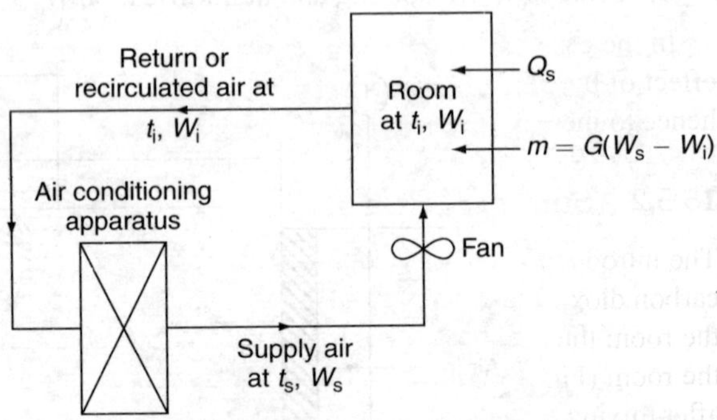

Fig. 15.24 *Simple air-conditioning system*

$$\text{RTH} = \text{RSH} + \text{RLH}$$

In the case of an air-conditioning apparatus, the room or recirculated air from the room is conditioned to a supply DBT t_s and a specific humidity W_s. The processed air is supplied to the room by a fan. Assuming steady-state conditions and the supply air flow rate as (cmm), we have:

$$\text{RSH} = \dot{Q}_s = Gc_p\,(t_i - t_s) = 0.0204(\text{cmm})(t_i - t_s)$$

$$\text{RLH} = \dot{Q}_L = G\left(h_{fgo}\right)(W_i - W_s) = 2500G = 50(\text{cmm})(W_i - W_s)$$

15.5.1 Summer Air-Conditioning—Apparatus Dew Point

In summer, the outside air temperature and humidity are both high. The room, therefore gains heat as well as moisture. It is thus required to cool and dehumidify the recirculated room air in the air – conditioning apparatus either by the use of a cooling coil or by an air washer in which chilled water is sprayed. The process follows the room sensible heat factor (RSHF) line. The room sensible heat factor is the ratio of the room sensible heat to the room total heat, i.e.,

$$\text{RSHF} = \frac{\text{RSH}}{\text{RSH} + \text{RLH}} = \frac{\text{RSH}}{\text{RTH}}$$

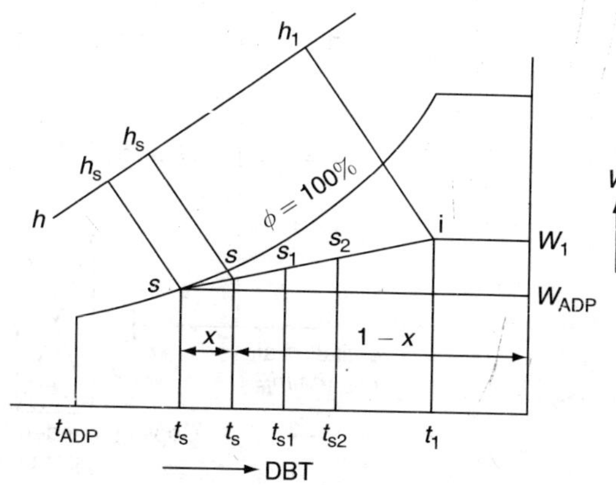

Fig. 15.25 *Locus of supply air states for cooling and dehumidifying*

In a cooling and dehumidifying process, the temperature at which the RSFH or condition line intersects the saturation curve is called the room apparatus dew point (RADP). Thus, t_{ADP} in Fig. 15.25 denotes the effective surface temperature t_s. The condition line *i-s* represents the locus of all possible supply air states like s, s_1, s_2, etc. The minimum supply air quantity will be

$$(\text{cmm})_{\text{min.}} = \frac{\text{RSH}}{0.0204(t_i - t_{ADP})}$$

$$= \frac{\text{RLH}}{50(W_i - W_{ADP})} = \frac{\text{RTH}}{0.02(h_i - h_{ADP})}$$

In the case of an actual coil with a bypass factor of x, the leaving air state will be at s. It is seen that the effect of bypass factor is to decrease the difference in temperature between the room air and supply air, and hence to increase the supply air quantity over its minimum value.

15.5.2 Summer Air-Conditioning System with Ventilation Air–Bypass Factor X

The introduction of the fresh outside air for the ventilation of the conditioned space is necessary to dilute the carbon dioxide and adours and other air contaminants for maintaining the purity of room air. The supply air to the room thus comprises fresh air and recirculated room air. An amount equal to the fresh air is ejected from the room (Fig. 15.26). In the figure, *o* and *i* represent the outside and inside air states and 1 is the state of air after mixing of recirculated room air (m_i) and ventilation air (m_o). The room sensible heat factor (RSHF) line is drawn from *i* to intersect the saturation curs *c* at room ADP at 2. Point 2 indicates the supply air state for a minimum rate of air supply. The line 1–2 represents the condition line for the apparaturs called the grand sensible heat factor (GSHF) line. Figure 15.27 shows psychrometric processes with zero and finite bypass factors. The total load on the air-conditioning apparatus from state 1 to 2 is

$$\dot{Q} = \dot{m}(h_1 - h_2) = (\dot{m}_i h_i + \dot{m}_o h_o) - \dot{m}_s h_2$$

$$= (\dot{m}_i - \dot{m}_o) h_i + \dot{m}_o h_o - \dot{m}_s h_2$$

$$= \dot{m}_i (h_i - h_2) + \dot{m}_o (h_o - h_i)$$

$$= \text{Room load} + \text{Ventilation air load}$$

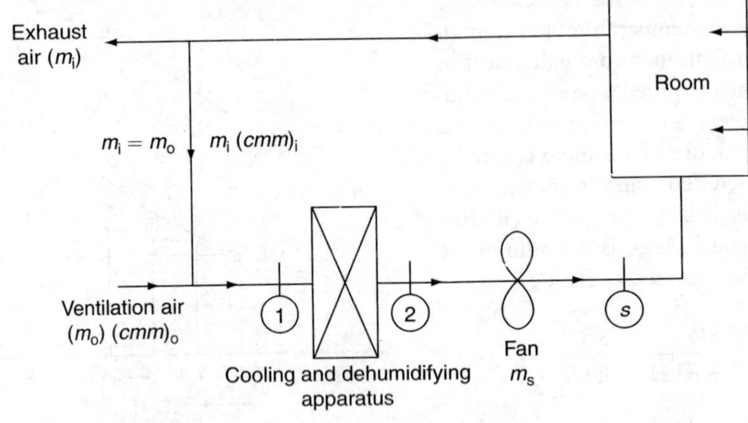

Fig. 15.26 *Summer air-conditioning scheme with ventilation air*

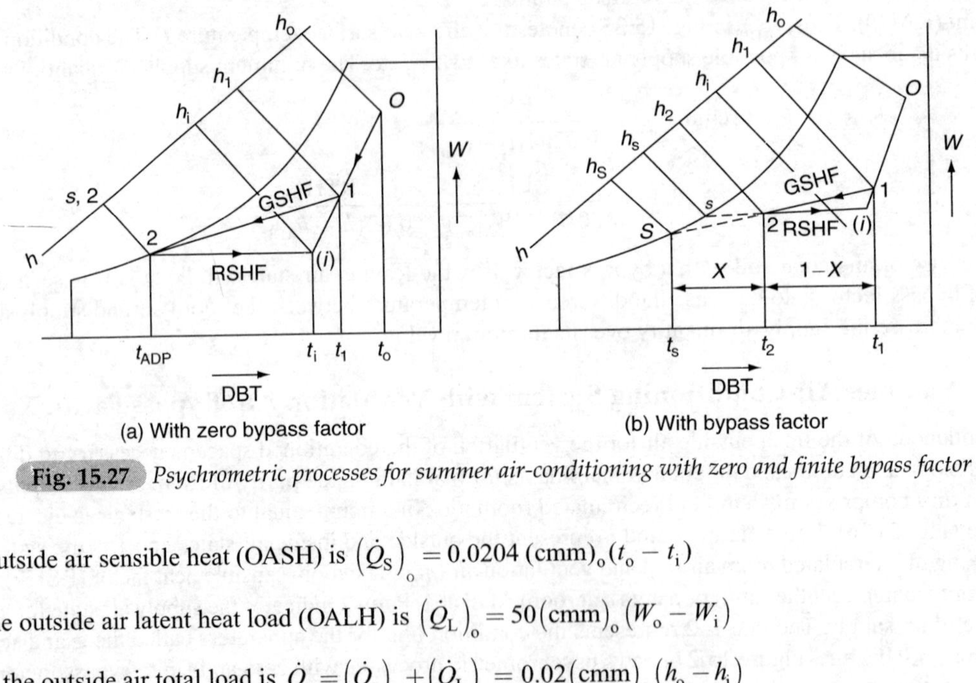

Fig. 15.27 *Psychrometric processes for summer air-conditioning with zero and finite bypass factor*

The outside air sensible heat (OASH) is $\left(\dot{Q}_S\right)_o = 0.0204\,(\text{cmm})_o(t_o - t_i)$

and the outside air latent heat load (OALH) is $\left(\dot{Q}_L\right)_o = 50(\text{cmm})_o(W_o - W_i)$

Thus, the outside air total load is $\dot{Q}_o = \left(\dot{Q}_s\right)_o + \left(\dot{Q}_L\right)_o = 0.02(\text{cmm})_o(h_o - h_i)$

15.5.3 Winter Air Conditioning

The processes in the conditioning apparatus for winter air conditioning for comfort usually involve heating and humidifying. Two typical processes are

1. Preheating the air with steam or hot water in a coil followed by adiabatic saturation and then reheat.
2. Heating and humidifying air in an air washer with pumped recirculation and external heating of water followed by reheat.

The processes for the two systems are shown in Fig. 15.28. The first system with preheating and adiabatic saturation follows processes 1–2 and 2–3 respectively. The second system replaces the two processes with heated water spray in the air washer and the process line is 1–3 (dotted). The reheating process 3-s is common to both.

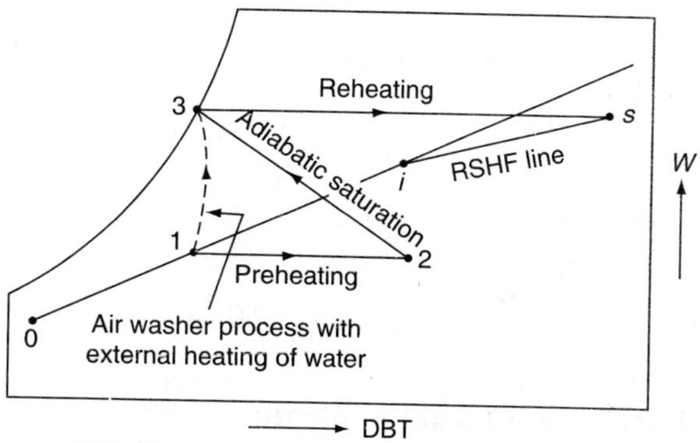

Fig. 15.28 *Winter air-conditioning processes*

15.6 LOAD CALCULATIONS

The major components of load in buildings are due to the direct solar radiation through the west glass, transmission through building structures and fresh air for ventilation, which depend on the *external environment*. The sensible and latent heat gains due to occupants, lights, appliances, machines, piping, etc. within the conditioned space, form the components of load of the *internal heat gains*.

15.6.1 Occupancy Load

The occupants in a conditioned space give out heat at a metabolic rate that depends on their rate of working. The typical values of sensible heat (S) and latent heat (L) given out by a person are given in Table 15.2.

15.6.2 Lighting Load

Electric lights generate sensible heat equal to the electric power consumed.

Incandescent light: $Q =$ Total watts

Fluorescent tube: $Q = 1.25 \times$ total watts

15.6.3 Appliance Load

Most appliances contribute both sensible and latent heats, as given in Table 15.3 below.

Table 15.2
Heat Liberated due to Occupancy

Activity	*Metabolic rate (W)*	*Heat liberated (W) Room DBT in °C*					
		20		24		26	
		S	*L*	*S*	*L*	*S*	*L*
Seated at rest	115	90	25	75	40	65	50
Office Work	140	40	90	80	60	70	70
Standing	150	105	45	82	68	72	78
Light work in factory	235	130	105	100	135	80	155
Dancing	265	140	125	105	160	90	175

Table 15.3
Appliance Load, W

Appliance	*Sensible heat*	*Latent heat*
Coffee brewer	265	65
Egg boiler	353	253
Food warmer/m^2 of plate	1150	1150
Toaster, 360 slices/h	1500	382

15.6.4 Piping, Tank, Steam

Heat is added to conditioned space from running pipes carrying hot or cold fluids. Open tanks containing warm water contribute both SHL and LHL to the space due to evaporation.

15.6.5 Product Load

In the case of cold storages, the enclosures are insulated with at least 10–15 cm of thermocole and are almost sealed. In addition to the heat to be removed from products at the time of initial loading, there is also the heat produced by the commodities during storage. The *heat of respiration* forms a sizeable product load even at a storage temperature of 0°C. At higher temperatures, it is more, as given below.

Table 15.4
Heat of Respiration of Products in J/kg per 24 hours

Product	*Storage temperature*	
	0°C	15.6°C
Apples	312–1500	2300–8215
Cabbage	1248	4265
Carrots	2183	8420
Cauliflower	4680	10,500
Grapes	416	2290
Lemon	520	2390
Orange	728	2495
Potato	1350	1560
Tomato	1040	5820

15.7 AIR-CONDITIONING SYSTEMS

Among many applications of air conditioning, the important ones are

(a) Cold storage
(b) Industrial air conditioning
(c) Comfort air conditioning

15.7.1 Cold Storage

Though cold storage is understood to be merely an application of refrigeration, it is in fact a complete air-conditioning system in which room air is cooled to much lower temperatures over a cooling coil and supplied back to the storage space. The conditions maintained inside the storage space depend on the nature of the product stored.

15.7.2 Industrial Air Conditioning

One category of industries requires constant temperature like metrology laboratories, precision-machine tools, computer centres, etc. In these, a variation of 10 to 20% in RH will not have much effect. The other category like paper and textile mills requires a constant RH. In a textile mill, e.g., an RH of 70 to 75% is required.

15.7.3 Comfort Air Conditioning

There is a problem of measuring comfort in terms of a single parameter which would include all three governing factors, namely, air temperature, humidity and air velocity in addition to air purity. Often, a single parameter called the *effective temperature* is used as an index of comfort.

Effective temperature (ET) is defined as that temperature of saturated air at which a person would experience the same feeling of comfort as experienced in the actual unsaturated environment. Based on the concept of effective temperatures, some *comfort charts* have been developed. The general practice is to recommend the following optimum inside design conditions for comfort for summer air conditioning:

ET 21.7°C
DBT 25 ± 1°C
RH 50 ± 5%

The corresponding room air velocity is 0.4 m/s.

During winter, the body gets acclimatized to withstand lower temperatures. Consequently, a DBT of 21°C at 50% RH and 0.15 – 0.2 m/s air velocity is quite comfortable. In addition to the maintenance of temperature, humidity and air velocity it is also important to maintain the purity of room air. To achieve this, it is necessary to introduce fresh air or *ventilation air*. In the case of theatres and auditoriums, ventilation air requirement is very large due to very large occupancy.

Solved Examples

Example 15.1

Atmospheric air at 1.0132 bar has a dbt of 32°C and a wbt of 26°C. Compute (a) the partial pressure of water vapour, (b) the specific humidity, (c) the dew point temperature, (d) the relative humidity, (e) the degree of saturation, (f) the density of the air in the mixture, (g) the density of the vapour in the mixture and (h) the enthalpy of the mixture.

Solution The state of air is shown on the DBT-W plot in Fig. Ex. 15.1. The path 1–2 represents the constant wbt and enthalpy of the air, which also holds good approximately for an adiabatic saturation process. From Eq. (15.1), the specific humidity at state 2 is given by

$$W_2 = 0.622 \frac{p_s}{p - p_s}$$

The saturation pressure p_s at the wbt of 26°C is 0.03363 bar.

$$\therefore W_2 = 0.622 \frac{0.03363}{1.0132 - 0.3363}$$

$$= 0.021148 \text{ kg vap./kg dry air}$$

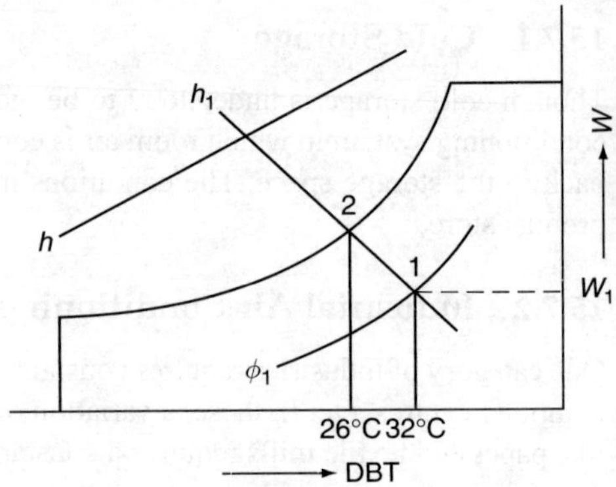

Fig. Ex. 15.1

From Eq. (15.3), for adiabatic saturation

$$W_1 = \frac{c_{pa}(T_2 - T_1) + W_2 \cdot h_{fg2}}{h_{w1} - h_{f2}}$$

From the steam tables, at 26°C

$h_{fg2} = 2439.9$ kJ/kg, $h_{f2} = 109.1$ kJ/kg

At 32°C, $h_{w1} = h_g = 2559.9$ kJ/kg

(a) $$W_1 = \frac{1.005(26 - 32) + 0.021148\,(2439.9)}{2559.9 - 109.1}$$

$$= 0.0186 \text{ kg vap./kg dry air} \qquad \textit{Ans.}$$

(b) $$W_1 = 0.622 \frac{p_w}{p - p_w} = 0.0186$$

$$\frac{p_w}{p - p_w} = \frac{0.622}{0.0186} = 33.44$$

$$p_w = 0.03 \text{ bar} \qquad \textit{Ans.}$$

(c) Saturation temperature at 0.03 bar, dpt = 24.1°C *Ans.*

(d) Relative humidity, $\phi = \dfrac{p_w}{p_{sat}}$

At 32°C, $p_{sat} = 0.048$ bar

$$\phi = \frac{0.03}{0.048} = 0.625 \text{ or } 62.5\% \qquad \textit{Ans.}$$

(e) Deg. of saturation $\mu = \dfrac{W}{W_s} = \dfrac{p_w}{p_s}\dfrac{p - p_s}{p - p_w} = \dfrac{0.03\,(1.0132 - 0.048)}{0.048\,(1.0132 - 0.03)} = 0.614$

(f) Partial pressure of dry air

$$p_a = p - p_w = 1.0132 - 0.03 = 0.9832 \text{ bar}$$

$\therefore$ Density of dry air

$$\rho_a = \frac{p_a}{R_a T_{db}} = \frac{0.9832 \times 100}{0.287 \times (273 + 32)}$$

$$= 1.12 \text{ kg/m}^3 \text{ dry air} \qquad \textit{Ans.}$$

(g) Density of water vapour

$$\rho_w = 0.0186 \frac{\text{kg vap.}}{\text{kg dry air}} \times 1.12 \frac{\text{kg dry air}}{\text{m}^3 \text{ dry air}}$$

$$= 0.021 \text{ kg vap/m}^3 \text{ dry air}$$ *Ans.*

(h) Enthalpy of the mixture

$$h = h_a + Wh_w = c_p t_a + W[h_g + 1.88\,(t_{db} - t_{dp})]$$

$$= 1.005 \times 32 + 0.0186\,[2559.9 + 1.88\,(32 - 24.1)]$$

$$= 80.55 \text{ kJ/kg}$$ *Ans.*

Example 15.2

An air-water vapour mixture enters an adiabatic saturator at 30°C and leaves at 20°C, which is the adiabatic saturation temperature. The pressure remains constant at 100 kPa. Determine the relative humidity and the humidity ratio of the inlet mixture.

Solution The specific humidity at the exit

$$W_2 = 0.622 \frac{p_s}{p - p_s} = 0.622 \left(\frac{2.339}{100 - 2.339}\right) = 0.0149 \frac{\text{kg vap.}}{\text{kg dry air}}$$

The specific humidity at the inlet (Eq. 15.3)

$$W_1 = \frac{c_{pa}(T_2 - T_1) + W_2 h_{fg2}}{h_{w1} - h_{f2}} = \frac{1.005\,(20 - 30) + 0.0149 \times 2454.1}{2556.3 - 83.96}$$

$$= 0.0107 \text{ kg vap./kg dry air}$$

$$W_1 = 0.622 \left(\frac{p_{w1}}{100 - p_{w1}}\right) = 0.0107$$

$$p_{w_1} = 1.691 \text{ kPa}$$

$$\therefore \quad \phi_1 = \frac{p_{w1}}{p_{s1}} = \frac{1.691}{4.246} = 0.398 \text{ or } 39.8\%$$ *Ans.*

Example 15.3

Saturated air at 2°C is required to be supplied to a room where the temperature must be held at 20°C with a relative humidity of 50%. The air is heated and then water at 10°C is sprayed in to give the required humidity.Determine the temperature to which the air must be heated and the mass of spray water required per m³ of air at room conditions. Assume that the total pressure is constant at 1.013 bar ana neglect the fan power.

Solution The process is shown in Fig. Ex. 15.3. From the steam tables, at 20°C, $p_{sat} = 2.339$ kPa

$$\phi_3 = \frac{p_{w3}}{(p_{sat})_{t3}} = \frac{p_{w3}}{2.339} = 0.50$$

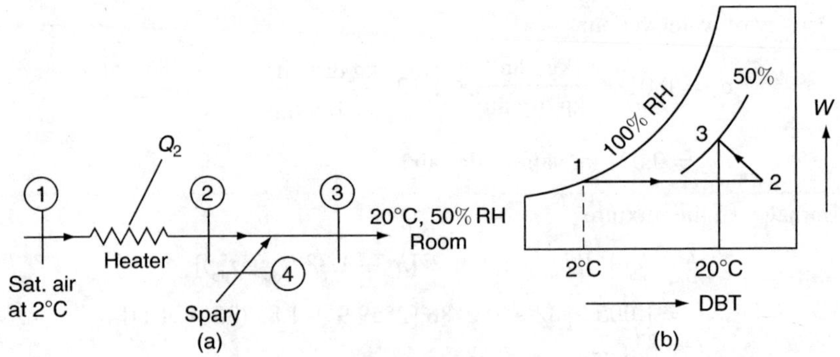

Fig. Ex. 15.3

$$\therefore \quad p_{w3} = 1.17 \text{ kPa}$$

$$\therefore \quad p_{a3} = 101.3 - 1.17 = 100.13 \text{ kPa}$$

$$W_3 = 0.622 \frac{p_{w3}}{p_{a3}} = 0.622 \times \frac{1.17}{100.13} = 0.00727$$

$$\phi_1 = \frac{p_{w1}}{(p_{sat})_{2°C}} = 1.00$$

At $2°C,\ p_{sat} = 0.7156$ kPa

$$\therefore \quad p_{w1} = 0.7156 \text{ kPa}$$

$$p_{a1} = 101.3 - 0.7156 = 100.5844 \text{ kPa}$$

$$W_1 = 0.622 \frac{0.7156}{100.5844} = 0.00442$$

$$W_3 - W_1 = 0.00727 - 0.00442 = 0.00285 \text{ kg vap./kg dry air}$$

$$v_{a3} = \frac{R_a T_3}{p_{a3}} = \frac{0.287 \times 293}{100.13} = 0.84 \text{ m}^3\text{/kg dry air}$$

$$\therefore \quad \text{Spray water} = 0.00285 \frac{\text{kg vap.}}{\text{kg dry air}} \times \frac{\text{kg dry air}}{0.84 \text{ m}^3}$$

$$= 0.00339 \text{ kg moisture/m}^3$$ *Ans.*

$$G_2 h_2 + m_{w4} h_4 = G_3 h_3$$

$$\therefore \quad h_2 + (W_3 - W_2) h_4 = h_3$$

$$h_{a2} + W_2 h_{w2} + (W_3 - W_2) h_4 = h_{a3} + W_3 h_{w3}$$

$$\therefore \quad c_p (t_3 - t_2) + W_3 h_{w3} - W_2 h_{w2} - (W_3 - W_2) h_4 = 0$$

From the steam tables, at $p_w = 1.17$ kPa

$$h_g = 2518 \text{ kJ/kg and } t_{sat} = 9.65°\text{C}$$

$$1.005\,(20 - t_2) + 0.00727\,[2518 + 1.884\,(20 - 9.65)]$$
$$- 0.00442\,[2518 + 1.884\,(t_2 - 9.65)] - 0.00285 \times 10 = 0$$

$\therefore$ $$t_2 = 27.2°\text{C}$$ *Ans.*

Example 15.4

An air conditioning system is designed under the following conditions:
Outdoor conditions—30°C dbt, 75% R.H.
Required indoor condition—22°C dbt, 70% R.H.
Amount of free air circulated—3.33 m³/s
Coil dew point temperature—14°C

The required condition is achieved first by cooling and dehumidification and then by heating. Estimate (a) the capacity of the cooling coil in tonnes, (b) the capacity of the heating coil in kW, and (c) the amount of water vapour removed in kg/s.

Solution The processes are shown in Fig. Ex. 15.4. The property values, taken from the psychrometric chart, are

$$h_1 = 82,\ h_2 = 52,\ h_3 = 47,\ h_4 = 40 \text{ kJ/kg dry air}$$
$$W_1 = 0.020,\ W_2 = W_3 = 0.0115 \text{ kg vap./kg dry air}$$
$$v_1 = 0.887 \text{ m}^3\text{/kg dry air}$$
$$G = \frac{3.33}{0.887} = 3.754 \text{ kg dry air/sec}$$

$\therefore$ $$\text{Cooling coil capacity} = G(h_1 - h_3) = 3.754\,(82 - 47) \text{ kJ/s}$$
$$= \frac{3.754 \times 35 \times 3600}{14{,}000} = 33.79 \text{ tonnes}$$ *Ans.* (a)

$$\text{Capacity of the heating coil} = G(h_2 - h_3) = 3.754\,(52 - 47) \text{ kJ/s}$$
$$= 3.754 \times 5 = 18.77 \text{ kW}$$ *Ans.* (b)

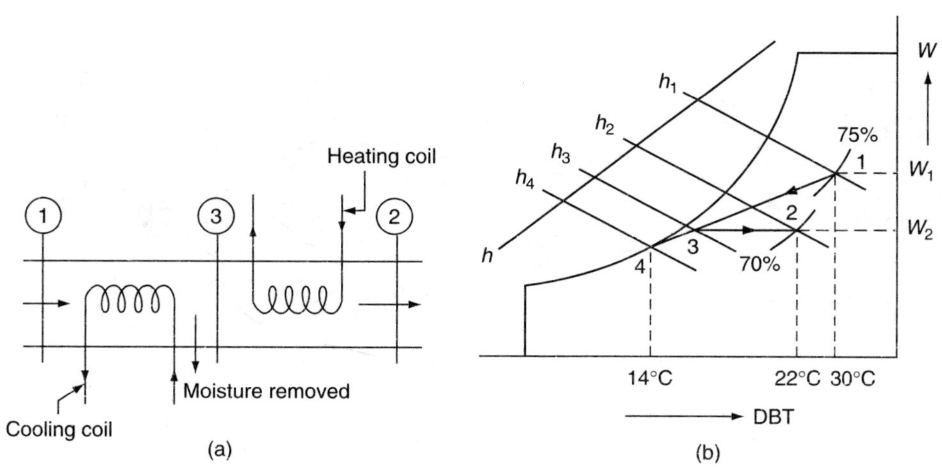

Fig. Ex. 15.4

Rate of water vapour removed $= G(W_1 - W_3) = 3.754 \times (0.0200 - 0.0115)$

$= 0.0319$ kg/s *Ans.* (c)

Example 15.5

Air at 20°C, 40% RH is mixed adiabatically with air at 40°C, 40% RH in the ratio of 1 kg of the former with 2 kg of the latter (on dry basis). Find the final condition of air.

Solution Fig. Ex. 15.5 shows the mixing process of two air streams. The equations

$$G_1 + G_2 = G_3$$

$$G_1 W_1 + G_2 W_2 = G_3 W_3$$

$$G_1 h_1 + G_2 h_2 = G_3 h_3$$

result in

$$\frac{W_2 - W_3}{W_3 - W_1} = \frac{h_2 - h_3}{h_3 - h_1} = \frac{G_1}{G_2}$$

From the psychrometric chart

$$W_1 = 0.0058,\ W_2 = 0.0187 \text{ kg vap./kg dry air}$$

$$h_1 = 35,\ h_2 = 90 \text{ kJ/kg dry air}$$

$$\therefore \quad \frac{0.0187 - W_3}{W_3 - 0.0058} = \frac{G_1}{G_2} = \frac{1}{2}$$

$$\therefore \quad W_3 = \frac{2}{3} \times 0.187 + \frac{1}{3} \times 0.0058 = 0.0144 \text{ kg vap./kg dry air}$$

Again

$$\frac{h_2 - h_3}{h_3 - h_1} = \frac{G_1}{G_2} = \frac{1}{2}$$

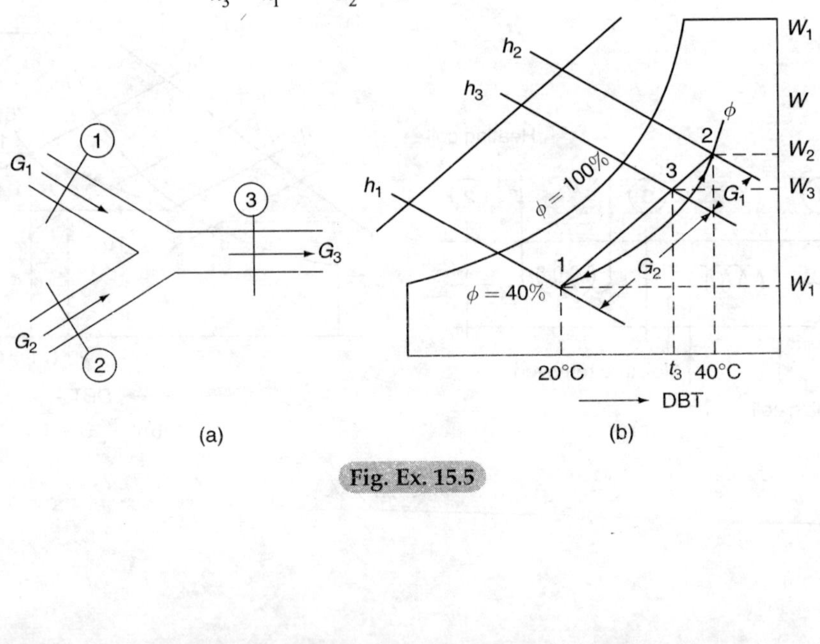

Fig. Ex. 15.5

$$\therefore \quad h_3 = \frac{2}{3} h_2 = \frac{1}{3} h_1 = \frac{2}{3} \times 90 + \frac{1}{3} \times 35 = 71.67 \text{ kJ/kg dry air}$$

$\therefore$ Final condition of air is given by

$$W_3 = 0.0144 \text{ kg vap./kg dry air}$$

$$h_3 = 71.67 \text{ kJ/kg dry air}$$

Example 15.6

Saturated air at 21°C is passed through a drier so that its final relative humidity is 20%. The drier uses silica gel adsorbent. The air is then passed through a cooler until its final temperature is 21°C without a change in specific humidity. Find out (a) the temperature of air at the end of the drying process, (b) the heat rejected in kJ/kg dry air during the cooling process, (c) the relative humidity at the end of the cooling process, (d) the dew point temperature at the end of the drying process, and (e) the moisture removed during the drying process in kg vap./kg dry air.

Solution From the psychrometric chart (Fig. Ex. 15.6)

$T_2 = 38.5°C$ *Ans.* (a)

$h_1 - h_3 = 60.5 - 42.0 = 18.5$ kJ/kg dry air *Ans.* (b)

$\phi_3 = 53\%$ *Ans.* (c)

$t_4 = 11.2°C$ *Ans.* (d)

$W_1 - W_2 = 0.0153 - 0.0083 = 0.0070$ kg vap/kg dry air *Ans.* (e)

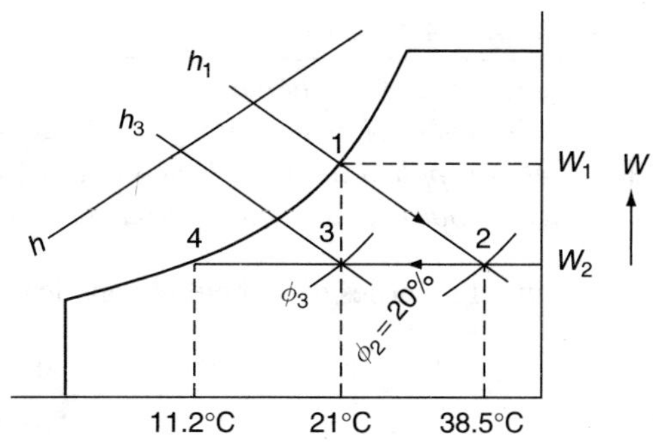

Fig. Ex. 15.6

Example 15.7

For a hall to be air-conditioned, the following conditions are given
Outdoor condition—40°C dbt, 20°C wbt
Required comfort condition—20°C dbt, 60% RH
Seating capacity of hall—1500
Amount of outdoor air supplied—0.3 m³/min per person

If the required condition is achieved first by adiabatic humidification and then by cooling, estimate (a) the capacity of the cooling coil in tonnes, and (b) the capacity of the humidifier in kg/h.

Solution From the psychrometric chart (Fig. Ex. 15.7)

$h_1 = h_2 = 57.0$, $h_3 = 42.0$ kJ/kg dry air

$W_1 = 0.0065$, $W_2 = W_3 = 0.0088$ kg vap./kg dry air

$t_2 = 34.5°C$, $v_1 = 0.896$ m³/kg dry air

Amount of dry air supplied

$$G = \frac{1500 \times 0.3}{0.896} = 502 \text{ kg/min}$$

∴ Capacity of the cooling coil $= G(h_2 - h_3)$
$= 502\ (57 - 42)$ kJ/min

$$= \frac{502 \times 15 \times 60}{14,000} = 32.27 \text{ tonnes} \qquad \textit{Ans. (a)}$$

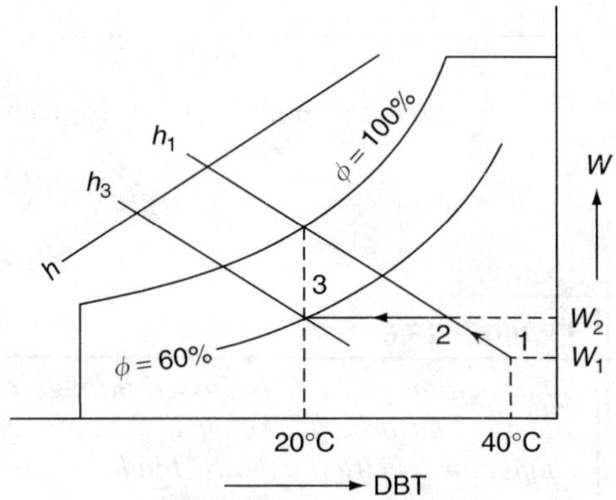

Fig. Ex. 15.7

Capacity of the humidifier $= G\ (W_2 - W_1)$
$= 502\ (0.0088 - 0.0065)$ kg/min

$$= 502 \times 23 \times 10–4 \times 60 = 69.3 \text{ kg/h} \qquad \textit{Ans. (b)}$$

Example 15.8

Water at 30°C flows into a cooling tower at the rate of 1.15 *kg per kg of air. Air enters the tower at a dbt of* 20°C *and a relative humidity of* 60% *and leaves it at a dbt of* 28°C *and* 90% *relative humidity. Make-up water is supplied at* 20°C*. Determine:* (i) *the temperature of water leaving the tower,* (ii) *the fraction of water evaporated, and* (iii) *approach and range of the cooling tower.*

Solution Properties of air entering and leaving the tower (Fig. 15.13) are

$$t_{wb1} = 15.2°C$$

$$t_{wb2} = 26.7°C$$

$$h_1 = 43 \text{ kJ/kg dry air}$$

$$h_2 = 83.5 \text{ kJ/kg dry air}$$

$$W_1 = 0.0088 \text{ kg water vapour/kg dry air}$$

$$W_2 = 0.0213 \text{ kg water vapour/kg dry air}$$

Enthalpies of the water entering the tower and the make-up water are

$$h_{w3} = 125.8 \text{ kJ/kg} \quad h_m = 84 \text{ kJ/kg}$$

From the energy balance Eq. (15.5),

$$h_{w3} - h_{w4} = \frac{G}{m_w}[(h_2 - h_1) - (W_2 - W_1)\,h_w]$$

$$= \frac{1}{1.15}[(83.5 - 43) - (0.0213 - 0.0088)\ 84] = 34.2 \text{ kJ/kg}$$

Temperature drop of water

$$t_{w3} - t_{w4} = \frac{34.2}{4.19} = 30 - t_{w4}$$

$\therefore \quad t_{w4} = 21.8°C$

$\therefore \quad \text{Approach} = t_{w4} = t_{wb1} = 21.8 - 15.2 = 6.6°C$

$$\text{Range} = t_{w3} = t_{w4} = 30 - 21.8 = 8.2°C$$

Fraction of water evaporated, $x = G(W_2 - W_1) = 1(0.0213 - 0.0088) = 0.0125$ kg/kg dry air

Example 15.9

Air having DBT of 40°C and RH 40% has to be cooled and dehumidified such that its DBT is 25°C and RH 70%. Determine the amount of moisture removed and the tons of refrigeration required when the air flow rate is 30 cmm. Estimate the bypass factor of the cooling coil.

Solution

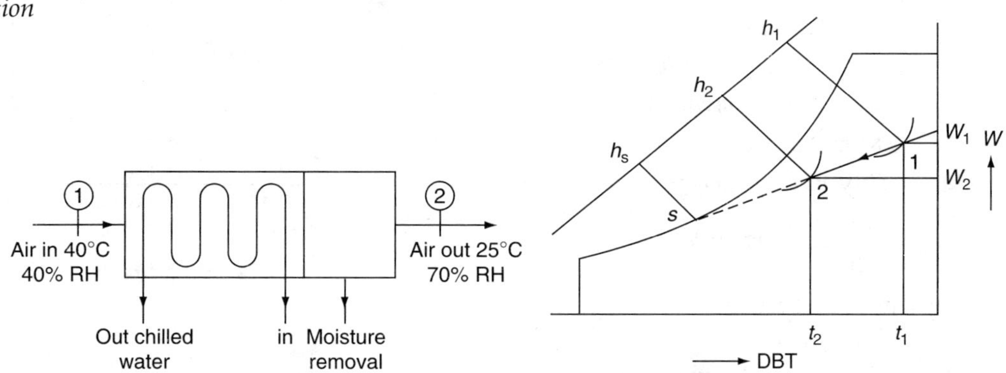

Fig. Ex. 15.9 *Cooling and dehumidification*

From psychrometric chart (fig. above)

$$\upsilon_1 = 0.9125 \text{ m}^3\text{/kg d.a. (at 40°C, 40\% RH)}$$

Mass of dry air entering the cooling coil

$$G = \frac{30 \text{ m}^3/\text{min}}{0.9125} = 32.88 \text{ kg/min}$$

$$W_1 = 0.0182 \text{ kg vapour/kg d.a., } h_1 = 86 \text{ kJ/kg d.a.}$$
$$W_2 = 0.0136 \text{ kg/kg d.a., } h_2 = 60 \text{ kJ/kg d.a.}$$

$$\text{Cooling load} = G(h_1 - h_2) = \frac{32.88(86 - 60)}{60 \times 3.5} = 4.07 \text{ tonnes of refrigeration}$$ *Ans.*

$$\text{Rate of moisture removal} = G(W_1 - W_2)$$
$$= 32.88 (0.0182 - 0.0136)$$
$$= 0.151 \text{ kg/min}$$ *Ans.*

$$\text{Bypass factor} = \frac{h_2 - h_5}{h_1 - h_5}$$
$$= \frac{60 - 41.5}{86 - 41.5} = 0.468$$ *Ans.*

Example 15.10

A classroom with seating capacity of 75 persons has to be air conditioned, first by heating and then humidifying, during winter. The outdoor conditions are: DBT 10°C, WBT 8°C. The indoor conditions: DBT 20°C, RH 50%. An amount of 0.3 cmm per person is circulated Calculate the capacity of the heating coil and its surface temperature when the bypass factor is 0.5 and the capacity of the humidifier.

Solution In Fig. Ex.15.10 above, line 1–4 is sensible heating and line 2–3 is adiabatic saturation when $h_2 = h_3$. From the chart

$W_1 = 0.0058$ kg moisture/kg d.a.,

$h_1 = 24.5$ kJ/kg d.a.

$h_2 = h_3 = 39.5$ kJ/kg d.a, $t_db_2 = 25$°C.

$W_3 = 0.0074$ kg moisture/kg d.a.

$\upsilon_1 = 0.81$ m3/kg d.a.

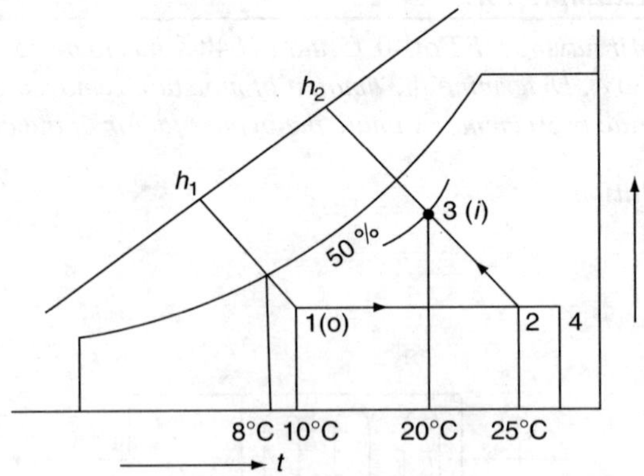

Fig. Ex. 15.10

Mass of dry air circulated per min

$$G = \frac{0.3 \times 75}{0.81} = 27.78 \text{ kg d.a./min}$$

Capacity of the heating coil $= G(h_2 - h_1)$
$= 27.78\ (39.5 - 24.5)/60 = 6.94$ kW *Ans.*

Bypass factor of the heating coil

$$= \frac{t_4 - t_2}{t_4 - t_1} = \frac{t_4 - 25}{t_4 - 10} = 0.5$$

$\therefore\ t_4 = 40°C = t_s$ *Ans.*

Capacity of the humidifier $= G(W_3 - W_1) = 27.78\ (0.0074 - 0.0058) \times 60$
$= 2.667$ kg/h *Ans.*

Example 15.11

A mixture of recirculated room air and outdoor air enters a cooling coil at 31°C DBT and 18.5°C WBT at the rate of 39.6 cmm. The effective surface temperature of the coil is 4.4°C. The surface area of the coil is such as would gives 12.5 kW of refrigeration. Determine the DBT and WBT of the air leaving the coil and the coil bypass factor.

Solution From Fig. Ex. 15.11, at the ADP, of 4.4°C,

$W_s = 5.25$g/kg d.a.

$h_s = 17.7$ kJ/kg.d.a.

State of entering air: $\upsilon_1 = 0.872$ m^3/kg d.a.

$h_1 = 52.5$ kJ/kg d.a.

$w_1 = 8.2$g/kg d.a.

Mass flow rate of dry air,

$$G = \frac{39.6}{0.872} = 44.4 \text{ kg d.a./min}$$

Cooling load, $\dot{Q} = G\,(h_1 - h_2)$

$$\therefore \quad h_1 - h_2 = \frac{12.5 \times 60}{44.4} = 16.89 \text{ kJ/kg d.a.}$$

$$h_2 = 52.5 - 16.89 = 35.61 \text{ kJ/kg d.a.}$$

For the condition line 1–5,

$$\frac{h_1 - h_2}{h_1 - h_s} = \frac{w_1 - w_2}{w_1 - w_s}$$

$$\frac{52.5 - 25.61}{52.5 - 17.7} = \frac{8.2 - w_2}{8.2 - 5.25}$$

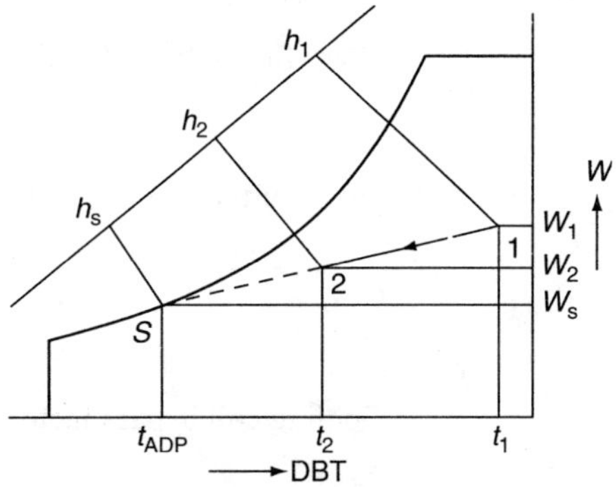

Fig. Ex. 15.11

$\therefore W_2 = 6.77$ g water vapour/kg d.a.

From the psychrometric chart, for $h_2 = 35.61$ kJ/kg d.a. and $W_2 = 6.77$g water vapour/kg d.a.,

$$t_2 = 18.6°\text{C},\ t' = 12.5°\text{C (WBT)}. \qquad \textit{Ans.}$$

Coil bypass factor

$$X = \frac{h_2 - h_s}{h_1 - h_s} = \frac{35.61 - 17.7}{52.5 - 17.7} = 0.515 \qquad \textit{Ans.}$$

***Example* 15.12**

An auditorium has to be air-conditioned first by cooling and dehumidifying and then heating for summer when the outdoor conditions are: DBT 35°C, RH 70%. The desired indoor conditions are: DBT 20°C, RH 60%. The cooling coil dew point temperature is 10°C. The amount of free air to be circulated is 300 cmm. Estimate (a) the capacity of cooling coil and its bypass factor, (b) the capacity of the heating coil and its surface temperature when the bypass factor is 0.25, and (c) the mass of water vapour removed.

Solution

Point 1(*o*): DBT 35°C RH 70%

$h_1 = 98$ kJ/kg d.a.

$w_1 = 0.0246$ kg vap./kg d.a.

$v_1 = 0.907$ m^3/kg d.a.

Point 3(*c*): DBT 20°C, RH 60%,

$h_3 = 42$ kJ/kg d.a., $W_3 = 0.0088$ kg vap./kg d.a.

Point(*s*) $t_s = 10$°C.

$h_2 = 34$ kJ/kg d.a., $h_s = 30$ kJ/kg d.a.,

$t_2 = 12$°C.

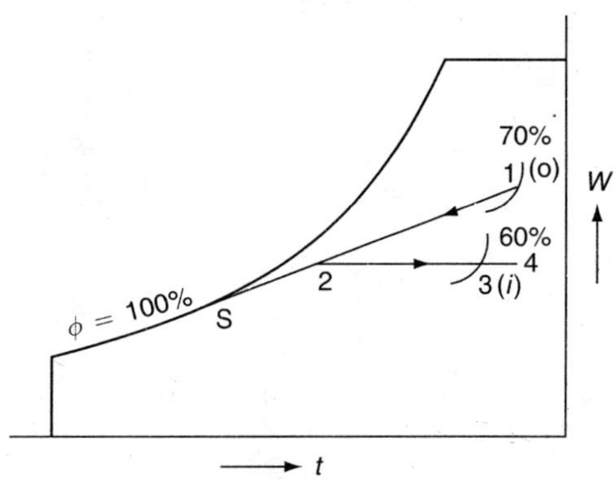

Fig. Ex. 15.12

$$G = \text{mass flow rate of dry air} = \frac{300}{0.907}$$

$$= 330.76 \text{ kg/min}$$

$$\text{Capacity of the cooling coil} = G(h_1 - h_2)$$

$$= 330.76\ (98 - 34)/60 \times 3.5$$

$$= 100.8 \text{ tonnes of refrigeration} \qquad \textit{Ans.}$$

$$\text{Cooling coil bypass factor} = \frac{h_2 - h_s}{h_1 - h_s} = \frac{34 - 30}{98 - 30} = 0.059 \qquad \textit{Ans.}$$

$$\text{Capacity of the heating coil} = G(h_3 - h_2) = \frac{330.76}{60}(42 - 34) = 44.1 \text{ kW}$$

$$\text{Bypass factor of the heating coil} = 0.25 = \frac{t_4 - t_3}{t_4 - t_2} = \frac{t_4 - 20}{t_4 - 12} \qquad \textit{Ans.}$$

$$\therefore \quad t_4 = 22.67°\text{C} = \text{surface temperature of the heating coil} \qquad \textit{Ans.}$$

$$\text{Mass of water vapour removed} = G(W_1 - W_3) = 330.76\ (0.0246 - 0.0088)$$

$$= 5.23 \text{ kg/min.} \qquad \textit{Ans.}$$

Example 15.13

Water from a cooling system is itself to be cooled in a cooling tower at a rate of 2.78 kg/s. The water enters the tower at 65°C and leaves a collecting tank at the base at 38°C. Air flows through the tower, entering the base at 15°C, 0.1 MPa, 55% RH, and leaving the top at 35°C, 0.1 MPa, saturated. Make-up water enters the collecting tank at 14°C. Determine the air flow rate into the tower in m³/s and the make-up water flow rate in kg/s.

Solution Figure Ex. 15.13 shows the flow diagram of the cooling tower. From the steam tables.

$$\text{at } 15°\text{C}, p_{sat} = 0.01705 \text{ bar}, h_g = 2528.9 \text{ kJ/kg}$$

$$\text{at } 35°\text{C}, p_{sat} = 0.05628 \text{ bar}, h_g = 2565.3 \text{ kJ/kg}$$

$$\phi_1 = \frac{p_w}{(p_{sat})_{15°C}} = 0.55$$

$$\therefore \quad p_{w1} = 0.55 \times 0.01705 = 0.938 \times 10^{-2} \text{ bar}$$

$$\phi_2 = \frac{p_w}{(p_{sat})_{35°C}} = 1.00$$

$$\therefore \quad p_{w2} = 0.05628 \text{ bar}$$

$$W_1 = 0.622 \frac{p_w}{p - p_w} = 0.622 \times \frac{0.938 \times 10^{-2}}{1.00 - 0.00938} = 0.00589 \text{ kg vap./kg dry air}$$

$$W_2 = 0.622 \times \frac{0.05628}{1.00 - 0.05628} = 0.0371 \text{ kg vap./dry air}$$

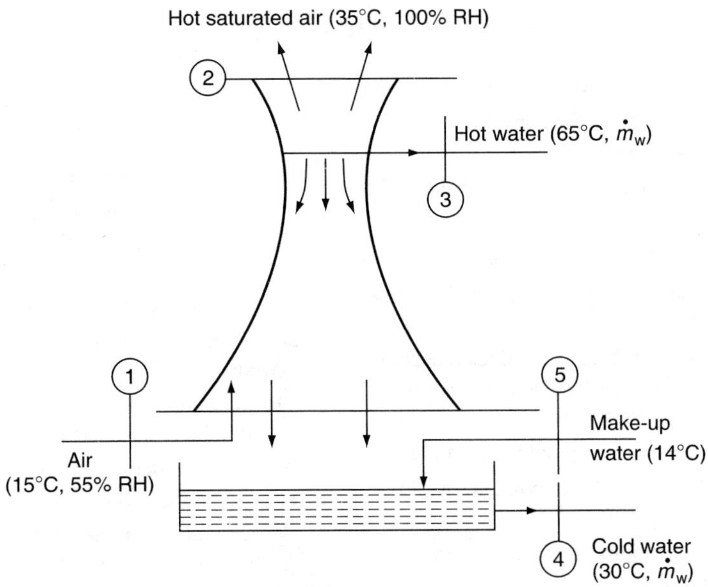

Fig. Ex. 15.13

$\therefore$ Make-up water $= W_2 - W_1 = 0.0371 - 0.00589 = 0.03121$ kg vap./kg dry air

Energy balance gives

$$H_2 + H_4 - H_1 - H_3 - H_5 = 0$$

For 1 kg of dry air

$$c_{pa}(t_2 - t_1) + W_2 h_2 - W_1 h_1 + \dot{m}_w (h_4 - h_3) - (W_2 - W_1) h_5 = 0$$

$$\therefore \quad 1.005\,(35 - 15) + 0.0371 \times 2565.3 - 0.00589 \times 2528.9$$

$$+ \dot{m}_w (-35)\, 4.187 - 0.03121 \times 4.187 \times 14 = 0$$

or

$$146.55\ \dot{m}_w = 98.54$$

$$\therefore \quad \dot{m}_w = 0.672 \text{ kg water/kg dry air}$$

Since water flow rate is 2.78 kg/s

$$\therefore \text{ Rate of dry air flow} = \frac{2.78}{0.672} = 4.137 \text{ kg/s}$$

$\therefore$ Make-up water flow rate $= 0.03121 \times 4.137 = 0.129$ kg/s *Ans.*

Rate of dry air flow $= 4.137$ kg/s

Rate of wet air flow $= 4.137\,(1 + W_1) = 4.137 \times 1.00589 = 4.16$ kg/s

$$\therefore \text{ Volume flow rate of air} = \frac{\dot{m}_a RT}{p} = \frac{4.16 \times 0.287 \times 288}{100} = 3.438 \text{ m}^3\text{/s}$$ *Ans.*

Review Questions

15.1 What do you mean by air conditioning?

15.2 What is moist air? Name the constituents of dry air and its composition.

15.3 What is Gibbs theorem? State the molecular mass of atmospheric air.

15.4 What do you understand by psychrometry?

15.5 What do you mean by saturated air and unsaturated air?

15.6 What is specific humidity? When does it become maximum? Show that specific humidity is a function of partial pressure of water vapour.

15.7 What do you mean by dew point temperature?

15.8 What is degree of saturation? What are its limiting values?

15.9 What is relative humidity? How is it defined as a ratio of two partial pressures?

15.10 Show that the relative humidity of air given by

$$\phi = \frac{\mu}{1-(1-\mu)p_s/p}$$

15.11 What is enthalpy of moist air? Show that it is given by

$$h = 1.005\,t + W\,(2500 + 1.88\,t)\text{ kJ/kg d.a.}$$

15.12 What do your mean by wet bulb temperature? Is it a thermodynamic property?

15.13 When do the DBT, DPT and WBT become equal?

15.14 What is a sling psychrometer?

15.16 Explain the adiabatic saturation process. What is thermodynamic wet-bulb temperature?

15.17 Show that the enthalpy of air-vapour mixture remains constant during an adiabatic saturation process.

15.18 Explain with a neat sketch the constant property lines in a psychrometric chart.

15.19 What do you mean by enthalpy deviation?

15.20 What are the psychrometric processes encountered in airconditioning practice?

15.21 What is sensible heating or cooling?

15.22 Explain SHL, LHL and total heat load.

15.23 Define sensible heat factor (SHF).

15.24 What is humid specific heat?

15.25 Show that

$$\text{SHF} = \frac{0.0204\,(\text{cmm})\,\Delta t}{0.0204\,(\text{cmm})\,\Delta t + 50\,(\text{cmm})\Delta W} e^{i\theta}$$

where cmm is m^3/min.

15.26 What do you mean by cooling and dehumidification? Explain it with flow diagram and psychrometric charts.

15.27 Explain the process of heating and humidification.

15.28 Explain the process of adiabatic mixing of two air streams.

15.29 What do you mean by chemical dehumidification? Why does the DBT of air increase?

15.30 What is adiabatic evaporative cooling?

15.31 Explain the operation of a cooling tower. What is the minimum temperature to which warm water can be cooled in a cooling tower?

15.32 How is a cooling tower specified? Define the approach and range.

15.33 What do you mean by bypass factor? What is contact factor?

15.34 Explain the operation of an air washer with sketches. How can it be used for year-round airconditioning?

15.35 Explain with a sketch a simple airconditioning system. Define RSH and RLH.

15.36 Explain with sketches summer air-conditioning schemes with and without ventilation air.

15.37 Explain the terms GSHF and RSHF. What are OASH and OALH?

15.38 Explain with sketches the schemes of winter air conditioning.

15.39 Explain the external and internal heat gains. What is occupancy load?

15.40 What is product load? What do you mean by heat of respiration?

15.41 Explain the applications of air conditioning in cold storages and typical industries.

15.42 What is comfort air conditioning? What do you mean by effective temperature? What are comfort charts?

15.43 State the optimum inside design conditions of air recommended for summer and winter air conditioning.

15.44 Explain the importance of ventilation air in a conditioned space.

Problems

15.1 An air-water vapour mixture at 0.1 MPa, 30°C, 80% RH has a volume of 50 m^3. Calculate the specific humidity, dew point, wbt, mass of dry air, and mass of water vapour.

If the mixture is cooled at constant pressure to 5°C, calculate the amount of water vapour condensed.

15.2 A sling psychrometer reads 40°C dbt and 36°C wbt. Find the humidity ratio, relative humidity, dew point temperature, specific volume, and enthalpy of air.

15.3 Calculate the amount of heat removed per kg of dry air if the initial condition of air is 35°C, 70% RH, and the final condition is 25°C, 60% RH.

15.4 Two streams of air 25°C, 50% RH and 25°C, 60% RH are mixed adiabatically to obtain 0.3 kg/s of dry air at 30°C. Calculate the amounts of air drawn from both the streams and the humidity ratio of the mixed air.

15.5 Air at 40°C dbt and 27°C wbt is to be cooled and dehumidified by passing it over a refrigerant-filled coil to give a final condition of 15°C and 90% RH. Find the amounts of heat and moisture removed per kg of dry air.

15.6 An air-water vapour mixture enters a heater-humidifier unit at 5°C, 100 kPa, 50% RH. The flow rate of dry air is 0.1 kg/s. Liquid water at 10°C is sprayed into the mixture at the rate of 0.002 kg/s. The mixture leaves the unit at 30°C, 100 kPa. Calculate (a) the relative humidity at the outlet, and (b) the rate of heat transfer to the unit.

15.7 A laboratory has a volume of 470 m^3, and is to be maintained at 20°C, 52.5% RH. The air in the room is to be completely changed once every hour and is drawn from the atmosphere at 1.05 bar, 32°C, 86% RH, by a fan absorbing 0.45 kW. This air passes through a cooler which reduces its temperature and causes condensation, the condensate being drained off at 8°C. The resulting saturated air is heated to room condition. The total pressure is constant throughout. Determine (a) the temperature of the air leaving the cooler, (b) the rate of condensation, (c) the heat transfer in the cooler, and (d) the heat transfer in the heater.

Ans. (a) 10°C, (b) 10.35 kg/h, (c) 11.33 kW, (d) 1.63 kW

15.8 In an air conditioning system, air is to be cooled and dehumidified by means of a cooling coil. The data are as follows:

Initial condition of the air at inlet to the cooling coil: dbt = 25°C, partial pressure of water vapour = 0.019 bar, absolute total pressure = 1.02 bar

Final condition of air at exit of the cooling coil: dbt = 15°C, RH = 90%, absolute total pressure = 1.02 bar.

Other data are as follows:

Characteristic gas constant for air = 287 J/kg K

Characteristic gas constant for water vapour = 461.5 J/kg K

Saturation pressure for water at 15°C = 0.017 bar

Enthalpy of dry air = 1.005 t kJ/kg

Enthalpy of water vapour = (2500 + 1.88 t) kJ/kg where t is in °C

Determine (a) the moisture removed from air per kg of dry air, (b) the heat removed by the cooling coil per kg of dry air. *Ans.* (a) 0.0023 kg/kg d.a. (b) 16.1 kJ/kg d.a.

15.9 Air at 30°C, 80% RH is cooled by spraying in water at 12°C. This causes saturation, followed by condensation, the mixing being assumed to take place adiabatically and the condensate being drained off at 16.7°C. The resulting saturated mixture is then heated to produce the required conditions of 60% RH at 25°C. The total pressure is constant at 101 kPa. Determine the mass of water supplied to the sprays to provide 10 m^3/h of conditioned air. What is the heater power required? *Ans.* 2224 kg/h, 2.75 kW

15.10 An air-conditioned room requires 30 m^3/min of air at 1.013 bar, 20°C, 52.5% RH. The steady flow conditioner takes in air at 1.013 bar, 77% RH, which it cools to adjust the moisture content and reheats to room temperature. Find the temperature to which the air is cooled and the thermal loading on both the cooler and heater. Assume that a fan before the cooler absorbs 0.5 kW, and that the condensate is discharged at the temperature to which the air is cooled. *Ans.* 10°C, 25 kW, 6.04 kW

15.11 An industrial process requires an atmosphere having a RH of 88.4% at 22°C, and involves a flow rate of 2000 m^3/h. The external conditions are 44.4% RH, 15°C. The air intake is heated and then humidified by water spray at 20°C. Determine the mass flow rate of spray water and the power required for heating, if the pressure throughout is 1 bar. *Ans.* 23.4 kg/h, 20.5 kW

15.12 Cooling water enters a cooling tower at a rate of 1000 kg/h and 70°C. Water is pumped from the base of the tower at 24°C and some make-up water is added afterwards. Air enters the tower at 15ºC, 50% RH, 1.013 bar, and is drawn from the tower saturated at 34°C, 1 bar. Calculate the flow rate of the dry air in kg/h and the make-up water required per hour. *Ans.* 2088 kg/h, 62.9 kg/h

15.13 A grain dryer consists of a vertical cylindrical hopper through which hot air is blown. The air enters the base at 1.38 bar, 65°C, 50% RH. At the top, saturated air is discharged into the atmosphere at 1.035 bar, 60°C.

Estimate the moisture picked up by 1 kg of dry air, and the total enthalpy change between the entering and leaving streams expressed per unit mass of dry air. *Ans.* 0.0864 kJ/kg air, 220 kJ/kg air

15.14 Air enters a counterflow cooling tower at a rate of 100 m^3/s at 30°C dbt and 40% relative humidity. Air leaves at the top of the tower at 32°C and 90% relative humidity. Water enters the tower at 35°C and the water flow rate is 1.2 times the mass flow rate of air. Make-up water is supplied at 20°C. What are the range and approach of the tower? At what rate is heat absorbed from the load by the stream of water on its way back to the top of the tower? What percentage of the water flow rate must be supplied as make-up water to replace the water evaporated into the air stream? *Ans.* Range = 8.7°C, Approach = 6.3°C, Q = 5005 kW, % make-up = 1.39%

15.15 Moist air at standard atmospheric pressure is passed over a cooling coil the inlet conditions are: DBT 30°C, 50% RH. The outlet conditions are: DBT 15°C, RH 80% show the process on a psychrometric chart. Estimate the amount of heat and moisture removed per kg of dry air. *Ans.* (–27.23 kJ/kg d.a., –0.00482 kg/kg d.a.)

15.16 Moist air enters a chamber at 5°C DBT and 2.5°C WBT at a rate of 90 cmm. While passing through the chamber, the air absorbs sensible heat at the rate of 40.7 KW and picks up 40 kg/h of saturated steam at 110°C. Determine the DBT and WBT of leaving air. *Ans.* 26.5°C 18.1°C

15.17 An air-conditioned room is maintained at 27°C DBT and 50% RH. The ambient conditions are 40°C DBT and 27°C WBT. The room has a sensible heat gain of 14 kW. Air is supplied to the room at 7°C saturated. Determine (a) mass of moist air supplied to the room is kg/h, (b) latent heat gain of room in KW, (c) cooling load of the air washer in KW if 30% of the air supplied to the room is fresh, the remainder being recirculated.
Ans. (a) 2482 kg/h, (b) 8.57 kw, (c) 28.71 kw

15.18 The outdoor summer design condition for a bank for 100 persons at a place is DBT 310 K and WBT 300 K. The required inside conditions are DBT 295 K and 60% RH. The room sensible heat gain is 400,000 kJ/h. and the latent heat gain is 200,000 kJ/h. Ventilation air required is 0.0047 m^3/s per person. Estimate (a) the grand total heat load, (b) SHF, (c) apparatus dew point, (d) volume flow rate of dehumidified air
Ans. (a) 5,70,480 kJ/h, (b) 0.6622, (c) 282.5K, (d) 532.7 m^3/min.

15.19 A class room of seating capacity 50 persons (Sensible heat gain per person 100 W, latent hear gain per person 25 W, other heat gains are 3000W) is to be maintained at DBT 15°C, 50% RH. The ambient conditions are 40°C DBT and 27°C WBT. Air is supplied to the room at 7°C, 100% RH. Calculate (a) the mass of moist air supplied to the room, (b) the cooling load of the air washer in kW, if 30% of the air supplied to the room is fresh and the remainder being recirculated. *Ans.* (a) 85.38 kg/min, (b) 6.97 tonnes.

C H A P T E R

16

Reactive Systems

In this chapter we shall study the thermodynamics of mixtures that may be undergoing chemical reaction. With every chemical reaction is associated a chemical equation which is obtained by balancing the atoms of each of the atomic species involved in the reaction. The initial constituents which start the reaction are called the *reactants*, and the final constituents which are formed by chemical reaction with the rearrangement of the atoms and electrons are called the *products*. The reaction between the reactants, hydrogen and oxygen, to form the product water can be expressed as

$$H_2 + \frac{1}{2} O_2 \rightleftharpoons H_2O \tag{16.1}$$

The equation indicates that one mole of hydrogen and half a mole of oxygen combine to form one mole of water. The reaction can also proceed in the reverse direction. The coefficients 1, $\frac{1}{2}$, 1 in the chemical Eq. (16.1) are called *stoichiometric coefficients*.

16.1 DEGREE OF REACTION

Let us suppose that we have a mixture of four substances, A_1, A_2, A_3 and A_4, capable of undergoing a reaction of the type

$$\nu_1 A_1 + \nu_2 A_2 \rightleftharpoons \nu_3 A_3 + \nu_4 A_4$$

where the ν's are the stoichiometric coefficients.

Starting with arbitrary amounts of both initial and final constituents, let us imagine that the reaction proceeds completely to the right with the disappearance of at least one of the initial constituents, say, A_1. Then the original number of moles of the initial constituents is given in the form

$$n_1 \text{ (original)} = n_0 \nu_1$$
$$n_2 \text{ (original)} = n_0 \nu_2 + N_2$$

where n_0 is an arbitrary positive number, and N_2 is the residue (or excess) of A_2, i.e., the number of moles of A_2 which cannot combine. If the reaction is assumed to proceed completely to the left with the disappearance of the final constituent A_3, then

$$n_3 \text{ (original)} = n'_0 \nu_3$$
$$n_4 \text{ (original)} = n'_0 \nu_4 + N_4$$

where n'_0 is an arbitrary positive number and N_4 is the excess number of moles of A_4 left after the reaction is complete from right to left.

For a reaction that has occurred completely to the left, there is a maximum amount possible of each initial constituent, and a minimum amount possible of each final constituent, so that

$$n_1(\max) = \underset{\text{(Original number of moles of } A_1)}{n_0 \nu_1} + \underset{\text{(Number of moles of } A_1 \text{ formed by chemical reaction)}}{n'_0 \nu_1}$$

$$(n'_0 \nu_3 A_3 + n'_0 \nu_4 A_4 \rightarrow n'_0 \nu_1 A_1 + n'_0 \nu_2 A_2)$$

$$= (n_0 + n'_0)\, \nu_1$$

$$n_2\,(\text{max}) = (n_0\, \nu_2 + N_2) \quad + \quad n'_0\, \nu_2$$

(Original number of moles of A_2) (Number of moles of A_2 formed by chemical reaction)

$$= (n_0 + n'_0)\, \nu_2 + N_2$$

n_3 (min) = 0 (The constituent A_3 completely disappears by reaction)

n_4 (min) = N_4 (The excess number of moles of A_4 that are left after the reaction is complete to the left)

Similarly, if the reaction is imagined to proceed completely to the right, there is a minimum amount of each initial constituent, and a maximum amount of each final constituent, so that

$$n_1\,(\text{min}) = 0$$

$$n_2\,(\text{min}) = N_2$$

$$n_3\,(\text{max}) = n'_0\, \nu_3 \quad + \quad n_0\, \nu_3$$

(Original amount) (Amount formed by chemical reaction)

$$(n_0\, \nu_1\, A_1 + n_0\, \nu_2\, A_2 \rightarrow n_0\, \nu_3\, A_3 + n_0\, \nu_4\, A_4)$$

$$= (n_0 + n'_0)\nu_3$$

$$n_4\,(\text{max}) = (n_0 + n'_0)\, \nu_4 + N_4$$

Let us suppose that the reaction proceeds partially either to the right or to the left to the extent that there are n_1 moles of A_1, n_2 moles of A_2, n_3 moles of A_3, and n_4 moles of A_4. The *degree* (or *advancement*) *of reaction* ε is defined in terms of any one of the initial constituents, say, A_1, as the fraction

$$\varepsilon = \frac{n_1(\text{max}) - n_1}{n_1(\text{max}) - n_1(\text{min})}$$

It is seen that when $n_1 = n_1$ (max), $\varepsilon = 0$, the reaction will start from left to right. When $n_1 = n_1$ (min), $\varepsilon = 1$, reaction is complete from left to right.

The degree of reaction can thus be written in the form

$$\varepsilon = \frac{(n_0 + n'_0)\nu_1 - n_1}{(n_0 + n'_0)\nu_1}$$

Therefore
$$n_1 = (n_0 + n'_0)\, \nu_1 - (n_0 + n'_0)\, \nu_1 \varepsilon$$

$= n$ (at start) $- \; n$ (consumed)

= Number of moles of A_1 at start − number of moles of A_1 consumed in the reaction

$$= (n_0 + n'_0)\, \nu_1 (1 - \varepsilon)$$

$$n_2 = n\,(\text{at start}) - n\,(\text{consumed})$$

$$= (n_0 + n'_0)\, \nu_2 + N_2 - (n_0 + n'_0)\, \nu_2 \varepsilon$$

$$= (n_0 + n'_0)\, \nu_2\, (1 - \varepsilon) + N_2$$

$$n_3 = n\,(\text{at start}) + n\,(\text{formed})$$

$$= 0 + (n_0 + n'_0)\, \nu_3 \varepsilon$$

$$= (n_0 + n'_0)\, \nu_3 \varepsilon$$

$$n_4 = n\,(\text{at start}) + n\,(\text{formed})$$

$$= N_4 + (n_0 + n'_0)\, \nu_4 \varepsilon$$

$$= (n_0 + n'_0)\, \nu_4 \varepsilon + N_4 \qquad (16.2)$$

The number of moles of the constituents change during a chemical reaction, not independently but restricted by the above relations. These equations are the *equations of constraint*. The n's are functions of ε only. In a homogeneous system, in a given reaction, the mole fraction x's are also functions of ε only, as illustrated below.

Let us take the reaction

$$H_2 + \frac{1}{2} O_2 \rightleftharpoons H_2O$$

in which n_0 moles of hydrogen combine with $n_0/2$ moles of oxygen to form n_0 moles of water. The n's and x's as functions of ε are shown in the table given below.

Table 16.1

A	ν	n	x
$A_1 = H_2$	$\nu_1 = 1$	$n_1 = n_0 (1 - \varepsilon)$	$x_1 = \frac{n_1}{\sum n} = \frac{2(1-\varepsilon)}{3-\varepsilon}$
$A_2 = O_2$	$\nu_2 = \frac{1}{2}$	$n_2 = \frac{n_0}{2} (1 - \varepsilon)$	$x_2 = \frac{1-\varepsilon}{3-\varepsilon}$
$A_3 = H_2O$	$\nu_3 = 1$	$n_3 = n_0\varepsilon$	$x_3 = \frac{2\varepsilon}{3-\varepsilon}$
		$\sum n = \frac{n_0}{2} (3 - \varepsilon)$	

If the reaction is imagined to advance to an infinitesimal extent, the degree of reaction changes from ε to $\varepsilon + d\varepsilon$, and the various n's will change by the amounts

$$dn_1 = -(n_0 + n'_0)\, \nu_1\, d\varepsilon$$

$$dn_2 = -(n_0 + n'_0)\, \nu_2\, d\varepsilon$$

$$dn_3 = (n_0 + n'_0)\, \nu_3\, d\varepsilon$$

$$dn_4 = (n_0 + n'_0)\, \nu_4\, d\varepsilon$$

or

$$\frac{dn_1}{-\nu_1} = \frac{dn_2}{-\nu_2} = \frac{dn_3}{\nu_3} = \frac{dn_4}{\nu_4} = (n_0 + n'_0)\, d\varepsilon$$

which shows that the dn's are proportional to the ν's.

16.2 REACTION EQUILIBRIUM

Let us consider a homogeneous phase having arbitrary amounts of the constituents, A_1, A_2, A_3 and A_4, capable of undergoing the reaction

$$\nu_1 A_1 + \nu_2 A_2 \rightleftharpoons \nu_3 A_3 + \nu_4 A_4$$

The phase is at uniform temperature T and pressure p. The Gibbs function of the mixture is

$$G = \mu_1 n_1 + \mu_2 n_2 + \mu_3 n_3 + \mu_4 n_4$$

where the n's are the number of moles of the constituents at any moment, and the μ's are the chemical potentials.

Let us imagine that the reaction is allowed to take place at constant T and p. The degree of reaction changes by an infinitesimal amount from ε to $\varepsilon + d\varepsilon$. The change in the Gibbs function is

$$dG_{T,\,p} = \Sigma \mu_k \, dn_k = \mu_1 \, dn_1 + \mu_2 \, dn_2 + \mu_3 \, dn_3 + \mu_4 \, dn_4$$

The equations of constraint in differential form are

$$dn_1 = -\,(n_0 + n'_0)\,\nu_1 \, d\varepsilon, \quad dn_3 = (n_0 + n'_0)\,\nu_3 \, d\varepsilon$$

$$dn_2 = -\,(n_0 + n'_0)\,\nu_2 \, d\varepsilon, \quad dn_4 = (n_0 + n'_0)\,\nu_4 \, d\varepsilon$$

On substitution

$$dG_{T,\,p} = (n_0 + n'_0)(-\,\nu_1\mu_1 - \nu_2\mu_2 + \nu_3\mu_3 + \nu_4\mu_4)\, d\varepsilon \tag{16.3}$$

When the reaction proceeds *spontaneously* to the right, $d\varepsilon$ is positive, and since $dG_{T,\,p} < 0$

$$(\nu_1\mu_1 + \nu_2\mu_2) > (\nu_3\mu_3 + \nu_4\mu_4)$$

If $\Sigma \nu_k \mu_k = (\nu_3\mu_3 + \nu_4\mu_4) - (\nu_1\mu_1 + \nu_2\mu_2)$, then it is negative for the reaction to the right.

When the reaction proceeds spontaneously to the left, $d\varepsilon$ is negative

$$(\nu_1\mu_1 + \nu_2\mu_2) < (\nu_3\mu_3 + \nu_4\mu_4)$$

i.e., $\Sigma \nu_k \mu_k$ is positive.

At equilibrium, the Gibbs function will be minimum, and

$$\nu_1\mu_1 + \nu_2\mu_2 = \nu_3\mu_3 + \nu_4\mu_4 \tag{16.4}$$

which is called the *equation of reaction equilibrium*.

Therefore, it is the value of $\Sigma \nu_k \mu_k$ which causes or forces the spontaneous reaction and is called the '*chemical affinity*'.

16.3 LAW OF MASS ACTION

For a homogeneous phase chemical reaction at constant temperature and pressure, when the constituents are *ideal gases*, the chemical potentials are given by the expression of the type

$$\mu_k = \bar{R}T\,(\phi_k + \ln p + \ln x_k)$$

where the ϕ's are functions of temperature only (Article 10.11).

Substituting in the equation of reaction equilibrium (16.4)

$$\nu_1\,(\phi_1 + \ln p + \ln x_1) + \nu_2\,(\phi_2 + \ln p + \ln x_2)$$

$$= \nu_3\,(\phi_3 + \ln p + \ln x_3) + \nu_4\,(\phi_4 + \ln p + \ln x_4)$$

On rearranging

$$\nu_3 \ln x_3 + \nu_4 \ln x_4 - \nu_1 \ln x_1 - \nu_2 \ln x_2 + (\nu_3 + \nu_4 - \nu_1 - \nu_2) \ln p$$

$$= -\,(\nu_3\phi_3 + \nu_4\phi_4 - \nu_1\phi_1 - \nu_2\phi_2)$$

$$\therefore \quad \ln \left[\frac{x_3^{\nu_3} \cdot x_4^{\nu_4}}{x_1^{\nu_1} \cdot x_2^{\nu_2}}\right] p^{\nu_3 + \nu_4 - \nu_1 - \nu_2}$$

$$= -\,(\nu_3\phi_3 + \nu_4\phi_4 - \nu_1\phi_1 - \nu_2\phi_2)$$

Denoting

$$\ln K = -\,(\nu_3\phi_3 + \nu_4\phi_4 - \nu_1\phi_1 - \nu_2\phi_2)$$

where K, known as the *equilibrium constant*, is a function of temperature only

$$\left[\frac{x_3^{\nu_3} \cdot x_4^{\nu_4}}{x_1^{\nu_1} \cdot x_2^{\nu_2}}\right]_{\varepsilon = \varepsilon_e} p^{\nu_3 + \nu_4 - \nu_1 - \nu_2} = K \tag{16.5}$$

This equation is called the *law of mass action*. K has the dimension of pressure raised to the $(\nu_3 + \nu_4 - \nu_1 - \nu_2)$ power. Here the x's are the values of mole fractions at equilibrium when the degree of reaction is ε_e.

The law of mass action can also be written in this form

$$\frac{p_3^{\nu_3} \cdot p_4^{\nu_4}}{p_1^{\nu_1} \cdot p_2^{\nu_2}} = K$$

where the p's are the partial pressures.

16.4 HEAT OF REACTION

The equilibrium constant K is defined by the expression

$$\ln K = -\,(\nu_3\phi_3 + \nu_4\phi_4 - \nu_1\phi_1 - \nu_2\phi_2)$$

Differentiating ln K with respect to T

$$\frac{d}{dT}\ln K = -\left(\nu_3\frac{d\phi_3}{dT} + \nu_4\frac{d\phi_4}{dT} - \nu_1\frac{d\phi_1}{dT} - \nu_2\frac{d\phi_2}{dT}\right)$$

Now, from Eq. (10.73) $$\phi = \frac{h_0}{\bar{R}T} - \frac{1}{\bar{R}}\int\frac{\int c_p dT}{T^2}\cdot dT - \frac{s_0}{\bar{R}}$$

Therefore $$\frac{d\phi}{dT} = -\frac{h_0}{\bar{R}T^2} - \frac{\int c_p dT}{\bar{R}T^2}$$

$$= -\frac{1}{\bar{R}T^2}\left(h_0 + \int c_p dT\right) = -\frac{h}{\bar{R}T^2}$$

Therefore $$\frac{d}{dT}\ln K = \frac{1}{\bar{R}T^2}\,(\nu_3 h_3 + \nu_4 h_4 - \nu_1 h_1 - \nu_2 h_2)$$

where the h's refer to the same temperature T and the same pressure p. If ν_1 moles of A_1 and ν_2 moles of A_2 combine to from ν_3 moles of A_3 and ν_4 moles of A_4 at constant temperature and pressure, the heat transferred would be, as shown in Fig. 16.1, equal to the final enthalpy $(\nu_3 h_3 + \nu_4 h_4)$ minus the initial enthalpy $(\nu_1 h_1 + \nu_2 h_2)$. This is known as the *heat of reaction*, and is denoted by ΔH.

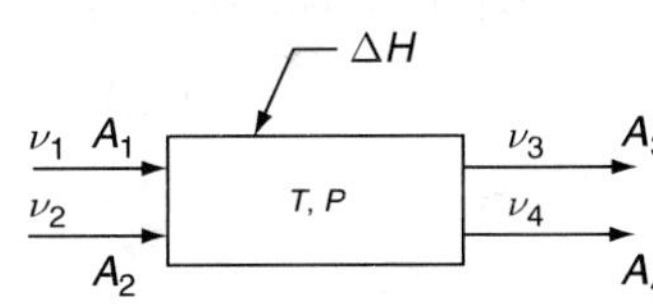

Fig. 16.1 *Heat of reaction*

$$\Delta H = \nu_3 h_3 + \nu_4 h_4 - \nu_1 h_1 - \nu_2 h_2 \qquad (16.6)$$

Therefore $$\frac{d}{dT}\ln K = \frac{\Delta H}{\bar{R}T^2} \qquad (16.7)$$

This is known as the *van't Hoff equation*. The equation can be used to calculate the heat of reaction at any desired temperature or within a certain temperature range.

By rearranging. Eq. (16.7)

$$\frac{\dfrac{d\ln K}{dT}}{T^2} = \frac{\Delta H}{\bar{R}}$$

$$\frac{d\ln K}{d(1/T)} = -\frac{\Delta H}{\bar{R}}$$

Therefore, $$\Delta H = -\,2.303\bar{R}\frac{d\log K}{d\left(\dfrac{1}{T}\right)} = -\,19.148\,\frac{d\log K}{d\left(\dfrac{1}{T}\right)} \text{ kJ/kg mol}$$

If K_1 and K_2 are the equilibrium constants evaluated at temperatures T_1 and T_2 respectively

$$\Delta H = -\ 19.148\ \frac{\log K_1 - \log K_2}{\dfrac{1}{T_1} - \dfrac{1}{T_2}}$$

or

$$\Delta H = 19.148\ \frac{T_1 T_2}{T_1 - T_2} \log \frac{K_1}{K_2}$$

If ΔH is positive, the reaction is said to be *endothermic*. If ΔH is negative, the reaction is *exothermic*.

16.5 TEMPERATURE DEPENDENCE OF THE HEAT OF REACTION

$$\Delta H = \nu_3 h_3 + \nu_4 h_4 - \nu_1 h_1 - \nu_2 h_2$$

$$h = h_0 + \int c_p \mathrm{d}T$$

Therefore

$$\Delta H = \nu_3 h_{03} + \nu_4 h_{04} - \nu_1 h_{01} - \nu_2 h_{02} + \int (\nu_3\, c_{p3} + \nu_4\, c_{p4} - \nu_1\, c_{p1} - \nu_2\, c_{p2}) \mathrm{d}T$$

Denoting

$$\Delta H_0 = \nu_3 h_{03} + \nu_4 h_{04} - \nu_1 h_{01} - \nu_2 h_{02}$$

$$\Delta H = \Delta H_0 + \int (\nu_3 c_{p3} + \nu_4 c_{p4} - \nu_1 c_{p1} - \nu_2 c_{p2}) \mathrm{d}T$$

If c_p is known as a function of temperature and if at any temperature ΔH is known, then at any other temperature, ΔH can be determined for a certain chemical reaction from the above relation.

Some chemical reaction may be expressed as the result to two or more reactions. If ΔH_0 is known for each fo the separate reactions, then the ΔH_0 of the resultant reaction may be calculated. For example,

$$H_2O \rightleftharpoons H_2 + \frac{1}{2} O_2 \qquad \Delta H_0 = 239{,}250 \text{ J/g mol}$$

$$CO_2 \rightleftharpoons CO + \frac{1}{2} O_2 \qquad \Delta H_0 = 279{,}890 \text{ J/g mol}$$

$$CO_2 + H_2O \rightleftharpoons CO_2 + H_2 \qquad \Delta H_0 = -40{,}640 \text{ J/g mol}$$

16.6 TEMPERATURE DEPENDENCE OF THE EQUILIBRIUM CONSTANT

$$\ln K = -(\nu_3 \phi_3 + \nu_4 \phi_4 - \nu_1 \phi_1 - \nu_2 \phi_2)$$

where

$$\phi = \frac{h_0}{\bar{R}T} - \frac{1}{\bar{R}} \int \frac{\int c_p \mathrm{d}T}{T^2} \mathrm{d}T - \frac{s_0}{\bar{R}}$$

On substitution

$$\ln K = \frac{1}{\bar{R}T} (\nu_3 h_{03} + \nu_4 h_{04} - \nu_1 h_{01} - \nu_2 h_{02}) + \frac{1}{\bar{R}} \int \frac{\int (\nu_3 c_{p3} + \nu_4 c_{p4} - \nu_1 c_{p1} - \nu_2 c_{p2}) \mathrm{d}T}{T^2} \cdot \mathrm{d}T + \frac{1}{\bar{R}} (\nu_3 s_{03} + \nu_4 s_{04} - \nu_1 s_{01} - \nu_2 s_{02})$$

If
$$\Delta H_0 = \nu_3 h_{03} + \nu_4 h_{04} - \nu_1 h_{01} - \nu_2 h_{02}$$
$$\Delta S_0 = \nu_3 s_{03} + \nu_4 s_{04} - \nu_1 s_{01} - \nu_2 s_{02}$$
$$\ln K = -\frac{\Delta H_0}{\bar{R}T} + \frac{1}{\bar{R}}\int \frac{\int (\nu_3 c_{p3} + \nu_4 c_{p4} - \nu_1 c_{p1} - \nu_2 c_{p2})\mathrm{d}T}{T^2}\mathrm{d}T + \frac{\Delta S_0}{\bar{R}} \tag{16.8}$$

This equation is sometimes known as the *Nernst's equation.*

16.7 THERMAL IONIZATION OF A MONATOMIC GAS

One interesting application of Nernst's equation was made by Dr. M.N. Saha to the thermal ionization of a monatomic gas. If a monatomic gas is heated to a high enought temperature (5000 K and above), some ionization occurs, with the electrons in the outermost orbit being shed off, and the atoms, ions, and electrons may be regarded as a mixture of three ideal monatomic gases, undergoing the reaction

$$A \rightleftharpoons A^+ + e$$

Starting with n_0 moles of atoms, it is shown in Table 16.2.
The equilibrium constant is given by

$$\ln K = \ln \left\{\frac{x_3^{\nu_3} \cdot x_4^{\nu_4}}{x_1^{\nu_1} \cdot x_2^{\nu_2}}\right\}_{\varepsilon_e} \cdot p^{\nu_3+\nu_4-\nu_1-\nu_2}$$

$$\ln K = \frac{\dfrac{\varepsilon_e}{1+\varepsilon_e} \cdot \dfrac{\varepsilon_e}{1+\varepsilon_e}}{\dfrac{1-\varepsilon_e}{1+\varepsilon_e}} \cdot p^{\nu_3+\nu_4-\nu_1}$$

or,
$$\ln K = \ln \frac{\varepsilon_e^2}{1-\varepsilon_e^2} \cdot p$$

Since the three gases are monatomic, $c_p = \frac{5}{2}\bar{R}$ which, on being substituted in the Nernst's equation, gives

$$\ln \frac{\varepsilon_e^2}{1-\varepsilon_e^2} \cdot p = -\frac{\Delta H_0}{\bar{R}T} + \frac{5}{2}\ln T + \ln B \tag{16.9}$$

Table 16.2

A	ν	n	x
$A_1 = A$	$\nu_1 = 1$	$n_1 = n_0(1 - \varepsilon_e)$	$x_1 = \dfrac{1-\varepsilon_e}{1+\varepsilon_e}$
$A_3 = A^+$	$\nu_3 = 1$	$n_3 = n_0\varepsilon_e$	$x_3 = \dfrac{\varepsilon_e}{1+\varepsilon_e}$
$A_4 = e$	$\nu_4 = 1$	$n_4 = n_0\varepsilon_e$	$x_4 = \dfrac{\varepsilon_e}{1+\varepsilon_e}$
		$\Sigma n = n_0(1 + \varepsilon_e)$	

where
$$\frac{\Delta S_0}{\bar{R}} = \ln B$$

$$\ln \frac{\varepsilon_e^2}{1-\varepsilon_e^2} \cdot \frac{p}{T^{5/2} \cdot B} = -\frac{\Delta H_0}{\bar{R}T}$$

$$\therefore \quad \frac{\varepsilon_e^2}{1-\varepsilon_e^2} = Be^{-\Delta H_0/\bar{R}T} \cdot \frac{T^{5/2}}{p} \tag{16.10}$$

where ε_e is the equilibrium value of the degree of ionization. This is known as the *Saha's equation*. For a particular gas the degree of ionization increases with an increase in temperature and a decrease in pressure.

It can be shown that ΔH_0 is the amount of energy necessary to ionize 1 gmol of atoms. If we denote the ionization potential ΔH_0 of the atom in volts by E, then

$$\Delta H_0 = \text{E(volts)} \times 1.59 \times 10^{-19} \frac{\text{coulomb}}{\text{electron}} \times 6.06 \times 10^{23} \frac{\text{electron}}{\text{g mol}}$$

$$= 9.6354 \times 10^4 \text{ E J/g mol}$$

Equation (16.9) becomes

$$\text{In} \frac{\varepsilon_e}{1-\varepsilon_e^2} p = -\frac{96354\,E}{\bar{R}T} + \frac{5}{2} \text{In}\, T + \text{In}\, B \tag{16.11}$$

Expressing p in atmospheres changing to common logarithms and introducing the value of B from statistical mechanics, Saha finally obtained the equation:

$$\log \frac{\varepsilon_e^2}{1-\varepsilon_e^2} p(\text{atm}) = -\frac{96{,}354\,E}{19.148\,T} + \frac{5}{2} \log T + \log \frac{\omega_i \omega_e}{\omega_a} - 6.491 \tag{16.12}$$

where ω_i, ω_e and ω_a are constants that refer to the ion, electron and atom respectively. The value ω_e for an electron is 2. The value of E. ω_i and ω_a for a few elements are given below:

Table 16.3
Values for E and w

Element	*E, volts*	ω_a	ω_i
Na	5.12	2	1
Cs	3.87	2	1
Ca	6.09	1	2
Cd	8.96	1	2
Za	9.36	1	2

For alkali metals like Cs, Na, K etc., the ionization potential is less. It means less energy is required to ionize one gmol of atoms. So these are used as *seed for magnetohydrodynamic power generation.* Saha applied his Equation (16.12) to the determination of temperature of a stellar atmosphere. The spectrum of a star contains lines which originate from atoms (arc lines) and those which originate from ions (spark lines). A comparison of the intensities of an arc line and a spark line from the same element gives a measure of ε_e. Considering a star as a sphere of ideal gas, the pressure of a star can be estimated. Thus, knowing ε_e, *p, E* and the ω's, the temperature of the star can be calculated.

16.8 GIBBS FUNCTION CHANGE

Molar Gibbs function of an ideal gas at temperature T and pressure p is equal to (Article 10.11)

$$g = \bar{R}T\ (\phi + \ln p)$$

For the reaction of the type $\nu_1 A_1 + \nu_2 A_2 \rightleftharpoons \nu_3 A_3 + \nu_4 A_4$

the Gibbs function change of the reaction ΔG is defined by the expression

$$\Delta G = \nu_3\, g_3 + \nu_4\, g_4 - \nu_1\, g_1 - \nu_2\, g_2 \tag{16.13}$$

where the g's refer to the gases completely separated at T, p. ΔG is also known as the *free energy change*. Substituting the values of the g's

$$\Delta G = \bar{R}T(\nu_3\phi_3 + \nu_4\phi_4 - \nu_1\phi_1 - \nu_2\phi_2) + \bar{R}T \ln p^{\nu_3+\nu_4-\nu_1-\nu_2}$$

But $$\ln K = -(\nu_3\phi_3 + \nu_4\phi_4 - \nu_1\phi_1 - \nu_2\phi_2)$$

$\therefore$ $$\Delta G = -\bar{R}T\ \ln K + \bar{R}T \ln p^{\nu_3+\nu_4-\nu_1-\nu_2}$$

If p is expressed in atmospheres and ΔG is calculated from each gas is at a pressure of 1 atm., the second term on the right drops out. Under these conditions ΔG is known as the *standard Gibbs function change* and is denoted by $\Delta G°$

$$\Delta G° = \bar{R}T\ \ln K \tag{16.14}$$

This is an important equation which relates the standard Gibbs function change with temperature and the equilibrium constant. From this the equilibrium constant can be calculated from changes in the standard Gibbs function, or vice versa.

For dissociation of water vapour, $H_2O \rightleftharpoons H_2 + \frac{1}{2}O_2$, the value of ln K_{298} is found to be – 93.7. Therefore,

$$\Delta G°_{298} = -8.3143 \times 298 \times (-93.7) = 232{,}157 \text{ J/gmol.}$$

Substituting ln K from Nernst's equation:

$$\Delta G° = \Delta H_0 - T\,\frac{(\nu_3 c_{p_3} + \nu_4 c_{p_4} - \nu_1 c_{p_1} - \nu_2 c_{p_2})\mathrm{d}T}{T^2}\,\mathrm{d}T - T\Delta S_0$$

from which also $\Delta G°$ may be calculated directly. Values of ΔH_0, ΔS_0 and $\Delta G°_{298}$ for fundamental ideal gas reactions are given in Table 16.4.

Both ΔS_0 and $\Delta G°_{298}$ may be added and subtracted in the same manner as ΔH_0. For example.

H_2O	$\rightleftharpoons H_2 + \frac{1}{2}O_2$	$\Delta G°_{298} = 232{,}545$ J/gmol
CO_2	$\rightleftharpoons CO + \frac{1}{2}O_2$	$\Delta G°_{298} = 258{,}940$ J/gmol
$CO + H_2O$	$\rightleftharpoons CO_2 + H_2$	$\Delta G°_{298} = -26{,}395$ J/gmol

$$\ln K_{298} = \frac{\Delta G°_{298}}{\bar{R}T} = \frac{26{,}395}{8.3143 \times 298} = 10.653$$

$$K_{298} = 42{,}330$$

From this, the value of the degree of reaction at equilibrium ε_e may be calculated.

Table 16.4
Fundamental Idea Gas Reactions

Reaction	ΔH_0 J/gmol	ΔS_0 J/gmol-K	ΔG°_{298} J/gmol
$H_2 \rightleftharpoons 2H$	427,380	4.90	404,335
$HCl \rightleftharpoons \frac{1}{2}H_2 + \frac{1}{2}Cl_2$	91,760	– 22.25	95,110
$HBr \rightleftharpoons \frac{1}{2}H_2 + \frac{1}{2}Br_2$	50,280	– 24.05	54,050
$HI \rightleftharpoons \frac{1}{2}H_2 + \frac{1}{2}I_2$	5,320	– 21.00	8,380
$NO \rightleftharpoons \frac{1}{2}N_2 + \frac{1}{2}O_2$	– 90,500	– 10.48	– 87,570
$H_2O \rightleftharpoons H_2 + \frac{1}{2}O_2$	239,250	14.70	232,545
$H_2S \rightleftharpoons H_2 + \frac{1}{2}S_2$	80,450	6.90	71,230
$CO_2 \rightleftharpoons CO + \frac{1}{2}O_2$	279,890	18.69	258,940
$NO_2 \rightleftharpoons NO + \frac{1}{2}O_2$	59,500	11.44	37,290
$SO_2 \rightleftharpoons \frac{1}{2}S_2 + O_2$	349,025	3.77	329,330
$NH_3 \rightleftharpoons \frac{1}{2}N_2 + \frac{3}{2}H_2$	39,800	40.27	16,380
$SO_3 \rightleftharpoons SO_2 + \frac{1}{2}O_2$	94,690	89.50	67,880
$N_2O_4 \rightleftharpoons 2NO_2$	56,980	174.30	5,030

From Eq. (16.3), if $n_0 = 1$ and $n'_0 = 0$

$$\left(\frac{\partial G}{\partial \varepsilon}\right)_{T,p} = (\nu_3\mu_3 + \nu_4\mu_4 - \nu_1\mu_1 - \nu_2\mu_2)$$

Since $\mu_k = \bar{R}T\ (\phi_k + \ln p + \ln x_k)$

and $g_k = \bar{R}T\ (\phi_k + \ln p)$

so $\mu_k = g_k\ \bar{P}T\ \ln x_k$

Therefore
$$\left(\frac{\partial G}{\partial \varepsilon}\right)_{T,p} = \nu_3 g_3 + \nu_4 g_4 - \nu_1 g_1 - \nu_2 g_2 + \bar{R}T\ \ln \frac{x_3^{\nu_3}\cdot x_4^{\nu_4}}{x_1^{\nu_1}\cdot x_2^{\nu_2}}$$
$$= \Delta G + \bar{R}T\ \ln \frac{x_3^{\nu_3}\ x_4^{\nu_4}}{x_1^{\nu_1}\ x_2^{\nu_2}}$$

At $\varepsilon = 0, x_3 = 0, x_4 = 0$

$$\left(\frac{\partial G}{\partial \varepsilon}\right)_{T,p} = -\infty$$

and at $\varepsilon = 1, x_1 = 0, x_2 = 0$

$$\left(\frac{\partial G}{\partial \varepsilon}\right)_{T,p} = +\infty$$

At $\varepsilon = \frac{1}{2}, x_1 = \frac{\nu_1}{\Sigma\nu}, x_2 = \frac{\nu_2}{\Sigma\nu}$

$$x_3 = \frac{\nu_3}{\Sigma\nu}, x_4 = \frac{\nu_4}{\Sigma\nu}$$

where $\Sigma v = \nu_3 + \nu_4 + \nu_1 + \nu_2$

$$\left(\frac{\partial G}{\partial \varepsilon}\right)_{\substack{T,p\\ \varepsilon=\frac{1}{2}}} = \Delta G + \bar{R}T \ \ln \frac{\left(\frac{\nu_3}{\Sigma v}\right)^{\nu_3}\left(\frac{\nu_4}{\Sigma v}\right)^{\nu_4}}{\left(\frac{\nu_1}{\Sigma v}\right)^{\nu_1}\left(\frac{\nu_2}{\Sigma v}\right)^{\nu_2}}$$

If $p = 1$ atm., $T = 298$ K

$$\left(\frac{\partial G}{\partial \varepsilon}\right)_{\substack{p=1\,\text{atm}\\ T=298\,\text{K}\\ \varepsilon=\frac{1}{2}}} = \Delta G^\circ_{298}$$

because the magnitude of the second term on the right hand side of the equation is very small compared to ΔG°. The slope $\left(\frac{\partial G}{\partial \varepsilon}\right)_{T,p}$ at $\varepsilon = \frac{1}{2}$ is called the 'affinity' of the reaction, and it is equal to ΔG° at the standard reference state. The magnitude of the slope at $\varepsilon = 1/2$ (Fig. 16.2) indicates the direction in which the reaction will proceed. For water vapour reaction, ΔG°_{298} is a large positive number, which indicates the equilibrium point is far to the left of $\varepsilon = 1/2$, and therefore, ε_e is very small. Again for the reaction $NO \rightleftharpoons \frac{1}{2} N_2 + \frac{1}{2} O_2$, ΔG°_{298} is a large negative value, which shows that the equilibrium point is far to the right of $\varepsilon = 1/2$, and so ε_e is close to unity.

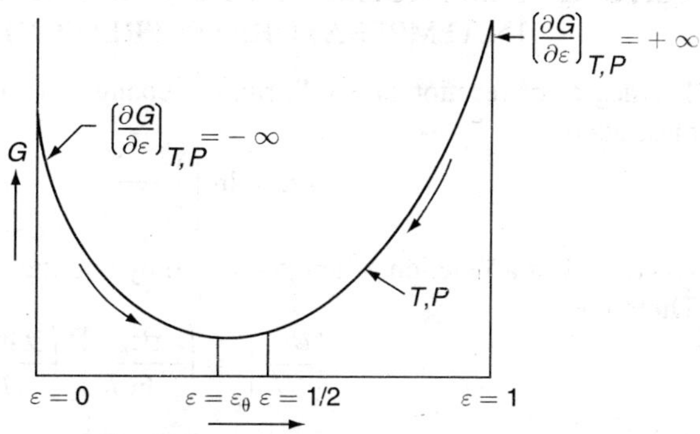

Fig. 16.2 *Plot of G against ε at constants T and p*

16.9 FUGACITY AND ACTIVITY

The differential of the Gibbs function of an ideal gas undergoing an isothermal process is

$$dG = Vdp = \frac{n\bar{R}T}{p}\ dp = n\bar{R}T\ d(\ln p)$$

Analogously, the differential of the Gibbs function for a real gas is

$$dG = n\bar{R}T\ d(\ln f) \quad (16.15)$$

where f is called the *fugacity*, first used by Lewis. The value of fugacity approaches the value of pressure as the latter tends to zero, i.e., when ideal gas conditions apply. Therefore

$$\lim_{p \to 0} \frac{f}{p} = 1$$

For an ideal gas $f = p$. Fugacity has the same dimension as pressure. Integrating Eq. (16.15)

$$G - G^0 = n\bar{R}T \ \ln \frac{f}{f^0}$$

where G^0 and f^0 refer to the reference state when $p_0 = 1$ atm. The ratio f/f^0 is called the *activity*.

Therefore
$$G - G^0 = n\bar{R}T \ \ln a \tag{16.16}$$

For ideal gases, the equilibrium constant is given by $K = \dfrac{p_3^{\nu 3} \cdot p_4^{\nu 4}}{p_1^{\nu 1} \cdot p_2^{\nu 2}}$

For real gases
$$K_{\text{real}} = \frac{f_3^{\nu 3} \cdot f_4^{\nu 4}}{f_1^{\nu 1} \cdot f_2^{\nu 2}}$$

Similarly, it can be shown that
$$\Delta G^0 = - n\bar{R}T \ \ln K_{\text{real}}$$

and
$$\frac{\text{d} \ln K_{\text{real}}}{\text{d}T} = \frac{\Delta H^0}{\bar{R}T^2}$$

16.10 DISPLACEMENT OF EQUILIBRIUM DUE TO A CHANGE IN TEMPERATURE OR PRESSURE

The degree of reaction at equilibrium ε_e changes with temperature and also with pressure. From the law of mass action

$$\ln K = \ln \left[\frac{x_3^{\nu_3} \cdot x_4^{\nu_4}}{x_1^{\nu_1} \cdot x_2^{\nu_2}}\right]_{\varepsilon = \varepsilon_e} + (\nu_3 + \nu_4 - \nu_1 - \nu_2) \ln p$$

where $\ln K$ is a function of temperature only and the first term on the right hand side is function of ε_e only. Therefore

$$\left(\frac{\partial \varepsilon_e}{\partial T}\right)_p = \left(\frac{\partial \varepsilon_e}{\partial \ln K}\right)_p \left(\frac{\partial \ln K}{\partial T}\right)_p$$

$$\left(\frac{\partial \varepsilon_e}{\partial T}\right)_p = \frac{\left(\dfrac{\partial \ln K}{\partial T}\right)_p}{\left(\dfrac{\partial \ln K}{\partial \varepsilon_e}\right)_p} = \frac{\Delta H}{\bar{R}T^2 \ \dfrac{\text{d}}{\text{d}\varepsilon_e} \ln \left[\dfrac{x_3^{\nu_3} \cdot x_4^{\nu_4}}{x_1^{\nu_1} \cdot x_2^{\nu_2}}\right]_{\varepsilon = \varepsilon_e}} \tag{16.17}$$

It can be shown (Example 16.4) that

$$\frac{\text{d}}{\text{d}\varepsilon} \ln \frac{x_3^{\nu_3} \cdot x_4^{\nu_4}}{x_1^{\nu_1} \cdot x_2^{\nu_2}} = \frac{n_0}{\Sigma n_k} \cdot \frac{(\nu_1 + \nu_2)(\nu_3 + \nu_4)}{\varepsilon (1 - \varepsilon)}$$

which is always positive.

Therefore, for endothermic reaction, when ΔH is positive, $\left(\dfrac{\partial \varepsilon_e}{\partial T}\right)_p$ is positive, and for exothermic reaction, when ΔH is negative, $\left(\dfrac{\partial \varepsilon_e}{\partial T}\right)_p$ is negative.

Again,
$$\left(\frac{\partial \varepsilon_e}{\partial p}\right)_T = -\left(\frac{\partial \varepsilon_e}{\partial \ln K}\right)_p \left(\frac{\partial \ln K}{\partial p}\right)_{\varepsilon_e}$$
$$= \frac{-\left(\dfrac{\partial \ln K}{\partial p}\right)_{\varepsilon_e}}{\left(\dfrac{\partial \ln K}{\partial \varepsilon_e}\right)_p}$$

Using the law of mass action
$$\left(\frac{\partial \varepsilon_e}{\partial p}\right)_T = -\frac{\nu_3 + \nu_4 - \nu_1 - \nu_2}{p \dfrac{d}{d\varepsilon_e} \ln \left[\dfrac{x_3^{\nu_3} \cdot x_4^{\nu_4}}{x_1^{\nu_1} \cdot x_2^{\nu_2}}\right]_{\varepsilon=\varepsilon_e}} \tag{16.18}$$

If $(\nu_3 + \nu_4) > (\nu_1 + \nu_2)$, i.e., the number of moles increase or the volume increases due to reaction, $\left(\dfrac{\partial \varepsilon_e}{\partial p}\right)_T$ is negative. If $(\nu_3 + \nu_4) < (\nu_1 + \nu_2)$, i.e. volume decreases in an isothermal reaction, $\left(\dfrac{\partial \varepsilon_e}{\partial p}\right)_T$ is positive.

16.11 HEAT CAPACITY OF REACTING GASES IN EQUILIBRIUM

For a reaction of four ideal gases, such as
$$\nu_1 A_1 + \nu_2 A_2 \rightleftharpoons \nu_3 A_3 + \nu_4 A_4$$
the enthalpy of mixture at equilibrium is
$$H = \Sigma n_k h_k$$
where $n_1 = (n_0 + n'_0)\, \nu_1 (1 - \varepsilon_e)$, $n_2 = (n_0 + n'_0)\, \nu_2\, (1 - \varepsilon_e) + N_2$

$n_3 = (n_0 + n'_0)\, \nu_3\, \varepsilon_e$, and $n_4 = (n_0 + n'_0)\, \nu_4\, \varepsilon_e + N_4$

Let us suppose that an infinitesimal change in temperature takes place at constant pressure in such a way that equilibrium is maintained.

ε_e will change to the value $\varepsilon_e + d\varepsilon_e$, and the enthalpy will change by the amount
$$dH_p = \Sigma n_k\, dh_k + \Sigma h_k\, dn_k$$
where $\qquad dh_k = c_{pk}\, dT,\ dn_k = \pm\, (n_0 + n'_0)\, \nu_k\, d\varepsilon_e$

Therefore $\qquad dH_p = \Sigma n_k\, c_{pk}\, dT + (n_0 + n'_0)\, (\nu_3 h_3 + \nu_4 h_4 - \nu_1 h_1 - \nu_2 h_2)\, d\varepsilon_e$

The heat capacity of the reacting gas mixture is
$$C_p = \left(\frac{\partial H}{\partial T}\right)_p = \Sigma n_k\, c_{pk} + (n_0 + n'_0)\, \Delta H \left(\frac{\partial \varepsilon_e}{\partial T}\right)_p$$

Using Eq. (16.14)
$$C_p = \Sigma n_k\, c_{pk} + (n_0 + n'_0)\, \frac{(\Delta H)^2}{\bar{R}T^2 \dfrac{d}{d\varepsilon} \ln \left[\dfrac{x_3^{\upsilon_3} \cdot x_4^{\upsilon_4}}{x_1^{\upsilon_1} \cdot x_2^{\upsilon_2}}\right]_{\varepsilon=\varepsilon_e}} \tag{16.19}$$

16.12 COMBUSTION

Combustion is a chemical reaction between a fuel and oxygen which proceeds at a fast rate with the release of energy in the form of heat. In the combustion of methane, e.g.

$$\underset{\text{Reactants}}{CH_4 + 2O_2} \rightarrow \underset{\text{Products}}{CO_2 + 2H_2O}$$

One mole of methane reacts with 2 moles of oxygen to form 1 mole of carbon dioxide and 2 moles of water. The water may be in the liquid or vapour state depending on the temperature and pressure of the products of combustion. Only the initial and final products are being considered without any concern for the intermediate products that usually occur in a reaction.

Atmospheric air contains 21% oxygen, 78% nitrogen, and 1% argon by volume. In combustion calculations, however, the argon is usually neglected, and air is assumed to consist of 21% oxygen and 79% nitrogen by volume (or molar basis). On a mass basis, air contains 23% oxygen and 77% nitrogen.

For each mole of oxygen taking part in a combustion reaction, there are 79.0/21.0 = 3.76 moles of nitrogen. So for the combustion of methane, the reaction can be written as

$$CH_4 + 2O_2 + 2(3.76)N_2 \rightarrow CO_2 + 2H_2O + 7.52N_2$$

The minimum amount of air which provides sufficient oxygen for the complete combustion of all the elements like carbon, hydrogen, etc., which may oxidize is called the *theoretical or stoichiometric air*. There is no oxygen in the products when complete combustion (oxidation) is achieved with this theoretical air. In practice, however, more air than this theoretical amount is required to be supplied for complete combustion. Actual air supplied is usually expressed in terms of percent theoretical air; 150% theoretical air means that 1.5 times the theoretical air is supplied. Thus, with 150% theoretical air, the methane combustion reaction can be written as

$$CH_4 + 2(1.5)\,O_2 + 2(3.76)\,(1.5)N_2$$

$$\rightarrow CO_2 + 2H_2O + O_2 + 11.28N_2$$

Another way of expressing the actual air quantity supplied is in terms of excess air. Thus 150% theoretical air means 50% excess air.

With less than theoretical air supply, combustion will remain incomplete with some CO present in the products. Even with excess air supply also, there may be a small amount of CO present, depending on mixing and turbulence during combustion, e.g., with 115% theoretical air

$$CH_4 + 2(1.15)O_2 + 2(1.15)\,3.76N_2 \rightarrow$$

$$\rightarrow 0.95\,CO_2 + 0.05CO + 2H_2O + 0.325O_2 + 8.65N_2$$

By analyzing the products of combustion, the actual amount of air supplied in a combustion process can be computed. Such analysis is often given on the 'dry' basis, i.e., the fractional analysis of all the components, except water vapour. Following the principle of the conservation of mass of each of the elements, it is possible to make a carbon balance, hydrogen balance, oxygen balance, and nitrogen balance from the combustion reaction equation, from which the actual air-fuel ratio can be determined. It has been illustrated in Ex. 16.4 later.

16.13 ENTHALPY OF FORMATION

Let us consider the steady state steady flow combustion of carbon and oxygen to from CO_2 (Fig. 16.3). Let the carbon and oxygen each enter the control volume at 25°C and 1 atm. pressure, and the heat transfer be such that the product CO_2 leaves at 25°C, 1 atm. pressure. The measured value of heat transfer is −393,522 kJ per kg mol of CO_2 formed. If H_R and H_P refer to the total enthalpy of the reactants and products respectively, then the first law applied to the reaction $C + O_2 \rightarrow CO_2$ gives

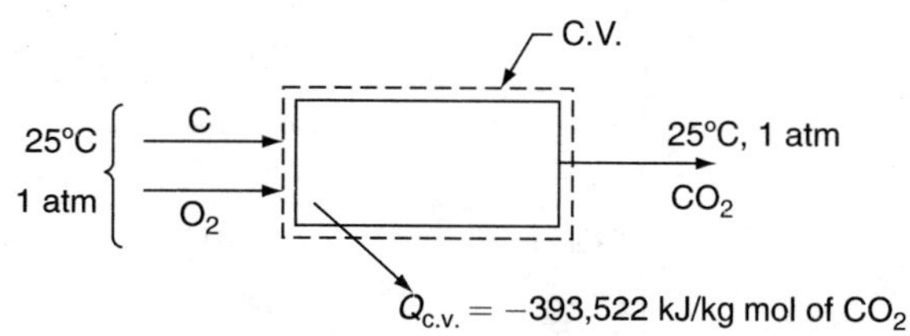

Fig. 16.3 *Enthalpy of formation*

$$H_R + Q_{C.V.} = H_P$$

For all the reactants and products in a reaction, the equation may be written as

$$\sum_R n_i \bar{h}_i + Q_{C.V.} = \sum_P n_e \bar{h}_e$$

where R and P refer to the reactants and products respectively.

The enthalpy of all the elements at the standard reference state of 25°C, 1 atm. is assigned the value of zero. In the carbon-oxygen reaction, $H_R = 0$. So the energy equation gives

$$Q_{C.V.} = H_p = -393{,}522 \text{ kJ/kg mol}$$

This is what is known as the *enthalpy of formation of* CO_2 at 25°C, 1 atm., and designated by the symbol, $\bar{h}_f^0$. So

$$\left(\bar{h}_f^0\right)_{CO_2} = -393{,}522 \text{ kJ/kg mol}$$

In most cases, however, the reactants and products are not at 25°C, 1 atm. Therefore, the change of enthalpy (in the case of constant pressure process or S.S.S.F. process) between 25°C, 1 atm. and the given state must be known. Thus the enthalpy at any temperature and pressure, $\bar{h}_{T,p}$ is

$$\bar{h}_{T,p} = \left(\bar{h}_f^0\right)_{298\text{ K}, 1\text{ atm}} + \left(\Delta\bar{h}\right)_{298\text{ K}, 1\text{ atm} \rightarrow T, p}$$

For convenience, the subscripts are usually dropped, and

$$\bar{h}_{T,p} = \bar{h}_f^0 + \Delta\bar{h}$$

where $\Delta\bar{h}$ represents the difference in enthalpy between any given state and the enthalpy at 298.15 K, 1 atm.

Table 16.5 gives the values of the enthalpy of formation of a number of substances in kJ/kg mol.

Table C in the appendix gives the values of $\Delta\bar{h} = \bar{h}^0 - \bar{h}_{298}^0$ in kJ/kg mol for various substances at different temperatures.

Table 16.5

Enthalpy of Formation, Gibbs Function of Formation, and Absolute Entropy of Various Substances at 25°C, 1 atm. pressure

Substance	*Molecular weight, M*	$\bar{h}_f^0$ kJ/kg mol	$\bar{g}_f^0$ kJ/kg mol	s^0 kJ/kg mol K
CO(g)	28.011	– 110529	– 137150	197.653
CO_2(g)	44.001	– 393522	– 394374	213.795
H_2O(g)	18.015	– 241827	– 228583	188.833
H_2O(l)	18.015	– 285838	– 237178	70.049
CH_4(g)	16.043	– 74873	– 50751	186.256
C_2H_2(g)	26.038	+ 226731	+ 209234	200.958
C_2H_4(g)	28.054	+ 52283	+ 68207	219.548
C_2H_6(g)	30.070	– 84667	– 32777	229.602
C_3H_8(g)	44.097	– 103847	– 23316	270.019
C_4H_{10}(g)	58.124	– 126148	– 16914	310.227
C_8H_{18}(g)	114.23	– 208447	+ 16859	466.835
C_8H_{18}(l)	114.23	– 249952	+ 6940	360.896

16.14 FIRST LAW FOR REACTIVE SYSTEMS

For the S.S.S.F. process as shown in Fig. 16.4 the first law gives

$$H_R + Q_{C.V.} = H_P + W_{C.V.}$$

or

$$\sum_R n_i \bar{h}_i + Q_{C.V.} = \sum_P n_e \bar{h}_e + W_{C.V.}$$

$$\sum_R n_i \left[\bar{h}_f^0 + \Delta\bar{h}\right]_i + Q_{C.V.} = \sum_P n_e \left[\bar{h}_f^0 + \Delta\bar{h}\right]_e + W_{C.V.} \tag{16.20}$$

When the states of reactants and products are not in the standard reference state (298K, 1 atm), then, as shown in Fig. 16.5,

$$Q_{C.V.} = H_P - H_R = (H_P - H_{P_0}) + (H_{P_0} - H_{R_0}) + (H_{R_0} - H_R)$$

$$= \sum_P n_e \left(\bar{h}_P - \bar{h}_{P_0}\right) + \Delta\bar{h}_{RP}^0 - \sum_R n_i \left(\bar{h}_R - \bar{h}_{R_0}\right) \tag{16.21}$$

where $\Delta\bar{h}_{PR}^0$ is the enthalpy of reaction at the standard temperature (298 K). The variation of enthalpy with pressure is not significant.

For a constant volume reaction,

$$Q_{C.V.} = U_P - U_R = (U_P - U_{P_0}) + (U_{P_0} - U_{R_0}) + (U_{R_0} - U_R)$$

$$= \sum_P n_e \left(\bar{u}_P - \bar{u}_{P_0}\right) + \Delta\bar{u}_{RP}^0 - \sum_R n_i \left(\bar{u}_R - \bar{u}_{R_0}\right) \tag{16.22}$$

where $\Delta\bar{u}_{RP}^0$ is the internal energy of reaction at 298 K.

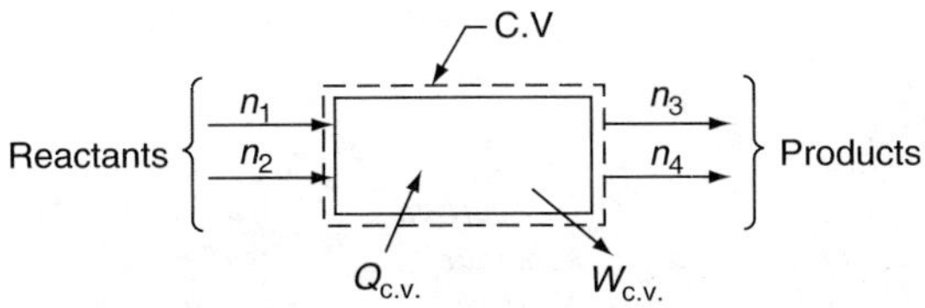

Fig. 16.4 *First law for a reactive system*

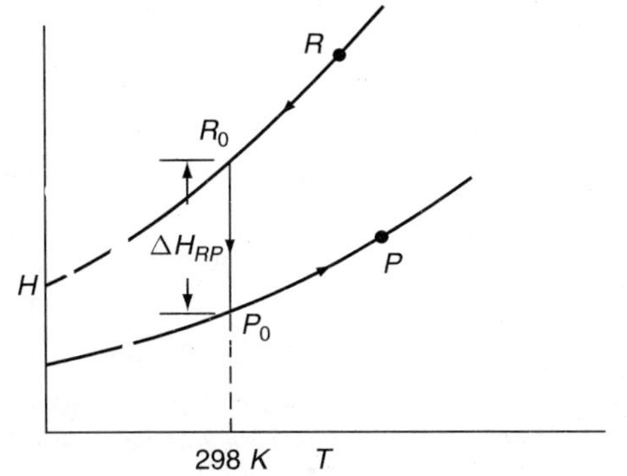

Fig. 16.5 *Enthalpy of reactants and products varying with temperature*

16.15 ADIABATIC FLAME TEMPERATURE

If a combustion process occurs adiabatically in the absence of work transfer or changes in K.E. and P.E. then the energy equation becomes

$$H_R = H_P$$

or

$$\sum_R n_i \bar{h}_i = \sum_p n_e \bar{h}_e$$

or

$$\sum_R n_i \left[\bar{h}_f^0 + \Delta \bar{h}\right]_i = \sum_P n_e \left[\bar{h}_f^0 + \Delta \bar{h}\right]_e \tag{16.23}$$

For such a process, the temperature of the products is called the *adiabatic flame temperature* which is the maximum temperature achieved for the given reactants. The adiabatic flame temperature can be controlled by the amount of excess air supplied; it is the maximum with a stoichiometric mixture. Since the maximum permissible temperature in a gas turbine is fixed from metallurgical considerations, close control of the temperature of the products is achieved by controlling the excess air.

For a given reaction the adiabatic flame temperature is computed by trial and error. The energy of the reactants H_R being known, a suitable temperature is chosen for the products so that the energy of the products at that temperature becomes equal to the energy of the reactants.

16.16 ENTHALPY AND INTERNAL ENERGY OF COMBUSTION: HEATING VALUE

The *enthalpy of combustion* is defined as the difference between the enthalpy of the products and the enthalpy of the reactants when complete combustion occurs at a given temperature and pressure.

Therefore

$$\bar{h}_{RP} = H_P - H_R$$

or

$$\bar{h}_{RP} = \sum_P n_e [\bar{h}_f^0 + \Delta \bar{h}]_e - \sum_R n_i [\bar{h}_f^0 + \Delta \bar{h}]_i \tag{16.24}$$

where $\bar{h}_{RP}$ is the enthalpy of combustion (kJ/kg or kJ/kg mol) of the fuel.

The values of the enthalpy of combustion of different hydrocarbon fuels at 25°C, 1 atm. are given in Table 16.6.

The *internal energy of combustion*, u_{RP}, is defined in a similar way.

Table 16.6

Enthalpy of Combustion of Some Hydrocarbons at 25°C

Hydrocarbon (1)	Formula (2)	Liquid H_2O in Products (Negative of Higher Heating Value): Liquid Hydrocarbon kJ/kg fuel (3)	Liquid H_2O in Products (Negative of Higher Heating Value): Gaseous Hydrocarbon kJ/kg fuel (4)	Vapour H_2O in Products (Negative of Lower Heating Value): Liquid Hydrocarbon kJ/kg fuel (5)	Vapour H_2O in Products (Negative of Lower Heating Value): Gaseous Hydrocarbon kJ/kg fuel (6)
Paraffin Family					
Methane	CH_4		– 55496		– 50010
Ethane	C_2H_6		– 51875		– 47484
Propane	C_3H_8	– 49975	– 50345	– 45983	– 46353
Butane	C_4H_{10}	– 49130	– 49500	– 45344	– 45714
Pentane	C_5H_{12}	– 48643	– 49011	– 44983	– 45351
Hexane	C_6H_{14}	– 48308	– 48676	– 44733	– 45101
Heptane	C_7H_{16}	– 48071	– 48436	– 44557	– 44922
Octane	C_8H_{18}	– 47893	– 48256	– 44425	– 44788
Decane	$C_{10}H_{22}$	– 47641	– 48000	– 44239	– 44598
Dodecane	$C_{12}H_{26}$	– 47470	– 47828	– 44100	– 44467
Olefin Family					
Ethene	C_2H_4		– 50296		– 47158
Propene	C_3H_6		– 48917		– 45780
Butene	C_4H_8		– 48453		– 45316
Pentene	C_5H_{10}		– 48134		– 44996
Hexene	C_6H_{12}		– 47937		– 44800
Heptene	C_7H_{14}		– 47800		– 44662
Octene	C_8H_{16}		– 47693		– 44556
Nonene	C_9H_{18}		– 47612		– 44475
Decene	$C_{16}H_{20}$		– 47547		– 44410
Alkylbenzene Family					
Benzene	C_6H_6	– 41831	– 42266	– 40141	– 40576
Methylbenzene	C_7H_8	– 42473	– 42847	– 40527	– 40937
Ethylbenzene	C_8H_{10}	– 42997	– 43395	– 40924	– 41322
Propylbenzene	C_9H_{12}	– 43416	– 43800	– 41219	– 41603
Butylbenzene	$C_{10}H_{14}$	– 43748	– 44123	– 41453	– 41828

$$\bar{u}_{\mathrm{RP}} = U_{\mathrm{P}} - U_{\mathrm{R}} = \sum_{p} n_{\mathrm{e}}\left[\bar{h}_{\mathrm{f}}^{0} + \Delta\bar{h} - p\bar{v}\right]_{\mathrm{e}} - \sum_{\mathrm{R}} n_{\mathrm{i}}\left[\bar{h}_{\mathrm{f}}^{0} + \Delta\bar{h} - p\bar{v}\right]_{\mathrm{i}}$$

If all the gaseous constituents are considered ideal gases and the volume of liquid and solid considered is assumed to be negligible compared to gaseous volume

$$\bar{u}_{\mathrm{RP}} = \bar{u}_{\mathrm{RP}} - \bar{R}T(n_{\text{gaseous products}} - n_{\text{gaseous reactants}}) \tag{16.25}$$

In the case of a constant pressure or steady flow process, the negative of the enthalpy of combustion is frequently called the *heating value at constant pressure*, which represents the heat transferred from the chamber during combustion at constant pressure.

Similarly, the negative of the internal energy of combustion is sometimes designated as the *heating value at constant volume* in the case of combustion, because it represents the amount of heat transfer in the constant volume process.

The *higher heating value* (HHV) or higher calorific value (HCV) is the heat transferred when H_2O in the products is in the liquid state. The *lower heating value* (LHV) or lower calorific value (LCV) is the heat transferred in the reaction when H_2O in the products is in the vapour state.

Therefore

$$\mathrm{LHV} = \mathrm{HHV} - m_{\mathrm{H_2O}} \cdot h_{\mathrm{fg}}$$

where $m_{\mathrm{H_2O}}$ is the mass of water formed in the reaction.

16.17 ABSOLUTE ENTROPY AND THE THIRD LAW OF THERMODYNAMICS

So far only the first law aspects of chemical reactions have been discussed. The second law analysis of chemical reactions needs a base for the entropy of various substances. The entropy of substances at the absolute zero of temperature, called absolute entropy, is dealt with by the third law of thermodynamics formulated in the early twentieth century primarily by W.H. Nernst (1864–1941) and Max Planck (1858–1947). The third law states that the entropy of a perfect crystal is zero at the absolute zero of temperature and it represents the maximum degree of order. A substance not having a perfect crystalline structure and possessing a degree of randomness such as a solid solution or a glassy solid, has a finite value of entropy at absolute zero. The third law (see Chapter 11) provides an absolute base from which the entropy of each substance can be measured. The entropy relative to this base is referred to as the absolute entropy. Table 16.3 gives the absolute entropy of various substances at the standard state 25°C, 1 atm. For any other state

$$\bar{s}_{\mathrm{T,\,p}} = \bar{s}_{\mathrm{T}}^{0} + (\Delta\bar{s})_{\mathrm{T,\,1\,atm,}\to\mathrm{T,\,p}}$$

where $\bar{s}_{\mathrm{T}}^{0}$ refers to the absolute entropy at 1 atm. and temperature T, and $(\Delta\bar{s})_{\mathrm{T,\,1\,atm,}\to\mathrm{T,\,p}}$ refers to the change of entropy for an isothermal change of pressure from 1 atm. to pressure p (Fig. 16.6). Table C in the appendix gives the values of $\bar{s}^{0}$ for various substances at 1 atm. and at different temperatures. Assuming ideal gas behaviour $(\Delta\bar{s})_{\mathrm{T,\,1\,atm,}\to\mathrm{T,\,p}}$ can be determined (Fig. 16.6)

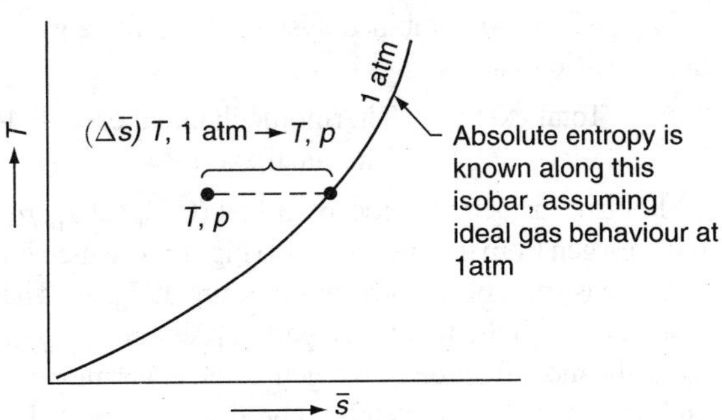

Fig. 16.6 *Absolute entropy*

$$\bar{s}_2 - \bar{s}_1 = -\bar{R}\ln\frac{p_2}{p_1}$$

or

$$(\Delta\bar{s})_{\mathrm{T,\,1\,atm,}\to\mathrm{T,\,p}} = -\bar{R}\ln p$$

where p is in atm.

16.18 SECOND LAW ANALYSIS OF REACTIVE SYSTEMS

The reversible work for a steady state steady flow process, in the absence of changes in K.E. and P.E., is given by

$$W_{\text{rev}} = \sum n_i (h_i - T_0 s_i) - \sum n_e (h_e - T_0 s_e)$$

For an S.S.S.F. process involving a chemical reaction

$$W_{\text{rev}} = \sum_{R} n_i [\bar{h}_f^0 + \Delta\bar{h} - T_0 \bar{s}]_i - \sum_{P} n_e [\bar{h}_f^0 + \Delta\bar{h} - T_0 \bar{s}]_e \tag{16.26}$$

The irreversibility for such a process is

$$I = \sum_{P} n_e T_0 \bar{s}_e - \sum_{R} n_i T_0 \bar{s}_i - Q_{\text{C.V.}}$$

The availability, ψ, in the absence of K.E. and P.E. changes, for an S.S.S.F. process is

$$\psi = (h - T_0 s) - (h_0 - T_0 s_0)$$

When an S.S.S.F. chemical reaction takes place in such a way that both the reactants and products are in temperature equilibrium with the surroundings, the reversible work is given by

$$W_{\text{rev}} = \sum_{R} n_i \bar{g}_i - \sum_{P} n_e \bar{g}_e \tag{16.27}$$

where the $\bar{g}$'s refer to the Gibbs function. The *Gibbs function for formation*, $\bar{g}_f^0$, is defined similar to enthalpy of formation, $\bar{h}_f^0$. The Gibbs function of each of the elements at 25°C and 1 atm. pressure is assumed to be zero, and the Gibbs function of each substance is found relative to this base. Table 16.1 gives $\bar{g}_f^0$ for some substances at 25°C, 1 atm.

16.19 CHEMICAL EXERGY

In Chapter 8, it was stated that when a system is at the dead state, it is in thermal and mechanical equilibrium with the environment, and the value of its exergy is zero. To state it more precisely, the *thermomechanical* contribution to exergy is zero. However, the contents of a system at the dead state may undergo chemical reaction with environmental components and produce additional work.

We will here study a combined system formed by an environment and a system having a certain amount of fuel at T_0, p_0. The work obtainable by allowing the fuel to react with oxygen from the environment to produce the environmental components of CO_2 and H_2O is evaluated. The *chemical exergy* is thus defined as the maximum theoretical work that could be developed by the combined system. Thus for a given system at a specific state:

Total exergy = Thermomechanical exergy + Chemical exergy

Let us consider a hydrocarbon fuel (C_aH_b) at T_0, p_0 reacting with oxygen from the environment (Fig. 16.7) which is assumed to be consisting of an ideal gas mixture at T_0, p_0. The oxygen that reacts with the fuel is at a partial pressure of x_{O_2}, p_0, where x_{O_2} is the mole fraction of oxygen in the environment. The fuel and oxygen react completely to produce CO_2 and H_2O, which exit in separate streams at T_0 and respective partial pressures of $x_{CO_2} \cdot p_0$ and $x_{H_2O} \cdot p_0$. The reaction is given by:

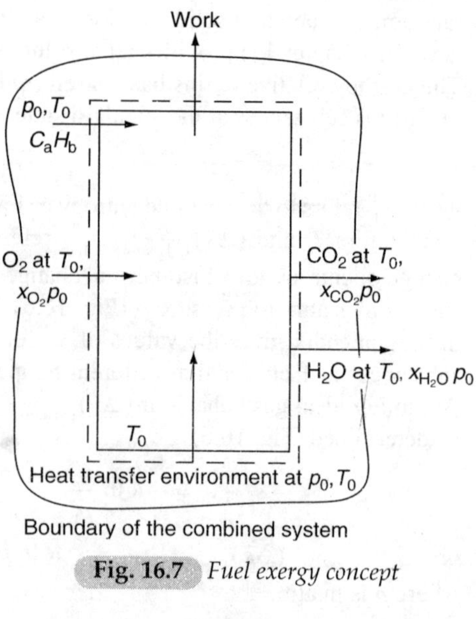

Fig. 16.7 *Fuel exergy concept*

$$\mathrm{C_aH_b} + \left(a + \frac{b}{4}\right)\mathrm{O_2} \longrightarrow a\mathrm{CO_2} + \frac{b}{2}\mathrm{H_2O}$$

At steady state, the energy balance gives

$$\dot{Q}_{\text{C.V.}} + \dot{n}\,H_{\text{R}} = \dot{W}_{\text{C.V.}} + \dot{n}\,H_{\text{P}}$$

or,
$$\frac{\dot{W}_{\text{C.V.}}}{\dot{n}} = \frac{\dot{Q}_{\text{C.V.}}}{\dot{n}} + H_{\text{R}} - H_{\text{P}}$$

$$= \frac{\dot{Q}_{\text{C.V.}}}{n} + \left(\bar{h}_{\text{f}}^{0} + \Delta\bar{h}\right)\mathrm{C_aH_b} + \left(a + \frac{b}{4}\right)\bar{h}_{\mathrm{O_2}} - a\bar{h}_{\mathrm{CO_2}} - \frac{b}{2}\bar{h}_{\mathrm{H_2O}} \tag{16.28}$$

where $\dot{n}$ is the rate of fuel flow in moles, and K.E. and P.E. effects are neglected.

An entropy balance for the control volume gives:

$$0 = \frac{\dot{Q}_{\text{C.V.}}/\dot{n}}{T_0} + \bar{s}_{\mathrm{C_aH_b}} + \left(a + \frac{b}{4}\right)\bar{s}_{\mathrm{O_2}} - a\bar{s}_{\mathrm{CO_2}} - \frac{b}{2}\bar{s}_{\mathrm{H_2O}} + \frac{s_{\text{gen}}}{\dot{n}} \tag{16.29}$$

Eliminating $\dot{Q}_{\text{C.V.}}$ between Eqs (16.28) and (16.29),

$$\frac{\dot{W}_{\text{C.V.}}}{\dot{n}} = \left[\bar{h}_{\mathrm{C_aH_b}} + \left(a + \frac{b}{4}\right)\bar{h}_{\mathrm{O_2}} - a\bar{h}_{\mathrm{CO_2}} - \frac{b}{2}\bar{h}_{\mathrm{H_2O}}\right]$$
$$-T_0\left[\bar{s}_{\mathrm{C_aH_b}} + \left(a + \frac{b}{4}\right)\bar{s}_{\mathrm{O_2}} - a\bar{s}_{\mathrm{CO_2}} - \frac{b}{2}\bar{s}_{\mathrm{H_2O}}\right] - \frac{T_0\,\dot{s}_{\text{gen}}}{\dot{n}} \tag{16.30}$$

The specific enthalpies in Eq. (16.30) can be determined knowing only the temperature T_0 and the specific entropies can be determined knowing T_0, p_0 and the composition of the environment.

For maximum work, $\dot{I} = \dfrac{T_0\,\dot{s}_{\text{gen}}}{n} = 0$

Therefore, the chemical exergy $\bar{a}_{\text{ch}}$ can be expressed as

$$\bar{a}_{\text{ch}} = \left[\bar{h}_{\mathrm{C_aH_b}} + \left(a + \frac{b}{4}\right)\bar{h}_{\mathrm{O_2}} - a\bar{h}_{\mathrm{CO_2}} - \frac{b}{2}\bar{h}_{\mathrm{H_2O}}\right]$$
$$-T_0\left[\bar{s}_{\mathrm{C_aH_b}} + \left(a + \frac{b}{4}\right)\bar{s}_{\mathrm{O_2}} - a\bar{s}_{\mathrm{CO_2}} - \frac{b}{2}\bar{s}_{\mathrm{H_2O}}\right] \tag{16.31}$$

The specific entropies for $\mathrm{O_2}$, $\mathrm{CO_2}$ and $\mathrm{H_2O}$ are written from (Eq. 16.22),

$$\bar{s}_{\text{i}}\,(T_0, x_{\text{i}}\,p_0) = \bar{s}_{\text{i}}\,(T_0, p_0) - \bar{R}\,\ln x_{\text{i}} \tag{16.32}$$

where the first term on the right is the absolute entropy at T_0 and p_0, and x_{i} is the mole fraction of component i in the environment. Therefore, Eq. (16.31) becomes,

$$\bar{a}_{\text{ch}} = \left[\bar{h}_{\mathrm{C_aH_b}} + \left(a + \frac{b}{4}\right)\bar{h}_{\mathrm{O_2}} - a\bar{h}_{\mathrm{CO_2}} - \frac{b}{2}\bar{h}_{\mathrm{H_2O}}\right] \text{ (at } T_0, p_0)$$
$$-T_0\left[\bar{s}_{\mathrm{C_aH_b}} + \left(a + \frac{b}{4}\right)\bar{s}_{\mathrm{O_2}} - a\bar{s}_{\mathrm{CO_2}} - \frac{b}{2}\bar{s}_{\mathrm{H_2O}}\right] \text{ (at } T_0, p_0)$$
$$+RT_0 \ln \frac{(x_{\mathrm{O_2}})^{a+b/4}}{\left(x_{\mathrm{CO_2}}\right)^{a}\left(x_{\mathrm{H_2O}}\right)^{b/2}} \tag{16.33}$$

In terms of Gibbs functions of respective substances,

$$\bar{a}_{\text{ch}} = \left[\bar{g}_{\mathrm{C_aH_b}} + \left(a + \frac{b}{4}\right)\bar{g}_{\mathrm{O_2}} - a\bar{g}_{\mathrm{CO_2}} - \frac{b}{2}\bar{g}_{\mathrm{H_2O}}\right] \text{ (at } T_0, p_0)$$

$$+\bar{R}T_0 \ln \frac{\left(x_{O_2}\right)^{a+b/4}}{\left(x_{CO_2}\right)^{a}\left(x_{H_2O}\right)^{b/2}} \tag{16.34}$$

where $\bar{g}(T_0, p_0) = \bar{g}_f^0 + \Delta\bar{g}_{T_0, p_0 \to T_{ref}, P_{ref}}$

For the special case when T_0 and p_0 are the same as T_{ref} and p_{ref}, $\Delta\bar{g}$ will be zero.
The chemical exergy of pure CO at T_0, p_0 where the reaction is given by:

$$CO + \frac{1}{2}O_2 \longrightarrow CO_2$$

$$(\bar{a}_{ch})_{CO} = \left[\bar{g}_{CO} + \frac{1}{2}\bar{g}_{O_2} - \bar{g}_{CO_2}\right] \text{(at } T_0, p_0\text{)}$$

$$+\bar{R}T_0 \ln \frac{\left(x_{O_2}\right)^{1/2}}{x_{CO_2}}$$

Water is present as a vapour within the environment, but normally is a liquid at T_0, p_0. The chemical exergy of liquid water is

$$H_2O(l) \longrightarrow H_2O(g)$$

$$(\bar{a}_{ch})_{H_2O(l)} = \left[\bar{g}_{H_2O(l)} - \bar{g}_{H_2O(g)}\right] \text{(at } T_0, p_0\text{)}$$

$$+\bar{R}T_0 \ln \frac{1}{x_{H_2O(g)}}$$

The specific exergy of system is

$$a = a_{thermo\text{-}mech} + a_{chem}$$

$$= (u - u_0) + p_0(v - v_0) - T_0(s - s_0) + \frac{V^2}{2} + gz + a_{ch} \tag{16.35}$$

and the specific flow exergy is given by:

$$a = (h - h_0) - T_0(s - s_0) + \frac{V^2}{2} + gz + a_{ch} \tag{16.36}$$

Between two states of a system of constant composition, a_{ch} cancels, leaving just the thermomechanical contribution. For example, to find the chemical exergy of liquid octane, the reaction is:

$$C_8H_{18}(l) + 12.5O_2 \longrightarrow 8CO_2 + 9H_2O(g)$$

Assume the composition of the environment on molar basis be:
N_2 75.67%, O_2 20.35%, H_2O 3.12%, CO_2 0.03%, others 0.83%. Then,

$$\bar{a}_{ch} = \left[\bar{g}_{C_8H_{18}(l)} + 12.5\bar{g}_{O_2} - 8\bar{g}_{CO_2} - 9\bar{g}_{H_2O(g)}\right] \text{(at } T_0, p_0\text{)}$$

$$+\bar{R}T_0 \ln \frac{\left(x_{O_2}\right)^{12.5}}{\left(x_{CO_2}\right)^{8}\left(x_{H_2O}\right)^{9}}$$

Using the values given in Table 16.1

$$(\bar{a}_{ch})\, C_8H_{18}(l) = 6610 + 12.5(0) - 8(-394, 380) - 9(-228, 590)$$

$$+ 8.3143\ (298.15) \ln \frac{(0.2035)^{12.5}}{(0.0003)^{8}\ (0.0312)^{9}}$$

$$= 5,407,843 \text{ kJ/kg mol} = 47,346 \text{ kJ/kg}$$

16.20 SECOND LAW EFFICIENCY OF A REACTIVE SYSTEM

For a fuel at T_0, p_0, the chemical exergy is the maximum theoretical work that could be obtained through reaction with environmental substances. However, due to various irreversibilities like friction and heat loss, the actual work obtained is only a fraction of this maximum theoretical work. The second law efficiency may thus be defined as the ratio of:

$$\eta_{II} = \frac{\text{Actual work done}}{\text{Maximum theoretical work}} = \frac{W_{\text{C.V.}}}{m_{\text{fuel}} \times a_{ch}} \tag{16.37}$$

The associated irreversibilities and the consequent exergy losses require to be reduced to enhance the second law efficiency, which in turn, reduces the fuel consumption and also increases the cost. The trade off between the fuel savings and the additional costs must be carefully weighed.

16.21 FUEL CELLS

Fuel cells are receiving prominent attention as clean, efficient and affordable alternative energy technologies for the future. A *fuel cell* is a direct energy conversion device in which fuel and an oxidizer undergo a controlled chemical reaction, producing products and supplying electrical current directly to an external circuit. The fuel and oxidizer react in stages on two separate electrodes, a positive electrode called *cathode*, and a negative electrode called *anode*. The two electrodes are separated by an electrolyte. The rates of reaction are limited by the time it takes for diffusion of chemical species through the electrodes and the electrolyte and by the reaction kinetics.

Fuel cells are of various kinds. One type, the hydrogen-oxygen fuel cell, is shown schematically in Fig. 16.8. The fuel is hydrogen, which diffuses through the porous anode and reacts on the anode surface with OH^- ions, forming water and yielding free electrons according to:

$$H_2 + 2OH^- \longrightarrow 2H_2O + 2e^-$$

The electrons flow into the external circuit, and the water passes on to the electrolyte. On the cathode surface, oxygen fed into the cell combines with water from the electrolyte and electrons from the external circuit to produce OH^- ions and water according to:

$$\frac{1}{2}O_2 + 2H_2O + 2e^- \longrightarrow 2(OH)^- + H_2O$$

The electrolyte separating the electrodes transports the OH^- ions, completing, the circuit, and the water (products) is removed from the cell. The overall reaction is:

$$H_2 + \frac{1}{2}O_2 \longrightarrow H_2O$$

which is the same as the equation for the highly exothermic combustion reaction. However, in a fuel cell, only a relatively small amount of heat transfer between the cell and its surroundings takes place, and the temperature rise is also relatively much smaller.

Energy is removed from the fuel cell as electrical energy, whereas energy is removed from a combustion reaction as heat or as heat and work together. Because the fuel cell operates almost isothermally and continuously, the extent of its conversion of chemical energy to electrical energy is not limited by second law of thermodynamics.

In a fuel cell, there is a continuous supply of the reactants. The overall reaction, as stated, is divided into two reactions that occur on separate electrodes. The fuel and the oxidizer do not come directly into contact with each other, because direct contact would generally involve a non-isothermal (exothermic) reaction as in a normal combustion process.

One reaction, occuring on the surface of one electrode, ionizes the fuel and sends released electrons into an external electric circuit. On the surface of the other electrode, a reaction occurs that accepts electrons from

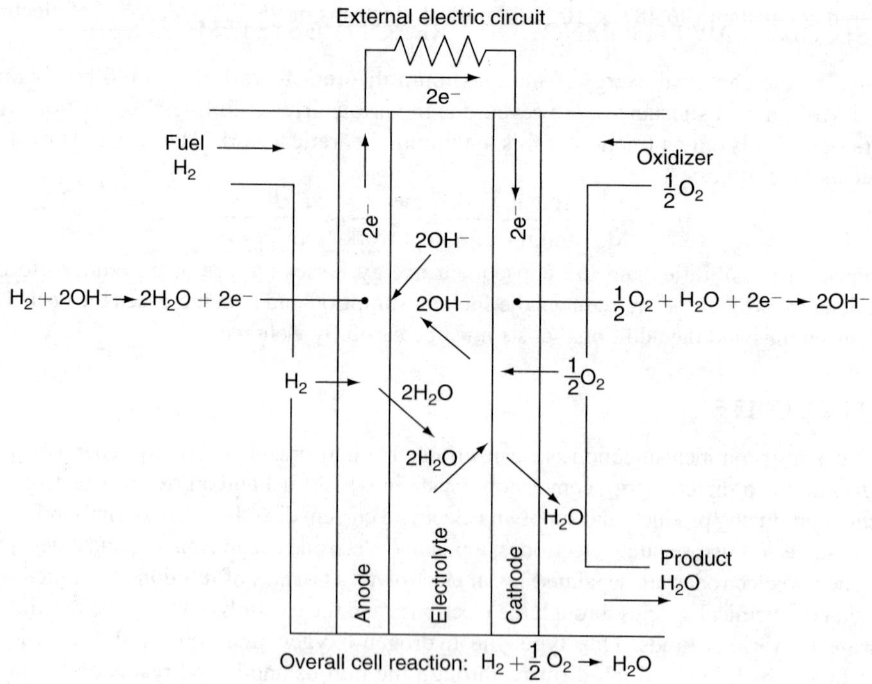

Fig. 16.8 *Hydrogen-oxygen fuel cell*

the external circuit and when combined with the oxidizer creates ions. The ions from each reaction are combined in the electrolyte to complete the overall reaction. The electrolyte between the electrodes is necessary to transport ions, and it is not electrically conductive, thus, not allowing the flow of electrons through it.

The maximum work obtainable in a fuel all is given by Eq. (16.27),

$$W_{max} = -\Delta G = \sum_R n_i \bar{g}_i - \sum_P n_e \bar{g}_e$$

where

$$\bar{g} = g_f^0 + \Delta \bar{g}$$

Also, from Eq. (16.2),

$$W_{max} = W_{rev} = \sum_R n_i \left[\bar{h}_f^0 + \Delta \bar{h} - T_0 \bar{s}\right]_i - \sum_P n_e \left[\bar{h}_f^0 + \Delta \bar{h} - T_0 \bar{s}\right]_e$$

$$= \Delta H - T_0 \left[\sum_R n_i \bar{s}_i - \sum_P n_e \bar{s}_e\right]$$

The fuel cell efficiency is defined as:

$$\varepsilon = \Delta G / \Delta H \tag{16.38}$$

which can be less than, equal to, or greater than unity. The work done in a fuel cell can be written as:

$$W_{max} = \text{moles of reactant} \times \frac{\text{molecules}}{\text{kgmol}} \times \frac{\text{electrons}}{\text{molecule}} \times \frac{\text{charge}}{\text{electron}} \times \text{voltage}$$

$$= n \times (6.023 \times 10^{26})\text{j}\,(1.6022 \times 10^{19})V = \mathcal{F} n\, j\, V,$$

where $\mathcal{F}$ is Faraday constant (96.487 × 10⁶ C/k mol) of electrons or 96,487 kJ/V-kmol of electrons, n is the number of moles of reactant having a valence j, and V is the cell terminal voltage. The ideal cell terminal voltage is given by:

$$V_i = \frac{\Delta G}{\mathcal{F} n j} \tag{16.39}$$

$$CH_4 + 2O_2 \longrightarrow CO_2 + 2H_2O$$

Solved Examples

Example 16.1

Starting with n_0 moles of NH_3, which dissociates according to the equation

$$NH_3 \rightleftharpoons \frac{1}{2} N_2 + \frac{3}{2} H_2,$$

show that at equilibrium

$$K = \frac{\sqrt{27}}{4} \frac{\varepsilon_e^2}{1-\varepsilon_e^2} \cdot p$$

Solution

$$NH_3 \rightleftharpoons \frac{1}{2} N_2 + \frac{3}{2} H_2$$

Comparing the equation with

$$\nu_1 A_1 + \nu_2 A_2 \rightleftharpoons \nu_3 A_3 + \nu_4 A_4$$

we have

$$A_1 = NH_3,\ \nu_1 = 1,\ n_1 = n_0 (1 - \varepsilon),\ x_1 - \frac{1-\varepsilon}{1+\varepsilon}$$

$$A_3 = N_2,\ \nu_3 = \frac{1}{2},\ n_3 = \frac{n_0 \varepsilon}{2},\ x_3 = \frac{\varepsilon}{2(1+\varepsilon)}$$

$$A_4 = H_2,\ n_4 = \frac{3}{2},\ \nu_4 = \frac{3 n_0 \varepsilon}{2},\ x_4 = \frac{3\varepsilon}{2(1+\varepsilon)}$$

$$\Sigma n = n_0(1 + \varepsilon)$$

Substituting in the law of mass action

$$K = \left[\frac{x_3^{\nu_3} \cdot x_4^{\nu_4}}{x_1^{\nu_1} \cdot x_2^{\nu_2}}\right]_{\varepsilon=\varepsilon_e} p^{\nu_3+\nu_4-\nu_1-\nu_2}$$

$$= \frac{\left[\dfrac{\varepsilon_e}{2(1+\varepsilon_e)}\right]^{1/2} \left[\dfrac{3\varepsilon_e}{2(1+\varepsilon_e)}\right]^{3/2}}{\left(\dfrac{1-\varepsilon_e}{1+\varepsilon_e}\right)} \cdot p^{\frac{3}{2}+\frac{1}{2}-1}$$

$$= \frac{\sqrt{27}\, \varepsilon^2}{4(1+\varepsilon_e)^2} \cdot \frac{1+\varepsilon_e}{1-\varepsilon_e} \cdot p = \frac{\sqrt{27}}{4} \cdot \frac{\varepsilon_e^2}{1-\varepsilon_e^2} \cdot p$$

Proved.

Example 16.2

At 35°C and 1 atm. the degree of dissociation of N_2O_4 at equilibrium is 0.27. (a) Calculate K. (b) Calculate ε_e at the same temperature when the pressure is 100 mm Hg. (c) The equilibrium constant for the dissociation of N_2O_4 has the values 0.664 and 0.141 at temperatures 318 and 298 K respectively. Calculate the average heat of reaction within this temperature range.

Solution Nitrogen tetroxide dissociates according to the equation

$$N_2O_4 \rightleftharpoons 2NO_2$$

Comparing this equation with

$$\nu_1 A_1 + \nu_2 A_2 \rightleftharpoons \nu_3 A_3 + \nu_4 A_4$$

$$A_1 = N_2O_4, \nu_1 = 1, n_1 = n_0(1-\varepsilon), x_1 = \frac{1-\varepsilon}{1+\varepsilon}$$

$$A_3 = NO_2, \nu_3 = 2, n_3 = 2n_0\,\varepsilon, x_3 = \frac{2\varepsilon}{1+\varepsilon}$$

$$\Sigma n = n_0(1+\varepsilon)$$

$$\nu_2 = 0, \nu_4 = 0$$

Substituting in the law of mass action

$$K = \left[\frac{x_3^{\nu_3} \cdot x_4^{\nu_4}}{x_1^{\nu_1} \cdot x_2^{\nu_2}}\right]_{\varepsilon_e} \cdot p^{\nu_3+\nu_4-\nu_1-\nu_2}$$

$$= \frac{\left(\frac{2\varepsilon_e}{1+\varepsilon_e}\right)^2}{\frac{1-\varepsilon_e}{1+\varepsilon_e}} \cdot p^{2+0-1+0} = \frac{4\varepsilon_e^2 p}{1-\varepsilon_2^2}$$

(a) When $\varepsilon_e = 0.27$ and $p = 1$ atm.

$$K = \frac{4(0.27)^2 \cdot 1}{1-(0.27)^2} = 0.3145 \text{ atm.}$$ *Ans.*

(b) Taking the value of K to be the same at the same temperature

$$0.3145 \text{ atm.} = \frac{4\varepsilon_e^2}{1-\varepsilon_e^2} \cdot \frac{100}{760}$$

$$\therefore \quad \varepsilon_e = 0.612$$ *Ans*

(c) The heat of reaction is given by

$$\Delta H = -2.30\,\bar{R}\,\frac{d \log K}{d\frac{1}{T}}$$

$$= 2.30\,\bar{R}\,\frac{T_1 T_2}{T_1 - T_2} \log \frac{K_1}{K_2}$$

$$= 2.30 \times 8.3143 \times \frac{318 \times 298}{318 - 298} \log \frac{0.664}{0.141}$$

$$= 90700 \log 4.7 = 61{,}000 \text{ kJ/kg mol}$$ *Ans.*

Example **16.3**

For the chemical reaction

$$CO_2 + H_2 \rightleftharpoons CO + H_2O$$

the equilibrium value of the degree of reaction at 1200 *K is* 0.56. *Determine the equilibrium constant and the Gibbs function change.*

Solution $$CO_2 + H_2 \rightleftharpoons CO + H_2O$$

Here, $\nu_1 = \nu_2 = \nu_3 = \nu_4 = 1$

At equilibrium, $n_1 = 1 - \varepsilon,\ n_2 = 1 - \varepsilon,$

$n_3 = \varepsilon$ and $n_4 = \varepsilon$

$\therefore$ $\sum n = 2$

Therefore, $$x_1 = \frac{1-\varepsilon}{2} \quad x_2 = \frac{1-\varepsilon}{2} \quad x_3 = \frac{\varepsilon}{2} \quad x_4 = \frac{\varepsilon}{2}$$

Substituting $\varepsilon = 0.56,$

$$x_1 = 0.22 \quad x_2 = 0.22 \quad x_3 = 0.28 \quad x_4 = 0.28$$

$$K = \frac{x_3^{\nu_3}\, x_4^{\nu_4}}{x_1^{\nu_1}\, x_2^{\nu_2}} \times p^{\nu_3+\nu_4-\nu_1-\nu_2} = \frac{0.28 \times 0.28}{0.22 \times 0.22} = 1.62$$

$$\Delta G = -\bar{R}T \ \ln K + \bar{R}T \ln p^{(\nu_3+\nu_4-\nu_1-\nu_2)}$$
$$= -\bar{R}T \ \ln K = -8.3143 \times 1200 \ln 1.62 = -4813.2 \text{ J/g mol}$$

Example **16.4**

Prove that for a mixture of reacting ideal gases,

$$\frac{d}{d\varepsilon} \ln \frac{x_3^{\nu_3}\, x_4^{\nu_4}}{x_1^{\nu_1}\, x_2^{\nu_2}} = \frac{(n_0 + n_0')}{\Sigma n_k} \frac{(\nu_1 + \nu_2)(\nu_3 + \nu_4)}{\varepsilon(1-\varepsilon)}$$

which is always positive.

Solution From the law of mass action, the equilibrium constant is given by:

$$K = \frac{x_3^{\nu_3}\, x_4^{\nu_4}}{x_1^{\nu_1}\, x_2^{\nu_2}} p^{\nu_3+\nu_4-\nu_1-\nu_2}$$

$$= \frac{[n_3/\Sigma n]^{\nu_3}\, [n_4/\Sigma n]^{\nu_4}}{[n_1/\Sigma n]^{\nu_1}\, [n_2/\Sigma n]^{\nu_2}} p^{\nu_3+\nu_4-\nu_1-\nu_2}$$

$$= \frac{n_3^{\nu_3}\, n_4^{\nu_4}}{n_1^{\nu_1}\, n_2^{\nu_2}} [p/\Sigma n]^{\Delta\nu}$$

where $\Sigma n = n_1 + n_2 + n_3 + n_4$ and $\Delta\nu = \nu_3 + \nu_4 - \nu_1 - \nu_2$.

By logarithmic differentiation,

$$\frac{dK}{K} - \Delta\nu \frac{dp}{p} + \Delta\nu \frac{d\Sigma n}{\Sigma n} = \nu_3 \frac{dn_3}{n_3} + \nu_4 \frac{dn_4}{n_4} - \nu_1 \frac{dn_1}{n_1} - \nu_2 \frac{dn_2}{n_2} \quad (1)$$

Now,
$$n_1 = (n_0 + n'_0)\,\nu_1\,(1 - \varepsilon)$$
$$n_2 = (n_0 + n'_0)\,\nu_2\,(1 - \varepsilon)$$
$$n_3 = (n_0 + n'_0)\,\nu_3\varepsilon$$
$$n_4 = (n_0 + n'_0)\,\nu_4\varepsilon$$
$$\sum n = (n_0 + n'_0)\,(\nu_1 + \nu_2 + \varepsilon\Delta\nu) \quad (2)$$

Again,
$$-\frac{dn_1}{\nu_1} = -\frac{dn_2}{\nu_2} = \frac{dn_3}{\nu_3} = \frac{dn_4}{\nu_4} = (n_0 + n'_0)\,d\varepsilon$$
$$d\sum n = dn_1 + dn_2 + dn_3 + dn_4 = (n_0 + n'_0)\,\Delta\nu\, d\varepsilon \quad (3)$$

From Eqs. (1), (2) and (3),
$$\frac{dK}{K} - \Delta\nu\,\frac{dp}{p} + \Delta\nu\,(n_0 + n'_0)\,\Delta\nu\, d\varepsilon$$
$$= \nu_3\frac{(n_0 + n'_0)\,\nu_3\, d\varepsilon}{n_3} + \nu_4\frac{(n_0 + n'_0)\,\nu_4\, d\varepsilon}{n_4}$$
$$-\nu_1\frac{(n_0 + n'_0)\,\nu_1\, d\varepsilon}{n_1} - \nu_2\frac{-(n_0 + n'_0)\,\nu_2\, d\varepsilon}{n_2}$$
$$\frac{dK}{K} - \Delta\nu\,\frac{dp}{p} = (n_0 + n'_0)\left[\frac{\nu_3^2}{n_3} + \frac{\nu_4^2}{n_4} + \frac{\nu_1^2}{n_1} + \frac{\nu_2^2}{n_2} - \frac{(\Delta\nu)^2}{\Sigma n}\right] d\varepsilon$$
$$= \frac{n_0 + n'_0}{\Sigma n_K}\left[\frac{\nu_1^2}{x_1} + \frac{\nu_2^2}{x_2} + \frac{\nu_3^2}{x_3} + \frac{\nu_4^2}{x_4} - (\Delta\nu)^2\right] d\varepsilon$$
$$\frac{d(\ln K - \Delta\nu \ln p)}{d\varepsilon} = \frac{n_0 + n'_0}{\Sigma n_K}\,\frac{1}{\Psi}$$

where
$$\frac{1}{\Psi} = \frac{\nu_1^2}{x_1} + \frac{\nu_2^2}{x_2} + \frac{\nu_3^2}{x_3} + \frac{\nu_4^2}{x_4} - (\Delta\nu)^2$$

L.H.S.
$$= \frac{d}{d\varepsilon}\ln Kp^{-\Delta\nu} = \frac{d}{d\varepsilon}\ln \frac{x_3^{\nu_3}\, x_4^{\nu_4}}{x_1^{\nu_1}\, x_2^{\nu_2}}\, p^{\Delta\nu}\, p^{-\Delta\nu}$$
$$\frac{d}{d\varepsilon}\ln \frac{x_3^{\nu_3}\, x_4^{\nu_4}}{x_1^{\nu_1}\, x_2^{\nu_2}} = \frac{n_0 + n'_0}{\Sigma n_K}\,\frac{1}{\Psi} \quad (4)$$

where x's are the equilibrium values.

Now,
$$\frac{1}{\Psi} = \frac{\nu_1^2}{x_1} + \frac{\nu_2^2}{x_2} + \frac{\nu_3^2}{x_3} + \frac{\nu_4^2}{x_4} - (\Delta\nu)^2$$
$$= \left[\frac{\nu_1^2}{n_1} + \frac{\nu_2^2}{n_2} + \frac{\nu_3^2}{n_3} + \frac{\nu_4^2}{n_4}\right] \Sigma n - (\Delta\nu)^2$$
$$= \left[\frac{\nu_1^2}{\nu_1(1-\varepsilon)} + \frac{\nu_2^2}{\nu_2(1-\varepsilon)} + \frac{\nu_3^2}{\nu_3\varepsilon} + \frac{\nu_4^2}{\nu_4\varepsilon}\right] \frac{\Sigma n}{n_0 + n'_0} - (\Delta\nu)^2$$

$$= \left[\frac{\nu_1\varepsilon + \nu_2\varepsilon + \nu_3 - \nu_3\varepsilon + \nu_4 - \nu_4\varepsilon}{\varepsilon(1-\varepsilon)}\right](\nu_1 + \nu_2 + \varepsilon\Delta\nu) - (\Delta\nu)^2$$

$$= \frac{(\nu_3 + \nu_4) - \varepsilon\Delta\nu}{\varepsilon(1-\varepsilon)}[(\nu_1 + \nu_2) + \varepsilon\Delta\nu] - (\Delta\nu)^2$$

$$= \frac{(\nu_1 + \nu_2)(\nu_3 + \nu_4) - \varepsilon\Delta\nu(\nu_1 + \nu_2) + (\nu_3 + \nu_4) - \varepsilon^2(\Delta\nu)^2}{\varepsilon(1-\varepsilon)} - (\Delta\nu)^2$$

$$= \frac{(\nu_1 + \nu_2)(\nu_3 + \nu_4) + \varepsilon(\Delta\nu)^2 - \varepsilon^2(\Delta\nu)^2}{\varepsilon(1-\varepsilon)} - (\Delta\nu)^2$$

$$= \frac{(\nu_1 + \nu_2)(\nu_3 + \nu_4)}{\varepsilon(1-\varepsilon)} + \frac{\varepsilon(1-\varepsilon)(\Delta\nu)^2}{\varepsilon(1-\varepsilon)} - (\Delta\nu)^2 = \frac{(\nu_1 + \nu_2)(\nu_3 + \nu_4)}{\varepsilon(1-\varepsilon)}$$

$$\therefore \frac{d}{d\varepsilon}\ln\left[\frac{x_3^{\nu_3}\, x_4^{\nu_4}}{x_1^{\nu_1}\, x_2^{\nu_2}}\right]_{\varepsilon=\varepsilon_e} = \frac{(n_0 + n'_0)}{\sum n_K}\cdot\frac{(\nu_1 + \nu_2)(\nu_3 + \nu_4)}{\varepsilon(1-\varepsilon)}$$ Proved.

Example 16.5

For the dissociation of nitrogen tetraoxide according to the equation

$$N_2O_4 \rightleftharpoons 2NO_2$$

Show that the degree of dissociation at equilibrium is

$$e = \frac{V_e}{V_0} - 1$$

where V_0 = initial volume and V_e volume at equilibrium. At 50°C and 0.124 atm, there is a 77.7% increase in volume when equilibrium is reached. Find the value of the equilibrium constant.

Solution $N_2O_4 \rightleftharpoons 2NO_2$

Starting with n_0 moles of N_2O_4 at temperature T and pressure p, the initial volume V_0 is

$$V_0 = n_0\frac{\bar{R}T}{p}$$

If V_e denotes the volume at equilibrium, the temperature and pressure remaining the same, then

$$V_e = [n_0(1 - \varepsilon_e) + 2n_0\varepsilon_e]\frac{\bar{R}T}{p}.$$

where ε_e is the value of the degree of dissociation at equilibrium. This can be written:

$$V_e = (1 + \varepsilon_e)V_0$$

or,

$$\varepsilon_e = \frac{V_e}{V_0} - 1$$ Proved.

Given $V_e/V_0 = 1.777,\ \varepsilon_e = 0.777.$

Now, $n_1 = n_0\,\nu_1(1 - \varepsilon_e) = n_0(1 - \varepsilon_e)$

$$n_3 = n_0\, \nu_3, \varepsilon_e = n_0 \cdot 2\varepsilon_e$$

$$\Sigma n = n_0(1 + \varepsilon_e)$$

$$x_1 = \frac{1-\varepsilon_e}{1+\varepsilon_e}, x_3 = \frac{2\varepsilon_e}{1+\varepsilon_e}$$

$$K = \left[\frac{x_3^{\nu_3} \cdot x_4^{\nu_4}}{x_1^{\nu_1} \cdot x_2^{\nu_2}}\right]_{\varepsilon=\varepsilon_e} p^{\nu_3+\nu_4-\nu_1-\nu_2}$$

$$= \frac{\left[\dfrac{2\varepsilon_e}{1+\varepsilon_e}\right]^2}{\left[\dfrac{1-\varepsilon_e}{1+\varepsilon_e}\right]} p^{2-1} = \frac{4\varepsilon_e^2}{1-\varepsilon_e^2} p$$

$$= \frac{4\times(0.777)^2}{1-(0.777)^2}\, 0.124 = \frac{0.2994}{0.3963} = 0.755 \text{ atm}$$ *Ans.*

Example 16.6

Consider the equilibrium mixture of H_2O vapour, H_2 and O_2 caused by the dissociation of 1 g mol of H_2O at 1 atm and 1900 K. If $\Delta H = 250, 560$ J/g mol, $\varepsilon_e = 3.2 \times 10^{-3}$, estimate $C_P - \sum n_k C_{P_k}$.

Solution $H_2O \rightleftharpoons H_2 + \frac{1}{2} O_2$

$$\nu_1 = 1, \nu_2 = 0, \nu_3 = 1, \nu_4 = \frac{1}{2}$$

$$n_1 = n_0\nu_1 (1 - \varepsilon_e) = n_0(1 - \varepsilon_e),$$

$$n_3 = n_0\, \nu_3\, \varepsilon_e = n_0\, \varepsilon_e, n_4 = n_0\, \nu_4\, \varepsilon_e = n_0\, \frac{\varepsilon_e}{2}$$

$$\Sigma n = n_0\left(1+\frac{\varepsilon_e}{2}\right)$$

Now,

$$\frac{d}{d\varepsilon_e}\ln\frac{x_3^{\nu_3}\, x_4^{\nu_4}}{x_1^{\nu_1}\, x_2^{\nu_2}} = \frac{n_0}{\Sigma n}\frac{(\nu_1+\nu_2)(\nu_3+\nu_4)}{\varepsilon_e(1-\varepsilon_e)}$$

$$C_P - \sum n_k c_{P_k} = \frac{(\Delta H)^2 (1+\varepsilon_e/2)\,\varepsilon_e\,(1+\varepsilon_e)}{\bar{R}T^2(\nu_1+\nu_2)(\nu_3+\nu_4)}$$

$$= \frac{(250{,}560)^2 \times 3.2\times 10^{-3}}{8.3143\times(1900)^2 \times \frac{3}{2}} = 4.462 \text{ J/g mol K}$$ *Ans.*

Example 16.7

The products of combustion of an unknown hydrocarbon $C_x H_y$ have the following composition as measured by an Orsat apparatus:

CO_2 8.0%, CO 0.9%, O_2 8.8% and N_2 82.3%

Determine: (a) the composition of the fuel, (b) the air-fuel ratio, and (c) the percentage excess air used.

Solution Let 'a' moles of oxygen be supplied per mole of fuel. The chemical reaction can be written as follows.

$$C_xH_y + aO_2 + 3.76\,aN_2 \rightarrow 8CO_2 + 0.9CO + 8.8O_2 + 82.3N_2 + \frac{y}{2}\,H_2O$$

Oxygen balance gives

$$2a = 16 + 0.9 + 17.6 + \frac{y}{2} \qquad (1)$$

Nitrogen balance gives

$$3.76a = 82.3$$

$$\therefore \quad a = 21.89$$

By substituting the value of a in Eq. (1),

$$y = 18.5$$

Carbon balance gives

$$x = 8 + 0.9 = 8.9$$

Therefore, the chemical formula of the fuel is $C_{8.9}H_{18.5}$.

(a) The composition of the fuel is

$$\%\text{ carbon} = \frac{8.9 \times 12}{8.9 \times 12 + 18.5 \times 1} \times 100 = 58.23\%$$

$$\therefore \quad \%\text{ hydrogen} = 14.77\%$$

(b) $$\text{Air fuel ratio} = \frac{32a + 3.76a \times 28}{12x + y}$$

$$= \frac{21.89(32 + 3.76 \times 28)}{12 \times 8.9 + 18.5} = \frac{3005}{125.3} = 24$$

(c) % Excess air used

$$= \frac{8.8 \times 32}{21.89 \times 32 - 8.8 \times 32} \times 100 = 67.22\%$$

Example 16.8

Determine the heat transfer per kg mol of fuel for the following reaction

$$CH_4 + 2O_2 \rightarrow CO_2 + 2H_2O\ (l)$$

The reactants and products are each at a total pressure of 100 *kPa and* 25°*C.*

Solution By the first law

$$Q_{C.V.} + \sum_R n_i\bar{h}_i = \sum_P n_e\bar{h}_e$$

From Table C in the appendix

$$\sum_R n_i\bar{h}_i = (\bar{h}_f)_{CH_4} = -74{,}874 \text{ kJ}$$

$$\sum_P n_e\bar{h}_e = (\bar{h}_f^0)_{CO_2} + 2(\bar{h}_f^0)_{H_2O(l)}$$

$$= -393{,}522 + 2(-285{,}838) = -965{,}198 \text{ kJ} \qquad \textit{Ans.}$$

$$\therefore \quad Q_{C.V.} = -965{,}198 - (-74{,}873) = -890{,}325 \text{ kJ} \qquad \textit{Ans.}$$

Example 16.9

A gasoline engine delivers 150 kW. The fuel used is C_8H_{18} (1) and it enters the engine at 25°C. 150% theoretical air is used and it enters at 45°C. The products of combustion leave the engine at 750 K, and the heat transfer from the engine is 205 kW. Determine the fuel consumption per hour, if complete combustion is achieved.

Solution The stoichiometric equation (Fig. Ex. 16.9) gives

$$C_8H_{18}\ (l) + 12.5\ O_2 + 3.76 \times 12.5N_2 \rightarrow 8CO_2 + 9H_2O + 3.76 \times 12.5N_2$$

With 150% theoretical air $\quad C_8H_{18} + 18.7O_2 + 70N_2 \rightarrow 8CO_2 + 9H_2O + 6.25O_2 + 70N_2$

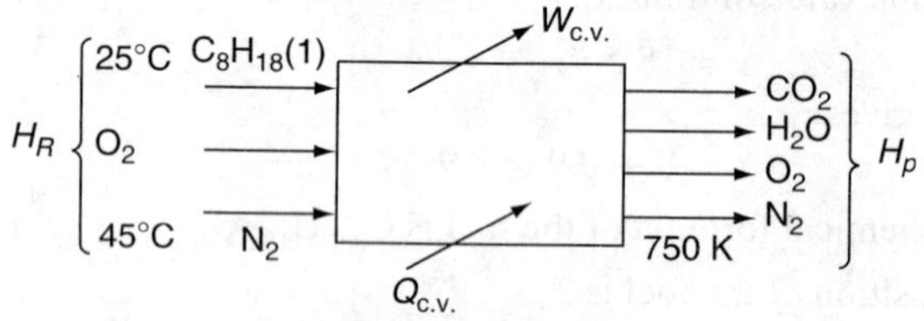

Fig. Ex. 16.9

$$H_R = \sum_R n_i[\bar{h}_f^0 + (\bar{h}_T^0 - \bar{h}_{298})]_i$$

$$= 1(\bar{h}_f)_{C_8H_{18}(l)} + 18.7[\bar{h}_f^0 + (\bar{h}_T - \bar{h}_{298})]_{O_2} + 70\ [\bar{h}_f^0 + (\bar{h}_T - \bar{h}_{298})]_{N_2}$$

$$= (-\ 249{,}952) + 18.7 \times 560 + 70 \times 540 = -\ 201{,}652 \text{ kJ/kg mol fuel}$$

$$H_P = 8[-\ 393{,}522 + 20288] + 9\ [-\ 241{,}827 + 16087] + 6.25\ [14171] + 70\ [13491]$$

$$= 8\ (-\ 373{,}234) + 9\ [-\ 225{,}740] + 6.25\ (14171) + 70\ [13491]$$

$$= -\ 3{,}884{,}622 \text{ kJ/kg mol fuel}$$

Energy output from the engine, $W_{C.V.} = 150$ kW

$$Q_{C.V} = -\ 205 \text{ kW}$$

Let n kg mol of fuel be consumed per second.

By the first law

$$n(H_R - H_P) = W_{C.V.} - Q_{C.V.}$$

$$n\ [-\ 201{,}652 + 3{,}884{,}662] = 150 - (-\ 205) = 355 \text{ kW}$$

$$\therefore \quad n = \frac{355 \times 3600}{3683010} \text{ kg mol/h} = 0.346 \text{ kg mol/h}$$

$\therefore$ Fuel consumption rate $= 0.346 \times 114 = 39.4$ kg/h *Ans.*

Example 16.10

Determine the adiabatic flame temperature when liquid octane at 25°C is burned with 300% theoretical air at 25°C in a steady flow process.

Solution The combustion equation with 300% theoretical air is given by

$$C_8H_{18}(l) + 3(12.5)O_2 + 3(12.5)\ (3.76)\ N_2 \rightarrow$$
$$\rightarrow 8CO_2 + 9H_2O + 25O_2 + 141N_2$$

By the first law

$$H_R = H_P$$

$$\sum_R n_i[\bar{h}_f^0 + \Delta\bar{h}]_i = \sum_P n_e[\bar{h}_f^0 + \Delta\bar{h}]_e$$

Now $H_R = (\bar{h}_{f^0})_{C_8H_{18}(l)} = -\ 249{,}952$ kJ/kg mol fuel

$$H_P = \sum_P n_e[\bar{h}_{f^0} + \Delta\bar{h}]_e = 8\,[-\ 393{,}522 + \Delta\,\bar{h}_{CO_2}]$$

$$+\ 9[-\ 241{,}827 + \Delta\bar{h}_{H_2O}] + 25\ \Delta\bar{h}_{O_2} + 141\ \Delta\bar{h}_{N_2}$$

The exit temperature, which is the adiabatic flame temperature, is to be computed by trial and error satisfying the above equation.

Let $T_e = 1000$ K, then

$$H_P = 8(-\ 393{,}522 + 33{,}405) + 9(-\ 241{,}827 + 25{,}978)$$

$$+\ 25(22.707) + 141(21{,}460)$$

$$= -\ 1226{,}577 \text{ kJ/kg mol}$$

If $T_e = 1200$ K

$$H_P = 8(-\ 393{,}522 + 44{,}484) + 9(-241{,}827 + 34{,}476)$$

$$+\ 25(29{,}765) + 141(28{,}108)$$

$$= +\ 46{,}537 \text{ kJ/kg mol}$$

If $T_e = 1100$ K,

$$H_P = 8(-\ 393{,}522 + 38{,}894) + 9(-\ 241{,}827 + 30{,}167)$$

$$+\ 25(26{,}217) + 141(24{,}757)$$

$$= -\ 595{,}964 \text{ kJ/kg mol}$$

Therefore, by interpolation, the adiabatic flame temperature is 1182 K *Ans.*

Example 16.11

(a) Propane (g) at 25°C and 100 kPa is burned with 400% theoretical air at 25°C and 100 kPa. Assume that the reaction occurs reversibly at 25°C, that the oxygen and nitrogen are separated before the reaction takes place (each at 100 kPa, 25°C), that the constituents in the products are separated, and that each is at 25°C, 100 kPa. Determine the reversible work for this process. (b) If the above reaction occurs adiabatically, and each constituent in the products is at 100 kPa pressure and at the adiabatic flame temperature, compute (a) the increase is entropy during combustion, (b) the irreversibility of the process, and (c) the availability of the products of combustion.

Solution The combustion equation (Fig. Ex. 16.11) is

$$C_3H_8(g) + 5(4)O_2 + 5(4)\,(3.76)N_2 \rightarrow$$

$$3CO_2 + 4H_2O(g) + 15O_2 + 75.2N_2$$

(a) $$W_{rev} = \sum_R n_i\bar{g}_i - \sum_P n_e\bar{g}_e$$

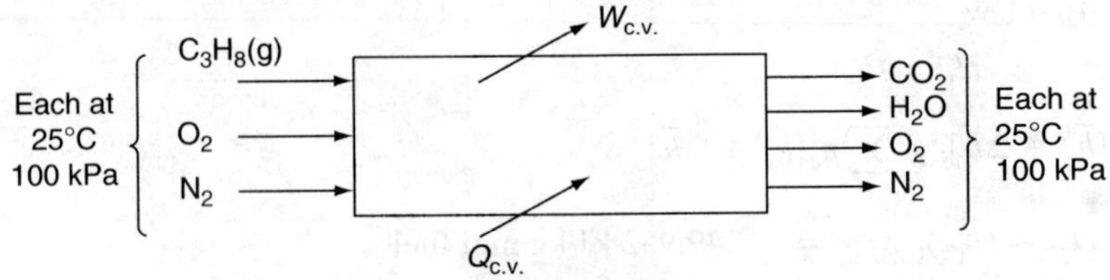

Fig. Ex. 16.11

From Table 16.6

$$W_{rev} = (\bar{g}_{f^0})_{C_3H_8(g)} - 3(\bar{g}_{f^0})_{CO_2} - 4(\bar{g}_{f^0})_{H_2O(g)}$$

$$= -23{,}316 - 3(-394{,}374) - 4(-228{,}583)$$

$$= 2{,}074{,}128 \text{ kJ/kg mol} \qquad \textit{Ans.}$$

$$= \frac{2{,}074{,}128}{44.097} = 47{,}035.6 \text{ kJ/kg} \qquad \textit{Ans.}$$

(b) $$H_R = H_P$$

$$(\bar{h}_{f^0})_{C_3H_8(g)} = 3(\bar{h}_{f^0} + \Delta\bar{h})_{CO_2} + 4(\bar{h}_{f^0} + \Delta\bar{h})_{H_2O(g)} + 15\,\Delta\bar{h}_{O_2} + 75.2\,\Delta\bar{h}_{N_2}$$

From Table 16.1

$$-103{,}847 = 3(-393{,}522 + \Delta\bar{h})_{CO_2} + 4(-241{,}827 + \Delta\bar{h})_{H_2O(g)} + 15\,\Delta\bar{h}_{O_2} + 75.2\,\Delta\bar{h}_{N_2}$$

Using Table C in the appendix, and by trial and error, the adiabatic flame temperature is found to be 980 K.

The entropy of the reactants

$$S_R = \sum_R \left(n_i \bar{s}_i^0\right)_{298} = \left(\bar{s}^0_{C_3H_8(g)} + 20\bar{s}^0_{O_2} + 75.2\bar{s}^0_{N_2}\right)_{298}$$

$$= 270.019 + 20\,(205.142) + 75.2\,(191.611)$$

$$= 18{,}782.01 \text{ kJ/kg mol K}$$

The entropy of the products

$$S_P = \sum_P \left(n_e \bar{s}_e^0\right)_{980}$$

$$= \left(3\bar{s}^0_{CO_2} + 4\bar{s}^0_{H_2O(g)} + 15\bar{s}^0_{O_2} + 75.2\bar{s}^0_{N_2}\right)_{980}$$

$$= 3(268.194) + 4(231.849) + 15(242.855) + 75.2(227.485)$$

$$= 22{,}481.68 \text{ kJ/kg mol K}$$

∴ The increase in entropy during combustion

$$S_P - S_R = 3{,}699.67 \text{ kJ/kgmol K} \qquad \textit{Ans.}$$

The irreversibility of the process

$$I = T_0\left[\sum_{P} n_e \bar{s}_e - \sum_{R} n_i \bar{s}_i\right]$$

$$= 298 \times 3699.67 = 1{,}102{,}501.66 \text{ kJ/kgmol}$$

$$= \frac{1{,}102{,}501.66}{44.097} = 25{,}001 \text{ kJ/kg}$$

The availability of combustion products

$$\psi = W_{rev} - I$$

$$= 47{,}035.6 - 25{,}001$$

$$= 22{,}034.6 \text{ kJ/kg}$$ *Ans.*

Example 16.12

Determine the chemical exergy of (a) carbon, (b) hydrogen, (c) methane, (d) carbon monoxide, (e) liquid methanol, (f) nitrogen, (g) oxygen, (h) carbon dioxide and (i) water in kJ/kg, in respect of the environment in which the gas phase obeys the ideal gas model.

Environment:

$T_0 = 298.15$ K, $p_0 = 1$ atm, N_2 75.67%, O_2 20.35%, $H_2O(g)$ 3.12%, CO_2 0.03% and others 0.83%, by volume.

Solution (a) Carbon: $C + O_2 \rightarrow CO_2$

$$\bar{a}_{ch} = \left[\bar{g}_c + \bar{g}_{O_2} - \bar{g}_{CO_2}\right]_{T_0,\, p_0} + \bar{R}T_0 \ln \frac{x_{O_2}}{x_{CO_2}}$$

$$= 0 + 0 - (-394{,}380) + 8.3143 \times 298.15 \ln \frac{0.2035}{0.0003}$$

$$= 410{,}541 \text{ kJ/k mol} = 34{,}212 \text{ kJ/kg}$$

(b) Hydrogen: $H_2 + \frac{1}{2} O_2 \rightarrow H_2O$

$$\bar{a}_{ch} = \left[\bar{g}_{H_2} + \frac{1}{2}\bar{g}_{O_2} - \bar{g}_{H_2O}\right]_{p_0,\, T_0} + \bar{R}T_0 \ln \frac{(x_{O_2})^{1/2}}{x_{H_2O(g)}}$$

$$= 0 + 0 - (-228590) + 8.3143 \times 298.15 \ln \frac{(0.2035)^{1/2}}{0.0312}$$

$$= 235{,}212 \text{ kJ/kg mol} = 116{,}557 \text{ kJ/kg.}$$ *Ans.*

(c) Methane: $CH_4 + 2O_2 \rightarrow CO_2 + 2H_2O$

$$\bar{a}_{ch} = \left[\bar{g}_{CH_4} + 2\bar{g}_{O_2} - \bar{g}_{CO_2} - 2\bar{g}_{H_2O(g)}\right]$$

$$+ \bar{R}T_0 \ln \frac{x_{O_2}^2}{(x_{CO_2})^1 (x_{H_2O})^2}$$

$$= -50{,}790 + 0 - (-394{,}380) - 2(-228{,}590)$$

$$+ 8.3143 \times 298.15 \ln \frac{(0.2035)^2}{(0.0003)(0.0312)^2}$$

$$= 830{,}174 \text{ kJ/kg mol} = 51{,}756 \text{ kJ/kg}$$ *Ans.*

(d) Carbon monoxide: $CO + \frac{1}{2} O_2 \rightarrow CO_2$

$$\bar{a}_{ch} = \left[\bar{g}_{CO} + \frac{1}{2}\bar{g}_{O_2} - \bar{g}_{CO_2}\right] - \bar{R}T_0 \ln \frac{x_{O_2}^{1/2}}{x_{CO_2}}$$

$$= -137{,}150 + 0 - (-394{,}380) + 8.3143 \times 298.15 \ln \frac{(0.2035)^{1/2}}{0.0003}$$

$$= 275{,}364 \text{ kJ/kg mol} = 9831 \text{ kJ/kg}$$ *Ans.*

(e) Liquid methanol: $CH_3OH(l) + 1.5O_2 \rightarrow CO_2 + 2H_2O$

$$\bar{a}_{ch} = \left[\bar{g}_{CH_3OH(l)} + 1.5\bar{g}_{O_2} - \bar{g}_{CO_2} - 2\bar{g}_{H_2O(g)}\right] + \bar{R}T_0 \ln \frac{x_{O_2}^{1.5}}{(x_{CO_2})(x_{H_2O})^2}$$

$$= -166{,}240 + 0 - (-394{,}380) - 2(-228{,}590)$$

$$+ 8.3143 \times 298.15 \ln \frac{(0.2035)^{1.5}}{(0.0003)(0.0312)^2}$$

$$= 716{,}647 \text{ kJ/kg mol} = 22{,}360 \text{ kJ/kg}$$ *Ans.*

(f) Nitrogen: $N_2 \rightarrow N_2$

$$\bar{a}_{ch} = \bar{R}T_0 \ln \frac{1}{x_{N_2}} = 8.3143 \times 298.15 \ln \frac{1}{0.7567}$$

$$= 691.1 \text{ kJ/k mol} = 24.7 \text{ kJ/kg}$$ *Ans.*

(g) Oxygen: $O_2 \rightarrow O_2$

$$\bar{a}_{ch} = \bar{R}T_0 \ln \frac{1}{x_{O_2}} = 8.3143 \times 298.15 \ln \frac{1}{0.2035}$$

$$= 3947 \text{ kJ/k mol} = 123.3 \text{ kJ/kg}$$ *Ans.*

(h) Carbon dioxide:

$$\bar{a}_{ch} = \bar{R}T_0 \ln \frac{1}{x_{CO_2}} = 8.3143 \times 298.15 \ln \frac{1}{0.0003}$$

$$= 20{,}108 \text{ kJ/kg mol} = 456.9 \text{ kJ/kg}$$ *Ans.*

(i) Water: $H_2O(l) \rightarrow H_2O(g)$

$$\bar{a}_{ch} = \bar{g}_{H_2O(l)} - \bar{g}_{H_2O(g)} + \bar{R}T_0 \ln \frac{1}{x_{H_2O}}$$

$$= -237{,}180 - (-228{,}590) + 8.3143 \times 298.15 \ln \frac{1}{0.0312}$$

$$= 4.9 \text{ kJ/kg mol} = 0.272 \text{ kJ/kg}$$ *Ans.*

Example 16.13

Liquid octane (C_8H_{18}) at 25°C, 1 atm and a mass flow rate of 0.57 kg/h enters a small internal combustion engine operating at steady state. The fuel burns with air entering the engine in a separate stream at 25°C, 1 atm. Combustion products exit at 670 K, 1 atm with a dry molar analysis of 11.4% CO_2, 2.9% CO, 1.6% O_2 and 84.1% N_2. If the engine develops power at the rate of 1 kW, determine (a) the rate of heat transfer from the engine in kW, (b) the second law efficiency of the engine. Use the environment as given below and neglect KE and PE effects.

Environment:

1 atm, 298.15 K and N_2 75.67%, O_2 20.25%, H_2O(g) 3.12%, CO_2 0.03% and other 0.83%, by volume.

Solution

$$C_8H_{18}(1) + a(O_2 + 3.76N_2) \rightarrow b\,[0.114CO_2 + 0.029\,CO + 0.016O_2 + 0.841N_2] + CH_2O$$

By carbon balance,

$$b(0.114 = 0.029) = 8$$

$$b = 55.9$$

By hydrogen balance, $18 = 2C \quad C = 9$

By oxygen balance, $a = b \times 0.114 + \frac{b}{2} \times 0.029 + b \times 0.016 + \frac{C}{2}$

$$= b\,(0.114 + 0.0145 + 0.016) + \frac{9}{2} = 12.58$$

The combustion equation becomes,

$$C_8H_{18}(1) + 12.58(O_2 + 3.76N_2) \rightarrow 55.9[0.114\,CO_2 + 0.029\,CO + 0.016\,O_2 + 0.841N_2] + 9H_2O$$

By first law, $Q + H_R = W + H_P$

$$\frac{Q_{CV}}{\dot{n}_{fuel}} = \frac{\dot{W}_{CV}}{\dot{n}_{fuel}} - (-249{,}910) + 55.9\,[0.114(-393{,}520 + 25284) + 0.029$$

$$(-110{,}530 + 11{,}089) + 0.016(1151.5) + 0.841(11016) + 9(-241{,}820 + 13064)$$

$$= \frac{\dot{W}_{CV}}{\dot{n}_{fuel}} - 3{,}845{,}872 \text{ kJ/kg mol } C_8H_{18}(1)$$

$$\dot{Q}_{CV} = \dot{W}_{CV} - \dot{n}_{fuel} \times 3{,}845{,}872 \text{ kW}$$

$$= 1 - 3{,}845{,}872\,\frac{\text{kJ}}{\text{kg mol}} \times \frac{0.57 \text{ kg/h}}{3600 \text{ s/h}} \times \frac{1 \text{ kg mol}}{114.22 \text{ kg}}$$

$$= 1 - 5.33 = -4.33 \text{ kW}$$

Ans. (a)

(b) Chemical exergy of C_8H_{18} was found to be 5,407,843 kJ/kg mol (Art. 16.19).

$$\eta_{II} = \frac{\dot{W}_{CV}}{\dot{n}_{fuel} \times \bar{a}_{ch}}$$

$$= \frac{1 \text{ kW}}{5407843\,\frac{\text{kJ}}{\text{kg mol}} \times \frac{0.57 \text{ kg}}{3600 \text{ s}} \times \frac{1 \text{ kg mol}}{114.22 \text{ kg}}}$$

$$= \frac{1}{7.496} = 0.133 \text{ or } 13.3\%$$

Ans. (b)

Review Questions

16.1 What are stoichiometric coefficients?

16.2 What is the degree of reaction? What are its limiting values?

16.3 When does a chemical reaction reach equilibrium?

16.4 What do you understand by the equilibrium constant of a chemical reaction?

16.5 What is the law of mass action?

16.6 What is the heat of reaction? When is it positive and when negative?

16.7 What is the van't Hoff equation?

16.8 Give Nernst's equation.

16.9 What is thermal ionization?

16.10 What is the significance of Saha's equation?

16.11 What do you understand by the standard Gibbs function change?

16.12 What is chemical affinity?

16.13 Define fugacity and activity.

16.14 How can the heat capacity of reacting gases in equilibrium be estimated?

16.15 What is stoichiometric air?

16.16 What do you understand by the enthalpy of formation?

16.17 Define adiabatic flame temperature. How is it estimated?

16.18 What is enthalpy of combustion? What is internal energy of combustion?

16.19 What do you understand by higher heating value and lower heating value of a fuel?

16.20 State Planck's formulation of the third law of thermodynamics.

16.21 What is the Gibbs function of formation?

Problems

16.1 Starting with n_0 moles of NO, which dissociates according to the equation

$$NO \rightleftharpoons \frac{1}{2}N_2 + \frac{1}{2}O_2$$

show that at equilibrium

$$K = \frac{1}{2}\cdot\frac{\varepsilon_e}{1-\varepsilon_e}$$

16.2 A mixture of $n_0\,\nu_1$ moles of A_1 and $n_0\,\nu_2$ moles of A_2 at temprature T and pressure p occupies a volume V_0. When the reaction

$$\nu_1 A_1 + \nu_2 A_2 \rightleftharpoons \nu_3 A_3 + \nu_4 A_4$$

has come to equilibrium at the same T and p, the volume is V_e. Show that

$$\varepsilon_e = \frac{V_e - V_0}{V_0}\cdot\frac{\nu_1+\nu_2}{\nu_3+\nu_4-\nu_1-\nu_2}$$

16.3 The equilibrium constant of the reaction

$SO_3 \rightleftharpoons SO_2 + \frac{1}{2}O_2$ has the following values

T	800 K	900 K	1000 K	1105 K
K	0.0319	0.153	0.540	1.59

Determine the average heat of dissociation graphically. *Ans.* 94 MJ/k mol

16.4 At high temperature the potassium atom is ionized according to the equation $K \rightleftharpoons K^+ + e^-$. The values of the equilibrium constant at 3000 K and 3500 K are 8.33×10^{-6} and 1.33×10^{-4} respectively. Compute the average heat of reaction in the given temperature range. *Ans.* 483,819 J/g mol

16.5 (a) Show that $\Delta G = \Delta H + T\left(\frac{\partial \Delta G}{\partial T}\right)_p$

(b) Show that $\Delta G = -\bar{R}T\ln\frac{x_3^{\nu_3}\cdot x_4^{\nu_4}}{x_1^{\nu_1}\cdot x_2^{\nu_2}}$

where the x's are equilibrium values.

16.6 When 1 kg mol of HI dissociates according to the reaction

$$HI \rightleftharpoons \frac{1}{2}H_2 + \frac{1}{2}I_2$$

at $T = 675$ K, $K = 0.0174$, and $\Delta H = 5910$ kJ/kg mol.

Calculate $(\partial\varepsilon_e/\partial T)_p$ at this temperature. *Ans.* $0.505 \times 10^{-4}\ K^{-1}$

16.7 Calculate the degree of ionization of cesium vapour at 10^{-6} atm at the two temperatures of 2260 and 2520 K. *Ans.* 0.393, 0.808

16.8 In the case of ionization of a monatomic gas, show that:

(a) $$\frac{G - G(\text{mia})}{\overline{R}T} = \varepsilon\left[\ln\frac{\varepsilon^2}{1-\varepsilon^2} - \ln\frac{\varepsilon_e^2}{1-\varepsilon_e^2}\right] + \ln\frac{1-\varepsilon}{1+\varepsilon} - \ln\frac{1-\varepsilon_e}{1+\varepsilon_e}$$

(b) At $\varepsilon = 0$, $$\frac{G_0 - G\,(\text{main})}{\overline{R}T} = -\ln\frac{1-\varepsilon_e}{1+\varepsilon_e}$$

(c) At $\varepsilon = 1$, $$\frac{G_1 - G\,(\text{main})}{\overline{R}T} = \ln\frac{1}{4} - \ln\frac{\varepsilon_e^2}{(1+\varepsilon_e)^2}$$

(d) Plot $\left[G - G(\min)\right]/(230\overline{R}T)$ against ε for the ionization of cesium vapour at 2260 K and 10^{-6} atm.

16.9 Prove that, for a mixture of reacting ideal gases in equilibrium

(a) $$\left(\frac{\partial V}{\partial T}\right)_T = -\frac{V}{T} - \frac{(n_0 + n_0')\overline{R}T(\Delta\nu)^2}{p^2 \cdot \dfrac{d}{d\varepsilon_e}\ln\dfrac{x_3^{\nu_3}\cdot x_4^{\nu_4}}{x_1^{\nu_1}\cdot x_2^{\nu_2}}}$$

(b) $$\left(\frac{\partial V}{\partial T}\right)_p = \frac{V}{T} + \frac{(n_0 + n_0')\Delta\nu\Delta H}{pT\dfrac{d}{d\varepsilon_e}\ln\dfrac{x_3^{\nu_3}\cdot x_4^{\nu_4}}{x_1^{\nu_1}\cdot x_2^{\nu_2}}}$$

(c) $$\left(\frac{\partial p}{\partial T}\right)_{\varepsilon_e} = \frac{p\Delta H}{\overline{R}T^2\Delta\nu}$$

16.10 Prove that, for a mixture of reacting ideal gases in equilibrium,

$$dS = \sum n_k\left[\sum x_k c_{pk} + \frac{\psi(\Delta H)^2}{\overline{R}T^2}\right]\frac{dT}{T} - \overline{R}\sum n_k\left[1 + \frac{\psi\Delta H\Delta\gamma}{\overline{R}T^2}\right]\frac{dp}{p}$$

16.11 Oxygen dissociates according to the relation $O_2 \rightarrow 2O$. Show that the equilibrium constant is given by

$$K = \frac{4\varepsilon_e^2 p}{1-\varepsilon_e^2}$$

where ε_e is the degree of dissociation. Obtain an expression for the rate of variation of ε_e with temperature, $\left(\frac{\partial \varepsilon_e}{\partial T}\right)_p$, in terms of ε_e, $\frac{dK}{dT}$ and the total pressure p. At a temperature of 3800 K, $K = 1$ atm and $\frac{\Delta H}{\overline{R}} = 59$ K, calculate the value of ε and $\left(\frac{\partial \varepsilon_e}{\partial T}\right)_p$ at a pressure of 1 atm.

16.12 (a) Starting with n_0 moles of water vapour which dissociates according to the equation $H_2O \rightleftharpoons H_2 + \frac{1}{2}O_2$, show that at equilibrium

$$K = \frac{\varepsilon_e^{3/2}}{(2+\varepsilon_e)^{1/2}(1-\varepsilon_e)} \cdot p^{1/2}$$

(b) At an average temperature of 1900 K, the slope of the graph of log K against $1/T$ for the dissociation of water vapour is found to be −13,000. Determine the heat of dissociation. Is it exothermic or endothermic?

Ans. 248,885 kJ/kg mol, endothermic

16.13 The volumetric composition of the 'dry' products of combustion of an unknown hydrocarbon fuel, C_xH_y, gives: CO_2 12.1%, O_2 3.8%, CO 0.9% and N_2 83.2%. Determine (a) the chemical formula of the fuel, (b) the air fuel ratio, and (c) the percentage of excess air used.

Ans. (a) $C_{13}H_{23}$, (b) 17.03, (c) 18%

16.14 Liquid octane is burned with air in a combustion test, and the dry volumetric analysis of the products reveals the following composition: CO_2 11%, O_2 3.6%, CO 1.5% and N_2 83.9%. Compute the actual air-fuel ratio used during the test and the percentage excess air. *Ans.* 17.267, 14.19%

16.15 Propane (C_3H_8) is reacted with air in such a ratio that an analysis of the products of combustion gives CO_2 11.5%, O_2 2.7% and CO 0.7%. What is the percent theoretical air used during the test?

Ans. 111%

16.16 Carbon monoxide and 300% theoretical air enter a steady flow combustor at 400 K and a low pressure. The energy released by the reaction heats the product gases to 1400 K. If the combustion is complete, estimate the heat gained or lost through the walls of the combustor.

Ans. −17,360 kJ/kg mol

16.17 Liquid octane enters the combustion chamber of a gas turbine at 25°C and air enters from the compressor at 227°C. It is determined that 98% of the carbon in the fuel burns to form CO_2 and the remaining 2% burns to form CO. The temperature of the products is limited to 827°C. Estimate the air-fuel ratio used and the percentage excess air.

Ans. 66,338%

16.18 A mixture of methane and oxygen, in the proper ratio for complete combustion and at 25°C and 1 atm, reacts in a constant volume calorimeter bomb. Heat is transferred until the products of combustion are at 400 K. Determine the heat transfer per mole of methane.

Ans. –794414 kJ/kg mol

16.19 Liquid hydrazine (N_2H_4) and oxygen gas, both at 25°C, 0.1 MPa are fed to a rocket combustion chamber in the ratio of 0.5 kg O_2/kg N_2H_4. The heat transfer from the chamber to the surroundings is estimated to be 100 kJ/kg N_2H_4. Determine the temperature of the products, assuming only H_2O, H_2, and N_2 to be present. The enthalpy of the formation of N_2H_4 (l) is + 50,417 kJ/kg mol.

Ans. 2855 K

If saturated liquid oxygen at 90 K is used instead of 25°C oxygen gas in the combustion process, what will the temperature of the products be?

16.20 Liquid ethanol (C_2H_5OH) is burned with 150% theoretical oxygen in a steady flow process. The reactants enter the combustion chamber at 25°C, and the products are cooled and leave at 65°C, 0.1 MPa. Calculate the heat transfer per kg mol of ethanol. The enthalpy of formation of C_2H_5OH (l) is –277,634 kJ/kg mol.

16.21 A small gas turbine uses C_8H_{18}(l) for fuel and 400% theoretical air. The air and fuel enter at 25°C and the combustion products leave at 900 K. If the speicific fuel consumption is 0.25 kg/s per MW output, determine the heat transfer from the engine per kg mol of fuel, assuming complete combustion.

Ans. –48.830 kJ/kg mol.

16.22 Hydrogen peroxide (H_2O_2) enters a gas generator at the rate of 0.1 kg/s, and is decomposed to steam and oxygen. The resulting mixture is expanded through a turbine to atmospheric pressure, as shown in Fig. P. 16.22. Determine the power output of the turbine and heat transfer rate in the gas generator. The enthalpy of formation of H_2O_2(l) is –187,583 kJ/kg mol.

Ans. 38.66 kW, –83.3 kW

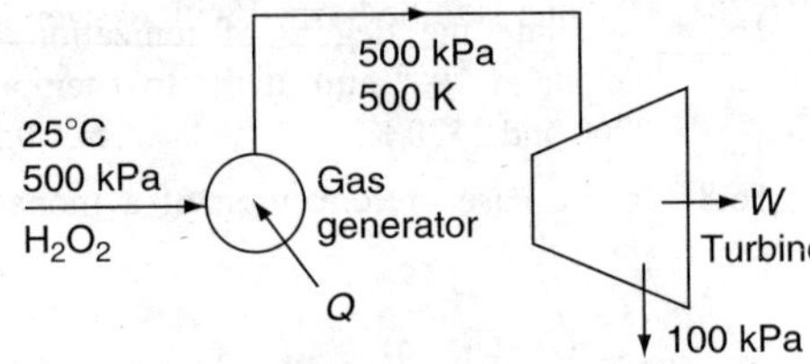

Fig. P. 16.22

16.23 An internal combustion engine burns liquid octane and uses 150% theoretical air. The air and fuel enter at 25°C, and the products leave the engine exhaust ports at 900 K. In the engine 80% of the carbon burns to CO_2 and the remainder burns to CO. The heat transfer from this engine is just equal to the work done by the engine. Determine (a) the power output of the engine if the engine burns 0.006 kg/s of fuel, and (b) the composition and the dew point of the product of combustion.

16.24 Gaseous butane at 25°C is mixed with air at 400 K and burned with 400% theoretical air. Determine the adiabatic flame temperature.

16.25 A mixture of butane and 150% theoretical air enters a combustion chamber of 25°C, 150 kPa, and the products of combustion leave at 1000 K, 150 kPa. Determine the heat transfer from the combustion chamber and the irreversibility for the process.

16.26 The following data are taken from the test of a gas turbine:

Fuel—C_4H_{10}(g) 25°C, 0.1 MPa
Air—300% theoretical air at 25°C, 0.1 MPa
Velocity of inlet air—70 m/s
Velocity of products at exit—700 m/s
Temperature and pressure of products—900 K, 0.1 MPa

Assuming complete combustion, determine (a) the net heat transfer per kg mol of fuel, (b) the net increase of entropy per kg mol of fuel, and (c) the irreversibility for the process.

16.27 Calculate the equilibrium composition if argon gas is heated in an arc to 10,000 K, 1 kPa, assuming the plasma to consist of A, A^+, e^-. The equilibrium constant for the reaction

$$A \rightleftharpoons A^+ + e^-$$

at this temperature is 0.00042.

16.28 Methane is to be burned with oxygen in an adiabatic steady-flow process. Both enter the combustion chambar at 298 K, and the products leave at 70 kPa. What percent of excess O_2 should be used if the flame temperature is to be 2800 K? Assume that some of the CO_2 formed dissociates to CO and O_2, such that the products leaving the chamber consist of CO_2, CO, H_2O and O_2 at equilibrium.

16.29 (a) Octane burns with the stoichiometric amount of air. Determine the air fuel ratio and the partial pressure of CO_2 in the dry products of combustion when the total pressure is 1 atm. (b) If 25% excess air is used in burning octane and combustion is complete determine the dry volumetric analysis of the products of combustion and the dew point temperature of the products.
Ans. (a) 15.11, 0.1455 atm, (b) CO_2 11.45%, O_2 4.47%, N_2 84.0 8; 47.356°C

16.30 Methane gas at 25°C, 1 atm and a volumetric flow rate of 27 m^3/h enters a furnace operating at steady state. The methane burns completely with 140% theoretical air, entering at 127°C at 1 atm. Products of combustion leave at 427°C, 1 atm. Determine (a) the volumetric flow rate of air in m^3/h, (b) the rate of heat transfer from the furnace in kW. Neglect KE and PE changes.
Ans. (a) 483.03 m^3/h, (b) −202.5 kW

16.31 Octane gas at 25°C, 1 atm enters a jet engine and burns completely with 300% theoretical air at 25°C, 1 atm. Products of combustion leave at 990 K, 1 atm. If the fuel and air enter with relatively low air velocities, determine the velocity of the combustion products at exit. Neglect PE effects and heat transfer between the engine and surroundings.
Ans. 641.84 m/s

16.32 Liquid octane (C_8H_{18}) at 25°C, 1 atm enters the combustior of a simple open gas turbine power plant and burns completely with 400% theoretical air entering the compressor at 25°C, 1 atm. Products of combustion exit the turbine at 627°C, 1 atm. If the rate of heat transfer from the gas turbine is estimated as 15% of the net power developed, determine the net power developed in kJ per kg mol of fuel. Kinetic and potential energy effects are negligible. *Ans.* 439,750 kJ/kg mol

16.33 Gaseous ethanol (C_6H_5OH) at 25°C, 1 atm enters a reactor operating at steady state and burns completely with 130% theoretical air entering at 25°C, 1 atm. Products of combustion exit at 127°C, 1 atm. If the rate of heat transfer from the reactor is 900 kW, determine the mass flow rate of the fuel, in kg/s, neglecting the KE and PE effects.
Ans. 0.034 2 kg/s

16.34 Carbon monoxide at 25°C, 1 atm enters an insulated reactor operating at steady state and reacts completely with the theoretical amount of air entering in a separate steam at 25°C, 1 atm. The products exit as mixture at 1 atm. Determine in kJ/kg mol of carbon (a) the availability entering with CO, (b) the availability exiting with the products, (c) the irreversibility rate and (d) the second law efficiency. Assume the environment as given in Example 16.13.
Ans. (a) 275,364 kJ/k mol CO, (b) 226,105 kJ/k mol CO, (c) 49,259 kJ/mol CO, (d) 82.1%

16.35 Liquid octane (C_8H_{18}) at 25°C, 1 atm enters an adiabatic reactor operated at steady state and burns completely with air entering at 227°C, 1 atm. If the combustion products exit at 1127°C, determine the percent excess air used, neglecting KE and PE effects. *Ans.* 168.5%

C H A P T E R 17

Compressible Fluid Flow

A fluid is defined as a substance which continuously deforms under the action of shearing forces. Liquids and gases are termed as fluids. A fluid is said to be *incompressible* if its density (or specific volume) does not change (or changes very little) with a change in pressure (or temperature or velocity). Liquids are incompressible. A fluid is said to be *compressible* if its density changes with a change in pressure or temperature or velocity. Gases are compressible. The effect of compressibility must be considered in flow problems of gases. Thermodynamics is an essential tool in studying compressible flows, because of which Theodore von Karman suggested the name 'Aerothermodynamics' for the subject which studies the dynamics of compressible fluids.

The basic principles in compressible flow are:

(a) Conservation of mass (continuity equation)
(b) Newton's second law of motion (momentum principle)
(c) Conservation of energy (first law of thermodynamics)
(d) Second law of thermodynamics (entropy principle)
(e) Equation of state.

For the first two principles, the student is advised to consult a book on fluid mechanics, and the last three principles have been discussed in the earlier chapters of this book.

17.1 VELOCITY OF PRESSURE PULSE IN A FLUID

Let us consider an infinitesimal pressure wave initiated by a slight movement of a piston to the right (Fig. 17.1) in a pipe of uniform cross-section. The pressure wave front propagates steadily with a velocity c, which is known as the velocity of sound, sonic velocity or acoustic velocity. The fluid near the piston will have a slightly increased pressure and will be slightly more dense, than the fluid away from the piston.

To simplify the analysis, let the observer be assumed to travel with the wave front to the right with the velocity c. Fluid flows steadily from right to left and as it passes through the wave front, the velocity is reduced from c to c-dV. At the same time, the pressure rises from p to $p + \mathrm{d}p$ and the density from ρ to $\rho + \mathrm{d}\rho$.

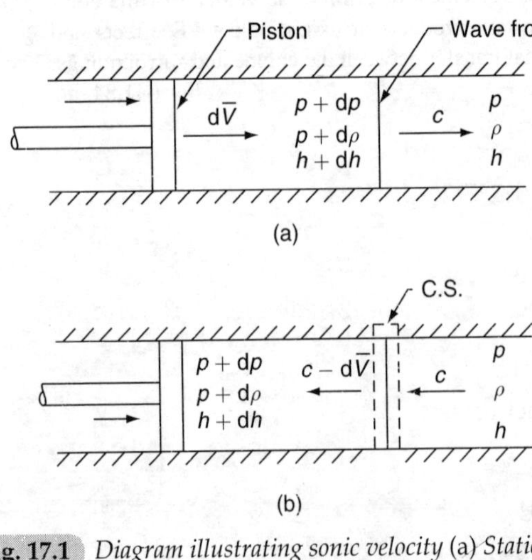

Fig. 17.1 *Diagram illustrating sonic velocity* (a) *Stationary observer* (b) *Observer travelling with the wave front*

The continuity equation for the control volume gives

$$\rho A c = (\rho + \mathrm{d}\rho)\, A(c - \mathrm{dV})$$

$$\rho c = \rho c - \rho\, \mathrm{dV} + c\mathrm{d}\rho - \mathrm{d}\rho \cdot \mathrm{dV}$$

Neglecting the product $\mathrm{d}\rho \cdot \mathrm{dV}$, both being very small

$$\rho\, \mathrm{dV} = c\mathrm{d}\rho \tag{17.1}$$

The momentum equation for the control volume gives

$$[p - (p + \mathrm{d}p)]\, A = w\, [(c - \mathrm{dV}) - c]$$

$$-\, \mathrm{d}p\, A = \rho A c\, (c - \mathrm{dV} - c)$$

$$\mathrm{d}p = \rho\, c\mathrm{dV} \tag{17.2}$$

From Eqs (17.1) and (17.2)

$$\frac{\mathrm{d}p}{c} = c\mathrm{d}\rho$$

$\therefore$

$$c = \sqrt{\frac{\mathrm{d}p}{\mathrm{d}\rho}}$$

Since the variations in pressure and temperature are negligibly small and the change of state is so fast as to be essentially adiabatic, and in the absence of any internal friction or viscosity, the process is reversible and isentropic. Hence, the sonic velocity is given by

$$c = \sqrt{\left(\frac{\partial p}{\partial \rho}\right)_s} \tag{17.3}$$

No fluid is truly incompressible, although liquids show little change in density. The velocity of sound in common liquids is of the order of 1650 m/s.

17.1.1 Velocity of Sound in an Ideal Gas

For an ideal gas, in an isentropic process

$$pv^{\gamma} = \text{constant} \quad \text{or} \quad \frac{p}{\rho^{\gamma}} = \text{constant}$$

By logarithmic differentiation (i.e., first taking logarithm and then differentiating)

$$\frac{\mathrm{d}p}{p} - \gamma \frac{\mathrm{d}\rho}{\rho} = 0$$

$\therefore$

$$\frac{\mathrm{d}p}{\mathrm{d}\rho} = \gamma \frac{p}{\rho}$$

Since

$$c^2 = \frac{\mathrm{d}p}{\mathrm{d}\rho} \quad \text{and} \quad p = \rho\, RT$$

$$c^2 = \gamma\, RT$$

or

$$c = \sqrt{\gamma RT} \tag{17.4}$$

where

$$R = \text{characteristic gas constant} = \frac{\text{Universal gas constant}}{\text{Molecular weight}}$$

The lower the molecular weight of the fluid and higher the value of γ, the higher is the sonic velocity at the same temperature. c is a thermodynamic property of the fluid.

17.1.2 Mach Number

The Mach number, M, is defined as the ratio of the actual velocity V to the sonic velocity c.

$$\mathrm{M} = \frac{\mathrm{V}}{c}$$

When $M > 1$, the flow is supersonic, when $M < 1$, the flow is subsonic, and when $M = 1$, the flow is sonic.

17.2 STAGNATION PROPERTIES

The isentropic *stagnation state* is defined as the state a fluid in motion would reach if it were brought to rest isentropically in a steady-flow, adiabatic, zero work output device. This is a reference state in a compressible fluid flow and is commonly designated with the subscript zero. The stagnation enthalpy h_0 (Fig. 17.2) is related to the enthalpy and velocity of the moving fluid by

$$h_0 = h + \frac{V^2}{2} \tag{17.5}$$

For an ideal gas, $h = h(T)$ and c_p is constant. Therefore

$$h_0 - h = c_p (T_0 - T) \tag{17.6}$$

From Eqs (17.5) and (17.6)

$$c_p (T_0 - T) = \frac{V^2}{2}$$

$$\frac{T_0}{T} = 1 + \frac{V^2}{2c_p T}$$

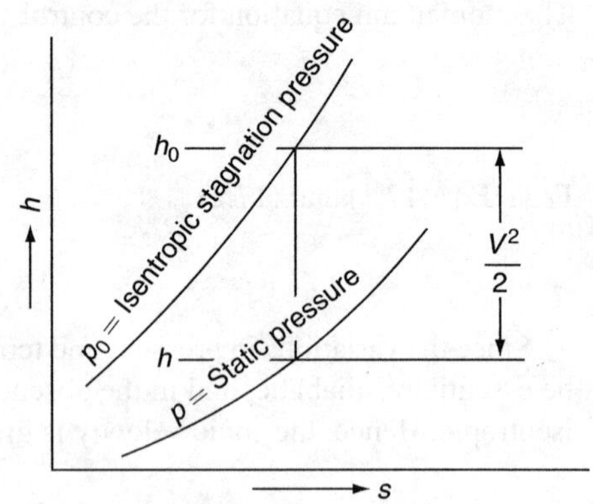

Fig. 17.2 *Stagnation state*

The properties without any subscript denote static properties.

Since
$$c_p = \frac{\gamma R}{\gamma - 1}$$

$$\frac{T_0}{T} = 1 + \frac{V^2(\gamma - 1)}{2\gamma RT}$$

Using Eq. (17.4) and the Mach number

$$\frac{T_0}{T} = 1 + \frac{\gamma - 1}{2} M^2 \tag{17.7}$$

The stagnation pressure p_0 is related to the Mach number and static pressure in the case of an ideal gas by the following equation

$$\frac{p_0}{p} = \left(\frac{T_0}{T}\right)^{\gamma/(\gamma-1)} = \left(1 + \frac{\gamma - 1}{2} M^2\right)^{\gamma/(\gamma-1)} \tag{17.8}$$

Similarly,
$$\frac{\rho_0}{\rho} = \left(1 + \frac{\gamma - 1}{2} M^2\right)^{1/(\gamma-1)} \tag{17.8a}$$

17.3 ONE-DIMENSIONAL STEADY ISENTROPIC FLOW

From a one dimensional point of view, the three most common factors which tend to produce continuous changes in the state of a flowing stream are

(a) Changes in cross-sectional area
(b) Wall friction
(c) Energy effects, such as external heat exchange, combustion, etc.

A study will first be made of the effects of area change in the absence of friction and energy effects. The process, which has been called isentropic flow, might aptly be termed as simple area change.

A nozzle is any duct which increases the kinetic energy of a fluid at the expense of its pressure. A diffuser is a passage through which a fluid loses kinetic energy and gains pressure. The same duct or passage may be either a nozzle or diffuser depending upon the end conditions across it. A nozzle or diffuser with both a converging and a diverging section is shown in Fig. 17.3. The minimum section is known as the throat. For the control volume shown in Fig. 17.3, since stagnation enthalpy and stagnation temperature do not change in adiabatic flow

$$h_0 = h + \frac{V^2}{2}$$

$$dh = -V\,dV \qquad (17.9)$$

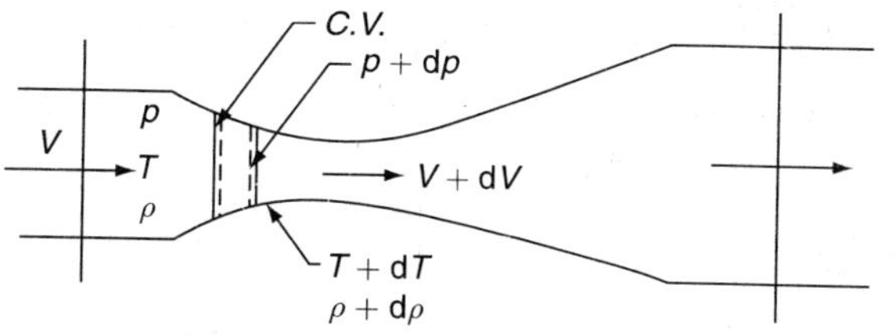

Fig. 17.3 *Reversible adiabatic flow through a nozzle*

From the property relation $\quad Tds = dh - v\,dp$

For isentropic flow

$$dh = \frac{dp}{\rho} \qquad (17.10)$$

From Eqs (17.9) and (17.10)

$$dp = -\rho V\,dV \qquad (17.11)$$

or

$$\frac{dp}{dV} < 0 \qquad (17.12)$$

As pressure decreases, velocity increases, and vice versa.

The continuity equation gives

$$w = \rho A V$$

By logarithmic differentiation

$$\frac{d\rho}{\rho} + \frac{dA}{A} + \frac{dV}{V} = 0$$

∴

$$\frac{dA}{A} = \frac{dV}{V} - \frac{d\rho}{\rho}$$

Substituting from Eq. (17.11)

$$\frac{dA}{A} = \frac{dp}{\rho V^2} - \frac{d\rho}{\rho} = \frac{dp}{\rho V^2}\left[1 - V^2\frac{d\rho}{dp}\right]$$

or

$$\frac{dA}{A} = \frac{dp}{\rho V^2}(1 - M^2) \qquad (17.13)$$

Also

$$\frac{dA}{A} = (M^2 - 1)\frac{dV}{V} \qquad (17.14)$$

When $M < 1$, i.e., the inlet velocity is subsonic, as now area A decreases, the pressure decreases and velocity increases, and when flow area A increases, pressure increases and velocity decreases. So for subsonic flow, a convergent passage becomes a nozzle (Fig. 17.4a) and a divergent passage becomes a diffuser (Fig. 17.4b).

When $M > 1$, i.e., when the inlet velocity is supersonic, as now area A decreases, pressure increases and velocity decreases, and as flow area A increases, pressure decreases and velocity increases. So for supersonic flow, a convergent passage is a diffuser (Fig. 17.4c) and a divergent passage is a nozzle (Fig. 17.4d).

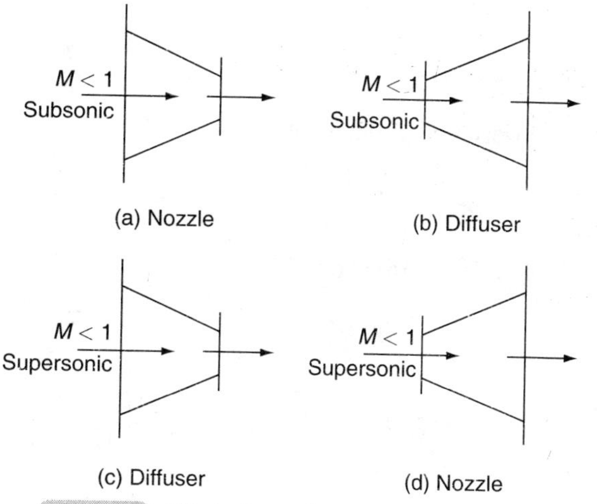

Fig. 17.4 *Effect of area change in subsonic and supersonic flow at inlet to duct*

17.4 CRITICAL PROPERTIES–CHOKING IN ISENTROPIC FLOW

Let us consider the mass rate of flow of an ideal gas through a nozzle. The flow is isentropic

$$w = \rho A \mathrm{V}$$

or $$\frac{w}{A} = \frac{p}{RT} \cdot cM = \frac{p}{RT}\sqrt{\gamma RT} \cdot M = \frac{p}{p_0} \cdot p_0 \sqrt{\frac{T_0}{T}} \cdot \sqrt{\frac{1}{T_0}}\sqrt{\frac{\gamma}{R}} \cdot M$$

$$= \left(\frac{T_0}{T}\right)^{-\gamma/\gamma-1}\left(\frac{T_0}{T}\right)^{1/2} \cdot \frac{p_0}{\sqrt{T_0}}\sqrt{\frac{\gamma}{R}} \cdot M = \sqrt{\frac{\gamma}{R}}\frac{p_0 M}{\sqrt{T_0}} \cdot \frac{1}{\left(1+\frac{\gamma-1}{2}M^2\right)^{(\gamma+1)/2(\gamma-1)}} \tag{17.15}$$

Since p_0, T_0, γ and R are constant, the discharge per unit area w/A is a function of M only. There is a particular value of M when w/A is a maximum. Differentiating Eq. (17.15) with respect to M and equating it to zero,

$$\frac{\mathrm{d}(w/A)}{\mathrm{d}M} = \sqrt{\frac{\gamma}{R}} \cdot \frac{p_0}{\sqrt{T_0}} \frac{1}{\left(1+\frac{\gamma-1}{2}M^2\right)^{(\gamma+1)/2(\gamma-1)}}$$

$$+\sqrt{\frac{\gamma}{R}} \cdot \frac{p_0 M}{\sqrt{T_0}}\left[-\frac{\gamma+1}{2(\gamma-1)}\right]\left(1+\frac{\gamma-1}{2}M^2\right)^{-\frac{\gamma+1}{2(\gamma-1)}-1}\left(\frac{\gamma-1}{2} \cdot 2M\right) = 0$$

$$1 - \frac{M^2(\gamma+1)}{2\left(1+\frac{\gamma-1}{2}M^2\right)} = 0$$

$$M^2(\gamma+1) = 2 + (\gamma-1)M^2$$

$$M^2 = 1 \quad \text{or} \quad M = 1$$

So, the discharge w/A is maximum when $M = 1$.

Since $\mathrm{V} = cM = \sqrt{\gamma RT} \cdot M$, by logarithmic differentiation

$$\therefore \quad \frac{\mathrm{dV}}{\mathrm{V}} = \frac{\mathrm{d}M}{M} + \frac{1}{2}\frac{\mathrm{d}T}{T} \tag{17.16}$$

and $$\frac{T}{T_0} = \left(1+\frac{\gamma-1}{2}M^2\right)^{-1}$$

by logarithmic differentiation

$$\therefore \quad \frac{\mathrm{d}T}{T} = -\frac{(\gamma-1)M^2}{1+\frac{\gamma-1}{2}M^2}\frac{\mathrm{d}M}{M} \tag{17.17}$$

From Eqs (17.16) and (17.17)

$$\therefore \quad \frac{\mathrm{dV}}{\mathrm{V}} = \frac{1}{1+\frac{\gamma-1}{2}M^2} \cdot \frac{\mathrm{d}M}{M} \tag{17.18}$$

From Eqs (17.14) and (17.18) $$\frac{\mathrm{d}A}{A}\left(\frac{1}{M^2-1}\right) = \frac{1}{1+\frac{\gamma-1}{2}M^2}\frac{\mathrm{d}M}{M}$$

$$\therefore \quad \frac{\mathrm{d}A}{A} = \frac{(M^2-1)\mathrm{d}M}{M\left(1+\frac{\gamma-1}{2}M^2\right)} \tag{17.19}$$

By substituting $M = 1$ in any one of the Eqs (17.13), (17.14) or (17.19), $dA = 0$ or A = constant. So $M = 1$ *occurs only at the throat and nowhere else, and this happens only when the discharge is the maximum.*

If the convergent-divergent duct acts as a nozzle, in the divergent part also, the pressure will fall continuously to yield a continuous rise in velocity. The velocity of the gas is subsonic before the throat, becomes sonic at the throat, and then becomes supersonic till its exit in isentropic flow, provided the exhaust pressure is low enough. The reverse situation prevails when the inlet velocity is supersonic. The whole duct then becomes a diffuser. The transition from subsonic flow to supersonic flow and vice versa can occur only when the compressible fluid flows through a throat and the exit pressure is maintained at the appropriate value.

When $M = 1$, the discharge is maximum and the nozzle is said to be choked. The properties at the throat are then termed as critical properties and these are designated by a superscript asterisk (*). Substituting $M = 1$ in Eq. (17.7)

$$\frac{T_0}{T^*} = 1 + \frac{\gamma-1}{2}M^2 = 1 + \frac{\gamma-1}{2} = \frac{\gamma+1}{2}$$

$$\therefore \quad \frac{T^*}{T_0} = \frac{2}{\gamma+1} \tag{17.20}$$

The *critical pressure ratio* p^*/p_0 is then given by

$$\frac{p^*}{p_0} = \left(\frac{2}{\gamma+1}\right)^{\gamma/\gamma-1} \tag{17.20a}$$

For diatomic gases, like air, $\gamma = 1.4$

$$\frac{p^*}{p_0} = \left(\frac{2}{2.4}\right)^{1.4/0.4} = 0.528$$

The critical pressure ratio for air is 0.528. For superheated steam, $\gamma = 1.3$ and p^*/p_0 is 0.546.

For air,
$$\frac{T^*}{T_0} = 0.833$$

and
$$\frac{\rho^*}{\rho_0} = \left(\frac{2}{\gamma+1}\right)^{1/(\gamma-1)} = 0.634$$

By substituting $M = 1$ in Eq. (17.15)

$$\frac{w}{A^*} = \sqrt{\frac{\gamma}{R}} \cdot \frac{p_0}{\sqrt{T_0}} \cdot \frac{1}{\left(\frac{\gamma+1}{2}\right)^{(\gamma+1)/2(\gamma-1)}} \tag{17.21}$$

Dividing Eq. (17.21) by Eq. (17.15)

$$\frac{A}{A^*} = \left[\left(\frac{2}{\gamma+1}\right)\left(1+\frac{\gamma-1}{2}M^2\right)\right]^{(\gamma+1)/2(\gamma-1)} \frac{1}{M} \tag{17.22}$$

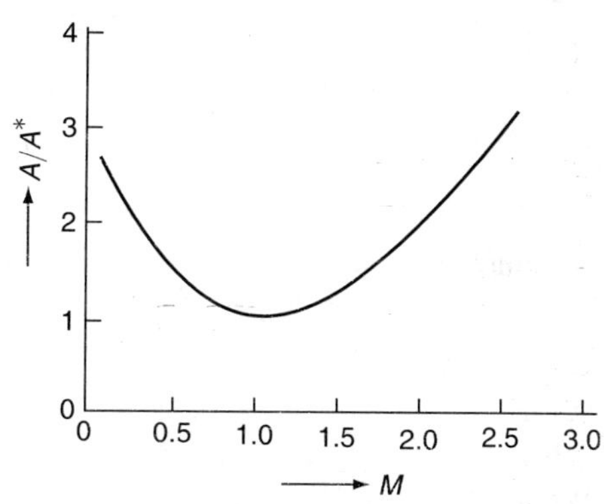

Fig. 17.5 *Area ratio as a function of mach number in an isentropic nozzle*

The area ratio A/A^* is the ratio of the area at the point where the Mach number is M to the throat area A^*. Figure 17.5 shows a plot of A/A^* vs. M, which shows that a subsonic nozzle is converging and a supersonic nozzle is diverging.

17.4.1 Dimensionless Velocity, M*

Since the Mach number M is not proportional to the velocity alone and it tends towards infinity at high speeds, one more dimensionless parameter M^* is often used, which is defined as

$$M^* = \frac{V}{c^*} = \frac{V}{V^*} \tag{17.23}$$

where $c^* = \sqrt{\gamma RT^*} = V^*$

$$\therefore \quad M^{*2} = \frac{V^2}{c^{*2}} = \frac{V^2}{c^2} \cdot \frac{c^2}{c^{*2}}$$

$$= M^2 \cdot \frac{c^2}{c^{*2}}$$

For the adiabatic flow of an ideal gas

$$\frac{V^2}{2} + c_p T = \text{constant} = h_0 = c_p T_0$$

$$\frac{V^2}{2} + \frac{\gamma RT}{\gamma - 1} = \frac{\gamma RT_0}{\gamma - 1}$$

$$\frac{V^2}{2} + \frac{c^2}{\gamma - 1} = \frac{c_0^2}{\gamma - 1}$$

Since

$$\frac{c_0}{c^*} = \sqrt{\frac{\gamma RT_0}{\gamma RT^*}} = \sqrt{\frac{T_0}{T^*}} = \sqrt{\left(\frac{\gamma + 1}{2}\right)}$$

$$\therefore \quad \frac{V^2}{2} + \frac{c^2}{\gamma - 1} = \frac{\gamma + 1}{2} c^{*2} \frac{1}{\gamma - 1}$$

$$\frac{V^2}{c^{*2}} + \frac{2}{\gamma - 1} \frac{c^2}{c^{*2}} = \frac{\gamma + 1}{\gamma - 1}$$

$$M^{*2} + \frac{2}{\gamma - 1} \frac{M^{*2}}{M^2} = \frac{\gamma + 1}{\gamma - 1}$$

On simplification

$$M^{*2} = \frac{\frac{\gamma + 1}{2} M^2}{1 + \frac{\gamma - 1}{2} M^2} \tag{17.24}$$

When $M < 1, \quad M^* < 1$

When $M > 1, \quad M^* > 1$

When $M = 1, \quad M^* = 1$

When $M = 0, \quad M^* = 0$

When $M = \infty, \quad M^* = \sqrt{\frac{\gamma + 1}{\gamma - 1}}$

17.4.2 Pressure Distribution and Choking in a Nozzle

Let us first consider a convergent nozzle as shown in Fig. 17.6, which also shows the pressure ratio p/p_0 along the length of the nozzle. The inlet condition of the fluid is the stagnation state at p_0, T_0, which is assumed to be constant. The pressure at the exit plane of the nozzle is denoted by p_E and the back pressure is p_B which can be varied by the valve. As the back pressure p_B is decreased, the mass flow rate w and the exit plane pressure p_E/p_0 vary, as shown in Fig. 17.7.

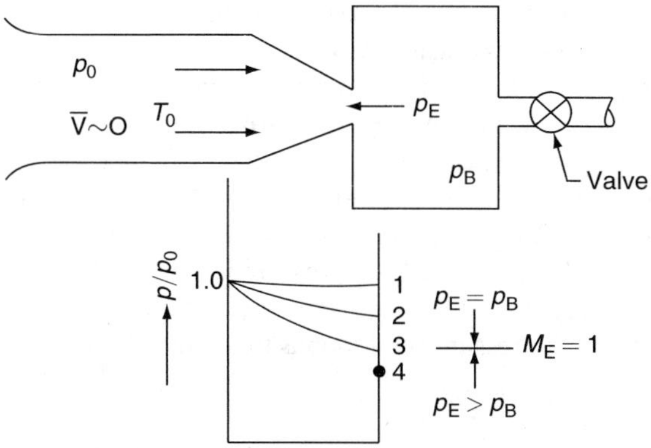

Fig. 17.6 *Compressible flow through a converging nozzle*

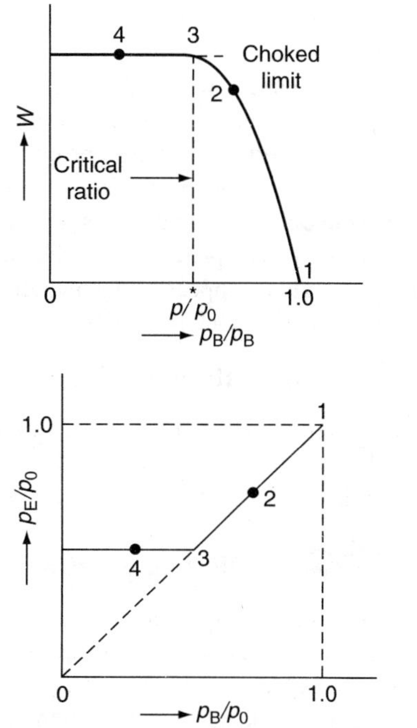

Fig. 17.7 *Mass flow rate and exit pressure as a function of back pressure in a converging nozzle*

When $p_B/p_0 = 1$, there is no flow, and $p_E/p_0 = 1$, as designated by point 1. If the back pressure p_B is now decreased to a value as designated by point 2, such that p_B/p_0 is greater than the critical pressure ratio, the mass flow rate has a certain value, and $p_E = p_B$. The exit Mach number M_E is less than 1. Next the back pressure p_B is lowered to the critical pressure, denoted by point 3. The exit Mach number M_E is now unity, and $p_E = p_B$. When p_B is increased below the critical pressure, indicated by point 4, there is no increase in the mass flow rate, and p_E remains constant at a value equal to critical pressure, and $M_E = 1$. The drop in pressure from p_E to p_B occurs outside the nozzle exit. This is choking limit which means that for given stagnation conditions the nozzle is passing the maximum possible mass flow.

Let us next consider a convergent-divergent nozzle, as shown in Fig. 17.8. Point 1 designates the condition when $p_B = p_0$, and there is no flow. When p_B is lowered to the pressure denoted by point 2, so that p_B/p_0 is less than 1 but greater than the critical pressure ratio, the velocity increases in the convergent section, but $M < 1$ at the throat. The divergent section acts as a subsonic diffuser in which the pressure increases and velocity decreases. Point 3 indicates the back pressure at which $M = 1$ at the throat, but the diverging section acts as a subsonic diffuser in which the pressure increases and velocity decreases. Point 4 indicates one other back pressure for which the flow is isentropic throughout and the diverging section acts as a supersonic nozzle with a continuous decrease in pressure and a continuous increase in velocity, and $p_{E4} = p_{B4}$. This condition of supersonic flow past the throat with the isentropic

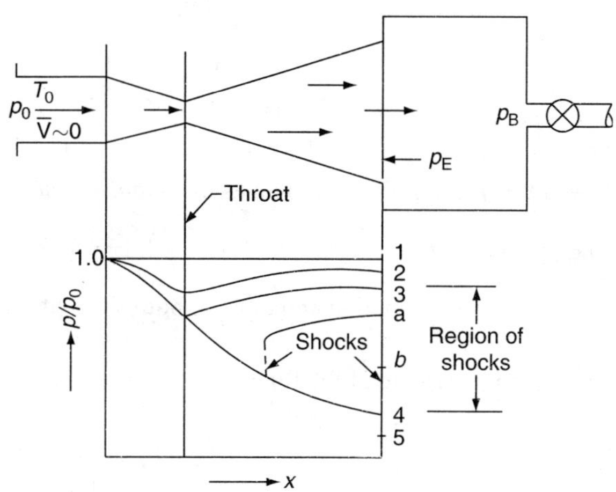

Fig. 17.8 *Compressible flow through a convergent divergent nozzle*

conditions is called the *design pressure ratio of the nozzle*. If the pressure is lowered to 5, no further decrease in exit pressure occurs and the drop in pressure from p_E to p_B occurs outside the nozzle.

Between the back pressures designated by points 3 and 4, flow is not isentropic in the diverging part, and it is accompanied by a highly irreversible phenomenon, known as *shocks*. Shocks occur only when the flow is supersonic, and after the shock the flow becomes subsonic, when the rest of the diverging portion acts as a diffuser. Properties very discontinuously across the shock. When the back pressure is as indicated by point *b* (Fig. 17.8), the flow throughout the nozzle is isentropic, with pressure continuously decreasing and velocity increasing, but a shock appears just at the exit of the nozzle. When the back pressure is increased from *b* to *a*, the shock moves upstream, as indicated. When the back pressure is further increased, the shock moves further upstream and disappears at the nozzle throat where the back pressure corresponds to 3. Since flow throughout is subsonic, no shock is possible.

17.4.3 Gas Tables for Isentropic Flow

The values of M^*, A/A^*, p/p_0, ρ/ρ_0, and T/T_0 computed for an ideal gas having $\gamma = 1.4$ for various values of Mach number M from the Eqs (17.24), (17.22), (17.8), and (17.7) respectively are given in Table D.1 in the appendix. These may be used with advantage foı computations of problems of isentropic flow.

17.5 NORMAL SHOCKS

Shock waves are highly localized irreversibilities in the flow. Within the distance of a mean free path of a molecule, the flow passes from a supersonic to a subsonic state, the velocity decreases abruptly, and the pressure rises sharply. Figure 17.9 shows a control surface that includes such a normal shock. Normal shocks may be treated as shock waves perpendicular to the flow. The fluid is assumed to be in thermodynamic equilibrium upstream and downstream of the shock, the properties of which are designated by the subscripts *x* and *y* respectively. For the control surface,

Continuity equation $\dfrac{w}{A} = \rho_x V_x = \rho_y V_y = G$ (17.25)

where G is the mass velocity (kg/m²s).

Momentum equation $p_x - p_y = \dfrac{w}{A}(V_y - V_x)$

$$= \rho_y V_y^2 - \rho_x V_x^2$$

$$\therefore \quad p_x + \rho_x V_x^2 = p_y + \rho_y V_y^2 \quad (17.26)$$

or

$$F_x = F_y \quad (17.26a)$$

where $F = pA + \rho A V^2$ is known as the *impulse function*.

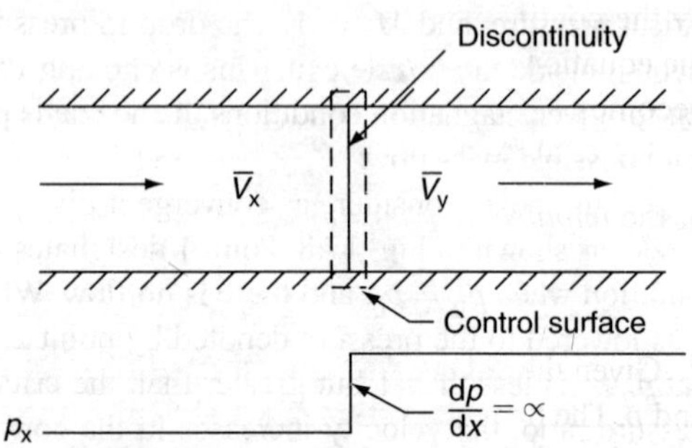

Fig. 17.9 *One dimensional normal shock*

Energy equation

$$h_x + \frac{V_x^2}{2} = h_y + \frac{V_y^2}{2} = h_{0x} + h_{0y} = h_0 \quad (17.27)$$

where h_0 is the stagnation enthalpy on both sides of the shock. Second law states

$$s_y - s_x \geq 0 \quad (17.28)$$

The equation of state of the fluid may be written implicitly in the form

$$h = h(s, \rho) \quad (17.29a)$$

$$dh = \left(\frac{\partial h}{\partial s}\right)_\rho ds + \left(\frac{\partial h}{\partial \rho}\right)_s d\rho$$

or

$$s = s(p, \rho) \quad (17.29b)$$

Combining the continuity and energy equations and dropping the subscripts x and y

$$h_0 = h + \frac{G^2}{2\rho^2} = \text{constant} \tag{17.30}$$

or

$$h = h_0 - \frac{G^2}{2\rho^2}$$

Given the values of G and h_0, this equation relates h and ρ (Fig. 17.10). The line representing the locus of points with the same mass velocity and stagnation enthalpy is called a *Fanno line*. The end states of the normal shock must lie on the Fanno line.

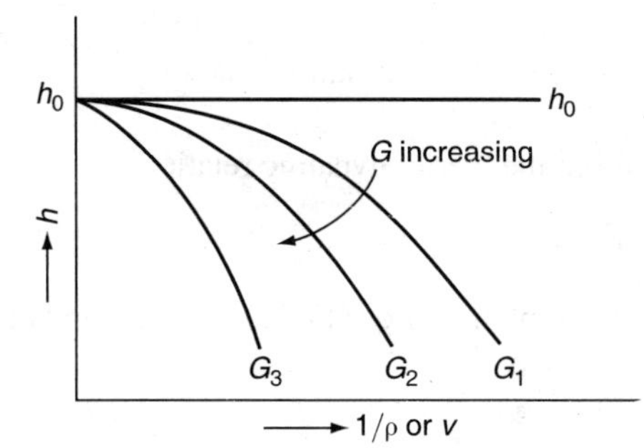

Fig. 17.10 *Fanno line on h-1/ρ coordinate*

The Fanno line may also be represented in the *h-s* diagram. The upstream properties h_x, ρ_x, p_x and V_x are known. Let a particular value of V_y be chosen, then ρ_y may be computed from the continuity Eq. (17.25), h_y from the energy Eq. (17.27), and s_y from Eq. (17.29a). By repeating the calculation for various values of V_y the Fanno line may easily be constructed (Fig. 17.11). Since the momentum equation has not been introduced, the Fanno line represents states with the same mass velocity and stagnation enthalpy, but not the same value of the impulse function.

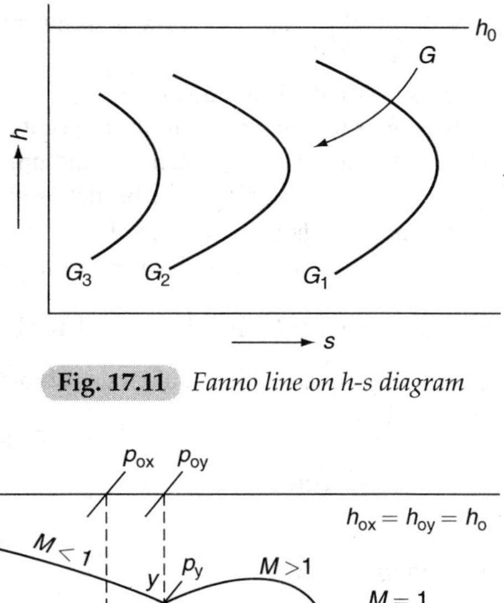

Fig. 17.11 *Fanno line on h-s diagram*

Adiabatic flow in a constant area duct with friction, in a one dimensional model, has both constant G and constant h_0, and hence must follow a Fanno line.

Let us next consider the locus of states which are defined by the continuity Eq. (17.25), the momentum Eq. (17.26) and the equation of state (17.29). The impulse function in this case becomes

$$F = pA + \rho A V^2$$

or the *impulse pressure* I is given by

$$= \frac{F}{A} = p + \rho V^2 = p + \frac{G^2}{\rho} \tag{17.31}$$

Given the values for I and G, the equation relates p and ρ. The line representing the locus of states with the same impulse pressure and mass velocity is called the *Rayleigh line*. The end states of the normal shock must lie on the Rayleigh line, since $I_x = I_y$, and $G_x = G_y$.

The Rayleigh line may also be drawn on the *h-s* plot. The properties upstream of the shock are all known. The downstream properties are to be known. Let a particular value of V_y be chosen. Then ρ_y may be computed from the continuity Eq. (17.25) and p_y from the momentum Eq. (17.26), and s_y from Eq. (17.29b) may be found. By repeating the calculations for various values of V_y, the locus of possible states reachable from, say, state x and may be plotted, and this is the Rayleigh line (Fig. 17.12).

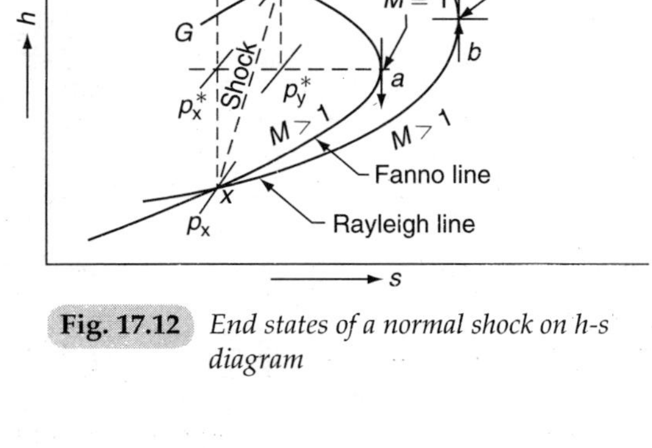

Fig. 17.12 *End states of a normal shock on h-s diagram*

Since the normal shock must satisfy Eqs (17.25), (17.26), (17.27), and (17.29) simultaneously, the end states x and y of the shock must lie at the intersections of the Fanno line and the Rayleigh line for the same G (Fig. 17.12).

The Rayleigh line is also a model for flow in a constant area duct with heat transfer, but without friction.

For an infinitesimal process in the neighbourhood of the point of maximum entropy (point a) on the Fanno line, from the energy equation

$$dh + \mathbf{V}d\mathbf{V} = 0 \tag{17.32}$$

and from the continuity equation

$$\rho\, dV + V d\,\rho = 0 \tag{17.33}$$

From the thermodynamic relation

$$Tds = dh - Vdp$$

or

$$dh = \frac{dp}{\rho} \tag{17.34}$$

By combining Eqs (17.32), (17.33), and (17.34)

$$\frac{dp}{d\rho} + V\left(-\frac{Vd\rho}{\rho}\right) = 0$$

$$\therefore \qquad \frac{dp}{d\rho} = V^2$$

or $V = \sqrt{\left(\frac{\partial p}{\partial \rho}\right)_s}$, since the flow is isentropic.

This is the local sound velocity.

So the Mach number is unity at point a. Similarly, it can be shown that at point b on the Rayleigh line, $M = 1$. It may also be shown that the upper branches of the Fanno and Rayleigh lines represent subsonic speeds ($M < 1$) and the lower branches represent supersonic speeds ($M > 1$).

The normal shock always involves a change from supersonic to subsonic speed with a consequent pressure rise, and never the reverse. By the second law, entropy always increases during irreversible adiabatic change.

17.5.1 Normal Shock in an Ideal Gas

The energy equation for an ideal gas across the shock becomes

$$c_p T_x + \frac{V_x^2}{2} = c_p T_y + \frac{V_y^2}{2} = c_p T_0$$

Now

$$h_{0x} = h_{0y} = h_0, \text{ and } T_{0x} = T_{0y} = T_0$$

Substituting

$$c_p = \frac{\gamma R}{\gamma - 1}, c_x = \sqrt{\gamma R T_x}, \text{ and } c_y = \sqrt{\gamma R T_y}$$

$$\frac{T_0}{T_x} = 1 + \frac{\gamma - 1}{2} M^2{}_x, \text{ and } \frac{T_0}{T_y} = 1 + \frac{\gamma - 1}{2} M^2{}_y$$

$$\therefore \qquad \frac{T_y}{T_x} = \frac{1 + \frac{\gamma - 1}{2} M_x^2}{1 + \frac{\gamma - 1}{2} M_y^2} \tag{17.35}$$

Again

$$\rho_x V_x = \rho_y V_y$$

$$\therefore \qquad \frac{p_x}{RT_x} V_x = \frac{p_y}{RT_y} \cdot V_y$$

or
$$\frac{T_y}{T_x} = \frac{p_y}{p_x}\cdot\frac{V_y}{V_x} = \frac{p_y}{p_x}\cdot\frac{M_y c_y}{M_x c_x} = \frac{p_y}{p_x}\cdot\frac{M_y}{M_x}\sqrt{\frac{T_y}{T_x}}$$

or
$$\frac{T_y}{T_x} = \left(\frac{p_y}{p_x}\right)^2\left(\frac{M_y}{M_x}\right)^2 \tag{17.36}$$

From Eqs (17.35) and (17.36)
$$\frac{p_y}{p_x} = \frac{M_x}{M_y}\sqrt{\frac{1+\frac{\gamma-1}{2}M_x^2}{1+\frac{\gamma-1}{2}M_y^2}} \tag{17.37}$$

Also
$$p_x + \rho_x V_x^2 = p_y + \rho_y V_y^2$$

or
$$p_x + \frac{\gamma p_x}{\gamma R T_x} V_x^2 = p_y + \frac{\gamma p_y}{R T_y \gamma} V_y^2$$

$$p_x(1+\gamma M_x^2) = p_y(1+\gamma M_y^2)$$

$\therefore$
$$\frac{p_y}{p_x} = \frac{1+\gamma M_x^2}{1+\gamma M_y^2} \tag{17.38}$$

From Eqs (17.37) and (17.38), upon rearrangement
$$M_y^2 = \frac{M_x^2 + \frac{2}{\gamma-1}}{\frac{2\gamma}{\gamma-1}M_x^2 - 1} \tag{17.39}$$

Then from Eqs (17.37), (17.38), and (17.39)
$$\frac{p_y}{p_x} = \frac{2\gamma}{\gamma+1}M_y^2 - \frac{\gamma-1}{\gamma+1} \tag{17.40}$$

and from Eqs (17.35) and (17.39)
$$\frac{T_y}{T_x} = \frac{\left(1+\frac{\gamma-1}{2}M_x^2\right)\left(\frac{2\gamma}{\gamma-1}M_x^2 - 1\right)}{\frac{(\gamma+1)^2}{2(\gamma-1)}M_x^2} \tag{17.41}$$

Then
$$\frac{\rho_y}{\rho_x} = \frac{p_y}{T_y}\cdot\frac{T_x}{p_x} = \frac{\left(\frac{2\gamma}{\gamma+1}M_x^2 - \frac{\gamma-1}{\gamma+1}\right)\left(\frac{(\gamma+1)^2}{2(\gamma-1)}M_x^2\right)}{\left(1+\frac{\gamma-1}{2}M_x^2\right)\left(\frac{2\gamma}{\gamma-1}M_x^2 - 1\right)} \tag{17.42}$$

The ratio of the stagnation pressures is a measure of the irreversibility of the shock process. Now
$$\frac{p_{oy}}{p_{ox}} = \frac{p_{oy}}{p_y}\cdot\frac{p_y}{p_x}\cdot\frac{p_x}{p_{ox}}$$

and
$$\frac{p_{oy}}{p_y} = \left(1+\frac{\gamma-1}{2}M_y^2\right)^{\gamma/(\gamma-1)}$$

$$\frac{p_{ox}}{p_x} = \left(1+\frac{\gamma-1}{2}M_x^2\right)^{\gamma/(\gamma-1)}$$

$$\therefore \qquad \frac{p_{oy}}{p_{ox}} = \left\{\frac{\frac{\gamma+1}{2}M_x^2}{1+\frac{\gamma-1}{2}M_x^2}\right\}^{\gamma(\gamma-1)} \Bigg/ \left[\frac{2\gamma}{\gamma+1}M_x^2 - \frac{\gamma-1}{\gamma+1}\right]^{1/(\gamma-1)} \qquad (17.43)$$

$$\frac{p_{oy}}{p_x} = \frac{p_{oy}}{p_y}\cdot\frac{p_y}{p_x} = \left(1+\frac{\gamma-1}{2}M_y^2\right)^{\gamma/(\gamma-1)}\left[\frac{2\gamma}{\gamma+1}M_x^2 - \frac{\gamma-1}{\gamma+1}\right] \qquad (17.44)$$

For different values of M_x, and for $\gamma = 1.4$, the values of M_y, p_y/p_x, T_y/T_x, ρ_y/ρ_x, p_{oy}/p_{ox}, and p_{oy}/p_x computed from Eqs (17.39), (17.40), (17.42), (17.43), and (17.44) respectively, are given in Table D.2 in the appendix.

To evaluate the entropy change across the shock, for an ideal gas

$$ds = c_p \frac{dT}{T} - R\frac{dp}{p}$$

$$\therefore \qquad s_y - s_x = c_p \ln\frac{T_y}{T_x} - R\ln\frac{p_y}{p_x} = c_p\left[\ln\frac{T_y}{T_x} - \ln\left(\frac{p_y}{p_x}\right)^{(\gamma-1)/\gamma}\right]$$

since $$R = \frac{c_p(\gamma-1)}{\gamma}$$

$$\therefore \qquad s_y - s_x = c_p \ln\frac{T_y/T_x}{(p_y/p_x)^{(\gamma-1)/\gamma}} = c_p \ln\frac{T_{oy}/T_{ox}}{(p_{oy}/p_{ox})^{(\gamma-1)/\gamma}} = -R\ln\frac{p_{oy}}{p_{ox}} \qquad (17.45)$$

The strength of a shock wave, P, is defined as the ratio of the pressure increase to the initial pressure, i.e.

$$P = \frac{p_y - p_x}{p_x} = \frac{p_y}{p_x} - 1$$

Substituting from Eq. (17.40) $$P = \frac{2\gamma}{\gamma+1}M_y^2 - \frac{\gamma-1}{\gamma+1} - 1$$

$$= \frac{2\gamma}{\gamma+1}(M_y^2 - 1) \qquad (17.46)$$

17.6 ADIABATIC FLOW WITH FRICTION AND DIABATIC FLOW WITHOUT FRICTION

It was stated that the Fanno line representing the states of constant mass velocity and constant stagnation enthalpy also holds for adiabatic flow in a constant area duct with friction. For adiabatic flow the entropy must increase in the flow direction. Hence a Fanno process must follow its Fanno line to the right, as shown in Fig. 17.13. Since friction will tend to move the state of the fluid to the right on the Fanno line, the Mach number of subsonic flows increases in the downstream section (Fig. 17.13), and in supersonic flows friction acts to decrease the Mach number. Hence, friction tends to drive the flow to the sonic point.

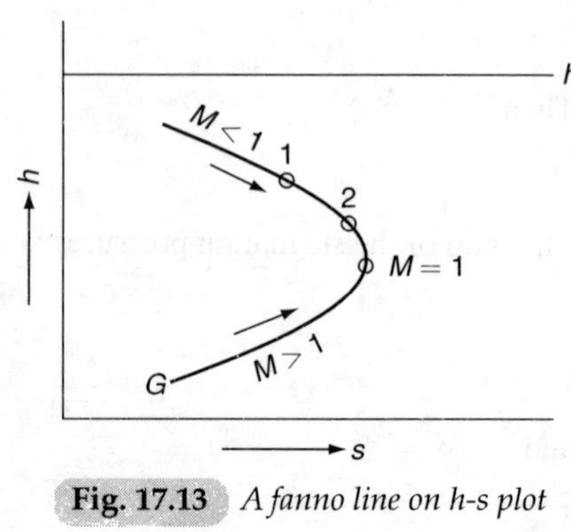

Fig. 17.13 *A fanno line on h-s plot*

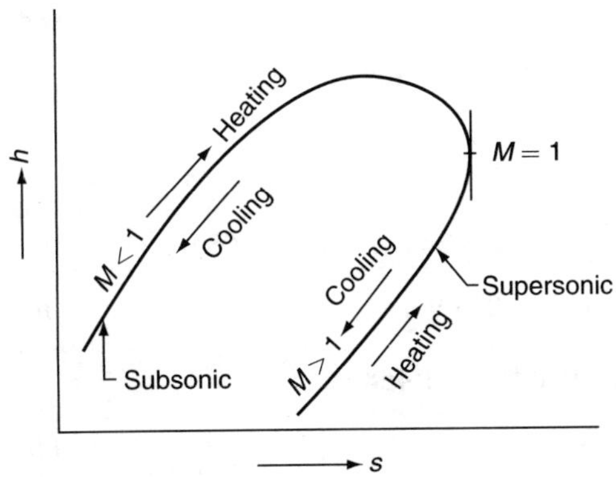

Fig. 17.14 *A rayleigh line on h-s plot*

Let us consider a short duct with a given h_0 and G, i.e. a given Fanno line, with a given subsonic exit Mach number represented by point 1 in Fig. 17.13. If some more length is added to the duct, the new exit Mach number will be increased due to friction, as represented by, say, point 2. The length of the duct may be further increased till the exit Mach number is unity. Any further increase in duct length is not possible without incurring a reduction in the mass flow rate. Hence subsonic flows can be choked by friction. There is a maximum flow rate that can be passed by a pipe with given stagnation conditions. Choking also occurs in supersonic flow with friction, usually in a very short length. It is thus difficult to use such flows in applications.

Diabatic flows, i.e., flows with heating or cooling, in a constant area duct, in the absence of friction, can be treated by the Rayleigh process (Fig. 17.14). The process is reversible, and the direction of entropy change is determined by the sign of the heat transfer. Heating a compressible flow has the same effect as friction, and the Mach number goes towards unity. Therefore, there is a maximum heat input for a given flow rate which can be passed by the duct, which is then choked. Although the cooling of the fluid increases the flow stagnation pressure with a decrease in entropy, a nonmechanical pump is not feasible by cooling a compressible flow, because of the predominating effect of friction.

Solved Examples

Example 17.1

A stream of air flows in a duct of 100 mm diameter at a rate of 1 kg/s. The stagnation temperature is 37°C. At one section of the duct the static pressure is 40 kPa. Calculate the Mach number, Velocity, and stagnation pressure at this section.

Solution

$$T_0 = 37 + 273 = 310 \text{ K}, p = 40 \text{ kPa}, \gamma = 1.4$$

The mass flow rate per unit area is

$$\frac{w}{A} = \rho V = \frac{p}{RT}\sqrt{\gamma RT}\cdot M = \sqrt{\frac{\gamma}{R}}\cdot\frac{pM}{\sqrt{T_0}}\sqrt{\frac{T_0}{T}} = \sqrt{\frac{\gamma}{R}}\cdot\frac{pM}{\sqrt{T_0}}\left(1+\frac{\gamma-1}{2}M^2\right)^{1/2}$$

$$\frac{1}{\frac{\pi}{4}(0.1)^2}\text{ kg/m}^2\text{s} = \sqrt{\frac{1.4}{0.287\text{ kJ/kg K}}}\frac{40\text{ kN/m}^2\times M}{\sqrt{310\text{ K}}}(1+0.2M^2)^{1/2}$$

$$127.39 = \sqrt{\frac{1.4}{0.287\times10^3\times310}}\,.\,40\times10^3 M(1+0.2M^2)^{1/2}$$

$$\therefore \quad M(1+0.2M^2)^{1/2} = 0.803$$

$$M^2(1+0.2M^2) = 0.645$$

$$M^4 + 5M^2 - 3.225 = 0$$

$$M^2 = \frac{-5\pm\sqrt{25+1.29}}{2} = \frac{-5+6.16}{2} = 0.58$$

$$\therefore \quad M = 0.76 \qquad \textit{Ans.}$$

$$\frac{T_0}{T} = 1 + \frac{\gamma - 1}{2} M^2 = 1 + 0.2 \times (0.76)^2 = 1.116$$

$$\therefore \quad T = \frac{310}{1.116} = 277.78 \text{ K}$$

$$c = \sqrt{\gamma RT} = \sqrt{1.4 \times 0.287 \times 277.78 \times 10^3} = 334.08 \text{ m/s} \qquad \textit{Ans.}$$

$$\therefore \quad V = c\, M = 334.08 \times 0.76 = 253.9 \text{ m/s} \qquad \textit{Ans.}$$

$$\frac{p_0}{p} = \left(\frac{T_0}{T}\right)^{\gamma/(\gamma-1)} = (1.116)^{1.4/0.4} = 1.468$$

$$\therefore \quad p_0 = 40 \times 1.468 = 58.72 \text{ kPa} \qquad \textit{Ans.}$$

Example 17.2

A conical air diffuser has an intake area of 0.11 m² and an exit area of 0.44 m². Air enters the diffuser with a static pressure of 0.18 MPa, static temperature of 37°C, and Velocity of 267 m/s. Calculate (a) the mass flow rate of air through the diffuser, (b) the Mach number, static temperature, and static pressure of air leaVing the diffuser, and (c) the net thrust acting upon the diffuser, assuming that its outer surfaces are wetted by atmospheric pressure at 0.1 MPa.

Solution The conical diffuser is shown in Fig. Ex. 17.2. The mass flow rate of air through it

$$\omega = \rho A V$$

$$= \frac{p_1}{RT_1} \cdot A_1 V_1 = \frac{0.18 \times 10^3}{0.287 \times 310} \times 0.11 \times 267 = 59.42 \text{ kg/s} \quad \textit{Ans. (a)}$$

$$c_1 = \sqrt{\gamma RT_1}$$

$$= 20.045 \sqrt{T_1} = 20.045 \sqrt{310} = 352 \text{ m/s}$$

$$\therefore \quad M_1 = \frac{V_1}{c_1} = \frac{267}{352} = 0.76$$

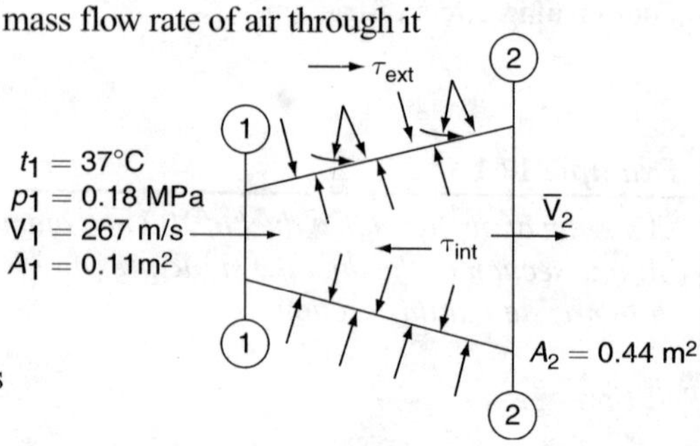

Fig. Ex. 17.2

From the gas tables for the isentropic flow of air ($\gamma = 1.4$), given in the appendix, when $M_1 = 0.76$

$$\frac{A_1}{A^*} = 1.0570, \quad \frac{p_1}{p_{01}} = 0.68207, \quad \frac{T_1}{T_{01}} = 0.89644$$

$$\frac{F_1}{F^*} = 1.0284$$

Now

$$\frac{A_2}{A_1} = \frac{0.44 \text{ m}^2}{0.11 \text{ m}^2} = 4 = \frac{A_2 / A^*}{A_1 / A^*}$$

$$\therefore \quad \frac{A_2}{A^*} = 4 \times 1.0570 = 4.228$$

Again, from the isentropic flow tables, when $\frac{A_2}{A^*} = 4.228$

$$M_2 \simeq 0.135, \quad \frac{p_2}{p_{02}} = 0.987, \quad \frac{T_2}{T_{02}} = 0.996, \quad \frac{F_2}{F^*} = 3.46$$

$$\frac{p_2}{p_1} = \frac{p_2/p_{02}}{p_1/p_{01}} = \frac{0.987}{0.682} = 1.447$$

$$\frac{T_2}{T_1} = \frac{T_2/T_{02}}{T_1/T_{01}} = \frac{0.996}{0.89644} = 1.111$$

$$\frac{F_2}{F_1} = \frac{F_2/F^*}{F_1/F^*} = \frac{3.46}{1.0284} = 3.364$$

$\therefore$ $p_2 = 1.447 \times 0.18 = 0.26$ MPa

$T_2 = 1.111 \times 310 = 344.4 \text{ K} = 71.3°\text{C}$ *Ans.*

Impulse function at inlet

$$F_1 = p_1 A_1 + \rho_1 A_1 V^2_1 = p_1 A_1 \left(1 + \frac{1}{RT_1} V_1^2\right) = p_1 A_1 (1 + \gamma M^2_1)$$

$$= 0.18 \times 10^3 \times 0.11 (1 + 1.4 \times 0.76^2) = 35.82 \text{ kN}$$

Internal thrust τ will be from right to left, as shown in Fig. Ex. 17.2.

$$\tau_{int} = F_2 - F_1 = 3.364 F_1 - F_1$$

$$= 2.364 \times 35.82 = 84.68 \text{ kN}$$

External thrust is from left to right

$$\tau_{ext} = p_0(A_2 - A_1) = 0.1 \times 10^3 (0.44 - 0.11) = 33 \text{ kN}$$

Net thrust $= \tau_{int} - \tau_{ext}$

$= 84.68 - 33 = 51.68$ kN *Ans.*(c)

Example 17.3

A conVergent-diVergent nozzle has a throat area 500 mm² and an exit area of 1000 mm². Air enters the nozzle with a stagnation temperature of 360 K and a stagnation pressure of 1 MPa. Determine the maximum flow rate of air that the nozzle can pass, and the static pressure, static temperature, Mach number, and Velocity at the exit from the nozzle, if (a) the diVergent section acts as a nozzle, and (b) the diVergent section acts as a diffuser.

Solution

$$\frac{A_2}{A^*} = \frac{1000}{500} = 2$$

From the isentropic flow tables, when $A_2/A^* = 2$ there are two values of the Mach number, one for supersonic flow when the divergent section acts as a nozzle, and the other for subsonic flow when the divergent section acts as a diffuser, which are $M_2 = 2.197, 0.308$ (Fig. Ex. 17.3).

(a) When $M_2 = 2.197, \frac{p_2}{p_0} = 0.0939, \frac{T_2}{T_0} = 0.5089$

$\therefore$ $p_2 = 0.0939 \times 1000 = 93.9$ kPa

$T_2 = 0.5089 \times 360 = 183.2$ K

$c_2 = \sqrt{\gamma R T_2} = 20.045 \sqrt{183.2} = 271.2$ m/s

$\therefore$ $V_2 = 271.2 \times 2.197 = 596$ m/s *Ans.*

Mass flow rate $w = A^* \rho^* V^* = \rho_2 A_2 V_2 = \rho_1 A_1 V_1$

For air $\frac{p^*}{p_0} = 0.528$ and $\frac{T^*}{T_0} = 0.833$

$$\rho^* = \frac{p^*}{RT^*} = \frac{0.528 \times 1000}{0.287 \times 0.833 \times 360}$$
$$= 6.13 \text{ kg/m}^3$$

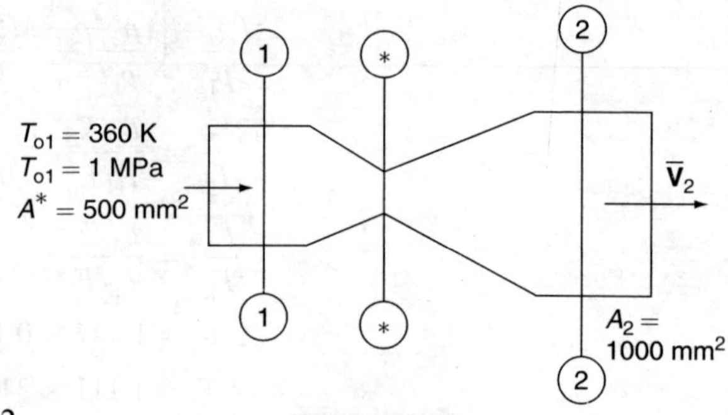

Fig. Ex. 17.3

$$T^* = 360 \times 0.833 = 300 \text{ K}$$
$$V^* = \sqrt{\gamma RT^*}$$
$$= 20.045\sqrt{300} = 347.2 \text{ m/s}$$
$$\therefore \quad w = (500 \times 10^{-6}) \times 6.13 \times 347.2$$
$$= 1.065 \text{ kg/s}$$

Ans.

(b) When $M = 0.308$, $\frac{p_2}{p_0} = 0.936$, $\frac{T_2}{T_0} = 0.9812$

$$p_2 = 0.936 \times 1000 = 936 \text{ kPa}$$
$$T_2 = 0.9812 \times 360 = 353.2 \text{ K}$$
$$c_2 = \sqrt{\gamma RT_2} = 20.045\sqrt{353.2} = 376.8 \text{ m/s}$$
$$V_2 = 376.8 \times 0.308 = 116 \text{ m/s}$$
$$w = 1.065 \text{ kg/s}$$

Ans.

Example 17.4

When a Pitot tube is immersed in a supersonic stream, a curVed shock waVe is formed ahead of the Pitot tube mouth. Since the radius of the curVature of the shock is large, the shock may be assumed to be a normal shock. After the normal shock, the fluid stream decelerates isentropically to the total pressure p_{0y} at the entrance to the Pitot tube.

A pitot tube traVelling in a supersonic wind-tunnel giVes Values of 16 kPa and 70 kPa for the static pressure upstream of the shock and the pressure at the mouth of the tube respectiVely. Estimate the Mach number of the tunnel. If the stagnation temperature is 300°C, calculate the static temperature and the total (stagnation) pressure upstream and downstream of the tube.

Solution With reference to the Fig. Ex. 17.4

$$p_x = 16 \text{ kPa}, p_{oy} = 70 \text{ kPa}$$
$$\therefore \quad \frac{p_{oy}}{p_x} = \frac{70}{16} = 4.375$$

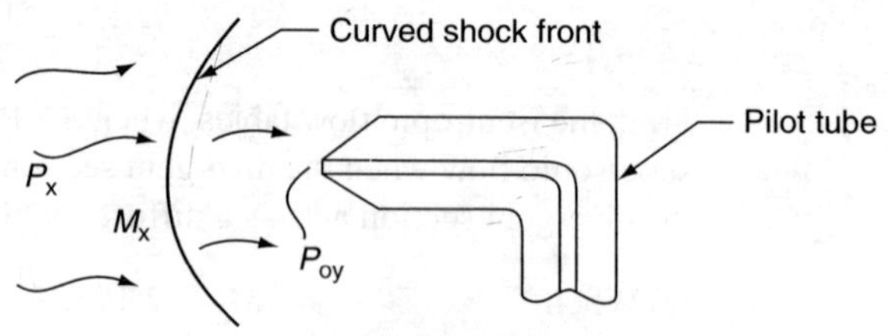

Fig. Ex. 17.4

From the gas tables for normal shock

When

$$\frac{p_{oy}}{p_x} = 4.375, M_x = 1.735, \frac{p_y}{p_x} = 3.34, \frac{\rho_y}{\rho_x} = 2.25$$
$$\frac{T_y}{T_x} = 1.483, \frac{p_{oy}}{p_{ox}} = 0.84, M_y = 0.631$$
$$T_{ox} = T_{oy} = 573 \text{ K}$$

$$T_{ox} = \left(1 + \frac{\gamma - 1}{2} M_x^2\right) T_x = (1 + 0.2 \times 3) T_x = 1.6\ T_x$$

$$T_x = \frac{573}{1.6} = 358 \text{ K}$$

$$T_y = 358 \times 1.483 = 530 \text{ K} = 257°\text{C}$$

$$p_{ox} = \frac{p_{oy}}{0.84} = \frac{70}{0.84} = 83.3 \text{ kPa}$$

$$M_x = 1.735$$

Ans.

Example 17.5

A converGent-diVergent nozzle operates at off-design condition while conducting air from a high pressure tank to a large container. A normal shock occurs in the diVergent part of the nozzle at a section where the cross-sectional area is 18.75 cm². If the stagnation pressure and stagnation temperature at the inlet of the nozzle are 0.21 MPa and 36°C respectiVely, and the throat area is 12.50 cm² and the exit area is 25 cm² estimate the exit Mach number, exit pressure, loss in stagnation pressure, and entropy increase, during the flow between the tanks.

Solution With reference to Fig. Ex. 17.5, at shock section

$$\frac{A_x}{A^*} = \frac{18.75}{12.50} = 1.5$$

Upto the shock, the flow is isentropic. From the isentropic flow tables, when

$$\frac{A}{A^*} = 1.5, M_x = 1.86$$

$$\frac{p_x}{p_{ox}} = 0.159$$

∴ Static pressure upstream of the shock

$$p_x = 0.159 \times 0.21 \times 10^3 = 33.4 \text{ kPa}$$

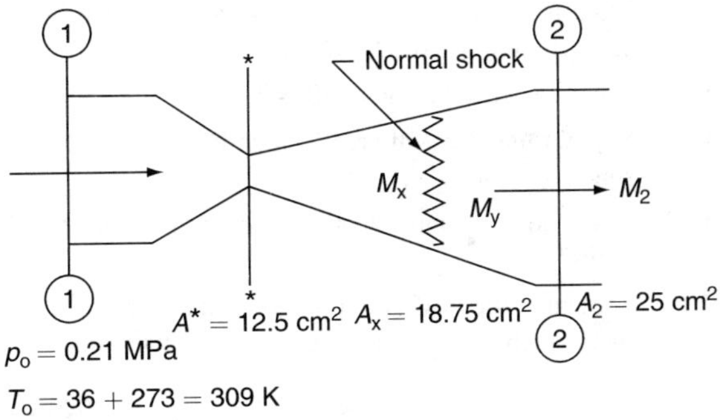

Fig. Ex. 17.5

From the gas tables on normal shocks

When $M_x = 1.86, M_y = 0.604, \frac{p_y}{p_x} = 3.87, \frac{p_{oy}}{p_x} = 4.95, \frac{p_{oy}}{p_{ox}} = 0.786$

∴
$$p_y = 3.87 \times 33.4 = 129.3 \text{ kPa}$$

$$p_{oy} = 4.95 \times 33.4 = 165.3 \text{ kPa}$$

From the shock section to the exit of the nozzle, the flow is again isentropic.

When $M_y = 0.604$, from the isentropic flow tables

$$\frac{A_y}{A^*} = 1.183$$

∴
$$\frac{A_2}{A^*} = \frac{A_2 A_y}{A_y A^*} = \frac{25}{18.75} \times 1.183 = 1.582$$

When $A_2/A^* = 1.582$, from the isentropic flow tables, $M_2 = 0.402$ *Ans.*

$$\frac{p_2}{p_{oy}} = 0.895$$

$$\therefore \quad p_2 = 0.895 \times 165.3 = 147.94 \text{ kPa}$$ *Ans.*

Loss in stagnation pressure occurs only across the shock

$$p_{ox} - p_{oy} = 210 - 165.3 = 44.7 \text{ kPa}$$ *Ans.*

$$\text{Entropy increase, } s_y - s_x = -R \ln \frac{p_{oy}}{p_{ox}} = 0.287 \ln \frac{210}{165.3}$$

$$= 0.287 \times 0.239 = 0.0686 \text{ kJ/kg K}$$ *Ans.*

Review Questions

17.1 What is a compressible fluid?

17.2 What are the basic laws in compressible flow?

17.3 How is sonic velocity defined in terms of pressure and density of the fluid?

17.4 Show that the sonic velocity in an ideal gas depends on the temperature and the nature of the gas.

17.5 What is Mach number?

17.6 What is a stagnation state? What do you mean by stagnation properties?

17.7 What are a nozzle and a diffuser?

17.8 Explain the effect of area change in subsonic and supersonic flows.

17.9 What do you understand by choking in nozzle flows?

17.10 Show that the discharge through a nozzle is maximum when there is a sonic condition at its throat.

17.11 What do you understand by critical pressure ratio? What is its value for air?

17.12 Explain the effect of area ratio as a function of Mach number in an isentropic nozzle?

17.13 What is M^*?

17.14 What is a shock? Where does it occur in a nozzle?

17.15 What is the impulse function?

17.16 What is a Fanno line? Why do the end states of a normal shock lie on the Fanno line?

17.17 What is a Rayleigh line? Why do the end states of a normal shock also lie on the Rayleigh line?

17.18 Where does the local sound velocity occur on the Fanno line and on the Rayleigh line?

17.19 How is the strength of a shock defined?

17.20 Explain the occurrence of choking for adiabatic flow with friction and for diabatic flow without friction.

Problems

17.1 Air in a reservoir has a temperature of 27°C and a pressure of 0.8 MPa. The air is allowed to escape through a channel at a rate of 2.5 kg/s. Assuming that the air velocity in the reservoir is negligible and that the flow through the channel is isentropic, find the Mach number, the velocity, and the area at a section in the channel where the static pressure is 0.6 MPa.

17.2 A supersonic wind tunnel nozzle is to be designed for $M = 2$, with a throat section, 0.11 m^2 in area. The supply pressure and temperature at the nozzle inlet, where the velocity is negligible, are 70 kPa and 37°C respectively. Compute the mass flow rate, the exit area, and the fluid properties at the throat and exit. Take $\gamma = 1.4$.

17.3 An ideal gas flows into a convergent nozzle at a pressure of 0.565 MPa, a temperature of 280°C, and negligible velocity. After reversible adiabatic expansion in the nozzle the gas flows directly into a large vessel. The gas in the vessel may be maintained at any specified state while the nozzle supply state is held constant. The exit area of the nozzle is 500 mm^2. For this gas $\gamma = 1.3$ and $c_p = 1.172$ kJ/kg K. Determine (a) the pressure of the gas leaving the nozzle when its temperature is 225°C, and (b) the gas mass flow rate when the pressure in the vessel is 0.21 MPa. *Ans.* 0.36 MPa, 0.48 kg/s

17.4 Air flows adiabatically through a pipe with a constant area. At point 1, the stagnation pressure is 0.35 MPa and the Mach number is 0.4. Further downstream the stagnation pressure is found to be 0.25 MPa. What is the Mach number at the second point for subsonic flow?

17.5 The intake duct to an axial flow air compressor has a diameter of 0.3 m and compresses air at 10 kg/s. The static pressure inside the duct is 67 kPa and the stagnation temperature is 40°C. Calculate the Mach number in the duct. *Ans.* 0.526

17.6 Show that for an ideal gas the fractional change in pressure across a small pressure pulse is given by

$$\frac{dp}{p} = \gamma \frac{dV}{c}$$

and that the fractional change in absolute temperature is given by

$$\frac{dT}{T} = (\gamma - 1)\frac{dV}{c}$$

17.7 An airplane flies at an altitude of 13,000 m (temperature – 55°C, pressure 18.5 kPa) with a speed of 180 m/s. Neglecting frictional effects, calculate (a) the critical velocity of the air relative to the aircraft, and (b) the maximum possible velocity of the air relative to the aircraft.

17.8 A stream of air flowing in a duct is at a pressure of 150 kPa, has a Mach number of 0.6, and flows at a rate of 0.25 kg/s. The cross-sectional area of the duct is 625 mm^2. (a) Compute the stagnation temperature of the stream in °C. (b) What is the maximum percentage reduction in area which could be introduced without reducing the flow rate of the stream? (c) For the maximum area reduction of part (b), find the velocity and pressure at the minimum area, assuming no friction and heat transfer.

17.9 A rocket combustion chamber is supplied with 12 kg/s of hydrogen and 38 kg/s of oxygen. Before entering the nozzle all the oxygen is consumed, the pressure is 2.3 MPa, and the temperature is 2800°C. Neglecting dissociation and friction, find the throat area of the nozzle required. Assume $\gamma = 1.25$.

17.10 A gas with specific heat ratio $\gamma = 1.4$ passes through a plane normal shock Immediately before the shock the Mach number is 2, and the stagnation pressure and temperature are 600 kPa and 400 K respectively. Find the static pressure and static temperature immediately downstream of the shock. *Ans.* 345 kPa, 376 K

17.11 Air flowing through a nozzle encounters a shock. The Mach number upstream of the shock is 1.8, and the static temperature downstream of the shock is 450 K. How much has the velocity changed across the shock? *Ans.* 378 m/s

17.12 The stagnation temperature and stagnation pressure of air in a reservoir supplying a convergent-divergent nozzle are 450 K and 400 kPa respectively. The nozzle throat area is 625 mm^2 and the nozzle exit area is 1875 mm^2. A shock is noted at a position in the divergent section where the area is 1250 mm^2. (a)What are the exit pressure, temperature, and velocity? (b) What value of the back pressure would cause the flow through the nozzle to be completely supersonic? (c) What value of back pressure would result in completely isentropic flow interior and exterior to the nozzle? *Ans.* (a) 234 kPa, 440 K, 141 m/s, (b) 150 kPa, (c) 18.8 kPa

17.13 Assume that a flow through the nozzle has been established by setting the back pressure at 200 kPa. Because a back pressure greater than 150 kPa leads to a shock in the nozzle, find the area ratio where the shock occurs.

17.14 Air flows through a frictionless convergent-divergent nozzle. The area of the exit section is three times the area of the throat section, and the ratio of stagnation pressure at the entrance to the static pressure in the exit section is 2.5. Calculate (a) the Mach number M_x where the shock occurs, (b) the area A_x where the shock occurs, and (c) the entropy increase.

17.15 A nozzle is designed assuming reversible adiabatic flow with an exit Mach number of 2.6. Air flows through it with a stagnation pressure and temperature of 2 MPa and 150°C respectively. The mass rate of flow is 5 kg/s. (a) Determine the exit pressure, temperature, area, and throat area. (b) If the back pressure at the nozzle exit is raised to 1.35 MPa, and the flow remains isentropic except for a normal shock wave, determine the exit Mach number and temperature, and the mass flow rate through the nozzle.

17.16 A jet plane travels through the air with a speed of 1000 km per hour at an altitude of 6000 m, where the pressure is 40 kPa and the temperature is – 12°C. Consider the diffuser of the engine. The air leaves the diffuser with a velocity of 100 m/s. Determine the pressure and temperature leaving the diffuser, and the ratio of the inlet to the exit area of the diffuser, assuming isentropic flow. *Ans.* 61 kPa, 295 K, 0.487

17.17 A normal shock occurs in the diverging section of a nozzle under steady flow conditions at a point where the air is at 150 kPa and 300 K travelling at 1000 m/s. What are the pressure and temperature on the subsonic side of the front? If the surroundings are at 20°C, what is the irreversibility caused by the shock process?

Ans. 1426.5 kPa, 762.1 K, 84.74 kJ/kg

C H A P T E R 18

Elements of Heat Transfer

18.1 BASIC CONCEPTS

Energy balances by first law have been made in a variety of physical situations, say, in a feedwater heater or a cooling coil. However, no indication has been given regarding the size of the heat exchanger, for heating or cooling of a fluid. If we consider a steel block heated in a furnace, to be allowed to cool in room air, we can estimate the amount of heat lost by the block in cooling by energy balance. But how long the cooling process will take place cannot be answered by thermodynamics. It is the science of heat transfer which is concerned with the estimation of the rate at which heat is transferred, the duration of heating or cooling for a certain heat duty and the surface area required to accomplish that heat duty.

There are three modes in which heat may be transferred: (a) conduction, (b) convection and (c) radiation.

Conduction refers to the transfer of heat between two bodies or two parts of the same body through molecules which are more or less stationary. In liquids and gases conduction results from the transport of energy by molecular motion near the wall and in solids it takes place by a combination of lattice vibration and electron transport. In general, good electrical conductors are also good thermal conductors.

Convection heat transfer occurs because of the motion of a fluid past a heated surface-the faster the motion, the greater the heat transfer. The convection heat transfer is usually assumed to be proportional to the surface area in contact with the fluid and the difference in temperature of the surface and fluid. Thus,

$$Q = hA\,[T_w - T_f]$$

where h is called the convection heat transfer coefficient, which is a strong function of both fluid properties and fluid velocity (W/m^{2}K).

Radiation heat transfer is the result of electromagnetic radiation emitted by a surface because of the temperature of the surface. This differs from other forms of electromagnetic radiation such as radio, television, *X*-rays and γ-rays which are not related to temperature.

18.2 CONDUCTION HEAT TRANSFER

Fourier's law of heat conduction states that the rate of heat flux is linearly proportional to the temperature gradient. For one dimensional or unidirectional heat conduction,

$$\mathbf{q} \propto \frac{dt}{dx} \quad \text{or} \quad \mathbf{q} = -K\frac{dt}{dx} \tag{18.1}$$

where $\mathbf{q}$ is the rate of heat flux in W/m^2, dt/dx is the temperature gradient in x-direction, and K is the constant of proportionality which is a property of the material through which heat is being conducted and is known as *thermal conductivity*. $\mathbf{q}$ is a vector quantity. The negative sign is being used because heat flows from a high to a low temperature region and the slope dt/dx is negative.

For a finite temperature difference $(t_1 - t_2)$ across a wall of thickness x (Fig. 18.1)

$$\mathbf{q} = -K \frac{t_2 - t_1}{x} = K \frac{t_1 - t_2}{x} \text{ W/m}^2$$

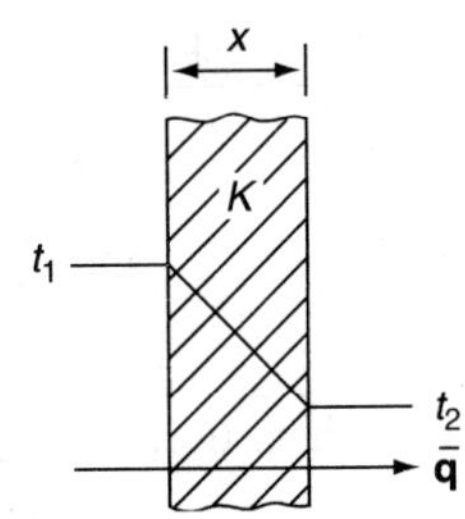

Fig. 18.1 *Heat conduction through a wall*

If A is the surface area normal to heat flow, then the rate of heat transfer

$$Q = \mathbf{q} \cdot A = -KA \frac{t_2 - t_1}{x} \quad \text{or} \quad Q = KA \frac{t_1 - t_2}{x} \text{ Watts} \qquad (18.2)$$

The dimension of thermal conductivity is W/mK. Since $dt/dx = \mathbf{q}/K$ for the same $\mathbf{q}$, if K is low (i.e., for an insulator), dt/dx will be large, i.e., there will be a large temperature difference across the wall, and if K is high (i.e., for a conductor), dt/dx will be small, or there will be a small temperature difference across the wall.

18.2.1 Resistance Concept

Heat flow has an analogy in the flow of electricity. Ohm's law states that the current I flowing through a wire (Fig. 18.2) is proportional to the potential difference E, or

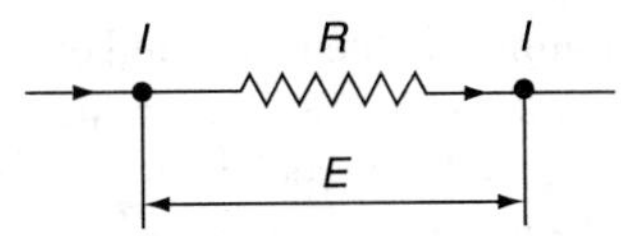

Fig. 18.2 *Electrical resistance concept*

$$I = \frac{E}{R}$$

where $1/R$ is the constant of proportionality, and R is known as the resistance of the wire, which is a property of the material. Since the temperature difference and heat flux in conduction are similar to the potential difference and electric current respectively, the heat conduction rate through the wall may be written as

$$Q = -KA \frac{t_2 - t_1}{x} = \frac{t_1 - t_2}{x/KA} = \frac{t_1 - t_2}{R}$$

where $R = x/KA$ is the thermal resistance to heat flow offered by the wall (Fig. 18.3). For a composite wall, as shown in Fig. 18.4, there are two resistances in series. The slope of the temperature profile depends on the thermal conductivity of the material. t_i is the interface temperature. The total thermal resistance

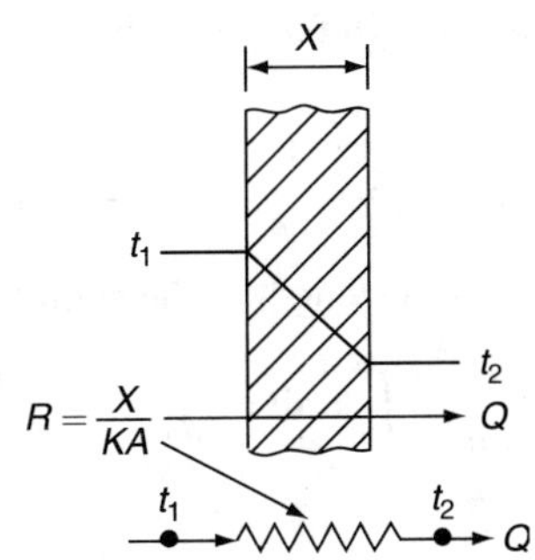

Fig. 18.3 *Thermal resistance offered by a wall*

$$R = R_1 + R_2 = \frac{x_1}{K_1 A} + \frac{x_2}{K_2 A}$$

and the rate of heat flow $\qquad Q = \dfrac{t_1 - t_2}{R}$

Again, $t_1 - t_i = Q \cdot R_1 = Q \cdot \dfrac{x_1}{K_1 A}$, from which t_i can be evaluated.

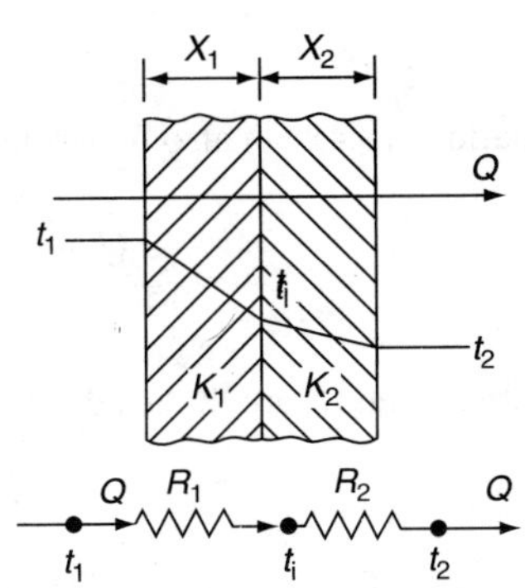

Fig. 18.4 *Heat conduction through resistance in series*

For two resistances in parallel (Fig. 18.5), the total resistance R is given by

$$\frac{1}{R}=\frac{1}{R_1}+\frac{1}{R_2} \quad \text{or} \quad R=\frac{R_1R_2}{R_1+R_2}$$

where $R_1 = \dfrac{x}{K_1 A_1}$ and $R_2 = \dfrac{x}{K_2 A_2}$

and the rate of heat flow $\quad Q = \dfrac{t_1 - t_2}{R}$

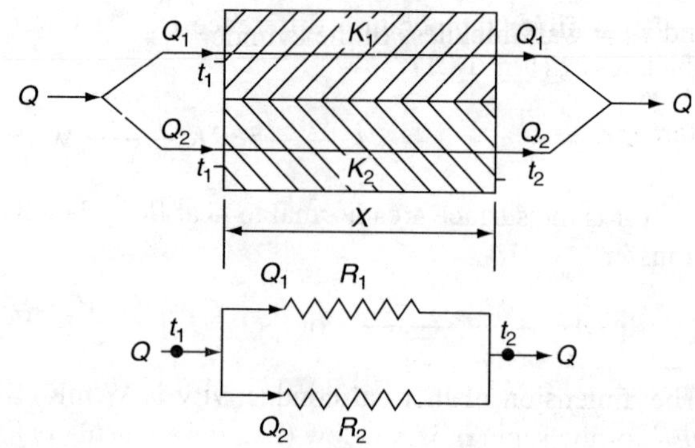

Fig. 18.5 *Heat conduction through resistance in parallel*

18.2.2 Heat Conduction through a Cylinder

Let us assume that the inside and outside surfaces of the cylinder (Fig. 18.6) are maintained at temperatures t_1 and t_2 respectively, and t_1 is greater than t_2. We will also assume that heat is flowing, under steady state, only in the radial direction, and there is no heat conduction along the length or the periphery of the cylinder. The rate of heat transfer through the thin cylinder of thickness dr is given by

$$Q = -KA\,\frac{dt}{dr} = -K\,2\,\pi\, rL\,\frac{dt}{dr} \qquad (18.3)$$

where L is the length of the cylinder.

or $$\int_{t=t_1}^{t=t_2} dt = \int_{r=r_1}^{r=r_2} -\frac{Q}{2\pi KL}\frac{dr}{r}$$

$\therefore$ $$t_2 - t_1 = -\frac{Q}{2\pi KL}\ln\frac{r_2}{r_1}$$

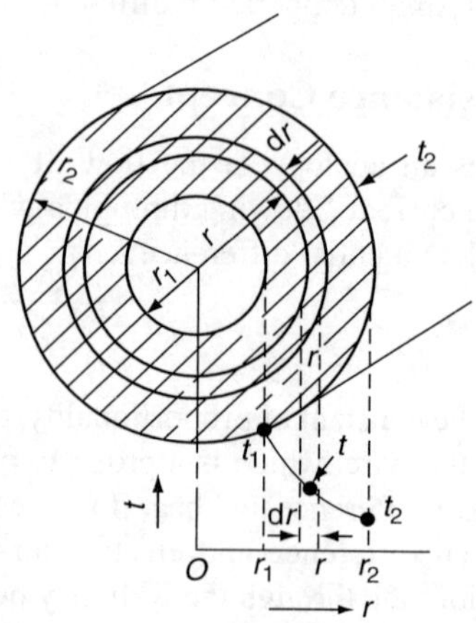

Fig. 18.6 *Heat conduction through a cylindrical wall*

$\therefore$ $$Q = \frac{2\pi KL(t_1 - t_2)}{\ln\dfrac{r_2}{r_1}} \qquad (18.4)$$

Equation (18.4) can also be written in the following form

$$Q = \frac{2\pi L(r_2 - r_1)\,K\,(t_1 - t_2)}{(r_2 - r_1)\ln\dfrac{2\pi r_2 L}{2\pi r_1 L}} = \frac{K\,(A_2 - A_1)\,(t_1 - t_2)}{(r_2 - r_1)\ln\dfrac{A_2}{A_1}}$$

where A_1 and A_2 are the inside and outside surface areas of the cylinder.

$\therefore$ $$Q = -KA_{\text{l.m.}}\frac{t_2 - t_1}{r_2 - r_1} \text{ or } Q = -KA_{\text{l.m.}}\frac{t_2 - t_1}{x_w} \qquad (18.5)$$

where $A_{\text{l.m.}}$ = log-mean area $= \dfrac{A_2 - A_1}{\ln\dfrac{A_2}{A_1}}$

and x_w = wall thickness of the cylinder

$$= r_2 - r_1$$

Here the thermal resistance offered by the cylinder wall to radial heat conduction is

$$R = \frac{x_w}{K \cdot A_{1\cdot m}}$$

From Eq. (18.3) $$dt = -\frac{Q}{2\pi KL}\frac{dr}{r} = C_1 \frac{dr}{r} \quad \text{or} \quad t = C_1 \ln r + C_2 \qquad (18.6)$$

where C_1 and C_2 are the constants to be evaluated from the conditions: when

$$r = r_1, t = t_1 \quad r = r_2, t = t_2$$

The temperature across the wall of the cylinder thus varies logarithmically.

For two cylindrical resistances in series (Fig. 18.7)

$$R = R_1 + R_2 = \frac{x_{w1}}{k_1 A_{1\cdot m1}} + \frac{x_{w2}}{k_2 A_{1\cdot m2}}$$

where $$x_{w1} = r_2 - r_1 \quad x_{w2} = r_3 - r_2$$

$$A_{1\cdot m1} = \frac{A_2 - A_1}{\ln \frac{A_2}{A_1}} = \frac{2\pi (r_2 - r_1) L}{\ln \frac{r_2}{r_1}}$$

and $$A_{1\cdot m2} = \frac{A_3 - A_2}{\ln \frac{A_3}{A_2}} = \frac{2\pi (r_3 - r_2) L}{\ln \frac{r_3}{r_2}}$$

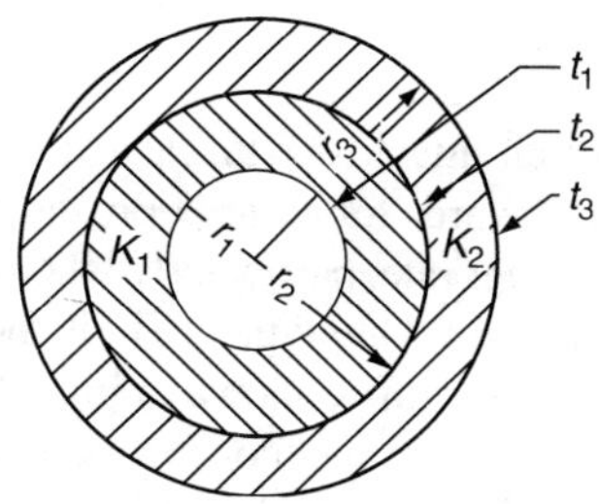

Fig. 18.7 *Heat conduction through two cylindrical thermal resistances in series*

The rate of heat transfer will be $$Q = \frac{t_1 - t_3}{R} = \frac{t_1 - t_2}{R_1} = \frac{t_2 - t_3}{R_2}$$

from which the interface temperature t_2 can be evaluated.

18.2.3 Heat Conduction through a Sphere

Heat flowing through the thin spherical strip (Fig. 18.8) at radius r, of thickness dr, is

$$Q = -KA\frac{dt}{dr}$$

where A is the spherical surface at radius r normal to heat flow,

$$\therefore \quad Q = -K4\pi r^2 \frac{dt}{dr}$$

$$\int_{t=t_1}^{t=t_2} dt = \int_{r=r_1}^{r=r_2} -\frac{Q}{4\pi K}\frac{dr}{r^2}$$

$$t_2 - t_1 = -\frac{Q}{4\pi K}\left(\frac{1}{r_1} - \frac{1}{r_2}\right)$$

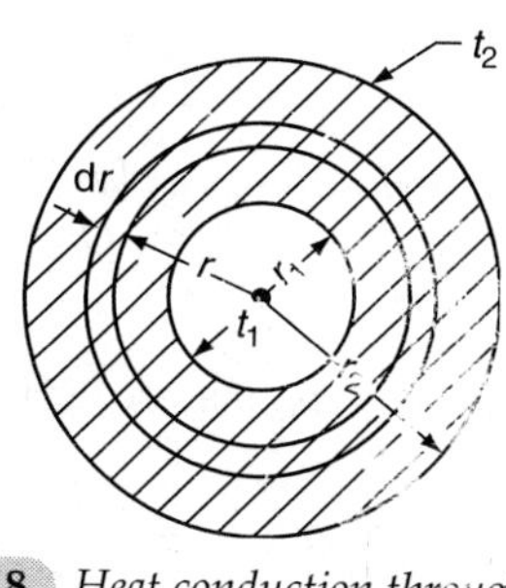

Fig. 18.8 *Heat conduction through a sphere*

$$\therefore \qquad Q = \frac{4\pi K(t_1 - t_2) r_1 r_2}{r_2 - r_1} \quad \text{or} \quad Q = -KA_{\text{g.m.}} \frac{t_2 - t_1}{x_w} \tag{18.7}$$

where $A_{\text{g.m.}}$ = geometrical mean area

$$= \sqrt{A_1 A_2} = \sqrt{4\pi r_1^2 \, 4\pi r_2^2} = 4\pi r_1 \, r_2$$

and x_w = wall thickness of the sphere $= r_2 - r_1$

Here the thermal resistance offered by the wall to heat conduction is $R = \dfrac{x_w}{KA_{\text{g.m.}}}$

Thus similar expressions of thermal resistance hold good for flat plate, cylinder, and sphere, which are

$$R_{\text{plate}} = \frac{x_w}{KA}, \; R_{\text{cylinder}} = \frac{x_w}{KA_{\text{l.m.}}} \quad \text{and} \quad R_{\text{sphere}} = \frac{x_w}{KA_{\text{g.m.}}}$$

where K is the thermal conductivity of the wall material.

18.2.4 Fins

Fins are often attached with heat transfer surfaces to increase the area for convection and thus increase the heat transfer rate, particularly when heat transfer coefficient is low. One type of fin is the straight fin of a rectangular profile shown in Fig. 18.9. By combining a conduction and convection energy balance on the fin and neglecting heat loss from the tip, it is possible to derive an expression for the temperature distribution at any x position as (see "Heat and Mass Transfer," by P.K. Nag, Tata McGraw-Hill, 2007)

$$\frac{t - t_\infty}{t_1 - t_\infty} = \frac{\cosh m(l - x)}{\cosh ml} \tag{18.8}$$

where t = temperature at the base of the fin, t_α = free-stream temperature, l = length of fin, $m = [(hp)/(KA)]^{1/2}$, h = heat transfer coefficient, P = perimeter of fin, A = cross-sectional area of fin and K = thermal conductivity.

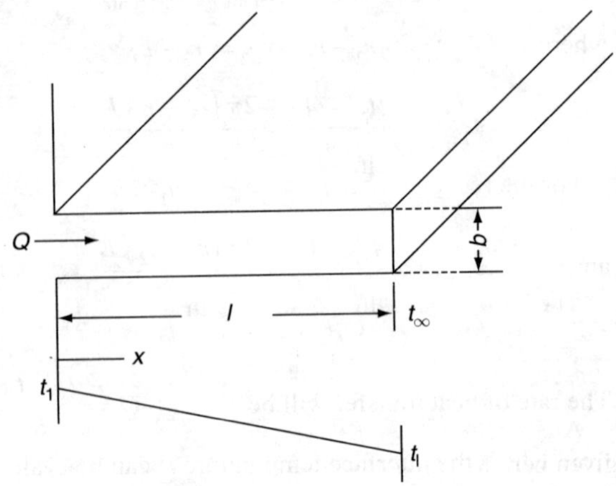

Fig. 18.9 *Rectangular fin*

If heat loss from the tip is neglected, the tip temperature t_t is given by:

$$\frac{t_t - t_\infty}{t_1 - t_\infty} = \frac{1}{\cosh ml}$$

The heat transfer from one fin is

$$Q = m\,K\,A\,\theta_1 \tan h\, ml \tag{18.9}$$

where $\qquad \theta_1 = t_1 - t_\infty.$

A fin efficiency (η_f) may be defined as:

$$\eta_f = \frac{\text{actual heat transfer}}{\text{heat which would be transferred if entire fin area were at base temperature}} \tag{18.10}$$

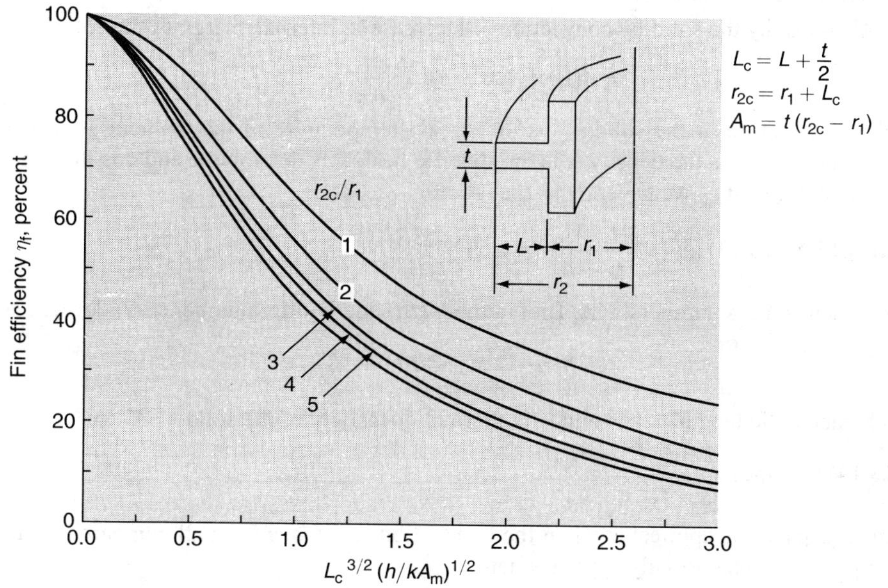

Fig. 18.10 *Efficiencies of circumferential fins of rectangular profile*

For the rectangular fin, the fin efficiency may be calculated from:

$$\eta_f = \frac{\tan h\, ml}{ml} \tag{18.11}$$

The hyperbolic functions are defined as;

$$\cosh x = \frac{e^x + e^{-x}}{2}, \quad \tanh x = \frac{e^x + e^{-x}}{e^x - e^{-x}}$$

Analytical expressions for circular fins may also be derived but are very complicated, so the results are given here in graphical form (Fig. 18.10). The top curve for $r_{cr}/r_1 = 1.0$ is also the efficiency for the straight fin of rectangular profile. From the definition of η_f, the actual heat transfer is thus:

$$Q = \eta_f\, h\, a_f(t_1 - t_\infty) + h(a - a_f)(t_1 - t_\infty) \tag{18.12}$$

where a_f = surface area of fin and a = total area of finned and unfinned surface area.

18.2.5 Transient Heat Conduction

We have so far considered only steady-state conduction where the temperatures do not vary with time. We will now consider a simple case of transient conduction, say, cooling of a solid by convection. Figure 18.11 shows a thermal "lump" which has a low internal resistance compared to the external convective resistance at the surface. Thus, there will be small temperature gradients in the solid material, so that at any instant of time the material remains essentially uniform in temperature.

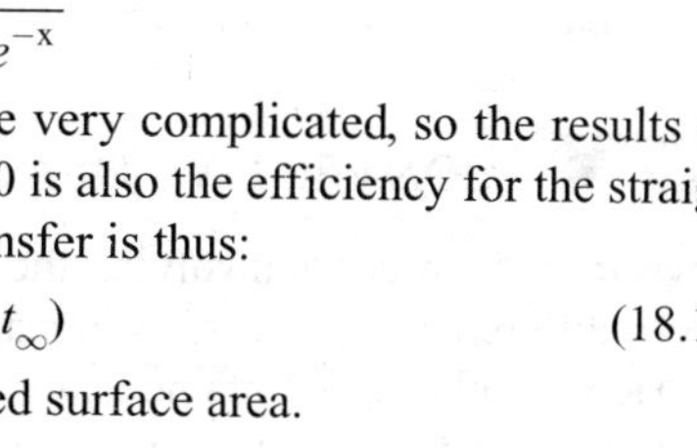

Fig. 18.11 *Cooling of a solid by convection*

Therefore, Heat lost by the solid by convection = Decrease in internal energy of the solid

$$h\,A(t - t_\infty) = -\rho\,c\,V\,\frac{dt}{d\tau} \qquad (18.13)$$

where t is the temperature of the solid, t_∞ is the stream temperature of the ambient atmosphere, A is the surface area of the solid, ρ is the density, c is the specific heat, V is the volume and τ is the time.

If the solid is initially at t_0, we have, $t = t_0$ at $\tau = 0$

Integration of Eq. (18.13) gives $\dfrac{t - t_\infty}{t_0 - t_\infty} = e^{-(h A \tau)/\rho c V}$ (18.14)

If we use dimensionless numbers, viz., Biot number (Bi) and Fourier number (FO) defined as:

$$Bi = \frac{h\,L}{K} \text{ and } FO = \frac{\alpha\tau}{L^2}$$

where L = characteristic length = V/A and α = thermal diffusivity of the solid = $K/\rho c$.

Equation (18.14) becomes $\dfrac{t - t_\infty}{t_0 - t_\infty} = e^{-\text{Bi Fo}}$ (18.15)

This lumped-capacity is applicable when the conduction resistance is small compared to the convection resistance. In practice, this normally applies when

$$\frac{h\,L}{K} \text{ or } Bi < 0.1 \qquad (18.16)$$

Equation (18.14) can also be expressed in terms of a thermal resistance for convection, $R_{th} = 1/hA$, and a thermal capacitance, $C_{th} = \rho\,cV$, so that

$$\frac{h\,A}{\rho c V} = \frac{1}{R_{th}\,C_{th}}$$

and the system behaves as an electric capacitance discharging through a resistor.

The transient heat conduction in a lumped capacity system is thus characterized by Biot number and Fourier number. Fourier number can be regarded as dimensionless time.

18.3 CONVECTIVE HEAT TRANSFER

Convection is a process involving the mass movement of fluids. When a temperature difference produces a density difference which results in mass movement (Fig. 18.12), the process is called *free or natural convection*. Here the plate is maintained isothermal at temperature t_w, which is higher than the surrounding fluid temperature t_f. The fluid near the wall, on getting heated, moves up due to the effect of buoyancy, and is replaced by the cold fluid moving towards the wall. Thus a circulation current is set up.

When the mass motion of the fluid is caused by an external device like a pump, compressor, blower or fan, the process is called *forced convection* (Fig. 18.13). Here the fluid is made to flow along the hot surface and heat is transferred from the wall to the fluid.

Whether the convection process is natural of forced, there is always a fluid film immediately adjacent to the wall where the temperature varies from t_w to t_f (Fig. 18.14). Heat is first conducted through this fluid film and then it is transported by fluid motion.

Rate of heat transfer through the film

$$Q = -K_f \cdot A\,\frac{t_f - t_w}{\delta}$$

where δ is the film thickness and K_f is the thermal conductivity of the film.

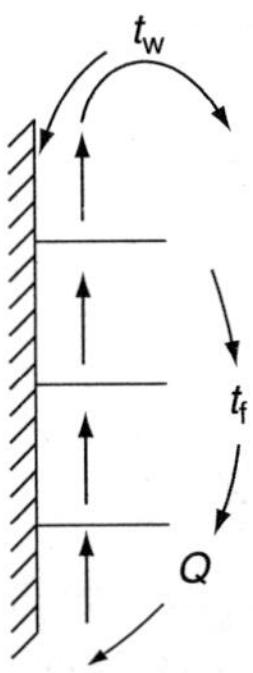

Fig. 18.12 *Natural of free convection*

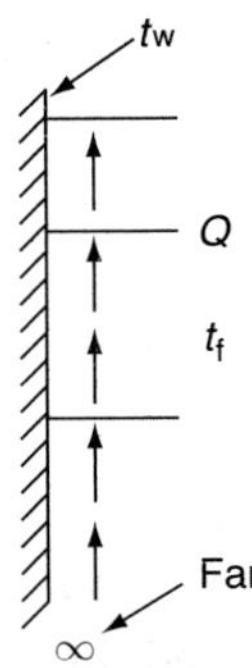

Fig. 18.13 *Forced convection*

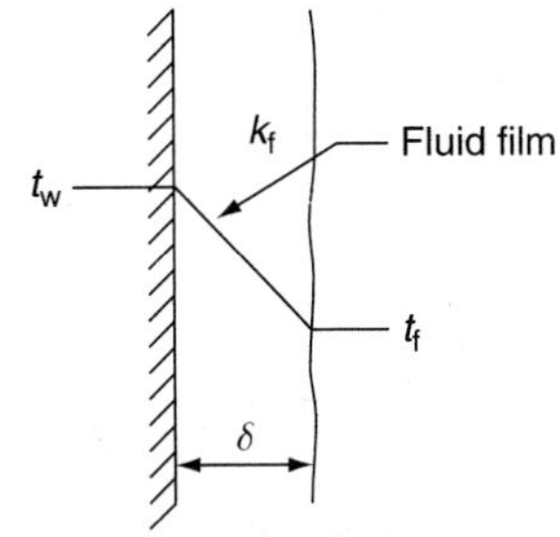

Fig. 18.14 *Heat conduction through a fluid film at the wall*

The film coefficient of heat transfer, or heat transfer coefficient h, is defined as

$$h = \frac{K_f}{\delta} \text{ (W/m}^2\text{ K)}$$

$$\therefore \qquad Q = hA\,(t_w - t_f) \tag{18.17}$$

The rate of heat transfer Q increases with the increase in the value of heat transfer coefficient h. The higher the value of K_f and the lower the value of film thickness δ, the higher will be the value of h. As the velocity of the fluid increases, the film thickness decreases. For gases, K_f is low, so the rate of heat transfer from the solid wall to a gas is small compared to a liquid.

The above equation is known as *Newton's law of cooling*. Strictly speaking, convection applies only to fluid motion. The mechanism of heat transfer is by conduction.

The thermal resistance offered by the fluid film

$$R = \frac{t_w - t_f}{Q} = \frac{1}{hA}$$

For heat transfer from a hot fluid to a cold fluid through a wall (Fig. 18.15), there are three resistances in series

$$R = R_1 + R_2 + R_3 = \frac{1}{h_1 A} + \frac{x}{KA} + \frac{1}{h_2 A}$$

$$Q = \frac{t_h - t_c}{R}$$

or

$$Q = \frac{t_h - t_c}{\dfrac{1}{h_1 A} + \dfrac{x}{KA} + \dfrac{1}{h_2 A}} = UA\,(t_h - t_c)$$

where U is known as the overall heat transfer coefficient (W/m² K) and is given by

$$\frac{1}{UA} = \frac{1}{h_1 A} + \frac{x}{KA} + \frac{1}{h_2 A} = R$$

or

$$\frac{1}{U} = \frac{1}{h_1} + \frac{x}{K} + \frac{1}{h_2}$$

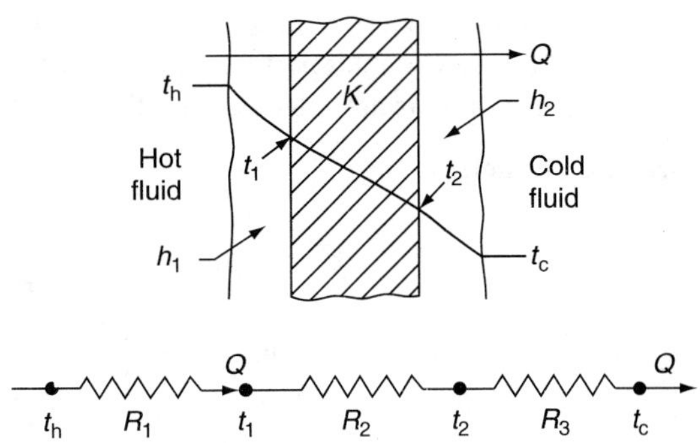

Fig. 18.15 *Heat transfer from hot to cold fluid through a plane wall*

For heat transfer from a hot fluid inside a cylinder to the cold fluid outside (Fig. 18.16)

$$Q = \frac{t_h - t_c}{R_1 + R_2 + R_3} = \frac{t_h - t_c}{\dfrac{1}{h_i A_i} + \dfrac{x_w}{K_w A_{l.m.}} + \dfrac{1}{h_o A_o}}$$

or $\quad Q = U_o A_o (t_h - t_c)$

where $$\frac{1}{U_o A_o} = \frac{1}{h_i A_i} + \frac{x_w}{K_w A_{l.m.}} + \frac{1}{h_o A_o} \qquad (18.18)$$

U_o being the overall heat transfer coefficient based on outside area A_o, h_i the inside heat transfer coefficient, and h_o the outside heat transfer coefficient.

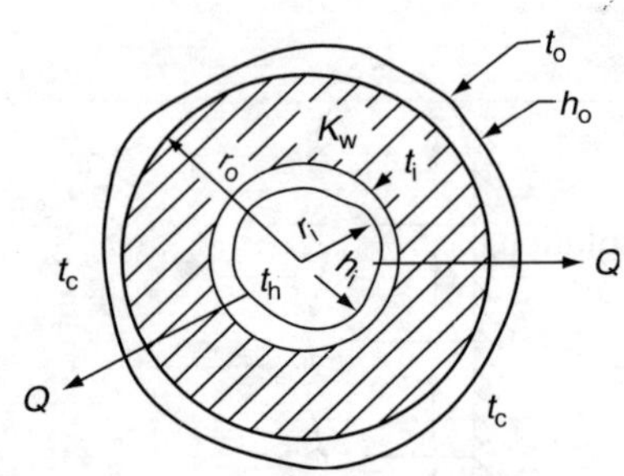

Fig. 18.16 *Radial heat transfer from hot to cold fluid through a cylinder*

When the wall thickness x_w is small, $A_o \cong A_{l.m.} \cong A_i$, and

$$\frac{1}{U_o} = \frac{1}{h_i} + \frac{x_w}{k_w} + \frac{1}{h_o}$$

18.3.1 Correlations in Convective Heat Transfer

Imagine there to be a curve in a fluid system in motion. If the tangent at every point of this curve indicates the direction of the velocity of the fluid particle, then the curve is known as a *streamline*. When one streamline slides over the other, the flow is laminar. When the streamlines are interwoven with one another and there is transverse flow of fluid particles, the flow is *turbulent*. The transition from laminar to turbulent flow in a tube depends on the mean velocity u_m, the diameter of the tube D, the density ρ, and the viscosity μ of the fluid. These variables are grouped together in a dimensionless parameter, called Reynolds number (Re), which is

$$\text{Re} = \frac{u_m D \rho}{\mu} = \frac{u_m D}{\nu}$$

where v = kinematic viscosity $\quad = \dfrac{\mu}{\rho}$ (m²/s)

If Re < 2100, the flow is laminar, and if Re > 2100, the flow is turbulent. The flow is most often turbulent and hardly ever laminar.

The heat transfer coefficient may be evaluated from correlations developed by dimensional analysis. In this method, all the variables pertinent to the phenomenon are to be enlisted, either by intuition or experience. Four fundamental units are selected, and these are mass M, length L, time T, and temperature θ.

Forced Convection

$$h = f(D, u, \rho, \mu, K, c_p)$$

Let $$h = B \cdot D^a \cdot u^b \cdot \rho^c \cdot \mu^d \cdot K^e \cdot c_p^f$$

Expressing the variables in terms of their dimensions

$$MT^{-3}\,\theta^{-1} = B \cdot L^a \cdot (LT^{-1})^b \cdot (ML^{-3})^c \cdot (ML^{-1}\,T^{-1})^d \cdot (MLT^{-3}\,\theta^{-1})^e \cdot (L^2 \cdot T^{-2}\,\theta^{-1})^f$$

$$= B \cdot L^{a+b-3c-d+e+2f} \cdot T^{-b-d-3e-2f} \cdot M^{c+d+e} \cdot \theta^{-e-f}$$

$\therefore \quad a + b - 3c - d + e + 2f = 0$

$$-b - d - 3e - 2f = -3$$
$$c + d + e = 1$$
$$-e - f = -1$$

The solution of these equations gives

$$a = c - 1, b = c, d = -c + f, e = 1 - f$$

$$\therefore \quad h = B \cdot D^{c-1} \cdot u^c \cdot \rho^c \cdot \mu^{-c+f} \cdot k^{1-f} \cdot c_p^f$$

$$\frac{hD}{K} = B\left(\frac{uD\rho}{\mu}\right)^c \cdot \left(\frac{\mu c_p}{K}\right)^f$$

$$Nu_d = B(Re_d)^c \cdot (Pr)^f \tag{18.19}$$

where

$$Nu_d = \text{Nusselt number} = \frac{hD}{K}$$

$$Re_d = \text{Reynolds number} = \frac{uD\rho}{\mu}$$

$$Pr = \text{Prandtl number} = \frac{\mu c_p}{K}$$

The constants B, c, and f are evaluated from experimental data. For fully developed turbulent flow inside tubes, the following equation, attributed to Dittus and Boelter, may be used

$$Nu_d = 0.023\ \mathrm{Re}_d^{0.8}\ \mathrm{Pr}^n \tag{18.20}$$

where $n = 0.4$, when the fluid is heated

$n = 0.3$, when the fluid is cooled

For fully developed laminar flow in a tube, (Fig. 18.17a), $R_{ed} < 2100$,

$$Nu_d = 4.364 \text{ for constant wall heat flux}$$
$$Nu_d = 3.66 \text{ for constant wall temperature} \tag{18.21}$$

Properties (μ, c_p, k, ρ) of the fluid are to be evaluated at the mean fluid temperature:

$$t_f = \frac{t_{f_1} + t_{f_2}}{2}$$

where t_{f_1} and t_{f_2} are the inlet and exit bulk or energy-average temperatures of the fluid.

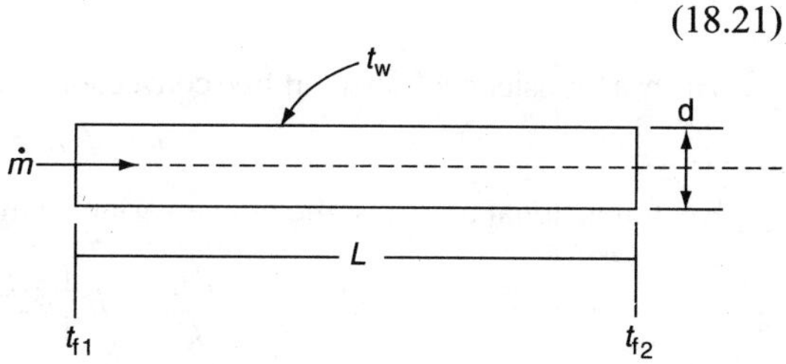

Fig. 18.17 (a) *Flow in a circular tube*

For flow over a flat plate with constant wall temperature (Fig. 18.17b)

$$Q = hA\,(t_w - t_\infty)$$

Laminar flow, $Re_L < 5 \times 10^5$

$$Nu_x = 0.332\ Re_x^{1/2}\ Pr^{1/3}$$

$$\overline{Nu_L} = 0.664\ Re_L^{1/2}\ Pr^{1/3} \tag{18.22}$$

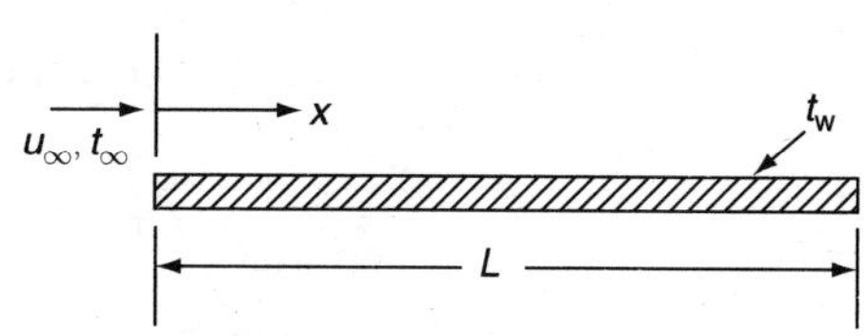

Fig. 18.17 (b) *Flow over a flat plate*

Turbulent flow, $Re_L > 5 \times 10^5$

$$Nu_x = 0.0296\, Re_x^{0.8}\, Pr^{1/3}$$

$$Nu_L = 0.037\, [Re_L^{0.8} - 850]\, Pr^{1/3} \qquad (18.23)$$

Properties are to be evaluated at $t_f = \dfrac{t_w + t_\infty}{2}$

For flow across a cylinder at constant wall temperature (Fig. 18.17c)

$$Q = h\,A\,[t_w - t_\infty]$$

$$Nu_d = C\, Re_d^n\, Pr^{1/3}$$

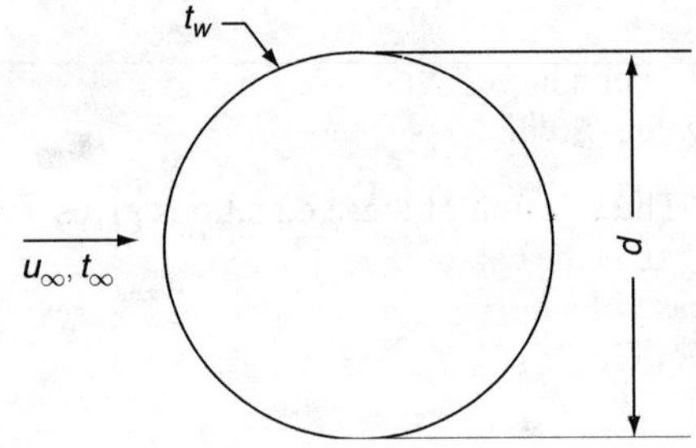

Fig. 18.17 (c) *Flow across a cylinder*

Re_d	c	n
0.4–4	0.989	0.330
4–40	0.911	0.385
40–4000	0.683	0.466
4000–40000	0.193	0.618
40000–400000	0.0266	0.805

For flow across a sphere, constant wall temperature:

Gases: $Nu_d = 0.37\, Re_d^{0.6}$

Water and Oil: $Nu_d = [1.2 + 0.53\, Re_d^{0.54}]\, Pr^{0.3}$ (18.24)

Free Convection

Let a fluid at T_o, with density ρ_o, change to temperature T with density ρ.

Then the buoyancy force, $F = \dfrac{(\rho_0 - \rho)\,g}{\rho}$

Now, let β = coefficient of volume expansion

then $\dfrac{1}{\rho} = \dfrac{1}{\rho_0} + \beta(T_0 - T)$ or $\rho_0 = \rho\,(1 + \beta \cdot \Delta T)$

$\therefore$ $$F = \beta \cdot g \cdot \Delta T$$

where $$\Delta T = T_0 - T$$

For an ideal gas $$\beta = \frac{1}{v}\left(\frac{\partial v}{\partial T}\right)_p = \frac{1}{v}\cdot\frac{R}{p} = \frac{1}{T}\ (K^{-1})$$

The heat transfer coefficient in free convection may be assumed to be a function of the variables as given below

$$h = f(L, K, c_p, \rho, \mu, g\,\beta\,\Delta T)$$

By dimensional analysis, the above variables may be arranged in three non-dimensional groups

$$\frac{hL}{K} = B\left(\frac{g\beta\Delta T \cdot L^3 \cdot \rho^2}{\mu^2}\right)^a \cdot \left(\frac{\mu c_p}{K}\right)^b$$

or $\text{Nu} = B \cdot \text{Gr}^a \cdot \text{Pr}^b$

where Nu = Nusselt number $= \dfrac{hL}{K}$

Gr = Grashoff number $= \dfrac{g\beta\Delta T L^3 \rho^2}{\mu^2}$

Pr = Prandtl number $= \dfrac{\mu c_p}{K}$

For a large number of experiments made on fluids it has been found that exponents a and b are of the same value. So the expression reduces to

$$\mathrm{Nu} = B \cdot (\mathrm{Gr} * \mathrm{Pr})^{a} \qquad (18.25)$$

L is the characteristic length, which is the length in the case of a vertical plate and cylinder, diameter in the case of a horizontal cylinder, and radius in the case of a sphere.
For Gr. Pr $< 10^9$, the flow is laminar, and

$$\mathrm{Nu} = 0.59\,(\mathrm{Gr.\ Pr})^{1/4} \qquad (18.26)$$

and for Gr. Pr $> 10^9$, the flow is turbulent, and

$$\mathrm{Nu} = 0.13\,(\mathrm{Gr.\ Pr})^{1/3} \qquad (18.27)$$

18.4 HEAT EXCHANGERS

A heat exchanger is a device in which heat is transferred between two moving fluids.

Heat exchangers may be parallel flow, counterflow or crossflow, depending upon the direction of the motion of the two fluids. If both the fluids move in the same direction, it is a parallel flow heat exchanger. If the fluids flow in the opposite directions, it is a counterflow heat exchanger. If they flow normal to each other, it is a crossflow heat exchanger.

18.4.1 Parallel Flow Heat Exchanger

Let us assume that the cold fluid (subscript c) is flowing through the inner tube and the hot fluid (subscript h) is flowing through the annulus. The hot fluid enters at t_{h1} and leaves at t_{h2}, while the cold fluid enters at t_{c1} and leaves at t_{c2}. Let us consider a differential length dL of the heat exchanger, as shown in Fig. 18.18, where the hot fluid is at t_h and the cold fluid is at t_c, and the temperature difference between the two fluids is Δt $(= t_h - t_c)$. Δt varies from Δt_i, at the inlet to Δt_e at the exit of the heat exchanger. Let dQ be the rate of heat transfer in that differential length. Then by energy balance

$$dQ = U_o \cdot dA_o \cdot \Delta t = -\, \dot{m}_h c_h\, dt_h = \dot{m}_c c_c\, dt_c$$

Here dt_h is negative, because there is a decrease in t_h along the flow, where dt_c is positive. Symbols $\dot{m}$ and c represent the mass flow rate and specific heat respectively.

$$\Delta t = t_h - t_c$$

$$\therefore \quad d(\Delta t) = dt_h - dt_c = -\frac{dQ}{\dot{m}_h c_h} - \frac{dQ}{\dot{m}_c c_c} = -\,dQ\left(\frac{1}{\dot{m}_h c_h} + \frac{1}{\dot{m}_c c_c}\right) = -\,\mu_p \cdot dQ$$

where μ_p is used for the expression in parenthesis.

$$\int_{\Delta t_i}^{\Delta t_e} d(\Delta t) = \int_i^e \mu_p\, dQ \quad \text{or} \quad \mu_p = \frac{\Delta t_i - \Delta t_e}{Q}$$

Also

$$dQ = U_o\, dA_o\, \Delta t \quad \text{or} \quad -\frac{d(\Delta t)}{\mu_p} = U_o\, dA_o\, \Delta t$$

$$\therefore \quad -\int_{\Delta t_i}^{\Delta t_e} \frac{d(\Delta t)}{\Delta t} = \int_i^e \mu_p\, U_o\, dA_o$$

$$\therefore \quad \ln \frac{\Delta t_i}{\Delta t_e} = \mu_p\, U_o A_o$$

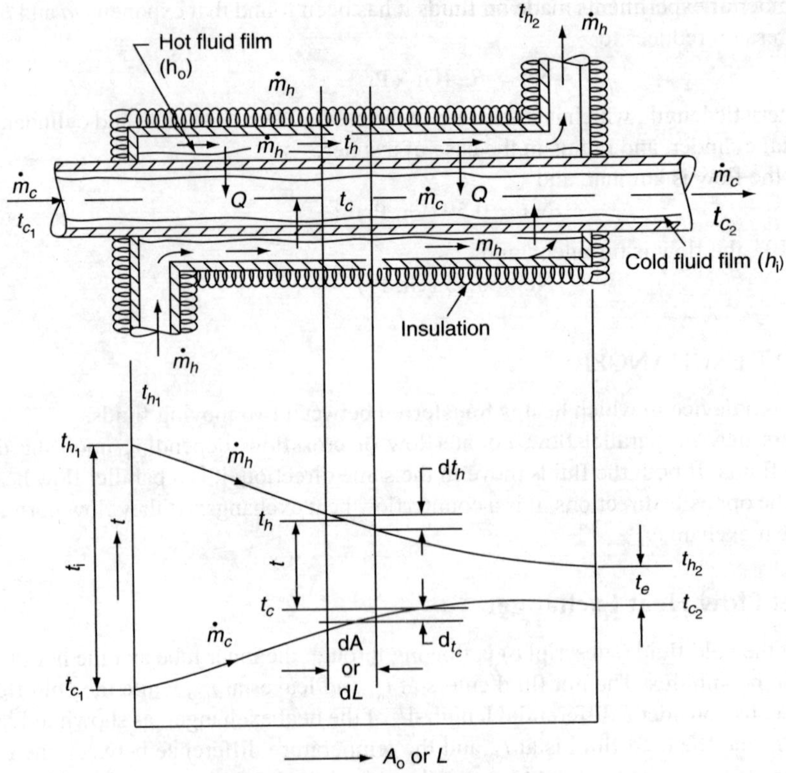

Fig. 18.18 Parallel flow heat exchanger

Substituting the expression for μ_p

$$\ln \frac{\Delta t_i}{\Delta t_e} = \frac{\Delta t_i - \Delta t_e}{Q} \cdot U_o A_o$$

$$\therefore \quad Q = \frac{\Delta t_i - \Delta t_e}{\ln \dfrac{\Delta t_i}{\Delta t_e}} U_o A_o$$

where $\Delta t_i = t_{h1} - t_{c1}$ and $\Delta t_e = t_{h2} - t_{c2}$

$$\therefore \quad Q = U_o A_o \Delta t_{l.m.} = \dot{m}_h c_h (t_{h1} - t_{h2}) = \dot{m}_c c_c (t_{c2} - t_{c1}) \tag{18.28}$$

where $\Delta t_{l.m.}$ = log-mean temperature difference (LMTD)

$$= \frac{\Delta t_i - \Delta t_e}{\ln \dfrac{\Delta t_i}{\Delta t_e}}$$

and
$$\frac{1}{U_o A_o} = \frac{1}{h_i A_i} + \frac{x_w}{k_w A_{l.m.}} + \frac{1}{h_o A_o}$$

18.4.2 Counterflow Heat Exchanger

The two fluids flow in opposite directions (Fig. 18.19). In the differential length, the rate of heat transfer

$$dQ = -\,\dot{m}_\mathrm{h}\, c_\mathrm{h} \mathrm{d}t_\mathrm{h} = -\,\dot{m}_\mathrm{c}\, c_\mathrm{c}\, \mathrm{d}t_\mathrm{c} = U_\mathrm{o} \cdot \mathrm{d}A_\mathrm{o} \cdot \Delta t$$

where both $\mathrm{d}t_\mathrm{h}$ and $\mathrm{d}t_\mathrm{c}$ are negative for positive x direction (towards the right). Now

$$\Delta t = t_\mathrm{h} - t_\mathrm{c}$$

$$\mathrm{d}(\Delta t) = \mathrm{d}t_\mathrm{h} - \mathrm{d}t_\mathrm{c} = -\frac{\mathrm{d}Q}{\dot{m}_\mathrm{h} c_\mathrm{h}} + \frac{\mathrm{d}Q}{\dot{m}_\mathrm{c} c_\mathrm{c}} = -\,\mathrm{d}Q \left(-\frac{1}{\dot{m}_\mathrm{h} c_\mathrm{h}} - \frac{1}{\dot{m}_\mathrm{h} c_\mathrm{c}} \right) = -\,\mathrm{d}Q\, \mu_\mathrm{c}$$

where

$$\mu_\mathrm{c} = \frac{1}{\dot{m}_\mathrm{h} c_\mathrm{h}} - \frac{1}{\dot{m}_\mathrm{c} c_\mathrm{c}}$$

$$\int_{\Delta t_\mathrm{i}}^{\Delta t_\mathrm{e}} \mathrm{d}(\Delta t) = \int_i^e -\,\mathrm{d}Q \cdot \mu_\mathrm{c}$$

$$\Delta t_\mathrm{i} - \Delta t_\mathrm{e} = \mu_\mathrm{c} \cdot Q \quad \text{or} \quad \mu_\mathrm{c} = \frac{\Delta t_\mathrm{i} - \Delta t_\mathrm{e}}{Q}$$

Again

$$\mathrm{d}Q = U_\mathrm{o}\, \mathrm{d}A_\mathrm{o}\, \Delta t$$

$$-\frac{\mathrm{d}(\Delta t)}{\mu_\mathrm{c}} = U_\mathrm{o} \cdot \mathrm{d}A_\mathrm{o} \cdot \Delta t$$

$$\int_{\Delta t_\mathrm{i}}^{\Delta t_\mathrm{e}} -\frac{\mathrm{d}(\Delta t)}{\Delta t} = \int_i^e U_\mathrm{o}\, \mathrm{d}A_\mathrm{o} \cdot \mu_\mathrm{c}$$

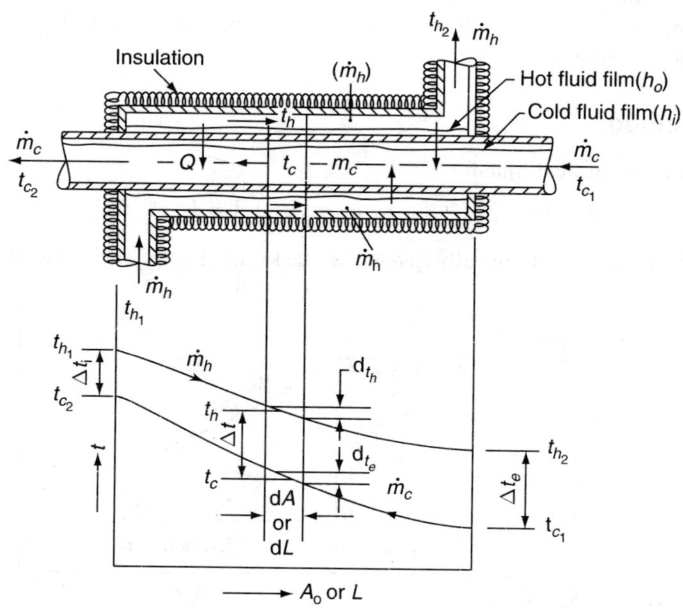

Fig. 18.19 *Counterflow heat exchanger*

$$\ln \frac{\Delta t_i}{\Delta t_e} = U_o \cdot A_o \cdot \mu_c = U_o A_o \cdot \frac{\Delta t_i - \Delta t_e}{Q}$$

$\therefore$
$$Q = U_o A_o \Delta t_{l.m.}$$

where
$$\Delta_{tl.m.} \cdot \frac{\Delta t_i - \Delta t_e}{\ln \dfrac{\Delta t_i}{\Delta t_e}}$$

$$\Delta t_i = t_{h1} - t_{c2},\ \Delta t_e = t_{h2} - t_{c1}$$

$$\frac{1}{U_o A_o} = \frac{1}{h_i A_i} + \frac{x_w}{K_w A_{l.m.}} + \frac{1}{h_0 A_o}$$

and when x_w is small,

$$\frac{1}{U_o} \cong \frac{1}{h_i} + \frac{x_w}{K_w} + \frac{1}{h_o}$$

$\therefore$
$$Q = U_o A_o \Delta t_{l.m.}$$
$$= \dot{m}_c c_h (t_{h1} - t_{h2}) = \dot{m}_c c_c [t_{c1} - t_{c2}] \quad (18.29)$$

For the same rate of heat transfer, and inlet and exit temperatures, $(\Delta t_{l.m.})$ counterflow is greater than $(\Delta t_{l.m.})$ parallel flow. So the surface area required is less for counterflow operation. For parallel flow, $t_{h2} > t_{c2}$, i.e., the hot fluid cannot be cooled below t_{c2} or the cold fluid cannot be heated above t_{h2}. But for counterflow operation, t_{h2} may be less than t_{c2} which means that the hot fluid can be cooled below t_{c2} or the cold fluid heated above t_{h2}. For these reasons, counterflow heat exchangers are much more common in practice.

When one of the two fluids undergoes phase change (at constant temperature and pressure), e.g. condensation and evaporation, $\Delta t_{l.m.}$ is the same for parallel flow and the counterflow (Fig. 18.20), and the heating surface required is also the same.

18.4.3 ε-NTU Method

In a heat exchanger, the rate of heat transfer

$$Q = \dot{m}_h c_h [t_{h1} - t_{h2}] = \dot{m}_c m_c [t_{c_2} t_{c_1}] = U_o A_o \Delta t_{l.m.}$$

where $\dot{m}_h$, c_h, $\dot{m}_c$, c_c, t_{h_1}, t_{c_1} and U_o are usually given. Two tasks are mostly encountered:

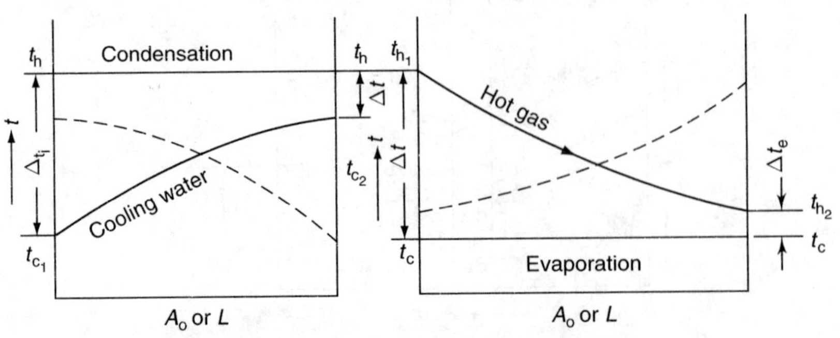

Fig. 18.20 Heat transfer with phase change

1. To estimate the surface area required (A_o). Then either t_{c_2} or t_{h_2} is given. We can use the LMTD method to find A_o.
2. For a given heat exchanger (i.e. A_o), to estimate the exit temperature t_{h_2} and t_{c_2}.

We cannot find t_{h_2} and t_{c_2} directly by the LMTD method. We have to use trial-and-error method. Assume t_{c_2}, find Q, t_{h_2}, $\Delta T_{\text{l.m.}}$ and then Q'. If $Q' \neq Q$, assume another t_{c_2} and repeat calculations till $Q' = Q$. ε-NTU method can here be easily used. We introduce three terms in this regard.

(a) Heat capacity ratio, $R = \dfrac{(\dot{m}c)_s}{(\dot{m}c)_l} = \dfrac{C_{\min}}{C_{\max}}$, (< 1)

where $(\dot{m}c)_s$ = smaller value of $\dot{m}_h c_h$ and $\dot{m}_c c_c = C_{\min}$
and $(\dot{m}c)_l$ = larger value of the two $\dot{m}_h c_h$ and $\dot{m}_c c_c = C_{\max}$

(b) Effectiveness, $\varepsilon = \dfrac{\text{Actual heat transfer}}{\text{Maximum possible heat transfer}}$

In a counterflow heat exchanger, e.g., $(t_{h_2})_{\min} = t_{c_1}$ or $(t_{c_2})_{\max} = t_{h_1}$.

$$\therefore \qquad \varepsilon = \frac{\dot{m}_h c_h \left(t_{h_1} - t_{h_2}\right)}{(\dot{m}c)_s \left(t_{h_1} - t_{c_1}\right)} = \frac{\dot{m}_c c_c \left(t_{c_2} - t_{c_1}\right)}{(\dot{m}c)_s \left(t_{h_1} - t_{c_1}\right)} (< 1)$$

If we consider $(\dot{m}c)_l$ in the denominator, then the other fluid should undergo a temperature change greater than the maximum available temperature difference, $t_{h_1} - t_{c_1}$. Therefore,

$$Q = \varepsilon\ (\dot{m}c)_s (t_{h_1} - t_{c_1}) \tag{18.30}$$

(c) NTU = Number of transfer units = $\dfrac{U_o A_o}{(\dot{m}c)_s} = \dfrac{U_o A_o}{c_{\min}}$

It gives the size of the heat exchanger. For a *parallel-flow* heat exchanger.

$$\ln \frac{\Delta t_e}{\Delta t_i} = -\mu_p U_o A_o$$

$$1 - \frac{t_{h_2} - t_{c_2}}{t_{h_1} - t_{c_1}} = 1 - e^{-\mu_p U_o A_o} = \frac{t_{h_1} - t_{c_1} - t_{h_2} + t_{c_2}}{t_{h_1} - t_{c_1}}$$

Let $\dot{m}_h c_h < \dot{m}_c c_c$. Then $R = \dfrac{\dot{m}_h c_h}{\dot{m}_c c_c} = \dfrac{t_{c_2} - t_{c_1}}{t_{h_1} - t_{h_2}}$ and $\varepsilon = \dfrac{t_{h_1} - t_{h_2}}{t_{h_1} - t_{c_1}}$

$$\varepsilon(1 + R) = 1 - \exp\left[-U_o A_o \left(\frac{1}{\dot{m}_h c_h} + \frac{1}{\dot{m}_c c_c}\right)\right]$$

$$\varepsilon_p = \frac{1 - \exp[-\text{NTU}(1 + R)]}{1 + R} \tag{18.31}$$

Similarly, for a counterflow heat exchanger,

$$\ln \frac{\Delta t_e}{\Delta t_i} = -U_o A_o \mu_c$$

$$\frac{t_{h_2} - t_{c_1}}{t_{h_1} - t_{c_2}} = e^{-U_o A_o \mu_c}$$

or, $$\frac{(t_{h_1} - t_{c_1}) - (t_{h_1} - t_{h_2})}{(t_{h_1} - t_{c_1}) - (t_{c_2} - t_{c_1})} = \frac{1 - \dfrac{t_{h_1} - t_{h_2}}{t_{h_1} - t_{c_1}}}{1 - \dfrac{t_{c_2} - t_{c_1}}{t_{h_1} - t_{c_1}}} = \frac{1 - \varepsilon}{1 - \varepsilon R} = K \text{ (say)} \quad \text{or,} \quad K - \varepsilon KR = 1 - \varepsilon$$

$$\varepsilon_c = \frac{1 - K}{1 - KR} = \frac{1 - e^{-U_o A_o \mu_c}}{1 - Re^{-U_o A_o \mu_c}}$$

$$= \frac{1 - \exp[-\text{NTU}(1 - R)]}{1 - R\exp[-\text{NTU}(1 - R)]} \tag{18.32}$$

When one of the two fluids undergoes phase change, $R = 0$. Then,

$$\varepsilon_p = \varepsilon_c = 1 - e^{-\text{NTU}} \tag{18.33}$$

For a *balanced* heat exchanger, $\dot{m}_h c_h = \dot{m}_c c_c$,

$$\therefore \quad R = 1, \Delta t_i = \Delta t_e = \Delta t_{l.m.}$$

$$t_{h_1} - t_{h_2} = t_{c_2} - t_{c_1} \text{ or, } t_{h_2} - t_{c_1} = t_{h_1} - t_{c_2}$$

$$U_o A_o \Delta t_{l.m.} = U_o A_o (t_{h_1} - t_{c_2}) = \dot{m}_h c_h (t_{h_1} - t_{h_2}) = (\dot{m}c)_s (t_{h_1} - t_{h_2})$$

$$\text{NTU} = \frac{U_o A_o}{(\dot{m}c)_s} = \frac{t_{h_1} - t_{h_2}}{t_{h_1} - t_{c_2}}$$

$$\varepsilon = \frac{t_{h_1} - t_{h_2}}{t_{h_1} - t_{c_1}} = \frac{t_{h_1} - t_{h_2}}{(t_{h_1} - t_{h_2}) + (t_{h_2} - t_{c_1})}$$

$$= \frac{\text{NTU}(t_{h_1} - t_{c_2})}{\text{NTU}(t_{h_1} - t_{c_2}) + (t_{h_1} - t_{c_2})} = \frac{\text{NTU}}{\text{NTU} + 1} \tag{18.34}$$

18.5 RADIATION HEAT TRANSFER

All bodies radiate heat. The phenomenon is identical to the emission of light. Two similar bodies isolated together in a vacuum radiate heat to each other, but the colder body will receive more heat than the hot body and thus become heated.

If Q is the total radiant energy incident upon the surface of a body some part of it (Q_a) will be absorbed, some (Q_r) will be reflected, and some (Q_t) will be transmitted through the body. Therefore,

$$Q = Q_a + Q_r + Q_t$$

or $$\frac{Q_a}{Q} + \frac{Q_r}{Q} + \frac{Q_t}{Q} = 1 \quad \text{or} \quad \alpha + \rho + \tau = 1$$

where α is known as *absorptivity*, ρ as *reflectivity*, and τ as transmissivity. For an *opaque* body, $\tau = 0$ and $\alpha + \rho = 1$. Most solids are opaque.

A body which absorbs all the incident radiation is called a black body. A black body is also the best radiator. Most radiating surfaces are grey and have an emissivity factor ε less than unity, where

$$\varepsilon = \frac{\text{Actual radiation of gray body at } T\text{ K}}{\text{Radiation of a black body at } T\text{ K}}$$

It can be shown that the *emissivity* or ability to radiate heat is equal to the *absorptivity* or ability to absorb heat (Kirchhoff's law), which justifies the statement that good absorbers are also good emitters. A brightly polished surface will have a low absorptivity and low emissivity.

The rate at which energy is radiated by a black body at temperature T (K) is given by the *Stefan-Boltzmann law*

$$Q = \sigma A T^4$$

where Q = rate of energy radiation, W

A = surface area radiating heat, m^2

and σ = Stefan-Boltzmann constant = 5.67×10^{-8} W/m^2 K^4

If the radiation from a heated body is dispersed into a spectrum by a prism, it is found that the radiant energy is distributed among various wave lengths. The *total emissive power* of a body, E, is defined as the total energy emitted by the body at a certain temperature per unit time and per unit surface area at all wavelengths. The *monochromatic emissive power*, E_λ, is defined as the radiant energy emitted by a body per unit time and per unit surface area at a particular wavelength and temperature. The variation of E_λ and λ and T is shown in Fig. 18.21. At a certain temperature,

$$E = \int_0^\infty E_\lambda \, d_\lambda \tag{18.35}$$

$$= \text{Area under the curve at that temperature.}$$

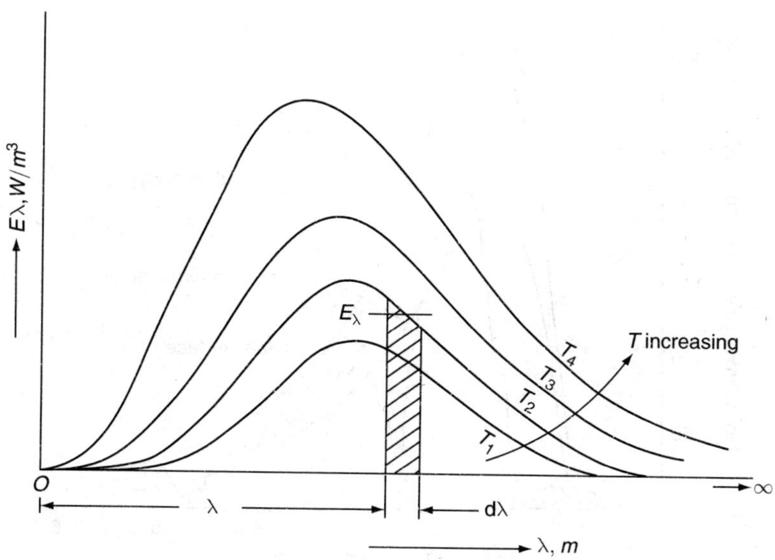

Fig. 18.21 *Radiation intensity $E\lambda$ varying with λ and T*

Thermal radiation extends over a spectral range of wavelengths from 0.1 μm to 100 μm and the spectral energy distribution of a black body is given by Planck's law:

$$E_{\lambda B} = \frac{C_1 \lambda^{-5}}{e^{C_2/\lambda T} - 1} \tag{18.36}$$

where λ = wavelength, μm; T = temperature, K;

$C_1 = 3.743 \times 10^8$ W $\times$ (μm)^4/m^2; $C_2 = 1.4387 \times 10^4$ μm.K.

$E_{\lambda B}$ is called the monochromatic emissive power of a black body. The *emissitivity* of a surface is then:

$$\varepsilon = \frac{E}{E_B} \tag{18.37}$$

where E_B is the total emissive power of a black body. A *gray body* has the *monochromatic emissivity*, ε_λ, constant over all wavelengths.

$$\varepsilon_\lambda = \frac{E_\lambda}{E_{\lambda B}} = \text{constant for a gray body} \tag{18.38}$$

Real surfaces are not gray and have a jagged emissive power distribution as shown in Fig. 18.22.

The actual radiant energy transfer between two bodies depends upon the (i) two surface temperatures, (ii) the surface emissivities, and (iii) the geometric orientation of the surfaces, i.e., how they view each other. *A radiation shape factor* F_{12} (or view factor) is defined as the fraction of energy leaving surface 1 and reaching surface 2. Similarly, F_{21} is the fraction of energy leaving 2 and reaching 1. It can be shown that

$$A_1 F_{12} = A_2 F_{21} \tag{18.39}$$

which is known as the *reciprocity theorem*.

The total emissive power of a black body is given by:

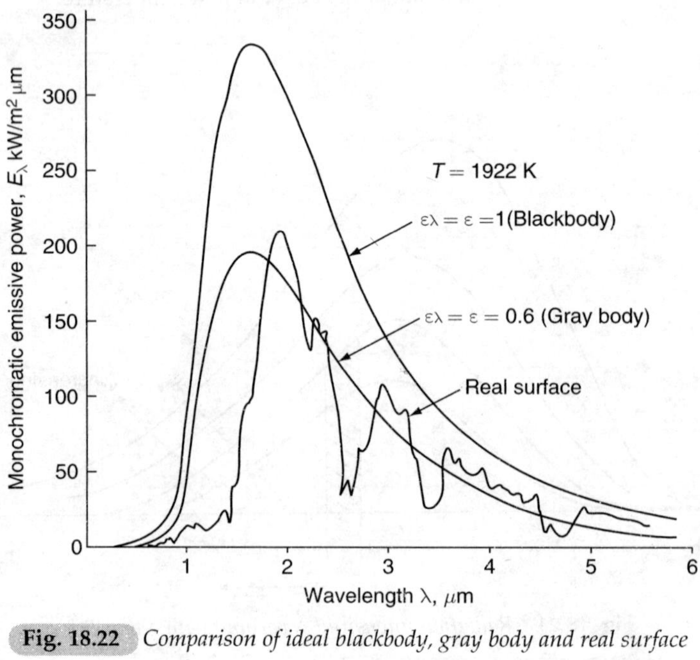

Fig. 18.22 *Comparison of ideal blackbody, gray body and real surface*

$$E_B = \int_0^\infty E_{\lambda B}\, d\lambda = \int_0^\infty \frac{C_1\lambda^{-5}}{e^{C_2/\lambda T} - 1}\, d\lambda = \sigma T^4$$

which is the Stefan-Boltzmann law, as stated earlier.

Charts of F_{12} for three geometries are shown in Figs 18.23, 18.24 and 18.25.

In estimating radiant heat transfer from gray surfaces, two terms will be introduced.

Radiosity (J) = total energy leaving a surface per unit area and per unit time (sum of emitted and reflected energies)

Irradiation (G) = total energy incident on a surface per unit area and per unit time.

The energy balance for the gray body (Fig. 18.26) assumed to be opaque ($\tau = 0$) gives:

$$J = \varepsilon E_B + \rho G = \varepsilon E_B + (1 - \varepsilon)G$$

$$G = [J - \varepsilon E_B]/[1 - \varepsilon]$$

The net energy leaving the surface is:

$$\frac{Q}{A} = J - G = \varepsilon E_B + (1 - \varepsilon)\, G - G$$

$$= \varepsilon(E_B - G) = \varepsilon \left[E_B - \frac{J - \varepsilon E_B}{1 - \varepsilon}\right] = \varepsilon\, \frac{E_B - J}{1 - \varepsilon}$$

$$Q_{net} = \frac{E_B - J}{(1 - \varepsilon)/A} \qquad (18.40)$$

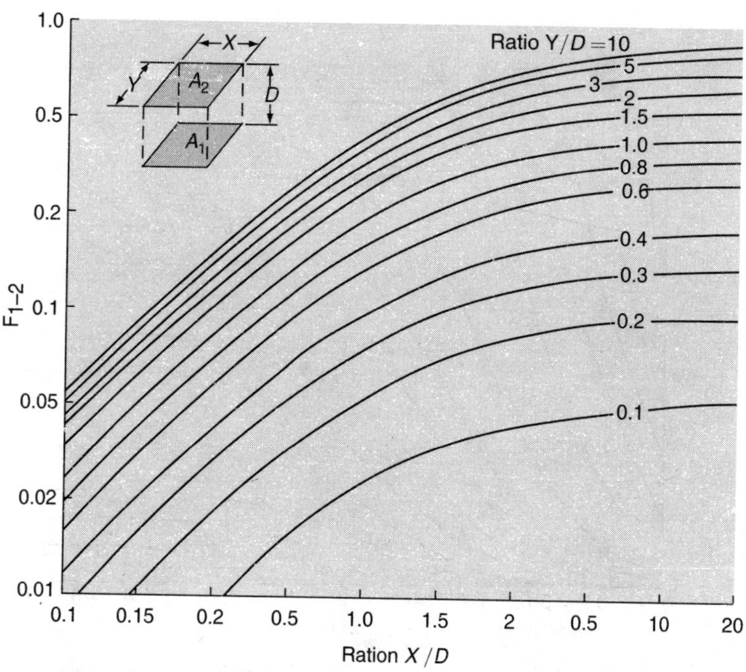

Fig. 18.23 *Radiation shape factor for radiation between parallel rectangles*

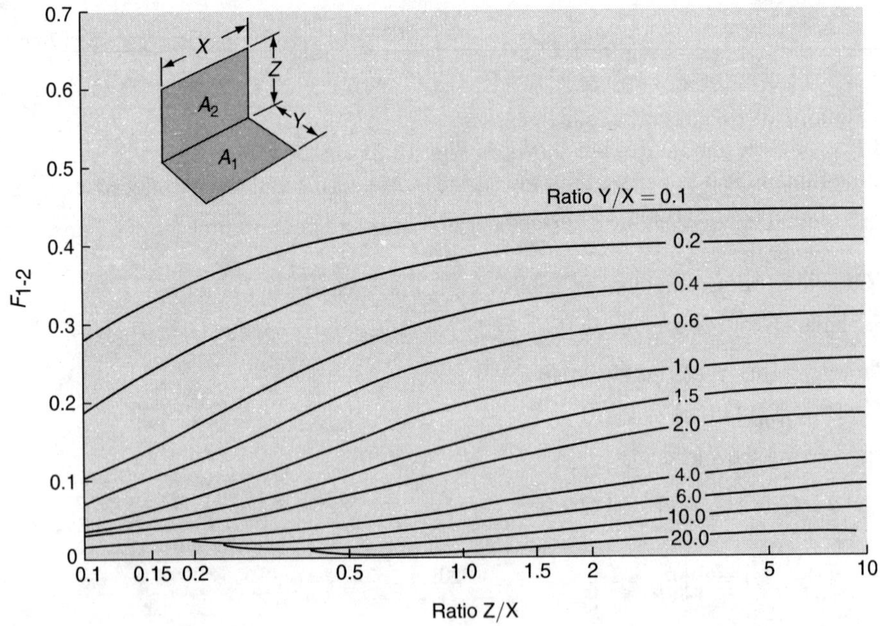

Fig. 18.24 *Radiation shape factor for radiation between perpendicular rectangles with a common edge*

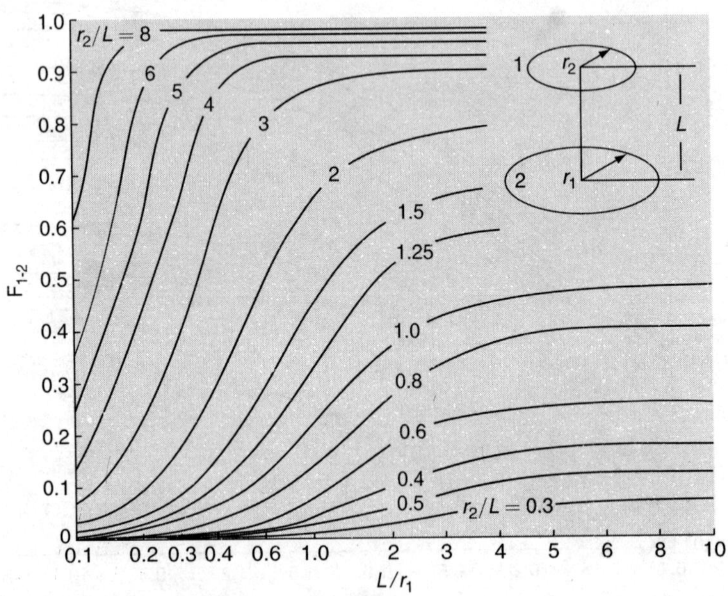

Fig. 18.25 *Radiation shape factor for radiation between two parallel concentric disks*

If the numerator is considered as the potential difference, the denominator as the "*surface resistance*" to radiation, and the heat flow as current, then a network element could be drawn as shown in Fig. 18.27.

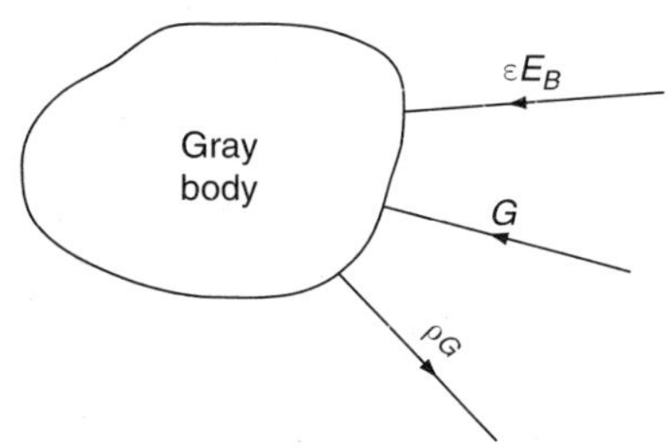

Fig. 18.26 *Radiosity and irradiation in a gray body*

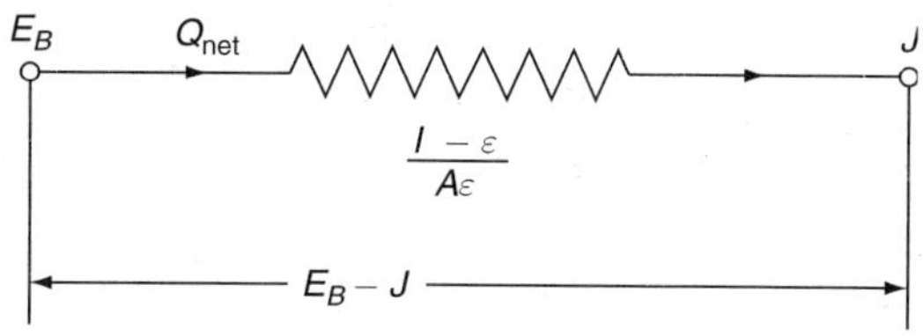

Fig. 18.27 *Surface resistance in radiation network*

Let us now consider the exchange of radiant energy by two surfaces A_1 and A_2 (Fig. 18.28). Of the total radiation which leaves the surface 1, the amount that reaches surface 2 is $J_1 A_1 F_{12}$, and of the total energy leaving surface 2, the amount that reaches surface 1 is $J_2 A_2 F_{21}$.

The net energy interchange between two surface is:

$$Q_{1-2} = J_1 A_1 F_{12} - J_2 E_{21} A_2$$

$$= (J_1 - J_2) A_1 F_{12} = \frac{J_1 - J_2}{1/(A_1 F_{12})} \tag{18.41}$$

The denominator $1/(A_1F_{12})$ is called the "space resistance" and the numerator $(J_1 - J_2)$ is the potential difference, as shown in Fig. 18.29.

Figure 18.30 shows a network which represents two surfaces exchanging radiative energy with each other.

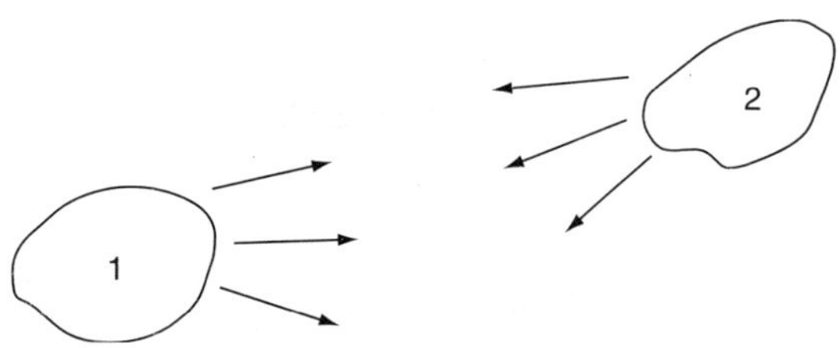

Fig. 18.28 *Radiation interaction between two gray surfaces*

The net heat transfer is:

$$(Q_{1-2})_{net} = \frac{E_{B_1} - E_{B_1}}{\frac{1-\varepsilon_1}{A_1\varepsilon_1} + \frac{1}{A_1 F_{12}} + \frac{1-\varepsilon_2}{A_2\varepsilon_2}}$$

$$= \sigma A_1 F_{12} [T_1^4 - T_2^4] \tag{18.42}$$

where $\frac{1}{A_1F_{12}} = \frac{1-\varepsilon_1}{A_1\varepsilon_1} + \frac{1}{A_1F_{12}} + \frac{1-\varepsilon_2}{A_2\varepsilon_2}$ or, $F_{12} = \frac{1}{\left(\frac{1}{\varepsilon_1} - 1\right) + \frac{1}{F_{12}} + \frac{A_1}{A_2}\left(\frac{1}{\varepsilon_2} - 1\right)}$ (18.43)

For a three-body problem, the radiation network is shown in Fig. 18.31.

For two gray surfaces connected by a nonconducting and re-radiating wall in an enclosure, the radiation network is shown in Fig. 18.32.

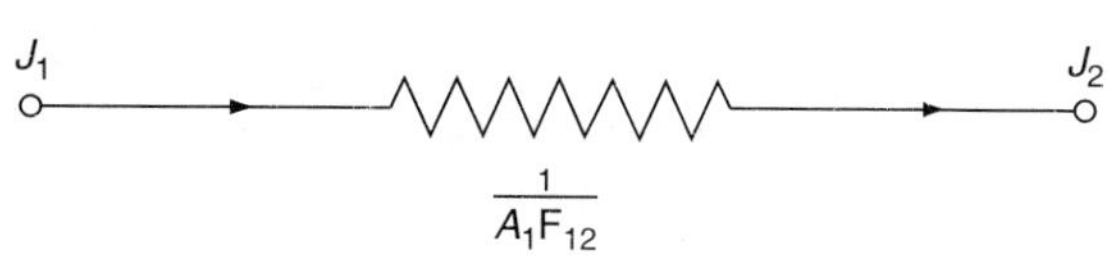

Fig. 18.29 *Space resistance in radiation network*

Node J_R is not connected to any surface resistance, since the surface R has no source of its own and it only reflects energy. It is called a *floating node*. Now,

$$F_{11} + F_{12} + F_{IR} = 1$$

$$F_{11} = 0$$

$$F_{IR} = 1 - F_{12}$$

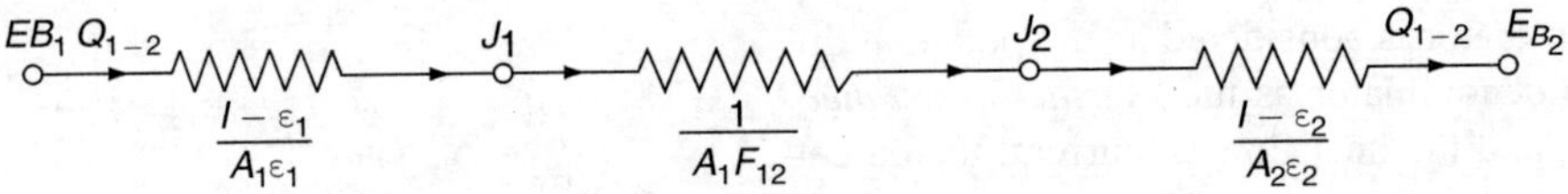

Fig. 18.30 *Radiation network for two gray surfaces*

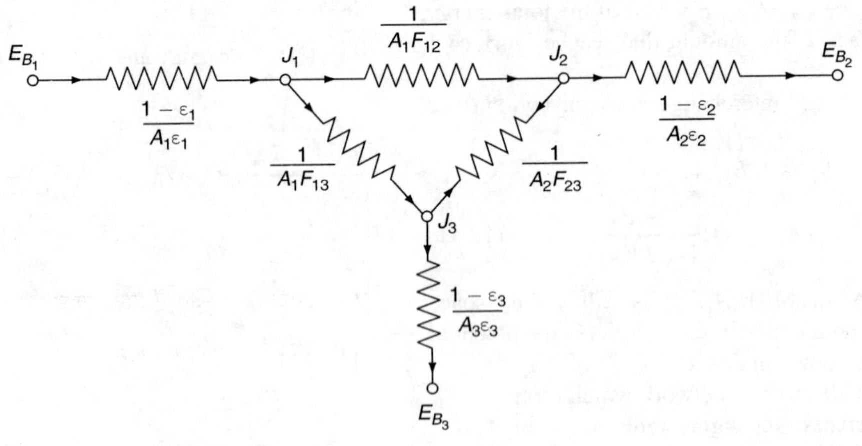

Fig. 18.31 *Radiation network for three gray surfaces*

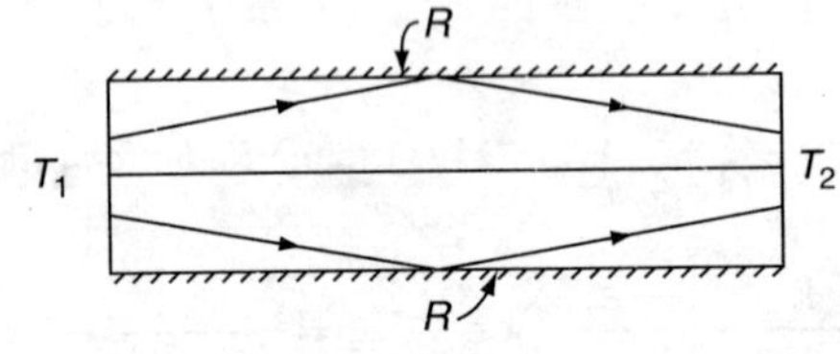

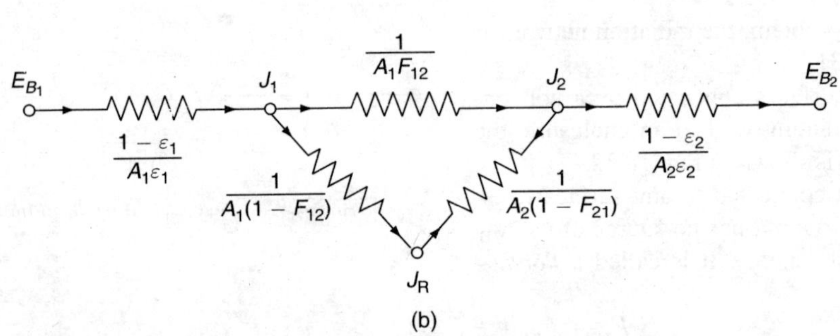

Fig. 18.32 *Radiation network for two gray surfaces enclosed by a third surface (R) which is nonconducting and re-radiating*

Similarly, $F_{2R} = 1 - F_{21}$

The network is a simple series-parallel system and may be solved to compute the heat flow. From Eq. (18.43) some simple situations can be explained.

(a) For two black surfaces, $\varepsilon_1 = \varepsilon_2 = 1, F_{12} = F_{12}$ (18.44)

(b) For two infinite parallel planes, $A_1 = A_2, F_{12} = 1$

$$F_{12} = \frac{1}{\dfrac{1}{\varepsilon_1} + \dfrac{1}{\varepsilon_2} - 1} \tag{18.45}$$

(c) For two concentric cylinders or spheres, A_1 = surface area of the inner cylinder or sphere,

$$F_{12} = \frac{1}{\dfrac{1}{\varepsilon_1} + \dfrac{A_1}{A_2}\left(\dfrac{1}{\varepsilon_2} - 1\right)} \tag{18.46}$$

(d) When the enclosed body (area A_1) is very small compared to the enclosure surface, $A_2 >> A_1$, then

$$F_{12} = \varepsilon_1 \tag{18.47}$$

18.5.1 Combined Convection and Radiation

Heat is transferred from a hot body both by natural convection and radiation. Rate of heat transfer by natural convection

$$Q_c = h_c A(t_w - t_f)$$

where h_c is the convective heat transfer coefficient.
Rate of heat transfer by radiation

$$Q_r = \sigma A_1 F_{1-2} (T_w^4 - T_f^4) = h_r A_1(t_w - t_f)$$

where h_r is known as the radiation heat transfer coefficient.

$$\therefore \quad h_r = \sigma F_{1-2} (T_w + T_f)(T_w^2 - T_f^2) \tag{18.48}$$

$\therefore$ Total rate of heat transfer

$$Q = Q_c + Q_r = (h_c + h_r) A_1(t_w - t_f) \tag{18.49}$$

Solved Examples

***Example* 18.1**

A cold storage room has walls made of 0.23 *m of brick on the outside,* 0.08 *m of plastic foam, and finally* 1.5 *cm of wood on the inside. The outside and inside air temperatures are* 22°C *and* – 2°C *respectively. If the inside and outside heat transfer coefficients are respectively* 29 *and* 12 *W/m² K, and the thermal conductivities of brick, foam, and wood are* 0.98, 0.02, *and* 0.17 *W/mK respectively, determine (a) the rate of heat removed by refrigeration if the total wall area is* 90 *m², and (b) the temperature of the inside surface of the brick.*

Solution Figure Ex. 18.1 shows the wall of the cold storage

$$\frac{1}{U} = \frac{1}{h_0} + \frac{x_1}{k_1} + \frac{x_2}{k_2} + \frac{x_3}{k_3} + \frac{1}{h_i} = \frac{1}{12} + \frac{0.23}{0.98} + \frac{0.08}{0.02} + \frac{1.5}{100 \times 0.17} + \frac{1}{29}$$

$$\frac{1}{U} = 0.0833 + 0.2347 + 4.0 + 0.0882 + 0.0345 = 4.4407 \text{ m}^2 \text{ K/W}$$

$$\therefore \quad U = 0.2252 \text{ W/m}^2 \text{ K}$$

Rate of heat transfer $Q = UA\,(t_0 - t_i) = 0.2252 \times 90\,[22 - (-2)] = 486.4$ W *Ans.*

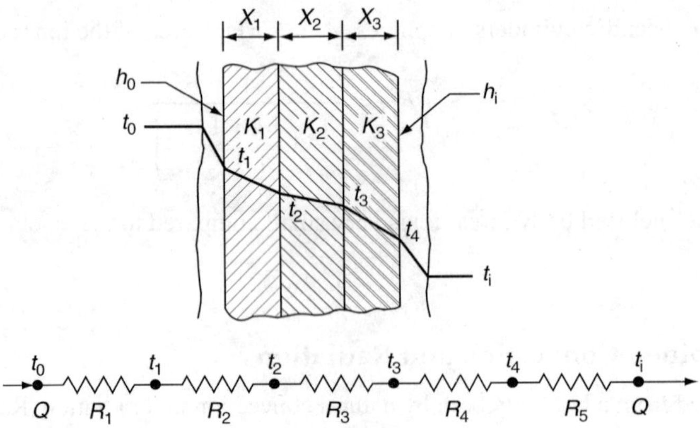

Fig. Ex. 18.1

Now

$$R_1 + R_2 = \frac{1}{h_0} + \frac{x_1}{k_1} = \frac{1}{12} + \frac{0.23}{0.98} = 0.318$$

$$\therefore \quad t_0 - t_2 = \frac{486.4 \times 0.318}{90} = 1.72\text{°C}$$

$$\therefore \quad t_2 = 22 - 1.72 = 20.28\text{°C}$$ *Ans.*

Example 18.2

Hot air at a temperature of 60°C is flowing through a steel pipe of 10 cm diameter. The pipe is covered with two layers of different insulating materials of thicknesses 5 cm and 3 cm, and their corresponding thermal conductivities are 0.23 and 0.37 W/mK. The inside and outside heat transfer coefficients are 58 and 12 W/m²K. The atmosphere is at 25°C. Find the rate of heat loss from a 50 m length of pipe. Neglect the resistance of the steel pipe.

Solution As shown in Fig. Ex. 18.2

$$r_1 = 5 \text{ cm}, \quad r_2 = 10 \text{ cm}, \quad r_3 = 13 \text{ cm}$$

$$K_1 = 0.23 \text{ and } K_2 = 0.37 \text{ W/mK}$$

$$h_i = 58 \text{ and } h_0 = 12 \text{ W/m}^2\text{/K}$$

$$t_i = 60°C,\ t_0 = 25°C$$

Rate of heat transfer

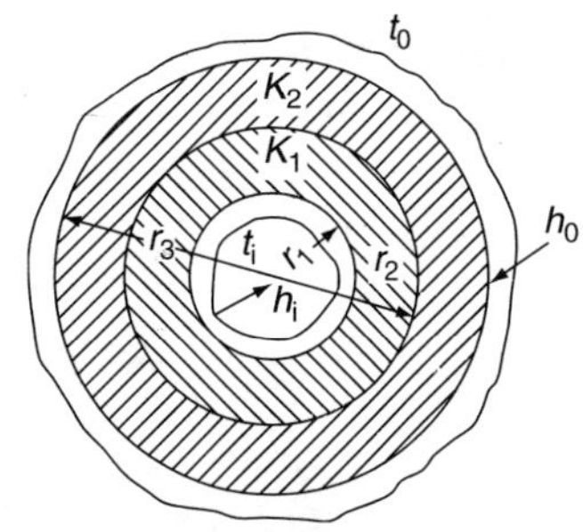

Fig. Ex. 18.2

$$Q = \frac{2\pi L (t_i - t_0)}{\frac{1}{h_i r_i} + \frac{1}{K_1} \ln \frac{r_2}{r_1} + \frac{1}{K_2} \ln \frac{r_3}{r_2} + \frac{1}{h_0 r_3}}$$

$$= \frac{2\pi \times 50 (60 - 25)}{\frac{1}{58 \times 0.05} + \frac{1}{0.23} \ln \frac{10}{5} + \frac{1}{0.37} \ln \frac{13}{10} + \frac{1}{12 \times 0.13}}$$

$$= \frac{6.28 \times 50 \times 35}{0.3448 + 0.7092 + 0.3448 + 0.6410} = 2334 \text{ W}$$

$$= 2.334 \text{ kW}$$ *Ans.*

Example 18.3

Three 10 mm dia. rods A, B and C protrude from a steam path at 100°C to a length of 0.25 m into the atmosphere at 20°C. The temperatures of the other ends are found to be 26.76°C for A, 32.00°C for B and 36.93°C for C. Neglecting the effects of radiation and assuming the surface film coefficient of 23 W/m² K, evaluate their thermal conductivities.

Solution If the tip loss is neglected, the tip temperature t_1 is given by

$$\frac{t_1 - t_\infty}{t_1 - t_\infty} = \frac{1}{\cosh ml}$$

For rod *A*,

$$\frac{26.76 - 20}{100 - 20} = \frac{6.76}{80} = \frac{1}{\cosh ml}$$

$$\cosh ml = 11.8$$

$$\text{or,} \quad ml = 3.16$$

$$m = \frac{3.16}{0.25} = 12.64 \text{ m}^{-1} = \left[\frac{hp}{KA}\right]^{1/2}$$

Here,

$$P = \pi \text{ d and } A = \frac{\pi}{4} d^2$$

$$\frac{23 \times \pi \times 0.01 \times 4}{K \times \pi \times 0.01 \times 0.01} = (12.64)^2$$

$$K_A = 57.58 \text{ W/mK}$$ *Ans.*

For rod *B*,

$$\frac{32 - 20}{100 - 20} = \frac{12}{80} = \frac{1}{\cosh ml}$$

$$\cosh ml = 6.67$$

$$ml = 2.6$$

or,

$$m = \frac{2.6}{0.25} = 10.4 = \left[\frac{23 \times \pi \times 0.01 \times 4}{K \times \pi \times 0.01 \times 0.01}\right]^{1/2}$$

$$K_B = \frac{9200}{(10.4)^2} = 85.2 \text{ W/mK} \qquad \textit{Ans.}$$

For rod C,

$$\frac{1}{\cosh ml} = \frac{36.93 - 20}{100 - 20} = \frac{16.93}{80}$$

$$\cosh ml = 4.73 \quad \text{or,} \quad ml = 2.23 \quad \text{or,} \quad m = 2.23/0.25 = 8.92$$

$$K_c = \frac{9200}{(8.92)^2} = 116 \text{ W/mK} \qquad \textit{Ans.}$$

Example 18.4

A 50 mm × 50 mm square iron bar 0.4 m long is connected to the walls of two heated reservoirs, each at 120°C. The ambient air temperature is 35°C and the convective heat transfer coefficient is 17.4 W/m² K. If the thermal conductivity of the bar is 52.2 W/mK, calculate the rate of heat loss from the bar and the temperature of the bar midway between the reservoirs.

Solution At the central plane of the bar (Fig. Ex. 18.4) $dT/dx = 0$. Taking one-half of the bar,

$$\frac{\theta_l}{\theta_1} = \frac{1}{\cosh ml}, \quad \frac{Q_1}{2} = m\,K\,A\,\theta_1 \tanh m_1$$

Here $m = \left[\frac{hp}{KA}\right]^{1/2} = \left[\frac{17.4 \times 4 \times 0.05}{52.2 \times 0.05 \times 0.05}\right]^{1/2} = 5.16 \text{ m}^{-1}$

$$ml = 5.16 \times 0.2 = 1.032$$

$$\cosh ml = 1.58, \ \tanh ml = 0.776$$

$$\frac{t_l - 35}{120 - 35} = \frac{1}{1.58}$$

$t_1 = 88.8°\text{C}$ = Midway temperature of the rod *Ans.*

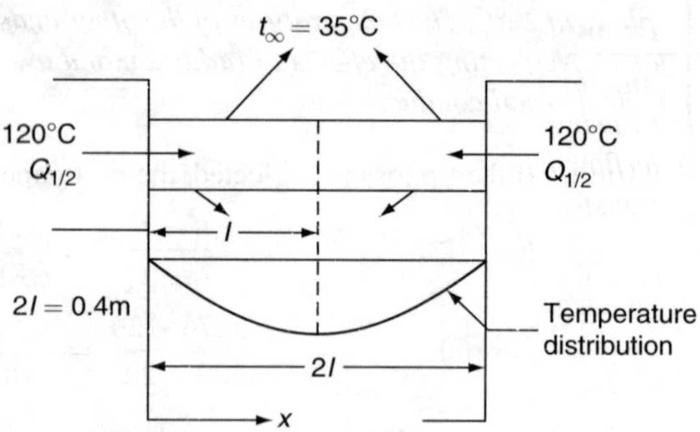

Fig. Ex. 18.4

$$Q_1/2 = 5.12 \times 52.2 \times 0.0025 \times 85 \times 0.776 = 44.2 \text{ W}$$

$$Q_1 = 88.4 \text{ W} \qquad \textit{Ans.}$$

Example 18.5

A load of peas at a temperature of 25°C is to be cooled down in a room at a constant air flow of 1°C. (a) How long will the peas require to cool down to 2°C when the surface heat transfer coefficient of the peas is 5.8 W/m² K? (b) What is the temperature of the peas after 10 min from the start of cooling? (c) What air temperature must be used if the peas were to be cooled down to 5°C in 30 min? The peas are supposed to have an average dia. of 8 mm. Their density is 750 kg/m³ and specific heat 3.35 kJ/kg K.

Solution From Eq. (18.14)

$$\frac{t - t_\infty}{t_0 - t_\infty} = e^{-(hA\tau)/(\rho c V)}$$

Here, $$\frac{\rho V}{A} = \frac{750 \times 4\pi (d/2)^3}{3 \times 4\pi (d/2)^2} = \frac{750}{3} \times \frac{d}{2} = \frac{750}{3} \times \frac{0.008}{2} = 1$$

$$\ln \frac{2-1}{25-1} = -\frac{hA\tau}{\rho c V} = -\frac{h\tau}{c} = -\frac{5.8 \times 10^{-3}\tau}{3.35}$$

(a) $$\ln 24 = \frac{5.8 \times 10^{-3} \times 3600\,\tau}{3.35}$$

$$\tau = 0.51\ h = 31\ \text{min}$$ *Ans.* (a)

(b) $$\frac{t-1}{25-1} = e^{-(5.8 \times 10^{-3} \times 10 \times 60)/3.5}$$

$$\tau = 9.5°\text{C}$$ *Ans.* (b)

(c) $$\frac{5-t_\infty}{25-t_\infty} = e^{-(5.8 \times 10^{-3} \times 30 \times 60)/3.35}$$

$$\tau = 4.1°\text{C}$$ *Ans.* (c)

Example 18.6

An oil cooler for a lubrication system has to cool 1000 kg/h of oil (c_p = 2.09 kJ/kg K) from 80°C to 40°C by using a cooling water flow of 1000 kg/h available at 30°C. Give your choice for a parallel flow or counterflow heat exchanger, with reasons. Estimate the surface area of the heat exchanger, if the overall heat transfer coefficient is 24 W/m² K (c_p of water = 4.18 kJ/kg K).

Solution Rate of heat transfer

$$Q = \dot{m}_h c_h (t_{h1} - t_{h2}) = \dot{m}_c c_c (t_{c2} - t_{c1})$$

$$1000 \times 2.09\ (80 - 40) = 1000 \times 4.18\ (t_{c2} - 30)$$

$$\therefore \quad t_{c2} = 50°\text{C}$$

Since $t_{c2} > t_{h2}$, counterflow arrangement must be used (Fig. Ex. 18.6).

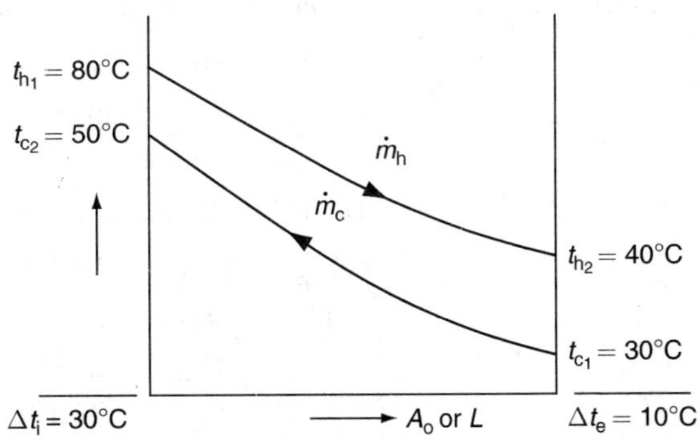

Fig. Ex. 18.6

$$(\Delta t)_{\text{l.m.}} = \frac{30-10}{\ln \frac{30}{10}} = 18.2°\text{C}$$

$$Q = 1000 \times 2.09 \times 40$$
$$= 83.6 \times 10^3 = U_o A_o \Delta t_{\text{l.m.}}$$

$$A_o = \frac{83{,}600}{24 \times 18.2 \times 3.6} = 53.16\ \text{m}^2$$ *Ans.*

Example 18.7

Water is evaporated continuously at 100°C in an evaporator by cooling 500 kg/h of air from 260°C to 150°C. Calculate the heat transfer surface area required and the steam evaporation rate per hour if the liquid enters at 100°C. Take U_0 = 46 W/m² K and c_p of air = 1.005 kJ/kg K.

Solution At 100°C, h_{fg} = 2257 kJ/kg. If $\dot{m}$ is the rate of evaporation

$$Q = \dot{m}\, h_{fg} = \dot{m}_a\, c_{pa}\, (t_{a1} - t_{a2})$$

$$\therefore\ \dot{m} \times 2257 = 500 \times 1.005\ (260 - 150)$$

$$\therefore\quad \dot{m} = 24.49 \text{ kg/h} \qquad \textit{Ans.}$$

As shown in Fig. Ex. 18.7

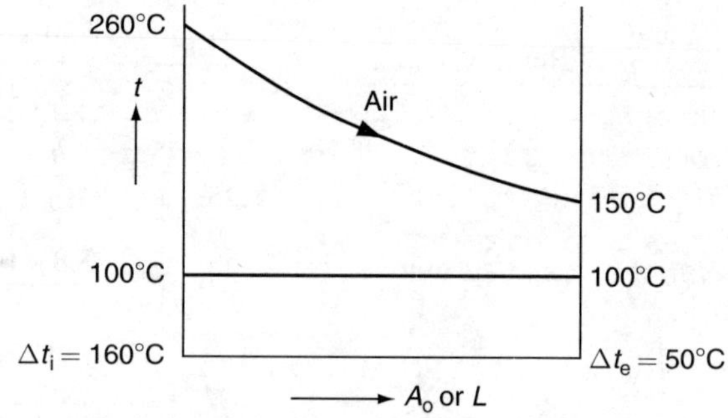

Fig. Ex. 18.7

$$\Delta t_{l.m.} = \frac{160 - 50}{\ln \dfrac{160}{50}} = 94.5°\text{C}$$

$$\therefore\quad Q = U_o\, A_o\, \Delta t_{l.m.} = 500 \times 1.005 \times 110$$

$$\therefore\quad A_o = \frac{500 \times 1.005 \times 110}{46 \times 94.5 \times 3.6} = 3.53 \text{ m}^2 \qquad \textit{Ans.}$$

Example 18.8

An oil fraction at 121°C is to be cooled at the rate of 20.15 kg/s in a simple counterflow heat exchanger using 5.04 kg/s of water initially at 10°C. The exchanger contains 200 tubes each 4.87 m long and 1.97 cm o.d., with U_o = 0.34 kW/m² K. If the specific heat of oil is 2.094 kJ/kg K, calculate the exit temperature of the oil and the rate of heat transfer.

Solution

$$\dot{m}_h = 20.15 \text{ kg/s},\ c_h = 2.094 \text{ kJ/kg K},\ t_{h_1} = 121°\text{C},$$

$$\dot{m}_c = 5.04 \text{ kg/s},\ c_c = 4.2 \text{ kJ/kg K},\ t_{c_1} = 10°\text{C},$$

$$U_o = 0.34 \text{ kW/m}^2 \text{ K}$$

$$A_o = n\, d_0\, l = 200 \times \pi \times 0.0197 \times 4.87 = 60.28 \text{ m}^2$$

$$(\dot{m}c)_{oil} = 20.15 \times 2.094 = 42.19 \text{ kW/K}$$

$$(\dot{m}c)_{water} = 5.04 \times 4.2 = 21.09 \text{ kW/K}$$

$$C_{min} = 21.09 \text{ kW};\ C_{max} = 42.19 \text{ kW/K}$$

$$R = \frac{C_{min}}{C_{max}} = \frac{21.09}{42.19} = 0.5$$

$$\text{NTU} = \frac{U_o A_o}{C_{min}} = \frac{0.34 \times 60.28}{21.09} = 0.972$$

$$\varepsilon = \frac{1 - \exp\left[-\text{NTU}\,(1 - R)\right]}{1 - R\exp\left[-\text{NTU}\,(1 - R)\right]} = \frac{0.3849}{0.6925} = 0.5558 = \frac{\Delta t_{larger}}{t_{h_1} - t_{c_1}}$$

$$\Delta t_{larger} = (\Delta T)_{water} = 0.5558\ (121 - 10) = 61.69°\text{C}$$

$$\Delta t_{oil} = \frac{\Delta t_{water} \times (\dot{m}c)_{water}}{(\dot{m}c)_{oil}} = 30.84°\text{C} = t_{h_1} - t_{h_2}$$

$$t_{h_2} = 121 - 30.84 = 90.16°\text{C} \qquad \textit{Ans.}$$

$= \text{exit temperature of oil.}$

$$Q = \dot{m}_h c_h [t_{h_1} - t_{h_2}] = 42.19 \times 30.84 = 1308 \text{ kW}$$ *Ans.*

Example 18.9

Water flows inside a tube 5 cm in diameter and 3 m long at a velocity 0.8 m/s. Determine the heat transfer coefficient and the rate of heat transfer if the mean water temperature is 50°C and the wall is isothermal at 70°C. For water at 60°C, take $K = 0.66$ W/mK, $v = 0.478 \times 10^{-6}$ m²/s, and $Pr = 2.98$.

Solution Reynolds number,

$$\text{Re} = \frac{u_m D}{v} = \frac{0.8 \times 0.05}{0.478 \times 10^{-6}} = 83{,}700$$

The flow is turbulent. Prandtl number, Pr = 2.98

From the Dittus and Boelter equation

$$\text{Nu} = 0.023\,(\text{Re})^{0.8}\,(\text{Pr})^{0.4}$$

$$\frac{hD}{K} = 0.023\,(83{,}700)^{0.8}\,(2.98)^{0.4}$$

$$\frac{h \times 0.05}{0.06} = 0.023 \times 8673.6 \times 1.5478$$

$$\therefore \quad h = 4075 \text{ W/m}^2\text{ K}$$ *Ans.*

$$\therefore \quad Q = hA\,(t_w - t_f) = 4075 \times \pi \times 0.05 \times 3 \times 20 = 38387 \text{ W}$$

$$= 38.39 \text{ kW}$$ *Ans.*

Example 18.10

An electrically heated plate 15 cm high and 10 cm wide is maintained at 140°C. Estimate the rate of heat dissipation from both sides of the plate in an atmosphere at 20°C, if the radiation heat transfer coefficient is 8.72 W/m² K. For air at the mean temperature of 80°C, take $v = 21.09 \times 10^{-6}$ m²/s, $Pr = 0.692$, and $K = 0.0305$ W/mK.

Solution

$$Q = (h_c + h_r)\,A(t_w - t_f)$$

where

$$A = 0.1 \times 0.15 \times 2 = 0.03 \text{ m}^2$$

$$h_r = 8.72 \text{ W/m}^2\text{ K}, \; t_w - t_f = 140 - 20 = 120°\text{C}$$

$$\beta = \frac{1}{v}\left(\frac{\partial v}{\partial T}\right)_p = \frac{1}{T_{mean}} = \frac{1}{273 + 80} = \frac{1}{353} \;(K)^{-1}$$

$$L = 0.15 \text{ m}$$

$$\therefore \text{ Grashoff number,} \quad G_r = \frac{g\beta\,\Delta T L^3}{v^2} = \frac{9.81 \times 120 \times (0.15)^3}{353 \times (21.09 \times 10^{-6})^2} = 25.4 \times 10^6$$

$$\text{Gr. Pr} = 25.4 \times 10^6 \times 0.692 = 17.6 \times 10^6$$

Since the product Gr. Pr is less than 10^9

$$\text{Nu} = 0.59\ (\text{Gr. Pr})^{1/4} = 0.59\ (17.6 \times 10^6)^{1/4} = 38.2$$

$$\therefore \quad h_c = \frac{38.2 \times 0.0305}{0.15} = 7.77 \text{ W/m}^2 \text{ K}$$

$$\therefore \quad Q = (8.72 + 7.77) \times 0.03 \times 120$$

$$= 59.36 \text{ W}$$

Ans.

Example 18.11

A long steel rod, 2 cm in diameter, is to be heated from 427°C to 538°C. It is placed concentrically in a long cylindrical furnace which has an inside diameter of 16 cm. The inner surface of the furnace is at a temperature of 1093°C, and has an emissivity of 0.85. If the surface of the rod has an emissivity of 0.6, find the time required for the heating operation. Take for steel, $\rho = 7845$ kg/m³ and $c = 0.67$ kJ/kg K.

Solution The surface area of the rod (Fig. Ex. 18.11). $A_1 = \pi \times 2 \times L$ cm², and the surface area of the furnace, $A_2 = \pi \times 16 \times L$ cm²

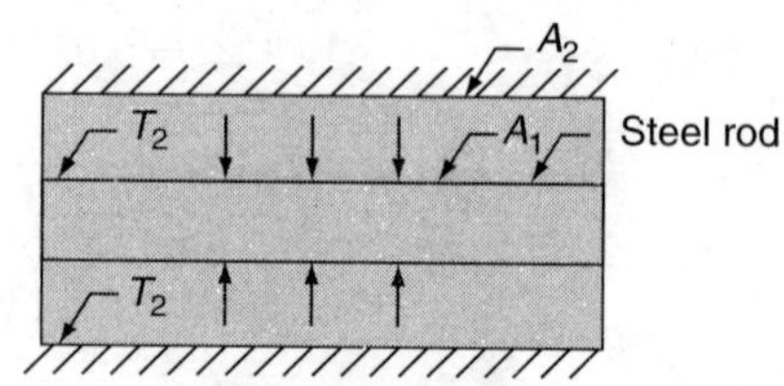

Fig. Ex. 18.11

$$\therefore \quad \frac{A_1}{A_2} = \frac{1}{8}$$

Now $$F_{1-2} = \frac{1}{\frac{1}{\varepsilon_1} + \frac{A_1}{A_2}\left(\frac{1}{\varepsilon_2} - 1\right)} = \frac{1}{\frac{1}{0.6} + \frac{1}{8}\left(\frac{1}{0.85} - 1\right)}$$

$$= \frac{1}{1.684} = 0.592$$

Initial rate of heat absorption by radiation, when the rod is at 427°C or 700 K

$Q_i = \sigma A_1 F_{1-2} (T_1^4 - T_2^4) = 5.67 \times 10^{-8} \times \pi \times 2 \times 10^{-2} \times 1 \times 0.592\ (700^4 - 1366^4) = -6332$ W/m

Rate of heat absorption at the end of the heating process, when the rod is at 538°C.

$$Q_e = 5.67 \times 10^{-8} \times \pi \times 2 \times 10^{-2} \times 1 \times 0.592\ (811^4 - 1366^4) = -5937 \text{ W/m}$$

$\therefore$ Average rate of heat absorption during the heating process

$$Q_{a.v.} = \frac{6322 + 5937}{2} = 6130 \text{ W/m}$$

Time required for heating, τ is obtained from the equation

$$m\, c_p\, \Delta T = Q_{av} \times \tau = 6130\, \tau$$

$$\tau = \frac{7845 \times \pi/4 \times 10^{-4} \times 1 \times 0.67 \times 111}{6130 \times 10^{-3}} \text{ s}$$

$$= 29.88 \text{ s}$$

Ans.

Example 18.12

Two long concentric cylinders exchange heat by radiation. The inner cylinder having a diameter of 10 cm has $T_1 = 1000$ K and $\varepsilon_1 = 0.65$ while the outer cylinder having a diameter of 20 cm has $T_2 = 500$ K and $\varepsilon_2 = 0.40$. Calculate the net heat transfer between the two cylinders per unit length.

Solution Referred to Fig. Ex. 18.12

$$A_1 = \pi d_1 = \pi \times 0.10 = 0.314 \text{ m}$$

$$A_2 = \pi d_2 = \pi \times 0.20 = 0.628 \text{ m}$$

$$\frac{1-\varepsilon_1}{A_1\varepsilon_1} = \frac{1-0.65}{0.314 \times 0.65} = 1.715$$

$$\frac{1-\varepsilon_2}{A_2\varepsilon_2} = \frac{1-0.4}{0.628 \times 0.4} = 2.39$$

$$\frac{1}{A_2F_{12}} = \frac{1}{0.314 \times 0.4} = 3.19$$

$$\sum R = \frac{1-\varepsilon_1}{A_1\varepsilon_1} + \frac{1}{A_1F_{12}} + \frac{1-\varepsilon_2}{A_2\varepsilon_2}$$

$$= 1.715 + 2.39 + 3.19 = 7.295$$

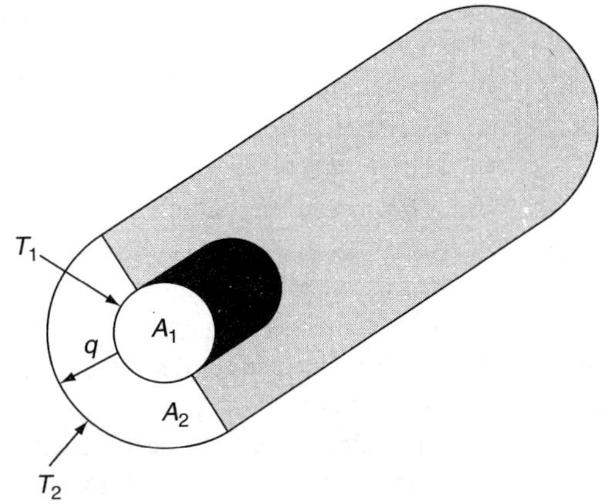

Fig. Ex. 18.12

$$E_{B_1} = \sigma T_1^4 = 5.67 \times 10^{-8} \times (1000)^4 = 5.67 \times 10^4 \text{ W/m}^2$$

$$E_{B_2} = \sigma T_2^4 = 5.67 \times 10^{-8} \times (500)^4 = \frac{5.67 \times 10^4}{16} = 0.35 \times 10^4 \text{ W/m}^2$$

$$Q = \frac{E_{B_1} - E_{B_2}}{\sum R} = \frac{5.32 \times 10^4}{7.295} = 0.729 \times 10^4 = 7290 \text{ W/m length}$$ *Ans.*

Review Questions

18.1 How is the subject of Heat Transfer different from the subject of Thermodynamics?

18.2 What are the three basic modes in which heat is transferred?

18.3 Why are good electrical conductors also good thermal conductors?

18.4 What is Fourier's law of heat conduction?

18.5 How does the slope of the temperature profile in a wall depend on its thermal conductivity?

18.6 Show that, for estimating radial heat conduction through a cylindrical wall, the log-mean area of the inner and outer surfaces has to be considered.

18.7 Show that for estimating radial heat conduction through a spherical wall, the geometric mean area of the inner and outer surfaces should be considered.

18.8 How do fins affect the heat transfer rate?

18.9 How is fin efficiency defined?

18.10 What is meant by transient heat conduction?

18.11 What is lumped-capacity analysis?

18.12 What are Biot number and Fourier number? What is their physical significance?

18.13 What do you understand by natural convection and forced convection?

18.14 How is heat transfer coefficient defined? What is its dimension?

18.15 What are the three resistances offered to heat transfer from one fluid to another through a clean wall?

18.16 What is Reynolds number? What is its critical value when the flow through a tube becomes turbulent?

18.17 What are Prandtl number and Nusselt number?

18.18 For fully developed laminar flow in a tube, what are the values of Nusselt number (a) for constant wall temperature, (b) for constant wall heat flux?

18.19 What are the expressions of Nusselt number for (a) laminar flow and (b) turbulent flow, over a flat plate?

18.20 What is Dittus-Boelter equation? Where is it used?

18.21 What is Grashof number? When does it become significant?

18.22 What is a heat exchanger?

18.23 Why are counterflow heat exchangers superior to parallel flow heat exchangers?

18.24 What is log-mean temperature difference?

18.25 Why do the directions of flows of the two fluids in a heat exchanger become immaterial when one of the two fluids undergoes phase change?

18.26 When is ε-NTU method convenient to use in heat exchanger analysis?

18.27 Define (a) effectiveness, (b) heat capacity ratio and (c) NTU, in regard to a heat exchanger.

18.28 Derive the expression for effectiveness in a (a) parallel flow heat exchanger, (b) counter-flow heat exchanger.

18.29 What is the expression for effectiveness when one of the fluids undergoes phase change?

18.30 Find the expression for effectiveness of a balanced heat exchanger with equal heat capacities.

18.31 Define absorptivity, reflectivity and transmissivity.

18.32 What is emissivity? What is Kirchhoff's law?

18.33 What is a black body?

18.34 What is the Stefan-Boltzmann law?

18.35 What is the view factor? Why is it significant in radiant heat exchange between two bodies?

18.36 What is the reciprocity theorem?

18.37 What is a gray body?

18.38 What are meant by (a) monochromatic emissive power, (b) total emissive power?

18.39 What is Planck's law of thermal radiation? Explain its importance.

18.40 Define radiosity and irradiation for a gray body.

18.41 Explain (a) surface resistance and (b) space resistance in radiant energy change between two gray bodies.

18.42 Give the radiation network for two gray surfaces and derive the view factor.

18.43 What do you mean by "floating node"?

18.44 How would you define the heat transfer coefficient for combined convection and radiation?

Problems

18.1 A room has a brick wall 25 cm in thickness. The inside air is at 25°C and the outside air is at – 15°C. The heat transfer coefficients on the inside and outside are 8.72 and 28 W/m^2 K respectively. The thermal conductivity of brick is 0.7 W/mK. Find the rate of heat transfer through the wall and the inside surface temperature.

18.2 For the wall in the above problem, it is proposed to reduce the heat transfer by fixing an insulating board (K = 0.05 W/mK), 2.5 cm in thickness, to the inside surface. Find the rate of heat transfer and the inside surface temperature.

18.3 Sheets of brass and steel, each 1 cm thick, are placed in contact. The outer surface of brass is kept at 100°C and the outer surface of steel is kept at 0°C. What is the temperature of the common interface? The thermal conductivities of brass and steel are in the ratio 2:1. *Ans.* 66.7°C

18.4 In a pipe carrying steam, the outside surface (15 cm OD) is at 300°C. The pipe is to be covered with insulation (K = 0.07 W/m K) such that the outside surface temperature does not exceed 60°C. The atmosphere is at 25°C and the heat transfer coefficient is 11.6 W/m^{2}K. Find the thickness of insulation required and the rate of heat loss per m length of pipe.

18.5 The passenger compartment of a jet transport is essentially a cylindrical tube of diameter 3 m and length 20 m. It is lined with 3 cm of insulating material (K = 0.04 W/mK), and must be maintained at 20°C for passenger comfort although the average outside temperature is – 30°C at its operating height. What rate of heating is required in the compartment, neglecting the end effects?

18.6 A hollow sphere (K = 35 W/mK), the inner and outer diameters of which are 28 cm and 32 cm respectively, is heated by means of a 20 ohm coil placed inside the sphere. Calculate the current required to keep the two surfaces at a constant temperature difference of 50°C, and calculate the rate of heat supply.

18.7 (a) Develop an expression for the steady state heat transfer rate through the walls of a spherical container of inner radius r_i and outer radius

r_0. The temperatures are t_0 and t_1 at radii r_0 and r_1 respectively. Assume that the thermal conductivity of the wall varies as

$$K = K_0 + (K_1 - K_0)\ \frac{t - t_0}{t_1 - t_0}$$

(b) Estimate the rate of evaporation of liquid oxygen from a spherical container, 1.8 m ID, covered with 30 cm of asbestos insulation. The temperatures at the inner and outer surfaces of the insulation are – 183°C and 0°C respectively. The boiling point of oxygen is – 183°C and the latent heat of vaporization is 215 kJ/kg. The thermal conductivities of the insulation are 0.16 and 0.13 W/mK at 0°C and – 183°C respectively. *Ans.* 19.8 kg/h

18.8 A counterflow double-pipe heat exchanger using superheated steam is used to heat water at the rate of 10,500 kg/h. The steam enters the heat exchanger at 180°C and leaves at 130°C. The inlet and exit temperatures of water are 30°C and 80°C respectively. If the overall heat transfer coefficient from steam to water is 814 W/m^{2}K, calculate the heat transfer area. What would the increase in area be if the fluid flows were parallel?
Ans. 7.49 m^2, 8.23 m^2, Increase = 0.74 m^2

18.9 An oil cooler consists of a straight tube of ID 1.25 cm, wall thickness 0.125 cm, enclosed within a pipe and concentric with it. The external surface of the pipe is well insulated. Oil flows through the tube at the rate of 250 kg/h and cooling water flows in the annuals at the rate of 300 kg/h in the direction opposite to that of oil. The oil enters the tube at 180°C and is cooled to 66°C. The cooling water enters at 10°C. Estimate the length of tube required, given that the heat transfer coefficient from the oil to the tube surface is 1628 and that from the tube surface to the water is 3722 W/m^2 K. Neglect the temperature drop across the tube wall, c_p of oil = 1.675 kJ/kg K. *Ans.* 2.81 m

18.10 A marine steam turbine has a condenser flow (of steam) at a full load of 12,000 kg/h. The quality of steam at the turbine exhaust is 0.85. The condenser pressure is 0.07 bar. Sea water used for circulation is at 21°C. The terminal temperature difference between the steam and outgoing water is 5.7°C. The circulating water velocity is maintained at 1.7 m/s. The condenser tubes are of 1.3 cm ID and 0.122 cm thickness. Determine (a) the rate of flow of cooling water, (b) the length of tubes, and (c) the number of tubes, in the condenser. Take U_o = 3256 W/m^2 K.
Ans. 490*t*/*h*, 6.8 m, 604

18.11 Fifty kg of water per min is heated from 30°C to 50°C by passing through a pipe of 2 cm in diameter. The pipe is heated by condensing steam on its surface maintained at 100°C. Find the length of the pipe required. Take the following properties of water at the mean bulk temperature of 40°C: ρ = 992.2 kg/m^3, ν = 0.659 × 10^{-6} m^2/s, K = 0.63 W/mK and Pr = 4.31.

18.12 Air flows at a velocity of 0.3 m/s through the annular space between two concentric tubes. The outer tube is 5 cm ID and the inner tube is 3.125 cm OD. The air enters at 16°C and leaves at 32°C. The temperature of the outside surface of the inner tube is 50°C. Determine the heat transfer coefficient between the air and inner tube. Properties of air at 24°C: ρ = 1.18 kg/m^3, K = 0.03 W/mK, Pr = 0.7 and ν = 15.5 × 10^{-6} m^2/s.

18.13 A rectangular duct, 30 cm × 20 cm in cross-section, carries cold air. The temperature of the outer surface of the duct is 5°C and the surrounding air temperature is 25°C. Estimate the rate of heat gain by the duct, assuming that the duct, one metre in length, is exposed to the air in the vertical position. Properties of air at 15°C: ρ = 1.22 kg/m^3, ν = 14.6 × 16^{-6} m^2/s, K = 0.03 W/mK, and Pr = 0.7. *Ans.* 0.186 W

18.14 A hot square plate, 50 cm × 50 cm at 100°C, is exposed to atmospheric air at 20°C. Compute the rate of heat loss from both the surfaces of the plate if the plate is kept in the vertical plane. Properties of air at 60°C:

ρ = 1.06 kg/m^3, ν = 18.97 × 10^{-6} m^2/s,

K = 0.03 W/mk, and Pr = 0.696

18.15 Determine the heat lost by radiation per metre length of 7.5 cm diameter oxidized steel pipe at 300°C if (a) located in a large room with red brick walls at a temperature of 25°C, and (b) enclosed in a 25 cm × 25 cm red brick conduit at a temperature of 25°C. The emissivity of oxidized steel is 0.79 and that of red brick is 0.93.
Ans. 1.052, 1.035 kW

18.16 Two concentric spheres, 21 cm and 30 cm in diameter, with the space between them evacuated are to be used to store liquid air (– 153°C) in a room at 27°C. The surfaces of the spheres are flushed with aluminium ($\varepsilon = 0.03$) and the latent heat of vaporization of liquid air is 209.35 kJ/kg. Calculate the rate of evaporation of liquid air.

Ans. 0.0216 kg/h

18.17 Estimate the net radiant heat exchange per square metre for two very large parallel planes at temperatures 560°C and 300°C respectively. Assume that the emissivities of the hot and cold planes are 0.8 and 0.6 respectively. *Ans.* 11.28 kW

18.18 The inner sphere of a Dewar flask is of 30 mm dia. and outer sphere is of 360 mm dia. Both the spheres are plated for which $\varepsilon = 0.05$. The space between them is evacuated. Determine the rate at which liquid oxygen would evaporate at – 183°C when the outer sphere temperature is 20°C. The latent heat of vaporization of liquid oxygen is 214.2 kJ/kg. [*Ans.* 1.6466×10^{-5} kg/s]

18.19 A metal plate 0.609 m in height forms the vertical wall of an oven and is at a temperature of 171°C. Within the oven is air at a temperature of 93.4°C and atmospheric pressure. Assuming that natural convection conditions hold near the plate, for which

$$\text{Nu} = 0.548\ (\text{Gr Pr})^{0.25},$$

find the mean heat transfer coefficient, and the heat taken up by the air per sec. per metre width. For air at 132.2°C, $k = 32.2 \times 10^{-6}$ kW/mK, $\mu = 0.232 \times 10^{-4}$ kg/m–s, $c_p = 1.005$ kJ/kg K. Assume air as an ideal gas with $R = 0.287$ kJ/kg K.

[*Ans.* $h = 4.18$ W/m^2 K, $Q = 197.6$ W/m]

18.20 The heat transfer coefficients for the flow of air at 28°C over a 12.5 mm dia. sphere are measured by observing the temperature-time history of a copper ball of the same dimension. The temperature of the copper ball was measured by two thermocouples, one located in the centre, the other near the surface. Both the thermocouples, measured the same temperature at an instant. In one test the initial temperature of the ball was 65°C and in 1.15 min the temperature decreased by 11°C. Calculate the heat transfer coefficient for this case.

[*Ans.* 37.41 W/m^2 K]

18.21 A cubical piece of aluminium 1 cm on a side is to be heated from 50°C to 300°C by a direct flame. How long should the aluminium remain in the flame, if the flame temperature is 800°C and the convective heat transfer coefficient is 190 W/m^2 K? For aluminium, take $\rho = 2719$ kg/m^3 and $c = 0.871$ kJ/kg K. [*Ans.* 8.36 s]

18.22 The cooling system of an electronic package has to dissipate 0.153 kW from the surface of an aluminium plate 100 mm × 150 mm. It is proposed to use 8 fins, each 150 mm long and 1 mm thick. The temperature difference between the plate and the surroundings is 50 K, the thermal conductivity of the plate and fins is 0.15 kW/mK and the convective coefficient is 0.04 kW/m^{2}K. Calculate the height of fins required. [*Ans.* 30 cm]

18.23 Oil ($c_p = 2$ kJ/kgK) is cooled from 110°C to 70°C by a flow of water in a counterflow heat exchanger. The water ($c_p = 4.18$ kJ/kgK) flows at the rate of 2 kg/s and is heated from 35°C to 65°C. The overall heat transfer coefficient is 0.37 kW/m^{2}K. Determine the exit temperatures of oil and water, if the water flow rate drops to 1.5 kg/s at the same oil flow rate. *Ans.* 72.5°C, 72.5°C

18.24 A tank contains 272 kg of oil which is stirred so that its temperature is uniform. The oil is heated by an immersed coil of pipe 2.54 cm dia. in which steam condenses at 149°C. The oil of specific heat 1.675 kJ/kgK is to be heated from 32.2°C to 121°C in 1 hour. Calculate the length of pipe in the coil if the surface coefficient is 0.653 kW/m^{2}K.

[*Ans.* 3.47 m]

CHAPTER 19

Gas Compressors

A gas compressor is a device in which work is done on the gas to raise its pressure, with an appreciable increase in its density. The compression of gases is an important process in many power plants, refrigeration plants, and industrial plants. Industrial uses occur in connection with compressed air motors for tools, air brakes for vehicles, servo-mechanisms, metallurgical and chemical processes, conveying of materials through ducts, transporting of natural gas, and production of bottled gases.

Compressors can be (a) positive displacement machines like reciprocating compressors, Root's blower and vane-sealed machines and (b) turbine type like centrifugal and axial flow compressors.

19.1 COMPRESSION PROCESSES

A gas compression process may be either adiabatic or involving heat transfer, depending upon the purpose for which the gas is compressed. If the compressed gas is to be used promptly in an engine or in a combustion process, adiabatic compression is desirable so as to obtain the maximum energy in the gas at the end of compression. In many applications, however, the gas is not used promptly but is stored in a tank or receiver for use later when needed. The gas in the tank transfer heat to the surroundings so that when finally used it is at room temperature. In such cases the objective of the compression and storage process is simply to increase the pressure of the gas without change of temperature. It will be shown below that if the gas is cooled during compression, the work required will be less than for adiabatic compression. Another advantage of cooling is the reduction in volume of gas and less pipe friction losses. Since cooling during the compression process is not very effective, *after-coolers* are often used to cool the gas leaving the compressor.

19.2 WORK OF COMPRESSION

The steady flow energy equation for a compressor (Fig. 19.1) gives

$$h_1 + Q = h_2 + W_x \tag{19.1}$$

neglecting the changes of potential and kinetic energy. From the property relation

$$T\mathrm{d}s = \mathrm{d}h - v\mathrm{d}p \tag{19.2}$$

For a reversible process,

$$Q = \Delta h - \int v\mathrm{d}p \tag{19.3}$$

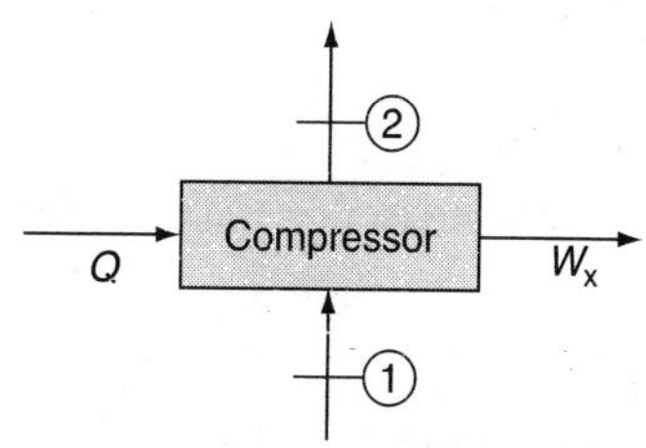

Fig. 19.1

Let us consider two particular idealized cases, namely, reversible adiabatic and reversible isothermal, as well as a general case of a reversible polytropic process (pv^n = constant). The paths of such processes are plotted in Fig. 19.2 for an ideal gas compressed from state 1 to state 2.

As shown in Sec. 7.14 or from Eqs (19.1 to 19.3) for adiabatic compression of unit mass,

$$W_x = -\int v\mathrm{d}p \tag{19.4}$$

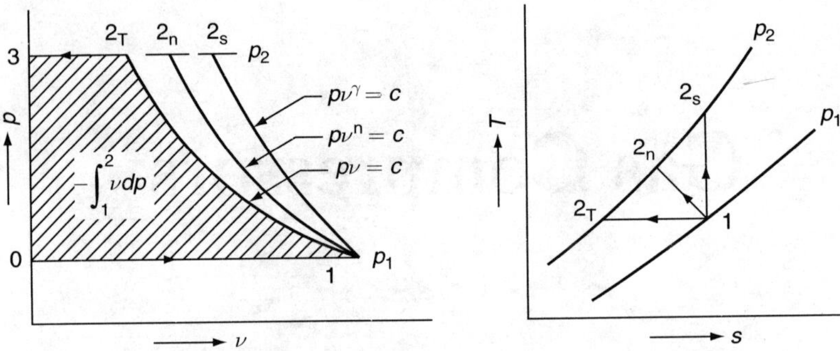

Fig. 19.2 *Reversible compression processes*

For a reversible adiabatic process, $pv^\gamma = C$,

$$v = \left(\frac{pv^\gamma}{p}\right)^{\frac{1}{\gamma}} = \frac{p_1^{1/\gamma} v_1}{p^{1/\gamma}} = \frac{p_2^{1/\gamma} v_2}{p^{1/\gamma}}$$

and

$$W_x = -\int_1^2 \frac{p_1^{1/\gamma} v_1}{p^{1/\gamma}} dp = -p_1^{1/\gamma} \cdot v_1 \cdot \frac{1}{1-\frac{1}{\gamma}} \left[p_2^{1-\frac{1}{\gamma}} - p_1^{1-\frac{1}{\gamma}} \right]$$

$$= -\frac{\gamma}{\gamma-1} p_1 v_1 \left[\left(\frac{p_2}{p_1}\right)^{\frac{\gamma-1}{\gamma}} - 1 \right]$$

The *work of compression* or the steady flow work input to the gas is the negative of the shaft work W_x, i.e. for reversible adiabatic compression,

$$W_s = \frac{\gamma}{\gamma-1} p_1 v_1 \left[\left(\frac{p_2}{p_1}\right)^{\frac{\gamma-1}{\gamma}} - 1 \right] \tag{19.5}$$

Similarly, for reversible polytropic compression

$$W_n = \frac{n}{n-1} p_1 v_1 \left[\left(\frac{p_2}{p_1}\right)^{\frac{n-1}{n}} - 1 \right] \tag{19.6}$$

For reversible isothermal compression of an ideal gas,

$$W_t = -\int_1^2 v dp = -\int_1^2 \frac{p_1 v_1}{p} dp = p_1 v_1 \ln \frac{p_2}{p_1} \tag{19.7}$$

In the p-v plot of Fig. 19.2 the work of compression for each type of process is represented by the area between the path of that process and the axis of pressures. For the general process $pv^n = C$, taking logarithm

$$\ln p + n \ln v = \ln C$$

On differentiation,

$$\frac{dp}{p} + n\frac{dv}{v} = 0$$

∴ The slope at any state (say 1) is given by

$$\frac{dp}{dv} = -n\frac{p_1}{v_1} \tag{19.8}$$

For $\gamma > n > 1$ and for the same pressure ratio p_2/p_1, the isothermal compression needs the minimum work, whereas adiabatic compression needs the maximum work, while the polytropic compression needing work in between the two. In isothermal compression all the work done on the gas during the process is transformed into an internal energy increase where $Q - W = u_2 - u_1 = O$ (for an ideal gas) or $W = Q$. This heat is taken away from the gas by cooling. Thus in isothermal compression, considered to be ideal, no energy would be imparted to the gas, since its function is simply to raise the pressure of the gas (and not its temperature).

The efficiency of a compressor working in a steady flow process may be defined as

$$\eta_c = \frac{h_{2S} - h_1}{W_c} = \frac{W_s}{W_c} \tag{19.9}$$

where W_c is the shaft work supplied to the compressor per unit mass. For the idealized reversible isothermal process the compressor efficiency is sometimes defined as

$$\eta_c = \frac{W_t}{W_c} \tag{19.10}$$

where W_t is the work of reversible isothermal compression. The two efficiencies of Eqs (19.9) and (19.10) are called respectively the *adiabatic efficiency* and *isothermal efficiency*. Because of the effects of cooling, the adiabatic efficiency of a real compressor may be greater than unity.

Many turbine-type compressors are essentially adiabatic machines due to their high speeds. For an adiabatic machine the work of compression is equal to the enthalpy rise of the gas,

$$W_c = h_2 - h_1$$

Then for an adiabatic compressor the efficiency is

$$\eta_c = \frac{h_{2S} - h_1}{h_2 - h_1} \tag{19.11}$$

For a reciprocating machine the compressor efficiency may be on an indicated work basis or a brake work basis, depending upon where the work input is measured (See Sec. 3.3).

19.3 SINGLE-STAGE RECIPROCATING AIR COMPRESSOR

Figure 19.3 shows the arrangement of a single-stage reciprocating air compressor, together with a typical indicator diagram. The compressor operates on a two-stroke cycle as follows:

Stroke 1 *(a-c)* The piston withdraws, causing the air in the clearance volume to expand, and when the pressure in the cylinder falls below atmospheric pressure (at *b*) the inlet valve opens and air is drawn into the cylinder for the remainder of the stroke.

Stroke 2 *(c-a)* The piston moves inwards, compressing the air in the cylinder, and the inlet valve closes when the cylinder pressure reaches atmospheric pressure. Further compression follows as the piston moves towards the top of its stroke until, when the pressure in the cylinder is more than that in the receiver, the delivery valve opens and air is delivered to the receiver for the remainder of the stroke.

The intercept V on the indicator diagram represents the volume of air taken in per cycle. It is seen that the effect of clearance air is to reduce the quantity of air drawn in during the suction stroke, so that in practice the clearance space is made as small as possible.

The total area of the indicator diagram represents the actual work of the compressor on the gas or air. The areas above p_2 and below p_1 represent work done because of pressure drop through the valves and port

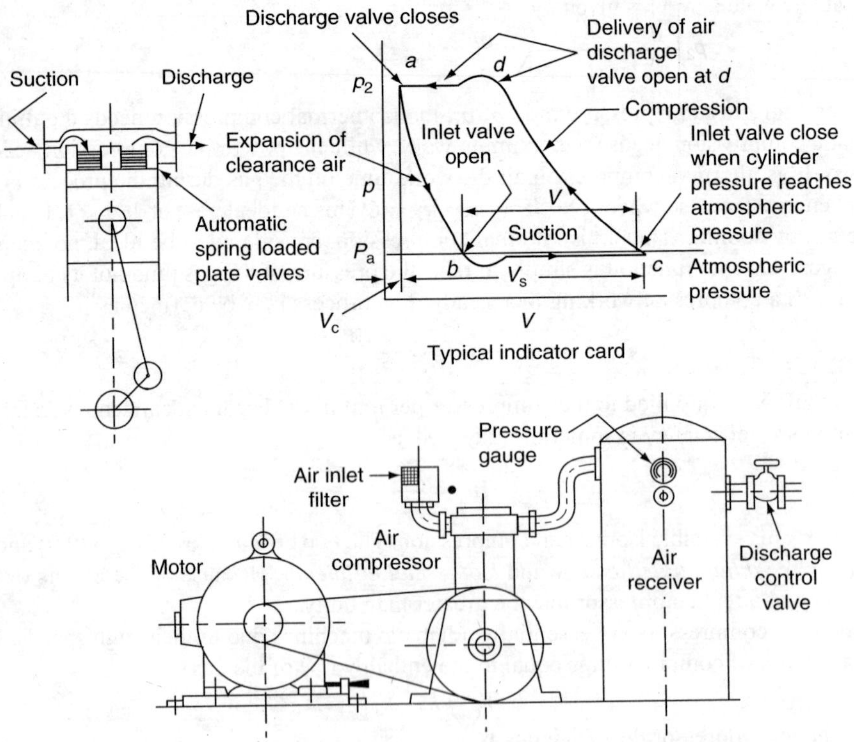

Fig. 19.3 *Arrangement of single-stage reciprocating air compression*

passages; this work is called the valve loss. The idealized machine to which the actual machine is compared has an indicator diagram like Fig. 19.4 Both expansion and compression are supposed to follow the same law $pv^n = C$. The small quantity of high pressure air in the clearance volume expands to V_a and air drawn in during the suction stroke is $V_b - V_a$. From the diagram:

Work done on the air during the cycle = Enclosed area *abcd* = Area *dcfh* + Area *cbef* – Area *hgad* – Area *geba*

$$\therefore \quad W = p_2(V_c - V_d) + \frac{p_2V_c - p_1V_b}{n-1} - \frac{p_2V_d - p_1V_a}{n-1} - p_1(V_b - V_a)$$

$$= \frac{n}{n-1}\left[(p_2V_c - p_1V_b) - (p_2V_d - p_1V_a)\right]$$

Now, $p_2 V_c = m_c R T_2$, $p_1 V_b = m_b RT_1$ and $m_c - m_b$ = mass of air taken during compression.

Also, $p_2 V_d = m_d RT_2$, $p_1 V_a = m_a RT_1$ and $m_a - m_d$ = mass of air present in the clearance volume after delivery.

$$\therefore \quad \text{Work done/cycle} = \frac{n}{n-1}R[m_b(T_2 - T_1) - m_d(T_2 - T_1)]$$

$$= \frac{n}{n-1}(m_b - m_d)R(T_2 - T_1)$$

$$= \frac{n}{n-1}(m_b - m_d)RT_1\left[\left(\frac{p_2}{p_1}\right)^{\frac{n-1}{n}} - 1\right] \quad (19.12)$$

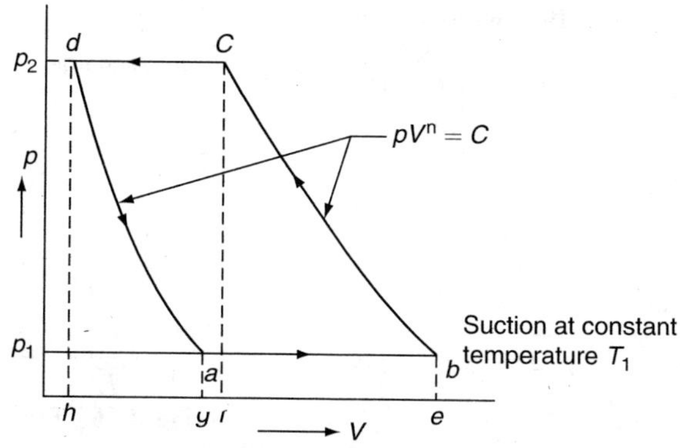

Fig. 19.4 *Indicator diagram*

where $m_b - m_d$ is the difference between the mass of air present at the end of suction and that present at the end of delivery.

Thus the expression Eq. (19.12) is the same as obtained from the steady flow energy equation given in Eq. (19.6) where the clearance volume was neglected. Thus the mass of gas in the clearance volume does not have any effect on the work of compression.

19.4 VOLUMETRIC EFFICIENCY

The amount of air dealt with in a given time by an air compressor is often referred to at free air conditions, i.e. the temperature and pressure of the environment, which may be taken as 15°C and 101.325 kPa, if not mentioned. It is known as free air delivery (FAD). The ratio of the actual volume of gas taken into the cylinder during suction stroke to the piston displacement volume (PD) or the swept volume (V_s) of the piston is called the *volumetric efficiency*, or

$$\eta_{vol} = \frac{\dot{m}v_1}{PD} = \frac{\dot{m}v_1}{V_s}$$

where $\dot{m}$ is the mass flow rate of the gas and v_1 is the specific volume of the gas at inlet to the compressor. With reference to Fig. 19.5.

$$\eta_{vol} = \frac{V_2 - V_1}{V_s} = \frac{V_c + V_s - V_1}{V_s} = 1 + \frac{V_c}{V_s} - \frac{V_1}{V_s}$$

$$\text{Let } C = \text{clearance} = \frac{\text{Clearance volume}}{\text{P.D. or } V_s} = \frac{V_c}{V_s}$$

$$\text{Since} \quad p_1 V_1^n = p_2 V_4^n, \therefore V_1 = V_4 \left(\frac{p_2}{p_1}\right)^{\frac{1}{n}} = V_c \left(\frac{p_2}{p_1}\right)^{\frac{1}{n}}$$

$$\therefore \quad \eta_{vol} = 1 + C - \frac{V_c}{V_s}\left(\frac{p_2}{p_1}\right)^{\frac{1}{n}} = 1 + C - C\left(\frac{p_2}{p_1}\right)^{\frac{1}{n}} \qquad (19.13)$$

Equation (19.13) is plotted in Fig. 19.6. Since (p_2/p_1) is always greater than unity, it is evident that the volumetric efficiency decreases as the clearance increases and as the pressure ratio increases.

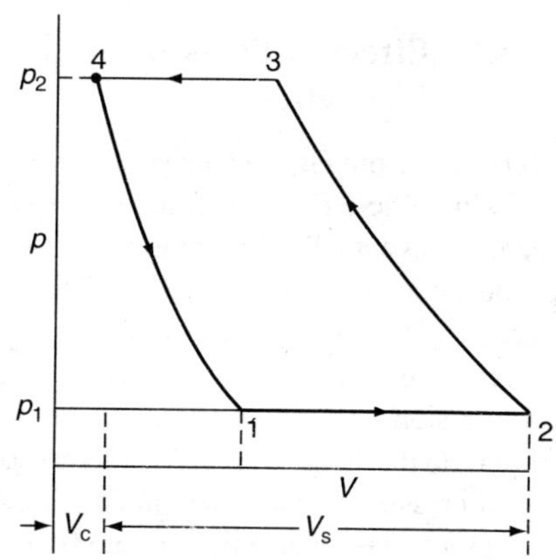

Fig. 19.5 *Volumetric efficiency*

For air compressors, the prescribed inlet conditions are often atmospheric conditions in which the volume of air induced is often called the free air delivery (F.A.D). If the ambient air is at pressure p_a and the temperature T_a, then the volumetric efficiency

$$\eta_{vol} = \frac{V_2 - V_1}{V_2 - V_4} \times \frac{p_1}{p_a} \times \frac{T_a}{T_1}$$

$$= \left[1 + C - C\left(\frac{p_2}{p_1}\right)^{\frac{1}{n}}\right] \times \frac{p_1}{p_a} \times \frac{T_a}{T_1} \quad (19.13a)$$

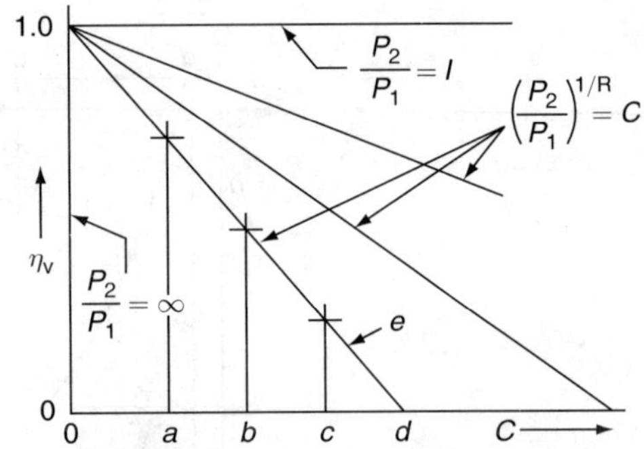

Fig. 19.6 *Effect of clearance on volumetric efficiency*

In order to get maximum flow capacity, compressors are built with the minimum practical clearance. Sometimes, however, the clearance is deliberately increased $\left(\dot{m} = \frac{V_s n_v}{v_1}\right)$ as a means of controlling the flow through a compressor driven by a constant speed motor. The compressor cylinder of Fig. 19.7 is fitted with a clearance pocket which can be opened at will by a valve. Let us suppose that this machine is operating at conditions corresponding to line e in Fig. 19.6. If the clearance volume is at minimum value a the volumetric efficiency and the flow through the machine are maximum. If the clearance pocket is then opened to increase the clearance to b, the volumetric efficiency and the flow are reduced. By increasing the clearance in steps, as indicated by points c and d, the flow may be reduced in steps to zero. The work per kg of gas compressed is, however, not affected by the clearance volume in an idealized compressor.

For a given pressure ratio, η_{vol} is zero when the maximum clearance is

$$C_{max} = \frac{1}{\left(\frac{p_2}{p_1}\right)^{\frac{1}{n}} - 1} \quad (19.14)$$

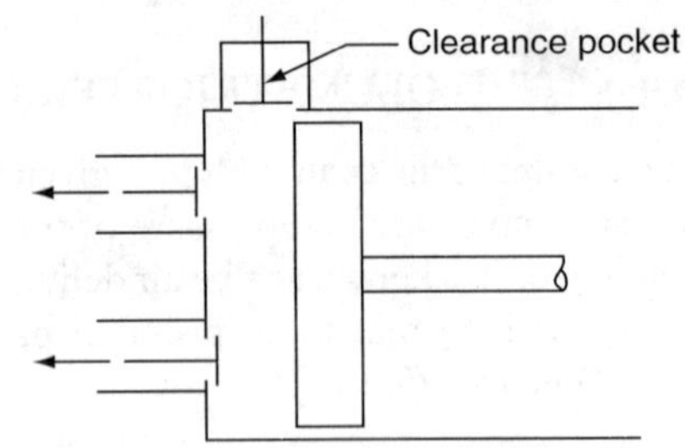

Fig. 19.7 *Clearance pocket for capacity control*

19.4.1 Effect of Pressure Ratio on Volumetric Efficiency

It is evident from Fig. 19.6 that as the pressure ratio is increased the volumetric efficiency of a compressor of fixed clearance decreases eventually becoming zero. This can also be seen in an indicator diagram, Fig. 19.8. As the discharge pressure is increased from p_{2a} to p_{2b}, the volume V_{1a} taken at pressure p_1 decreases to V_{1b}, and η_{vol} decreases. At some pressure p_{2c} the compression line intersects the line of clearance volume and there is no discharge of gas. An attempt to pump to p_{2c} (or any higher pressure) would result in compression an re-expansion of the same gas repeatedly, with no flow in or out.

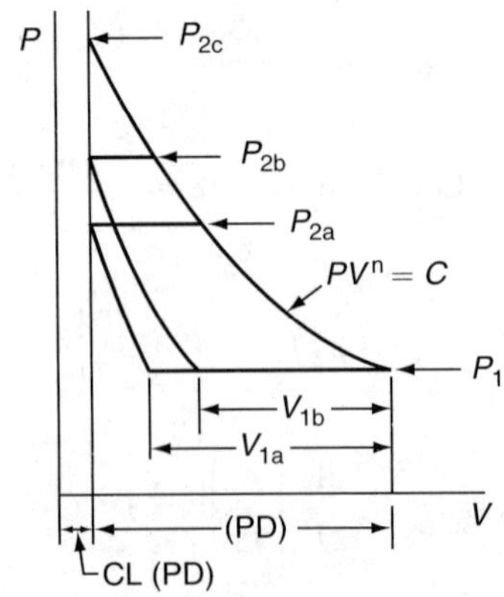

Fig. 19.8 *Effect of pressure-ratio on capacity*

The maximum pressure ratio attainable with a reciprocating compressor cylinder is then seen to be limited by the clearance, which is given from Eq. (19.15), as

$$\frac{p_{2\max}}{p_1} = \left(1 + \frac{1}{C}\right)^n \tag{19.15}$$

19.5 MULTI-STAGE COMPRESSION

When compressing a gas (or air) to high pressure it is advantageous to do it in stages. The condition for minimum work requires the compression to be isothermal. Since the temperature after compression is given by $T_2 = T_1(p_2/p_1)^{(n-1)/n}$ the delivery temperature, T_2, increases with the pressure ratio. Also the volumetric efficiency as given by Eq. (19.13) decreases as the pressure ratio increases, as mentioned earlier.

The volumetric efficiency can be improved by carrying out the compression in two stages. After the first stage of compression (L.P.) from the state p_1, T_1 to the state p_x, T_x, the fluid is passed into a smaller cylinder (H.P.) in which the gas is compressed to the required final pressure p_2 at temperature T_2 (Fig. 19.9).

The gas after being compressed in the L.P. (low pressure) cylinder ($a - b$) is passed on to an intercooler for getting cooled. After leaving the intercooler the gas enters the H.P. (high pressure) cylinder for further compression (c–d). Figure 19.10 shows the p-V diagram for two-stage compression. Complete or perfect intercooling (b-c) means that the exiting gas from the intercooler at temperature T_x is cooled completely to the original (inlet) temperature T_1.

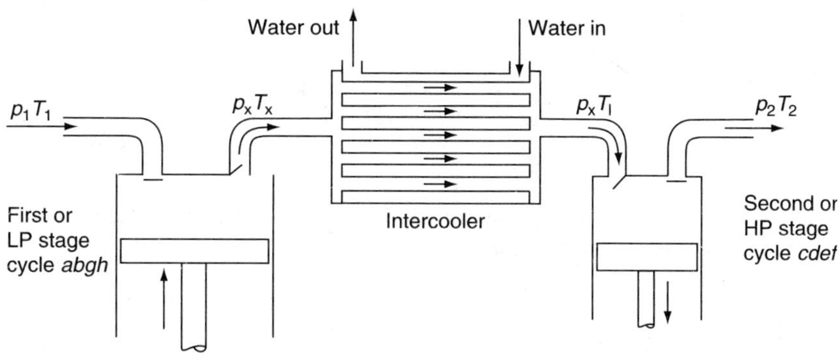

Fig. 19.9 *Plan showing intercooling between compressor stages*

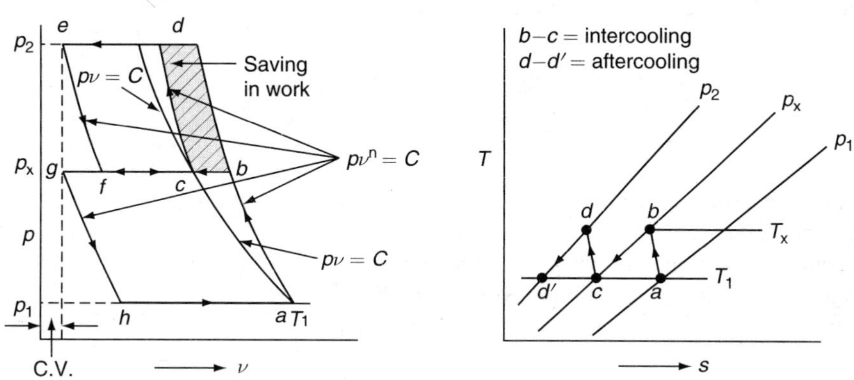

Fig. 19.10 *p-V and T-s diagrams for two-stage compression with perfect intercooling, showing the work saved*

Similarly, perfect aftercooling (d–d') makes the gas, leaving the H.P. compressor, cooled also to the inlet temperature T_1. This aftercooling reduces the volume of the gas leaving, and thus the size of the receiver becomes smaller. The clearance volume in both the cylinders has here been assumed to be the same.

19.5.1 Ideal Intermediate Pressure

The intermediate pressure p_x (Fig. 19.10) has an optimum value for minimum work of compression. The total work per kg of gas is given by

$$W_c = \frac{n}{n-1} RT_1 \left[\left(\frac{p_x}{p_1} \right)^{\frac{n-1}{n}} - 1 \right] + \frac{n}{n-1} RT_1 \left[\left(\frac{p_2}{p_x} \right)^{\frac{n-1}{n}} - 1 \right]$$

$$= \frac{n}{n-1} RT_1 \left[\left(\frac{p_x}{p_1} \right)^{\frac{n-1}{n}} + \left(\frac{p_2}{p_x} \right)^{\frac{n-1}{n}} - 2 \right] \tag{19.16}$$

Here p_1, T_1 and p_2 are fixed and p_x is the only variable. Differentiating Eq. (19.16) with respect to p_x and making it equal to zero,

$$\frac{dW_c}{dp_x} = \frac{n}{n-1} RT_1 \left[\frac{n-1}{n} \left(\frac{1}{p_1} \right)^{\frac{n-1}{n}} (p_x)^{1-\frac{1}{n}-1} + p_2^{\frac{n-1}{n}} \cdot p_x^{-1+\frac{1}{n}-1} \left(-\frac{n-1}{n} \right) \right] = 0$$

$$\therefore \quad p_x^{-\frac{1}{n}+2-\frac{1}{n}} = (p_2 p_1)^{\frac{n-1}{n}} \quad \text{or,} \quad p_x^{2\left(\frac{n-1}{n}\right)} = (p_1 p_2)^{\frac{n-1}{n}}$$

$$\therefore \quad p_x = \sqrt{p_1 p_2} \tag{19.17}$$

Thus, for minimum work of compression, the intermediate pressure is the geometric mean of the suction and discharge pressures for a two-stage compressor.

From Eq. (19.17)

$$\frac{p_x}{p_1} = \frac{p_2}{p_x} = \left(\frac{p_2}{p_1} \right)^{\frac{1}{2}} \tag{19.18}$$

Pressure ratio in L.P. stage = Pressure ratio in H.P. stage

Also,

$$\frac{T_x}{T_1} = \left(\frac{p_x}{p_1} \right)^{\frac{n-1}{n}} \quad \text{and} \quad \frac{T_2}{T_1} = \left(\frac{p_2}{p_x} \right)^{\frac{n-1}{n}}$$

$$\therefore \quad T_2 = T_x \tag{19.19}$$

From Fig. 19.6 it is seen that work required in L.P. compressor = Work required in H.P. compressor.

Thus, the intermediate pressure that produces minimum work will also result in equal pressure ratios in the two stages of compression, equal discharge temperatures, and equal work for the two stages.

For two stage-compression, the minimum work, using Eq. (19.16) becomes

$$W_c = \frac{2nRT_1}{n-1} \left[\left(\frac{p_x}{p_1} \right)^{\frac{n-1}{n}} - 1 \right] \tag{19.20}$$

where p_x/p_1 is the pressure ratio in each stage. In terms of overall pressure ratio, it becomes

$$W_c = \frac{2nRT_1}{n-1}\left[\left(\frac{p_2}{p_1}\right)^{\frac{n-1}{2n}} - 1\right] \quad (19.21)$$

Heat rejected in the intercooler,

$$Q_{bc} = c_p\,(T_x - T_1)\ \text{kJ/kg}$$

Let us now consider compression efficiencies and imperfect intercooling. As shown in Fig. 19.11, an ideal gas is compressed from the initial state p_1, T_1 to p_x. It is then cooled at constant pressure to T_y and then compressed from p_x, T_y to p_2. Given p_1, T_1, T_y and p_2, it is desired to find the value of p_x which gives minimum work. Let the adiabatic compression efficiencies of the two stages be respectively η_{C_1} and η_{C_2}; then the work of compression is

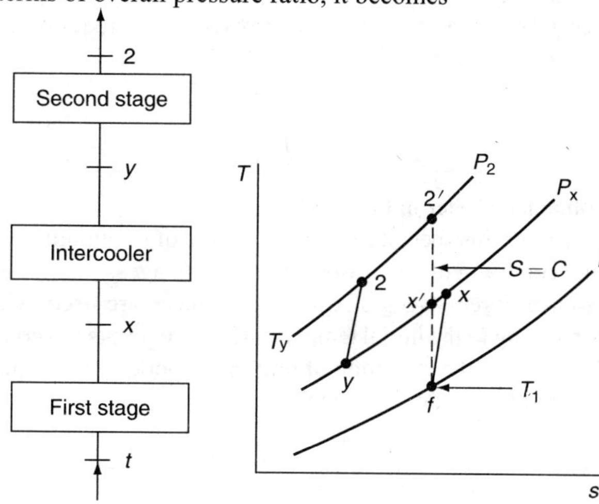

Fig. 19.11 *T-s plot of two-stage compression process*

$$W_c = W_1 + W_2 = \frac{1}{\eta_{C_1}}\frac{\gamma}{\gamma-1}RT_1\left[\left(\frac{p_x}{p_1}\right)^{\frac{\gamma-1}{\gamma}} - 1\right] + \frac{1}{\eta_{C_2}}\frac{\gamma}{\gamma-1}RT_y\left[\left(\frac{p_2}{p_y}\right)^{\frac{\gamma-1}{\gamma}} - 1\right]$$

But

$$\left(\frac{p_x}{p_1}\right)^{\frac{\gamma-1}{\gamma}} = \frac{T_x'}{T_1} \quad \text{and} \quad \left(\frac{p_2}{p_y}\right)^{\frac{\gamma-1}{\gamma}} = \left(\frac{p_2}{p_x}\right)^{\frac{\gamma-1}{\gamma}} = \frac{T_2'}{T_x'}$$

$$W_c = \frac{\gamma R}{\gamma-1}\left[\frac{T_1}{\eta_{C_1}}\left(\frac{T_x'}{T_1} - 1\right) + \frac{T_y}{\eta_{C_2}}\left(\frac{T_2'}{T_x'} - 1\right)\right]$$

Taking the derivative with respect to T_x' and setting it equal to zero (noting that T_1, T_2', and T_y are constant),

$$\frac{dW_c}{dT_x'} = 0 = \frac{\gamma R}{\gamma-1}\left[\frac{1}{\eta_{C_1}} + \frac{T_y T_2'}{\eta_{C_2}}\left(-\frac{1}{(T_x')^2}\right)\right]$$

Then,

$$(T_x')^2 = \frac{\eta_{C_1}}{\eta_{C_2}} T_y\, T_2' \quad \text{and} \quad \frac{T_x'}{T_1} = \sqrt{\frac{\eta_{C_1}}{\eta_{C_2}} \cdot \frac{T_y}{T_1} \cdot \frac{T_2'}{T_1}}$$

for minimum work. Now

$$\frac{T_x'}{T_1} = \left(\frac{p_x}{p_1}\right)^{\frac{\gamma-1}{\gamma}} \quad \text{and} \quad \frac{T_2'}{T_1} = \left(\frac{p_2}{p_1}\right)^{\frac{\gamma-1}{\gamma}}$$

Therefore, for minimum work in two-stage compression of an ideal gas with intercooling to a fixed temperature T_y,

$$\frac{p_x}{p_1} = \sqrt{\left(\frac{\eta_{C_1}}{\eta_{C_2}}\frac{T_y}{T_1}\right)^{\frac{\gamma}{\gamma-1}} \cdot \frac{p_2}{p_1}} \quad (19.22)$$

For the special case of $T_y = T_1$ and $\eta_{C_1} = \eta_{C_2}$, which is often taken as a standard of comparison, the requirement for minimum work is

$$\frac{p_x}{p_1} = \sqrt{\frac{p_2}{p_1}} \tag{19.23}$$

as obtained earlier in Eq. (19.18).

Also for this special case the condition of minimum work is the condition of equal work in the two stages.

When, three stages of equal efficiencies are used, with intercooling to the initial temperature at two points as shown in Fig. 19.12, the condition of minimum work, and of equal division of work among stages is

$$\frac{p_{x1}}{p_1} = \frac{p_{x2}}{p_{x1}} = \frac{p_2}{p_{x2}} = \left(\frac{p_2}{p_1}\right)^{1/3} \tag{19.24}$$

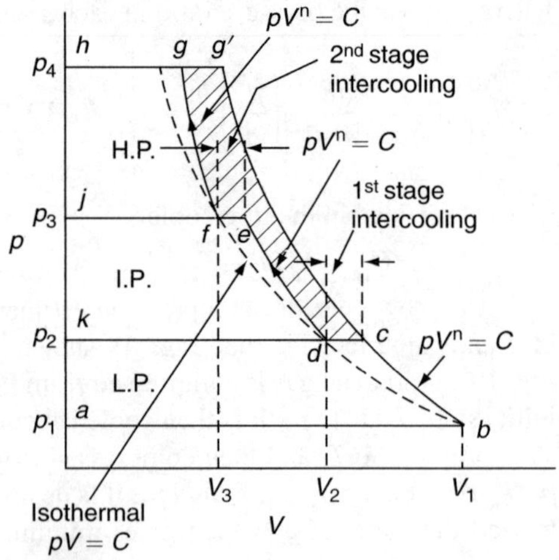

Fig. 19.12 *Three-stage compression with perfect intercooling*

Thus for 3-stage compressor the optimum pressure ratio per stage can be written as

$$\frac{p_x}{p_1} = \left(\frac{p_d}{p_s}\right)^{\frac{1}{3}}$$

where p_d and p_x are the discharge and suction pressure respectively.

For N-stage compressor, the optimum pressure ratio per stage is

$$\frac{p_x}{p_1} = \left(\frac{p_d}{p_s}\right)^{\frac{1}{N}} \tag{19.25}$$

The minimum work of compression for N-stages is then

$$W_c = \frac{NnRT_1}{n-1}\left[\left(\frac{p_d}{p_s}\right)^{\frac{n-1}{Nn}} - 1\right] \tag{19.26}$$

In the case of gas compression, the desirable idealized process is, however, a reversible isothermal process, and the isothermal efficiency is given by

$$\eta_t = \frac{W_t}{W_c} = \frac{p_1 v_1 ln \frac{p_2}{p_1}}{W_c} \tag{19.27}$$

The indicator diagram for a three-stage compressor with perfect intercooling is shown is Fig. 19.12. The air delivered from the first stage at *c* is cooled to point *d* on the curve $pV = C$ (show dotted) before admission to the second stage cylinder. Again the air is cooled from *e* to *f* after compression in the second stage and before admission to the third stage cylinder. The shaded area represents the saving of work resulting from such intercooling since without intercooling the compression curve would have followed the path bcg'.

Advantages of multi-stage compression are:

(i) Improved overall volumetric efficiency. If all compression were done in one cylinder the gas in the clearance volume would expand to a large volume before the new intake could begin. This results in a very low volumetric efficiency. By cooling the gas between the stages a much higher efficiency can be obtained.
(ii) A reduction in work required per stroke, and therefore the total driving power.
(iii) Size and strength of cylinders can be adjusted to suit volume and pressure of gas.
(iv) Multi-cylinders give more uniform torque and better mechanical balance thus needing smaller flywheel.
(v) Since the maximum temperature reached during the compression process is greatly reduced by intercooling lubrication difficulties and explosion hazards are lessened.
(vi) Leakage losses are reduced considerably.

Practice appears to indicate that the economical value of pressure ratio per stage is in the range of 3 to 5. For compression of atmospheric air, single-stage machines are often used up to 550 kPa discharge pressure, two-stage from 350 kPa to 2.1. MPa, three stage from 2.1 MPa to 7.0 MPa, and four or more stages for higher pressures.

19.6 AIR MOTORS

Compressed air is used in a wide variety of applications where the use of electric motors or S.I. engines is not permissible and where high safety is required to be met as in mining applications. Pneumatic breakers, picks, spades, rammers, vibrators, riveters, etc. form a range of hand tools having wide use in construction work.

Basically the cycle in the reciprocating expander is the reverse of that in the reciprocating compressor (Fig.19.13). Air is supplied from an air receiver (process 5-1) in which air is at approximately ambient air temperature. The state 1 is the point of cut-off. Air expands in the cylinder according to the law pv^n = constant (where $n \cong 1.3$) pushing the piston out and doing work against the surroundings till the point of release at 2. There is blowdown of air from 2 to 3 and air is exhausted from 3 to 4. The compression of trapped or cushion air begins at 4 and ends a 5 according to pV^n = constant and the compressed air is again admitted to repeat the cycle.

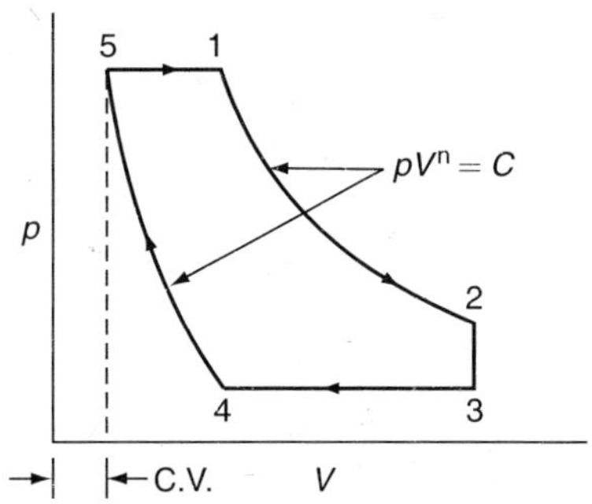

Fig. 19.13 *p-V diagram for an air motor*

19.7 ROTARY COMPRESSOR

Rotary compressors are used where large quantities of air or gas are needed at releatively low pressure. They may be classified into two main types.

(a) **Displacement compressors** in which air is compressed by being trapped in the reducing space formed by two sets of engaging surfaces.
(b) **Steady-flow compressors** in which compression occurs by the transfer of kinetic energy from a rotor.

Rotary positive displacement machines are generally uncooled and the compression is largely adiabatic. There are two types: Roots blower and vane-type compressor.

Roots blower is an extension of the idea of a gear pump, popular in engines for pumping oil. There are two lobes on each rotor, and their shape is of cycloidal or involute form (Fig.19.14). One of the lobes is connected to the drive and the second is gear driven from the first, the two rotating in opposite directions. Very small clearances between the lobes and between the casing and the lobes are provided to prevent leakages and to reduce wear. Four times the volume between the casing and one side of the rotor will be displaced in each revolution of the driving shaft. As each side of each lobe faces its side of the casing a volume of gas V, at pressure p_1 is displaced towards the delivery side at constant pressure. A further rotation of the rotor opens this volume to the receiver, and the gas flows back from the receiver, since this gas is at a higher pressure. The gas induced is compressed irreversibly by that from the receiver to the pressure p_2, and then delivery begins. This process is carried out four times per revolution of the driving shaft.

The p-V diagram for this machine is shown in Fig. 19.14 (c), in which the pressure rises irreversibly from p_1 to p_2 at constant volume.

For pressure ratios of 1.2,1.6 and 2.0, the Roots blower efficiency becomes 0.945,0.84 and 0.765 respectively, which show that the efficiency decreases as the pressure ratio increases.

Vane type compressor (Fig. 19.15) consists of a rotor mounted eccentrically in the body and supported by ball-and-roller bearings at the ends. The rotor is slotted to take the blades which are of a non-metallic material, usually fibre or carbon. As each blade moves past the inlet passage, compression begins due to decreasing volume between the rotor and casing. Delivery begins with the arrival of each blade at the delivery passage. The p-V diagram is shown in Fig. 19.15 (b) where V_s is the induced volume at state p_1, T_1. Compression occurs to the pressure p_i isentropically. At this pressure the displaced gas is opened to the receiver and the gas flowing back from the receiver raises the pressure irreversibly to p_2. The work input is given by the sum of the areas A and B. Comparing the areas of Figs19.14 (c) and 19.15 (b) it is seen that for a given air flow and given pressure ratio the vane type requires less work input than the Roots blower.

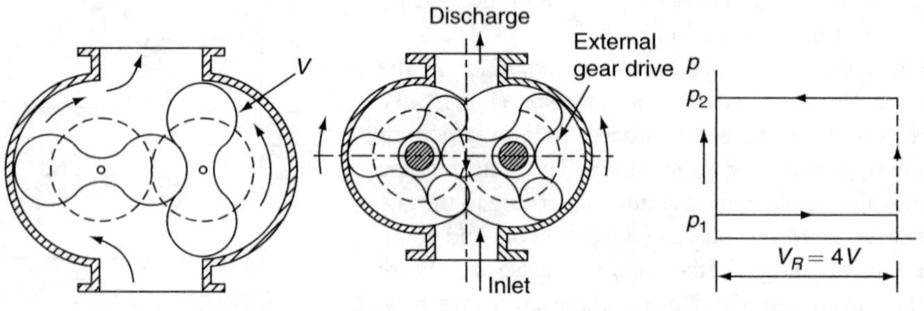

Fig. 19.14 *Roots blower with (a) two lobe rotor, (b) three lobe rotor, and (c) P-V diagram*

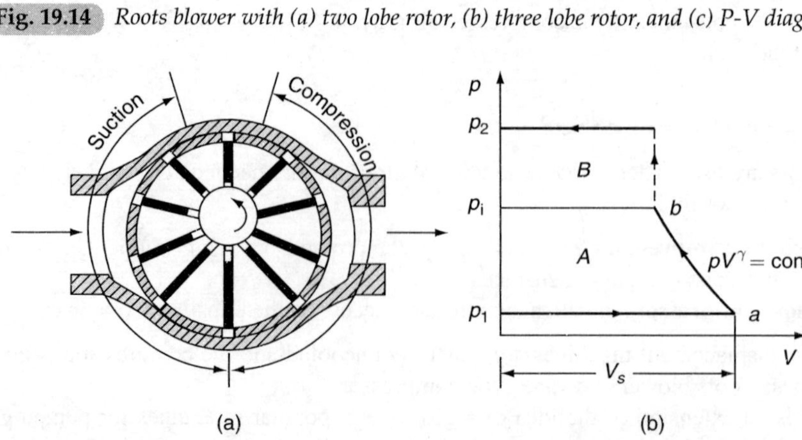

Fig. 19.15 *Vane-type positive displacement compressor with p-v diagram*

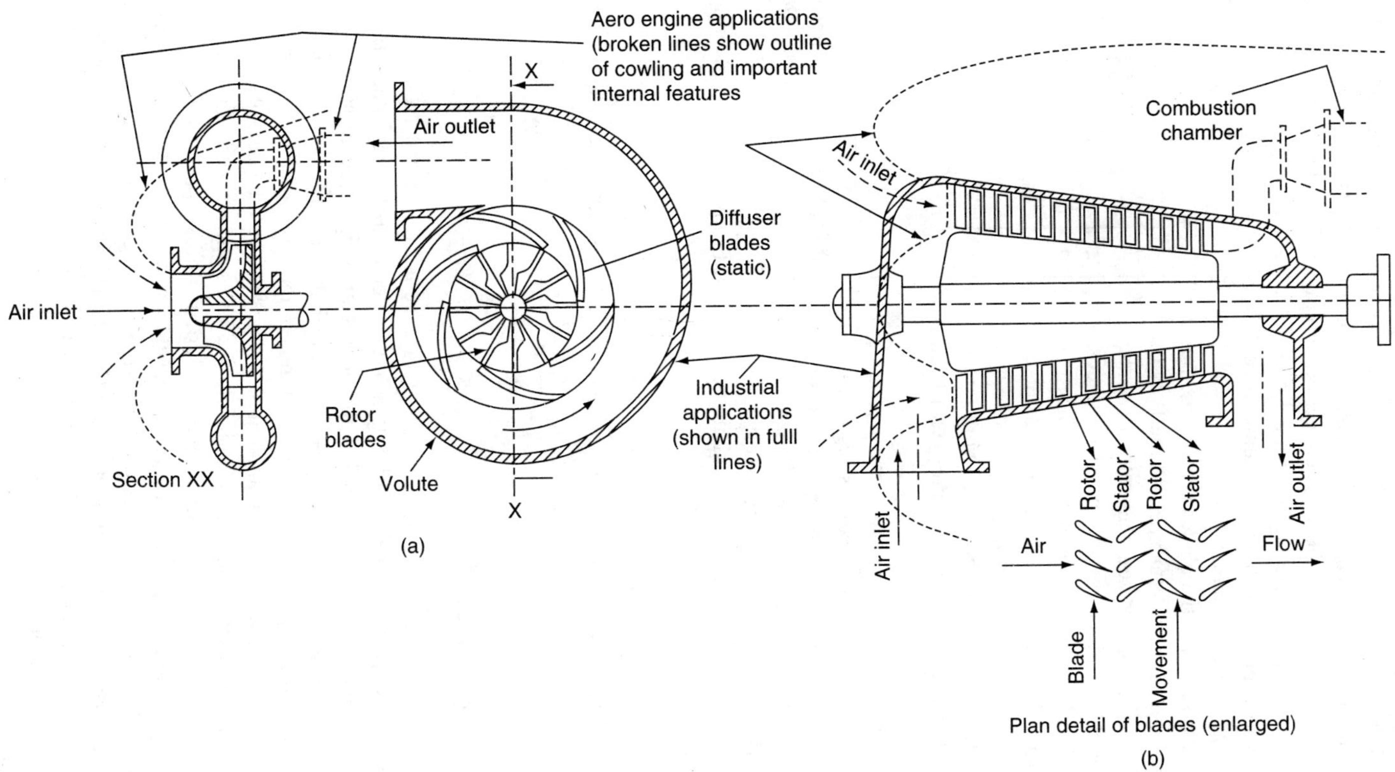

Fig. 19.16 *(a) Centrifugal compressor, (b) axial-flow compressor*

Steady flow compressors are of two types: centrifugal and axial flow, the diagrammatic sketches of which are shown in Fig. 19.16. In the **centrifugal compressor**, rotation of the impeller causes the air to be compressed by centrifugal action into the reduced volume at the tips of the compressor. A diffuser is fitted in which a part of the K.E. (Velocity) of the air is converted into internal energy so that there is a rise of pressure in both the impeller and the diffuser. The typical pressure ratio is about 1.4 to 1, but multistage arrangements can provide delivery pressures upto 10 atm.

In the **axial-flow compressor** the blades are arranged in much the same way as in a reaction turbine and the flow of air is along the axis of the compressor. The velocity of the air is changed during its passage through the sets of blades, and pressure lift occurs both in the moving blade passages and in the fixed blade passages which act as a diffuser. One set of moving and fixed blades represent a stage and a typical pressure ratio per stage is about 1.2. Multi-stage arrangements achieve higher delivery pressures with high efficiency.

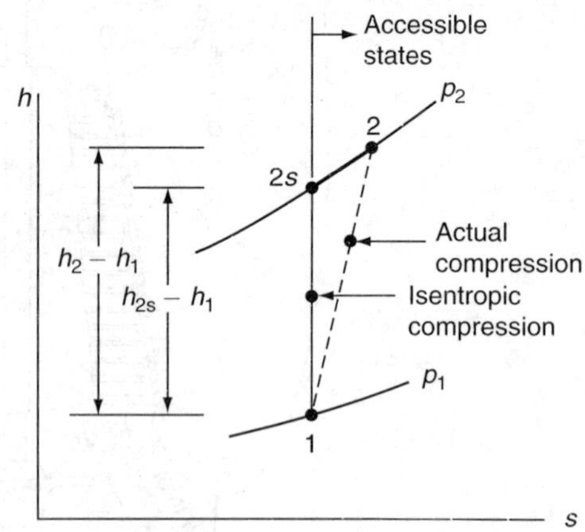

Fig. 19.17 *Comparison of actual and isentropic compressions*

For uncooled rotary compressors having a necessarily high-speed action, the compression process is regarded as adiabatic while the ideal process is isentropic (reversible and adiabatic) (Fig.19.17). In an irreversible adiabatic process the extra work which flows into the system to overcome friction increases the temperature and hence enthalpy of the gas. From S.F.E.E. neglecting P.E. and K.E. terms, the work input is

$$W_c = h_2 - h_1 = c_p (T_2 - T_1) \tag{19.28}$$

where T_2 is the final temperature for the actual compression and the isentropic efficiency η_s is given by

$$\eta_s = \frac{W_s}{W_c} = \frac{h_{2s} - h_1}{h_2 - h_1} = \frac{T_{2s} - T_1}{T_2 - T_1} \tag{19.29}$$

19.7.1 Centrifugal Compressor

The centrifugal compressor consists of an impeller with a series of curved radial vanes (Fig. 19.16a) housed in a stationary casing. The impeller imparts a high velocity to the air which flows through fixed divergent passages, in which air is decelerated with a consequent increase in static pressure. The main components of the compressor are (i) inlet pipe, (ii) impeller, (iii) diffuser (vaneless or with vanes), (iv) volute casing, and (v) outlet pipe. The air is sucked in near the hub called the eye through the inlet pipe in the axial direction and is turned through an angle in the impeller. The impeller rotates at a very high speed (15000 to 30000 rpm) and imparts the air a very high velocity. The static pressure of air increases from the eye to the tip of the impeller. Air leaving the impeller tip flows through diffuser passages (scroll) which convert the kinetic energy to pressure energy. The delivery pressure developed in a single stage compressor is of the order of 10 to 150 kPa gauge. Multi-stage compressors can develop discharge pressures as high as 1000 kPa.

The compressor may have single inlet or double inlet (Fig. 19.18). In a double inlet impeller having an eye on either side, air is drawn in on both sides (Fig. 19.18b). The impeller is subjected to approximately equal forces in the axial direction. About half the pressure rise occurs in the impeller vanes and the other half occurs in the diffuser passages.

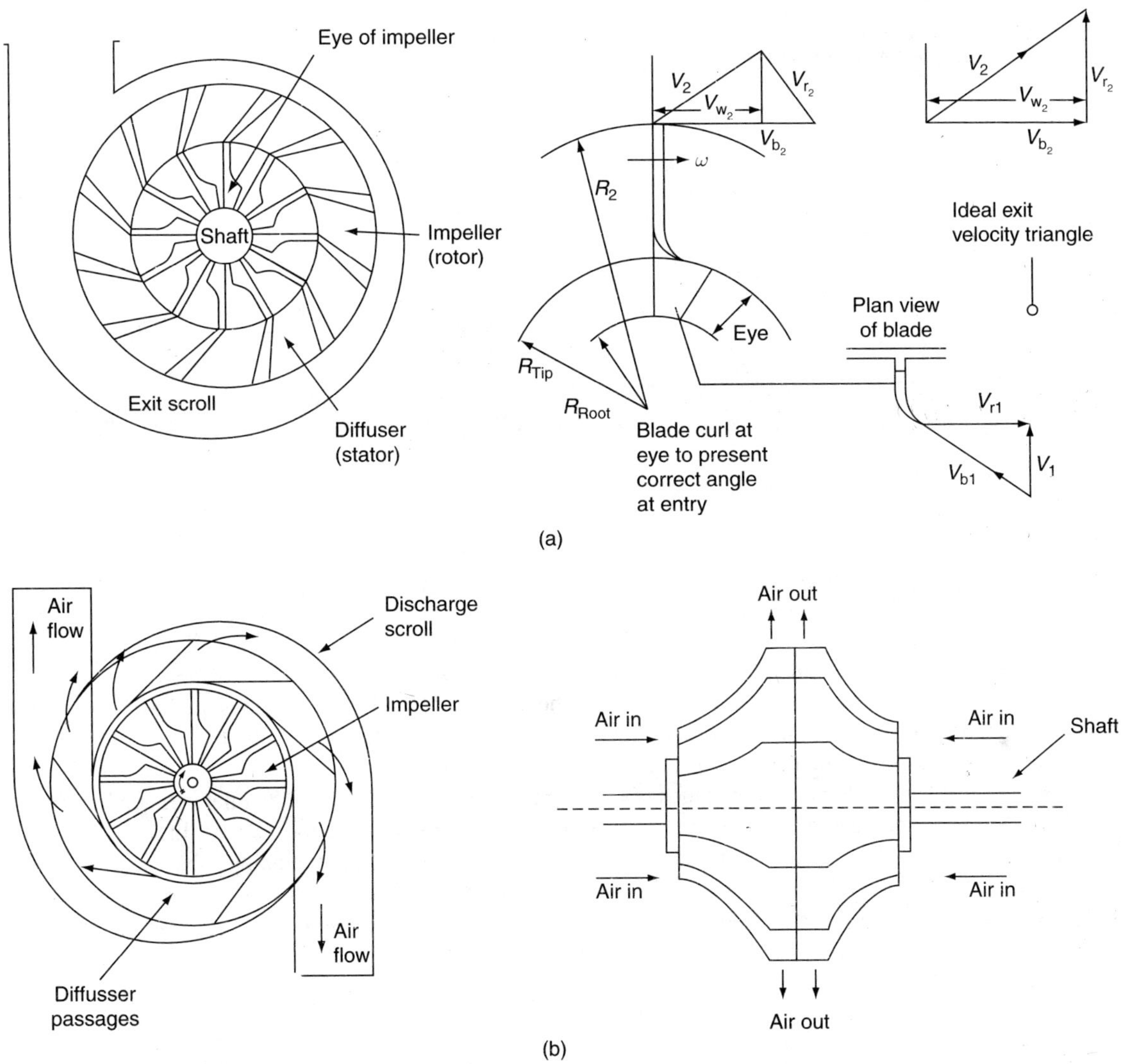

Fig. 19.18 *(a) The centrifugal compressor (b) Doulble-sided impeller of a centrifugal compressor*

The velocity diagrams at inlet and outlet of the impeller are shown in Fig. 19.19. The air enters the impeller with an absolute velocity $V_1 = V_{f1}$ in the axial direction and V_{b1} is the peripheral velocity at the inlet. Thus, the air will be entering the blades at an angle β_1, the blade angle at inlet with relative velocity V_{r1}. The air will be leaving the blades in the radial direction ($\beta_2 = 90°$) with relative velocity V_{r2}, and with a n absolute velocity at outlet V_2. There is no velocity of whirl at inlet ($V_{\omega 1} = 0$) (Fig.19.19a), the theoretical work required to drive the compressor would be

$$W_c = \rho Q\, V_{\omega_2}\, V_{b_2} = \rho Q\, V_{\omega_2}^2 = \rho V_{b_2}^2 \tag{19.30}$$

where Q is the volume flow rate of air (m^3/s).

The fluid leaving the impeller is directed into the fixed diffuser vanes.

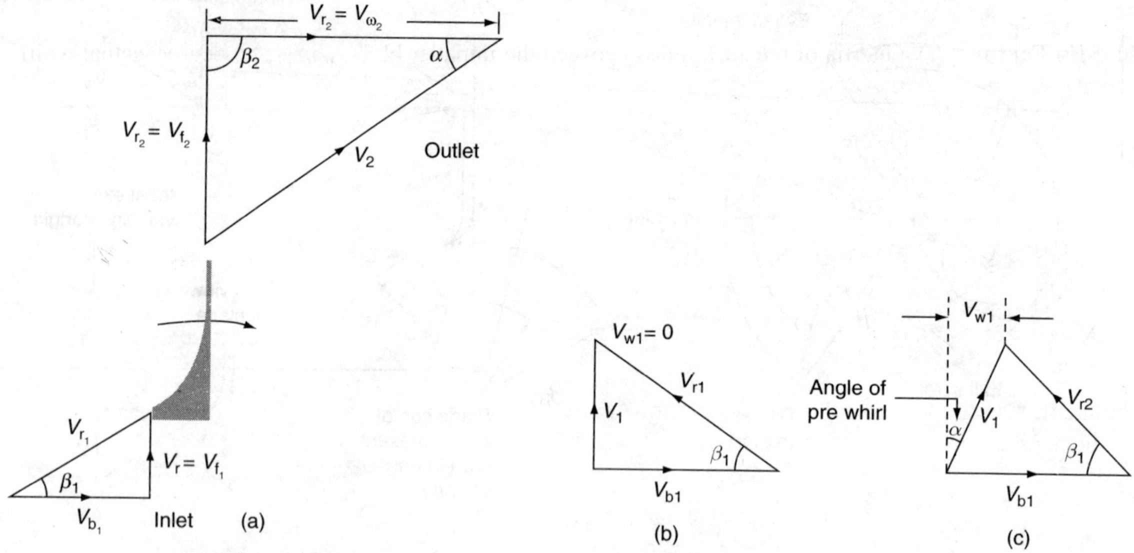

Fig. 19.19 *Blade velocity diagrams at blade inlet of a (a) Velocity triangles centrifugal compressor (b) without, and (c) with pre-whirl*

(a) Pre-whirl The rise in the static pressure of the fluid flowing through the centrifugal compressor is caused by the impeller running at a very high speed. The profiles of the impeller blades are designed in such a way that the absolute velocity of the fluid at the inlet is axial and the fluid enters the blades without shock. The velocity triangle at the inlet is a right-angled triangle ($V_{\omega 1} = 0$).

When the peripheral velocity of the blades is very high, the relative velocity at the inlet is much higher and the flow through the passage may be supersonic leading to the formation of shock waves. In order to reduce the relative velocity at the inlet, the fluid is given a pre-whirl so that the flow remains subsonic through the flow passage, shock waves do not form and the losses are minimum. The pre-whirl is achieved by providing guide vanes at the impeller inlet, without reducing the mass-flow rate. The velocity triangle at the inlet gets modified (Fig. 19.19b), and the work input to the compressor is given by

$$W_c = \rho Q V_{b_2} \left(V_{\omega_2} V_{b_2} - V_{\omega_1} V_{b_1} \right) \tag{19.31}$$

That is, the pre-whirl reduces the work input of the compressor by $V_{\omega_1} V_{b_1}$. The inlet velocity of the impeller eye is inclined by an angle, known as *pre-whirl*, α.

At exit from the impeller, the flow is in the radial direction and the blade velocity V_{b_2} is larger, since the radius of the impeller is larger at the outlet.

The blade velocity diagram is shown in Fig. 19.20 (a) being the case of radially inclined blades, and (b) being that of blades inclined backward at an angle β_2.

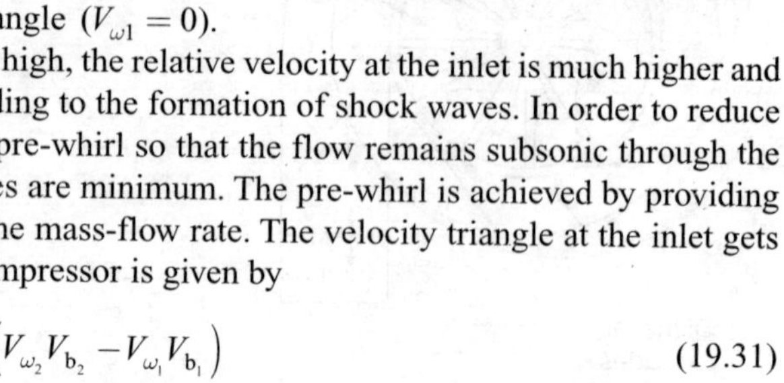

Fig. 19.20 *Blade velocity diagrams at blade outlet of a centrifugal compressor for (a) radially inclined and (b) backward inclined blading*

(b) *Slip Factor* The inertia of the air trapped between the impeller blades, however, causes, actual whirl velocity v'_{ω_2} to be less than v_{ω_2}. It is known as *slip* $(v_{\omega_2} - v'_{\omega_2})$.

$$\text{Slip factor, } \sigma = \frac{V'_{\omega_2}}{V_{\omega_2}} = \frac{V'_{\omega_2}}{V_{b_2} - V_{f2}\cot\beta_2} \tag{19.32}$$

$$\text{Power input, } W_c = \dot{m}\left(V_{b2}\cdot V'_{\omega 1} - V_{b_1}V_{\omega_1}\right) \tag{19.33}$$

The *slip factor* σ is defined as the ratio of work done (when the number of blades are finite) to the work done under ideal conditions (when the number of blades are infinite). Thus

$$\sigma = \frac{V'_{\omega_2}V_{b_2}}{V_{\omega_2}V_{b_2}} = \frac{V'_{\omega 2}}{V_{\omega 2}}$$

The smaller the slip factor, the lower the pressure ratio achieved. The slip factor may be increased by increasing the number of vanes in the impeller, but this has the adverse effect of making the eye more solid.

Because of skin friction, heat losses and leakage around the vanes, the actual work input is greater than the theoretical value.

The work input to the compressor is then

$$W_c = Z\dot{m}\sigma\left(V'_{\omega_2}V_{b_2} - V_{\omega_1}V_{b_1}\right) \tag{19.34}$$

where Z in greater than 1.

The steady-flow energy equation applied to a control volume around the adiabatic compressor gives

$$\dot{W}_c = \dot{m}c_p(T_2 - T_1) \tag{19.35}$$

The *isentropic efficiency* of the compressor is defined by (Fig. 19.21)

$$\eta_s = \frac{(\dot{W}_c)_{\text{ideal}}}{(W_c)_{\text{actual}}} = \frac{T_{2s} - T_1}{T_2 - T}$$

and

$$\frac{T_{2s}}{T_1} = \left(\frac{p_2}{p_1}\right)^{\frac{\gamma-1}{\gamma}} = r_p^{\frac{\gamma-1}{\gamma}},$$

r_p being the pressure ratio.

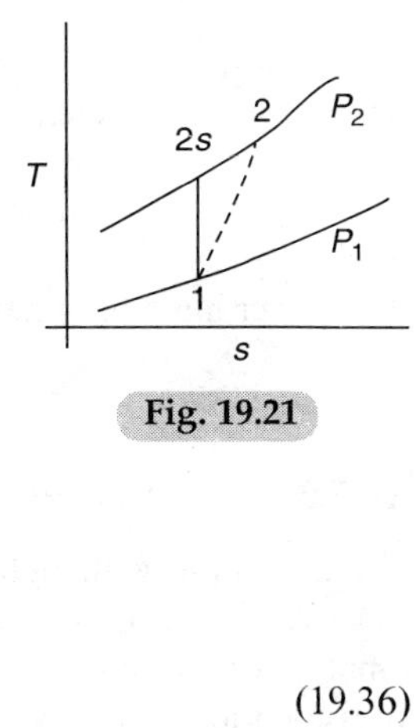

Fig. 19.21

$$\therefore \qquad W_c = \frac{\dot{m}c_pT_1}{\eta_s}\left[(r_p)^{\frac{\gamma-1}{\gamma}} - 1\right]$$

From Eq. (19.34) if $V_{\omega_1} = 0, Z = 1, V_{b_2} = V_{\omega_2}$,

$$r_p = 1 + \eta_s\left[1 + \eta_s\left(\frac{\sigma V_{b_2}^2}{c_pT_1}\right)\right] \tag{19.36}$$

For values of $T_1 = 289$ K, $\eta_s = 0.8$ and V_{b_2} limited to 440 m/s by stress conditions, it is found that the pressure ratio r_p achieved by a centrifugal air compressor will not exceed 4 to 1. This means that the centrifugal compressor has limited application and when high pressure ratios are required axial flow machines must be used.

(c) *Influence of Impeller Blade Shape* The various shapes of blades are shown in Fig.19.22. It is seen from the velocity triangles that V_{ω_2} is maximum for forward curved blades ($\beta_2 > 90°$) and V_{ω_2} is minimum for backward curved blades ($\beta_2 < 90°$). Since the power required is directly proportional to V_{ω_2}, the power input to the radial blades will be between the two cases. From the efficiency point of view, backward curved blades are the best, requiring the minimum work input.

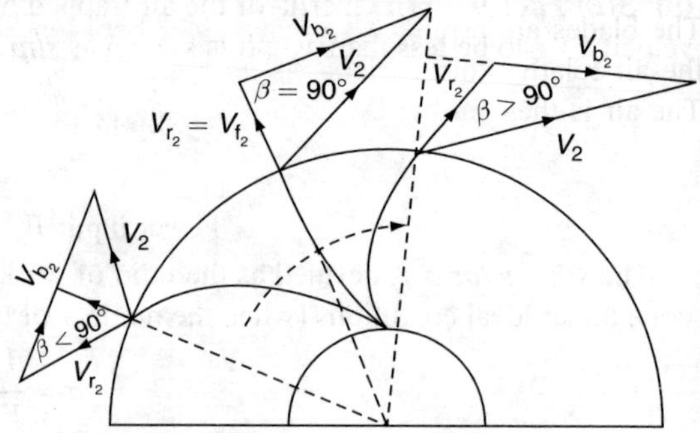

Fig. 19.22 *Influence of blade shape*

(d) Stagnation Values in Centrifugal Compressors Since very high velocity is imparted to the air flowing through the compressor, the total or stagnation properties of air are significant. Let us assume that air is entering the compressor with velocity V_1, pressure p_1 and temperature T_1. The stagnation temperature is

$$T_{01} = T_1 + \frac{V_1^2}{2c_p}$$

and stagnation pressure is

$$p_{01} = p_1\left(\frac{T_{01}}{T_1}\right)^{\frac{\gamma}{\gamma-1}}$$

For isentropic compression from p_{01} to p_{02} (Fig. 19.23)

$$\frac{T'_{02}}{T_{01}} = \left(\frac{p_{02}}{p_{01}}\right)^{\frac{\gamma-1}{\gamma}}$$

The isentropic efficiency of the compressor is

$$\eta_c = \eta_s = \frac{T'_{02} - T_{01}}{T_{02} - T_{01}}$$

The power input becomes

$$W_c = \dot{m}\, c_p \left(T_{02} - T_{01}\right)$$

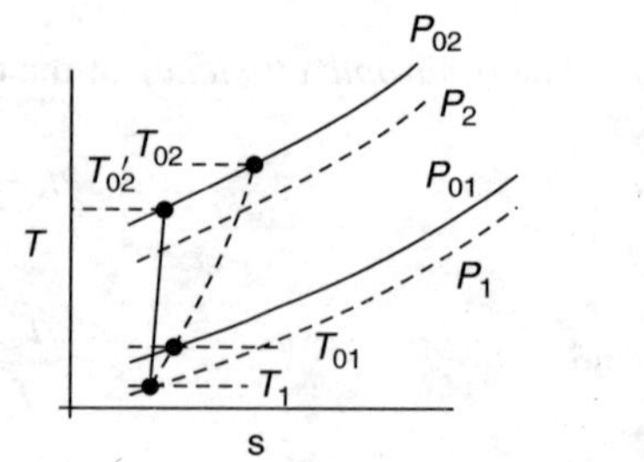

Fig. 19.23 *Static and stagnation properties*

19.7.2 Axial Flow Compressor

For larger units with high pressure ratios the axial flow compressor is more efficient and is usually preferred. For industrial and large gas turbine plants axial compressors are normally used, although some units may employ two or more centrifugal compressors with intercooling between stages. In aircraft units with higher pressure ratios, the advantage of the smaller diameter axial flow compressor can offset the disadvantage of the increased length and weight compared with an equivalent centrifugal compressor. However, centrifugal compressors are cheaper to produce, more robust, less prone to icing problems at high altitudes, and have a wider operating range than the axial flow type.

An axial flow compressor (Figs 19.16b, 19.24a) is similar to an axial flow turbine with a succession of moving blades on the rotor shaft and fixed blades arranged around the stator casing (Fig. 19.24b). Air flows axially through the moving and fixed blades, with stationary guide blades provided at entry to the first row of moving blades.

The work input to the rotor shaft is transferred by the moving blades to the air, thus accelerating it. The blades are arranged so that spaces between blades form diffuser passages, and hence the velocity of the air relative to the blades is decreased as the air passes through them, and there is a rise in pressure. The air is then further diffused in the stator blades which also form diffusing passages (Fig. 19.24a). One

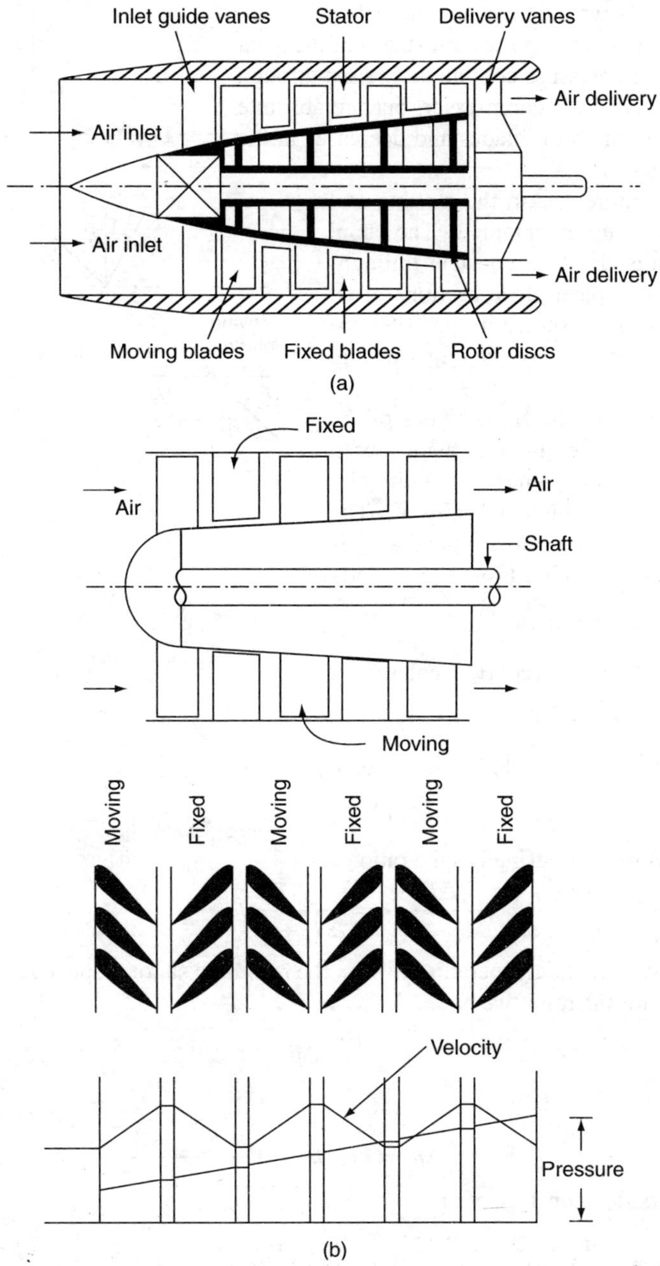

Fig. 19.24 *(a) Axial flow compressor (b) velocity and pressure variation in axial flow compressor*

row of moving blades along with a row of fixed blades constitutes a stage, and the pressure ratio developed in a stage is about 1.2. There can be a large number of stages (5 to 14) with a constant work input per stage.

Figure 19.24(b) shows the pressure and velocity variation of the fluid flowing through several stages of the axial flow compressor. The pressure continuously rises through the successive stages of the compressor, but the absolute velocity increases in the rotor blades and decreases in the stator blade passages.

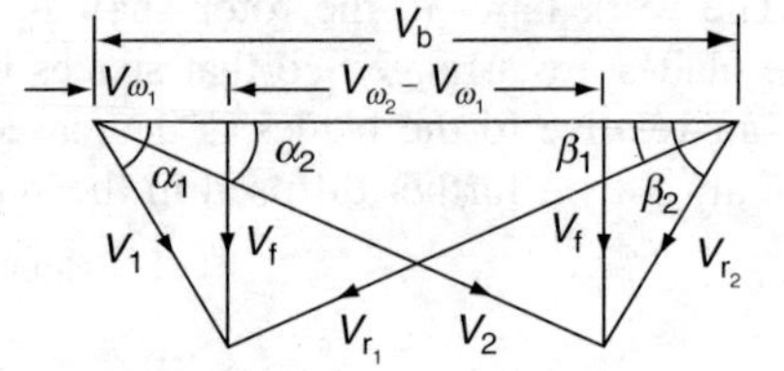

Fig. 19.25 *Velocity triangles for the rotor and stator of an axial flow compressor*

An equal temperature rise in the moving and fixed blades is usually maintained. The axial velocity of air (V_f) is also kept constant throughout the compressor. Typical blade sections of an axial flow compressor are shown in Fig. 19.26(a) and the corresponding velocity diagrams in Fig. 19.26(b).

The air coming out of the fixed blades of the previous stage enters the moving blades with relative velocity V_{r1} and comes out with relative velocity V_{r2} and absolute velocity V_2. The peripheral speed V_b and the velocity of flow V_f is constant at the inlet and outlet. From the velocity triangles (Fig. 19.25).

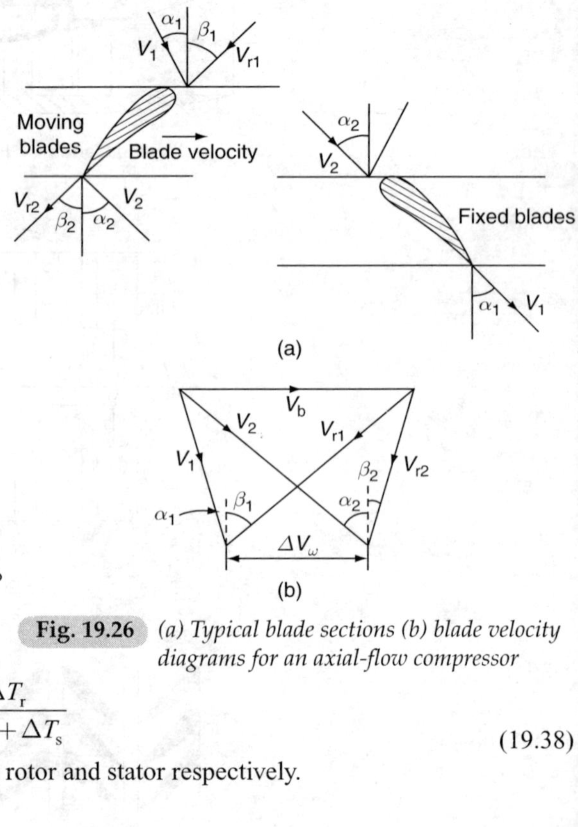

Fig. 19.26 *(a) Typical blade sections (b) blade velocity diagrams for an axial-flow compressor*

$$V_b = V_f(\cot\beta_1 + \cot\alpha_1) = V_f(\cot\beta_2 + \cot\alpha_2)$$

Power input for a mass flow rate of $\dot{m}$ kg/s,

$$W = \Delta V_\omega V_b = (V_{\omega_2} - V_{\omega_1})V_b = V_f(\cot\alpha_2 - \cot\alpha_1)V_b \tag{19.37}$$

The degree of reaction R is defined as the ratio

$$R = \frac{\text{Enthalpy rise in rotor}}{\text{Enthalpy rise in the stage}} = \frac{\Delta h_r}{\Delta h_r + \Delta h_s} = \frac{\Delta T_r}{\Delta T_r + \Delta T_s} \tag{19.38}$$

where ΔT_r and ΔT_s are the temperature rises in the rotor and stator respectively.

From the SFEE for the rotor, we have

$$h_1 + \frac{V_1^2}{2} + W = h_2 + \frac{V_2^2}{2}$$

neglecting any heat loss,

$$\therefore \quad \Delta h_r = h_2 - h_1 = W - \frac{V_2^2 - V_1^2}{2}$$

where W is the work done on 1 kg of air.

$$\therefore \quad R = \frac{W - (V_2^2 - V_1^2)/2}{W} = 1 - \frac{V_2^2 - V_1^2}{2W}$$

$$=1-\frac{V_2^2-V_1^2}{2V_b(\cot\alpha_2-\cot\alpha_1)V_f}=1-\frac{V_f^2(\text{cosec}^2\alpha_2-\text{cosec}^2\alpha_1)}{2V_bV_f(\cot\alpha_2-\cot\alpha_1)}$$

$$=1-\frac{V_f}{2V_1}\cdot\frac{(\cot^2\alpha_2-\cot^2\alpha_1)}{(\cot\alpha_2-\cot\alpha_1)}$$

$$=1-\frac{V_f}{2V_b}(\cot\alpha_1+\cot\alpha_2) \qquad (19.39)$$

Also, $$R=\frac{V_f}{2V_b}(\tan\beta_1+\tan\beta_2)$$

For $$R=0.5,\ \frac{V_b}{V_f}=\cot\beta_1+\cot\alpha_1$$

$\therefore$ $$\beta_2=\alpha_1,\ \alpha_2=\beta_1$$

and $$V_1=V_{r_2},\ V_2=V_{r_1}$$

For 50% reaction blading, with equal enthalpy or temperature rise in the rotor and stator, the velocity triangles are symmetrical.

Blades are usually of twisted section designed according to free vortex theory (see Cohen et al.).

Due to non-uniformity of velocity profiles in the blade passages, the work that can be put into a given blade is less than that given by the ideal diagram. It is taken care of by introducing a *work done factor, y*, defined as

Work done factor, $$y=\frac{\text{Actual power input}}{\dot{m}\Delta V_\omega\cdot V_b}$$

which is about 0.85 for a compressor stage.

(c) Polytropic Efficiency The compression process between the stages of an axial flow compressor is never isentropic, through it can be nearly adiabatic due to its high speed of rotation.

In Fig. 19.27, for a 4-stage compression, the pressure ratios for the stages are p_2/p_1, p_3/p_2, p_4/p_3 and p_5/p_4. If the entire process of compression would have been isentropic, the state points will follow the line 1–6. But for polytropic compression, the actual path followed is 1-2′-3′-4′-5′. The isentropic efficiency for the second stage, e.g., is

$$\eta_{st}=\frac{T_3-T_2'}{T_3'-T_2'}$$

while the *polytropic or internal* efficiency

$$\eta_{int}=\eta_{pol.}=\frac{T_6-T_1}{T_5'-T_1}$$

The polytropic efficiency will always be less than the average stage efficiency because the sum $\left[(h_2-h_1)+(h_3-h_{2'})h+(h_4-h_{3'})+(h_5-h_{4'})\right.$ is greater than (h_6-h_1).

For a polytropic process,

$$\frac{T_2}{T_1}=\left(\frac{p_2}{p_1}\right)^{\frac{n-1}{n}} \quad \text{or} \quad T=\text{const.}\times p^{\frac{n-1}{n}}$$

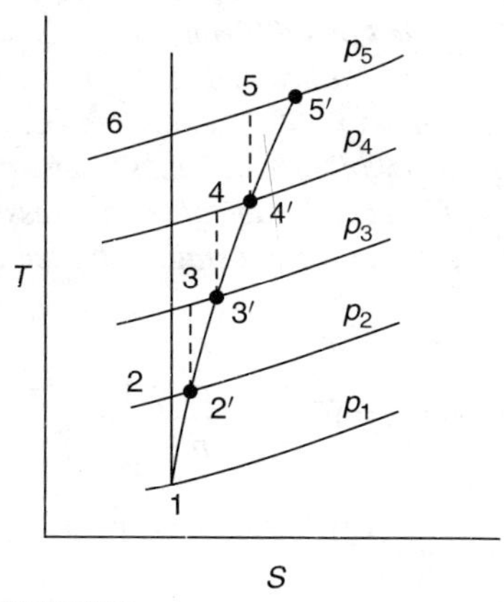

Fig. 19.27 *Polytropic compression in a 4-stage axial compression*

or,
$$\frac{dT}{T} = \frac{n-1}{n}\frac{dp}{p}.$$

For an isentropic process,
$$\frac{dT}{T} = \frac{\gamma-1}{\gamma}\frac{dp}{p}.$$

Let us assume that when the compression process is isentropic the temperature rise is dT and dT' is the temperature rise for a polytropic process, then the stage or isentropic efficiency

$$\eta_s = \frac{dT}{dT'} = \frac{\gamma-1}{\gamma} \times \frac{n}{n-1}$$

$$\therefore \quad \frac{n-1}{n} = \frac{\gamma-1}{\eta\gamma}$$

For a given compressor, let the total pressure rise be given by $p_r = p_5/p_1$, and the actual temperature rises from T_1 to $T_{5'}$, then internal or polytropic efficiency is given by

$$\eta_{int.} = \eta_{pol.} = \frac{T_6 - T_1}{T_{5'} - T_1} = \frac{(T_6/T_1 - 1)}{(T_{5'}/T_1 - 1)} = \frac{(p_r)^{\frac{\gamma-1}{\gamma}} - 1}{(p_r)^{\frac{\gamma-1}{\gamma\eta_s}} - 1}$$

19.8 BLOWERS AND FANS

Fans are used to handle large volume flows of air with a *pressure ratio varying between* 1.0 *and* 1.1. They run at slow speed, compared to compressors, and with or without casing. They are either centrifugal or axial flow type.

Blowers have *pressure ratios varying between* 1.1 to 4.0 and their speeds are higher than fans. Blowers may have more than one stage of compression and are widely used in power plants, workshops and for ventilation. The principles of operation of blowers and fans are similar to those of compressors.

Solved Examples

Example 19.1

A single cylinder reciprocating compressor has a bore of 120 *mm and a stroke of* 150 *mm, and is driven at a speed of* 1200 *rpm. It is compressing CO_2 gas from a pressure of* 120 *kPa and a temperature of* 20°C *to a temperature of* 215°C. *Assuming polytropic compression with n =* 1.3. *no clearance and volumetric efficiency of* 100%, *calculate (a) pressure ratio, (b) indicated power, (c) shaft power, with a mechanical efficiency of* 80%, *(d) mass flow rate.*

If a second stage of equal pressure ratio were added, calculate (e) the overall pressure ratio and (f) the bore of the second stage cylinder, if the same stroke was maintained.

Solution (a)
$$\frac{p_2}{p_1} = \left(\frac{T_2}{T_1}\right)^{\frac{n}{n-1}} = \left(\frac{488}{298}\right)^{\frac{1.3}{0.3}} = 8.48$$

(b)
$$V_l = V_S = \frac{\pi}{4}(0.12)^2 \times 0.15 = 0.0017 \text{ m}^3$$

$$W = \frac{n}{n-1} p_1 V_1 \left[\left(\frac{p_2}{p_1}\right)^{\frac{n-1}{n}} - 1\right]$$

$$\therefore \qquad W = \frac{1.3}{0.3} \times 120 \times 10^3 \times 0.0017 \left[(8.48)^{\frac{0.3}{1.3}} - 1 \right] = 563.6 \text{ J}$$

$$\text{Indicated power} = \frac{563.6 \times 1200}{60} \times 10^{-3} = 11.27 \text{ kW}$$

$$\text{(c) Shaft power} = \frac{11.27}{0.8} = 14.1 \text{ kW}$$

$$\text{(d) Volume flow rate } \dot{V} = 0.0017 \times \frac{1200}{60}$$

$$= 0.034 \text{ m}^3/\text{s}$$

$$\therefore \qquad \dot{m} = \frac{p_1 \dot{V}_1}{RT_1}$$

$$\frac{120 \times 10^3 \times 0.034}{(8314/44) \times 298} = 0.0725 \text{ kg/s}$$

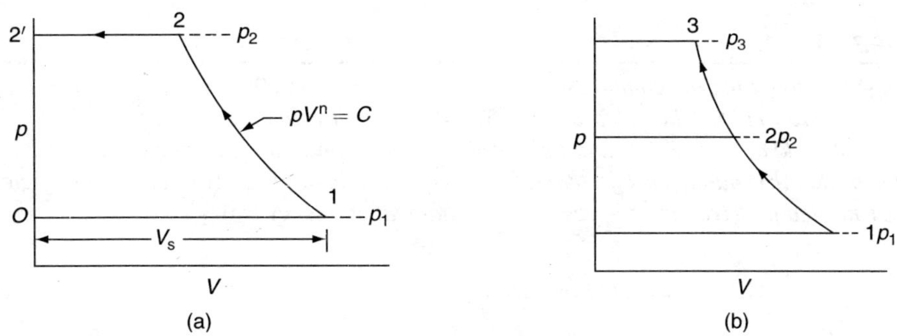

Fig. Ex. 19.1

(e) If a second stage were added with the same pressure ratio, the overall pressure ratio would be

$$\frac{p_3}{p_1} = \frac{p_3}{p_2} \times \frac{p_2}{p_1} = 8.48 \times 8.48 = 71.9$$

(f) Volume delivered per cycle is V_2

$$p_1 V_1^n = p_2 V_2^n$$

$$\therefore \qquad V_2 = \left(\frac{p_1}{p_2} \right)^{\frac{1}{n}} \cdot V_1$$

$$= \left(\frac{1}{8.48} \right)^{\frac{1}{1.3}} \times 0.0017 = 0.00033 \text{ m}^3$$

The second stage would have a swept volume of 0.00033 m^3. Since stroke is the same,

$$\frac{\pi}{4} d^2 \times 0.15 = 0.00033 \text{ m}^3$$

$$\therefore \qquad d = 0.053 \text{ m} = 53 \text{ mm}$$

Example 19.2

A single stage reciprocating compressor takes in air at a pressure of 96 kPa and a temperature of 305 K. The air is compressed to a pressure of 725 kPa and delivered to a reservoir. The clearance volume of the compressor is 5 percent of the swept volume. Both the compression and expansion processes may be represented by a reversible relation $pv^{1.3}$ = constant. Determine the volumetric efficiency referred to atmospheric conditions of 101.3 kPa and 292 K.

Solution

$$\eta_{\text{vol.}} = \left[1 + C - C\left(\frac{p_2}{p_1}\right)^{\frac{1}{n}}\right] \times \frac{p_1}{p_a} \times \frac{T_a}{T_1}$$

$$= \left[1 + 0.05 - 0.05\left(\frac{725}{96}\right)^{\frac{1}{1.3}}\right] \times \frac{96}{101.3} \times \frac{292}{305}$$

Example 19.3

A single stage reciprocating air compressor has a swept volume of 2000 cm³ and runs at 800 rpm. It operates on a pressure ratio of 8, wih a clearance of 5% of the swept volume. Assume NTP room conditions and at inlet (p = 101.3 kPa, t = 15°C), and polytropic compression and expansion with n = 1.25. Calculate (a) indicated power, (b) volumetric efficiency, (c) mass flow rate, (d) FAD, (e) isothermal efficiency, (f) the actual power needed to drive the compressor, if mechanical efficiency is 0.85.

$$= 0.737 \text{ or } 73.7\%$$ *Ans.*

Solution $p_1 = 101.3 \text{ kPa}, p_2 = 8p_1 = 810.4 \text{ kPa}$

$$T_1 = 288 \text{ K}, V_S = 2000 \text{ cm}^3$$

$$V_3 = V_c = 0.05\, V_S = 100 \text{ cm}^3$$

$$V_1 = V_c + V_s = 2100 \text{ cm}^3$$

$$p_3 V_3^n = p_4 V_4^n$$

$$\therefore \quad V_4 = \left(\frac{p_3}{p_4}\right)^{\frac{1}{n}} \cdot V_3 = (8)^{\frac{1}{1.25}} \times 100 = 528 \text{ cm}^3$$

$$V_1 - V_4 = 2100 - 528 = 1572 \text{ cm}^3$$

$$W = \frac{n}{n-1} p_1 (V_1 - V_4)\left[\left(\frac{p_2}{p_1}\right)^{\frac{n-1}{n}} - 1\right]$$

$$= \frac{1.25}{0.25} \times 101.3 \times 10^3 \times 1572 \times 10^{-6}\left[(8)^{\frac{0.25}{1.25}} - 1\right] = 411 \text{ J}$$

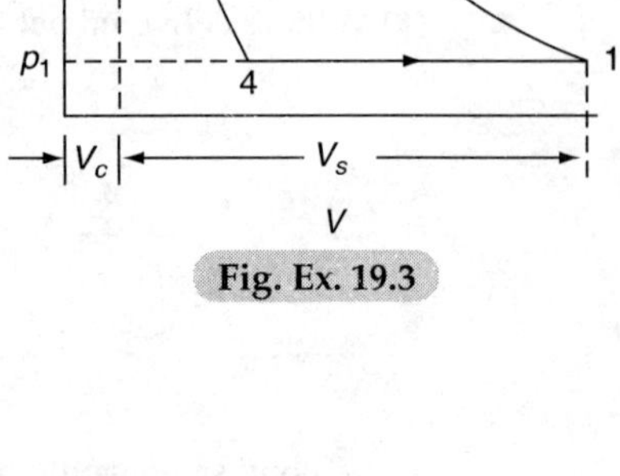

Fig. Ex. 19.3

(a) Indicated power $= \dfrac{411 \times 800 \times 10^{-3}}{60} = 5.47$ kW

(b) Volumetric efficiency $= \dfrac{1572}{2000} \times 100 = 78.6\%$

(c) Mass of air compressed per cycle

$$m = \frac{pV}{RT} = \frac{101.3 \times 10^3 \times 1572 \times 10^{-6}}{287 \times 288} = 1.93 \times 10^{-3}\,\text{kg}$$

$\therefore$ Mass flow rate $= 1.93 \times 10^{-3} \times 800 = 1.54$ kg/min

(d) FAD = Free air delivery $= 1572 \times 10^{-6} \times 800 = 1.26\ \text{m}^3/\text{min}$

(e) $W_t = p_1 (V_1 - V_4) \ln \dfrac{p_2}{p_1} = 101.3 \times 10^3 \times 1572 \times 10^{-6} \ln 8 = 331$ J

$$\eta_{\text{isothermal}} = \frac{0.331 \times 800}{5.47 \times 60} = 80.7\%$$

(f) Input power $= \dfrac{5.47}{0.85} = 6.44$ kW

Example 19.4

A single cylinder double-acting reciprocating compressor compresses 3 kg/min of air from 1 bar, 300 K to 6 bar. The clearance volume is such that the maximum pressure at the end of compression cannot exceed 15 bar. Calculate the power input, the volumetric efficiency and the cylinder dimensions, when the compressor runs at 300 rpm, the index of compression and expansion is equal to 1.3 and stroke/bore ratio is 1.5.

Solution As shown in Fig. Ex. 19.4, as the pressure ratio (p_2/p_1) increases, the volume of air trapped in the clearance space reduces the actual quantity of air taken inside the cylinder and the volumetric efficiency decreases. When the delivery pressure is 15 bar, the volumetric efficiency would be zero. The power input is

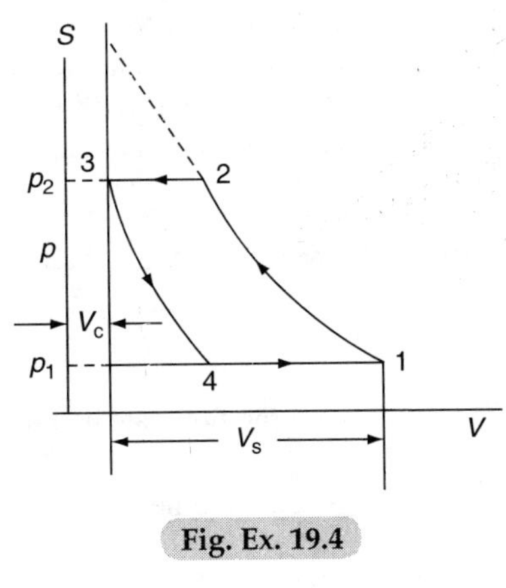

Fig. Ex. 19.4

$$W_c = \frac{n}{n-1} \dot{m} R T_1 \left[\left(\frac{p_2}{p_1} \right)^{\frac{n-1}{n}} - 1 \right]$$

$$= \frac{1.3}{0.3} \times \frac{3}{60} \times 0.287 \times 300 \times \left[(6)^{\frac{0.3}{1.3}} - 1 \right]$$

$= 9.55$ kW *Ans.*

Now, $p_1 V_1^n = p_5 V_5^n$

$$\therefore \quad V_5 = V_c = \left(\frac{p_1}{p_5} \right)^{\frac{1}{n}} V_1 = \left(\frac{1}{15} \right)^{\frac{1}{1.3}} V_1 = 0.1245\, V_1$$

$$\frac{V_1}{V_c} = \frac{1}{0.1245} = 8.032$$

$$\frac{V_s}{V_c} = 8.032 - 1 = 7.032$$

$$p_3 V_3^{\,n} = p_4 V_4^{\,n}$$

$$\therefore \quad V_4 = V_3 \left(\frac{p_3}{p_4}\right)^{\frac{1}{n}} = V_c (6)^{\frac{1}{1.3}} = 3.968\, V_c$$

Volume of air taken in (F.A.D.) at 1 bar, 15°C

$$= V_1 - V_4 = (8.032 - 3.968) V_c \times \frac{288}{300} = 3.9\, V_c$$

$$\eta_{vol} = \frac{V_1 - V_4}{V_s} = \frac{3.9\, V_c}{7.032\, V_c} = 0.5548 \text{ or } 55.48\%$$ *Ans.*

The compressor handles 3 kg/min and it is double-acting and runs at 300 rpm.

Volume of air taken in on one side of the piston per cycle at 1 bar, 288 K,

$$= \frac{3 \times 287 \times 288}{2 \times 10^5 \times 300} = 4.1328 \times 10^{-3} \text{ m}^3$$

$$V_s = \frac{4.1328 \times 10^{-3}}{0.5548} = 7.45 \times 10^{-3} \text{ m}^3 = \frac{\pi}{4} \text{d}^2 \times 1.5\text{d}$$

$$\therefore \quad d = 0.1848 \text{ m} \quad \text{and} \quad l = 0.2772 \text{ m}$$ *Ans.*

Example 19.5

A single acting reciprocating compressor with cylinder of 15 cm diameter and 18 cm stroke has a clearance volume of 4% of swept volume. It takes in air at 1 bar, 25°C and delivers at 8 bar while running at 1200 rpm. The actual power input is 18 kW. Estimate (a) the power required to drive the unit, (b) the isothermal efficiency, and (c) the mechanical efficiency when the mass flow rate in 4 kg/min.

Solution Specific volume of air entering the compressor at 1 bar, 25°C is

$$v = \frac{RT}{p} = \frac{0.287 \times 298}{100} = 0.855 \text{ m}^3/\text{kg}$$

Volume flow rate handled, $\dot{V} = 0.855 \times 4 = 3.42 \text{ m}^3/\text{min}$

$$\text{Swept volume} = \frac{\pi}{4} \text{d}^2 \, \text{lN} = \frac{\pi}{4} \times (0.15)^2 \times 0.18 \times 1200 = 3.84 \text{ m}^3/\text{min}$$

$$\therefore \quad \eta_{vol} = \frac{3.42}{3.84} = 0.89$$

Now,

$$\eta_{vol} = 1 + C - C \left(\frac{p_2}{p_1}\right)^{\frac{1}{n}}$$

$$1 + 0.04 - 0.04 \times 8^{1/n} = 0.89$$

$$8^{1/n} = \frac{0.15}{0.04} = 3.75$$

$$\frac{1}{n}\ln 8 = \ln 3.75$$

$\therefore$

$$n = 1.573$$

$$\frac{T_2}{T_1} = \left(\frac{p_2}{p_1}\right)^{\frac{n-1}{n}} = (8)^{\frac{0.573}{1.573}} = \frac{T_2}{298} = 8^{0.364}$$

$$T_2 = 298 \times 2.13 = 635.2 \text{ K.}$$

Power required

$$W_c = \frac{n}{n-1}\dot{m}R(T_2 - T_1) = \frac{1.573}{0.573} \times \frac{4}{60} \times 0.287 \times (635.2 - 298)$$

$$= 17.71 \text{ kW}$$

$\therefore$

$$\eta_{\text{mech}} = \frac{17.71}{18} = 0.984 \text{ or } 98.4\%$$ *Ans.*

$$W_{\text{isothermal}} = \dot{m}\,RT_1 \ln\frac{p_2}{p_1} = \frac{4}{60} \times 0.287 \times 298 \ln 8 = 11.85 \text{ kW}$$

$$\eta_{\text{iso.}} = \frac{11.85}{18} = 0.6585 \quad \text{or} \quad 65.85$$ *Ans.*

Example 19.6

A single-stage reciprocating compressor has two double-acting cylinders each having 40-cm diameter and 50-cm stroke. The piston rod diameter is 5-cm and the speed is 300 rpm. The inlet condition of air is at 1 bar, 15°C. The delivery pressure is 7.5 bar. If the volumetric efficiency is 80%, mechanical efficiency is 95% and the isothermal efficiency is 70%, determine the power required to drive the compressor.

Solution Swept volume of the cylinder (head end) $= \frac{\pi}{4} \times (0.4)^2 \times 0.5 = 0.063 \text{ m}^3$

Swept volume on the crank side $= \frac{\pi}{4}\left[(0.4)^2 - (0.05)^2\right] \times 0.5 = 0.0619 \text{ m}^3$

Swept volume/min for both cylinders = (0.063 + 0.0619) × 2 × 300 = 74.94 m^3/min

Actual volume handled by the compressor $= 74.94 \times \eta_{\text{vol}} = 74.94 \times 0.8 = 59.95 \text{ m}^3/\text{min}$

Isothermal work $= p_1 \dot{V}_1 \ln\frac{p_2}{p_1}$

$$= 59.95 \times 10^5 \times \ln 7.5 = 120.79 \text{ J/min}$$

Input power $= \frac{120.79}{0.7} = 172.56 \text{ kW}$ *Ans.*

Power required to drive the compressor $= \frac{\text{power input}}{\eta_{\text{mech}}} = \frac{172.56}{0.95} = 181.64 \text{ kW}$ *Ans.*

Example 19.7

A two-stage air compressor with perfect intercooling takes in air at 1 bar pressure and 27°C. The law of compression in both the stages is $pv^{1.3}$ = constant. The compressed air is delivered at 9 bar from the H.P. cylinder to an air receiver. Calculate, per kilogram of air, (a) the minimum work done and (b) the heat rejected to the intercooler.

Solution The minimum work required in a two-stage compressor is given by Eq. (19.20),

$$W_c = \frac{2nRT_1}{n-1}\left[\left(\frac{p_2}{p_1}\right)^{n-1/n} - 1\right]$$

$$= \frac{2\times1.3\times0.287\times300}{0.3}\left[(3)^{0.3/1.3} - 1\right]$$

$$= 26 \times 0.287 \times 100 \times 0.287 = 214.16 \text{ kJ/kg}$$

$$p_2 = \sqrt{p_1 p_3} = \sqrt{1\times9} = 3 \text{ bar}$$

$$\frac{T_2}{T_1} = \left(\frac{p_2}{p_1}\right)^{n-1/n} = 3^{0.3/1.3} = 1.28856$$

$$T_2 = 386.56 \text{ K}$$

Heat rejected to the intercooler

$$= 1.005\ (386.56 - 300) = 86.99 \text{ kJ/kg}$$

Example 19.8

A single-acting two-stage air compressor deals with 4 m³/min of air at 1.013 bar and 15°C with a speed of 250 rpm. The delivery pressure is 80 bar. Assuming complete intercooling, find the minimum power required by the compressor and the bore and stroke of the compressor. Assume a piston speed of 3 m/s, mechanical efficiency of 75% and volumetric efficiency of 80% per stage. Assume the polytropic index of compression in both the stages to be n = 1.25 and neglect clearance.

Solution $p_2 = \sqrt{p_1 p_4} = \sqrt{1.013\times80} = 9$ bar

Minimum power required by the compressor

$$\dot{W} = \frac{2n}{n-1} p_1 \dot{V}_1 \left[\left(\frac{p_2}{p_1}\right)^{\dot{n}-1/n} - 1\right] \times \frac{1}{\eta_{mech}}$$

$$= \frac{2\times1.25}{0.25} \times \frac{1.013\times100}{0.75} \times \frac{4}{60}\left[\left(\frac{9}{1.013}\right)^{0.25/1.25} - 1\right]$$

$$= \frac{1013\times4}{45} \times 0.548 = 49.34 \text{ kW}$$

If L be the stroke length of the piston.

$$2L\frac{N}{60} = 3 \text{ m/s}$$

$$L = \frac{90 \times 100}{250} = 36 \text{ cm}$$

Effective LP swept volume = 4/250 = 0.016 m^3

$$\frac{\pi}{4}(D_{LP})^2 \times 0.36 \times \eta_{vol} = 0.016$$

$$D_{LP} = \sqrt{\frac{0.016 \times 4}{\pi \times 0.36 \times 0.8}}$$

$$= 0.266 \text{ m} = 26.6 \text{ cm}$$

$$\frac{p_1 V_1}{T_1} = \frac{p_3 V_3}{T_3} \qquad \therefore \qquad \frac{V_3}{V_1} = \frac{p_1}{p_3}$$

$\therefore$

$$\frac{\frac{\pi}{4} D^2{}_{HP} L}{\frac{\pi}{4} D^2{}_{LP} L} = \frac{1.013}{9}$$

$$D_{HP} = 0.266\sqrt{\frac{1.013}{9}} = 0.892 \text{ m} = 8.92 \text{ cm}$$

Example 19.9

A gas is compressed in a two-stage reciprocating compressor from 1 bar, 300 K to 9 bar. Assuming perfect intercooling, estimate the compressor work and the total heat transfer. Take R = 0.287 kJ/kg K, c_p = 1.042 kJ/kg K and n = 1.3.

Solution Inter cooler pressure, $p_i = \sqrt{p_1 p_4} = \sqrt{1 \times 9} = 3$ bar

$$T_2 = T_1\left(\frac{p_2}{p_1}\right)^{\frac{n-1}{n}} = 300(3)^{0.286} = 386.57 \text{ K}$$

$$W_c = \frac{2n}{n-1} mR(T_2 - T_1)$$

$$= \frac{2 \times 1.3}{0.3} \times 1 \times 0.287(386.67 - 300) = 222.83 \text{ kJ/kg}$$

For eacy cylinder, W_c = 111.415 kJ/kg

Assuming perfect intercooling, heat transfer in the intercooler,

$$Q = mc_p (T_2 - T_1) = 1 \times 1.042\ (386.57 - 300) = 90.206 \text{ kJ/kg}$$

LP compressor: SFEE gives

$$h_1 + W_c = h_2 + Q$$

$$Q = h_1 - h_2 + W_c$$

$$= 1.042\ (300 - 386.57) + 111.415$$

$$= 21.209 \text{ kJ/kg}$$

Similary, for HP compressor, $Q' = 21.209$ kJ/kg

$\therefore$ total heat transfer to the surroundings

$$= 90.206 + 2 \times 21.209 = 132.624 \text{ kJ/kg} \qquad \textit{Ans.}$$

Example 19.10

The intake conditions of a single-acting two-stage air compressor running at 300 rpm are 0.98 bar and 305 K. The delivery pressure is 20 bar. The intermediate pressure is 5 bar and the clearance volume of the low-pressure compressor is 4% of the stroke volume. The compressor delivers 3 m³/min at 1 bar, 25°C. Determine (a) the power required to drive the compressor, (b) the low pressure cylinder dimensions if L = D, and (c) isothermal efficiency when the intercooling is perfect and the index n = 1.3 for compression and expansion in both the cylinders.

Solution The P-V diagram has been shown in Fig. Ex. 19.10. For the LP cylinder,

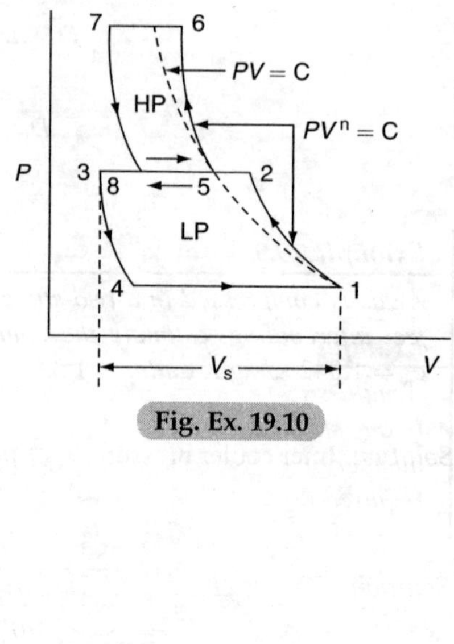

Fig. Ex. 19.10

$$\frac{V_3}{V_s} = 0.04$$

$$V_3 = 0.04\, V_s$$

$$V_1 = 1.04\, V_s$$

$$V_4 = V_3\left(\frac{p_3}{p_4}\right)^{\frac{1}{n}} = 0.04\, V_s\left(\frac{5}{0.98}\right)^{\frac{1}{1.3}} = 0.14\, V_s$$

Volume of air taken in at 0.98 bar, 305 K $= V_1 - V_4$

$$= 1.04\, V_s - 0.14\, V_s$$

$$= 0.90\, V_s$$

$\therefore$ Volume of air taken at free air conditions

$$= 0.90\, V_s \times \frac{298}{305} \times \frac{0.98}{1.00} = 0.861\, V_s$$

$$\therefore \text{ Volumetric efficiency} = \frac{V_1 - V_4}{V_s} = 0.861 \text{ or } 86.1\%.$$

$$\text{Mass of air delivered, } \dot{m} = \frac{p\dot{V}}{RT} = \frac{3 \times 10^5}{287 \times 298 \times 60} = 5.846 \times 10^{-2} \text{ kg/s}$$

$$\frac{T_2}{T_1} = \left(\frac{p_2}{p_1}\right)^{\frac{n-1}{n}}$$

$$\therefore \qquad T_2 = 305\left[\frac{5}{0.98}\right]^{\frac{0.3}{1.3}} = 444.24 \text{ K}$$

For perfect intercooling, $T_1 = T_5 = 305$ K

$$T_6 - T_5\left(\frac{p_6}{p_5}\right)^{\frac{n-1}{n}} = 305\left(\frac{20}{5}\right)^{\frac{0.3}{1.3}} = 420 \text{ K}$$

Power required to drive the compressor

$$\dot{W}_c = \frac{n}{n-1}\dot{m}R\left[(T_2 - T_1) + (T_6 - T_5)\right]$$

$$= \frac{1.3}{0.3} \times 5.846 \times 10^{-2} \times 0.287\left[(444.24 - 305) + (420 - 305)\right]$$

$$= 18.484 \text{ kW}$$ *Ans.* (a)

$$\text{Mass of air delivered per stroke} = 5.846 \times 10^{-2} \times \frac{60}{300} = 1.1692 \times 10^{-2} \text{ kg}$$

$$\text{Volume of free air per stroke} = \frac{1.1692 \times 10^{-2} \times 287 \times 298}{10^5} = 10^{-2} \text{ m}^3$$

$$\text{Stroke volume} = \frac{\pi}{4} d^2 l = \frac{10^{-2}}{0.861} = 1.61 \times 10^{-2} \text{ m}^3$$

$$\therefore \quad d^3 = \frac{1.61 \times 10^{-2} \times 4}{\pi}$$

$$\therefore \quad d = l = 0.2735 \text{ m} = 27.35 \text{ cm}$$ *Ans.* (b)

Isothermal power required

$$= \dot{m}RT_1 \ln\frac{p_2}{p_1} = 5.846 \times 10^{-2} \times 0.287 \times 305 \ln\frac{20}{0.98} = 15.433 \text{ kW}$$

$$\therefore \text{ isothermal efficiency} = \frac{15.433}{18.484} = 0.835 \text{ or } 83.5\%$$ *Ans.* (c)

Example 19.11

A multi-stage air compressor takes in air at 1 bar, 298 K and delivers at 36 bar. The maximum temperature in any stage is not to exceed 390 K. If the law of compression and expansion is $pv^{1.3}$ = constant, find the number of stages for minimum power input. Estimate the power required. What would have been the power required for single-stage compression? What would be the maximum temperature in any-stage?

Solution The optimum pressure ratio per stage given by Eq. (19.25),

$$\frac{p_2}{p_1} = \left(\frac{p_d}{p_s}\right)^{\frac{1}{N}},$$

where N is the number of stages.

Since the maximum temperature should not exceed 390 K,

$$\frac{p_2}{p_1} = \left(\frac{T_2}{T_1}\right)^{\frac{n}{n-1}} = \left(\frac{390}{298}\right)^{\frac{1.3}{0.3}} = 3.205 \text{ or } 4 \text{ stages}$$ *Ans.*

Corresponding intermediate pressures are

$$p_2 = (36)^{\frac{1}{4}} = 2.45 \text{ bar}$$

$$P_3 = (36)^{0.5} = 6.0 \text{ bar}$$

$$P_4 = (36)^{0.75} = 14.7 \text{ bar}$$

Total power required (Eq. 19.26)

$$W_c = \frac{N_n R T_1}{n-1}\left[\left(\frac{p_d}{p_s}\right)^{\frac{n-1}{N_n}} - 1\right]$$

$$= \frac{4 \times 1.3 \times 0.287 \times 298}{0.3}\left[\left(\frac{36}{1}\right)^{\frac{0.3}{4\times 1.3}} - 1\right]$$

$$= 1482.45\left[(2.45)^{0.23} - 1\right]$$

$$= 339.3 \text{ kW/kg air.}$$

For a single-stage compressor, the power required

$$W_c = \frac{n}{n-1} mRT_1\left[\left(\frac{p_d}{p_s}\right)^{\frac{n-1}{n}} - 1\right]$$

$$= \frac{1.3}{0.3} \times 1 \times 0.287 \times 298\left[(36)^{\frac{0.3}{1.3}} - 1\right]$$

$$= 370.61\ (2.28 - 1) = 474.4 \text{ kW}$$ *Ans.*

Maximum temperature in any stage

$$T = T_1(2.45)^{\frac{0.3}{1.3}} = 298 \times (2.45)^{0.23} = 366.2 \text{ K.}$$

Example 19.12

The pressure and temperature of air supplied to an air engine are 700 kPa 38°C respectively. Cut-off takes place at 0.4 of the stroke. Expansion follows the law $pv^{1.3} = c$ (constant) to the release point, which is at the end of the outstroke. The pressure then falls to the constant back pressure of 112 kPa. Neglect the effect of clearance and assuming that the area of the actual indicator diagram is 0.85 of that outlined above, determine the indicated output if the air mass is 1.25 kg.

Solution $V_1 = 0.4\, V_2, \frac{T_1}{T_2} = \left(\frac{V_2}{V_1}\right)^{n-1}$

$$= (2.5)^{0.3} = 1.316$$

$$\therefore \quad T_2 = \frac{273 + 38}{1.316} = 236.3 \text{ K}$$

$$p_1 V_1^{1.3} = p_2 V_2^{1.3}$$

$$\therefore \quad p_2 = p_1\left(\frac{V_1}{V_2}\right)^{1.3} = 700(0.4)^{1.3} = 212.7 \text{ kPa}$$

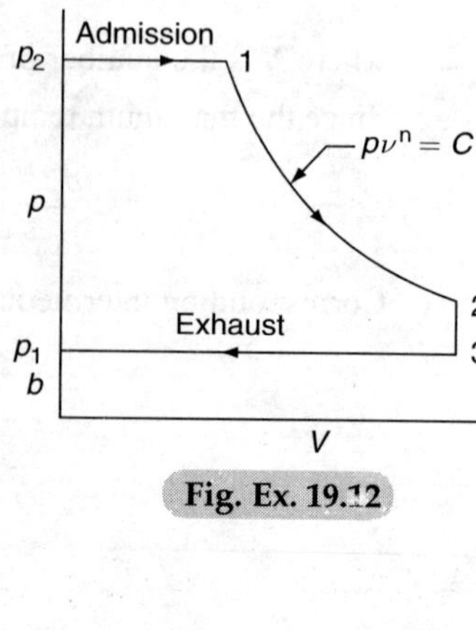

Fig. Ex. 19.12

$$v_2 = \frac{mRT_2}{p_2} = \frac{1.25 \times 0.287 \times 236.3}{212.7} = 0.3986 \text{ m}^3$$

$$v_2 = \frac{V_2}{m} = \frac{0.3986}{1.25} = 0.3188 \text{ m}^3/\text{kg}$$

Area of indicator diagram

$$= p_1 v_1 + \frac{p_1 v_1 - p_2 v_2}{n-1} - p_3 v_3 = RT_1 + \frac{R(T_1 - T_2)}{n-1} - p_3 v_3$$

$$= 0.287 \times 311 + \frac{0.287(311 - 236.3)}{0.3} - 112 \times 0.3188 = 125 \text{ kJ/kg}$$

Indicated output $= 125 \times 0.85 = 106.26$ kJ/kg $= 106.26 \times 1.25 = 132.82$ kJ

Example 19.13

A three-stage single-acting reciprocating air compressor has a low-pressure (L.P.) cylinder of 450 mm bore and 300 mm stroke. The clearance volume of the l.p. cylinder is 5% of the swept volume. Intake pressure and temperature are 1 bar and 18°C respectively, while the delivery pressure is 15 bar. Intermediate pressures are ideal and intercooling is perfect. The compression and expansion index can be taken as 1.3. Estimate (a) the intermediate pressures, (b) the effective swept volume of L.P. cylinder, (c) the temperature and volume of air delivered per stroke at 15 bar, and (d) the work done per kg of air. Take R = 0.29 kJ/kg K.

Solution

(a) $$\frac{p_2}{p_1} = \frac{p_3}{p_2} = \frac{p_4}{p_3} = \left(\frac{p_4}{p_1}\right)^{1/3} = (15)^{1/3} = 2.466$$

$p_2 = 2.466$ bar

$p_3 = 2.466 \times 2.466 = 6.081$ bar

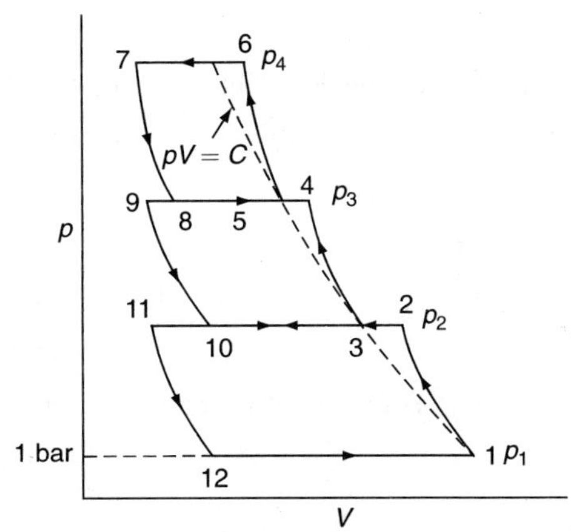

Fig. Ex. 19.13

(b) Swept volume of L.P. cylinder (Fig. Ex. 19.13)

$$= V_1 - V_{11} = \frac{\pi}{4} \times (0.45)^2 \times 0.3 = 0.0477 \text{ m}^3$$

$$V_{11} = 0.05 \times 0.0477 = 0.00239 \text{ m}^3$$

$$\therefore \quad V_1 = 0.0477 + 0.00239 = 0.05009 \text{ m}^3$$

$$p_{11} V_{11}^{1.3} = p_{12} V_{12}^{1.3}$$

$$V_{12} = V_{11} \times \left(\frac{p_{11}}{p_{12}}\right)^{1/1.3} = 0.00239 \times (2.466)^{1/1.3} = 0.00478 \text{ m}^3$$

Effective swept volume of the L.P. cylinder is

$$V_1 - V_{12} = 0.05009 - 0.00478 = 0.04531 \text{ m}^3$$

$$T_5 = T_3 = T_1 = 291 \text{ K}$$

$$\frac{T_6}{T_5} = \left(\frac{p_4}{p_3}\right)^{\frac{n-1}{n}}$$

$$\therefore \quad T_6 = 291\times(2.466)^{\frac{0.3}{1.3}} = 358.5 \text{ K}$$

$$\therefore \quad t_6 = \text{delivery temperature} = 85.5°\text{C}$$

$$\frac{p_4(V_6 - V_7)}{T_6} = \frac{p_1(V_1 - V_{12})}{T_1}$$

$$\therefore \quad V_6 - V_7 = \frac{p_1}{p_6}\cdot\frac{T_6}{T_1}\cdot(V_1 - V_{12}) = \frac{1}{15}\times\frac{358.5}{291}\times 0.04531 = 0.00372 \text{ m}^3$$

$$\text{Work per kg air} = \frac{3nRT_1}{n-1}\left[\left(\frac{p_2}{p_1}\right)^{\frac{n-1}{n}} - 1\right]$$

$$= \frac{3\times1.3\times0.29\times291}{0.3}\left[(2.466)^{\frac{0.3}{1.3}} - 1\right] = 254.3 \text{ kJ}$$

Example 19.14

For a Roots blower, the inlet pressure is 1.013 bar and the pressure ratio is 1.5 to 1. The induced volume of air is 0.03 m³/rev. Estimate the work input. What would be the work input for a vane-type compressor if the internal compression takes place through half the pressure range.

Solution For the Roots blower (Fig. Ex. 19. 14(a))

$$p_1 = 1.013 \text{ bar}$$

$$p_2 = 1.5\times1.013 = 1.52 \text{ bar}$$

$$V_s = 0.03 \text{ m}^3/\text{rev.}$$

$$\therefore \quad \text{Work done} = (p_2 - p_1)V_s$$

$$= (1.52 - 1.013) \times 100 \times 0.03 = 1.52 \text{ kJ/rev}$$

For the vane-type compressor (Fig. Ex. 19.14(b))

$$p_i = \frac{(1.5\times1.013) + 1.013}{2} = 1.266 \text{ bar}$$

$$\text{Work required} = \text{Area } A + \text{Area } B$$

$$\text{Area } A = \frac{\gamma}{\gamma - 1} p_1 V_s \left[\left(\frac{p_i}{p_1}\right)^{\frac{\gamma-1}{\gamma}} - 1\right]$$

$$= \frac{1.4}{0.4}\times1.013\times100\times0.03\left[\left(\frac{1.266}{1.013}\right)^{\frac{0.4}{1.4}} - 1\right] = 0.70 \text{ kJ/rev}$$

$$\text{Area } B = (p_2 - p_i)\, V_b$$

$$p_1 V_a^\gamma = p_i V_b^\gamma$$

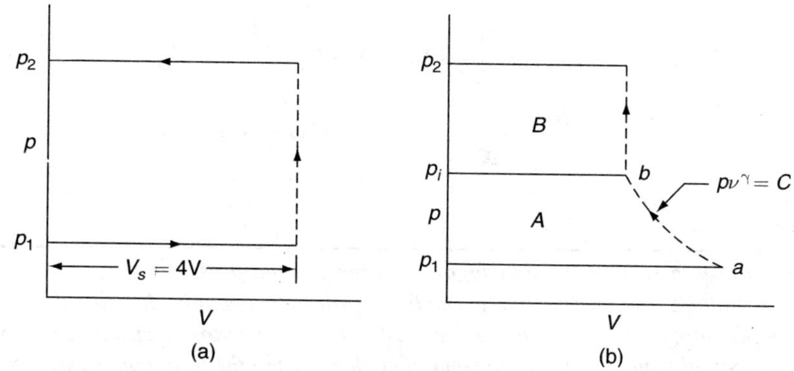

Fig. Ex. 19.14

$$V_b = V_a \left(\frac{p_1}{p_i}\right)^{\frac{1}{\gamma}} = 0.03\left(\frac{1.013}{1.266}\right)^{\frac{1}{1.4}} = 0.0256 \text{ m}^3$$

Area $B = (1.52 - 1.266) \times 100 \times 0.0256 = 0.65$ kJ/rev

∴ Worked required = 0.70 + 0.65 = 1.35 kJ/rev.

(Compared to 1.52 kJ/rev. for the Roots blower)

Example 19.15

A Roots blower supplies air at the rate of 1 *kg/s. The pressure ratio of the blower is* 2:1 *with an intake pressure and temperature of* 1 *bar and* 70°C *respectively. Find the power required to drive the blower. Take* $R = 0.29$ *kJ/kg K.*

If a vane pump having the same air flow, pressure ratio and intake conditions has the volume reduced to 0.7 *of the intake volume before delivering the air, estimate the power required.*

Solution $$\dot{V} = \frac{\dot{m} RT_1}{p_1} = \frac{1 \times 0.29 \times 343}{100} = 0.995 \text{ m}^3/\text{s}$$

∴ Power required by the Roots blower

$$= \dot{V}(p_2 - p_1) = 0.995 \times 100 = 99.5 \text{ kW}$$

For the vane compressor, $p_1 V_1^\gamma = p_2 V_2^\gamma$

$$\therefore \quad p_2 = p_1 \left(\frac{V_1}{V_2}\right)^\gamma = 1 \times \left(\frac{1}{0.7}\right)^{1.4} = 1.65 \text{ bar}$$

$$\dot{V}_2 = 0.7\, \dot{V}_1 = 0.7 \times 0.995 = 0.696 \text{ m}^3/\text{s}$$

$$\text{Power required} = \frac{\gamma}{\gamma - 1} p_1 \dot{V}_1 \left[\left(\frac{p_2}{p_1}\right)^{\frac{\gamma-1}{\gamma}} - 1\right] + \dot{V}_2 (p_3 - p_2)$$

$$= \frac{1.4}{0.4} \times 100 \times 0.995 \left[\left(\frac{1.65}{1} \right)^{\frac{0.4}{1.4}} - 1 \right]$$

$$+ 0.696\,(2 - 1.65) \times 100 = 78\text{kW}$$

Example 19.16

A gas turbine utilizes a two-stage centrifugal compressor. The pressure ratios for the first and second stages are 2.5 to 1 and 2.1 to 1 respectively. The flow of air is 5 kg/s, this air being drawn at 1.013 bar, and 10°C. If the temperature drop in the intercooler is 50°C and the isentropic efficiency is 85% for each stage, calculate (a) the actual temperatures at the end of each stage (b) the total compressor power. Take $\gamma = 1.4$ and $c_p = 1.005$ kJ/kg K.

Solution Stage I

$$\frac{T_{2s}}{T_1} = \left(\frac{p_2}{p_1} \right)^{\frac{\gamma-1}{\gamma}} \quad \text{(Fig. Ex. 19.16)}$$

$$T_{2s} = 283(2.5)^{0.4/1.4} = 367.7 \text{ K}$$

$$T_2 - T_3 = 50°\text{C}$$

$$T_{2s} - T_1 = \eta_s (T_2 - T_1)$$

$$T_2 = T_1 + \frac{T_{2s} - T_1}{n_s}$$

$$= 283 + \frac{367.7 - 283}{0.85} = 382.65 \text{ K}$$

Stage II $T_3 = T_2 - 50 = 382.65 - 50 = 332.65$ K

$$\frac{T_{4s}}{T_3} = \left(\frac{p_3}{p_2} \right)^{\frac{\gamma-1}{\gamma}}$$

$$\therefore \quad T_{4s} = 332.65 \times (2.1)^{0.4/1.4} = 411.19 \text{ K}$$

$$T_4 - T_3 = \frac{T_{4s} - T_3}{n_s}$$

$$= \frac{411.19 - 322.65}{0.85} = 104.17 \text{ K}$$

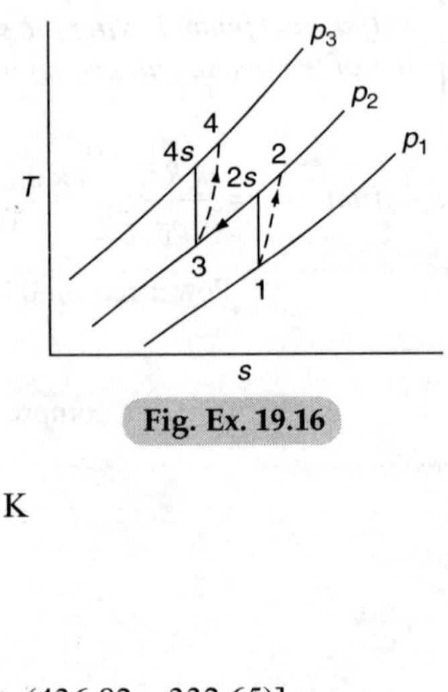

Fig. Ex. 19.16

$$\therefore \quad T_4 = 104.17 + 332.65 = 436.82 \text{ K}$$

$$\text{Total compressor power} = \dot{m}_a c_p [(T_2 - T_1) + (T_4 - T_3)]$$

$$= 5 \times 1.005\,[(382.65 - 283) + (436.82 - 332.65)]$$

$$= 1024.2 \text{ kW}$$

***Example* 19.17**

A rotary compressor is used to supercharge a petrol engine. The static pressure ratio across the rotor is 2.5 : 1*. Static inlet pressure and temperature are* 0.6 *bar and* 5°C *respectively. The air-fuel ratio is* 13:1 *and the engine consumes* 0.04 *kg fuel/s. For the air-fuel mixture take* $\gamma = 1.39$ *and* $c_p = 1.005$ *kJ/kg K. The isentropic efficiency of the compressor is* 84%*. Estimate the power required to drive the compressor. Taking the exit velocity from the compressor as* 120 *m/s and assuming that the mixture is adiabatically brought to rest in the engine cylinders, estimate the stagnation temperature and pressure of the mixture at the beginning of the compression stroke in the engine cylinder.*

Solution

$$\frac{T_{2s}}{T_1} = \left(\frac{p_2}{p_1}\right)^{\frac{\gamma-1}{\gamma}} = (2.5)^{\frac{0.39}{1.39}} = 1.294$$

$$T_{2s} = 278 \times 1.294 = 359.7 \text{ K}$$

$$T_2 - T_1 = \frac{T_{2s} - T_1}{\eta_s} = \frac{359.7 - 278}{0.84} = 97.3 \text{ K}$$

$$\therefore \quad T_2 = 97.3 + 278 = 375.3 \text{ K}$$

Mass of air-fuel mixture = 0.04 × (13+1) = 0.56 kg/s

Power to drive the compressor = $\dot{m}_g c_p (T_2 - T_1)$

$$= 0.56 \times 1.005 \times 97.3 = 54.8 \text{ kW}$$

Stagnation temperature

$$T_{02} = T_2 + \frac{V_2^2}{2c_p} = 375.3 + \frac{120^2}{2 \times 1.005 \times 1000}$$

$$= 382.46 \text{ K} = 109.46°\text{C}$$

The temperature in the engine cylinder is 109.46°C

$$p_{02} = p_2 \times \left(\frac{T_{02}}{T_2}\right)^{\frac{\gamma}{\gamma-1}} = 0.6 \times 2.5 \times \left(\frac{382.46}{375.3}\right)^{0.36} = 1.605 \text{ bar or } 160.5 \text{ kPa}$$

***Example* 19.18**

A centrifugal compressor running at 10,000 *rpm delivers* 1.2 m^3/s *of free air. The pressure and temperature at the inlet are* 1 *bar and* 27°C*. The pressure ratio is* 5*, the blades are radial at outlet. The velocity of flow is* 60 *m/s and is constant throughout, and the slip factor is* 0.9*. Calculate* (*i*) *the pressure of air at outlet,* (*ii*) *power input,* (*iii*) *the impeller diameter and blade angle at inlet, and* (*iv*) *diffuser inlet angle. Take the isentropic efficiency as* 0.85 *and the impeller diameter at inlet as half of that at outlet.*

Solution

$$T_{2s} = T_1\left(\frac{p_2}{p_1}\right)^{\frac{\gamma-1}{\gamma}} = (273 + 27)5^{\frac{0.4}{1.4}} = 475.36 \text{ K}$$

$$\eta_s = \frac{T_2 - T_1}{T_{2s} - T_1} = 0.85$$

$$\therefore \quad T_2 = T_1 + \frac{T_{2s} - T_1}{\eta_s} = 300 + \frac{175.36}{0.85} = 506.3\,\text{K} = 233.3\ \text{C} \qquad \textit{Ans. (i)}$$

$$\dot{m} = \frac{p\dot{v}}{RT} = \frac{100 \times 1.2}{0.287 \times 288} = 1.452\,\text{kg/s}$$

$$\text{Power input, } \dot{W}_c = \dot{m}c_p(T_2 - T_1)$$

$$= 1.452 \times 1.005 \times (506.3 - 300) = 301\ \text{kW} \qquad \textit{Ans. (ii)}$$

Since the blades are radial at outlet

$$\dot{W}_c = \dot{m}\,\sigma\,V_{b_2}^2$$

$$V_{b_2} = \left[\frac{301 \times 10^3}{1.452 \times 0.9}\right]^{\frac{1}{2}} = 480\,\text{m/s} = \frac{\pi DN}{60}$$

$\therefore$ Diameter of the impeller

$$D = \frac{480 \times 60}{\pi \times 10{,}000} = 0.9166\,\text{m} \qquad \textit{Ans. (iii)}$$

$$V_{b_1} = \frac{V_{b_2}}{2} = \frac{480}{2} = 240\,\text{m/s}$$

$$\tan\beta_1 = \frac{V_f}{V_{b_1}} = \frac{60}{240} = 0.25$$

$$\therefore \quad \beta_1 = \text{blade inlet angle} = 14° \qquad \textit{Ans. (iii)}$$

From the outlet velocity triangle (Fig. 19.19),

$$\tan\alpha = \frac{V_f}{\sigma V_{b_2}} = \frac{60}{0.9 \times 480}$$

$$\therefore \quad \alpha = 7.9° \qquad \textit{Ans. (iv)}$$

Example 19.19

The following data comprise the design specification for a single-sided centrifugal air compressor. Calculate the total head pressure ratio that should be achieved, the required power at the input shaft and the inlet angles of the blades at the root and tip of the impeller eye.

Data: Rotational speed = 264 rps; slip factor = 0.91; Impeller diameter = 0.482 m; Impeller eye tip diameter = 0.306 m; Impeller eye root diameter = 0.153 m; Uniform axial inlet velocity = 138 m/s; Air mass-flow rate = 9.1 kg/s; Inlet air stagnation temperature = 294 K; Total head isentropic efficiency = 0.80; Mechanical efficiency = 0.98.

The compression process is adiabatic, and take air as an ideal gas with $c_p = 1.006$ kJ/kg K and $\gamma = 1.4$.

Solution Power input, $\dot{W}_c = \dot{m}\sigma V_{b_2}^2$

where $V_{b_2} = 2\pi R_2 N$

$$\therefore \qquad \dot{W}_c = \frac{9.1 \times 0.91}{10^3}\left(2\pi \times \frac{0.482}{2} \times 264\right) = 1320 \text{ kW}$$

External power required $= \dfrac{1320}{0.98} = 1350$ kW *Ans.*

Now, $\dot{m}c_p \Delta T_t = \dot{W}_c$

$\therefore$ Total head temperature rise, $\Delta T_t = \dfrac{1320}{9.1 \times 1.006} = 144.5$ K

$\therefore$ Ideal temperature rise $= 144.5 \times \eta_s = 144.5 \times 0.8 = 115.5$ K

$$\therefore \qquad (T_2)_i = 294 + 115.5 = 409.5 \text{ K} = T_{02}$$

Now,
$$\frac{p_{02}}{p_{01}} = \left(\frac{T_{02}}{T_{01}}\right)^{\frac{\gamma}{\gamma-1}} = \left(\frac{409.5}{294}\right)^{3.5} = 3.18$$
Ans.

Blade angles: At the eye, tip velocity $= 2\pi \times 264 \times \dfrac{0.306}{2} = 254$ m/s

Eye root velocity = 127 m/s

From velocity triangle, $\beta_1 = \tan^{-1}\left(\dfrac{V_f}{V_b}\right) = \tan^{-1}\left(\dfrac{138}{254}\right) = 28°31'$

= blade inlet angle at the eye tip

$$\beta_2 = \tan^{-1}\left(\frac{138}{127}\right) = 47°23'$$

= blade inlet angle at the eye root *Ans.*

Example 19.20

A centrifugal compressor running at 16000 rpm takes in air at 17°C and compresses it through a pressure ratio of 4 with an isentropic efficiency of 82%. The blades are radially inclined and the slip factor is 0.85. Guide vanes at inlet give the air an angle of pre-whirl of 20° to the axial direction. The mean diameter of the impeller eye is 200 mm and the absolute air velocity at inlet is 120 m/s. Calculate the impeller tip diameter. Take c_p = 1.005 kJ/kg K and γ = 1.4.

Solution With reference to Fig. Ex. 19.20(a),

$$\frac{T_{2s}}{T_1} = (4)^{\frac{1.4-1}{1.4}} = 4^{0.286} = 1.487$$

$$T_{2s} = 290 \times 1.487 = 431 \text{ K}$$

$$\Delta T_s = T_{2s} - 1 = 431 - 290 = 141 \text{ K}$$

$$\Delta T = 141/0.82 = 171.95 \text{ K}$$

$$W_c = mc_p \Delta T = 1 \times 1.005 \times 171.95 = 172.81 \text{ kJ/kg}$$

Absolute air velocity at inlet (Fig. Ex. 19.20(b)),

$$V_1 = 120 \text{ m/s}$$

$$V_{b_1} = \frac{\pi d_1 N}{60} = \frac{\pi \times 0.2 \times 16{,}000}{60} = 167.55 \text{ m/s}$$

Pre-whirl angle = 20°

$$V_{\omega_1} = V_1 \sin 20° = 120 \sin 20° = 41.04 \text{ m/s}$$

At exit of the vanes (Fig. Ex.19.20(c)),

$$V_{\omega_2} = V_{b_2} \text{ (since the blades are radially inclined)}$$

$$\text{slip factor } \sigma = \frac{V'_{\omega_2}}{V_{\omega_2}} = 0.85$$

$$V'_{\omega_2} = 0.85 V_{b_2}$$

$$\text{Power input per kg } V'_{\omega_2} V_{b_2} - V_{\omega_1} V_{b_1}$$

$$172.81 \times 10^3 = 0.85\, V_{b_2}^2 - 167.55 \times 41.05$$

$$\therefore \quad V_{b_2} = 459.78 \text{ m/s} = \frac{\pi d_2 \times 16000}{60}$$

Tip diameter $d_2 = 0.5488 \text{ m} = 549 \text{ mm}$ *Ans.*

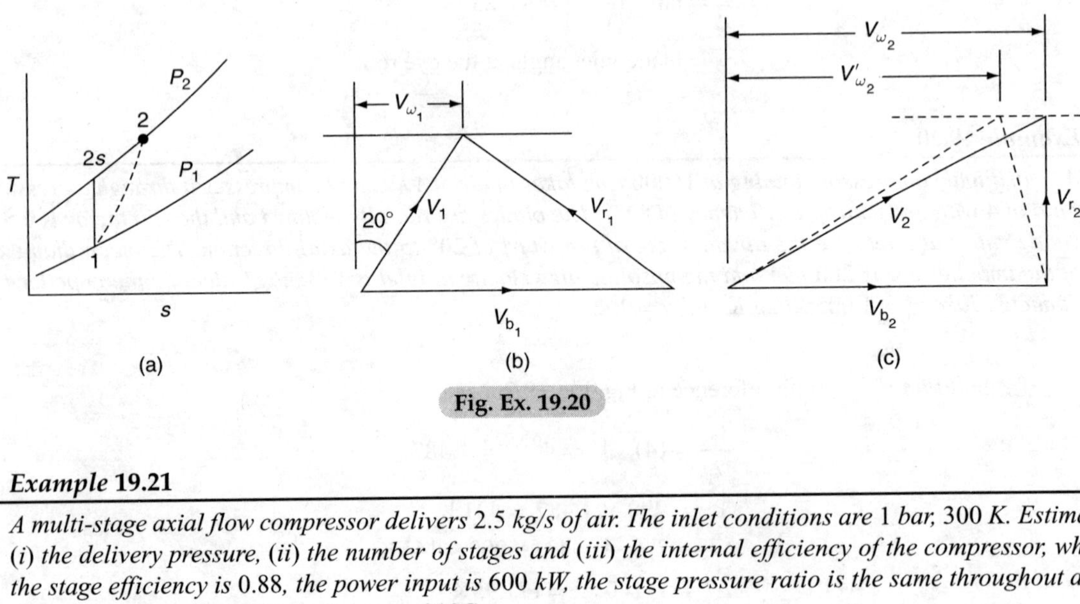

Fig. Ex. 19.20

Example 19.21

A multi-stage axial flow compressor delivers 2.5 *kg/s of air. The inlet conditions are* 1 *bar,* 300 *K. Estimate (i) the delivery pressure, (ii) the number of stages and (iii) the internal efficiency of the compressor, when the stage efficiency is* 0.88*, the power input is* 600 *kW, the stage pressure ratio is the same throughout and the temperature rise in the first stage is* 21°C.

Solution The stage efficiency for polytropic compression is

$$\eta_s = \frac{\gamma-1}{\gamma}\times\frac{n}{n-1}$$

$$0.88 = \frac{1.4-1}{1.4}\times\frac{n}{n-1}$$

$$\therefore \quad \frac{n-1}{n} = 3.08$$

Now, $W_c = mc_p(T - T_1)$,

where T is the actual temperature of air at the end of compression Fig. Ex. 19.21.

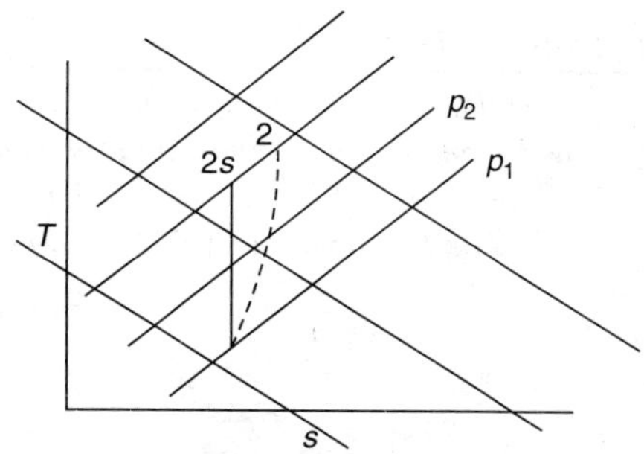

Fig. Ex. 19.21

$$\therefore \quad T = \frac{W_c}{mc_p} + T_1 = \frac{600}{2.5\times1.005} + 300 = 538.81 \text{ K}$$

If the delivery pressure is denoted by p, then

$$\frac{p}{p_1} = \left(\frac{T}{T_1}\right)^{\frac{n}{n-1}} = \left(\frac{538.81}{300}\right)^{3.08} = 6.07$$

$\therefore$ delivery pressure, $p = 6.07 \times 1 = 6.07$ bar *Ans.* (i)

The stage efficiency is given by (for the first stage)

$$\eta_s = \frac{T_{2s}-T_1}{T_2-T_1}$$

$$T_{2s} - T_1 = \eta_s\left(T_2 - T_1\right) = 0.88\times21 = 18.48$$

$$\therefore \quad T_2 = 318.48 \text{ K}$$

$$\frac{p_2}{p_1} = \left(\frac{T_2}{T_1}\right)^{\frac{\gamma}{\gamma-1}} = \left(\frac{318.48}{300}\right)^{3.5} = 1.2327$$

p_1 = Pressure ratio for the first stage

Taking the same pressure ratio for each stage,

$$\left(\frac{p_2}{p_1}\right)^{N} = 6.07$$

where N is the number of stages.

$$N \ln\left(\frac{p_2}{p_1}\right) = \ln 6.07$$

$$N = \frac{\ln 6.07}{\ln 1.2327} = 8.619$$

i.e., 9 stages *Ans.* (ii)

Temperature at the end of isentropic compression

$$T_s = T_1\left(\frac{p}{p_1}\right)^{\frac{\gamma-1}{\gamma}} = 300(6.07)^{0.286} = 502.47 \text{ K}$$

$$\therefore \quad \text{Internal efficiency} = \frac{T_s - T_1}{T_1 - T_1} = \frac{502.47-300}{538.81-300} = 0.85$$ *Ans.* (iii)

Example 19.22

An axial flow compressor having an impeller of mean diameter 0.5 m rotates at 15000 rpm. The velocity of flow is constant at 230 m/s, and the velocity of whirl at inlet is 80 m/s. The inlet pressure and temperature are 1 bar and 300 K. The stage efficiency is 0.88. The pressure ratio through the stage is 1.5. Calculate the (i) fluid deflection angle, (ii) the power input, and (iii) the degree of reaction.

Solution

$$V_b = \text{peripheral speed} = \frac{\pi DN}{60}$$

$$= \frac{\pi \times 0.5 \times 15000}{60} = 392.75 \text{ m/s}$$

Temperature at the end of isentropic compression

$$T_s = T_1 \left(\frac{p}{p_1}\right)^{\frac{\gamma-1}{\gamma}} = 300 \times (1.5)^{0.286} = 336.85 \text{ K}$$

Actual temperature $T = T_1 + \dfrac{T_s - T_1}{\eta_s} = 300 + \dfrac{36.85}{0.88} = 341.87$ K

Work input, $W_c = c_p(T - T_1) = 1.005 \times 41.87 = 42.08$ kJ/kg *Ans.* (ii)

$\therefore$ $\left(V_{\omega_2} - V_{\omega_1}\right)V_b = 42.08$

$$V_{\omega_2} = 80 + \frac{42.08 \times 10^3}{392.75} = 187.14 \text{ m/s}$$

From velocity triangles at inlet and outlet of blades

$$\tan \beta_1 = \frac{V_f}{V_b - V_{\omega_1}} = \frac{220}{392.75 - 80}$$

$\therefore$ $\beta_1 = 35.12°$

and $\tan \beta_2 = \dfrac{V_f}{V_b - V_{\omega_2}} = \dfrac{220}{392.75 - 187.14} = \dfrac{220}{205.61}$

$$\beta_2 = 46.94°$$

$\therefore$ fluid deflection angle $= \beta_2 - \beta_1 = 46.94 - 35.12 = 11.82°$ *Ans.* (i)

Degree of reaction, $R = 1 - \dfrac{V_1(\cot \alpha_1 + \cot \alpha_2)}{2V_b} = 1 - \dfrac{V_{\omega_1} + V_{\omega_2}}{2V_b}$

$$= 1 - \frac{80 + 187.14}{2 \times 392.75} = 0.66 \text{ or } 66\%$$ *Ans.* (iii)

Example 19.23

An axial flow fan delivering 5m^3/s has the mean impeller diameter of 1.0 m and the hub diameter of 0.6 m. It rotates at 600 rpm and develops a theoretical head equal to 35 mm of water. Determine the blade angles at the tip and at the hub. Assume that the velocity of flow is independent of radius and the energy input per unit. The length of the blade is constant. Take the density of air as 1.2 kg/m^3 and the density of water as 1000 kg/m^3.

Solution Velocity of flow, $V_f = \dfrac{5\times4}{\pi\left(1^2-0.6^2\right)} = 9.95$ m/s

$$V_b \text{ at blade tip} = \frac{\pi\times1.0\times600}{60} = 31.42 \text{ m/s}$$

$$V_b \text{ at the hub} = \frac{\pi\times0.6\times600}{60} = 18.85 \text{ m/s}$$

From Fig. Ex. 19.23,

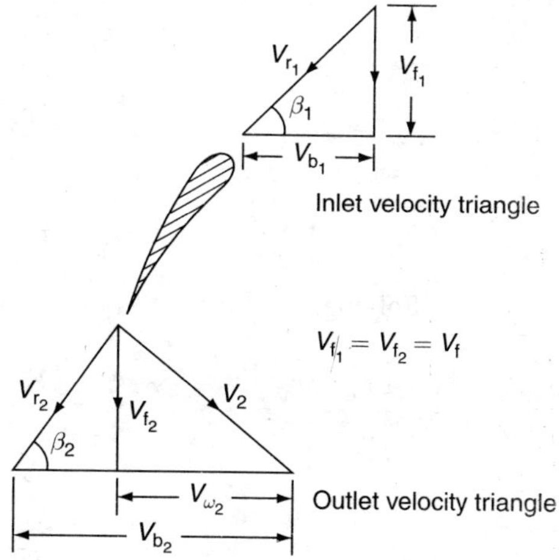

Fig. Ex. 19.23

$$\tan\beta_2 = \frac{V_{f_2}}{V_{b_2}-V_{\omega_2}}; \text{ Head developed by the fan}$$

$$= \frac{V_{b_2}.V_{\omega_2}}{g}$$

Now, H = 35 mm of water

$$= \frac{0.035\times10^3}{1.2} = 29.17 \text{ m of air}$$

$$V_{\omega_2} \text{ at the tip} = \frac{29.17\times9.81}{31.42} = 9.11 \text{ m/s}$$

$$V_{\omega_2} \text{ at the hub} = \frac{29.17\times9.81}{18.85} = 15.18 \text{ m/s}$$

$$\beta_{tip} = \tan^{-1}\frac{V_f}{V_{b_2}-V_{\omega_2}} = \tan^{-1}\frac{9.95}{31.42-9.11} = 24°$$

$$\beta_{hub} = \tan^{-1}\frac{9.95}{18.85-15.18} = 69.75°$$ *Ans.*

Example 19.24

A centrifugal blower running at 9000 rpm compresses 6 m³/s of air from 1 bar, 25°C to 2.2 bar. The compression index is 1.33. The velocity of flow is 75 m/s and is constant throughout the impeller. The impeller blade angles at the inlet and outlet are 30° and 55° respectively. Estimate the speed of the impeller and its width at inlet and outlet. The outlet diameter of the impeller is 0.75 m.

Solution The outlet temperature of air of the impeller

$$T_2 = T_1\left(\frac{p_2}{p_1}\right)^{\frac{n-1}{n}} = 298\times\left(\frac{2.2}{1}\right)^{\frac{0.33}{1.33}} = 362.39 \text{ K}$$

Power input to the impeller

$$V_{\omega_2} V_{b_2} = c_p(T_2 - T_1) = 1.005\,(362.39 - 298) = 64.71 \text{ kJ/kg}$$

From outlet velocity triangle

$$\tan\beta_2 = \frac{V_f}{V_{b_2}-V_{\omega_2}} = \frac{75}{V_{b_2}-V_{\omega_2}} = \tan 55° = 1.428$$

$$\therefore \quad V_{b_2} - V_{\omega_2} = 52.51 \text{ m/s}$$

$$V_{\omega_2} + V_{b_2} = \left[\left(V_{b_2} - V_{\omega_2} \right)^2 + 4 V_{b_2} \cdot V_{\omega_2} \right]$$

$$= \left[(52.51)^2 + 4 \times 64.71 \times 10^3 \right] = 511.43$$

$$V_{b_2} - V_{\omega_2} = 52.41$$

Solving,

$$V_{b_2} = 281.97 \text{ m/s}, \; V_{\omega_2} = 229.46 \text{ m/s}$$

$$V_{b_2} = \frac{\pi D_2 N}{60} = 281.97$$

$$\therefore \quad N = \frac{281.97 \times 60}{\pi \times 0.75} = 7180 \text{ rpm} \qquad \textit{Ans.}$$

$$V_{b_1} = \frac{\pi D_1 \times 7180}{60} = \frac{V_f}{\tan 30^\circ} = \frac{75}{0.5773} = 129.9$$

$$\therefore \quad D_1 = \text{impeller diameter at inlet} = 0.345 \text{ m} \qquad \textit{Ans.}$$

Discharge at inlet

$$Q = \pi D_1 \, b_1 \, V_f = 6 \text{ m}^3/\text{s}$$

$$\therefore \quad b_1 = \frac{6}{\pi \times 0.345 \times 75} = 7.38 \text{ cm} \qquad \textit{Ans.}$$

Discharge at outlet when the pressure is 2.2 bar and temperature 362.39 K,

$$Q = 6 \times \frac{1}{2.2} \times \frac{362.39}{298} = 3.316 \text{ m}^3/\text{s} \; = \pi D_2 \, b_2 \, V_f$$

$$\therefore \quad b_2 = \frac{3.316}{\pi \times 0.75 \times 75} = 1.88 \text{ cm} \qquad \textit{Ans.}$$

Review Questions

19.1 What is the function of a compressor? What are the different types of compressors?

19.2 What are the main applications of compressors?

19.3 Derive the expressions for the reversible work of compression if the compression process is (a) adiabatic, (b) polytropic, and (c) isothermal.

19.4 Which of the compression processes needs minimum work and which the maximum work?

19.5 Define (a) the adiabatic efficiency and (b) the isothermal efficiency.

19.6 Show that for a single-stage reciprocating compressor the work of compression remains the same irrespective of its derivation on the basis of (a) a steady flow system and (b) a closed system.

19.7 Define the volumetric efficiency of a compressor. On what factors does it depend?

19.8 Discuss how the volumetric efficiency varies with the clearance and the pressure ratio. For a given pressure ratio and the polytropic index, find the maximum clearance when the volumetric efficiency is reduced to zero.

19.9 What is the maximum pressure ratio attainable with a reciprocating compressor for a given clearance?

19.10 What is the need of staging the compression process?

19.11 Show that the optimum intermediate pressure of a two-stage reciprocating compressor for minimum work is the geometric mean of the suction and discharge pressures.

19.12 Explain how does the use of intermediate pressure for minimum work result in equal pressure ratios in the two-stages of compression, equal discharge temperatures, and equal work for the two stages.

19.13 What do you mean by perfect intercooling? What is the amount of heat rejected in the intercooler?

19.14 What is the function of an aftercooler?

19.15 Explain the advantages of multistage compression?

19.16 Explain the operation of an air motor.

19.17 What are rotary compressors? How does the Roots blower operate? What is the power input to it? Define Roots blower efficiency.

19.18 What is a vane type compressor? Briefly explain its operation.

19.19 How does the pressure rise in a centrifugal compressor? Where is it used?

19.20 How does an axial flow compressor operate?

19.21 How does the pressure rise occur in a centrifugal compressor?

19.22 Why is pre-whirl provided at the vanes of a centrifugal compressor? Draw the velocity triangles at inlet and exit of the vanes and explain.

19.23 Explain what you understand by slip factor.

19.24 Explain the mechanism of flow in an axial flow compressor. Show the pressure and velocity profiles in a multi-stage compressor.

19.25 What is the degree of reaction? Draw the velocity triangles for R=50%.

19.26 Define work done factor. What is polytropic efficiency?

19.27 What is the difference between a blower and a fan? How does a compressor differ?

Problems

19.1 A 4-cylinder single stage air compressor has a bore of 200 mm and a stroke of 300 mm and runs at 400 rpm. At a working pressure of 620 kPa (g) it delivers 3.1 m^3 of air per min. at 270°C. Calculate (a) mass flow rate, (b) free air delivery (FAD), (c) effective swept volume, (d) volumetric efficiency. Take free air conditions at inlet as 101.3 kPa, 21°C. *Ans.* (a) 0.239 kg/s, (b) 0.199 m^3s, (c) 0.0299 m^3 (d) 79.2 %

19.2 0.2 m^3 of air at 20°C and 100 kPa is compressed according to the relation $pv^{1.3}$ = constant by the piston in an engine cylinder that has a compression ratio of 6. Heat is then added while the pressure remains the same, until the piston returns to its original position. Calculate (a) mass of air, (b) pressure at the end of compression, (c) final temperature, (d) network transfer in the combined process. *Ans.* (a) 0.238 kg, (b) 1.027 MPa, (c) 2736°C, (d) 123.7 kJ

19.3 A single stage single-acting air compressor deals with 90 m^3/h of air at 101.325 kPa and 15°C. The pressure and temperature during the suction stroke remain constant at 98 kPa and 40°C respectively, n= 1.22. The air is delivered at 735 kPa, R_a = 0.287 kJ/kg K. Find (a) the power needed to drive the compressor if the mechanical efficiency is 0.85, (b) the swept volume if the speed is 120 rpm. Take the volumetric efficiency as 0.78.

19.4 Find the stroke, piston diameter, and indicated power of a single-acting air compressor, which operates under the following conditions: volume of F.A.D. at 101.325 kPa and 15°C = 105 m^3/min; pressure and temperature at the beginning of compression, 98 kPa and 30°C; discharge pressure is 40 kPa; speed 220 rpm: n = 1.25; stroke = bore and the clearance volume is 6% of the swept volume.

19.5 A two-stage single-acting compressor compresses air from 101.325 kPa and 15°C to a pressure of 162 kPa. Calculate the work done per kg of air delivered, and the heat transferred to the intercooler if the intercooling is ideal.

If the volumetric efficiency is expected to be 85% for the L.P. stage and the speed is 400 rpm, calculate a suitable diameter for the L.P. cylinder for an air delivery of 2.5 kg/min if the stroke-bore ratio is unity. Assume the polytropic index to be 1.3 and state all the assumptions made. For air, R = 0.287 kJ/kg K and c_p = 1.005 kJ/kg K.

19.6 Air flows steadily into a compressor at a temperature of 17°C and a pressure of 1.05 bar and leaves at a temperature of 247°C and a pressure of 6.3 bar. There is no heat transfer to or from the air as it flows through the compressor; changes in elevation and velocity are negligible. Evaluate the external work done per kg of air, assuming air as an ideal gas for which R = 0.287 kJ/kg K and

$\gamma = 1.4$. Evaluate the minimum external work required to compress the air adiabatically from the same initial state to the same final pressure and the isentropic efficiency of the compressor.
Ans. –225 kJ/kg, –190 kJ/kg, 84.4%

19.7 A slow-speed reciprocating air compressor with a water jacket for cooling approximates a quasi-static compression process following a path $pv^{1.3}$ = const. If air enters at a temperature of 20°C and a pressure of 1 bar, and is compressed to 6 bar at a rate of 1000 kg/h, determine the discharge temperature of air, the power required and the heat transferred per kg. *Ans.* 443 K, 51.82 kW, 36 kJ/kg

19.8 A single-acting two-stage reciprocating air compressor with complete intercooling delivers 6 kg/min at 15 bar pressure. Assume an intake condition of 1 bar and 15°C and that the compression and expansion processes are polytropic with $n = 1.3$. Calculate: (a) the power required, (b) the isothermal efficiency. *Ans.* (a) 26.15 kW (b) 85.6%

19.9 A two-stage air compressor receives 0.238 m³/s of air at 1 bar and 27°C and discharges it at 10 bar. The polytropic index of compression is 1.35. Determine (a) the minimum power necessary for compression, (b) the power needed for single-stage compression to the same pressure, (c) the maximum temperature for (a) and (b), and (d) the heat removed in the intercooler.
Ans. (a) 63.8 kW, (b) 74.9 kW, (c) 404.2 K, 544.9 K, (d) 28.9 kW

19.10 A single-acting air compressor has a cylinder of bore 15 cm and the piston stroke is 25 cm. The crank speed is 600 rpm. Air taken from the atmosphere (1 atm, 27°C) is delivered at 11 bar. Assuming polytropic compression $pv^{1.25}$ = const., find the power required to drive the compressor, when its mechanical efficiency is 80%. The compressor has a clearance volume which is 1/20th of the stroke volume. How long will it take to deliver 1 m³ of air at the compressor outlet conditions. Find the volumetric efficiency of the compressor.
Ans. 12.25 kW, 3.55 min, 72%

19.11 A multistage air compressor compresses air from 1 bar to 40 bar. The maximum temperature in any stage is not to exceed 400 K. (a) If the law of compression for all the stages is $pv^{1.3}$ = const., and the initial temperature is 300 K, find the number of stages for the minimum power input. (b) Find the intermediate pressures for optimum compression as well as the power needed. (c) What is the heat transfer in each of the intercooler?
Ans. (a) 3 (b) 3.48 bar, 12.1 bar, 373.1 kJ/kg (c) 100.5 kJ/kg

19.12 A single stage reciprocating air compressor has a swept volume of 2000 cm³ and runs at 800 rpm. It operates on a pressure ratio of 8, with a clearance of 5% of the swept volume. Assume NTP room conditions, and at inlet, and polytropic compression and expansion with $n = 1.25$. Calculate (a) the indicated power, (b) volumetric efficiency, (c) mass flow rate, (d) the free air delivery FAD, (e) isothermal efficiency, (f) actual power needed to drive the compressor, if the mechanical efficiency is 0.85.
Ans. (a) 5.47 kW (b) 78.6% (c) 1.54 kg/min (d) 1.26 m³/min, (e) 80.7% (f) 6.44 kW

19.13 A two-stage single-acting reciprocating compressor takes in air at the rate of 0.2 m³/s. Intake pressure and temperature are 0.1 MPa and 16°C respectively. The air is compressed to a final pressure of 0.7 MPa. The intermediate pressure is ideal, and intercooling is perfect. The compression index is 1.25 and the compressor runs at 10 rps. Neglecting clearance, determine (a) the intermediate pressure, (b) the total volume of each cylinder, (c) the power required to drive the compressor, (d) the rate of heat absorption in the intercooler.
Ans. (a) 0.264 MPa (b) 0.0076 m³ h.p. and 0.02 m³ l.p. cylinder (c) 42.8 kW (d) 14.95 kW

19.14 A 3-stage single-acting air compressor running in an atmosphere at 1.013 bar and 15°C has a free air delivery of 2.83 m³/min. The suction pressure and temperature are 0.98 bar and 32°C respectively. The delivery pressure is to be 72 bar. Calculate the indicated power required, assuming complete intercooling, $n = 1.3$ and that the compressor is designed for minimum work. What will be the heat loss to the intercoolers? *Ans.* 25.568 kW, 13.78 kW

19.15 A two-stage single-acting air compressor delivers 0.07 m³ of free air per sec (free air condition 101.325 kN/m² and 15°C). Intake conditions are 95 kN/m² and 22°C. Delivery pressure from the compressor is 1300 kN/m². The intermediate pressure is ideal and there is perfect intercooling. The compression index is 1.25 in both cylinders. The overall mechanical and electrical efficiency is 75%. Neglecting clearance, determine (a) the energy input to the driving motor, (b) the heat transfer rate in the intercooler, (c) the percentage saving in work by using a two-stage intercooled compressor instead of a single-stage compressor. Take $R = 0.287$ kJ/kg K and $c_p = 1.006$ kJ/kg K.
Ans. (a) 29.1 kW, (b) 6.99 kW, (c) 13%

19.16 Air at 1.013 bar and 15°C is to be compressed at the rate of 5.6 m^3/min to 1.75 bar. Two machines are considered: (a) the Roots blower, and (b) a sliding vane rotary compressor. Compare the powers required, assuming for the vane type that internal compression takes place through 75% of the pressure rise before delivery takes place, and that the compressor is an ideal uncooled machine.

Ans. 6.88 kW, 5.71 kW

19.17 A rotary air compressor has an inlet static pressure and temperature of 100 kN/m^2 and 20°C respectively. The compressor has an air mass flow rate of 2 kg/s through a pressure ratio of 5:1. The isentropic efficiency of compression is 85%. Exit velocity from the compressor is 150 m/s. Neglecting change of velocity through the compressor, determine the power required to drive the compressor. Estimate the stagnation temperature and pressure at exit from the compressor.

Ans. 404 kW, 232.2°C, 541.5 kN/M^2

19.18 The cylinder of an air motor has a bore of 63.5 mm and stroke of 114 mm. The supply pressure is 6.3 bar, the supply temperature 24°C, and the exhaust pressure is 1.013 bar. The clearance volume is 5% of the swept volume and the cut-off ratio is 0.5. The air is compressed by the returning piston after it has travelled through 0.95 of its stroke. The law of compression and expansion is $pv^{1.3}$ = constant. Calculate (a) the temperature at the end of expansion, (b) the indicated power of the motor which runs at 300 rpm, and (c) the air supplied per minute.

Ans. (a) – 28.4°C, (b) 0.749 kW, (c) 0.42 kg/min.

19.19 A single-acting two-stage air compressor runs at 300 rpm and delivers 510 m^3/h with intake air at 1 bar, 15°C. The delivery pressure is 40 bar. Calculate for minimum work input (i) the optimum pressure ratio for each stage, (ii) the theorectical power input of each stage when the intercooling is perfect and the index of compression is $n = 1.3$, (iii) the amount of heat rejected from each cylinder and the intercooler, and (iv) the swept volumes of the L.P. and H.P. cylinders, if their volumetric efficiencies are 0.9 and 0.85 respectively.

Ans. (i) 6.324 bar, (ii) 65.14 kW, (iii) 6.358 kW, 26.214 kW, (iv) 0.026 m^3, 0.0045 m^3

19.20 A centrifugal compressor has a pressure ratio of 4 with an isentropic efficiency 0.82 when running at 16000 rpm. It takes in air at 1 bar, 17°C. Guide vanes at the inlet give the air a pre-whirl of 20° to the axial direction and the mean diameter of the eye is 200 mm, the absolute air velocity at the inlet is 120 m/s. At the exit, the blades are radially inclined and the impeller tip diameter is 550 mm. Calculate the slip factor of the compressor.

Ans. 0.847

19.21 A centrifugal compressor compresses at the rate of 2k g/s from 1 bar, 20°C to a total pressure at outlet equal to 5 bar. The velocity of air at inlet is 150 m/s and the compressor runs at 20,000 rpm. The isentropic efficiency is 0.8 and the slip factor is 0.9. Calculate (i) the change in total temperature, (ii) the impeller diameter at outlet and inlet if the hub diameter is 10 cm, and (iii) the power input.

Ans. (i) $T_{02} - T_{01} = 200$ K, (ii) $D_2 = 0.45$ m, $D_1 = 0.156$ m, (iii) 402 kW

19.22 The mean diameter of the rotor of an axial flow compressor is 0.55 m and it rotates at 16000 rpm. The velocity of flow, 230 m/s, is constant and the velocity of whirl at the inlet is 85 m/s. The air inlet condions are 1 bar, 305 K. The stage efficiency is 0.9. The pressure ratio through the stage is 1.55. Calculate (i) the fluid deflection angle, (ii) the work input, and (iii) the degree of reaction.

Ans. (i) 8.23°, (ii) 45.48 kJ/kg, (iii) 0.708

19.23 An axial-flow compressor stage has a mean blade velocity of 210 m/s, stagnation temperature rise 28 K, the axial velocity of flow is constant and is equal to 150 m/s. The degree of reaction is 0.5. Determine the appropriate air angles, the stagnation pressure ratio of the stage, if the isentropic efficiency of the stage is 0.85 and the stagnation temperature at the inlet is 520 K.

Ans. Outlet angle 75.78°, Inlet angle 41.09°; Pressure ratio =1.169

19.24 A centrifugal fan delivers 9 m^3/s of air while running at 1200 rpm. The impeller diameter at the inlet and outlet are 0.6 m and 0.8 m respectively. The blades are curved backwards and the blade outlet angle is 70°. The width of the blade at the outlet is 0.12 m. The volute casing recovers 35% of the outlet velocity head, while the impeller losses are 20% of the outlet velocity head. Determine (i) the pressure rise at outlet, (ii) manometric efficiency, when the air enters the impeller in the axial direction at 16 m/s.

Ans. (i) Net pressure rise = 109.13 m of air (ii) 54%

C H A P T E R

20 Internal Combustion Engines

20.1 INTERNAL AND EXTERNAL COMBUSTION ENGINES

The mechanism in which the energy released by the combustion of fuel is converted to mechanical or shaft work is called a heat engine. The heat engine can be an internal or external combustion engine. In an *internal combustion engine* (*IC engine*), the combustion of fuel occurs inside the engine cylinder itself and the products of combustion are the working fluid which by virtue of its high internal energy with high temperature and pressure pushes the piston outward to do work. In an *external combustion engine,* on the other hand, the fuel is burnt outside the engine or turbine and the heat released is utilized to heat an intermediate fluid like water as in a steam power plant, which by virtue of its high enthalpy, does work on the turbine blades to produce mechanical or shaft work.

An IC engine has high overall efficiency, is compact with a high ratio of maximum output to its weight and bulk, has greater mechanical simplicity to operate, and lower initial cost. Its wide ranging applications include (i) transport vehicles in roads (cars, trucks, buses, bikes etc), rail locomotive, aircraft, marine propulsion, (ii) as a prime mover for electric generator, welding sets, grinders, pumps, compressors and blowers, fork lifts etc., (iii) agricultural machinery like harvesters, thrashers, pump sets, (iv) earth moving equipment like dumpers, excavators and so on. Because of their multifarious activities, IC engines play a very important role in industry.

20.2 CLASSIFICATION OF IC ENGINES

IC engines are of two types—reciprocating and rotary. Rotary IC engines are the open-cycle gas turbines which have been discussed in the next chapter. *Reciprocating IC engines* can be classified on the following basis (Fig. 20.1):

(a) Thermodynamic Cycle

(i) Otts or Constant Volume Cycle The energy released during the combustion of fuel occurs at constant volume.

(ii) Diesel or Constant Pressure Cycle Energy is released by combustion at constant pressure.

(iii) Dual or Limited Pressure Cycle Energy is released by combustion of fuel partly at constant volume and partly at constant pressure.

Energy is rejected at constant volume in all the three cycles.

(b) Number of Strokes Per Cycle Reciprocating IC engines use a piston which moves back and forth inside a cylinder. The distance travelled by the piston inside the cylinder from one extreme end (top dead centre or tdc) to the other (bottom dead centre or bdc) is called one *stroke.*

(i) 4-stroke Engine The engine cycle is completed in four strokes of the piston

(ii) 2-stroke Engine The engine cycle is completed in two strokes of the piston

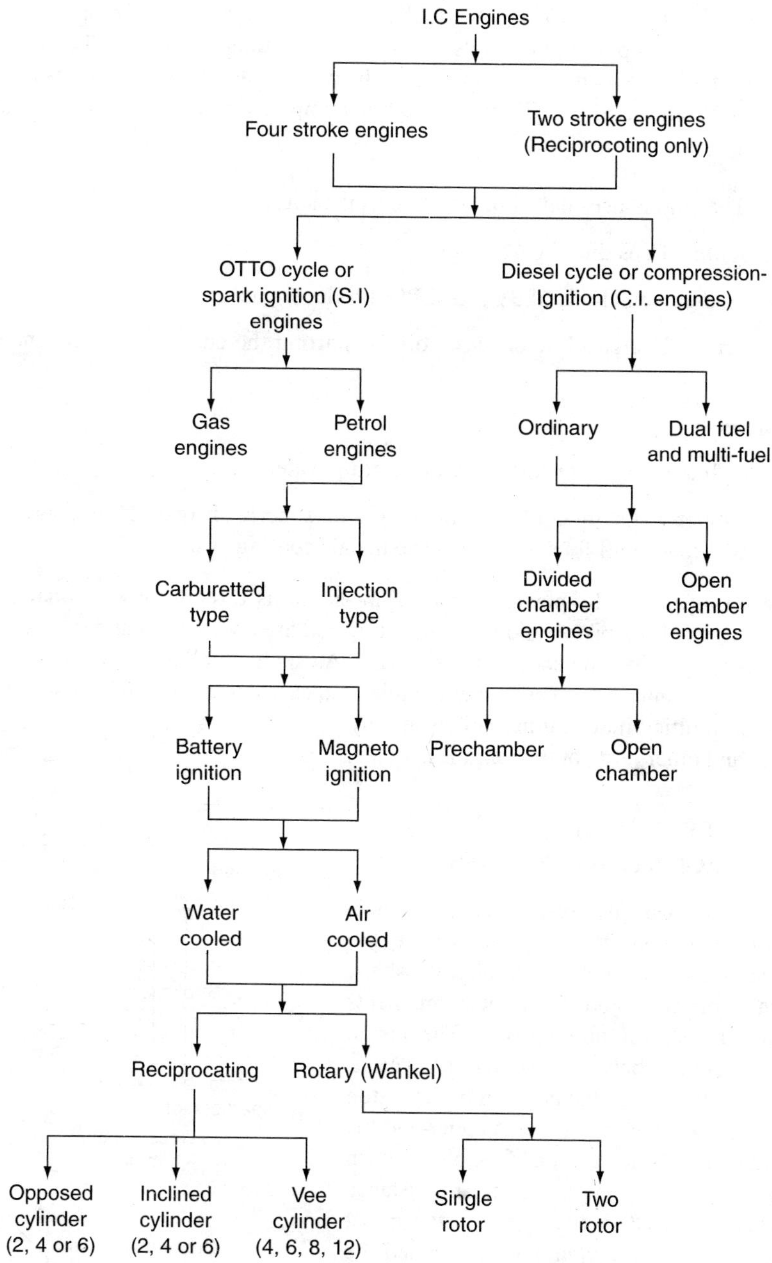

Fig. 20.1 *Classification of IC engines*

(c) Ignition System

(*i*) *Spark Ignition (S.I.) Engines* A homogeneous mixture of air and fuel vapour is supplied to the engine and the combustion is initiated by a spark plug.

(ii) Compression Ignition (C.I.) Engines Air sucked inside the cylinder is compressed to a higher pressure and temperature (this temperature exceeds the self-ignition temperature, S.I.T., of the fuel), the fuel is injected into the cylinder in the form of fine spray and the mixture automatically burns inside the cylinder.

The engine cycle of both the SI and CI engines can be completed in 2 strokes or 4 strokes of the piston.

(d) Fuels Used

(i) Petrol Engine The engine uses petrol or gasoline as the source of energy

(ii) Oil or Diesel Engine Uses diesel oil as fuel

(iii) Gas Engines Use gaseous fuels like CNG, LPG, producer gas, etc.

(iv) Multi-fuel Engines Use gasoline or diesel oil for starting the engine and kerosene or biogas as their primary fuel

(e) Cooling Systems

(i) Water Cooled Cylinder walls are cooled by circulating water

(ii) Air Cooled Cylinder walls are cooled by blowing atmospheric air over the hot surfaces. Motor cycles, Scooters, aircraft and a few small four wheelers have the air-cooling system.

(f) Multicylinder Engines The power output of an engine is directly proportional to its speed. When the reciprocating masses of the piston and the connecting rod are accelerated and decelerated, inertia forces develop and they put a limit on the speed of the engine. As such, smaller cylinders are used to reduce the inertia forces per cylinder and the forces in one cylinder can easily be balanced by a suitable arrangement of other cylinders. Thus, multicylinder engines are invariably used in 4-wheelers and other large output engines.

20.3 AN OVERVIEW OF RECIPROCATING ENGINES

The reciprocating engine, basically a piston-cylinder device, has a wide range of applications. It is the powerhouse of the vast majority of automobiles, trucks, light aircraft, ships, electric power generators and so on. The basic components of such an engine are shown in Fig. 20.2. The piston reciprocates in the cylinder between two fixed positions called the top dead centre (TDC)-the position of the piston when it forms the smallest volume in the cylinder-and the **bottom dead centre (BDC)**-the position of the piston when if forms the largest volume in the cylinder. The distance between the TDC and the BDC is the largest distance that the piston can travel in one direction, and it is called the **stroke** of the engine. The diameter of the piston is called the **bore**. The air or air-fuel mixture is drawn into the cylinder through the **intake valve,** and the combustion products are expelled from the cylinder through the **exhaust valve.**

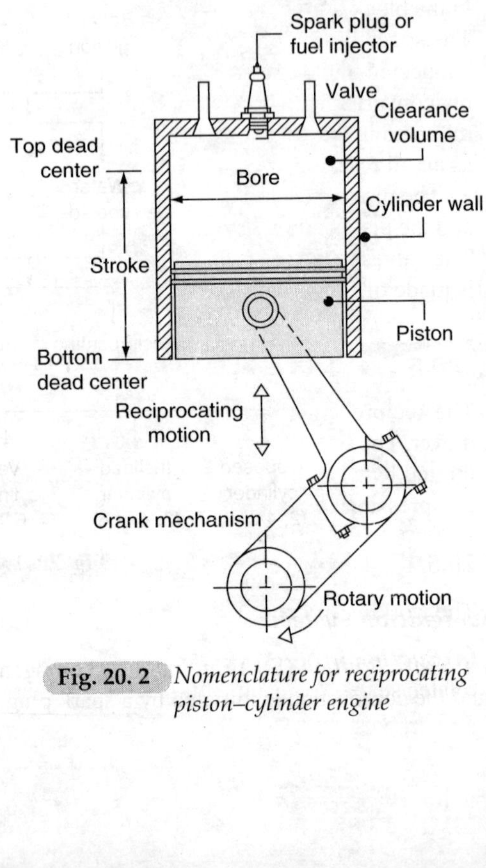

Fig. 20. 2 *Nomenclature for reciprocating piston–cylinder engine*

The minimum volume formed in the cylinder when the piston is at TDC is called the **clearance volume**. The volume displaced by the piston as it moves between TDC and BDC is called the **displacement volume.** The ratio of the maximum volume formed in the cylinder to the minimum (clearance) volume is called the **compression ratio** r_k of the engine

$$r_k = V_{max}/V_{min} = V_{BDC}/V_{TDC}$$

Note that r_k is a **volume ratio** and should not be confused with the pressure ratio.

Another term frequently used in regard to reciprocating engines is the **mean effective pressure (mep)**. It is a fictitions pressure that, if it acted on the piston during the entire power stroke, would produce the same amount of net work as that produced during the actual cycle.

$$W_{net} = \text{m.e.p.} \times \text{piston area} \times \text{stroke} = \text{m.e.p.} \times \text{displacement volume}$$

or,
$$\text{m.e.p., } p_m = (W_{net})/[V_{max} - V_{min}] \text{ (k Pa)}$$

The mean effective pressure can be used as a parameter to compare the performance of reciprocating engines of equal size. The engine with a larger value of m.e.p. will deliver more net work per cycle and thus will perform better.

Reciprocating engines are classified as **spark-ignition** (SI) **engines** or **compression-ignition** (CI) **engines**, depending on how the combustion process in the cylinder in initiated. In SI engines, also called petrol or gasoline engines, the combustion of the air-fuel mixture is initiated by a spark plug. In CI engines, also called diesel engines, the air-fuel mixture is self-ignited as a result of compressing the mixture above its self-ignition temperature.

20.4 DESCRIPTION OF AN SI ENGINE

The different parts of an SI engine including its main components are shown in Fig. 20.3. The cylinder block is made of cast iron because of its good wear resistance and low cost. The cooling water passages are provided in the block during casting. The piston is made of aluminum in small engines and cast steel in bigger engines. The space between the cylinder head and the piston crown is called the *combustion chamber*. The piston is connected to the crankshaft through a connecting rod made of forged steel. The connecting rod-crank assembly converts the reciprocating motion of the piston to rotary motion. The crankshaft is supported in bearings mounted in the crank case which is sealed at the bottom with a pressed-steel or cast-aluminium oil pan acting as an oil reservoir for the lubrication system.

The piston rings do not permit the high pressure gases to escape through the gap between the cylinder wall and the piston and also scraps the excess oil from the cylinder walls and allows the oil to return to the oil pan. The valves made of chrome – nickel alloy are usually poppet valves and are operated by a cam. The camshaft is made of forged steel and is driven by the crankshaft through gears.

20.5 OPERATING PRINCIPLES

The reciprocating internal combustion engine remain by far the most common form of engine or prime mover. The two main types of internal combustion engines are: Spark Ignition (SI) engines where the fuel is ignited by a spark; and compression ignition (CI) engines where the rise in temperature and pressure during compression is sufficient to cause spontaneous ignition of the fuel.

20.5.1 4-Storke Spark Ignition Engine

The working principles of a typical 4-stroke spark ignition or petrol engine can be described as follows:

(a) Suction Process Let us consider that the piston is at the top dead centre (tdc) and the crankshaft is being rotated so that the piston moves downwards (from tdc to bdc) and the inlet valve opens to admit a homogeneous

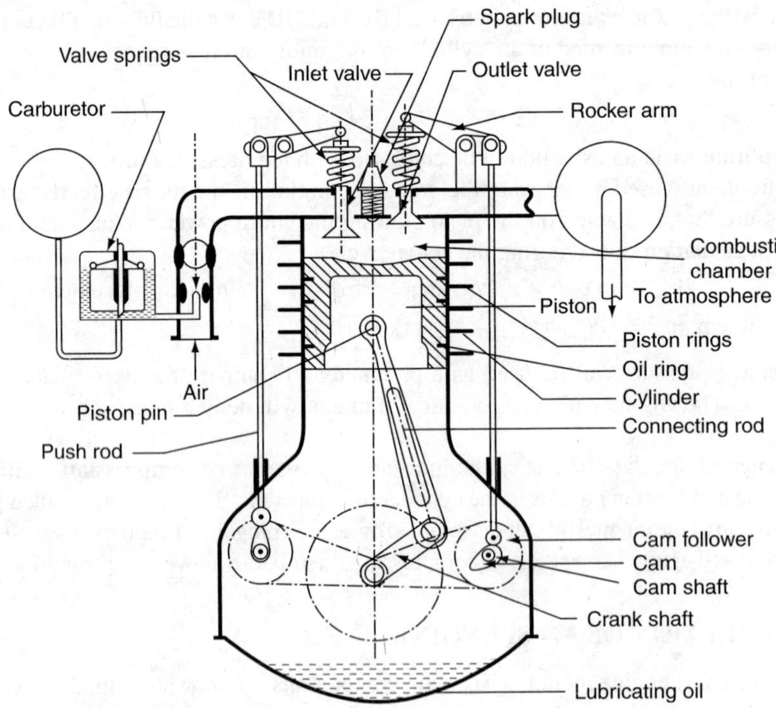

Fig. 20.3 *Main components of a spark ignition engine*

mixture of air and fuel inside the engine cylinder, Fig 20.4(a). During the suction process, the pressure inside the cylinder is lower than the ambient pressure by an amount that depends upon the speed of the engine.

(b) Compression Process The fresh charge admitted during the suction process mixes with the residual gases present inside the cylinder and the mixture is compressed to a higher pressure when the piston describes the return stoke, i.e., moves from the bdc to tdc, while the valves, inlet and outlet, are closed. The pressure at the end of the compression process, Fig 20.4(b), depends upon the compression ratio, varying between 6 and 11, and its magnitude is about 0.6 to 0.9 MPa.

(c) Ignition and Expansion Process Near the end of the compression stroke, there is an electric discharge across the spark plug, between 10 and 40 crank angle degrees before tdc, and that initiates the combustion process. The fuel burns and the combustion process is completed within a few milliseconds. It is assumed that the combustion process takes place at constant volume and the liberated heat energy rapidly increases the pressure and temperature of the working fluid present inside the cylinder. The products of combustion exert pressure at the piston crown and the piston is forced to descend downwards, from tdc to bdc, Fig. 20.4(c). During this expansion process, both the valves are closed.

(d) Exhaust Process The exhaust valve opens near the end of the expansion process and the burned gases are forced out of the cylinder by the piston moving from bdc to tdc, Fig. 20.4(d). The exhaust valve communicates with the muffler or the silencer through which the burned gases are released to the atmosphere. During the exhaust process, the pressure inside the engine cylinder is a little more than the ambient pressure. The cycle is completed in four strokes of the piston or two revolutions of the crankshaft. The theoretical and actual pressure-volume diagram is shown in Fig. 20.5.

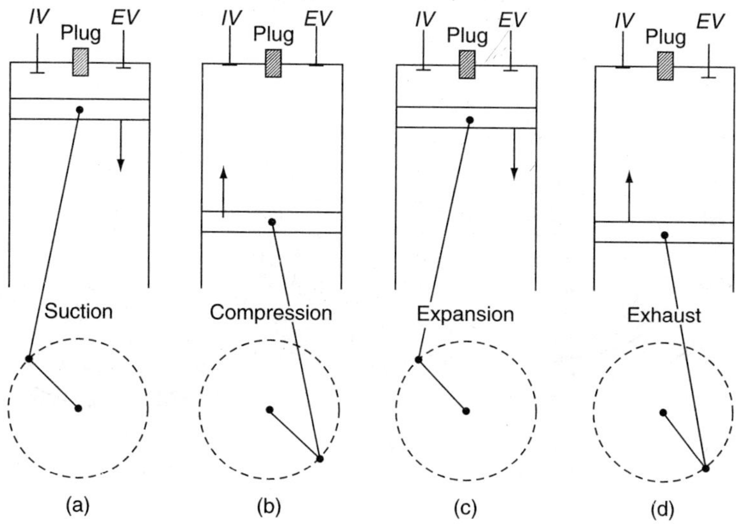

Fig. 20.4 *Sequence of operation on a 4-stroke SI engine*

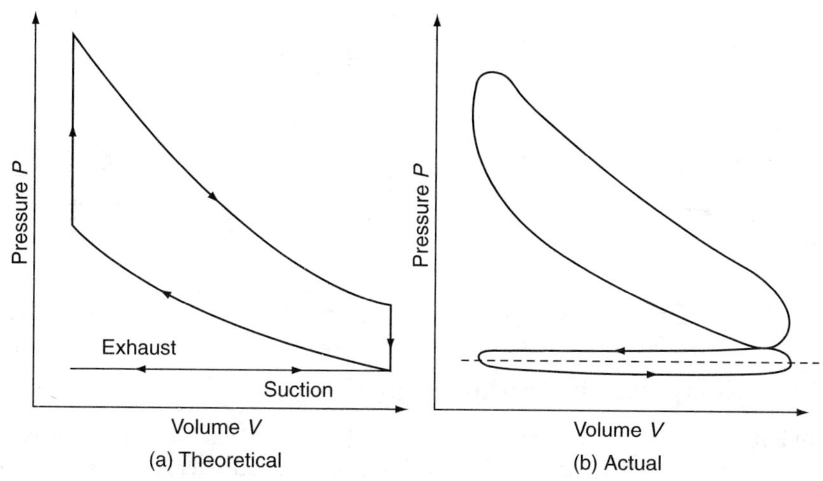

Fig. 20.5 *Typical pressure-volume diagram for SI engine*

Otto, a German engineer, using the principles of Alphonse Beau de Rochas, described above, built a 4-stroke engine that became highly successful and is known as Otto Cycle.

The inlet and outlet valves are opened and closed by cam mechanism. The clearance between the cam, tappet and valve has to be taken up slowly and in order to avoid noise and wear, the valve is lifted slowly. Similarly the valve cannot be closed suddenly because it will bounce on its seat. Therefore, the opening and closing periods of valves are spread over a few crank angles. Figure 20.6(a) shows the valve-timing diagram for a 4-stroke SI engine.

Crank angle is also an independent and useful variable for analyzing the performance of an IC engine because the engine processes occupy almost 180 crank-angle intervals over a wide range of engine speeds. The sequence of events that take place inside the engine cylinder of 4-stroke SI engine is shown in Fig. 20.6(b). This figure also shows the valve timing.

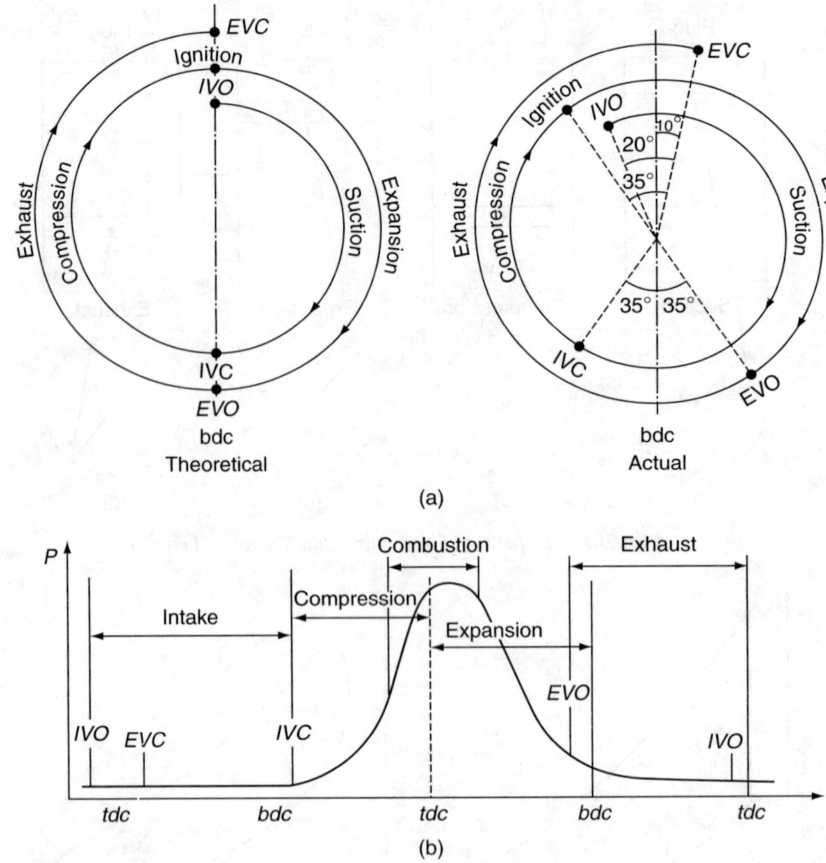

Fig. 20.6 *(a) Valve timing diagram for a 4-stroke SI engine (b) Sequence of events in 4-stroke SI engine-pressure vs crank angle*

20.5.2 4-Stroke Compression Ignition Engine

In 1892, Dr Rudolf Diesel, a German engineer, designed and fabricated an engine capable of running at higher compression ratios and using cheaper fuels. The cycle named after him is known as 'Diesel cycle'. The different processes are:

(a) Suction Process Air alone is admitted inside the cylinder through the inlet valve while the piston moves from tdc to bdc. The inlet valve starts opening at about 30 before tdc so that it is fully open before the piston describes the suction process.

(b) Compression Process The piston compresses the air admitted earlier while moving from bdc to tdc and during this process both the valves should remain closed. The compression ratio varies from 12:1 to 24:1 and the air is compressed to a very high pressure, about 4 MPa, and as such the temperature at the end of the compression stroke is about 800 K, much higher than the self ignition temperature of the fuel. The fuel usually a light fuel oil, is injected directly into the engine cylinder nearly at the end of compression stroke. The injection starts at about 20 deg before tdc and continues during a part of the expansion process.

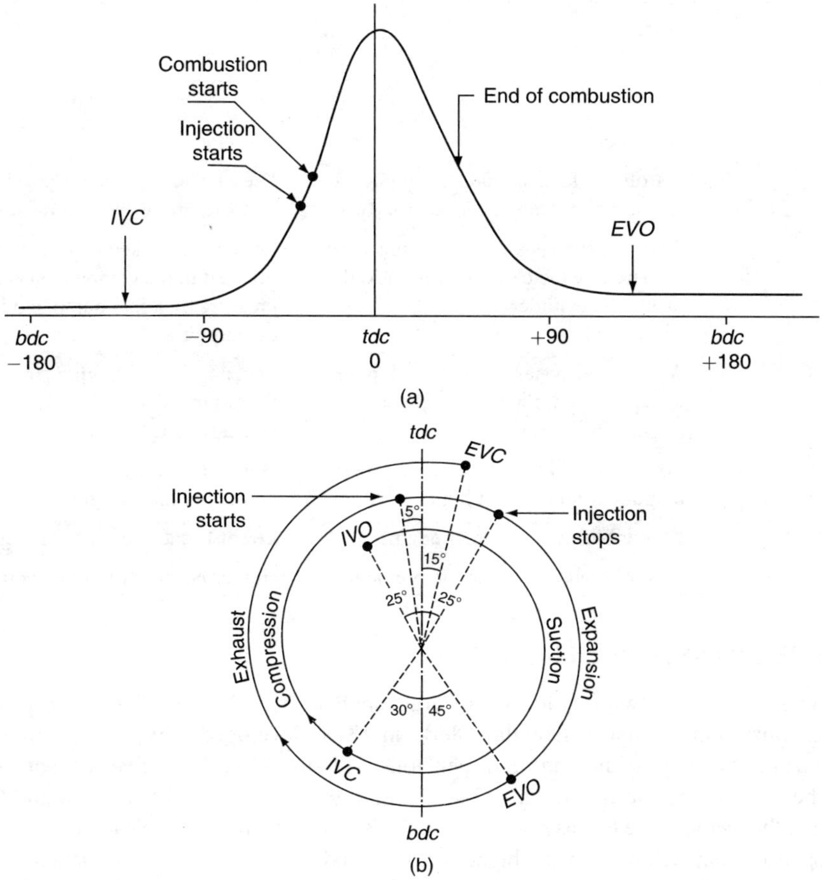

Fig. 20.7 *(a) Sequence of events during compression, combustion and expansion in 4-stroke CI engine: pressure-crank angle (b) Valve timing diagram for a diesel engine*

(c) Combustion and Expansion Process The liquid fuel enters the engine cylinder in the form of a fine spray and after entering the cylinder, it atomises, evaporates and mixes with heated air to form a combustible mixture. Since the temperature of air is much more than the self ignition temperature of the fuel, the fuel autoignites and the flame rapidly spreads through the combustion chamber. The products of combustion push the piston downwards from tdc to bdc, and the expansion stroke completed.

(d) Exhaust Process The exhaust valve starts opening at about 30° before the bdc (before the completion of the expansion process) and a substantial part of the working fluid leaves the cylinder during the expansion process itself. During the exhaust process, the piston moves from the bdc to tdc and the gases are pushed out of the cylinder and the exhaust valve close at about 30 deg after tdc. The sequence of events that take place inside the engine cylinder of a 4-stroke CI engine is shown in Fig. 20.7.

The basic difference between the operation of the two types of engines lies in the method of injecting the fuel inside the engine and the system that initiates the combustion process and these lead to considerable variations in their operation. These are given in a tabular form, Table 20.1.

Table 20.1

Comparison of SI and CI Engines

Criteria	SI engine	CI engine
1. Working cycle	Otto cycle	Diesel cycle
2. Fuel used	Petrol/gasoline, should have higher self ignition temperature for better efficiency	Diesel, should have lower self ignition temperature for better efficiency
3. Suction process	A homogeneous air-fuel mixture prepared by the carburetor is sucked inside the cylinder	Only air is sucked inside, fuel is directly injected in the combustion chamber at high pressure by a fuel pump and injector. No carburetor
4. Ignition & combustion process	An ignition system with a spark plug is required, combustion takes place at constant volume	Fuel autoignites at high temperature inside the engine. Combustion at constant pressure, no spark plug required
5. Compression ratio and engine weight	Lower (6 to 11) lighter engine, power output per unit weight high	Higher (12 to 24), heavier engine, power output per unit weight low
6. Speed	Higher because of lighter engine	Lower because of heavier engine
7. Efficiency	Lower because of lower compression ratio	Higher because of higher compression ratio

20.6 THE TWO-STROKE ENGINE

A four-stroke engine requires two revolutions of crankshaft to complete the four basic processes-suction, compression, expansion and exhaust. Douglas Clerk, in 1878, developed a two-stroke engine to produce a greater output without changing the engine dimensions. He simplified the valve mechanism, reduced the proportion of the idle strokes and found that a 2-stroke engine can be applied to both SI and CI engines.

Let us consider the piston at the tdc, as shown in Fig. 20.8. For a SI engine, the clearance volume will be occupied by a homogeneous air-fuel mixture at a higher pressure and the spark plug will initiate the combustion process. For a CI engine, the clearance volume will be occupied by compressed air only and the fuel would be sprayed

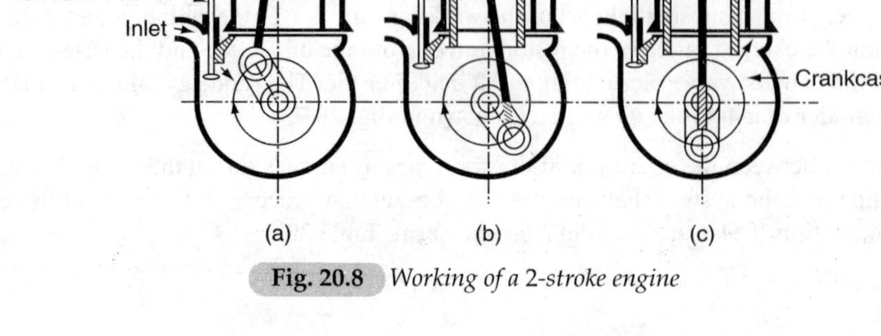

Fig. 20.8 *Working of a 2-stroke engine*

inside the cylinder. After combustion, the released heat energy will raise the pressure and temperature of the working fluid and the piston would be pushed downwards describing the power stroke. Near the end of the downward motion, the piston uncovers an opening/port, provided in the cylinder wall at B and most of the products of combustion would escape to the atmosphere through the exhaust manifold.

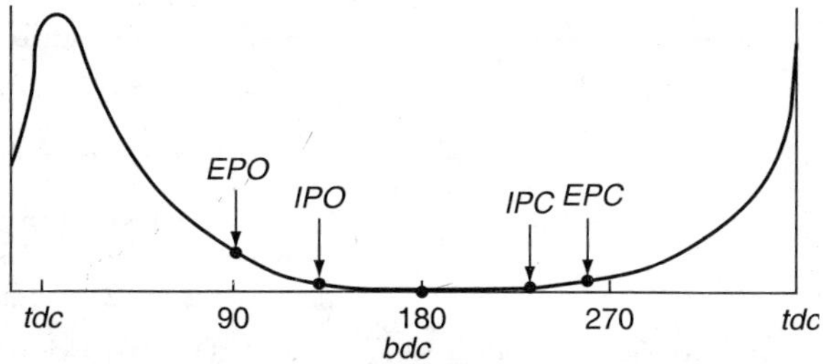

Fig. 20.9 *Sequence of operation in 2-stroke engine pressure—crank angle*

While the piston continues to descend downwards, it uncovers the transfer/inlet port A and either air (in CI engines) or air-fuel mixture (in SI engines) enters the engine cylinder. The entering fluid drives the exhaust gases out of the cylinder through the exhaust port B. This is an example of cross-scavenging. The deflector provided at the piston crown prevents the incoming charge from passing straight across the cylinder to the exhaust manifold.

During the upward movement of the piston (from bdc to tdc), the two ports A and B are covered and the charge admitted inside the cylinder is compressed whereas the crankcase sucks in the fresh charge through the reed spring inlet valve. This charge is compressed during the downward movement of the piston and is transferred to the main cylinder through the transfer port A during the last part of the downward stroke and in the beginning of the upward stroke. Thus, the cycle is completed in one revolution of the crankshaft or two-strokes of the piston. The sequence of operations that takes place inside the engine cylinder of a 2-stroke engine is shown in Fig. 20.9, and Fig. 20.10 shows the ideal and actual p-v diagrams for a two stroke engine along with the port opening diagram.

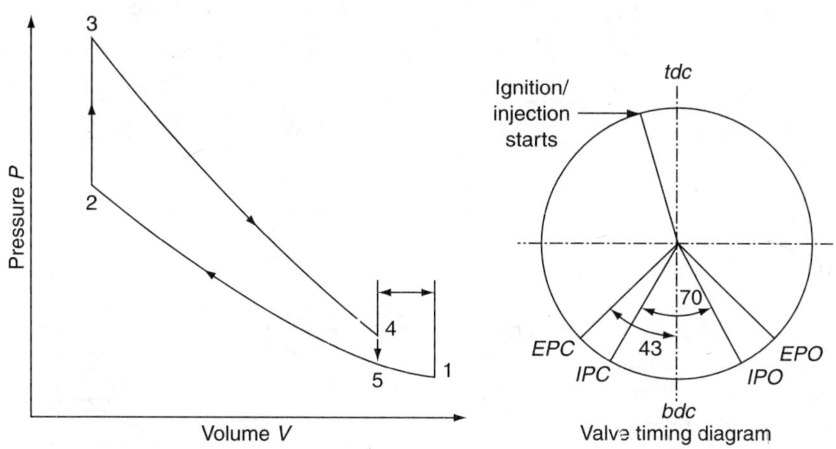

Fig. 20.10 *Pressure—volume and valve timing diagram for 2-stroke engine*

It is clear from the working of a 2-stroke engine that some freshly inducted working fluid is lost through the exhaust port during the scavenging process and therefore, the power output of a 2-stroke engine per unit displaced volume is less than twice the power output of an equivalent 4-stroke engine running at the same speed. The relative merits and demerits of 4-stroke and 2-stroke engines have been put in a tabular form in Table 20.2.

20.7 ENGINE PERFORMANCE

The power output of an engine is measured with the help of a brake or dynamometer, and is called '*brake power*' or '*shaft power*', which is given by

Table 20.2

Comparison of 4-Stroke and 2-Stroke Engines

Criteria	*Four-stroke engine*	*Two-stroke engine*
1. Thermodynamic cycle	Cycle is completed in 4-stroke or 2 revolutions of the crankshaft. One power stroke in 2 revolutions	Cycle is completed in 2-stroke or 1 revolution of the crankshaft. One power stroke in one revolution
2. Thermal efficiency	Higher at normal and part loads 3	Lower at normal and part loads
3. Volumetric efficiency	Higher because of more time for induction	Lower because less time for induction
4. Power output by weight ratio	Lower, a heavier engine is required to produce a given output because of one power stroke in two revolutions	Higher, a lighter and compact engine produces the given output because of one power stroke in each revolution
5. Turning moment diagram	Less uniform because of one power stroke in 2 revolution, heavier fly-wheel is required	More uniform because of one power stroke in each revolution, lighter fly-wheel is required
6. Cooling & lubrication	Normally water cooled, except in 2-or 3-wheelers, less wear and tear, less lubricant required	Normally air cooled, wear and tear more, usually mobil oil is mixed with fuel, more lubrication required
7. Valve mechanism	Required, design gets complicated	Not required, design is simplified
8. Initial cost	High	Low
9. Use	Heavy duty vehicles where efficiency is more important	Light vehicles where compactness and weight are more important
10. Effect on enviroment	Less	More polluting gases

$$\text{BP (or bp)} = \frac{2\pi TN}{60} \qquad (20.1)$$

where T is torque developed in Nm and N is the r.p.m. of the shaft.

The power output developed inside the engine cylinder is measured by obtaining the p–v diagram with the help of an indicator and is called '*indicated power*' which is given by

$$ip \text{ or IP} = \frac{\text{IMEP} \times L \times A \times \left(N \text{ or } \dfrac{N}{2} \right)}{60}$$

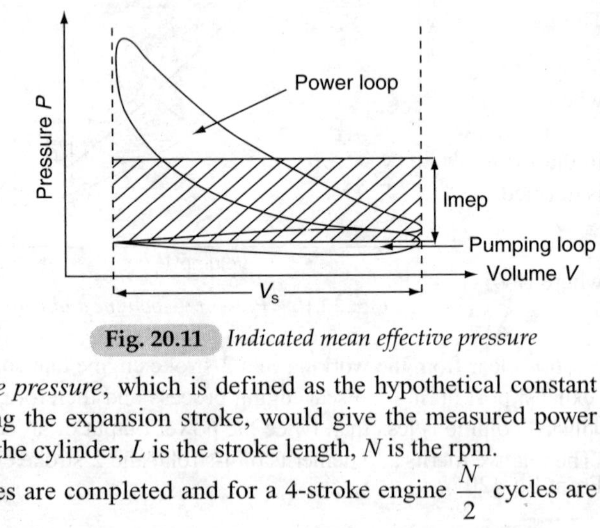

Fig. 20.11 *Indicated mean effective pressure*

where IMEP (or imep) = *indicated mean effective pressure*, which is defined as the hypothetical constant pressure which when applied to each piston during the expansion stroke, would give the measured power (Fig. 20.11). Here, A is the cross-sectional area of the cylinder, L is the stroke length, N is the rpm.

In the expression, for a 2-stroke engine N cycles are completed and for a 4-stroke engine $\dfrac{N}{2}$ cycles are completed per minute.

The difference between the IP and BP is the power absorbed in mechanical friction, in driving the auxiliaries of the engine and the pumping power. This difference is called the *friction power* or

$$\text{IP} - \text{BP} = \text{FP} \tag{20.2}$$

The *mechanical efficiency* is defined as

$$\eta_{\text{mech}} = \frac{\text{BP}}{\text{IP}} \tag{20.3}$$

The brake power can also be expressed as

$$\text{BP} = \frac{\text{BMEP} \times L \times A \times \left(N \text{ or } N/2\right)}{60} \tag{20.4}$$

where BMEP is the *brake mean effective pressure.*

The BMEP may be considered as that mep acting on the pistons which would give the measured bp. (or BP) if the engine were frictionless.

The volumetric efficiency, η_{vol} is a measure of the effectiveness of the suction and exhaust process of the engine. It is defined as the ratio of the volume of air sucked inside per cycle per cylinder at the ambient condition to the stroke volume of the cylinder, or

$$\eta_{\text{vol}} = \frac{\text{volume of air inducted at ambient condition}}{\text{stroke volume}}$$

$$= \frac{\text{mass of air inducted per cycle (or minute)}}{\text{mass of air occupying the stroke volume at ambient conditions}}$$

$$= \frac{\dot{m}_a \text{ kg/min}}{\rho_a \times V_s \times \left(N \text{ or } \frac{N}{2}\right)} \tag{20.5}$$

where V_s is the stroke volume.

$$\dot{m}_a = \eta_{\text{vol}} \times \rho_a \times \frac{\pi}{4} D^2 L \left(N \text{ or } \frac{N}{2}\right)$$

The fuel burning role

$$\dot{m}_f = \dot{m}_a \times \frac{F}{A}$$

where F/A is the fuel-air ratio.

The thermal efficiency of an IC engine is expressed as the ratio of power output and rate of energy input in the form of fuel. It is defined on the basis of indicated or brake output. The *indicated thermal efficiency* is defined as

$$\eta_{\text{i.th}} = \frac{\text{IP}}{\dot{m}_f \times \text{HV}} \tag{20.6}$$

where HV is the heating value of the fuel. Similarly, the *brake thermal* efficiency is given by

$$\eta_{\text{br.th}} = \frac{\text{BP}}{\dot{m}_f \times \text{HV}} \tag{20.7}$$

Thus,

$$\eta_{\text{mech}} = \frac{\text{BP}}{\text{IP}} = \frac{\text{BMEP}}{\text{IMEP}} = \frac{\eta_{\text{i.th}}}{\eta_{\text{br.th}}} \tag{20.8}$$

From Eq. (20.6),

$$\text{I.P.} = \dot{m}_f \times \text{HV} \times \eta_{\text{i.th}}$$

or,

$$\text{B.P.} = \eta_{\text{vol}} \times \rho_a \times \frac{\pi}{4} D^2 L n \times \left(N \text{ or } \frac{N}{2}\right) \times \text{HV} \times \eta_{\text{i.th}} \times \eta_{\text{mech}}$$

or, $$\text{B.P.} = \eta_{vol} \times \eta_{mech} \times \eta_{i.th} \times \rho_a \times \frac{\pi}{4} D^2 L n \times \left(N \text{ or } \frac{N}{2} \right) \times \frac{F}{A} \times \text{HV} \tag{20.9}$$

where n = number of cylinders.

Thus, $$\text{B.P.} \propto \rho_a$$

Power developed by the engine is directly proportional to the density of air. As density of air decreases, power output of the engine decreases. At high altitudes the air density is less, so the power output is also less. A supercharger is often used to supply compressed air to the engine to improve the engine output at high altitudes.

The power output of an IC engine depends greatly upon the amount of charge which can be induced into the cylinder. This is called the breathing capacity of the engine and is expressed by the *volumetric efficiency*. The volumetric efficiency with normal aspiration is seldom above 80% and to improve it, supercharging is used. The η_{vol} depends on compression ratio, valve timing, induction and port design, air-fuel ratio, enthalpy of vaporization of fuel, cylinder temperature and the atmospheric conditions.

The specific fuel consumption (s.f.c) of an engine can be expressed either on indicated power or brake power basis. Indicated specific fuel consumption (isfc) is the fuel consumption per unit IP or

$$isfc = \dot{m}_f / \text{IP kg/kwh}$$

Similarly, $$bsfc = \dot{m}_f / \text{BP kg/kwh} \tag{20.10}$$

The *bsfc* is the mass-flow rate of fuel consumed per unit shaft output and is the criterion of economical power production.

The rate of work done by the gas on the piston is evaluated from the indicator diagram obtained from the engine (Fig. 20.11). The i_{mep} is given by

$$imep = \frac{\text{net area of indicator diagram}}{\text{length of diameter}} \times \text{Constant}$$

where the constant depends on the scales of the recorder and net area = area of power loop-area of pumping loop.

Then $$\text{IP or } ip = \frac{imep \times LA(N/2 \text{ or } N)n}{60}$$

where N = rpm and n = number of cylinders.

Brake power (*bp*) is the output of the engine at the shaft measured by a dynamometer. Absorption dynamometers which are more common can be (a) friction-type like prony brake, (b) hydraulic, and (c) electrical, where

$$bp = \frac{2\pi TN}{60}$$

where T is the torque measured.

20.8 AIR STANDARD CYCLES

Internal combustion engines are noncyclic heat engines. The working fluid, the fuel–air mixture, undergoes permanent chemical change due to combustion, and the products of combustion after doing work are thrown out of the engine, and a fresh charge is taken. So the working fluid does not undergo a complete thermodynamic cycle.

To simplify the analysis of IC engines, air standard cycles are conceived. In an air standard cycle, a certain mass of air operates in a complete thermodynamic cycle where heat is added and rejected with external heat reservoirs, and all the processes in the cycle are reversible. Air is assumed to behave as an ideal gas, and its specific heats are constant.

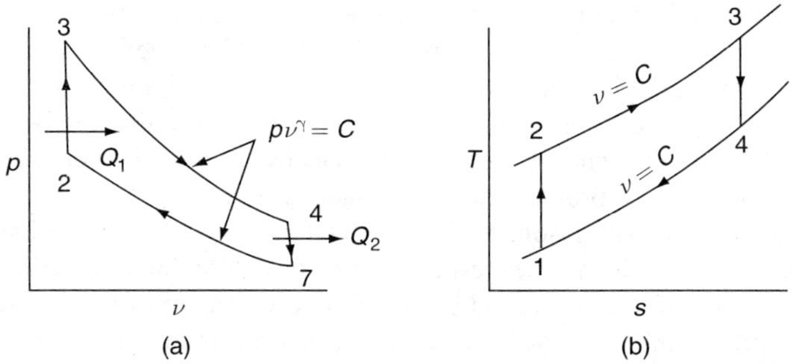

Fig. 20.12 *Otto cycle*

20.8.1 Otto Cycle

The Otto cycle is the air standard cycle of the SI engine, the operation of which has been explained in Fig. 20.5. It consists of two reversible constant volume processes in which heat is supplied and rejected and two reversible adiabatics in which the working fluid, air, is expanded and compressed (Fig. 20.12).

$$\eta_{\text{Otto}} = 1 - \frac{Q_2}{Q_1} = 1 - \frac{mc_v(T_4 - T_1)}{mc_v(T_3 - T_2)} = 1 - \frac{T_4 - T_1}{T_3 - T_2}$$

$$\text{Process } 1-2, \quad \frac{T_2}{T_1} = \left(\frac{v_1}{v_2}\right)^{\gamma-1}, \quad \text{Process } 3\text{–}4, \quad \frac{T_3}{T_4} = \left(\frac{v_4}{v_3}\right)^{\gamma-1} = \left(\frac{v_1}{v_2}\right)^{\gamma-1}$$

$$\therefore \quad \frac{T_2}{T_1} = \frac{T_3}{T_4} \quad \text{or,} \quad \frac{T_3}{T_2} = \frac{T_4}{T_1}, \quad \text{or,} \quad \frac{T_4 - T_1}{T_3 - T_2} = \left(\frac{v_2}{v_1}\right)^{\gamma-1}$$

$$\therefore \quad \eta_{\text{Otto}} = 1 - \left(\frac{v_2}{v_1}\right)^{\gamma-1} = 1 - \frac{1}{r_k^{\gamma-1}} \tag{20.11}$$

where $r_k = \dfrac{v_1}{v_2}$ is called the compression ratio.

The efficiency of the air standard Otto cycle is thus a function of the compression ratio only. The higher the compression ratio, the higher the efficiency. It is independent of the temperature levels at which the cycle operates. The compression ratio cannot, however, be increased beyond a certain limit, because of a noisy and destructive combustion phenomenon, known as detonation. It also depends upon the fuel, the engine design, and the operating conditions.

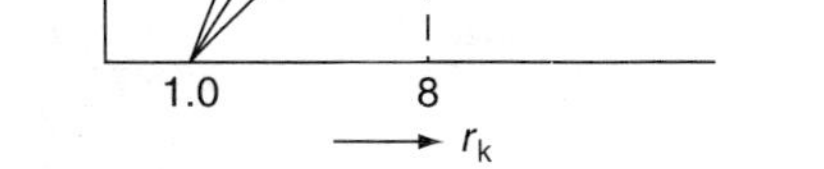

Fig. 20.13 *Effect of r_k and γ on Otto cycle Efficiency*

Figure 20.13 shows the effect of compression ratio and the specific heat ratio on the efficiency of Otto cycle. For an air standard cycle, air is the working fluid, $\gamma = 1.4$.

We can observe that the thermal efficiency curve is rather steep at low compression ratios but flattens our starting with a r_k of about 8. Therefore, the increase in thermal efficiency with the compression ratio is not that pronounced at high compression ratios. Also, when high compression ratios are used, the temperature of the air fuel mixture rises above the self-ignition temperature of the fuel when the mixture ignites without the spark, causing an early and rapid combustion of the fuel ahead of the flame front, followed by almost instantaneous burning of the remaining mixture. This premature ignition of the fuel, called **autoignition**, produces an audible noise, which is called **engine knock or detonation.** This auto-ignition hurts performance and can cause engine damage, thus setting the upper limit of the compression ratio that can be used in SI engines.

Improvement of the thermal efficiency of gasoline engines by utilizing higher compression ratios (upto 12) without facing the auto-ignition problem has been made possible by using gasoline blends that have good antiknock characteristics, such as gasoline mixed with tetraethyl lead. Tetraethyl lead has been added to gasoline since the 1920s because it is the cheapest method of raising the **octane rating**, which is a measure of the engine knock resistance of a fuel. Leaded gasoline, however, has a very undesirable side effect as it forms compounds during the combustion process that are hazardous to health and pollute the environment. Most cars made since 1975 have been designed to use unleaded gasoline, and the compression ratios had to be lowered to avoid engine knock. As a result the thermal efficiency of car engines has somewhat decreased. However, owing to improvement in other areas like reduction in overall automobile weight, improved aerodynamic design etc, today's cars have better fuel economy.

The second parameter affecting the thermal efficiency of the Otto cycle is the specific heat ratio γ. For a given r_k, the use of a monatomic gas such as argon or helium as the working fluid yields the highest thermal efficiency (Fig. 20.13). The specific heat ratio γ and thus the thermal efficiency of the Otto cycle decreases as the molecules of the working fluid get larger. At room temperature it is 1.4 for air, 1.3 for carbon dioxide and 1.2 for ethane. The working fluid in actual engines contains larger molecules such as CO_2, and γ decreases with temperature, because of which the thermal efficiencies of the actual engines are lower than those of the Otto cycle and vary from 25 to 30 percent.

The ratio of brake thermal efficiency and air standard efficiency is called *relative efficiency.*

$$\eta_{rel} = \frac{\eta_{br.th}}{\eta_{air\ std.}}$$

20.8.2 Work Output

The net work output for an Otto cycle (Fig. 20.13) can be expressed

$$W_{net} = \frac{p_3V_3 - p_4V_4}{\gamma-1} - \frac{p_2V_2 - p_1V_1}{\gamma-1}$$

Now,

$$\frac{v_1}{v_2} = \frac{V_1}{V_2} = r_k, \quad \text{or} \quad V_1 = V_2 r_k = V_4$$

$$\frac{p_2}{p_1} = \frac{p_3}{p_4} = \left(\frac{V_1}{V_2}\right)^{\gamma} = r_k^{\gamma}$$

$$\frac{p_3}{p_2} = \frac{p_4}{p_1} = r_p \text{ (say)}$$

$$W_{net} = \frac{p_1V_1}{\gamma-1}\left(\frac{p_3V_3}{p_1V_1} - \frac{p_4V_4}{p_1V_1} - \frac{p_2V_2}{p_1V_1} + 1\right)$$

$$= \frac{p_1V1}{\gamma-1}\left(\frac{r_p r_k^{\gamma}}{r_k} - r_p - \frac{r_k^{\gamma}}{r_k} + 1\right) = \frac{p_1V_1}{\gamma-1}\left(r_p r_k^{\gamma-1} - r_p - r_k^{\gamma-1} + 1\right)$$

$$W_{net} = \frac{p_1V_1}{\gamma-1}\left(r_p - 1\right)\left(r_k^{\gamma-1} - 1\right) \tag{20.12}$$

20.8.3 Mean Effective Pressure

The mean effective pressure (m.e.p) of the cycle is given by

$$p_m = \frac{\text{Net work output}}{\text{Swept volume}}$$

where swept volume $= V_1 - V_2 = V_2\,(r_k - 1)$

$$\therefore \qquad p_m = \frac{\dfrac{p_1 V_1}{\gamma - 1}(r_p - 1)(r_k^{\gamma-1} - 1)}{V_2(r_k - 1)}$$

$$= \frac{p_1 r_k (r_p - 1)(r_k^{\gamma-1} - 1)}{(\gamma - 1)(r_k - 1)} \qquad (20.13)$$

Thus, it is seen that the net work output is directly proportional to the pressure ratio r_p. For given values of r_k and γ, p_m increases with r_p. For an Otto cycle, an increase in r_k leads to an increase in p_m, W_{net} and cycle efficiency.

20.8.4 Diesel Cycle

The limitation on compression ratio in the S.I. engine can be overcome by compressing air alone, instead of the fuel-air mixture, and then injecting the fuel into the cylinder in spray form when combustion is desired. The CI engine, first proposed by Rudolph Diesel in the 1890s, is very similar to the SI engine, differing mainly in the method of initiating combustion. In SI engines, a mixture of air fuel is compressed during compression stroke, and the compression ratios are limited by the onset of autoignition or engine knock. In CI engines, only air is compressed during the compression stroke. Therefore, diesel engines can operate at much higher compression ratios, typically between 12 and 24. The spark plug and carburettor (for mixing fuel and air) are replaced by a fuel injector in diesel engines. The temperature of air after compression must be high enough so that the fuel sprayed into the hot air burns spontaneously. The rate of burning can, to some extent, be controlled by the rate of injection of fuel. An engine operating in this way is called a *compression ignition (C.I.) engine.* The sequence of processes in the elementary operation of a C.I. engine, shown in Fig. 20.14 is as follows.

Process 1–2 intake The air valve is open. The piston moves out admitting air into the cylinder at constant pressure.

Process 2–3 Compression The air is then compressed by the piston to the minimum volume with all the valves closed.

Process 3–4 Fuel injection and combustion The fuel valve is open, fuel is sprayed into the hot air, and combustion takes place at constant pressure.

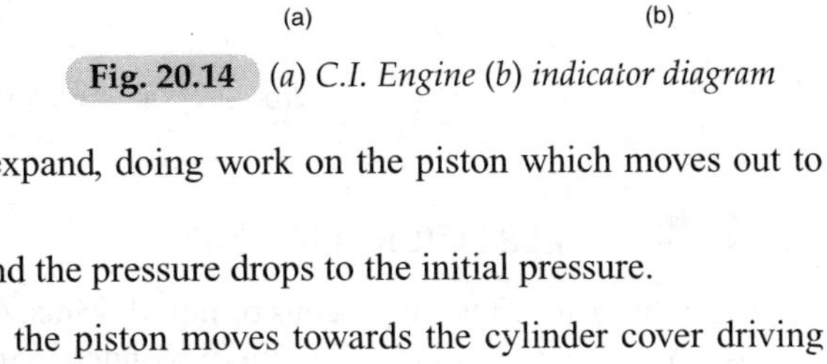

Fig. 20.14 (*a*) *C.I. Engine* (*b*) *indicator diagram*

Process 4–5 Expansion The combustion products expand, doing work on the piston which moves out to the maximum volume.

Process 5–6 Blowdown The exhaust valve opens, and the pressure drops to the initial pressure.

Process 6–1 Exhaust With the exhaust valve open, the piston moves towards the cylinder cover driving away the combustion products from the cylinder at constant pressure.

The above processes constitute an engine cycle, which is completed in four strokes of the piston or two revolutions of the crankshaft.

Figure 20.15 shows the air standard cycle called the Diesel cycle, corresponding to the CI engine as described above. The cycle is composed of two reversible adiabaties, one reversible isobar and one reversible isochore.

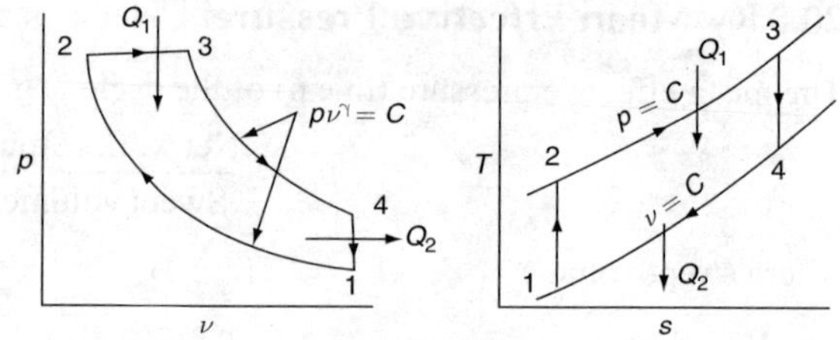

Fig. 20.15 *Diesel cycle*

Air is compressed reversibly and adiabatically in process 1–2. Heat is then added to it from an external source reversibly at constant pressure in process 2–3. Air then expands reversibly and adiabatically in process 3–4. Heat is then rejected reversibly at constant volume in process 4–1 and the cycle repeats itself.

For m kg of air,

heat supplied $$Q_1 = Q_{2-3} = mc_p(T_3 - T_2)$$

heat rejected $$Q_2 = Q_{4-1} = mc_v(T_4 - T_1)$$

$$\therefore \quad \eta = 1 - \frac{Q_2}{Q_1} = 1 - \frac{1}{\gamma}\cdot\frac{T_4 - T_1}{T_3 - T_2} \tag{20.14}$$

Compression ratio, $r_k = \dfrac{V_1}{V_2} = \dfrac{v_1}{v_2}$; Expansion ratio, $r_e = \dfrac{V_4}{V_3} = \dfrac{v_4}{v_3}$

Cut-off ratio, $$r_c = \frac{V_3}{V_2} = \frac{v_3}{v_2}$$

Now, $$r_k = r_e \cdot r_c$$

Also, $$\frac{T_4}{T_3} = \left(\frac{v_3}{v_4}\right)^{\gamma-1} = \frac{1}{r_e^{\gamma-1}}; \quad T_4 = T_3\frac{r_c^{\gamma-1}}{r_k^{\gamma-1}} \tag{20.15}$$

$$\frac{T_2}{T_3} = \frac{p_2v_2}{p_3v_3} = \frac{1}{r_c}; \quad T_2 = \frac{T_3}{r_c} \tag{20.16}$$

$$\frac{T_1}{T_2} = \left(\frac{v_2}{v_1}\right)^{\gamma-1} = \frac{1}{r_k^{\gamma-1}}; \quad \therefore \quad T_1 = \frac{T_3}{r_c}\cdot\frac{1}{r_k^{\gamma-1}} \tag{20.17}$$

$\therefore$ substituting T_1, T_2 and T_4 in Eq. (20.14)

$$\eta_{\text{Diesel}} = 1 - \frac{1}{\gamma}\cdot\frac{1}{r_k^{\gamma-1}}\cdot\frac{r_c^{\gamma} - 1}{r_c - 1} \tag{20.18}$$

As $r_c > 1$, $\eta_{\text{Diesel}} < 1 - \dfrac{1}{r_k^{\gamma-1}}$, i.e., $\eta_{\text{Diesel}} \nless \eta_{\text{Otto}}$ for the same compresion ratio, r_k.

20.9 FUELS FOR IC ENGINES

IC engine fuels are either in gaseous or liquid forms. *Gaseous fuels* can mix with air and distribute homogeneously to the various cylinders in a multicylinder engine. They can burn completely and leave relatively little combustion deposits compared with other fuels.

However, in vehicular engines, large containers are required for storing and carrying the fuel in the vehicles. The gases are thus compressed to high pressures and stored in thick containers. In stationary engines, they can be located close to the gas source. The gaseous fuels mostly used are natural gas (CNG), coke oven gas, blast furnace gas, refinery gas, producer gas and liquefied petroleum gas (LPG).

Liquid fuels are mostly derived from crude petroleum which consists of a large number of hydrocarbons having different molecular structures and boiling range. Generally, the compounds are classified into five groups — paraffins ($C_n H_{2n+2}$), olefins (C_nH_{2n}), diolefins (C_nH_{2n-2}), all being chain structure, hapthenes (C_nH_{2n} ring), and aromatics (C_nH_{2n-6} ring). No individual fuel belongs to a single group only, for example, petrol is generally a blend of paraffin, napthene and aromatic. A fuel tends to exhibit the characteristics of the hydrocarbon which forms a major part of it: The anti-knock quality of an SI engine fuel is worst in the paraffin group, improves gradually with the groups stated in the order above, and is best in the aromatic group. But with CI engines the suitability of fuels are in the reverse order, i.e., paraffin is the best and aromatic the worst. Boiling point temperature of a fuel depends on the number of atoms in the molecular structure and increases with the number of atoms. Fuels with fewer atoms in the molecule tend to be more volalile. The heating value of a fuel increases as the ratio of hydrogen to carbon atoms in the molecule increases. Therefore, paraffin has the greatest and aromatic has the least calorific value.

Crude petroleum, as such, is rarely used as fuel, and it is processed in a refinery to get the desirable products. Fractional distillation (separation of compounds depending on their boiling points) is used to separate crude petroleum, and fractions separated with increasing boiling points are in the order: petrol, naptha, kerosene, fuel and gas oil, lubricating oil, residue.

SI engines need liquid fuels with high volatility (vaporizing ability) for carburetion (mixing the fuel with air before admission to the engine), but CI engines fuels are from the fractions of kerosene, and fuel and gas oil range. CI engine fuels are less volatile, more viscous and heavier than petrol.

Fuel-Air/Air-Fuel Ratio The relative proportions of fuel and air in the engine are very important for combustion and efficiency of the engine. This is expressed either as a ratio of the mass of mass to that of air (F/A) or vice versa (A/F). A mixture that contains just enough air for complete combuation of the fuel in the mixture is called chemically correct or stoichiometric fuel–air ratio. A mixture having more fuel than this is a rich mixture and a mixture that contains less fuel is called a lean mixture. The ratio actual F/A ratio to stoichiometrice F/A ratio is called the *equivalence radio*, ϕ so that

$$\phi = \frac{\text{Actual F/A ratio}}{\text{Stoichiometric F/A ratio}}$$

When $\phi > 1$ it is a rich mixture. If $\phi = 1$, it is chemically correct mixture.

20.9.1 SI Engine Fuels

The important qualities to be considered for a *SI* engine fuel are the following:

(i) Volatility It is one of the most important properties of *SI* engine fuel and it influences the operation and maintenance of the engine. For ease in engine starting, good acceleration, proper distribution of air-fuel mixture in the cylinders, avoiding crankcase dilution (i.e. dilution of the lubricating oil in the crankcase by liquid petrol droplets) and washing of the cylinder lubricating oil, it is necessary to have high volatility, i.e. low boiling-off temperature of the fuel. Again, a high volatility can stop the flow of fuel to the engine by forming vapour locks in the fuel passages. Therefore, actual selection of a fuel is a compromise between requirements.

(ii) Gum Deposits Some components (hydrocarbons) of a fuel have a tendency to form gum due to oxidation during storage. This gum causes undesirable deposits in the inlet valves, piston rings, carburetor and other engine parts. Therefore, for smooth operation of *SI* engines, the gum content and the tendency to form gum during storage of a fuel should be within reasonable limit.

(iii) Sulphur Content Any sulphur present in the fuel tends to form corrosive compounds (H_2SO_4) which will attack the engine parts and damage them. Therefore, sulphur content in a fuel must be within permissible limit.

(iv) Anti-knock Quality Under certain conditions, *SI* engine fuel (petrol) tends to cause engine knock due to sudden rise in pressure during combustion, accompanied by a hammer like sound (detonation). This kind of engine knock occurs towards the end of the pressure rise *after ignition*, and depends on the chemical composition and molecular structure of the fuel, and other factors such as engine load, speed, spark ignition timing, air-fuel ratio, temperature in the later stages of combustion (this temperature depends upon compression ratio and a given fuel will have increasing tendency to knock with increase in compression ratio).

The anti-knock quality of a fuel is expressed by a number (*Octane number*). The usual method used to determine the anti-knock quality of a *SI* engine fuel consists in testing the given fuel in a *CFR* (Co-operative Fuel Research) test engine with variable compression, and comparing it with a certain mixture whose knocking property is known. Iso-octane practically does not have any knocking tendency and it is arbitrarily assigned a rating of 100 octane number; normal heptane knocks worse than any known petrol and is given a rating of 0 octane number. These two can be mixed to match the knocking characteristic of any given fuel, and such a mixture is designated by the percentage of iso-octane (by volume) in the mixture. The octane number of a fuel is the percentage by volume of iso-octane in a mixture of iso-octane and normal heptane which produces the same knock intensity as the given fuel. Knocking (detonation) limits the power output and efficiency of *SI* engines. The higher the octane number of a fuel the greater will be its resistance to knock. Ordinary petrol can be made less likely to knock by adding small quantities of certain compound such as tetraethyl lead. However, now its use is discouraged due to severe environmental pollution.

20.9.2 CI Engine Fuels

Commonly used fuels in *CI* engines are high-speed diesel, light-diesel and heavy-diesel oils. The main qualities to be considered in the selection of *CI* engine fuels are:

(i) Ignition It is a measure of the ability of the fuel to ignite promptly after it is injected into the cylinder. The ideal fuel should ignite immediately after injection, but in reality there is a certain lapse of time between the beginning of injection and ignition of the fuel. This time lag is known as *ignition lag* or *delay*, and if the ignition lag is large, the whole amount of the fuel may be injected before it ignites and when it ignites it burns suddenly with excessive pressure rise, producing a *combustion knock*. This combustion knock will be less with fuels having short ignition lag, because, the fuel will begin to burn soon after injection and hence there is less chance of the fuel to accumulate and burn suddenly.

The ignition quality or ignition lag of *CI* engine fuels is measured in terms of *Cetane number*. The ignition quality of a given fuel is compared in a *CFR* test engine with a reference fuel. Cetane $C_{16}H_{34}$ has very good ignition quality and is arbitrarily given a rating of 100 cetane number, and it is mixed with alphamethyl naphthalene $C_{11}H_{10}$ which has very poor ignition quality and assigned 0 cetane number. The percentage of cetane (by volume) in a mixture of cetane and alphamethyl naphthalene, which has the same ignition quality as the given fuel, is called the cetane number. Thus a fuel with cetane number 44 has the same ignition quality as a mixture of 44% of cetane and 56% of alphamethyl naphthalene. Adding certain compounds like amyl nitrate improves the ignition quality of a fuel.

(ii) Volatility Though *CI* engine fuels are less volatile than *SI* engine fuels, it should be volatile enough to promote good mixing and combustion to ensure a clean exhaust.

(iii) Viscosity *CI* engine fuels are more viscous than *SI* engine fuels, but they should be fluid enough to flow through the fuel system and filters.

(iv) Impurities Sulphur and solid particle contents in the fuel should be within permissible limit to prevent formation of corrosive compounds, engine wear, and clogging of fuel injectors.

(v) Flash point The flash point, i.e. the temperature at which the fuel gives off sufficient vapour to form a combustible mixture with air, should be sufficiently low to prevent fire hazard.

20.10 COMBUSTION IN SI ENGINES

Chemically correct or stoichiometric air-fuel ratio (*AFR*) is the ratio as obtained from the chemical reaction equation. The actual *AFR* during operation is, however, likely to be different from this ratio. *SI* engines generally operate on a limited range of *AFR* (8–20), beyond which the mixture is either too rich or too lean. There is no single *AFR* which is best for all conditions; for example the *AFR* for maximum power is not the same as the *AFR* for maximum economy (i.e. the power for which fuel consumption per kWh is minimum). The best power *AFR* is richer and the best economy *AFR* is leaner than the stoichiometric ratio (Fig. 20.16).

Combustion in *SI* engines may be of two types-normal combustion, and abnormal combustion. During *normal combustion* the spark plug starts the burning of the mixture and a flame-front spreads in all directions from the spark through compressed charge until the whole mixture (charge) is burned. The speed with which the flame spreads is called the rate of flame propagation or flame speed and it influences the combustion and knocking in the engine. The flame speed depends on the following factors:

(i) Turbulence Increased turbluence in the charge increases the flame speed due to increased mixing of the burning and unburned particles in the charge. Again, increase in engine speed increases turbulence.

(ii) Air-fuel Ratio Highest flame speed occurs at an *AFR* somewhat richer than the stoichiometric *AFR*.

(iii) Other Factors Pressure, temperature, and humidity of the intake charge, amount of residual gas (charge dilution), compression ratio, spark timing; the effect of these factors are, however, small.

The rate at which the pressure rises in the cylinder, during the combustion process, is of primary interest to both the designer and operator. This *rate of pressure rise* exerts considerable influence on the peak pressures encountered, the power produced, and the smoothness with which the forces are transmitted to the piston. The rate of pressure rise is dependent upon the *mass rate of combustion of the mixture* in the cylinder.

A convenient and generally used graphical presentation depicting the pressure rise in the cylinder is known as the **pressure-crank angle** or the **pressure-time** (*p-t*) diagram. Such a diagram is shown in Fig. 20.17. In addition to showing the rate of pressure rise, the *p-t* diagram indicates, more clearly than the *p-V* diagram, the events occurring near TDC.

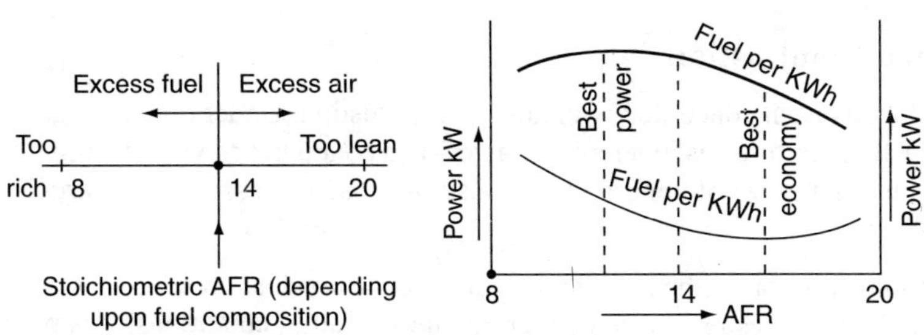

Fig. 20.16

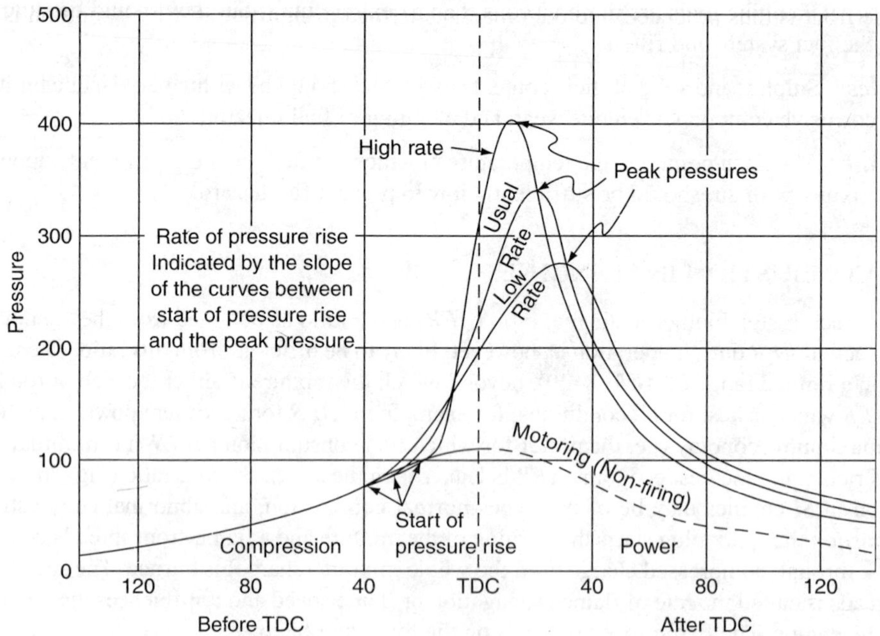

Fig. 20.17 *Illustration of various combustion rates*

Figure. 20.17 shows the relationship between pressure and crank angle for three different rates of combustion, namely, a high, a normal, and a low rate. Note that with the lower rates of combustion, it becomes necessary to initiate burning at an earlier point on the compression stroke because of the longer time necessary to complete combustion. Also note that higher rates of pressure rise, as a result of the higher rates of combustion, generally produce higher peak pressures at a point closer to TDC, which is a generally desirable feature. Higher peak pressures closer to TDC produce a greater force acting through a larger portion of the power stroke, and hence, increase the power output. Practical operating considerations, however, place a limit on the rate of pressure rise. If the rate is too high, the forces exerted on the piston tend to become "impactive," causing rough, or "jerky" operation. Also, if the peak pressures become excessive, they tend to create a situation conducive to an undesirable occurrence known as detonation. A compromise between these opposing factors is obviously necessary. This is accomplished by designing and operating the engine in such a manner that approximately one-half of the pressure rise has taken place as the piston reaches TDC. The result is a peak pressure reasonably close to the beginning of the power stroke, yet maintaining smooth engine operation.

20.10.1 Abnormal Combustion

The discussion thus far has been concerned with normal combustion. Under certain engine operating conditions, however, the air-fuel mixture has inherent characteristics conducive to various forms of abnormal combustion. The more important types of abnormal combustion, to be discussed in this article, are pre-ignition and detonation.

(1) *Preignition* If some portion of the boundary of the combustion chamber, such as a spark plug, exhaust valve, or carbon particle, becomes overheated under certain operating conditions, it is possible for this part to act in the same manner as the regular spark, and ignite the adjacent fresh combustible charge. An entirely distinct flame front is thus produced, and the process is termed **preignition**. Such a condition is undesirable since combustion becomes both erratic and uncontrollable. Moreover, preignition tends to raise the temperatures and pressures in the chamber which cause the temperature of the "hot spot" to rise further, and encourage still earlier preignition on succeeding cycles. The cumulative effect not only tends to raise peak pressures and encourage the possibility of detonation, but also tends to cause the peak pressures to occur progressively earlier in the cycle. In fact, preignition may advance these peak pressures to such a point that they occur before the piston reaches TDC on the compression stroke. In such a case, the peak pressure in those cylinders which are preigniting will oppose piston movement during the last part of the compression stroke, thus decreasing total output as well as causing rough engine operation. Preignition may also cause damage, through burning, to those engine parts which are subjected to the extreme temperatures.

(2) *Detonation* A combustible mixture of fuel and air, under certain conditions of temperature, pressure, and density, has the faculty of igniting without the assistance of an initiating flame or spark. Such an event is known as auto-ignition. It is comparable to the more familiar layman's term of "spontaneous combustion."

In SI engines, the main "actor" in the auto-ignition phenomenon is *the last portion of the unburned charge* in the combustion chamber. As the normal flame front proceeds across the chamber, it raises the pressure and temperature of the remaining portion of the unburned charge. Under certain conditions of pressure, temperature, and density of the unburned charge, this charge may auto-ignite and burn almost instantaneously, thus releasing energy at a much greater rate than during the normal combustion process. The extremely rapid release of energy causes pressure differentials of considerable magnitude in the combustion chamber which give rise to radical vibrations of the gaseous products, producing an audible knock. This condition is known as detonation.

Detonation is a most important aspect in the operation of SI engines, since it is *the major factor limiting the compression ratio* of an engine. The violent pressure fluctuations accompanying detonation can cause severe damage to the engine, and *sustained detonation cannot be tolerated.* In the operation of automobile engines, sonic warnings are given in the form of an audible "pinging" sound, and operating conditions can be eased to prevent damage. In a SI aircraft engine, however, the engine noises override any detonation sounds and thus increase the inherent danger of detonation damage.

Although little is actually known concerning the processes which occur during detonation, it is desirable to follow through the events which are believed to take place.

It is known that, in order to auto-ignite, the last unburned portion of the charge must reach and remain for a definite amount of time above a certain critical temperature which is dependent upon conditions of pressure and density of the unburned charge. Once these conditions are reached, a "preparation" phase commences, followed by the "actual burning" phase. The "Preparation" phase is known as the ignition delay.

Figure 20.18(a) represents a normal flame front travelling across a combustion chamber from A toward D, and increasing the pressure, temperature, and density of the unburned charge (area $BB'D$). If this unburned charge does not reach its critical temperature for auto-ignition, it will not auto-ignite, and the flame front BB' will proceed on through the unburned charge to point D in an orderly manner. The *p-t* diagram for such normal combustion is illustrated in Fig. 20.18(b).

Refer now to Fig. 20.18(c) If the unburned charge (area $BB'D$) reaches and remains above its critical conditions for auto-ignition, there is a possibility of detonation. In essence, a "race" develops between the flame front and the ignition delay. If the flame front can proceed from BB' to D and consume the unburned charge in a normal manner, prior to completion of the ignition delay period, there will be no detonation. If, however, the flame front is able to proceed only as far as, say CC', during the ignition delay period, then the remaining portion of the unburned charge (area $CC'D$) will detonate. Figure 20.18(d) is a *p-t* diagram showing the extreme pressure fluctuations occurring during detonation. Note that *detonation occurs, in a SI engine, near the end of combustion.*

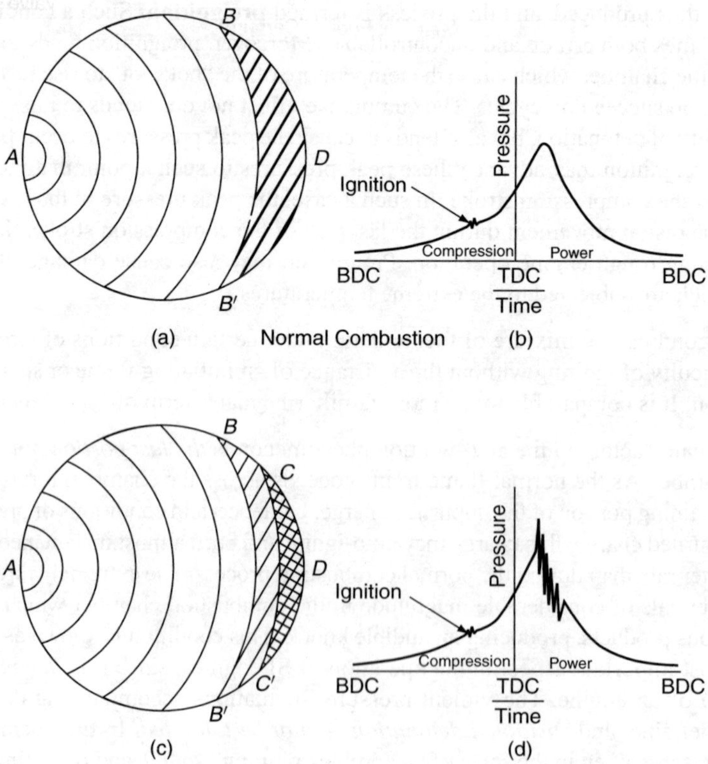

Fig. 20.18 *Schematic presentation of the principles of normal and detonating combustion processes*

In summary, if the unburned charge does not reach its critical temperature, there can be no detonation. If the ignition delay period is longer than the time required for the flame front to burn through the unburned charge, there can be no detonation. But if the critical temperature is reached and maintained, and the ignition delay is shorter than the time it takes for the flame front to burn through the unburned charge, then the charge will detonate. Consequently, in order to inhibit detonation, *a high critical temperature for auto-ignition, and a long ignition delay, are desirable qualities in SI engine fuels.*

20.10.2 SI Engine combustion chamber

Figure 20.19 shows some combustion chambers with valve and spark plug locations. The main considerations in designing a good combustion chamber for SI engines are the following:

(i) *High volumetric efficiency* The incoming charge should enter with a high velocity with minimum pressure loss.

(ii) Combustion be completed near the TDC, which would give high pressure to the piston for high output

(iii) *Turbulence*-Correct degree of turbulence should be generated to promote rapid combustion

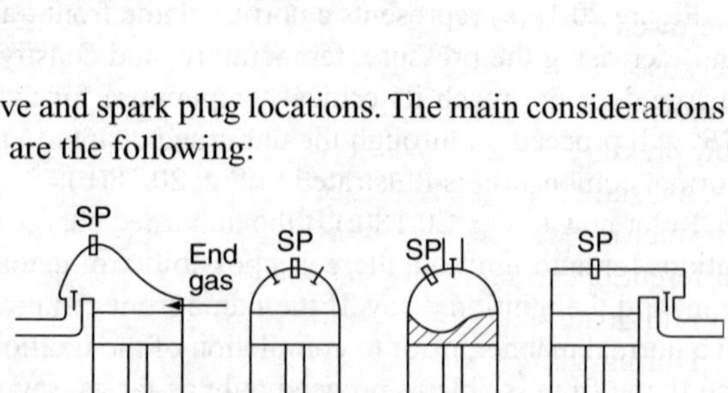

Fig. 20.19 *SI engine combustion chambers*

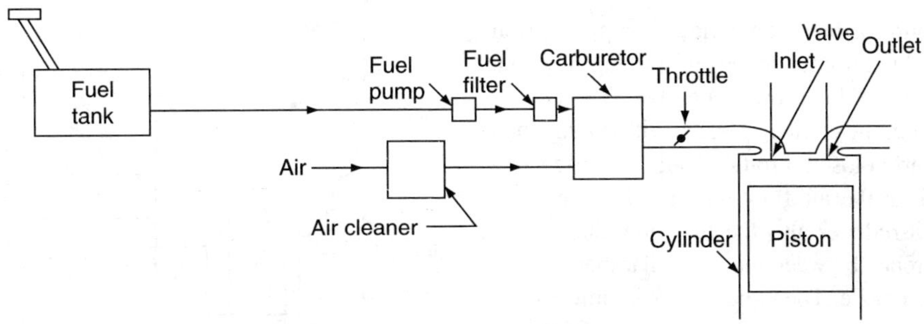

Fig. 20.20 *Fuel and air induction system*

(iv) Satisfactory cooling of exhaust valve head
(v) Good scavenging of hot exhaust gases.

SI engines now having compression ratios of 9 or more use overhead valves as shown.

20.10.3 Ignition Systems in SI Engines

In SI engines the fuel system consists of a fuel supply tank, fuel pump, filters, lines, carburetor and intake pipe or manifold as shown in Fig. 20.20. The fuel pump may not be used if the fuel tank is placed at a higher level. The carburetor is the most important component; it prepares the correct air–fuel mixture and delivers this mixture to each cylinder through a ducting called intake manifold.

20.11 SIMPLE FLOAT CARBURETOR

Figure 20.21 shows a simple carburetor which consists of a venturi, a fuel nozzle with a metering orifice, a fuel tank with a float, a throttle valve and a choke. The fuel is pumped or flows by gravity to the float chamber. The fuel level is maintained at a constant level in the chamber with the help of a fuel supply valve controlled by the float. When air flows through the intake pipe, a vacuum is created at the venturi throat, which draws the fuel out through the discharge nozzle like a stream of fuel droplets, because of the pressure difference between the atmospheric pressure in the float chamber and the low pressure at the throat.

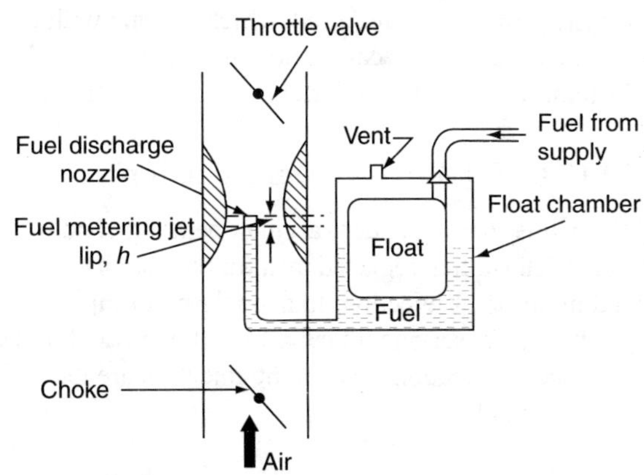

Fig. 20.21 *Simple float-type carburetor*

As stated earlier, the carburetor must provide air–fuel mixture with appropriate air–fuel ratio for different operating conditions of the engine, which are (i) starting and warming, (ii) idling, (iii) part load, (iv) full load, and (v) acceleration.

During starting and warming up, the engine speed is low, the quantity of air flowing through the venturi is low and the vacuum created at the venturi throat is insufficient to draw fuel into the air stream. When the speed of the engine increases, the quantity of air flowing through the venturi increases and this creates a large vacuum at the throat. Therefore, a proportionately larger amount of fuel is sprayed into the air stream. Thus, a simple carburetor has a tendency to supply a rich mixture ($\phi > 1$) at higher speeds of the engine and a weak mixture ($\phi < 1$) at lower speeds.

During idling, the engine operates without any load and with a nearly closed throttle. As the throttle valve is almost closed, the air flow through the venturi is reduced, and it reduces also the fuel flow due to less pressure difference between the float chamber and the fuel discharge nozzle. The exhaust gas is drawn back into the cylinder along with a fraction of fresh charge when the throttle is closed. The overall mixture will contain a high percentage of exhaust gas, and in order to have combustion inside the engine, the carburetor must supply an increasingly rich mixture. There is an air bleed arrangement so that fuel can flow to the intake manifold when the throttle is almost closed (Fig. 20.22).

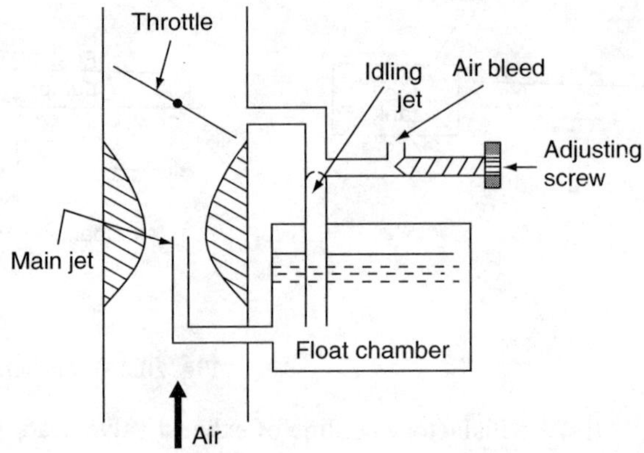

Fig. 20.22 *Idling jet in a carburetor*

When the engine is operating at part load with the throttle partly opened, maximum economy is the objective and therefore the air–fuel ratio for maximum efficiency should be selected, which would require a lean mixture by supplying extra air. When the throttle is fully opened, the carburetor must supply a rich mixture for maximum output with the supply of extra fuel.

In order to obtain a smooth and rapid acceleration of the engine, extra quantity of fuel is momentarily required to supply a richer mixture. An accelerating pump discharges a jet of fuel through a special jet in the carburetor. It is actuated by a lever from the throttle linkage.

When the engine is started under cold conditions by cranking, the speed of the engine is very low and a very rich mixture is required, which is done with the help of choke closing the air passage.

Thus a complete carburetor must have the following systems: (i) idling system, (ii) the fuel metering system, (iii) power/fuel enrichment system, (iv) accelerating system, and (v) choke.

20.11.1 S.I. Engine Fuel Injection

An engine running under varying operating conditions with a good exhaust control requires a complex carburetor, which is often replaced by a fuel injection system. The latter can be of two types, viz, single-point injection and multi-point injection. In a single-point injection system, a simple injector injects fuel just upstream of the throttle valve, serving all cylinders. It is difficult to achieve proper distribution of fuel to all the cylinders. In a multi-point injection system, the injectors are located immediately before the inlet valve accommodated in the cylinder head.

20.11.2 Flow Equations in a Carburetor

(a) Air Flow By applying steady flow energy equation between section 1 (atmospheric condition) and venturi throat, section 2, we have

$$h_1 + \frac{V_1^2}{2} = h_2 + \frac{V_2^2}{2}$$

Since $V_2 << V_1$ and $h = c_p T$ for air,

$$V_2 = \sqrt{2(h_1 - h_2)} = \sqrt{2c_p(T_1 - T_2)} = \sqrt{2c_p T_1 \left[1 - \left(\frac{p_2}{p_1}\right)^{\frac{\gamma-1}{\gamma}}\right]},$$

assuming isentropic expansion.

The mass flow late of air $\dot{m}_a$ is obtained as

$$\dot{m} = \rho_2 A_2 V_2 = \frac{p_2}{RT_2} \cdot A_2 \times \sqrt{2c_p T_1 \left[1 - \left(\frac{p_2}{p_1}\right)^{\frac{\gamma-1}{\gamma}}\right]}$$

(b) Fuel Flow Since gasoline is incompressible, Bernoulli's equation is used between sections 1 and 2, which gives

$$\frac{p_1}{\rho_f g} = \frac{p_2}{\rho_f g} + \frac{V_2^2}{2g} + h$$

where h is the lip.

and

$$V_f = \sqrt{2g \frac{p_1 - p_2}{\rho_f} - h}$$

where ρ_f is the density of fuel.

The mass flow rate of fuel $\dot{m}_f$ is given by $\dot{m}_f = C_d \rho_f A_f V_f$

where C_d is the coefficient of discharge of the fuel orifice.

20.11.3 Ignition System in SI Engines

The ignition system is an essential element in the operation of *SI* engines. The two commonly used systems are

(i) Battery ignition system
(ii) Magneto ignition system

Figure 20.23 shows the essential components of a battery ignition system for a four cylinder engine. A voltage of the order of 10000 to 20000 volts is required at the spark plug to overcome the resistance at the spark gap and to release sufficient energy to start the ignition of the charge. The ignition coil, which is essentially an induction coil, consists of two separate insulated windings (primary and secondary) on an iron core. The primary coil consists of a small number of thick wire while the secondary coil consists of a large number of thin wire. The primary coil is energized by a 6–12 V storage battery and the high voltage current is induced in the secondary coil at each break at the contact breaker which is operated by a special cam. The number of lobes in this cam is equal to the number of cylinders in the engine. The high voltage current from the secondary is led to the spark plugs in turn at correct times by the distributor. Both the cam and the distributor are driven by the engine cam shaft. The purpose of the condenser is to facilitate the collapse of the magnetic field and to prevent arcing (sparking) at the breaker points.

The magneto ignition system works on the same principle as the battery ignition system except that no battery is used; the magnetic field in the core of the primary and secondary coils is produced by a rotating permanent magnet. The magneto ignition system is light, compact and requires less maintenance; therefore, it is suitable for applications where storage batteries prove heavy and bulky, e.g. in aircraft and scooter engines. The disadvantage with magneto system is that at starting and at low speeds the voltage developed (which depends upon engine speed) is low.

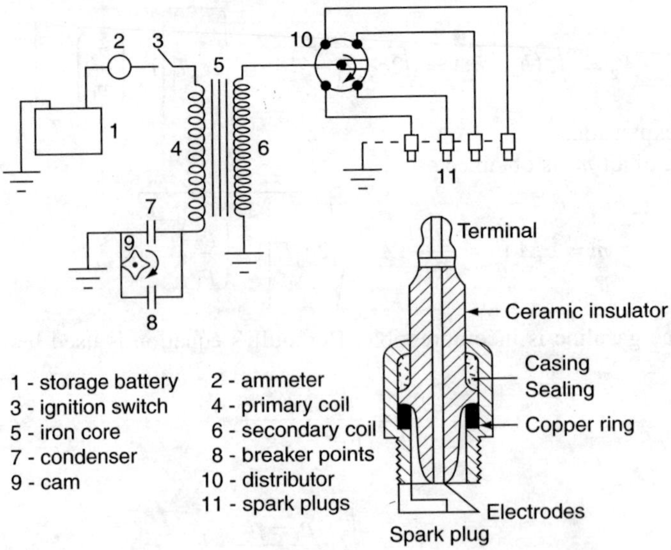

Fig. 20.23 *Battery ignition system and spark plug*

20.12 FUEL INJECTION IN CI ENGINES

The fuel injection in CI engines consists of a fuel tank, filters, lines, fuel pump and fuel injector. Its function is to supply correct quantity of fuel and inject it at the correct time without after dribbling, atomize the fuel properly, and ensure that the fuel spray penetrates the desired areas of the combustion chambers. CI engine fuel injection system may be of two types: air injection and airless or solid injection.

In air-injection, the fuel is injected into the cylinder by means of compressed air at about 7 MPa. Though this system was used in early years, it is seldom used now. Advantages of this system are good atomization and distribution, and possibility of using high viscosity, less expensive fuel. Disadvantages are complication of the engine with high pressure multi-stage air compressor which absorbs a part of engine power.

Now-a-days the airless or solid injection is used, and this system consists of two main parts-a high pressure fuel pump (15–30 MPa), and a fuel injector. Depending upon the arrangement of the fuel pumps and injectors, solid injection system may be classified as: *common rail system, unit injector system, individual pump and injector system* (Fig. 20.24). Of these, the latter two are mostly used.

In common rail system a single pump supplies fuel to a high pressure header or common rail from which the high pressure fuel goes to injectors in each cylinder. The injectors are operated by cams through push rods and rocker arms at correct sequence and time. In unit injector system the fuel pump and the injector or nozzle are combined into a single unit. Each cylinder is provided with one unit injector which is operated by cams as shown. In individual pump and nozzle system each cylinder has one pump and one nozzle or injector, but they are separated from each other. The injector is placed on the cylinder and the pump is placed on the side of the engine. The fuel pumps may be placed separately near each cylinder or in a cluster as shown.

20.12.1 Fuel Pump

The fuel pump plunger is actuated by a cam and produces the required fuel pressure to open the injector valve at correct time. The amount of fuel injected depends upon the *effective stroke* of the plunger which is controlled by a rack. The fuel injector contains a delivery valve which is actuated by the fuel pressure. Figure 20.25 shows a schematic diagram of a fuel pump.

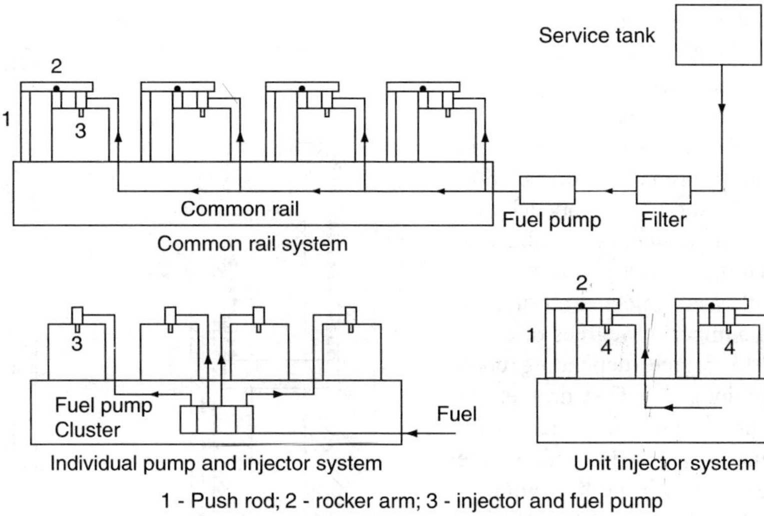

Fig. 20.24 *Fuel injection system*

The pump plunger has a constant stroke as determined by the cam lift. The plunger can be rotated inside the barrel with the rack to control the amount of fuel delivered by the pump to the injector. The vertical groove *C* is connected to the helical groove *D* and the space *S* as shown. During upward stroke of the plunger it covers the ports *A* and *B*, and fuel above the plunger and in the space *S* is raised to high pressure. This high pressure fuel is delivered to the injector until the edge *F* (on the helical edge) of the plunger uncovers the port *B*; at that instant the pressure developed is released and the fuel goes out through the overflow port *B*. The effective stroke of the pump is *EF*, because only during this movement of the plunger the fuel is delivered, although the plunger stroke may be more (constant and equal to cam lift). The effective stroke *EF* of the pump can be changed with the rack through the fuel control lever, and hence the amount of fuel injected to match the engine load.

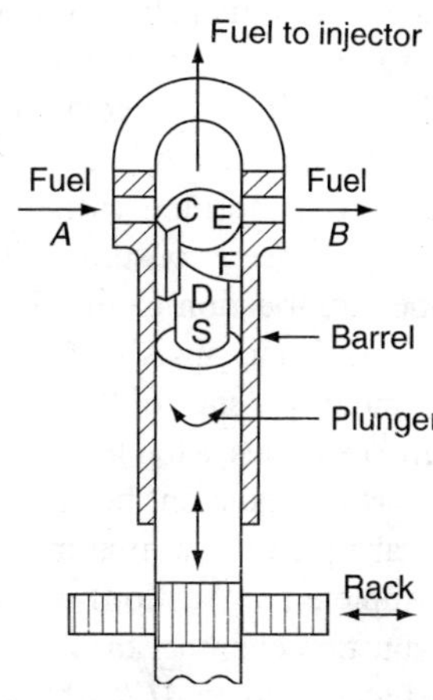

Fig. 20.25 *Fuel pump*

20.12.2 Fuel Injector

Figure 20.26(a) shows a schematic diagram of a fuel injector. Injection of fuel begins when the valve *V* is lifted from its seat against the spring force by the high pressure oil from the fuel pump acting on the annular area *A*.

Fuel injectors may be of single-orifice (20.26b), multi-orifice (20.26c) and pintle (20.26d) type. A single-orifice nozzle or injector gives good penetration but requires high injection pressure for fine atomization. A multi-orifice nozzle gives better atomization but get easily clogged. Pintle type nozzles can be designed to produce various spray-cone angles and are less likely to be clogged.

20.12.3 Combustion in CI Engines

In *SI* engines a homogenous mixture of air and fuel (produced by the carburettor) within the combustible range maintains the combustion. In *CI* engines the air-fuel mixture is not homogeneous and the *AFR* in the various parts of the combustion chamber is different. As the combustion chamber contains compressed air at a temperature above the ignition temperature of the fuel, combustion occurs at many points within the chamber. As the fuel is injected, each minute droplet produced after atomization by the injector is enveloped by its own vapour, and after a small interval, combustion begins at the surface of this envelope. As stated in Section 20.5, the fuel is not injected all at once but continues over a number of degrees of crank angles (upto about 35 degree, depending upon speed and size of engine). The first droplets of fuel entering the cylinder come in contact with air whose temperature is only a little above the ignition temperature and hence the beginning of the burning process takes a little time. The droplets which enter later find the air already heated to a temperature much above the ignition temperature due to burning of the earlier droplets and begin to burn almost immediately as they enter the cylinder, but the last droplets to enter find some difficulty in burning as much of the oxygen in the air has been consumed. Therefore, to ensure proper combustion of the fuel, sufficient mixing of the fuel and air is necessary by dispersion of fuel and turbulence of the air.

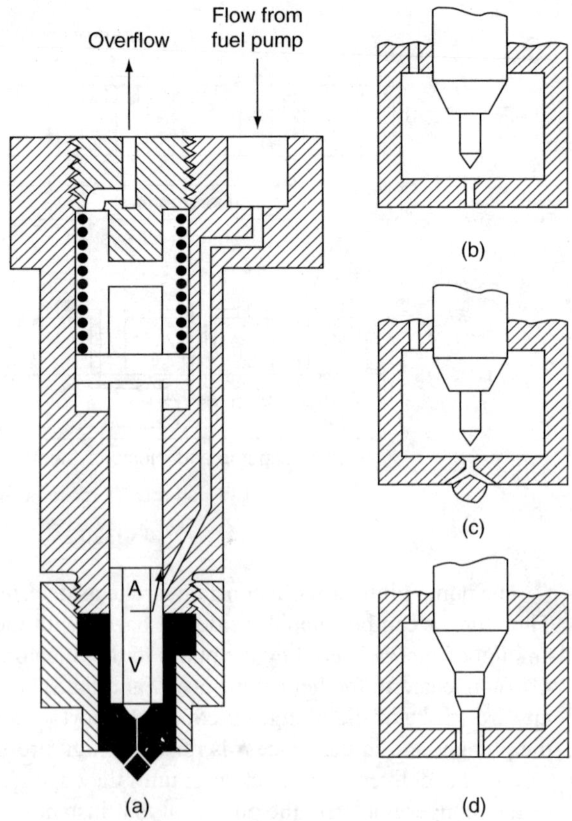

Fig. 20.26 *Fuel injector*

In *SI* engines, a too lean air-fuel mixture cannot support combustion, but in a *CI* engine burning can take place in a mixture which is infinitely lean. A *CI* engine begins to give out black exhaust (due to incomplete combustion) when the *AFR* is close to the stoichiometric ratio, because, in certain parts of the combustion chamber the *AFR* is too rich for complete combustion, and therefore, it is necessary to operate a *CI* engine with an *AFR* which is more (i.e. leaner) than the stoichiometric ratio.

As stated earlier, in *CI* engines the fuel does not ignite immediately on injection. The period between start of injection and start of combustion (ignition) is called *ignition delay* or *ignition lag,* and during this period there is no pressure rise in the cylinder due to combustion (Fig. 20.27).

Fuel injection in *CI* engines takes place over a certain time interval and as the initial droplets go through the delay period, additional droplets continue to enter the cylinder. If the ignition delay of the fuel is short, the first droplets will begin to burn soon after injection and a very small amount of fuel will accumulate in the combustion chamber.

This will give normal combustion with smooth pressure variation in the cylinder. But if the ignition delay of the fuel is long, a considerable part of the fuel injected will accumulate and when ignition begins it will suddenly burn, causing rapid pressure rise and pressure fluctuations. This will result in vibrations and audible knocks in a phenomenon similar to *SI* engine detonation (knocks); but, while in *SI* engines knocks due to detonation occur near the end of combustion, in *CI* engines knocks (called combustion or diesel knocks) due to detonation occur at the beginning of combustion.

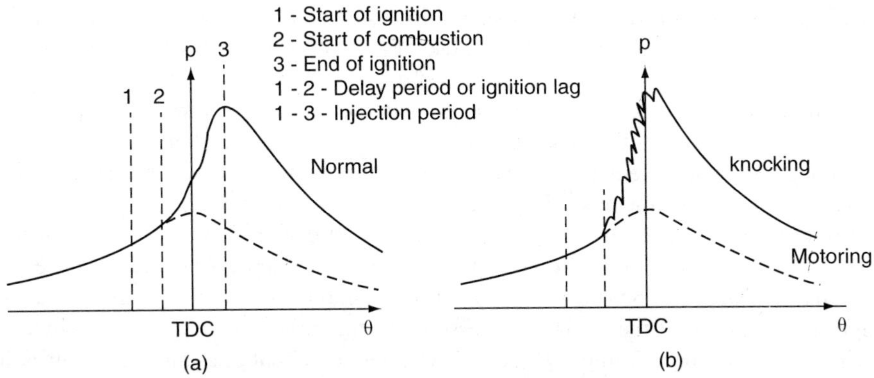

Fig. 20.27 *Normal and abnormal combustion in CI engines*

From the above we see that ignition delay of a fuel plays an important part in smooth combustion in a *CI* engine. The ignition delay of a *CI* engine fuel may be reduced by adding certain additives to the fuel. Factors which influence ignition delay are:

(i) Compression Ratio Increase in compression ratio reduces ignition lag; a higher pressure increases density resulting in closer contact of the molecules which reduces the time of reaction when fuel is injected.

(ii) Inlet Air Temperature An increase in inlet temperature increases the air temperature after compression and hence decreases the ignition lag.

(iii) Coolant Temperature Increase in coolant temperature increases cylinder air temperature and thus reduces ignition lag.

(iv) Engine Speed Increase in speed increases turbulence and this reduces the ignition lag.

20.12.4 CI Engine Combustion Chamber

In *SI* engine the mixing of the fuel and air is done in the carburetor outside the cylinder, but in a *CI* engine this mixing is done in the combustion chamber. Therefore, in *CI* engines the combustion chamber must be designed to provide for this mixing of fuel and air. In general, combustion chambers in *CI* engines may be of two types-non-turbulent type, and turbulent type (Fig. 20.28).

In non-turbulent type combustion chambers the mixing of the fuel and air depends only to a small extent upon the turbulence created by the chamber. Mixing is mainly achieved by using orifice type nozzles working under high pressure. The fuel is highly atomized and spreads throughout the combustion chamber at high velocities and thereby promotes good mixing. In this type of combustion

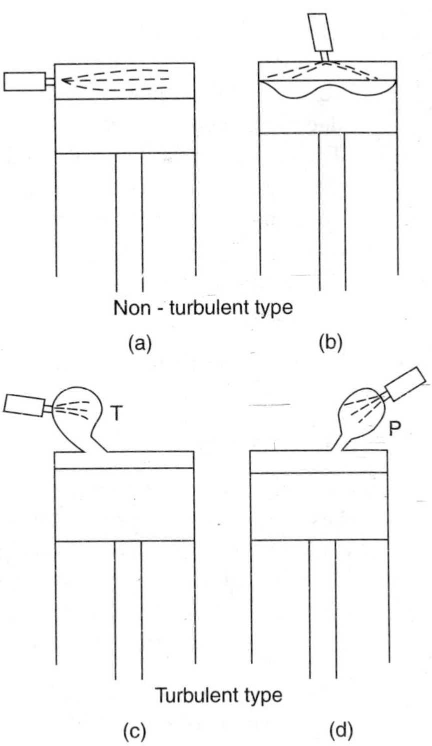

Fig. 20.28 *CI Engine combustion chambers*

chamber (Fig. 20.28a and b), the surface to volume ratio is small and hence heat loss through the walls is also small, and consequently, starting is easier. But the use of orifice type nozzles with small openings increases the possibility for frequent clogging. Use of air swirl produced by designed cavities in the piston crown improves performance. This type of combustion chambers are suitable for low speed stationary engines.

In turbulent type, the combustion chamber produces the desired air-fuel mixing through turbulence (Fig. 20.28c and d). With turbulent type combustion chambers, pintle type nozzles working under lower pressure are used, and therefore, the possibility of clogging is less. Due to greater turbulence and larger surface to volume ratio, heat loss is more with consequent starting difficulties. This type of combustion chamber is suitable for variable speed operation due to greater turbulence, which enables ready response to any change in the fuel injection rate. Figure 20.28c shows a turbulent type combustion chamber with a turbulence chamber T. During compression, almost all the air is compressed into the turbulence chamber T. Figure 20.28d shows a chamber with pre-combustion chamber P; during compression about one-third of the air is forced into the small pre-combustion chamber. Combustion produces high pressure in T and P, which force the burning mixture out, and thereby producing high turbulence to promote good mixing and combustion.

20.13 COOLING SYSTEM OF IC ENGINES

The temperature of the gases inside the cylinder may be as high as 2750°C. If there is no external cooling, the cylinder walls and piston will tend to assume the average temperature of the gases which may be of the order of 1000° to 1500°C. The cooling of the engine is necessary for the following reasons.

(a) The lubricating oil used determines the maximum engine temperature that can be used. This temperature varies from 160°C to 200°C. Above these temperatures the lubricating oil deteriorates very rapidly and may evaporate and burn damaging the piston and cylinder surfaces. Piston seizure due to overheating may also occur.
(b) The strength of the materials used for various engine parts decreases with increase in temperature. Local thermal stresses can develop due to uneven expansion of various parts, often resulting in cracking.
(c) High engine temperatures may result in very hot exhaust valve, giving rise to pre-ignition and detonation or knocking.
(d) Due to high cylinder head temperature, the volumetric efficiency and hence power output of the engine are reduced.

Following are the two methods of cooling the engine.

(i) Air cooling
(ii) Water cooling

Air cooling is used in small engines, where fins are provided to increase heat transfer surface area.

Big diesel engines are always water cooled. The cylinder and its head are enclosed in a water jacket which is connected to a radiator. Water flowing in the jacket carries away the heat from the engine and becomes heated. The hot water then flows into the radiator and gets cooled by rejecting heat to air from the radiator walls. Cooled water is again circulated in the water jacket.

Various methods used for circulating the water around the cylinder are the following.

(a) *Thermosiphon cooling* In this method water flow is caused by density difference (Fig. 20.29). The rate of circulation is however slow and insufficient.
(b) *Forced cooling by pump* In this method a pump, taking power from the engine, forces water to circulate, ensuring engine cooling under all operating conditions. There may be overcooling which may cause low temperature corrosion of metal parts due to the presence of acids.

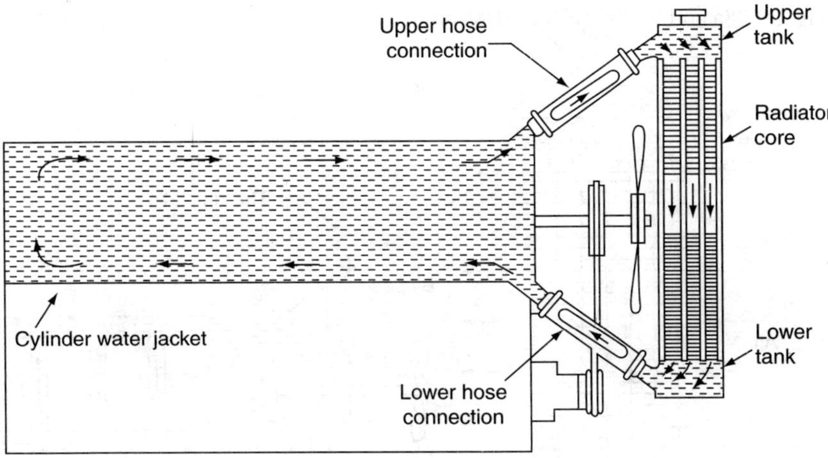

Fig. 20.29 *Thermosiphon cooling*

(c) *Thermostat cooling* This is a method in which a thermostat maintains the desired temperature and protects the engine from getting overcooled (Fig. 20.30).

(d) *Pressurized water cooling* In this method a higher water pressure, 1.5 to 2 bar, is maintained to increase heat transfer in the radiator. A pressure relief valve is provided against any pressure drop or vacuum.

(e) *Evaporative cooling* In this method water is allowed to evaporate absorbing the latent heat of evaporation from the cylinder walls. The cooling circuit is such that the coolant is always liquid and the steam flashes in a separate vessel (Fig. 20.31).

20.14 LUBRICATING SYSTEM

Lubrication is the flow of oil between two surfaces having relative motion. The following are the important functions of a lubricating system.

(a) *Lubrication* To keep moving parts sliding freely past each other, thus reducing engine friction and wear.

(b) *Cooling* To keep the surfaces cool by taking away a part of the heat caused by friction.

(c) *Cleaning* To keep the bearings and piston rings clean of the products of wear as well as of combustion by washing them away.

(d) *Sealing* To form a good seal between the piston rings and cylinder walls.

(e) *Reducing noise* To reduce the noise of the engine by absorbing vibration.

Various lubrication systems used for IC engines may be classified in the following manner.

1. Mist lubrication system
2. Wet sump lubricating system
3. Dry sump lubricating system

1. *Mist Lubrication System* This system is used for two-stroke cycle engines which employ crankcase compression. Crankcase lubrication is thus not suitable in these engines.

2. *Wet Sump Lubricating System* The bottom part of the crankcase, called sump, contains the lubricating oil which is pumped to various parts of the engine. There can be three types of wet sump lubrication system.

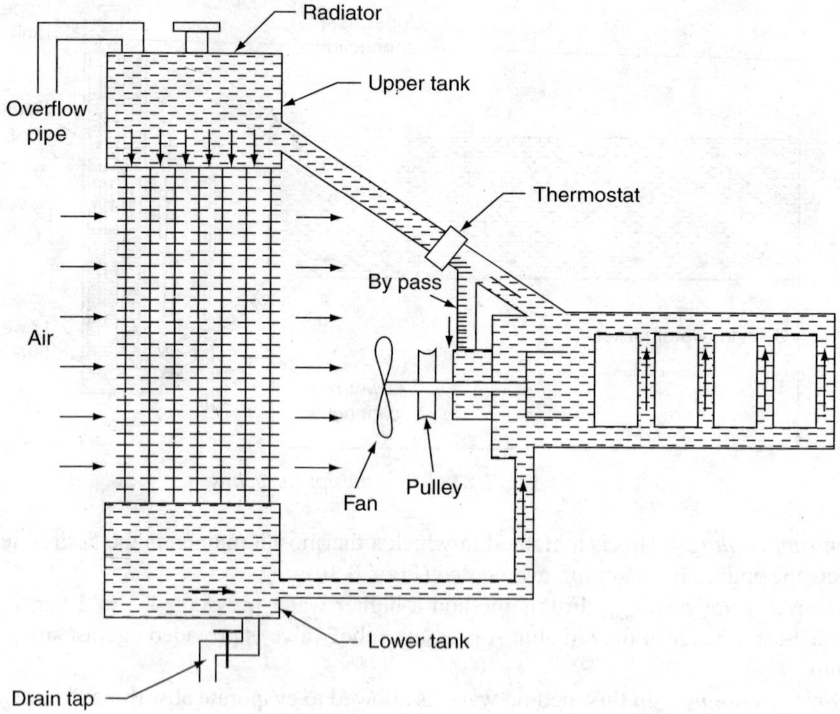

Fig. 20.30 *Thermostat cooling*

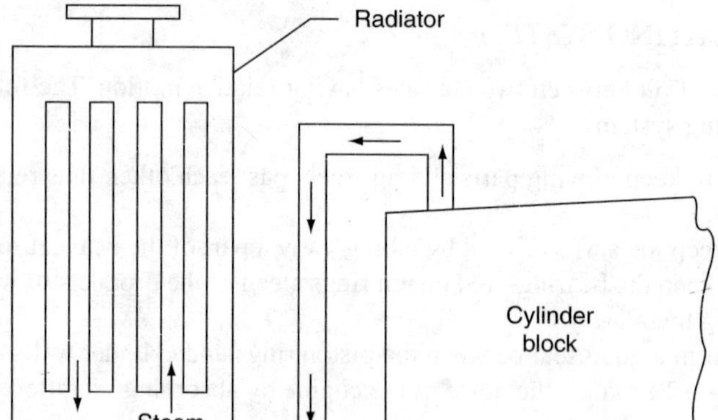

Fig. 20.31 *Evaporative cooling*

Splash System It is used for small four-stroke stationary engines (Fig. 20.32). The oil level in the sump is maintained in a way that when the connecting rod's big end is at its lowest position the drippers at that end strike the oil in the troughs which are supplied with oil from the sump by an oil pump. Due to striking of the drippers, oil splashes over various engine parts like crankpin bearings, piston rings, piston pins, etc. Excess oil drips back to the sump.

Modified Splash System The splash system is not sufficient if the bearing loads are high. For such cases, the modified splash system is used, where the main and camshaft bearings are lubricated by oil under pressure pumped by an oil pump. The other engine parts are lubricated by splash as shown in Fig. 20.32.

Full Pressure System An oil pump is used to lubricate all parts of the engine (Fig. 20.33). Oil is pumped to the main bearings of crankshaft and camshaft at pressures varying between 1.5 and 4 bar. Drilled passages are used to lubricate connecting rod end bearings.

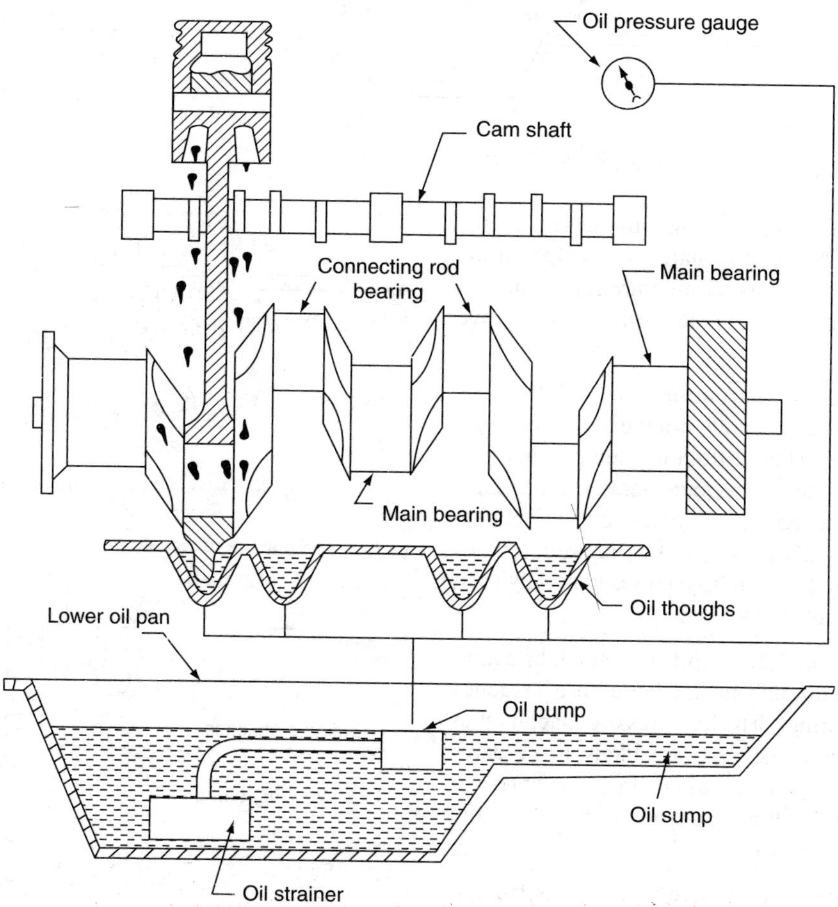

Fig. 20.32 *Splash lubrication system*

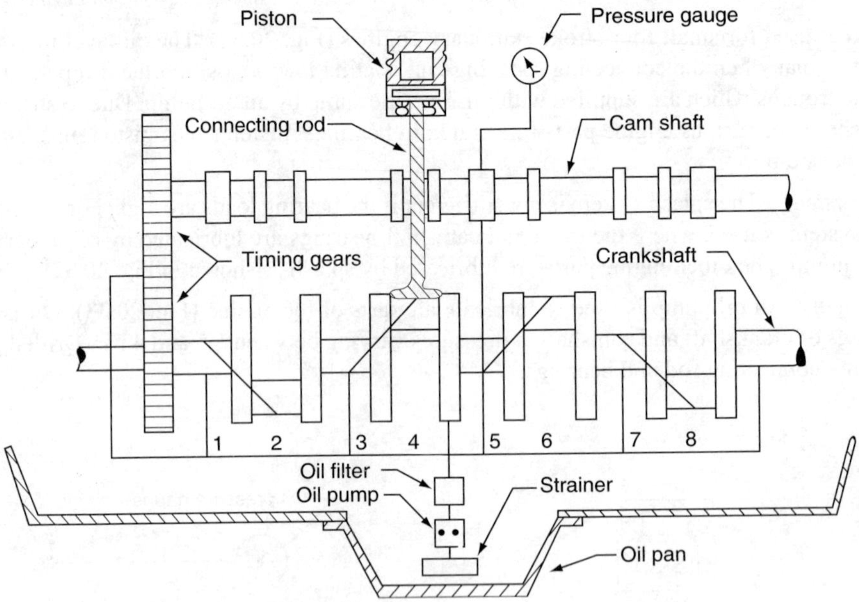

Fig. 20.33 *Full pressure lubrication system*

A gear pump (Fig. 20.34) submerged in the oil and driven by the camshaft draws oil from the sump through a strainer. A pressure relief valve is provided on the delivery side to prevent excessive pressure.

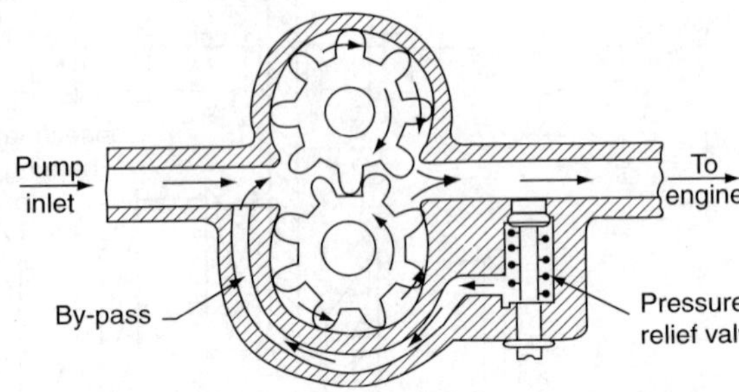

Fig. 20.34 *Gear oil pump with relief valve*

3. Dry Sump Lubrication System Oil from the sump is carried to a separate storage tank outside the engine cylinder. The oil from the sump is pumped through filters to the storage tank (Fig. 20.35). Oil from the storage tank is pumped to the engine cylinder through oil cooler. Oil pressure varies from 3 to 8 bar. Dry sump system is generally used for high capacity engines.

From the pump all the oil used for lubrication usually passes through an oil filter before it reaches the engine bearings. The bearings are machined to a very close tolerance and are likely to be damaged if any foreign material is allowed to enter the lubrication line. Filter arrangement may be of the following two types.

(i) Full flow type, where all the oil is filtered before it is fed to the bearings, as in dry sump lubrication (Fig. 20.35).

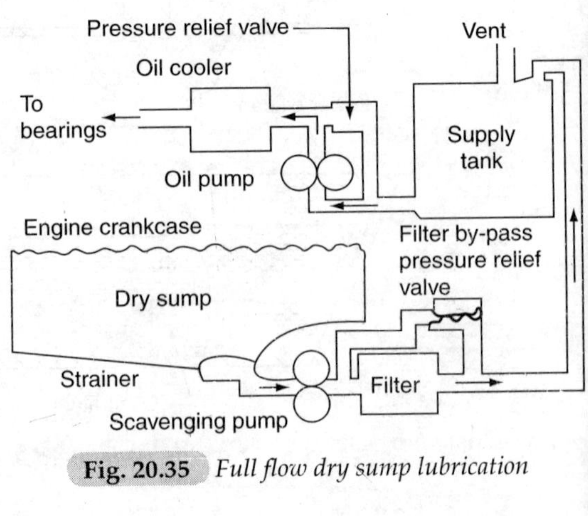

Fig. 20.35 *Full flow dry sump lubrication*

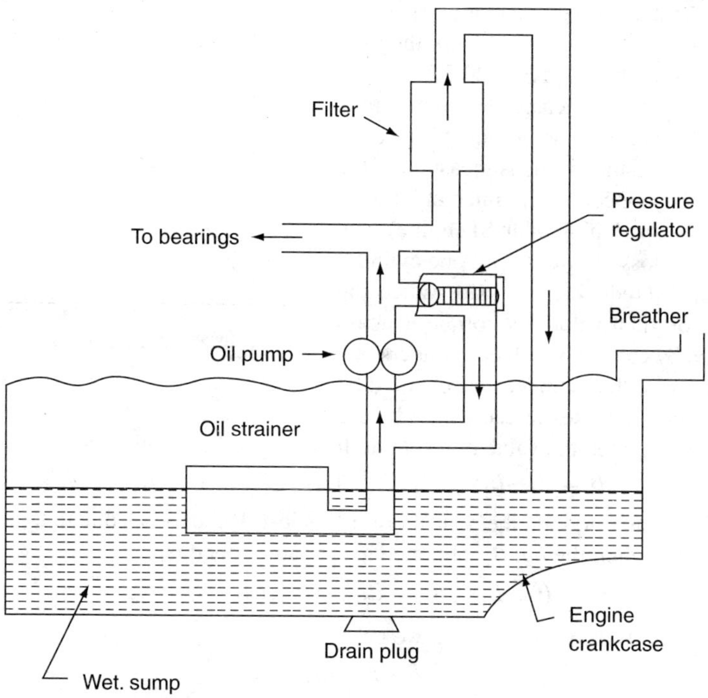

Fig. 20.36 *By-pass type wet sump lubrication*

(ii) By-pass type, where only a small part of the oil is passed through the filter and the rest is directly supplied to the bearings, as in wet sump lubrication system (Fig. 20.36).

20.15 STARTING OF ENGINE

Following are the three common methods of starting an engine.

(i) By an auxiliary engine, which is mounted close to the main engine and drives the latter through a clutch and gears.

(ii) By using an electric motor, in which a storage battery of 12 to 36 volts is used to supply power to an electric motor that derives the engine.

(iii) By compressed air system, in which compressed air at about 17 bar supplied from an air tank is admitted to a few engine cylinders making them work like reciprocating air motors to run the engine shaft. Fuel is admitted to the remaining cylinders and ignited in the normal way causing the engine to start. The compressed air system is commonly used for starting large diesel engines employed for stationary power plant service.

20.16 PERFORMANCE CHARACTERISTICS

It was stated in Art. 20.7 that

$$fp = ip - bp \quad \text{and} \quad \eta_M = bp/ip.$$

The *fp* is very nearly constant at a given engine speed. If the load is decreased giving lower values of *bp,* then the variation in η_M with *bp* is as shown in Fig. 20.37.

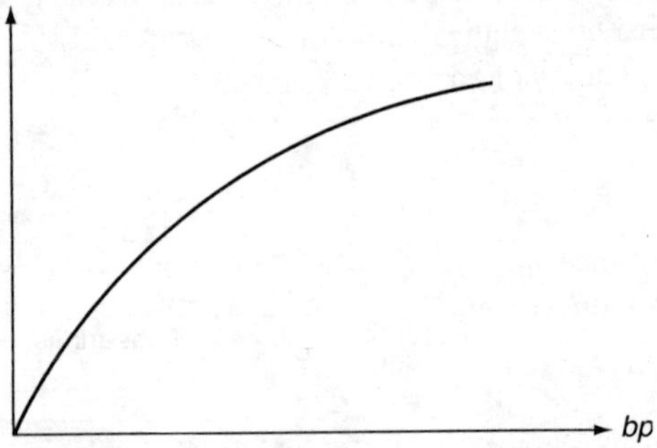

Fig. 20.37 *Variation of mechanical efficiency*

The Morse test can be used to measure the *ip* of multi-cylinder engines. The engine, say having four cylinders, is run at the required speed and the torque is measured. One cylinder is cut out by disconnecting the injector of a CI engine (or by shorting the spark plug of an SI engine). The speed falls because of the loss of power with one cylinder cut out, but is restored by reducing the load. When the speed has reached the original value, the torque is again measured. It is repeated by cutting out other cylinders one by one. If the values of *ip* of the cylinders are denoted by I_1, I_2, I_3 and I_4 and the power losses in each cylinder are denoted by L_1, L_2, L_3 and L_4, then the value of *bp*, *B*, at the test speed with all cylinders firing is given by

$$B = (I_1 - L_1) + (I_2 - L_2) + (I_3 - L_3) + (I_4 - L_4) \tag{20.19}$$

If number 1 cylinder is cut out, then the contribution I_1 is lost. If the losses due to that cylinder remain the same as when it was firing, then the *bp*, B_1, obtained at the same speed is

$$B_1 = (0 - L_1) + (I_2 - L_2) + (I_3 - L_3) + (I_4 - L_4) \tag{20.20}$$

By subtracting Eq. (20.20) from Eq. (20.19), we get

$$B - B_1 = I_1$$

By cutting out each cylinder in turn the values of I_2, I_3 and I_4, can be similarly obtained. Therefore, the total indicated power of the engine is

$$I = I_1 + I_2 + I_3 + I_4$$

At a constant engine speed if the load is decreased in steps and the corresponding *bp* and fuel consumption are measured and plotted, a graph called the "Willan's Line" is obtained as shown in Fig. 20.38.

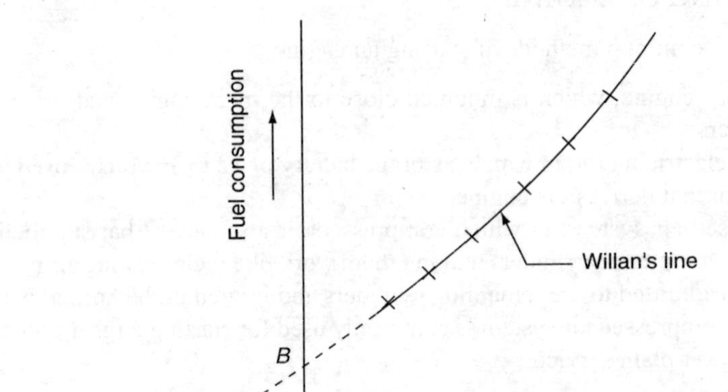

Fig. 20.38 *Fuel consumption against brake power for a CI engine*

The extrapolated reading *OA* is the power loss of the engine at that speed and the fuel consumption *OB* at zero *bp* is equivalent to the power loss *OA*.

The *bp* of an engine can be obtained accurately and conveniently using a dynamometer, which is

$$bp = \eta_M \times ip = \eta_M \times \frac{P_i AL \frac{N}{2} n}{60}$$

Since η_M and p_i are difficult to obtain they may be combined and replaced by a brake mean effective pressure (bmep), p_b, i.e.

$$bp = \frac{p_b AL \frac{N}{2} n}{60} \tag{20.21}$$

where $p_b = \eta_M \times p_i$

The bmep may be considered as that mep acting on the pistons which would give the measured *bp* if the engine were frictionless ($fp = 0$). Again,

$$\frac{p_b AL \frac{N}{2} n}{60} = \frac{2\pi TN}{60}$$

$$p_b = KT$$

where *K* is a constant.

Therefore, bmep is directly proportional to the engine torque and is independent of the engine speed.

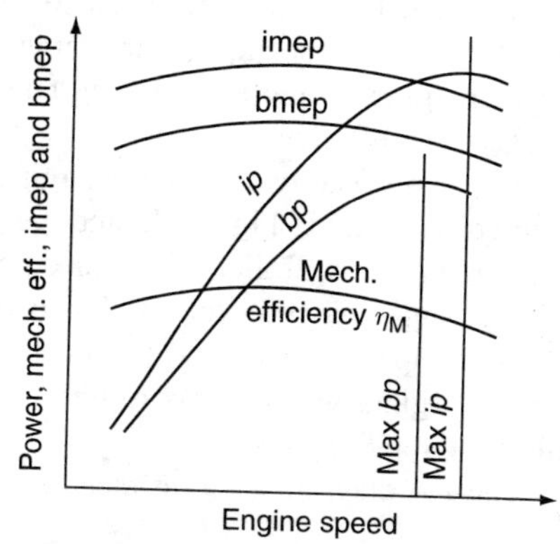

Fig. 20.39 *Engine characteristics of power, imep, bmep, and mechanical efficiency against engine speed*

Energy balance Energy supplied by the fuel

$\dot{m}_f \times CV = bp +$ the heat transferred to cooling water (Q_c)
+ the energy of the exhaust referred to inlet condition (Q_{ex})
+ unaccounted energy losses which include losses by radiation and convection.

Energy to exhaust is $(Q_{ex}) = (\dot{m}_a + \dot{m}_f) h_e - \dot{m}_a h_a$

where $\dot{m}_a$ and $\dot{m}_f$ are the air and fuel mass flow rates, h_e is the enthalpy of the exhaust gas (dry exhaust + steam), reckoned from 0°C, and h_a is the enthalpy of air at inlet, reckoned from 0°C.

For a diesel engine at full load, typical values are: to *bp* 35%, to cooling water 20%, to exhaust 35%, and to radiation, etc. 10%. The heat to the jacket cooling water and exhaust can be utilized in industries which have heating loads as space heating and hot water systems.

Engine characteristics of power, imep, bmep and mechanical efficiency against engine speed are shown in Fig. 20.39. The difference between the *ip* and the *bp* at any speed is the *fp* which increases with speed. Both *ip* and *bp* curves show maximum values, but they occur at different speeds. After the maximum both *ip* and *bp* fall because of the reduction in volumetric efficiency with speed (Fig. 20.40).

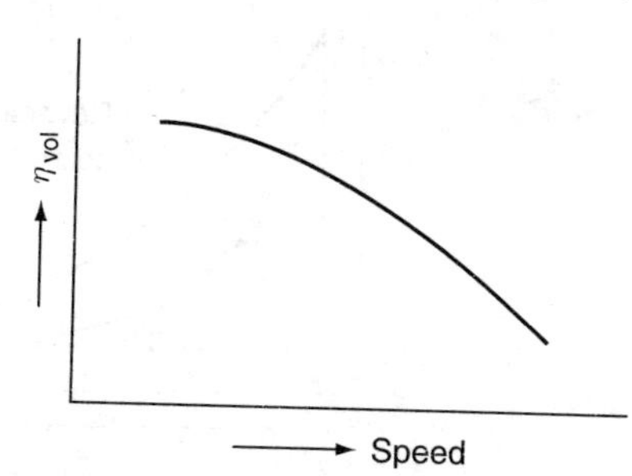

Fig. 20.40 *Variation of volumetric efficiency with speed*

The limiting condition of fuel supply to a CI engine is the smoke limit when black smoke appears in the exhaust. It occurs at air-fuel ratios of about 16/1. Figure 20.41 shows sfc varying with bmep. A minimum sfc and hence a maximum brake thermal efficiency occur at part load (i.e. less than the maximum bmep).

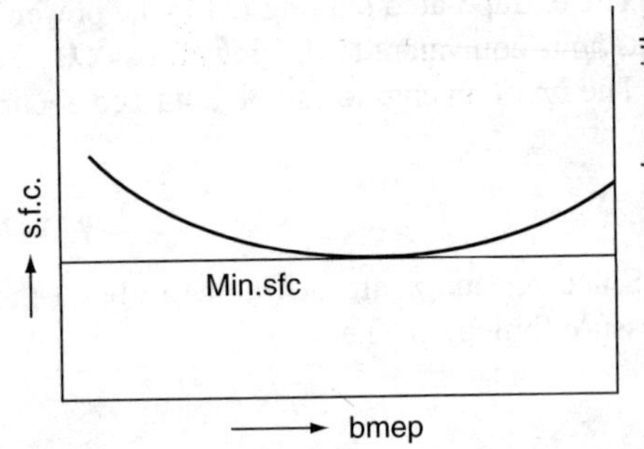

Fig. 20.41 *Specific fuel consumption against bmep for a CI engine*

20.17 SUPERCHARGING

The power output of an engine is affected by the reduction in volumetric efficiency at increased engine speed (Fig. 20.40). The object of supercharging is to increase the volumetric efficiency above that obtained with normal aspiration. Supercharging of air by a compressor or blower increases its density and also its mass flow rat which permits the burning of more fuel and thus augments the output of the engine.

The increase in pressure and temperature of the intake air in diesel engine reduces ignition delay, lowe the rate of pressure rise and makes the combustion process better, quieter and smoother. There is a decreas in the exhaust gas temperature due to high expansion ratio as well as *A/F* ratio. Increased reliability, durabili and better fuel consumption are some other benefits of supercharging. In a spark ignition engine, howeve supercharging promotes knocking tendency, which leads to the use of lower compression ratio and henc leads to less efficiency.

The main features of supercharging are illustrated in *p–V* diagrams for the idealized constant volum four-stroke cycle in Fig. 11.42 and the plant line diagrams in Fig. 20.43. Figure 20.42(a) shows the normal aspirated cycle with line 5–1 representing both the inlet and exhaust strokes at about ambient air pressu p_a. The supercharged cycle is shown in Fig. 20.42(b), where p_i, T_i refer to the engine inlet condition. Tw ways of supercharging, viz. (a) mechanical supercharging with a blower driven by the engine itself and (turbocharging, where a gas turbine driven by the engine exhaust drives the compressor (centrifugal), a illustrated in Fig. 20.43.

The *T–s* diagram for a turbocharger is shown in Fig. 20.44. The compressor power input is,

$$\dot{W}_c = \dot{m}_a C p_a T_a \left[(p_i / p_a)^{(\gamma_a - 1)/\gamma_a} - 1 \right] \eta_c$$

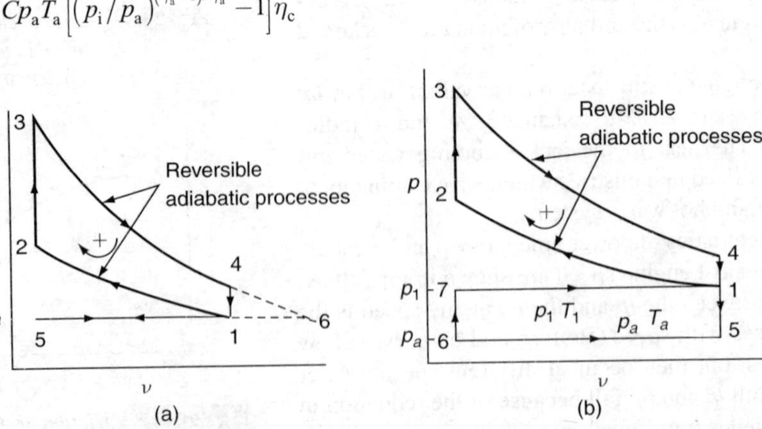

Fig. 20.42 *Pressure-volume diagram for a four-stroke CI engine (a) Without supercharging and (b) With supercharging*

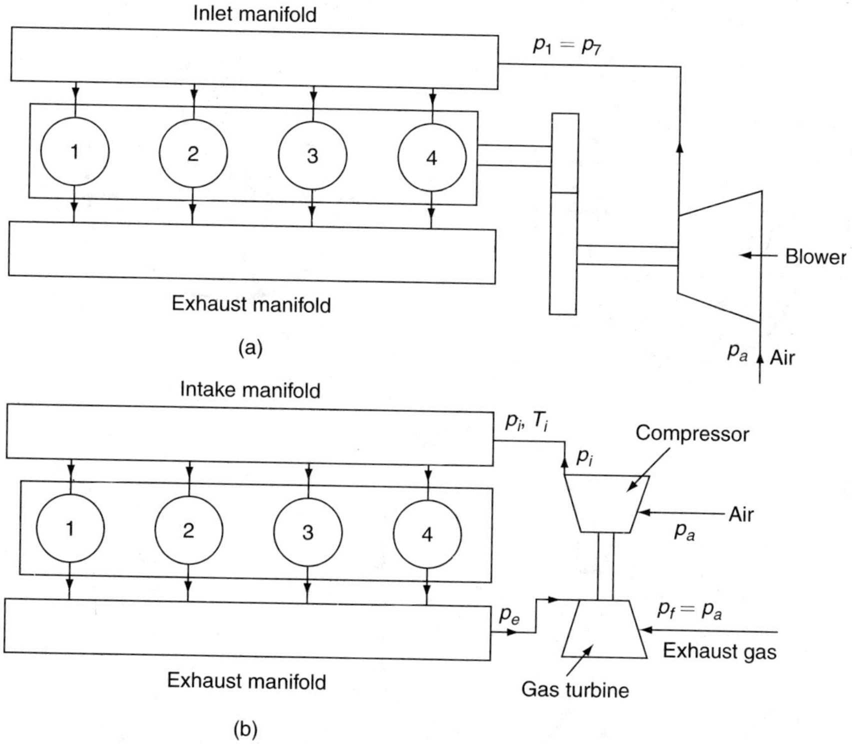

Fig. 20.43 *Diagrams of a four-stroke four-cylinder CI engine with (a) Mechanical supercharging and (b) turbocharging*

The turbine power output is,

$$\dot{W}_{\mathrm{T}} = \dot{m}_{\mathrm{g}} c p_{\mathrm{e}} T_{\mathrm{e}} \left[1 - (p_{\mathrm{a}}/p_{\mathrm{e}})^{(\gamma_{\mathrm{a}}-1)\gamma_{\mathrm{a}}}\right] \times \eta_{\mathrm{T}}$$

where η_c and η_T ar e the isentropic efficiencies of the compressor and turbine respectively. Now,

$$\dot{m}_{\mathrm{e}} = \text{rate of flow of exhaust gas from the engine}$$

$$= \dot{m}_{\mathrm{a}} + \dot{m}_{\mathrm{f}}$$

$$\dot{m}_{\mathrm{e}}/\dot{m}_{\mathrm{a}} = 1 + (\dot{m}_{\mathrm{f}}/\dot{m}_{\mathrm{a}}) = 1 + F/A$$

where F/A is the fuel air ratio.

Also, $W_{\mathrm{c}} = W_{\mathrm{T}} \times \eta_{\mathrm{M}}$

where, η_M is the mechanical efficiency.

$$\left[(p_{\mathrm{i}}/p_{\mathrm{a}})^{(\gamma_{\mathrm{a}}-1)/\gamma_{\mathrm{a}}} - 1\right] = \left[1 - (p_{\mathrm{a}}/p_{\mathrm{e}})^{(\gamma_{\mathrm{e}}-1)/\gamma_{\mathrm{e}}}\right]\left[c_{\mathrm{p_e}}/c_{\mathrm{p_a}}\right]\left[T_{\mathrm{e}}/T_{\mathrm{a}}\right]$$

$$[1 + F/A] \times \eta_0$$

where, $\eta_0 = \eta_{\mathrm{M}} \times \eta_{\mathrm{T}} \times \eta_{\mathrm{c}}$ = Overall efficiency of the supercharger.

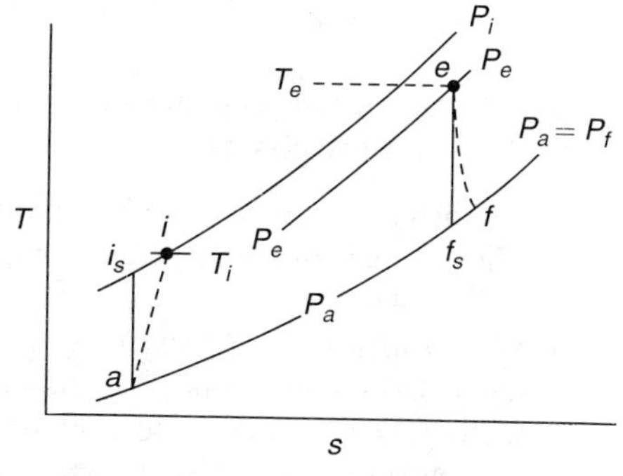

Fig. 20.44 *T-s diagram for a turbocharger of Fig.* 20.43

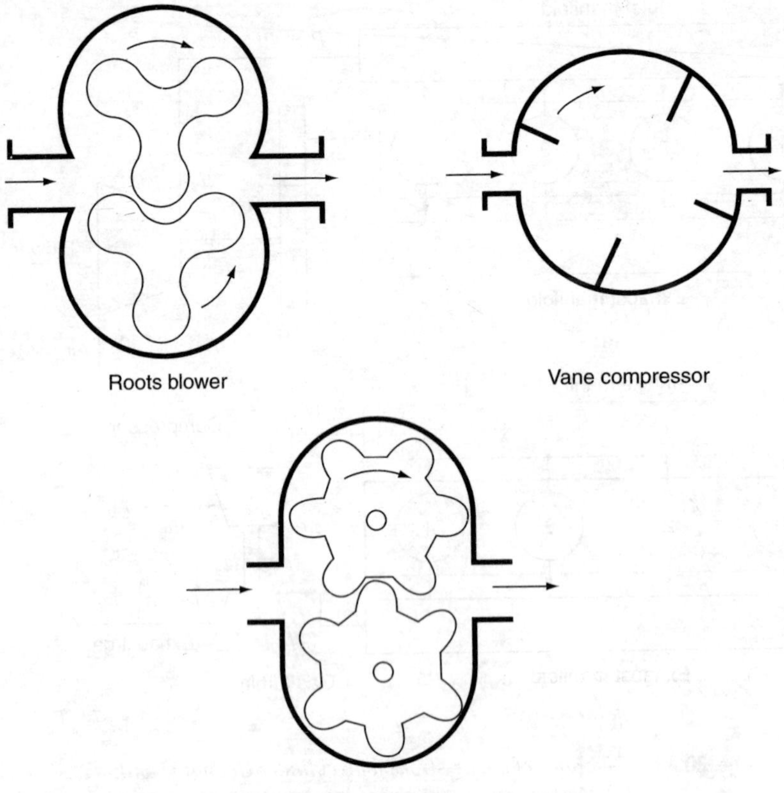

Fig. 20.45 *Types of positive displacement compressor*

Compressors for turbocharging can be divided into the following two classes.

(*i*) *Positive displacement* type like roots blower, vane compressor and screw compressor (Fig 20.45),

(*ii*) *Non-positive displacement* type like axial and radial compressors (Fig. 20.46). Turbochargers are superior to superchargers, since the former use the exhaust gas energy during blowdown and the latter consume a part of the engine output.

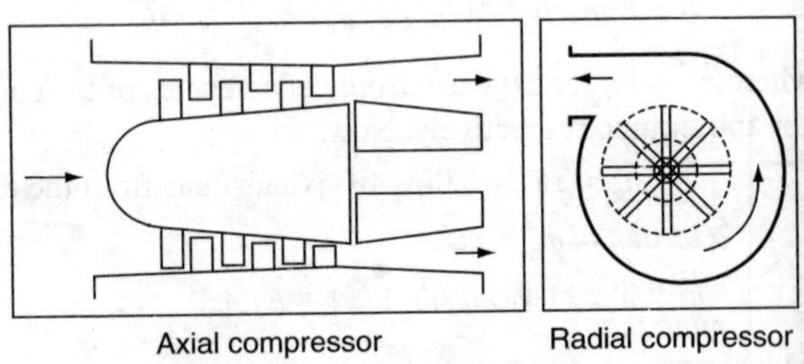

Fig. 20.46 *Types of dynamic or non-positive displacement compressor*

Figure 20.47 shows the continuous running performance characteristics of a diesel engine in three modes, i.e. (a) normally aspirated (b) turbocharged and (c) turbocharged with intercooling for a specific engine as given by Eastop and McConkey. It is seen that intercooling of air has a very good influence on the engine performance. Typical performances for a CI engine at full load and part load are shown in Fig. 20.48.

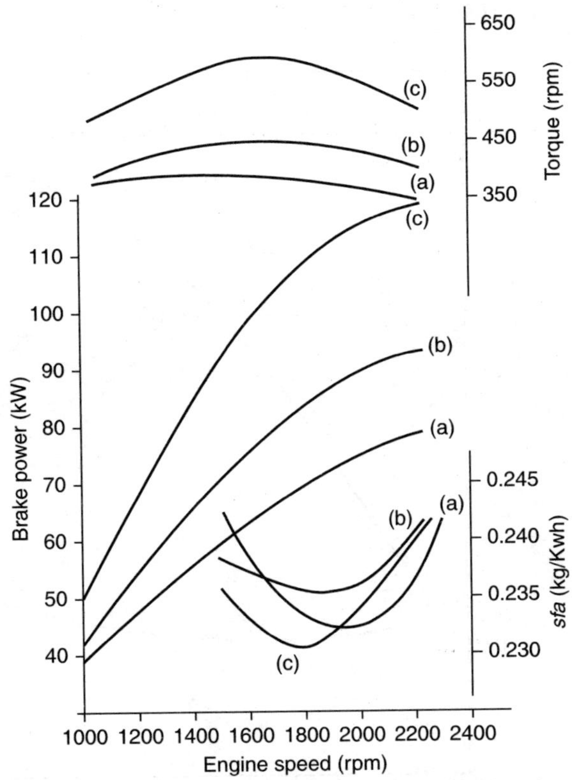

Fig. 20.47 *Continuous running performance characteristics of a CI engine in three modes: (a) Normally aspirated, (b) Turbocharged and (c) Turbocharged with intercooling*

Solved Examples

Example 20.1

A four cylinder four-stroke gasoline engine has a 65-mm diameter and 95-mm stroke. On test, it developed a torque of 64 Nm when running at 3000 rpm. If the clearance volume in each cylinder is 63 cm³, the brake efficiency ratio based on air standard efficiency is 0.5 and calorific value of gasoline is 42 MJ/kg, determine the fuel consumption in kg/h and the bmep.

Solution Swept volume per cylinder, V_s

$$= \frac{\pi}{4} d^2 L = \frac{\pi}{4}(6.5)^2 \times 9.5 = 315.24 \text{ cm}^3$$

Compression ratio, $$r_k = \frac{V_s + V_c}{V_c} = \frac{315.24 + 63}{63} = 6$$

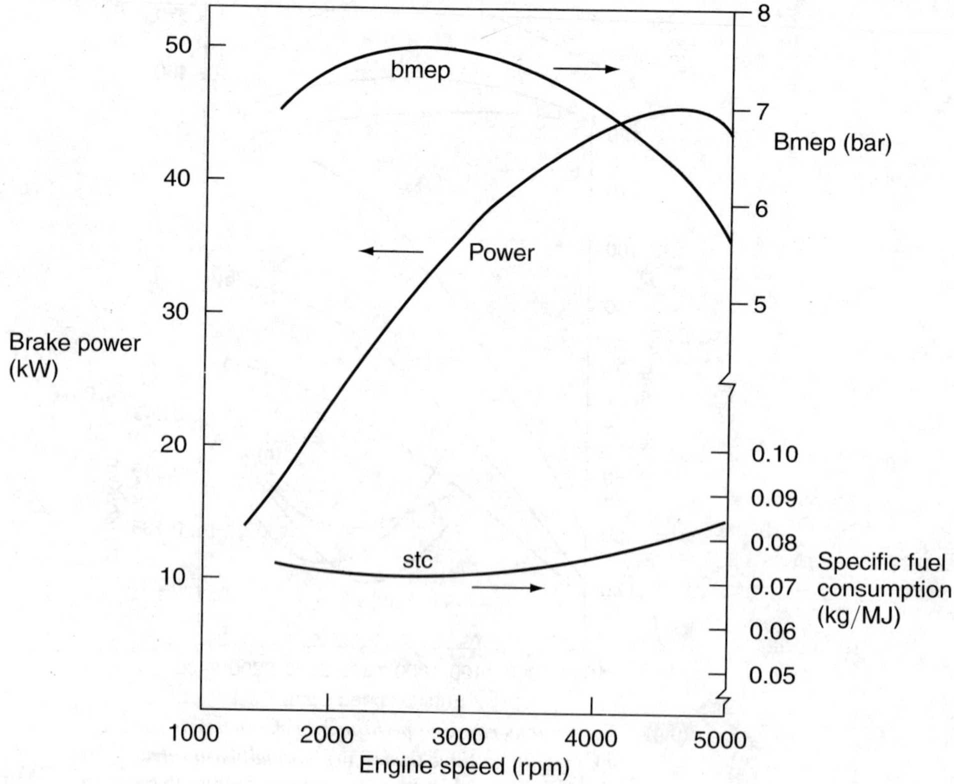

(a)

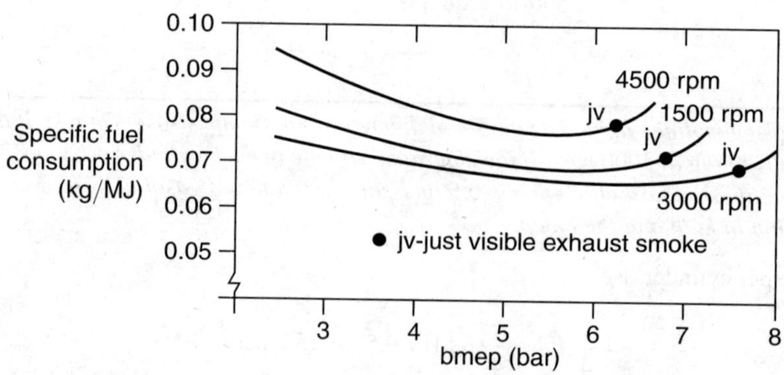

(b)

Fig. 20.48 *Typical performance for a CI engine at (a) Full load and (b) Part load at different speeds*

$$\text{Air standard efficiency, } \eta_{as} = 1 - \frac{1}{r_k^{\gamma-1}} = 1 - \frac{1}{6^{0.4}} = 0.5125$$

$$\text{Relative efficiency} = \frac{\text{Brake thermal efficiency}}{\text{Air standard efficiency}} = 0.5$$

$$\text{Brake thermal efficiency} = 0.5 \times 0.5125 = 0.2562$$

$$\text{BP} = \frac{2\pi TN}{60} = \frac{2\pi \times 3000 \times 64}{60 \times 1000} = 20.1 \text{ kW}$$

$$\eta_{br.\,th} = \frac{\text{BP}}{m_f \times cv} = \frac{20.1}{\dot{m}_f \times 42000} = 0.2562$$

$$\therefore \quad m_f = \frac{20.1 \times 3600}{0.2562 \times 42000} = 6.726 \text{ kg/h}$$ *Ans.*

$$\text{BP} = \frac{\text{BMEP}\frac{\pi}{4} D^2 L N}{60 \times 2} = \frac{20.1}{4}$$

$$\text{BMEP} = \frac{5.025 \times 2 \times 60 \times 1000 \times 4}{0.095 \times \pi \times (0.065)^2 \times 3000}$$

$$= 627887 \cdot 4 \frac{\text{N}}{\text{m}^2} = 627 \cdot 8874 \text{ kN/m}^2$$ *Ans.*

Example 20.2

A four-cylinder two-stroke cycle petrol engine develops 30 kW at 2500 rpm. The mean effective pressure of each is 800 kN/m² and mechanical efficiency is 80%. Calculate the diameter and stroke of each cylinder if the stroke-to-bore ratio is 1.5. Also calculate the brake specific fuel consumption of the engine, if brake thermal efficiency is 28%. The heating value of petrol is 44 MJ/kg.

Solution No. of cylinders = 4, B.P = 30 kW

$$N = 2500 \text{ RPM}, \; P_m = 800 \text{ kN/m}^2, \; \eta_{mech} = 80$$

$$L/d = 1 \cdot 5, \; \eta_{br \cdot th.} = 28\%, \; \text{C.V} = 44 \text{ MJ/kg.}$$

$$\text{I} \cdot \text{P} = \frac{\text{B.P.}}{\eta_{mech}} = \frac{30}{0 \cdot 8} = 37.5 \text{ kW}$$

$$\text{I} \cdot \text{P} = \frac{p_m \text{ LAN} n}{60}$$

$$37 \cdot 5 = \frac{800 \times 10^3 \times 1 \cdot 5d \times \frac{\pi}{4} d^2 \times 2500 \times 4}{60 \times 1000}$$

$$d^3 = 2 \cdot 3873 \times 10^{-4} \text{ m}^3$$

$$\therefore \quad d = 0.062\,\text{m} = 6.2\,\text{cm, L} = 1.5\times6.2 = 9.3\,\text{cm}$$

$$\eta_{\text{br.th}} = \frac{\text{BP}}{\dot{m}_f \times cv}$$

$$0.28 = \frac{\text{B.P.}}{\dot{m}_f \times 44{,}000}$$

$$\text{bsfc} = \frac{\dot{m}_f}{\text{BP}} = \frac{1\times3600}{0.28\times44{,}000} = 0.2922\,\text{kg/kWh}$$ *Ans.*

Example 20.3

The following particulars refer to a 2-stroke diesel engine: bore = 10cm, stroke = 15cm, piston speed = 300 m/min, torque developed = 58 Nm, mechanical efficiency = 80%, indicated thermal efficiency = 40%, calorific value of fuel = 44 MJ/kg. Determine (a) indicated power, (b) indicated mean effective pressure, and (c) fuel consumption per kWh on brake power output.

Solution Given: $d = 10$ cm $= 0.1$m, $L = 15$ cm $= 0.15$ m.

Piston Speed = $2LN$=300 m/min, $T = 58$ Nm.

Type of engine = 2-Stroke diesel engine.

$$\text{C.V} = 44\ \text{MJ/kg},\ \eta_{\text{mech}} = 0\cdot80,\ \eta_{\text{i.th.}} = 0.40$$

$$\text{Now, } 2\text{LN} = 300,\ \therefore N = \frac{300}{2\times0.15} = 1000\ \text{rpm.}$$

$$\eta_{\text{mech}} = \text{BP/IP. Also, B.P.} = \frac{2\pi\text{TN}}{60} = \frac{2\pi\times58\times1000}{60\times1000} = 6.073\,\text{kW}$$

$$\text{I.P.} = \frac{6.073}{0\cdot8} = 7\cdot592\,\text{kW}$$ *Ans.*(a)

$$\text{I.P.} = \frac{\text{p}_\text{m}\,\text{L}\frac{\pi}{4}\text{d}^2\text{Nn}}{60}$$

$$p_\text{m} = i_{\text{mep}} = \frac{60\times7.592\times4\times1000}{0.15\times\frac{\pi}{4}(0.1)^2\times1000\times1}$$

$$= 386666.67\ \text{N/m}^2 = 386.67\ \text{kN/m}^2$$ *Ans.*(a)

$$\eta_{\text{i.th}} = \frac{\text{IP}}{\dot{\text{m}}_\text{f}\times\text{C.V.}} = \frac{\text{B.P.}}{\dot{\text{m}}_\text{f}\times\text{C.V.}\times\eta_{\text{mech}}} = \frac{6.053\times3600}{\dot{\text{m}}_\text{f}\times44000\times0.8}$$

$$\therefore \quad \dot{\text{m}}_\text{f} = \frac{7.5921\times3600}{0\cdot4\times44000} = 1.5529\ \text{kg/h}.$$

$$\therefore \quad \text{bsfc} = 1.5529/6.073 = 0.2557\,\text{kg/kWh}$$ *Ans.*

Example 20.4

Following are the observations made for a 20 minute trial of a 2-stroke diesel engine:

Net brake load = 680*N, mep* = 3 *bar, N* = 360 *rpm, Fuel consumption* = 1.56 *kg, Cooling water* = 160 *kg, Water inlet temperature* = 57°*C, Air used/kg fuel* = 30 *kg, Room temperature* = 27°*C, Exhaust gas temperature* = 310°*C, cylinder dimensions* = 210-*mm bore* × 290-*mm stroke, Brake diameter* = 1*m, Calorific value of fuel* = 44 *MJ/kg, Steam formed per kg fuel in the exhaust* = 1.3 *kg, specific heat of steam in the exhaust* = 2.093 *kJ/kg k. Specific heat of dry exhaust gases* =1.01 *kJ/kg k.*

Calculate the indicated power and the brake power, and make an energy balance of the engine

Solution For a two-stroke engine,

$$i_p = \frac{p_m \, \text{LANn}}{60} = \frac{3\times100\times0\cdot29\times\frac{\pi}{4}(0\cdot21)^2\times360\times1}{60}$$

$$= 18\cdot08\,\text{kW}$$ *Ans.*

$$b_p = \frac{2\pi TN}{60} = \frac{2\pi\times\left(680\times\frac{1}{2}\right)\times360}{60} = 12.818 \text{ kW}$$

$$\eta_M = bp/ip = 0.7089 \text{ or } 70.89\%$$

Heat supplied during the trial = 1.56×44,000 = 68,640 kJ.

Energy equivalent of *ip* in trial period = 18.08 kJ/s × (20 × 60) s = 21,696 kJ

Energy carried away by cooling water = 160 × 4.187 × (57 – 32) = 16,748 kJ

Total mass of exhaust gas = 1.56 × 30 + 1.56 = 48.36 kg

Mass of steam formed = 1.3 × 1.56 = 2.028 kg

Mass of dry exhaust gas = 48.36 – 2.028 = 46.332 kg

Energy carried away by dry exhaust gases

= 46.332 × 1.01 × (310 – 27) = 13,243 kJ

Energy carried away by steam

= 2.028 [4.187 (100 – 27) + 2257.9 + 2.093(310 – 100)] = 6090 kJ

Total energy carried away by exhaust gases = 19.333 kJ

Energy Balance Sheet		
Energy released by combustion of fuel	68,640 kJ	
1. Energy equivalent of ip	21,696 kJ	31.61%
2. Energy carried away by cooling water	16,748 kJ	24.40%
3. Energy carried away by exhaust gases	19,333 kJ	28.17%
4. Unaccounted energy loss (by difference)	10,863 kJ	15.82%
	68,640 kJ	100.00%

Example 20.5

The following particulars refer to a 2-Stroke oil engine: bore = 20 cm, stroke = 30 cm, speed = 350 rpm, i.m.e.p. = 275 kN/m², net brake load = 610 N, diameter of brake drum = 1 m, oil consumption = 4.25 kg/h, calorific value of fuel = 44 MJ/kg. Determine (a) IP, (b) B.P., (c) η_{mech} (d) indicated thermal efficiency, and (e) brake thermal efficiency.

Solution

$$\text{IP} = \frac{\text{imep LAN}}{60 \times 1000}\,\text{kW} = \frac{275 \times 10^3 \frac{\text{KN}}{\text{m}^2} \times 0 \cdot 3\,\text{m} \times \frac{\pi}{4}(0 \cdot 2^2\ \text{m}^2 \times 350\ \text{rpn}}{60 \times 1000}$$

$$= 15 \cdot 1189\ \text{kW} \qquad \textit{Ans.}\text{(a)}$$

$$\text{BP} = \frac{2\pi \text{N}(W - S)R}{60 \times 1000}\,\text{kW} = \frac{2\pi \times 350 \times 610 \times 0 \cdot 5}{60 \times 1000}$$

$$= 11.1788\ \text{kW} \qquad \textit{Ans.}\text{(b)}$$

$$\eta_{\text{mech}} = \frac{\text{BP}}{\text{IP}} = 0.7393 \text{ or } 73.93\% \qquad \textit{Ans.}\text{(c)}$$

$$\eta_{\text{i.th.}} = \frac{IP \times 3600}{\dot{m}_f \times CV} = \frac{15 \cdot 1189 \times 3600}{4.25 \times 44000} = 0.291 \text{ or } 29.1\% \qquad \textit{Ans.}\text{(d)}$$

$$\eta_{\text{br.th.}} = 0.291 \times 0.7393 = 0.2151 \text{ or } 21.51\% \qquad \textit{Ans.}\text{(e)}$$

Example 20.6

A 4-stroke 6-cylinder gas engine with a stroke volume of 1.75 litres develops 26.25 kW at 506 rpm. The m.e.p. is 600 kN/m². Find the average number of times each cylinder misfired in one minute.

Solution Given : $V_s = 1.75 \times 10^{-3}\,\text{m}^3$, IP = 26.25 kW, N = 506 rpm, imep = 600 kN/m², $n = 6$

$$\text{IP/cylinder} = \frac{\text{imep} \times \text{LA}\frac{N}{2}}{60}$$

$$\frac{26.25}{6} = \frac{600 \times 10^3 \times 1.75 \times 10^{-3} \times n}{60{,}000}$$

where n = number of explosions

$$n = \frac{26.25}{6 \times 1 \cdot 75 \times 10^{-3}} = 250$$

As it is a 4-stroke, the number of explosions per min. should be 506/2 = 253.

∴ no. of misfiring per min per cylinder = 253 – 250 = 3 *Ans.*

Example 20.7

An engine is used on a job requiring 110 kW brake power. The mechanical efficiency of the engine is 80% and the engine used 50 kg/h of fuel under the conditions of operation. A design improvement is made which reduces the engine friction by 5 kW. Assuming that the indicated thermal efficiency remains the same, how many kg of fuel per hour will be saved?

Solution BP = 110 kW, $\eta_{mech} = 80\%$

$\therefore$ IP = 110/0.8 = 137.5 kW

Frictional power = IP – BP = 137.5 – 110 = 27.5 kW

New frictional power = 27.5 – 5 = 22.5 kW

New IP = 110 + 22.5 = 132.5 kW

As the indicated thermal efficiency remains the same in both cases,

$$\frac{\text{New IP}}{\dot{m}_{f_1} \times \text{C.V.}} = \frac{\text{IP}}{\dot{m}_f \times \text{CV}}, \quad \text{where } m_{f_1} = \text{new fuel consumption per hour.}$$

$$\therefore \quad \frac{132.5}{\dot{m}_{f_1}} = \frac{137.5}{50}$$

$$\therefore \quad \dot{m}_{f_1} = \frac{132.5 \times 50}{137.5} = 48.18 \text{ kg/h.}$$

$\therefore$ Saving in fuel = 50 – 48.18 = 1.82 kg/h *Ans.*

Example 20.8

During a trail of a four-cylinder four-stroke petrol engine coupled to a hydraulic dynamometer at constant speed, the following readings were obtained:

B.P. with all cylinders operating = 14.7 *kW*
B.P. with cylinder no. 1 *cut out* = 10.14 *kW*
B.P. with cylinder no. 2 *cut out* =10.3 *kW*
B.P. with cylinder no. 3 *cut out* = 10.36 *kW*
B.P. with cylinder no. 4 *cut out* = 10.21 *kW*
Fuel consumption = 5.5 *kg/h*
Calorific value of fuel = 42 *MJ/kg*
Diameter of cylinder = 8 *cm*
Stroke of piston = 10 *cm*
Clearance volume = 0.1 *litre*

Calculate (*a*) *mechanical efficiency and* (*b*) *relative efficiency on indicated power basis.*

Solution $IP_1 = 14.7 - 10.14 = 4.56$

$IP_2 = 14.7 - 10.3 = 4.4$

$IP_3 = 14.7 - 10.36 = 4.34$

$IP_4 = 14.7 - 10.21 = 4.49$

Total IP = 4.56 + 4.4 + 4.34 + 4.49 = 17. 79 kW

$$\text{Mechanical efficiency} = \frac{\text{BP}}{\text{IP}} = \frac{14.7}{17.79} = 0.8263 \quad \text{or} \quad 82.63\%$$ *Ans.* (*a*)

$$V_s = \frac{\pi}{4} d^2 l = \frac{\pi}{4}(0.08)^2 \times (0.1) = 5.0265 \times 10^{-4} \text{ m}^3$$

$V_c = 0.1 \text{litre} = 0.1 \times 10^{-3} = 10^{-4} \text{m}^2$

Compression ratio, $r_k = \frac{V_s + V_c}{V_c} = \frac{5.0265 \times 10^{-4} + 10^{-4}}{10^{-4}} = 6.0265$

Air standard efficiency, $\eta_{\text{A.S.E.}} = 1 - \frac{1}{r_k^{\gamma-1}} = 1 - \frac{1}{(6.0265)^{0.4}} = 0.5215$

Indicated thermal efficiency $= \frac{\text{IP} \times 3600}{\dot{m}_f \times \text{C.V.}} = \frac{17.79 \times 3600}{5.5 \times 42000} = 0.2772$

$\therefore$ Relative efficiency $= \frac{0.2772}{0.5215} = 0.5316$ or 53.16% *Ans.* (*b*)

Example 20.9

A four-cylinder four-stroke petrol engine of 82-*mm bore*, 130-*mm stroke develops* 28.35 *kW brake power while running at* 1500 *rpm and using a* 20% *rich mixture. If the volume of the air into the cylinder when measured at* 15.5°C *and* 760 *mm of mercury is* 70% *of the swept volume, the theoretical air fuel ratio is* 14.8, *the heating value of petrol is* 44 *MJ/kg and the mechanical efficiency of the engine is* 90%. *Find* (*a*) *indicated thermal efficiency, and* (*b*) *brake mean effective pressure.*

Solution Given: $n = 4$, 4-stroke, $d = 0.082$ mm, L = 0.13 m, BP = 28.35 kW, $N = 1500$ rpm, Mixture = 20% rich.

Volume of air at 15.5°C and 760 mm Hg = 70% of swept volume

Theoretical A/F ratio = 14.8, CV = 44000 kJ/kg

$$\eta_{\text{mech}} = 0.9, \quad R = 0.287 \text{ kJ/kgK.}$$

$$\text{IP} = \frac{\text{BP}}{\eta_{\text{mech}}} = \frac{28.35}{0.9} = 31.5\text{kW}$$

760 mm Hg = 1.01325 kN/m^2

Volume of air sucked into the cylinder in m^3/min = 70% of swept volume × 4

$$= 0.7 \times \frac{\pi}{4}(0.082)^2 \times 0.13 \times \frac{1500}{2} \times 4 = 1.4417 \text{ m}^3/\text{min}$$

Now, $P\dot{V} = \dot{m}RT$

$$\dot{m} = \frac{1.01325 \times 1.4417}{0.287(15.5 + 273)} = 1.764 \text{ kg/min}$$

Theoretically, 14.8 kg of air are required for 1 kg fuel, but the mixture is 20% rich,

Actual fuel consumption $= \frac{1.764}{14.8} \times 1.2 = 0.143 \text{ kg/min}$

Indicated thermal efficiency $= \frac{31.5 \times 3600}{0.143 \times 44000 \times 60} = 0.30$ or 30% *Ans.*

$$\text{Brake power} = bmep \times \frac{\frac{\pi}{4} d^2 l \times \frac{N}{2} \times n}{60}$$

$$28.35 = \frac{bmep \times \frac{\pi}{4}(0.082)^2 \times 0.13 \times \frac{1500}{2} \times 4}{60} = bmep \times 0.0343$$

$$\therefore \quad bmep = 825.53\,\text{kN/m}^2$$ *Ans.* (b)

Example 20.10

In a simple carburetor, the throat diameter is 25 *mm and the main jet diameter is* 1.2 *mm. The coefficients of discharge for the venturi and the fuel jet are* 0.85 *and* 0.65 *respectively. The gasoline surface is* 6 *mm below the throat. Estimate the minimum velocity of air required to start the flow of fuel. The ambient conditions are pressure* 1 *bar,* $T_1 = 300$ *K and* $\rho_f = 760$ *kg/m*^{3}.

Solution The lip is provided in the carburetor to avoid the fuel wastage. The fuel will start flowing when the depression at the throat overcomes the effect of the lip. Thus,

$$\Delta p = h\rho_f g = 760 \times 6 \times 10^{-3} \times 9.81$$

$$= 44.75\ \text{N/m}^2$$

$$p_2 = p_1 - \Delta p = 1.0 - 44.75 \times 10^{-5}\ \text{bar}$$

Since the depression is very small, we can consider air as an incompressible fluid, and we have

$$p_1 = p_2 + \rho \frac{v^2}{2}$$

$$\therefore \quad \rho \frac{v^2}{2} = p_1 - p_2 = \Delta p = 44.75 \times 10^{-5}\ \text{bar} = 44.75\ \text{N/m}^2$$

$$V = \left(\frac{2 \times 44.75}{1.16}\right)^{\frac{1}{2}}, \text{ since } \rho_{\text{air}} = \frac{p}{RT} = \frac{10^5}{287 \times 300} = 1.168\ \text{kg/m}^3$$

$$= 8.75\ \text{m/s}$$ *Ans.*

Example 20.11

A six-cylinder four-stroke S.I. engine develops 40 *kW. During a Morse test conducted on the engine at* 2000 *rpm, the power output with each cylinder made inoperative turn by turn was* 32.2, 32.0, 32.5, 32.4, 32.1 *and* 32.3 *kW respectively. Estimate the mechanical efficiency, bmep, air standard efficiency when the bore* = 100 *mm, stroke* = 125 *mm, clearance volume* = $1.23 \times 10^{-4} m^4$, *the brake thermal efficiency when fuel consumption is* 9 *kg/h and the heating value is* 40 *MJ/kg, and the relative efficiency.*

Solution Morse test is conducted to determine the IP of multi-cylinder engine. The engine is made to run at constant speed and after steady-state conditions are reached, one cylinder, at one time, is made inoperative (by short circuiting the spark plug or by disconnecting the injector). Under these conditions, all other cylinders 'motor' this cut-out cylinder. The output is measured by keeping the speed constant at its original value. The difference in the output gives the indicated power of the cut-out cylinder. Thus, the IP produced by the engine is

$$IP = (40 - 32.2) + (40 - 32) + (40 - 32.5) + (40 - 32.4) + (40 - 32.1) + (40 - 32.3) = 46.5 \text{ kW}$$

$$\therefore \quad \eta_{mech} = 40/46.5 = 0.86 \quad \text{or} \quad 86\%$$ *Ans.*

$$bmep = \frac{BP \times 2 \times 60}{LAN} = \frac{40 \times 2 \times 60}{0.125 \times \frac{\pi}{4}(0.1)^2 \times 2000} = 305.54 \text{ kPa}$$

$$= 3.055 \text{ bar}$$ *Ans.*

Stroke volume, $V_s = \frac{\pi}{4} \times (0.1)^2 \times 0.125 = 9.819 \times 10^{-4} \text{ m}^2$

Compression ratio, $r_k = \frac{V_c + V_s}{V_c} = 1 + \frac{9.819 \times 10^{-4}}{1.23 \times 10^{-4}} = 8.99$

$\therefore$ air-standard efficience (ASE) $= 1 - \frac{1}{r_k^{\gamma - 1}} = 0.5835$ or 58.35% *Ans.*

Brake thermal efficiency $= \frac{40 \times 3600}{9 \times 40,000} = 0.4$ or 40% *Ans.*

Relative efficiency $= \frac{0.4}{0.5835} = 0.686$ or 68.6% *Ans.*

Example 20.12

Air is sucked into a diesel engine fitted with a turbocharger at a pressure of 0.95 bar and a temperature of 25°C. The delivery pressure is 2 bar. The air-fuel ratio is 18:1. The temperature of the gases leaving the engine is 600°C. and the pressure in 1.8 bar. The gases come out of the turbine at 1.04 bar. The efficiency of the compresor is 0.75 and that of turbine is 0.85. Calculate the power lost in the supercharger expressed as a percentage of the power developed by the turbine. For air, take $c_p = 1.005$ *and for gases* $c_p = 1.15$ *kJ/kgk,* $\gamma = 1.4$.

Solution The compressor in driven by the turbine (turbocharging) and both of them are mounted on the same shaft (Fig. Ex. 20.12). The energy available in the exhaust gases from the engine is converted into useful work by the turbine which drives the compressor. The T-s diagram is shown in the figure. For isentropic compression,

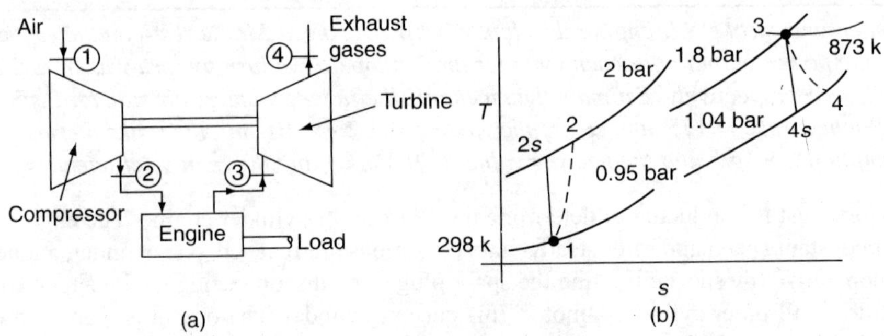

Fig. Ex. 20.12

$$\frac{T_{2s}}{T_1} = \left(\frac{P_2}{P_1}\right)^{\frac{\gamma-1}{\gamma}}$$

$$T_{2s} = 298\left(\frac{2}{0.95}\right)^{0.286} = 368.7\,\text{K}$$

$$\eta_c = \frac{T_{2s} - T_1}{T_2 - T_1} = 0.75 = \frac{368.7 - 298}{T_2 - 298}$$

$$\therefore \quad T_2 = 392.28\text{ K}$$

Power required by the compressor, $W_c = mc_p\,(T_2 - T_1)$

$$= 1 \times 1.005\,(392.28 - 298) = 94.75\text{ kJ/kg}$$

For isentropic expansion in the turbine

$$\eta_T = \frac{T_3 - T_4}{T_3 - T_{4s}}$$

$$\frac{T_3}{T_{4s}} = \left(\frac{p_3}{p_4}\right)^{\frac{\gamma-1}{\gamma}} = \left(\frac{1.8}{1.04}\right)^{0.286} = 1.17$$

$$T_{4s} = \frac{873}{1.17} = 746.23\,\text{K}$$

$$T_4 = T_3 - (T_3 - T_{4s})\eta_T = (873 - 746.23) \times 0.85 = 765.2\text{ K}$$

$$\frac{A}{F} = \frac{18}{1},\quad m_g = m_a\left(1 + \frac{1}{18}\right) = \frac{19}{18}m_a.$$

Power developed by the turbine, $W_T = \dfrac{19}{18} \times 1 \times 1.15(873 - 765.2) = 130.86\,\text{kJ/kg air}$

Power lost as a percentage of the power produced by the turbine $= \dfrac{130.86 - 94.75}{130.86}$

$= 0.276$ or 27.6% *Ans.*

Example 20.13

A four-stroke 6-cylinder diesel engine develops 250 kW at 2000 rpm. The bsfc is 0.20 kg/kWh. At the beginning of injection, pressure is 35 bar and the maximum cylinder pressure is 55 bar. The injection is expected to be 180 bar and the maximum pressure at the injector is set to be about 520 bar. Assuming $c_d = 0.78$, specific gravity of fuel oil = 0.85, atmospheric pressure 1 bar, estimate the total arifice area per injector if the injection takes place over 18° crank angle.

Solution

BP per cylinder $= \dfrac{250}{6} = 41.67\,\text{kW}$

$$\dot{m}_f = 41.67 \times 0.2 = 8.33\text{ kg/h} = 0.139\text{ kg/min}$$

$$\text{Fuel injected per cycle} = 0.139 \times \frac{2}{N} = \frac{0.139}{1000} = 0.139 \times 10^{-3}\ \text{kg}$$

$$\text{Time for injection} = \frac{\theta}{360 \times \frac{N}{60}} = \frac{18}{360 \times \frac{2000}{60}} = 1.5 \times 10^{-3}\ \text{s}$$

Pressure difference at the beginning = 180 − 35 = 145 bar

Pressure difference at the end = 520 − 55 = 465 bar

$$\text{Average pressure difference} = \frac{465 + 145}{2} = 305\ \text{bar}$$

$$\text{Velocity of injection} = c_d \sqrt{\frac{2\Delta^{2}}{\rho_f}} = 0.78\sqrt{\frac{2 \times 305 \times 10^5}{850}}$$

$$= 208.96\ \text{m/s}$$

$$\text{Volume of fuel injected per cycle} = \frac{0.139 \times 10^{-3}}{850}$$

$$\therefore \quad \text{area of orifice} = \frac{\text{Volume of fuel injected per cylcle}}{\text{Injection velocity} \times \text{Injection timing}}$$

$$= \frac{0.139 \times 10^{-3}}{850 \times 208.96 \times 1.5 \times 10^{-3}} = 0.522\ \text{mm}^2 \qquad \textit{Ans.}$$

Example 20.14

The compression curve on the indicator digram from a gas engine follows the law $pv^{1.3}$ = constant. At two points on the curve at ¼ stroke and ¾ stroke, the pressures are 140 kN/m² and 360 kN/m² respectively. Determine the compression ratio of the engine. Calculate the thermal efficiency and the gas consumption per kWh on indicated power basis if the relative efficiency is 0.4 and gas has the calorific value of 18840 KJ/m³.

Solution Let the point 1 is at $\frac{1}{4}$ th stroke and point 2 is at $\frac{3}{4}$ th stroke of the compression curve AB (Fig. Ex. 20.14).

Let

$$V_s = \text{stroke volume}$$

$$V_C = \text{clearance volume}$$

$$\therefore \quad V_1 = V_C + V_S - \frac{V_S}{4} = V_C + 0.75 V_S$$

$$\text{Volume at point 2,} \quad V_2 = V_C + V_S - \frac{3}{4}V_S = V_C + 0.25 V_S$$

$$\frac{V_1}{V_2} = \left(\frac{P_2}{P_1}\right)^{\frac{1}{1.3}} = \left(\frac{360}{140}\right)^{\frac{1}{1.3}} = 2.067 = \frac{V_C + 0.75V_S}{V_C + 0.25V_S}$$

$$\therefore \quad \frac{V_S}{V_C} = 4.5735$$

$$\text{Compression ratio, } \gamma_k = \frac{V_S + V_C}{V_C} = \frac{V_S}{V_C} + 1 = 5.5735 \qquad \textit{Ans.}$$

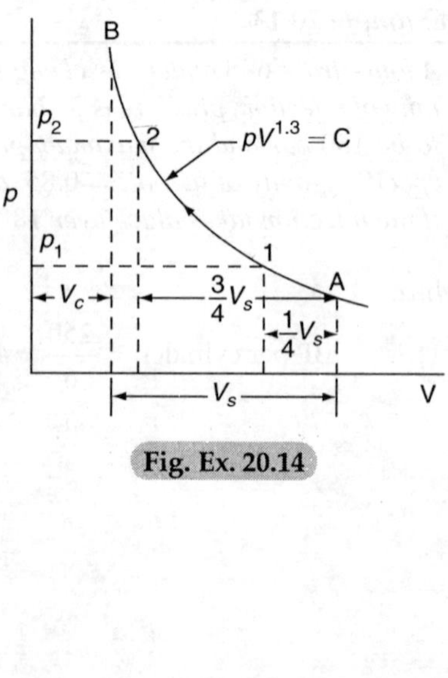

Fig. Ex. 20.14

$$\eta_{ASE} = 1 - \frac{1}{\gamma_k^{\gamma-1}} = 1 - \frac{1}{(5.5735)^{0.4}} = 0.4971$$

Relative efficiency $= \dfrac{\eta_{i.th.}}{\eta_{ASE}} = 0.4$

$\therefore \quad \eta_{i.th.} = 0.4 \times 0.4971 = 0.19884$ or 19.884% *Ans.*

$$\therefore \quad 0.19884 = \frac{3600}{18840 \times \dot{V}_f}$$

$\therefore \quad \dot{V}_f = 0.961\ \text{m}^3/\text{kWh}$ *Ans.*

Example 20.15

During an engine trial on a 6-cylinder 4-stroke diesel engine, bore of 180 mm, stroke of 200 mm, the following observations are recorded. BP = 245 kW, speed = 1500 rpm, mep = 8 bar, fuel consumption 70 kg/h, HV of fuel = 42 MJ/kg, hydrogen content of fuel = 12% by mass; air consumption 26 kg/min, mass of cooling water 82 kg/min, cooling water temperature rise 44°C, cooling oil circulated through the engine = 50 kg/min, temperature rise of cooling oil 24°C, specific heat of cooling oil = 2.1 kJ/kgK, room temperature 30°C, exhaust gas temperature 400°C, c_p of dry exhaust gas, 1.045 kJ/kgK, partial pressure of steam in exhaust gas is 0.035 bar, estimate the mechanical efficiency and draw up an energy balance sheet.

Solution

$$\text{Heat energy supplied by fuel} = \frac{70 \times 42{,}000}{3600} = 816.67\,\text{kW}$$

$$\text{Indicated power} = \frac{p_m LAN}{2 \times 60} = \frac{8 \times 10^5 \times 0.2 \times \frac{\pi}{4}(0.18)^2 \times 1500 \times 6}{2 \times 60 \times 1000} = 305.4\ \text{kW}$$

$$\eta_{mech} = \frac{\text{BP}}{\text{IP}} = \frac{245}{305.4} = 0.802 \text{ or } 80.2\%$$ *Ans.*

$$\text{Heat carried away by cooling water} = \frac{82}{60} \times 4.187 \times 44 = 251.78\ \text{kW}$$

$$\text{Heat carried away by cooling oil} = \frac{50}{60} \times 2.1 \times 24 = 42\ \text{kW}$$

Mass of exhaust gas $= \dot{m}_f + \dot{m}_a$

$$= \frac{70}{60} + 26 = 27.167\ \text{kg/min}$$

$$\text{Mass of vapour in the exhaust gases} = 0.12 \times 9 \times \frac{70}{60} = 1.26\ \text{kg/min}$$

($2H_2 + O_2 = 2H_2O$

4 kg 32 36 kg

1 kg 8 9 Kg)

Mass of dry exhaust gas = (27.167 – 1.26)/60 = 0.4318 kg/s

From steam table, the enthalpy of steam at the exhaust condition (0.035 bar) = 3060 kJ/kg

Heat carried away by steam = 0.21 × 3060 = 64.28 kW

Heat carried away by dry exhaust gases

$$= 0.4318 \times 1.045 \times (400 - 30) = 166.96 \text{ kW}$$

Energy Balance

	Input	*Output*
Heat supplied by fuel	816.67 kW	-
Useful work (BP)	-	245 kW
Heat carried by cooling water	-	251.78 kW
Heat carried away by steam	-	64.28 kW
Heat carried away by cooing oil	-	42.0 kW
Heat carried away by dry exhaust gas	-	166.96 kW
Heat transferred to surroundings	-	46.65 kW
	816.67 kW	816.67 kW

Example 20.16

A 4-stroke 4-cylinder S.I. engine running at 3000 rpm produces a torque of 66.5 Nm. It has a bore of 60 mm and a stroke of 100 mm. The clearance volume in each cylinder is 60 cc and the relative efficiency with respect to brake efficiency is 0.5. If the calorific value of fuel is 42 MJ/kg, determine the fuel consumption in kg/h and the bmep.

Solution Swept volume, $V_s = \frac{\pi}{4} \times (0.60)^2 \times 0.1 = 2.83 \times 10^{-4}$ m³

Compression ratio, $$\gamma_k = \frac{2.83 \times 10^{-4} + 0.6 \times 10^{-4}}{0.6 \times 10^{-4}} = 5.71$$

$$\therefore \quad \eta_{\text{air std}} = 1 - \frac{1}{5.71^{1.4-1}} = 0.50$$

$$\eta_{br} = 0.50 \times \eta_{rel} = 0.50 \times 0.50 = 0.25$$

$$\text{BP} = \frac{2\pi \text{TN}}{60}$$

$$= \frac{2\pi \times 66.5 \times 3000}{60} \times 10^{-3} = 20.89 \text{ kW}$$

$$\eta_{br} = \frac{\text{bp}}{\dot{m}_f \times \text{C.V.}} = \frac{20.89}{\dot{m}_f \times 42 \times 10^3} = 0.25$$

$$\therefore \quad \text{Fuel consumption, } \dot{m}_f = \frac{20.89}{42 \times 10^3 \times 0.25} \times 3600 = 7.16 \text{ kg/h}$$ *Ans.*

$$bp = \frac{\text{bmep} \times \frac{\pi}{4} D^2 L \times \frac{N}{2}}{60}$$

$$20.89 = \frac{bmep \times 2.83 \times 10^{-4} \times \frac{3000}{2}}{60} \times 10^{-3}$$

Ans.

$$bmep = 7.38 \times 10^5 \text{ Pa} = 7.38 \text{ bar}$$

Review Questions

20.1 What are the differences between internal and external combustion engines?

20.2 What are the advantages and applications of IC engines?

20.3 Describe how IC engines are classified.

20.4 Give with a neat sketch an overview of a reciprocating IC engine.

20.5 Explain the differences between S.I. and C.I. engines.

20.6 Describe with a neat sketch the main components of an S.I. engine.

20.7 Explain the working principle of a 4-stroke S.I. engine with the help of *p-v* and valve timing diagrams.

20.8 Explain with sketches the operation of 4-stroke C.I. engine.

20.9 How does the operation of a 2-stroke S.I. engine differ from that of a 4-stroke S.I. engine?

20.10 Explain the valve-timing diagram of a 2-stroke S.I. engine.

20.11 Define the terms: (a) brake power, (b) indicated power, (c) mechanical efficiency, (d) friction power, (e) imep, (f) bmep, (g) isfc, and (h) bsfc.

20.12 What do you mean by volumetric efficiency? On what factors does it depend?

20.13 Show that the brake power of an IC engine depends on the density of air used.

20.14 How are these measured: (a) bp, (b) ip, (c) bmep, (d) imep and (e) bsfc?

20.15 Why are air-standard cycles conceived? Show that the efficiency of the Otto cycle depends on the compression ratio only.

20.16 Explain the effects of compression ratio and specific heat ratio on Otto cycle efficiency.

20.17 Derive the expressions for net work output and m.e.p. of the Otto cycle in terms of r_k, r_p, and γ.

20.18 Explain the air-standard cycle of a C.I. engine. Show that the cycle efficiency depends on r_k and r_c.

20.19 Describe the types of fuels used for I.C. Engines.

20.20 What are the important qualities of S.I. engine fuels?

20.21 Explain the octane number rating of S.I. engine fuels. Why is unleaded petrol used in automobiles?

20.22 Explain the main qualities of C.I. engine fuels. What do you mean by cetane number rating of diesel oil?

20.23 Describe the combustion mechanism in an S.I. engine. What is the relationship between pressure and crank angle for different rates of combustion?

20.24 What is abnormal combustion? Explain the phenomena of preigntion and detonation. What is auto-ignition?

20.25 Explain with a sketch the ignition system of an S.I. engine. How does the spark plug operate?

20.26 What is the function of a carburetor? Describe the operation of a simple float carburetor with a neat sketch. What are idling jet and compensating jet? When is the accelerating pump used?

20.27 Describe the fuel injection systems in C.I. engines. Explain the operation of a fuel pump. What are the types of fuel injectors?

20.28 Explain the combustion mechanism in C.I. engines. What is ignition delay? What are the factors which influence ignition delay?

20.29 Explain the C.I. engine combustion chambers.

20.30 Describe the cooling system of I.C. engines with neat sketches.

20.31 What are the various components to be lubricated in an engine? Explain how it is accomplished.

20.32 Explain the various mechanisms of lubrication in an I.C. engine.

20.33 What are the common methods of starting an engine?

20.34 What are the distinctive features in combustion in a C.I. engine as compared to an S.I. engine?

20.35 Explain the performance characteristics of an I.C. engine.

20.36 Describe the Morse test in measuring the I.P. of a multi-cylinder engine.

20.37 What do you mean by volumetric efficiency? What are the variables which affect the volumetric efficiency of an engine?

20.38 Give an energy balance of an IC engine.

20.39 What is the objective of supercharging? Why is it more beneficial in an IC engine compared to a SI engine?

20.40 Explain the main features of supercharging with the help of a p–V diagram. What do you mean by mechanical supercharging and turbocharging? What is the effect of intercooling in turbocharging?

20.41 What are the different types of compressors used for supercharging? Why are turbochargers superior to superchargers?

20.42 Explain the running performance characteristics of a diesel engine operating in various modes.

Problems

20.1 Calculate the air flow rate, fuel flow rate, bmep and imep for a 4-stroke engine developing 300 kW at 2500 rpm when the stroke volume is 2000 cc, thermal efficiency is 0.30 and air/fuel ratio is 15.
Ans. 0.075 kg/s, 0.9 kg/s, 7.2 MPa, 9 MPa.

20.2 A petrol engine having a compression ratio of 8, has a brake thermal efficiency which is 40% of air standard efficiency. The calorific value of fuel is 44 MJ/kg. Calculate the fuel consumption in kg/h, if the engine delivers 7.5 kW. *Ans.* 2.7165 kg/h

20.3 A four-cylinder engine running at 1200 rpm gave 18.6 kW brake power. The average torque when one cylinder was cut off was 105 Nm. Determine the indicated thermal efficiency if the heating value of fuel is 42000 kJ/kg and the engine uses 0.34 kg of petrol per kW brake power.
Ans. 29.3%

20.4 A 4-stroke engine uses natural gas and has a stroke of 400 mm, bore of 350 mm. It runs at 1000 rpm and the air-fuel ratio 5:1 by volume and the volumetric efficiency of 0.80. Estimate the bmep when the brake thermal efficiency is 0.28 and the HV of the gas is 8 MJ/m^3 at standard conditions.
Ans. 298.6 kPa.

20.5 In a simple carburetor, the gasoline in the float chamber stands 5-mm below the nozzle opening. The engine consumes 6.1 kg of fuel per hour. The fuel jet diameter is 1.25mm and the discharge coefficient for fuel orifice is 0.64. Estimate the air velocity at the throat, the throat diameter if the air-fuel ratio is 16:1. C_d for air is 0.87 and the ambient conditions are: p_1 = 1 bar, T_1 = 300 K, ρ_f = 775 kg/m^3. *Ans.* d_2 = 2.19 cm

20.6 A single cylinder 4-stroke diesel engine running at 2000 rpm uses 2.8 kg of fuel (sp. gravity 0.88) per hour. It has an injector with a single orifice nozzle and the injection period is 22° crank angle. If the injection pressure is 150 bar and the average pressure inside the cylinder is 30 bar, estimate the fuel orifice diameter. Take C_d for the nozzle = 0.88. *Ans.* 0.503 mm

20.7 A Diesel engine has a compression ratio of 14:1 and fuel is cut off at 0.08 of stroke. Calculate the mass of fuel used per kWh if the calorific value is 42 MJ/kg and the relative efficiency 0.54.
Ans. 0.269 kg/kWh.

20.8 A 4-cylinder 4-stroke single acting S.I. engine gave the following results:

Petrol consumption =5 kg/h
Speed =750 rpm
Air-fuel ratio = 11:1
Temperature at the end of compression=375°C
Pressure at the end of compression = 1200 kN/m^2
Relative efficiency = 0.50
Diameter of cylinder = 8.5 cm
Stroke of piston = 12 cm
Calorific value of fuel = 42 MJ/kg
R for mixture=0.285 kJ/kgK.

Estimate specific fuel consumption, (isfc)
Ans. 0.308 kg/kWh

20.9 Find the diameter of the cylinder of a single acting diesel engine, working on four-stroke cycle, which is required to give 38 kW indicated power at 200 rpm from the following data: compression ratio 14:1, fuel cut-off 5% of the stroke, index of compression curve 1.4; index of expansion curve 1.3, pressure at the beginning of compression 100 kPa, ratio of stroke to bore 1.5:1.

Ans. 33.76 cm

20.10 A four-stroke diesel engine of 2000 cc produces 12.1 kW per m^3 of free air inducted per minute. The speed is 3000 rpm and the volumetric efficiency is 0.84. When the ambient conditions are: $p_1 = 1$ bar, $T_1 = 300$ K, a supercharger is introduced in the system which raises the pressure to 1.8 bar with an isentropic efficiency of 0.8. Estimate the increase in the output of the engine.

Ans. 55.72 kW

20.11 A single cylinder 4- stroke gas engine has a bore of 180 mm, stroke of 330 mm and is governed to run at 750 rpm. At full load, indicator cards are taken which give a working loop mep of 7.8 bar and a pumping loop mep 0.41 bar. The dead cycle mep is 0.68 far. The number of firing strokes were 125 per min at no load conditions. Estimate the brake power and the mechanical efficiency.

Ans. 28.25 kW, 72.8%

20.12 A 4-stroke 3000 cc 6 cylinder engine has a maximum torque of 240 Nm at 2000 rpm. The specific fuel consumption is 0.09 kg/MJ and the air-flow rate is 4.08 m^3/min. The ambient conditions are p_1=1bar, T_1=300 K. Estimate the volumetric efficiency, air-fuel ratio and the thermal efficiency if the heating value of fuel is 44 MJ/kg.

Ans. 0.907, 11.64, 0.2525

20.13 Estimate the cylinder dimensions of 4-stroke 3 cylinder engine producing 25 kW at 3500 rpm, when the air-fuel ratio is 15, the heating value of fuel is 40 MJ/kg, the volumetric efficiency is 0.85, brake thermal efficiency is 0.30.

Ans. d =74.8 mm, L = 82.32 mm

20.14 Find the air velocity and the corresponding venturi throat diameter of a carburetor when the fuel jet cross-sectional area is 1.95 mm^2 and the fuel flow rate is 4.98 kg/h. The intake conditions are p = 1 bar, T = 300 K, Air fuel ratio = 14.5:1, C_d for fuel jet = 0.6, C_d for venturi = 0.84, ρ_f =780 kg/m^3.

Ans. 25.6 m/s, 32 mm.

20.15 A 6-cylinder 4-stroke SI engine having a bore of 90 mm and a stroke of 100 mm has a compression ratio of 7. The relative efficiency is 0.55 when the indicated specific fuel consumption is 300 g/kWh. Calculate (i) the calorific value of the fuel, (ii) the fuel consumption in kg/h, when the imep is 8.5 bar and speed is 2500 rpm.

Ans. (i) 40.4 MJ/kg, (ii) 20.28 kg/h.

20.16 A 4-cylinder gasoline engine having a bore of 75 mm and a stroke of 90 mm, and running at 3000 rpm develops a brake power of 20.9 kW. A Morse Test was conducted on this engine, and the brake power (kW) obtained when each cylinder was made inoperative by short circuiting the spark plug are 14.9, 14.3, 14.8 and 14.5 respectively. The test was conducted at constant speed Estimate the indicated power, mechanical efficiency and the bmep, when all the cylinders are firing.

Ans. 25.1 kW, 83.3%, 5.25kW

20.17 The following observations were made during a trial of a single-cylinder four stroke cycle gas engine having a bore of 180 mm and stroke 240 mm:

Duration of trial=30 min
Total number of revolutions = 9000
Total number of explosions = 4450
Mean effective pressure = 5 bar
Net load on the brake wheel = 40 kg
Effective diameter of the brake wheel = 1m
Total gas used at NTP = 2.4 m^3
Calorific value of gas at NTP = 19 MJ/m^3
Total air used = 36 m^3
Pressure of air = 720 mm Hg
Temperature of air = 17°C
Density of air at WTP = 1.29 kg/m^3
Temperature of exhaust gas = 350°C
Room temperature = 17°C
Specific heat of exhaust gas = 1 kJ/kgk.
Cooling water circulated = 80 kg
Temperature rise of cooling water = 30°C.

Draw up an energy balance of the engine and estimate the mechanical and indicated efficiencies of the engine.

Ans. 81.6%, 20.8%

20.18 A six-cylinder 4-stroke gasoline engine having a bore of 80 mm and a stroke of 100 mm for each cylinder consumes fuel at the rate of 20 kg/h and develops a torge of 150 Nm at the speed of 4000 rpm. The clearance volume in each cylinder is 70 cc. Estimate (i) the brake power, (ii) the brake mean effective pressure, (iii) the brake efficiency if the calorific value of fuel is 43 MJ/kg, and (iv) the relative efficiency based on brake power.

Ans. (i) 62.8 kW, (ii) 6.25 bar, (iii) 26.3%, (iv) 46.3%

CHAPTER 21

Gas Turbines and Propulsion Systems

The economics of power generation by gas turbines is now quite attractive due to low capital cost and its high reliability and flexibility in operation. Another outstanding feature is its capability of quick starting and using a wide variety of fuels from natural gas to residual oil or powdered coal. Its consumption of lubricating oil is quite low because of the absence of rubbing and reciprocating parts, and the balancing does not pose a major problem. Due to better materials being made available and with the use of adequate blade cooling, the inlet gas temperature to the gas turbine (GT) blades can now exceed 1200°C, as a result of which the overall efficiency of a GT plant can be about 35%, almost the same as that of a conventional steam power plant.

Because of its low weight per unit power, gas turbine is exclusively used to drive aviation system of all kinds of aircraft. It is also being increasingly used in land vehicles like buses and trucks and also to drive locomotives and marine ships. In oil and gas industries, the gas turbine is widely employed to drive auxiliaries like compressors blowers and pumps.

21.1 CLOSED CYCLE AND OPEN CYCLE PLANTS

The essential components of a gas turbine (GT) power plant are the compressor, combustion chamber and the turbine. The air standard cycle of a GT plant is the Brayton cycle.

A GT plant can either be open or closed. Figure 21.1 shows the arrangement of an open-cycle plant which is more common. The compressor takes in ambient air and raises its pressure. The temperature of air is increased when it flows through a combustion chamber where a fossil fuel is burned or a heat exchanger (nuclear fuel being used as a source of energy) is present. The high-pressure, high-temperature working fluid, mostly a gas, enters a turbine where it expands to a low-pressure (equal to or a little above the atmospheric pressure) fluid. In an open unit, the gas is released from the turbine to the surroundings and in a closed unit, the working fluid is cooled in a cooler after the exhaust from the turbine and is returned to the compressor (Fig. 21.2). A significant part of the power developed by the turbine is utilized to drive the compressor and any other auxiliaries, and the remainder is available as useful work.

21.2 ADVANTAGES OF A GT PLANT

The advantages of a GT plant for power generation are being enumerated below:

(1) *Warm-up Time* Once the turbine is brought up to the rated speed by the starting motor and the fuel is ignited, the GT will accelerate from cold start to full load without warm-up time.

(2) *Low Weight and Size* The weight of the plant per kW output is low, which is a favourable feature in all vehicles (land, air and sea). In utilities also, the foundation of the plant is lighter.

(3) *Fuel Flexibility* Any hydrocarbon fuel from high octane gasoline to heavy diesel oil and pulverized coal can be used effectively.

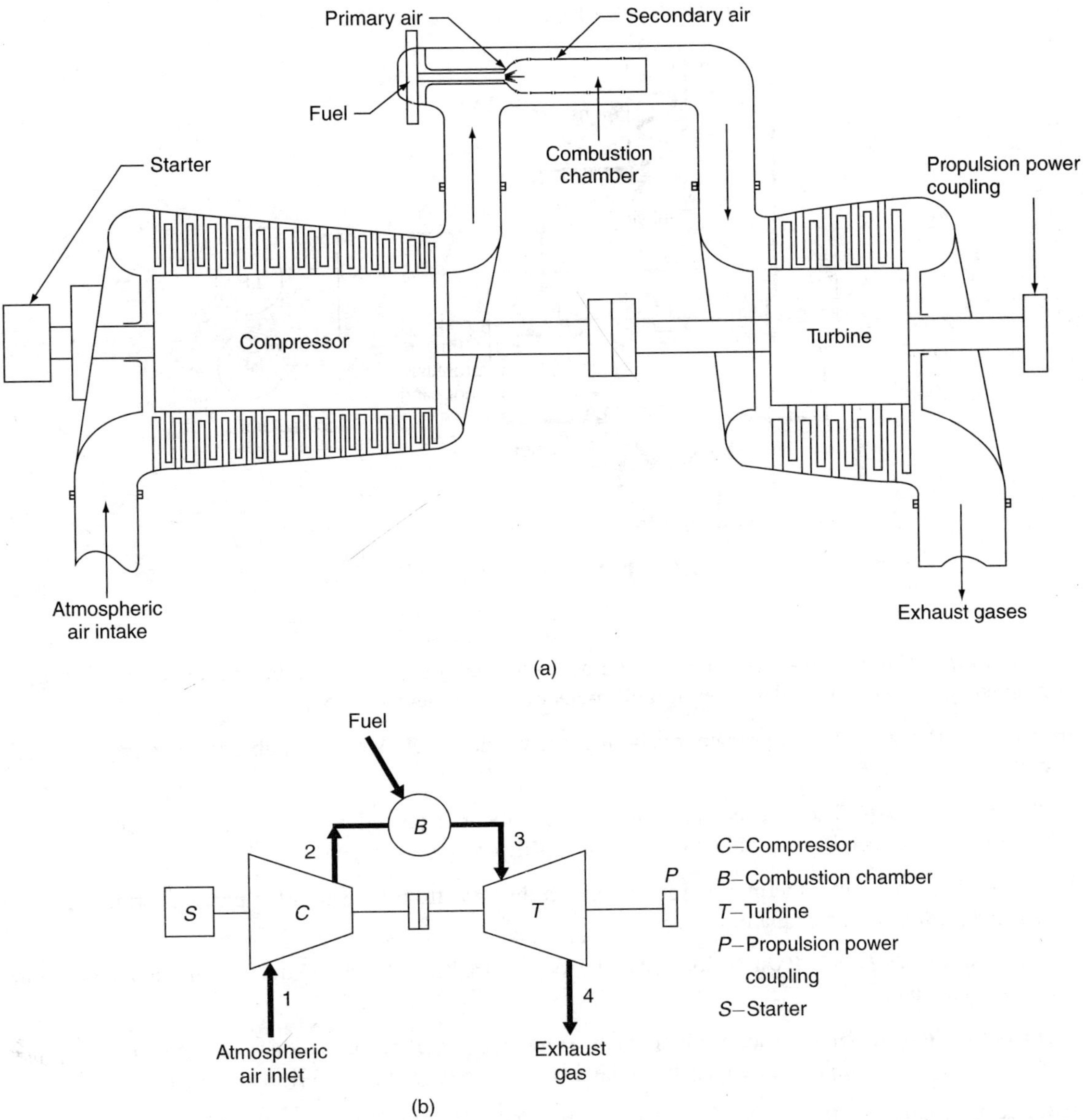

Fig. 21.1 *(a) Component parts of a simple open cycle (constant-pressure combustion) gas turbine, (b) Simple open cycle gas turbine*

(4) *Floor Space* Because of its smaller size, the floor space required for its installation is less.

(5) *Start-up and Shut-down* A GT plant can be started up as well as shut down quickly, like a diesel engine. Thus it is eminently suitable to meet the peak load demand of a certain region.

(6) *High Efficiency* Suitable blade cooling permits the use of high GT inlet temperature (as high as 1300°C) yielding a high thermal efficiency (on the order of 37%).

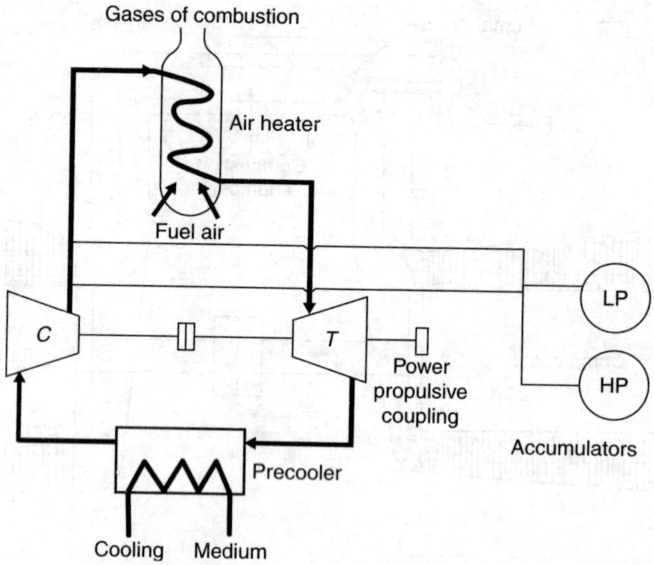

Fig. 21.2 *Simple closed-cycle GT plant*

(7) *Combined Cycle Mode* A GT plant can be used in conjunction with a bottoming steam plant in the combined cycle mode to yield an overall fuel-to-electricity efficiency of 55%.

(8) *Cooling Water* The requirement of cooling water is not much. Water availability is not a restriction for installing a GT plant.

(9) *Lubricating Oil* The absence of rubbing and reciprocating parts does not warrant high consumption of lubricating oil and there is no problem of balancing as in a reciprocating engine.

(10) *Ash Disposal* In a thermal power station ash disposal from the site often poses a serious problem. This is not so in a GT power plant.

(11) *Transmission Loss* It can be located at the load centre itself. Therefore, the transmission loss is minimal in such a plant.

(12) *Cost of Installation* The installation cost is much less compared to a thermal plant. Only a foundation is required. The whole plant comes from the factory to the site, almost fully assembled.

(13) *Scope of Cogeneration* A GT exhaust can be used to produce process heat for various uses.

(14) *Low Capital Cost* GT plants are available in standard sizes. The capital cost per kW is considerably less than a thermal plant.

21.3 DISADVANTAGES OF A GT PLANT

These are enlisted below:

1. Part load efficiency is low
2. Highly sensitive to component efficiency like η_c and η_T

3. The efficiency depends on the ambient condition (p_a and T_a)
4. High air rate is required to limit the maximum GT inlet temperature, as a result of which the exhaust losses are high, unless the waste heat in it is utilized
5. Compressor work required is quite large, which tells upon the efficiency of the plant
6. Air and gas filters have to be of very high quality so that no dust enters to erode and corrode the turbine blades

21.4 ANALYSIS OF A GT PLANT

The analysis of Brayton cycle, the air standard cycle of the GT plant, has been given in Chapter 13. The salient features of the cycle (Fig. 21.3) are being given below:

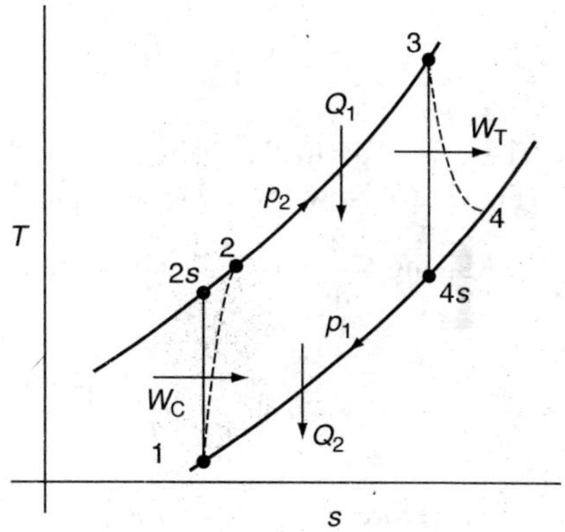

Fig. 21.3 Brayton cycle

Heat supplied, $Q_1 = m_a c_p (T_3 - T_2)$

Heat rejected, $Q_2 = m_a c_p (T_4 - T_1)$

$$\frac{T_{2s}}{T_1} = \frac{T_3}{T_{4s}} = \left(\frac{p_2}{p_1}\right)^{\frac{\gamma-1}{\gamma}} = r_p^{\frac{\gamma-1}{\gamma}}$$

where m_a = mass of air and r_p = pressure ration, p_2/p_1.

The compressor efficiency, $\eta_c = \dfrac{T_{2s} - T_1}{T_2 - T_1}$

and turbine efficiency, $\eta_T = \dfrac{T_3 - T_4}{T_3 - T_{4s}}$

For the ideal cycle, 1–2*s* – 3– 4*s* – 1,

$$\eta_{cycle} = 1 - \frac{1}{r_p^{\frac{\gamma-1}{\gamma}}} \tag{21.1}$$

As r_p increases, η_{cycle} increases till Carnot cycle is reached (Fig. 21.4). With the increase or r_p, the mean temperature of heat addition T_{m_1} increases, and the mean temperature of heat rejection T_{m_2} decreases (Fig. 21.5). When in the limit $T_{m_1} \to T_3$ and $T_{m_2} \to T_1$, the Carnot efficiency is obtained.

$$\eta_{cycle} = 1 - \frac{1}{r_p^{\frac{\gamma-1}{\gamma}}} = 1 - \frac{T_1}{T_3} = 1 - \frac{T_{min}}{T_{max}}$$

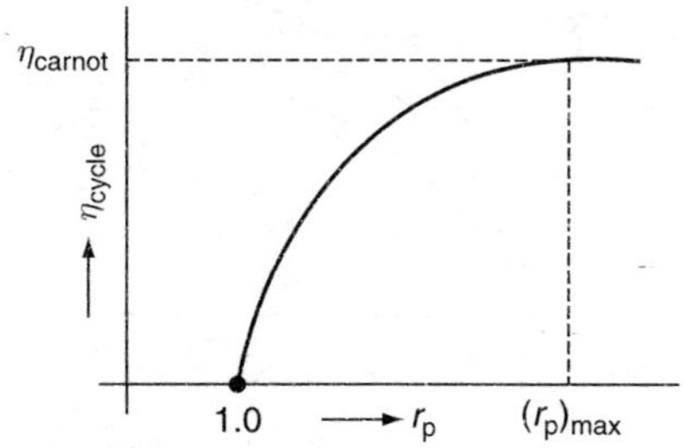

Fig. 21.4 Variation of η_{cycle} with r_p

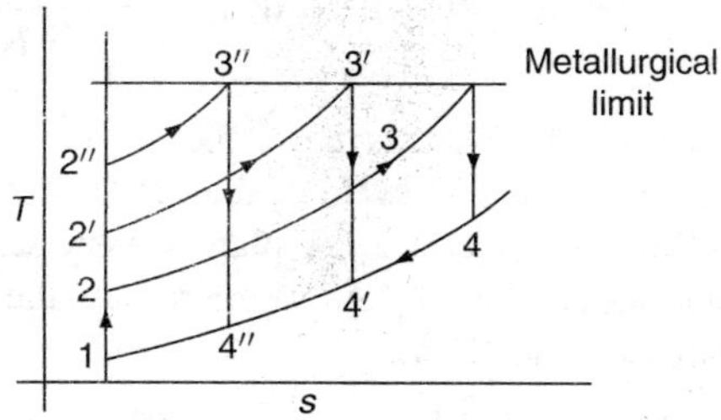

Fig. 21.5 Effect of r_p on Brayton cycle

$$\therefore \quad (r_p)_{max} = \left(\frac{T_{max}}{T_{min}}\right)^{\frac{\gamma}{\gamma-1}} = \left(\frac{T_3}{T_1}\right)^{\frac{\gamma}{\gamma-1}} \tag{21.2}$$

For a particular value of r_p,

$$W_{net} = Q_1 - Q_2 = m_a c_p [T_3 - T_{2s} - T_{4s} + T_1]$$

$$= m_a c_p \left[T_3 - T_1 r_p^{\frac{\gamma-1}{\gamma}} - T_3 r_p^{-\frac{\gamma-1}{\gamma}} + T_1\right]$$

There is a particular value of r_p when W_{net} is maximum (Fig. 21.6). Now, the values of T_1 and T_3 are known.

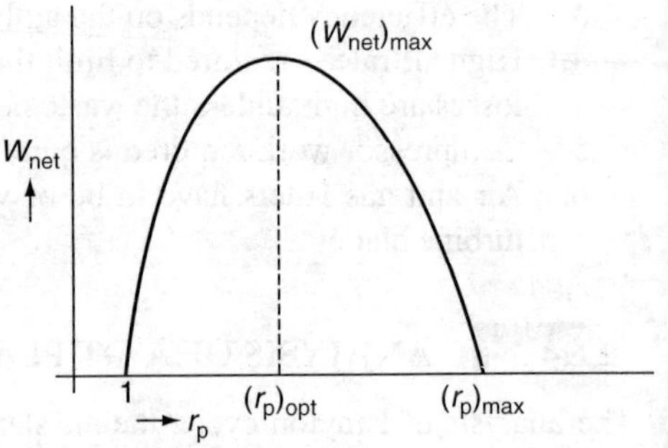

Fig. 21.6 *Effect of r_p on W_{net}*

Making $\frac{dW_{net}}{dr_p} = 0$, the optimum value of r_p becomes

$$(r_p)_{opt} = \left(\frac{T_{max}}{T_{min}}\right)^{\frac{\gamma}{2(\gamma-1)}}$$

Therefore, $(r_p)_{opt} = \sqrt{(r_p)_{max}}$

$$(W_{net})_{max} = m_a c_p \left[T_3 - 2\sqrt{T_3 T_1} + T_1\right]$$

$$= m_a c_p \left(\sqrt{T_3} - \sqrt{T_1}\right)^2 \tag{21.3}$$

and
$$\eta_{cycle} = 1 - \sqrt{\frac{T_{min}}{T_{max}}} \tag{21.4}$$

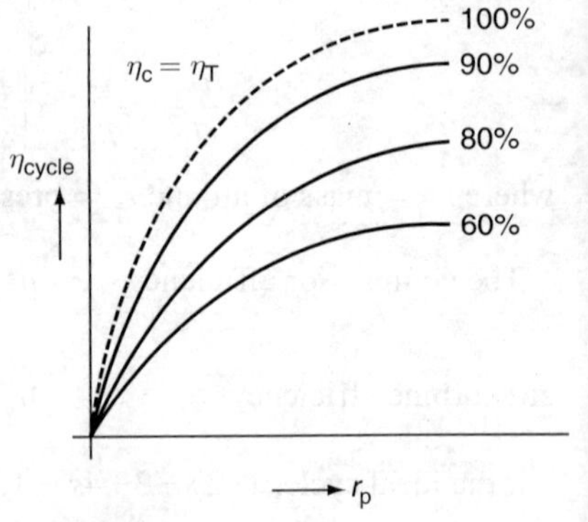

Fig. 21.7 *Effect of η_c and η_T on η_{cycle}*

If the compressor and turbine efficiencies are considered, it can be shown

$$(r_p)_{opt} = \left(\eta_c \eta_T \frac{T_3}{T_1}\right)^{\frac{\gamma}{2(\gamma-1)}} \tag{21.5}$$

The work ratio r_w is defined as the ratio of net work to work done by the turbine.

$$\therefore \quad r_w = \frac{W_{net}}{W_T} = (W_T - W_c)/W_T = 1 - \frac{T_1}{T_3} r_p^{\frac{\gamma-1}{\gamma}} \tag{21.6}$$

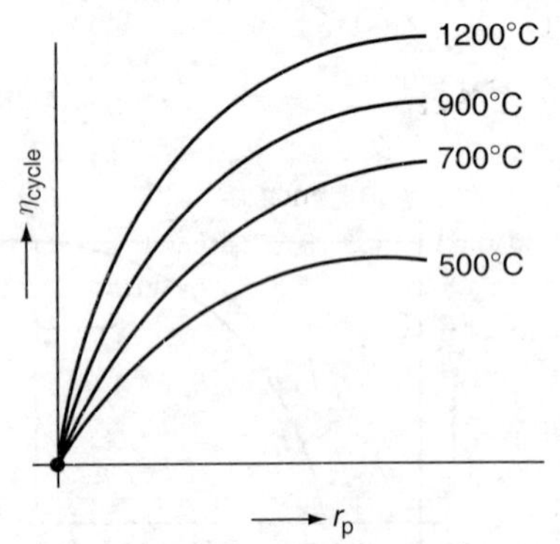

Fig. 21.8 *Effect of turbine inlet temperature on $\eta_{thermal}$ of a GT plant*

Figure 21.7 shows the effect of turbine and compressor efficiencies on cycle efficiency. The thermal efficiency of Brayton cycle or a GT plant is very sensitive to turbine inlet temperature T_3. As T_3 increases, the cycle efficiency increases (Fig. 21.8).

(a) Effect of Regeneration The thermal efficiency of a simple open-cycle gas turbine may be improved by the utilization of the energy available in the turbine exhaust gases in a regeneration process. Inspection of Fig. 21.9 reveals that the temperature of the exhaust gases leaving the turbine at the state 4 is higher than the temperature of compressed air at the state 2. This difference in temperatures makes the regeneration possible. The recovery of a part of the thermal energy of the exhaust gases is accomplished by installing a heat exchanger called a *regenerator* in the flow system as shown in Fig. 21.10. The exhaust gases at a high temperature enter the hot side of the regenerator and are circulated around tubes containing the cold compressed air in the cold side of the regenerator. In this system, the temperature of the compressed air is increased before it reaches the combustion chamber (B) and therefore, less fuel is required to raise the air to the specified turbine inlet temperature. The effectiveness of regenerator is defined as

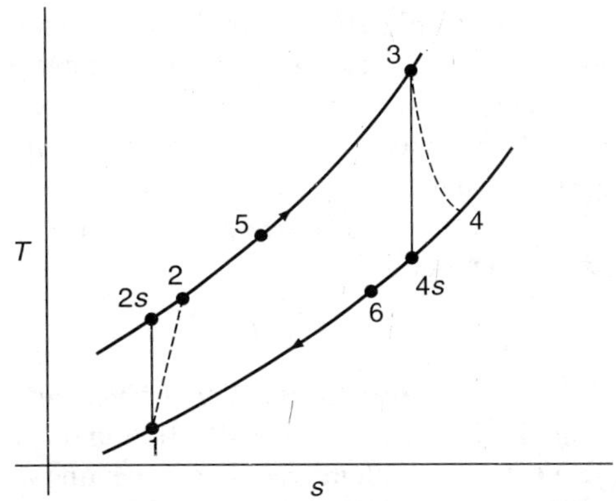

Fig. 21.9 *T-s diagram of an open-cycle gas turbine with regenerator*

$$\varepsilon = \frac{\text{Actual temperature rise of air}}{\text{Maximum temperature rise possible}} = \frac{T_5 - T_2}{T_4 - T_2} \tag{21.7}$$

Here, $Q_1 = m_a c_p (T_3 - T_5)$ and $Q_2 = m_a c_p (T_6 - T_1)$, both of which decrease, whereas W_{net}, i.e., $(W_T - W_c)$ remains unchanged. Therefore, the efficiency of the cycle, W_{net}/Q_1, increases. In order to obtain an ideal regenerator, i.e., one having a heat exchanger effectiveness or regenerator efficiency of 100 per cent, the temperature of the compressed air, T_2, must be raised to the temperature of the exhaust gases, T_4, entering the regenerator. This could only be accomplished by having a heat-transfer surface of infinite area. Since the regenerators are restricted in size due to weight and space limitation, they have a maximum effectiveness of about 75 per cent.

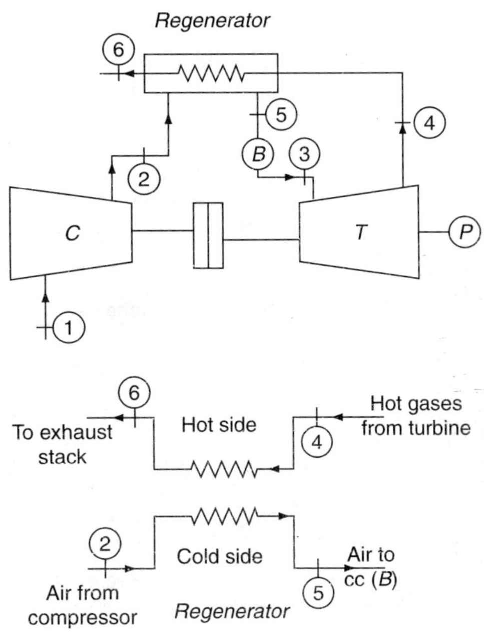

Fig. 21.10 *Open-cycle gas turbine with regenerator*

(b) Effect of Intercooling By staging the compression process (1–2 and 3–4) with perfect intercooling (2–3), the cycle efficiency decreases, as shown in Fig. 21.11, where the small cycle 1–2–3–4–4′–1 is added to the basic cycle 1–4′–5–6–1 without intercooling. However, it permits more heat recovery from hot gases exiting the turbine at the state 6 by heating air leaving the compressor at the state 4. For minimum work of compression, the intercooler pressure $p_i = (p_1 p_2)^{1/2}$, where p_1 and p_2 are suction and discharge pressures, respectively.

(c) Effect of Reheating Similarly, by staging the heat supply process with a combustor and a reheater, the cycle efficiency decreases, but it permits more heat recovery from the turbine exhaust gases (Fig. 21.12) (since $T_6 > T_4'$) with the result that reheating along with regeneration may bring about an improvement in cycle efficiency.

It can be shown that the optimum reheat pressure for maximum net work output is

$$p_i = \sqrt{p_1 \, p_2} \tag{21.8}$$

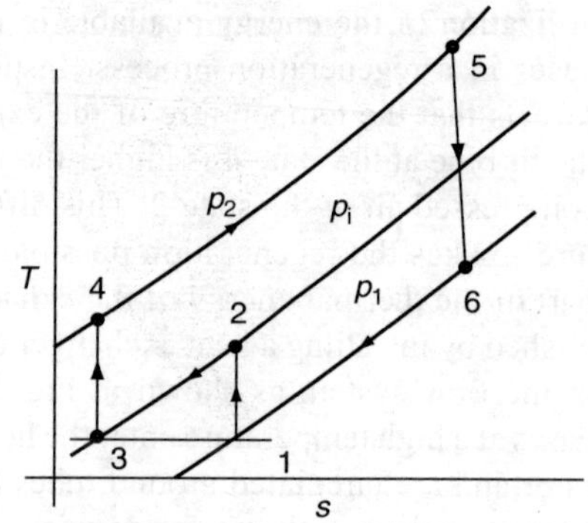

Fig. 21.11 *Effect of intercooling on Brayton cycle*

(d) Effect of Intercooling, Reheating and Regeneration Figure 21.13(a) and (b) shows the flow and *T-s* diagrams of a closed-cycle GT plant with intercooling, reheating and regeneration.

The net work of the GT plant is given by

$$W_{net} = W_T - W_c$$

$$= (m_a + m_f)\, c_{p_g} \left[(T_6 - T_7) + (T_8 - T_9)\right] - m_a c_{p_a} \left[(T_2 - T_1) + (T_4 - T_3)\right] \tag{21.9}$$

Heat supply,

$$Q_1 = (m_a + m_f)\, c_{p_g} \left[(T_6 - T_5) + (T_8 - T_7)\right]$$

$$= \dot{m}_f \times \text{C.V.}$$

Therefore, the overall plant efficiency

$$\eta_0 = \frac{W_{net}}{\dot{m}_f \times \text{C.V.}} \tag{21.10}$$

where C.V. = calorific value of the fuel.

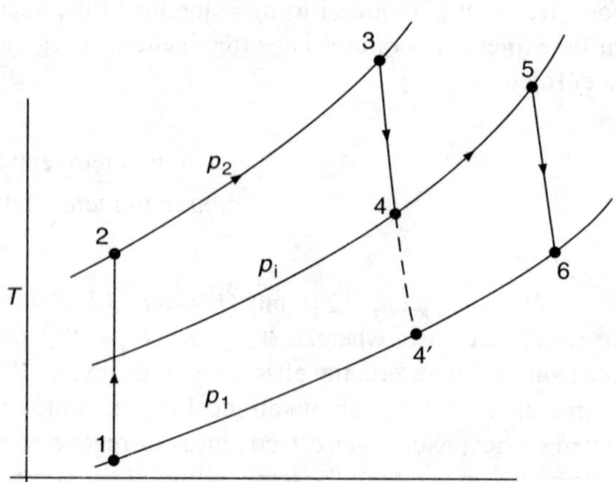

Fig. 21.12 *Effect of reheat on Brayton cycle*

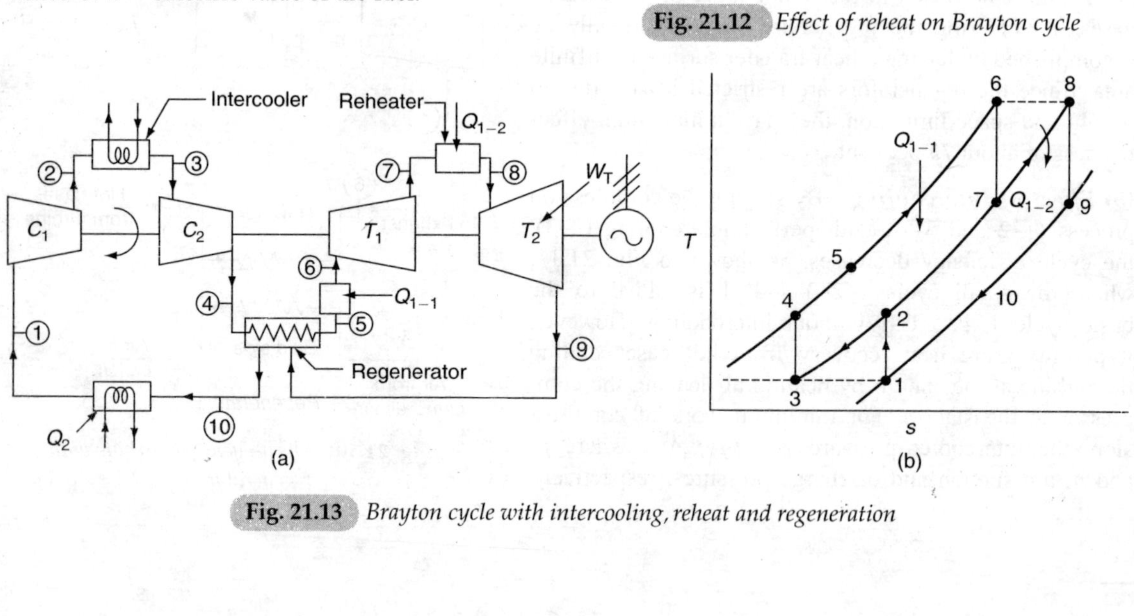

Fig. 21.13 *Brayton cycle with intercooling, reheat and regeneration*

Figure 21.14 shows an open-cycle gas turbine with a regenerator, intercooler and reheater.

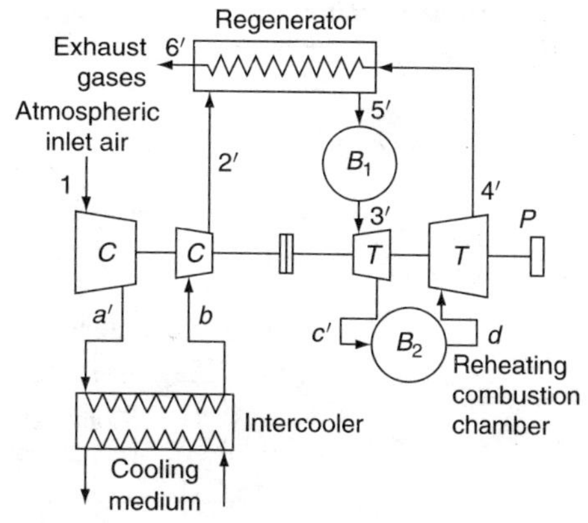

Fig. 21.14 *Open-cycle gas turbine with regenerator, intercooler, and reheater*

21.5 CLOSED-CYCLE GAS TURBINE

The engines discussed so far have been the open-cycle gas turbines. In a closed-cycle GT, the same working medium is continuously circulated (Fig. 21.2). The heat supply in the cycle takes place through a heat exchanger where a fuel may be burnt and heat rejection also occurs through a cooling medium in another heat exchanger. The performance characteristics, the effect of different variables and the component elements on the performance, and the equations developed for the open cycle apply equally as well to the closed cycle. The advantages of the closed cycle over that of the open cycle are (1) reduced size, (2) improved part-load efficiency (3) fuel flexibility. The disadvantages are (1) dependent system (cooling water availability) (2) complexity and cost, (3) air heater (not efficient in heat transfer).

21.6 SEMI-CLOSED CYCLE GT PLANT

The advantages of the open cycle plant, viz. quick and easy starting and the closed cycle plant, viz. constant efficiency at all loads and higher unit rating permitting the use of higher back pressure, are combined in a semi-closed cycle gas turbine power plant. Here, part of the compressed air is heated by the gases exiting the combustion chamber (CC) and then expanded in an air turbine which drives the compressor, thus operating in a closed cycle. The remaining air is used in the CC to burn fuel, and the combustion products after heating the air expand in a gas turbine to drive the generator before exhausting to the atmosphere (Fig. 21.15 a). Figure 21.15 (b) shows a combined combustion chamber and a heat exchanger, where hot gases of combustion leave to expand in the gas turbine in the open cycle and the heated air flows to the air turbine in the closed cycle.

21.7 PERFORMANCE OF GAS TURBINE POWER PLANTS

The gas turbine plant works under variable load conditions. It is thus necessary to study the effect of load on the cycle efficiency which is directly concerned with the running cost of the plant.

It is necessary to study the effect of pressure ratio on the thermal efficiency, air mass flow and specific fuel consumption with regenerative reheat and intercooled cycle, because smaller mass flow rate for the given output reduces the component sizes and the plant capital costs. Lower fuel consumption reduces the running cost of the plant. Some of these characteristics are represented graphically and also discussed.

(a) Part Load Efficiency The part load efficiencies for open cycle, closed cycle and semi-closed cycle are shown in Fig. 21.16. The part load performance of the semi-closed cycle is seen to be the best.

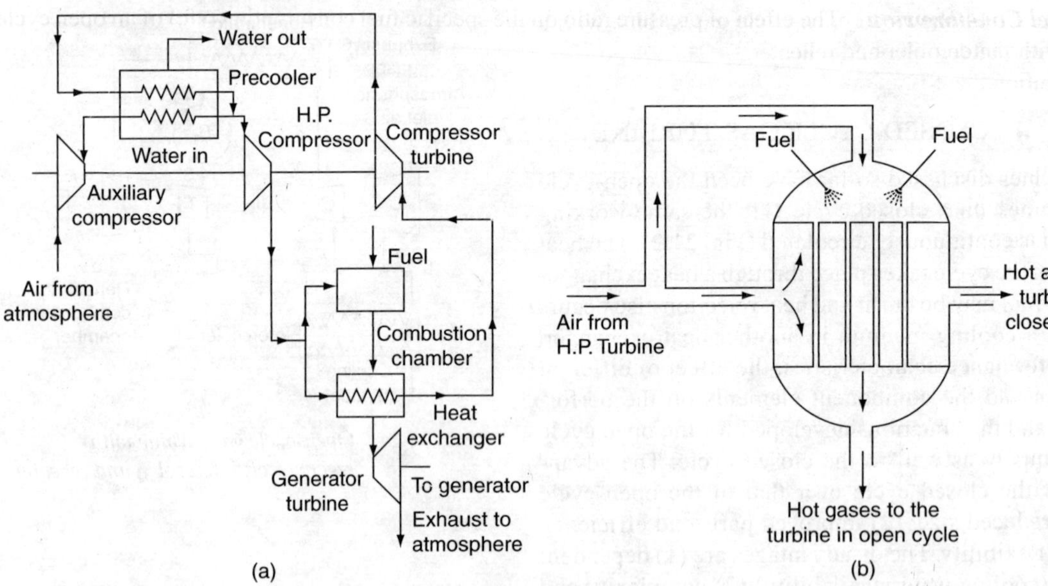

Fig. 21.15 *(a) Semi-closed cycle gas turbine plant, (b) Combined combustion chamber and air heater*

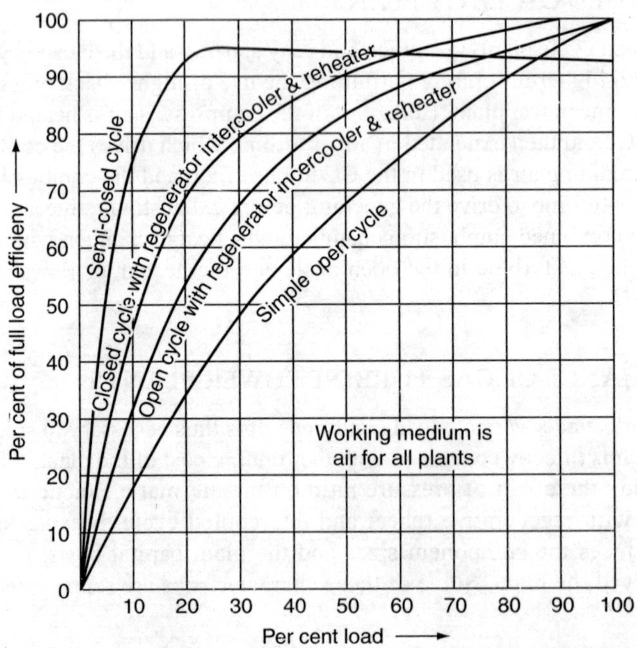

Fig. 21.16 *Part load efficiencies of different plants*

(b) Fuel Consumption The effect of pressure ratio on the specific fuel consumption (sfc) of an open cycle plant with the degree of regeneration as a parameter is shown in Fig. 21.17. It shows that for each degree of regeneration there is an optimum pressure ratio for minimum sfc.

(c) Air Rate In the addition to the thermal efficiency which is a measure of the fuel economy, the size of the plant is equally important in many applications, particularly in the field of aviation. For a given duty, the size of a plant is dependent on the air flow rate in relationship to the useful shaft output. The air rate is defined as

$$AR = \frac{w_a \text{ kg/s} \times 3600}{W_{net} \text{ (kW)}}, \quad \text{i.e. kg/kWh}.$$

Air rate is a criterion of the size of the plant, i.e., the lower the air rate, the smaller the plant. From the mechanical and metallurgical standpoint, the lowering of the air rate results in turbines of smaller physical dimensions with a more uniform temperature distribution. Any means by which the physical dimensions can be reduced and the inherent distortions minimized are steps toward greater reliability of the gas turbine . The effect of pressure ratio on air rate for an open cycle plant with the turbine inlet temperature as a parameter is shown in Fig. 21.18. It indicates optimum pressure ratio for different turbine inlet temperatures requiring minimum air rates.

(d) Thermal Efficiency The effect of pressure ratio of a simple open-cycle plant with turbine inlet temperature as a parameter is shown in Fig. 21.19 and with compressor inlet temperature as a parameter in Fig. 21.20. As the turbine inlet temperature increases for a particular pressure ratio, thermal efficiency increases, and for each temperature, there is an optimum value of r_p when efficiency is maximum. An increase in the compressor inlet air temperature increases the compressor work. W_{net} is decreased, air rate increases and thermal efficiency decreases.

(e) Regeneration The effect of regeneration on thermal efficiency of a simple cycle, taking pressure ratio and turbine inlet air temperature as parameters, is shown in Fig. 21.20(a) and Fig. 21.20(b) respectively.

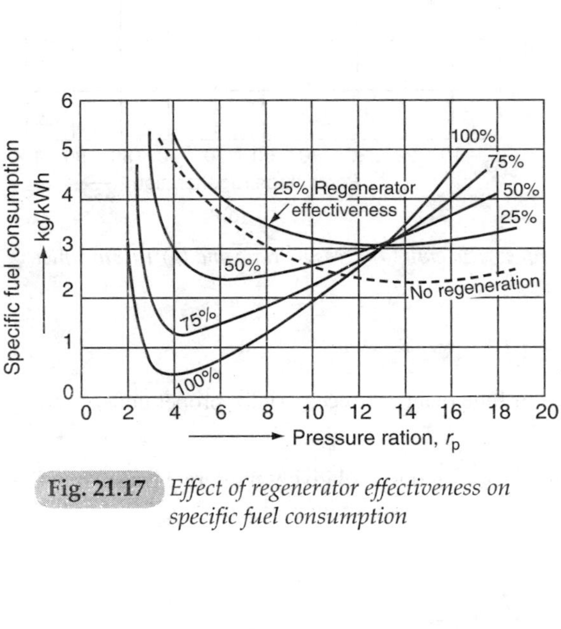

Fig. 21.17 *Effect of regenerator effectiveness on specific fuel consumption*

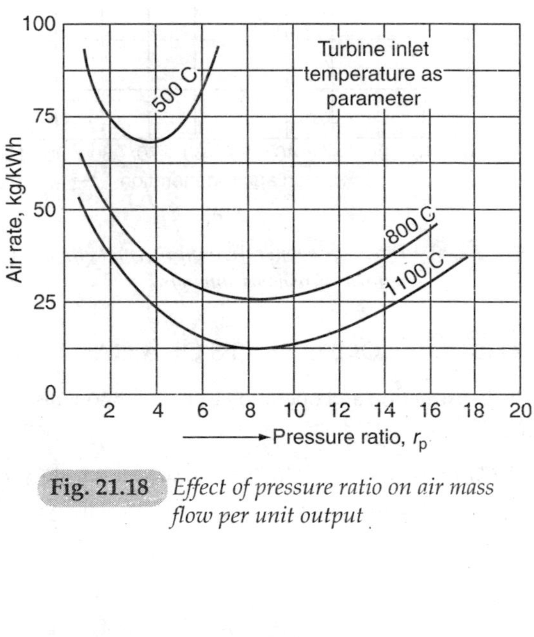

Fig. 21.18 *Effect of pressure ratio on air mass flow per unit output*

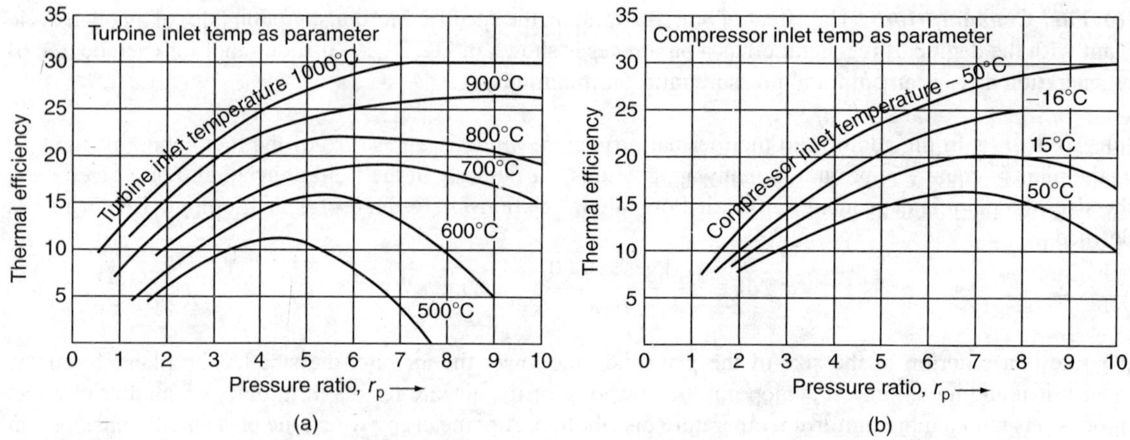

Fig. 21.19 *Effect of pressure ratio on thermal efficiency of a simple open cycle plant with (a) Turbine inlet temperature and (b) Compressor inlet temperature as parameters*

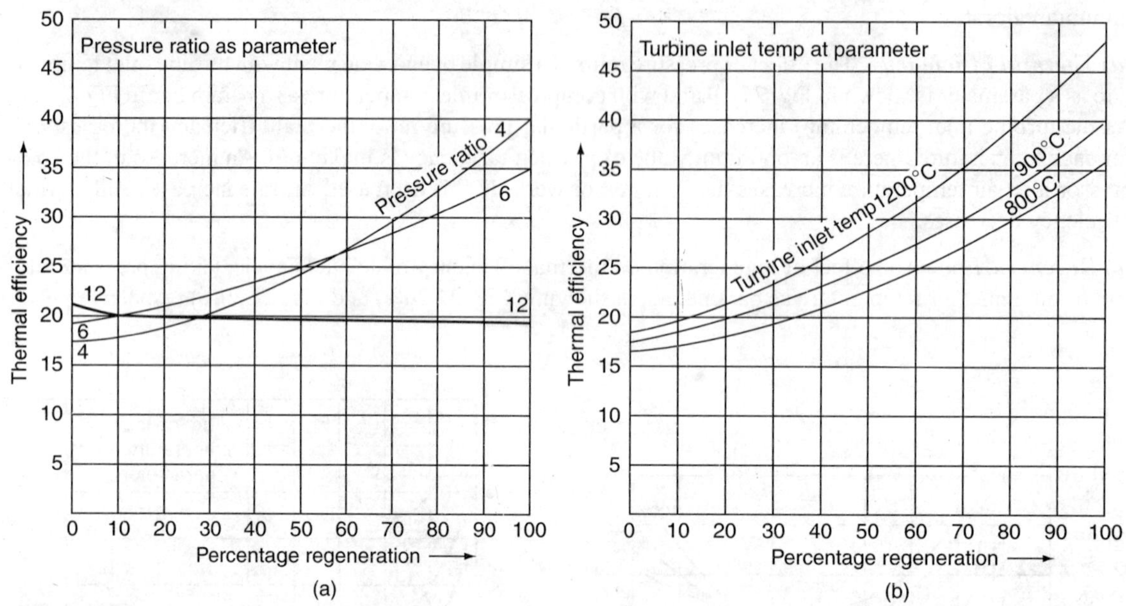

Fig. 21.20 *Effect of regeneration on thermal efficiency of simple cycle with (a) Pressure ratio and (b) Turbine inlet temperature as parameters*

21.8 COMPONENTS OF A GAS TURBINE PLANT

The construction and operation of the components of a gas turbine plant are necessary for proper understanding and design.

(a) Compressor The high flow rates of air through the turbine and the relatively moderate pressure ratios necessiate the use of rotary compressors. The types of compressors commonly used are

1. Centrifugal compressors
2. Axial flow compressors

These have been discussed in Chapter 19. The centrifugal compressor is comprised of two major parts, the impeller, or rotating component and the diffusor. The air enters the compressor at the hub and it then moves radially outward through the impeller and into the diffusor. The impeller converts the mechanical energy, available to the compressor, into kinetic energy, plus heat due to friction, in the working media. The diffusor then transforms the kinetic energy in the air into pressure energy in accordance with Bernoulli's principle. The flow through the diffusor is subject to frictional losses as well. Also, because the air leaves the impeller radially, it must normally be turned 90° to enter the combustion chamber or regenerator, involving more frictional losses. The choice of the blade shape (i.e., bent backward, forward or straight radialers) and the compressor rpm depend on stress limits and manufacturing costs.

In general, the *centrifugal compressor*, as compared to axial flow, is more rugged, simpler, relatively insensitive to surface deposits, has a wider stability range, is less expensive, and attains a higher pressure ratio per stage. However, the efficiency is lower, the diameter larger, and it is not readily adaptable to multi-staging. The single-stage compressors for use in industry may obtain efficiencies from 80 to 84% at pressure ratios between 2.5 and 3, while for aircraft use, pressure ratios are between 4 and 4.5 with efficiencies in the range of 76 to 81%.

The important characteristics of the *axial flow compressor* are its high peak efficiencies, adaptability to multistaging to obtain higher overall pressure ratios, high flow-rate capabilities, and relatively small diameter. However, the axial flow compressor is sensitive to changes in air flow and rpm, which result in a rapid drop off in efficiencies, i.e., the stability range of speeds for good efficiencies is small.

The axial flow compressor consists of a series of rotor-stator stages. The rotor comprises a series of blades that move relative to a series of stationary blades called the stator. The blades transmit the mechanical energy into kinetic energy in the air. Compression is accomplished in both the rotor and stator blades into pressure energy (i.e., continually diffusing it from a high velocity to a lower velocity with a corresponding rise in static pressure). The details of the flow diagram and the velocity triangles as well as the power input and efficiency of the compressor are given in Chapter 19.

(b) Combustion Chamber (CC) Its characteristic feature is that combustion of fuel has to take place with high air velocities (50 to 140 m/s) to limit the size of the cc and with high air fuel ratios (50:1 to 250:1) to keep the turbine inlet temperature to permissible limits.

In an open cycle GT plant combustion may be arranged to take place in one or two large cylindrical can-type combustion chambers (CC) with ducting to convey the hot gases to the turbine. Combustion is initiated by an electric spark and once the fuel starts burning, the flame is required to be stabilized. A pilot or recirculated zone is created in the main flow to establish a stable flame which helps to sustain combustion continuously. The common methods of flame stabilization are by swirl flow and by bluff body.

Figure 21.21 shows a can-type combustor with swirl flow flame stabilization. About 20 per cent of the total air from the compressor is directly fed through a swirler to the burner as primary air, to provide a rich fuel-air mixture in the primary zone, which continuously burns, producing high temperature gases. Air flowing through the swirler produces a vortex motion creating a low pressure zone along the axis of the CC to cause reversal of flow. About 30 per cent of total air is supplied through dilution holes in the secondary zone through the annulus round the flame tube to complete the combustion. The secondary air must be admitted at right points in the CC, otherwise the cold injected air may chill the flame locally thereby reducing he rate of reaction. The secondary air not only helps to complete the combustion process but also helps to cool the flame tube. The remaining 50 per cent of air is mixed with burnt gases in the tertiary zone to cool the gases down to the temperature suited to the turbine blade materials.

Figure 21.22 shows a can-type combustor with a bluff body stabilizing the flame. The fuel is injected upstream into the air flow and a sheet metal cone and perforated baffle plate ensure the necessary mixing of

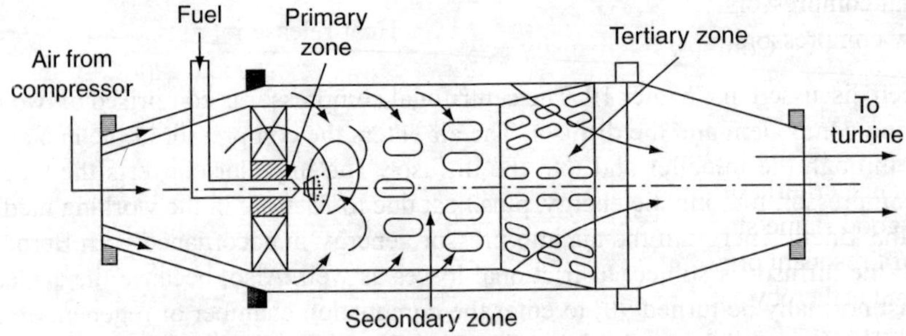

Fig. 21.21 *Can-type combustor with swirl flow flame stabilizer*

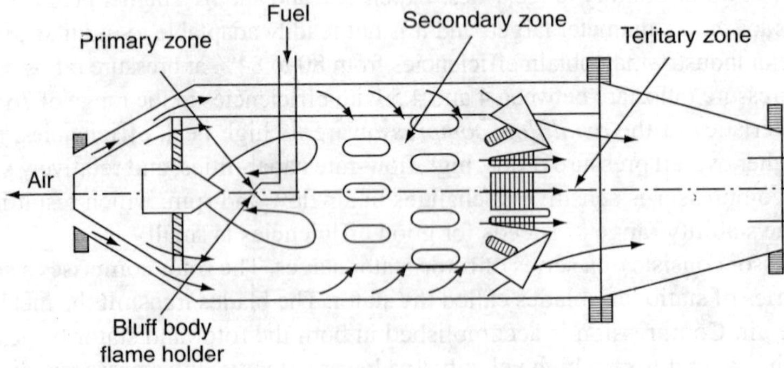

Fig. 21.22 *Can-type combustor with bluff-body flame stabilizer*

fuel and air. The low pressure zone created downstream side causes the reversal of flow along the axis of the CC to stabilize the flame. Sufficient turbulence is produced in all three zones of the CC for uniform mixing and good combustion.

The air-fuel ratio in a GT plant varies from 60/1 to 120/1 and the air velocity at entry to the CC is usually not more than 75 m/s. There is a rich and a weak limit of flame stability and the limit is usually taken at flame blowout. Instability of the flame results in rough running with consequent effect on the life of the CC.

Because of the high air-fuel ratio used, the gases entering the HP turbine contain a high percentage of oxygen and therefore if reheating is performed, the additional fuel can be burned satisfactorily in HP turbine exhaust, without needing further air for oxygen.

A team "combustion efficiency" is often used in this regard, which is defined as follows.

$$\text{Combustion efficiency} = \frac{\text{Theoretical fuel–air ratio for actual temperature rise}}{\text{Actual fuel air ratio for actual temperature rise}}$$

Theoretical temperature rise depends on the calorific value of the fuel used, the fuel-air ratio and the initial temperature of air. To evaluate the combustion efficiency, the inlet and outlet temperatures and the fuel and air mass flow rates are measured. The fuel used in aircraft gas turbine is a light petroleum distillate or kerosene of gross calorific value of 46.4 MJ/kg. For gas turbines used in power production or in cogeneration plants, the fuel used can be natural gas.

In order to give a comparison of combustion chambers operating under different ambient conditions, combustion intensity is defined as the following.

$$\text{Combustion intensity} = \frac{\text{Heat release rate}}{\text{Volume of CC} \times \text{inlet pressure}}$$

The lower the combustion intensity, the better the design. In aircraft a figure of about 23 kW/(m^{3}atm) is normal, whereas in large industrial plants it is about 0.2 kW/(m^3 atm).

The performance criteria of the combustion chamber are (1) low pressure loss, (2) high combustion efficiency, and (3) good flame stability. Flame stability implies steady and continuous flame. This is serious to the phenomena of resonant or pulsating combustion, and due to blowouts, where the flame is blown out of the exit of the CC and is thereby extinguished, which may happen in aircraft applications. The main requirements for a good CC are (1) low carbon deposit in the CC, turbine, and regenerator, (2) low weight and frontal area, (3) reliability and serviceability, and (4) thorough mixing of cold air with the hot products of combustion to give uniform temperature distribution.

(c) *Gas Turbines* Like steam turbines, gas turbines are also of the axial-flow type (Fig. 21.23). The basic requirements of the turbines are light weight, high efficiency, reliability in operation and long working life. Large work output can be obtained per stage with high blade speeds when the blades are designed to sustain higher stresses. More stages are always preferred in gas turbine power plants, because it helps to reduce the stresses in the blades and increases the overall life of the turbine. The cooling of gas turbine blades is essential for long life as it is continuously subjected to high temperature gases.

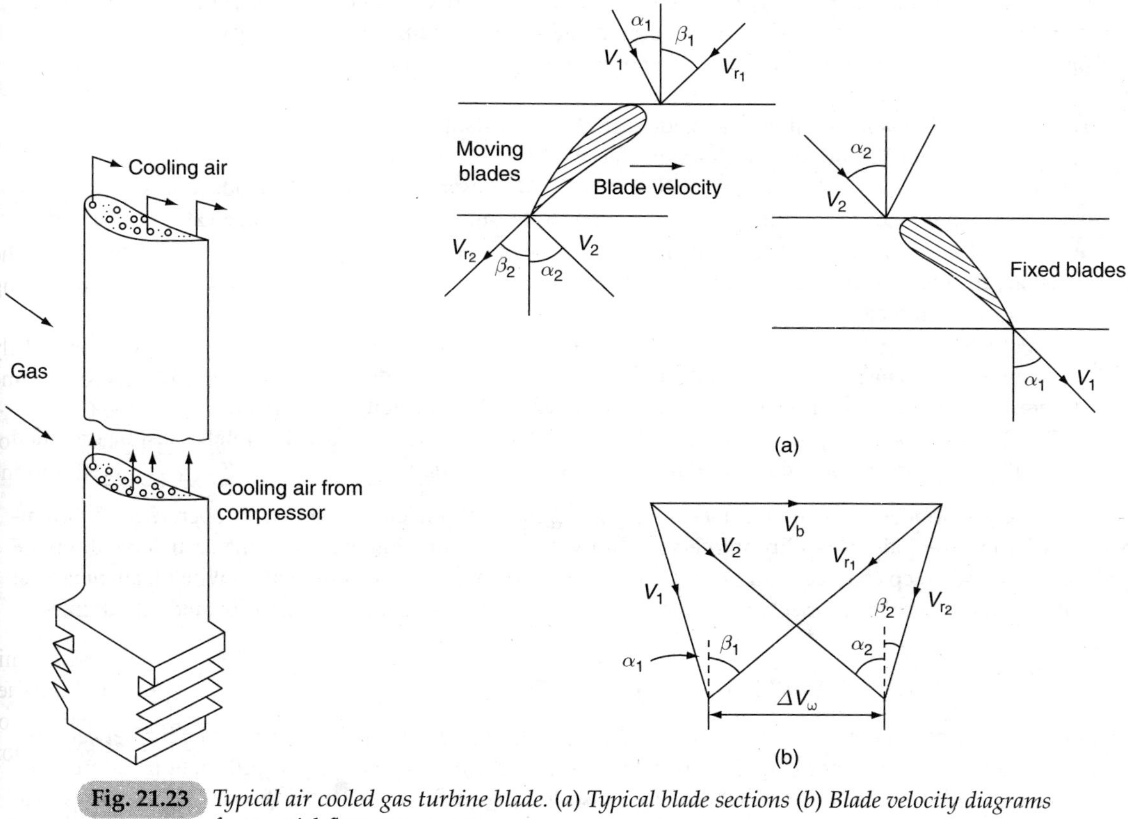

Fig. 21.23 *Typical air cooled gas turbine blade. (a) Typical blade sections (b) Blade velocity diagrams for an axial-flow compressor*

Blade angles of gas turbines follow the axial-flow compressor blading (Fig. 21.23(a)), where the degree of reaction is not 50 per cent. It is usually assumed for any stage that the absolute velocity at inlet to each stage (V_2) is equal to the absolute velocity at exit from the moving blades (i.e. V_2) and that the same flow velocity V_f is constant throughout the turbine.

The degree of reaction, R, as defined for a steam turbine, is valid for gas turbines also. It is the ratio of the enthalpy drop in the moving blades to the enthalpy drop in the stage. As shown in Fig. 21.23(a), we have

$$R = \frac{V_{r_2}^2 - V_{r_1}^2}{2V_b \Delta V_w} = \frac{V_f^2(\sec^2\beta_2 - \sec^2\beta_1)}{2V_b(V_f \tan\beta_2 + V_f \tan\beta_1)}$$

$$= \frac{V_f(\tan^2\beta_2 - \tan^2\beta_1)}{2V_b(\tan\beta_2 + \tan\beta_1)} = \frac{V_f}{2V_b}(\tan\beta_2 - \tan\beta_1) \qquad (21.11)$$

Putting $R = 0.5$ in Eq. (21.11), we get $\quad V_f(\tan\beta_2 - \tan\beta_1) = V_b$

or $\quad V_b + V_f \tan\alpha_2 - V_f \tan\beta_1 = V_b \quad \alpha_2 = \beta_1$

It also follows that $\alpha_1 = \beta_2$. The fixed and moving blades have the same cross-section and the diagram is symmetrical.

Vortex Blading is the name given to the twisted blades which are designed by using three dimensional flow equations with a view to decrease fluid flow losses. A radial equilibrium equation can be derived (*see the book of Cohen et al.*) and it can be shown that one set of conditions which satisfies this equation is as follows.

(a) Constant axial velocity along the blades, i.e. V_f = constant.
(b) Constant specific work over the annulus, i.e. $V_b \Delta V_w$ = constant.
(c) Free vortex at entry to the moving blades. i.e. $V_{w1} r$ = constant, where r is the blade radius at any point.

Since the specific work output is constant over the annulus, it can be calculated at the mean radius, and multiplied by the mass flow rate it becomes the power for the stage. Since the fluid density varies along the blade height, the density at the mean radius can be used, so that $\dot{m} = \rho_m V_f A$, where A is the blade annular area.

(d) *Duct work* The duct work consists of ducts between the compressor and the combustion chamber, combustion chamber to the turbine, and the exhaust duct. The ducts must be sized to minimize the pressure losses, as the loss in pressure directly reduces the capacity of the plant.

Ducts should be supported from the floor to reduce vibrations. Expansion joints must be provided to allow for dimensional changes due to temperature variation.

The basic requirements for the turbines are light weight, high efficiency, ability to operate at high temperatures for long periods, reliability and serviceability. The determination of blading material depends on the stress-rupture and creep characteristics of the various blading materials, in combination with mechanical and thermal stresses, resistance to mechanical and thermal shock, and resistance to corrosion and vibration.

21.9 GAS TURBINE WITH WATER INJECTION

One method used to improve the performance of the gas turbine is to inject water into the air between the compressor and the regenerator (Fig. 21.24). The quantity of water injected is just sufficient to saturate the compressed air. Any excess water injected may cause fouling of the regenerator. Increased mass flow rate flowing through the turbine increases the turbine output without increasing the compressor work input. Water

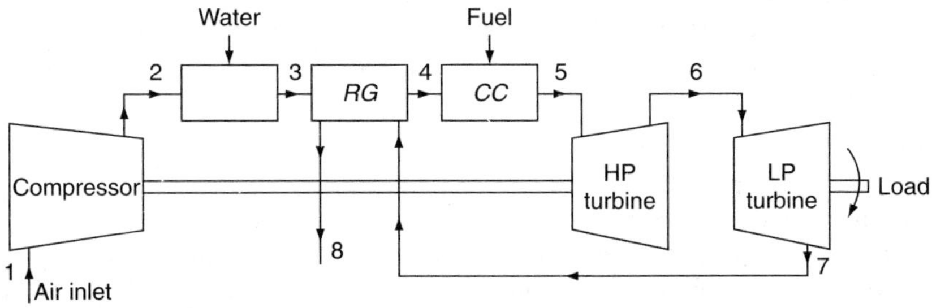

Fig. 21.24 *Schematic layout of a two-shaft open cycle gas turbine with water injection and regenerator*

injection is most commonly used as a power boost for take-off and emergency requirement with jet propelled aircraft (see later). However, if water has impurities, it may cause corrosion or deposits on the blades, which will have a detrimental effect on the performance and maintenance.

21.10 GAS TURBINE FUELS

Gas turbines are basically designed to operate on petroleum-based fuels like natural gas, kerosene, aviation fuel and residual fuel oil. Other fuels like powdered coal, sewage gas, etc. are also being actively considered.

(a) Natural Gas It generally contains a major percentage of methane and a small percentage of ethane propane and butane. The sulphur compound (H_2S) is kept below 0.1 per cent by volume. It is obtained from wells in oil-fields and used for auxiliary power production within the oil-fields.

(b) Liquid Fuels The liquid fuels range from gas oil to residual fuel oil. The major consideration in selecting the fuel is cost. The cheapest fuels are always preferred.

Distillate fuels in the gas oil range (boiling point varies between 200°C to 370°C) may be used in a gas turbine without any difficulty. Residual fuels include fuel oils, furnace oils, boiler fuel oils, etc. If the viscosity of oil is high, some heating arrangement need be provided. Caution should be observed against corrosion of blades and other components by sulphur compounds and vanadium.

(c) Solid Fuels The use of coal as fuel for closed cycle gas turbine plant is universally accepted, but its use in open cycle plant is now in active development. Coal is burnt in two modes, viz. (i) integrated gasification, where coal is completely or partially gasified and the fuel gas produced is consumed in the gas turbine combustor and (ii) pressurized bubbling or circulating fluidized bed, where the fuel gas, after it is adequately filtered, expands in the gas turbine. Coal is normally considered as a gas turbine fuel in combined cycle power generation. The development of proper filters, ceramic or others, is the key to its use.

21.11 GAS TURBINE MATERIALS

The combustion chamber of a diesel engine is subjected to the highest temperature of the cycle, 2000°C, to 2500°C, for a very short period during only one stroke of the cycle. During the remaining three strokes, the engine gets time to be cooled. Therefore, special materials are not required for diesel engine plants.

In gas turbine plants, however, the components are continuously exposed to the hot gases and are made of special materials with necessary arrangements for cooling. Blades are also subjected to high centrifugal stresses due to high rotative speeds, in addition to thermal stress. These have a high creep rate, due to which the blades increase in length gradually. Contact with the casing can thus occur with resultant failure. Blade materials should possess the following properties.

(i) Materials must withstand high temperature and high stress.
(ii) It must have low creep rate.
(iii) It must have high resistance to oxidation, corrosion and erosion.
(iv) It must neither be brittle at ordinary temperature nor plastic like when hot.
(v) It must have good castability or forgeability characteristics, depending on the process of manufacture.
(vi) It must have good machinability to achieve precise dimension.
(vii) It must have high resistance to fatigue failure.
(viii) It must maintain structural stability when exposed to varying temperature.

All these required properties cannot be obtained in one material. Therefore, the selection of material for each component is a difficult job.

1. *Metals for Turbine Rotor Discs* The turbine rotor disc is subjected to centrifugal and thermal stresses. The thermal stresses (due to temperature gradient) can be reduced by using an alloy of high conductivity.

The disc hub stresses tend to cause tensile deformation. This can be minimised by using a material of low expansion coefficient. Austenitic steels with 12 to 18 per cent chromium, 8 to 12 per cent nickel and small percentages of tungsten, molybdenum and titanium are used for turbine rotor discs.

These days the turbine discs are cooled by tapping compressed air from the compressor. Therefore, less expensive materials can be used. Ferritic steels having higher creep strength at low temperature (up to 600°C) can be used for the central portion, whereas austenitic steel is used on the outer surface of the ferritic rotor disc.

2. *Material for Turbine Rotor Blade* Blades are subjected to the highest stresses and temperatures. Most satisfactory materials for blades are the stainless steel alloys and 8-20 nickel chromium alloys, known as Nimonic alloys. These alloys have high resistance to oxidation, scaling with ceramics (silicon carbide, silicon nitride, aluminium nitride, etc.) on the blades of nimonic alloys provides better mechanical properties. Blades are cooled by compressed air taken by a bleed from the compressor.

3. *Material for Combustion Chamber* The gas turbine combustion chamber is generally made of Nimonic 75 alloy. This alloy has an excellent creep resistance, capacity to withstand heavy thermal shocks, and high resistance to oxidation.

4. *Material for Compressor* The impeller of centrifugal compressor is subjected to high centrifugal and thermal stresses, the latter being due to the temperature difference between the air inlet and air discharge temperatures. To minimise centrifugal stresses, lighter materials like aluminium alloys are used. These alloys suffer from high thermal expansion, for which allowance is provided.

The axial flow compressor blades are now made of titanium alloys, which are of low density, possess good strength at high temperatures (400–500°C) and are strongly resistant to corrosion. Light weight, good creep strength and fatigue resistance are attractive features of titanium alloys.

21.12 JET PROPULSION SYSTEM

Jet propulsion, like all means of propulsion, is based on Newton's second and third laws of motion. Newton's second law states that *the rate of change of momentum in any direction is proportional to the force acting in that direction*. Newton's third law states that *for every action there is an equal and opposite reaction*.

With regard to vehicles operating entirely in a fluid, the reaction principle is based on imparting momentum to a mass of fluid in such a manner that the reaction of the imparted momentum furnishes a propulsive force. Peculiar to jet propulsion, however, this mass of fluid, whose velocity has been increased, is rejected from the vehicle in a jet stream. The jet aircraft draws in air and expels it to the rear at a markedly increased velocity; the rocket greatly changes the velocity of the fuel which it ejects rearward in the form of products of combustion. In each case, the *action* of accelerating the mass of fluid in a given direction creats a *reaction* in the opposite direction in the form of a propulsive force. The magnitude of this propulsive force is defined as *thrust*.

Aircraft propulsion may be achieved by using a heat engine to drive an airscrew or propeller, or by allowing a high-energy fluid to expand and leave the aircraft in a rearward direction as a high-velocity jet. In the propeller type of aircraft engine, the propeller takes a large mass flow and gives it a moderate velocity backwards relative to the aircraft. In the jet engine, the aircraft induces a relatively small air flow and gives it a high velocity backwards relative to the aircraft. In both cases the rate of change of momentum of the air provides a reactive forward thrust which propels the aircraft. The propeller-type engine can be driven by a petrol engine or by a gas turbine unit.

If the velocity of the jet backwards relative to the aircraft is V_j and the velocity of the aircraft is V_o, then the atmospheric air, initially at rest, is given a velocity of $(V_j - V_o)$ (Fig. 21.25). The thrust available for propulsion is solely due to the rate of change of momentum of the air stream.

$$\text{Thrust per unit mass-flow rate} = V_j - V_o$$

$$\text{propulsive power is then} = V_o(V_j - V_o)$$

This is the rate at which work must be done in order to keep the aircraft moving at the constant velocity V_o against the frictional resistance or drag.

$$\text{Net work output from the engine} = \text{Increase in kinetic energy of air} = V_j^2 - V_o^2$$

It is used in the two ways: (i) it provides the thrust and (ii) it gives the air, previously at rest, an absolute velocity $(V_j - V_o)$ and KE $(V_j - V_o)^2/2$. Therefore,

$$V_o(V_j - V_o) + \frac{(V_j - V_o)^2}{2} = V_o V_j - V_o^2 + \frac{V_j^2 - 2V_jV_o + V_o^2}{2}$$

$$= \frac{V_j^2 - V_o^2}{2}$$

∴ Propulsive efficienty,

$$\eta_p = \frac{2V_o(V_j - V_o)}{V_j^2 - V_o^2} = \frac{2V_o}{V_j + V_o}$$

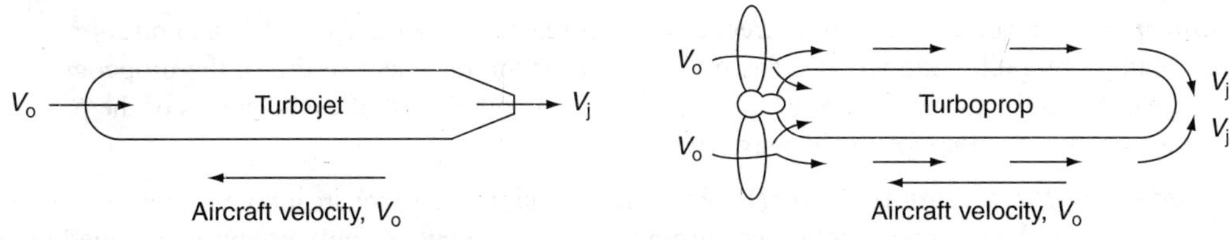

Fig. 21.25 *Flow through a turbojet and turboprop*

21.13 PROPULSIVE DEVICES FOR AIRCRAFTS AND MISSILES

Various propulsive devices used for aircrafts and missiles are given below:

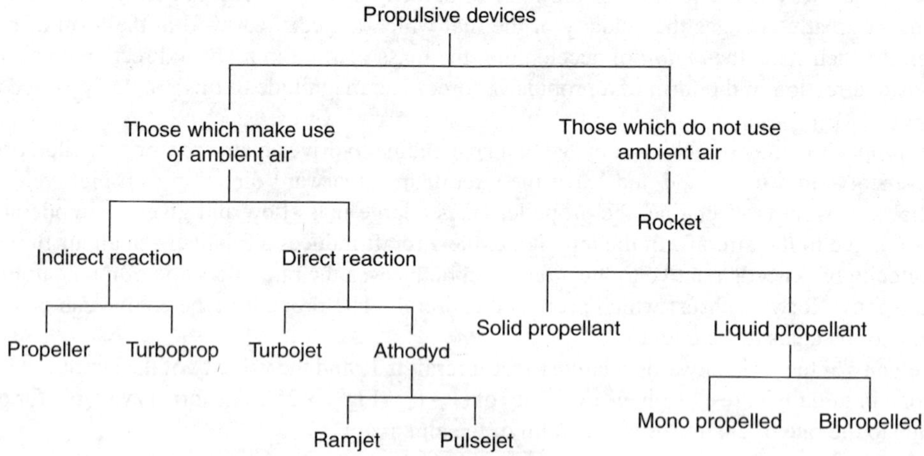

Propulsive devices are basically of two types: those which make use of atmospheric air as the main working medium supplying oxygen for combustion of fuel, *the atmospheric jet engines*, and those which carry oxygen required for combustion of fuel, the *rockets*. The performance of jet engines depends on the forward speed of the engine and upon the atmospheric pressure and temperature. The rocket engine carries its own oxidizer for the combustion of fuel and is, therefore, independent of the atmospheric air as well as the forward speed.

The devices which make use of the ambient air are further subdivided into indirect reaction and direct reaction devices. The main propulsive devices are the following:

1. *Propeller* It is an indirect reaction device. Earlier, it used to be driven by the reciprocating internal combustion engine. A propeller handles relatively a large mass of air and accelerates it rearwards at low speeds. It is the reaction of the rate of change of momentum of the air, called the thrust, which propels the aircraft. The function of the engine is only to revolve the propeller at the desired speed. Piston engines are, however, now used only for small aircrafts.

2. *Turbojet* A turboject is the most important direct reaction device. It utilizes a gas turbine power plant. In a turbine, partial expansion takes place to produce just sufficient power to drive the compressor. The exhaust of the turbine which is at a pressure higher than the atmospheric pressure is expanded in a nozzle given a high-velocity jet. Compared to propeller units, in turbojet units a small mass of air flows through the unit, but has a high rearward velocity. Turbojets are very efficient at high speed and high altitude, and inefficient at low speed and low altitude.

3. *Turboprop* It is a combination of indirect and direct reaction devices (propeller and turbojet). Thrust is produced both by propeller and jet. Besides the compressor the turbine also drives the propeller through a reduction gear. It has the thermal advantage of a turbojet, combined with the advantages of the propeller for efficient take-off, particularly for the heavily loaded aircraft.

4. *Athodyd (Aero-thermodynamic Duct)* The ram jet and the pulse jet are athodyds, i.e., a straight duct-type of jet engine without a compressor and turbine wheels. The entire compression is obtained by a ram, eliminating the need of a turbine. Athodyds are used for pilotless aircraft, helicopter rotor and missiles.

5. ***Rocket*** It does not use ambient air for propulsion. Both the fuel and oxidizer are carried with the power plant, and are accelerated from zero velocity to a high velocity at nozzle outlet. A rocket is the only propulsion device suitable for space travel.

21.13.1 Turbojet Engine

The basic components of a turbojet engine are shown in Fig. 21.26 are:

(*a*) *Diffuser* The ambient air enters with a velocity equal to the velocity of the aircraft and this velocity is slowed down in the diffuser. The kinetic energy of the airstream is converted to pressure energy. This is called 'ram compression'.

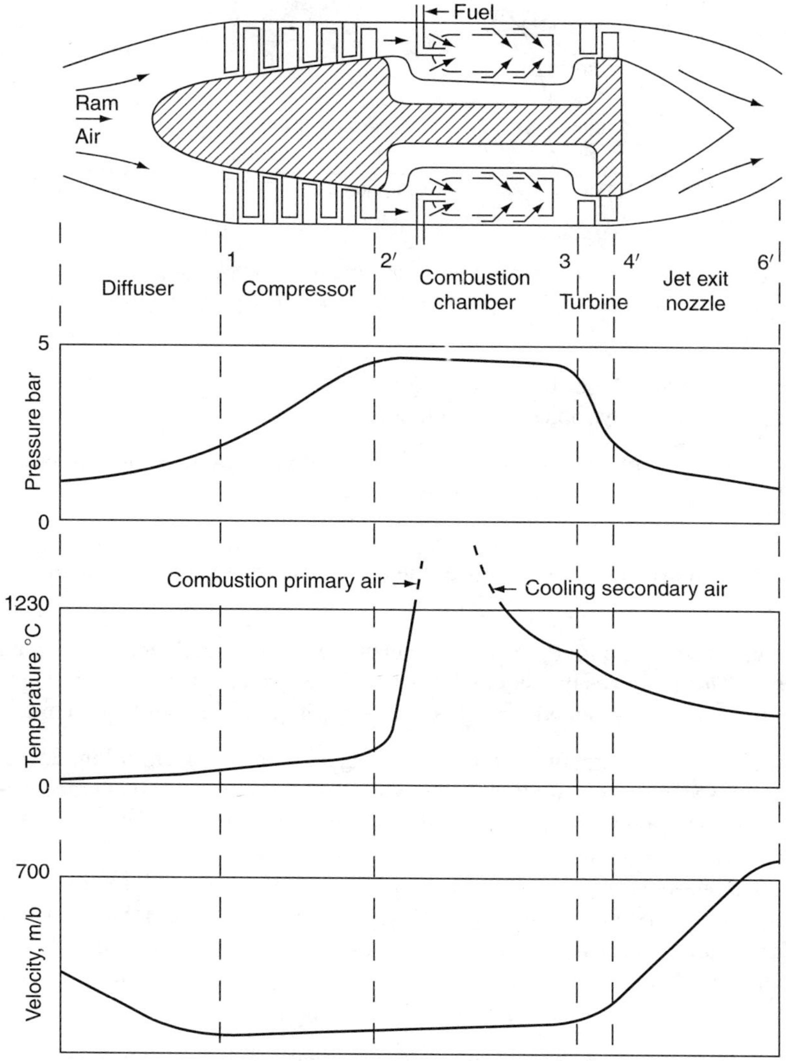

(a)

Fig. 21.26 *Continued*

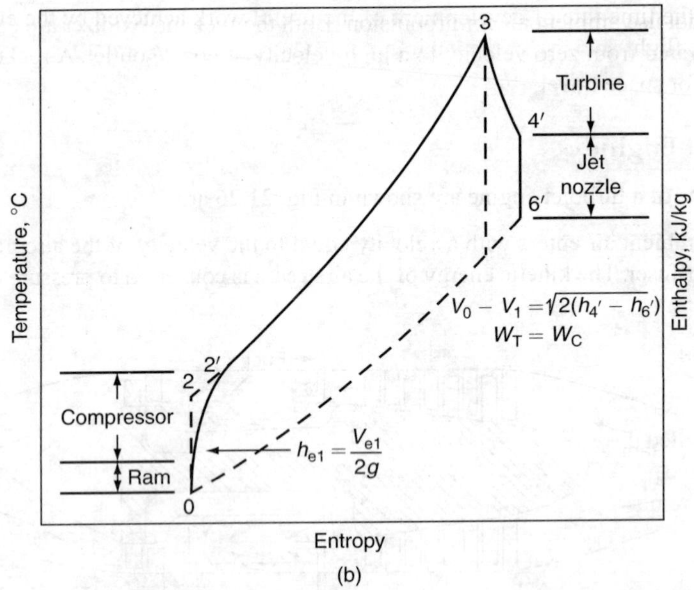

Fig. 21.26 *(a) General representation of the pressure, temperature, and velocities of the air and gases of combustion as they pass through the engine, (b) Temperature-entropy diagram of a typical turbojet engine cycle*

(b) Compressor Air leaving the diffuser with negligible velocity enters the compressor (usually an axial flow type) and is compressed polytropically to high pressure.

(c) Combustion Chamber (cc) Compressed air enters the combustion chamber where the fuel is sprayed and it is assumed that the combustion takes place at constant pressure.

(d) Turbine The products of combustion of high pressure and temperature undergo polytropic expansion to an intermediate pressure such that the output of the turbine is just sufficient to run the compressor and the auxiliaries of the unit.

(e) Nozzle The gases coming out of the turbine expand down to the ambient pressure and a high-velocity jet leaves the nozzle. This produces the required thrust and the aircract is propelled in the forward direction. This method of propulsion is best suited for aircrafts flying with a speed of 800 km/h or more.

The temperature – entropy diagram for a typical turbojet engine is given in Fig. 21.26(b). The entering atmospheric air is diffused isentropically from velocity V_0 down to zero ($V_1 = 0$) in process 0–1. The hot gases leaving the turbine are assumed to expand isentropically (process 4′–6′), and the turbine work W_T is equal to the compressor work ($W_T = W_c$). The rise in ram pressure ratio p_1/p_o increases with Mach number. At a Mach number of 2, the ideal ram-pressure ratio is 8. The thrust T is the magnitude of propulsive force created by the jet engine depending on the rate of change of air flowing through the engine. Since the weight rate of flow of fuel through the engine is normally in the vicinity of 1% of the rate of flow of air, it will not introduce any appreciable error if it is assumed that the working medium is comprised of air only. It can be expressed as

$$T = w_a(V_j - V_o) \text{ newtons}$$

where w_a = mass flow rate of air, kg/s; V_j = exit velocity of gases leaving the nozzle, m/s; and V_o = vehicle velocity through the air, m/s. Since the atmospheric air is assumed to be at rest, the velocity of air entering the engine is the velocity of the vehicle V_o.

Thrust power, TP, the time rate of development of the useful work achieved by the engine, is the product of the thrust times the flight velocity of the vehicle, or

$$\text{TP} = TV_o = w_a(V_j - V_o)\, V_o \tag{21.12}$$

Propulsive power, PP, representing the energy required to change the momentum of the mass flow of air, may be expressed as the difference between the kinetic energies of the entering air and the exit gases, or

$$\text{PP} = \frac{w_a\,(V_j^2 - V_o^2)}{2} \tag{21.13}$$

Therefore, the *propulsive efficiency*, η_p, may be expressed as

$$\eta_p = \frac{\text{TP}}{\text{pp}} = \frac{2(V_j - V_o)V_o}{V_j^2 - V_o^2} = \frac{2V_o}{V_j + V_o} = \frac{2}{1 + \dfrac{V_j}{V_o}} \tag{21.14}$$

This is also often called the Froude efficiency. As $V_o \to V_j$, η_p approaches maximum value. But as this occurs, the thrust and propulsive power approach zero. Thus, the ratio of velocities (V_j/V_o) for maximum efficiency and for maximum power are not the same.

An alternate way of defining propulsive efficiency is to express propulsive power in terms of thrust power and kinetic energy losses, i.e.,

$$\text{PP} = \text{TP} + \text{K.E. losses}.$$

The propulsive efficiency then becomes

$$\eta_p = \frac{\text{TP}}{\text{PP}} = \frac{\text{TP}}{\text{TP} + \text{K.E. losses}} \tag{21.15}$$

The main components of a turbojet unit and of on open-cycle gas turbine unit are the same and as such, the performance of a turbojet unit depends on the pressure ratio of the compressor, efficiency of the individual constituents and the turbine inlet temperature. Since the turbojet unit propels an aircraft at different altitudes with varying speed, the performance of these units is a function of the flight speed and altitude. Figure 21.27(a) shows the variation of propulsive efficiency with flight speed and Fig. 21.27(b) shows the variation of thrust power with flight speed at different altitudes. It is seen that as the aircraft velocity

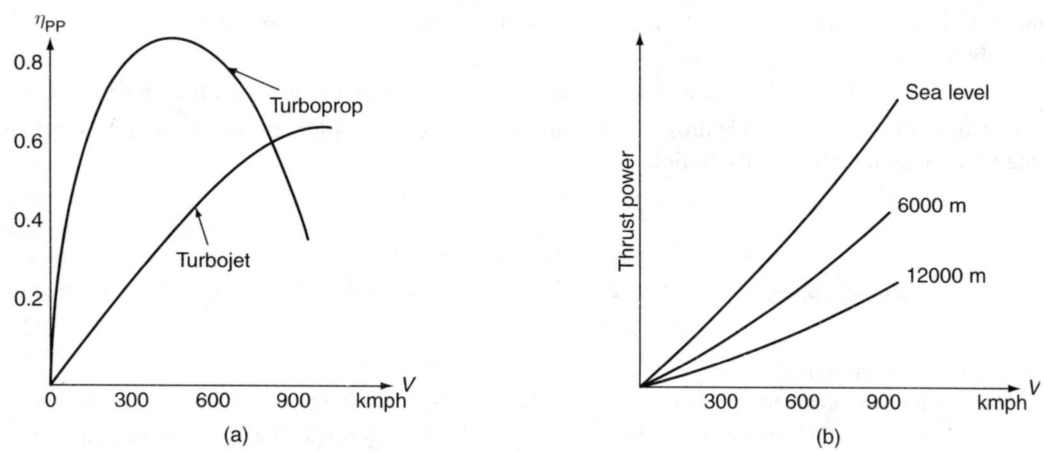

Fig. 21.27 *(a) Variation of the propulsive efficiency with flight speed, (b) Variation of thrust power with flight velocity at different altitude for a turbojet engine*

V_o increases, the propulsive efficiency increases. For a propeller driven aircraft the change of η_p is greater initially, but at speeds at which the propeller tip attains the sonic velocity, η_p falls off rapidly and Eq. (21.14) is no longer valid. For aircraft speeds upto about 850 km/h, the propeller is the more efficient means of propulsion, but for speeds above this the jet engine is superior.

The thrust power decreases with increasing altitude because the density of the ambient air decreases with increasing altitude, But it has been found that the drag on the aircraft decreases at a greater rate than the air density and therefore, it is possible to fly the aircraft at a higher speed and with better economy.

21.13.2 Pressure Thrust

In aircraft gas turbine work it becomes important to use stagnation conditions, since velocity changes through the unit is no longer negligible. Also, in general, temperature - measuring instruments such as thermocouples, measure stagnation temperature and not static temperature: using stagnation conditions, the isentropic efficiencies of the compressor and turbine, and intake duct and nozzle or jet pipe efficiency are redifined.

Referring to Fig. 21.28(b) of a typical jet engine, we have

$$(\eta_s)_{\text{intake duct(ram)}} = \frac{T_{01s} - T_0}{T_{01} - T_0}, \qquad (\eta_s)_{\text{compressor}} = \frac{T_{02s} - T_{01}}{T_{02} - T_{01}}$$

$$(\eta_s)_{\text{turbine}} = \frac{T_{03} - T_{04}}{T_{03} - T_{04s}}, \qquad (\eta_s)_{\text{nozzle}} = \frac{T_{04} - T_{05}}{T_{04} - T_{5s}}$$

For adiabatic flow, the total temperature remains constant and $T_0 = T_{01}$. There is a loss of pressure in the combustion chamber from 2 to 3.

It was assumed that the gases expand down to atmospheric pressure in the jet nozzle. In the case of a convergent nozzle, the back pressure will normally be lower than the nozzle exit pressure. This phenomenon is called *underexpansion*, which is explained in Chapter 17.

Due to the difference in pressure between the nozzle exit and the atmosphere in which the aircraft is flying, there will be an additional thrust, called the *pressure thrust.* In the case of supersonic aircraft, the pressure at the air intake is higher than the atmospheric pressure because of compression through the shock wave formed, which will reduce the net thrust calculated purely from momentum considerations.

If we consider an aircraft like the turbojet in Fig. 21.28(a) with an air intake of area A_1, inlet air pressure p_1, and a nozzle exit area A_2, exit pressure p_2, and the atmospheric pressure p_0, we have from Newton's second law of motion,

$$F + p_1 A_1 - p_2 A_2 = \text{rate of change of air in the direction of motion of the fluid}$$

where F is the net force due to the hydrostatic pressure and friction exerted by the inside of the aircraft on the working fluid in the direction of its motion,

$$\therefore \quad F + p_1 A_1 - p_2 A_2 = \dot{m}(V_j - V_o)$$

$$\therefore \quad F = \dot{m}(V_j - V_a) - p_1 A_1 + p_2 A_2.$$

There is an equal and opposite force, R; exerted by the working fluid on the inside of the aircraft engine,

$$R = \dot{m}(V_j - V_a) - p_1 A_1 + p_2 A_2$$

in the direction of motion of the aircraft.

Let us consider the forces acting on the aircraft. There is the force R, there is the total drag D due to the air resistance, and there is a pressure force due to the atmospheric pressure acting on the projected area in the direction of flight. In flight there is considerable pressure variation over the aircraft surfaces causing lift and drag forces, the total drag force D and the form drag due to the vortices formed.

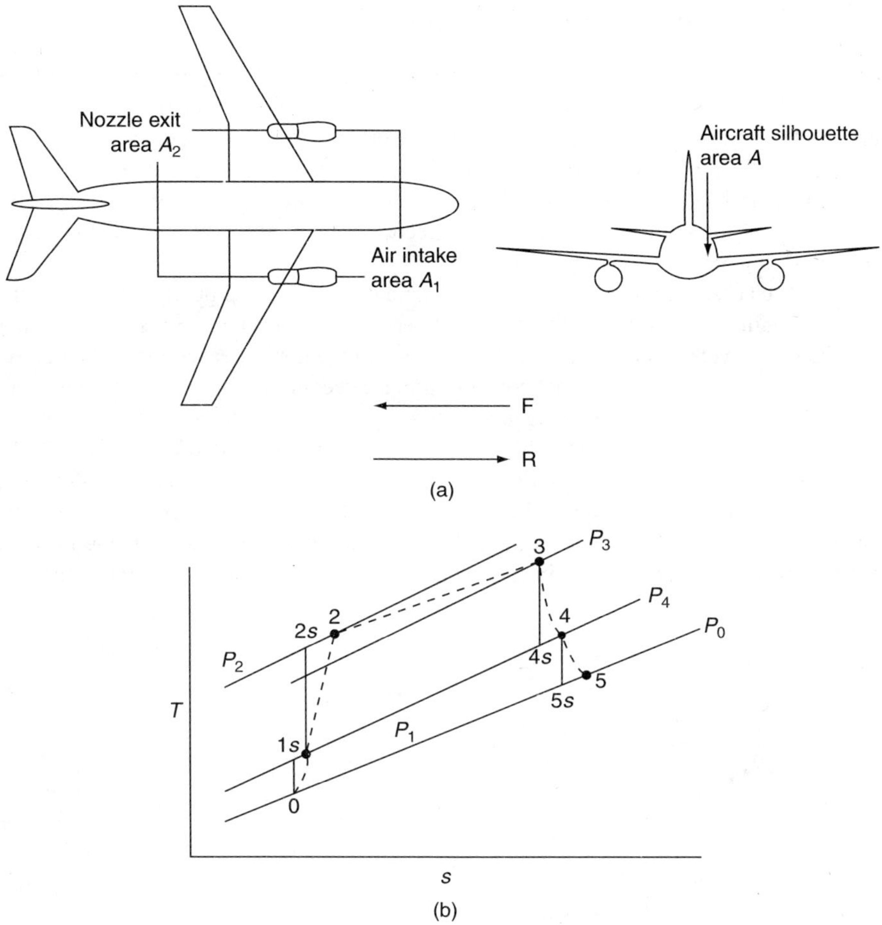

Fig. 21.28 *(a) Plan view of aircraft and aircraft silhouette (b) T–s diagram of a jet engine*

Assuming that the aircraft silhouette (profile) area in the direction of flight is *A,* then the net pressure force in the direction of flight is given by

$$p_0(A - A_2) - p_0(A - A_1) = p_0(A_1 - A_2).$$

Since the aircraft is flying at constant velocity, the net force acting is zero, i.e,

$$R - D + p_0(A_1 - A_2) = 0$$

Therefore, the total thrust required to overcome the total drag force is given by

$$\text{Total thrust } (T) = D = R + p_0\,(A_1 - A_2)$$

$$= \dot{m}(V_j - V_o) - p_1A_1 + p_2A_2 + p_0(A_1 - A_2)$$

or,

$$T = \dot{m}(V_j - V_o) + A_2(p_2 - p_0) - A_1(p_1 - p_0)$$

For subsonic aircraft, however, the last term is zero, since then $p_1 = p_o$.

$$\therefore \qquad \text{Total thrust} = \dot{m}\left(V_j - V_0\right) + A_2(p_2 - p_0) = \text{Momentum thrust} + \text{Pressure thrust.}$$

21.14 TURBOJET ENGINE WITH AFTERBURNER

A turbojet engine cannot produce the extra thrust necessary during take-off, for high and for increased maneuvering of military aircraft. So these units are equipped with additional devices for augmenting the thrust. The thrust can be increased (i) by increasing the mass flow rate of the working fluid, or (ii) by increasing the jet velocity.

The mass flow rate of the working fluid is increased by injecting a mixture of water and alcohol (methanol) into the ram air stream at the compressor inlet. The evaporation of the liquid mixture during the compression process cools the air and the compressor work input decreases. The increase in mass flow rate through the turbine and the nozzle increases the jet velocity.

The most widely used method for achieving thrust augmentation in the turbojet engine is the use of an after-burner. This system requires burning of additional fuel between the turbine and nozzle and this is equivalent to a reheater of the open-cycle gas turbine unit (Fig. 21.29). Since the increased thrust is required for a short period of time, an increase in the specific fuel consumption is acceptable while the unit is developing the additional thrust. Thus, the working fluid enters the nozzle at a higher temperature and the jet velocity at exit would increase. Since the velocity of the working fluid at inlet to the burner should be sufficiently low for stable combustion and minimum pressure losses, a diffuser is provided between the turbine outlet and burner inlet. Moreover, the exit area of the nozzle can also be varied so that the engine may operate as a simple turbojet.

Another diagram of a turbojet engine equipped with after burning or tail pipe burning is shown in Figure 21.30. Tail-pipe burning consists of introducing and burning fuel between the turbine and the nozzle,

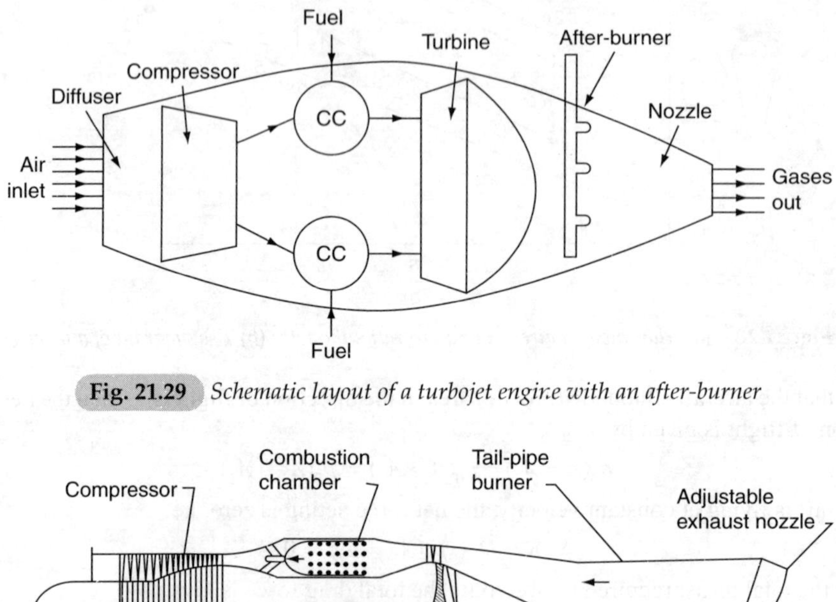

Fig. 21.29 *Schematic layout of a turbojet engine with an after-burner*

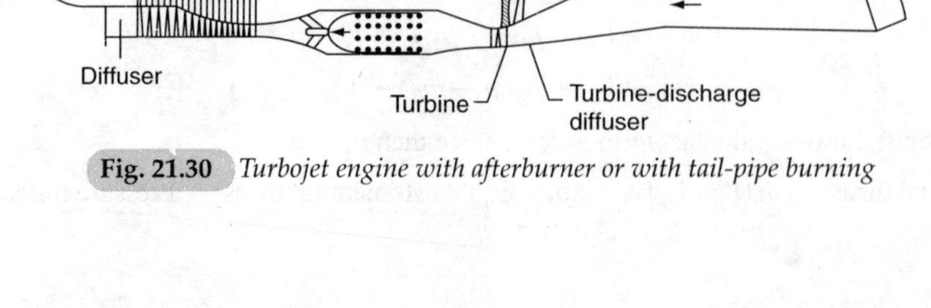

Fig. 21.30 *Turbojet engine with afterburner or with tail-pipe burning*

raising the gas exit velocity and producing an increased thrust. Since the turbojet with tail-pipe burning consists of a turbojet and essentially a ram jet(explained later), the engine is sometimes designated a turbo-ram-jet. Tail-pipe burning is not only an augmentation device for improving the take-off and high speed performance of an airplane, but it also may be considered a distinct type of engine for flight at supersonic speed.

21.15 TURBO-PROP ENGINE

A turbojet engine is quite efficient at flight sppeds above 800 km/h. However, at lower speeds it is not so efficient. This led to the development of turbo-prop engines. A schematic layout of a turbo-prop engine is shown in Fig. 21.31. It comprises of a geared propeller connected to a turbojet engine. The turbine of the turboprop engine is bigger than that of the turbojet engine as it drives both the compressor and the propeller. The propeller consumes about 80 to 90 per cent of the net power available from the turbine and the remaining 10 to 20% of turbine power is left to produce the jet thrust.

Turbo prop engines having two independent turbines have operating convenience for control. One of the turbines drives the compressor, while the other drives the propeller through a reduction gear (Fig. 21.31). The turbine speeds are from 11,000 to 40,000 rpm and propeller speeds are one-tenth to one-twentieth of it. The turboprop engine cycle, Fig. 21.32 is the same as that of the turbojet engine cycle, except that the turbine expansion process is greater. The energy supplied to the propeller, with no losses in the reduction gear, can be written as

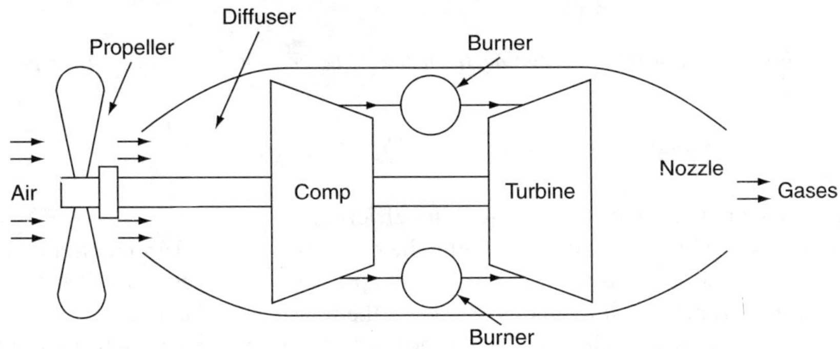

Fig. 21.31 *Schematic layout of a turbo-prop jet engine*

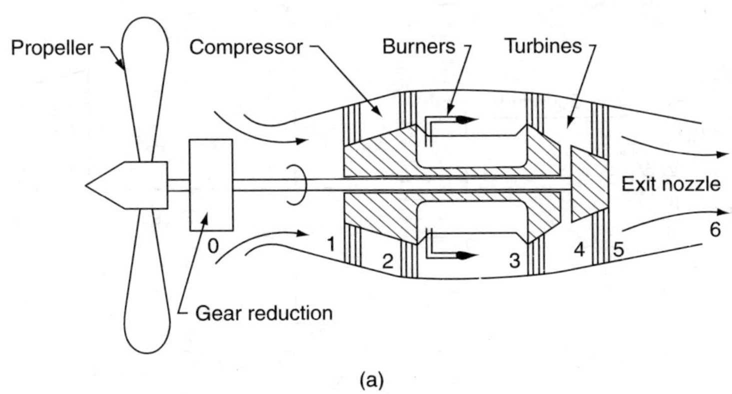

Fig. 21.32 *Continued*

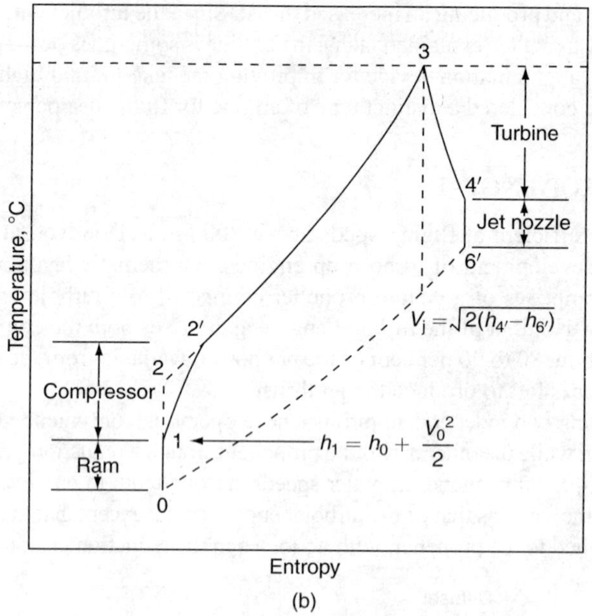

Fig. 21.32 *(a) Two-shaft turbo-prop engine, (b) Temperature-entropy diagram for a turboprop engine*

$$\text{Propeller power} = \eta_g \times \eta_p \times \left[(h_3 - h_{4'}) - (h_{2'} - h_1)\right] \frac{\text{kJ}}{\text{kg}} \tag{21.16}$$

where η_g = reduction gear efficiency and η_p= propeller efficiency.

A propeller is used to produce a thrust by changing the momentum of fluid around it. Figure 21.33 shows the flow through a propeller. It is assumed that the flow upstream at the section 1 is undisturbed and the fluid moves towards the propeller with velocity V_1. Since the rotation of blades causes a reduction in pressure on the upstream side of the propeller, the fluid is accelerated towards the propeller and it flows over the

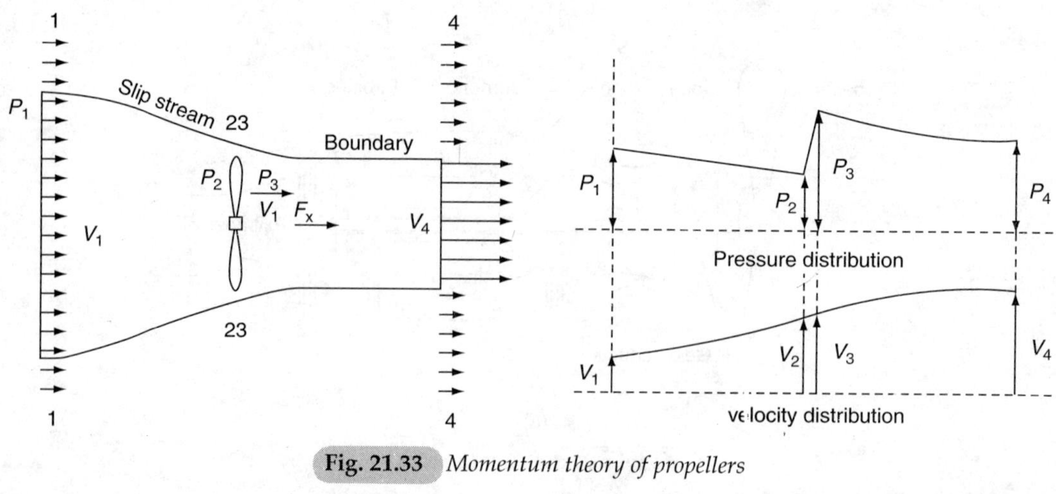

Fig. 21.33 *Momentum theory of propellers*

propeller with velocity V, and the pressure increases considerably. The velocity at the downstream section 4 is V_4. The body of the fluid affected by the propeller is called 'slipstream', and it the pressure is assumed constant all along the slipstream boundary and at sections 1 and 4. Thus the thrust exerted by the propeller is given by

$$F_x = \rho Q (V_4 - V_1) = (p_3 - p_2) A \tag{21.17}$$

where A = cross-sectional area swept by the propeller $V = \dfrac{V_1 + V_4}{2}$, the average velocity of the upstream and downstream fluids

$$Q = \text{volume flow rate} = A \times V.$$

The useful workdone by the propeller (rate) $P = F \times V = \rho Q (V_4 - V_1) V_1$

$$\text{Power input to the propeller} = \frac{1}{2} \rho Q \left(V_4^2 - V_1^2\right)$$

$$\therefore \text{ Propeller efficiency, } \quad \eta_p = \frac{\rho Q (V_4 - V_1) V_1}{\frac{1}{2} \rho Q \left(V_4^2 - V_1^2\right)} = \frac{2 V_1}{V_4 + V_1} = \frac{V_1}{V} \tag{21.18}$$

Figure 21.34 shows the variation in the thrust produced by a turbo-prop engine with flight velocity. Since the thrust produced at lower flight velocity is considerably higher, the turbo-prop engine has a good take-off characterstics. From Fig. 21.27(a), it is evident that the propulsive efficiency of a turbo-prop engine is higher than that of a turbojet engine at lower flight velocity (up to 800 km/h). The present trend requires higher flight velocities ($M > 1$) and this would require considerable research and experimentation before a turbo-prop will reach its peak performance and service operation.

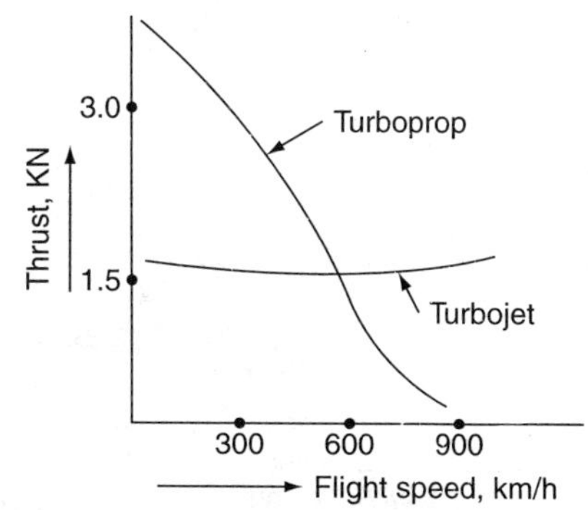

Fig. 21.34 *Variation of thrust with flight speed for a turbojet and a turboprop engine*

21.16 BYPASS TURBO-JET ENGINE

A bypass turbojet engine increases the thrust without adversely affecting the propulsive efficiency and fuel economy (Fig. 21.35). There is a fan at the front intake driven by the main shaft. The part of the air drawn by the fan is sent over the combustion chamber (CC) through suitable ducting, to the exhaust unit, thus bypassing the engine. A portion of air is sent to the engine compressor with an added advantage of creating a supercharging effect. The bypass ratio is selected according to the aircraft operational requirement.

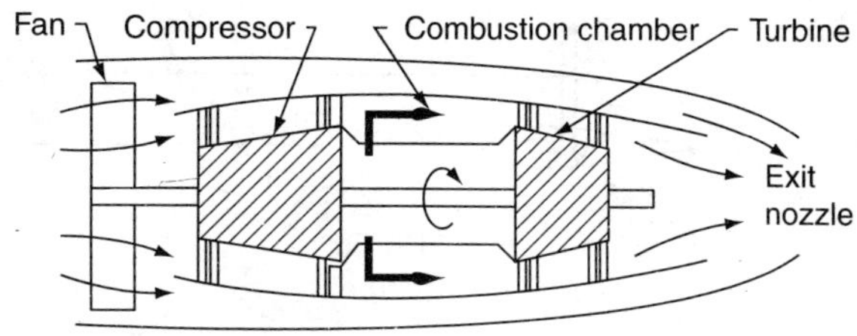

Fig. 21.35 *Bypass turbojet engine*

Its advantage is that the lower the velocity of jet efflux, the better the propulsive efficiency and fuel economy, than a straight jet for a given thrust with less K.E. lost to atmosphere. This applies for aircrafts flying at subsonic speeds and for long range. Installation and fire protection are also assisted by the air-cooled duct in the bypass surrounding the hot parts of the engine.

21.17 ATHODYDS-RAMJETS AND PULSEJETS

It is earlier mentioned that at high aircraft speeds, turbojet is designed to take advantage of ram compression. At a Mach number of 2, the ideal ram pressure ratio is nearly 8. At such high ram compression there is no necessity of a mechanical compressor, and so no turbine is required to drive it. Ramjet and pulsejet power plants, called *athodyds* utilize this principle. The word 'athodyd' is derived from the words *aero-thermodynamicduct.*

21.17.1 Ramjet

A ram jet, also known as Lorin tube, is the simplest propulsion device among the air-breathing engine category. It consists of a convergent-divergent diffuser, a combustion chamber and an exit nozzle or tail pipe (Fig. 21.36). In the absence of compressor the compression of air is obtained by the diffuision of the

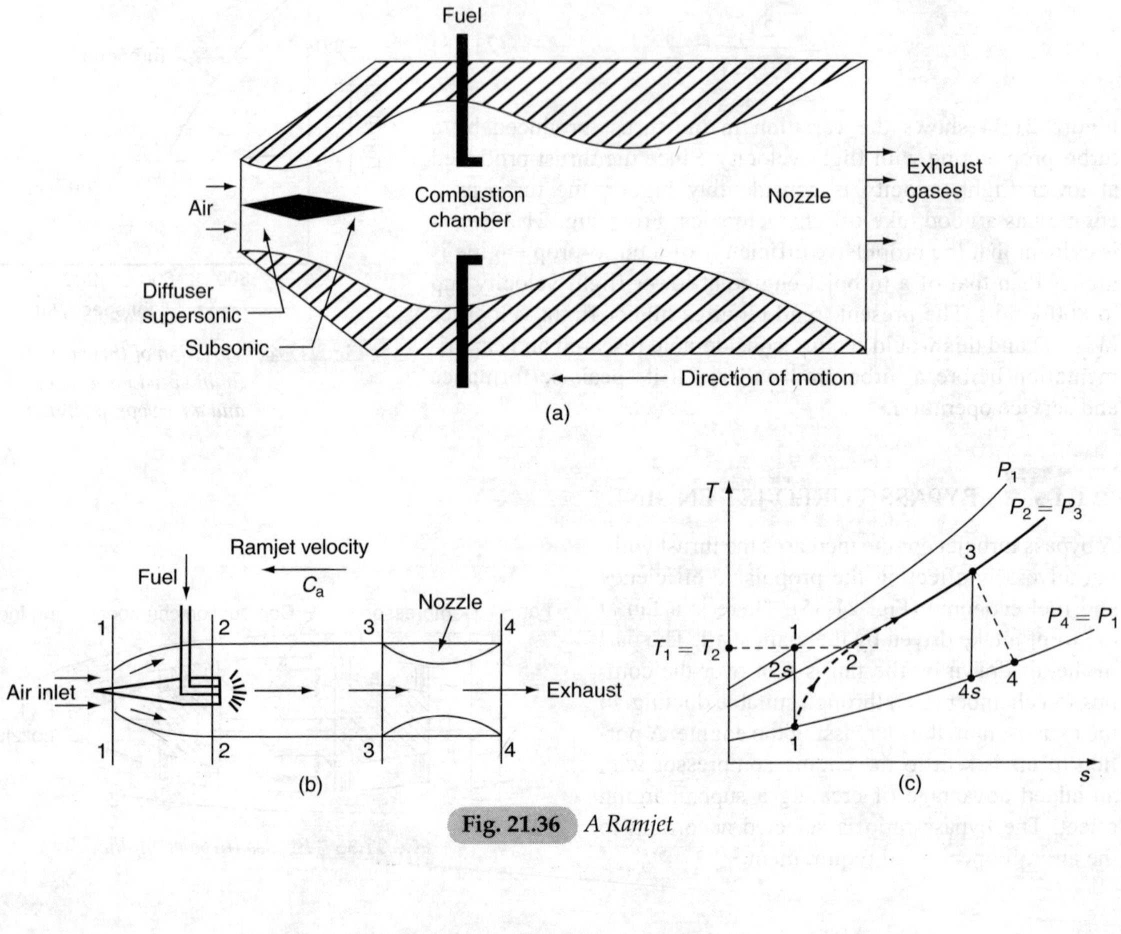

Fig. 21.36 *A Ramjet*

high-velocity air stream approaching the diffuser inlet. The velocity is supersonic and the stream is first compressed adiabatically while passing through the normal and oblique shocks. The velocity after the shocks is subsonic while the diffusion takes place in the diverging section of the diffuser, the Mach number being limited to about 0.25 at inlet to the combustion chamber (cc) so that the flame in the CC is stable. The ideal ram pressure ratio increases with the increase in Mach number. (Fig. 21.37) At $M = 2.5$, the pressure ratio p_2/p_i is about 17 which is very high. However, there will be pressure losses due to shocks, wall friction and flow separation as a result of which this pressure ratio is not obtained.

The fuel is sprayed into the CC with injection nozzles and a spark plug initiates the combustion process. The air–fuel ratio is about 15:1 and the temperature after combustion is very high, about 2000 K, much higher than that in the turbojet engines. The products of combustion expand through the nozzle producing the required thrust. In the *T–s* diagram of the ram jet (Fig. 21.38), the combustion process takes place at constant pressure.

For ideal compression and expansion,

$$p_{02}/p_1 = \left(1+\frac{\gamma-1}{2}M_i^2\right)^{\frac{\gamma}{\gamma-1}}$$

where p_{02} is the total pressure after isentropic compression, and

$$p_{03}/p_4 = \left(1+\frac{\gamma-1}{2}M_e^2\right)^{\frac{\gamma}{\gamma-1}}$$

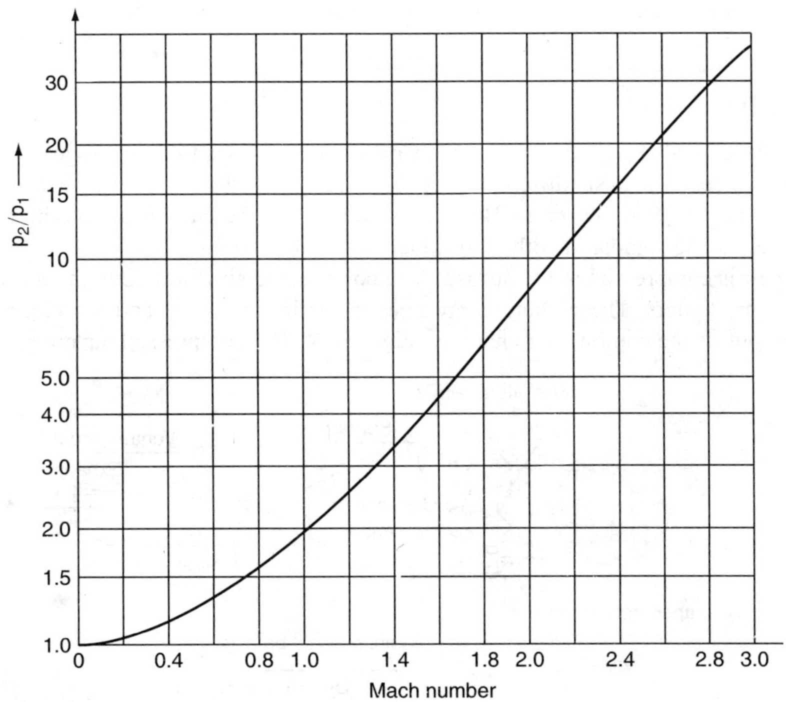

Fig. 21.37 *Ideal ram pressure ratio vs. Mach number of vehicles for sea level conditions*

Since the compression and expansion ratios are equal,

$$1+\frac{\gamma-1}{2}M_i^2 = 1+\frac{\gamma-1}{2}M_e^2, \text{ or } M_i = M_e$$

$$\frac{V_1}{\sqrt{\gamma RT_1}} = \frac{V_e}{\sqrt{\gamma RT_e}}, \text{ or } V_e = V_1\left(\frac{T_4}{T_1}\right)^{1/2}$$

Thrust, $$T = \dot{m}_g V_e - \dot{m}_a V_1 \quad (21.19)$$

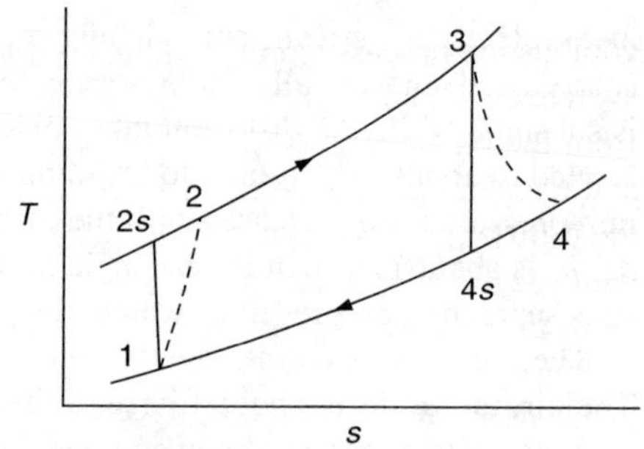

Fig. 21.38 *T–s plot for Ramjet*

21.17.2 Ram-jet Characteristics

1. The system does not contain any moving parts, is light in weight, is less costly and requires no maintenance.
2. The fuel consumption is much lower at higher speeds but is much higher at low and moderate speeds.
3. The system can accept a large variety of fuels.
4. The design of diffuser is quite difficult because the successful operation of ram jet is solely dependent on this convergent-divergent passage.
5. The device is a continuous burning duct with a continuous flow of atmospheric air at high speeds and this requires an elaborate device for fuel injection.
6. The device produces greater thrust per unit weight than any other propulsion device at supersonic speeds except the rockets.
7. The system is not self starting, it cannot accelerate from a rest position and therefore requires a launching device.

21.17.3 Pulse Jet

The pulse jet was developed in Germany during the world war II and is similar to ram jet. It also does not have a compressor and turbine. The basic difference between a ram jet and a pulse jet is that the former is a continuously operating engine working on the Brayton cycle whereas the latter is an intermittent firing engine, the operating cycle may be compared with Otto cycle and is self starting.

The incoming air is compressed in the diffuser. The compressed air flows through mechanically operated non-return valves or shutters. These shutters are opened for incoming air and are closed by the expanding gases coming out of the combustion chamber, Fig. 21.39. The compressed air mixes with fuel and the

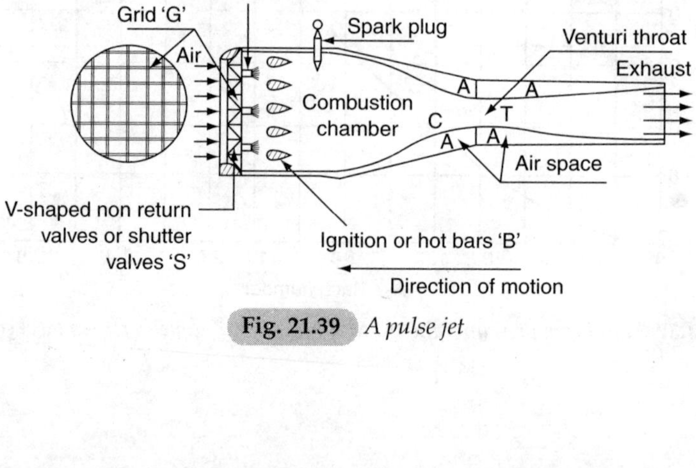

Fig. 21.39 *A pulse jet*

combustion process is initiated by a spark plug. Once the engine starts operating normally, the spark plug is switched off and the residual flame inside the combustion chamber and hot ignition bars which retain heat from the previous explosion is used for ignition in the succeeding cycles. During combustion the pressure and temperature of the products of combustion increase and when its value is greater than the pressure at the diffuser outlet, the mechanical valves are closed and the gases expand through the nozzle producing the required thrust. After the gases leave the combustion chamber, the pressure decreases and the valves open to allow the compressed air again. The air mixes with fuel and the cycle repeats.

Pulse jet engines are much cheaper than turbo-jet engines and are self starting (compressed air is supplied to the system) but its propulsive efficiency is lower than the turbo-jet engines.

21.18 ROCKET PROPULSION

A rocket engine does not draw its oxidiser from the surrounding air. The fuel and the oxidizer, called the *propellant*, is carried by the propelling unit. The function of the rocket is thus independent of speed and has an unlimited operating altitude.

The rocket motors using a chemical fuel are generally divided into two categories: *Solid propellant rocket* and *liquid propellant rocket*. A hybrid rocket combines solid fuel with a liquid oxidizer.

The schematic diagram of a solid propellant rocket is shown in Fig. 21.40. It consists of a seamless tube, usually made of steel, closed at one end. The tube is filled with solid propellant called the '*grain*'. The grain which contains both oxidizer and fuel is electrically ignited and burns fully as there is no method to stop the burning. The open end holds the convergent – divergent nozzle, through which the gases are ejected out at supersonic speed. The reaction to the ejection of the high velocity gases produces the thrust of the rocket motor.

The grain configuration for solid propellant rocket motor varies according to the required thrust-time program for the engine, on which depends the rate of burning. There are two principal types of solid propellants: (a) heterogeneous or composite in which inorganic oxidizer such as potassium perchlorate or sodium nitrate is dispersed in a fuel matrix like organic plastic, asphalt or oil mixtures (75% oxidizer and 25% fuel), and (b) homogeneous propellant consisting of the colloid of nitroglycerine-nitrocellulose: $C_3H_6\ (NO_3)_3$ – $C_6H_7O_2(NO_3)_3$. The main advantage of solid propellant rockets is its simplicity, having no moving parts and any fuel supply system. The disadvantages are that it has to be large enough to store the entire amount of propellant and is strong enough to withstand high pressures (40 to 140 bar) and temperatures (1600 to 3000°C). There is no provision of cooling. These rockets are suitable for producing thrust for short durations. They are used to power rocket projectiles, guided missiles and as boosters for aircraft as well as spacecraft. The use of additional thrust by rocket motors at take-off is termed jet-assisted take-off (JATO) or rocket-assisted take-off (RATO).

As stated earlier the solid propellant rocket differs from other engines in that total mass of fuel is stored and burned within the combustion chamber. There is no fuel supply system.

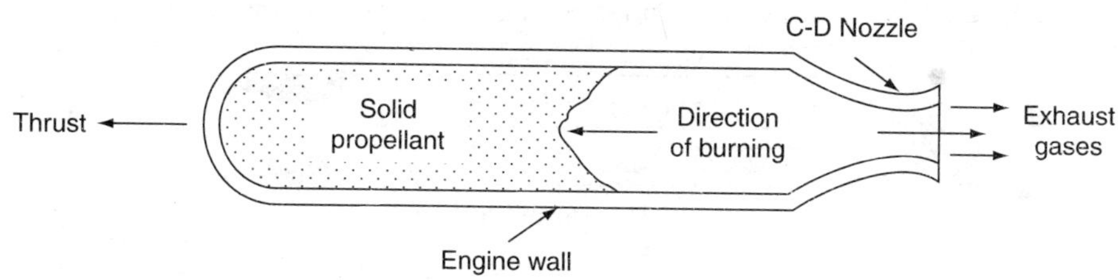

Fig. 21.40 *Schematic diagram of a solid propellent rocket*

The solid propellant rocket motor consists of a seamless tube, usually made of steel, closed solidly at one end. The open end holds the nozzle which may be a single or multi-orifice type. Usually, small projectile rockets (less than 10 cm in diameter) use single orifice nozzles, while larger ones use multi-orifice nozzles. The single orifice nozzles have their axis parallel to the axis of the motor, while the axis of the orifices of the multi-orifice nozzle may be at an angle to the axis in order to rotate the projectile, thus providing spin stabilization. It should be noted, however, that JATO and large missile booster units are often single nozzle type engines. The solid propellant rockets are divided into two main types according to the amount of surface area exposed to burning. These two types are restricted burning and unrestricted burning.

The restricted burning rocket is one in which the propellant is constrained to burn on only one surface, Fig. 21.41(a). The manner in which a cigarette burns is similar to the type of burning of a restricted burning rocket. For example, the charge in a JATO unit is poured in while liquid and fits the chamber tightly so that it is only free to burn on one end. The unrestricted burning rocket is essentially free to burn on all surfaces at the same time, Fig. 21.41(b). The restricted burning rocket delivers a small thrust for a relatively long period while the unrestricted burning rocket delivers a relatively large thrust for a short period. Table 18-1 shows an approximate comparison between the two types of solid propellant rockets.

Liquid propellant rockets utilize propellant stores in the container outside the combustion chamber. The liquid propellants are of two types: monopropellant and bipropellant. A monopropellant is a liquid that requires no auxiliary material (oxidizer) for bringing about the release of the thermochemical energy. Nitromethane, propyl nitrate and hydrogen peroxide (H_2O_2) are some monopropellants with only one storage tank. A liquid propellant rocket consists of three major components, viz, (1) rocket motor, (2) propellant system, and (3) controls. An advantage of a liquid propellant is its ability to discontinue combustion at any time.

The *rocket motor* consists of an exit nozzle, Fig. 21.42, a combustion chamber, propellant injectors and an ignition system. As it was stated previously, most liquid fuel rocket motors are cooled by circulating fuel around the engine walls. The cooling fluid may move axially, or may circulate in a helical path around the motor. Normally, *axial flow* is only used for large size motors, because the amount of fuel used for propulsion of small motors is not sufficient to fill the practical size passages that can be manufactured. *Helical flow,* on

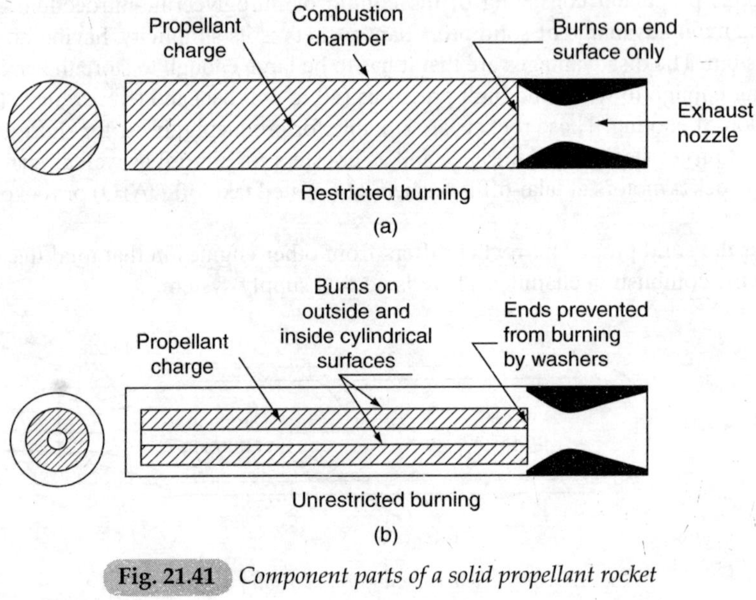

Fig. 21.41 *Component parts of a solid propellant rocket*

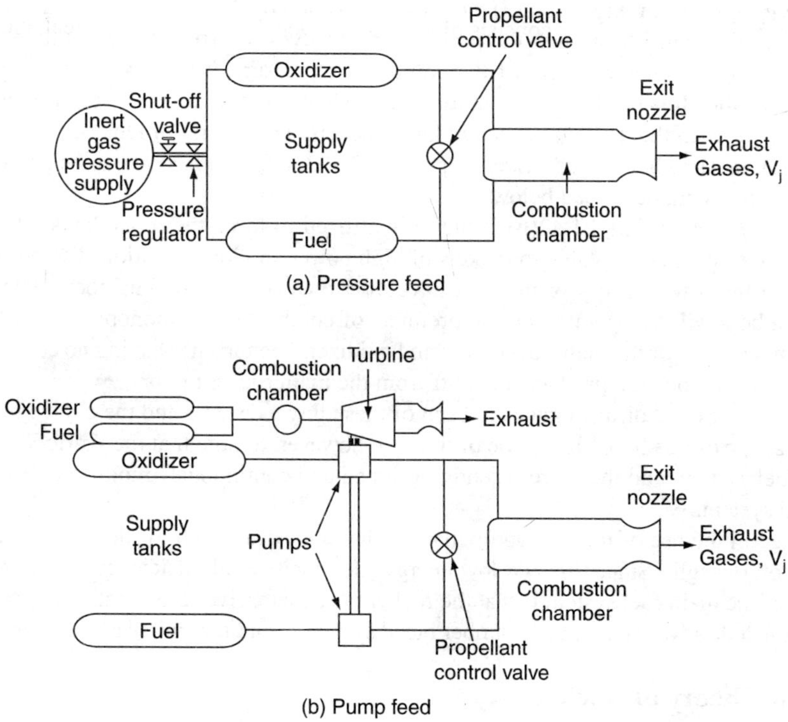

Fig. 21.42 *Liquid bipropellant rocket systems*

the other hand, is usually used in small motors, since it produces too much pressure drop to be practical when considerable quantities of fuel are used as in large size motors.

Film cooling is another method of cooling in which a thin layer of fuel covers the exposed wall surfaces from excessive heat. The thin film of fuel may enter the combustion chamber and nozzle through a series of holes in the motor walls or walls of a special porous material that may be used. This method, however, reduces the efficiency of the engine, because some of the fuel used for cooling is not burned. The best way to arrange for the maintenance of the protective film is not yet known.

The combustion chamber is usually of cylindrical shape (Fig. 21.42) with one end closed and the other end terminating at the entrance to the exit nozzle, which is usually of the DeLaval type. There are no obstructions on the walls between the combustion chamber and the nozzle. In the liquid propellant rocket, the combustion chamber pressures vary from 20 to 50 bar and combustion chamber temperatures from (1650 to 3350)°C. The exhaust gas velocities vary for 1800 to 4000 m/s.

Injection of the fuel and oxidizer into the combustion chamber is accomplished through injectors, which have the same function as those in a compression ignition engine, i.e., atomize and mix the propellants so that a fuel-oxidizer mixture results which can be readily vaporized and burned.

To start a rocket motor an electrical igniter, may be necessary for some propellants. However, propellants which ignite upon contact with oxidizer are more desirable. These self-igniting propellants are called *hypergolic* propellants.

The *propellant system* employs either a pressure feed or a pump feed to transfer fuel from a storage tank to the combustion chamber.

In the *pressure feed* system, Fig. 21.42(a) the pressure exerted by the inert gas stowed under high pressure forces the fuel and the oxidizer through the proportioning valves or orifices that regulate the fuel-oxidizer ratio into the combustion chamber against the combustion pressure, The pressure feed system is simple, inexpensive and reliable. It is limited, however, to small or short duration rockets, because the weight of tanks becomes prohibitive when this system is used in large rockets. The weight of tanks capable of carrying several tons of propellants and being pressurized to pressures of 300 to 750 psia would be greater than the more complex pump feed system described below.

In the *pump feed* system, Fig. 21.42(b), a pump is utilized to force the propellants into the combustion chamber. This feed system is adaptable to rockets of high power and long duration. The pumps are driven by relatively small turbines which may or may not have their own combustion chambers. The turbines having combustion chambers will be operated by the products of combustion of monopropellant (explained later) such as hydrogen peroxide or the main rocket fuel and oxidizer. The turbines having no combustion chambers can be driven by the combustion products bled off from the main rocket motor.

There are advantages and disadvantages to each of these three systems and the choice is based upon factors beyond the scope of this book. It may be of interest, however, to note that the World War II designed V-2 used the monofuel system, and the more recently designed and continuously modified "Viking" rocket uses a chamber bleed system.

A disadvantage in the use of the monopropellant is that the extra weight of the turbine fuel may be from 8 to 5% of the total weight. Since this is a low energy fuel, the over-all efficiency of the rocket is reduced. A disadvantage of the main fuel system is that the rocket fuels themselves are usually too powerful to be used by a turbine, while a disadvantage of the chamber bleed system is that it requires an auxiliary starting system.

21.18.1 Basic Theory of Rockets

Figure 21.43 shows a schematic diagram of the basic components of a liquid bipropellant rocket engine. It consists of an injection system, a combustion chamber, and an exit nozzle. The oxidizer and fuel burn in the CC producing a high pressure. The pressure produced is governed by the weight rate of flow of the propellants, the chemical characteristics of the propellants, and the cross-sectional area of the nozzle throat. The gases are ejected to the atmosphere at supersonic speeds through the nozzle. The nozzle converts the pressure energy of the gases into kinetic energy. The reaction to the ejection of the high-velocity gases produces the thrust of the rocket engine.

The thrust developed is a resultant of the pressure forces acting upon the inner and outer surfaces of the rocket motor. The internal forces acting on the engine are

$$\text{Resultant internal forces} = w_p \cdot V_j + p_j A_j \text{ (newtons)}$$

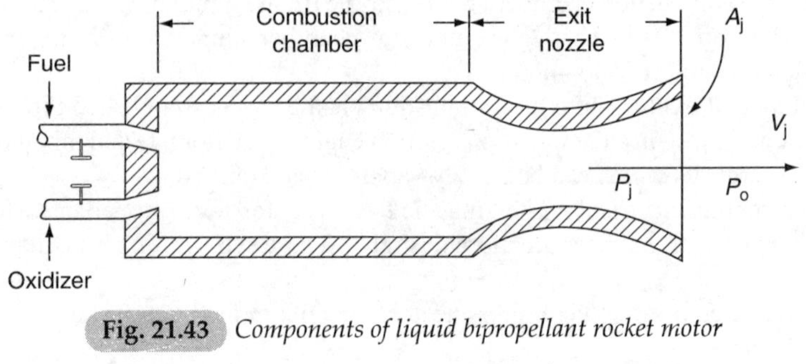

Fig. 21.43 *Components of liquid bipropellant rocket motor*

where w_p = weight rate of flow, kg/s, V_j = exit jet velocity relative to nozzle, m/s, P_j = exit static pressure, N/m^2 and A_j = exit area, m^2.

The resultant external forces acting on the rocket motor are $p_0 A_j$, where p_0 is the atmospheric pressure, N/m^2. The thrust which is a resultant of the total pressure force, becomes

$$T = w_p V_j + A_j\left(p_j - p_o\right) \text{ newtons} \quad (21.20)$$

The above equation shows the effect of atmospheric pressure on the thrust of a rocket engine. The thrust is maximum when $p_0 = 0$, i.e., operating in vacuum.

In testing a rocket engine, the thrust, the total propellant consumption, and the total time are readily measured. The thrust can be expressed as

$$T = w_p \cdot V_{je} \text{ newtons} \quad (21.21)$$

where V_{je} is the effective jet exit velocity defined by

$$V_{je} = V_j + \frac{A_j(p_j - p_o)}{w_p} \text{ m/s}$$

The thrust power, TP, developed by the rocket engine is

$$\text{TP} = \text{TV}_o = w_p V_{je} V_o \text{ (watts)}$$

The propulsive efficiency, η_p is given by

$$\eta_p = \frac{\text{TP}}{\text{PP}}$$

where the propulsive power is the thrust power plus the kinetic energy lost in the exhaust.

$$\text{K.E. loss} = w_p(V_{je} - V_o)^2/2$$

$$\therefore \quad \eta_\phi = \frac{\text{TP}}{\text{TP} + \text{KE loss}} = \frac{w_p V_{je} V_0}{w_p V_{je} V_0 + \left[w_p (V_{je} - V_o)^2/2\right]}$$

$$= \frac{2\,V_o V_j}{1 + (V_o/V_{je})^2} \quad (21.22)$$

Specific impulse, I_{sp}, is the thrust produced per unit weight of propellant consumption

$$I_{sp} = \frac{I}{w_p} = V_{je}$$

It is desirable to use propellants with the greatest possible I_{sp}, since this allows a greater useful load to be carried for a given overall rocket weight.

Let us consider an accelerating rocket as shown in Fig. 21.44. It is assumed that R = air resistance, m_1 = initial mass of rocket, w_p = rate at which the propellant burns and m = mass of the rocket and propellant at a given time. Assuming that the pressure at the nozzle exit is the same as the ambient pressure and the momentum of the system does not change with time, the equation of linear momentum can be written as

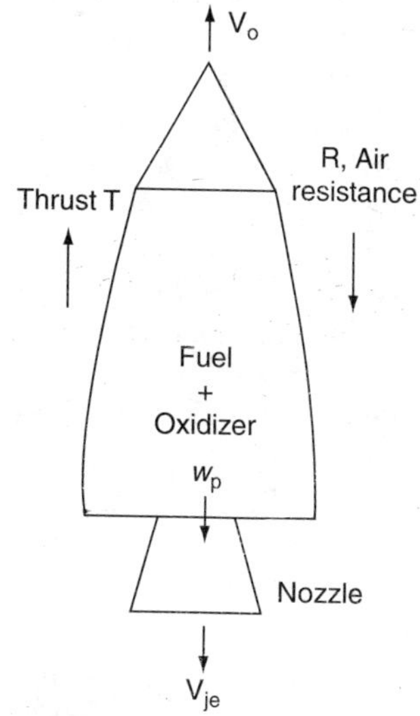

Fig. 21.44 *Analysis of rocket acceleration*

$$-R - mg + T = ma, \quad \text{where } a = \text{acceleration of the rocket.} \quad (21.23)$$

21.19 NUCLEAR ROCKETS

The main components of a nuclear rocket engine are shown in Fig. 21.45. A nuclear reactor replaces the combustion chamber. Liquid hydrogen is used as working fluid and is heated by the reactor before it expands through the thrust-producing nozzle. A small fraction of the heated gas, propellant, is extracted for driving the turbine which drives the propellant pump.

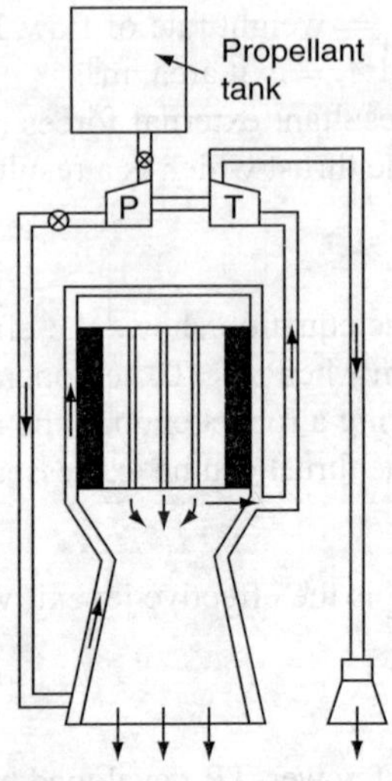

Fig. 21.45 *Main components of a nuclear rocket engine*

Solved Examples

Example 21.1

In a gas turbine plant, air enters the L.P. compressor having a pressure ratio of 3.5 and efficiency 0.85 at 1 bar and 300 K. It then enters the intercooler where it is cooled down to 310 K. The cooled air is further compressed in the H.P. compressor also having a pressure ratio of 3.5 and an efficiency of 0.85. It enters the regenerator having an effectiveness of 0.8. The gases coming out of the combustion chamber enter the H.P. turbine of efficiency 0.88 at 1100 K. The H.P. turbine drives the compressor and there is a reheater between the two turbines. The gases enter the L.P. turbine at 1050 K and the exhaust gases coming out of L.P. turbine of 0.88 efficiency are used to heat the air in the regenerator before leaving to the atmosphere. Determine (a) the power output, and (b) the overall efficiency of the plant. Take c_p for air as 1.005 and for gases as 1.15 kJ/kgK, and γ for air as 1.4 and for gases as 1.33.

Solution The flow and *T–s* diagrams are shown in Fig. Ex. 21.1.

$$p_1 = 1 \text{ bar}, T_1 = 300 \text{ K}, p_2 = 3.5 \text{ bar}, T_3 = 310 \text{ K}, p_4 = 12.25 \text{ bar}, T_6 = 1100 \text{ K}, T_8 = 1050 \text{ K}$$

$$\frac{T_{2s}}{T_1} = \left(\frac{p_2}{p_1}\right)^{\frac{\gamma-1}{\gamma}}$$

$$\therefore \quad T_{2s} = 300\,(3.5)^{0.286} = 429.26 \text{ K}$$

$$T_2 = T_1 + \frac{T_{2s} - T_1}{\eta_c} = 300 + \frac{429.26 - 300}{0.85} = 452.076 \text{ K}$$

$$T_{4s} = T_3\left(p_4/p_3\right)^{\frac{\gamma-1}{\gamma}} = 310(3.5)^{0.286} = 443.5$$

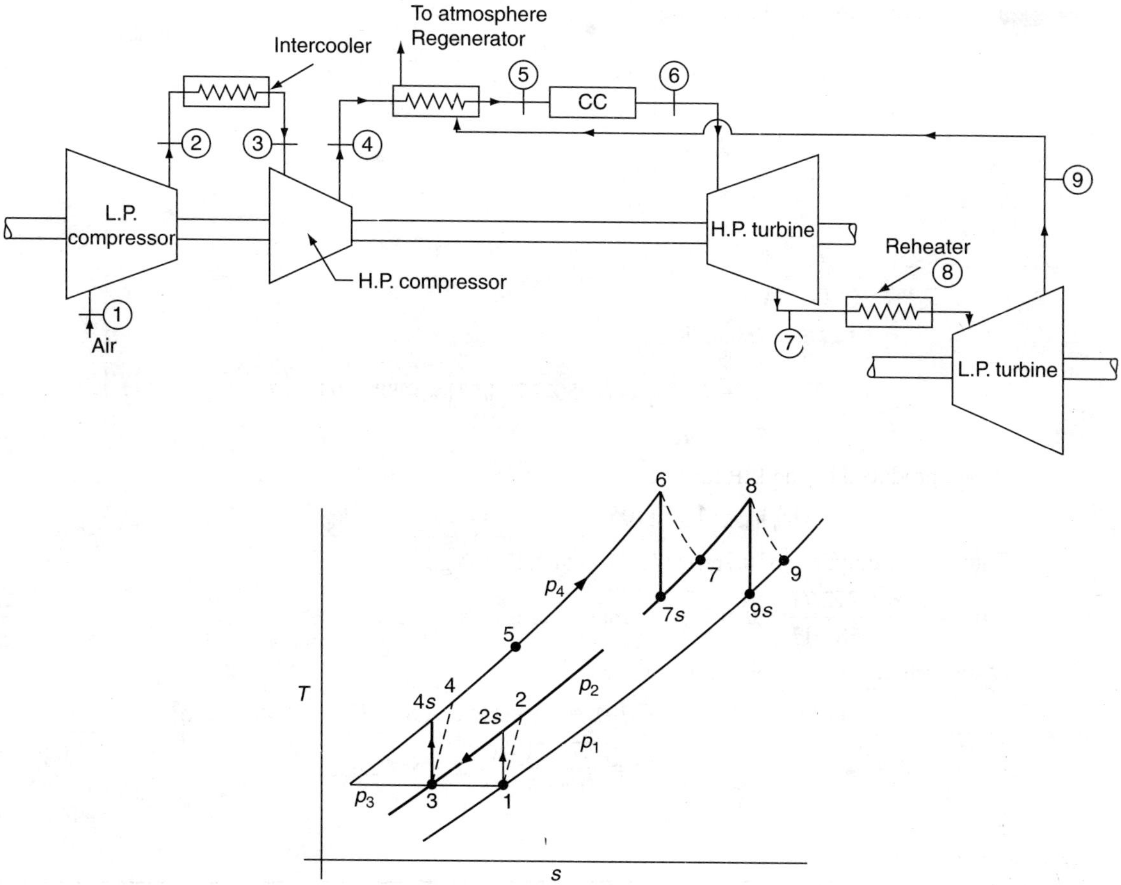

Fig. Ex. 21.1

$$T_4 = T_3 + \frac{T_{4s} - T_3}{\eta_c} = 310 + \frac{443.57 - 310}{0.85} = 467.15$$

Power input to the compressor

$$W_c = 1.005\,[(452.076 - 300) + (467.15 - 310)] = 310.76 \text{ kJ/kg}$$

Power output of the HP turbine

$$(W_T)_{HP} = 310.76 = 1.15\,(1100 - T_7)$$

$$\therefore \quad T_7 = 829.77 \text{ K}$$

$$\frac{T_6 - T_7}{T_6 - T_{7s}} = \eta_T = 0.88$$

$$T_{7s} = T_6 - (T_6 - T_7)/\eta_T = 1100 - (1100 - 829.77)/0.88 = 792.92 \text{ K}.$$

Pressure ratio for the H.P. turbine

$$\frac{p_6}{p_7} = \left(\frac{T_6}{T_7}\right)^{\frac{\gamma}{\gamma-1}} = \left(\frac{1100}{792.92}\right)^{\frac{1.33}{0.33}} = 3.74$$

$$\therefore \quad p_7 = \frac{12.25}{3.74} = 3.275 \text{ bar}$$

$$\frac{T_8}{T_{9s}} = \left(\frac{p_7}{p_1}\right)^{\frac{\gamma-1}{\gamma}} = (3.275)^{\frac{0.33}{1.33}}$$

$$\therefore \quad T_{9s} = 780.52 \text{ K}$$

$$T_8 - T_9 = (T_8 - T_{9s})\eta_T = (1050 - 780.52) \times 0.88 = 050 - T_9$$

$$\therefore \quad T_9 = 812.86 \text{ K}$$

Power produced by the L.P. turbine

$$(W_T)_{LP} = 1.15\,(1050 - 812.86) = 272.71 \text{ kJ/kg}$$

Total power output = 310.76 + 272.71 = 583.47 kJ/kg

$$\text{Work ratio} = \frac{272.71}{583.47} = 0.467 \text{ (with intercooling)}$$

Q_1 to system having intercooling

$$= (1.15 \times 1100 - 1.005 \times 467.14) + 1.15\,(1050 - 829.77) = 1048.5 \text{ kJ/kg}$$

$$\therefore \quad \eta_{plant} = \frac{272.71}{1048.5} = 0.26 \quad \text{or} \quad 26\%$$ *Ans.*

Example 21.2

The blade velocity at the mean diameter of a gas turbine stage is 360 m/s. The blade angles at inlet and exit are 20° and 52° respectively and the blades at this section are designed to have a degree of reaction of 50 percent. The mean diameter of the blades is 0.450 m and the mean blade height is 0.08 m. Assuming that the blades are designed according to vortex theory, calculate (a) the flow velocity (b) the blade angles at the tip and the root (c) the degree of reaction at the tip and at the root of the blades.

Solution Given: $\beta_1 = 20° = \alpha_2, \beta_2 = 52° = \alpha_1, V_{b_m} = 360$ m/s, $D_m = 0.450$ m, $h_b = 0.08$ m (Fig. Ex. 21.2(a)).

$$V_f \tan\beta_2 - V_f \tan\beta_1 = V_{b_m} = 360$$

$$V_f(\tan 52° - \tan 20°) = 360$$

$$V_f = \frac{360}{1.2799 - 0.364} = 393 \text{ m/s}$$

The flow velocity at the tip and root is also the same, i.e. 393 m/s. *Ans.* (a)

Blade tip: Using free vortex theory,

$$V_{w_1} r = \text{contant}$$

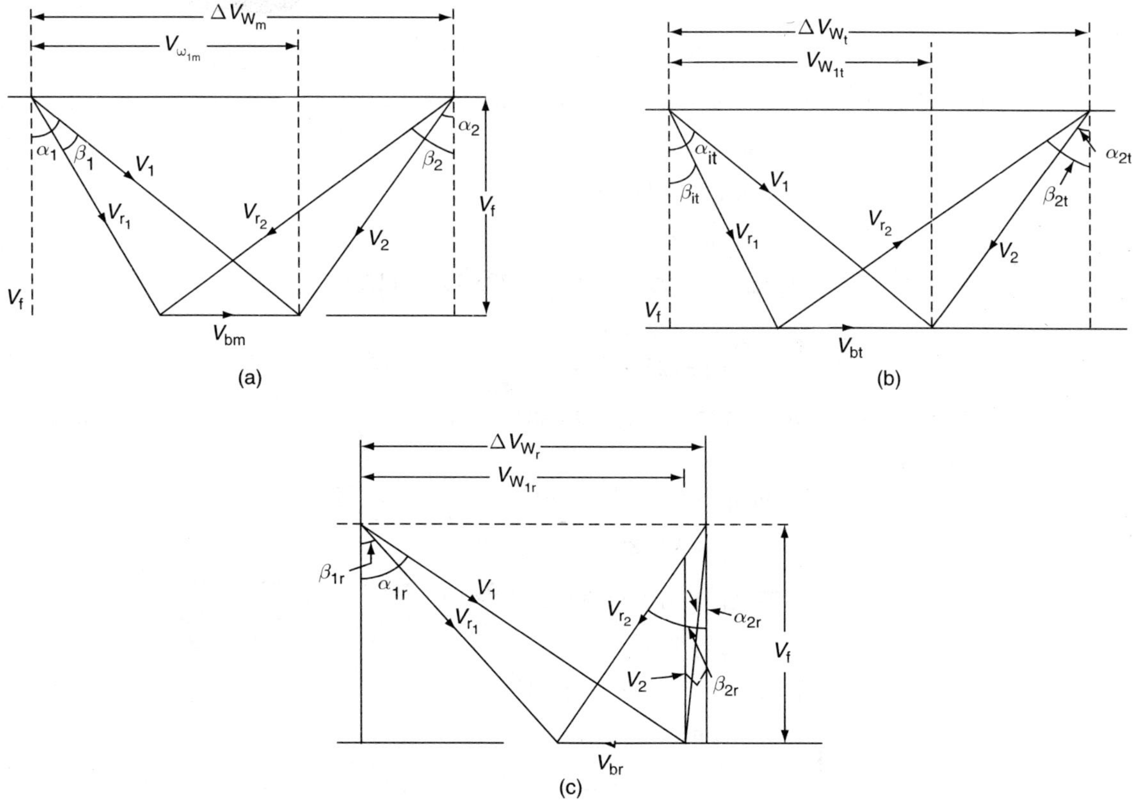

Fig. Ex. 21.2

$$r_t = \frac{0.45}{2} + 0.04 = 0.265 \text{ m}$$

$$V_{b_t} = \frac{\pi D_t N}{60} = \frac{\pi D_m N}{60} \times \frac{D_t}{D_m}$$

$$= 360 \times \frac{r_t}{r_m} = 360 \times \frac{0.265}{0.225} = 424 \text{ m/s}$$

$$V_{w_{1m}} = V_f \tan \alpha_1 = V_f \tan \beta_2 = 393 \tan 52^\circ$$

$$= 503 \text{ m/s}$$

$$V_{w_{1m}} r_m = V_{w_{1t}} r_t$$

$$V_{w_{1t}} = \frac{503 \times 0.225}{0.265} = 427 \text{ m/s}$$

Using the condition of constant specific work,

$$\Delta V_{w_t} = \frac{\Delta V_{w_m} V_{b_m}}{V_{b_t}} = \frac{V_f \left(\tan \beta_1 + \tan \beta_2\right) V_{bm}}{V_{b_t}}$$

$$= \frac{393\left(\tan 20^\circ + \tan 52^\circ\right)\times 360}{424} = 548.57 \text{ m/s}$$

Blade velocity diagram at the blade tip is shown in Fig. Ex. 21.2(b).

$$V_f \tan\alpha_{1_r} = V_{w_{1r}}$$

$$393\tan\alpha_{1r} = 611.76$$

$$\alpha_{1r} = 57.28^\circ$$

$$V_f \tan\alpha_{2r} = \Delta V_{w_r} - V_{w_{1r}} = 785.79 - 611.76$$

$$= 174.03$$

$$\alpha_{2r} = 23.88^\circ$$

Fixed blades (root): $\alpha_{1r} = 57.28^\circ$, $\alpha_{2r} = 23.88^\circ$ *Ans.* (b)

For moving blades

$$V_f \tan\beta_{1r} = V_{w_{1r}} - V_{b_r} = 611.7 - 296 = 315.76$$

$$\beta_{1r} = 38.78^\circ$$

$$V_f \tan\beta_{2r} = V_{br} + V_f \tan\alpha_{2r} = 296 + 174.03 = 470.03$$

$$\beta_{2r} = 50.1^\circ$$

Moving blades (root): $\beta_{1r} = 38.78^\circ$, $\beta_{2r} = 50.1^\circ$ *Ans.* (b)

$$V_f \tan\alpha_{1t} = V_{w_{1t}}$$

$$\alpha_{1t} = 47.37^\circ$$

$$V_f \tan\alpha_{2_t} = \Delta V_{w_t} - V_{w_{1t}} = 548.57 - 427 = 121.57$$

$$\tan\alpha_{2t} = \frac{121.57}{393}$$

$$\alpha_{2t} = 17.19^\circ$$

Fixed blades (tip) $\alpha_{1t} = 47.37^\circ, \alpha_{2t} = 17.19^\circ$

Similarly for moving blades,

$$V_f \tan\beta_{1t} = V_{w_{1t}} - V_{b_t}$$

$$393\tan\beta_{1t} = 427 - 424$$

$$\beta_{1t} = 0.44^\circ$$

$$V_f \tan\beta_{2t} = 424 + 121.57 = 545.57$$

$$\beta_{2t} = 54.23^\circ$$

Moving blades (tip) $\beta_{1t} = 0.44^\circ$, $\beta_{2t} = 54.23^\circ$

Blade root

$$r_r = 0.225 - 0.04 = 0.185 \text{ m}$$

$$V_{b_r} = 360 \times \frac{0.185}{0.225} = 296 \text{ m/s}$$

$$V_{w_{1m}} r_m = V_{w_{1r}} r_r$$

$$V_{w_{1r}} = -\frac{503 \times 0.225}{0.185} = 611.76 \text{ m/s}$$

$$\Delta V_{w_r} = \frac{\Delta V_{w_m} V_{b_m}}{V_{b_r}} = \frac{232593.68}{296} = 7285.79 \text{ m/s}$$

Blade velocity diagram at the blade root is shown in Fig. Ex. 21.2(c)

The degree of reaction is given by

$$R = \frac{V_f(\tan\beta_2 - \tan\beta_1)}{2V_b}$$

$$\text{At the tip, } R = \frac{393(\tan 54.23° - \tan 0.44°)}{2 \times 424} = 0.64 \quad \text{or} \quad 64\%$$ *Ans.*

$$\text{At the root, } R = \frac{393(\tan 50.1° - \tan 38.78°)}{2 \times 296} = 0.26 \quad \text{or} \quad 26\%$$ *Ans.*

***Example* 21.3**

The products of combustion enter an axial flow gas turbine at 8 bar, 1125 K and leave at 1.5 bar. There are 11 stages, each developing the same specific work with the same stage efficiency. The axial velocity of flow is constant at 110 m/s and the polytropic efficiency is 0.85. At a particular stage, the mean blade velocity is 140 m/s, the stage has 50% reaction at the mean blade height and the specific work output is constant across the stage at all radii. Assuming that $\gamma = 1.33$ and $c_p = 1.15$ kJ/kgK for gases, estimate (a) the blade angles at inlet and exit, (b) the overall isentropic efficiency of the turbine, and (c) the stage efficiency.

Solution Polytropic efficiency η_p is defined by

$$\eta_p = \frac{dh}{dh_s} = \frac{c_p dT}{vp} \cong \frac{c_p dT}{\frac{RT}{p} dp}$$

$$\eta_p \int_1^2 \frac{dp}{p} = \frac{c_p}{R} \int_1^2 \frac{dT}{T}$$

$$\eta_p \ln \frac{p_2}{p_1} = \frac{\gamma}{\gamma - 1} \ln \frac{T_2}{T_1}$$

$$\therefore \quad \frac{T_2}{T_1} = \left(\frac{p_2}{p_1}\right)^{\frac{(\gamma-1)\eta_p}{\gamma}}$$

$$T_2 = 1125\left(\frac{1.5}{8}\right)^{0.2125} = 788.25 \text{ K}$$

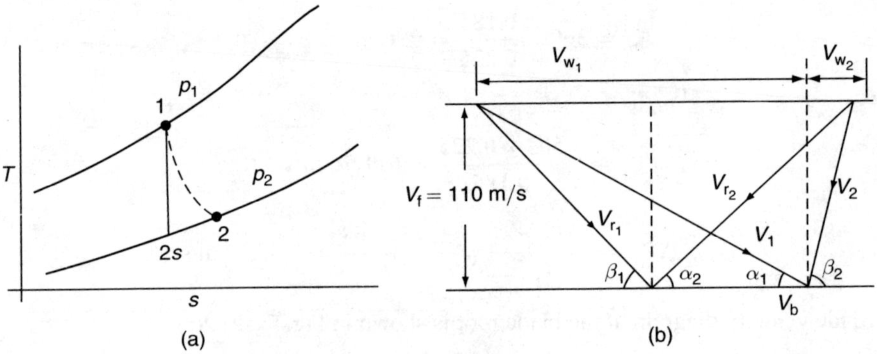

Fig. Ex. 21.3

For isentropic expansion,

$$\frac{T_{2s}}{T_1} = \left(\frac{p_2}{p_1}\right)^{\frac{\gamma-1}{\gamma}} = \left(\frac{1.5}{8}\right)^{0.25} = 0.658$$

$$\therefore \qquad T_{2s} = 1125 \times 0.658 = 740.29 \text{ K}$$

$\therefore$ overall isentropic efficiency of the turbine

$$\eta_s = \frac{T_1 - T_2}{T_1 - T_{2s}} = \frac{1125 - 788.25}{1125 - 740.29} = 0.875 \quad \text{or} \quad 87.5\%$$ *Ans.* (b)

Work output of the turbine

$$W_T = c_p(T_1 - T_2) = 1.15\,(1125 - 788.25) = 387.26 \text{ kJ/kg}$$

$\therefore$ work output per stage $\dfrac{387.26}{11} = 35.2$ kJ/kg

From the velocity triangles at inlet and exit of the blades,

$$V_1 = V_{r_2}, V_{r_1} = V_2$$

$$\text{Work output per stage} = \left(V_{w_1} + V_{w_2}\right)V_b = 35.2 \times 10^3 \text{ J/kg}$$

$$\therefore \qquad V_{w_1} + V_{w_2} = \frac{35.2 \times 10^3}{140} = 251.43$$

$$V_{w_2} = V_{w_1} - V_b$$

$$V_{w_1} + V_{w_1} - V_b = 251.43$$

$$\therefore \qquad V_{w_1} = \frac{251.43 + 140}{2} = 195.71 \text{ m/s}$$

$$\alpha_1 = \tan^{-1}\frac{110}{195.71} = 29.33^\circ = \alpha_2$$

These are the blade outlet angles measured with the direction of rotation.

$$\beta_1 = \tan^{-1}\frac{110}{195.71-140} = 63.14° \text{ (blade angle from direction of rotation)}$$ *Ans.* (a)

Enthalpy drop per stage = 35.20 kJ/kg

$$\therefore \quad \text{Temperature drop per stage } = \frac{35.2}{1.15} = 30.7 \text{ K}$$

$$\therefore \quad T_1' = 1125 - 30.7 = 1094.3 \text{ K} = T_1 - T_1'$$

Now $$\frac{T_1'}{T_1} = \left(\frac{p_1'}{p_1}\right)^{\frac{(\gamma-1)\eta_p}{\gamma}} = \left(\frac{p_1'}{p_1}\right)^{0.2125}$$

$$\therefore \quad \frac{p_1'}{p_1} = \left(\frac{1094.3}{1125}\right)^{\frac{1}{0.2125}} = 0.878$$

Isentropic temperature ratio

$$\frac{T_{1s}}{T_1} = (0.878)^{\frac{\gamma-1}{\gamma}} = 0.968$$

$$T_{1s} = 1125 \times 0.968 = 1089.1 \text{ K}$$

$$\therefore \quad \text{Stage efficiency, } \eta_{st} = \frac{1125-1094.3}{1125-1089.1} = 0.853 \quad \text{or} \quad 85.3\%$$ *Ans.* (c)

Example 21.4

A turbojet aircraft is flying at 800 km/h at 10,700 m where the pressure and temperature of the atmosphere are 0.24 bar and − 50°C respectively. The compressor pressure ratio is 10/1 and the maximum cycle temperature is 1093 K. Calculate the thrust developed and the specific fuel consumption using the following particulars: entry duct efficiency = 0.9, isentropic efficiency of the compressor = 0.9, stagnation pressure loss in the combustion chamber = 0.14 bar, calorific value of fuel = 43.3 MJ/kg, combustion efficiency = 0.98, isentropic efficiency of turbine = 0.92, mechanical efficiency of drive = 0.98, jet pipe efficiency = 0.92, nozzle outlet area = 0.08 m², c_p = 1.005 KJ/kg K and γ = 1.4 for air, c_p = 1.15 kJ/kg K and γ = 1.333. Assume that the nozzle is convergent.

Solution K.E. of air at inlet $= \frac{1}{2} \times \left(\frac{800 \times 1000}{3600}\right)^2 = \frac{1}{2} \times (222.2)^2 \text{ Nm/kg} = 24.7 \text{ kJ/kg}$

$$T_{01} - T_0 = \frac{24.7}{1.005} = 24.6 \text{ K}$$

$$\therefore \quad T_{01} = [(-50) + 273] + 24.6 = 247.6 \text{ K}$$

Intake efficiency, $0.9 = \dfrac{T_{01s} - T_0}{T_{01} - T_0}$

$$T_{01s} - T_0 = 0.9 \times 24.6 = 22.1 \text{ K}$$

$$T_{01s} = 22.1 + (-50 + 273) = 245.1 \text{ K}$$

$$\frac{p_{01}}{p_0} = \left(\frac{T_{01s}}{T_0}\right)^{\frac{\gamma}{\gamma-1}} = \left(\frac{245.1}{223}\right)^{\frac{1.4}{0.4}} = (1.1)^{3.5} = 1.393$$

$$\therefore \qquad p_{01} = 1.393 \times 0.24 = 0.334 \text{ bar}$$

For the compressor,

$$\frac{T_{02s}}{T_{01}} = \left(\frac{p_{02}}{p_{01}}\right)^{\frac{\gamma}{\gamma-1}} = (10)^{\frac{0.4}{1.4}} = 1.931$$

$$\therefore \qquad T_{02s} = 247.6 \times 1.931 = 478 \text{ K}$$

$$\eta_c = \frac{T_{02s} - T_{01}}{T_{02} - T_{01}} = 0.9$$

$$T_{02} - T_{01} = \frac{478 - 247.6}{0.9} = 256 \text{ K}$$

$$T_{02} = 247.6 + 256 = 503.6 \text{ K}$$

$$p_{02} = 10 \times p_{01} = 10 \times 0.334 = 3.34 \text{ bar}$$

$$p_{03} = 3.34 - 0.14 = 3.2 \text{ bar}$$

Since $W_c = W_T$, by energy balance

$$T_{03} - T_{04} = \frac{1.005\,(503.6 - 247.6)}{1.15 \times 0.98} = 228.3 \text{ K}$$

$$T_{04} = 1093 - 228.3 = 864.7 \text{ K}$$

$$\eta_T = \frac{T_{03} - T_{04}}{T_{03} - T_{04s}} = 0.92$$

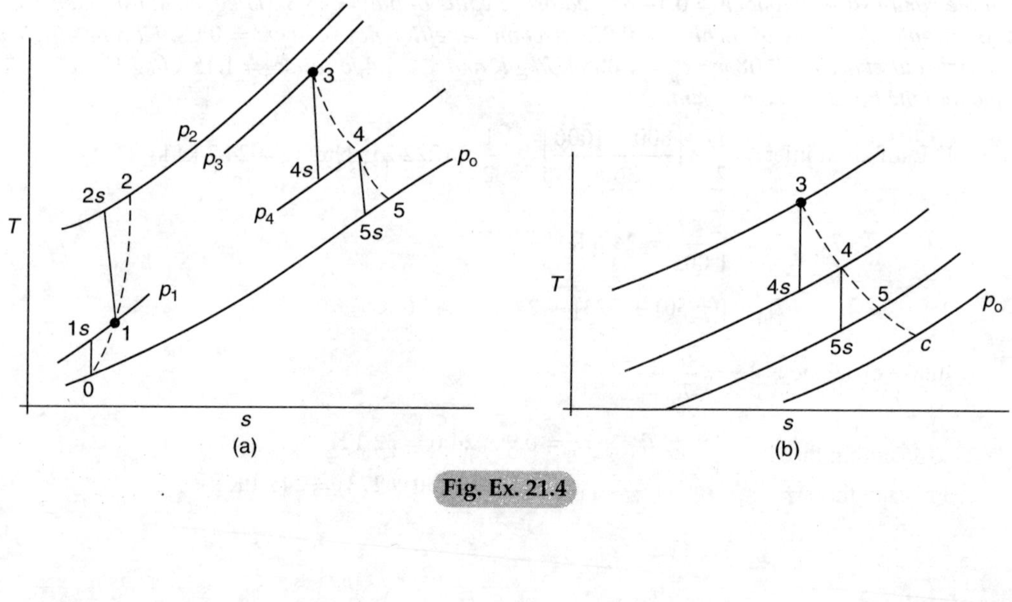

Fig. Ex. 21.4

$$T_{03} - T_{04s} = \frac{228.3}{0.92} = 248.2 \text{ K}$$

$$T_{04s} = 1093 - 248.2 = 844.8 \text{ K}$$

$$\frac{p_{03}}{p_{04}} = \left(\frac{T_{03}}{T_{04s}}\right)^{\frac{\gamma}{\gamma-1}} = \left(\frac{1093}{844.8}\right)^{\frac{1.333}{0.333}} = 2.803$$

$$\therefore \quad p_{04} = \frac{3.2}{2.803} = 1.156 \text{ bar}$$

For choked flow in the nozzle, the critical pressure ratio is given by

$$\frac{p_c}{p_{04}} = \left(\frac{2}{\gamma+1}\right)^{\frac{\gamma}{\gamma-1}} = \left(\frac{2}{2.333}\right)^{\frac{1.333}{0.333}} = 0.54$$

$$p_{cr} = 0.54 \times 1.156 = 0.624 \text{ bar}$$

Since the atmospheric pressure is 0.24 bar, the nozzle is choked and hence the nozzle exit velocity is sonic (Fig. Ex. 21.4(b)).

$$\frac{T_5}{T_{04}} = \frac{2}{\gamma+1} = \frac{2}{2.333} = \text{T}_5 = 864.77 \times 0.857 = 741.33 \text{ K}$$

$$\text{Jet pipe efficiency} = \frac{T_{04} - T_{05}}{T_{04} - T_{05s}} = 0.92$$

$$T_{05s} = 864.7 - \frac{864.7 - 741.33}{0.92} = 730.6 \text{ K}$$

$$\frac{p_{05}}{p_5} = \left(\frac{T_{05}}{T_{05s}}\right)^{\frac{\gamma}{\gamma-1}} = \left(\frac{864.7}{730.6}\right)^{\frac{1.333}{0.333}} = 1.963$$

$$p_5 = \frac{1.156}{1.963} = 0.589 \text{ bar}$$

$$R = \frac{c_p(\gamma-1)}{\gamma} = \frac{1.15 \times 0.333}{1.333} = 0.2873 \text{ kJ/kg K}$$

$$v_5 = \frac{RT_5}{p_5} = \frac{0.2873 \times 741.3}{0.589 \times 10^5} = 3.616 \text{ m}^3\text{/kg}$$

Jet velocity, $V_j = \sqrt{\gamma R T_5} = \sqrt{1.333 \times 287.3 \times 741.3} = 532.8$ m/s

Mass flow rate, $\dot{m} = \dfrac{0.08 \times 532.8}{3.616} = 11.79$ kg/s

Momentum thrust $= \dot{m}\left(V_j - V_o\right) = 11.79\ (532.8 - 222.2) \doteq 3661$ N

Pressure thrust $= (p_5 - p_0)\, A = (0.589 - 0.24) \times 0.08 \times 10^5 = 2792$ N

$\therefore$ total thrust $= 3661 + 2792 = 6453$ N *Ans.*

Heat supplied, $Q_1 = \dot{m}c_{p_g}(T_{03} - T_{02})$

$$= 11.79 \times 1.15\,(1093 - 503.6) = 7990 \text{ kW}$$

$$Q_1 = \dot{m}_f \times C.V. \times \eta_{cc}$$

$\therefore$ Fuel burning rate, $\dot{m}_f = \dfrac{7990}{43{,}300 \times 0.98} = 0.188$ kg/s

$\therefore$ Specific fuel consumption $= \dfrac{0.188 \times 10^3}{6453} = 0.0291 \dfrac{\text{kg}}{\text{KNs}}$ *Ans.*

Because of the low value of fuel – air ratio, one can assume that the mass flow rate of air is equal to the mass flow rate of gases.

Example 21.5

A turbojet flying at 850 km/h has an air mass flow rate of 50 kg/s. The enthalpy drop across the nozzle is 200 kJ/kg and the nozzle efficiency is 0.9. The air fuel ratio is 80 and the heating value of the fuel is 40 MJ/kg. Estimate the propulsive power, thrust power, propulsive and thermal and overall efficiency of the unit.

Solution Velocity of aircraft, $V_0 = \dfrac{850 \times 1000}{3600} = 236.11$ m/s

A/F ratio $= 80$.

$\therefore$ $\dot{m}_f = \dfrac{50}{80} = 0.625$ kg/s

Velocity of gases at exit from the nozzle, V_e.

$$= \sqrt{2c_p(\Delta h)\eta_n} = \sqrt{2 \times 1005 \times 200 \times 0.9} = 601.5 \text{ m/s}$$

Thrust, $T = \dot{m}_g V_e - \dot{m}_a V_0 = 50.625 \times 601.5 - 50 \times 236.11 = 18.645$ kN

$\therefore$ Thrust power (TP) $= TV_o = 18.645 \times 236.11 = 4402.37$ kW *Ans.*

Propulsive power (pp) is the energy required to change the momentum of the fluid and is expressed by the difference in K.E. as given by

$$\text{PP} = \frac{1}{2}(\dot{m}_a + \dot{m}_f)V_e^2 - \frac{1}{2}m_a V_o^2$$

$$= \frac{1}{2}(50.625) \times (601.5)^2 - \frac{1}{2} \times 50 \times (236.11)^2$$

$$= 7.72 \text{ MW}$$ *Ans.*

Propulsive efficiency, $\eta_p = \dfrac{\text{TP}}{\text{PP}} = \dfrac{4.402}{7.72} = 0.567$ or 56.7 *Ans.*

Thermal efficiency $= \dfrac{\text{PP}}{\dot{m}_f \times \text{CV}} = \dfrac{7.72}{0.625 \times 40} = 0.31$ or 31% *Ans.*

Overall efficiency $= \eta_{th} \times \eta_p = 0.31 \times 0.567 = 0.176$ or 17.6% *Ans.*

Example 21.6

Air at the ambient conditions of 0.56 bar and 260 K enters the compressor having a pressure ratio of 6 and an efficiency of 0.85 of a turboprop aircraft flying at a speed of 360 km/h. The propeller diameter is 3 m, propeller efficiency 0.8 and gear reduction efficiency is 0.95. The products of combustion enter the gas turbine having an expansion ratio of 5 and an efficiency of 0.88 at 1100 K. Determine the air – fuel ratio, thrust produced by the nozzle with an efficiency of 0.9, thrust by the propeller and the mass flow rate of air flowing through the compressor. Given: C. V. of fuel = 40 MJ/kg.

Solution Velocity of aircraft, $V_0 = \dfrac{360 \times 10^3}{3600} = 100$ m/s

Compressor: $p_1 = 0.56$ bar, $p_2 = 6 \times 0.56 = 3.36$ bar,

$$\frac{T_{2s}}{T_1} = 6^{0.286} = 1.716, \quad T_{2s} = 260 \times 1.716 = 446.16 \text{ K}$$

$$\eta_c = \frac{T_{2s} - T_1}{T_2 - T_1} = 0.85, \quad T_2 = 260 + \frac{446.16 - 260}{0.85} = 479.01 \text{ K}$$

Power input to compressor

$W_c = c_p (T_2 - T_1)$

$= 1.003\,(479.01 - 260) = 220.1$ kJ/kg

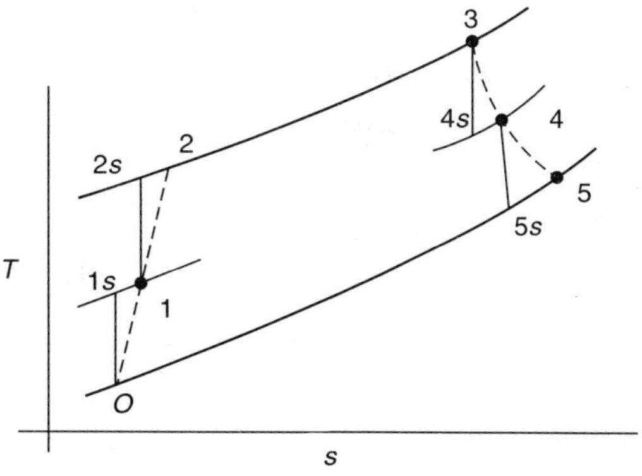

Fig. Ex. 21.6

Combustion chamber

By making an energy balance,

$$C_{p_a} T_2 + \dot{m}_f \times \text{C.V.} = C_{p_g} T_3 (1 + m_f)$$

Assuming $C_{p_a} = C_{p_g}$,

$$479.01 + \dot{m}_f \times \frac{40{,}000}{1.005} = 1100(1 + \dot{m}_f)$$

$\therefore$ $\dot{m}_f = 0.016$ kg/s

$\therefore$ $\text{air – fuel ratio} = \dfrac{1}{\dot{m}_f} = \dfrac{1}{0.016} = 62.23$ *Ans.*

Turbine

$$T_3 = 1100 \text{ K}, p_3 = 6 \times 0.56 = 3.36 \text{ bar}$$

$$p_4 = \frac{3.36}{5} = 0.672 \text{ bar}$$

Temperature at the end of isentropic expansion

$$T_{4s} = T_3 / 5^{0.286} = 644.44 \text{ K}$$

$$T_4 = T_3 - (T_3 - T_{4s})/\eta_T = 1100 - (1100 - 644.44)/0.88$$
$$= 699.11 \text{ K}$$

Power output of turbine, $W_T = (1 + m_f)C_p(T_3 - T_4)$

$= 1.016 \times 1.005\,(1100 - 699.11)$

$= 409.34$ kJ/kg air

Power available for propeller $= 409.34 - 220.1 = 189.24$ kJ/kg

Nozzle

$$\frac{T_4}{T_{5s}} = \left(\frac{p_4}{p_5}\right)^{0.286} = \left(\frac{0.672}{0.56}\right)^{0.286} = 1.053$$

$$T_{5s} = \frac{699.11}{1.053} = 613.36 \text{ K}$$

Velocity of gases at nozzle exit

$$V_j = \sqrt{2 \times 1.005 \times 0.9(699.11 - 663.36)} = 254.31 \text{ m/s}$$

Thrust by jet $= m_g V_j - m_a V_\phi = 1.016 \times 254.31 - 1 \times 100$

$= 158.38$ N/kg air

Propeller

Propeller efficiency $= 0.8 = \dfrac{V_0}{V}$

$$\therefore \quad V = \frac{100}{0.8} = 125 \text{ m/s}$$

$$V = \frac{V_4 + V_0}{2} = 125$$

$$\therefore \quad V_4 = 250 - 100 = 150 \text{ m/s}$$

$$Q = \frac{\pi}{4} d^2 \times V = \frac{\pi}{4}(3)^2 \times 125 = 883.7 \text{ m}^3/\text{s}$$

Thrust power of the propeller

$$= \frac{1}{2} PQ\left(V_4^2 - V_0^2\right) = \frac{1}{2} \times \frac{0.56 \times 10^5}{287 \times 260} \times 883.7\left(150^2 - 100^2\right)$$

$$= 4142.3 \frac{\text{kg}}{\text{m}^3} \times \frac{\text{m}^3}{\text{s}} \times \frac{\text{m}^2}{\text{s}^2} = 4142.3 \text{ kJ/s}$$ *Ans.*

Reduction gear efficiency $= 0.95$

Power to be supplied by the turbine $= \dfrac{4142.3}{0.95} = 4360.3$ kW

Mass flow rate through the compressor

$$\dot{m}_a = \frac{4360.3 \text{ kW}}{189.23 \text{ kJ/kg}} = 23.04 \text{ kg/s}$$ *Ans.*

Thrust produced by the propeller $= \dfrac{4142.3 \times 0.8}{100} = 33.14$ kN *Ans.*

Example 21.7

A rocket is fixed vertically starting from rest. The initial mass of the rocket is 15000 kg and the stored propellant burns at the rate of 125 kg/s. The gases come out at a velocity of 2000 m/s relative to the rocket. Neglecting air resistance, calculate (a) the velocity it will attain in 70 seconds, and (b) the maximum height that the rocket will attain.

Solution Neglecting air resistance, the linear momentum equation (Eq. 21.23) can be written as

$$T - mg = m\mathrm{d}V/\mathrm{d}t \text{ and } m = m_i - \dot{m}_p t$$

where m_i = initial mass of rocket and $\dot{m}_p$ = rate of burning fuel.

$$\mathrm{d}V = \frac{T - (m_i - \dot{m}_p t)g}{m_i - \dot{m}_p t}\mathrm{d}t$$

$$= \frac{T}{m_i - \dot{m}_p t}\mathrm{d}t - g\mathrm{d}t$$

On integration,

$$V = -\frac{T}{\dot{m}_p}\ln\left(1 - \frac{\dot{m}_p t}{m_i}\right) - gt$$

since $\quad T = \dot{m}_p V_e, \quad V_e = T/\dot{m}_p = 2000$ m/s

After 70 seconds, the velocity of the rocket will be

$$V = -2000\ \ln\left(1 - \frac{125 \times 70}{15000}\right) - 9.81 \times 70$$

$$= 1064.24 \text{ m/s} \qquad \textit{Ans. (a)}$$

The height travelled during the first 70 seconds,

$$h_1 = \int_0^t v\mathrm{d}t = \int_0^{70}\left[-2000\ \ln\left(1 - \frac{125t}{15000}\right) - gt\right]\mathrm{d}t$$

$$= 28.4 \text{ km}$$

The rocket will be assumed to continue its flight upward until its K.E. is converted to P.E.

$$\frac{1}{2}mV^2 = mgh_2$$

$$\therefore \quad h_2 = (1064.24)^2/2 \times 9.81$$

$$= 57.73 \text{ km.}$$

$$\therefore \quad \text{Maximum height} = 28.4 + 57.73$$

$$= 86.13 \text{ km.} \qquad \textit{Ans.}$$

Example 21.8

The pressure and temperature developed in the combustion chamber of a chemical rocket engine are 2.4 MPa and 3170 K respectively. The atmosperic pressure is 55 kPa. The pressure at the nozzle exit is 85 kPa and the cross-sectional area at the nozzle throat is 0.06 m². The nozzle efficiency is 0.91, the coefficient of discharge is 0.98, $\gamma = 1.25$, $R = 0.693$ kJ/kgk, the half angle of divergence is 12: Determine the thrust and the specific impulse.

Solution Jet velocity, $V_j = \sqrt{2(h_c - h_j)} = \sqrt{2c_p(T_c - T_j)}$

$$= \sqrt{2c_pT_c\left[1-\left(\frac{p_j}{p_\gamma}\right)^{\frac{\gamma-1}{\gamma}}\right]}$$

$$= \sqrt{\frac{2\gamma RT_c}{\gamma-1}\left[1-\left(\frac{p_j}{p_c}\right)^{\frac{\gamma-1}{\gamma}}\right]}$$

$$= \sqrt{\frac{2\times1.25\times0.693\times3170}{0.25}\left[1-\left(\frac{55}{2400}\right)^{\frac{0.25}{1.25}}\right]}$$

$= 3271.94$ m/s.

The thrust developed by the nozzle is directly proportional to the axial velocity, a correction factor is used to account for the nonaxial flow of gases. The correction factor is given by $k = (1+\cos\alpha)/2$, where α is the semidivergence angle. The actual velocity of gases is

$$(V_j)_{act} = \sqrt{\eta_n}\,.\,k \times \text{theoretical velocity}$$

$$\therefore \quad (V_j)_{act} = \sqrt{0.91}\times(1+\cos12^\circ)/2 \times 3271.94$$

$$= 3086.9 \text{ m/s.}$$

The velocity at the throat is sonic.
The mass flow rate, $\dot{m}$

$$= P_tA_tV_t = A_tP_c\left[\frac{\gamma}{RT_c}\left(\frac{2}{\gamma+1}\right)^{\frac{\gamma+1}{\gamma-1}}\right]^{\frac{1}{2}}$$

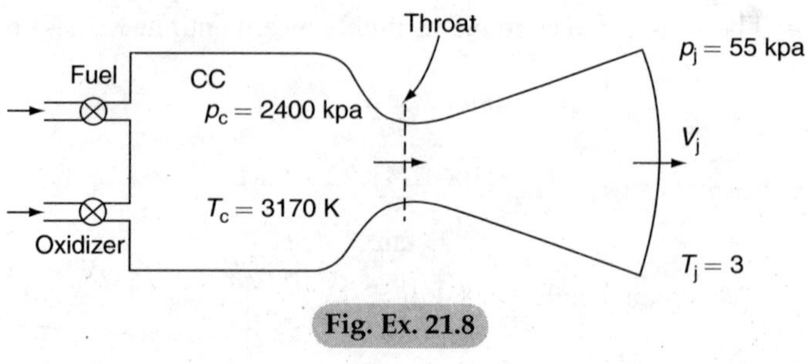

Fig. Ex. 21.8

(see chapter 17)

$$= 0.06 \times 2.4 \times 10^6 \left[\frac{1.25}{693 \times 3170} \left(\frac{2}{2.25} \right)^{\frac{2.25}{0.25}} \right]^{\frac{1}{2}} = 63.93 \text{ kg/s}$$

Actual mass flow rate = 0.98 × 63.93 = 62.66 kg/s

$$(\dot{m})_{act} = P_e A_e V_j$$

which gives $A_e = 0.2589 \text{ m}^2$

$$\text{Thrust produced} = \dot{m} V_j + A_e (p_e - p_o)$$

$$= 62.66 \times 3086.9 + 0.2589\,(85 - 55) \times 10^3$$

$$= 201.18 \text{ kN}$$ *Ans.*

$$\text{Specific impulse} = \frac{201.18 \times 10^3}{62.66} = 3210 \text{N/kg/s}$$ *Ans.*

Review Questions

21.1 What are the applications of gas turbines?

21.2 How does the open-cycle GT plant compare with the closed-cycle GT plant?

21.3 Explain the advantages of a GT plant.

21.4 Give an analysis of the GT plant. What is the effect of pressure ratio on GT plant output and efficiency?

21.5 Derive the expression of optimum pressure ratio for a GT plant having compressor efficiency η_c and turbine efficiency η_T.

21.6 What is work ratio of a GT plant ? Show that it is

$$r_w = 1 - \frac{T_1}{T_3} r_p^{\frac{\gamma-1}{\gamma}}$$

21.7 Explain the effect of regeneration on the performance of a GT plant with sketches. What is its effectiveness?

21.8 Explain the effects of intercooling and reheating in a GT plant.

21.9 Draw the schematic diagram of a GT plant with intercooling, reheating and regeneration and define the overall plant efficiency.

21.10 What is a semi-closed cycle GT plant?

21.11 Discuss the GT plant performance under variable loads.

21.12 Compare centrifugal compressor with an axial flow compressor.

21.13 Explain with suitable sketches the operation of combustion chambers in a GT plant.

21.14 Define combustion efficiency. What are the performance criteria of a CC?

21.15 Explain the cooling of GT blades. What is degree of reaction ? Draw the velocity triangles at inlet and exit of the blades.

21.16 Explain vortex blading. What are the basic requirements for a gas turbine?

21.17 Explain the effect of water injection on gas turbine performance.

21.18 Discuss the use of different fuels in a GT plant.

21.19 Discuss on the different materials used for GT plant construction.

21.20 On which laws does the jet propulsion system depend? Explain.

21.21 What is the propulsive force? What is thrust?

21.22 Enlist the various propulsive devices for aircrafts and missiles.

21.23 How does a turboprop differ from a turbojet engine? Explain propulsive power and propulsive efficiency.

21.24 Explain the operation of a propeller. What is an athodyd ? What is a rocket ?

21.25 Explain with a neat sketch the operation of a turbojet engine. Draw the *T–s* diagram.

21.26 Define propulsive power, thrust power and propulsive efficiency.

21.27 Explain the effect of flight speed on propulsive efficiency and the thrust power.

21.28 What is the effect of afterburner in a turbojet engine? Explain with a neat sketch.

21.29 Explain the operation of a turboprop engine with neat sketches.

21.30 Explain the momentum theory of propellers. What is the effect of flight speed on the thrust for a turbojet and a turboprop engine?

21.31 What is a bypass turbojet engine? What is its advantage?

21.32 Explain the operation of a ramjet engine. What is the thrust produced?

21.33 Explain with a sketch the operation of a pulse-jet engine.

21.34 What is the principle of operation of rocket propulsion? What is 'grain'?

21.35 What do you mean by JATO and RATO?

21.36 What are restricted and unrestricled burning of a solid propellant rocket?

21.37 Explain the operation of liquid propellant rocket with sketches. What is a hypergolic propellant ?

21.38 Derive the expression for thrust produced in a rocket. What is propulsive efficiency? What is specific impulse?

21.39 Explain a nuclear rocket engine with a sketch.

21.40 What do you mean by je thrust and pressure thrust in aircrafts and missiles ? Derive the expression for total thrust in a subsonic aircraft.

Problems

21.1 Estimate the pressure ratio at which the specific output is maximum if the temperature at the compressor inlet is 300 K and that at the turbine inlet is 1000 K. The isentropic efficiencies of the compressor and turbine are 0.85 and 0.88 relatively. Also, obtain the mass flow rate per kWh of air and fuel if the calorific value of fuel is 40 MJ/kg.

$\left(Ans.\ \gamma_p = 5, \dot{m}_a = 30.27 \text{ kg/kwh}, \dot{m}_f = 0.375 \text{ kg/kwh}\right)$

21.2 The pressure ratio of the compressor in a GT plant is 7. Air enters the compressor at 1 bar, 25°C, and the temperature at the turbine inlet is 1050 K. The gases leave the turbine at 1 bar. The combustion efficiency is 0.89 and the pressure drop in the combustion chamber is 0.11 bar. Taking $\eta_c =$ 0.82 and $\eta_T = 0.85$, estimate (a) the thermal efficiency of the plant, (b) the work ratio, and (c) the specific fuel consumption. The heating value of the fuel is 42 MJ/kg. Take C_p for air as 1.005 and C_p for gases as 1.11 kJ/kgK.

(*Ans.* (a) 22.53%, (b) 0.3633, (c) 0.38 kg/kwh)

21.3 Air at 1 bar, 300 K enters the compressor of a gas turbine unit and comes out at 8 bar. The compressor is driven by an H.P. turbine, and the L.P. turbine drive a generator. The isentropic efficiencies of the compressor, H.P. turbine and L.P. turbine are 0.82, 0.85 and 0.85 respectively. Determine the net power developed, the work ratio and the overall efficiency of the unit if the air flow rate is 5 kg/s, the air – fuel ratio is 62:1, the maximum temperature of gases at the H.P. turbine inlet is 1000 K. Take C_p of air as 1.005 and C_p for gases as 1.11 kJ/kgK. The heating value of fuel is 40 mJ/kg.

(*Ans.* 698.53 kW, 0.318, 21.65 %)

21.4 Air enters the compressor of a gas turbine equipped with a regenerator at 1 bar, 300 K. The pressure ratio of the compressor is 6.6, its efficiency is 0.82. The regenerator effectiveness is 0.85, the combustion chamber efficiency is 0.8 and the turbine efficiency is 0.85. The pressure drop in the regenerator on the air and gas side is 0.06 and 0.05 bar respectively. The heating value of the fuel is 40 mJ/kg and the air – fuel ratio is 65:1. Estimate the efficiency of the plant with and without a regenerator.

(*Ans.* 28.95 % and 19.15 %, with and without regenerator.)

21.5 In a closed-cycle gas turbine cycle, helium at 15 bar, 300 K enters the compressor ($\eta_c = 0.8$) and comes out at 60 bar. The gas is heated in a heat exchanger to 1000 K before entering the turbine ($\eta_T = 0.85$). The exhaust from the turbine is cooled to the initial temperature in a cooler. Calculate the cycle efficiency and the net output for a flow rate of 75 kg/s. Take $C_p = 5$ kJ/kg K and $\gamma = 1.5$. (*Ans.* 19.6 %, 35.54 mw.)

21.6 A simple open-cycle gas turbine plant works between the pressures of 1 bar and 6 bar and temperatures of 300 K and 1023 K. The calorific value of fuel is 44 MJ/kg. If the mechanical efficiency and the generator efficiency are 95% and 96% respectively, and for an air-flow rate of 20 kg/s, calculate (a) the air-fuel ratio, (b) the thermal efficiency, and (c) the power output.

21.7 A turbojet aircraft is travelling at 925 km/h in atmospheric conditions of 0.45 bar and –26°C. The compressor pressure ratio is 8, the air mass flow rate is 45 kg/s and the maximum allowable cycle temperature is 800°C. The compressor, turbine and jet pipe stagnation isentropic efficiencies are 0.85, 0.89 and 0.9 respectively, the mechanical efficiency of the drive is 0.98 and the combustion efficiency is 0.99. Assuming a convergent propulsion nozzle, loss of stagnation pressure in the combustion chamber of 0.2 bar, and a fuel of calorific value of 43.3 MJ/kg, calculate (a) the required nozzle exit area, (b) the net thrust developed, (c) the air–fuel ratio, and (d) the specific fuel consumption. For gases in the turbine and propulsion nozzle, take $\gamma = 1.333$ and $C_p = 1.15$ kJ/kgK.

Ans. ((a) 0.216 m^2 (b) 19.94 kN, (c) 70.87, (d) 0.0319 kg/kNs)

21.8 In a turboprop engine, the compressor pressure ratio is 6 and the maximum cycle temperature is 760°C. The stagnation isentropic efficiencies of the compressor and turbine are 0.85 and 0.88 respectively, the mechanical efficiency is 0.99. The aircraft is flying at 725 km/h at an altitude where the ambient temperature is –7°C. Taking an intake duct efficiency of 0.9, neglecting and pressure loss in the combustion chamber, and assuming that the gases in the turbine expand down to atmospheric pressure, leaving the aircraft at 725 km/h relative to the aircraft, calculate (a) the specific power output, and (b) the cycle efficiency. For the gases in the turbine, take $\gamma = 1.333$ and $c_p = 1.15$ kJ/kgK.

(*Ans.* (a) 170.2 kW per kg/s (b) 28.4%)

21.9 Afterburning is used in the aircraft of Problem 21.8 to obtain an increase in thrust. The stagnation temperature after the afterburner is 700°C and the pressure loss in the afterburning process is 0.07 bar. Calculate the nozzle exit area now required to pass the same mass flow rate as in Problem 21.8 and the new net thrust.

(*Ans.* 0.24 m^2, 22 kN.)

21.10 A turbojet unit flies at 225 m/s in air where $\phi =$ 0.62 bar and T = 265 K. The unit has diffuser ($\eta = 0.9$), compressor ($\eta = 0.85$, pressure ratio 6), combustion chamber (pressure loss 0.15 bar, $\eta = 0.9$), turbine ($\eta = 0.88$, inlet temperature 1100 K, and nozzle ($\eta = 0.9$) Calculate the thrust and the air–fuel ratio, if the heating value of the fuel is 40 MJ/kg. (*Ans.* 392.32 N/kg air, 59.73.)

21.11 The drag for a turbojet aircraft is 7 kN and the propulsive efficiency is 0.55. When the aircraft is flying with a speed of 800 km/h, estimate (a) the diameter of the jet if the air density is 0.17 kg m^3, (b) the air–fuel ratio when the overall efficiency of the unit is 20% and (c) the specific fuel consumption if the heating value of the fuel is 40 MJ/kg.

(*Ans.* (a) 4.96 cm, (b) 98.97, (c) 0.45 kg/kWh)

21.12 A propeller of 3.5 m, diameter-produces a thrust of 20 kN while flying at a speed of 100 m/s where the pressure is 1 bar and the temperature is 280 K. Estimate the velocity in the final wake, the pressure rise through the propellerer, and the propeller efficiency.

(*Ans.* $V_4 = 115.5$ m/s, $\Delta p = 2.0786$ kPa, $\eta_p = 92.8\%$)

21.13 An aircraft is flying at a speed of 250 m/s and the air mass flow rate is 45 kg/s. The air–fuel ratio is 60 and the heating value is 40 MJ/kg. If the gases expand to ambient pressure in the nozzle, determine (a) the jet velocity when thrust power is maximum, (b) the thrust and thrust power, and (c) the propulsive and overall efficiency of the system.

(*Ans.* (a) 500 m/s, (b) 11.625 kN and 2.91 Mw, (c) 0.674 and 0.0968.)

21.14 The combustion gases from a rocket motor have negligible velocity when leaving the CC at a pressure of 25 bar and a temperature of 3000 K. The gases expand through a convergent–divergent nozzle to a pressure of 1.5 bar in the exit plane of the nozzle. Assuming isentropic expansion aı d taking for gases $\gamma = 1.2$ and a motor mass of 32 kg/kgmol and neglecting dissociation effects, calculate the ratio of exit area A_e to throat area A_t. Also, calculate the thrust developed per unit throat area when the rocket is operating in outer space.

Ans. 3.19, 11.415 MN/m^2 throat area.

C H A P T E R

22 Transport Processes in Gases

Collisions between molecules were not considered in Chapter 21 while deriving the expressions for pressure and temperature of an ideal gas in terms of its molecular properties. Intermolecular collisions will now be considered.

22.1 MEAN FREE PATH AND COLLISION CROSS-SECTION

Let us single out one particular molecule represented by the black circle and trace its path among the other molecules, which would be assumed to be frozen in their respective positions (Fig. 22.1). The distance traversed by a molecule between successive collisions is called the free path, denoted by x, and the average length of these paths is called the *mean free path*, denoted by λ. The molecules are assumed to be perfectly elastic spheres of radius r. As two molecules collide, the centre-to-centre distance is $2r$, which would remain the same if the radius of the moving molecule is increased to $2r$ and the stationary molecules are shrunk to geometrical points, as shown in Fig. 22.1 The cross-sectional area of the moving molecule is called the *collision cross-section* σ, and it is given by $\sigma = 4\,\pi r^2$

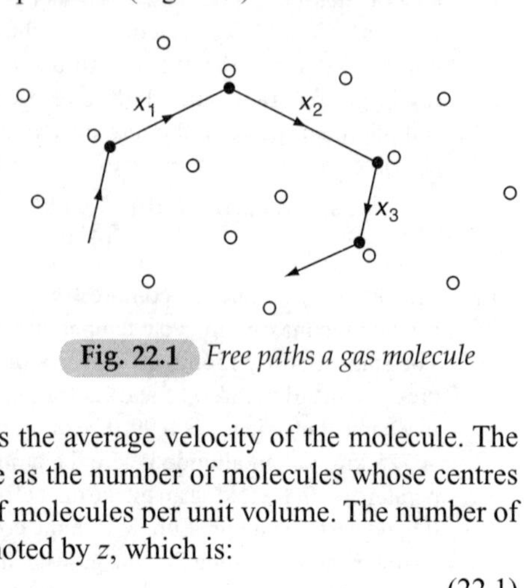

Fig. 22.1 *Free paths a gas molecule*

The moving molecule sweeps out in time t, a cylindrical volume of cross-sectional area σ and length $\bar{v}t$, where $\bar{v}$ is the average velocity of the molecule. The number of collisions it makes during this time, will be the same as the number of molecules whose centres lie within this volume, which is $\sigma n\bar{v}t$, where n is the number of molecules per unit volume. The number of collisions per unit time is known as the *collision frequency*, denoted by z, which is:

$$z = \sigma n \bar{v} \tag{22.1}$$

The mean free path of the molecules is given by:

$$\lambda = \frac{\text{Distance travelled in time } t}{\text{Number of collisions in time } t} = \frac{\bar{v}t}{\sigma n \bar{v} t} = \frac{1}{\sigma n} \tag{22.2}$$

On an average, the diameter (d) of the molecules is (2 to 3) $\times\ 10^{-10}$ m, the distance between molecules 3×10^{-9} m (or 10d), and the mean free path is about 3×10^{-8} m (or $100d$).

If motion of all the molecules if considered and all the molecules move with the same speed, a correction is required and λ is obtained as:

$$\lambda = 0.75/\sigma n \tag{22.3}$$

If the Maxwellian velocity distribution is assumed for the molecules, $\lambda = 0.707/\sigma n$ (22.4)

For an electron moving among molecules of a gas, the radius of the electron is so small compared to that of a molecule that in a collision the electron may be treated as a point and the centre-to-centre distance becomes

r, instead of $2r$, where r is radius of the molecule. Also, the velocity of the electron is so much greater than the velocities of the molecules that the latter can be considered stationary. As a result, no correction is required, and the electronic mean free path λ_e is given by:

$$\lambda_e = 4/\sigma n \tag{22.5}$$

where $\sigma = 4\pi r^2$

22.2 DISTRIBUTION OF FREE PATHS

The distance travelled by a molecule between successive collisions or the free path x varies widely. It may be greater or less than λ, or equal to it. Just like distribution of molecular velocities, we will now determine how many molecules will have free paths in a certain range, say between x and $x + dx$.

Let us consider a large number of molecules N_0 at a certain instant (Fig. 22.2). If the molecules collide, they will be assumed to get removed from the group. Let N represent the number of molecules left in the group after travelling a distance x. Then these N molecules have free paths larger than x. In the next short distance dx, let dN number of molecules make collisions and get removed from the group. So, these dN molecules which have free paths lying between x and x + dx are proportional to N and to dx. Since N is always decreasing, dN is negative and it is given by:

$$dN = -P_c N\, dx \tag{22.6}$$

where P_c is the constant of proportionality, known as the *collision probability*.

Then

$$dN/N = -P_c\, dx$$

$$\ln N = -P_c x + A$$

Where A is constant. When $x = 0$, $N = N_0$, and so $A = \ln N_0$. Therefore,

$$N = N_0 e^{-P_c x} \tag{22.7}$$

The number of molecules that remains in the group falls off exponentially with x. From Eq. (22.6),

$$dN = -P_c N_0 e^{-P_c x}\, dx \tag{22.8}$$

Using this expression for dN, the mean free path λ becomes

$$\lambda = \frac{\int x\, dN}{\int dN} = \frac{\int_0^\infty -x P_c N_0 e^{-P_c x}\, dx}{N_0} = \frac{1}{P_c}$$

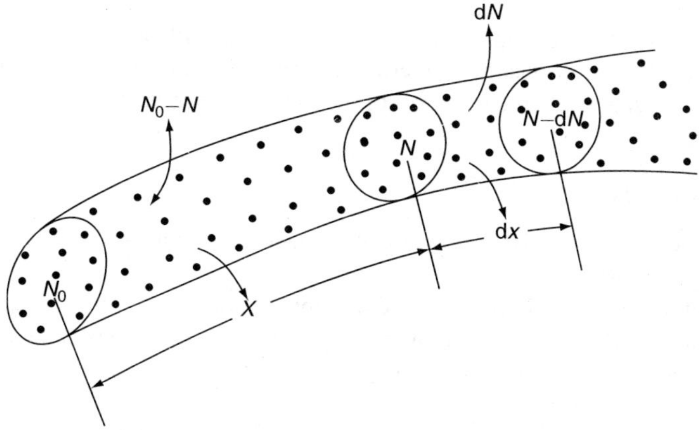

Fig. 22.2 *Molecules colliding and getting removed from the group*

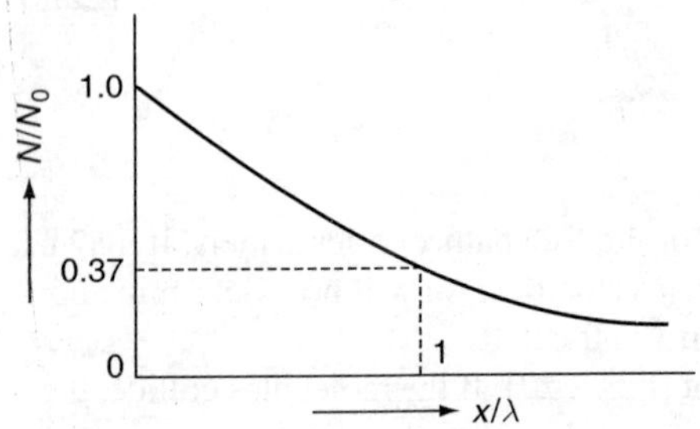

Fig. 22.3 *Plot of survival equation*

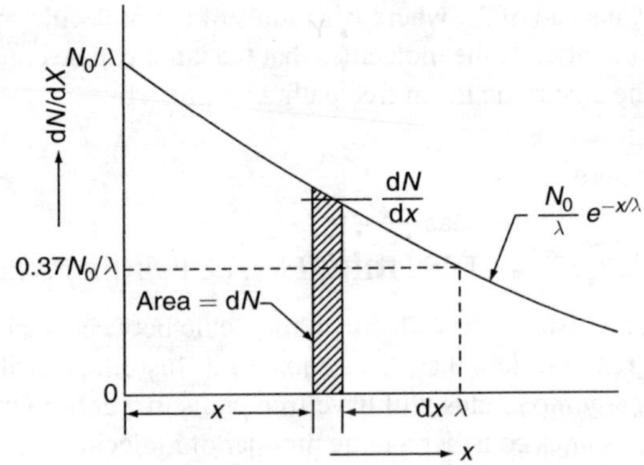

Fig. 22.4 *Distribution of free paths*

Since $\lambda = 1/\sigma n$, $P_c = \sigma n$. The collision probability is thus proportional to the collision cross-section and the number of molecules per unit volume. The Eq. (22.7) can thus be written as

$$N = N_0 e^{-x/\lambda} \tag{22.9}$$

It is know as the *survival equation* which indicates the number of molecules N, out of N_0, which survive collision and have free paths longer than x. A plot of N/N_0 vs x/λ is shown in Fig. 22.3. If $x/\lambda = 1$, ie., $x = \lambda$, $N/N_0 = 0.37$. The fraction of free paths longer than λ is, therefore, 37% and the fraction shorter than λ is 63%.

Differentiating Eq. (22.9)

$$dN = \frac{N_0}{\lambda} e^{-x/\lambda} dx$$

or,

$$dN/dx = -\frac{N_0}{\lambda} e^{-x/\lambda} \tag{22.10}$$

This equation represents the distribution of free paths. It is plotted in Fig. 22.4. The area of the narrow vertical strip of thickness dx at a distance x from the origin represents dN, the number of molecules with free paths of lengths between x and x + dx.

22.3 TRANSPORT PROPERTIES

A Simple treatment based on the concept of the mean free path will now be given for four transport properties of a gas, viz., coefficient of viscosity, thermal conductivity, coefficient of diffusion and electrical conductivity, which govern respectively the transport of momentum, energy, mass, and electric charge within the gas by molecular motion.

22.3.1 Coefficient of Viscosity

Let us consider a gas flowing over a flat stationary plate. Due to viscous effect there is the growth of a boundary layer over the plate surface.

The velocity of fluid at the surface will be zero, and it gradually increases to free stream velocity as shown in Fig. 22.5, drawn for laminar flow. Let us imagine a surface *P-P* within the gas at an arbitrary height from the plate, where the fluid velocity is u and the velocity gradient du/dy. The velocity u is superposed on the random thermal motion of the molecules.

Let us consider a volume element dV at a distance r from an element of area dA in the plane *P-P*, making an angle θ with normal to dA (Fig. 22.6), the plane *P-P* being the same as shown in Fig. 22.5. The volume

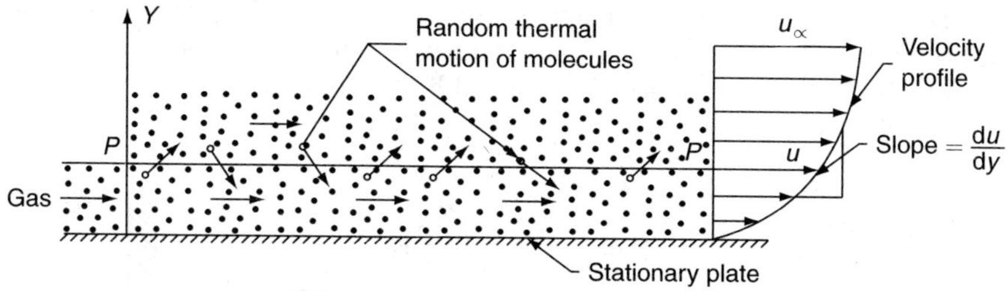

Fig. 22.5 *Flow of a gas over a flat plate*

element is very small when compared with the physical dimensions of the system, but large enough to contain many molecules. The total number of molecules in dV is ndV, and the total number of collisions within dV in time dt is $\frac{1}{2}Zn\cdot$ d$V \cdot$ dt, where z is the collision frequency of a molecule, n is the number of molecules per unit volume, and the factor $\frac{1}{2}$ is required since two molecules are involved in each collision. Since two new free paths originate at each collision, the total number of new free paths, or molecules, originating in dV is ZndVdt. If we assume that these molecules are uniformly distributed in direction throughout the solid angle 4π, then the number headed towards the elemental area dA is:

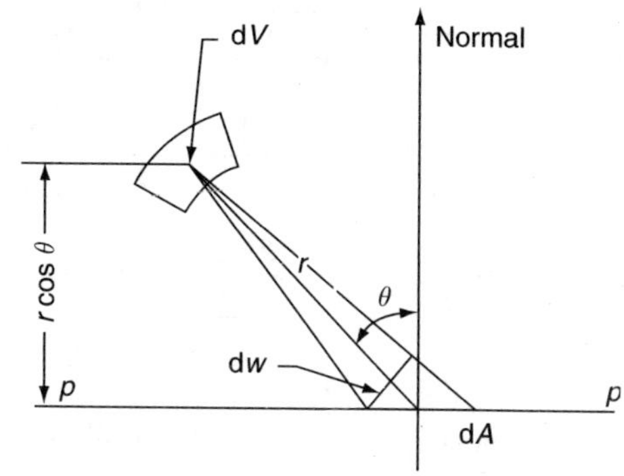

Fig. 22.6 *Transfer of momentum across the plane P-P by molecules in random thermal motion*

$$\frac{zn\mathrm{d}V\,\mathrm{d}t}{4\pi}\mathrm{d}w$$

where dw is the solid angle subtended at the centre of dV by the area dA and is equal to (dA cos θ)/r^2.

The number of molecules that leave dV and reach dA without having made a collision may be found from the survival equation, Eq. (22.9), as given below:

$$\frac{zn\,\mathrm{d}V\mathrm{d}t}{4\pi}\mathrm{d}w\,e^{-r/\lambda}$$

Since d$V = r^2 \sin\theta \cdot \mathrm{d}\theta \cdot \mathrm{d}\phi \cdot \mathrm{d}r$, the number of molecules leaving dV in time dt and crossing dA without any collision is

$$\frac{1}{4\pi}\frac{\mathrm{d}A\cos\theta}{r^2}Zn\,r^2\sin\theta\,\mathrm{d}\theta\,\mathrm{d}\phi\,\mathrm{d}r\,\mathrm{d}t\,e^{-r/\lambda}$$

The total number of molecules crossing dA in time dt from the top is

$$N_{\text{total}} = \frac{Zn\,\mathrm{d}A\,\mathrm{d}t}{4\pi}\int_{\theta=0}^{\pi/2}\sin\theta\cos\theta\,\mathrm{d}\theta\int_{\phi=0}^{2\pi}\mathrm{d}\phi\int_{r=0}^{\infty}e^{-r/\lambda}\mathrm{d}r$$

$$= \frac{Zn\,\mathrm{d}A\,\mathrm{d}t}{4\pi}\frac{1}{2}2\pi\cdot\lambda = \frac{1}{4}Zn\,\lambda\,\mathrm{d}A\,\mathrm{d}t \tag{22.11}$$

Since the dimensions of the physical system are very much larger than the molecular free path, the integral over r has been extended to infinity.

But $z = \bar{v}/\lambda$, so the number of molecules crossing the plane P-P from the top (or bottom) per unit area and per unit time is $\frac{1}{4}n\bar{v}$. This is the same result obtained earlier in Sec. 21.4 without considering any intermolecular collision.

These molecules crossing the plane *P-P* may be visualized as carrying properties characteristic of an average distance $\bar{y}$, either above or below the plane at which they made their last collisions before crossing. To find $\bar{y}$, each molecule crossing from dV is multiplied by its distance $r\cos\theta$ from the *P-P* plane, it is integrated over θ, ϕ and r and then divided by N_{total} crossing the plane.

$$\bar{y} = \frac{\int y\, dN}{\int dN} = \frac{\int r\cos\theta\, dN}{N_{\text{total}}}$$

$$= \frac{1}{N_{\text{total}}}\left[\int_r \int_\phi \int_\theta \frac{1}{4\pi} Znd\,A\,dt.\,e^{-r/\lambda}\sin\theta.\cos\theta.\,d\theta\,d\phi\,dr\,r\cos\theta\right]$$

$$= \frac{1}{N_{\text{total}}}\left[\frac{ZndA\,dt}{4\pi}\int_{\theta=0}^{\pi/2}\sin\theta\cos^2\theta\,d\theta\int_{\phi=0}^{2\pi}d\phi\int_{r=0}^{\infty}re^{-r/\lambda}dr\right]$$

$$= \frac{\frac{Zn\,dA\,dt}{4\pi}\frac{1}{3}2\pi\cdot\lambda^2}{\frac{1}{4}Zn\,\lambda\,dA\,dt} = \frac{2}{3}\lambda \tag{22.12}$$

The velocity of gas at a height $\bar{y}$ above *PP* is,

$$u + \frac{2}{3}\lambda\frac{du}{dy}$$

if the velocity gradient is considered constant over distances of the order of a free path.

The net momentum in the direction of flow carried across the plane by the molecules crossing *PP* from above per unit area and per unit time is:

$$\frac{1}{4}n\,\bar{v}\,m\left[u + \frac{2}{3}\lambda\frac{du}{dy}\right]$$

Similarly, the net momentum transfer from below is:

$$\frac{1}{4}n\,\bar{v}\,m\left[u - \frac{2}{3}\lambda\frac{du}{dy}\right]$$

The difference between the above two quantities is the net rate of transport of momentum per unit area and per unit time, given by:

$$\frac{1}{3}nm\bar{v}\lambda\frac{du}{dy}$$

From Newton's law of viscosity, this is the viscous force per unit area $\tau = \mu\frac{du}{dy}$, where μ is the coefficient of viscosity.

Therefore,

$$\mu = \frac{1}{3}nm\bar{v}\lambda \tag{22.13}$$

Putting $\sigma = 1/\lambda n$ from Eq. (22.2),

$$\mu = \frac{1}{3}\frac{m\bar{v}}{\sigma} \tag{22.14}$$

Where σ is the collision cross-section.

For a gas with a Maxwellian velocity distribution, $\bar{v} = [8\,KT/\pi m]^{1/2}$, $\lambda = 0.707/\sigma n$

Therefore, from Eq. (22.13),
$$\mu = \frac{2}{3\sigma}[mKT/\pi]^{1/2} \tag{22.15}$$

Putting
$$\sigma = \pi d^2$$
$$\mu = \frac{2}{3\pi^{3/2}} \frac{(mKT)^{1/2}}{d^2} \tag{22.16}$$

A significant conclusion from this equation is that the viscosity of a gas is independent of the pressure or the density and *depends only on the temperature*. As the temperature of the gas increases, its viscosity increases.

22.3.2 Thermal Conductivity

It is possible to derive an expression for the thermal conductivity of a gas by repeating the argument leading up to the Eq. (22.12). For a system, not being isothermal, the molecules moving from the warmer region to the colder region carry with them more energy than those moving in the opposite direction, resulting in a net transfer of energy.

Let us consider a gas confined between two stationary plates (Fig. 22.7) maintained at different temperatures. Let T be the temperature at the plane P-P and dT/dY the temperature gradient. The mean energy of a molecule at a temperature T is given by $\frac{f}{2}KT$, where f is the number of degrees of freedom. It was concluded from Eq. (22.12) that the molecules crossing the plane from either direction carry property values characteristic of the planes $(2/3)\lambda$ distance from the plane.

Therefore, the energy carried across the plane per unit area and per unit time, by the molecules crossing the plane from above:
$$\frac{1}{4} n \bar{v} \frac{f}{2} K \left[T + \frac{2}{3} \lambda \frac{dT}{dy} \right]$$
and the energy carried by the molecules crossing from below:
$$\frac{1}{4} n \bar{v} \frac{f}{2} K \left[T - \frac{2}{3} \lambda \frac{dT}{dy} \right]$$

The net rate energy transfer per unit area is the difference of the above two quantities, which gives:
$$\frac{1}{6} n \bar{v} f K \lambda \frac{dT}{dy}$$

By Fourier's law, this is given by:
$$q = K \frac{dT}{dy}$$

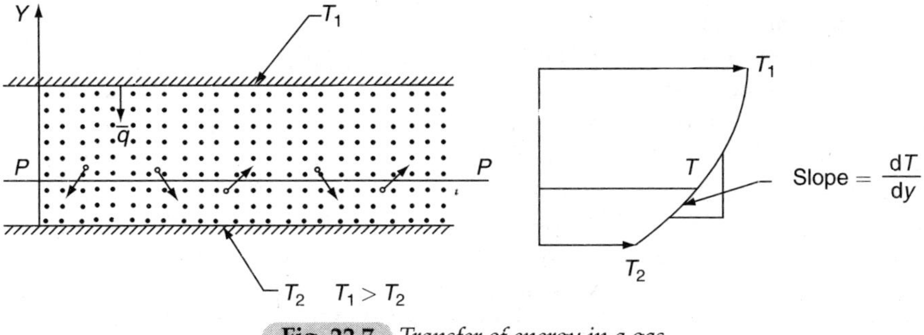

Fig. 22.7 *Transfer of energy in a gas*

where k is the thermal conductivity of the gas. Therefore,

$$k = \frac{1}{6} n \bar{v} f K \lambda \tag{22.17}$$

or,

$$k = \frac{1}{6} \frac{\bar{v} f K}{\sigma} \tag{22.18}$$

For a gas with a Maxwellian velocity distribution,

$$\bar{v} = [8KT/\pi m]^{1/2} \text{ and } \lambda = 0.707/\sigma n$$

$$k = \frac{1}{3} \frac{f K}{\sigma} [KT/\pi m]^{1/2} \tag{22.19}$$

The above equation predicts that the thermal conductivity of a gas, like the viscosity, is independent of pressure or density, and *depends only on temperature*. It increases as the temperature increases.

For a monatomic gas, $f = 3$ and putting $\sigma = \pi d^2$,

$$k = \frac{1}{\pi^{3/2} d^2} \left[\frac{K^3 T}{m} \right]^{1/2}$$

Dividing Eq. (22.15) by Eq. (22.19),

$$\mu / k = 2m / f K \tag{22.20}$$

But

$$m = M/N_0,\ K = \bar{R}/N_0,\ c_v = \frac{f}{2} R = \frac{f}{2} \frac{\bar{R}}{M},$$

where M is the molecular weight and N_0 is the Avogadro's number. Therefore, on substitution in Eq. (22.20)

$$\mu c_v / k = 1 \tag{22.21}$$

or, Prandtl number,

$$Pr = \mu c_p / k = \gamma \tag{22.22}$$

The results given by the Eqs (22.21) and (22.22) agree with the experimental values only as regards order of magnitude.

22.3.3 Coefficient of Diffusion

In a gaseous mixture, diffusion results from random molecular motion whenever there is a concentration gradient of any molecular species. Let us consider two different gases A and B at the same temperature and pressure on the two opposite sides of the partition in a vessel (Fig. 22.8). The number of molecules per unit volume (p/KT) is, therefore, the same on both sides. When the partition is removed, both the gases diffuse into each other, and after a lapse of time both the gases are uniformly distributed throughout the entire volume. The diffusion process is often superposed by the hydrodynamic flow resulting from pressure differences, and the effects of molecules rebounding from the walls of the vessel. When more than one type of molecule is present, the rates of diffusion of one species into another are different. To simplify the problem we assume: (1) the molecules of a single species diffusing into others of the same species (self-diffusion), (2) the containing vessel very large compared with the mean free path so that collisions with the walls can be neglected in comparison with collisions

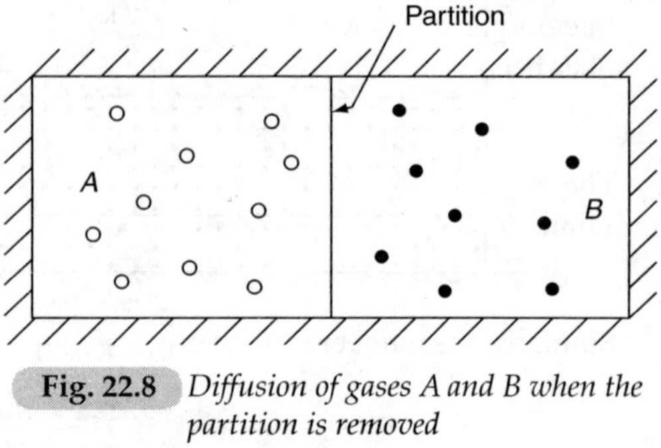

Fig. 22.8 *Diffusion of gases A and B when the partition is removed*

with other molecules, and (3) a uniform pressure maintained so that there is no hydrodynamic flow. Of course, if all the molecules are exactly alike, there would be no way experimentally to identify the diffusion process. However, the diffusion of molecules that are isotopes of the same element is a practical example of the self-diffusion process.

Let n denote the number of molecules per unit volume of one gas, blackened for identification (Fig. 22.9). Let us consider diffusion across an imaginary vertical plane y-y in the vessel. Let us also assume that n increases from left to right in the positive x-direction. The number of black molecules crossing the plane from right to left exceeds the number crossing in the opposite direction. The net transfer of black molecules from left to right per unit area and per unit time, denoted by Γ, is given by:

$$\Gamma = -D\frac{dn}{dx} \tag{22.23}$$

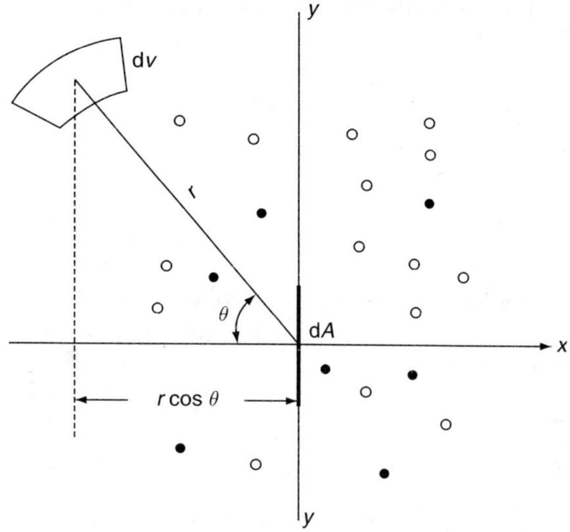

Fig. 22.9 *Diffusion across an imaginary plane*

where D is the coefficient of diffusion.

Let us consider a volume element dV at a distance r from an element of area dA in the plane y-y, making an angle θ with normal to dA. The concentration of black molecules in the plane of dV at a distance x from the vertical plane,

$$n = n_0 + x\frac{dn}{dx} = n_0 - r\cos\theta\frac{dn}{dx} \tag{22.24}$$

Where n_0 is the concentration at the vertical plane. The total number of free paths originating in dV in time dt is $Zn_t dV dt$, where n_t is the total number of molecules per unit volume. The number of free paths of black molecules will be $\frac{n}{n_t}(Zn_t dV dt)$, or $Zn dV dt$. The number of black molecules crossing dA without making a collision, as found in Section 22.3.1 is $\frac{1}{4\pi} Zn\, dA\, dt \sin\theta \cos\theta\, e^{-r/\lambda}\, d\theta\, d\phi\, dr$. By substituting for n from Eq. (22.24), it is:

$$\frac{1}{4\pi} Zn_0\, dA\, dt \sin\theta \cos\theta\, e^{-r/\lambda} d\theta\, d\phi\, dr$$

$$-\frac{1}{4\pi} Z\frac{dn}{dx} dA\, dt \sin\theta \cos^2\theta\, r\, e^{-r/\lambda} d\theta\, d\phi\, dr$$

Integrating the above expressions over θ from 0 to $\pi/2$, over ϕ from 0 to 2π and over r from 0 to ∞, the number becomes

$$\frac{1}{4} Z\, n_0 \lambda\, dA\, dt - \frac{1}{6} Z\lambda^2 \frac{dn}{dx} dA\, dt$$

Therefore, the number of black molecules crossing the plane y-y from left to right per unit area and per unit time is:

$$\underset{\rightarrow}{\Gamma} = \frac{1}{4} Z\, n_0 \lambda - \frac{1}{6} Z\lambda^2 \frac{dn}{dx} \tag{22.25}$$

Similarly, the number crossing from right to left is

$$\underset{\leftarrow}{\Gamma} = \frac{1}{4} Z\, n_0 \lambda + \frac{1}{6} Z\, \lambda^2 \frac{dn}{dx} \tag{22.26}$$

By subtracting Eq. (22.26) from Eq. (22.25), the net number crossing from left to right becomes

$$\Gamma = -\frac{1}{3} Z\,\lambda^2 \frac{dn}{dx} \tag{22.27}$$

From Eqs (22.23) and (22.27)

$$D = \frac{1}{3} Z\lambda^2 \tag{22.28}$$

Putting $z = \bar{v}/\lambda$,

$$D = \frac{1}{3}\,\bar{v}\lambda \tag{22.29}$$

For Maxwellian velocity distribution of the molecules,

$$\bar{v} = [8KT/\pi m]^{1/2} \quad \text{and} \quad \lambda = 0.707/\sigma n_t$$

$$\therefore \quad D = \frac{2}{3\sigma n_t}\,[KT/\pi m]^{1/2} \tag{22.30}$$

or,

$$D = \frac{2}{3\pi^{3/2} d^2 n_t}[KT/m]^{1/2} \tag{22.31}$$

The equation applies to diffusion in a binary mixture of almost identical gases. Dividing Eq. (22.15) by Eq. (22.30),

$$\mu/D = n_t \,.\, m = \rho$$

or, Schmidt number

$$Sc = \mu/\rho D = 1 \tag{22.32}$$

Measured values of Schmidt number for the diffusion of isotopic tracer molecules yield values between 1.3 and 1.5, which indicate qualitative agreement of theory with measured data.

22.3.4 Electrical Conductivity

Conduction of electricity in a gas arises as a result of motion of the free electrons present in the gas. When high-energy atoms in the gas collide, some collisions cause ionization when an electron is separated from its atom, so that a negatively charged electron and a positively charged ion are produced from the neutral atom. Most gases at room temperature do not have many such high-energy molecules and thus have very few free electrons. At high temperatures, however, an appreciable number of electrons may be liberated and the gases may become highly conductive. In absence of external electrical fields, the electrons will be distributed uniformly throughout the gas volume. As an electrical field is impressed on the gas, the charged particles are accelerated with a force

$$F = q_e E = m_e \frac{dv_e}{dt} \tag{22.33}$$

where q_e is the charge on the electron and E is the electric field strength (volts per metre). Integrating Eq. (22.33)

$$v_e = \frac{dx_e}{dt} = \frac{q_e Et}{m_e}$$

At $t = 0$, $v_e = 0$, $x_e = 0$. Similarly, the velocity of a single charged ion is: $v_i = \dfrac{q_i Et}{m_i}$ (22.34)

where q_i is the charge on an ion and m_i is the mass of the ion. Since, $m_i/m_e = 1840\,M$, where M is the atomic weight of the ion, $v_i/v_e = 1/1840\,M$. Therefore, the velocity of the ions is very small compared to the velocity of electrons and is neglected.

The electron velocity v_e is randomly oriented in absence of an electric field with no preferred direction of motion. The electron would attain the velocity v_e (in Eq. 22.33) after t seconds of impressing the electric field, assuming it had to collision. To consider the effects of collision, it is assumed that the electrons are brought to rest after each collision, reaccelerated by the electric field, and then undergo another collision. The distance traveled between collision x_e is obtained by integrating Eq. (22.33).

$$x_e = (q_e E t^2)/2m_e$$

where t is the collision time.

The average distance travelled between collisions is

$$\bar{x}_e = \frac{1}{N_e}\int x_e \mathrm{d}N_e$$

where N_e is the total number of electrons in the gas.

The distribution of collision times is identical to distribution of free paths given by Eq. (22.10), which is

$$\mathrm{d}N_e = \frac{N_e}{\tau_e} e^{-t/\tau_e} \mathrm{d}t \tag{22.35}$$

where τ_e is the mean collision time for electrons.

The average distance travelled in the electric field by the electrons between collisions, $\bar{x}_e$, divided by the mean collision time τ_e is called the *electron drift velocity*, v_{e_d}. Therefore

$$v_{e_d} = \bar{x}_e/\tau_e = \frac{1}{\tau_e}\frac{1}{N_e}\int x_e \mathrm{d}N_e$$

$$= \frac{1}{\tau_e}\frac{1}{N_e}\int_0^\infty \frac{q_e E t^2}{2m_e}\frac{N_e}{\tau_e} e^{-t/\tau_e}\mathrm{d}t$$

$$= \frac{1}{\tau_e^2}\frac{q_e E}{2m_e}\int_0^\infty t^2 e^{-t/\tau_e}\mathrm{d}t = \frac{q_e E \tau_e}{m_e} \tag{22.36}$$

From Eq. (22.5), the electronic mean free path is

$$\lambda_e = 4/\sigma n$$

Also, $\lambda_e = \bar{v}_e \tau_e$, where $\bar{v}_e$ is the average velocity of the electrons, given by $[8KT/\pi m_e]^{1/2}$ for Maxwellian velocity distribution. Therefore,

$$v_{e_d} = \frac{q_e E}{m_e}\frac{4}{\sigma n}\left[\frac{\pi m_e}{8KT}\right]^{1/2}$$

Using

$$n = p/KT,$$

$$u_{e_d} = \frac{4 q_e E}{\sigma p}\left[\frac{\pi KT}{8 m_e}\right]^{1/2} \tag{22.37}$$

where σ is the atomic cross-section.

The flux of charge across unit area per unit time is called the *current density*, J. The current density is defined with respect to the average drift velocity by the following equation,

$$J = n_e q_e v_{e_d} \tag{22.38}$$

where n_e is the number density of electrons. The motion of the ions is neglected. From Eqs (22.36) and (22.38),

$$J = \frac{q_e^2 n_e \tau_e}{m_e} E \tag{22.39}$$

The current density is proportional, to the electric field and the constant of proportionality is called the *electrical conductivity*, σ_e $\quad J = \sigma_e E$

From Eq. (22.39), σ_e is given by

$$\sigma_e = \frac{q_e^2 n_e \tau_e}{m_e}$$

Putting

$$\lambda_e = \bar{v}\,\tau_e = 4/\sigma n,$$

$$\sigma_e = \frac{q_e^2 n_e}{m_e}\frac{4}{\sigma n}\left[\frac{\pi m_e}{8KT}\right]^{1/2}$$

$$= \frac{q_e^2 n_e}{\sigma n}\left[\frac{2\pi}{m_e KT}\right]^{1/2} \qquad (22.40)$$

The electron drift velocity v_{e_d} is also proportional to the electric field and the proportionality constant is called the electron mobility, μ_e

$$v_{e_d} = \mu_e E$$

From Eq. (22.36)

$$\mu_e = (q_e/\tau_e)/m_e$$

Again, using

$$\lambda_e = \bar{v}_e\,\tau_e = 4/\sigma n,$$

$$\mu_e = \frac{q_e}{m_e}\frac{4}{\sigma n}\left[\frac{\pi m_e}{8KT}\right]^{1/2} = \frac{q_e}{\sigma n}\left[\frac{2\pi}{m_e KT}\right]^{1/2} \qquad (22.41)$$

Solved Examples

Example 22.1

Calcuate the mean free path for oxygen molecules at 1 atm pressure and 300 K. What fraction of molecules have free paths longer than 2λ?. The effective diameter of the oxygen molecule is 3.5 Angstroms.

Solution The collision cross section of the oxygen molecule is:

$$\sigma = \pi d^2 = \pi(3.5\times10^{-10})^2 = 3.84\times10^{-19}\,\text{m}^2$$

Number of molecules per unit volume

$$n = p/KT = (1.013\times10^5)/(1.38\times10^{-23}\times300)$$

$$= 2.45\times10^{25}\,\text{molecules/m}^3$$

The mean free path is given by Eq. (22.4)

$$\lambda = 0.707/\sigma n = 0.707/(3.84\times10^{-19}\times2.45\times10^{25})$$

$$= 7.25\times10^{-8}\,\text{m}$$ *Ans.*

From the Eq. (22.9), putting $x = 2\lambda$

$$N/N_0 = e^{-x/\lambda} = e^{-2} = 0.135$$

i.e., 13.5% of the molecules have free paths greater than 2λ.

Example 22.2

The mean free path of the molecules of a certain gas at a temperature of 25°C is 2.63×10^{-5} m. The radius of the molecules is 2.56×10^{-10} m.
(a) Find the pressure of the gas. (b) Calculate the number of collisions made by a molecule per metre of path.

Solution The collision cross-section, $\sigma = 4\pi r^2 = 4\pi \times (2.56 \times 10^{-10})^2 = 82.36 \times 10^{-20}\,\text{m}^2$

Number of molecules per unit volume of the gas is:

$$n = \frac{0.707}{\sigma\lambda} \quad \text{(assuming Maxwell-Boltzmann velocity distribution)}$$

$$= \frac{0.707}{82.36 \times 10^{-20} \times 2.63 \times 10^{-5}} = 0.326 \times 10^{23}\,\text{molecules m}^3$$

From the ideal gas equation of state,

$$P = nKT$$

$$= 0.326 \times 10^{23}\,\frac{\text{molecules}}{m^3} \times 1.38 \times 10^{-23}\,\frac{\text{J}}{\text{molecule - K}} \times 298\text{ K}$$

$$= 0.326 \times 1.38 \times 298\text{ N/m}^2 = 134.06\text{ Pa}$$ *Ans.* (a)

Number of collisions made by a molecule:

$$= \frac{1}{\lambda} = \frac{1}{2.63 \times 10^{-5}} = 3.8 \times 10^4$$ *Ans.* (b)

Example 22.3

The mean free path in a certain gas is 10 cm. If there are 10,000 free paths, how many are longer than (a) 10 cm, (b) 20 cm, (c) 50 cm, (d) How many are longer than 5 cm, but shorter than 10 cm. (e) How many are between 9.5 cm and 10.5 cm in length? (f) How many are between 9.9 cm and 10.1 cm in length? (g) How many are exactly 10 cm in length?

Solution From the survival equation, $N = N_0 e^{-x/\lambda}$

where $N_0 = 10{,}000$ and $\lambda = 10$ cm (given)

$$N = 10{,}000\, e^{-x/10}$$

is the governing equation,

(a) For $x > 10$ cm, $N = 10{,}000\, e^{-1} = 10{,}000/2.718 = 3680$ *Ans.*

(b) For $x > 20$ cm, $N = 10{,}000\, e^{-2} = 1354$ *Ans.*

(c) For $x > 50$ cm, $N = 10{,}000\, e^{-5} = 67.4$ or 68 *Ans.*

(d) Fox x between 5 cm and 10 cm, we have

$$dN = \frac{N_0}{\lambda} e^{-x/\lambda} dx$$

$$N = \int_5^{10} -\frac{N_0}{\lambda} e^{-x/\lambda} dx = -\frac{N_0}{\lambda}[e^{-x/10}(-\lambda)]_5^{10}$$

$$= -N_0[e^{-x/10}]_5^{10} = N_0[e^{-1} - e^{-0.5}]$$
$$= N_0[0.3679 - 0.6066] = -0.2387 \times 10{,}000$$
$$= -2387$$ *Ans.*

The negative sign implies than N is decreasing with x.

(e) For x between 9.5 cm and 10.5 cm

$$N = -10{,}000[e^{-9.5/10} - e^{-10,5/10}]$$
$$= -10{,}000(0.3870 - 0.3500) = -370$$ *Ans.*

(f) For x between 9.9 cm and 10.1 cm

$$N = -10{,}000[e^{-9.9/10} - e^{-10.1/10}]$$
$$= -10{,}000(0.3716 - 0.3642) = -74$$ *Ans.*

(g) For $x = 10$ cm exactly

$$N = -\frac{N_0}{\lambda}\int_{10}^{10} e^{-x/\lambda} dx = \text{zero}$$ *Ans.*

Example 22.4

Calculate the coefficient of viscosity of oxygen at 1 atm pressure and 300 K.

Solution From previous examples, we have

$$m = 5.31 \times 10^{-26} \text{ kg/molecule}$$
$$\bar{v} = 445 \text{ m/s}$$
$$\sigma = 3.84 \times 10^{-19} \text{ m}^2$$

Therefore,

$$\mu = \frac{1}{3}\frac{m\bar{v}}{\sigma}$$
$$= \frac{(5.31 \times 10^{-26}) \text{ kg/molecule} \times 445 \text{ m/s}}{3 \times 3.84 \times 10^{-19} \text{ m}^2/\text{molecule}}$$
$$= 2.05 \times 10^{-5} \text{ kg/ms}$$ *Ans.*
$$= 2.05 \times 10^{-5} \text{ Ns/m}^2$$ *Ans.*

Example 22.5

Calculate the thermal conductivity of oxygen at 1 atm, 300 K.

Solution For oxygen, a diatomic gas, the degree of freedom $f = 5$.

$$\bar{v} = [8\, KT/\pi m]^{1/2} = 445 \text{ m/s}$$
$$\sigma = \pi d^2 = 3.84 \times 10^{-19} \text{ m}^2$$
$$k = \frac{1}{6}\frac{\bar{v} f K}{\sigma}$$
$$= \frac{1}{6}\frac{445 \text{ m/s} \times 5 \times 1.38 \times 10^{-23} \text{ J/molecule K}}{3.84 \times 10^{-19} \text{ m}^2/\text{molecule}}$$
$$= 0.0133 \text{ W/mK}$$ *Ans.*

If the gas has Maxwellian velocity distribution,

$$k = \frac{1}{3}\frac{f\,K}{\sigma}[KT/\pi m]^{1/2}$$

$$= \frac{1}{3}\frac{5 \times 1.38 \times 10^{-23}\ \text{J/molecule K}}{3.84 \times 10^{-19}\ \text{m}^2/\text{molecule}} \times \left[\frac{1.38 \times 10^{-23}(\text{J/molecule K}) \times 300}{\pi \times 5.31 \times 10^{-26}\ \text{kg/molecule}}\right]$$

$$= \frac{5 \times 1.38 \times 10^{-23}}{3 \times 3.84 \times 10^{-19}}\frac{J}{m^2\text{K}}\left[\frac{1.38 \times 10^{-23} \times 300}{\pi \times 5.31 \times 10^{-26}}\right]^{1/2} \text{m/s}$$

$$= 0.0095\ \text{W/mK}$$

Ans.

Example 22.6

Determine the pressure in a cathode-ray tube such that 90% of the electrons leaving the cathode ray reach the anode 20 cm away without making a collision. The diameter of an ion is 3.6 × 10^{-10} m and the electron temperature is 2000 K. Use the electronic mean free path $\lambda_e = 4/\sigma n$, where σ is the cross-section of the ion.

Solution The several equation is $N = N_0 e^{-x/\lambda}$

where $N = 0.9\,N_0$ and $x = 0.2$ m

$$0.9 = e^{-x/\lambda}$$

$$x/\lambda = 0.1053$$

$$\lambda = 0.2/0.1053 = 1.9\ \text{m}$$

$$\sigma = 4\pi r^2 = 4\pi \times (1.8 \times 10^{-10})^2 = 40.715 \times 10^{-20}\ \text{m}^2$$

$$\lambda_e = \frac{4}{\sigma n} = \frac{4}{40.715 \times 10^{-20} \times n} = 1.9\ \text{m}$$

$$n = \frac{4}{40.715 \times 10^{-20} \times 1.9} = 5.17 \times 10^{18}\ \text{molecules/m}^3$$

Pressure in the cathode ray tube

$$P = nKT = 5.17 \times 10^{18} \times 1.38 \times 10^{-23} \times 2000\ \text{N/m}^2$$

$$= 14.27 \times 10^{-2} = 0.1427\ \text{Pa}$$

Ans.

Example 22.7

Oxygen gas is contained in a one-litre flask at atmospheric pressure and 300 K. (a) How many collisions per second are made by one molecule with the other molecules? (b) How many molecules strike one sq. cm of the flask per second? (c) How many molecules are there in the flask? Take radius of oxygen molecule as 1.8 × 10^{-10} m.

Solution (a) Number of molecules in the flask at

$$1\ \text{atm},\ 300\ \text{K} = N/V = n = p/KT$$

$$= \frac{101.325 \times 1000\ \text{N/m}^2}{1.38 \times 10^{-23}\ \text{J/molecule - K} \times 300\ \text{K}}$$

$$= 2.45 \times 10^{25} \text{ molecules/m}^3$$
$$= 2.45 \times 10^{22} \text{ molecules/litre}$$
$$\sigma = 4\pi r^2 = 4\pi \times (1.8 \times 10^{-10})^2 = 40.71 \times 10^{-20} \text{ m}^2$$

Collision frequency, $z = \sigma n \bar{v}$,

where $\bar{v} = [8KT/\pi m]^{1/2}$

Mass of an oxygen molecule, $m = 5.31 \times 10^{-26}$ kg/molecule

$$\bar{v} = \left[\frac{8 \times 1.38 \times 10^{-23} \times 300}{\pi \times 5.31 \times 10^{-26}}\right]^{1/2} = 445.58 \text{ m/s}$$

$$z = 40.71 \times 10^{-20} \times 2.45 \times 10^{25} \times 445.58$$
$$= 4.44 \times 10^9 \text{ collisions/s}$$ *Ans.* (a)

(b) Number of collisions $= \frac{1}{4} n\bar{v}$

$$= \frac{1}{4} \times 2.45 \times 10^{25} \times 445.58$$
$$= 272.92 \times 10^{25} \text{ collisions/m}^2\text{s}$$
$$= 2.73 \times 10^{23} \text{ collisions/cm}^2\text{s}$$ *Ans.* (b)

(c) Number of molecules in the flask

$$= 2.45 \times 10^{22}$$ *Ans.* (c)

Example 22.8

A group of oxygen molecules start their free paths at the same instant. The pressure is such that the mean free path is 2 cm. After how long a time will half of the group still remain, i.e., half will not as yet have made a collision? Assume that all particles have a speed equal to the average speed. The temperature is 300 K.

Solution The survival equation in $N = N_0 e^{-x/\lambda}$

The group is reduced to half after travelling a distance x, i.e., half of the group has a free path longer than x. The mean free path, λ, is 2 cm.

$$N/N_0 = 0.5 = e^{-x/\lambda}$$
$$x/\lambda = \ln 2 = 0.693$$
$$x = 1.386 \text{ cm}$$

For oxygen at 300K, $\bar{v} = 445.58$ m/s $= 4.456 \times 10^4$ cm/s

Time, $t = 1.386/(4.456 \times 10^4) = 3.11 \times 10^{-5}$ s *Ans.*

Example 22.9

To what pressure, in mm of Hg, must a cathode ray tube be evacuated in order that 90% of the electrons leaving the cathode shall reach the anode, 20 cm away, without making a collision? Take for ion $\sigma = 4.07 \times 10^{-19}$ m² and $T = 2000$ K.

Solution The survival equation is $N = N_0 e^{-x/\lambda}$

Here, $N = 0.9 N_0$, $x = 0.2$ m

$$0.9 = e^{-x/\lambda}$$

$$e^{x/\lambda} = 1.111$$
$$x/\lambda = \ln 1.111$$
$$= 0.2/0.105 \cong 2 \text{ m}$$

Electronic mean free path $\lambda_e = \dfrac{1}{\sigma n} = \dfrac{1}{4.07 \times 10^{-19} \times n} = 2 \text{ m}$

$$n = 1.23 \times 10^{18} \text{ molecules/m}^3$$

Now, pressure $p = nKT$

$$= 1.23 \times 10^{18} \frac{\text{molecules}}{\text{m}^3} \times 1.38 \times 10^{-23} \frac{J}{\text{molecule K}} \times 2000 \text{ K}$$

$$= 3.395 \times 10^{-2} \text{ N/m}^2$$ *Ans.*

Example 22.10

A tube 2 m long and 10^{-4} m^2 in cross-section contain CO_2 at atmospheric pressure and 0°C. The carbon atoms in one-half of the CO_2 molecules are radioactive isotope C^{14}. At time t = 0, all the molecules at the extreme left end of the tube contain radioactive carbon, and the number of such molecules per unit volume decreases uniformly to zero at the other end of the tube. (a) What is the initial concentration gradient of radioactive molecules? (b) Initially, how many radioactive molecules per sec cross a cross-section at the mid-point of the tube from left to right? (c) How many cross from right to left? What is the initial net rate of diffusion of radioactive molecules across the cross-section? Take $\sigma = \pi r^2 = 4 \times 10^{-19}$ m^2.

Solution (a) Number of molecules/m^3 at 1 atm, 273 K

$$n = p/KT = \frac{101.325 \times 1000}{1.38 \times 10^{-23} \times 273} = 2.69 \times 10^{25} \text{ molecules/m}^3$$

Concentration gradient, $dn/dx = (-2.69 \times 10^{25})/2$

$$= -1.345 \times 10^{25} \text{ molecules/m}^4$$ *Ans.*

(b)

$$\bar{v} = [2.55\, KT/m]^{1/2}$$

$$= \left[\frac{2.55 \times 1.38 \times 10^{-23} \times 273 \times 6.023 \times 10^{26}}{46}\right]^{1/2}$$

$$\bar{v} = 355 \text{ m/s}$$

$$\lambda = \frac{1}{\sigma n} = \frac{1}{2.69 \times 10^{25} \times 4 \times 10^{-19}} = 9.3 \times 10^{-6} \text{ m}$$

Number of molecules crossing from left to right per unit area per unit time:

$$\underset{\rightarrow}{\Gamma} = \frac{1}{4} z n_0 \lambda - \frac{1}{6} z \lambda_2 \frac{dn}{dx}$$

$$= \frac{1}{4} \bar{v} n_0 - \frac{1}{6} \bar{v} \lambda \frac{dn}{dx}$$

$$= \frac{1}{4} \times 355 \times 2.69 \times 10^{25} - \frac{1}{6} \times 355 \times 9.3 \times 10^{-6} \times (-1.345 \times 10^{25})$$

$$= 2.39 \times 10^{27} + 7.4 \times 10^{21} \text{ molecules/m}^2\text{s}$$ *Ans.*

(c) Number of molecules crossing from right to left per unit area unit time:

$$\overleftarrow{\Gamma} = \frac{1}{4} v n_0 + \frac{1}{6} z\lambda \frac{dn}{dx}$$

$$= 2.39 \times 10^{27} - 7.4 \times 10^{21} \text{ molecules/m}^2\text{s}$$ *Ans.*

Net rate of diffusion:

$$= 7.4 \times 10^{21} \times 2 = 14.8 \times 10^{2} \text{ molecules/m}^2\text{s}$$

$$= \frac{14.8 \times 10^{21} \times 46}{6.023 \times 10^{26}} \frac{\text{molecules}}{\text{m}^2\text{s}} \times \frac{\text{kg}}{\text{kgmol}} \times \frac{\text{kgmol}}{\text{molecules}}$$

$$= 113 \times 10^{-5} = 11.3 \times 10^{-4} \text{ kg/m}^2\text{s}$$

$$= 1.13 \text{ g/m}^2\text{s}$$ *Ans.*

Review Questions

22.1 Define mean free path, collision cross-section and collision frequency.

22.2 Show that $\lambda = 1/\sigma n$. What is electronic mean path? Why is it is equal to $4/\sigma n$?

22.3 What is collision probability? Show that it is reciprocal of the mean free path.

22.4 Derive the survival equation: $N = N_0 e^{-x/\lambda}$ and explain its significance.

22.5 Show that 37% of the molecules in a gas have free paths longer than λ.

22.6 Explain graphically the distribution of free paths of gas molecules.

22.7 What are transport properties? What do they signify?

22.8 Show that the number of molecules crossing a plane in a gas per unit area and per unit time is equal to $\frac{1}{4} n\bar{v}$.

22.9 Show that the average distance from a plane in a gas where the molecules made their last collisions before crossing that plane is equal to $\frac{2}{3}\lambda$.

22.10 Show that the coefficient of viscosity of a gas is equal to $\frac{1}{3} nm\bar{v}\lambda$. With a Maxwellian velocity distribution of gas molecules, show that

$$\mu = \frac{2}{3\sigma}\left[\frac{mKT}{\pi}\right]^{1/2}.$$

and hence, assert that the viscosity of a gas depends only on the temperature, and is independent of pressure or density of the gas.

22.11 Show that the thermal conductivity of a gas is given by $k = \frac{1}{6}\frac{\bar{v} f K}{\sigma}$. With a Maxwellian velocity distribution, show that:

$$k = \frac{1}{3}\frac{f K}{\sigma}\left[\frac{KT}{\pi m}\right]^{1/2}$$

Hence, assert that the thermal conductivity of a gas is independent of pressure or density, and depends only on the temperature.

22.12 Show that for a monatomic gas,

$$k = \frac{1}{\pi^{3/2} d^2}\left[\frac{K^3 T}{m}\right]^{1/2}$$

22.13 Show that the theoretical value of Prandtl number of a gas is equal to the specific heat ratio of the gas.

22.14 What do you understand by a self-diffusion process? What is the coefficient of diffusion?

22.15 Derive the expression for the coefficient of diffusion in a gas from molecular theory as given by:

$$D = \frac{1}{3}\bar{v}\lambda$$

With Maxwellian velocity distribution, show that

$$D = \frac{2}{3\pi^{1/2} d^2 n_t}\left[\frac{K\,T}{m}\right]^{1/2}$$

Where n_t is the total number of molecules per unit volume.

22.16 Define Schmidt number. What is its physical significance? Show that for a gas the Schmidt number is unity.

22.17 Why do all collisions of gas molecules not cause ionization?

22.18 Explain why at high temperatures a gas can be highly conductive of electricity.

22.19 Show that the ratio of velocities of an ion and a free electron is given by:

$$v_i/v_e = 1/1840M$$

where M is the atomic weight of the ion.

22.20 Explain the distribution of collision times of electrons.

22.21 What do you mean by electron drift velocity? Show that it is given by:

$$v_{e_d} = \frac{q_e E\tau_e}{m_e} = \frac{4q_e E}{\sigma p}\left[\frac{\pi KT}{8m_e}\right]^{1/2}$$

22.22 Show that the electrical conductivity of a gas is given by:

$$\sigma_e = \frac{q_e^2 n_e}{\sigma n}\left[\frac{2\pi}{m_e KT}\right]^{1/2}$$

22.23 Define electron mobility. Show that it is given by:

$$\mu_e = \frac{q_e}{\sigma n}\left[\frac{2\pi}{m_e KT}\right]^{1/2}$$

Problems

22.1 Calculate the collision frequency of a nitrogen molecule (a) at 300 K and 1 atm pressure, (b) at 300 K and 1 micron Hg abs. pressure. The radius of nitrogen molecule is 1.88×10^{-10}m.

Ans. (a) 7.35×10^9 collisions/s
(b) 9.63×10^3 collisions/s

22.2 Calculate the collision rate of a molecule in a Maxwellian gas. *Ans.* $\sqrt{2}\,\sigma n[8\,\text{KT}/\pi m]^{1/2}$

22.3 The mean free path of a certain gas in 12 cm. If there are 10,000 free paths, how many are longer than (a) 5 cm, (b) 15 cm, (c) 50 cm? (d) How many are longer than 6 cm, but shorter than 12 cm? (e) How many are between 11.5 cm and 12.5 cm in length? (f) How many are between 11.9 and 12.1 cm in length? (g) How many have free paths exactly equal to 12cm?

22.4 The mean free path of the molecules of a certain gas at 20°C is 3×10^{-5} m. (a) If the radius of the molecule is 3×10^{-10} m, find the pressure of the gas. (b) Calculate the number of collisions made by a molecule per metre of path.

22.5 The mean free path of the molecules of a certain gas at 298 K is 2.63×10^{-5} mm, the radius of each molecule is 2.56×10^{-10} m. Compute the number of collisions made by a typical particle in moving a distance of 1 m, and also the pressure of the gas.

22.6 Determine the pressure in a cathode ray tube such that 95 per cent of the electrons leaving the cathode ray reach the anode 25 cm away without making a collision. The diameter of an ion is 3.6×10^{-10} m and the electron temperature is 2000 K. Use the electronic mean free path $\lambda_e = 4/\sigma n$, where σ is the cross-section of the ion.

22.7 A beam of electrons is projected from an electron gun into a gas at a pressure p, and the number remaining in the beam at a distance x from the gun is determined by allowing the beam to strike a collecting plate and measuring the current to the plate. The electron current emitted by the gun is 100 μa, and the current to the plate when x = 10 cm and p = 1 mmHg is 37 μa. Determine (a) the electron mean free path, and (b) the current at 500 μ Hg pressure. *Ans.* (a) 10 cm, (b) 60.7 μa

22.8 A singly charged oxygen ion starts a free path in a direction at right angles to an electric field of intensity 100 volts/cm. The pressure is one atmosphere and the temperature 300 K. Calculate (a) the distance moved in the direction of the field in a time equal to that required to traverse one mean free path, (b) the ratio of the mean free path to this distance, (c) the average velocity in the direction of the field, (d) the ratio of the thermal velocity to this velocity, and (e) the ratio of the energy of thermal agitation to the energy gained from the field in one mean free path.

Ans. (a) 3.87×10^{-10} m, (c) 340 m/s, (e)10^4

22.9 A spherical satellite d metre in diameter moves through the earth's atmosphere with a speed of $\bar{v}$ m/s at an altitude where the number density is n molecules/m^3. How many molecules strike the satellite in 1 second? Derive an expression for the drag experienced by the satellite, assuming that all molecules which strike the sphere adhere to it.

Appendix A

TABLE A.1 STEAM TABLES*

Table A.1.1
Saturated Steam: Temperature Table

		Specific Volume, m^3/kg		Internal Energy, kJ/kg			Enthalpy, kJ/kg			Entropy, kJ/kg K		
Temp. °C T	Pressure kPa, MPa P	Sat. Liquid v_f	Sat. Vapour v_g	Sat. Liquid u_f	Evap. u_{fg}	Sat. Vapour u_g	Sat. Liquid h_f	Evap. h_{fg}	Sat. Vapour h_g	Sat. Liquid s_f	Evap. s_{fg}	Sat. Vapour s_g
0.01	0.6113	0.001000	206.132	0.00	2375.3	2375.3	0.00	2501.3	2501.3	0.0000	9.1562	9.1562
5	0.8721	0.001000	147.118	20.97	2361.3	2382.2	20.98	2489.6	2510.5	0.0761	8.9496	9.0257
10	1.2276	0.001000	106.377	41.99	2347.2	2389.2	41.99	2477.7	2519.7	0.1510	8.7498	8.9007
15	1.7051	0.001001	77.925	62.98	2333.1	2396.0	62.98	2465.9	2528.9	0.2245	8.5569	8.7813
20	2.3385	0.001002	57.790	83.94	2319.0	2402.9	83.94	2454.1	2538.1	0.2966	8.3706	8.6671
25	3.1691	0.001003	43.359	104.86	2304.9	2409.8	104.87	2442.3	2547.2	0.3673	8.1905	8.5579
30	4.2461	0.001004	32.893	125.77	2290.8	2416.6	125.77	2430.5	2556.2	0.4369	8.0164	8.4533
35	5.6280	0.001006	25.216	146.65	2276.7	2423.4	146.66	2418.6	2565.3	0.5052	7.8478	8.3530
40	7.3837	0.001008	19.523	167.53	2262.6	2430.1	167.54	2406.7	2574.3	0.5724	7.6845	8.2569
45	9.5934	0.001010	15.258	188.41	2248.4	2436.8	188.42	2394.8	2583.2	0.6386	7.5261	8.1647

* Adapted from Joseph H. Keenan, Frederick G. Keyes, Philip G. Hill, and Joan G. Moore, *Steam Tables*, John Wiley and Sons, New York, 1969.

(*Continued*)

Table A.1.1
(*Continued*)

Temp. °C T	Pressure kPa, MPa P	Specific Volume, m³/kg Sat. Liquid v_f	Sat. Vapour v_g	Internal Energy, kJ/kg Sat. Liquid u_f	Evap. u_{fg}	Sat. Vapour u_g	Enthalpy, kJ/kg Sat. Liquid h_f	Evap. h_{fg}	Sat. Vapour h_g	Entropy, kJ/kg K Sat. Liquid s_f	Evap. s_{fg}	Sat. Vapour s_g
50	12.350	0.001012	12.032	209.30	2234.2	2443.5	209.31	2382.7	2592.1	0.7037	7.3725	8.0762
55	15.758	0.001015	9.568	230.19	2219.9	2450.1	230.20	2370.7	2600.9	0.7679	7.2234	7.9912
60	19.941	0.001017	7.671	251.09	2205.5	2456.6	251.11	2358.5	2609.6	0.8311	7.0784	7.9095
65	25.033	0.001020	6.197	272.00	2191.1	2463.1	272.03	2346.2	2618.2	0.8934	6.9375	7.8309
70	31.188	0.001023	5.042	292.93	2176.6	2469.5	292.96	2333.8	2626.8	0.9548	6.8004	7.7552
75	38.578	0.001026	4.131	313.87	2162.0	2475.9	313.91	2321.4	2635.3	1.0154	6.6670	7.6824
80	47.390	0.001029	3.407	334.84	2147.4	2482.2	334.88	2308.8	2643.7	1.0752	6.5369	7.6121
85	57.834	0.001032	2.828	355.82	2132.6	2488.4	355.88	2296.0	2651.9	1.1342	6.4102	7.5444
90	70.139	0.001036	2.361	376.82	2117.7	2494.5	376.90	2283.2	2660.1	1.1924	6.2866	7.4790
95	84.554	0.001040	1.982	397.86	2102.7	2500.6	397.94	2270.2	2668.1	1.2500	6.1659	7.4158
100	0.10135	0.001044	1.6729	418.91	2087.6	2506.5	419.02	2257.0	2676.0	1.3068	6.0480	7.3548
105	0.12082	0.001047	1.4194	440.00	2072.3	2512.3	440.13	2243.7	2683.8	1.3629	5.9328	7.2958
110	0.14328	0.001052	1.2102	461.12	2057.0	2518.1	461.27	2230.2	2691.5	1.4184	5.8202	7.2386
115	0.16906	0.001056	1.0366	482.28	2041.4	2523.7	482.46	2216.5	2699.0	1.4733	5.7100	7.1832
120	0.19853	0.001060	0.8919	503.48	2025.8	2529.2	503.69	2202.6	2706.3	1.5275	5.6020	7.1295
125	0.2321	0.001065	0.77059	524.72	2009.9	2534.6	524.96	2188.5	2713.5	1.5812	5.4962	7.0774
130	0.2701	0.001070	0.66850	546.00	1993.9	2539.9	546.29	2174.2	2720.5	1.6343	5.3925	7.0269
135	0.3130	0.001075	0.58217	567.34	1977.7	2545.0	567.67	2159.6	2727.3	1.6869	5.2907	6.9777
140	0.3613	0.001080	0.50885	588.72	1961.3	2550.0	589.11	2144.8	2733.9	1.7390	5.1908	6.9298
145	0.4154	0.001085	0.44632	610.16	1944.7	2554.9	610.61	2129.6	2740.3	1.7906	5.0926	6.8832
150	0.4759	0.001090	0.39278	631.66	1927.9	2559.5	632.18	2114.3	2746.4	1.8417	4.9960	6.8378

155	0.5431	0.001096	0.34676	653.23	1910.8	2564.0	653.82	2098.6	2752.4	1.8924	4.9010	6.7934
160	0.6178	0.001102	0.30706	674.85	1893.5	2568.4	675.53	2082.6	2758.1	1.9426	4.8075	6.7501
165	0.7005	0.001108	0.27269	696.55	1876.0	2572.5	697.32	2066.2	2763.5	1.9924	4.7153	6.7078
170	0.7917	0.001114	0.24283	718.31	1858.1	2576.5	719.20	2049.5	2768.7	2.0418	4.6244	6.6663
175	0.8920	0.001121	0.21680	740.16	1840.0	2580.2	741.16	2032.4	2773.6	2.0909	4.5347	6.6256
180	1.0022	0.001127	0.19405	762.08	1821.6	2583.7	763.21	2015.0	2778.2	2.1395	4.4461	6.5857
185	1.1227	0.001134	0.17409	784.08	1802.9	2587.0	785.36	1997.1	2782.4	2.1878	4.3586	6.5464
190	1.2544	0.001141	0.15654	806.17	1783.8	2590.0	807.61	1978.8	2786.4	2.2358	4.2720	6.5078
195	1.3978	0.001149	0.14105	828.36	1764.4	2592.8	829.96	1960.0	2790.0	2.2835	4.1863	6.4697
200	1.5538	0.001156	0.12736	850.64	1744.7	2595.3	852.43	1940.7	2793.2	2.3308	4.1014	6.4322
205	1.7230	0.001164	0.11521	873.02	1724.5	2597.5	875.03	1921.0	2796.0	2.3779	4.0172	6.3951
210	1.9063	0.001173	0.10441	895.51	1703.9	2599.4	897.75	1900.7	2798.5	2.4247	3.9337	6.3584
215	2.1042	0.001181	0.09479	918.12	1682.9	2601.1	920.61	1879.9	2800.5	2.4713	3.8507	6.3221
220	2.3178	0.001190	0.08619	940.85	1661.5	2602.3	943.61	1858.5	2802.1	2.5177	3.7683	6.2860
225	2.5477	0.001199	0.07849	963.72	1639.6	2603.3	966.77	1836.5	2803.3	2.5639	3.6863	6.2502
230	2.7949	0.001209	0.07158	986.72	1617.2	2603.9	990.10	1813.8	2803.9	2.6099	3.6047	6.2146
235	3.0601	0.001219	0.06536	1009.88	1594.2	2604.1	1013.61	1790.5	2804.1	2.6557	3.5233	6.1791
240	3.3442	0.001229	0.05976	1033.19	1570.8	2603.9	1037.31	1766.5	2803.8	2.7015	3.4422	6.1436
245	3.6482	0.001240	0.05470	1056.69	1546.7	2603.4	1061.21	1741.7	2802.9	2.7471	3.3612	6.1083
250	3.9730	0.001251	0.05013	1080.37	1522.0	2602.4	1085.34	1716.2	2801.5	2.7927	3.2802	6.0729
255	4.3195	0.001263	0.04598	1104.26	1496.7	2600.9	1109.72	1689.8	2799.5	2.8382	3.1992	6.0374
260	4.6886	0.001276	0.04220	1128.37	1470.6	2599.0	1134.35	1662.5	2796.9	2.8837	3.1181	6.0018
265	5.0813	0.001289	0.03877	1152.72	1443.9	2596.6	1159.27	1634.3	2793.6	2.9293	3.0368	5.9661
270	5.4987	0.001302	0.03564	1177.33	1416.3	2593.7	1184.49	1605.2	2789.7	2.9750	2.9551	5.9301
275	5.9418	0.001317	0.03279	1202.23	1387.9	2590.2	1210.05	1574.9	2785.0	3.0208	2.8730	5.8937

(*Continued*)

Table A.1.1
(Continued)

Temp. °C T	Pressure kPa, MPa P	Specific Volume, m³/kg Sat. Liquid v_f	Sat. Vapour v_g	Internal Energy, kJ/kg Sat. Liquid u_f	Evap. u_{fg}	Sat. Vapour u_g	Enthalpy, kJ/kg Sat. Liquid h_f	Evap. h_{fg}	Sat. Vapour h_g	Entropy, kJ/kg K Sat. Liquid s_f	Evap. s_{fg}	Sat. Vapour s_g
280	6.4117	0.001332	0.03017	1227.43	1358.7	2586.1	1235.97	1543.6	2779.5	3.0667	2.7903	5.8570
285	6.9094	0.001348	0.02777	1252.98	1328.4	2581.4	1262.29	1511.0	2773.3	3.1129	2.7069	5.8198
290	7.4360	0.001366	0.02557	1278.89	1297.1	2576.0	1289.04	1477.1	2766.1	3.1593	2.6227	5.7821
295	7.9928	0.001384	0.02354	1305.21	1264.7	2569.9	1316.27	1441.8	2758.0	3.2061	2.5375	5.7436
300	8.5810	0.001404	0.02167	1331.97	1231.0	2563.0	1344.01	1404.9	2748.9	3.2533	2.4511	5.7044
305	9.2018	0.001425	0.01995	1359.22	1195.9	2555.2	1372.33	1366.4	2738.7	3.3009	2.3633	5.6642
310	9.8566	0.001447	0.01835	1387.03	1159.4	2546.4	1401.29	1326.0	2727.3	3.3492	2.2737	5.6229
315	10.547	0.001472	0.01687	1415.44	1121.1	2536.6	1430.97	1283.5	2714.4	3.3981	2.1821	5.5803
320	11.274	0.001499	0.01549	1444.55	1080.9	2525.5	1461.45	1238.6	2700.1	3.4479	2.0882	5.5361
330	12.845	0.001561	0.012996	1505.24	993.7	2498.9	1525.29	1140.6	2665.8	3.5506	1.8909	5.4416
340	14.586	0.001638	0.010797	1570.26	894.3	2464.5	1594.15	1027.9	2622.0	3.6593	1.6763	5.3356
350	16.514	0.001740	0.008813	1641.81	776.6	2418.4	1670.54	893.4	2563.9	3.7776	1.4336	5.2111
360	18.651	0.001892	0.006945	1725.19	626.3	2351.5	1760.48	720.5	2481.0	3.9146	1.1379	5.0525
370	21.028	0.002213	0.004926	1843.84	384.7	2228.5	1890.37	441.8	2332.1	4.1104	0.6868	4.7972
374.14	22.089	0.003155	0.003155	2029.58	0	2029.6	2099.26	0	2099.3	4.4297	0	4.4297

Table A.1.2

Saturated Water: Pressure Table

Pressure kPa P	Temp. °C T	Specific Volume, m^3/kg: Sat. Liquid v_f	Sat. Vapour v_g	Internal Energy, kJ/kg: Sat. Liquid u_f	Evap. u_{fg}	Sat. Vapour u_g	Enthalpy, kJ/kg: Sat. Liquid h_f	Evap. h_{fg}	Sat. Vapour h_g	Entropy, kJ/kg K: Sat. Liquid s_f	Evap. s_{fg}	Sat Vapour s_g
0.6113	0.01	0.001000	206.132	0	2375.3	2375.3	0.00	2501.3	2501.3	0	9.1562	9.1562
1.0	6.98	0.001000	129.208	29.29	2355.7	2385.0	29.29	2484.9	2514.2	0.1059	8.8697	8.9756
1.5	13.03	0.001001	87.980	54.70	2338.6	2393.3	54.70	2470.6	2525.3	0.1956	8.6322	8.8278
2.0	17.50	0.001001	67.004	73.47	2326.0	2399.5	73.47	2460.0	2533.5	0.2607	8.4629	8.7236
2.5	21.08	0.001002	54.254	88.47	2315.9	2404.4	88.47	2451.6	2540.0	0.3120	8.3311	8.6431
3.0	24.08	0.001003	45.665	101.03	2307.5	2408.5	101.03	2444.5	2545.5	0.3545	8.2231	8.5775
4.0	28.96	0.001004	34.800	121.44	2293.7	2415.2	121.44	2432.9	2554.4	0.4226	8.0520	8.4746
5.0	32.88	0.001005	28.193	137.79	2282.7	2420.5	137.79	2423.7	2561.4	0.4763	7.9187	8.3950
7.5	40.29	0.001008	19.238	168.76	2261.7	2430.5	168.77	2406.0	2574.8	0.5763	7.6751	8.2514
10.0	45.81	0.001010	14.674	191.79	2246.1	2437.9	191.81	2392.8	2584.6	0.6492	7.5010	8.1501
15.0	53.97	0.001014	10.022	225.90	2222.8	2448.7	225.91	2373.1	2599.1	0.7548	7.2536	8.0084
20.0	60.06	0.001017	7.649	251.35	2205.4	2456.7	251.38	2358.3	2609.7	0.8319	7.0766	7.9085
25.0	64.97	0.001020	6.204	271.88	2191.2	2463.1	271.90	2346.3	2618.2	0.8930	6.9383	7.8313
30.0	69.10	0.001022	5.229	289.18	2179.2	2468.4	289.21	2336.1	2626.3	0.9439	6.8247	7.7686
40.0	75.87	0.001026	3.993	317.51	2159.5	2477.0	317.55	2319.2	2636.7	1.0258	6.6441	7.6700
50.0	81.33	0.001030	3.240	340.42	2143.4	2483.8	340.47	2305.4	2645.9	1.0910	6.5029	7.5939

(Continued)

Table A.1.2
(Continued)

Pressure kPa P	Temp. °C T	Specific Volume, m³/kg Sat. Liquid v_f	Sat. Vapour v_g	Internal Energy, kJ/kg Sat. Liquid u_f	Evap. u_{fg}	Sat. Vapour u_g	Enthalpy, kJ/kg Sat. Liquid h_f	Evap. h_{fg}	Sat. Vapour h_g	Entropy, kJ/kg K Sat. Liquid s_f	Evap. s_{fg}	Sat. Vapour s_g
75.0	91.77	0.001037	2.217	384.29	2112.4	2496.7	384.36	2278.6	2663.0	1.2129	6.2434	7.4563
MPa												
0.100	99.62	0.001043	1.6940	417.33	2088.7	2506.1	417.44	2258.0	2675.5	1.3025	6.0568	7.3593
0.125	105.99	0.001048	1.3749	444.16	2069.3	2513.5	444.30	2241.1	2685.3	1.3739	5.9104	7.2843
0.150	111.37	0.001053	1.1593	466.92	2052.7	2519.6	467.08	2226.5	2693.5	1.4335	5.7897	7.2232
0.175	116.06	0.001057	1.0036	486.78	2038.1	2524.9	486.97	2213.6	2700.5	1.4848	5.6868	7.1717
0.200	120.23	0.001061	0.8857	504.47	2025.0	2529.5	504.68	2202.0	2706.6	1.5300	5.5970	7.1271
0.225	124.00	0.001064	0.7933	520.45	2013.1	2533.6	520.69	2191.3	2712.0	1.5705	5.5173	7.0878
0.250	127.43	0.001067	0.7187	535.08	2002.1	2537.2	535.34	2181.5	2716.9	1.6072	5.4455	7.0526
0.275	130.60	0.001070	0.6573	548.57	1992.0	2540.5	548.87	2172.4	2721.3	1.6407	5.3801	7.0208
0.300	133.55	0.001073	0.6058	561.13	1982.4	2543.6	561.45	2163.9	2725.3	1.6717	5.3201	6.9918
0.325	136.30	0.001076	0.5620	572.88	1973.5	2546.3	573.23	2155.8	2729.0	1.7005	5.2646	6.9651
0.350	138.88	0.001079	0.5243	583.93	1965.0	2548.9	584.31	2148.1	2732.4	1.7274	5.2130	6.9404
0.375	141.32	0.001081	0.4914	594.38	1956.9	2551.3	594.79	2140.8	2735.6	1.7527	5.1647	6.9174
0.40	143.63	0.001084	0.4625	604.29	1949.3	2553.6	604.73	2133.8	2738.5	1.7766	5.1193	6.8958
0.45	147.93	0.001088	0.4140	622.75	1934.9	2557.6	623.24	2120.7	2743.9	1.8206	5.0359	6.8565
0.50	151.86	0.001093	0.3749	639.66	1921.6	2561.2	640.21	2108.5	2748.7	1.8606	4.9606	6.8212
0.55	155.48	0.001097	0.3427	655.30	1909.2	2564.5	655.91	2097.0	2752.9	1.8972	4.8920	6.7892
0.60	158.85	0.001101	0.3157	669.88	1897.5	2567.4	670.54	2086.3	2756.8	1.9311	4.8289	6.7600
0.65	162.01	0.001104	0.2927	683.55	1886.5	2570.1	684.26	2076.0	2760.3	1.9627	4.7704	6.7330

0.70	164.97	0.001108	0.2729	696.43	1876.1	2572.5	697.20	2066.3	2763.5	1.9922	4.7158	6.7080
0.75	167.77	0.001111	0.2556	708.62	1866.1	2574.7	709.45	2057.0	2766.4	2.0199	4.6647	6.6846
0.80	170.43	0.001115	0.2404	720.20	1856.6	2576.8	721.10	2048.0	2769.1	2.0461	4.6166	6.6627
0.85	172.96	0.001118	0.2270	731.25	1847.4	2578.7	732.20	2039.4	2771.6	2.0709	4.5711	6.6421
0.90	175.38	0.001121	0.2150	741.81	1838.7	2580.5	742.82	2031.1	2773.9	2.0946	4.5280	6.6225
0.95	177.69	0.001124	0.2042	751.94	1830.2	2582.1	753.00	2023.1	2776.1	2.1171	4.4869	6.6040
1.00	179.91	0.001127	0.19444	761.67	1822.0	2583.6	762.79	2015.3	2778.1	2.1386	4.4478	6.5864
1.10	184.09	0.001133	0.17753	780.08	1806.3	2586.4	781.32	2000.4	2781.7	2.1791	4.3744	6.5535
1.20	187.99	0.001139	0.16333	797.27	1791.6	2588.8	798.64	1986.2	2784.8	2.2165	4.3067	6.5233
1.30	191.64	0.001144	0.15125	813.42	1777.5	2590.9	814.91	1972.7	2787.6	2.2514	4.2438	6.4953
1.40	195.07	0.001149	0.14084	828.68	1764.1	2592.8	830.29	1959.7	2790.0	2.2842	4.1850	6.4692
1.50	198.32	0.001154	0.13177	843.14	1751.3	2594.5	844.87	1947.3	2792.1	2.3150	4.1298	6.4448
1.75	205.76	0.001166	0.11349	876.44	1721.4	2597.8	878.48	1918.0	2796.4	2.3851	4.0044	6.3895
2.00	212.42	0.001177	0.09963	906.42	1693.8	2600.3	908.77	1890.7	2799.5	2.4473	3.8935	6.3408
2.25	218.45	0.001187	0.08875	933.81	1668.2	2602.0	936.48	1865.2	2801.7	2.5034	3.7938	6.2971
2.50	223.99	0.001197	0.07998	959.09	1644.0	2603.1	962.09	1841.0	2803.1	2.5546	3.7028	6.2574
2.75	229.12	0.001207	0.07275	982.65	1621.2	2603.8	985.97	1817.9	2803.9	2.6018	3.6190	6.2208
3.00	233.90	0.001216	0.06668	1004.76	1599.3	2604.1	1008.41	1795.7	2804.1	2.6456	3.5412	6.1869
3.25	238.38	0.001226	0.06152	1025.62	1578.4	2604.0	1029.60	1774.4	2804.0	2.6866	3.4685	6.1551
3.50	242.60	0.001235	0.05707	1045.41	1558.3	2603.7	1049.73	1753.7	2803.4	2.7252	3.4000	6.1252
4.0	250.40	0.001252	0.049778	1082.28	1520.0	2602.3	1087.29	1714.1	2801.4	2.7963	3.2737	6.0700
5.0	263.99	0.001286	0.039441	1147.78	1449.3	2597.1	1154.21	1640.1	2794.3	2.9201	3.0532	5.9733

(Continued)

Table A.1.2
(Continued)

Pressure kPa P	Temp. °C T	Specific Volume, m³/kg: Sat. Liquid v_f	Sat. Vapour v_g	Internal Energy, kJ/kg: Sat. Liquid u_f	Evap. u_{fg}	Sat. Vapour u_g	Enthalpy, kJ/kg: Sat. Liquid h_f	Evap. h_{fg}	Sat. Vapour h_g	Entropy, kJ/kg K: Sat. Liquid s_f	Evap. s_{fg}	Sat. Vapour s_g
6.0	275.64	0.001319	0.032440	1205.41	1384.3	2589.7	1213.32	1571.0	2784.3	3.0266	2.8625	5.8891
7.0	285.88	0.001351	0.027370	1257.51	1323.0	2580.5	1266.97	1505.1	2772.1	3.1210	2.6922	5.8132
8.0	295.06	0.001384	0.023518	1305.54	1264.3	2569.8	1316.61	1441.3	2757.9	3.2067	2.5365	5.7431
9.0	303.40	0.001418	0.020484	1350.47	1207.3	2557.8	1363.23	1378.9	2742.1	3.2857	2.3915	5.6771
10.0	311.06	0.001452	0.018026	1393.00	1151.4	2544.4	1407.53	1317.1	2724.7	3.3595	2.2545	5.6140
11.0	318.15	0.001489	0.015987	1433.68	1096.1	2529.7	1450.05	1255.5	2705.6	3.4294	2.1233	5.5527
12.0	324.75	0.001527	0.014263	1472.92	1040.8	2513.7	1491.24	1193.6	2684.8	3.4961	1.9962	5.4923
13.0	330.93	0.001567	0.012780	1511.09	985.0	2496.1	1531.46	1130.8	2662.2	3.5604	1.8718	5.4323
14.0	336.75	0.001611	0.011485	1548.53	928.2	2476.8	1571.08	1066.5	2637.5	3.6231	1.7485	5.3716
15.0	342.24	0.001658	0.010338	1585.58	869.8	2455.4	1610.45	1000.0	2610.5	3.6847	1.6250	5.3097
16.0	347.43	0.001711	0.009306	1622.63	809.1	2431.7	1650.00	930.6	2580.6	3.4760	1.4995	5.2454
17.0	352.37	0.001770	0.008365	1660.16	744.8	2405.0	1690.25	856.9	2547.2	3.8078	1.3698	5.1776
18.0	357.06	0.001840	0.007490	1698.86	675.4	2374.3	1731.97	777.1	2509.1	3.8713	1.2330	5.1044
19.0	361.54	0.001924	0.006657	1739.87	598.2	2338.1	1776.43	688.1	2464.5	3.9387	1.0841	5.0227
20.0	365.81	0.002035	0.005834	1785.47	507.6	2293.1	1826.18	583.6	2409.7	4.0137	0.9132	4.9269
21.0	369.89	0.002206	0.004953	1841.97	388.7	2230.7	1888.30	446.4	2334.7	4.1073	0.6942	4.8015
22.0	373.80	0.002808	0.003526	1973.16	108.2	2081.4	2034.92	124.0	2159.0	4.3307	0.1917	4.5224
22.09	374.14	0.003155	0.003155	2029.58	0	2029.6	2099.26	0	2099.3	4.4297	0	4.4297

Table A.1.3
Superheated Vapour

	P = 10 kPa (45.81)				P = 50 kPa, (81.33)				P = 100 kPa (99.62)			
T	v	u	h	s	v	u	h	s	v	u	h	s
Sat.	14.674	2437.9	2584.6	8.1501	3.240	2483.8	2645.9	7.5939	1.6940	2506.1	2675.5	7.3593
50	14.869	2443.9	2592.6	8.1749	—	—	—	—	—	—	—	—
100	17.196	2515.5	2687.5	8.4479	3.418	2511.6	2682.5	7.6947	1.6958	2506.6	2676.2	7.3614
150	19.513	2587.9	2783.0	8.6881	3.889	2585.6	2780.1	7.9400	1.9364	2582.7	2776.4	7.6133
200	21.825	2661.3	2879.5	8.9037	4.356	2659.8	2877.6	8.1579	2.1723	2658.0	2875.3	7.8342
250	24.136	2736.0	2977.3	9.1002	4.821	2735.0	2976.0	8.3555	2.4060	2733.7	2974.3	8.0332
300	26.445	2812.1	3076.5	9.2812	5.284	2811.3	3075.5	8.5372	2.6388	2810.4	3074.3	8.2157
400	31.063	2968.9	3279.5	9.6076	6.209	2968.4	3278.9	8.8641	3.1026	2967.8	3278.1	8.5434
500	35.679	3132.3	3489.0	9.8977	7.134	3131.9	3488.6	9.1545	3.5655	3131.5	3488.1	8.8341
600	40.295	3302.5	3705.4	10.1608	8.058	3302.2	3705.1	9.4177	4.0278	3301.9	3704.7	9.0975
700	44.911	3479.6	3928.7	10.4028	8.981	3479.5	3928.5	9.6599	4.4899	3479.2	3928.2	9.3398
800	49.526	3663.8	4159.1	10.6281	9.904	3663.7	4158.9	9.8852	4.9517	3663.5	4158.7	9.5652
900	54.141	3855.0	4396.4	10.8395	10.828	3854.9	4396.3	10.0967	5.4135	3854.8	4396.1	9.7767
1000	58.757	4053.0	4640.6	11.0392	11.751	4052.9	4640.5	10.2964	5.8753	4052.8	4640.3	9.9764
1100	63.372	4257.5	4891.2	11.2287	12.674	4257.4	4891.1	10.4858	6.3370	4257.3	4890.9	10.1658
1200	67.987	4467.9	5147.8	11.4090	13.597	4467.8	5147.7	10.6662	6.7986	4467.7	5147.6	10.3462
1300	72.603	4683.7	5409.7	11.5810	14.521	4683.6	5409.6	10.8382	7.2603	4683.5	5409.5	10.5182
	P = 200 kPa (120.23)				P = 300 kPa (133.55)				P = 400 kPa (143.63)			
Sat.	0.88573	2529.5	2706.6	7.1271	0.60582	2543.6	2725.3	6.9918	0.46246	2553.6	2738.5	6.8958
150	0.95964	2576.9	2768.8	7.2795	0.63388	2570.8	2761.0	7.0778	0.47084	2564.5	2752.8	6.9299
200	1.08034	2654.4	2870.5	7.5066	0.71629	2650.7	2865.5	7 3115	0.53422	2646.8	2860.5	7.1706
250	1.19880	2731.2	2971.0	7.7085	0.79636	2728.7	2967.6	7.5165	0.59512	2726.1	2964.2	7.3788

(Continued)

Table A.1.3
(Continued)

T	v	u	h	s	v	u	h	s	v	u	h	s
	P = 200 kPa (120.23)				*P* = 300 kPa (133.55)				*P* = 400 kPa (143.65)			
300	1.31616	2808.6	3071.8	7.8926	0.87529	2806.7	3069.3	7.7022	0.65484	2804.8	3066.7	7.5661
400	1.54930	2966.7	3276.5	8.2217	1.03151	2965.5	3275.0	8.0329	0.77262	2964.4	3273.4	7.8984
500	1.78139	3130.7	3487.0	8.5132	1.18669	3130.0	3486.0	8.3250	0.88934	3129.2	3484.9	8.1912
600	2.01297	3301.4	3704.0	8.7769	1.34136	3300.8	3703.2	8.5892	1.00555	3300.2	3702.4	8.4557
700	2.24426	3478.8	3927.7	9.0194	1.49573	3478.4	3927.1	8.8319	1.12147	3477.9	3926.5	8.6987
800	2.47539	3663.2	4158.3	9.2450	1.64994	3662.9	4157.8	9.0575	1.23722	3662.5	4157.4	8.9244
900	2.70643	3854.5	4395.8	9.4565	1.80406	3854.2	4395.4	9.2691	1.35288	3853.9	4395.1	9.1361
1000	2.93740	4052.5	4640.0	9.6563	1.95812	4052.3	4639.7	9.4689	1.46847	4052.0	4639.4	9.3360
1100	3.16834	4257.0	4890.7	9.8458	2.11214	4256.8	4890.4	9.6585	1.58404	4256.5	4890.1	9.5255
1200	3.39927	4467.5	5147.3	10.0262	2.26614	4467.2	5147.1	9.8389	1.69958	4467.0	5146.8	9.7059
1300	3.63018	4683.2	5409.3	10.1982	2.42013	4683.0	5409.0	10.0109	1.81511	4682.8	5408.8	9.8780
	P = 500 kPa (151.86)				*P* = 600 kPa (158.85)				*P* = 800 kPa (170.43)			
Sat.	0.37489	2561.2	2748.7	6.8212	0.31567	2567.4	2756.8	6.7600	0.24043	2576.8	2769.1	6.6627
200	0.42492	2642.9	2855.4	7.0592	0.35202	2638.9	2850.1	6.9665	0.26080	2630.6	2839.2	6.8158
250	0.47436	2723.5	2960.7	7.2708	0.39383	2720.9	2957.2	7.1816	0.29314	2715.5	2950.0	7.0384
300	0.52256	2802.9	3064.2	7.4598	0.43437	2801.0	3061.6	7.3723	0.32411	2797.1	3056.4	7.2372
350	0.57012	2882.6	3167.6	7.6328	0.47424	2881.1	3165.7	7.5463	0.35439	2878.2	3161.7	7.4088
400	0.61728	2963.2	3271.8	7.7937	0.51372	2962.0	3270.2	7.7078	0.38426	2959.7	3267.1	7.5715
500	0.71093	3128.4	3483.8	8.0872	0.59199	3127.6	3482.7	8.0020	0.44331	3125.9	3480.6	7.8672
600	0.80406	3299.6	3701.7	8.3521	0.66974	3299.1	3700.9	8.2673	0.50184	3297.9	3699.4	8.1332
700	0.89691	3477.5	3926.0	8.5952	0.74720	3477.1	3925.4	8.5107	0.56007	3476.2	3924.3	8.3770
800	0.98959	3662.2	4157.0	8.8211	0.82450	3661.8	4156.5	8.7367	0.61813	3661.1	4155.7	8.6033
900	1.08217	3853.6	4394.7	9.0329	0.90169	3853.3	4394.4	8.9485	0.67610	3852.8	4393.6	8.8153
1000	1.17469	4051.8	4639.1	9.2328	0.97883	4051.5	4638.8	9.1484	0.73401	4051.0	4638.2	9.0153
1100	1.26718	4256.3	4889.9	9.4224	1.05594	4256.1	4889.6	9.3381	0.79188	4255.6	4889.1	9.2049
1200	1.35964	4466.8	5146.6	9.6028	1.13302	4466.5	5146.3	9.5185	0.84974	4466.1	5145.8	9.3854
1300	1.45210	4682.5	5408.6	9.7749	1.21009	4682.3	5408.3	9.6906	0.90758	4681.8	5407.9	9.5575

Sat.	0.19444	2583.6	2778.1	6.5864	0.16333	2588.8	2784.8	6.5233	0.14084	2592.8	2790.0	6.4692
200	0.20596	2621.9	2827.9	6.6939	0.16930	2612.7	2815.9	6.5898	0.14302	2603.1	2803.3	6.4975
250	0.23268	2709.9	2942.6	6.9246	0.19235	2704.2	2935.0	6.8293	0.16350	2698.3	2927.2	6.7467
300	0.25794	2793.2	3051.2	7.1228	0.21382	2789.2	3045.8	7.0316	0.18228	2785.2	3040.4	6.9533
350	0.28247	2875.2	3157.7	7.3010	0.23452	2872.2	3153.6	7.2120	0.20026	2869.1	3149.5	7.1359
400	0.30659	2957.3	3263.9	7.4650	0.25480	2954.9	3260.7	7.3773	0.21780	2952.5	3257.4	7.3025
500	0.35411	3124.3	3478.4	7.7621	0.29463	3122.7	3476.3	7.6758	0.25215	3121.1	3474.1	7.6026
600	0.40109	3296.8	3697.9	8.0289	0.33393	3295.6	3696.3	7.9434	0.28596	3294.4	3694.8	7.8710
700	0.44779	3475.4	3923.1	8.2731	0.37294	3474.5	3922.0	8.1881	0.31947	3473.6	3920.9	8.1160
800	0.49432	3660.5	4154.8	8.4996	0.41177	3659.8	4153.9	8.4149	0.35281	3659.1	4153.0	8.3431
900	0.54075	3852.5	4392.9	8.7118	0.45051	3851.6	4392.2	8.6272	0.38606	3851.0	4391.5	8.5555
1000	0.58712	4050.5	4637.6	8.9119	0.48919	4050.0	4637.0	8.8274	0.41924	4049.5	4636.4	8.7558
1100	0.63345	4255.1	4888.5	9.1016	0.52783	4254.6	4888.0	9.0171	0.45239	4254.1	4887.5	8.9456
1200	0.67977	4465.6	5145.4	9.2821	0.56646	4465.1	5144.9	9.1977	0.48552	4464.6	5144.4	9.1262
1300	0.72608	4681.3	5407.4	9.4542	0.60507	4680.9	5406.9	9.3698	0.51864	4680.4	5406.5	9.2983
		P = 1.60 *MPa* (201.40)				*P* = 1.80 *MPa* (207.15)				*P* = 2.00 *MPa* (212.42)		
Sat.	0.12380	2595.9	2794.0	6.4217	0.11042	2598.4	2797.1	6.3793	0.09963	2600.3	2799.5	6.3408
225	0.13287	2644.6	2857.2	6.5518	0.11673	2636.6	2846.7	6.4807	0.10377	2628.3	2835.8	6.4146
250	0.14184	2692.3	2919.2	6.6732	0.12497	2686.0	2911.0	6.6066	0.11144	2679.6	2902.5	6.5452
300	0.15862	2781.0	3034.8	6.8844	0.14021	2776.8	3029.2	6.8226	0.12547	2772.6	3023.5	6.7663
350	0.17456	2866.0	3145.4	7.0693	0.15457	2862.9	3141.2	7.0099	0.13857	2859.8	3137.0	6.9562
400	0.19005	2950.1	3254.2	7.2373	0.16847	2947.7	3250.9	7.1793	0.15120	2945.2	3247.6	7.1270
500	0.22029	3119.5	3471.9	7.5389	0.19550	3117.8	3469.7	7.4824	0.17568	3116.2	3467.6	7.4316
600	0.24998	3293.3	3693.2	7.8080	0.22199	3292.1	3691.7	7.7523	0.19960	3290.9	3690.1	7.7023
700	0.27937	3472.7	3919.7	8.0535	0.24818	3471.9	3918.6	7.9983	0.22323	3471.0	3917.5	7.9487
800	0.30859	3658.4	4152.1	8.2808	0.27420	3657.7	4151.3	8.2258	0.24668	3657.0	4150.4	8.1766

(*Continued*)

Table A.1.3
(Continued)

T	P = 1.60 MPa (201.40)				P = 1.80 MPa (207.15)				P = 2.00 MPa (212.42)			
	v	u	h	s	v	u	h	s	v	u	h	s
900	0.33772	3850.5	4390.8	8.4934	0.30012	3849.9	4390.1	8.4386	0.27004	3849.3	4389.4	8.3895
1000	0.36678	4049.0	4635.8	8.6938	0.32598	4048.4	4635.2	8.6390	0.29333	4047.9	4634.6	8.5900
1100	0.39581	4253.7	4887.0	8.8837	0.35180	4253.2	4886.4	8.8290	0.31659	4252.7	4885.9	8.7800
1200	0.42482	4464.2	5143.9	9.0642	0.37761	4663.7	5143.4	9.0096	0.33984	4463.2	5142.9	8.9606
1300	0.45382	4679.9	5406.0	9.2364	0.40340	4679.4	5405.6	9.1817	0.36306	4679.0	5405.1	9.1382
	P = 2.50 MPa (223.99)				P = 3.00 MPa (233.90)				P = 3.50 MPa (242.60)			
Sat.	0.07998	2603.1	2803.1	6.2574	0.06668	2604.1	2804.1	6.1869	0.05707	2603.7	2803.4	6.1252
225	0.08027	2605.6	2806.3	6.2638	—	—	—	—	—	—	—	—
250	0.08700	2662.5	2880.1	6.4084	0.07058	2644.0	2855.8	6.2871	0.05873	2623.7	2829.2	6.1748
300	0.09890	2761.6	3008.8	6.6437	0.08114	2750.0	2993.5	6.5389	0.06842	2738.0	2977.5	6.4460
350	0.10976	2851.8	3126.2	6.8402	0.09053	2843.7	3115.3	6.7427	0.07678	2835.3	3104.0	6.6578
400	0.12010	2939.0	3239.3	7.0147	0.09936	2932.7	3230.8	6.9211	0.08453	2926.4	3222.2	6.8404
450	0.13014	3025.4	3350.8	7.1745	0.10787	3020.4	3344.0	7.0833	0.09196	3015.3	3337.2	7.0051
500	0.13998	3112.1	3462.0	7.3233	0.11619	3107.9	3456.5	7.2337	0.09918	3103.7	3450.9	7.1571
600	0.15930	3288.0	3686.2	7.5960	0.13243	3285.0	3682.3	7.5084	0.11324	3282.1	3678.4	7.4338
700	0.17832	3468.8	3914.6	7.8435	0.14838	3466.0	3911.7	7.7571	0.12699	3464.4	3908.8	7.6837
800	0.19716	3655.3	4148.2	8.0720	0.16414	3653.6	4146.0	7.9862	0.14056	3651.8	4143.8	7.9135
900	0.21590	3847.9	4387.6	8.2853	0.17980	3846.5	4385.9	8.1999	0.15402	3845.0	4384.1	8.1275
1000	0.23458	4046.7	4633.1	8.4860	0.19541	4045.4	4631.6	8.4009	0.16743	4044.1	4630.1	8.3288
1100	0.25322	4251.5	4884.6	8.6761	0.21098	4250.3	4883.3	8.5911	0.18080	4249.1	4881.9	8.5191
1200	0.27185	4462.1	5141.7	8.8569	0.22652	4460.9	5140.5	8.7719	0.19415	4459.8	5139.3	8.7000
1300	0.29046	4677.8	5404.0	9.0291	0.24206	4676.6	5402.8	8.9442	0.20749	4675.5	5401.7	8.8723

Sat.	0.04978	2602.3	2801.4	6.0700	0.04406	2600.0	2798.3	6.0198	0.03944	2597.1	2794.3	5.9733
275	0.05457	2667.9	2886.2	6.2284	0.04730	2650.3	2863.1	6.1401	0.04141	2631.2	2838.3	6.0543
300	0.05884	2725.3	2960.7	6.3614	0.05135	2712.0	2943.1	6.2827	0.04532	2697.9	2924.5	6.2083
350	0.06645	2826.6	3092.4	6.5820	0.05840	2817.8	3080.6	6.5130	0.05194	2808.7	3068.4	6.4492
400	0.07341	2919.9	3213.5	6.7689	0.06475	2913.3	3204.7	6.7046	0.05781	2906.6	3195.6	6.6458
450	0.08003	3010.1	3330.2	6.9362	0.07074	3004.9	3323.2	6.8745	0.06330	2999.6	3316.1	6.8185
500	0.08643	3099.5	3445.2	7.0900	0.07651	3095.2	3439.5	7.0300	0.06857	3090.9	3433.8	6.9758
600	0.09885	3279.1	3674.4	7.3688	0.08765	3276.0	3670.5	7.3109	0.07869	3273.0	3666.5	7.2588
700	0.11095	3462.1	3905.9	7.6198	0.09847	3459.9	3903.0	7.5631	0.08849	3457.7	3900.1	7.5122
800	0.12287	3650.1	4141.6	7.8502	0.10911	3648.4	4139.4	7.7942	0.09811	3646.6	4137.2	7.7440
900	0.13469	3843.6	4382.3	8.0647	0.11965	3842.1	4380.6	8.0091	0.10762	3840.7	4378.8	7.9593
1000	0.14645	4042.9	4628.7	8.2661	0.13013	4041.6	4627.2	8.2108	0.11707	4040.3	4625.7	8.1612
1100	0.15817	4248.0	4880.6	8.4566	0.14056	4246.8	4879.3	8.4014	0.12648	4245.6	4878.0	8.3519
1200	0.16987	4458.6	5138.1	8.6376	0.15098	4457.4	5136.9	8.5824	0.13587	4456.3	5135.7	8.5330
1300	0.18156	4674.3	5400.5	8.8099	0.16139	4673.1	5399.4	8.7548	0.14526	4672.0	5398.2	8.7055
		P = 6.00 *MPa* (275.64)				*P* = 7.00 *MPa* (285.88)				*P* = 8.00 *MPa* (295.06)		
Sat.	0.03244	2589.7	2784.3	5.8891	0.02737	2580.5	2772.1	5.8132	0.02352	2569.8	2757.9	5.7431
300	0.03616	2667.2	2884.2	6.0673	0.02947	2632.1	2838.4	5.9304	0.02426	2590.9	2785.0	5.7905
350	0.04223	2789.6	3043.0	6.3334	0.03524	2769.3	3016.0	6.2282	0.02995	2747.7	2987.3	6.1300
400	0.04739	2892.8	3177.2	6.5407	0.03993	2878.6	3158.1	6.4477	0.03432	2863.8	3138.3	6.3633
450	0.05214	2988.9	3301.8	6.7192	0.04416	2977.9	3287.0	6.6326	0.03817	2966.7	3272.0	6.5550
500	0.05665	3082.2	3422.1	6.8802	0.04814	3073.3	3410.3	6.7974	0.04175	3064.3	3398.3	6.7239
550	0.06101	3174.6	3540.6	7.0287	0.05195	3167.2	3530.9	6.9486	0.04516	3159.8	3521.0	6.8778
600	0.06525	3266.9	3658.4	7.1676	0.05565	3260.7	3650.3	7.0894	0.04845	3254.4	3642.0	7.0205
700	0.07352	3453.2	3894.3	7.4234	0.06283	3448.6	3888.4	7.3476	0.05481	3444.0	3882.5	7.2812
800	0.08160	3643.1	4132.7	7.6566	0.06981	3639.6	4128.3	7.5822	0.06097	3636.1	4123.8	7.5173

(Continued)

Table A.1.3
(Continued)

T	P = 6.00 MPa (275.64)				P = 7.00 MPa (285.88)				P = 8.00 MPa (295.06)			
	v	u	h	s	v	u	h	s	v	u	h	s
900	0.08958	3837.8	4375.3	7.8727	0.07669	3835.0	4371.8	7.7991	0.06702	3832.1	4368.3	7.7350
1000	0.09749	4037.8	4622.7	8.0751	0.08350	4035.3	4619.8	8.0020	0.07301	4032.8	4616.9	7.9384
1100	0.10536	4243.3	4875.4	8.2661	0.09027	4240.9	4872.8	8.1933	0.07896	4238.6	4870.3	8.1299
1200	0.11321	4454.0	5133.3	8.4473	0.09703	4451.7	5130.9	8.3747	0.08489	4449.4	5128.5	8.3115
1300	0.12106	4669.6	5396.0	8.6199	0.10377	4667.3	5393.7	8.5472	0.09080	4665.0	5391.5	8.4842
	P = 9.00 MPa (303.40)				P = 10.00 MPa (311.06)				P = 12.50 MPa (327.89)			
Sat.	0.02048	2557.8	2742.1	5.6771	0.01803	2544.4	2724.7	5.6140	0.01350	2505.1	2673.8	5.4623
325	0.02327	2646.5	2855.9	5.8711	0.01986	2610.4	2809.0	5.7568	—	—	—	—
350	0.02580	2724.4	2956.5	6.0361	0.02242	2699.2	2923.4	5.9442	0.01613	2624.6	2826.2	5.7117
400	0.02993	2848.4	3117.8	6.2853	0.02641	2832.4	3096.5	6.2119	0.02000	2789.3	3039.3	6.0416
450	0.03350	2955.1	3256.6	6.4843	0.02975	2943.3	3240.8	6.4189	0.02299	3912.4	3199.8	6.2718
500	0.03677	3055.1	3386.1	6.6575	0.03279	3045.8	3373.6	6.5965	0.02560	3021.7	3341.7	6.4617
550	0.03987	3152.2	3511.0	6.8141	0.03564	3144.5	3500.9	6.7561	0.02801	3124.9	3475.1	6.6289
600	0.04285	3248.1	3633.7	6.9588	0.03837	3241.7	3625.3	6.9028	0.03029	3225.4	3604.0	6.7810
650	0.04574	3343.7	3755.3	7.0943	0.04101	3338.2	3748.3	7.0397	0.03248	3324.4	3730.4	6.9218
700	0.04857	3439.4	3876.5	7.2221	0.04358	3434.7	3870.5	7.1687	0.03460	3422.9	3855.4	7.0536
800	0.05409	3632.5	4119.4	7.4597	0.04859	3629.0	4114.9	7.4077	0.03869	3620.0	4103.7	7.2965
900	0.05950	3829.2	4364.7	7.6782	0.05349	3826.3	4361.2	7.6272	0.04267	3819.1	4352.5	7.5181
1000	0.06485	4030.3	4613.9	7.8821	0.05832	4027.8	4611.0	7.8315	0.04658	4021.6	4603.8	7.7237
1100	0.07016	4236.3	4867.7	8.0739	0.06312	4234.0	4865.1	8.0236	0.05045	4228.2	4858.8	7.9165
1200	0.07544	4447.2	5126.2	8.2556	0.06789	4444.9	5123.8	8.2054	0.05430	4439.3	5118.0	8.0987
1300	0.08072	4662.7	5389.2	8.4283	0.07265	4660.4	5387.0	8.3783	0.05813	4654.8	5381.4	8.2717

Sat.	.010338	2455.4	2610.5	5.3097	.0079204	2390.2	2528.8	5.1418	.0058342	2293.1	2409.7	4.9269
350	.011470	2520.4	2692.4	5.4420	.0017139	1632.0	1662.0	3.7612	.0016640	1612.3	1645.6	3.7280
400	.015649	2740.7	2975.4	5.8810	.0124477	2685.0	2902.8	5.7212	.0099423	2619.2	2818.1	5.5539
450	.018446	2879.5	3156.2	6.1403	.0151740	2844.2	3109.7	6.0182	.0126953	2806.2	3060.1	5.9016
500	.020800	2996.5	3308.5	6.3442	.0173585	2970.3	3274.0	6.2382	.0147683	2942.8	3238.2	6.1400
550	.022927	3104.7	3448.6	6.5198	.0192877	3083.8	3421.4	6.4229	.0165553	3062.3	3393.5	6.3347
600	.024911	3208.6	3582.3	6.6775	.0210640	3191.5	3560.1	6.5866	.0181781	3174.0	3537.6	6.5048
650	.026797	3310.4	3712.3	6.8223	.0227372	3296.0	3693.9	6.7356	.0196929	3281.5	3675.3	6.6582
700	.028612	3410.9	3840.1	6.9572	.0243365	3398.8	3824.7	6.8736	.0211311	3386.5	3809.1	6.7993
800	.032096	3611.0	4092.4	7.2040	.0273849	3601.9	4081.1	7.1245	.0238532	3592.7	4069.8	7.0544
900	.035457	3811.9	4343.8	7.4279	.0303071	3804.7	4335.1	7.3507	.0264463	3797.4	4326.4	7.2830
1000	.038748	4015.4	4596.6	7.6347	.0331580	4009.3	4589.5	7.5588	.0289666	4003.1	4582.5	7.4925
1100	.042001	4222.6	4852.6	7.8282	.0359695	4216.9	4846.4	7.7530	.0314471	4211.3	4840.2	7.6874
1200	.045233	4433.8	5112.3	8.0108	.0387605	4428.3	5106.6	7.9359	.0339071	4422.8	5101.0	7.8706
1300	.048455	4649.1	5375.9	8.1839	.0415417	4643.5	5370.5	8.1093	.0363574	4638.0	5365.1	8.0441
		$P = 25$ MPa				$P = 30$ MPa				$P = 35$ MPa		
375	.001973	1798.6	1847.9	4.0319	.001789	1737.8	1791.4	3.9303	.001700	1702.9	1762.4	3.8721
400	.006004	2430.1	2580.2	5.1418	.002790	2067.3	2151.0	4.4728	.002100	1914.0	1987.5	4.2124
425	.007882	2609.2	2806.3	5.4722	.005304	2455.1	2614.2	5.1503	.003428	2253.4	2373.4	4.7747
450	.009162	2720.7	2949.7	5.6743	.006735	2619.3	2821.4	5.4423	.004962	2498.7	2672.4	5.1962
500	.011124	2884.3	3162.4	5.9592	.008679	2820.7	3081.0	5.7904	.006927	2751.9	2994.3	5.6281
550	.012724	3017.5	3335.6	6.1764	.010168	2970.3	3275.4	6.0342	.008345	2920.9	3213.0	5.9025
600	.014138	3137.9	3491.4	6.3602	.011446	3100.5	3443.9	6.2330	.009527	3062.0	3395.5	6.1178
650	.015433	3251.6	3637.5	6.5229	.012596	3221.0	3598.9	6.4057	.010575	3189.8	3559.9	6.3010
700	.016647	3361.4	3777.6	6.6707	.013661	3335.8	3745.7	6.5606	.011533	3309.9	3713.5	6.4631
800	.018913	3574.3	4047.1	6.9345	.015623	3555.6	4024.3	6.8332	.013278	3536.8	4001.5	6.7450

(*Continued*)

Table A.1.3
(Continued)

T	P = 25 MPa v	u	h	s	P = 30 MPa v	u	h	s	P = 35 MPa v	u	h	s
900	.021045	3783.0	4309.1	7.1679	.017448	3768.5	4291.9	7.0717	.014883	3754.0	4274.9	6.9886
1000	.023102	3990.9	4568.5	7.3801	.019196	3978.8	4554.7	7.2867	.016410	3966.7	4541.1	7.2063
1100	.025119	4200.2	4828.2	7.5765	.020903	4189.2	4816.3	7.4845	.017895	4178.3	4804.6	7.4056
1200	.027115	4412.0	5089.9	7.7604	.022589	4410.3	5079.0	7.6691	.019360	4390.7	5068.4	7.5910
1300	.029101	4626.9	5354.4	7.9342	.024266	4616.0	5344.0	7.8432	.020815	4605.1	5333.6	7.7652
	P = 40 MPa				P = 50 MPa				P = 60 MPa			
375	.0016406	1677.1	1742.7	3.8289	.0015593	1638.6	1716.5	3.7638	.0015027	1609.3	1699.5	3.1740
400	.0019077	1854.5	1930.8	4.1134	.0017309	1788.0	1874.6	4.0030	.0016335	1745.3	1843.4	3.9317
425	.0025319	2096.8	2198.1	4.5028	.0020071	1959.6	2060.0	4.2733	.0018165	1892.7	2001.7	4.1625
450	.0036931	2365.1	2512.8	4.9459	.0024862	2159.6	2283.9	4.5883	.0020850	2053.9	2179.0	4.4119
500	.0056225	2678.4	2903.3	5.4699	.0038924	2525.5	2720.1	5.1725	.0029557	2390.5	2567.9	4.9320
600	.0080943	3022.6	3346.4	6.0113	.0061123	2942.0	3247.6	5.8177	.0048345	2861.1	3151.2	5.6451
650	.0090636	3158.0	3520.6	6.2054	.0069657	3093.6	3441.8	6.0342	.0055953	3028.8	3364.6	5.8829
700	.0099415	3283.6	3681.3	6.3750	.0077274	3230.5	3616.9	6.2189	.0062719	3177.3	3553.6	6.0824
800	.0115228	3517.9	3978.8	6.6662	.0090761	3479.8	3933.6	6.5290	.0074588	3441.6	3889.1	6.4110
900	.0129626	3739.4	4257.9	6.9150	.0102831	3710.3	4224.4	6.7882	.0085083	3681.0	4191.5	6.6805
1000	.0143238	3954.6	4527.6	7.1356	.0114113	3930.5	4501.1	7.0146	.0094800	3906.4	4475.2	6.9126
1100	.0156426	4167.4	4793.1	7.3364	.0124966	4145.7	4770.6	7.2183	.0104091	4124.1	4748.6	7.1194
1200	.0169403	4380.1	5057.7	7.5224	.0135606	4359.1	5037.2	7.4058	.0113167	4338.2	5017.2	7.3082
1300	.0182292	4594.3	5323.5	7.6969	.0146159	4572.8	5303.6	7.5807	.0122155	4551.4	5284.3	7.4837

Table A.1.4
Compressed Liquid Water

T	P = 5.00 MPa (263.99) v	u	h	s	P = 10.00 MPa (311.06) v	u	h	s	P = 15.00 MPa (342.24) v	u	h	s
Sat.	.0012859	1147.78	1154.21	2.9201	.0014524	1393.00	1407.53	3.3595	.0016581	1585.58	1610.45	3.6847
0	.0009977	0.03	5.02	0.0001	.0009952	0.10	10.05	0.0003	.0009928	0.15	15.04	0.0004
20	.0009995	83.64	88.64	0.2955	.0009972	83.35	93.32	0.2945	.0009950	83.05	97.97	0.2934
40	.0010056	166.93	171.95	0.5706	.0010034	166.33	176.36	0.5685	.0010013	165.73	180.75	0.5665
60	.0010149	250.21	255.28	0.8284	.0010127	249.34	259.47	0.8258	.0010105	248.49	263.65	0.8231
80	.0010268	333.69	338.83	1.0719	.0010245	332.56	342.81	1.0687	.0010222	331.46	346.79	1.0655
100	.0010410	417.50	422.71	1.3030	.0010385	416.09	426.48	1.2992	.0010361	414.72	430.26	1.2954
120	.0010576	501.79	507.07	1.5232	.0010549	500.07	510.61	1.5188	.0010522	498.39	514.17	1.5144
140	.0010768	586.74	592.13	1.7342	.0010737	584.67	595.40	1.7291	.0010707	582.64	598.70	1.7241
160	.0010988	672.61	678.10	1.9374	.0010953	670.11	681.07	1.9316	.0010918	667.69	684.07	1.9259
180	.0011240	759.62	765.24	2.1341	.0011199	756.63	767.83	2.1274	.0011159	753.74	770.48	2.1209
200	.0011530	848.08	853.85	2.3254	.0011480	844.49	855.97	2.3178	.0011433	841.04	858.18	2.3103
220	.0011866	938.43	944.36	2.5128	.0011805	934.07	945.88	2.5038	.0011748	929.89	947.52	2.4952
240	.0012264	1031.34	1037.47	2.6978	.0012187	1025.94	1038.13	2.6872	.0012114	1020.82	1038.99	2.6770
260	.0012748	1127.92	1134.30	2.8829	.0012645	1121.03	1133.68	2.8698	.0012550	1114.59	1133.41	2.8575
280					.0013216	1220.90	1234.11	3.0547	.0013084	1212.47	1232.09	3.0392
300					.0013972	1328.34	1342.31	3.2468	.0013770	1316.58	1337.23	3.2259
320									.0014724	1431.05	1453.13	3.4246
340									.0016311	1567.42	1591.88	3.6545

(Continued)

Table A.1.4
(Continued)

T	P = 20 MPa (365.81)				P = 30 MPa				P = 50 MPa			
	v	u	h	s	v	u	h	s	v	u	h	s
Sat.	.0020353	1785.47	1826.18	4.0137	—	—	—	—	—	—	—	—
0	.0009904	0.02	20.00	0.0004	.0009856	0.25	29.82	0.0001	.0009766	0.20	49.03	−0.0014
20	.0009928	82.75	102.61	0.2922	.0009886	82.16	111.82	0.2898	.0009804	80.98	130.00	0.2847
40	.0009992	165.15	185.14	0.5646	.0009951	164.01	193.87	0.5606	.0009872	161.84	211.20	0.5526
60	.0010084	247.66	267.82	0.8205	.0010042	246.03	276.16	0.8153	.0009962	242.96	292.77	0.8051
80	.0010199	330.38	350.78	1.0623	.0010156	328.28	358.75	1.0561	.0010073	324.32	374.68	1.0439
100	.0010337	413.37	434.04	1.2917	.0010290	410.76	441.63	1.2844	.0010201	405.86	456.87	1.2703
120	.0010496	496.75	517.74	1.5101	.0010445	493.58	524.91	1.5017	.0010348	487.63	539.37	1.4857
140	.0010678	580.67	602.03	1.7192	.0010621	576.86	608.73	1.7097	.0010515	569.76	622.33	1.6915
160	.0010885	665.34	687.11	1.9203	.0010821	660.81	693.27	1.9085	.0010703	652.39	705.91	1.8890
180	.0011120	750.94	773.18	2.1146	.0011047	745.57	778.71	2.1024	.0010912	735.68	790.24	2.0793
200	.0011387	837.70	860.47	2.3031	.0011302	831.34	865.24	2.2892	.0011146	819.73	875.46	2.2634
220	.0011693	925.89	949.27	2.4869	.0011590	918.32	953.09	2.4710	.0011408	904.67	961.71	2.4419
240	.0012046	1015.94	1040.04	2.6673	.0011920	1006.84	1042.60	2.6489	.0011702	990.69	1049.20	2.6158
260	.0012462	1108.53	1133.45	2.8459	.0012303	1097.38	1134.29	2.8242	.0012034	1078.06	1138.23	2.7860
280	.0012965	1204.69	1230.62	3.0248	.0012755	1190.69	1228.96	2.9985	.0012415	1167.19	1229.26	2.9536
300	.0013596	1306.10	1333.29	3.2071	.0013304	1287.39	1327.80	3.1740	.0012860	1258.66	1322.95	3.1200
320	.0014437	1415.66	1444.53	3.3978	.0013997	1390.64	1432.63	3.3538	.0013388	1353.23	1420.17	3.2867
340	.0015683	1539.64	1571.01	3.6074	.0014919	1501.71	1456.47	3.5425	.0014032	1451.91	1522.07	3.4556
360	.0018226	1702.78	1739.23	3.8770	.0016265	1626.57	1675.36	3.7492	.0014838	1555.97	1630.16	3.6290
380	—	—	—	—	.0018691	1781.35	1837.43	4.0010	.0015883	1667.13	1746.54	3.8100

TABLE A.2 THERMODYNAMIC PROPERTIES OF REFRIGERANT-12*(DICHLORODIFLUOROMETHANE)

Table A.2.1
Saturated Refrigerant-12

Temperature °C t	Pressure MPa p	Specific Volume: Sat. Liquid v_f Cm^3/g	Specific Volume: Sat. Vapour v_g m^3/kg	Enthalpy: Sat. Liquid h_f	Enthalpy: Evap. h_{fg} kJ/kg	Enthalpy: Sat. Vapour h_g	Entropy: Sat. Liquid s_f kJ/kg K	Entropy: Sat. Vapour s_g
−90	0.0028	0.608	4.415545	−43.243	189.618	146.375	−0.2084	0.8268
−85	0.0042	0.612	3.037316	−38.968	187.608	148.640	−0.1854	0.8116
−80	0.0062	0.617	2.038345	−34.688	185.612	150.924	−0.1630	0.7979
−75	0.0088	0.622	1.537651	−30.401	183.625	153.224	−0.1411	0.7855
−70	0.0123	0.627	1.127280	−26.103	181.640	155.536	−0.1197	0.7744
−65	0.0168	0.632	0.841166	−21.793	179.651	157.857	−0.0987	0.7643
−60	0.0226	0.637	0.637910	−17.469	177.653	160.184	−0.0782	0.7552
−55	0.0300	0.642	0.491000	−13.129	175.641	162.512	−0.0581	0.7470
−50	0.0391	0.648	0.383105	−8.772	173.611	164.840	−0.0384	0.7396
−45	0.0504	0.654	0.302682	−4.396	171.558	167.163	−0.0190	0.7329
−40	0.0642	0.659	0.241910	−0.000	169.479	169.479	−0.0000	0.7269
−35	0.0807	0.666	0.195398	4.416	167.368	171.784	0.0187	0.7214
−30	0.1004	0.672	0.159375	8.854	165.222	174.076	0.0371	0.7165
−25	0.1237	0.679	0.131166	13.315	163.037	176.352	0.0552	0.7121
−20	0.1509	0.685	0.108847	17.800	160.810	178.610	0.0730	0.7082
−15	0.1826	0.693	0.091018	22.312	158.534	180.846	0.0906	0.7046
−10	0.2191	0.700	0.076646	26.851	156.207	183.058	0.1079	0.7014
−5	0.2610	0.708	0.064963	31.420	153.823	185.243	0.1250	0.6986

* Adapted from *Fundamentals of Classical Thermodynamics by*-G. J. Van Wylen and R. Sonntag. John Wiley, New York 1976, P. 667–673 (with the kind permission of the publishers, John Wiley & Sons, Inc, New York).

(Continued)

Table A.2.1
(Continued)

t	p	v_f	v_g	h_f	h_{fg}	h_g	s_f	s_g
0	0.3086	0.716	0.055389	36.022	151.376	187.397	0.1418	0.6960
5	0.3626	0.724	0.047485	40.659	148.859	189.518	0.1585	0.6937
10	0.4233	0.733	0.040914	45.337	146.265	191.602	0.1750	0.6916
15	0.4914	0.743	0.035413	50.058	143.586	193.644	0.1914	0.6897
20	0.5673	0.752	0.030780	54.828	140.812	195.641	0.2076	0.6879
25	0.6516	0.763	0.026854	59.653	137.933	197.586	0.2237	0.6863
30	0.7449	0.774	0.023508	64.539	134.936	199.475	0.2397	0.6848
35	0.8477	0.786	0.020641	69.494	131.805	201.299	0.2557	0.6834
40	0.9607	0.798	0.018171	74.527	128.525	203.051	0.2716	0.6820
45	1.0843	0.811	0.016032	79.647	125.074	204.722	0.2875	0.6806
50	1.2193	0.826	0.014170	84.868	121.430	206.298	0.3034	0.6792
55	1.3663	0.841	0.012542	90.201	117.565	207.766	0.3194	0.6777
60	1.5259	0.858	0.01111	95.665	113.443	209.109	0.3355	0.6760
65	1.6988	0.877	0.009847	101.279	109.024	210.303	0.3518	0.6742
70	1.8858	0.897	0.008725	107.067	104.255	211.321	0.3683	0.6721
75	2.0874	0.920	0.007723	113.058	99.068	212.126	0.3851	0.6697
80	2.3046	0.946	0.006821	119.291	93.373	212.665	0.4023	0.6667
85	2.5380	0.976	0.006005	125.818	87.047	212.865	0.4201	0.6631
90	2.7885	1.012	0.005258	132.708	79.907	212.614	0.4385	0.6585
95	3.0569	1.056	0.004563	140.068	71.658	211.726	0.4579	0.6526
100	3.3440	1.113	0.003903	148.076	61.768	209.843	0.4788	0.6444
105	3.6509	1.197	0.003242	157.085	49.014	206.099	0.5023	0.6319
110	3.9784	1.364	0.002462	168.059	28.425	196.484	0.5322	0.6064

Table A.2.2
Superheated Refrigerant-12

t °C	v m³kg	h kJ/kg	s kJ/kg K	v m³/kg	h kJ/kg	s kJ/kg K	v m/kg	h kJ kg	s kJ kg K
		0.05 MPa			*0.10 MPa*			*0.15 MPa*	
– 20.00	0.341857	181.042	0.7912	0.167701	179.861	0.7401			
– 10.00	0.356227	186.757	0.8133	0.175222	185.707	0.7628	0.114716	184.619	0.7318
00.00	0.370508	192.567	0.8350	0.182647	191.628	0.7849	0.119866	190.660	0.7543
10.00	0.384716	198.471	0.8562	0.189994	197.628	0.8064	0.124932	196.762	0.7763
20.00	0.398863	204.469	0.8770	0.197277	203.707	0.8275	0.129930	202.927	0.7977
30.00	0.412959	210.557	0.8974	0.204506	209.866	0.8482	0.134873	209.160	0.8186
40.00	0.427012	216.733	0.9175	0.211691	216.104	0.8684	0.139768	215.463	0.8390
50.00	0.441030	222.997	0.9372	0.218839	222.421	0.8883	0.144625	221.835	0.8591
60.00	0.455017	229.344	0.9565	0.225955	228.815	0.9078	0.149450	228.277	0.8787
70.00	0.468978	235.774	0.9755	0.233044	235.285	0.9269	0.154247	234.789	0.8980
80.00	0.482917	242.282	0.9942	0.240111	241.829	0.9457	0.159020	241.371	0.9169
90.00	0.496838	248.868	1.0126	0.247159	248.446	0.9642	0.163774	248.020	0.9354
		0.20 MPa			*0.25 MPa*			*0.30 MPa*	
00.0	0.088608	189.669	0.7320	0.069752	188.644	0.7139	0.057150	187.583	0.6984
10.00	0.092550	195.878	0.7543	0.073024	194.969	0.7366	0.059984	194.034	0.7216
20.00	0.096418	202.135	0.7760	0.076218	201.322	0.7587	0.062734	200.490	0.7440
30.00	0.100228	208.446	0.7972	0.079350	207.715	0.7801	0.065418	206.969	0.7658
40.00	0.103989	214.814	0.8178	0.082431	214.153	0.8010	0.068049	213.480	0.7869
50.00	0.107710	221.243	0.8381	0.085470	220.642	0.8214	0.070635	220.030	0.8075
60.00	0.111397	227.735	0.8578	0.088474	227.185	0.8413	0.073185	226.627	0.8276
70.00	0.115055	134.291	0.8772	0.091449	233.785	0.8608	0.075750	233.273	0.8473
80.00	0.118690	240.910	0.8962	0.094398	240.443	0.8800	0.078200	239.271	0.8665
90.00	0.122304	247.593	0.9149	0.097327	247.160	0.8987	0.080673	246.723	0.8853

(Continued)

Table A.2.2
(Continued)

t	v	h	s	v	h	s	v	h	s
100.0	0.125901	254.339	0.9332	0.100238	253.936	0.9171	0.83127	253.530	0.9038
110.0	0.129483	261.147	0.9512	0.103134	260.770	0.9352	0.085566	260.391	0.9220
		0.40 MPa			*0.50 MPa*			*0.60 MPa*	
20.0	0.045836	198.762	0.7199	0.035646	196.935	0.6999			
30.0	0.047971	205.428	0.7423	0.037464	203.814	0.7230	0.030422	202.116	0.7063
40.0	0.050046	212.095	0.7639	0.039214	210.656	0.7452	0.031966	209.154	0.7291
50.0	0.052072	218.779	0.7849	0.040911	217.484	0.7667	0.033450	216.141	0.7511
60.0	0.054059	225.488	0.8054	0.042565	224.315	0.7875	0.034887	223.104	0.7723
70.0	0.056014	232.230	0.8253	0.044184	232.161	0.8077	0.036285	230.062	0.7929
80.0	0.057941	239.012	0.8448	0.045774	238.031	0.8275	0.037653	237.027	0.8129
90.0	0.059846	245.837	0.8638	0.047340	244.932	0.8467	0.038995	244.009	0.8324
100.0	0.061731	252.707	0.8825	0.048886	251.869	0.8656	0.040316	251.016	0.8514
110.0	0.063600	259.624	0.9008	0.050415	258.845	0.8840	0.041619	258.053	0.8700
		0.70 MPa			*0.80 MPa*			*0.90 MPa*	
40.0	0.026761	207.580	0.7148	0.022830	205.924	0.7016	0.019744	204.170	0.6982
50.0	0.028100	214.745	0.7373	0.024068	213.290	0.7248	0.020912	211.765	0.7131
60.0	0.029387	221.854	0.7590	0.025247	220.558	0.7469	0.022012	218.212	0.7358
70.0	0.030632	228.931	0.7799	0.026380	227.766	0.7682	0.023062	226.564	0.7575
80.0	0.031843	235.997	0.8002	0.027477	234.941	0.7888	0.024072	233.856	0.7785
90.0	0.033027	243.066	0.8199	0.028545	242.101	0.8088	0.025051	141.113	0.7987

		0.70 MPa			*0.80 MPa*			*0.90 MPa*	
100.0	0.034189	250.146	0.8392	0.029588	249.260	0.8283	0.026005	248.355	0.8184
110.0	0.035332	257.247	0.8579	0.030612	256.428	0.8472	0.026937	255.593	0.8376
		1.00 MPa			*1.20 MPa*			*1.40 MPa*	
50.0	0.018366	210.162	0.7021	0.014483	206.661	0.6812			
60.0	0.019410	217.810	0.7254	0.015463	214.805	0.7060	0.012579	211.457	0.6876
70.0	0.020397	225.319	0.7476	0.016368	222.687	0.7293	0.013448	219.822	0.7123
80.0	0.021341	232.739	0.7689	0.017221	230.398	0.7514	0.014247	227.891	0.7355
90.0	0.022251	240.101	0.7895	0.018032	237.995	0.7727	0.014997	235.766	0.7575
100.0	0.023133	247.430	0.8094	0.018812	245.518	0.7931	0.015710	243.512	0.7785
110.10	0.023993	254.743	0.8287	0.019567	252.993	0.8129	0.016393	251.170	0.7988
		1.60 MPa			*1.80 MPa*			*2.00 MPa*	
70.0	0.011208	216.650	0.6959	0.009406	213.049	0.6794			
80.0	0.011984	225.177	0.7204	0.010187	222.198	0.7057	0.008704	218.859	0.6909
90.0	0.012698	233.390	0.7433	0.010884	230.835	0.7298	0.009406	228.056	0.7166
100.0	0.013366	241.397	0.7651	0.011526	239.155	0.7524	0.010035	236.760	0.7402
110.0	0.014000	249.264	0.7859	0.012126	2467.264	0.7739	0.010615	245.154	0.7624

TABLE A.3 THERMODYNAMIC PROPERTIES OF REFRIGERANT-22 (MONOCHLORODIFLUOROMETHANE)

Table A.3.1
Saturated Refrigerant-22

		Specific volume (m^3/kg)			Enthalpy (kJ/kg)			Entropy (kJ/kg K)		
Temp. °C	Abs. Press. MPa P	Sat. Liquid v_f	Evap. v_{fg}	Sat. Vapour v_g	Sat. Liquid h_f	Evap. h_{fg}	Vapour h_g	Sat. Liquid s_f	Evap. s_{fg}	Sat. Vapour s_g
– 70	0.0205	0.000670	0.940268	0.94093	– 30.607	249.425	218.180	– 0.1401	1.2277	1.0876
– 65	0.0280	0.000676	0.704796	0.705478	– 25.658	246.925	221.267	– 0.1161	1.1862	1.0701
– 60	0.0375	0.000682	0.536470	0.537152	– 20.652	244.354	223.702	– 0.0924	1.1463	1.0540
– 55	0.0495	0.000689	0.414138	0.414827	– 15.585	241.703	226.117	– 0.0689	1.1079	1.0390
– 50	0.0644	0.000695	0.323862	0.324557	– 10.456	238.965	228.509	– 0.0457	1.0708	1.0251
– 45	0.0827	0.000702	0.256288	0.256990	– 5.262	236.132	230.870	– 0.0227	1.0349	1.0122
– 40	0.1049	0.000709	0.205036	0.205745	0	233.198	233.197	0	1.0002	1.0002
– 35	0.1317	0.000717	0.165683	0.166400	5.328	230.156	235.484	0.0225	0.9664	0.9889
– 30	0.1635	0.000725	0.135120	0.135844	10.725	227.001	237.726	0.0449	0.9335	0.9784
– 25	0.2010	0.000733	0.111126	0.111859	16.191	223.727	239.918	0.0670	0.9015	0.9685
– 20	0.2448	0.000741	0.092102	0.092843	21.728	220.327	242.055	0.0890	0.8703	0.9593
– 15	0.2957	0.000750	0.076876	0.077625	27.334	216.798	244.132	0.1107	0.8398	0.9505
– 10	0.3543	0.000759	0.064581	0.065340	33.012	213.132	246.144	0.1324	0.8099	0.9422
– 5	0.4213	0.000768	0.054571	0.055339	38.762	209.323	248.085	0.1538	0.7806	0.9344
0	0.4976	0.000778	0.046357	0.047135	44.586	205.364	249.949	0.1751	0.7518	0.9269
5	0.5838	0.000789	0.039567	0.040356	50.485	201.246	251.731	0.1963	0.7235	0.9197

10	0.6807	0.000800	0.033914	0.034714	56.463	196.960	253.423	0.2173	0.6956	0.9129
15	0.7891	0.000812	0.029176	0.029987	62.523	192.495	255.018	0.2382	0.6680	0.9062
20	0.9099	0.000824	0.025179	0.026003	68.670	187.836	256.506	0.2590	0.6407	0.8997
25	1.0439	0.000838	0.021787	0.022624	74.910	182.968	257.877	0.2797	0.6137	0.8934
30	1.1919	0.000852	0.018890	0.019742	81.250	177.869	259.119	0.3004	0.5867	0.8871
35	1.3548	0.000867	0.016401	0.017269	87.700	172.516	260.216	0.3210	0.5598	0.8809
40	1.5335	0.000884	0.014251	0.015135	94.272	166.877	261.149	0.3417	0.5329	0.8746
45	1.7290	0.000902	0.012382	0.013284	100.982	160.914	261.896	0.3624	0.5058	0.8682
50	1.9423	0.000922	0.010747	0.011669	107.851	154.576	262.428	0.3832	0.4783	0.8615
55	2.1744	0.000944	0.009308	0.010252	114.905	147.800	262.705	0.4042	0.4504	0.8546
60	2.4266	0.000969	0.008032	0.009001	122.180	140.497	262.678	0.4255	0.4217	0.8472
65	2.6999	0.000997	0.006890	0.007887	129.729	132.547	262.276	0.4472	0.3920	0.8391
70	2.9959	0.001030	0.005859	0.006889	137.625	123.772	261.397	0.4695	0.3607	0.8302
75	3.3161	0.001069	0.004914	0.005983	145.986	113.902	259.888	0.4927	0.3272	0.8198
80	3.6623	0.001118	0.004031	0.005149	155.011	102.475	257.486	0.5173	0.2902	0.8075
85	4.0368	0.001183	0.003175	0.004358	165.092	88.598	253.690	0.5445	0.2474	0.7918
90	4.4425	0.001282	0.002282	0.003564	177.204	70.037	247.241	0.5767	0.1929	0.7695
95	4.8835	0.001521	0.001030	0.002551	196.359	34.925	231.284	0.6273	0.0949	0.7222
96.006	4.9773	0.001906	0	0.001906	212.546	0	212.546	0.6708	0	0.6708

Table A.3.2
Superheated Refrigerant-22

Temp. °C	v m³/kg	h kJ/kg	s kJ/kg K	v m³/kg	h kJ/kg	s kJ/kg K	v m³/kg	h kJ/kg	s kJ/kg K
		0.05 MPa			*0.10 MPa*			*0.15 MPa*	
– 40	0.440633	234.724	1.07616	0.216331	233.337	1.00523	—	—	—
– 30	0.460641	240.602	1.10084	0.226754	239.359	1.03052	0.148723	238.078	0.98773
– 20	0.480543	246.586	1.12495	0.237064	245.466	1.05513	0.155851	244.319	1.01288
– 10	0.500357	252.676	1.14855	0.247279	251.665	1.07914	0.162879	250.631	1.03733
0	0.520095	258.874	1.17166	0.257415	257.956	1.10261	0.169823	257.022	1.06116
10	0.539771	265.180	1.19433	0.267485	264.345	1.12558	0.176699	263.496	1.08444
20	0.559393	271.594	1.21659	0.277500	270.831	1.14809	0.183516	270.057	1.10721
30	0.578970	278.115	1.23846	0.287467	277.416	1.17017	0.190284	276.709	1.12952
40	0.598507	284.743	1.25998	0.297394	284.101	1.19187	0.197011	283.452	1.15140
50	0.618011	291.478	1.28114	0.307287	290.887	1.21320	0.203702	290.289	1.17289
60	0.637485	298.319	1.30199	0.317149	297.772	1.23418	0.210362	297.220	1.19402
70	0.656935	305.265	1.32253	0.326986	304.757	1.25484	0.216997	304.246	1.21479
80	0.676362	312.314	1.34278	0.336801	311.842	1.27519	0.223608	311.368	1.23525
90	0.695771	319.465	1.36275	0.346596	319.026	1.29524	0.230200	318.584	1.25540
		0.20 MPa			*0.25 MPa*			*0.30 MPa*	
– 20	0.115203	243.140	0.98184	—	—	—	—	—	—
– 10	0.120647	249.574	1.00676	0.95280	248.492	0.98231	0.078344	247.382	0.96170
0	0.126003	256.069	1.03098	0.099689	255.097	1.00695	0.082128	254.104	0.98677
10	0.131286	262.633	1.05458	0.104022	261.755	1.03089	0.085832	260.861	1.01106
		0.20 MPa			*0.25 MPa*			*0.30 MPa*	
20	0.136509	269.273	1.07763	0.108292	268.476	1.05421	0.089469	267.667	1.03468
30	0.141681	275.992	1.10016	0.112508	275.267	1.07699	0.093051	274.531	1.05771

		0.20 MPa			0.25 MPa			0.30 MPa	
40	0.146809	282.796	1.12224	0.116681	282.132	1.09927	0.096588	281.460	1.08019
50	0.151902	289.686	1.14390	0.120815	289.076	1.12109	0.100085	288.460	1.10220
60	0.156963	296.664	1.16516	0.124918	296.102	1.14250	0.103550	295.535	1.12376
70	0.161997	303.731	1.18607	0.128993	303.212	1.16353	0.106986	302.689	1.14491
80	0.167008	310.890	1.20663	0.133044	310.409	1.18420	0.110399	309.924	1.16569
90	0.171999	318.139	1.22687	0.137075	317.692	1.20454	0.113790	317.241	1.18612
100	0.176972	325.480	1.24681	0.141089	325.063	1.22456	0.117164	324.643	1.20623
110	0.181931	332.912	1.26646	0.145086	332.522	1.24428	0.120522	332.129	1.22603
		0.40 MPa			0.50 MPa			0.60 MPa	
0	0.060131	252.051	0.95359	—	—	—	—	—	—
10	0.063060	259.023	0.97866	0.049355	257.108	0.95223	0.040180	255.109	0.92945
20	0.065915	266.010	1.00291	0.051751	264.295	0.97717	0.042280	262.517	0.95517
30	0.068710	273.029	1.02646	0.054081	271.483	1.00128	0.044307	269.888	0.97989
40	0.071455	280.092	1.04938	0.056358	278.690	1.02467	0.046276	277.250	1.00378
50	0.074160	287.209	1.07175	0.058590	285.930	1.04743	0.048198	284.622	1.02695
60	0.076830	294.386	1.09362	0.060786	293.215	1.06963	0.050081	292.020	1.04950
70	0.079470	301.630	1.11504	0.062951	300.552	1.09133	0.051931	299.456	1.07149
80	0.082085	308.944	1.13605	0.065090	307.949	1.11257	0.053754	306.938	1.09298
90	0.084679	316.332	1.15668	0.067206	315.410	1.13340	0.055553	314.475	1.11403
100	0.087254	323.796	1.17695	0.069303	322.939	1.15386	0.057332	322.071	1.13466
110	0.089813	331.339	1.19690	0.071384	330.539	1.17395	0.059094	329.731	1.15492
120	0.092358	338.961	1.21654	0.073450	338.213	1.19373	0.060842	337.458	1.17482
130	0.094890	346.664	1.23588	0.075503	345.963	1.21319	0.062576	345.255	1.19441

(Continued)

Table A.3.2
(Continued)

Temp. °C	v m³/kg	h kJ/kg	s kJ/kg K	v m³/kg	h kJ/kg	s kJ/kg K	v m³/kg	h kJ/kg	s kJ/kg K
		0.70 MPa			*0.80 MPa*			*0.90 MPa*	
20	0.035487	260.667	0.93565	0.030366	258.737	0.91787	0.026355	256.713	0.90132
30	0.037305	268.240	0.96105	0.032034	266.533	0.94402	0.027915	264.760	0.92831
40	0.039059	275.769	0.98549	0.033632	274.243	0.96905	0.029397	272.670	0.95398
50	0.040763	283.282	1.00910	0.035175	281.907	0.99314	0.030819	280.497	0.97859
60	0.042424	290.800	1.03201	0.036674	289.553	1.01644	0.032193	288.278	1.00230
70	0.044052	298.339	1.05431	0.038136	297.202	1.03906	0.033528	296.042	1.02526
80	0.045650	305.912	1.07606	0.039568	304.868	1.06108	0.034832	303.807	1.04757
90	0.047224	313.527	1.09732	0.040974	312.565	1.08257	0.036108	311.590	1.06930
100	0.048778	321.192	1.11815	0.042359	320.303	1.10359	0.037363	319.401	1.09052
110	0.050313	328.914	1.13856	0.043725	328.087	1.12417	0.038598	327.251	1.11128
120	0.051834	336.696	1.15861	0.045076	335.925	1.14437	0.039817	335.147	1.13162
130	0.053341	344.541	1.17832	0.046413	343.821	1.16420	0.041022	343.094	1.15158
140	0.054836	352.454	1.19770	0.047738	351.778	1.18369	0.042215	351.097	1.17119
150	0.056321	360.435	1.21679	0.049052	359.799	1.20288	0.043398	359.159	1.19047
		1.00 MPa			*1.20 MPa*			*1.40 MPa*	
30	0.024600	262.912	0.91358	—	—	—	—	—	—
40	0.025995	271.042	0.93996	0.020851	267.602	0.91411	0.017120	263.861	0.89010
50	0.027323	279.046	0.96512	0.022051	276.011	0.94055	0.018247	272.766	0.91809
60	0.028601	286.973	0.98928	0.023191	284.263	0.96570	0.019299	281.401	0.94441
70	0.029836	294.859	1.01260	0.024282	292.415	0.98981	0.020295	289.858	0.96942
80	0.031038	302.727	1.03520	0.025336	330.508	1.01305	0.021248	298.202	0.99339
90	0.032213	310.599	1.05718	0.026359	308.570	1.03556	0.022167	306.473	1.01649
100	0.033364	318.488	1.07861	0.027357	316.623	1.05744	0.023058	314.703	1.03884

		1.00 MPa			*1.20 MPa*			*1.40 MPa*	
110	0.034495	326.405	1.09955	0.028334	324.682	1.07875	0.023926	322.916	1.06056
120	0.035609	334.360	1.12004	0.029292	332.762	1.09957	0.024775	331.128	1.08172
130	0.036709	342.360	1.14014	0.030236	340.871	1.11994	0.025608	339.354	1.10238
140	0.037797	350.410	1.15986	0.031166	349.019	1.13990	0.026426	347.603	1.12259
150	0.038873	358.514	1.17924	0.032084	357.210	1.15949	0.027233	355.885	1.14240
160	0.039940	366.677	1.19831	0.032993	365.450	1.17873	0.028029	364.206	1.16183
		1.60 MPa			*1.80 MPa*			*2.00 MPa*	
50	0.015351	269.262	0.89689	0.013052	265.423	0.87625	—	—	—
60	0.016351	278.358	0.92461	0.014028	275.097	0.90573	0.012135	271.563	0.88729
70	0.017284	287.171	0.95068	0.014921	284.331	0.93304	0.013008	281.310	0.91612
80	0.018167	295.797	0.97546	0.015755	293.282	0.95876	0.013811	290.640	0.94292
90	0.019011	304.301	0.99920	0.016546	302.046	0.98323	0.014563	299.697	0.96821
100	0.019825	312.725	1.02209	0.017303	310.683	1.00669	0.015277	308.571	0.99232
110	0.020614	321.103	1.04424	0.018032	319.239	1.02932	0.015960	317.322	1.01546
120	0.021382	329.457	1.06576	0.018738	327.745	1.05123	0.016619	325.991	1.03780
130	0.022133	337.805	1.08673	0.019427	336.224	1.07253	0.017258	334.610	1.05944
140	0.022869	346.162	1.10721	0.020099	344.695	1.09329	0.017881	343.201	1.08049
150	0.023592	354.540	1.12724	0.020759	353.172	1.11356	0.018490	351.783	1.10102
160	0.024305	362.945	1.14688	0.021407	361.666	1.13340	0.019087	360.369	1.12107
170	0.025008	371.386	1.16614	0.022045	370.186	1.15284	0.019673	368.970	1.14070
180	0.025703	379.869	1.18507	0.022675	378.738	1.17193	0.020251	377.595	1.15995
		2.50 MPa			*3.00 MPa*			*3.50 MPa*	
70	0.009459	272.677	0.87476	—	—	—	—	—	—
80	0.010243	283.332	0.90537	0.007747	274.530	0.86780	0.005765	262.739	0.82489
90	0.010948	293.338	0.93332	0.008465	286.042	0.89995	0.006597	277.268	0.86548
100	0.011598	302.935	0.95939	0.009098	296.663	0.92881	0.007257	289.504	0.89872
110	0.012208	312.261	0.98405	0.009674	306.744	0.95547	0.007829	300.640	0.92818

(Continued)

Table A.3.2
(Continued)

Temp. °C	v m³/kg	h kJ/kg	s kJ/kg K	v m³/kg	h kJ/kg	s kJ/kg K	v m³/kg	h kJ/kg	s kJ/kg K
		2.50 MPa			*3.00 MPa*			*3.50 MPa*	
120	0.012788	321.400	1.00760	0.010211	316.470	0.98053	0.008346	311.129	0.95520
130	0.013343	330.412	1.03023	0.010717	325.955	1.00435	0.008825	321.196	0.98047
140	0.013880	339.336	1.05210	0.011200	335.270	1.02718	0.009276	330.976	1.00445
150	0.014400	348.205	1.07331	0.011665	344.467	1.04918	0.009704	340.554	1.02736
160	0.014907	357.040	1.09395	0.012114	353.584	1.07047	0.010114	349.989	1.04940
170	0.015402	365.860	1.11408	0.012550	362.647	1.09116	0.010510	359.324	1.07071
180	0.015887	374.679	1.13376	0.012976	371.679	1.11131	0.010894	368.590	1.09138
190	0.016364	383.508	1.15303	0.013392	380.695	1.13099	0.011268	377.810	1.11151
200	0.016834	392.354	1.17912	0.013801	389.708	1.15024	0.011634	387.004	1.13115
		4.00 MPa			*5.00 MPa*			*6.00 MPa*	
90	0.005037	265.629	0.82544	—	—	—	—	—	—
100	0.005804	280.997	0.86721	0.003334	253.042	0.78005	—	—	—
110	0.006405	293.748	0.90094	0.004255	275.919	0.84064	0.002432	243.278	0.74674
120	0.006924	305.273	0.93064	0.004851	291.362	0.88045	0.003333	272.385	0.82185
130	0.007391	316.080	0.95778	0.005335	304.469	0.91337	0.003899	290.253	0.86675
140	0.007822	326.422	0.98312	0.005757	316.379	0.94256	0.004345	304.757	0.90230
150	0.008226	336.446	1.00710	0.006139	327.563	0.96931	0.004728	317.633	0.93310
160	0.008610	346.246	1.02999	0.006493	338.266	0.99431	0.005071	329.553	0.96094
170	0.008978	355.885	1.05199	0.006826	348.633	1.01797	0.005386	340.849	0.98673
180	0.009332	365.409	1.07324	0.007142	358.760	1.04057	0.005680	351.715	1.01098
190	0.009675	374.853	1.09386	0.007444	368.713	1.06230	0.005958	362.271	1.03402
200	0.010009	384.240	1.11391	0.007735	378.537	1.08328	0.006222	372.602	1.05609
210	0.010335	393.593	1.13347	0.008018	388.268	1.10363	0.006477	382.764	1.07734
220	0.010654	402.925	1.15259	0.008292	397.932	1.12343	0.006722	392.801	1.09790

TABLE A.4 THERMODYNAMIC PROPERTIES OF REFRIGERANT-134A (1, 1, 1, 2–TETRAFLUOROETHANE)

Table A.4.1
Saturated R-134a

		Specific Volume, m^3/kg			Enthalpy, kJ/kg			Entropy, kJ/kg K		
Temp. °C	Abs. Press. MPa P	Sat. Liquid v_f	Evap. v_{fg}	Sat. Vapour v_g	Sat. Liquid h_f	Evap. h_{fg}	Sat. Vapour h_g	Sat. Liquid s_f	Evap. s_{fg}	Sat. Vapour s_g
– 33	0.0737	0.000718	0.25574	0.25646	157.417	220.491	377.908	0.8346	0.9181	1.7528
– 30	0.0851	0.000722	0.22330	0.22402	161.118	218.683	379.802	0.8499	0.8994	1.7493
– 26.25	0.1013	0.000728	0.18947	0.19020	165.802	216.360	382.162	0.8690	0.8763	1.7453
– 25	0.0173	0.000730	0.17956	0.18029	167.381	215.569	382.950	0.8754	0.8687	1.7441
– 20	0.1337	0.000738	0.14575	0.14649	173.744	212.340	386.083	0.9007	0.8388	1.7395
– 15	0.1650	0.000746	0.11932	0.12007	180.193	209.004	389.197	0.9258	0.8096	1.7354
– 10	0.2017	0.000755	0.098454	0.099209	186.721	205.564	392.285	0.9507	0.7812	1.7319
– 5	0.2445	0.000764	0.081812	0.082576	193.324	202.016	395.340	0.9755	0.7534	1.7288
0	0.2940	0.000773	0.068420	0.069193	200.000	198.356	398.356	1.0000	0.7262	1.7262
5	0.3509	0.000783	0.057551	0.058334	206.751	194.572	401.323	1.0243	0.6995	1.7239
10	0.4158	0.000794	0.048658	0.049451	213.580	190.652	404.233	1.0485	0.6733	1.7218
15	0.4895	0.000805	0.041326	0.042131	220.492	186.582	407.075	1.0725	0.6475	1.7200
20	0.5728	0.000817	0.035238	0.036055	227.493	182.345	409.838	1.0963	0.6220	1.7183
25	0.6663	0.000829	0.030148	0.030977	234.590	177.920	412.509	1.1201	0.5967	1.7168
30	0.7710	0.000843	0.025865	0.026707	241.790	173.285	415.075	1.1437	0.5716	1.7153
35	0.8876	0.000857	0.022237	0.023094	249.103	168.415	417.518	1.1673	0.5465	1.7139
40	1.0171	0.000873	0.019147	0.020020	256.539	163.282	419.821	1.1909	0.5214	1.7123
45	1.1602	0.000890	0.016499	0.017389	264.110	157.852	421.962	1.2145	0.4962	1.7106

(Continued)

Table A.4.1
(Continued)

Temp. °C	Abs. Press. MPa P	Specific Volume, m^3/kg Sat. Liquid v_f	Evap. v_{fg}	Sat. Vapour v_g	Enthalpy, kJ/kg Sat. Liquid h_f	Evap. h_{fg}	Sat. Vapour h_g	Entropy, kJ/kg K Sat. Liquid s_f	Evap. s_{fg}	Sat. Vapour s_g
50	1.3180	0.000908	0.014217	0.015124	271.830	152.085	423.915	1.2381	0.4706	1.7088
55	1.4915	0.000928	0.012237	0.013166	279.718	145.933	425.650	1.2619	0.4447	1.7066
60	1.6818	0.000951	0.010511	0.011462	287.794	139.336	427.130	1.2857	0.4182	1.7040
65	1.8898	0.000976	0.008995	0.009970	296.088	132.216	428.305	1.3099	0.3910	1.7009
70	2.1169	0.001005	0.007653	0.008657	304.642	124.468	429.110	1.3343	0.3627	1.6970
75	2.3644	0.001038	0.006453	0.007491	313.513	115.939	429.451	1.3592	0.3330	1.6923
80	2.6337	0.001078	0.005368	0.006446	322.794	106.395	429.189	1.3849	0.3013	1.6862
85	2.9265	0.001128	0.004367	0.005495	332.644	95.440	428.084	1.4117	0.2665	1.6782
90	3.2448	0.001995	0.003412	0.004606	343.380	82.295	425.676	1.4404	0.2266	1.6670
95	3.5914	0.001297	0.002432	0.003729	355.834	64.984	420.818	1.4733	0.1765	1.6498
101.15	4.0640	0.001969	0	0.001969	390.977	0	390.977	1.5658	0	1.5658

Table A.4.2
Superheated R-134a

Temp. °C	v m³/kg	h kJ/kg	s kJ/kg K	v m³/kg	h kJ/kg	s kJ/kg K	v m³/kg	h kJ/kg	s kJ/kg K
		0.10 MPa			0.15 MPa			0.20 MPa	
– 25	0.19400	383.212	1.75058	—	—	—	—	—	—
– 20	0.19860	387.215	1.76655	—	—	—	—	—	—
– 10	0.20765	395.270	1.79775	0.13603	393.839	1.76058	0.10013	392.338	1.73276
0	0.21652	403.413	1.82813	0.14222	402.187	1.79171	0.10501	400.911	1.76474
10	0.22527	411.668	1.85780	0.14828	410.602	1.82197	0.10974	409.500	1.79562
20	0.23393	420.048	1.88689	0.15424	419.111	1.85150	0.11436	418.145	1.82563
30	0.24250	428.564	1.91545	0.16011	427.730	1.88041	0.11889	426.875	1.85491
40	0.25102	437.223	1.94355	0.16592	436.473	1.90879	0.12335	435.708	1.88357
50	0.25948	446.029	1.97123	0.17168	445.350	1.93669	0.12776	444.658	1.91171
60	0.26791	454.986	1.99853	0.17740	454.366	1.96416	0.13213	453.735	1.93937
70	0.27631	464.096	2.02547	0.18308	463.525	1.99125	0.13646	462.946	1.96661
80	0.28468	473.359	2.05208	0.18874	472.831	2.01798	0.14076	472.296	1.99346
90	0.29303	482.777	2.07837	0.19437	482.285	2.04438	0.14504	481.788	2.01997
100	0.30136	492.349	2.10437	0.19999	491.888	2.07046	0.14930	491.424	2.04614

(Continued)

Table A.4.2
(Continued)

Temp. °C	v m³/kg	h kJ/kg	s kJ/kg K	v m³/kg	h kJ/kg	s kJ/kg K	v m³/kg	h kJ/kg	s kJ/kg K
		0.25 MPa			*0.30 MPa*			*0.40 MPa*	
0	0.082637	399.579	1.74284	—	—	—	—	—	—
10	0.086584	408.357	1.77440	0.071110	407.171	1.75637	0.051681	404.651	1.72611
20	0.090408	417.151	1.80492	0.074415	416.124	1.78744	0.054362	413.965	1.75844
30	0.090139	425.997	1.83460	0.077620	425.096	1.81754	0.056926	423.216	1.78947
40	0.097798	434.925	1.86357	0.080748	434.124	1.84684	0.059402	432.465	1.81949
50	0.101401	443.953	1.89195	0.083816	443.234	1.87547	0.061812	441.751	1.84868
60	0.104958	453.094	1.91980	0.086838	452.442	1.90354	0.064169	451.104	1.87718
70	0.108480	462.359	1.94720	0.089821	461.763	1.93110	0.066484	460.545	1.90510
80	0.111972	471.754	1.97419	0.092774	471.206	1.95823	0.068767	470.088	1.93252
90	0.115440	481.285	2.00080	0.095702	480.777	1.98495	0.071022	479.745	1.95948
100	0.118888	490.955	2.02707	0.098609	490.482	2.01131	0.073254	489.523	1.98604
110	0.122318	500.766	2.05302	0.101498	500.324	2.03734	0.075468	499.428	2.01223
120	0.125734	510.720	2.07866	0.104371	510.304	2.06305	0.077665	509.464	2.03809
		0.50 MPa			*0.60 MPa*			*0.70 MPa*	
20	0.042256	411.645	1.73420	—	—	—	—	—	—
30	0.044457	421.221	1.76632	0.036094	419.093	1.74610	0.030069	416.809	1.72770
40	0.046557	430.720	1.79715	0.037958	428.881	1.77786	0.031781	426.933	1.76056
50	0.048581	440.205	1.82696	0.039735	438.589	1.80838	0.033392	436.895	1.79187
60	0.050547	449.718	1.85596	0.041447	448.279	1.83791	0.034929	446.782	1.82201

70	0.052467	459.290	1.88426	0.043108	457.994	1.86664	0.036410	456.655	1.85121
80	0.054351	468.942	1.91199	0.044730	467.764	1.89471	0.037848	466.554	1.87964
90	0.056205	478.690	1.93921	0.046319	477.611	1.92220	0.039251	476.507	1.90743
100	0.058035	488.546	1.96598	0.047883	487.550	1.94920	0.040627	486.535	1.93467
110	0.059845	498.518	1.99235	0.049426	497.594	1.97576	0.041980	496.654	1.96143
120	0.061639	508.613	2.01836	0.050951	507.750	2.00193	0.043314	506.875	1.98777
130	0.063418	518.835	2.04403	0.052461	518.026	2.02774	0.044633	517.207	2.01372
140	0.065184	529.187	2.06940	0.053958	528.425	2.05322	0.045938	527.656	2.03932
		0.80 MPa			*0.90 MPa*			*1.00 MPa*	
40	0.027113	424.860	1.74457	0.023446	422.642	1.72943	0.020473	420.249	1.71479
50	0.028611	435.114	1.77680	0.024868	433.235	1.76273	0.021849	431.243	1.74936
60	0.030024	445.223	1.80761	0.026192	443.595	1.79431	0.023110	441.890	1.78181
70	0.031375	455.270	1.83732	0.027447	453.835	1.82459	0.024293	452.345	1.81273
80	0.032678	465.308	1.86616	0.028649	464.025	1.85387	0.025417	462.703	1.84248
90	0.033944	475.375	1.89427	0.029810	474.216	1.88232	0.026497	473.027	1.87131
100	0.035180	485.499	1.92177	0.030940	484.441	1.91010	0.027543	483.361	1.89938
110	0.036392	495.698	1.94874	0.032043	494.726	1.93730	0.028561	493.736	1.92682
120	0.037584	505.988	1.97525	0.033126	505.088	1.96399	0.029556	504.175	1.95371
130	0.038760	516.379	2.00135	0.034190	515.542	1.99025	0.030533	514.694	1.98013
140	0.039921	526.880	2.02708	0.035241	526.096	2.01611	0.031495	525.305	2.00613
150	0.041071	537.496	2.05247	0.036278	536.760	2.04161	0.032444	536.017	2.03175

(Continued)

Table A.4.2
(Continued)

Temp. °C	v m³/kg	h kJ/kg	s kJ/kg K	v m³/kg	h kJ/kg	s kJ/kg K	v m³/kg	h kJ/kg	s kJ/kg K
		1.20 MPa			*1.40 MPa*			*1.60 MPa*	
50	0.017243	426.485	1.72373	—	—	—	—	—	—
60	0.018439	438.210	1.75837	0.015032	434.079	1.73597	0.012392	429.322	1.71349
70	0.019530	449.179	1.79081	0.016083	445.720	1.77040	0.013449	441.888	1.75066
80	0.020548	459.925	1.82168	0.017040	456.944	1.80265	0.014378	453.722	1.78466
90	0.021512	470 551	1.85135	0.017931	467.931	1.83333	0.015225	465.145	1.81656
100	0.022436	481.128	1.88009	0.018775	478.790	1.86282	0.016015	476.333	1.84695
110	0.023329	491.702	1.90805	0.019583	489.589	1.89139	0.016763	487.390	1.87619
120	0.024197	502.307	1.93537	0.020362	500.379	1.91918	0.017479	498.387	1.90452
130	0.025044	512.965	1.96214	0.021118	511.192	1.94634	0.018169	509.371	1.93211
140	0.025874	523.697	1.98844	0.021856	522.054	1.97296	0.018840	520.376	1.95908
150	0.026691	534.514	2.01431	0.022579	532.984	1.99910	0.019493	531.427	1.98551
160	0.027495	545.426	2.03980	0.023289	543.994	2.02481	0.020133	542.542	2.01147
170	0.028289	556.443	2.06494	0.023988	555.097	2.05015	0.020761	553.735	2.03702
		1.80 MPa			*2.0 MPa*			*2.50 MPa*	
70	0.011341	437.562	1.73085	0.009581	432.531	1.71011	—	—	—
80	0.012273	450.202	1.76717	0.010550	446.304	1.74968	0.007221	433.797	1.70180
90	0.013099	462.164	1.80057	0.011374	458.951	1.78500	0.008157	449.499	1.74567
100	0.013854	473.741	1.83202	0.012111	470.996	1.81772	0.008907	463.279	1.78311
110	0.014560	485.095	1.86205	0.012789	482.693	1.84866	0.009558	476.129	1.81709

120	0.015230	496.325	1.89098	0.013424	494.187	1.87827	0.010148	488.457	1.84886
130	0.015871	507.498	1.91905	0.014028	505.569	1.90686	0.010694	500.474	1.87904
140	0.016490	518.659	1.94639	0.014608	516.900	1.93463	0.011208	512.307	1.90804
150	0.017091	529.841	1.97314	0.015168	528.224	1.96171	0.011698	524.037	1.93609
160	0.017677	541.068	1.99936	0.015712	539.571	1.98821	0.012169	535.722	1.96338
170	0.018251	552.357	2.02513	0.016242	550.963	2.01421	0.012624	547.399	1.99004
180	0.018814	563.724	2.05049	0.016762	562.418	2.03977	0.013066	559.098	2.01614
190	0.019369	575.177	2.07549	0.017272	573.950	2.06494	0.013498	570.841	2.04177
		3.0 MPa			*3.50 MPa*			*4.0 MPa*	
90	0.005755	436.193	1.69950	—	—	—	—	—	—
100	0.006653	453.731	1.74717	0.004839	440.433	1.70386	—	—	—
110	0.007339	468.500	1.78623	0.005667	459.211	1.75355	0.004277	446.844	1.71840
120	0.007924	482.043	1.82113	0.006289	474.697	1.79346	0.005005	465.987	1.76415
130	0.008446	494.915	1.85347	0.006813	488.771	1.82881	0.005559	481.865	1.80404
140	0.008926	507.388	1.88403	0.007279	502.079	1.86142	0.006027	496.295	1.83940
150	0.009375	519.618	1.91328	0.007706	514.928	1.89216	0.006444	509.925	1.87200
160	0.009801	531.704	1.94151	0.008103	527.496	1.92151	0.006825	523.072	1.90271
170	0.010208	543.713	1.96892	0.008480	539.890	1.94980	0.007181	535.917	1.93203
180	0.010601	555.690	1.99565	0.008839	552.185	1.97724	0.007517	548.573	1.96028
190	0.010982	567.670	2.02180	0.009185	564.430	2.00397	0.007837	561.117	1.98766
200	0.011353	579.678	2.04745	0.009519	576.665	2.03010	0.008145	573.601	2.01432

TABLE A.5 THERMODYNAMIC PROPERTIES OF AMMONIA

Table A.5.1
Saturated Ammonia Table

T (°C)	P (kPa)	Specific volume (m^3/kg) Sat. liquid vt	t	Sat. vapour, vg	Enthalpy (kJ/kg) Sat. liquid, ht	Evap., htg	Sat. vapour, hg	Entropy (kJ/kg K) Sat. liquid, st	Sat. vapour sg
– 50	40.88	0.001424		2.6254	– 44.3	1416.7	1372.4	– 0.1942	6.1561
– 48	45.96	0.001429		2.3533	– 35.5	1411.3	1375.8	– 0.1547	6.1149
– 46	51.55	0.001434		2.1140	– 26.6	1405.8	1379.2	– 0.1156	6.0746
– 44	57.69	0.001439		1.9032	– 17.8	1400.3	1382.5	– 0.0768	6.0352
– 42	64.42	0.001444		1.7170	– 8.9	1394.7	1385.8	– 0.0382	5.9967
– 40	71.77	0.001449		1.5521	0.0	1389.0	1389.0	0.0000	5.9589
– 38	79.80	0.001454		1.4058	8.9	1383.3	1392.2	0.0380	5.9220
– 36	88.54	0.001460		1.2757	17.8	1377.6	1395.4	0.0757	5.8858
– 34	98.05	0.001465		1.1597	26.8	1371.8	1398.5	0.1132	5.8504
– 32	108.37	0.001470		1.0562	35.7	1365.9	1401.6	0.1504	5.8156
– 30	119.55	0.001476		0.9635	44.7	1360.0	1404.6	0.1873	5.7815
– 28	131.64	0.001481		0.8805	53.6	1354.0	1407.6	0.2240	5.7481
– 26	144.70	0.001487		0.8059	62.6	1347.9	1410.5	0.2605	5.7153
– 24	158.78	0.001492		0.7388	71.6	1341.8	1413.4	0.2967	5.6831
– 22	173.93	0.001498		0.6783	80.7	1335.6	1416.2	0.3327	5.6515
– 20	190.22	0.001504		0.6237	89.7	1329.3	1419.0	0.3684	5.6205
– 18	207.71	0.001510		0.5743	98.8	1322.9	1421.7	0.4040	5.5900
– 16	226.45	0.001515		0.5296	107.8	1316.5	1424.4	0.4393	5.5600
– 14	246.51	0.001521		0.4889	116.9	1310.0	1427.0	0.4744	5.5305
– 12	267.95	0.001528		0.4520	126.0	1303.5	1429.5	0.5093	5.5015

−10	290.95	0.001534	0.4185	135.2	1296.8	1432.0	0.5440	5.4730
−8	315.25	0.001540	0.3878	144.3	1290.1	1434.4	0.5785	5.4449
−6	341.25	0.001546	0.3599	153.5	1283.3	1436.8	0.6128	5.4173
−4	368.90	0.001553	0.3343	162.7	1276.4	1439.1	0.6469	5.3901
−2	398.27	0.001559	0.3109	171.9	1269.4	1441.3	0.6808	5.3633
0	429.44	0.001566	0.2895	181.1	1262.4	1443.5	0.7145	5.3369
2	462.49	0.001573	0.2698	190.4	1255.2	1445.6	0.7481	5.3108
4	497.49	0.001580	0.2517	199.6	1243.0	1447.6	0.7815	5.2852
6	531.51	0.001587	0.2351	208.9	1240.6	1449.6	0.8148	5.2599
8	573.64	0.001594	0.2198	218.3	1233.2	1451.5	0.8479	5.2350
10	614.95	0.001601	0.2056	227.6	1225.7	1453.3	0.8808	5.2104
12	658.52	0.001608	0.1926	237.0	1218.1	1455.1	0.9136	5.1861
14	704.44	0.001616	0.1805	246.4	1210.4	1456.8	0.9463	5.1621
16	752.79	0.001623	0.1693	255.9	1202.6	1458.5	0.9788	5.1385
18	803.66	0.001631	0.1590	265.4	1194.7	1460.0	1.0112	5.1151
20	857.12	0.001639	0.1494	274.9	1186.7	1461.5	1.0434	5.0920
22	913.27	0.001647	0.1405	284.4	1178.5	1462.9	1.0755	5.0692
24	972.19	0.001655	0.1322	294.0	1170.3	1464.3	1.1075	5.0467
26	1033.97	0.001663	0.1245	303.6	1162.0	1465.6	1.1394	5.0244
28	1098.71	0.001671	0.1173	313.2	1153.6	1466.8	1.1711	5.0023
30	1166.49	0.001680	0.1106	322.9	1145.0	1467.9	1.2028	4.9805
32	1237.41	0.001689	0.1044	332.6	1136.4	1469.0	1.2343	4.9589
34	1311.55	0.001698	0.0986	342.3	1127.6	1469.9	1.2656	4.9374
36	1369.03	0.001707	0.0931	352.1	1118.7	1470.8	1.2969	4.9161

(Continued)

Table A.5.1
(Continued)

T (°C)	P (kPa)	Specific volume (m³/kg) Sat. liquid vt	t	Sat. vapour, vg	Enthalpy (kJ/kg) Sat. liquid, ht	Evap., htg	Sat. vapour, hg	Entropy (kJ/kg K) Sat. liquid, st	Sat. vapour sg
38	1469.92	0.001716		0.0880	361.9	1109.7	1471.5	1.3281	4.8950
40	1554.33	0.001726		0.0833	371.7	1100.5	1472.2	1.3591	4.8740
42	1642.35	0.001735		0.0788	381.6	1091.2	1472.8	1.3901	4.8530
44	1734.09	0.001745		0.0746	391.5	1081.7	1473.2	1.4209	4.8322
46	1829.65	0.001756		0.0707	401.5	1072.0	1473.5	1.4518	4.8113
48	1929.13	0.001766		0.0669	411.5	1062.2	1473.7	1.4826	4.7905
50	2032.62	0.001777		0.0635	421.7	1052.0	1473.7	1.5135	4.7696

Table A.5.2
Superheated Ammonia Table

Abs. Press. (kPa) (Sat. temp., °C)		Temperature (°C) −20	−10	0	10	20	30	40	50	60	70	80	100
	v	2.4474	2.5481	2.6482	2.7479	2.8479	2.9464	3.0453	3.1441	3.2427	3.3413	3.4397	
50	h	1435.8	1457.0	1478.1	1499.2	1520.4	1541.7	1563.0	1584.5	1606.1	1627.8	1649.7	
(−46.54)	s	6.3256	6.4077	6.4865	6.5625	6.6360	6.7073	6.7766	6.8441	6.9099	6.9743	7.0372	
	v	1.6233	1.6915	1.7591	1.8263	1.8932	1.9597	2.0261	2.0933	2.1584	2.2244	2.2903	
75	h	1433.0	1454.7	1476.1	1497.5	1518.9	1540.3	1561.8	1583.4	1605.1	1626.9	1648.9	
(−39.18)	s	6.1190	6.2028	6.2828	6.3597	6.4339	6.5058	6.5756	6.6434	6.7096	6.7742	6.8373	
	v	1.2110	1.2631	1.3145	1.3654	1.4160	1.4664	1.5165	1.5664	1.6163	1.6659	1.7155	1.8145
100	h	1430.1	1452.2	1474.1	1495.7	1517.3	1538.9	1560.5	1582.2	1604.1	1626.0	1648.0	1692.6
(−33.01)	s	5.9695	6.0552	6.1366	6.2144	6.2894	6.3618	6.4321	6.5003	0.5668	6.6316	6.6950	6.8177
	v	0.9635	1.0059	1.0476	1.0889	1.1297	1.1703	1.2107	1.2509	1.2909	1.3309	1.3707	1.4501

125	*h*	1427.2	1449.8	1472.0	1493.9	1515.7	1537.5	1559.3	1581.1	1603.0	1625.0	1647.2	1691.8
(– 29.08)	*s*	5.8512	5.9389	6.0217	6.1006	6.1763	6.2494	6.3201	6.3887	6.4555	6.5206	6.5842	6.7072
	v	0.7984	0.8344	0.8697	0.9045	6.9388	0.9729	1.0068	1.0405	1.0740	1.1074	1.1408	1.2072
150	*h*	1424.1	1447.3	1469.8	1492.1	1514.1	1536.1	1558.0	1580.0	1602.0	1624.1	1646.3	1691.1
(– 25.23)	*s*	5.7526	5.8424	5.9266	6.0066	6.0831	6.1568	6.2280	6.2970	6.3641	6.4295	6.4933	6.6167
	v		0.6199	0.6471	0.6738	0.7001	0.7261	0.7519	0.7774	0.8029	0.8282	0.8533	0.9035
200	*h*		1442.0	1465.5	1448.1	1510.9	1533.2	1555.5	1577.7	1599.9	1622.2	1644.6	1689.6
(– 18.86)	*s*		5.6863	5.7737	5.8559	5.9342	6.0091	6.0813	6.1512	6.2189	6.2849	6.3491	6.4732
	v		0.4910	0.5135	0.5354	0.5568	0.5780	0.5989	0.6196	0.6401	0.6605	0.6809	0.7212
250	*h*		1436.6	1461.0	1484.5	1507.6	1530.3	1552.9	1575.4	1597.8	1620.3	1642.8	1688.2
(– 13.67)	*s*		5.5609	5.6517	5.7465	5.8165	5.8928	5.9661	6.0368	6.1052	6.1717	6.2365	6.3613
	v			0.4243	0.4430	0.4613	0.4792	0.4968	0.5113	0.5316	0.5488	0.5658	0.5997
300	*h*			1456.3	1480.6	1504.2	1527.4	1550.3	1573.0	1595.7	1618.4	1641.1	1686.7
(– 9.23)	*s*			5.5193	5.6366	5.7186	5.7963	5.8707	5.9423	6.0114	6.0785	6.1437	6.2693
	v			0.3605	0.3770	0.3929	0.4086	0.4239	0.4391	0.4541	0.4689	0.4837	0.5129
350	*h*			1451.5	1478.5	1590.7	1521.4	1547.6	1570.7	1593.6	1616.5	1639.3	1685.2
(– 5.35)	*s*			5.4600	5.5502	5.6312	5.7135	5.7800	5.8615	5.9314	5.9990	6.0647	6.1910
	v			0.3125	0.3274	0.3417	0.3556	0.3692	0.3826	0.3959	0.4090	0.4220	0.4478
400	*h*			1446.5	1472.4	1497.2	1521.3	1544.9	1568.3	1591.5	1614.5	1637.6	1683.7
(– 1.89)	*s*			5.3803	5.4735	5.5597	5.6405	5.7173	5.7907	5.8613	5.9296	5.9957	6.1228
	v			0.2752	0.2887	0.3017	0.3143	0.3266	0.3387	0.3506	0.3624	0.3740	0.3971
450	*h*			1441.3	1468.1	1493.6	1518.2	1542.2	1565.9	1589.3	1612.6	1635.8	1682.2
(1.26)	*s*			5.3078	5.4042	5492.6	5.5752	5.6532	5.7275	5.7989	5.8678	5.9345	6.0623
	v	0.2698	0.2813	0.2926	0.3036	0.3144	0.3251	0.3357	0.3565	0.3771	0.3975		
500	*h*	1489.9	1515.0	1539.5	1563.4	1587.1	1610.6	1634.0	1680.7	1727.5	1774.7		
(4.14)	*s*	5.4314	5.5157	5.5950	5.6704	5.7425	5.8120	5.8793	6.0079	6.1301	6.2472		
	v	0.2217	0.2317	0.2414	0.2508	0.2600	0.2691	0.2781	0.2957	0.3130	0.3302		
600	*h*	1482.4	1508.6	1533.8	1558.5	1582.7	1606.6	1630.4	1677.7	1724.9	1772.4		

(*Continued*)

Table A.5.2
(Continued)

Abs. Press. (kPa) (Sat. temp., °C)		Temperature (°C) 20	30	40	50	60	70	80	100	120	140	160	180
(9.29)	s	5.3222	5.4102	5.4923	5.5697	5.6436	5.7144	5.7826	5.9129	6.0363	6.1541		
	v	0.1874	0.1963	0.2048	0.2131	0.2212	0.2291	0.2369	0.2522	0.2672	0.2821		
700	h	1474.5	1501.9	1528.1	1553.4	1578.2	1602.6	1626.8	1674.6	1722.4	1770.2		
(13.81)	s	5.2259	5.3179	5.4029	5.4826	5.5582	5.6303	5.6997	5.8316	5.9562	6.0749		
	v	0.1615	0.1696	0.1773	0.1848	0.1920	0.1991	0.2060	0.2196	0.2329	0.2459	0.2589	
800	h	1466.3	1495.0	1522.2	1548.3	1573.7	1598.6	1623.1	1671.6	1719.8	1768.0	1816.4	
(17.86)	s	5.1387	5.2351	5.3232	5.4053	5.4827	5.5562	5.6268	5.7603	5.8861	6.0057	6.1202	
	v		0.1488	0.1559	0.1627	0.1693	0.1757	0.1820	0.1942	0.2061	0.2178	0.2294	
900	h		1488.0	1516.2	1543.0	1569.1	1594.4	1619.4	1668.5	1717.1	1765.7	1814.4	
(21.54)	s		5.1593	5.2508	5.3354	5.4147	5.4897	5.5614	5.6968	5.8237	5.9442	6.0594	
	v		0.1321	0.1388	0.1450	0.1511	0.1570	0.1627	0.1739	0.1847	0.1954	0.2058	0.2162
1000	h		1480.6	1510.0	1537.7	1564.4	1590.3	1615.6	1665.4	1714.5	1763.4	1812.4	1861.7
(24.91)	s		5.0889	5.1840	5.2713	5.3525	5.4292	5.5021	5.6392	5.7674	5.8888	6.0047	6.1159
	v			0.1129	0.1185	0.1238	0.1289	0.1338	0.1434	0.1526	0.1616	0.1705	0.1792
1200	h			1497.1	1526.6	1554.7	1581.7	1608.0	1659.2	1709.2	1758.9	1808.5	1858.2
(30.96)	s			5.0629	5.1560	5.2416	5.3215	5.3970	5.5379	5.6687	5.7919	5.9091	6.0214
	v			0.0944	0.0995	0.1042	0.1088	0.1132	0.1216	0.1297	0.1376	0.1452	0.1528
1400	h			1483.4	1515.1	1544.7	1573.0	1600.2	1652.8	1703.9	1754.3	1804.5	1854.7
(36.28)	s			4.9534	5.0530	5.1434	5.2270	5.3053	5.4501	5.5836	5.7087	5.8273	5.9406
	v				0.0851	0.0895	0.0937	0.0977	0.1053	0.1125	0.1195	0.1263	0.1330
1600	h				1502.9	1534.4	1564.0	1592.3	1646.4	1698.5	1749.7	1800.5	1851.2
(14.05)	s				4.9584	5.0543	5.1419	5.2232	5.3722	5.5084	5.6355	5.7555	5.8699

Table A.6
Variation of $\overline{c}_p$ with Temperature for Various Ideal Gases

$$\frac{\overline{c}_p}{R} = \alpha + \beta T + \gamma T2 + \delta T3 + \in T4$$

T is in K, equations valid from 300 to 1000 K

Gas	α	$\beta \times 10w^3$	$\gamma \times 10^6$	$\delta \times 10^9$	$\varepsilon \times 10^{12}$
CO	3.710	– 1.619	3.692	– 2.032	0.240
CO_2	2.401	8.735	– 6.607	2.002	0
H_2	3.057	2.677	– 5.180	5.521	– 1.812
H_2O	4.070	– 1.108	4.152	– 2.964	0.807
O_2	3.626	– 1.878	7.056	– 6.764	2.156
N_2	3.675	– 1.208	2.324	– 0.632	– 0.226
Air	3.653	– 1.337	3.294	– 1.913	0.2763
SO_2	3.267	5.324	0.684	– 5.281	2.559
CH_4	3.826	– 3.979	24.558	– 22.733	6.963
C_2H_2	1.410	19.057	24.501	16.391	– 4.135
C_2H_4	1.426	11.383	7.989	– 16.254	6.749
Monatomic gases*	2.5	0	0	0	0

* For monatomic gases, such as He, Ne, and Ar, $\overline{c}_p$ is constant over a wide temperature range and is very nearly equal to 5/2 $\overline{R}$.

Source: Adapted from K. Wark, *Thermodynamics, 4th ed. McGraw-Hill*, New York, 1983, as based on NASA SP-273, U.S. Government Printing Office, Washington, DC, 1971.

Table A.7
Ideal Gas Specific Heats of Some Common Gases (kJ/kg.K)

Temp. K	c_p (Air)	c_v (Air)	k (Air)	c_p (Nitrogen, N_2)	c_v (Nitrogen, N_2)	k (Nitrogen, N_2)	c_p (Oxygen, O_2)	c_v (Oxygen, O_2)	k (Oxygen, O_2)	Temp. K
250	1.003	0.716	1.401	1.039	0.742	1.400	0.913	0.653	1.398	250
300	1.005	0.718	1.400	1.039	0.743	1.400	0.918	0.658	1.395	300
350	1.008	0.721	1.398	1.041	0.744	1.399	0.928	0.668	1.389	350
400	1.013	0.726	1.395	1.044	0.747	1.397	0.941	0.681	1.382	400
450	1.020	0.733	1.391	1.049	0.752	1.395	0.956	0.696	1.373	450
500	1.029	0.742	1.387	1.056	0.759	1.391	0.972	0.712	1.365	500

(*Continued*)

Table A.7
(Continued)

Temp. K	c_p	c_v	k	c_p	c_v	k	c_p	c_v	k	Temp. K
	Air			Nitrogen, N_2			Oxygen, O_2			
550	1.040	0.753	1.381	1.065	0.768	1.387	0.988	0.728	1.358	550
600	1.051	0.764	1.376	1.075	0.778	1.382	1.003	0.743	1.350	600
650	1.063	0.776	1.370	1.086	0.789	1.376	1.017	0.758	1.343	650
700	1.075	0.788	1.364	1.098	0.801	1.371	1.031	0.771	1.337	700
750	1.087	0.800	1.359	1.110	0.813	1.365	1.043	0.783	1.332	750
800	1.099	0.812	1.354	1.121	0.825	1.360	1.054	0.794	1.327	800
900	1.121	0.834	1.344	1.145	0.849	1.349	1.074	0.814	1.319	900
1000	1.142	0.855	1.336	1.167	0.870	1.341	1.090	0.830	1.313	1000
Temp. K		Carbon Dioxide, CO_2			Carbon Monoxide, CO			Hydrogen, H_2		Temp. K
250	0.791	0.602	1.314	1.039	0.743	1.400	14.051	9.927	1.416	250
300	0.846	0.657	1.288	1.040	0.744	1.399	14.307	10.183	1.045	300
350	0.895	0.706	1.268	1.043	0.746	1.398	14.427	10.302	1.400	350
400	0.939	0.750	1.252	1.047	0.751	1.395	14.476	10.352	1.398	400
450	0.978	0.790	1.239	1.054	0.757	1.392	14.501	10.377	1.398	450
500	1.014	0.825	1.229	1.063	0.767	1.387	14.513	10.389	1.397	500
550	1.046	0.857	1.220	1.075	0.778	1.382	14.530	10.405	1.396	550
600	1.075	0.886	1.213	1.087	0.790	1.376	14.546	10.422	1.396	600
650	1.102	0.913	1.207	1.100	0.803	1.370	14.571	10.447	1.395	650
700	1.126	0.937	1.202	1.113	0.816	1.364	14.604	10.480	1.394	700
750	1.148	0.959	1.197	1.126	0.829	1.358	14.645	10.521	1.392	750
800	1.169	0.980	1.193	1.139	0.842	1.353	14.695	10.570	1.390	800
900	1.204	1.015	1.186	1.163	0.866	1.343	14.822	10.698	1.385	900
1000	1.234	1.045	1.181	1.185	0.888	1.335	14.983	10.859	1.380	1000

Source: Adapted from K. Wark, *Thermodynamics, 4th ed.*, McGraw-Hill, New York, 1983, as based on "Tables of Thermal Properties of Gases," NBS Circular 564, 1955.

TABLE A.8 LOGARITHM TO THE BASE 10 OF THE EQUILIBRIUM CONSTANT K

Table A.8

*Enthalpy of formation at 25°C, ideal gas enthalpy, and Absolute Entropy at 0.1 MPa (1 bar) Pressure**

					log 10 K				
Temp. K	$H_2 \rightleftharpoons 2H$	$O_2 \rightleftharpoons 2O$	$N_2 \rightleftharpoons 2N$	$\frac{1}{2} O_2 + \frac{1}{2} N_2 \rightleftharpoons NO$	$H_2O \rightleftharpoons H_2 + \frac{1}{2} O_2$	$H_2O \rightleftharpoons OH + \frac{1}{2} H_2$	$CO_2 \rightleftharpoons CO + \frac{1}{2} O_2$	$CO_2 + H_2 \rightleftharpoons CO + H_2O$	Temp. °R
298	– 71.224	– 81.208	– 159.600	– 15.171	– 40.048	– 46.054	– 45.066	– 5.018	537
500	– 40.316	– 45.880	– 92.672	– 8.783	– 22.886	– 26.130	– 25.025	– 2.139	900
1000	– 17.292	– 19.614	– 43.056	– 4.062	– 10.062	– 11.280	– 10.221	– 0.159	1800
1200	– 13.414	– 15.208	– 34.754	– 3.275	– 7.899	– 8.811	– 7.764	+ 0.135	2160
1400	– 10.630	– 12.054	– 28.812	– 2.712	– 6.347	– 7.021	– 6.014	+ 0.333	2520
1600	– 8.532	– 9.684	– 24.350	– 2.290	– 5.180	– 5.677	– 4.706	+ 0.474	2880
1700	– 7.666	– 8.706	– 22.512	– 2.116	– 4.699	– 5.124	– 4.169	+ 0.530	3060
1800	– 6.896	– 7.836	– 20.874	– 1.962	– 4.270	– 4.613	– 3.693	+ 0.577	3240
1900	– 6.204	– 7.058	– 19.410	– 1.823	– 3.886	– 4.190	– 3.267	+ 0.619	3420
2000	– 5.580	– 6.356	– 18.092	– 1.699	– 3.540	– 3.776	– 2.884	+ 0.656	3600
2100	– 5.016	– 5.720	– 16.898	– 1.586	– 3.227	– 3.434	– 2.539	+ 0.688	3780

(Continued)

Table A.8
(Continued)

Temp. K	log 10 K: $H_2 \rightleftharpoons 2H$	$O_2 \rightleftharpoons 2O$	$N_2 \rightleftharpoons 2N$	$\frac{1}{2} O_2 + \frac{1}{2} N_2 \rightleftharpoons NO$	$H_2O \rightleftharpoons H_2 + \frac{1}{2} O_2$	$H_2O \rightleftharpoons OH + \frac{1}{2} H_2$	$CO_2 \rightleftharpoons CO + \frac{1}{2} O_2$	$CO_2 + H_2 \rightleftharpoons CO + H_2O$	Temp. °R
2200	−4.502	−5.142	−15.810	−1.484	−2.942	−3.091	−2.226	+0.716	3960
2300	−4.032	−4.614	−14.818	−1.391	−2.682	−2.809	−1.940	+0.742	4140
2400	−3.600	−4.130	−13.908	−1.305	−2.443	−2.520	−1.679	+0.764	4320
2500	−3.202	−3.684	−13.070	−1.227	−2.224	−2.270	−1.440	+0.784	4500
2600	−2.836	−3.272	−12.298	−1.154	−2.021	−2.038	−1.219	+0.802	4680
2700	−2.494	−2.892	−11.580	−1.087	−1.833	−1.823	−1.015	+0.818	4860
2800	−2.178	−2.536	−10.914	−1.025	−1.658	−1.624	−0.825	+0.833	5040
2900	−1.882	−2.206	−10.294	−0.967	−1.495	−1.438	−0.649	+0.846	5220
3000	−1.606	−1.898	−9.716	−0.913	−1.343	−1.265	−0.485	+0.858	5400
3100	−1.348	−1.610	−9.174	−0.863	−1.201	−1.103	−0.322	+0.869	5580
3200	−1.106	−1.340	−8.664	−0.815	−1.067	−0.951	−0.189	+0.878	5760
3300	−0.878	−1.086	−8.186	−0.771	−0.942	−0.809	−0.054	+0.888	5940
3400	−0.664	−0.846	−7.736	−0.729	−0.824	−0.674	+0.071	+0.895	6120
3500	−0.462	−0.620	−7.312	−0.690	−0.712	−0.547	+0.190	+0.902	6300

Source: Based on data from the JANAF Thermochemical Tables, NSRDS–NBS–37, 1971.

Table A.9a

*Enthalpy of formation at 25°C, ideal gas enthalpy, and Absolute Entropy at 0.1 MPa (1 bar) Pressure**

Temp.	Nitrogen, Diatomic (N_2) $(h^\circ_f)_{298} = 0$ kJ/kmol $M = 28.013$		Oxygen, Diatomic (O_2) $(h^\circ_f)_{298} = 0$ kJ/kmol $M = 31.999$	
	$(\bar{h}^o - \bar{h}^o_{298})$	s°	$(\bar{h}^o - \bar{h}^o_{298})$	$\bar{s}^o$
K	kJ/kmol	kJ/kmol K	kJ/kmol K	kJ/kmol K
0	– 8669	0	– 8682	0
100	– 5770	159.813	– 5778	173.306
200	– 2858	179.988	– 2866	193.486
298	0	191.611	0	205.142
300	54	191.791	54	205.322
400	2971	200.180	3029	213.874
500	5912	206.740	6088	220.698
600	8891	212.175	9247	226.455
700	11937	216.866	12502	231.272
800	15046	221.016	15841	235.924
900	18221	224.757	19246	239.936
1000	21406	288.167	22707	243.585
1100	24757	231.309	26217	246.928
1200	28108	234.225	29765	250.016
1300	31501	236.941	33351	252.886
1400	34936	239.484	36966	255.564
1500	38405	241.878	40610	258.078
1600	41903	244.137	44279	260.446
1700	45430	246.275	47970	262.685
1800	48982	248.304	51689	264.810
1900	52551	250.237	55434	266.835
2000	56141	252.078	59199	268.764

* Adapted from Tables A. 11 (Pages 687 to 696) in *Fundamentals of Classical Thermodynamics* by G.J. Van Wylen and R. Sonntag, John Wiley, New York, 1976 (with the kind permission of the publishers, John Wiley & Sons, Inc., New York).

Table A.9b
(Continued)

Temp.	Carbon Dioxide (CO_2) $(h^\circ_f)_{288} = -393522$ kJ/kmol, M = 44.01		Carbon Monoxide (CO) $(h^\circ_f)_{298} = -110529$ kJ/kmol, M = 28.01	
K	$(\bar{h}^o - \bar{h}^o_{298})$ kJ/kmol	$\bar{s}^o$ kJ/kmol K	$(\bar{h}^o - \bar{h}^o_{298})$ kJ/kmol	$\bar{s}^o$ kJ/kmol K
0	– 9364	0	– 8669	0
100	– 6456	179.109	– 5770	165.850
200	– 3414	199.975	– 2858	186.025
298	0	213.795	0	197.653
300	67	214.025	54	197.833
400	4008	225.334	2975	206.234
500	8314	234.924	5929	212.828
600	12916	243.309	8941	218.313
700	17765	250.773	12021	223.062
800	22815	257.517	15175	227.271
900	28041	263.668	18397	231.006
1000	33405	269.325	21686	234.531
1100	38894	274.555	25033	237.719
1200	44484	279.417	28426	240.673
1300	50158	283.956	31865	243.426
1400	55907	288.216	35338	245.999
1500	61714	292.224	38848	248.421
1600	67580	296.010	42384	250.702
1700	73492	299.592	45940	252.861
1800	79442	302.993	49522	254.907
1900	85429	306.232	53124	256.852
2000	91450	309.320	56739	258.710

Table A.9c
(Continued)

Temp. K	Water (H_2O) $(\bar{h}_f^o)_{298} = -241827$ kJ/kmol, $M = 18.015$: $(\bar{h}^o - \bar{h}^o_{298})$ kJ/kmol	$\bar{s}^o$ kJ/kmol K	Hydrogen, Diatomic (H_2) $(\bar{h}_f^o)_{298} = 0$ kJ/kmol, $M = 2.016$: $(\bar{h}^o - \bar{h}^o_{298})$ kJ/kmol	$\bar{s}^o$ kJ/kmol K
0	– 9904	0	– 8468	0
100	– 6615	152.390	– 5293	102.145
200	– 3280	175.486	– 2770	119.437
298	0	188.833	0	130.684
300	63	189.038	54	130.864
400	3452	198.783	2958	139.215
500	6920	206.523	5883	145.738
600	10498	213.037	8812	151.077
700	14184	218.719	11749	155.608
800	17991	223.803	14703	159.549
900	21924	228.430	17682	163.060
1000	25978	232.706	20686	166.223
1100	30167	236.694	23723	169.118
1200	34476	240.443	26794	171.792
1300	38903	243.986	29907	174.281
1400	43447	247.350	33062	176.620
1500	48095	250.560	36267	178.833
1600	52844	253.622	39522	180.929
1700	57685	256.559	42815	182.929
1800	62609	259.372	46150	184.833
1900	67613	262.078	49522	186.657
2000	72689	264.681	52932	188.406

TABLE A.10 GAS TABLES

Table A.10.1
Sentropic Flow

M	M^*	$\frac{A}{A^*}$	$\frac{p}{p_0}$	$\frac{\rho}{\rho_0}$	$\frac{T}{T_0}$
0	0	∞	1.00000	1.00000	1.00000
0.10	0.10943	5.8218	0.99303	0.99502	0.99800
0.20	0.21822	2.9635	0.97250	0.98027	0.99206
0.30	0.32572	2.0351	0.93947	0.95638	098232
0.40	0.43133	1.5901	0.89562	0.92428	0.96899
0.50	0.53452	1.3398	0.84302	0.88517	0.95238
0.60	0.63480	1.1882	0.78400	0.84045	0.93284
0.70	0.73179	1.09437	0.72092	0.79158	0.91075
0.80	0.82514	1.03823	0.65602	0.74000	0.88652
0.90	0.91460	1.00886	0.59126	0.68704	0.86058
1.00	1.00000	1.00000	0.52828	0.63394	0.83333
1.10	1.08124	1.00793	0.46835	0.58169	0.80515
1.20	1.1583	1.03044	0.41238	0.53114	0.77640
1.30	1.2311	1.06631	0.36092	0.48291	0.74738
1.40	1.2999	1.1149	0.31424	0.43742	0.71839
1.50	1.3646	1.1762	0.27240	0.39498	0.68965
1.60	1.4254	1.2502	0.23527	0.35573	0.66138
1.70	1.4825	1.3376	0.20259	0.31969	0.63372
1.80	1.5360	1.4390	0.17404	0.28682	0.60680
1.90	1.5861	1.5552	0.14924	0.25699	0.58072
2.00	1.6330	1.6875	0.12780	0.23005	0.55556
2.10	1.6769	1.8369	0.10935	0.20580	0.53135
2.20	1.7179	2.0050	0.09352	0.18405	0.50813
2.30	1.7563	2.1931	0.07997	0.16458	0.48591
2.40	1.7922	2.4031	0.06840	0.14720	0.46468
2.50	1.8258	2.6367	0.05853	0.13169	0.44444
2.60	1.8572	2.8960	0.05012	0.11787	0.42517
2.70	1.8865	3.1830	0.04295	0.10557	0.40684
2.80	1.9140	3.5001	0.03685	0.09462	0.38941
2.90	1.9398	3.8498	0.03165	0.08489	0.37286

(Continued)

Table A.10.1
(Continued)

M	M^*	$\frac{A}{A^*}$	$\frac{p}{p_0}$	$\frac{\rho}{\rho_0}$	$\frac{T}{T_0}$
3.00	1.9640	4.2346	0.02722	0.07623	0.35714
3.50	2.0642	6.7896	0.01311	0.04523	0.28986
4.00	2.1381	10.719	0.00658	0.02766	0.23810
4.50	2.1936	16.562	0.00346	0.01745	0.19802
5.00	2.2361	25.000	$189(10)^{-5}$	0.01134	0.16667
6.00	2.2953	53.180	$633(10)^{-6}$	0.00519	0.12195
7.00	2.3333	104.143	$242(10)^{-6}$	0.00261	0.09259
9.00	2.3772	327.189	$474(10)^{-7}$	0.000815	0.05814
10.00	2.3904	535.938	$236(10)^{-7}$	0.000495	0.04762
∞	2.4495	∞	0	0	0

Table A.10.2
Normal Shocks

M_x	M_y	$\frac{p_y}{p_x}$	$\frac{\rho_y}{\rho_x}$	$\frac{T_y}{T_x}$	$\frac{p_{oy}}{p_{ox}}$	$\frac{p_{oy}}{p_x}$
1.00	1.00000	1.00000	1.0000	1.0000	1.00000	1.8929
1.10	0.91177	1.2450	1.1691	1.0649	0.99892	2.1328
1.20	0.84217	1.5133	1.3416	1.1280	0.99280	2.4075
1.30	0.78596	1.8050	1.5157	1.1909	0.97935	2.7135
1.40	0.73971	2.1200	1.6896	1.2547	0.95819	3.0493
1.50	0.70109	2.4583	1.8621	1.3202	0.92978	3.4133
1.60	0.66844	2.8201	2.0317	1.3880	0.89520	3.8049
1.70	0.64055	3.2050	2.1977	0.4583	0.85573	4.2238
1.80	0.61650	3.6133	2.3592	1.5316	0.81268	4.6695
1.90	0.58562	4.0450	2.5157	1.6079	0.76635	5.1417
2.00	0.57735	4.5000	2.6666	1.6875	0.72088	5.6405
2.10	0.56128	4.9784	2.8119	1.7704	0.67422	6.1655
2.20	0.54706	5.4800	2.9512	1.8569	0.62812	6.7163
2.30	0.53441	6.0050	3.0846	1.9968	0.58331	7.2937
2.40	0.52312	6.5533	3.2119	2.0403	0.54015	7.8969
2.50	0.51299	7.1250	3.3333	2.1375	0.49902	8.5262

(Continued)

Table A.10.2
(Continued)

M_x	M_y	$\frac{p_y}{p_x}$	$\frac{\rho_y}{\rho_x}$	$\frac{T_y}{T_x}$	$\frac{p_{oy}}{p_{ox}}$	$\frac{p_{oy}}{p_x}$
2.60	0.50387	7.7200	3.4489	2.2383	0.46012	9.1813
2.70	0.49563	8.3383	3.5590	2.3429	0.42359	9.8625
2.80	0.48817	8.9800	3.6635	2.4512	0.38946	10.569
2.90	0.48138	9.6450	3.7629	2.5632	0.35773	11.302
3.00	0.47519	10.333	3.8571	2.6790	0.32834	12.061
4.00	0.43496	18.500	4.5714	4.0469	0.13876	21.068
5.00	0.41523	29.000	5.0000	5.8000	0.06172	32.654
10.00	0.38757	116.50	5.7143	20.388	0.00304	129.217
∞	0.37796	∞	6.000	∞	0	∞

Table A.11
*Critical point data**

Substance	*Formula*	*Molecular Weight*	*Temperature K*	*Pressure MPa*	*Volume m³/kmol*
Ammonia	NH_3	17.03	405.5	11.28	.0724
Argon	Ar	30.948	151	4.86	.0749
Bromine	Br_2	159.808	584	10.34	.1355
Carbon Dioxide	CO_3	44.01	304.2	7.39	.0943
Carbon Monoxide	CO	28.001	133	3.50	.0930
Chlorine	CL_2	70.906	417	7.71	.1242
Deuterium (Normal)	D_2	4.00	38.4	1.66	—
Helium	He	4.003	5.3	0.23	.0578
Helium[3]	He	3.00	3.3	0.12	—
Hydrogen (Normal)	H_2	2.106	33.3	1.30	.0649
Krypton	Kr	83.80	209.4	5.50	.0924
Neon	Ne	20.183	44.5	2.73	.0417
Nitrogen	N_2	28.013	126.2	3.39	.0899
Nitrous Oxide	N_2O	44.013	309.7	7.27	.0961

(Continued)

Table A.11
(Continued)

Substance	*Formula*	*Molecular Weight*	*Temperature K*	*Pressure MPa*	*Volume $m^3/kmol$*
Oxygen	O_2	31.999	154.8	5.08	.0780
Sulfur Dioxide	SO_2	64.063	430.7	7.88	.1217
Water	H_2O	18.015	647.3	22.09	.0568
Xenon	Xe	131.30	289.8	5.88	.1186
Benzene	X_2H_6	78.115	562	4.92	.2603
n-Butane	C_4H_{10}	58.124	425.5	3.80	.2547
Carbon Tetra-chloride	CCL_4	153.82	556.4	4.56	.2759
Chloroform	$CHCL_3$	119.38	536.6	5.47	.2403
Dhichlorodi-fluoromethane	CCL_2F_2	120.91	384.7	4.01	.2179
Dichloro-fluoromethane	$CHCL_2F$	102.92	451.7	5.17	.1973
Ethane	C_2H_6	30.070	305.5	4.88	.1480
Ethyl Alcohol	C_2H_5OH	46.07	516	6.38	.1673
Ethylene	C_2H_{14}	28.054	282.4	5.12	.1242
n-Hexane	C_6H_4	86.178	507.9	3.03	.3677
Methane	CH_4	16.043	191.1	4.64	0.993
Methyl Alcohol	CH_3OH	32.042	513.2	7.95	.1180
Methyl Chloride	CH_3Cl	50.488	416.3	6.68	.1430
Propane	C_3H_8	44.097	370	4.26	.1998
Propene	C_3H_6	42.081	365	4.62	.1810
Propyne	C_3H_4	40.065	401	5.35	
Trichloro-fluoromethane	CCL_3F	137.37	471.2	4.38	.2478

* K.A. Kobe and R.E. Lynn, Jr., *Chem. Rev.*, 52: 117–236 (1953).

Appendix B

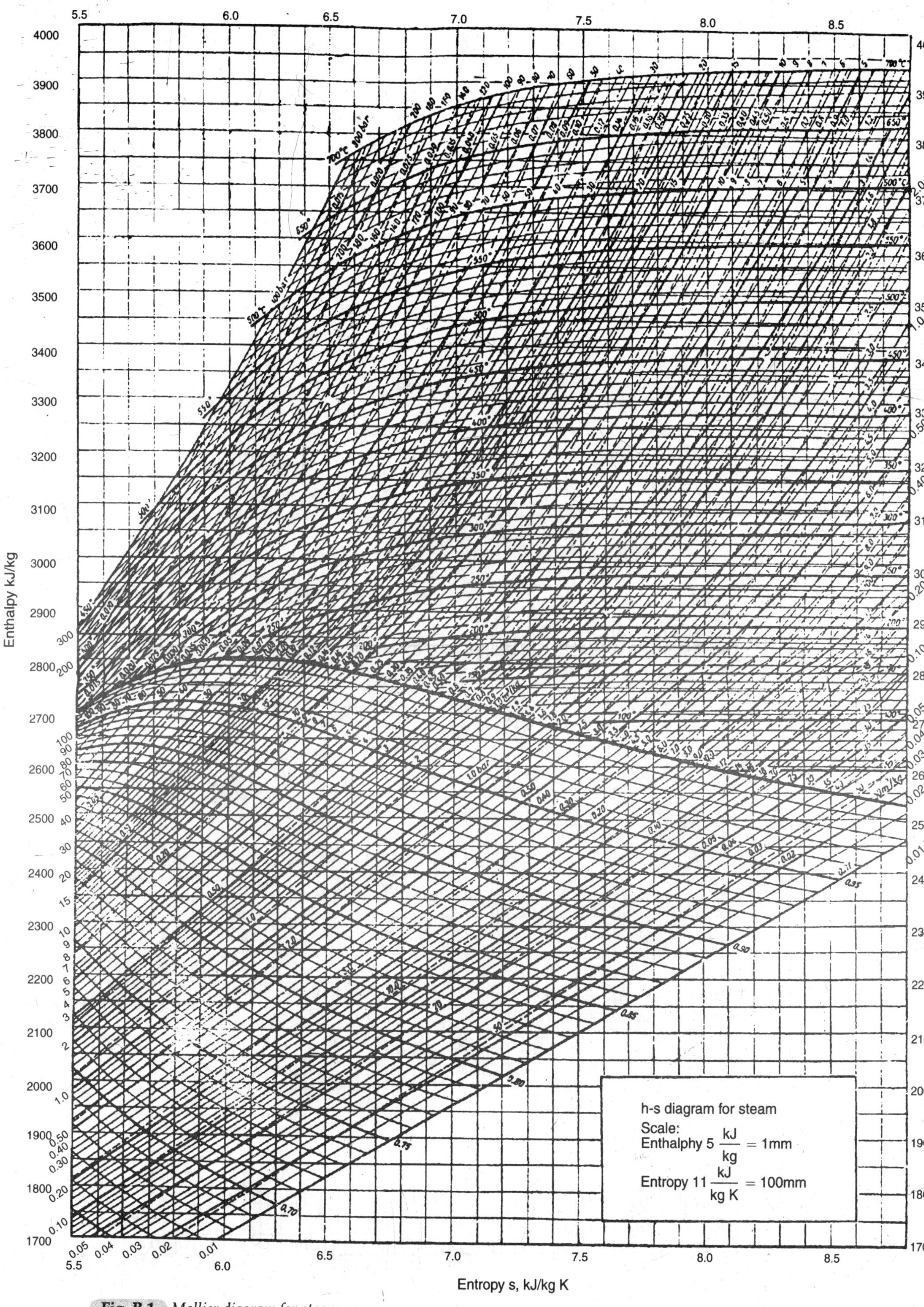

Fig. B.1 *Mollier diagram for steam*

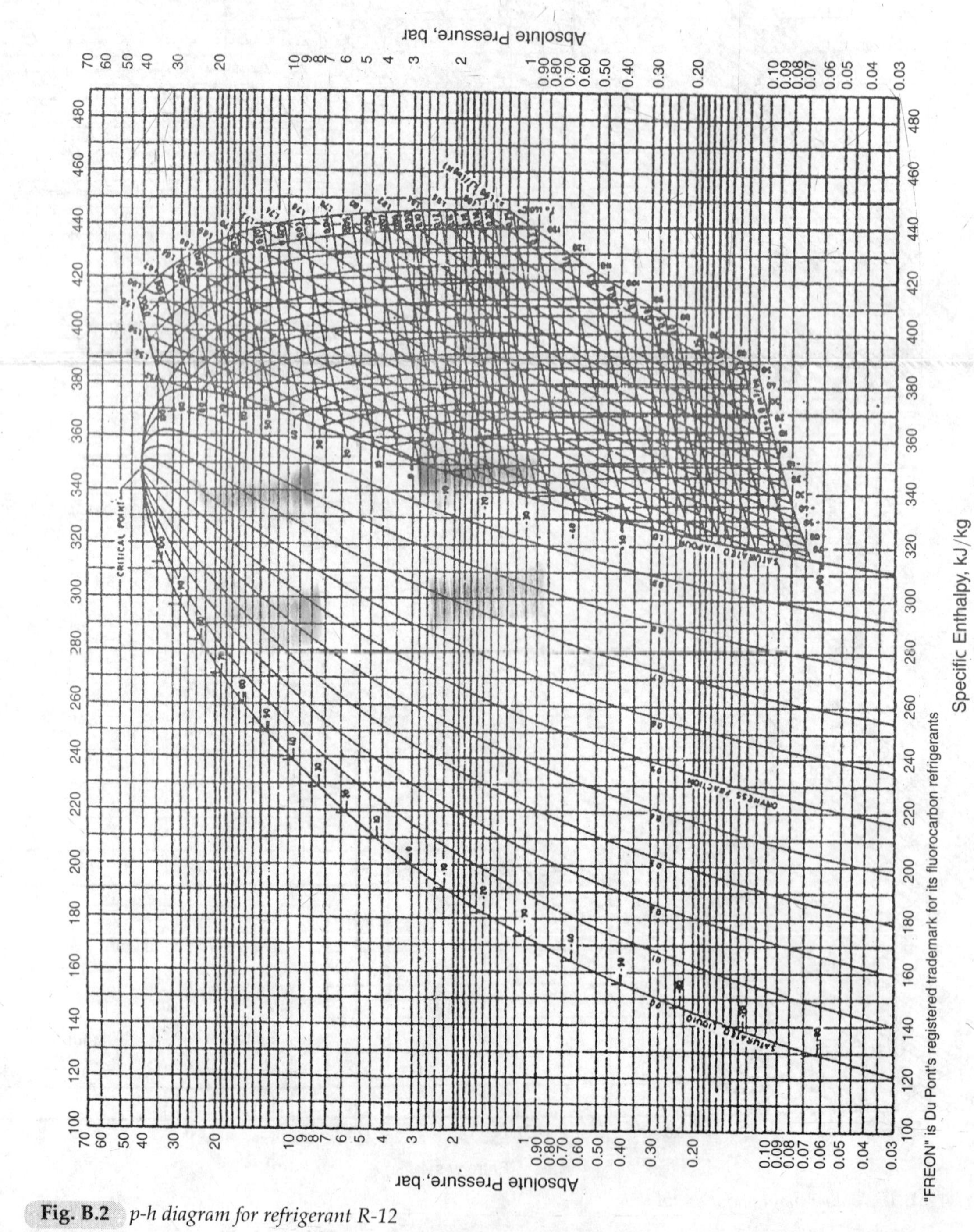

Fig. B.2 *p-h diagram for refrigerant R-12*

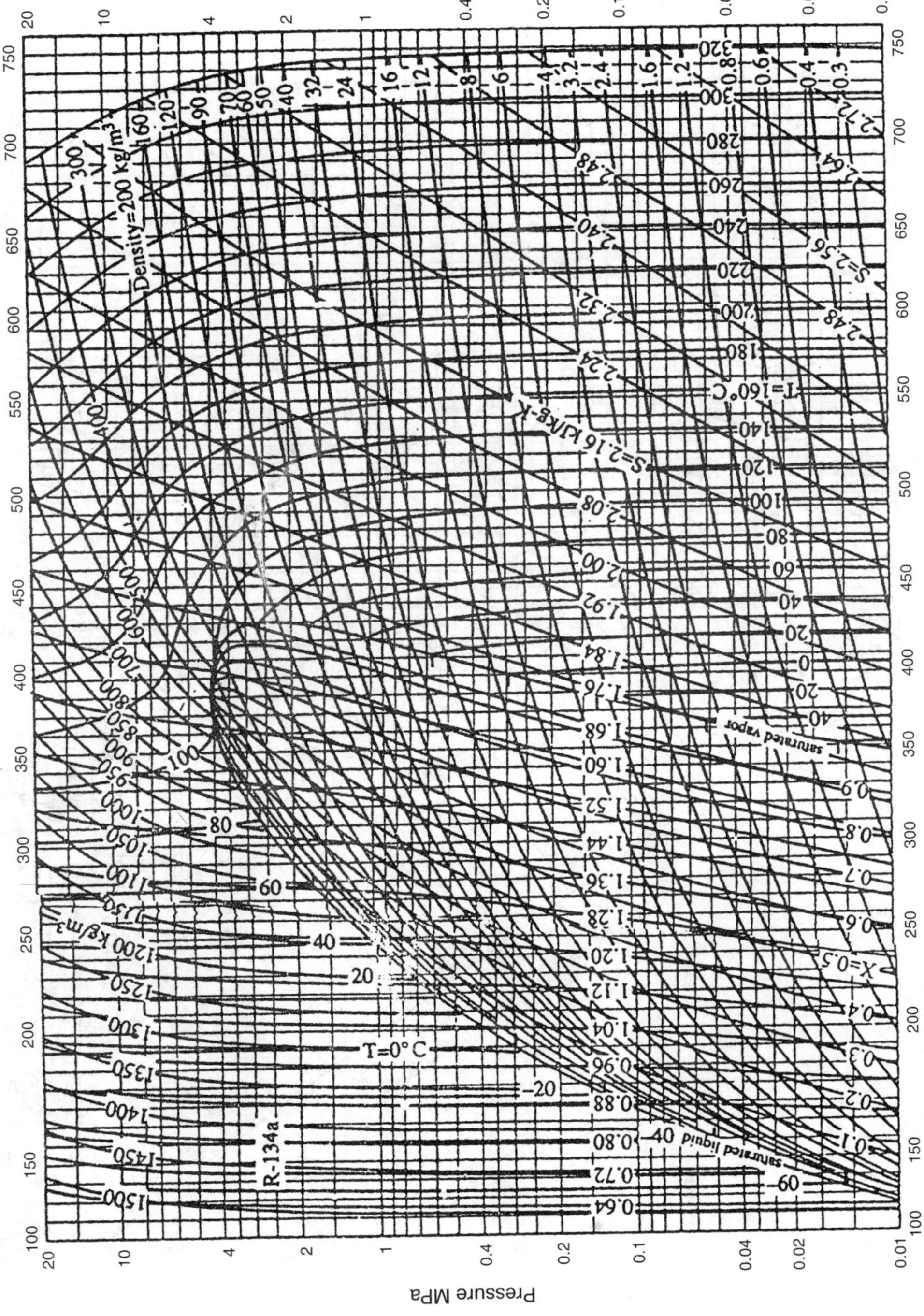

Fig. B.3 *p-h diagram for refrigerant R-134a*

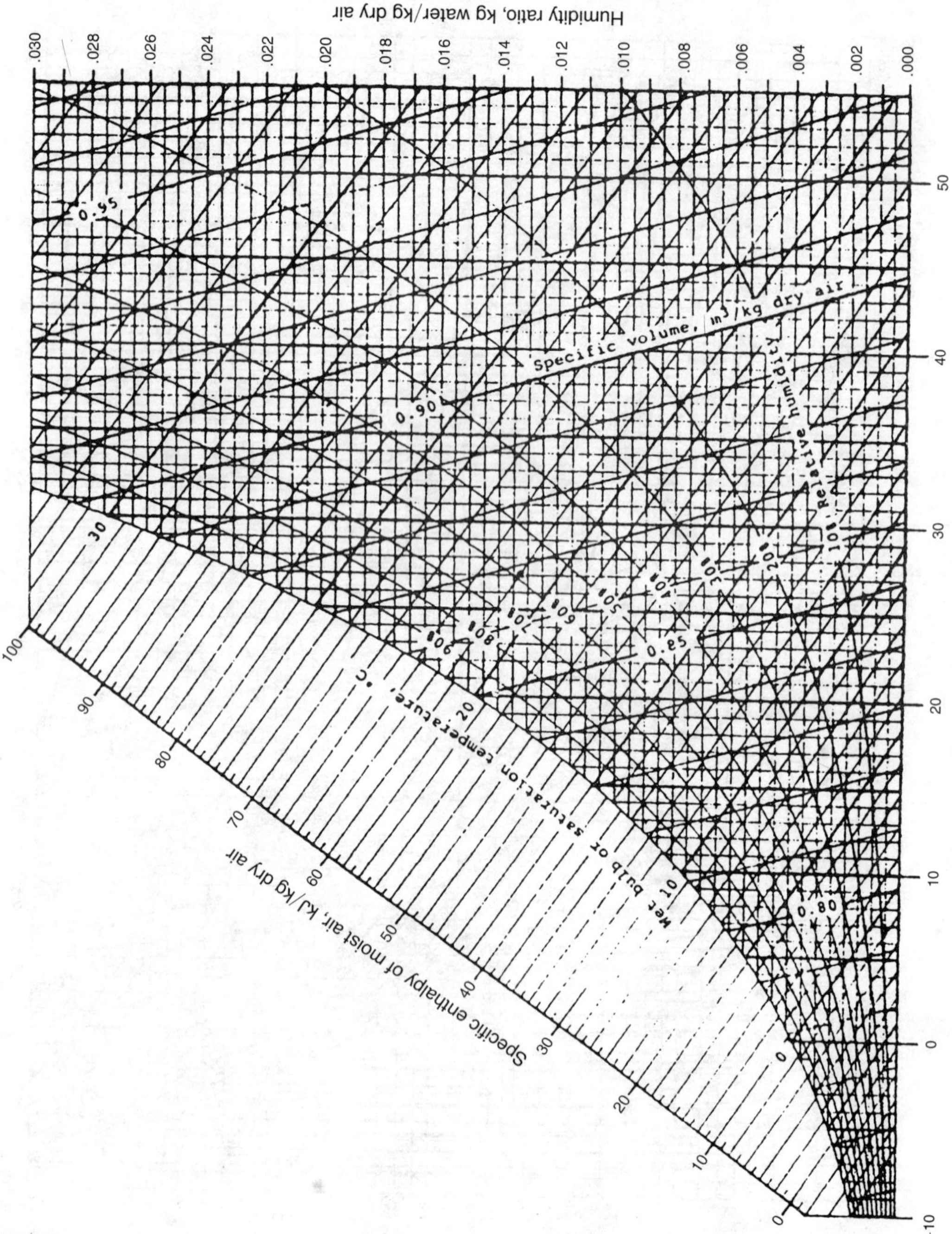

Fig. B.4 *Psychrometric chart for 1 atm (SI units)*
Source: Zhang, Z, and Pate, M.B., "A Methodology for Implementing a Psychrometric Chart in a Computer Graphics System," ASHRAE Transactions, Vol. 94,Pt. 1,1988

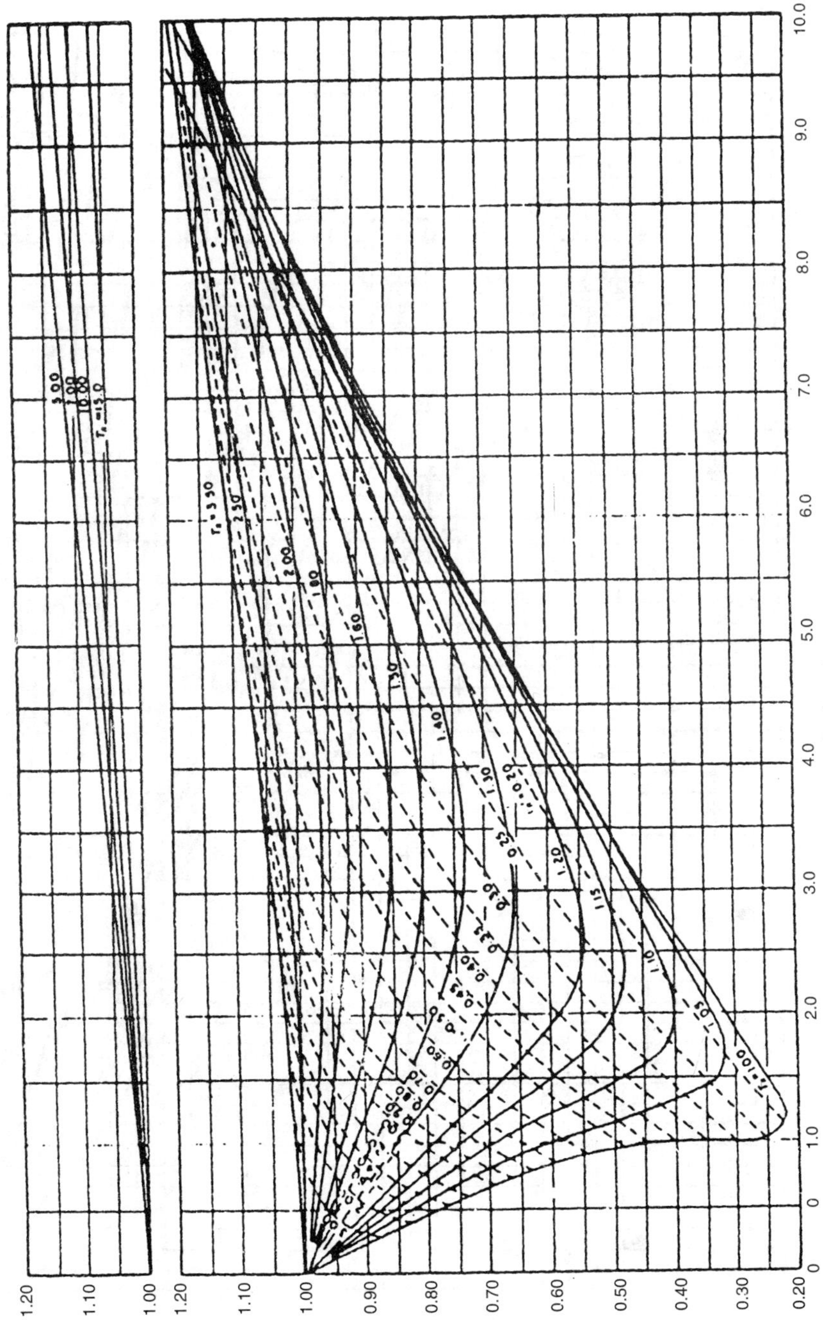

Fig. B.5 *Generalized compressibility chart, $p_R \le 10.0$*
Source: E.F. Obert, "Concepts of Thermodynamics, "McGraw-Hill, New York, 1960

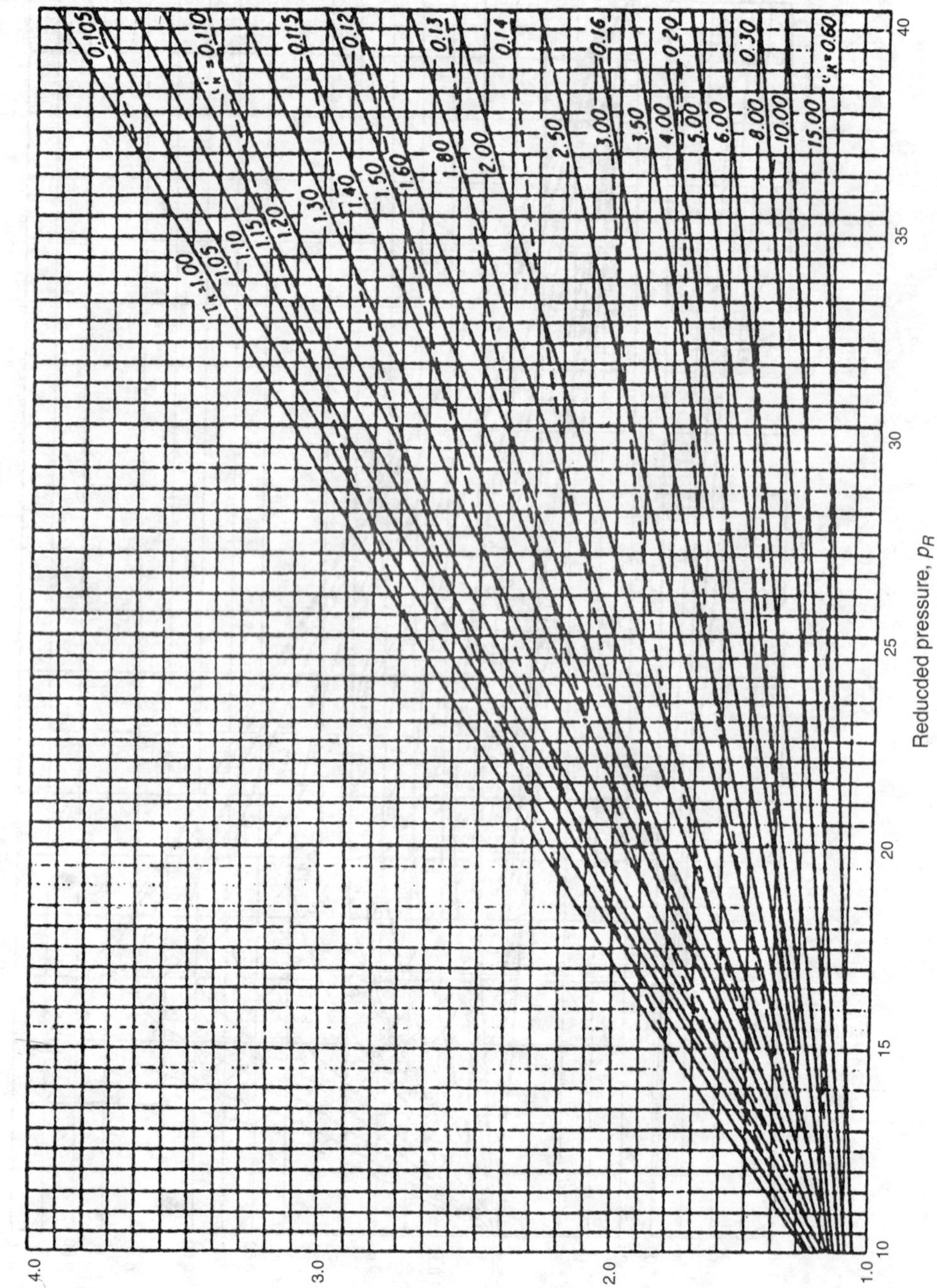

Fig. B.6 *Generalized compressibility chart,* $10 \le p_R \le 40$
Source: E.F. Obert, "Concepts of Thermodynamics, "McGraw-Hill, New York, 1960

Multiple-Choice Questions

1. Lead compounds were earlier used to be added to gasoline in S.I. engines so as to
   (a) reduce hydrocarbon emissions (b) reduce knocking
   (c) reduce exhaust gas temperature (d) increase power output

2. Decrease of air–fuel ratio in SI engines results in
   (a) an increase NO_x (b) a decrease of CO and unburnt hydrocarbon
   (c) an increase of CO and unburnt hydrocarbon (d) none of the above

3. If N is the rpm, the number of power strokes per minute in a four-stroke engine is
   (a) $2N$ (b) $N/2$ (c) N (d) $4N$

4. Volumetric efficiency is a measure of
   (a) speed of the engine (b) power of the engine
   (c) breathing capacity of the engine (d) pressure rise in the cylinder

5. If N is the rpm, the number of power strokes per minute in a two-stroke engine is
   (a) N (b) $N/2$ (c) $2N$ (d) $4N$

6. The volumetric efficiency of a well-designed engine is in the range of
   (a) 30 to 40% (b) 40 to 60% (c) 60 to 70% (d) 75 to 90%

7. If L is the stroke and N is the rpm, the mean piston speed of a two-stroke engine is
   (a) LN (b) $LN/2$ (c) $2LN$ (d) none of the above

8. Equivalence ratio is
   (a) $\dfrac{\text{actual fuel–air ratio}}{\text{stoichiometric fuel–air ratio}}$ (b) $\dfrac{\text{stoichiometric fuel–air ratio}}{\text{actual fuel–air ratio}}$
   (c) $\dfrac{\text{stoichiometric fuel–air ratio}}{\text{actual air–fuel ratio}}$ (d) $\dfrac{\text{stoichiometric air–fuel ratio}}{\text{actual air–fuel ratio}}$

9. The compression ratio in diesel engines is of the order of
   (a) 5–7 (b) 7–10 (c) 10–12 (d) 14–20

10. The compression ratio of an SI engine is in the range of
    (a) 4 to 6 (b) 6 to 8 (c) 6 to 10 (d) 10 to 14

11. The octane number of iso-octane is
    (a) 0 (b) 30 (c) 60 (d) 100

12. The ignition quality of diesel fuel is indicated by its
(a) octane number (b) cetane number (c) flash point (d) fire point

13. A simple carburetor supplies rich mixture during
(a) starting (b) idling (c) cruising (d) accelerating

14. Modern carburetors provide the correct quality of air–fuel mixture during
(a) starting (b) idling (c) cruising (d) all conditions

15. Fuel is injected in a four-stroke CI engine
(a) at the end of the suction stroke (b) at the end of the expansion stroke
(c) at the end of the compression stoke (d) at the end of the exhaust stroke

16. The advantage of fuel injection in SI engine is
(a) low initial cost (b) low maintenance requirement
(c) increased volumetric efficiency (d) none of the above

17. For engines operating with rich mixtures, the optimum spark timing
(a) must be advanced (b) must be retarded (c) must be at TDC (d) none of the above

18. By decreasing the cooling water temperature in SI engines, the knocking tendency
(a) increases (b) decreases (c) is not affected (d) none of the above

19. In CI engines knocking tendency increases with
(a) increase in compression ratio (b) increase in inlet temperature of air
(c) decrease in compression ratio (d) increasing coolant water temperature

20. In turbocharging, the supercharger is driven by
(a) a gas turbine using the exhaust gases (b) the engine itself
(c) a separate electric motor (d) none of the above

21. For a multi-stage compressor, the polytropic efficiency is
(a) the efficiency of all stages combined together (b) the isentropic efficiency of one stage
(c) constant throughout for all stages (d) a direct consequence of pressure ratio

22. When the flow of air through the compressor is parallel to the axis of the compressor, it is a
(a) rotary compressor (b) reciprocating compressor
(c) centrifugal compressor (d) axial flow compressor

23. The stip in a centrifugal compressor will
(a) increase with increasing number of vanes (b) decrease with increasing number of vanes
(c) remain the same with increasing number of vanes (d) have none of the above

24. **Assertion (A):** The polytropic efficiency of a compressor decreases with increasing pressure ratio.
Reasoning (R): Because more energy is required to compress air at higher pressures and temperatures.
(a) Both A and R are true, but R is not the correct answer of A
(b) both A and R are true and R is the correct answer of A
(c) A is true, R is false
(d) A is false, R is true

25. The maximum pressure ratio in an actual sing-stage axial flow compressor will be approximately
(a) 1:15 (b) 1:10 (c) 1:4 (d) 1: 1.2

26. The degree of reaction in a turbo compressor is defined as
(a) enthalpy drop in rotor/enthalpy drop in stage (b) enthalpy increase in the stage/enthalpy drop in rotor
(c) enthalpy drop in stage/enthalpy drop in rotor (d) enthalpy rise in rotor/enthalpy increase in stage

27. The volumetric efficiency of a reciprocating compressor having a given pressure ratio decreases if the index compression
(a) increases (b) decreases (c) remains constant (d) is equal to 1

28. For the same compression ratio, the efficiency of the Brayton cycle is
(a) equal to the Diesel cycle (b) equal to the Otto cycle
(c) equal to the dual cycle (d) greater than the Diesel cycle

29. If the temperature at the turbine inlet is kept constant, the net output of a simple gas turbine plant would
(a) increase with increasing pressure ratio
(b) decrease with increasing pressure ratio
(c) first increase and then decrease with increasing pressure ratio
(d) remain unaffected with changes in pressure ratio

30. The specific output of a gas turbine plant can be increased by water injection. The water should be injected between the
(a) regenerator and the combustion chamber (b) combustion chamber and the turbine
(c) compressor and the regenerator (d) turbine and compressor

31. The thermal efficiency of a gas turbine cycle can be increased by
(a) reheating (b) regeneration (c) intercooling (d) all the above

32. The maximum temperature in a gas turbine is approximately
(a) 300°C (b) 600°C (c) 900°C (d) 1200°C

33. **Assertion A:** A regenerator always increases the efficiency of a gas turbine unit.
Reasoning (R): Because it recovers the energy which was going as a waste and saves fuel consumption.
(a) Both A and R are false (b) A is false, R is true
(c) A is true, R is false (d) Both A and R are true

34. For an axial-flow gas turbine, the polytropic efficiency is equal to the
(a) isentropic efficiency (b) overall efficiency
(c) isentropic efficiency of an infinitely small stage (d) stage efficiency

35. In a gas turbine unit with a regenerator, perfect regeneration means
(a) $T_3 < T_4$ (b) $T_3 > T_4$ (c) $T_3 = T_4$ (d) none of the above
where T_3 is the temperature of air coming out of the regenerator and T_4 is the temperature of gases leaving the turbine.

36. **Assertion (A):** The output and thermal efficiency of a closed-cycle gas turbine can be greater than that of an open cycle gas turbine.
Reasoning (R): Because a closed cycle gas turbine can use helium as a working fluid.
(a) Both A and R are true and R is the correct explanation of A
(b) Both A and R are false
(c) A is false and R is true
(d) Both A and R are true and R is not the correct explanation of A

37. For a gas turbine unit, the pressure ratio for maximum output is given by

(a) $(T_3/T_1)^{\frac{\gamma}{2(\gamma+1)}}$ (b) $(T_3/T_1)^{\frac{2(\gamma-1)}{\gamma}}$ (c) $(T_3/T_1)^{\frac{\gamma}{2(\gamma-1)}}$ (d) $(T_3/T_1)^{\frac{2\gamma}{\gamma-1}}$

where T_1 is the temperature of air at the compressor inlet and T_3 is the temperature at the turbine inlet.

38. In a gas turbine, reheating is done mainly to
(a) increase the outlet temperature (b) reduce the size of the turbine
(c) increase the power output (d) reduce the peak temperature

39. Higher air–fuel ratio in a gas turbine would
(a) increase thermal efficiency (b) increase power output
(c) decrease the outlet temperature (d) make all of the above

40. The maximum thrust power is obtained in a turbojet engine when
(a) aircraft velocity is equal to the jet velocity (b) aircraft velocity is twice the jet velocity
(c) aircraft velocity is half the jet velocity (d) aircraft velocity is the square root of the jet velocity

41. Restricted burning in a solid propellant rocket is adopted when
(a) a large thrust is required for a small period of time (b) a small thrust is required for a small period of time
(c) a small thrust is required for a large period of time (d) a large thrust is required for a large period of time

42. **Assertion (A):** It is hot possible to attain 100% propulsive efficiency of a turbojet.
Reasoning (R): Because that would require the thrust and thrust power equal to zero.
(a) Both A and R are false (b) Both A and R are true
(c) A is true, but R is false (d) A is false and R is true

43. Propulsive efficiency is defined as
(a) thrust power/propulsive power (b) propulsive power/thrust power
(c) propulsive power × thrust power (d) none of the above

44. The maximum propulsive efficiency for maximum thrust power of a turbojet engine is
(a) 50% (b) 66.7% (c) 75% (d) 100%

45. The thrust of a jet-propulsion power input can be increased by
(a) burning fuel after gas turnbine (b) injecting water into the compressor
(c) injecting ammonia into the combustion chamber (d) none of the above

46. The propeller of a turboprop engine is driven by
(a) compressor (b) turbine (c) both 'a' and 'b' (d) none of the above

47. **Assertion (A)**: A turbojet employs a smaller gas turbine unit
Reasoning (R): Because aircrafts fly at comparatively lower speeds.
(a) Both A and R are false (b) Both A and R are true
(c) A is false, R is true (d) A is true, R is false

48. The efficiency of a jet engine is higher at
(a) high altitudes (b) low altitudes (c) low speeds (d) high speeds

49. In a refrigeration plant, if the condenser temperature increases, the power input to the compressor will
(a) decrease (b) increase (c) remain the same (d) be unpredictable

50. At a place where there is no electricity, we can use the following system to obtain refrigeration.
(a) Vapour compression (b) Vapour absorption (c) Steam jet refrigeration (d) Air cycle refrigeration

51. In an aircraft-refrigeration system, the pressure at the cooling turbine outlet is equal to
(a) ambient pressure
(b) cabin pressure
(c) pressure at inlet to compressor
(d) none of the above

52. Match list I and list II and select the answer from the code given below:

List I	*List II*
Equipment in a refrigeration system	Variation of properties
A. Compressor	1. Enthalpy remains constant
B. Evaporator	2. Enthalpy increases
C. Throttle valve	3. Enthalpy increases but pressure remains constant
D. Condenser	4. Enthalpy decreases but pressure remains constant

Code:

	A	*B*	*C*	*D*
(a)	3	2	1	4
(b)	2	3	4	1
(c)	2	3	1	4
(d)	4	2	3	1

53. In a 2-stage vapour compression system, a flash intercooler is installed between the
(a) HP compressor and condenser
(b) LP and HP compressor
(c) Evaporator and LP compressors
(d) none of the above

54. In a water-vapour refrigeration system, a steam ejector maintains a very low pressure in the
(a) condenser (b) absorber (c) flash chamber (d) none of the above

55. In a vapour compression system, the working fluid is a superheated vapour at inlet to
(a) evaporator (b) condenser (c) compressor (d) throttle valve

56. A good refrigerant should have a
(a) large latent heat at condensing pressure
(b) large latent heat at evaporator pressure
(c) condensing pressure close to critical pressure
(d) high critical pressure

57. The effects of superheating the vapour in the evaporator and subcooling in the condenser
(a) decrease of COP
(b) increase the COP
(c) superheating increases COP, but undercooling decreases COP
(d) superheating decreases COP, but subcooling increases COP

58. A humidification process means
(a) a decrease in relative humidity
(b) a decrease in specific humidity
(c) an increase in specific humidity
(d) an increase in temperature

59. In an adiabatic saturation process
(a) the enthalpy remains constant
(b) the temperature remains constant
(c) the absolute humidity remains constant
(d) the relative humidity remains constant

60. When air is adiabatically saturated, the temperature attained is
(a) dew point temperature
(b) dry bulb temperature
(c) wet bulb temperature
(d) triple point temperature

61. The saturation temperature at the partial pressure of water vapour in an air–water mixture is the
(a) dew point temperature (b) wet bulb temperature
(c) dry bulb temperature (d) adiabatic saturation temperature

62. The use of reheating of steam in a steam-power plant
(a) permits the use of higher boiler steam pressure to increase the net power output
(b) reduces the steam rate and the heat rate
(c) limits the permissible quality of steam at turbine exhaust
(d) accomplishes all the above

63. The ideal regenerative steam cycle which yields an efficiency equal to that of Carnot cycle is not practicable because
(a) reversible heat transfer cannot be obtained in finite time
(b) heat exchanger in the turbine is mechanically not practicable
(c) the moisture content of the steam in the turbine will be high
(d) of all the above reasons

64. If the number of feedwater heaters used in a steam power plant are seven, the maximum gain in cycle efficiency occurs when the overall temperature rise of feedwater is about x times the difference between the condenser and boiler saturation temperature (with dry saturated steam at turbine inlet),
Where
(a) $x = \frac{1}{2,}$ (b) $x = \frac{3}{4}$ (c) $x = \frac{7}{8}$ (d) $x = \frac{8}{9}$

65. In a steam power plant while the energy loss in the condenser is about 60%, the energy loss is nearly.
(a) zero (b) 4% (c) 60% (d) 100%

66. The overall efficiency of a steam power plant is expressed as the product of efficiencies of its components. If the boiler efficiency takes care of the energy loss in the boiler, the cycle efficiency takes care of the energy loss in the
(a) turbine (b) condenser (c) electric generator (d) the auxiliaries

67. Two Thermodynamic cycles are coupled in series where heat lost by one is absorbed by the other. If the topping cycle has an efficiency of 30% and the bottoming cycle has an efficiency of 20%, the overall efficiency of the combined cycle is
(a) 50% (b) 60% (c) 44% (d) 54%

68. This cycle is applicable to a pulse jet engine:
(a) Brayton cycle (b) Lenoir cycle (c) Atkinson cycle (d) Ericsson cycle

69. A gas turbine plant operates on an ideal Brayton cycle. If the minimum temperature is 300 K and the maximum temperature is 1200 K, the cycle efficiency corresponding to maximum net work is
(a) 0.30 (b) 0.40 (c) 0.50 (d) 0.75

70. In the expansion of steam is a turbine with friction and heat loss, the entropy decrease of steam due to heat loss happens to be equal to the entropy increase of steam due to friction. Then, the entropy of steam at the turbine exit minus that at turbine inlet is
(a) zero (b) positive (c) negative (d) more or less than zero

ANSWERS

1. (b)
2. (c)
3. (b)
4. (c)
5. (a)
6. (d)
7. (c)
8. (b)
9. (d)
10. (c)
11. (d)
12. (b)
13. (d)
14. (d)
15. (c)
16. (c)
17. (a)
18. (b)
19. (c)
20. (a)
21. (a)
22. (d)
23. (a)
24. (b)
25. (d)
26. (d)
27. (b)
28. (b)
29. (c)
30. (c)
31. (d)
32. (d)
33. (d)
34. (c)
35. (c)
36. (a)
37. (c)
38. (c)
39. (d)
40. (c)
41. (c)
42. (b)
43. (a)
44. (b)
45. (a)
46. (b)
47. (d)
48. (a)
49. (b)
50. (b)
51. (b)
52. (c)
53. (b)
54. (c)
55. (b)
56. (b)
57. (b)
58. (c)
59. (a)
60. (c)
61. (a)
62. (d)
63. (d)
64. (c)
65. (b)
66. (b)
67. (c)
68. (b)
69. (c)
70. (a)

Bibliography

1. Adrian Bejan, *Advanced Engineering Thermodynamics*, John Wiley, New York, 1988.
2. Adrian Bejan, *Entropy Generation through Heat and Fluid Flow*, Wiley Int. Science, New York, 1982.
3. R.W. Benson, *Advanced Engineering Thermodynamics*, Pergamon Press, London, 1977.
4. H.B. Callen, *Thermodynamics*, John Wiley, New York, 1960.
5. Y.A. Cengel and M.A. Boles, *Thermodynamics: An Engineering Approach*, McGraw-Hill, New York, 1988.
6. G.N. Hatsopoulos and J.H. Keenan, *Principles of General Thermodynamics*, John Wiley, New York, 1965.
7. Jack P. Holman, *Heat Transfer*, McGraw-Hill, New York, 1978.
8. J.R. Howell and R.O. Buckius, *Fundamentals of Engineering Thermodynamics,* McGraw-Hill, New York, 1992.
9. J.S. Hsieh, *Principles of Thermodynamics*, McGraw-Hill, 1975.
10. J.B. Jones and R.E. Dugan, *Engineering Thermodynamics*, Prentice-Hall, India, 1998.
11. J.H. Keenan, *Thermodynamics*, John Wiley, New York, 1941.
12. J.F. Lay, *Thermodynamics*, Merril Books, 1963.
13. J.F. Lee and F.W. Sears, *Thermodynamics*, Addison-Wesely, Cambridge, Mass, 1962.
14. David A. Mooney, *Mechanical Engineering Thermodynamics*, Prentice-Hall, New Jersey, 1953.
15. M.J. Moran and H.N. Shapiro, *Fundamentals of Engineering Thermodynamics*, John Wiley, New York, 1988.
16. P.K. Nag, *Heat and Mass Transfer*, 2e, Tata McGraw-Hill, New Delhi, 2007.
17. E.F. Obert, *Concepts of Thermodynamics*, McGraw-Hill, New York, 1960.
18. W.C. Reynolds and H.C. Perkins, *Engineering Thermodynamics*, McGraw-Hill, New York, 1977.
19. Michel A. Saad, *Thermodynamics for Engineers*, Prentice-Hall, New Delhi, 1969.
20. A.H. Shapiro, *The Dynamics and Thermodynamics of Compressible Fluid Flow*, The Ronald Press, New York, 1953.
21. G.J. Van Wylen, R.E. Sonntag and C. Borgnakke, *Fundamentals of Classical Thermodynamics*, John Wiley, New York, 1994.
22. D.B. Spalding and E.H. Cole, *Engineering Thermodynamics*, Edward Arnold, London, 1967.
23. Kenneth Wark, *Thermodynamics*, Fourth edition, McGraw-Hill, New York, 1985.
24. M.W. Zemansky, *Heat and Thermodynamics*, McGraw-Hill, New York, 1957.
25. M.W. Zemansky and H.C. Van Ness, *Basic Engineering Thermodynamics*, McGraw-Hill, New York, 1966.
26. ML Macheve and R.P. Sharma, A course in *Internal Combustion Engines,* Dhanpat Rai and Sons, Delhi, 1960.
27. P.W. Gill, J.H. Smith, E. J. Ziurys, Fundamentals of Internal Combustion Engines Fourth Edition, Oxford & IBH Publishing Co., Delhi, 1959.

28. T.D. Eastop and A. McKonkey *Applied Thermodynamics for Engineering* Technologists, Fifth Edition, Addison-Wesley Longman Ltd., UK, 1993.
29. H. Cohen, G.F.C. Rogers and H.L.H Saravanamutto, Gas Turbine Theory, Third Edition, Longman, UK, 1987.
30. G. M. Dusinberre and J.C. Lester Gas Turbine Power, International Textbook, 1958.

Index

A

A back pressure turbine 351
Abnormal combustion 670
Absolute entropy 132
Absolute entropy 525
Absolute pressure 8
Absolute temperature 238
Absolute thermodynamic temperature scale 94
Absolute zero 95
Absorption refrigeration cycle 436
Absorptivity 584
Activity 518
Adiabatic compressibility 287
Adiabatic efficiency 605
Adiabatic evaporative cooling 482
Adiabatic flame temperature 523
Adiabatic mixing of two streams 481
Adiabatic process 36
Adiabatic saturation process 473
Adiabatic saturation temperature 473
Adsorbents 482
Advantages of a gt plant 708
Aenergy 151
Affinity 517
Afterburning 397
After-coolers 603
Air conditioning 467
Air motors 613
Air standard cycles 380
Air washer 485
Air-conditioning systems 490
Aircraft propulsion 396
Amagat's law 468
Analyzer-rectifier combination 437
Anti-knock quality 668
Apparatus dew point 479
Approach 483
Aqua-ammonia absorption system. 436
As superheat horn 420
Athodyd 736
Athodyds 736
Autoignition 664
Availability 161
Availability function for a closed system 160
Availability function of a steady flow process 158
Availability in a nonflow process 161
Availability in a steady flow process 161
Available energy 150
Available energy referred to a cycle 150
Avogadro's law 236
Avogadro's number (a) 238
Axial flow compressor 620
Azeotropes 434

B

Balanced heat exchanger 584
Baro meter 8
Beattie-bridgeman equation 248
Bernoulli equation, 64
Binary vapour cycles 347
Biot number (bi) 574
Black body 585
Blowers 624
Bmep 687
Boiler efficiency 353
Boltzmann constant 238
Bootstrap system 446
Bottom dead centre 652
Boundary 1
Boyle temperature 254
Brake efficiency 353
Brake mean effective pressure 661
Brake power (BP) 32
Brake thermal efficiency 661
Brayton cycle 387
Brayton-rankine combined cycle 398
Breathing capacity 662
Bypass factor 484
Bypass ratio 397
Bypass turbo-jet engine 735

C

Caloric theory of heat 46
Capillary tube 428
Carnot cycle 113, 378
Carnot cyclea reversible cycle is an ideal hypothet 91
Carnot's theorem 93
Cascade systems 431
Causes of irreversibility 87
Celsius temperature scale 22
Centrifugal compressor 616
Cetane number 668
Characteristic gas constant 238
Chemical affinity 510
Chemical dehumidification 482
Chemical equilibrium 3
Chemical exergy 526
Chemical potential 298
Choked 553
Choking in a nozzle 554
Claude system 448
Clausius' statement of the second law 84
Clausius' theorem 109
Clausius-clapeyron equation 292
Closed system 1

Closed-cycle gas turbine 715
Coefficient of performance 85
Cogeneration plant 351
Com pressed liquid 209
Combustion in ci engines 677
Combustion knock 668
Comfort charts 491
Compressed liquid region 199
Compressibility factor 249
Compression ignition (C.I.) Engine 382, 652
Compression processes 603
Compression ratio 382, 383, 384, 388, 653
Compressor 428
Compressor efficiency 605
Condenser 427
Conditions for reversibility 91
Conduction heat transfer 568
Constant volume gas thermometer 20
Contact factor 485
Continuum 6
Control surface 2, 58
Control volume 2, 58
Convective heat transfer 574
Cooling and dehumidification 479
Cooling region 290
Cooling system of IC engines 680
Cooling tower 483
Corollary of carnot's theorem 93
Counterflow heat exchanger 581
Coupled cycles 349
Co-volume 248
Critical isotherm 200
Critical pressure 200
Critical pressure ratio 553
Critical properties 553
Critical state 199
Critical temperature 200
Critical volume 200
Cut-off ratio 383, 384
Cyclic heat engine 82

D

Dalton's law 467
Dalton's law of partial pressures 257
Dead state 160
Deaerator 344
Degradation 164
Degree (or advancement) of reaction 508
Degree of freedom 303
Degree of reaction 507, 622
Degree of saturation 470
Degree of subcooling 424
Degree of super-heat 209
Dehumidification and heating 481
Detonation 664, 671
Dew point temperature 469
Diather mic wall 36
Diesel cycle 382, 665
Diffuser 62, 550
Disadvantages of a gt plant 710
Discharging and charging a tank 68
Disorder number 131
Dissipation 164
Dissipative effects 89
Dittus and boelter 577
Dry air 467
Dry bulb temperature 468
Dry compression 420
Dual cycle 384

E

Effective temperature 491
Effectiveness of the regenerator 389
Effectiveness, ε 583
Efficiency of the reversible heat engine 96
Electric calorimeter 213
Electrical resistance thermometer 22
Electrical work 32
Electrolux refrigerator 439
Emissitivity 586
Emissivity 585
Energy 46
Energy balance 60, 687
Energy cascading 172
Energy converter 354
Energy equation 287
Energy interactions 27
Energy reservoirs 83
Energy–a property of the system 47
Engine indicator 30
Engine knock 664
Enthalpy 50
Enthalpy deviation 476
Enthalpy of combustion 523
Enthalpy of formation 521–522
Enthalpy of moist air 471
Entropy 109–110
Entropy generation in a closed system 124
Entropy generation number 166
Entropy generation 124
Entropy principle 116
Entropy transfer 122
Equation of continuity 60
Equations of state 248–236
Equilibrium constant 510
Equivalence of kelvin-planck and clausius statements 86
Equivalence radio 667
Ericsson cycle 379
Euler equation 64
Evaporator 429
Exact (or perfect) differential 282
Exergy 151, 526
Exergy balance for a closed system 167
Exergy balance for a steady flow system 168
Expansion device 427
Expansion ratio 383, 384
Extensive properties 3
External irreversibility 98

F

Fanno line 557
Fans 624
Feedwater heaters 343
Fin efficiency 572
Fins 572
First law for a closed system undergoing a 46
First law for a closed system undergoing a change of state 46
First law for a closed system undergoing
a cycle 45
First law for reactive systems 522
Fixed point 18
Flash point 669
Floating node 589
Flooded evaporators 429
Flow work 33
Force of cohesion 248
Forced convection 574
Fourier number 574

Fowler-guggenheim statement 96
Free expansion 35
Free or natural convection 574
Free-shaft turbine 395
Froude efficiency 729
Fuel cells 529
Fuel injection in ci engines 676
Fuel injector 677
Fuel pump 676
Fuel-air / Air-fuel ratio 667
Fugacity 517
Fusion curve 201

G

Gas cycle refrigeration 442
Gas tables 556
Gas turbine fuels 723
Gas turbine materials 723
Gas turbine with water injection 722
Gauge pressure 8
General energy equation 66
Generalized compressibility chart 251
Geometrical mean area 572
Gibb's theorem 468
Gibbs entropy equation 299
Gibbs function change 515
Gibbs function for formation 526
Gibbs function 163
Gibbs phase rule 303
Gibbs theorem 259
Gibbs-duhem equation 300
Global warming potential 435
Gouy-stodola theorem 164
Grashoff number 578
Grassmann diagram 165
Gray body 586

H

Heat capacity of reacting gases 519
Heat capacity ratio 583
Heat exchangers 579
Heat of reaction 511
Heat of respiration 490
Heat pump 85
Heat pump systems 441
Heat rate 331
Heat transfer 36
Heat transfer coefficient 575
Heating and humidification 480
Heating region 290
Heating value at constant pressure 525
Heating value at constant volume 525
Helmholtz function 163
Heterogeneous system 3
Higher heating value 525
Homogeneous system 3
Humid specific heat 478
Hypergolic propellants 741

I

Ideal gas 237
Ideal gas equation of state 20
Ideal gas temperature 21
Ideal regenerative cycle 337
Ignition delay 678
Ignition lag 668
Ignition systems in si engines 673
Imperfect differential 29
Impulse function 556
Impulse pressure 557
Indicated mean effective pressure 660
Indicated power (IP) 31, 660
Indicated specific fuel consumption 662
Indicated thermal efficiency 661
Indicator diagram 30
Intensive properties 3
Internal efficiency 353
Internal energy 46, 48
Internal energy of combustion 523
Internal irreversibility 98
International practical temperature scale 23
Inversion curve 290
Irradiation (g) 587
Irreversibility 163
Isentropic efficiency 619
Isolated system 2
Isothermal compressibility 286
Isothermal efficiency 605, 612

J

Jet propulsion cycle 396
Jet propulsion system 724
Joule's law 240
Joule-kelvin coefficient 290
Joule-kelvin effect 290

K

Kelvin scale 95
Kelvin-planck statement of second law 83
Kirchhoff's law 585

L

Laminar 576
Latent heat 37
Latent heat load 478
Latent heat load: humidification or dehumidification 477
Law of corresponding states 250, 252
Law of degradation of energy 154
Law of mass action 510–510
Ler 609
Limited pressure cycle 384
Linde–hampson system 447
Liquefaction of gases 446
Liquid-line heat exchanger 424
Load calculations 489
Log-mean area 570
Log-mean temperature difference 580
Lost work 125
Lower heating value 525
Lubricating system 681

M

Mach number 549
Macroscopic energy mode 47
Maximum inversion temperature 291
Maximum work in a reversible process 155
Maxwell's equations 283
Mean effective pressure 653
Mechanical efficiency 32, 353
Mechanical equilibrium 3
Mechanical irreversibility. 98
Mechanical stability 308
Metastable equilibrium 305
Microscopic energy mode 47
Minimum work of compression 610
Mole fraction 258
Mollier diagram 205

Monochromatic emissive power 585
Monochromatic emissivity 586
Morse test 686
Multi-evaporator systems 431
Multi-stage compression 609
Multi-stage vapour compression systems 429

N

Nernst's equation 513
Nernst-simon statement of third law 132
Newton's law of cooling 575
Nonequilibrium state 4
Normal boiling point 201
Normal combustion 669
Normal shocks 556
Nozzle 62, 550
Nozzle and diffusor 61
Ntu 583
Nuclear rockets 744
Number of transfer units 583
Nusselt number 577–578

O

Octane number 668
Octane rating 664
Of jumbo jets 398
Open system 2
Open-cycle plant 708
Otto cycle 384, 668
Overall efficiency 354
Overall heat transfer coefficient 575
Ozone depletion potential 435

P

Paddle-wheel work or stirring work 33
Parallel flow heat exchanger 579
Partial entropy of 260
Partial pressure 257
Partial volume 258
Pass-out turbines 352
Path 2
Path function and point function 28
Pdv–work or displacement work 28
Perfect differentials 29
Perfect intercooling 609
Perpetual motion machine of the first kind 51
Perpetual motion machine of the first kind–pmm1 51
Perpetual motion machine of the second kind 84
Perpetual motion of the third kind 86
Phase 3
Phase change of the first order 292
Pl 683
Planck's law 586
Polytropic efficiency 623
Polytropic process 245
Polytropic specific heat 247
Power or 'shaft power' 659
Prandtl number 577–578
Preignition 670
Pressure ratio 388
Pressure thrust 730
Pre-whirl 618
Process 3
Process heat 350
Production of solid ice 448
Propellant 739
Propeller 726
Properties 2
Propulsive devices 726
Propulsive efficiency 396, 725, 729
Propulsive power 396, 725, 729
Psychrometer 472
Psychrometric chart 474
Psychrometric processes 476
Psychrometric properties 468
Psychrometry 468
Pulse jet 738

Q

Qualitative difference between heat and work 81
Quality of energy 153
Quality or dryness fraction 206
Quasi-static process 4

R

Radiation heat transfer 584
Radiation shape factor 586
Radiosity (j) 587
Ram compression 727
Ramjet 398, 736
Range 483
Rankine cycle 329
Rayleigh line 557
Reaction equilibrium 509
Reciprocity theorem 586
Redlich-kwong equation 248
Reduced properties 250
Reflectivity 584
Refrigerant 419
Refrigerants 432
Refrigerating effect 422
Refrigerator 84
Regenerative cycle 338
Regenerator 713
Reheat cycle 335
Reheat-regenerative cycle 341
Relative efficiency 664
Relative humidity 470–471
Resistance concept 569
Reversed brayton cycle 443
Reversed heat engine 92
Reversed heat engine cycle 419
Reversible adiabatic process 242
Reversible isothermal process 244
Reversible process 87
Reversible work by an open system 156
Reversible work in a closed system 158
Reversible work in a steady flow process 158
Reynolds number 576–577
Rocket 727
Rocket motor 740
Rocket propulsion 739
Rockets 726
Room sensible heat factor 487
Roots blower 613
Rotary compressor 613

S

Saha's equation 514
Saha-bose equation of 248
Sankey diagram 165
Saturated liquid lines 199
Saturated vapour line 199
Saturation or vapour dome 199
Saturation pressure 201

Saturation states 198
Saturation temperature 201
Second law analysis of reactive systems 526
Second law efficiency 168
Second law efficiency of a reactive system 529
Semi-closed cycle gt plant 715
Sensible heat factor 479
Sensible heat load (qc) 477
Sensible heating or cooling 477
Separating and throttling calorimeter 212
Shaft work 32
Simple air-conditioning system 486
Simple float carburetor 673
Single-stage reciprocating air compressor 605
Sink 83
Slip factor 619
Sonic velocity 549
Source 84
Space resistance 589
Spark ignition (S.I.) Engine 380, 651
Specific heat 37
Specific heat at constant pressure 50
Specific heat at constant volume 49
Specific humidity humidity ratio 469
Stable equilibrium 304–305
Stagnation pressure p_0 550
Stagnation properties 550
Stagnation state 550
Starting of engine 685
State 2
Steady flow compressors 616
Steady flow energy equations 61
Steady flow process 59
Steady state of 59
Steam jet refrigeration system 440
Steam rate 331
Stefan-boltzmann constant 585
Stefan-boltzmann law 587
Stirling cycle 379
Stoichiometric air 521
Stoichiometric coefficients 507
Streamline 576
Strength of a shock wave 560
Subcooled 209
Subcooler 425
Sublimation curve 201
Summer air conditioning 487
Supercharging 688
Superheat 209
Surroundings 1
System 1

T

Tds equations 284
Temperature of the surroundings 150
Terminal temperature difference 343
The atmospheric jet engines 726
The inequality of clausius 114
The keenan function, 158
The two-stroke engine 658
Thermal conductivity 568
Thermal efficiency 83
Thermal equilibrium 3
Thermal irreversibility 98
Thermal stability 307
Thermocouple 22
Thermodynamic cycle 3
Thermodynamic equilibrium 3
Thermodynamic probability, 131
Thermomechanical 526
Thermometer 17
Thermometric property 17
Third law of thermodynamics 105
Third law of thermodynamic 525
Throat 550
Throttling calorimeter 211
Throttling device 62
Thrust 392, 725
Thrust power 729
Tonne of refrigeration 423
Top dead centre 652
Total emissive power 585
Total heat load 478
Transient heat conduction 573
Transmissivity 584
Triple point 199
Triple point of water 18
Trouton's rule 294
Turbo fan engine 397
Turbojet 726
Turbojet engine 396, 727
Turbojet engine with afterburner 732
Turboprop 726
Turboprop engine 398
Turbo-prop engine 733
Turbulent 576
Two-property rule 5
Types of irreversibility 98

U

Unavailable energy (u.E.) 150
Unit power consumption 423
Universal gas constant 237
Universe 1
Unsaturated 469
Unstable equilibrium 305
Useful work 159

V

Vacuum 8
Van der waals equation 248
Van't hoff equation 511
Vane type compressor 614
Vaporization curve 201
Vapour compression refrigeration cycle 420
Vapour pressure 201
Variable flow processes 65
Virial coefficients 249
Virial expansions 249
Volume expansivity 320
Volumetric efficiency 424, 429, 607, 661
Vortex blading 722

W

Wet compression 420
Wet-bulb depression 472
Wet-bulb temperature 472
Winter air conditioning 488
Work done factor 623
Work of compression 603–604
Work ratio 331, 388, 712
Work transfer 27

Z

Zeroth law of thermodynamics 17